D0086625

DNA REPLICATION

DNA REPLICATION

SECOND EDITION

ARTHUR KORNBERG
STANFORD UNIVERSITY

TANIA A. BAKER
STANFORD UNIVERSITY AND
NATIONAL INSTITUTES OF HEALTH

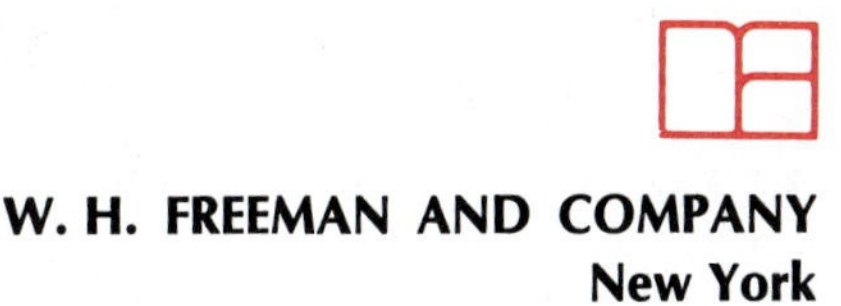

W. H. FREEMAN AND COMPANY
New York

COVER:

A looped rolling circle of the duplex replicative form of phage ϕX174, based on an electron micrograph provided by Dr Jack Griffith. See Fig. 8-13 for more detail. (Design by Robert Ishi)

Library of Congress Cataloging-in-Publication Data

Kornberg, Arthur, 1918–
DNA replication.—2d ed./by Arthur Kornberg and Tania A. Baker.
p. cm.
Includes bibliographical references and indexes.
ISBN 0-7167-2003-5
1. DNA—Synthesis. I. Baker, Tania A. II. Title.
QP624.K668 1991
574.87′3282—dc20 90-26456

Printed in the United States of America

2 3 4 5 6 7 8 9 RRD 9 8 7 6 5 4 3 2

Contents

4 DNA Polymerase I of *E. coli* 113

5 Prokaryotic DNA Polymerases Other Than *E. coli* Pol I 165

6 Eukaryotic DNA Polymerases 197

7 RNA Polymerases 227

8 Primases, Primosomes, and Priming 275

9 Ligases and Polynucleotide Kinases 307

10 DNA-Binding Proteins 323

11 DNA Helicases 355

12 Topoisomerases 379

13 Deoxyribonucleases 403

14 Inhibitors of Replication 439

15 Replication Mechanisms and Operations 471

16 Genome Origins 511

17 Bacterial DNA Viruses 553

18 Plasmids and Organelles 637

19 Animal DNA Viruses and Retroviruses 689

20 Regulation of Chromosomal Replication and Cell Division 731

21 Repair, Recombination, Transformation, Restriction, and Modification 771

22 Synthesis of Genes and Chromosomes 833

Preface
to *DNA Replication,* Second Edition

The rapid accretion of knowledge after the publication of *DNA Replication* in 1980 was the reason in 1982 for a *Supplement*. Because of its bulk (230 pages) and the exponential rate of research progress since then, further supplements would have been unwieldy. In this second edition, we have tried to limit the size of the volume by deletions and revisions and by severe selection of new material. Even so, with new chapters (Primerases, Primosomes, and Priming; DNA-Binding Proteins; DNA Helicases; Topoisomerases; Genome Origins; Plasmids and Organelles) and expansion of the others, growth has been unavoidable.

The plan of this book is the same — descriptions of the individual replication enzymes are followed by the genetic, physiologic, and structural features of the systems they serve. Beyond the obvious need to know the protein players to understand their genes, we hope that the emphasis on enzymology will illustrate the strength of this approach to the understanding of cell biology. As before, we include brief treatments of transcription, repair, and recombination, processes interwoven with replication.

The references have generally been selected to provide the most recent reviews and papers in which citations of the earlier work can be found. For economy of space and attention, a topical reference may be favored over a more venerable one.

Even more than in the first edition, we have been aided by expert reviews of most sections. We gratefully acknowledge the contributions of: Bruce Alberts, Stephen Benkovic, Kenneth Berns, Michael Botchan, Neal Brown, Patrick Brown, Richard Burgess, Richard Calendar, Michael Chamberlin, Gilbert Chu, Nicholas Cozzarelli, Harrison Echols, Gertrude Elion, Robert Fuller, Adayapalam Ganesan, Martin Gellert, Carol Gross, Richard Gumport, Masaki Hayashi, Kensuke Horiuchi, Joel Huberman, Kenneth Johnson, Mary Ellen Jones, Catherine Joyce, Roger Kornberg, Kenneth Kreuzer, Robert Lehman, Tomas Lindahl, Stuart Linn, Timothy Lohman, Kenneth Marians, Christopher Mathews, Steven Matson, Roger McMacken, Kiyoshi Mizuuchi, Mukund Modak, Peter Model, Paul Modrich, Gisela Mosig, Nancy Nossal, Leslie Orgel, Peter Reichard, Charles Richardson, Marjorie Russel, Margarita Salas, Robert Sauer, Kirsten Skarstad, Bruce Stillman, James Wang, Teresa Wang, Robert Webster, and Judith Zyskind.

It is with considerable guilt that we do not share coauthorship with Charlene Kornberg. The illustrations she designed are a major feature of the book. We also cannot adequately acknowledge our indebtedness to LeRoy Bertsch for his meticulous and insightful contributions to every aspect of the effort.

We are deeply grateful to Kelly Solis-Navarro for executing the complex illustrations, to Betty Bray for errorless typing, to Gloria Hamilton for extraordinary skill and concern in the editing and styling, and to Bob Ishi for his innovative book design. To Linda Chaput, President of W. H. Freeman, and to the staff, particularly Georgia Lee Hadler, we express our appreciation for their unstinting dedication to producing a book of high quality.

March 1991

Arthur Kornberg
Tania A. Baker

Preface
to *DNA Replication,* First Edition

I wrote *DNA Synthesis* in 1974 to review and record some of the early discoveries, which were receding from fashionable attention, and to discuss the advances and their wider ramifications. I anticipated that the pace of progress would require a major revision of the book in a few years. I did not anticipate how little I would retain five years later and how much would have to be added. But writing this virtually new and more massive book has been a pleasure rather than a chore. Besides extensive new information, recent advances in DNA synthesis have produced a qualitative change and justify a larger scope and a new title, *DNA Replication*.

The term DNA replication embraces biochemical, genetic, and physiological aspects, and also the numerous DNA transactions that determine the structure and function of genetic material. My goal is an up-to-date account of DNA replication and metabolism with a strong biochemical emphasis that can be used for orientation and as an information resource. I hope the book will serve students with a beginning interest in DNA synthesis as well as those working directly on the subject.

In this reorganization, I have devoted the first half of the book to the greatly expanded field of enzymology of DNA. I have chosen again to treat *E. coli* DNA polymerase I in detail as a prototypical polymerase. I have added discussions of supercoiling, binding, and twisting proteins, and expanded the accounts of ligases, nucleases, and inhibitors of replication. For a grasp of the variety of replication mechanisms, I have surveyed the replicative life cycles of bacterial and animal viruses and of plasmids and organelles.

Proper acknowledgment for help in preparing this book could easily fill a chapter. I want to express my gratitude to the Rockefeller Foundation for their hospitality at the Bellagio Study Center, where I was able to start this book in October 1977. Specialists in many areas generously gave me their points of view and the most recent information, often unpublished. These people have read and influenced the contents and style of one or more of the chapters. I hope they will feel rewarded with a useful book and to them my gratitude is unbounded: Bruce Alberts, Paul Berg, Maurice Bessman, Douglas Brutlag, Michael Chamberlin, Nicholas Cozzarelli, David Dressler, Adayapalam Ganesan, Mehran Goulian, Philip Hanawalt, Nicholas Hoogenraad, Dale Kaiser, David Korn, Roger Kornberg, Sylvy Kornberg, Gordon Lark, Robert Lehman, Stuart Linn, Robert Low, Mark Pearson, Peter Reichard, Charles Richardson, Joseph Shlomai, George Stark, Jean Thomas, and Olke Uhlenbeck.

I owe major debts to Charlene Levering who did the illustrations with artistic skill and unstinting devotion, to Patricia Brewer whose superb styling and editorial judgment smoothed a rough manuscript, and to LeRoy Bertsch, whose knowledgeable and meticulous final review reduced the errors to what I hope will be a forgivable level.

June 1979 *Arthur Kornberg*

Preface
to *DNA Synthesis*

The rapid flow of facts and ideas in biochemistry makes it difficult to write an article, let alone a book. But such turbulence submerges useful facts and ideas, which become too specialized for general textbooks and are lost sight of even in detailed annual reviews. That this was true of the biochemistry of DNA synthesis became clear during preparation of the Robbins Lectures, given at Pomona College in April 1972, and it prompted me to undertake this effort.

This book emphasizes biochemical rather than physiological aspects of DNA synthesis. The scope has been broadened beyond that of an earlier book (*Enzymatic Synthesis of DNA*, 1962) to include topics clearly pertinent to DNA synthesis: precursors, repair, recombination, restriction, and transcription. It is hoped that with an enlarged scope and simplified language the book will serve students with a beginning interest in DNA synthesis as well as those working directly on the subject. Citations of the literature favor reviews and recent papers, which will in turn give the interested reader more complete bibliographies.

At the conclusion of writing this book, I am surprised and embarrassed at the large number of people whom I have enlisted in the preparation of this relatively modest effort. I am most indebted to my wife, Sylvy, who helped me write this book and do the early work on DNA synthesis. I am also grateful to Fred Robbins for the initial stimulus, to Charlene Levering for the illustrations, to Inge Loper for the typing, to Stephanie Lee Rowen for a careful early reading of the text, to I. Robert Lehman, David S. Hogness, A. Dale Kaiser, and R. David Cole for a critical reading of the entire text, and to colleagues too numerous to mention who gave me information and advice in preparing many sections of the book. The best requital for all these contributions is a useful book, and this has been my primary goal.

February 1974 *Arthur Kornberg*

COLOR PLATES

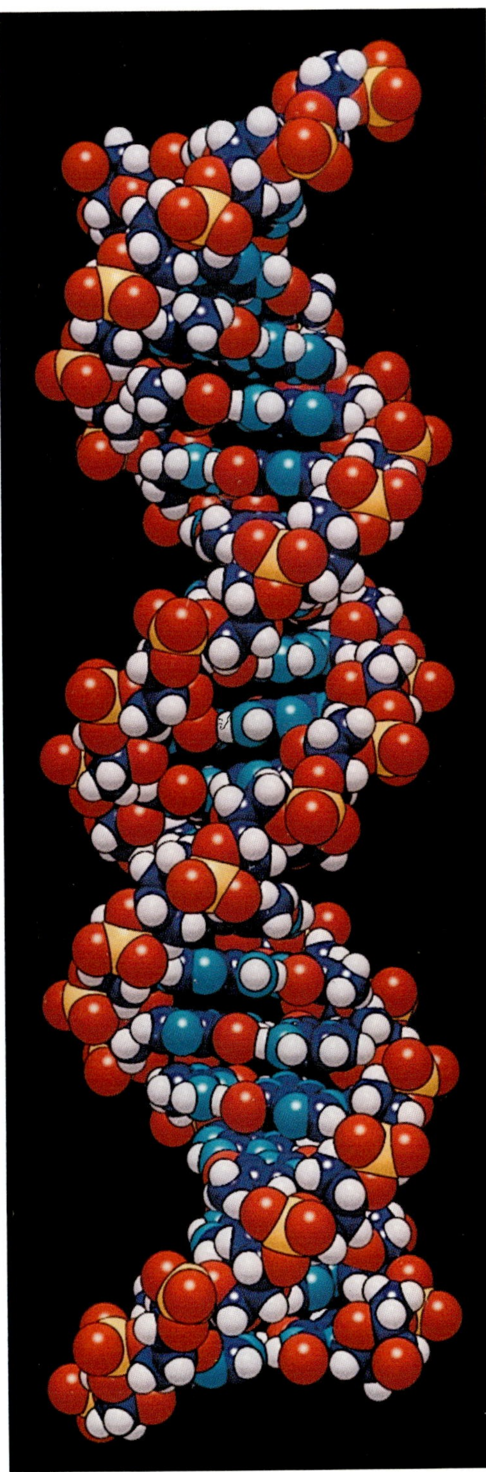

Plate 1
B-DNA. Space-filling atomic model of a DNA segment with two major grooves and one minor groove. Hydrogen, white; nitrogen, blue; oxygen, red; phosphorus, yellow. (Section 1-3; courtesy of Dr N Max, Lawrence Livermore National Labs)

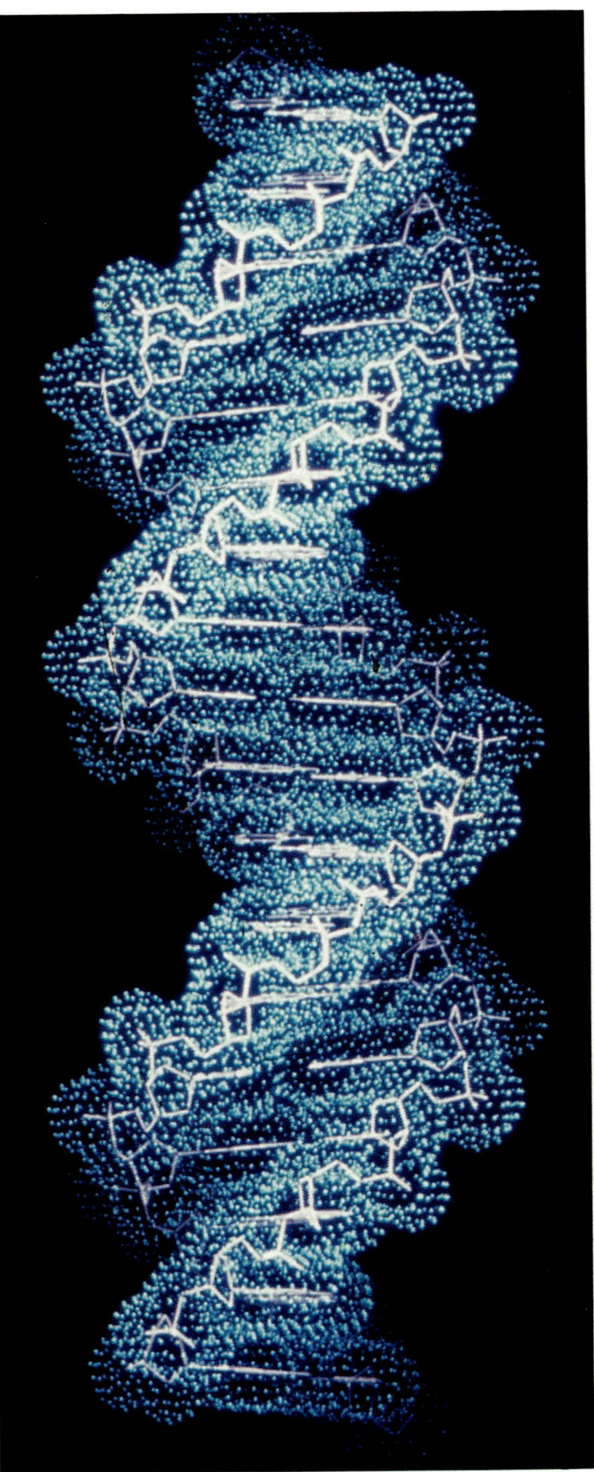

Plate 2
B-DNA. Computer-generated DNA structure with one major groove and two minor grooves. (Section 1-3; courtesy of Dr CA Frederick)

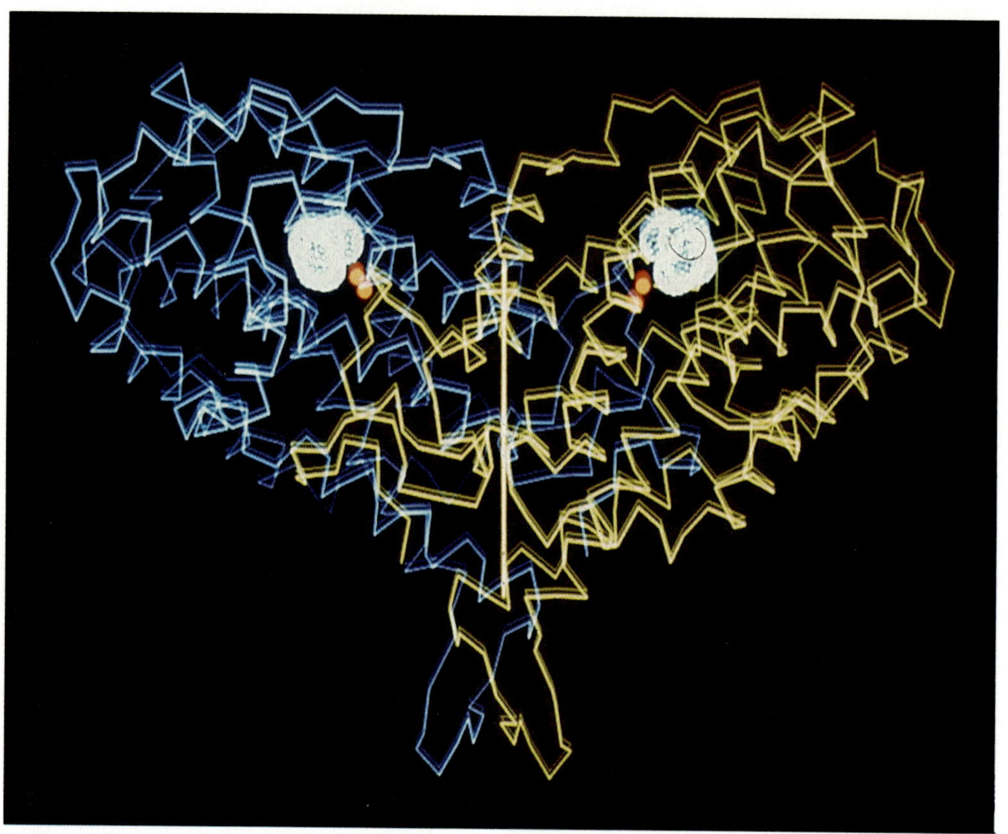

Plate 3

The R2 protein of *E. coli* ribonucleotide reductase. Buried among the many α-helices of the protein dimer (blue and yellow; axis of symmetry, vertical yellow line) are the iron centers (orange spheres) and the two Tyr 122 (van der Waals surface, white), the sites of the stable free radicals. (Section 2-7; courtesy of Drs P Nordlund, B-M Sjöberg, and H Eklund)

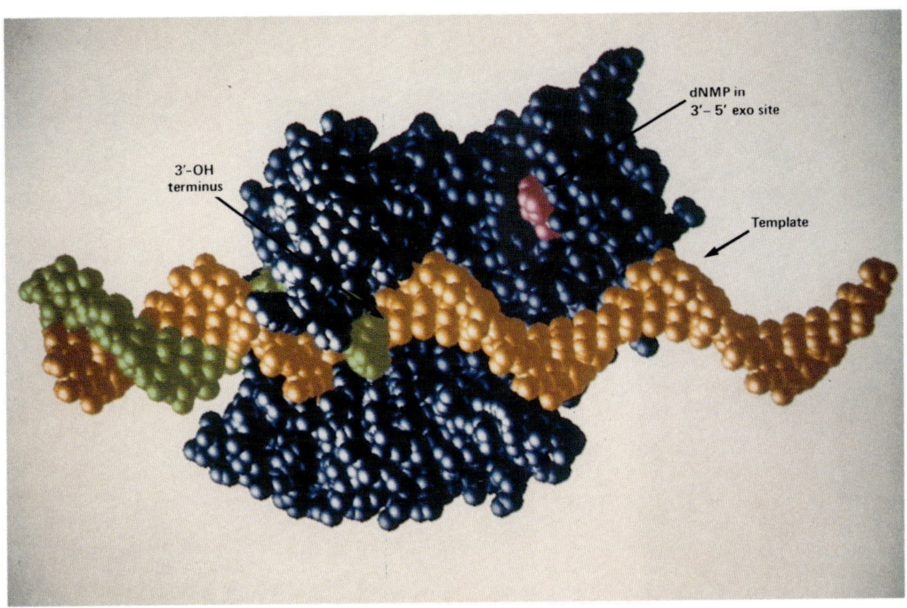

Plate 4
Space-filling model of the crystal structure of the large fragment of *E. coli* DNA pol I on the model-built template DNA strand with the 3′-OH terminus of the primer strand positioned adjacent to the proposed polymerase active site. (Section 4-4; courtesy of Dr TA Steitz)

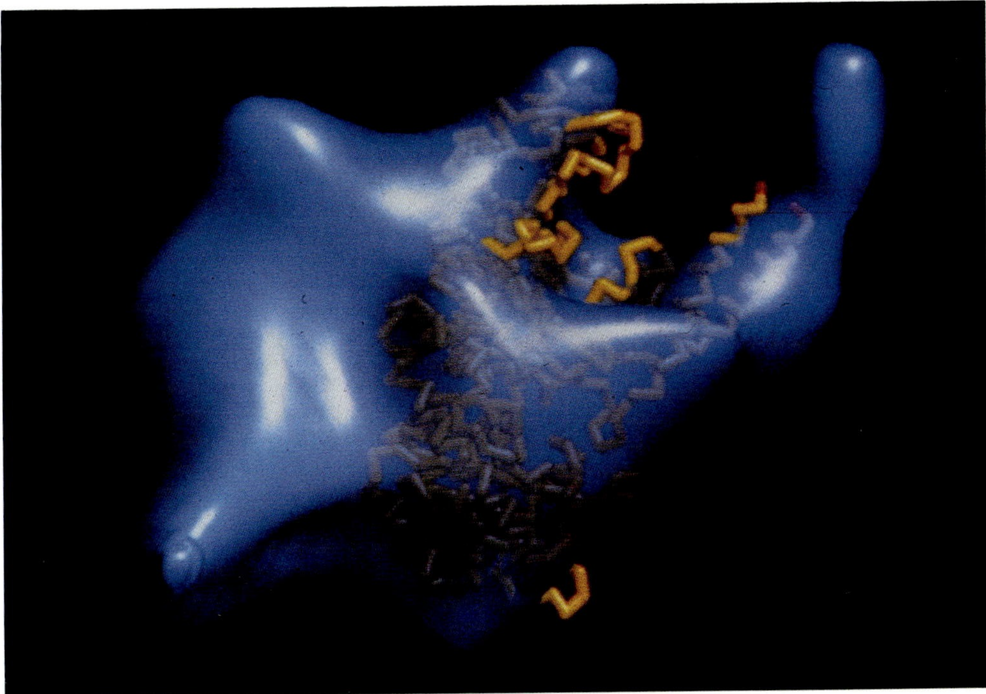

Plate 5
Three-dimensional structure of *E. coli* RNAP holoenzyme (blue solid figure) superimposed on the α-carbon backbone of *E. coli* DNA pol I (yellow). The enzymes clearly have a similar channel which can accommodate a strand of duplex DNA. (Section 7-2; courtesy of Dr R Kornberg)

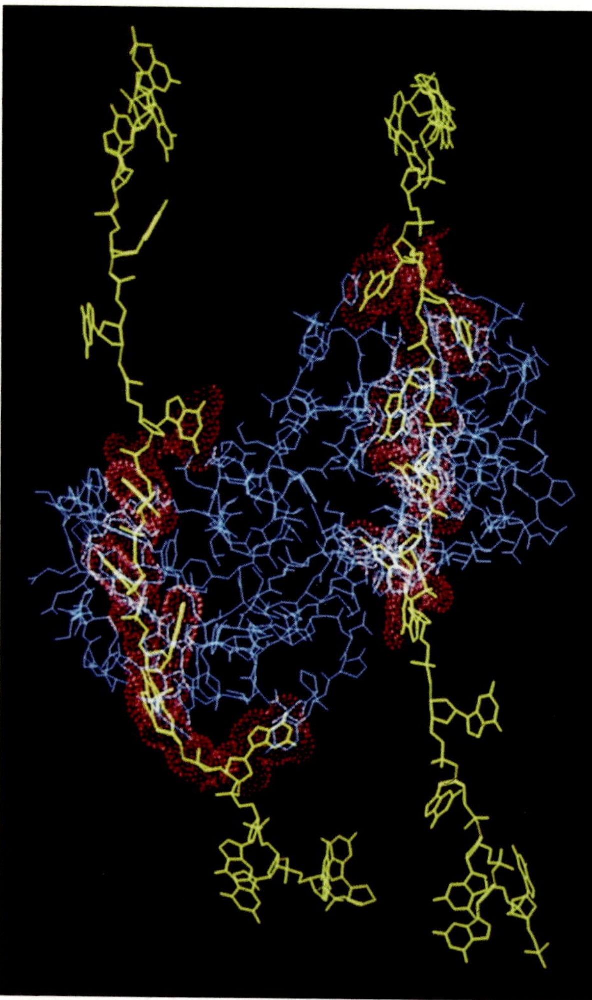

Plate 6

A general view of the complex of phage fd gp5 on fd ssDNA. Protein dimer, blue; fd ssDNA segments of opposite polarity, yellow; DNA-protein contact surface accessible to solvent, red. (Section 10-4; courtesy of Dr A McPherson)

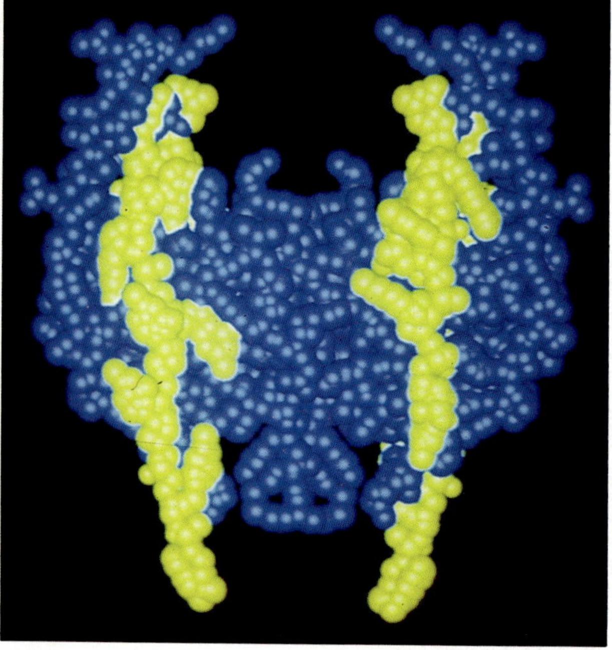

Plate 7

Atomic surface of phage sd gp5 (blue) associated with two antiparallel single strands of DNA (yellow). The dyad axis of the protein dimer is vertical. (Section 10-4; courtesy of Dr A McPherson)

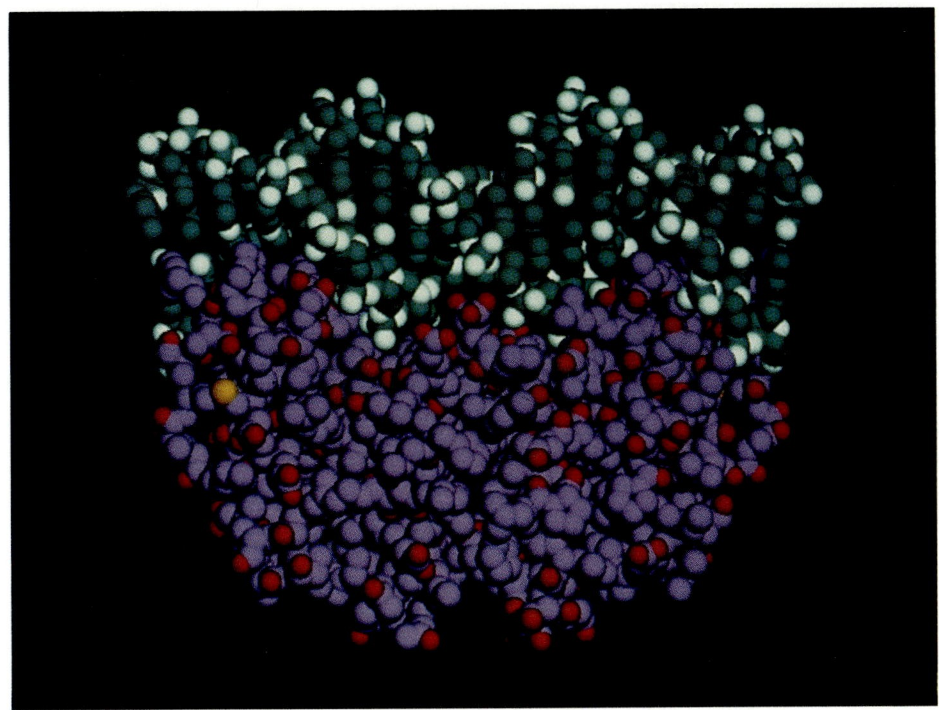

Plate 8
Atomic surface image of the *trp* repressor bound to its target DNA, the *trp* operator. Protein dimer, purple and red; DNA helix, blue and white. (Section 10-9; courtesy of Dr P Sigler)

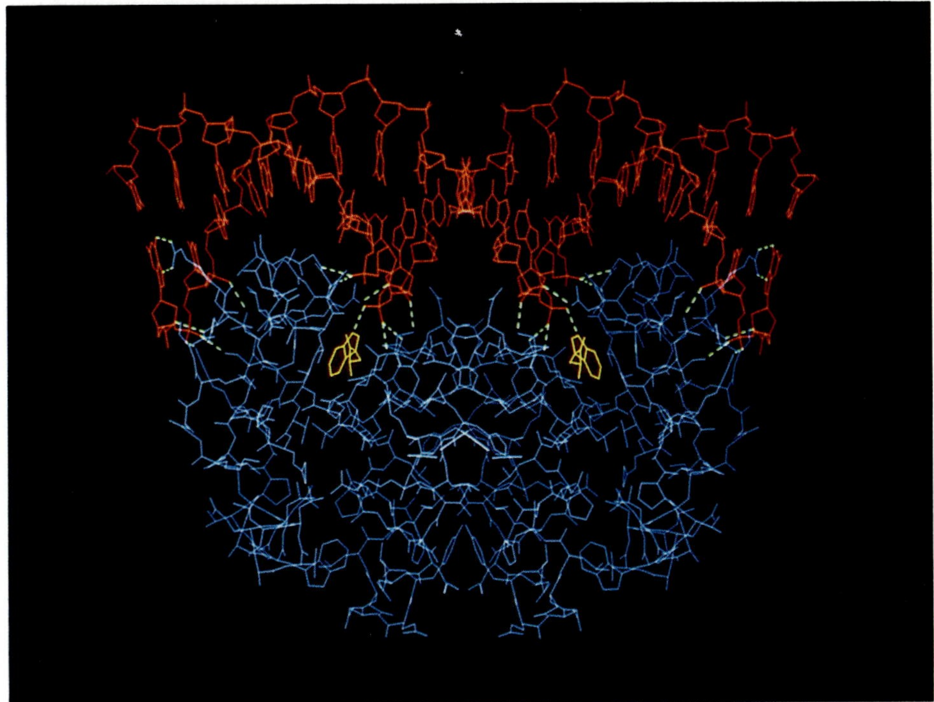

Plate 9
Model of *trp* repressor bound to *trp* operator DNA. Protein dimer, blue; DNA helix, red; tryptophan, yellow. (Section 10-9; courtesy of Dr P Sigler)

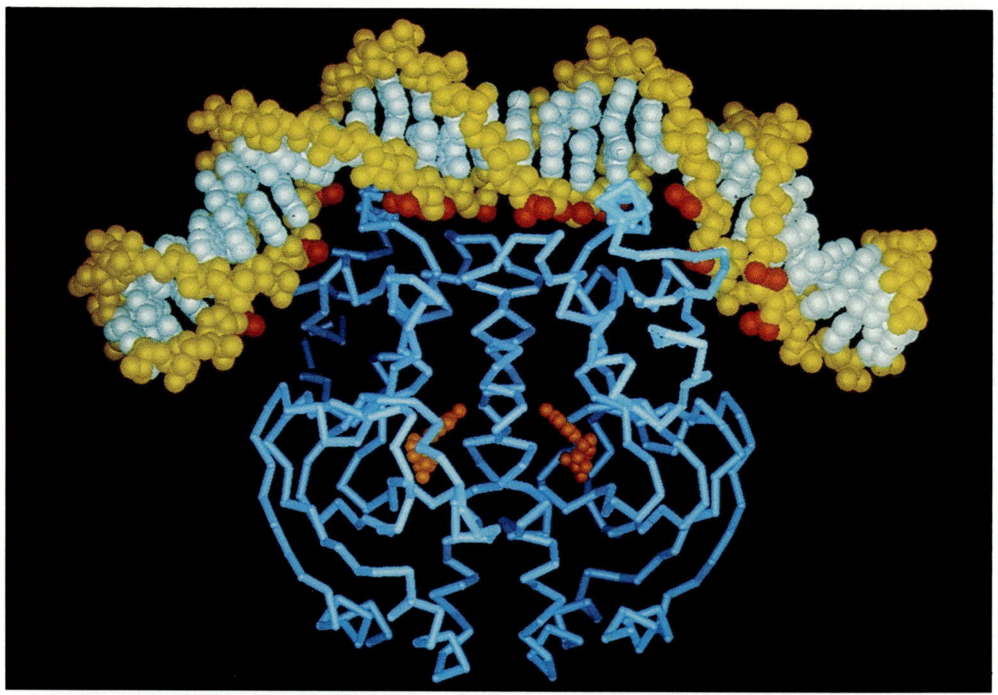

Plate 10
CAP protein bending ssDNA. Peptide backbone of the protein dimer, blue; cAMP, orange; DNA backbone, yellow; DNA base pairs, white; phosphates whose ethylation interferes with DNA binding, red. (Section 10-9; courtesy of Drs TA Steitz and S Schultz)

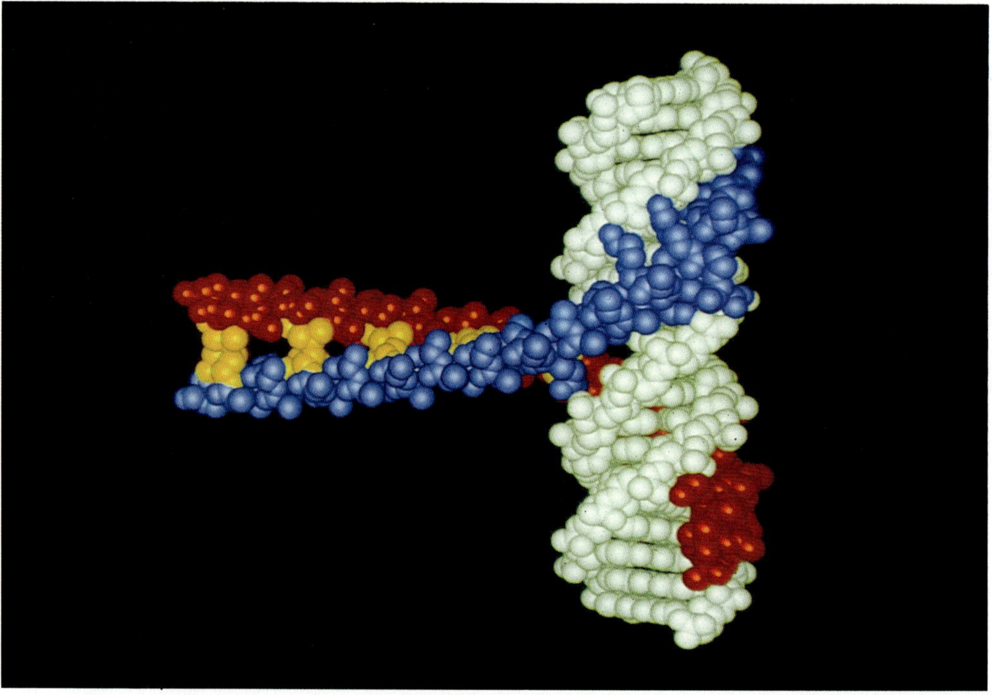

Plate 11
Model of a leucine zipper motif bound to B-DNA. DNA, white; polypeptides of the protein dimer, red or blue; interacting leucine side chains, yellow. (Section 10-9; courtesy of Dr S McKnight)

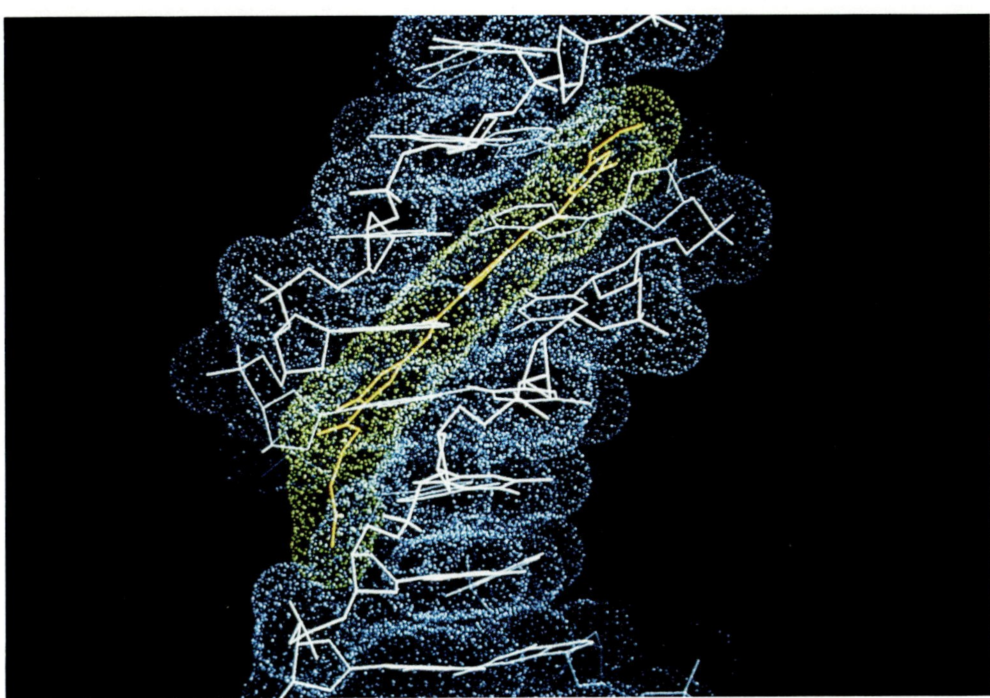

Plate 12
Single-crystal structure of distamycin (yellow) bound in the minor groove of B-DNA (blue). (Section 14-4; courtesy of Dr CA Frederick)

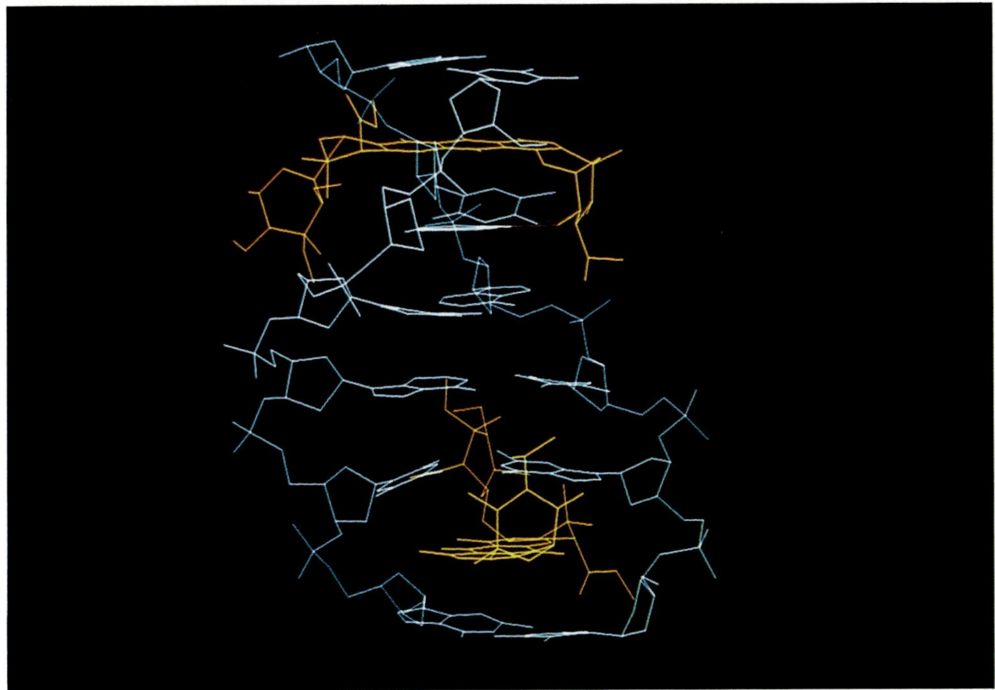

Plate 13
Single-crystal structure of nogalamycin (yellow) intercalated in B-DNA (blue). (Section 14-4; courtesy of Drs CA Frederick and LD Williams)

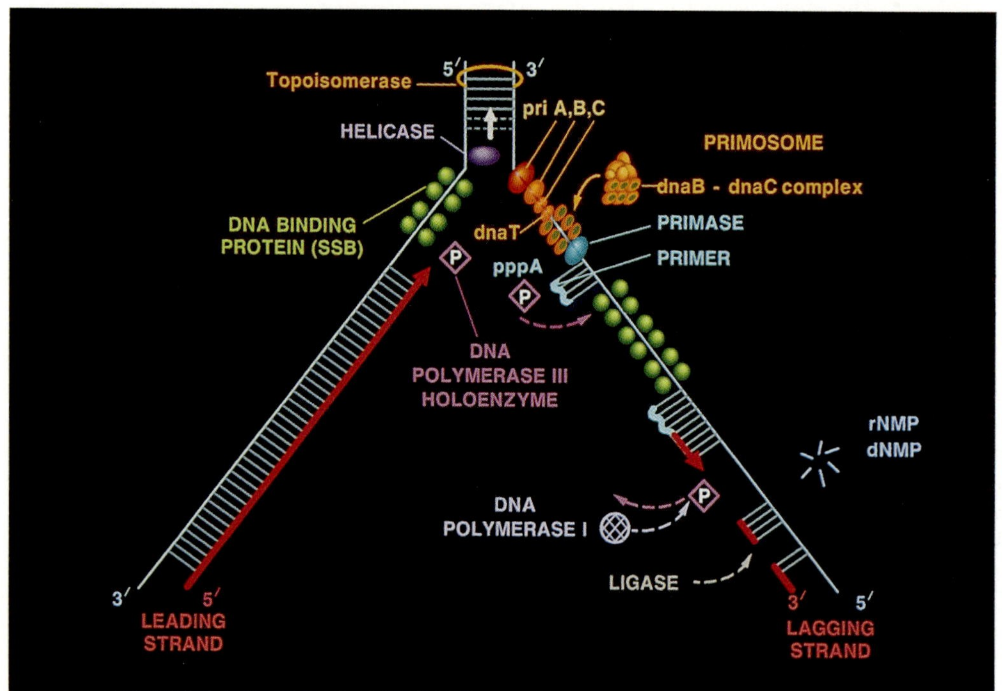

Plate 14
DNA replication fork. (Section 15-6)

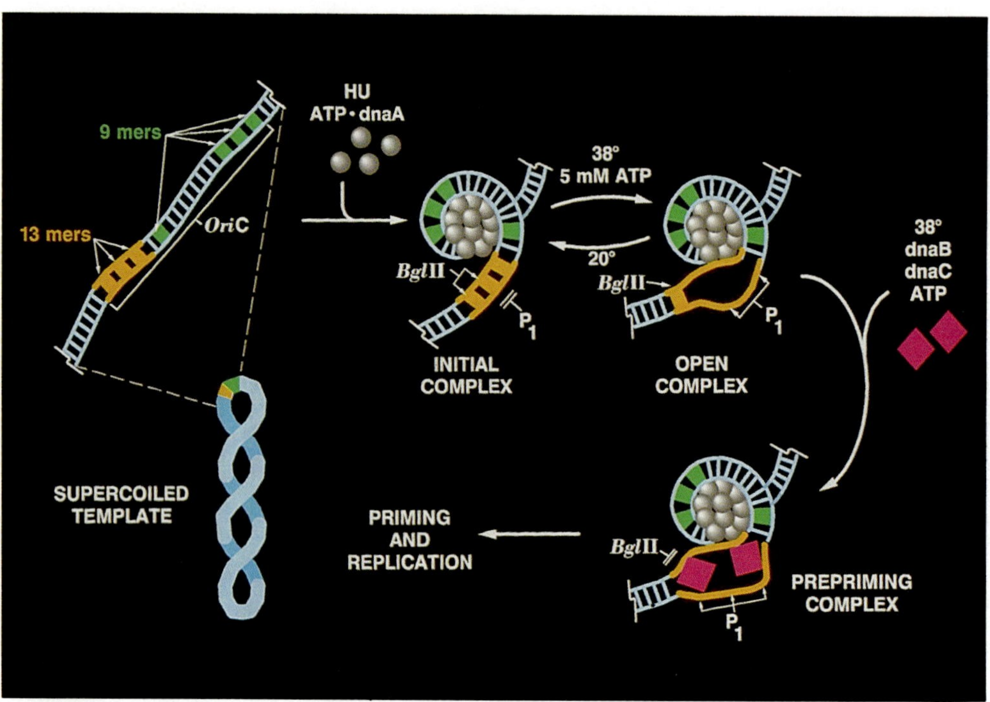

Plate 15
Mechanism of initiation of DNA replication at the *E. coli* origin. (Section 16-3)

CHAPTER 1

DNA Structure and Function

1-1 DNA: Past and Present[1]

1869–1943: The Discovery

With the discovery of a new organic phosphate compound in cells rich in nuclear substance, a new era in biology was born. This compound, at first called nuclein and later chromatin, was subsequently shown to consist of deoxyribonucleic acid (DNA) and protein. Other discoveries followed. Analysis of DNA showed it to contain four kinds of building blocks, called nucleotides. DNA was distinguished from ribonucleic acid (RNA) in having a different sugar, deoxyribose, in place of ribose, and a distinctive base, thymine, in place of uracil.

Although there was some reason to believe that DNA might be the genetic material, there was even more reason to assign this role to proteins. Each molecule of DNA was thought to be a repeating polymer of one kind of tetranucleotide unit. Proteins, since they are larger and are composed of twenty different amino acids, were thought more suitable for a genetic role. One must also recall that in the 1930s DNA was still called "thymus nucleic acid," and it was widely believed to occur only in animal cells. RNA had been isolated only from plant cells and was called "yeast nucleic acid." In fact, plant and animal cells were sometimes distinguished on the basis of this chemical feature.

1. Fruton JS (1972) *Molecules and Life*. Wiley, New York; McElroy WD, Glass B (eds.) (1957) *The Chemical Basis of Heredity*. Johns Hopkins Press, Baltimore; Mirsky AE (1968) *Sci Amer 218*(6):78; Olby R (1974) *The Path to the Double Helix*. Univ. Washington Press, Seattle.

1944 – 1960: The Genetic Substance

This "golden age"[2] began with the first important evidence that DNA is the genetic substance: the discovery, reported in 1944, that DNA prepared from one strain of pneumococcus could "transform" another strain.[3] The purified DNA carried a genetic message that could be assimilated and expressed by cells of another strain. DNA was further recognized to be a molecule far larger and more complex than a repeating tetranucleotide, varying in composition from organism to organism.[4] Yet, since traces of protein could not be ruled out as contaminants in the transforming DNA, doubt remained in some circles that DNA was the genetic material, and these findings had relatively little impact on geneticists.

Two persuasive discoveries were eventually made. The first was the demonstration in 1952 that infection of *Escherichia coli* by T2 bacteriophages involved injection of the DNA of the virus into the host cell.[5] The viral protein structures appeared to serve merely to inject the DNA into the bacterium and then to be largely discarded outside the cell. The DNA from the virus thus directed the bacterial cell to produce many identical copies of the infecting virus. This experiment dramatized the role of DNA as the carrier of information for producing the unique proteins of the virus and for duplicating its DNA many times over.

A second, remarkable event was the discovery by Watson and Crick, in 1953, of the complementary, double-stranded (duplex) structure of DNA and with it the recognition of how the molecule can be replicated.[6] Complementary pairing of the nucleotide constituents of one strand to those of the second strand was postulated to explain in a simple way how one DNA duplex can direct the assembly of two molecules identical to itself. In this model, each strand of the duplex serves as a template upon which the complementary strand is made.

These discoveries and other important ones that followed led to the realization that DNA has two major and discrete functions. One is to carry the genetic information that brings about the specific phenotype of the cell. DNA is transcribed into RNA, and the RNA is then translated into the amino acid language of the proteins. In the "central dogma" of molecular biology, information is transferred from nucleic acids to protein and never in the reverse direction.[7] The other major function of DNA is its own replication. For duplicating the genotype of the cell, DNA serves as a template for converting one chromosome into two identical chromosomes. An enzyme, named DNA polymerase,[8] was discovered to have the unprecedented property of taking instructions from this template and copying it by assembling activated nucleotides into long stretches of DNA, virtually error-free.

1960 – 1973: Consolidation

The beginning of this age was not marked by a specific event. It was an age in which the generally held conceptions of both the structure and the dual func-

---

2. Stent G (1969) *The Coming of the Golden Age*. Natural History Press, Garden City, New York; Stent G (1978) *Paradoxes of Progress*. WH Freeman, San Francisco.
3. Avery OT, MacLeod CM, McCarty M (1944) *J Exp Med 79*:137; Hotchkiss RD (1957) in *The Chemical Basis of Heredity* (McElroy WD, Glass B, eds.). Johns Hopkins Press, Baltimore, p. 321.
4. Chargaff E (1950) *Experientia 6*:201.
5. Hershey AD (1953) *CSHS 18*:135.
6. Watson JD, Crick FHC (1953) *Nature 171*:737.
7. Crick FHC (1970) *Nature 227*:561.
8. Kornberg A (1960) *Science 131*:1503.

tions of DNA were expanded. Without epochal discoveries, this age nevertheless brought a radical change in viewpoint toward DNA. Genetics and DNA became a branch of chemistry. Despite its chemical complexity, DNA was modified, dissected, analyzed, and synthesized in the test tube. There were insights into a metabolic dynamism of DNA that had not been anticipated. DNA, it was found, suffers lesions and is repaired. DNA molecules exchange parts with one another. DNA molecules are specifically modified and degraded, twisted and relaxed, and transcribed in reverse from RNA as well as directly into RNA. DNA functions not only in the nucleus but also in mitochondria and chloroplasts. These insights served as a stimulus to determine the total base sequence of DNA and to resynthesize it. There was a confidence, when *DNA Synthesis* appeared (in 1974), that the metabolic gyrations of DNA in the cell could be understood in as explicit detail as those of, say, glucose or glutamate.

1974–1980: Another Golden Age

Refinements in analytic methods uncovered unanticipated complexities and subtleties in the organization, replication, and expression of DNA. Revision of earlier concepts of chromosome organization, of expression of genes, and of replication and recombination made this era one of continuing drama and promise.

The concept that one stretch of DNA represents a single gene, expressed colinearly as one protein, was shaken. One stretch of bacteriophage ϕX174 DNA is used in coding five different proteins. One gene can code for a multifunctional protein in which several enzymatic activities are contained within a single polypeptide. The genetic sequence coding for a polypeptide in higher organisms is not necessarily continuous, as previously believed. Rather, it is often interspersed among regions (introns) that are functionally silent or have yet to be heard.

These discoveries were made with novel methods for dissecting, cloning, and amplifying genomes. A 5000-nucleotide-long genome could be sequenced in a few weeks. DNA insertion elements cropped up everywhere; they facilitate a dynamic transposition of DNA between plasmids, viruses, and chromosomes, merging their identities. The intricacy of enzymatic machinery operating on DNA was proving to be awesome: 20 polypeptides were shown to be engaged in the task of copying a small, single-stranded DNA circle to make it duplex. There was an expectation, when *DNA Replication* was published (in 1980), that with a few more clues the organization and control of replication and gene expression would be understood well enough to explain how cells develop, differentiate, and die.

1980–The Present: DNA in the Spotlight, Enzymes in the Wings

The ease of isolating, analyzing, synthesizing, and rearranging DNA sequences and genes, and the ability to insert these recombinant DNAs into cells, have fueled a stampede from all quarters of biology and have also titillated chemists and industrialists. Interest in DNA transactions such as replication, repair, transposition, and viral multiplication has increased but remains predominantly focused on intact cellular systems. More attention needs to be given to isolating the functional proteins that implement the genetic messages for cellular action. These proteins are the agents that are responsible for virtually all genetic events. Still, as this second edition of *DNA Replication* will illustrate, vigorous progress in almost all areas has provided insights into the nature of DNA replication that impress all of us who observe the advancing front of DNA knowledge.

1-2 Primary Structure[9]

The two kinds of nucleic acid, namely, ribonucleic acid (RNA) and deoxyribonucleic acid (DNA), are polymers of nucleotides.

Nucleotides

A *nucleotide* (Fig. 1-1) has three components: (1) a purine or pyrimidine *base*, linked through one of its nitrogens by an N-glycosidic bond to (2) a 5-carbon

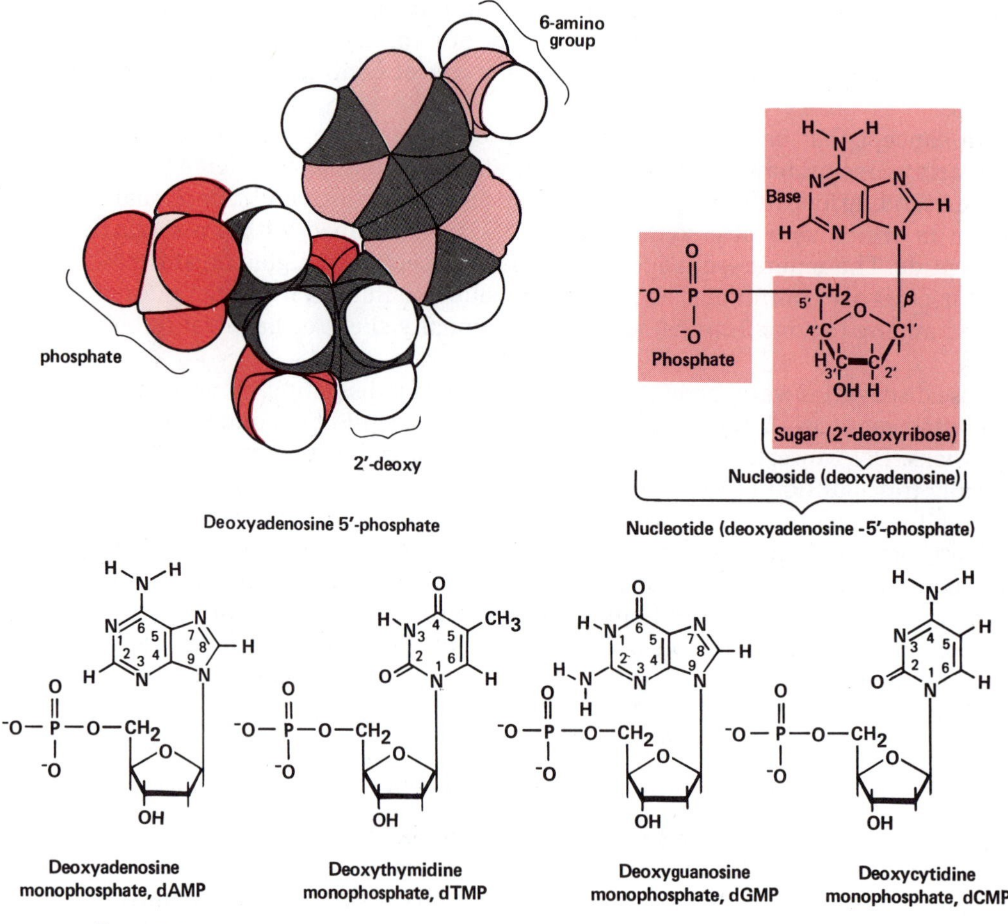

Figure 1-1
Space-filling model of dAMP, generic nucleotide structure, and deoxynucleotides of DNA.

9. Sober HA (ed.) (1970) *Handbook of Biochemistry: Selected Data for Molecular Biology*. 2nd ed. Chemical Rubber Co., Cleveland (Section G).

cyclic *sugar* (the combination of base and sugar is called a *nucleoside*), and (3) a *phosphate*, esterified to the hydroxyl on carbon 5 of the sugar. Nucleotides occur also in activated diphosphate and triphosphate forms, in which a phosphate or a pyrophosphate is linked to the nucleotide by a phosphoanhydride (pyrophosphate) bond.

In each of the two main kinds of nucleic acids there are generally four types of nucleotides. The bases and their nucleoside and nucleotide forms are listed in Table 1-1. Nucleotides are distinguished by their bases: in RNA these are adenine (A), guanine (G), uracil (U), and cytosine (C); in DNA, 5-methyluracil [thymine (T)] substitutes for uracil. Exceptions are the presence of thymine in transfer RNA and of uracil in the DNAs of certain phages, as well as aberrantly in other DNAs. RNA also contains a variety of other bases. The significance of the structures of the various bases becomes apparent when one considers the secondary structure of nucleic acids (Section 3). The truly major distinction between RNA and DNA is in their sugar–phosphate backbones: RNA contains only ribose; DNA contains only 2-deoxyribose.

Several advantages are apparent in the distinctive structure of DNA. The 2′-deoxy–containing backbone is more resistant to hydrolysis than is the ribo form (Section 22-1). Thymine ensures stability of the genetic message: chance deaminations of cytosine to uracil can be identified for repair by excision (Section 21-2); otherwise, retention of the uracil would result in mispairing and mutagenesis on subsequent replication. Finally, the 2′-deoxy also allows the sugar to pucker in a lower-energy (2′-*endo*) form, leading to a conformation of the deoxynucleotides different from that of the ribonucleotides. Proteins are thus able to distinguish DNA from RNA without interacting directly with the 2′ position (Section 8).

Table 1-1
Nucleic acid nomenclature

Base	Nucleoside[a]	Nucleotide[a]	Nucleic acid
PURINES			
Adenine	adenosine	adenylate	RNA
	deoxyadenosine	deoxyadenylate	DNA
Guanine	guanosine	guanylate	RNA
	deoxyguanosine	deoxyguanylate	DNA
{Hypoxanthine[b]	inosine	inosinate	precursor of adenylate and guanylate}
PYRIMIDINES			
Cytosine	cytidine	cytidylate	RNA
	deoxycytidine	deoxycytidylate	DNA
Thymine	thymidine or deoxythymidine	thymidylate or deoxythymidylate	DNA
Uracil	uridine	uridylate	RNA

[a] *Nucleoside* and *nucleotide* are generic terms which include both ribo and deoxyribo forms.
[b] Hypoxanthine and its derivatives are bracketed because they are precursors of the purine nucleotides but are found only in tRNA.

Nomenclature

Deoxyribonucleosides and deoxyribonucleotides are designated as *deoxynucleosides* and *deoxynucleotides*, respectively, to make the names less cumbersome. Designating deoxyribo-containing nucleosides and nucleotides with the prefix *deoxy-* or *deoxyribo-* has been accepted in all cases except for those containing thymine. Because thymine was originally thought to occur only in DNA, it seemed redundant to use the prefix, and the terms *thymidine, thymidine monophosphate,* and *thymidylate* were accepted. However, now that the natural occurrence of ribothymidylate has been recognized as one of the unusual nucleotides in transfer RNA and the thymine ribonucleotide and nucleoside have been made available by chemical and enzymatic synthesis, confusion in nomenclature does arise. In current practice, the terms *thymidine* and *deoxythymidine* are used interchangeably, as are *thymidylate* and *deoxythymidylate;* the ribonucleoside and ribonucleotides of thymine bear the prefix *ribo-*. Abbreviations for a deoxynucleoside mono-, di-, and triphosphate without a specified base are dNMP, dNDP, and dNTP, respectively. A polynucleotide sequence containing, for example, pCpGp, from left to right represents $p^{5'}C^{3'}p^{5'}G^{3'}p$. The capital letter denotes the base and the small *p* denotes the phosphate within the phosphodiester bond.

Polynucleotide Chains

Polynucleotide chains are long, almost always unbranched polymers formed by ester bonds between the 5′-phosphate (5′-P) of one nucleotide and the 3′-hydroxyl (3′-OH) of the sugar of the next (Fig. 1-2). The schematic diagram of nucleotide chains (Fig. 1-2, lower right) is frequently useful.

The important linkage is the 3′,5′-phosphodiester bridge. This linkage is especially vulnerable to hydrolytic cleavage, chemically and enzymatically. Depending on which of the phosphate ester bonds is broken, cleavage may yield either the biosynthetic intermediate 5′-phosphonucleoside or its 3′ isomer (Fig. 1-3).

Alkaline treatment hydrolyzes RNA to mononucleotides but does not affect the DNA backbone. Because of the proximity of the 2′-hydroxyl group to the phosphodiester bond in RNA, a cyclic nucleoside 2′:3′-phosphate intermediate is formed, which is then hydrolyzed to a mixture of nucleoside 2′- and 3′-phosphates. Breaks occur in mitochondrial DNA exposed to alkaline pH; these breaks have been attributed to interspersed ribonucleotides (Section 18-11). Evidence has been offered of alkali-sensitive linkers in nuclear DNA of kidney cells at a low, but significant, frequency of 1 per 150,000 base pairs.[10]

The polynucleotide chain was originally considered to be highly flexible and to assume essentially random conformations. However, detailed studies[11] indicate that the main degrees of rotational freedom are limited to the two O—P bonds in the phosphodiester linkage and to the glycosyl bond of base to sugar, and even here there are highly preferred conformations. To regard the single-strand polynucleotide chain as a completely random coil is therefore unwarranted (Section 3).

---

10. Filippidis E, Meneghini R (1977) *Nature* 269:445.
11. Levitt M (1972) *Polymerization in Biological Systems.* Ciba Foundation Symposium 7. Elsevier, New York, p. 147.

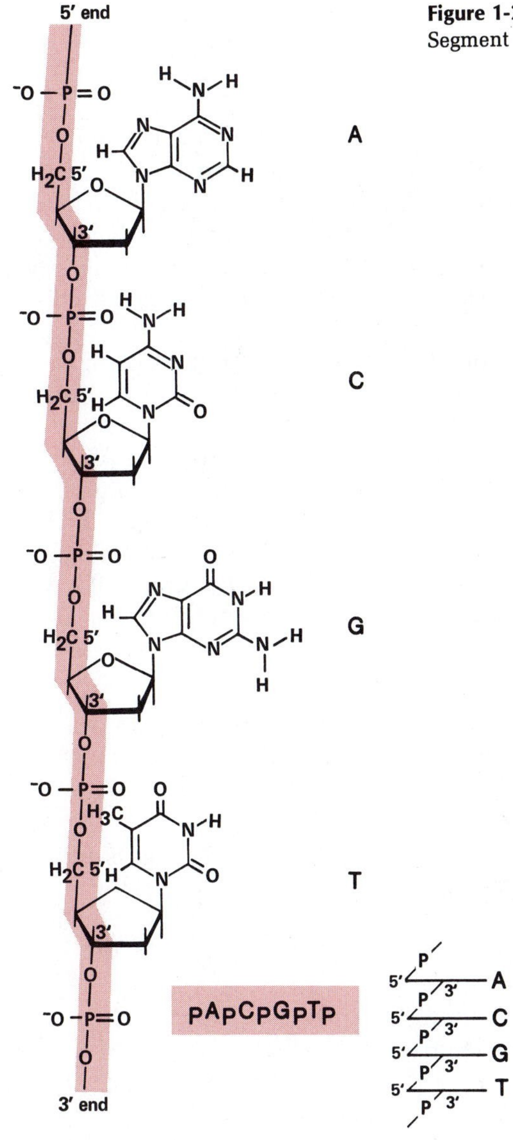

5′ end

A

C

G

T

3′ end

pApCpGpTp

Figure 1-2
Segment of a polydeoxynucleotide.

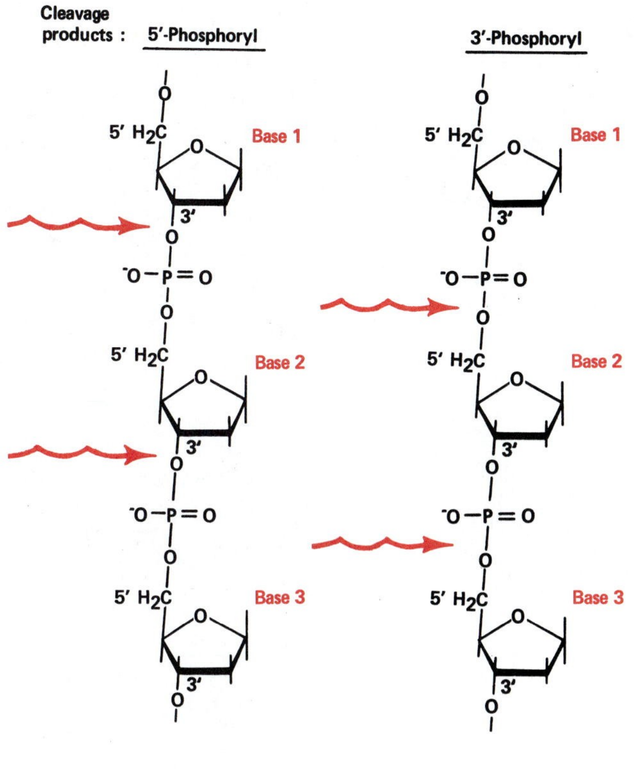

Cleavage products : 5′-Phosphoryl

3′-Phosphoryl

Base 1

Base 2

Base 3

Figure 1-3
Cleavages of polynucleotide chains yielding 5′-phosphoryl termini or 3′-phosphoryl termini.

1-3 A Double-Helical Structure[12]

Base Pairing

Two distinctions among the bases of the nucleotides are crucial to the secondary structure of nucleic acids. One rests on the presence of keto and amino groups that provide opportunities for hydrogen bonding. On this basis T or U, both keto compounds, can pair with A, an amino compound, by a hydrogen bond. G and C, each having both keto and amino groups, can form two hydrogen bonds. An additional hydrogen bond can be formed between the ring nitrogens in both the AT and GC pairs (Fig. 1-4).

The second important distinction among the bases is that they come in two sizes: the pyrimidines T, U, and C are smaller than the purines A and G. However, the *base pairs* (AT or GC) are nearly identical in size (by contrast, a pyrimidine pair would be much smaller, a purine pair much larger). The AT and GC base pairs have not only the same size but very similar dimensions. Thus, the two types of base pairs occupy the same amount of space, allowing a fairly uniform dimension throughout the DNA double helix. The AT and GC base pairs clearly differ, however, in the positions of their functional groups (O, N, and methyl groups) and in the accessibility of these groups in the major and

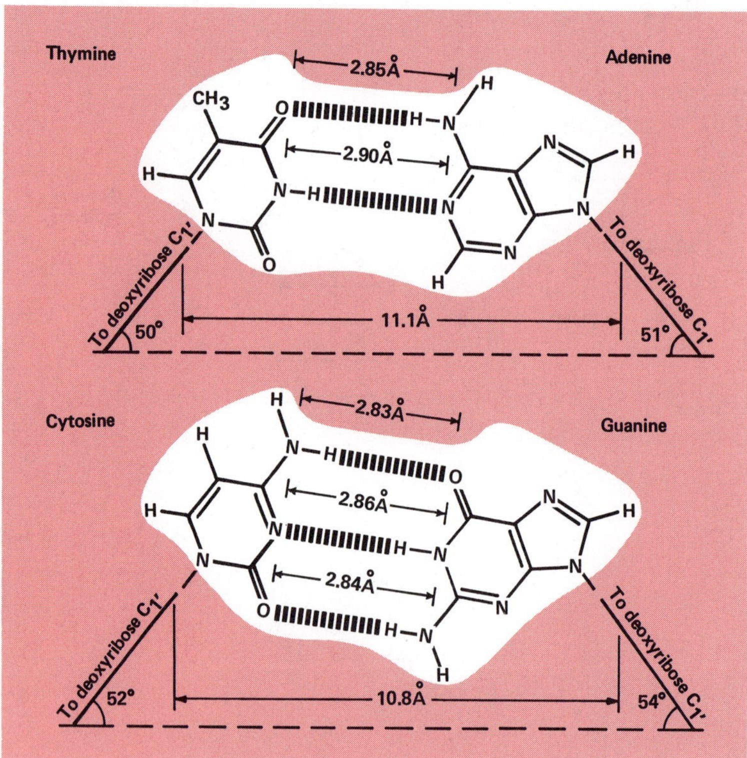

Figure 1-4
Hydrogen-bonded base pairs. Interatomic distances and angles are nearly the same for AT and GC pairs.

12. Watson JD, Crick FHC (1953) *Nature 171*:737; Arnott S, Wilkins MHF, Hamilton LD, Langridge R (1965) *JMB 11*:391; Saenger W (1984) *Principles of Nucleic Acid Structure*. Springer-Verlag, New York; Dickerson RE (1987) in *Unusual DNA Structures* (Wells RD, Harvy SC, eds.). Springer-Verlag, New York, p. 287.

minor grooves of the helix. These differences provide the means by which the nucleotide sequence can be specifically recognized by proteins interacting with the DNA.

Although the Watson-Crick AT and GC base pairs (Fig. 1-4) are most commonly found in double-helical DNA, other possible pairing configurations exist. Hoogsteen base pairs, which occur in triple-stranded RNA and DNA structures (see Fig. 1-31), represent alternative conformations of AT and GC pairs. Pairing also occurs between nucleotides other than AT and GC; these "mispaired" base pairs (see Fig. 1-18)—each contains two hydrogen bonds—are not dramatically different in geometry from the normal Watson-Crick structures, and can be accommodated by the double helix.

The base-pairing characteristics of the nucleotides are responsible for the pairing of two chains to form a strongly stabilized duplex (Plates 1 and 2). When two chains align themselves in this way, they assume a double-helical structure. The most important feature of the duplex model for DNA structure is the *base-pairing complementarity* of the two strands.[13] This property accounts for both the accurate replication (Fig. 1-5) of a very long chain and the transmission of information to form proteins via transcription into RNA. Complementarity is also the basis for the exchange of DNA segments between chromosomes in several varieties of recombination.

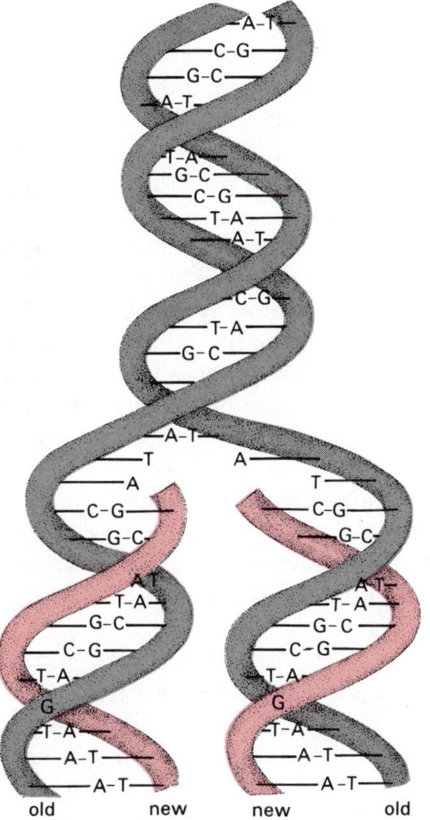

Figure 1-5
Replication of duplex DNA based on the complementarity of AT and GC base pairs. Parental strands are shaded; newly synthesized strands are colored. [After Watson JD, Crick FHC (1953) *CSHS 18*:123]

13. Watson JD, Crick FHC (1953) *Nature 171*:737.

The Double Helix

The basic structure of the now-familiar DNA duplex was determined from x-ray diffraction of DNA fibers[14] and from model building. The structures of homogeneous oligonucleotides have been determined at very high resolution (Section 8), and have revealed additional duplex forms. These sequence-dependent variations in the double helix had not been anticipated from the diffraction of DNA fibers in which the effects of the nucleotide sequences were averaged. The B form of DNA, the major form found in solution, will be described in this section; Section 8 describes the sequence-specific variations and alternative helix forms disclosed by crystallography of defined oligonucleotides.

The Standard B-Form Double Helix

This dominant form of duplex DNA, shown in Plates 1 and 2, has several remarkable features.

(1) The sugar–phosphate backbones of the two chains form a right-handed double helix with a diameter of 20 Å. It has a wide major groove and a relatively narrow minor groove.

(2) The sugar–phosphate backbones of the two chains are on the outside and are connected as in a ladder, the rungs being the bases connected by hydrogen bonds. The surface planes of the rungs are approximately perpendicular to the helix axis. The planes of neighboring base pairs are 3.4 Å apart. There are ten base pairs for each turn of the helix; thus each base pair is rotated 36° relative to its neighbor and each full turn has a length of 34 Å. These features apply to the B structure of the sodium salt of DNA found in hydrated fibers. DNA retains the B structure in physiologic solutions but with slightly different helix parameters, the number of base pairs per turn being about 10.5 rather than the 10.0 found in fibers.[15]

(3) Stacking interactions between the flat aromatic surfaces of the bases stabilize the helical structure against the electrostatic repulsive force between the negatively charged phosphate groups. This stabilizing energy may be equal to or greater than that of the hydrogen bonds connecting the bases between chains. The structure of the double helix adjusts at certain sequences to optimize the overlap between neighboring base pairs, thereby maximizing the stabilizing effects of stacking (Section 8). This adjustment is especially important for the large purine bases.

(4) The two chains are antiparallel. Chemically, they are arranged in opposite directions, with the structure . . . P—5′—sugar—3′—P . . . opposing the structure . . . P—3′—sugar—5′—P. . . . Thus, when each chain is traced from the 5′ end to the 3′ end (5′→3′), the direction is up the helix on one chain and down the helix on the other chain (Fig. 1-6).

(5) In general, the only base pairs allowed by the structure are AT and GC. [Mispaired bases do occur, however, and they perturb the geometry of the helix less than was initially thought (Section 8).] The base pairs have nearly the same shape and size, and the glycosidic bonds have the same positions and orientations in both types of pairs. This symmetry results in a double helix with a nearly uniform diameter throughout its length regardless of the base sequence.

---

14. Arnott S, Wilkins MHF, Hamilton LD, Langridge R (1965) *JMB 11*:391.
15. Wang JC (1979) *PNAS 76*:200.

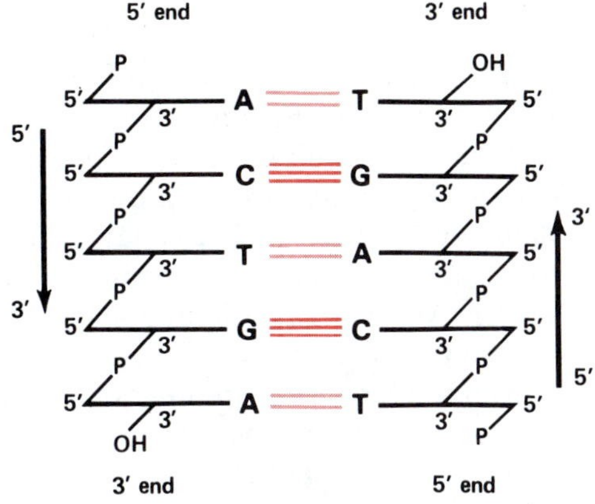

Figure 1-6
Segment of a DNA duplex showing the antiparallel orientation of the complementary chains.

Any sequence of bases is tolerated in a given chain, and DNAs of identical base composition can differ entirely in sequence. However, given the sequence and orientation of one chain, base-pairing rules and antiparallelism determine the sequence and orientation of the complementary chain. Modified bases (Section 2-13) can also enter into the double helix provided they form hydrogen bonds with a partner and stack properly with their neighbors.

(6) An important element of symmetry in the helix is a twofold (dyad) axis of rotation passing through the plane of each base pair and relating the N—C glycosidic bonds.[16] Rotation of a base pair by 180° allows the sugar and phosphate groups of the two antiparallel chains to have the same conformation. As a result, the connection between the glycosidic carbon atoms can be made by any one of the four nucleotide pairs: AT, TA, GC, and CG.

(7) Rotation about the N-glycosidic bond permits various geometric relationships of the base to the sugar. The conformations designated *anti* occur more frequently in solution than those conformations that are 180° opposite, designated *syn* (Fig. 1-7).[17] The nucleotides in B-form DNA are exclusively in the *anti* conformation.

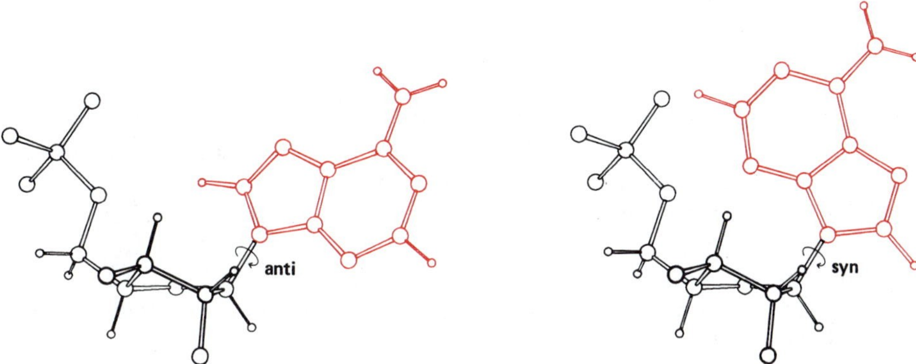

Figure 1-7
Rotation about the N—C glycosidic bond. The *anti* conformation is favored over the *syn*. (Courtesy of Professor M Sundaralingam)

16. Arnott S (1970) *Prog Biophys Mol Biol 21*:265.
17. Sundaralingam M (1969) *Biopolymers 7*:821.

Single-Stranded DNA

Although most biologically active DNAs are duplex, the genomes of several small viruses are single-stranded (Sections 17-3 and 17-4). Furthermore, characterization of replication enzymes, such as primases and polymerases, commonly involves using ssDNAs as substrates. Structurally, much less is known about ssDNA than about the duplex forms. However, ssDNA is clearly far from a random coil: base stacking, intrastrand hydrogen bonding between complementary sequences, and other types of pairing (non–Watson-Crick hydrogen bonds) are all available to the single strand. Even homopolymers, in which standard base pairing is impossible, may be helical in character at room temperature and neutral pH, indicating that rotational restrictions within the backbone and base stacking can be responsible for a helical conformation of ssDNA.

Hairpin-like regions are present in single-stranded phage DNAs, as judged by the insusceptibility of certain sequences to nucleases specific for single strands,[18] by their reactivity with cross-linking reagents specific for duplexes (e.g., psoralen),[19] and by their role as origins for replication (Sections 17-3 and 17-4). Parvovirus DNA has hairpin loops at each end, which may be important for the initiation of replication (Section 19-3). The vast conformational complexity taken on by ssRNAs, and the importance of these structural states to transcription, translation, and RNA processing, further argue that by analogy ssDNAs are ordered molecules with structures important for their function.

1-4 Melting and Reannealing

A most remarkable physical feature of the DNA helix, and one that is crucial to its functions in replication, transcription, and recombination, is the ability of the complementary chains to come apart and rejoin. Many techniques have been used to measure this melting (denaturation) and reannealing (renaturation) behavior. Nevertheless, important questions remain about the kinetics and thermodynamics of these processes and how they are influenced by other molecules in the cell and in vitro.

Melting

The chains of the DNA duplex come apart when the hydrogen bonds between the bases are broken. This melting of the secondary structure can be brought about in solution by increasing the temperature or by titrations with acid or alkali. Acid protonates the ring nitrogens of A, G, and C; alkali deprotonates ring nitrogens of G and T. Both treatments interfere with hydrogen bonding by generating charged groups close together inside the helix. The unusual lability of the purine glycosidic bonds at low pH limits the use of acid in melting procedures.

Stability of the duplex is a direct function of the number of triple-hydrogen-bonded GC base pairs it contains; the larger the mole fraction of GC pairs, the higher the temperature or pH of melting. Since DNA is usually experimentally

18. Schaller H, Voss H, Gucker S (1969) *JMB 44*:445.
19. Shen CKJ, Ikoku A, Hearst JE (1979) *JMB 127*:163.

manipulated in aqueous solution, where the potential for hydrogen bonding with water is readily available, it has been questioned why hydrogen bonds are so important to helix stability. A key reason why hydrogen bonding between the two DNA strands is more favorable than formation of hydrogen bonds between the bases and water lies in the difference in entropy of the two complexes. Once the first interstrand hydrogen bond is formed in the DNA, additional bonding does not generate any new complexes and thus has little unfavorable entropy associated with it. In contrast, association of free water molecules with the DNA strand, while allowing the same number of energetically favorable hydrogen bonds to form, is accompanied by an unfavorable decrease in entropy.

In thermal melting[20] (Fig. 1-8), unwinding of the chains begins in regions high in AT base pairs[21] and proceeds to regions of progressively higher GC content. Melting is conveniently monitored by an increase in absorbance at 260 nm (hyperchromic effect) that results from the disruption of base stacking. The midpoint melting temperature (T_m) of the DNA is commonly related to its GC content.

In addition to hydrogen bonds, other forces stabilizing the helix are base stacking, hydrophobic interactions, van der Waals forces, and neutralization of the electrostatic repulsive forces of the phosphate groups by cations. Although the primary sequence-specific effect on DNA melting is attributed to GC content, other stabilizing interactions, such as base stacking, also vary with the nucleotide sequence.[22] This variation gives rise to nonuniform melting behavior along different regions of the duplex (see Section 12).

Although melting is accomplished in vitro by changes in pH, solvent conditions, and temperature, the DNA helix is very stable in the cell. The melting that accompanies DNA replication requires enzymes specialized for this function: DNA helicases (Chapter 11) to disrupt the base pairs, topoisomerases (Chapter 12) to provide a swivel to unwind the strands, and single-strand binding proteins (SSBs) (Chapter 10) to coat and stabilize the unwound strands. Melting of the

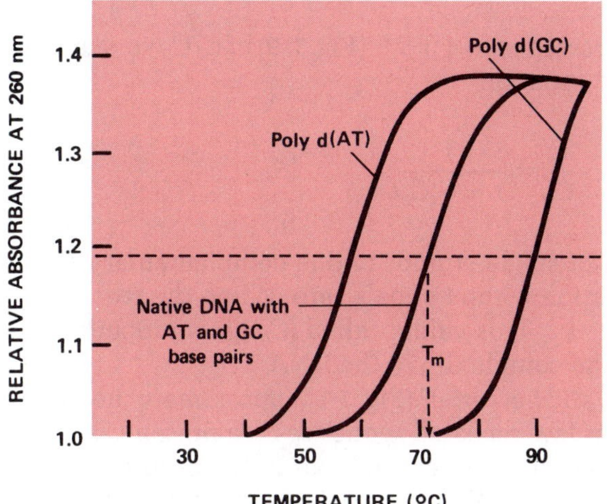

Figure 1-8
DNA melting curves. Note the dependence on GC content.

20. Marmur J, Rownd R, Schildkraut CL (1963) *Prog N A Res 1*:231.
21. Inman RB (1967) *JMB 28*:103.
22. Saenger W (1984) *Principles of Nucleic Acid Structure*. Springer-Verlag, New York.

duplex at a replication origin for initiation, and at the fork during elongation, requires an expenditure of energy, an investment justified by the functional and evolutionary benefits of maintaining the genome as duplex DNA.

Reannealing

Melting can be reversed even after the two chains have been completely separated. When complementary chains are incubated at a temperature below the T_m, they begin to reassociate and eventually reform the original helix. Reannealing is usually measured by a decrease in absorbance (hypochromic effect) or by insusceptibility to single-strand–specific nucleases.

The precision of reannealing is seen in the matching of one DNA chain of a wild-type phage λ to its complementary chain from a mutant carrying a deletion. The product looks normal in the electron microscope except for the area of the deletion, where the unmatched wild-type single strand forms a loop bridging the annealed duplex regions.[23] Such *heteroduplex* molecules make it possible to carry out refined physical mapping of the location of genes in chromosomes and introns within genes.

Measurement of Complementarity. The kinetics of reannealing provides an accurate and sensitive measure of complementarity (or diversity) in a population of DNA molecules. The two-step process begins with a slow nucleation event in which pairing of short sequences of matching bases is followed by a rapid "zippering" of the neighboring sequences to form the duplex. The rate of the overall process is governed by the second-order kinetics of the nucleation reaction. At a fixed total DNA concentration, the apparent rate constant, k_2, should be inversely proportional to the sum of the uniquely represented sequences N (in bp) in the DNA:[24]

$$k_2 \propto \frac{1}{N}$$

Rates of renaturation can be presented in a plot[25] (Fig. 1-9) of C/C_0 versus log C_0t according to the second-order equation

$$\frac{C}{C_0} = \frac{1}{1 + k_2C_0t}$$

where C is the concentration of single strands of one complementary type at time *t*, and C_0 is the concentration when *t* equals zero. When the reassociated fraction, C/C_0, is 50%, $C_0t = 1/k_2$. This value, called $(C_0t)_{1/2}$, is proportional to N and is a direct measure of the complexity of the DNA.

The "C_0t curves" for viruses and bacteria[26] (Fig. 1-9) approximate the second-order kinetics expected. When few repeated sequences are present, the value for N is commensurate with the total number of bases in the chromosome. By contrast, the reannealing kinetics of eukaryotic DNA usually shows a complex

---

23. Davis RW, Davidson N (1968) *PNAS 60*:243; Westmoreland BC, Szybalski W, Ris H (1969) *Science 163*:1343.
24. Wetmur JG, Davidson N (1968) *JMB 31*:349.
25. Britten RJ, Kohne DE (1968) *Science 161*:529.
26. Sober HA (ed.) (1970) *Handbook of Biochemistry: Selected Data for Molecular Biology.* 2nd ed. Chemical Rubber Co., Cleveland (Section H).

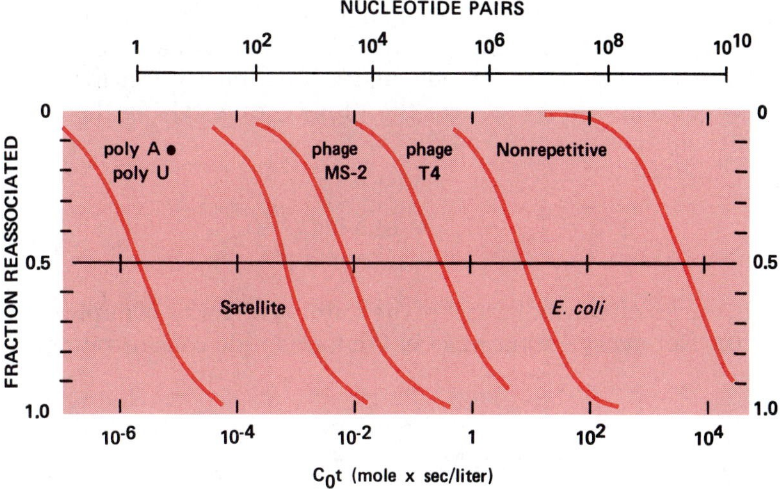

Figure 1-9
Reassociation (renaturation) of melted DNA and RNA as a function of time and concentration. The DNA was sheared into lengths of approximately 400 nucleotides before melting. Satellite DNA is from the mouse and nonrepetitive DNA is from the calf. [After Britten RJ, Kohne DE (1968) *Science 161*:529]

curve reflecting the presence of several classes of sequences. Part of the DNA, characterized by density-gradient centrifugation as a satellite of the main band (Section 5), has a simple sequence repeated 10^7 times or more and reanneals very rapidly. For most of the DNA, which is unique and reanneals slowly, the value of N is appropriately very large.

Hybridization. The reannealing of nucleic acids to identify sequences of interest is widely used. Hybridization takes advantage of the ability of a single-stranded DNA or RNA molecule to find its complement, even in the presence of large amounts of unrelated DNA. In the commonly used Southern blot technique,[27] total cellular DNA is immobilized on a support (such as nitrocellulose paper) and is then incubated with a radiolabeled nucleic acid (DNA or RNA) "probe" fragment under conditions that promote reannealing. The radiolabeled probe combines with its complement, forming a hybrid duplex that identifies the band on a gel, or the lysed colony on a petri plate, that contains the sequence of interest.

Hybridization conditions can be adjusted to different "stringencies" in order to change the proportion of complementarity in the nucleotide sequences required for efficient pairing. Relaxed conditions would be used to identify a human gene from a distantly related yeast probe, whereas the most demanding hybridization environment is required in order to distinguish cells carrying alleles of a gene differing by a single nucleotide.

Adherence to the principles of melting and reannealing is essential in generating the varying sensitivities for refined discriminations. Specific annealing is generally achieved at temperatures 12° below the T_m, which in turn depends on the GC percentage, the number of mismatches, and the solution conditions. For a perfectly base-paired fragment, at 0.2 M Na^+, the T_m can be calculated[28] as follows:

$$T_m = 69.3 + 0.41\ [(G + C)\%]$$

27. Southern E (1979) *Meth Enz 68*:152.
28. Marmur J, Doty P (1962) *JMB 5*:109.

This value is lowered by about 1° for every 1% to 1.5% difference in the sequence between the two hybridizing strands.[29] Ionic strength and formamide concentration are the two solution components most commonly used to vary the T_m. Higher salt concentrations stabilize the duplex, thus raising the T_m. In monovalent salt solutions below 0.3 M,[30]

$$\Delta T_m = T_{m2} - T_{m1} = 18.5 \log_{10} \frac{\mu_1}{\mu_2}$$

where μ_1 and μ_2 represent the molarities of the solutions. The T_m of duplex DNA is lowered by 0.7° for every 1% increase in the formamide concentration.[31]

1-5 Base Composition and Sequence

In the duplex DNA of all species, A is present in amounts equivalent to T, and G in amounts equivalent to C. The percentage content of amino bases matches that of the keto bases; purines equal pyrimidines. This striking feature of DNA base composition is explained by the double helix model of DNA, in which the complementarity of the two strands is based on the hydrogen-bonded pairing of A with T and of G with C.

The content (mole fraction, frequency) of any one of the four standard bases (A, T, G, and C) in duplex DNA establishes the content of each of the other three. Base compositions of DNAs are usually expressed by their G plus C content. In a duplex this is the mole fraction of GC pairs (i.e., GC pairs divided by the total base pairs). The rule of complementarity also dictates that the relative GC content of the duplex is equal to that of either strand.

Base composition can be determined by direct chemical analysis or by sedimentation to equilibrium in cesium chloride; the buoyant density measured is a direct function of GC content. Measurements of the stability of duplex DNA, such as midpoints of thermal melting, also correlate directly with GC content (Section 4). Individual strands with distinctive base compositions can be separated by the difference in their buoyant densities at alkaline pH.

The GC content is generally the same in all the cells of a species, but it varies considerably from one species to another, especially among bacteria, where the range is 0.3 to 0.7. The GC content in higher organisms is generally less than 0.5 (it is 0.40 in humans).

Methylation of Bases

An incompletely understood yet almost invariable aspect of base composition is the modification of certain bases, usually by methylation[32] (Sections 2-13 and 21-9). In most animal and plant DNAs, the C residues may bear a 5-methyl group, especially in the sequence CpG. The methylation pattern of CpG se-

---

29. Bonner TI, Brenner DJ, Neufeld BR, Britten RJ (1973) *JMB* 81:123.
30. Dove WF, Davidson N (1962) *JMB* 6:467.
31. Casey J, Davidson N (1977) *NAR* 4:1539.
32. Evans HH, Evans TE, Littman S (1973) *JMB* 74:563.

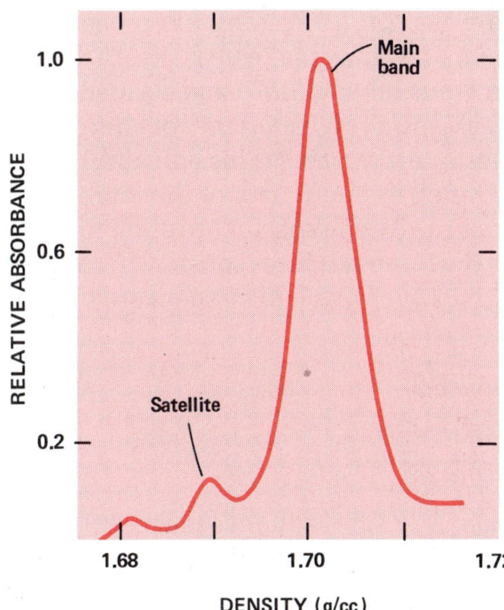

Figure 1-10
Embryonic nuclear DNA of *Drosophila melanogaster* sedimented to equilibrium in a cesium chloride density gradient.

quences has been correlated with the inactivation of genes during development.[33] There are also modifications of bases or substitutions by analogs in bacterial and phage DNAs (e.g., in phages T4, P1, and Mu) that protect the DNA against intracellular nucleases (Sections 17-7 and 17-8). In *E. coli* and related bacteria, methylation of the adenines within the sequence GATC marks the DNA for the postreplication mismatch repair enzymes. Methylation of this sequence by the *dam*-encoded methylase occurs shortly after synthesis of the new strand. This interval between replication and methylation, although brief, allows the newly synthesized and parental strands to be distinguished, ensuring that repair is directed to the unmethylated nascent sequence (Section 21-9). Methylation of GATC sites may serve additional functions in DNA replication and chromosome segregation (Sections 20-3 and 20-4).

Base composition appears to be uniform throughout prokaryotic DNA even when the genome is fragmented into several hundred 10-kb pieces. Sedimenting such fragments to equilibrium in a cesium chloride density gradient yields a single symmetrical peak. Eukaryotic DNA, by contrast, displays considerable nonuniformity in base composition, as reflected in the complex pattern of the reannealing curve of the DNA after it has been sheared, and in the asymmetry of DNA distribution in a buoyant-density gradient (Fig. 1-10). The DNA in the smaller peaks (called *satellite* bands), or in the shoulder of the main peak, may have a GC content considerably different from that in the main band.[34]

Heterogeneity of Eukaryotic Base Composition

Heterogeneity in base composition in eukaryotic DNA is due to three classes of DNA sequences: (1) highly repetitive, (2) moderately repetitive, and (3) unique.

---

33. Cedar H (1988) *Cell 53*:3.
34. Schachat FH, Hogness DS (1973) *CSHS 38*:371.

Highly Repetitive Sequences.[35] The sequences, 2 to 10 bp long, are repeated in tandem from 10^5 to 10^7 times in the genome. They may make up from 1% to 50% of the genome, averaging near 15%. An unusual satellite band formed from these sequences in several crab species is essentially a copolymer of alternating A and T, with only 3% of G and C residues interspersed.[36] This poly d(AT)-like satellite (Section 4-14) may constitute as much as one-fourth of the total DNA.

The sequences of different satellite bands in any one species are often related. For example, in one species of *Drosophila* the three satellites consist of heptanucleotides repeated some 10^7 times.[37] The sequences, as can be seen, are closely related:

5′-ACAAATT-3′

5′-ACAAACT-3′

5′-ATAAACT-3′

Satellite DNA may also have long repeats. A 172-bp sequence makes up 7% of a simian DNA.[38] One of the *Drosophila melanogaster* satellites has a 359-bp sequence repeat.[39] A bovine satellite consists of repetitive units 1400 bp in length, with each unit internally repetitious.[40]

Comparative analyses of the *Drosophila* satellite sequences, made possible by cloning and restriction nuclease techniques, have detected fine distinctions and heterogeneities. Models have been suggested for the evolution of these sequences by mutation and replication, starting from a single ancestral sequence.[41] The genetic function of the highly repeated sequences has not been determined. They are seldom, if ever, transcribed. Their well-defined locations, generally in the centromeric (heterochromatic) region of the chromosome, and the behavior of mutants with major lesions in this area suggest they may play a role in meiosis.

The ends of chromosomes (telomeric sequences) are also made up of a simple GC-rich sequence of DNA repeated 20 to 100 times (Section 15-12). The sequences presumably function to protect the genome by capping the ends of the linear chromosomes, and are synthesized by a special DNA polymerase that uses a short RNA, rather than the DNA, as a template (Section 15-12). In this way, telomeric sequences solve the inherent problem of successive shortening of the linear chromosomes during multiple rounds of replication.

Highly repetitive elements are also interspersed throughout the genome.[42] Such elements, designated *long* (6 to 7 kb) and *short* (79 to 300 bp) *interspersed repeated sequences*, are generated by reverse transcription of cellular RNAs and insertion of the resulting DNA at various chromosomal locations. The L1 family, the prototypic long repeat, is about 6.5 kb long and is present in some 20,000 copies in mammalian genomes.

The best-studied short interspersed element is the *Alu* repeat from primate DNA. The *Alu* family, composed of variants of a 281-bp sequence and present in 500,000 copies in the human genome, makes up 5% to 6% of the total DNA,

---

35. Brutlag D, Carlson M, Fry K, Hsieh TS (1978) *CSHS 42*:1137.
36. Swartz MN, Trautner TA, Kornberg A (1962) *JBC 237*:1961.
37. Gall J (1973) in *Symposium on Molecular Cytogenetics* (Hamkalo B, Papaconstantino J, eds.). Plenum Press, New York, p. 59.
38. Rosenberg H, Singer M, Rosenberg M (1978) *Science 200*:394.
39. Carlson M, Brutlag D (1977) *Cell 11*:371.
40. Botchan MR (1974) *Nature 251*:288.
41. Brutlag DL (1980) *ARG 14*:121; Singer MF (1982) *Int Rev Cytol 76*:67.
42. Weiner AM, Deininger PL, Efstratiadis A (1986) *ARB 55*:631.

occurring, on average, every 6 kb. *Alu* sequences were originally derived from 7SL RNA (a cellular RNA component of the signal recognition particle) by reverse transcription. Modern *Alu* repeats are clearly related in sequence to 7SL RNA, although modified by a duplication and a deletion. Elements related to *Alu* are found in other organisms; these include the B1 family in the mouse, similar but only half the length of *Alu*, and the *Alu*-equivalent in the Chinese hamster. Most other short interspersed repeated sequences are derived from reverse transcription of transfer RNAs rather than 7SL RNA.

These interspersed repetitive elements seem to have arisen quite recently in evolution, accounting for the relative lack of sequence divergence within the families. They serve no known function, except perhaps as raw material for natural selection.

Moderately Repetitive Sequences. These include the genes for ribosomal RNA (rRNA) and for histones, of which there are about 1000 copies per genome. The ribosomal DNA sequence is repeated many times in eukaryotic genomes, and is even present in multiple copies in prokaryotes (there are seven ribosomal DNA operons in *E. coli*), in order to achieve the high level of rRNA required to maintain production of ribosomes. The rRNA genes of *Xenopus laevis* (the South African clawed toad) are found in a satellite band in which tandem repeats of 18S and 28S RNA genes are separated by segments (spacers) of relatively high GC content.[43] Other moderately repetitive gene sequences are known to be distributed throughout the genome rather than in tandem.

Unique Sequences. Representing few or single copies of genes for enzymes and other specialized proteins, unique sequences make up 65% of the human genome and virtually all of bacterial, viral, and plasmid genomes.

Inverted Repetitive Sequences. These sequences form hairpin-like structures in denatured DNA by intrastrand folding of a single chain and are called "foldback," "snapback," or "hairpin" sequences. These sequences can be measured and isolated because of their very rapid reassociation during renaturation. As found in *Drosophila*, *Xenopus*, and human DNAs, they vary in length up to 1200 bp.[44] The ribosomal DNA of *Tetrahymena* macronuclei is part of a 20-kb inverted repeat.[45] Hairpin sequences comprise approximately 6% of the human genome, numbering about 2×10^6 per haploid set of chromosomes,[46] an abundance never encountered in prokaryotic DNA.

1-6 Size[47]

DNA is long and unbranched; its length has been measured directly in the electron microscope and by autoradiography (Table 1-2). An accurate estimate of the size of duplexes containing 40,000 base pairs has been obtained by ^{32}P-

43. Brown DD, Wensink PC, Jordan E (1972) *JMB 63*:57; Brown DD (1973) *Sci Amer 229*(2):21.
44. Cavalier-Smith T (1976) *Nature 262*:255.
45. Karrer KM, Gall JG (1976) *JMB 104*:421; Engberg J, Andersson P, Leick V, Collins J (1976) *JMB 104*:455.
46. Dott PJ, Chuang CR, Saunders GF (1976) *Biochemistry 15*:4120.
47. Sober HA (ed.) (1970) *Handbook of Biochemistry: Selected Data for Molecular Biology*. 2nd ed. Chemical Rubber Co., Cleveland (Section H).

Table 1-2
Approximate mass–length equalites of DNA

	Base pairs[a]	Length (μm)	Molecular weight
Duplex DNA[b] (Na^+, B form)	3000	1.0	2×10^6

[a] The average molecular weight of one base pair is 660.
[b] 1×10^{-12} g (one picogram) of duplex DNA contains 9.1×10^8 base pairs and is 30.9×10^4 μm long.

labeling of each of the 5′ ends.[48] Size has also been calculated from gel electrophoresis, sedimentation rate, viscosity, and light-scattering measurements.

The more complex organisms require more DNA (Table 1-3). The simplest chromosome, such as that of the polyoma or SV40 viruses, has only five or so genes in its 1.7-μm length. The DNA of larger viruses, phage λ for example, is 10 times as long; the DNA of vaccinia virus is 40 times as long. The length of DNA in more complex organisms such as bacteria or animals is correspondingly greater. The chromosome of *E. coli* is more than 1 mm long (1000 times the length of the

Table 1-3
Sizes of DNA molecules and genomes

	Size of genome		
Organism	Number of base pairs (thousands) (kb)	Total length[a] (μm)	Shape
VIRUSES			
Polyoma, SV40	5.1	1.7	circular
φX174	5.4	1.8	circular ss; ds replicative form[b]
M13 (fd, f1)	6.4	2.1	(circular ss; ds replicative form[b])
P4	11.6	3.6	linear
T7	39.9	13	linear
P22	43.2	14	linear
λ	48.6	16	linear
T2, T4, T6	166	55	linear
Vaccinia	190	63	linear
Fowlpox	280	93	linear
BACTERIA			
Mycoplasma hominis	760	260	circular
Escherichia coli	4,700	1,360	circular
EUKARYOTES			number of haploid chrosomes
Yeast (*S. cerevisiae*)	13,500	4,600	16
Drosophila	165,000	56,000	4
Human	2,900,000	990,000	23
South American lungfish	102,000,000	34,700,000	19

[a] Length = (kb) × 0.34 μm.
[b] ss = single-stranded; ds = double-stranded.

48. Weiss B, Richardson CC (1967) *JMB 23*:405.

bacterium) and contains 4.7 million base pairs. Assuming that 1000 bp is the average length of a gene (encoding a protein of about 40 kDa), that only one copy of each is present, that only one reading frame is used, and that appreciable intergenic regions are absent, then the number of genes in *E. coli* is near 4700.

Shearing stresses during isolation make it difficult to obtain and measure very long DNA strands. About 30 years ago, the size of DNA, whether isolated from phage, bacteria, or animal cells, was consistently estimated to be about 15 kb— a size delimited by the pressure of the thumb on a syringe delivering DNA through a needle (and, in the process, shearing it). When the shearing effect became known and care was taken to avoid it, the measured size of DNA shot up. Phage DNA 50 μm long was then seen in the electron microscope, and circular bacterial chromosomes 1 mm long were visualized by autoradiography. The longest stretch of animal DNA measured was also about 1 mm. However, human chromosomes contain from 48 to 240 million base pairs, and their DNA, if continuous, should measure from 1.6 to 8.32 cm.

Analysis that exploits the viscoelastic properties of DNA has revealed that the DNA in chromosomes of many animal and bacterial cells has an intact length that matches the size of the chromosome. DNA up to 4 cm long has been measured.[49] With this technique, there is remarkably good agreement between the DNA length of the largest *Drosophila* chromosome measured and the DNA mass determined cytologically (Fig. 1-11). The DNA chain runs the full length of the chromosome without discontinuity at the centromere, and the continuity is

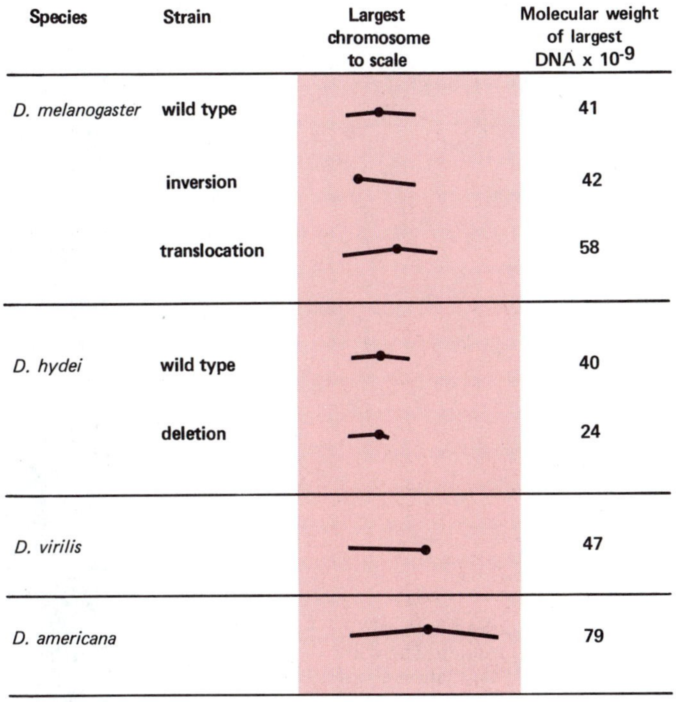

Species	Strain	Largest chromosome to scale	Molecular weight of largest DNA x 10^{-9}
D. melanogaster	wild type		41
	inversion		42
	translocation		58
D. hydei	wild type		40
	deletion		24
D. virilis			47
D. americana			79

Figure 1-11
Lengths of DNA in the largest chromosome of several *Drosophila* species, determined by viscometry. [After Kavenoff R, Zimm BH (1973) *Chromosoma 41*:1; Kavenoff R, Klotz LC, Zimm BH (1973) *CSHS 38*:1]

49. Kavenoff R, Zimm BH (1973) *Chromosoma 41*:1; Kavenoff R, Klotz LC, Zimm BH (1973) *CSHS 38*:1; Lauer GD, Roberts TM, Klotz LC (1977) *JMB 114*:507.

conserved throughout the cell cycle. These data provide the most compelling argument that in a higher organism, just as in a bacterium or virus, the *chromosome* is composed of *a single molecule of DNA.*

Conventional agarose gel electrophoresis is widely employed to measure the length of DNA molecules, but is severely limited in its ability to resolve molecules longer than 20 kb. Separation of molecules between 50 kb and 10 Mb (10 megabases; i.e., 10,000 kb) can be achieved by two improvements: (1) the avoidance of shearing forces by extracting the chromosomes from cells embedded in agarose and (2) the use of *pulsed-field gel electrophoresis.*[50] Development of this technology has had a wide-ranging impact on the ability to manipulate and analyze long pieces of DNA. Lengths approaching those defined by genetic distances (measured by recombination frequencies) can be visualized directly, bridging the gap between the physical and genetic maps of chromosomes.

Large DNA molecules (>50 kb) become distorted as they move through the pores of an agarose gel. Because long DNAs "snake" through the gel, their mobility is essentially independent of their length. In a pulsed-field gel, the direction of the applied electrical field is periodically changed, forcing the molecules to reorient repeatedly. The time between changes in the direction of the current, called the *pulse time,* determines the size range in which DNA molecules are effectively fractionated. Thus a pulse time of 0.1 second separates molecules smaller than 10 kb, whereas alternating 1-hour pulses are required for molecules 10 Mb long.

The chromosomes of yeasts and protozoa are linear molecules within the size range separable in a pulsed-field gel; the individual chromosomes form distinct bands corresponding to their molecular weights (Fig. 1-12). The circular chro-

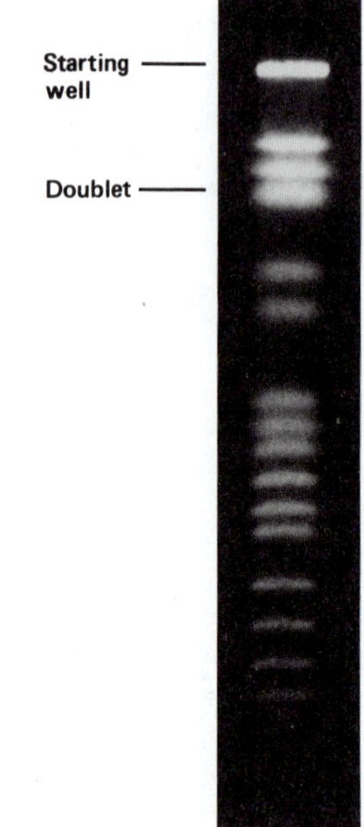

Figure 1-12
Whole chromosomes of yeast separated by pulsed-field gel electrophoresis and stained with ethidium bromide. There are 16 chromosomes; the largest is about 3 Mb, the smallest about 230 kb. (Courtesy of Dr R Davis)

---

50. Smith CL, Cantor CR (1987) *TIBS* 12:284; Cantor CR, Smith CL, Mathew MK (1988) *Ann Rev Biophys Biophys Chem* 17:287.

mosomes of bacteria can be divided into a simple pattern of linear fragments by use of a restriction endonuclease that cleaves the DNA infrequently; the lengths of these fragments add up to the total size of the genome.[51] Physical maps of even the human chromosomes can be constructed by dissecting them into fragments ranging from 100 kb to 1 Mb. DNA molecules separated by pulsed-field gel electrophoresis are visualized by staining them with ethidium bromide, or are transferred to a membrane support and probed by Southern blot hybridization (Section 4). The location of a gene on a large chromosome can thus be mapped in a single day. Furthermore, the length, location, and magnitude of genome rearrangements—insertions, deletions, translocations, and amplifications—can also be visualized directly.

1-7 Shape

The DNA in chromosomes may be linear or circular (closed-loop) in form (Fig. 1-13). The ends of linear molecules have special structural features essential for chromosome function. The image of DNA as an extended rod or a single untwisted circle stretched out on the electron microscope grid is misleading because it fails to portray the complex, contorted state of DNA condensed several thousandfold in the chromosome of a cell[52] or in the head of a virus.[53] Isolation procedures destroy or distort two salient features of the shape of DNA: the association with proteins or other complexing molecules and the path taken by DNA in its toroidal windings or supertwisted coilings (Section 9).

The complexing of DNA with proteins will be considered in Chapter 10.

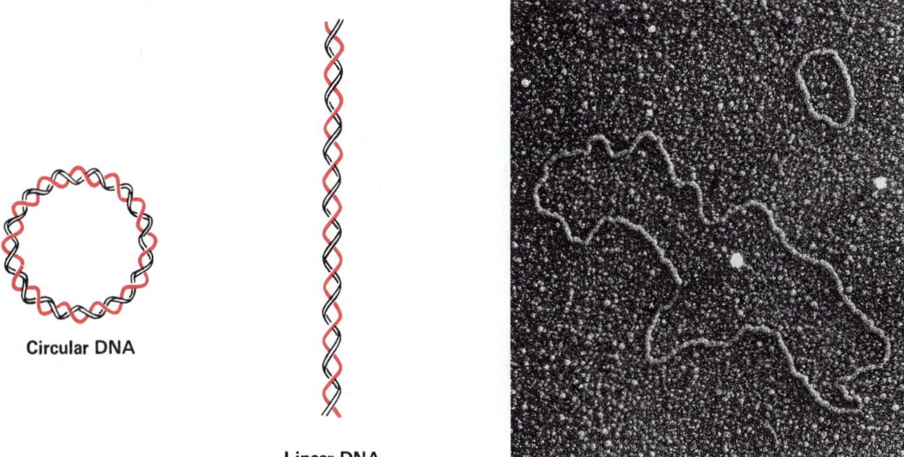

Figure 1-13
Diagrams of circular and linear forms of DNA and an electron micrograph of *E. coli* plasmids (λdv, the defective phage λ plasmid, below; the 15T-miniplasmid, above). (Micrograph courtesy of Dr T Broker)

51. Smith CL, Econome JG, Schutt A, Klco S, Cantor CR (1987) *Science 236*:1448.
52. Paulson JR, Laemmli UK (1977) *Cell 12*:817.
53. Richards KE, Williams RC, Calendar R (1973) *JMB 78*:255; Earnshaw WC, King J, Harrison SC, Eiserling FA (1978) *Cell 14*:559.

Histones profoundly influence the structure of the eukaryotic chromosome; nonhistone proteins[54] and transcriptional activity also have significant effects. The single molecule of DNA in the chromosome, complexed with these DNA-binding proteins in a chromatin fiber, is a beaded string coiled upon itself.[55] In the head of a polyhedral virus, DNA condensation, assisted by polyamines and metal ions, is achieved by tight and regular spool-like windings.[56]

Structural variations among circular genomes are found naturally and can also be generated in vitro. One of the strands of a covalently closed duplex may be nicked; such disruptions can lead to unraveling of the two strands. During the replication of duplex circular genomes, *catenanes* (interlocked rings) are formed (Fig. 1-14), and these are subsequently unlinked by a topoisomerase (Chapter 12). These forms are the normal products of recombination, as are knotted circular forms (Section 13), but are rarely present because they are untied rapidly by topoisomerases. Knots can also arise by the action of a topoisomerase (Chapter 12).

Chromosomes of multiple-genome length may be formed from single-stranded circles, duplex circles, or linear molecules. Especially striking are huge chains of phage λ or phage T7 genomes linked end to end *(concatemers)* in the infected cell. Concatemerization of λ DNA arises as a result of a rolling-circle mechanism of replication (Section 17-7). With T7, it may arise from terminally redundant ends (Section 17-5) generated by exonuclease digestion.

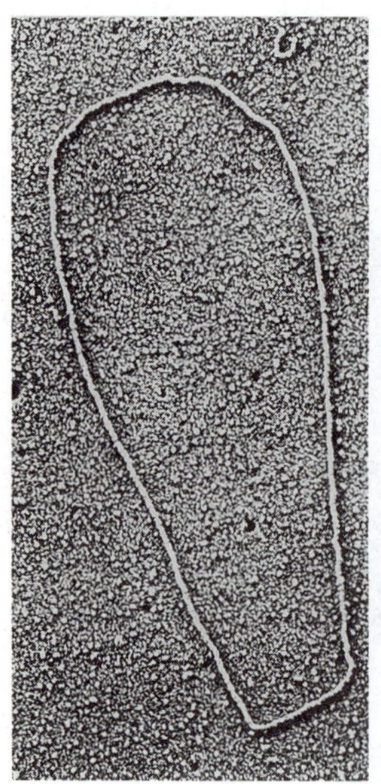

Monomer

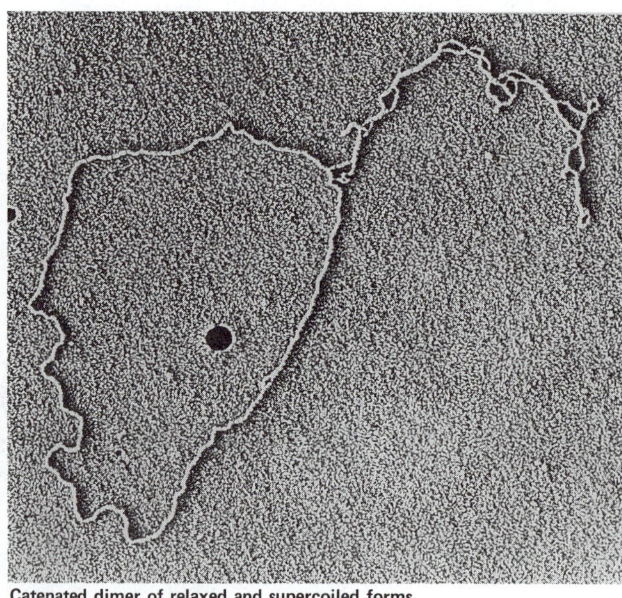

Catenated dimer of relaxed and supercoiled forms

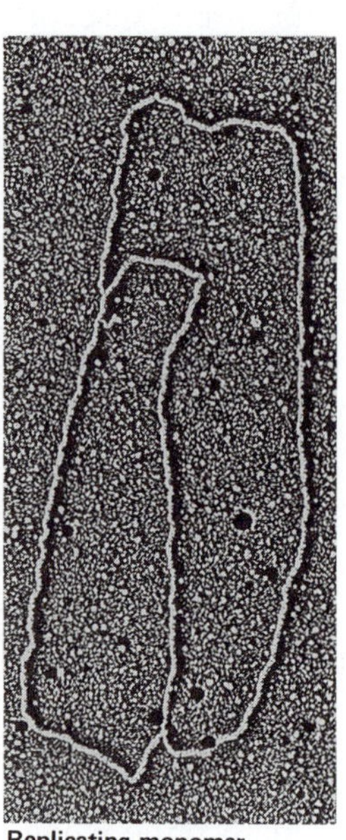

Replicating monomer

Figure 1-14
Relaxed and supercoiled forms of mitochondrial DNA from polyoma virus–transformed baby hamster kidney cells. The contour length of the monomer is about 5 μm. (Courtesy of the late Professor J Vinograd)

54. Earnshaw WC, King J, Harrison SC, Eiserling FA (1978) *Cell 14*:559.
55. Kornberg R (1977) *ARB 46*:931.
56. Earnshaw WC, King J, Harrison SC, Eiserling FA (1978) *Cell 14*:559.

With the advent of the capacity to synthesize large amounts of DNA of homogeneous sequence came the ability to observe the effects of nucleotide sequence on the geometry of the helix. In the earlier analysis of DNA fibers, sequence-specific perturbations in the structure were averaged (and hence unrecognized) due to the heterogeneous sequence of the long molecules used.

The most important newer technique, which has been widely applied, is x-ray diffraction of pure crystals. Structural information to greater than 2Å resolution has been obtained for several different sequences, and has clearly revealed the impact of the nucleotide sequence on the conformation of the helix.

A note of caution about the interpretation of crystal structures is, however, warranted. Conditions used for crystallization differ greatly from the dilute aqueous solutions used in studying the biochemistry of nucleic acids and certainly from the conditions present inside the cell. High concentrations of DNA and unusual solvents are often required to obtain crystals. As a result, the conformation of the oligonucleotides in the crystals may not represent the predominant structure that the same sequence will adopt in solution. A minor structural variant may also be dramatically stabilized by the crystal packing forces.

Combined use of a variety of techniques has helped to alleviate these problems. Spectroscopic analysis, especially nuclear magnetic resonance (NMR), can be very helpful. Enzymatic cleavage and chemical modification, two techniques that can be employed in dilute solution, may be used to confirm the presence of structural features of the DNA seen by other means.

Sequence-Dependent Alterations in B-Form DNA

Identifying the structures of homogeneous oligonucleotides by x-ray diffraction simultaneously confirmed the basic B-DNA conformation and established the role of the nucleotide sequence in the configuration of the helix. The fact that the base sequence is reflected in the structure of the DNA duplex has wide application to an understanding of DNA conformational dynamics (such as bending and melting—important to initiation of DNA replication and transcription). It also provides a chemical basis for the recognition of specific sequences by proteins.

Mechanisms of Sequence-Dependent Modifications. The basic geometry of the B-DNA helix has been outlined in Section 3. Three general mechanisms appear to determine when and how the nucleotide sequence modifies this structure. The double helix may distort in order to (1) maximize favorable base-stacking interactions, (2) minimize unfavorable clashes between the functional groups of neighboring nucleotides, and (3) form cross-strand hydrogen bonds, in addition to those of the standard Watson-Crick base pairs.

By enlarging the overlap of a base with the ones above and below it in the same polynucleotide chain, an increase in *propeller twist* (the twist of the two bases in a pair with respect to one another; Fig. 1-15) improves favorable base-stacking interactions and hence increases the stability of the helix. Base stacking is

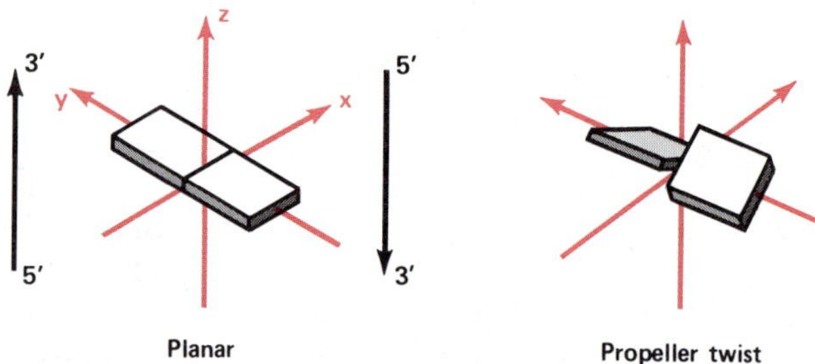

Figure 1-15
Schematic of propeller twist in a base pair. On the left, the two bases in the pair lie in the same plane; this planar form is the usual structure. On the right, a propeller twist has shifted each base out of the plane. [From Dickerson RE (1989) *NAR 17*:1797]

especially important for the large purine residues. In certain sequence contexts, propeller twist angles of 5° to 19° have been observed, compared to the average value of 2.1° derived from the x-ray fiber structure.

While maximizing base stacking, propeller twist may also introduce unfavorable interactions between neighboring bases, such as the clash between the amino groups of adjacent purines in opposite strands, which try to occupy the same space. *Purine clash* is especially pronounced in the minor groove at 5′-purine–pyrimidine-3′ steps, and must be avoided by either a decrease in the propeller twist or by another compensatory alteration in the helix structure. Increasing the *roll angles* of the base pairs to open a minor groove where clashing occurs or *sliding* the base pair out of the helix stack by altering the torsional angles of the sugars (Fig. 1-16) can avoid unfavorable clash. Finally, decreasing the *local twist angle* between base pairs can minimize unfavorable base contacts; substantial decreases in twist to angles as small as 28°, from the average of 36° per base-pair step, are observed in the helix in some sequence contexts. Other steps have twist angles larger than average, up to 42°.

Runs of AT Base Pairs. Structures of 12-bp oligonucleotides in which the internal base pairs are all AT and TA pairs [for example, d(CGCAAAAAAGCG),[57] d(CGCAAAAATGCG),[58] and d(CGCAAATTTGCG)[59]]* reveal additional sequence-specific alterations and explain some of the odd properties of the DNA homopolymer, poly dA·poly dT. Poly dA·poly dT is so resistant to bending that it cannot wrap around histones to form a nucleosome. Runs of A exceeding 10 bp are generally found only at the ends of natural nucleosomes, where bending of the DNA is at a minimum. Furthermore, runs of A, phased with the helix repeat,

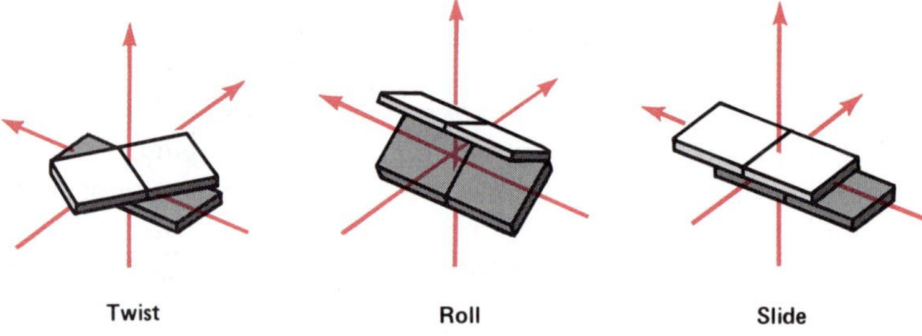

Figure 1-16
Schematic of twist, roll, and slide of two successive base pairs. [From Dickerson RE (1989) *NAR 17*:1797]

* Duplex DNAs were used but the sequence of only the top strand is shown. It is written 5′→3′ and the *d* denotes that all residues are deoxynucleotides.

57. Nelson HC, Finch JT, Luisi BF, Klug A (1987) *Nature 330*:221.
58. DiGabriele AD, Sanderson MR, Steitz TA (1989) *PNAS 86*:1816.
59. Coll M, Frederick CA, Wang AH-J, Rich A (1987) *PNAS 84*:8385.

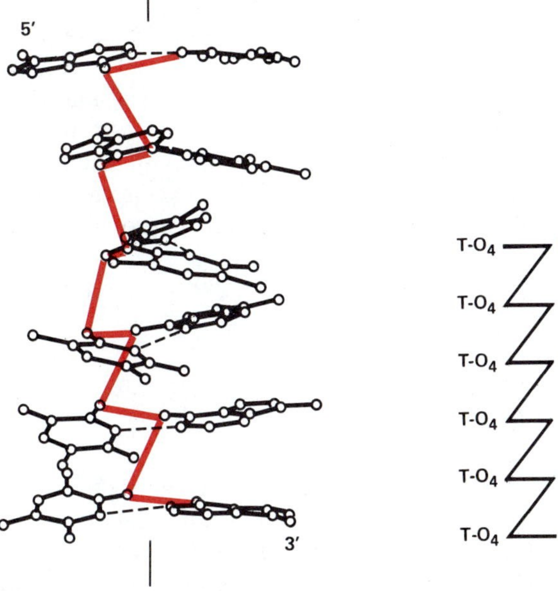

Figure 1-17
Structure of AT base pairs showing bifurcated hydrogen bonding (colored) and standard ring nitrogen–hydrogen bonds (dotted lines). [From Nelson HCM, Finch JT, Luisi BF, Klug A (1987) *Nature 330*:221]

introduce stable bending and are often found in regions identified with regulation of transcription and origins of replication (Section 10).

The large propeller twist (about 20°) of an AT base pair that is flanked by AT base pairs maximizes purine–purine stacking, thereby stabilizing the helix. This large angle between the base-paired A and T allows the formation of additional, forked (bifurcated) hydrogen bonds, linking each A not only with the T in the same base pair, but also with a T in an adjacent base pair (Fig. 1-17). (The N6 of adenine forms bonds with the O4 of each of two thymines.) This extra connection between the AT base pairs could explain the unusual rigidity of DNA helices containing runs of As.

Studies of poly dA · poly dT in solution,[60] employing enzymatic probes, Raman spectroscopy,[61] and NMR, argue for a structure of AT base pairs similar to that in the crystal. Poly dA · poly dT has a helical repeat of 10.0, instead of about 10.5 seen for most DNAs, consistent with the increased helical twist and decreased rise per base pair seen in crystals.

Runs of GC Base Pairs. Structures containing runs of GC base pairs, such as d(GGGGCCCC), d(GGATGGGAG), and d(GCCCGGGC), illustrate further how a favorable purine–purine stacking contributes to sequence-specific modifications of conformation. The GC base pairs are displaced (by sliding) from the central helix axis, creating an inner core similar to that in an A-form helix (see below). This similarity between the structure of duplex DNA containing runs of GC base pairs and the structure of duplex RNA and RNA–DNA hybrids (which form exclusively A-form helices) may explain the capacity of some proteins to bind DNA, RNA, and hybrid helices.

---

60. Poltev VI, Teplukhin AV (1987) *Mol Biol USSR 21*:102.
61. Patapoff TW, Thomas GA, Wang Y, Peticolas WL (1988) *Biopolymers 27*:493; Taillandier E, Ridoux J-P, Liquier J, Leupin W, Denny WA, Wang Y, Thomas GA, Peticolas WL (1987) *Biochemistry 26*:3361.

B-DNA Helices Containing Mismatched Base Pairs

Crystal structures of oligonucleotides containing GT, AC, and GA base pairs reveal that mispaired bases are surprisingly well accommodated in the B-form helix. Deformations in the conformational parameters (helical twist and rise; base-pair tilt, roll, and slide) are all within the same range as the alterations caused by interactions between standard base pairs (see above). Each mismatched base pair forms two hydrogen bonds and, with the exception of one conformation of GA, has a width (distance between the C1′ carbons) within 0.5Å of the width of the normal Watson-Crick base pairs (Fig. 1-18).

The very large GA base pair is accommodated by the helix in two different ways. In the sequence

5′CGC**G**AATT**A**GCG
3′GCG**A**TTAA**G**CGC

the As in the two GA base pairs adopt a *syn* conformation of the sugar—base bond, while the Gs remain in the standard *anti* orientation. The resulting G(*anti*)–A(*syn*) base pair is only 10.7Å wide, very close to the 10.8Å of a GC base pair, and much smaller than the 12.5Å of a G(*anti*)–A(*anti*) pair (Fig. 1-18). The *syn* conformation, previously seen only in Z-DNA, has surprisingly little effect on either the local or global helix structure and does not change the helix sense to left-handed, as observed in Z-DNA (see below). Of course, the positions of the functional groups on the adenine are substantially altered with respect to the rest of the helix, perhaps providing a means for recognition of this mismatched pair by the enzymes responsible for its repair.

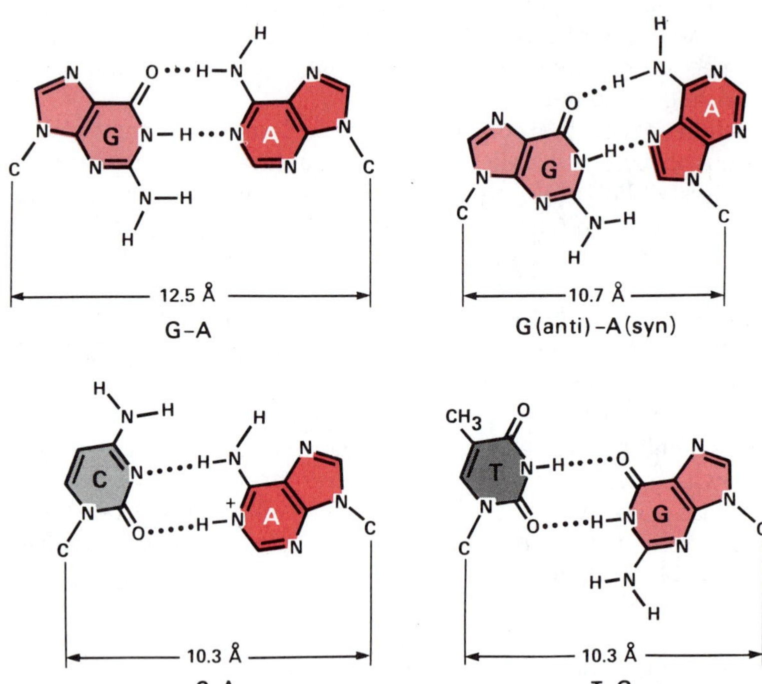

Figure 1-18
Hydrogen-bonding patterns and dimensions of mismatched base pairs. Compare with standard Watson-Crick base pairs in Fig. 1-4.

In the structure of

5′CCAA**GA**TTGG
3′GGTT**AG**AACC

both GA mismatches are in the wider G(*anti*)–A(*anti*) conformation, accommodated by a widening of the minor groove. The horizontal interphosphate distances adjust gradually over the central four base pairs. In addition to the two hydrogen bonds between G and A (Fig. 1-18), a high (i.e., wide-angle) propeller twist makes possible hydrogen bonding between the N2 of G and the O2 of the T in the adjacent AT base pair, stabilizing the structure.

A B-form helix containing unpaired bases has also been studied by crystallography.[62] The sequence

```
        A
       / \
d (C-G-C   G-A-A-T-T-C-G-C-G)
d (G-C-G-C-T-T-A-A-G   C-G-C)
                    \ /
                     A
```

contains two As that are unable to base-pair and are "looped out" of the B-form helix established by the rest of the molecule. Further evidence that such looped-out structures exist[63] in similar sequences is provided by NMR. However, extra bases can also stack into the helix by interacting with their neighbors, to yield a substantially different secondary structure near the insertion.

The accommodation of mismatched base pairs in the double helix, and the conformational changes they generate, clearly depend on the sequence environment in which they occur. Understanding the structure of mismatches in the helix context is needed for a proper understanding of the enzymology of proof reading during replication, postreplication mismatch repair, and recombination.

Non–B-Form Helices

In addition to B-DNA, two other families of DNA structure are recognized: the A and Z forms. These forms have also been crystallized and subjected to x-ray structure determinations (Fig. 1-19). Conversion of B-DNA to the A and Z forms usually attends a lowering of water activity—for example, by addition of ethanol, by a low humidity, or by a high salt concentration—yet some DNAs adopt the A and Z structures under physiologic conditions.

A-DNA. The A structure is induced in DNA in 70% to 75% ethanol,[64] is found in fibers of DNA in a dehydrated state, and is the principal conformation of RNA under all conditions. Polyribonucleotides, whether in an RNA–RNA duplex or a DNA–RNA heteroduplex, are unable to adopt the B conformation due to steric hindrance between the ribose 2′-hydroxyl group and the phosphate group of the adjacent nucleotide.

---

62. Joshua-Tor L, Rabinovich D, Hope H, Frolow F, Appella E, Sussman JL (1988) *Nature 334*:82.
63. Patel DJ, Shapiro L, Hare D (1987) *Q Rev Biophys 20*:5; Morden KM, Chu YG, Martin FH, Tinoco I (1983) *Biochemistry 22*:5557; Miller M, Harrison RW, Wlodawer A, Appella E, Sussman JL (1988) *Nature 334*:85.
64. Zimmerman SB, Pheiffer BH (1979) *JMB 135*:1023; Gray DM, Edmondson SP, Lang D, Vaughan M (1979) *NAR 6*:2089.

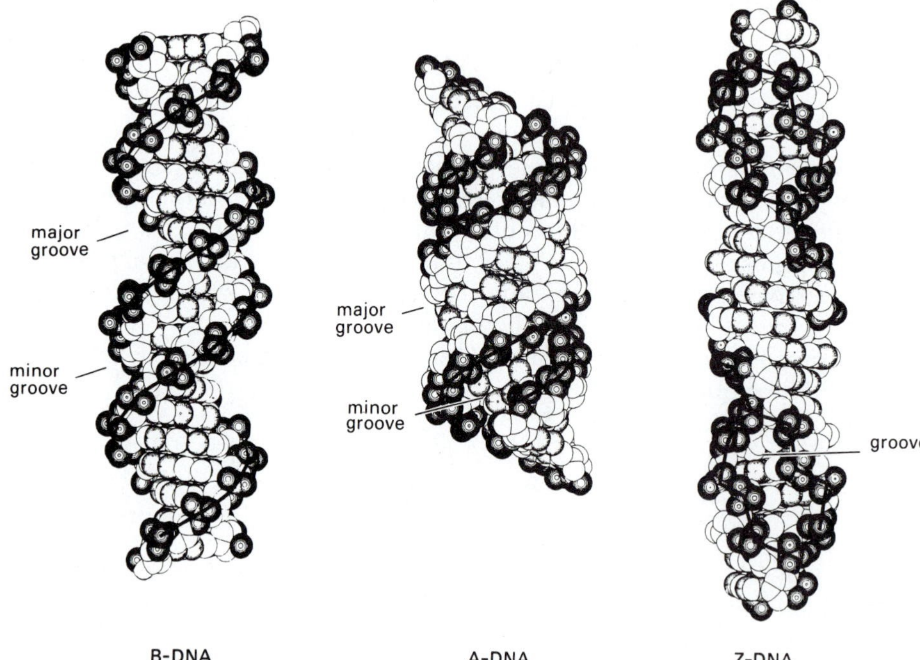

Figure 1-19
Structure of B-DNA (left), A-DNA (center), and Z-DNA (right).

The primary difference between A and B helices lies in the sugar ring conformation (pucker). The sugars are $C_{3'}$-*endo* in the A structure but $C_{2'}$-*endo* in the B form (Fig. 1-20). This altered sugar pucker shortens the distance between adjacent phosphates on one strand by about 1Å. Thus, A-DNA has between 11 and 12 base pairs per helix repeat, rather than the 10.5 typical of B-DNA.

The second major difference in A-DNA is that the base pairs are displaced from the central helix axis, toward the major groove. In B-DNA, the base pairs are essentially centered over the helix axis. In the A structure, by contrast, they are shifted (by slide; Fig. 1-16) nearly 5Å from the center, resulting in a ribbon-shaped helix with a cylindrical open core and a very deep but narrow major groove (Fig. 1-19). The shape of the DNA grooves and the exposure of the functional groups of the bases within these grooves undoubtedly render A and B forms clearly distinguishable by proteins interacting with the DNA. Base stacking, both intra- and interstrand, is the key element in stabilizing the A helix.

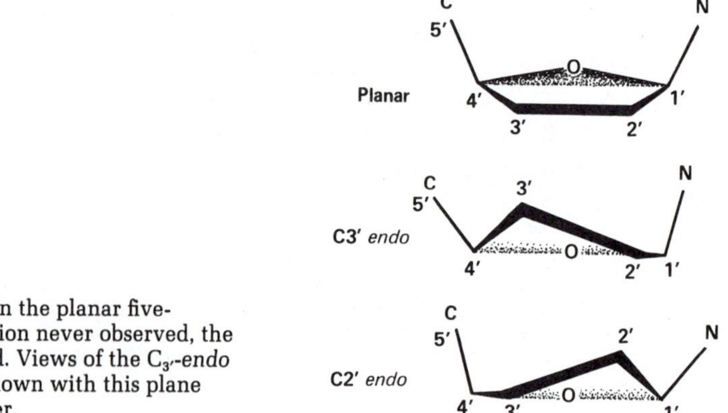

Figure 1-20
Sugar puckering modes. In the planar five-membered sugar, a situation never observed, the plane $C_{1'}$-$O_{4'}$-$C_{4'}$ is shaded. Views of the $C_{3'}$-*endo* and $C_{2'}$-*endo* forms are shown with this plane perpendicular to the paper.

A-DNA is quite uniform geometrically; the constraints of the A helix preclude the sequence-dependent structural diversity seen in B-DNA.

Z-DNA. Z-DNA, a new family of DNA structures, was so named because of the zigzag course of the backbone. The Z structure (Fig. 1-19) was found by x-ray crystallographic analysis of the oligonucleotides d(CGCGCG)[65] and d(CGCG).[66] Discovery of the left-handed Z-DNA sparked renewed interest in DNA structure and led to an appreciation of the influence of DNA sequence on helix conformation.

Z-DNA differs from the A and B forms not only in its left-handedness, but also in the orientation of the glycosyl bonds. The *syn* and *anti* orientations alternate, in contrast with the all-*anti* conformation in A- and B-DNA (Fig. 1-7). The pyrimidine nucleotides are in the standard *anti* conformation with a $C_{2'}$-*endo* sugar pucker, while the purine residues are *syn* and contain a $C_{3'}$-*endo* conformation. The crystal structures have a strong dinucleotide repeat unit due to the large alternation in helix twist angles between adjacent base pairs. The twist between the C and G in a CpG step is about 15°, while that between the G and C in the subsequent GpC is close to 45°.

The extent to which Z-DNA occurs in nature and its physiologic significance have not been established. The Z-DNA structure is more likely to occur in alternating purine–pyrimidine sequences, because steric repulsion makes it less favorable for a pyrimidine to adopt the *syn* conformation.[67] Moreover, a Z-compatible sequence may be forced into the B structure by a neighboring region of B-DNA. Factors that promote the Z structure include methylation,[68] bromination, specific DNA-binding proteins, and sufficient torsional stress as in negatively supercoiled DNA (Sections 11 and 12).

1-9 Supertwisting[69]

Definition

Supertwisting, supercoiling, and superhelicity are interchangeable terms that describe the twisting upon itself, or the spatial coiling, of a duplex DNA.[70] The term *super* is used because the helical coiling is over and above that of the twisting of the two strands of the DNA double helix about each other (see Fig. 1-21). When first observed, supertwisting was regarded as a special feature of small circular molecules such as viral chromosomes, bacterial plasmids, and mitochondrial DNA, and was even believed by some to be an artifact of isolation. Supertwisting is now known to be an integral and extraordinarily important feature of nearly all chromosomes, whether they be circular, as in bacteria, or linear, as in eukaryotic cells. The degree of supertwisting may be a crucial factor in stages of replication, transcription, and recombination.

---

65. Wang AH-J, Quigley GJ, Kolpak FJ, Crawford JL, van Boom JH, van der Marel G, Rich A (1979) *Nature 282*:680.
66. Drew H, Takano T, Tanaka S, Itakura K, Dickerson RE (1980) *Nature 286*:567; Drew HR, Dickerson RE (1981) *JMB 151*:535.
67. Drew H, Takano T, Tanaka S, Itakura K, Dickerson RE (1980) *Nature 286*:567.
68. Behe M, Felsenfeld G (1981) *PNAS 78*:1619; Möller A, Nordheim A, Nichols SR, Rich A (1981) *PNAS 78*:4777.
69. Crick FHC (1976) *PNAS 73*:2639; Wang JC, Peck LJ, Becherer K (1983) *CSHS 47*:93; Cozzarelli NR, Boles TC, White JH (1990) in *DNA Topology and Its Biological Effects* (Cozzarelli NR, Wang JC, eds.). CSHL, p. 139.
70. Vinograd J, Lebowitz J, Watson R (1968) *JMB 33*:173; Bauer W, Vinograd J (1970) *JMB 47*:419; Davidson N (1972) *JMB 66*:307.

The Topology of Supertwisting

For topological considerations, the DNA double helix can be represented as a twisted ribbon, with the edges of the ribbon representing the two DNA strands.[71] When the ends of the ribbon are joined, a simple ring-like structure results. In view of the antiparallel orientation of the two DNA strands, the edges of the ribbon have fixed and opposite direction. When the ends are united to form a ring, edges with the same direction must be joined together. When this circular structure lies flat on a planar surface without change in the global helix geometry, it is *relaxed* (Fig. 1-21, left). However, if, before sealing, one end of the ribbon is turned through one or more 360° revolutions in the direction of unwinding (untwisting the Watson-Crick helical turns) while the other is held fixed, the ring will no longer be relaxed, but instead will be *underwound*. This distortion of the DNA can be accommodated in several ways (Fig. 1-21):

(1) The strain may be taken up entirely within the structure of the double helix by a reduction in the number of times that one strand twists about the other. The axis of the double helix is unaffected, so that it can continue to be, on average, planar. The untwisting of the helix results in each turn of the double helix having more base pairs on average. This alteration may be smoothly

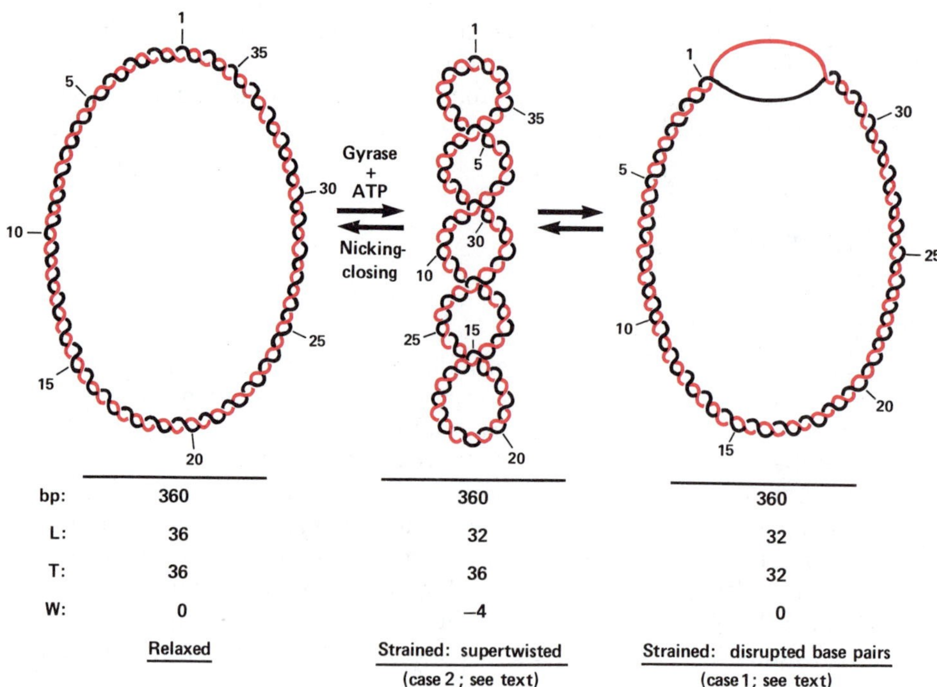

Figure 1-21
Isomeric topological states of a circular duplex DNA molecule. Conversion of the relaxed to the negatively supercoiled form is catalyzed by gyrase (Section 12-3); the reverse reaction is catalyzed by a nicking–closing enzyme (topoisomerase; Section 12-2). The strain in the supercoiled form may be taken up by supertwisting (case 2) or by local disruption of base pairing (case 1). (Adapted from a diagram supplied by Dr M Gellert)

---

71. Crick FHC (1976) *PNAS 73*:2639.

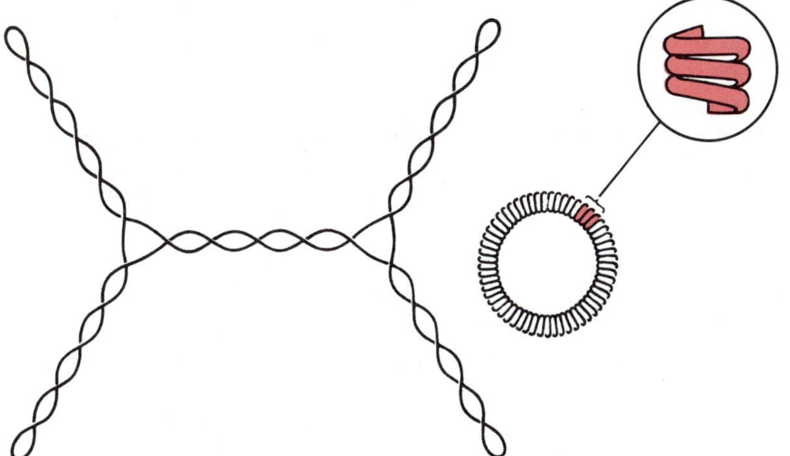

Figure 1-22
Plectonemic (left) and toroidal (right) supercoils of DNA of the same length. (Courtesy of Dr N R Cozzarelli)

distributed along the duplex molecule or concentrated in certain regions. In the extreme case, regional untwisting is complete and a stretch of ssDNA results, while the standard 10.5 bp per turn is maintained in the rest of the molecule.

(2) The strain may be taken up by *supertwisting*—twisting of the axis of the double helix upon itself. The number of base pairs per turn in the helix is little affected. However, the duplex molecule no longer lies in a plane but coils in three dimensions. When this supercoiling results from underwinding of the duplex, it is given a negative sign *(negative supertwisting)*. There are two types of supercoils, *plectonemic (intertwined)* and *toroidal* (Fig. 1-22). Plectonemic supercoils are the predominant form in negatively supercoiled DNA in solution, while toroidal coiling occurs when the DNA is wrapped around the surface of proteins, such as histones, to form nucleosomes. In both types of supercoiling a decrease in twist of the double helix is concomitant with the deformation of the helix axis: Part of the strain of underwinding is accommodated by unpairing and part by supertwisting.

(3) Underwinding of the DNA may induce the formation of an alternative structure (e.g., cruciform, triplex, or Z-DNA). This absorbs some of the invested energy, so that fewer superhelical turns accumulate (Sections 11 and 12).

If, before sealing, instead of being underwound, the duplex is given additional twists in the same direction (sense) as the helical turns, it becomes *overwound*. In this case, the average number of base pairs per helical repeat will have to be decreased to maintain a planar structure. If overwinding is taken up in superhelical turns, they will be opposite in sign to those described above, and thus the circle will be *positively supertwisted*.

Mathematical Description of Supertwisting

The reciprocal relationship between twisting of the strands of the double helix and supertwisting of the axis of the helix may be expressed in quantitative form through the following relationship:

$$Lk = Tw + Wr \tag{1}$$

where *Lk*, the *linking number*, is the total number of times one DNA strand crosses the other; *Tw*, the *twist* of one strand about the other, is the number of helical turns; and *Wr*, the *writhing number*, is the average number of times the helix axis crosses over itself when viewed from all directions—that is, the number of superhelical turns.

The linking number, an integer, is a topological property of the closed circular duplex molecule and remains unchanged, no matter what is done to the DNA, as long as the DNA backbone is not broken. *Lk can only be altered by breaking and resealing the DNA chains. Tw* and *Wr*, however, are geometric parameters and can vary freely, and they are not restricted to integral values. Molecules with the same linking number thus can have very different three-dimensional shapes depending on how the linking number is apportioned between twist and writhe. Molecules identical in sequence and length that differ only in topology are called *topoisomers.*

For relaxed DNA in the normal B form, the only intertwines in the molecule are due to the twist of the two DNA strands around one another in the helical turns. Therefore *Lk* equals the number of base pairs in the molecule divided by the number of base pairs per helical repeat (10.5 for standard B-DNA). Thus $Lk = Tw$ and $Wr = 0$ for relaxed DNA.

When strain is introduced by rotating the DNA through one or more turns before the ends are sealed to form a circular molecule (or by the action of a topoisomerase; see Chapter 12 and below), the value of *Lk* changes. The magnitude of this change, or the *linking number difference*, ΔLk, between the strained and relaxed molecules, is given by the number of turns introduced. The corresponding changes in *Tw* and *Wr* are related through equation (1) in the form

$$\Delta Lk = \Delta Tw + \Delta Wr \tag{2}$$

The magnitudes of ΔTw and ΔWr measure how the strain in the molecule is distributed between twisting of the strands of the helix and supertwisting of the helix axis. In underwinding, the strain could theoretically be taken up entirely by untwisting of the strands or entirely by supertwisting of the double helix. In reality, *a combination of changes in both helical twist and superhelical writhe accommodates the linking deficit.* When the structure of negatively supercoiled DNA in solution was analyzed, approximately 70% of ΔLk was apportioned to negative superhelical turns, while the remaining underwinding was reflected in a change in twist.[72]

Linking numbers of molecules with different lengths can be normalized to give a quantity, termed the *specific linking difference* or *superhelix density*, σ:

$$\sigma = \frac{\Delta Lk}{Lk^0} \tag{3}$$

where Lk^0 is the value of *Lk* when the DNA is relaxed. The superiority of using the superhelix density rather than *Lk* becomes clear when two molecules of different lengths are compared. *Lk*, the sum of both the superhelical and helical

72. Boles TC, White JH, Cozzarelli NR (1990) *JMB 213*:931.

turns, depends on both the length of the DNA and the topological state, whereas σ describes only the topological strain. The σ values of different-sized molecules can thus be compared directly. Values for the superhelix density for most molecules isolated from cells and virions lies between -0.05 and -0.07. The free superhelicity of DNA in *E. coli*, not constrained by proteins, may be about half this value.[73]

Methods of Measuring Superhelicity

The degree of supertwisting, its generation, and its removal can be measured in the following ways:

Agarose Gel Electrophoresis.[74] This method resolves DNA molecules on the basis of shape, a direct function of their superhelicity and a measure of *Lk* in equation (1). The technique is extraordinarily rapid, sensitive, and precise.

Circular molecules migrate in a gel at a rate proportional to their compactness. Molecules with superhelical turns migrate more rapidly than their relaxed counterparts. Thus, when a series of topoisomers are run on a gel, the relaxed molecule ($Wr = 0$) migrates most slowly; both negatively and positively supercoiled molecules migrate faster. A pattern is formed of bands that represent DNA molecules differing from each other by one unit of *Lk* (Fig. 1-23). All topoisomers between relaxed and native supercoiled can be resolved by using a series of overlapping "ladders," allowing the determination of ΔLk. Two-dimensional agarose gel electrophoresis (Section 11) increases the number of isomers that can be separated on a single gel and is thus very useful for measuring superhelicity.

Gel electrophoresis can be carried out in the presence of intercalating dyes, such as ethidium or chloroquine, to untwist the DNA. The presence of intercalating agents alters the mobility of circular molecules on gels by rendering them progressively less negatively supertwisted. Increasing the levels of intercalator relaxes the DNA and eventually generates positive supercoils. At the high concentrations of ethidium bromide routinely used during electrophoresis (0.5 μg/ml), most covalently closed circular forms are positively supercoiled and migrate together.

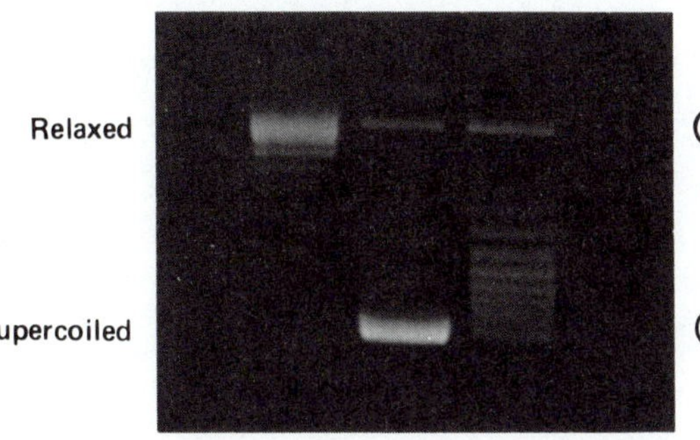

Figure 1-23
Agarose gel electrophoretic pattern of relaxed (left), supercoiled (center), and partially relaxed (right) DNA. Arrow = direction of migration. (Courtesy of Ms M O'Dea)

73. Bliska JB, Cozzarelli NR (1987) *JMB 194*:205.
74. Keller W (1975) *PNAS 72*:4876.

Sedimentation Methods.[75] The sedimentation velocity of DNA complexed with ethidium gives an estimate of all stages of supertwisting, negative and positive. Intercalation of one planar dye molecule between stacked base pairs of DNA untwists the DNA ribbon by 26°. Intercalation of ethidium changes a negatively supertwisted DNA circle into a more relaxed, open configuration; the DNA molecule is less compact and sediments less rapidly.

Binding of ethidium causes a negative ΔTw which is exactly compensated by a positive ΔWr because the value of Lk is constant (since there is no breakage and rejoining of DNA strands). When the positive ΔWr is numerically equal to the initial negative value of Wr in the negatively supercoiled molecule, $Wr = 0$ and the molecule adopts a fully relaxed configuration. The sedimentation coefficient passes through a minimum value because further binding of ethidium results in values of $Wr > 0$ and thus positive supercoiling, making the DNA again more compact and rapidly sedimenting. This response depends on the superhelical density of the DNA, rather than on Lk, and thus gives an estimate of σ.

Electrophoresis has largely replaced sedimentation of DNA as a means to measure the superhelical density. However, sedimentation is still used preparatively and for molecules that are very supercoiled, a condition in which resolving molecules by electrophoresis is difficult.

Electron Microscopy.[76] This technique clearly distinguishes the tightly supercoiled from the relaxed form (Fig. 1-24) and in a semiquantitative way identifies intermediate states.

Structural Dynamics of Supercoiled DNAs

Supertwisting occurs only when the free rotation of the ends of DNA is restrained. Such *topologically constrained* molecules include closed circles or linears with ends that are fixed or too remote to rotate rapidly. The intracellular circular DNAs of plasmids, viral replicative forms, and mitochondria are negatively supertwisted, as are the enormously larger eukaryotic chromosomes.

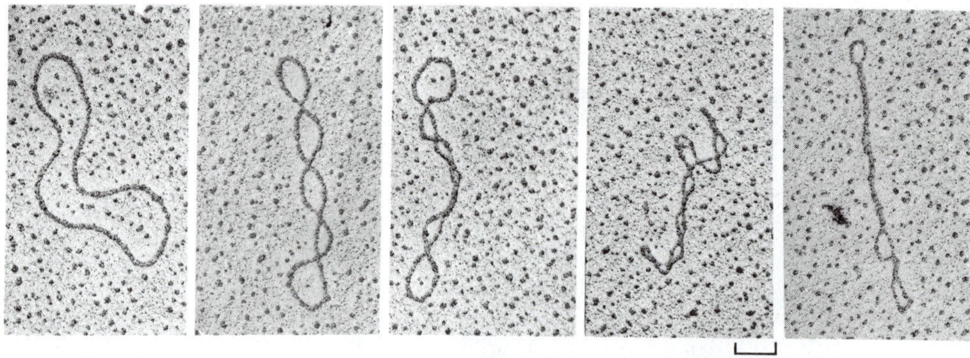

Figure 1-24
Electron micrographs of DNA (mini ColE1 plasmid dimer, 5 kb), selected to show degrees of supercoiling from fully relaxed (left) to tightly coiled (right). (Courtesy of Ms L Polder)

75. Wang JC (1974) *JMB 89*:783; Hsieh TS, Wang JC (1975) *Biochemistry 14*:527.
76. Germond JE, Hirt B, Oudet P, Gross-Bellard M, Chambon P (1975) *PNAS 72*:1843; Bellard M, Oudet P, Germond JE, Chambon P (1976) *EJB 70*:543.

How the strained state of the molecule is partitioned between twist and writhe is determined in part by interactions with DNA-binding proteins. The toroidal winding of DNA around globular histones, as occurs in chromatin (Section 10-7), accommodates negative supertwisting (−*Wr*), whereas ssDNA-binding proteins favor regions of ssDNA (−*Tw*). Topoisomerases directly change the superhelical state of DNA. Gyrase introduces negative supertwists, and relaxing enzymes remove them. The properties of these enzymes and how they work together with DNA-binding proteins, and the transcription and replication machinery to maintain the superhelical state of the DNA, are discussed further in Chapter 12.

1-10 Bending

Although long chains of DNA are quite flexible, pieces shorter than about 100 bp are relatively stiff, and the shorter they are the more energy is needed to bend them[77] (Fig. 1-25). However, certain sequences and structures in a DNA molecule generate an intrinsic bend or make the sequence susceptible to bending. Intrinsically bent molecules possess a degree and direction of curvature determined by the nucleotide sequence. "Bendable" DNA segments are inherently flexible enough that they can conform to a protein surface more easily than most sequences. Bent forms are important features in regions of DNA that control replication, transcription, and recombination. However, the molecular details defining the structural basis of bending need to be more thoroughly understood.

Each minicircle of the many thousands that make up the kinetoplast DNA (kDNA) of trypanosomes (Section 18-11) contains a small region with a prototypic bend. A 219-bp fragment of the minicircle migrates unusually slowly in polyacrylamide gel electrophoresis, appearing to be twice its actual size.[78] Such anomalous electrophoretic mobility is diagnostic of a bent DNA. Linear molecules of the same length travel through a gel at a rate determined by their

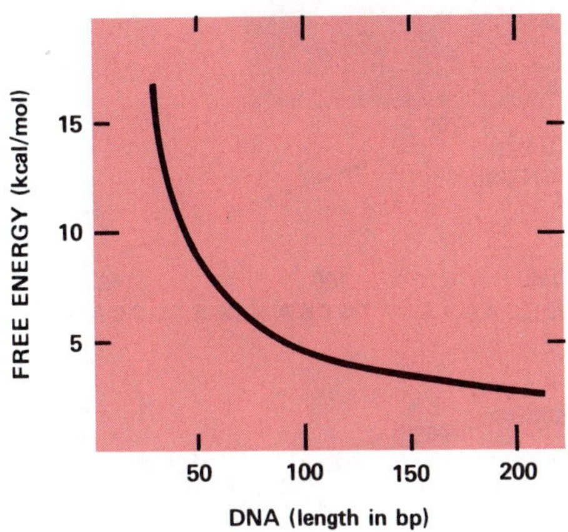

Figure 1-25
The free energy required for bending DNA segments of various lengths into a smooth semicircle.

77. Wang JC, Giaever GN (1988) *Science 240*:300.
78. Marini JC, Levene SD, Crothers DM, Englund PT (1983) *CSHS 47*:279.

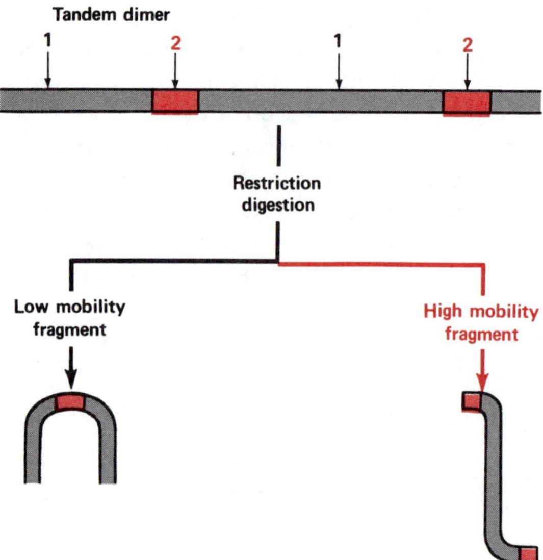

Figure 1-26
DNA bending. (Top) General scheme for identifying DNA with a bend (colored block). Cutting at site (1) produces DNA fragments with the bend in the center, while cutting at site (2) produces fragments with the bend at the ends. (Bottom) Scheme for identifying the location of a bend in a fragment of kinetoplast DNA (kDNA). [After Koo H-S, Wu H-M, Crothers DM (1986) *Nature 320*:501]

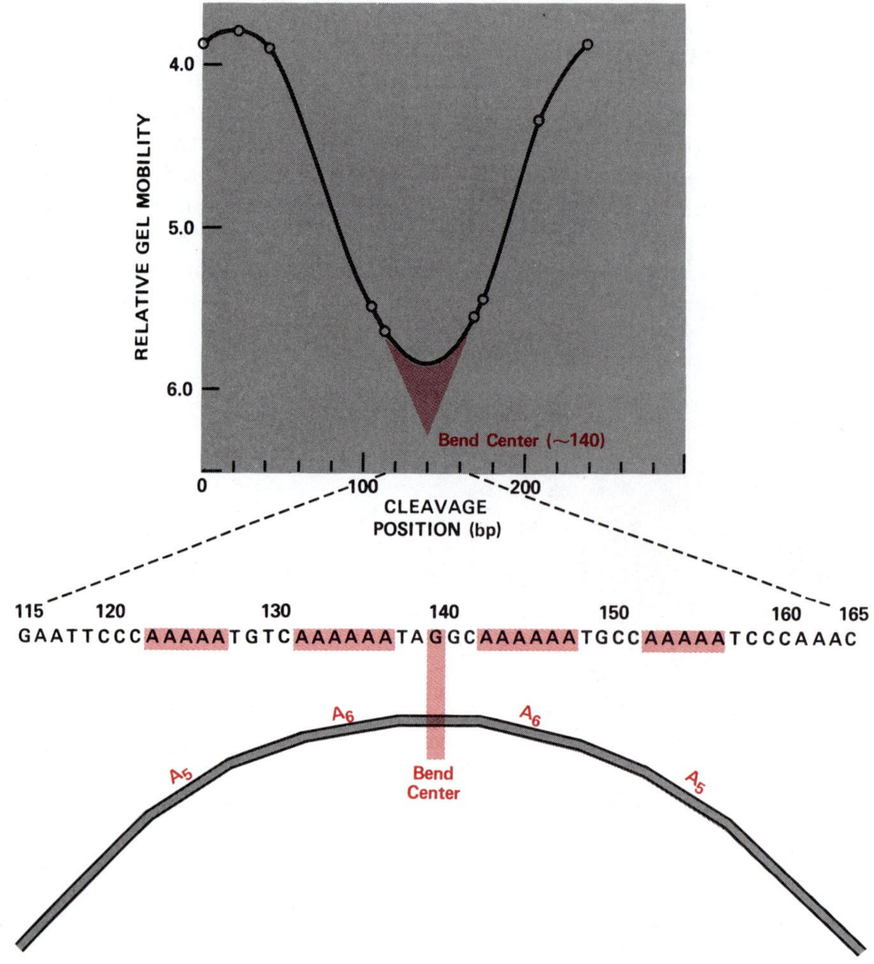

average end-to-end distance; a bend near the center of a fragment brings the ends closer together, impeding its mobility.

The locus of bending can be defined by electrophoresis of circularly permuted DNA fragments that are all identical in composition and length, but with the bend in different locations with respect to the ends[79] (Fig. 1-26). The bend distorts the electrophoretic mobility maximally when it is in the center and less so when near the ends.[80] Thus, a plot of the relative mobility of the fragments versus the position of cleavage, extrapolated to the maximum mobility, accurately locates the bend.

DNA bends are visible in the electron microscope.[81] The 219-bp bent fragment from the kDNA minicircle forms a nearly closed circle; when it is part of a longer molecule, the bend creates a tight loop (Fig. 1-27). Because a bend brings the ends of a molecule closer together, the probability of circularization by ligase is increased. The minimum length required to form a circle is therefore decreased and provides an estimate of the degree of curvature. Ligation kinetics has been used to show bending of the replication origins of the λ phage[82] and the R6K plasmid.[83]

Bending Induced by Adenine Repeats

Intrinsic bends can be caused by runs of A residues. For example, flanking the center of the kDNA bend are 18 runs of four to six A nucleotides, spaced at 5-bp intervals. To define the sequence requirements for bending, synthetic oligonucleotides have been useful.[84] Tracts of A nucleotides at least 3 bp long (A_3) are essential, with the maximum curvature observed for six sequential A residues (A_6). Phasing of the A tracts is also critical: [85] a 5-bp spacing between multiple A_5 segments is optimal; a 6-bp phasing also results in noticeable bending. How-

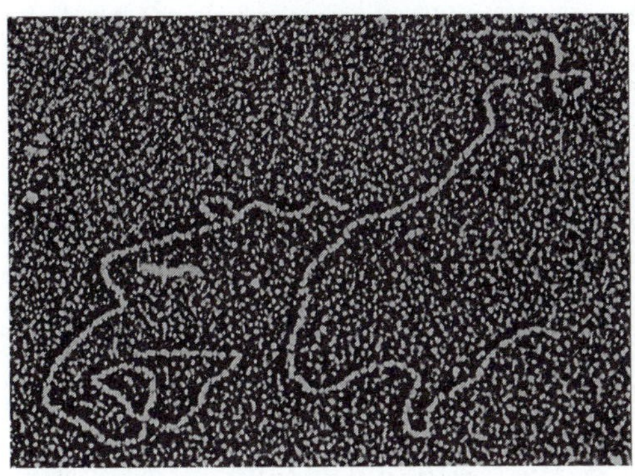
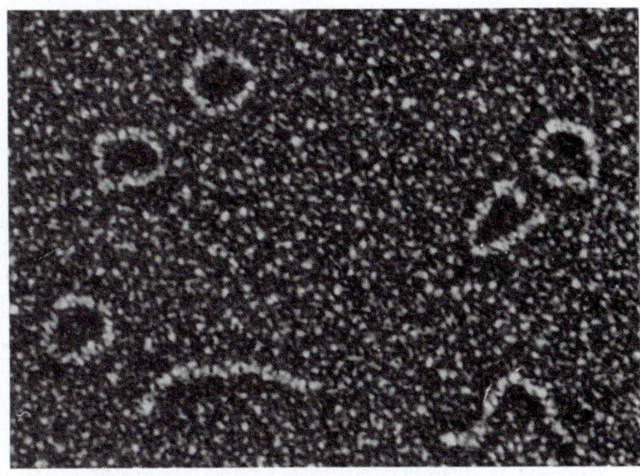

Figure 1-27
Electron micrographs of the kDNA bend in a linearized 2.5-kb minicircle (left) and in a 219-bp fragment (right). [From Griffith J, Bleyman M, Rauch CA, Kitchin PA, Englund PT (1986) *Cell* *46*:717]

79. Wu H-M, Crothers DM (1984) *Nature 308*:509.
80. Levene SD, Zimm BH (1989) *Science 245*:396.
81. Griffith J, Bleyman M, Rauch CA, Kitchin PA, Englund PT (1986) *Cell 46*:717.
82. Zahn K, Blattner FR (1987) *Science 236*:416.
83. Stenzel TT, Patel P, Bastia D (1987) *Cell 49*:709.
84. Koo H-S, Wu H-M, Crothers DM (1986) *Nature 320*:501.
85. Koo H-S, Wu H-M, Crothers DM (1986) *Nature 320*:501.

ever, a DNA fragment with identical A_5 elements spaced 4, 7, or 10 bp apart is not bent. Thus, maximum curvature occurs when the A tracts are phased at one-half the helical-repeat distance apart. By contrast, spacing by a full turn of the helix (i.e., 10 bp) negates the effect because individual A tracts all bend in the same direction, so that the 10-bp spacing causes DNA to adopt a zigzag conformation, rather than a bent one.

The additive behavior of multiple bends allows the direction of curvature to be determined. Changing the phasing between a bend of known direction and the kinetoplast minicircle segment demonstrates that the latter is curved toward the major groove at its center. Bending of each constituent A tract thus must be toward the minor groove. The pattern of chemical cleavage[86] and the ability of distamycin (which binds DNA in the minor groove) to straighten the curve further indicate that narrowing of the minor groove is responsible for bending.

Each A tract that is three or more nucleotides long contributes 17 to 22.5 degrees of curvature.[87] The minicircle fragment, containing 18 phased A tracts, is bent by 300° to 400°, consistent with the appearance in the electron micrograph of small circles (i.e., 360° bends; Fig. 1-27). Although bends caused by A tracts are the best characterized, thymine dimers,[88] cruciforms, and insertions of one or a few base pairs in one strand of the duplex[89] also bend DNA.

Protein-Induced Bends

Sequences that are not intrinsically curved may become bendable under the influence of a bound protein. Bending of DNA around a protein surface was first recognized in the nucleosome, in which a 145-bp segment is wrapped nearly twice around a histone octomer[90] and thus must bend more than 600°. To achieve this tight wrapping, the grooves of the DNA must be narrowed along the surface facing the protein and widened on the outside edges.

The variation in groove width on the inside and outside surfaces of nucleosomes correlates with the occurrence of short sequences.[91] Two or three consecutive AT base pairs are frequently located where the minor groove faces the protein surface (and must be narrowed), whereas where the major groove faces inward (and the minor groove outward), runs of two or three GC base pairs are favored. Repeating these preferred sequences, with the periodicity of the helix, imparts bendability to a fragment. Preferential positioning of the nucleosome occurs within this deformable region.

Many proteins other than histones bend DNA; among them are (1) transcription regulatory factors such as the catabolite activator protein (CAP) and the galactose repressor,[92] (2) *E. coli* RNA polymerase, (3) recombination enzymes such as Tn3 resolvase[93] and λ integrase, (4) the restriction enzyme EcoRI, (5) the small DNA-binding proteins IHF, HU,[94] and Fis, and (6) DNA gyrase.[95]

---

86. Burkhoff AM, Tullius TD (1987) *Cell 48*:935.
87. Koo H-S, Crothers DM (1988) *PNAS 85*:1763; Levene SD, Wu H-M, Crothers DM (1986) *Biochemistry 25*:3988; Zahn K, Blattner FR (1987) *Science 236*:416.
88. Husain I, Griffith J, Sancar A (1988) *PNAS 85*:2558.
89. Hsieh C-H, Griffith JD (1989) *PNAS 86*:4833.
90. Richmond TJ, Finch JT, Rushton B, Rhodes D, Klug A (1984) *Nature 311*:532.
91. Satchwell SC, Drew HR, Travers AA (1986) *JMB 191*:659; Drew HR, Travers AA (1985) *JMB 186*:773.
92. Kuhnke G, Theres C, Fritz H-J, Ehring R (1989) *EMBO J 8*:1247.
93. Hatfull GF, Noble SM, Grindley NDF (1987) *Cell 49*:103.
94. White SW, Appelt K, Wilson KS, Tanaka I (1989) *Proteins 5*:281; Hodges-Garcia Y, Hagerman PJ, Pettijohn DE (1989) *JBC 264*:14621.
95. Wahle E, Kornberg A (1988) *EMBO J 7*:1889.

The role of DNA sequence in protein-induced bending, established from nucleosome positioning, has been confirmed by studies with CAP.[96] Several consecutive AT base pairs are in the minor groove and GC runs are in the major groove of the DNA of the CAP binding site facing the protein. In addition to mutations that alter the core of the CAP binding site, nucleotide changes that disrupt this pattern also disrupt bending and lower the affinity of CAP for the DNA by as much as tenfold. Phasing of AT and GC sequences is observed in other DNA sequences bent by proteins,[97] such as sites bound by RNA polymerase, resolvase, and gyrase. Thus, dinucleotide phasing appears to be a general phenomenon governing the bending of DNA in protein complexes.

The magnitude of protein-induced bends is variable:[98] CAP and IHF both bend the DNA by greater than 140°; λ integrase, λ Xis, and Fis proteins induce bends ranging from 30° to 95°; EcoRI bends the DNA about 60° at its recognition site.

Functions of DNA Bending

Among the possible functions of bent DNA are (1) condensation and packaging of DNA, as in nucleosomes, (2) bringing together distant binding sites in linear DNA, (3) formation of special protein–DNA structures to align DNA sites, as in site-specific recombination, and (4) straining the DNA duplex to stimulate its cleavage[99] or melting.[100] The extraordinary bend in the kinetoplast minicircle is thought to mark the recently replicated molecules by adopting a conformation on nicked DNA distinct from that of the unreplicated, covalently closed circles (Section 18-11) and may also serve in packaging the huge number of minicircles.

1-11 Unusual Structures[101]

The Diversity of DNA Secondary Structures

In contrast to the novelty and variety of secondary and tertiary structures that emanate from the primary sequences of protein, those of DNA were once regarded as simple, uniform, and monotonous. Now it has become clear that the primary sequence of a DNA duplex endows it with considerable structural personality. In addition to bending (Section 10) and other consequences of the primary sequence (Section 8), the diversity of DNA can be expressed in unusual DNA forms that exploit the higher energy state of negatively supercoiled DNA. Such forms include Z-DNA segments, cruciforms, intramolecular triplexes, and ssDNA bubbles. These structures decrease the number of double helical crossovers ($-\Delta Tw$), causing a compensatory removal of negative writhe ($+\Delta Wr$). DNA breakage is not required during the conformational changes, and the linking number of the molecule is unaltered. In addition to these structures, the uptake by the duplex of an additional RNA or DNA strand, forming an R-loop or a D-loop, is also driven by the relaxation of negative supercoils. Enzyme-cata-

96. Gartenberg MR, Crothers DM (1988) *Nature 333*:824.
97. Travers A, Klug A (1987) *Nature 327*:280.
98. Thompson JF, Landy A (1988) *NAR 16*:9687.
99. Hatfull GF, Noble SM, Grindley NDF (1987) *Cell 49*:103.
100. Ramstein J, Lavery R (1988) *PNAS 85*:7231.
101. Wells RD, (1988) *JBC 263*:1095; Wells RD, Harvy SC (eds.) (1987) *Unusual DNA Structures*. Springer-Verlag, New York.

lyzed synapses of two duplexes may produce joints that are paranemic (with strands paired but not intertwined) or plectonemic (with strands intertwined) and Holliday structures (Section 13).

Physical, chemical, and enzymatic probes can be used to identify unusual DNA structures in vitro, but are difficult to apply in vivo. The problem is compounded by the lack of a way to measure the superhelical tension, the quantitative state of which is critical to the formation and stability of these structures.

Two-Dimensional Agarose Gel Electrophoresis[102]

This technique, useful for identifying alternative DNA structures in supercoiled circles, depends on the same principles that operate in separating DNA forms in one-dimensional gels (Section 9). Covalently closed circular molecules migrate as a function of their compactness, the molecules with the largest number of writhes (positive or negative supercoils) migrating the most rapidly.

To investigate the formation of an alternative DNA structure, a series of topoisomers ranging from fully relaxed to a superhelix density of −0.06 or more negative are generated by limited relaxation with a topoisomerase (Chapter 12). Electrophoresis is carried out first in the absence of an intercalating dye, with the ionic strength, pH, and temperature varied to stabilize the structure under study. The second dimension, in a direction perpendicular to the first, is run in the presence of ethidium or, more commonly, chloroquine. (Chloroquine has the same effect on DNA structure as ethidium, but binds DNA less tightly and is thus easier to use because the structural changes occur over a larger concentration range.)

The second dimension expands the range of molecules separable on one gel to as many as 40 resolvable topoisomer "spots." Furthermore, the presence of the intercalating dye eliminates the alternative DNA structure by removing the negative supercoils that stabilize it. Topoisomers of a plasmid lacking any alternative structure make a smooth curve or arc across the gel, with each isomer differing from its neighbors by one unit of linking number (Fig. 1-28). Although the presence of the dye in the second dimension changes the mobility of molecules by untwisting them (thus adding positive writhe), adjacent molecules differing by one in their linking number remain in the same relationship to each other. Thus, there is no discontinuity in the pattern.

In contrast, topoisomers of plasmids with sufficient superhelical density and the potential for forming an alternative structure [e.g., a $d(CG)_{16}$ insert, which forms Z-DNA] may undergo the particular conformational change, thereby reducing the number of supercoils. Migration of these molecules in the first dimension is retarded in proportion to the reduction in writhe. Upon addition of dye, the alternative conformation is destabilized, so that the topoisomers each have correspondingly more superhelical turns in the second dimension than they did in the first. The resulting discontinuity in the arc of the topoisomers in the gel clearly identifies the presence of the alternative DNA structure induced by supercoiling (Fig. 1-28).

Several characteristics of the alternative DNA conformation can be extracted from the electrophoresis pattern. (1) Visual inspection shows which topoisomers

102. Wang JC, Peck LJ, Becherer K (1983) *CSHS* 47:85.

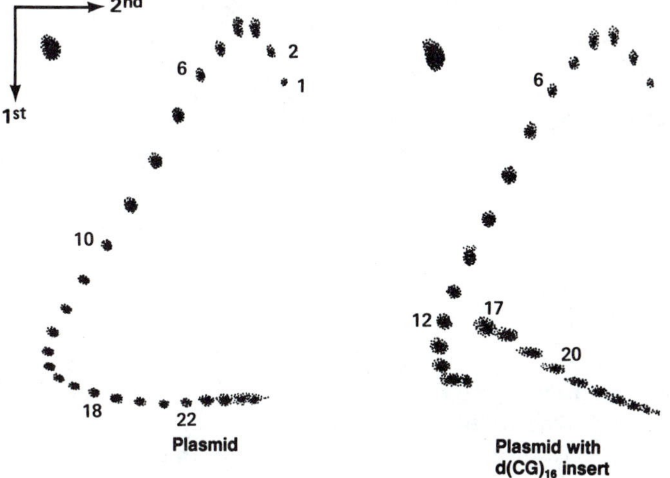

Figure 1-28
Two-dimensional agarose gel electrophoresis pattern of topoisomers of a plasmid without (left) and with (right) a Z-DNA segment. Spot in the upper left is the nicked circle; numbers label topoisomers with decreasing linking numbers. [From Peck LJ, Wang JC (1983) *PNAS 80*:620]

are able to form the structure. Thus, the superhelical density required for the structural change is readily apparent. (2) Since the free energy associated with different superhelicities is easily calculated, the energy invested in formation of the structure can be estimated. (3) The clarity of the spots, the number present at a given linking difference, and the effect of electrophoresis temperature on the pattern give indications of the kinetic stability of the structure. (4) Finally, from the number of negative supercoils lost, one can estimate the number of helical turns removed. In most cases, it is not the alternative structure itself that affects gel mobility, but rather the loss of negative supercoils. Correspondence between the anticipated loss of helical twist for a certain structure and its observed change in mobility is a diagnostic feature, inasmuch as each structure removes a characteristic number of turns.

1-12 Intramolecular Secondary Structures: Z-DNA; Cruciforms; Triplexes; Single-Stranded Bubbles

Z-DNA

Z-DNA, discovered by x-ray diffraction studies (Section 8), has the following characteristics: a left-handed sense of the double helix, a 12-bp helix repeat, and alternating *anti* and *syn* conformations about the base—sugar bond favored by alternating pyrimidine–purine (e.g., CG) sequences. The requirement for a repeating CG or TG sequence is not as strict as initially thought, for up to eight TA base pairs flanked by CG repeats has been shown to form a left-handed helix, although demanding more energy than a pure CG sequence. In fact, almost any sequence can adopt the Z conformation if provided enough energy.

An appropriate sequence adopts the Z conformation in response to superhelical tension. Because the sense of the Z helix is left- rather than right-handed, conversion of a stretch of B-DNA to the Z form reduces the total number of helical turns in the molecule. This decrease in twist number ($-\Delta Tw$) reduces the number of negative supercoils (Fig. 1-29). Thus, conversion of 1.2 right-

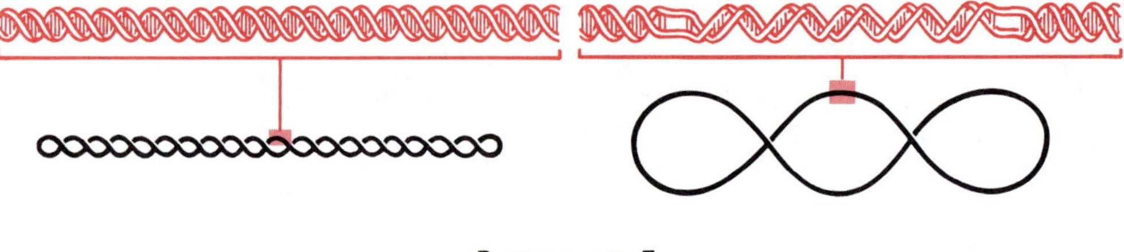

Figure 1-29
B to Z transition in a small segment of DNA and changes in the supercoiled state of the DNA. [After Wells RD (1988) *JBC 263*:1095]

handed double-helical turns (12 bp, $\Delta Tw = -1.2$) to 1 left-handed, double-helical turn ($\Delta Tw = -1$) removes about 2 negative supercoils.

The transition from B- to Z-DNA is cooperative, indicating an energy barrier (i.e., an unfavorable state) at the junctions between the two helix forms. As a result, the stability of a Z-DNA tract increases with its length. Conversion of a 32-bp segment decreases the number of negative supercoils by nearly six, whereas a 14-bp segment removes only about two turns. Yet both segments are flanked by identical, unfavorable B–Z junctions. Thus, for a segment to flip from B to Z, sufficient energy must be released by relaxing the molecule to overcome the unfavorable contributions of the junctions. To flip the 14-bp segment requires a more negative superhelix density (σ of -0.07) than for the 32-bp segment (σ of -0.05).[103] The rate of conversion from B to Z also depends on the superhelix density, being ten times slower on the more relaxed molecules.[104]

Although the levels of superhelicity that induce the B to Z transition in vitro are in the range found in plasmids isolated from cells, about half of this superhelical tension is unavailable in vivo, because of the constraint exerted by proteins associated with the chromosome. The presence of left-handed DNA in vivo is, however, implied by the failure of restriction endonucleases and methylases to act as they should within or near a $(CG)_n$ sequence in right-handed DNA[105] and by the recognition of cellular DNA by antibodies specific for the Z-DNA form.[106] The biological consequences of conversions to the Z-DNA form are still conjectural, although a high incidence of deletions near blocks of CG sequences suggests that Z-DNA facilitates recombination (see below).

Cruciforms

The cruciform structure consists of two exposed hairpins (see Fig. 1-30). Cruciforms may develop at the site of a *palindromic DNA sequence*, an inverted repeat with a twofold axis of symmetry. (The standard definition of a palindrome is a word or sentence that reads the same forward and backward.) Extrusion of the repeated sequence as a cruciform reduces the number of interstrand (helical) twists ($-\Delta Tw$); like a B → Z transition, this removes negative supercoils.[107] However, the magnitude of the reduction is smaller, with one superhelical turn removed for each ten base pairs involved in the structure.

---

103. Wang JC, Peck LJ, Becherer K (1983) *CSHS 47*:85; Nordheim A, Peck LJ, Lafer EM, Stollar BD, Wang JC, Rich A (1983) *CSHS 47*:93.
104. Peck LJ, Wang JC, Nordheim A, Rich A (1986) *JMB 190*:125.
105. Jaworski A, Hsieh W-T, Blaho JA, Larson JE, Wells RD (1987) *Science 238*:773.
106. Lafer EM, Sousa RJ, Rich A (1988) *JMB 203*:511; Wittig B, Dorbic T, Rich A (1989) *J Cell Biol 108*:755.
107. Mizuuchi K, Mizuuchi M, Gellert M (1982) *JMB 156*:229; Panayotatos N, Wells RD (1981) *Nature 289*:466.

Each hairpin in the cruciform structure is a completely base-paired B-form helix except for the loop at the tip of the stem, which contains at least three nucleotides. Hydrogen bonding is maintained at the four-way junction of the two hairpins, as judged by resistance of this region of DNA to enzymatic and chemical probes for single-strandedness. The junction region, however, appears to be either unusually flexible or to contain a stable bend or kink. This four-strand interchange is structurally similar to a Holliday junction (Section 13).

Two distinct pathways have been suggested for cruciform extrusion (Fig. 1-30). Most commonly, initiation involves the formation of little hairpins (*proto-cruciforms*) by melting of a few base pairs of the duplex followed by intrastrand pairing. Branch migration of the junction to elongate the arms is rapid and not energetically costly; for each base pair broken, an equivalent one is formed. In a less common pathway, extrusion of the cruciform is preceded by the melting of a large region of DNA to form a single-stranded DNA bubble, followed by intrastrand pairing to create the stems of the hairpins.[108] The latter pathway is favored at very low ionic strength, which destabilizes the duplex DNA, and by AT-rich inducing sequences as remote as 100 bp from the inverted repeat. Kinetic instability ("heavy breathing") in these inducing sequences may stimulate the initial melting phase of this pathway.[109] Certain sequences, such as a run of nine GC base pairs, block the effect of the inducing sequence if they lie between the AT-rich region and the inverted repeat.

Cruciforms can be easily detected in vitro by chemical, enzymatic, and structural methods. Sequences which form cruciforms efficiently in vitro do not always form them in vivo, presumably for lack of sufficient superhelical tension.[110] The major barrier to their formation is apparently kinetic. However,

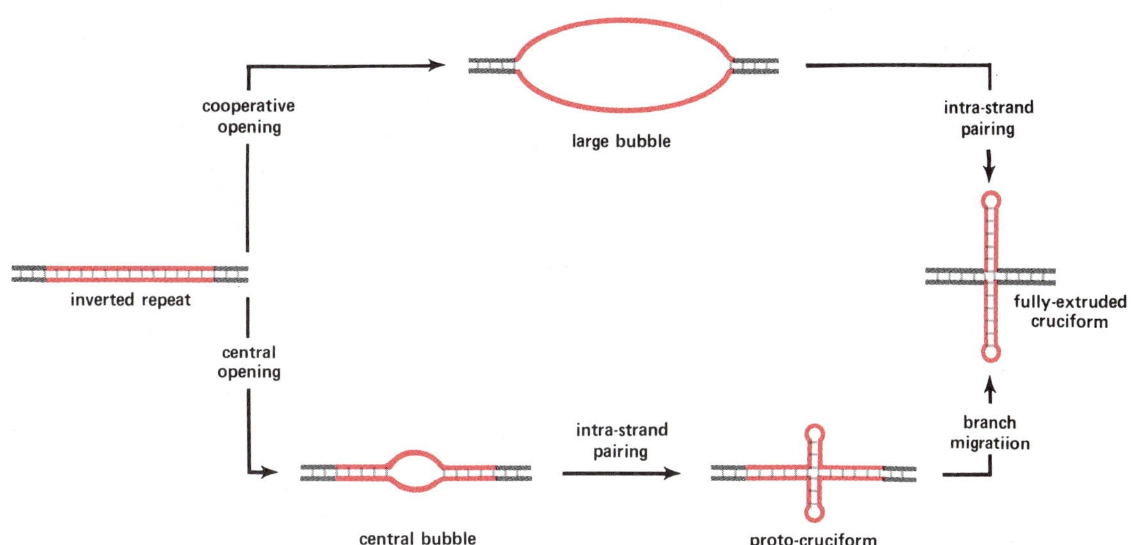

Figure 1-30
Two pathways for extrusion of an inverted repeat sequence into a cruciform. [After Lilley DMJ (1988) *TIG* 4:111]

108. Sullivan KM, Lilley DMJ (1986) *Cell 47*:817; Sullivan KM, Lilley DMJ (1988) *NAR 16*:1079.
109. Lilley DMJ (1988) *Trends Genet 4*:111.
110. Courey AJ, Wang JC (1983) *Cell 33*:817.

cruciforms do form at least transiently within the cell, as judged by psoralen cross-linking and the action of phage T7 endonuclease I, which specifically cleaves such forked molecules (Section 21-5).

Intramolecular Triplexes (H-DNA)

DNA containing repeated tracts of pyrimidines and purines, such as $(TC)_n$ paired with $(AG)_n$, can form a three-stranded helix in response to negative superhelicity or low pH (Fig. 1-31). The triplex is formed when half of the normal duplex is disrupted and the polypyrimidine strand folds back and inserts itself into the major groove of the remaining duplex. The donated strand associates by Hoogsteen base pairing with the purines in the duplex region without disrupting the conventional Watson-Crick pairs. Because formation of Hoogsteen base pairs requires protonation of the C residues, a low pH (e.g., a pH of 5) stabilizes the structure. (This requirement for H^+ ions gives H-DNA its name.) The triplex, a right-handed helix, has a loop of about four nucleotides in the polypyrimidine strand at its tip; the base forms a flexible hinge (hence the synonym *hinge DNA*), and the polypurine strand of the disrupted helix remains unpaired.

Formation of the triplex is driven by the removal of negative supercoils.[111] Long repeats remove more supercoils and require lower superhelix densities for triplex formation. Repeats longer than 30 bp remove about one superhelical turn for every 11 base pairs participating in the structure. Thus, both disruption of the duplex and formation of the triplex must contribute to unwinding. If

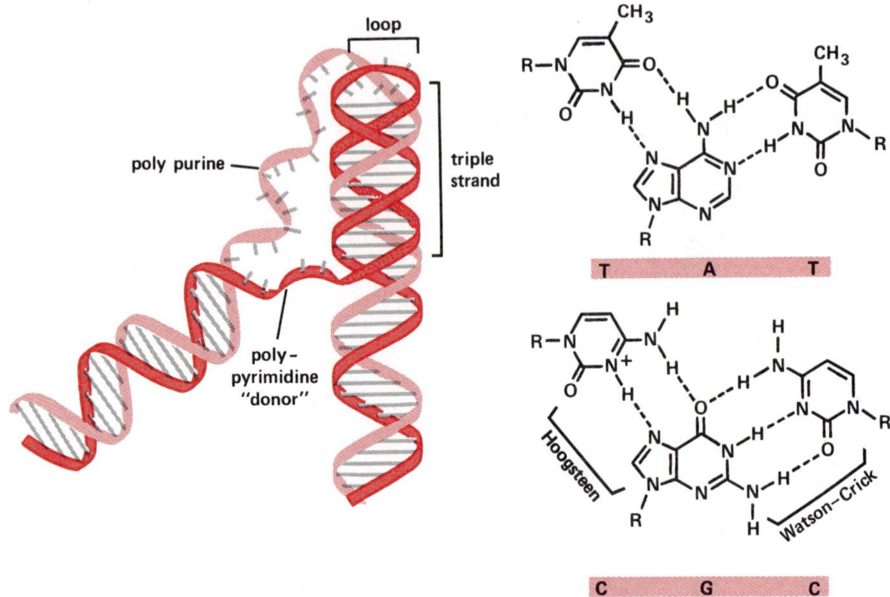

Figure 1-31
Creation of triple-stranded DNA (left) by Hoogsteen base pairs (right). [After Htun H, Dahlberg JE (1989) *Science 243*:1571]

111. Htun H, Dahlberg JE 1989) *Science 243*:1571.

melting of the duplex to generate the donor strand were the only source for relaxing the supercoils, then removal of only half this number of turns would be expected. The triplex region does not contribute to the linking number of the molecule because it has an equal number of strand crossings in both the positive and negative directions.

Model substrates and mutagenesis of existing triplexes have defined the sequence parameters, in addition to length, that stabilize triplex formation: (1) The exact alternating nucleotide sequence, rather than simply the one purine-rich and one pyrimidine-rich strand, is key, as hydrogen bonding between the strands is essential. (2) Some disruption in the repeats is tolerated: a 1- or 2-bp mutation will not block triplex formation providing enough repeat units are present to stabilize a structure containing the mismatches.[112] Alternatively, two shorter triplexes form as a result of a 3-bp disruption in a long repeat. Nucleotide insertions in the middle of the repeats are much less disruptive because they can be accommodated within the loop.[113] Up to ten additional base pairs in the center are compatible with triplex formation, although the structures formed are less stable compared to those with the optimal four-nucleotide loop. (3) A high GC content stabilizes the structure:[114] A_n-T_n repeats fail to form triplexes, but $(AG)_n-(CT)_n$ and G_n-C_n do so efficiently.

Inasmuch as triplexes can form at physiologic superhelix densities and neutral pH in vitro, they may do so in vivo as well. Sequences which form triplexes in vitro have been isolated from the regulatory regions of genes, suggesting a possible role in gene expression. However, direct evidence for the presence of triplexes in cells is still lacking.

Single-Stranded DNA Bubbles

Negatively supercoiled DNA has a more single-stranded character than does its relaxed counterpart. This property is readily apparent from its underwound nature (Section 9). The degree to which underwinding is partitioned into single-stranded regions, negative superhelical twists, or alternative structures depends on the DNA sequence, solution conditions, and the presence of proteins. However, some regions of DNA have an especially high propensity for adopting a single-strand, melted configuration.

Regions of DNA that are unusually stably unwound when present on a supercoiled circle have been identified by the sensitivity of an extensive region to single-strand–specific nucleases and by the removal of negative supercoils.[115] While the thermodynamically unwindable regions thus far described are very AT-rich, sequence contributions in addition to the AT content seem to be involved. The rules governing this behavior remain to be established.

Two such DNA unwinding elements are present on the commonly used plasmid, pBR322 (Section 18-2). Each region appears to extend over 100 to 120 nucleotides, as judged by nuclease digestion and by two-dimensional topoisomer gel analysis. No biological role for these sequences is known. Thus far these melted regions have been detected only at very low ionic strengths, which favor melting of the duplex; whether such regions occur under physiologic conditions

112. Hanvey JC, Shimizu M, Wells RD (1989) *JBC 264*:5950.
113. Shimizu M, Hanvey JC, Wells RD (1989) *JBC 264*:5944.
114. Hanvey JC, Shimizu M, Wells RD (1989) *JBC 264*:5950.
115. Kowalski D, Natale DA, Eddy MJ (1988) *PNAS 85*:9464.

has not yet been established. However, similar unwinding elements have been detected in the 13-bp repeated elements of the *E. coli* chromosomal origin sequence, *oriC* (Section 16-2), and in the autonomously replicating sequence (ARS) of the yeast 2μ plasmid (Section 16-6), where they may play a key role in destabilizing the duplex during initiation at the origin of chromosome replication.[116]

1-13 Intermolecular Structures: R-Loops; D-Loops; Joints; Holliday Junctions; Knots and Catenanes

R-Loops

R-loops form on duplex DNA when an RNA of complementary sequence stably base-pairs with one of the DNA strands. The other DNA strand is displaced as a single-stranded loop. The RNA–DNA hybrid duplex adopts the A-helix form (Section 8).

The association of a region of RNA with a negatively supercoiled duplex is stabilized by the removal of negative superhelical turns. R-loops form transiently during transcription, and may require RNA polymerase or some other enzyme to displace them. R-loop formation during transcription may be essential for DNA replication, to serve as a primer (e.g., for plasmid ColE1 and phage T7 replication) or as an activator (e.g., for ColE1 and *E. coli oriC*).

D-Loops

D-loops are generated when additional strands of DNA are taken up by the duplex. Formation of a D-loop is favored in negatively supercoiled DNA and can occur spontaneously. Although D-loops are formed in a high percentage of mammalian mitochrondial DNAs (Section 18-11), they are most prominent as intermediates in genetic recombination, in which their formation is catalyzed by strand-exchange enzymes; the best characterized of these is the RecA protein of *E. coli* (Section 21-5).

Paranemic Joints

Intermediates in the formation of stable D-loops, paranemic joints are protein-stabilized structures in which the incoming strand pairs with its complement in the duplex, but is *not* stably intertwined as in the classic helical duplex.[117] This lack of net topological interwinding enables a paranemic joint to be formed without the participation of a free end on either DNA molecule and to take place between a single-stranded and a duplex circle. The length of ssDNA that is able to pair with the duplex circle appears to be limited by the superhelicity of the duplex, which is relaxed by joint formation.[118]

The structure of the DNA strands within the paranemic joint that allows

---

116. Umek RM, Kowalski D (1988) *Cell 52*:559; Umek RM, Eddy MJ, Kowalski D (1988) in *Cancer Cells 6: Eukaryotic DNA Replication* (Kelly TJ, Stillman B, eds.). CSHL, p. 473.
117. Bianchi M, DasGupta C, Radding CM (1983) *Cell 34*:931; DasGupta C, Shibata T, Cunningham RP, Radding CM (1980) *Cell 22*:437; Cunningham RP, Wu AM, Shibata T, DasGupta C, Radding CM (1981) *Cell 24*:213; Riddles PW, Lehman IR (1985) *JBC 260*:165.
118. Christiansen G, Griffith J (1986) *PNAS 83*:2066.

pairing over several hundred base pairs in the absence of net helical intertwining has not been satisfactorily established and is of considerable interest. Two possibilities[119] are (1) unwinding of the duplex DNA so that one of its strands can pair side by side with the incoming strand; and (2) accommodating alternating segments of right- and left-handed helices across the joint region, involving no net change in the total number of helical turns introduced.

Plectonemic Joints

A free end on the DNA strand or the action of a topoisomerase[120] allows intertwining of the strands of a *paranemic* joint, converting it to the fully stabilized, helical *plectonemic* joint or D-loop. A standard B-form helix presumably forms between the incoming strand and its partner in the duplex. The length of the D-loop is determined by the available negative superhelicity. With completely homologous molecules in which the topological constraint is removed by nicking of the duplex or by the action of a topoisomerase, the size of the joint can increase to encompass the entire length of the molecule; the invading strand thus exchanges places with one of the strands in the original duplex.

Holliday Junctions

These junctions between four DNA strands are important intermediates in genetic recombination. Enzymes specific for these forked DNAs are encoded by phages T7 and T4, as well as by yeast;[121] the conformation of the fork determines how it will be recognized and cleaved by these enzymes and, as a consequence, the outcome of the recombination process (Section 21-5).

Although commonly represented in a flat, square, planar configuration with a fourfold axis of symmetry, Holliday junctions in fact exhibit twofold symmetry: the arms pair with each other, forming two sets of colinear helices[122] (Fig. 1-32).

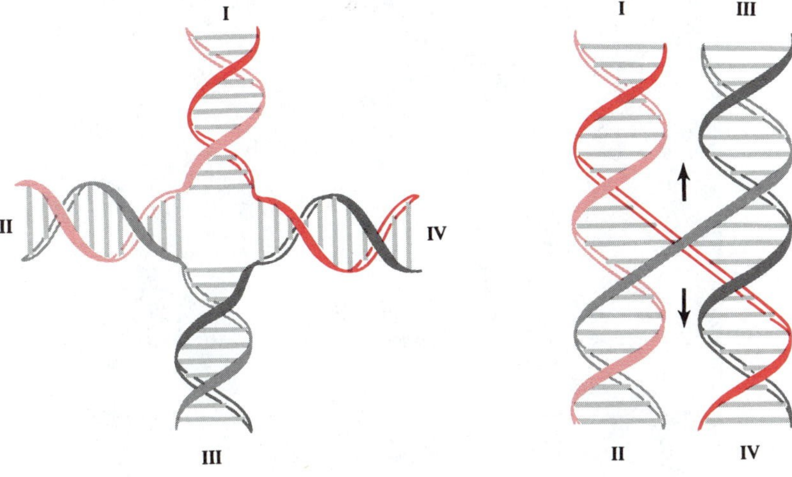

Figure 1-32
Structure of a Holliday junction. The commonly drawn square, planar conformation (left) and the probable stacked structure with twofold symmetry (right); arrows = directions of branch migration.

---

119. Riddles PW, Lehman IR (1985) *JBC 260*:165.
120. Cunningham RP, Wu AM, Shibata T, DasGupta C, Radding CM (1981) *Cell 24*:213.
121. West SC, Parsons CA, Picksley SM (1987) *JBC 262*:12752; Symington LS, Kolodner R (1985) *PNAS 82*:7247.
122. Duckett DR, Murchie AIH, Diekmann S, von Kitzing E, Kemper B, Lilley DMJ (1988) *Cell 55*:79; Churchill MEA, Tullius TD, Kallenbach NR, Seeman NC (1988) *PNAS 85*:4653.

This colinear conformation allows stacking between the bases across the junction of the arms. The identity of the final bases at the four-way junction of the arms determines which arms will pair with each other to provide the most favorable stacking energies and the most stable conformation. The angles between the two sets of colinear helices are not known, but electrophoretic analysis, analogous to that used to characterize DNA bends (Section 10), suggests that they are two acute and two obtuse angles.[123]

Knots and Catenanes

In knotted and catenated forms, the DNA duplex is interlinked; intramolecular linkage forms a knot (Fig. 1-33), while linkage between two circles generates a catenane[124] (Fig. 1-14). Interlocked DNA forms are often the products of replication and recombination on circular DNA molecules, although in most instances they are short-lived intermediates, quickly resolved by topoisomerases (Chapter 12). While easiest to imagine as arising from reactions on circular templates, these interlocked forms may occur as well in the topologically constrained domains of linear chromosomes. Analysis of the structure of these interlocked forms (the number, direction, and organization of the DNA crossovers) provides insight into the mechanisms by which they arise. The behavior of knots and catenanes as templates for recombination, replication, and transcription may in addition provide an experimental system for testing the structural requirements of the substrate.

Multiply intertwined catenated dimers are a common late intermediate in the replication of circular DNAs.[125] Helical intertwines between the two template strands can be converted into interlocks between the already replicated duplex

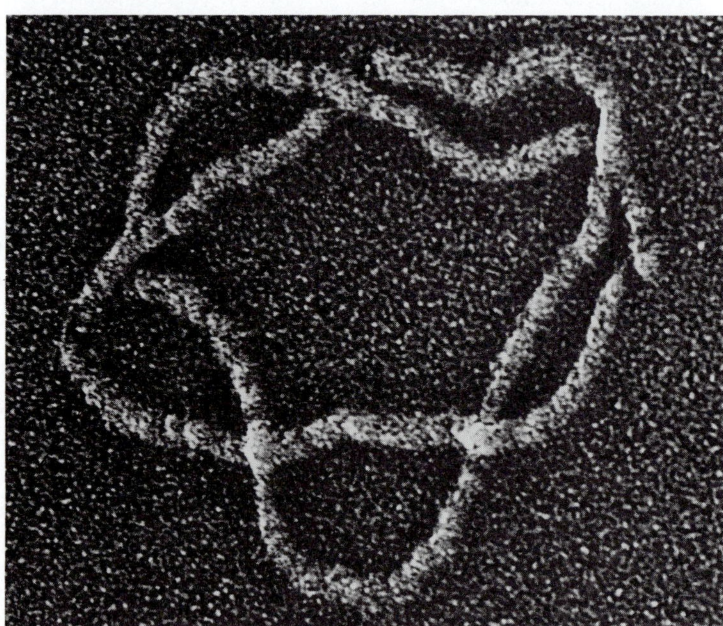

Figure 1-33
DNA knot formed by λ Int protein. (Courtesy of Dr J Griffith)

123. Diekmann S, Lilley DMJ (1987) *NAR 15*:5765.
124. Wasserman SA, Cozzarelli NR (1986) *Science 232*:951.
125. Sundin O, Varshavsky A (1980) *Cell 21*:103; Wasserman SA, Cozzarelli NR (1986) *Science 232*:951.

portions of the molecule, allowing replication to be completed (Section 12-1). For example, the final 100 base pairs of the template to be replicated could be converted into ten intermolecular intertwines between the daughter circles, which could be subsequently removed by a topoisomerase. Catenanes are clearly involved in the replication of many circular genomes, including the SV40 animal virus and the small *E. coli* plasmids. Furthermore, the behavior of topoisomerase mutants in yeast suggests that similar structures are encountered in the replication of large linear chromosomes.[126]

Knots and catenanes are useful in defining the mechanisms by which certain transactions of DNA occur. (1) Type I and type II topoisomerases can be unequivocally distinguished on the basis of their ability to untie knots in completely duplex DNA, since type II enzymes alone are active on such molecules (Section 12-3). (2) Interlocked DNA circles can also be used to determine whether a DNA-binding protein interacts with two distantly located sequences on a single circle by linear tracking along the DNA or by loop formation between the sites: looping can occur both intramolecularly and on catenated circles, whereas tracking is impossible between the two circles in a catenane.[127] (3) Determining whether or not catenate interlocks can replace the requirement for DNA superhelicity provides important insight into how the energy of supercoiling is exploited by a reaction.[128] (4) Finally, the structure of a knotted DNA product of a recombination reaction provides information about the organization of the synaptic complex[129] (Section 21-6).

A catalog of the characteristics of knots and catenanes important in biological systems has been delineated.[130] The study of DNA interlocks, in combination with more conventional biochemical approaches, represents an additional approach to an understanding of the dynamics of DNA transactions.

1-14 Functions of DNA

DNA structures need to be related to their functions, just as an understanding of DNA functions requires a knowledge of the variety of their structures. For some structures, such as the repetitive satellite DNAs of insects or the minicircle network of trypanosomes, the search for their true functions continues. However, the primary role of DNA, as the cell's genetic blueprint, is to be used for transcription of the encoded information and to serve as the template for replication. Furthermore, DNA must remain stable enough to allow only a modest level of change over many generations.

The structure of DNA can generally be rationalized as a "design" to fulfill these primary functions efficiently. The linear array of four nucleotides, arranged in a defined order, allows the molecule to encode an essentially endless and diverse amount of information. The duplex structure maintains two copies of the information at all times. If lost by errors in replication or by DNA damage, the information is often retrievable from an undamaged complementary strand

---

126. DiNardo S, Voelkel K, Sternglanz R (1984) *PNAS 81*:2616; Thrash C, Voelkel K, DiNardo S, Sternglanz R (1984) *JBC 259*:1375.
127. Krasnow MA, Cozzarelli NR (1983) *Cell 32*:1313.
128. Craigie R, Mizuuchi K (1986) *Cell 45*:793; Kanaar R, van de Putte P, Cozzarelli NR (1989) *Cell 58*:147.
129. Kanaar R, van de Putte P, Cozzarelli NR (1988) *PNAS 85*:752; Dröge P, Cozzarelli NR (1989) *PNAS 86*:6062.
130. White JH, MIllett KC, Cozzarelli NR (1987) *JMB 197*:585.

(Section 21-5). Complementarity in the nucleotide sequence provides the basis both for transcription into the temporary but directly usable messenger RNA and for replication prior to cell division.

Base complementarity, melting and annealing of the helix, and its ability to accommodate, at least transiently, a third nucleic acid strand provide the foundations for recombination between DNA molecules (Section 21-5). Recombination is essential for most pathways of damage repair, for replication of the ends of some linear molecules, and, in an increasingly recognized number of instances, for the initiation of replication. Finally, recombination provides for the "communication" between DNA molecules in a population, across species, and even across kingdom boundaries, affording a means for maintaining both genetic stability and diversity.

The shape of DNA molecules can also be understood from the viewpoint of efficiency of replication and transcription. Small genomes are often circular, avoiding the problems inherent in replicating the ends of linear molecules (Section 15-12). The latter, in turn, must have special structures at their ends, to ensure their replication. Finally, the bending and topological flexibility inherent in DNA play a key role in packing these long molecules into a cell that is only a tiny fraction of their length.

CHAPTER 2

Biosynthesis of DNA Precursors

2-1 The Building Blocks of DNA Synthesis

Elucidating the patterns of replication and understanding how they are controlled are predominant objectives in recent DNA research. Yet major developments in studies of DNA chain growth, as well as in the chemotherapy of cancer and in understanding autoimmune and viral diseases, have come from digging the tilled pathways of nucleotide biosynthesis.[1]

Interest in DNA precursors has unfortunately diminished—almost in proportion to their commercial availability. Nevertheless, the nature of DNA biosynthesis depends on the properties of the nucleotides and their structures, origins, and functions. An appreciation of the details of precursor biosynthesis, still unclear today, is fundamental for understanding the dynamics of DNA replication in the cell.

The building blocks of DNA synthesis are the *deoxyribonucleoside monophosphates* (deoxynucleotides, dNMPs), which, with an added pyrophosphoryl group, become the activated precursors, the *deoxyribonucleoside triphosphates* (dNTPs). Deoxyribonucleotides arise from the direct reduction of ribonucleotides (ribotides) at the level of the diphosphates (Section 7). Most of the nucleotides in DNA and RNA are supplied from pools which, except for ATP, are tiny in size (Table 2-1).[2]

1. Elion GB (1989) *Science 244*:41.
2. Neuhard J, Nygaard P (1987) in *Escherichia Coli and Salmonella Typhimurium* (Neidhardt FC, ed.). Am Soc Microbiol, Washington DC, Vol. 1 p. 445; Bochner BR, Ames BN (1982) *JBC 257*:9759.

Table 2-1
Nucleotide levels in cells[a]

DNA precursors	Amount[b] (mM)	RNA precursors	Amount[b] (mM)
dNMPs	43	rNMPs	270
dATP	0.18 (0.013)	ATP	3.0 (2.8)
dGTP	0.12 (0.005)	GTP	0.92 (0.48)
dCTP	0.07 (0.022)	CTP	0.52 (0.21)
dTTP	0.08 (0.023)	UTP	0.89 (0.48)

[a] Values were obtained from log phase *Escherichia coli* and *Salmonella typhimurium*, assuming that 1 g in bacterial dry weight corresponds to 2.33 g of intracellular water. Other values of interest: cAMP, 0.006 mM; ppGpp, 0.031 mM; PRPP, 0.47 mM. Data from Neuhard J, Nygaard P (1987) in *Escherichia Coli and Salmonella Typhimurium* (Neidhardt FC, ed.). Am Soc Microbiol, Washington DC, Vol. 1 p. 445; Bochner BR, Ames BN (1982) *JBC 257*:9759.

[b] Values in parentheses are averages of determinations from diverse mammalian (including human) cells and tissues, assembled from the literature by Professor Thomas W Traut.

Astonishingly, at any given moment the cellular supply of DNA building blocks for assembly of the complete chromosome is only 1% of the needed amount in *Escherichia coli* and is far less in mammalian cells. An inadequate supply of any one of the dNTPs is lethal and an oversupply is mutagenic. Beyond providing the right amount and balance of dNTPs, the cellular economy is aided by the efficient salvage of the available precursors and by intricate regulation of the synthetic pathways.

2-2 De Novo and Salvage Pathways of Nucleotide Synthesis[3]

In most cells, two fundamentally different pathways are used for the synthesis of nucleotides (Fig. 2-1). One is a *de novo pathway*, in which ribose phosphate, certain amino acids, CO_2, and NH_3 are combined in successive reactions to form the nucleotides. Neither the free purine bases (adenine and guanine), nor the pyrimidine bases (cytosine, uracil, and thymine), nor the corresponding nucleosides are intermediates in the de novo pathway. In other words, *nucleotides are synthesized by this pathway without passing through a stage or pool containing these free bases or nucleosides.*

In addition to the de novo pathway, cells also have various mechanisms for making use of the free bases and nucleosides released from the breakdown of nucleic acids (Fig. 2-1) by converting them back to nucleotides. These other sources have been called *salvage pathways*. The word *salvage* should not be taken in any sense that minimizes the importance of these biosynthetic routes. For example, a cell absolutely requires a salvage pathway if it lacks one of the enzymes required to support the de novo pathway at an adequate rate for survival. The genetic makeup of the cell or a deficiency resulting from disease, drugs, poisons, or physiologic stresses may contribute to inadequacy of the de

3. Neuhard J, Nygaard P (1987) in *Escherichia Coli and Salmonella Typhimurium* (Neidhardt FC, ed.). Am Soc Microbiol, Washington DC, Vol. 1 p. 445; Munch-Petersen A (ed.) (1983) *Metabolism of Nucleotides, Nucleosides, and Nucleobases in Microorganisms*. Academic Press, London; Reichard P (1988) *ARB* 57:349.

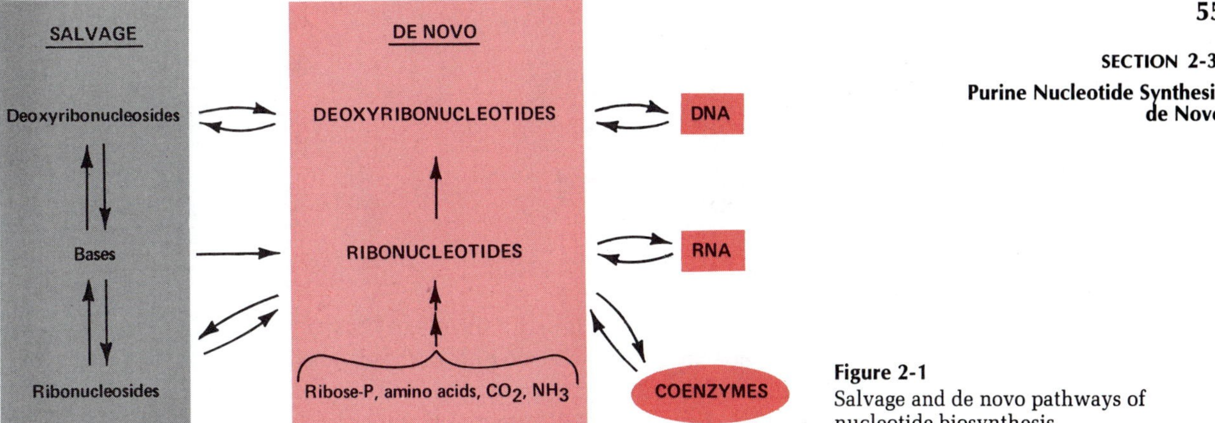

Figure 2-1
Salvage and de novo pathways of nucleotide biosynthesis.

novo pathway, making the recycling of bases and nucleosides from intracellular and extracellular sources essential.

The breakdown of nucleic acids *within cells* (bacterial, plant, and animal) is a major recycling source. This salvage pathway usually goes unnoticed because its efficiency often exceeds the capacity of extracellular components to enter the cell and compete with these pools of endogenously generated precursors.

Extracellular sources from which bases and nucleosides are recovered are the growth media for bacteria and the extracellular fluids for animals. In animals, the extracellular nucleotide precursors arise by digestion of foods in the alimentary tract and by release from neighboring cells or other tissues. Thus, a tissue which seems to lack the capacity for de novo synthesis of bases and nucleosides, such as brain, must salvage precursors synthesized elsewhere in the body (e.g., the liver).

Salvage pathways also serve important regulatory, disposal, and detoxification functions. Bacteria may also salvage nucleic acid bases and nucleosides as sources of carbon, nitrogen, and energy for growth. Whereas nucleoside degradation may furnish a purine for biosynthetic salvage in one situation, it may be essential for removing a toxic accumulation of a nucleoside in another. For lack of adenosine deaminase or a nucleoside phosphorylase in certain tissues, DNA synthesis is inhibited, leading to a severe immunodeficiency disease (Section 10). Protozoa, remarkably, lack de novo pathways and rely entirely on salvage, presenting numerous targets for chemotherapy of the pathogenic forms.[4]

2-3 Purine Nucleotide Synthesis de Novo[5]

The first indications of how the purine skeleton is synthesized came from experiments in which labeled nutrients were fed to pigeons. Birds excrete waste nitrogen as a purine, uric acid, in solid deposits. Chemical degradation of the uric acid revealed the ring positions of the labeled precursors. Glycine is the

4. Hammond DJ, Gutteridge WE (1984) *Mol Biochem Parasitol 13*:243; Ullman B (1984) *Pharm Res 1*:194.
5. Neuhard J, Nygaard P (1987) in *Escherichia Coli and Salmonella Typhimurium* (Neidhardt FC, ed.). Am Soc Microbiol, Washington DC, Vol. 1 p. 445.

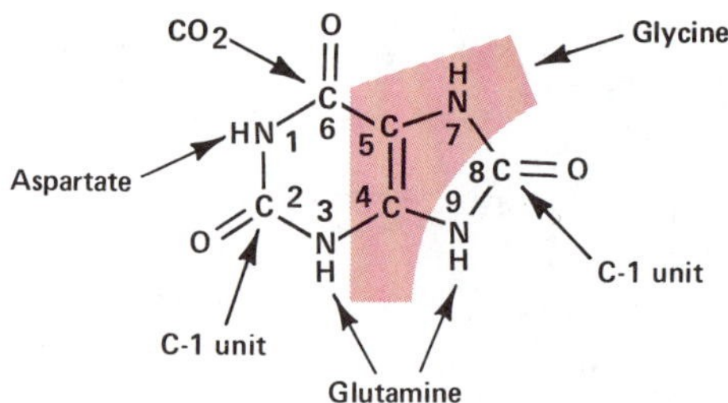

Figure 2-2
Sources of atoms in the biosynthesis of purines, shown in uric acid.
C-1 unit = one-carbon unit.

source of C4, C5, and N7, formate the source of C2 and C8, and CO_2 the source of C6 (Fig. 2-2). Aspartate is the source of N1 and glutamine of N3, but these were not established by nutritional experiments with intact cells and organisms because intracellular pools of NH_3 interfered with the ^{15}N labeling.

Cell-free extracts of animal and bacterial cells are remarkably active in catalyzing the whole sequence of reactions of purine biosynthesis. This fortunate circumstance has made it possible in most species to separate and characterize the enzymes involved, elucidating the details of every step in the pathway. Purine biosynthesis is now an exceedingly instructive chapter in the fundamentals of biochemistry in general and of biosynthetic metabolism in particular.

The Reactions of de Novo Purine Synthesis

The initial reactant in purine assembly is *5-phosphoribosyl pyrophosphate (PRPP)*. This activated form of ribose 5-phosphate is formed by PRPP synthase[6] by direct transfer of the β-γ pair of phosphates of ATP to C1 of the sugar in the α configuration (Fig. 2-3).[7] PRPP, first discovered as an intermediate in pyrimidine biosynthesis,[8] is a key reactant not only in purine and pyrimidine biosynthesis de novo, but also in the biosynthesis of the pyridine nucleotide coenzymes NAD and NADP, the biosynthesis of the amino acids histidine and tryptophan, and the utilization, or salvage, of preformed purine bases and nucleosides. PRPP synthase activity[9] depends on its state of aggregation, the concentrations of phosphate, and the end products of the many pathways it serves.

Condensation of PRPP with glutamine is catalyzed by PRPP amidotransferase;[10] this, followed by removal of glutamate from the product, fixes the first

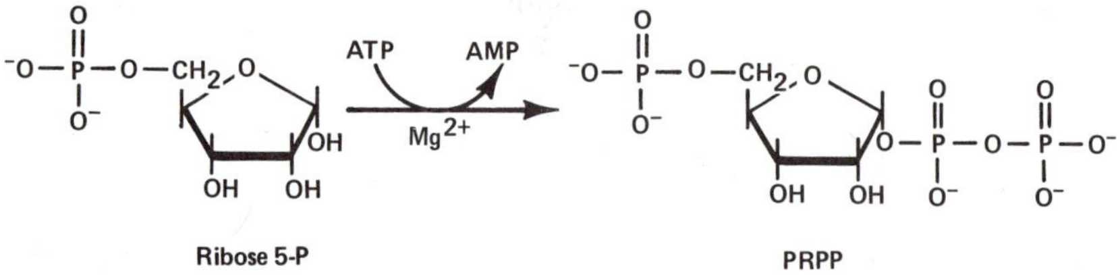

Ribose 5-P → PRPP (ATP → AMP, Mg^{2+})

Figure 2-3
Synthesis of PRPP from ribose 5-P and ATP by PRPP synthase.

6. Bower SG, Hove-Jensen B, Switzer RL (1988) *J Bact 170*:3243.
7. Khorana HG, Fernandes JF, Kornberg A (1958) *JBC 230*:941.
8. Kornberg A, Lieberman I, Simms ES (1955) *JBC 215*:389.
9. Switzer RL, Gibson KJ (1978) *Meth Enz 51*:3; Roth DG, White C, Deuel TF (1978) *Meth Enz 51*:12; Hove-Jensen B, Harlow KW, King CJ, Switzer RL (1986) *JBC 261*:6765.
10. Lewis JM, Hartman SC (1978) *Meth Enz 51*:171; Itakura M, Holmes EW (1979) *JBC 254*:333; Tso JY, Hermodson MA, Zalkin H (1982) *JBC 257*:3532.

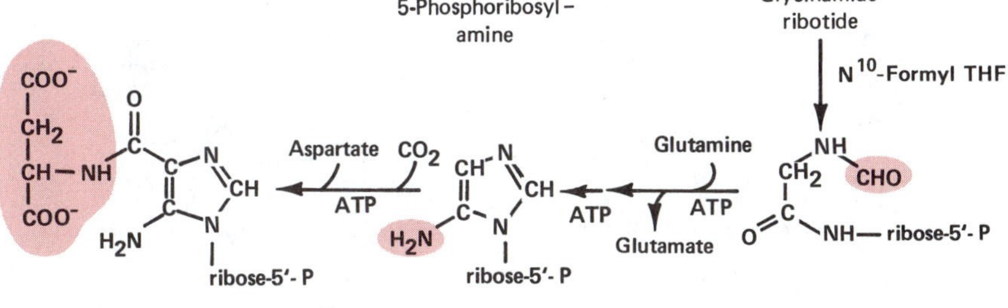
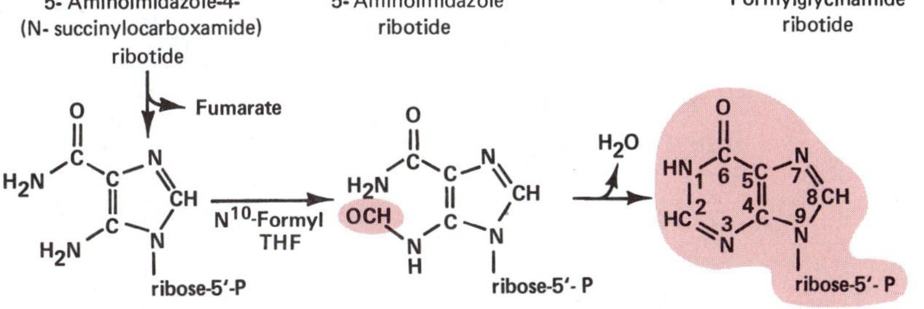

Figure 2-4
De novo pathway of purine nucleotide biosynthesis to inosinate. Addition of the purine ring component at each step is shown in color.

atom, N9, of the purine skeleton to the ribose phosphate (Fig. 2-4). This condensation involves an inversion in the configuration of the substituent on C1 of the sugar and thereby establishes the β configuration of the future nucleotide. An iron–sulfur cluster is at the active center of the enzyme in *Bacillus subtilis* and is a target for complex regulatory controls, particularly inactivation of the enzyme by molecular oxygen when cells enter the stationary phase.[11]

The succession of condensations and ring closures shown in Figure 2-4 produces the primary product of purine biosynthesis: the hypoxanthine nucleotide, inosinate (IMP), from which adenylate and guanylate are then derived (Fig. 2-5). (See Table 1-1 for a summary of the names of bases, nucleosides, and nucleotides.)

The individual steps[12] in the sequences illustrated in Figures 2-4 and 2-5 are the loci for various metabolic errors found in mutants or in other disordered states. Mutants that lack adenylosuccinase,[13] which is required for synthesis of adenylate (Fig. 2-5), also fail to convert the succinyl intermediate in purine biosynthesis (Fig. 2-4); the same enzyme is used for both reactions. Enzymes in

---

11. Wong JY, Bernlohr DA, Turnbough CL, Switzer RL (1981) *Biochemistry 20*:5669; Bernlohr DA, Switzer RL (1981) *Biochemistry 20*:5675; Souciet J-L, Hermodson MA, Zalkin H (1988) *JBC 263*:3323.
12. Nierlich DP (1978) *Meth Enz 51*:179; Buchanan JM, Lukens LN, Miller RW (1978) *Meth Enz 51*:186; Buchanan JM, Ohnoki S, Hong BS (1978) *Meth Enz 51*:193; Schrimsher JL, Schendel FJ, Stubbe J, Smith JM (1986) *Biochemistry 25*:4366; Smith JM, Daum HA III (1986) *JBC 261*:10632; Schendel FJ, Mueller E, Stubbe J, Shiau A, Smith JM (1989) *Biochemistry 28*:2459.
13. Woodward DO (1978) *Meth Enz 51*:202; Fischer HE, Muirhead KM, Bishop SH (1978) *Meth Enz 51*:207.

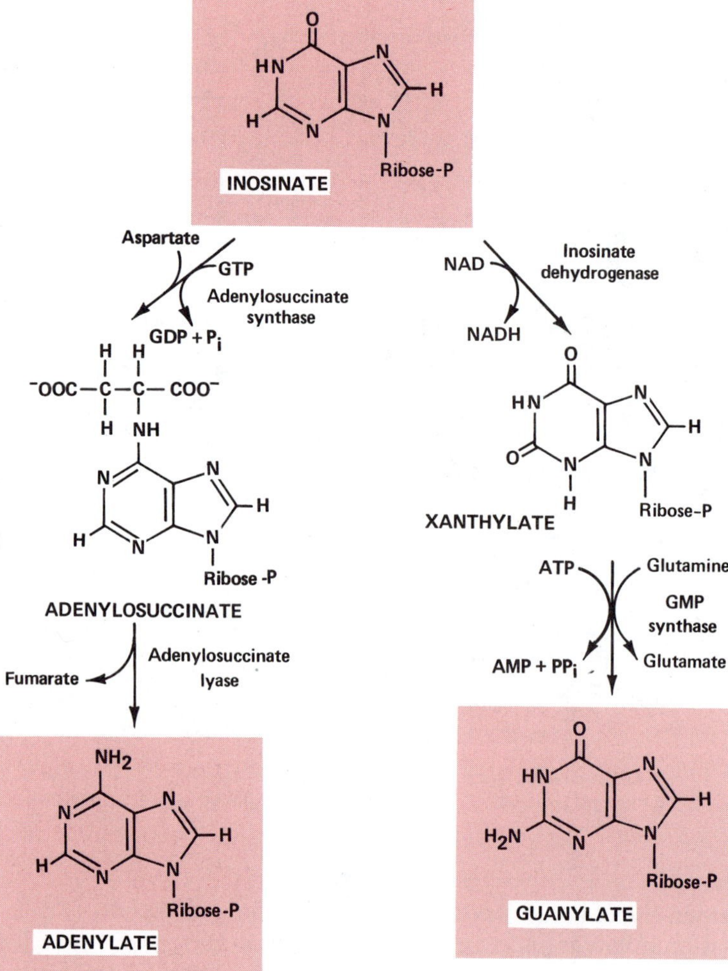

Figure 2-5
De novo pathway of purine nucleotide biosynthesis from inosinate to adenylate and guanylate.

these pathways may be involved in the fine regulation of rates of reactions important in growth and in cellular functions. These enzymes can be targets for inhibitors intended to block DNA synthesis (Section 14-2) in the treatment of proliferative diseases.

Both the GAR and AICAR formyltransferases (Table 2-2 and Fig. 2-4) have been isolated from chicken liver as a complex with the remarkable trifunctional folate enzyme that also includes serine transhydroxymethylase and synthesizes 10-formyl tetrahydrofolate and other folate cofactors (Section 14).[14] This proximity of the several enzyme catalytic centers provides a ready supply of the labile cofactor to the formylases. In addition, the integrity of this multienzyme complex is essential for activity of the GAR formyltransferase.[15] The AICAR formyltransferase is further linked in a single polypeptide with inosinicase,[16] ensuring the next and final step in the purine nucleotide synthetic pathway.

---

14. Caperelli CA, Benkovic PA, Chettur G, Benkovic SJ (1980) *JBC 255*:1885.
15. Smith GK, Mueller WT, Wasserman GF, Taylor WD, Benkovic SJ (1980) *Biochemistry 19*:4313.
16. Mueller WT, Benkovic SJ (1981) *Biochemistry 20*:337.

Table 2-2
Enzymes and genes of purine ribonucleotide biosynthesis in *E. coli*[a]

Enzyme	Reaction	Gene	Map position (minutes)
	ribose 5-phosphate + ATP		
PRPP synthase	↓	*prs*	26
	5-phosphoribosyl-α-pyrophosphate (PRPP)		
PRPP amidotransferase	↓	*purF*	50
	5-phosphoribosylamine		
GAR synthase	↓	*purD*	90
	glycinamide ribotide (GAR)		
GAR formyltransferase	↓	*purN*	54
	formylglycinamide ribotide (FGAR)		
FGAR amidotransferase	↓	*purL*	55
	formylglycinamidine ribotide (FGAM)		
AIR synthase	↓	*purM*	54
	5-aminoimidazole ribotide (AIR)		
AIR carboxylase	↓	*purE*	12
	carboxyaminoimidazole ribotide (CAIR)		
SACAIR synthase	↓	*purC*	53
	5-aminoimidazole-4-(N-succinylocarboxamide) ribotide (SACAIR)		
Adenylosuccinate lyase	↓	*purB*	25
	5-aminoimidazole-4-carboxamide ribotide (AICAR)		
AICAR formyltransferase	↓	*purH*	90
	5-formaminoimidazole-4-carboxamide ribotide (FAICAR)		
Inosinicase	↓	*purH*	90
	IMP		
	IMP		
Adenylosuccinate synthase	↓	*purA*	95
	Adenylosuccinate		
Adenylosuccinate lyase	↓	*purB*	25
	AMP		
Adenylate kinase	↓	*adk*	11
	ADP		
Nucleoside diphosphate kinase	↓	*ndk*	55
	ATP		
	IMP		
IMP dehydrogenase	↓	*guaB*	54
	XMP		
GMP synthase (xanthylate aminase)	↓	*guaA*	54
	GMP		
Guanylate kinase	↓	*gmk*	
	GDP		
Nucleoside diphosphate kinase	↓	*ndk*	55
	GTP		

[a] From Neuhard J, Nygaard P (1987) in *Escherichia Coli and Salmonella Typhimurium*, (Neidhardt FC, ed.). Am Soc Microbiol, Washington DC, Vol. 1 p. 445; Bachmann BJ (1990) *Microbiol Rev* 54:130; Masters M (1990) *J Bact* 172:1173.

Regulation of Purine Biosynthesis

There are two crucial stages in purine biosynthesis in *E. coli*, and both appear to be finely regulated. One, at PRPP amidotransferase, is the initial commitment to purine synthesis, and is under feedback regulation. The end products, IMP, AMP, and GMP, inhibit the pathway at the very first step. The PRPP amidotransferase that condenses PRPP with glutamine is inhibited by AMP, ADP, and ATP at one control site on the enzyme and by GMP, GDP, and GTP at another. A second stage of regulation occurs at the point in the pathway where inosinate goes either to adenylate or to guanylate[17] (Fig. 2-5). Adenylate synthesis by adenylosuccinate synthetase[18] requires GTP as the activating cofactor. An accumulation of GTP, reflecting a relative lack of ATP, accelerates production of adenylate, the ATP precursor. Conversely, the rate of inosinate conversion to guanylate depends on the level of ATP, the specific cofactor for this reaction. Furthermore, AMP inhibits adenylosuccinate synthetase, as does GMP in the initial reaction in the inosinate conversion to AMP. In addition to the important effects of nucleotides, the PRPP level exerts a major regulatory role at the PRPP amidotransferase step.

Genes Controlling Purine Nucleotide Biosynthesis

Genes are known in *Salmonella typhimurium* and in *E. coli*[19] for all but one of the twelve enzymes that convert PRPP to GMP and AMP (Table 2-2). Most of the genes, though physically unlinked, are transcriptionally coregulated by a single repressor. ATP and GTP may each act as corepressors with differential effects on reactions that precede the IMP branch point. What once appeared to be a simple and decisive regulation at IMP will probably prove to be more complex, acting both at this level and at earlier points in the de novo pathway.[20] In yeast, relatively little is known about the enzymes or their regulation, even though at least seven of the purine biosynthetic genes have long been identified.

Inhibitors of Purine Biosynthesis

An assortment of inhibitors can interrupt the pathway of purine biosynthesis at many points (Section 14-2). The antibiotic azaserine, which is isosteric with glutamine, competes with glutamine to block the reaction sequence by covalent bonding to a cysteine residue[21] at the active site of the PRPP amidotransferase. Amethopterin (methotrexate), aminopterin, and sulfonamides interrupt the purine pathway indirectly by reducing the supply of folate coenzymes: amethopterin and aminopterin prevent the regeneration of tetrahydrofolate from dihydrofolate produced in thymidylate synthesis (Sections 8 and 14).

Sulfonamides prevent the initial synthesis of folate coenzyme in microorganisms through competition with *p*-aminobenzoate, a precursor in the biosyn-

17. Sakamoto N (1978) *Meth Enz 51*:213; Spector T (1978) *Meth Enz 51*:219.
18. Wolfe SA, Smith JM (1988) *JBC 263*:19147.
19. Sampei G, Mizobuchi K (1989) *JBC 264*:21230; Aiba A, Mizobuchi K (1989) *JBC 264*:21239; Shen Y, Rudolph J, Stern M, Stubbe J, Flannigan KA, Smith JM (1990) *Biochemistry 29*:218.
20. Levine RA, Taylor MW (1981) *MGG 181*:313; Levine RA, Taylor MW (1982) *J Bact 149*:923.
21. Mizobuchi K, Buchanan JM (1968) *JBC 243*:4853; Thomas MS, Drabble WT (1985) *Gene 36*:45; Tiedeman AA, Smith JM (1985) *NAR 13*:1303; Tiedeman AA, Smith JM, Zalkin H (1985) *JBC 260*:8676.

thesis of folate; this is the mechanism of their antibacterial activity. The folate coenzyme is essential for the insertion of the one-carbon units at the C8 and C2 positions. Since the need for folate coenzyme for insertion at C2 is the more stringent, 5-aminoimidazole-4-carboxamide ribose phosphate accumulates in *E. coli* inhibited by sulfonamides. Catabolism of this intermediate to the free base, 5-aminoimidazole-4-carboxamide, and release of the free base into the culture medium by inhibited cells presented one of the earliest clues to the pattern of assembly of the purine skeleton.

Purine–Amino Acid Interrelationships

Figures 2-4, 2-5, and 2-6 illustrate the close interrelationships between purine biosynthesis and amino acid metabolism. Not only do certain amino acids (glutamine, glycine, aspartic acid) serve as precursors of the purines, but histidine is actually derived from ATP (Fig. 2-6). The N1 and C2 of the purine ring of ATP contribute to the imidazole skeleton, and what remains of the ATP, the 5-amino-4-imidazole carboxamide ribotide, is salvaged by insertion of C2 to form inosinate, thus restoring the purine nucleotide.

The Universality of Biochemical Processes

The purine biosynthetic pathway is essentially the same sequence of reactions in a wide range of species, including *E. coli*, yeast, pigeon, and human. The unity of biochemistry was dramatically illustrated in the first half of this century by the discovery that alcoholic fermentation in yeast is virtually identical with glycolysis in muscle, and this perception was reiterated with the delineation of purine pathways during the 1950s. The validity of this unity of biochemical mechanisms in nature is what justifies the confidence that studies of basic metabolic sequences in *E. coli* can be generally extended to the cells of more complicated, so-called "higher" organisms.

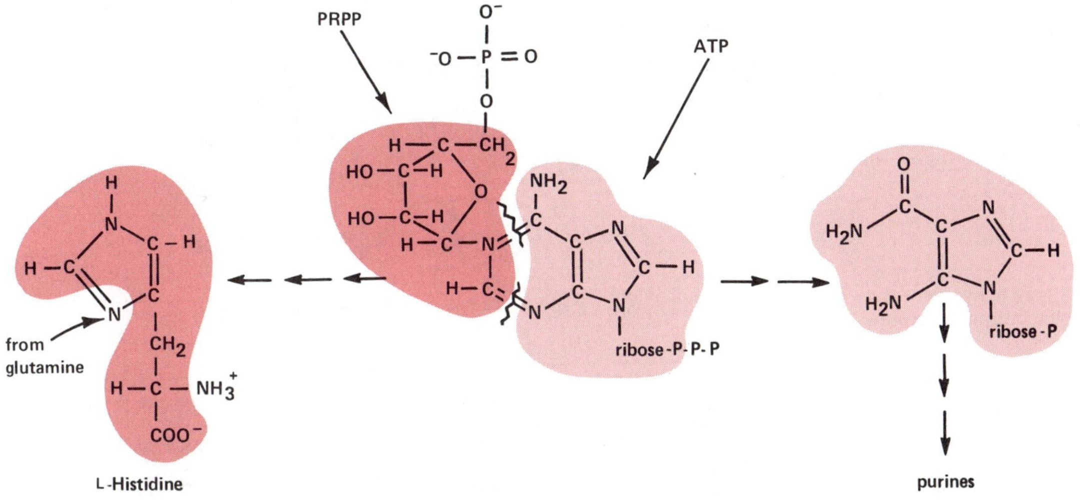

Figure 2-6
Histidine biosynthesis from PRPP and ATP.

2-4 Pyrimidine Nucleotide Synthesis de Novo[22]

The biosynthesis of pyrimidine nucleotides differs in pattern from that of purine nucleotides in that the pyrimidine skeleton is assembled first and is subsequently attached to ribose phosphate. The key pyrimidine intermediate is *orotate.*

The First Stage in Pyrimidine Biosynthesis

The pathway of orotate synthesis (Fig. 2-7) starts with the synthesis of carbamoyl phosphate,[23] which also serves arginine biosynthesis. (In mammalian cells, arginine synthesis and the urea cycle depend on a distinctive mitochondrial enzyme, carbamoyl phosphate synthetase.) In *E. coli*, transfer of the carbamoyl group to aspartate by aspartate carbamoyl transferase (transcarbamoylase) is the first unequivocal step of pyrimidine synthesis and is a key point for regulating the rate of commitment of carbamoyl phosphate and aspartate to the pathway. The enzyme is controlled through feedback inhibition by CTP, a terminal product of the pathway, and is activated by ATP. Studies of the control of aspartate carbamoyl transferase have pioneered and consistently advanced our general understanding of the mechanisms and regulation of enzyme activity.[24]

Dihydroorotase[25] catalyzes the cyclization of carbamoyl aspartate, by dehydration, to produce dihydroorotate, the reduced form of orotate. Oxidation of dihydroorotate then yields orotate. The oxidizing enzyme, initially isolated

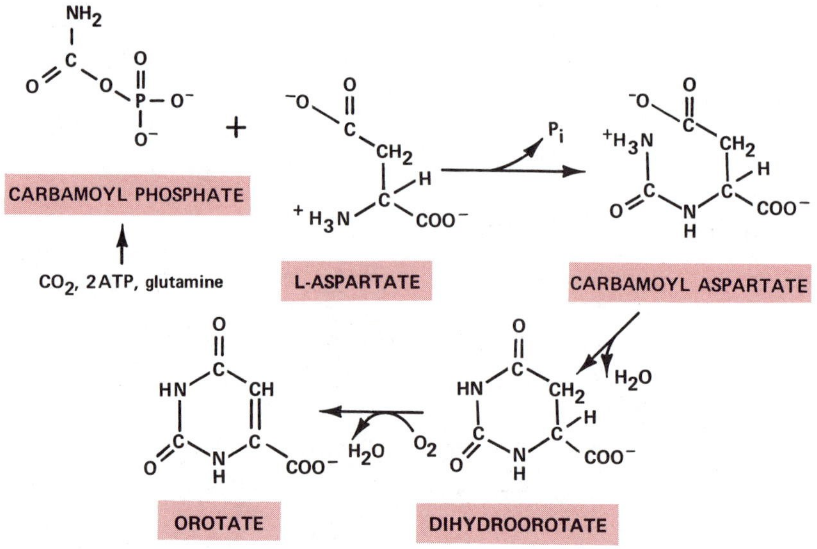

Figure 2-7
De novo pathway of pyrimidine nucleotide biosynthesis to orotate.

22. Neuhard J, Nygaard P (1987) in *Escherichia Coli and Salmonella Typhimurium* (Neidhardt FC, ed.). Am Soc Microbiol, Washington DC, Vol. 1 p. 445.
23. Makoff AJ, Radford A (1978) *Microbiol Rev 42*:307; Nyunoya H, Lusty CJ (1983) *PNAS 80*:4629; Nyunoya H, Lusty CJ (1984) *JBC 259*:9790.
24. Wente SR, Schachman HK (1987) *PNAS 84*:31; Kantrowitz ER, Lipscomb WN (1988) *Science 241*:669; Schachman HK (1988) *JBC 263*:18583.
25. Washabaugh MW, Collins KD (1984) *JBC 259*:3293; Neuhard J, Kelln RA, Stauning E (1986) *EJB 157*:335; Simmer JP, Kelly RE, Rinker AG Jr., Zimmerman BH, Scully JL, Kim H, Evans DR (1990) *PNAS 87*:174.

from anaerobic bacteria,[26] proved to be a flavoprotein, with only a feeble capacity to reoxidize the dihydroorotate to orotate. By contrast, another oxidase, subsequently isolated, catalyzes an extremely facile oxidation of dihydroorotate to orotate.[27]

The Independence of Biosynthesis and Energy Production

Genetic and biochemical studies make it clear that the bacterial flavoprotein is involved in the metabolism of orotate for energy needs and that the other enzyme is used for the biosynthesis of orotate. Thus, biosynthesis of orotate is managed by a set of enzymes that is largely or entirely distinct from the set that degrades orotate.

An appreciation in this instance of the duality of pathways of biosynthesis and degradation strengthens our belief that, in all cells, *the biochemical operations of energy production and of biosynthesis are distinct from one another.* This point needs emphasis because the early history of biochemistry focused on pathways of energy metabolism. And, when interest in biosynthesis began to develop, it was tacitly assumed that because individual degradative steps were catalyzed in both directions by the isolated enzymes, so must the entire pathway of degradation be reversible in a biosynthetic direction.

Biosynthesis of the Nucleotide

The second stage of pyrimidine biosynthesis involves condensation of orotate with PRPP and elimination of inorganic pyrophosphate (Fig. 2-8).[28] This reac-

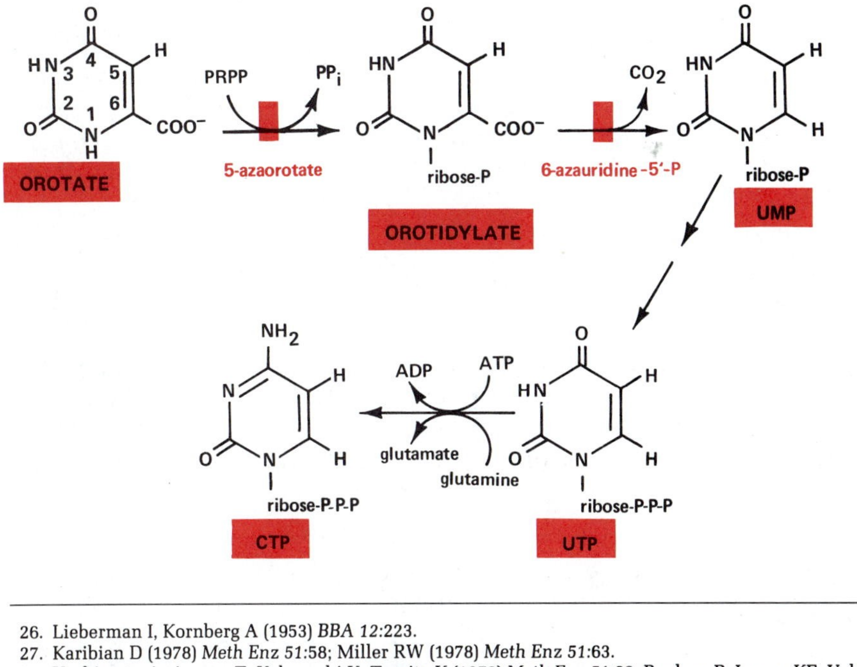

Figure 2-8
De novo pathway of pyrimidine nucleotide biosynthesis from orotate to UTP and CTP.

26. Lieberman I, Kornberg A (1953) *BBA 12*:223.
27. Karibian D (1978) *Meth Enz 51*:58; Miller RW (1978) *Meth Enz 51*:63.
28. Yoshimoto A, Amaya T, Kobayashi K, Tomita K (1978) *Meth Enz 51*:69; Poulsen P, Jensen KF, Valentin-Hansen P, Carlsson P, Lundberg LG (1983) *EJB 135*:223; Poulsen P, Bonekamp F, Jensen KF (1984) *EMBO J 3*:1783.

tion, analogous to the initial step of purine biosynthesis, fixes the pyrimidine in a β configuration; subsequent decarboxylation[29] produces uridylate.

The de novo route of thymidylate synthesis from deoxyuridylate is discussed in Section 8, following the description of conversion of ribonucleotides to deoxynucleotides.

Inhibitors of the Second Stage. Analogs of orotate and uridine inhibit the pathway (Fig. 2-8): 5-azaorotate competes with orotate in the condensation with PRPP, and 6-azauridylate blocks the decarboxylation of orotidylate, the condensation product. Inhibition of pyrimidine nucleotide synthesis by 5-azacytidine occurs at a particular stage of phage T4 DNA replication (Section 13).

The actions of inhibitors, so helpful in elucidating metabolic pathways in vivo and of such crucial importance as drugs in the treatment of disease, are discussed further in Chapter 14.

The Source of Cytidine Diphosphate. Cytosine nucleotides originate by amination of UTP, a reaction under refined allosteric control.[30] Inasmuch as nucleotides are reduced at the diphosphate level, the source of CDP is of great interest. A myokinase (adenylate kinase) type of enzyme for cytidylate (CTP:CMP phosphotransferase), detected in eukaryotic cell extracts,[31] may provide the major route from CTP to CDP and thence to dCDP.

Multifunctional Proteins in Pyrimidine Biosynthesis[32]

The contrast between the prokaryotic and eukaryotic systems of pyrimidine nucleotide synthesis demonstrates (1) the evolution of physical association between the enzyme units of this biosynthetic pathway and (2) the development of a miniorganelle.

In *E. coli*, six genes (Table 2-3) encode six enzymes that synthesize UMP from simple precursors (Figs. 2-7 and 2-8). The enzymes can be extracted in soluble and molecularly dispersed forms. Carbamoyl phosphate synthase and aspartate carbamoyl transferase, the first two enzymes in the chain, are each composed of two kinds of peptide chain, but none of the six is part of a multifunctional protein.[33]

Like the enzymes of fatty acid biosynthesis, several enzymes of the pyrimidine nucleotide system are linked in a single multifunctional unit in more complex organisms.[34]

Multifunctional Proteins of the First Stage. In *Drosophila*, the first three enzymes in the pathway are coded by a single complex genetic locus (the "rudimentary" locus), but physical evidence for a single gene or a multifunctional protein is lacking.[35] However, in cultures of rodent cells a 210-kDa protein contains the

29. Yoshimoto A, Umezu K, Kobayashi K, Tomita K (1978) *Meth Enz 51*:74; Donovan WP, Kushner SR (1983) *J Bact 156*:620.
30. Long C, Koshland DE Jr (1978) *Meth Enz 51*:79; Weinfeld H, Savage CR Jr, McPartland RP (1978) *Meth Enz 51*:84; Weng M, Makaroff CA, Zalkin H (1986) *JBC 261*:5568.
31. Chiba P, Cory JG (1988) *Cancer Biochem Biophys 9*:353.
32. Traut TW (1988) *Crit Rev B 23*:121; Coggins JR, Duncan K, Anton IA, Boocock MR, Chaudhuri S, Lambert JM, Lewendon A, Millar G, Mousdale DM, Smith DDS (1987) *Biochem Soc Trans 15*:754.
33. Yang YR, Kirschner MW, Schachman HK (1978) *Meth Enz 51*:35; Chang T-Y, Prescott LM, Jones ME (1978) *Meth Enz 51*:41; Adair LB, Jones ME (1978) *Meth Enz 51*:51.
34. Stark GR (1977) *TIBS 2*:64; Chaparian MG, Evans DR (1988) *FASEB J 2*:2982.
35. Rawls JM, Fristrom JW (1975) *Nature 255*:738.

three enzyme activities in a single polypeptide chain. Mutant cells selected for their resistance to N-phosphonacetyl-L-aspartate (PALA), an inhibitor of carbamoyl transferase, overproduced this enzyme 100-fold. The isolated enzyme showed a comparable increase in all three enzymatic activities but had only one PALA binding site, indicating that all the activities reside in a single poly-peptide.[36]

The trifunctional protein can be cleaved to peptide fragments, each with a single enzyme activity.[37] The carbamoyl transferase domain appears to be essential for the association of the 210-kDa polypeptide chains to form the trimers, hexamers, etc., of the native protein.[38] The domains of the aggregated protein are arranged in such a way that the two intermediates—carbamoyl phosphate and carbamoyl aspartate—are normally not released but are instead channeled to dihydroorotate, which is then made into UMP by the last three enzymes of the pathway.[39] Considerable homology is apparent between the aspartate carbamoyl transferase and dihydroorotase domains of the trifunctional protein and the respective discrete polypeptides in prokaryotes.

Table 2-3
Enzymes and genes of pyrimidine ribonucleotide biosynthesis in *E. coli*[a]

Enzyme	Reaction	Gene	Map position (minutes)
	glutamine, HCO_3^-, 2 ATP		
Carbamoyl phosphate synthase	↓	*carA, carB (pyrA)*	1
	carbamoyl phosphate		
Aspartate carbamoyl transferase	↓	*pyrB, pyrI*	97
	carbamoyl aspartate		
Dihydroorotase	↓	*pyrC*	23
	dihydroorotate		
Dihydroorotate oxidase	↓	*pyrD*	21
	orotate (+PRPP)		
Orotate phosphoribosyl transferase	↓	*pyrE*	82
	orotidine 5′-phosphate (OMP)		
OMP decarboxylase	↓	*pyrF*	28
	uridine 5′-phosphate (UMP)		
Uridylate kinase	↓	*pyrH*	5
	UDP		
Nucleoside diphosphate kinase	↓	*ndk*	55
	UTP		
CTP synthase	↓	*pyrG*	60
	CTP		

[a] From Neuhard J, Nygaard P (1987) in *Escherichia Coli and Salmonella Typhimurium* (Neidhardt FC, ed.). Am Soc Microbiol, Washington DC, Vol. 1 p. 445; Bachmann BJ (1990) *Microbiol Rev 54*:130; Sanderson KE, Hartman PE (1978) *Microbiol Rev 42*:471.

36. Padgett RA, Wahl GM, Coleman PF, Stark GR (1979) *JBC 254*:974.
37. Davidson JN, Rumsby PC, Tamaren J (1981) *JBC 256*:5220; Mally MI, Grayson DR, Evans DR (1981) *PNAS 78*:6647.
38. Dev IK, Harvey RJ (1978) *JBC 253*:4242; Smith GK, Benkovic PA, Benkovic SJ (1981) *Biochemistry 20*:4034; Coleman PF, Suttle DP, Stark GR (1977) *JBC 252*:6379.
39. Jones ME (1980) *ARB 49*:253; Christopherson RI, Jones ME (1980) *JBC 255*:11381; Mally MI, Grayson DR, Evans DR (1980) *JBC 255*:11372.

Multifunctional Proteins of the Second Stage. The conversion of orotate to UMP is effected by the bifunctional UMP synthetase, a 51-kDa polypeptide containing both orotate phosphoribosyl transferase and orotidylate decarboxylase.[40] As a monomer, the synthase lacks decarboxylase activity, but upon binding orotidylate (product of the transferase reaction), the enzyme becomes a bifunctional dimer. The barbituric acid analog of orotidylate binds the yeast decarboxylase domain 10^5 times more strongly than orotidylate does, suggesting a possible structure for the transition state.[41] Despite the plausibility of a bifunctional enzyme serving as a device for channeling an intermediate (i.e., orotidylate to UMP), available evidence indicates no such mechanism for UMP synthetase.[42]

When the overall pathway of pyrimidine biosynthesis functions in intact eukaryotic cells, the five intermediates—carbamoyl phosphate, carbamoyl aspartate, dihydroorotate, orotate, and orotidylate—are held at micromolar levels, in part because of the channeling by the multifunctional enzymes. Orotidylate can be degraded to orotidine by a pyrimidine nucleotidase or possibly by a specific nucleotidase;[43] orotidine is not converted back to orotidylate by kinase salvaging but rather by orotidine being converted first to orotate by uridine phosphorylase.[44] Humans afflicted with *orotic aciduria*,[45] an inborn metabolic error, are deficient in both the transferase and the decarboxylase; possibly, a single gene codes both enzymes in one polypeptide. These persons also accumulate and release carbamoyl aspartate, as do cultured cells with inhibited UMP synthetase.[46]

Advantages of Multifunctional Enzymes. Coordinate control of functionally linked enzymes in prokaryotes is achieved by regulation at the levels of transcription and translation. The individual polypeptides are so fashioned that they may later interact by specific associations to form a functional multiprotein complex. For the eukaryotic cell, however, the larger assortment of proteins in a greater volume may create problems in coordinate control and assembly. Thus, the advantages of a multifunctional enzyme include not only the covalent linkage of related enzymes but also a means of ensuring that they are made in equimolar amounts by a monocistronic message and a means of facilitating their developmental control.[47]

2-5 Nucleoside Monophosphate Conversion to Triphosphate[48]

Nucleoside monophosphates do not participate directly in nucleic acid biosynthesis but are first converted to diphosphates and then to triphosphates which

40. Floyd EE, Jones ME (1985) *JBC 260*:9443; Livingstone LR, Jones ME (1987) *JBC 262*:15726; Jacquet M, Guilbaud R, Garreau H (1988) *MGG 211*:441.
41. Levine HL, Brody RS, Westheimer FH (1980) *Biochemistry 19*:4993; Acheson SA, Bell JB, Jones ME, Wolfenden R (1990) *Biochemistry 29*:3198.
42. McClard RW, Shokat KM (1987) *Biochemistry 26*:3378; Traut TW (1989) *Arch B B 268*:108.
43. Traut TW (1980) *Arch B B 200*:590; el Kouni MH, Cha S (1981) *Fed Proc 40*:924.
44. Christopherson RI, Traut TW, Jones ME (1981) *Curr Top Cell Regul 18*:59; Traut TW (1989) *Arch B B 268*:108.
45. Suttle DP, Becroft DMO, Webster DR (1989) in *The Metabolic Basis of Inherited Disease* (Scriver CR, Beaudet AL, Sly WS, Valle D, eds.) 6th ed. McGraw-Hill, New York, p. 1095.
46. Jones ME (1980) *ARB 49*:253; Christopherson RI, Jones ME (1980) *JBC 25*:11381; Mally MI, Grayson DR, Evans DR (1980) *JBC 255*:11372.
47. Faure M, Kalekine M, Boy-Marcotte E, Jacquet M (1988) *Cell Differ 22*:159.
48. Ingraham JL, Ginther CL (1978) *Meth Enz 51*:371; Agarwal RP, Robison B, Parks RE Jr (1978) *Meth Enz 51*:376.

are incorporated into RNA and DNA. The only other metabolic fate for the nucleoside monophosphates is, as already mentioned, interconversion by amination and deamination reactions and by oxidations and reductions.

Conversion of the nucleoside monophosphates to diphosphates is catalyzed by kinases that, with two exceptions, are specific for each base but indifferent to whether the sugar is ribose or deoxyribose. The (d)CMP kinase also acts on UMP; the dTMP kinase also acts on dUMP but is specific for deoxyribose. The (deoxy)adenylate kinase,[49] for example, catalyzes the reversible reaction

$$\text{(d)AMP} + \text{ATP} \longleftrightarrow \text{(d)ADP} + \text{ADP}$$

The general reaction is

$$\text{(d)NMP} + \text{ATP} \longleftrightarrow \text{(d)NDP} + \text{ADP}$$

Synthesis of the diphosphate is favored by the regeneration of ATP through oxidative and other phosphorylations. ATP is the usual donor but may be replaced in some instances by dATP or other triphosphates.

An astonishing instance of the specificity of nucleoside monophosphate kinases is the enzyme encoded by T-even phages. This kinase is designed to furnish hydroxymethylcytosine deoxynucleoside triphosphate (HM·dCTP) in place of dCTP for incorporation into phage DNA at a high enough level to support the rate of phage DNA synthesis, which is tenfold greater than the rate of host DNA synthesis (Sections 13 and 17-6). The phage-induced kinase phosphorylates HM·dCMP, dTMP, and dGMP, but not dCMP. Furthermore, it ignores dAMP, which is phosphorylated at an adequate rate by adenylate kinase of the host cell.

The fate of nucleoside diphosphates is to become triphosphates. In only one puzzling instance, as substrates for the reduction of ribonucleotides to deoxyribonucleotides, do they participate directly in a proved biosynthetic sequence. Diphosphates are converted to triphosphates through the action of a powerful, ubiquitous, and *nonspecific* nucleoside diphosphate kinase.[50] The enzyme shows no preference either for purines or pyrimidines (including a large variety of synthetic analogs) or for ribose or deoxyribose. This lack of specificity applies to the donor triphosphate as well as the recipient diphosphate. The reaction may be formulated as

$$\text{(d)N}_X\text{PPP} + \text{(d)N}_Y\text{PP} \longleftrightarrow \text{(d)N}_X\text{PP} + \text{(d)N}_Y\text{PPP}$$

The donor is almost invariably ATP under aerobic conditions, but pyruvate kinase can provide an important route during anaerobiosis:

$$\text{phosphoenolpyruvate} + \text{(d)NDP} \longrightarrow \text{pyruvate} + \text{(d)NTP}$$

---

49. Reinstein J, Brune M, Wittenghofer A (1988) *Biochemistry 27*:4712.
50. Roisin MP, Kepes A (1978) *BBActa 526*:418; Muñoz-Dorado J, Inouye S, Inouye M (1990) *JBC 265*:2707.

2-6 Significance of Pyrophosphate-Releasing Reactions[51]

In the synthesis of purine and pyrimidine nucleotides, condensations with PRPP are key reactions. They are accompanied in each instance by the release of inorganic pyrophosphate (PP_i). The free energy change in these and other PP_i-releasing reactions is relatively small, and the synthetic direction of the reaction is demonstrably reversible upon addition of PP_i. However, PP_i is generally destined for hydrolysis due to the ubiquity of a potent inorganic pyrophosphatase. Exceptions are found in organisms that use PP_i in kinase reactions in place of ATP. As examples, the anaerobic parasite *Entamoeba histolytica* uses PP_i in the phosphofructokinase reaction, and *Propionibacterium shermanii* fixes CO_2 on pyruvate with the participation of PP_i.

When the released PP_i is hydrolyzed, reaction sequences such as nucleotide synthesis can be pulled far in the synthetic direction:

$$\begin{array}{c} \text{orotate} + \text{PRPP} \longleftrightarrow \text{orotidylate} + PP_i \\ PP_i + H_2O \longrightarrow 2\ P_i \\ \hline \text{Sum: orotate} + \text{PRPP} \longrightarrow \text{orotidylate} + 2\ P_i \end{array}$$

As mentioned already and to be discussed in Chapter 3, nucleoside triphosphates are used for nucleic acid synthesis. Polymerization involves repeated nucleotidyl additions, with the release of PP_i:

$$n\text{NTP} \longleftrightarrow (\text{NMP})_n + nPP_i$$

The removal of PP_i by hydrolysis drives the synthesis to completion, as in the biosynthesis of coenzymes (Fig. 3-1), proteins, lipids, and polysaccharides:

Coenzymes

$$\text{nicotinamide mononucleotide} + \text{ATP} \longleftrightarrow \text{NAD} + PP_i$$

Proteins

$$\text{amino acid} + \text{tRNA} + \text{ATP} \longleftrightarrow \text{aminoacyl tRNA} + \text{AMP} + PP_i$$

Lipids

$$\text{fatty acid} + \text{coenzyme A} + \text{ATP} \longleftrightarrow \text{fatty acyl CoA} + \text{AMP} + PP_i$$

Polysaccharides

$$\text{sugar–P} + \text{UTP} \longleftrightarrow \text{UDP–sugar} + PP_i$$

Because of the ready reversibility, enzymes catalyzing a PP_i-releasing reaction have been called *pyrophosphorylases*, by analogy with the phosphorylases. However, these enzymes are better regarded as transferases or *synthetases* because the generally low concentration of PP_i makes pyrophosphorolysis less likely. Phosphorylases, by contrast, usually serve in degradative pathways. They catalyze phosphorolytic cleavages, producing phosphorylated intermediates that conserve the energy of glycosidic and phosphodiester-linked com-

51. Kornberg A (1962) in *Horizons in Biochemistry* (Kasha M, Pullman B, eds.). Academic Press, New York, p. 251; Wood HG (1985) *Curr Top Cell Regul 26*:355.

pounds. But the main direction of the pathways involving phosphorylases is catabolic and toward energy production.

A PP_i-releasing synthetase coupled with an inorganic pyrophosphatase may seem an energetically wasteful process, but for the end achieved the expense is trivial. The de novo synthesis of a nucleotide consumes the equivalent of 65 to 70 ATPs. At the cost of only one additional ATP (lost by hydrolysis of PP_i), the integrity of a synthetic pathway is ensured against the vicissitudes of P_i or PP_i concentrations.

2-7 Ribonucleotide Reduction to a Deoxyribonucleotide[52]

The production of deoxyribonucleotides directly from ribonucleotides was suggested first by the observation[53] that in animals injected with cytidine labeled both in cytosine and in ribose, the ratio of labels was conserved in the deoxycytidine of DNA. An enzyme that reduces a ribonucleoside directly to the corresponding deoxyribonucleoside was subsequently discovered in extracts of E. *coli*. This enzyme, *ribonucleoside diphosphate reductase* (*ribonucleotide reductase*), catalyzes the reaction

$$\text{riboNDP} + \text{reductant-(SH)}_2 \longrightarrow \text{deoxyriboNDP} + \text{reductant-(S—S)}$$

The hydroxyl group at the 2′ position of the ribose moiety of a ribonucleoside diphosphate is replaced by a hydrogen (Fig. 2-9).

Thiol Coenzymes

The immediate source of reducing power comes from pairs of sulfhydryl groups on the enzyme,[54] which after oxidation are restored by a thiol coenzyme. The first such coenzyme to be identified was *thioredoxin*, a ubiquitous, heat-stable, 12-kDa protein with two sulfhydryl groups in nearby cysteine residues that are oxidized to cystine in the enzymatic reaction.[55] Thioredoxin, an efficient pro-

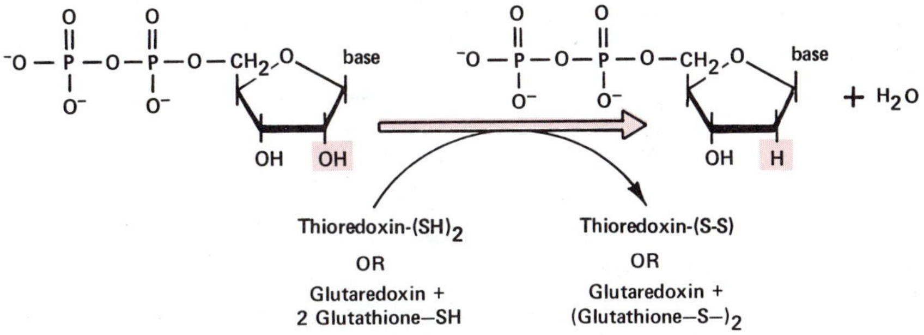

Thioredoxin-$(SH)_2$ OR Glutaredoxin + 2 Glutathione—SH → Thioredoxin-(S-S) OR Glutaredoxin + $(\text{Glutathione—S—})_2$ + H_2O

Figure 2-9
Ribonucleotide reduction to 2′-deoxyribonucleotide by ribonucleoside diphosphate reductase.

52. Reichard P (1988) *ARB 57*:349; Thelander L, Reichard P (1979) *ARB 48*:133; Reichard P (1987) *Biochemistry 26*:3245; Reichard P (1989) *BBActa 1000*:49.
53. Reichard P (1967) *The Biosynthesis of Deoxyribose*. Ciba Lectures in Biochemistry, Wiley, New York; Reichard P (1972) *Adv Enz Regul 10*:3.
54. Mao SS, Johnston MI, Bollinger JM, Stubbe J (1989) *PNAS 86*:1485.
55. Holmgren A (1985) *ARB 54*:, 237; Holmgren A (1989) *JBC 264*:2656; Katti SK, LeMaster DM, Eklund H (1990) *JMB 212*:167.

tein disulfide reductase, appears to function in oxidation-reductions to facilitate protein folding,[56] and serves in methionine sulfoxide and sulfate reductions. It is also essential for the function of phage T7 DNA polymerase (Section 5-9),[57] serves in the morphogenesis of filamentous phages (Section 17-3),[58] and, probably in a membranous location,[59] suppresses a lethal mutation in the gene for the *E. coli* dnaA initiator protein.[60]

When it was discovered that *E. coli* mutants totally lacking in thioredoxin[61] were viable, another hydrogen-donor system was recognized: glutathione (glutamylcysteinylglycine) and glutaredoxin, a heat-stable protein.[62] Reduction of oxidized thioredoxin and glutathione is achieved via specific flavoprotein reductases[63] employing NADPH as electron donor (Fig. 2-10). The further discovery of a double mutant, lacking both thioredoxin and glutaredoxin, indicates that still a third reducing system exists;[64] lipoic acid is a possible candidate.

Structure of Ribonucleotide Reductase

The ribonucleotide reductase of *E. coli* is the most thoroughly characterized of this class of enzymes. It consists of a one-to-one complex of two subunits, R1 and R2 (previously known as B1 and B2 for the bacterial enzyme and M1 and M2 for the mammalian enzyme), encoded by genes *nrdA* and *nrdB* (Table 2-4) within an essential operon. The subunits, which are separated during purification, constitute the active enzyme in the presence of Mg^{2+} (Fig. 2-11). Protein R1 is a dimer of α polypeptides (85.7 kDa) with two identical substrate binding sites and two different kinds of effector binding sites. One kind of effector binding site confers *substrate specificity*, and the other regulates *overall catalytic efficiency*. R1 also contains the thiol groups essential for activity. In the three-dimensional structure[65] of protein R2 (Plate 3), made up of two identical β polypeptides (43.4 kDa), two nonheme ferric irons are bridged by an oxygen atom and an organic free radical[66] is delocalized over the aromatic ring of a tyrosine residue (oxidized

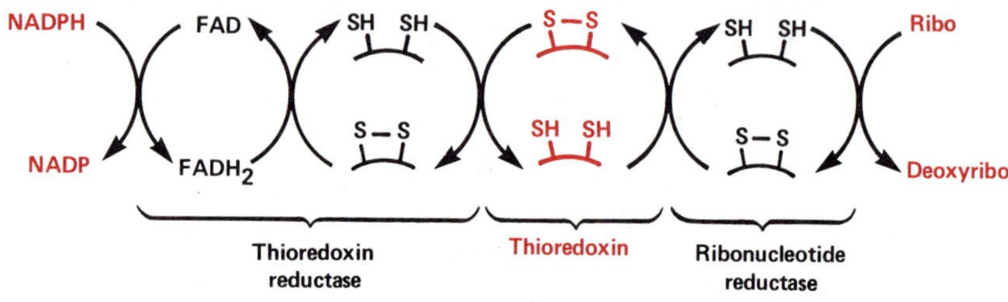

Figure 2-10
Electron transfer chain from NADPH, leading to reduction of ribonucleoside diphosphates. A similar scheme applies to glutaredoxin, which is reduced by a corresponding flavoprotein, glutathione reductase.

56. Pigiet VP, Schuster BJ (1986) *PNAS 83*:7643.
57. Conley RR, Pigiet V (1978) *JBC 253*:5568.
58. Lim C-J, Haller B, Fuchs JA (1985) *J Bact 161*:799; Russel M, Model P (1985) *PNAS 82*:29.
59. Bayer ME, Bayer MH, Lunn CA, Pigiet V (1987) *J Bact 169*:2659.
60. Hupp TR, Kaguni JM (1988) *MGG 213*:471.
61. Mark DF, Richardson CC (1976) *PNAS 73*:780; Holmgren A, Ohlsson I, Grankvist M (1978) *JBC 253*:430.
62. Luthman M, Eriksson S, Holmgren A, Thelander L (1979) *PNAS 76*:2158.
63. Holmgren A (1989) *JBC 264*:13963; Prongay AJ, Engelke DR, Williams CH Jr (1989) *JBC 264*:2656.
64. Russel M, Holmgren A (1988) *PNAS 85*:990.
65. Nordlund P, Sjöberg B-M, Eklund H (1990) *Nature 345*:593.
66. Larsson Å, Karlsson M, Sahlin M, Sjöberg B-M (1988) *JBC 263*:17780; Sahlin M, Gräslund A, Petersson L, Ehrenberg A, Sjöberg B-M (1989) *Biochemistry 28*:2618.

Table 2-4
Genes and enzymes of deoxyribonucleotide biosynthesis and salvage pathways of *E. coli*[a]

Gene	Map position (minutes)	Enzyme or function
apt	11	adenine phosphoribosyl transferase
add	36	adenosine deaminase
cdd	46	(d)cytidine deaminase
dcd	45	dCTP deaminase
deoA (tpp)	100	thymidine phosphorylase
deoB (drm)	100	deoxyribomutase
deoC (dra)	100	deoxyriboaldolase
deoD (pup)	100	purine nucleoside phosphorylase
deoR (nucR)	19	regulator of thymine required by *thy*⁻ strains
dut	82	deoxyuridine triphosphatase (dUTPase)
guaC	3	guanylate reductase
gpt	6	guanine phosphoribosyl transferase
grx	18.5	glutaredoxin
gsk	13	guanosine kinase
hpt	3	hypoxanthine–guanine phosphoribosyl transferase
nrdA	49	nucleoside diphosphate reductase
nrdB	49	nucleoside diphosphate reductase
tdk	27	thymidine kinase
thyA	61	thymidylate synthase
tmk		thymidylate kinase
trxA	85	thioredoxin
trxB		thioredoxin reductase
tsx (nup)	9	nucleoside uptake; phage T6 receptor; colicin K receptor
udp	86	uridine phosphorylase

[a] From Elion GB (1989) *Science 244*:41; Bachmann BJ (1990) *Microbiol Rev 54*:130; Sanderson KE, Hartman PE (1978) *Microbiol Rev 42*:471.

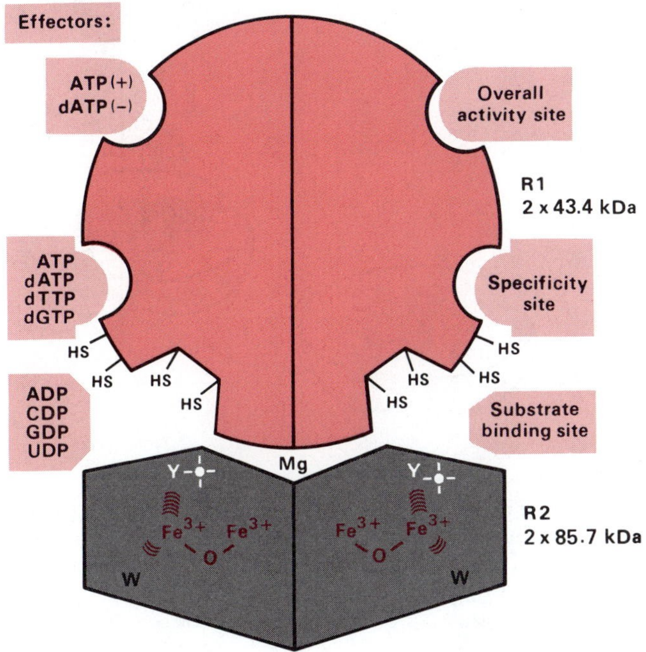

Figure 2-11
Ribonucleoside diphosphate reductase of *E. coli*: subunit structure, active site, and allosteric control sites. W = tryptophan, Y·⁺ = tyrosyl free radical. (Courtesy of Professor P Reichard)

for lack of a ring electron). The iron center helps to generate and stabilize the free radical. Hydroxyurea inhibits the enzyme by scavenging the free radical.

The catalytic site is formed from the interaction of R1 and R2: R1 contributes the thiol groups and binds the substrate; R2 supplies the free radical. The R1 subunit in ribonucleotide reductase is thus not strictly a regulatory subunit like those of other regulated enzymes, such as *E. coli* aspartate carbamoyl transferase; R1 is essential for catalytic activity as well as for binding effectors.

The mammalian reductase is essentially the same as the *E. coli* enzyme in structure, mechanism, and regulation.

Mechanism of Ribonucleotide Reductase Activity[67]

An attractive scheme (Fig. 2-12) proposes the formation of a very short-lived substrate radical, leading to the abstraction of a hydrogen atom from the 3′-carbon, possibly by the tyrosyl radical of the enzyme, and restoration of the enzyme radical following the loss of the 2′-hydroxyl. In favor of this mechanism is the fact that interaction between the reductase and the substrate analog, 2′-azido-CDP, results in the transient appearance of an electron paramagnetic resonance

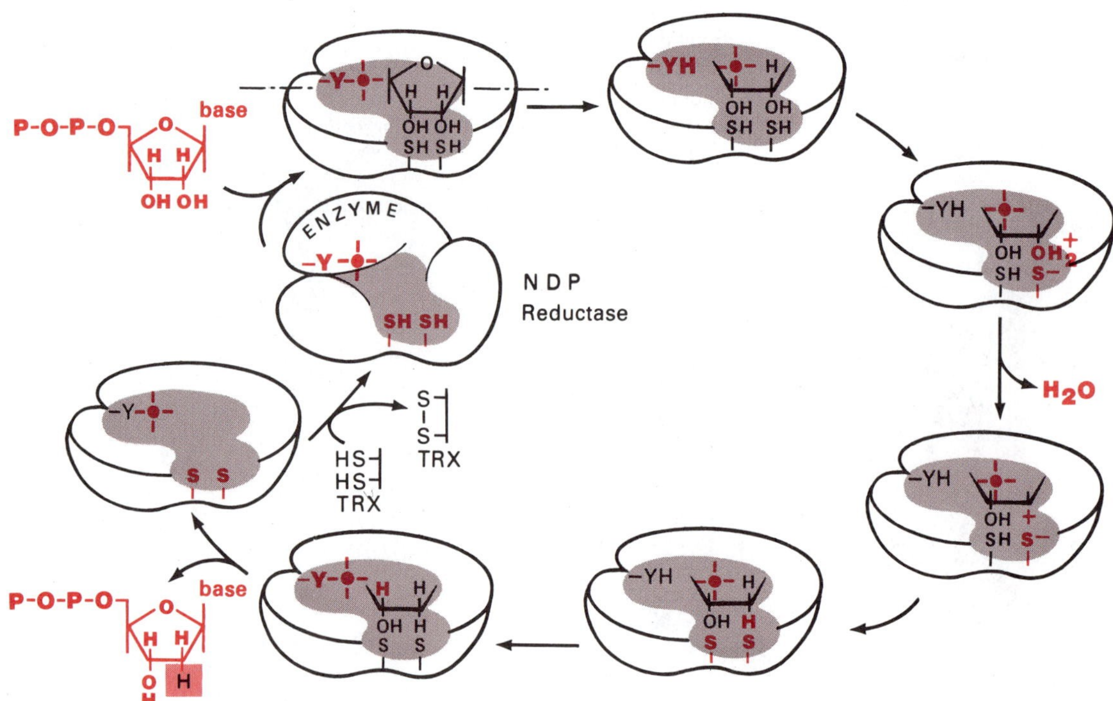

Figure 2-12
Mechanism of ribonucleotide reductase. Y·‡· = tyrosyl free radical; TRX = thioredoxin. [After a scheme proposed by Stubbe J, Ator M, Krenitsky T (1983) *JBC 258*:1625; Ashley GW, Stubbe J (1985) *Pharmacol Ther 30*:301. Two pairs of sulfhydryl groups, rather than one, are probably involved—see Mao SS, Johnston MI, Bollinger JM, Stubbe J (1989) *PNAS 86*:1485]

67. Stubbe J, Ator M, Krenitsky T (1983) *JBC 258*:1625; Ashley GW, Stubbe J (1985) *Pharmacol Ther 30*:301.

signal localized to the nucleotide. Presumably, the lifetime of the substrate radical is extended by the azido substituent. A similar abstraction of hydrogen from the 3′ position may occur with the *Lactobacillus* reductase (see below).

Treatment of R2 with a radical scavenger (e.g., hydroxyurea) or with an extract of *E. coli* under anaerobic conditions inactivates the reductase. Reactivation ("radicalization") of the enzyme can be achieved by a complex system which depends on NADPH, flavin, a dithiol, oxygen, and four enzymes, three of which have been identified as superoxide dismutase, catalase, and a pyridine–flavin oxidoreductase.[68] Anaerobically, *E. coli* must resort to an alternative way of producing deoxyribonucleotides, and, indeed, such an oxygen-independent system has been discovered in extracts from anaerobically grown cells.[69] The substrate for reduction is a ribonucleoside triphosphate (rNTP); the enzyme is sensitive to oxygen, is allosterically regulated by dNTPs, and requires S-adenosylmethionine as a cofactor.[70]

Regulation of Reductase Activity[71]

The levels of reductase can be modulated by allosteric control, by synthesis of the subunits, and possibly by a balance between radical formation and inactivation.

Allosteric Control. Regulation by nucleotide effectors of substrate specificity enables the reductase to attain a strikingly fine adjustment to the needs for DNA synthesis (Fig. 2-13). Binding of ATP to the overall activity sites makes the enzyme active; binding of dATP to these sites renders the enzyme inactive (Fig. 2-11). Binding of ATP, dATP, dTTP, or dGTP to the substrate specificity sites

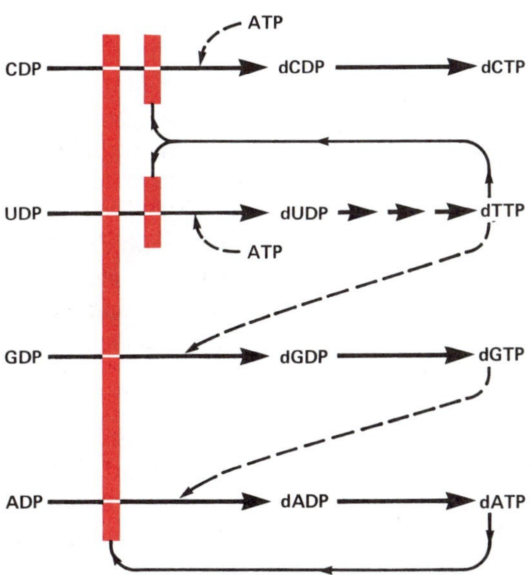

Figure 2-13
Allosteric control of ribonucleoside diphosphate reductase of *E. coli*. Bars indicate inhibitions; broken lines indicate activations. (Courtesy of Professor P Reichard)

68. Reichard P (1988) *ARB 57*:349. Fontecave M, Eliasson R, Reichard P (1987) *JBC 262*:12325; Fontecave M, Gräslund A, Reichard P (1987) *JBC 262*:12332.
69. Fontecave M, Eliasson R, Reichard P (1989) *PNAS 86*:2147.
70. Eliasson R, Fontecave M, Jörnvall H, Krook M, Pontis E, Reichard P (1990) *PNAS 87*:3314.
71. Thelander L, Reichard P (1979) *ARB 48*:133; Reichard P (1987) *Biochemistry 26*:3245.

determines the V_{max} and the K_m for the nucleoside diphosphates in the substrate binding sites. When ATP is bound in the substrate specificity sites, reductions of UDP and CDP are favored. Binding of dTTP stimulates GDP reduction, and binding of dGTP leads to ADP reduction. Binding of effectors must induce several distinct enzyme conformations.

The regulatory effects of the ribo- and deoxynucleoside triphosphates can be linked to DNA synthesis in the following way: ATP accumulation is a general signal for biosynthesis and ribonucleotide reduction. Piling up of dATP signifies an excess of one of the DNA building blocks: in small amounts, dATP calls for production of more pyrimidine deoxynucleotides; when dATP accumulates, it turns off the reductase entirely. Although dCTP shows no allosteric effects, dTTP is an adequate substitute; it inhibits further pyrimidine nucleotide reduction and stimulates GDP reduction. A build-up of dGTP calls for increased ADP reduction.

The pattern of regulation deduced from observations with the pure *E. coli* enzyme appears to apply to animal cells as well as to bacterial cells in vivo. Several examples illustrate this:

(1) Thymidine added to animal cells in culture at concentrations above 1 mM blocks DNA synthesis. Pools of dTTP, dGTP, and dATP are increased 25-fold, tenfold, and twofold, respectively, but that of dCTP is decreased tenfold.[72] Thus it appears that raising the dTTP level by thymidine administration stimulates the reduction of GDP and ADP and inhibits the reduction of CDP. The reduced dCTP level may block DNA synthesis; indeed, the addition of deoxycytidine overcomes the block.

(2) In humans, a deficiency of either (deoxy)adenosine deaminase or purine nucleoside phosphorylase can lead to immunodeficiency disease (Section 10). The lack of the salvage enzyme leads to large elevations of the levels of dATP or dGTP, with a consequent block in production of the other deoxynucleotides needed for DNA synthesis.

(3) Reductases encoded by some viruses (e.g., phages, vaccinia, and herpes) must meet the greater demand for precursors for synthesis of viral DNA.[73] The T4 phage reductase reacts specifically with a phage-encoded thioredoxin.[74] The phage system resembles that of the host cell except that both glutathione and thioredoxin reductase can reduce the phage thioredoxin, and dATP is the principal positive effector, rather than inhibitor, for reduction of pyrimidine nucleotides. Evidently, the nucleotide economy of the infected cell, geared to all-out phage DNA synthesis, requires a regulatory pattern unlike that of the uninfected cell. Curiously, the reductase encoded by vaccinia virus conforms more to feedback controls of the host cell.[75]

Regulation by Subunit Synthesis. The control of reductase synthesis in *E. coli* is effected by coordinate regulation of the genes for the R1 and R2 subunits within an operon. Regulation of the reductase levels in animal cells is more complex.[76] Quiescent mammalian cells lack the enzyme. Induction of DNA synthesis leads

---

72. Bjursell G, Reichard P (1973) *JBC 248*:3904.
73. Berglund O (1972) *JBC 247*:7270, 7276.
74. Sjöberg B, Söderberg B (1976) *JMB 100*:415; Söderberg B-O, Sjöberg B-M, Sonnerstam U, Bränden C-I (1978) *PNAS 75*:5827.
75. Slabaugh MB, Mathews CK (1986) *J Virol 60*:506; Slabaugh MB, Roseman N, Davis R, Mathews CK (1988) *J Virol 62*:519; Tengelsen LA, Slabaugh-MB, Bibler JK, Hruby DE (1988) *Virology 164*:121.
76. Engstrom Y, Rozell B (1988) *EMBO J 7*:1615.

to a steady synthesis of the R1 subunit throughout the growth cycle and a striking increase of R2 in each S phase of the cycle. A half-life of 3 hours for R2 contrasts with 15 hours for R1. During embryogenesis of the clam,[77] the unfertilized oocyte contains R1, but only the mRNA for R2; fertilization signals the translation of R2 mRNA, the appearance of active reductase, and a supply of dNTPs for DNA synthesis, all within a few minutes. In yeast, two genes encode R2, one of which responds to DNA damage.[78]

Reductase in Lactobacilli

The reductase system in lactobacilli differs in several ways from that in *E. coli*. The enzyme consists of a single polypeptide of 76 kDa.[79] Its cofactor is vitamin B_{12} coenzyme (5′-deoxyadenosylcobalamin) rather than nonheme iron, and the substrates are ribonucleoside triphosphates, not the diphosphates.[80] As in *E. coli*, the reductase uses thioredoxin as a reductant and employs a free radical mechanism.[81] Although it is a single polypeptide, the enzyme is under refined allosteric control,[82] a clear exception to the rule that subunit interactions are the basis for such regulation.

2-8 Origin of Thymine[83]

De Novo Synthesis of Thymidylate

Methylation of deoxyuridylate (dUMP), derived either by deaminase action on dCMP or by dUTPase (Section 9), is the de novo route for the synthesis of thymidylate (dTMP—see Section 1-2 on the optional use of "deoxy" as a prefix for thymidine, thymidylate, etc.). The dCMP deaminase supplies most of the dUMP in mammalian cells, yeast, and phage T2-infected *E. coli*. The enzyme, under allosteric control, is activated by dCTP (hydroxymethyl dCTP for T2) and is inhibited by dTTP.

Thymidylate synthase[84] catalyzes the transfer of a methylene group, a one-carbon unit at the hydroxymethyl (CH_2OH) oxidation state, from methylene tetrahydrofolate (THF) to deoxyuridylate, accompanied by reduction of the group to CH_3 (Fig. 2-14). Reduction is at the expense of THF to form dihydrofolate. Regeneration of THF, by a dihydrofolate reductase, is essential not only for sustained thymidylate synthesis but also for other biosynthetic events that

77. Standart N, Hunt T, Ruderman JV (1986) *J Cell Biol 103*:2129.
78. Elledge SJ, Davis RW (1987) *Mol Cell Biol 7*:2783.
79. Chen AK, Bhan A, Hopper S, Abrams R, Franzen JS (1974) *Biochemistry 13*:654; Blakley RL (1978) *Meth Enz 51*:246.
80. Ludwig W, Follman H (1978) *EJB 82*:393.
81. Mao SS, Johnston MI, Bollinger JM, Stubbe J (1989) *PNAS 86*:1485; Stubbe J, Ator M, Krenitsky T (1983) *JBC 258*:1625; Ashley GW, Stubbe J (1985) *Pharmacol Ther 30*:301.
82. Sjöberg B, Söderberg B (1976) *JMB 100*:415; Söderberg B-O, Sjöberg B-M, Sonnerstam U, Bränden C-I (1978) *PNAS 75*:5827.
83. Belfort M, Maley GF, Maley F (1983) *PNAS 80*:1858; Belfort M, Maley GF, Pedersen-Lane J, Maley F (1983) *PNAS 80*:4914; Blakley RL (1984) in *Folates and Pteridines* (Blakley RL, Benkovic SJ, eds.). Wiley, New York, Vol. 1, p. 191; Santi DV, Dannenberg PV (1984) in *Folates and Pteridines* (Blakley RL, Benkovic SJ, eds.). Wiley, New York, Vol. 1, p. 343; Hardy LW, Finer-Moore JS, Montfort WR, Jones MO, Santi DV, Stroud RM (1987) *Science 235*:448; Matthews DA, Appelt K, Oatley SJ (1989) *JMB 205*:449; Davisson VJ, Sirawaraporn W, Santi DV (1989) *JBC 264*:9145.
84. Climie S, Santi DV (1990) *PNAS 87*:633.

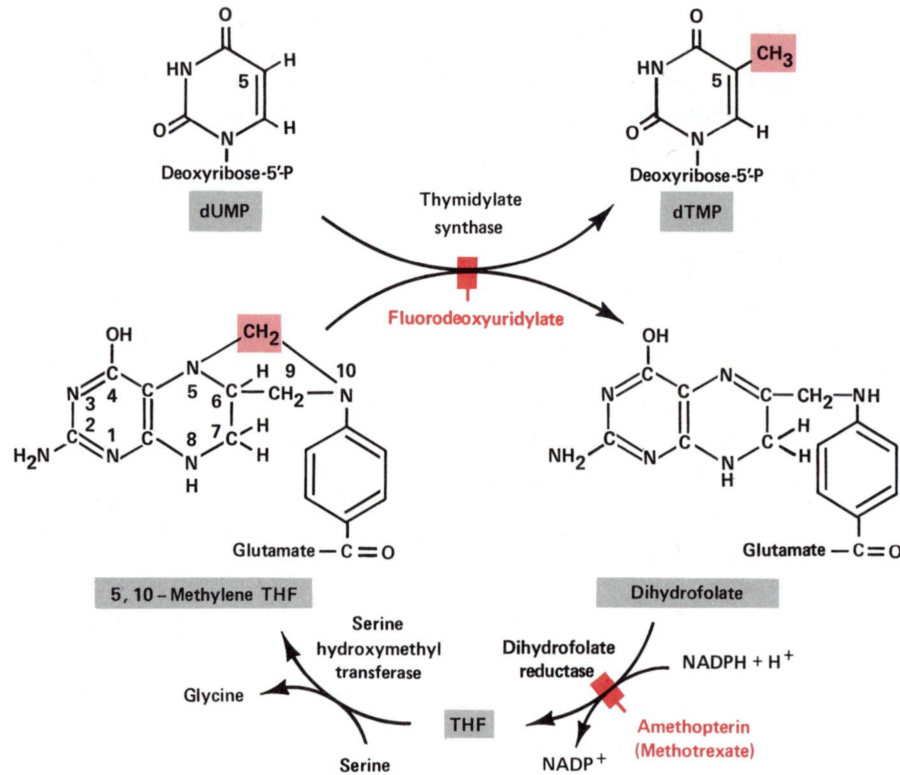

Figure 2-14
Thymidylate biosynthesis from deoxyuridylate (dUMP) by thymidylate synthase.

depend on THF coenzyme activity, such as the synthesis of purines, histidine, glycine, and methionine.

Dihydrofolate reductase served as the original object in studies of gene amplification[85] and is very much favored for protein structural studies.[86] A bifunctional thymidylate synthase–dihydrofolate reductase in a protozoan parasite (*Leishmania*)[87] resembles in its subunit size the sum of the synthase and reductase enzymes from a wide variety of other sources.[88] The synthase gene (*td*) in T4 phage is remarkable as one of only three examples (*nrdB* of T4 being another) of a prokaryotic gene containing an intron.[89] The active site of T4 synthase has been explored by site-directed mutagenesis.[90]

Inhibitors of Thymidylate Synthesis

The thymidylate synthesis step affords an opportunity to interrupt the synthesis of DNA and thus the growth of cells, especially rapidly proliferating cells (Sec-

85. Stark GR, Wahl GM (1984) *ARB 53*:447.
86. Garvey EP, Matthews CR (1989) *Biochemistry 28*:2083; Birdsall B, Feeney J, Tendler SJB, Hammond SJ, Roberts GCK (1989) *Biochemistry 28*:2297; Adams J, Johnson K, Matthews R, Benkovic SJ (1989) *Biochemistry 28*:6611; Bystroff C, Oatley SJ, Kraut J (1990) *Biochemistry 29*:3263.
87. Ferone R, Roland S (1980) *PNAS 77*:5802.
88. Meek TD, Garvey EP, Santi DV (1985) *Biochemistry 24*:678; Bzik DJ, Li W-B, Horii T, Inselburg J (1988) *PNAS 84*:8360.
89. Chu FK, Maley GF, Maley F, Belfort M (1984) *PNAS 81*:3049; Sjöberg B-M, Hahne S, Mathews CZ, Mathews CK, Rand KN, Gait MJ (1986) *EMBO J 5*:2031; Shub DA, Gott JM, Xu M-Q, Lang BF, Michel F, Tomaschewski J, Pedersen-Lane J, Belfort M (1988) *PNAS 85*:1151.
90. LaPat-Polasko L, Maley GF, Maley F (1990) *Biochemistry 29*:9561.

tion 14-2). Inhibitors of either thymidylate synthase or dihydrofolate reductase have been used clinically. Thymidylate synthase can be blocked by fluorodeoxyuridylate (Fig. 2-14), which is derived from administered fluorouracil or fluorodeoxyuridine by salvage reactions. Fluorodeoxyuridylate forms a ternary covalent complex with the synthase and methylene THF, immobilizing the enzyme.[91] Dihydrofolate reductase can be virtually totally inhibited by the folate antagonists amethopterin (methotrexate, 4-amino-10-methylfolate) and aminopterin (4-aminofolate), which have an affinity for the enzyme (K_d near 10^{-9} M) that is 100-fold greater than that of dihydrofolate.

In clinical treatments that use fluorodeoxyuridine and methotrexate together, their effects may be mutually antagonistic. (1) Depletion of THF by methotrexate may reduce the efficacy of fluorodeoxyuridylate, which requires the coenzyme for its inhibitory effect.[92] (2) The block of thymidylate synthase by fluorodeoxyuridylate halts the conversion of THF to dihydrofolate, and thus abolishes the cytotoxic effects of severe THF depletion.

The principles of methotrexate therapy in childhood leukemia are discussed in Section 14. Other aspects of the therapeutic application of methotrexate and mechanisms of cellular resistance to it are considered in Section 14-2.

Thymineless Death[93]

Depletion of dTTP due to lack of thymidylate synthase activity and inadequate salvage of precursors (Section 11) results in chromosome damage and death in bacteria and in eukaryotic cells. The mechanism may be due to excessive incorporation of uracil (Section 9), induction of an endonuclease, or both. Similar dire consequences are likely to result from the depletion of any dNTP pool and may affect DNA repair as well as replication.

A discussion of thymidylate synthesis by salvage of thymidine and a technique for selecting thymidylate synthase-defective mutants are presented in Section 11.

2-9 dUTPase in Thymidylate Biosynthesis[94]

An important source of dUMP (the precursor of dTMP) is by dUTPase action on dUTP (Fig. 2-15). This became apparent when it was realized that dUTP (derived from dUDP or by deamination of dCTP) is readily incorporated by DNA polymerase into DNA in place of dTTP. Inasmuch as uracil is not found in DNA, it followed that an enzyme must be eliminating dUTP as a substrate for DNA polymerase. Fulfilling this role, dUTPase was recognized in normal *E. coli* as well as in cells infected with phages which contain hydroxymethyluracil in

---

91. Pogolotti AL Jr, Santi DV (1977) *Enzyme action*. Bioorganic Chemistry Series Vol. 1. Academic Press, New York, p. 277.
92. Ullman B, Lee M, Martin DW Jr, Santi DV (1978) *PNAS 75*:980.
93. Seno T, Ayusawa D, Shimizu K, Koyama H, Takeishi K, Hori T (1985) in *Genetic Consequences of Nucleotide Pool Imbalance* (de Sernes FJ, ed.). Plenum Press, New York, p. 241; Kunz BA, Taylor GR, Haynes R (1986) *Genetics 114*:375.
94. Bertani LE, Häggmark A, Reichard P (1963) *JBC 238*:3407; Hochhauser S, Weiss B (1978) *J Bact 134*:157; Shlomai J, Kornberg A (1978) *JBC 253*:3305; Lundberg LG, Thoresson H-O, Karlstrom OH, Nyman P-O (1983) *EMBO J 2*:967; Hoffman I, Widström J, Zeppezauer M, Nyman P-O (1987) *EJB 164*:45.

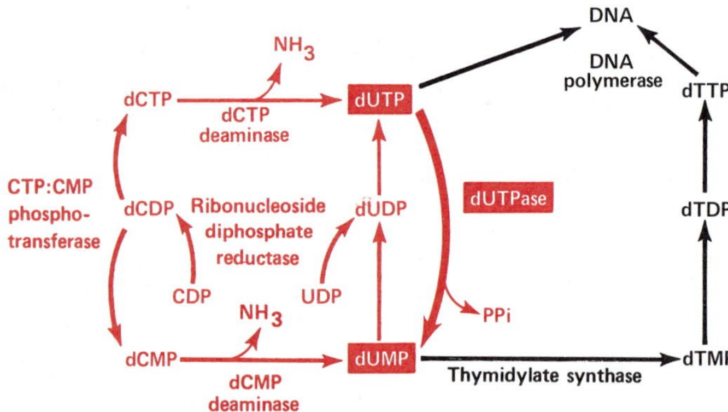

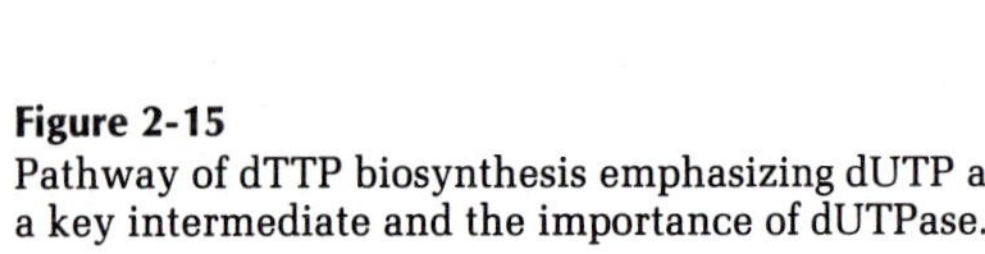
Figure 2-15
Pathway of dTTP biosynthesis emphasizing dUTP as a key intermediate and the importance of dUTPase.

place of thymine (Section 13); the latter cells must also eliminate dUTP from competition with hydroxymethyl dUTP for phage DNA synthesis.

The emphasis on the function of dUTPase in removing dUTP may be responsible for the long neglect in appreciating its biosynthetic role in generating dUMP as a precursor. Renewed interest in dUTPase has come from several directions:

(1) Recognition of an *E. coli* excision–repair system that detects uracil in DNA and removes it by N-glycosylase action (uracil N-glycosylase; Section 13-9). Following uracil removal, the apyrimidinic acid region is excised from the DNA backbone by an endonuclease (or by alkali in vitro). When uracil is found in DNA paired to adenine, it must have been incorporated directly into DNA; when paired to guanine, it must have been generated by the spontaneous deamination of cytosine in DNA.

(2) Identification of mutants, called *dut*⁻, with reduced levels of dUTPase.[95] These mutants were previously called *sof*, because they accumulate *s*hort Okazaki *f*ragments after a brief [^{3}H]thymine pulse. The fragmented nascent DNA is explained by uracil N-glycosylase excision of sites of uracil incorporation (Section 15-10).

(3) Identification of mutants called *ung*⁻ (lacking uracil *N*-glycosylase)[96] (Section 13-9), and *dut*⁻, *ung*⁻ double mutants, in which uracil accumulates in DNA with a frequency of about 1 per 100 thymines (Section 15-10).

(4) Purification and characterization of *E. coli* dUTPase, a tetramer of 16-kDa subunits.[97] Its abundance and kinetic properties suggest that cellular steady-state dUTP levels of about 0.5 μM would be needed were it the sole precursor for dTTP biosynthesis. However, at such a level of dUTP relative to a dTTP level near 100 μM, uracil residues would be incorporated into DNA with a frequency of 1 per 200 thymines.

(5) Observations from in vitro studies of phage ϕX174 DNA synthesis that purified replication enzymes freed of dUTPase, but still contaminated with

95. Tye BK, Nyman P-O, Lehman IR, Hochhauser S, Weiss B (1977) *PNAS 74*:154; Hochhauser S, Weiss B (1978) *J Bact 134*:157; el Hajj HH, Zhang H, Weiss B (1988) *J Bact 170*:1069.
96. Duncan BK, Rockstroh PA, Warner HR (1978) *J Bact 134*:1039.
97. Lundberg LG, Thoresson H-O, Karlstrom OH, Nyman P-O (1983) *EMBO J 2*:967; Hoffman I, Widström J, Zeppezauer M, Nyman P-O (1987) *EJB 164*:45; Shlomai J, Kornberg A (1978) *JBC 253*:3305.

uracil N-glycosylase and excision nucleases, produce fragmented DNA products. Addition of dUTPase, to reduce dUTP levels below 10^{-8} M (1000 times lower than the K_m for dUTPase), permitted the production of intact ϕX174 DNA circles (5386 residues) because uracil incorporation had been prevented.[98]

Although the studies cited have been derived from *E. coli*, the functions and significance of dUTPase, uracil incorporation, and excision–repair seem applicable to animal cells as well. Addition of dUTP to nuclei synthesizing polyoma viral DNA led to a reduced size of nascent DNA fragments, implying the action of a uracil N-glycosylase. Under normal circumstances, dUTPase operates very efficiently, leaving no measurable dUTP pools in animal cells.[99] However, dUTP was demonstrated after thymidylate synthesis in lymphoblast cells was inhibited with methotrexate.[100] The dUTP levels reached 20% of dTTP levels, and uracil incorporation and excision might well contribute to the toxicity of this drug.

From studies with *E. coli* mutants deficient in dCTP deaminase[101] it appears that three-fourths or more of the dUTP is normally derived via dCTP, and a small fraction comes from dUDP (Fig. 2-15). In *E. coli* infected with T-even phage (Section 13), and in *B. subtilis, S. typhimurium,*[102] and animal cells, the prominence of dCMP deaminase indicates a direct route to dUMP, bypassing dUTP.

2-10 Salvage Pathways of Nucleotide Synthesis[103]

Appreciation of the vital importance of various salvage pathways came from discoveries of diseases due to enzyme deficiencies, toxic effects of precursors and analogs, and instances in which deficiencies and toxic effects are combined. These discoveries illustrate the homeostatic function of salvage pathways—how they participate in a complex interplay of de novo, degradative, salvage, and excretory pathways to cope with changing environmental stresses, nutritional factors, and enzyme shortages.

Whereas de novo routes of purine and pyrimidine nucleotide synthesis, when present, are virtually the same in all cells, the salvage routes are far more diverse in nature and distribution. Information about the salvage enzymes, especially in animal cells, is fragmentary. Attempts to select animal cell mutants defective in nucleic acid metabolism, to understand changes in cancer cells, and to design drugs for the treatment of diseases are sometimes thwarted by the lack of such detail.

The principal salvage reactions are outlined in Figure 2-16 and in Table 2-5. The following discussion of salvage pathways parallels the order listed in Table 2-5.

---

98. Shlomai J, Kornberg A (1978) *JBC 253*:3305.
99. Nilsson S, Reichard P, Skoog L (1980) *JBC 255*:9552, Goulian M, Bleile B, Tseng BY (1980) *JBC 255*:10630.
100. Goulian M, Bleile B, Tseng BY (1980) *PNAS 77*:1956; Goulian M, Bleile B, Tseng BY (1980) *JBC 255*:10630.
101. Beck CF, Eisenhardt AR, Neuhard J (1975) *JBC 250*:609.
102. Møllgaard H, Neuhard J (1978) *JBC 253*:3536; Neuhard J (1978) *Meth Enz 51*:418.
103. Elion GB (1989) *Science 244*:41; Munch-Petersen A (ed.) (1983) *Metabolism of Nucleotides, Nucleosides, and Nucleobases in Microorganisms*. Academic Press, London; Reichard P (1988) *ARB 57*:349.

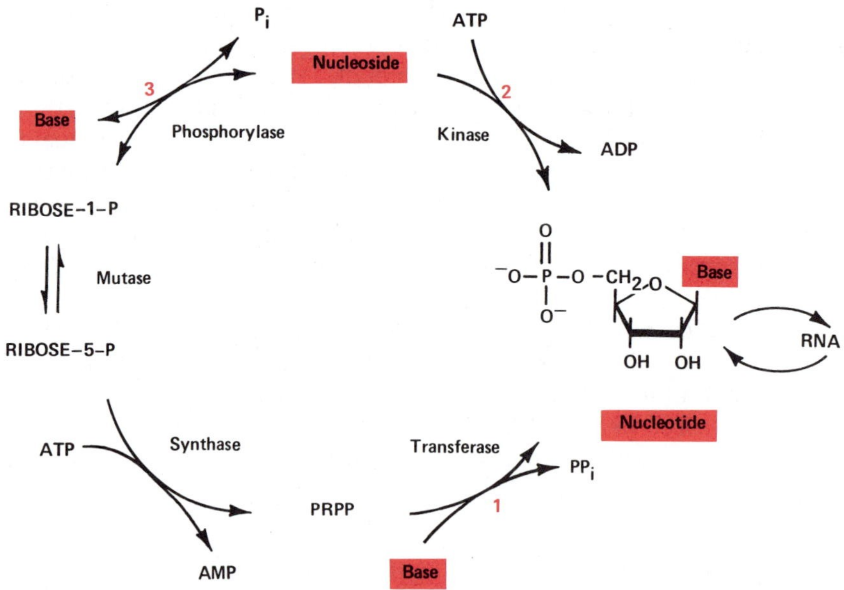

Figure 2-16
Principal pathways for salvage of bases and nucleosides. Numbers correspond to those in Table 2-5.

Table 2-5
Salvage pathways of nucleotide synthesis

Pathway	Reaction	Enzymes
(1)[a] Direct conversion of a base into a ribonucleotide	PRPP + base X → rNMP + PP_i	phosphoribosyl transferase
(2) Direct conversion of a nucleoside to a nucleotide	nucleoside N + ATP → NMP + ADP	nucleoside kinase
(3) Reversible conversion of bases to nucleosides	ribose 1-P + base X ↔ nucleoside X + P_i	nucleoside phosphorylase
	deoxyribose 1-P + base X ↔ deoxynucleoside X + P_i	
(4) Base exchange into deoxynucleosides	base X + deoxynucleoside Y ↔ base Y + deoxynucleoside X	nucleoside transglycosylases
(5) Interconversion by base alterations	adenine → hypoxanthine	deaminases
	cytosine → uracil	
	(d)adenosine → (d)inosine	
	(d)cytidine → (d)uridine	
(6) Reutilization of nucleotides	RNA → rNMP →→ rNTP DNA → dNMP →→ dNTP	nucleases plus kinases

[a] Numbers 1–3 correspond to those in Figure 2-16.

Direct Conversion of a Base into a Ribonucleotide[104]

Because this reaction is kinetically irreversible, the enzymes are designated as transferases—for example, hypoxanthine–guanine phosphoribosyl transferase (HGPRTase)—rather than pyrophosphorylases. They provide a direct route for the salvage of bases, particularly the purines. Transferases for the salvage of pyrimidines (uracil) have been identified in some bacterial cells.

Absence of HGPRTase in humans results in the severe Lesch-Nyhan syndrome (see below). Mutant cells defective in HGPRTase can be selected in media containing azaguanine. Normal cells incorporate the azaguanine via the enzyme and are blocked in purine nucleotide synthesis. Mutants that lack the transferase are spared because they cannot synthesize the inhibitory azaGMP. HGPRTase has been isolated from human erythrocytes, and its mutant forms have been characterized.[105]

Direct Conversion of a Nucleoside to a Nucleotide

The general kinase reaction is

$$\text{nucleoside N} + \text{ATP} \longrightarrow \text{NMP} + \text{ADP}$$

The distribution of kinases in cells and tissues is highly variable: mammalian cells possess kinases for all the deoxynucleosides, *E. coli* has one only for thymidine, and yeast has none. At an extreme, the protozoan parasite *Giardia lamblia* cannot synthesize purines de novo, lacks ribonucleotide reductase, and relies entirely on the salvage of ribo- and deoxyribonucleosides from the environment.[106]

Thymidine kinase, remarkable among the nucleoside kinases in its importance, will be described in Section 11. Abundant in some cells and not in others are specific kinases for (1) (d)adenosine, (2) cytidine or uridine, and (3) deoxyA, C, or G nucleoside (called deoxycytidine kinase because of its higher affinity for deoxycytidine). Like thymidine kinase, deoxycytidine kinase is allosterically controlled, figures prominently in substrate cycles (Section 15), and can be the basis of an immunodeficiency disease.[107] Inasmuch as a kinase for inosine is uncommon, this nucleoside is salvaged by the less direct routes described below (Fig. 2-18). Absence of a nucleoside kinase often accounts for the failure of a nucleoside to be used directly to correct a deficiency and explains why a nucleoside analog may be less effective as a drug.

In exponentially growing mouse fibroblasts, nearly a third of the dCDP produced by ribonucleotide reductase is excreted as the deoxynucleoside and is salvaged from the medium when needed. The balance between efflux and uptake, involving the opposing actions of a deoxynucleotidase and a deoxynucleoside kinase, is regulated by the supply of the dNTPs and the need for them in DNA synthesis.

---

104. Olsen AS, Milman G (1978) *Meth Enz 51*:543; Hochstadt J (1978) *Meth Enz 51*:549; Holden JA, Kelley WN (1978) *JBC 253*:4459; Caskey CT, Kruh GD (1979) *Cell 16*:1; Rasmussen UB, Mygind B, Nygaard P (1986) *BBA 881*:268.
105. Holden JA, Kelley WN (1978) *JBC 253*:4459; Johnson GG, Eisenberg LR, Migeon BR (1979) *Science 203*:174; Wilson JM, Baugher BW, Landa L, Kelley WN (1981) *JBC 256*:10306.
106. Baum KF, Berens RL, Marr J, Harrington JA, Spector T (1989) *JBC 264*:21087.
107. Bohman C, Eriksson S (1988) *Biochemistry 27*:4258.

Reversible Conversion of Bases to Nucleosides

The equilibrium catalyzed by a nucleoside phosphorylase, as with other phosphorolytic reactions, has a value near 1. Three nucleoside phosphorylases can be cited: one for pyrimidine ribonucleosides (e.g., uridine phosphorylase),[108] one for pyrimidine deoxynucleosides (e.g., thymidine phosphorylase),[109] and one for purine ribo- and deoxynucleosides (purine nucleoside phosphorylase).[110] The latter phosphorolyzes (deoxy)inosine and (deoxy)guanosine but not (deoxy)adenosine, which needs first to be deaminated to (deoxy)inosine.

Whereas these phosphorylases can be used on occasion to salvage preformed bases or sugar phosphates, the predominant direction, as with other phosphorylases, is degradative (Section 6). Failure to dispose of accumulating nucleosides, as in the example of a deficiency in purine nucleoside phosphorylase, can lead in humans to severe immunodeficiency disease (see below).

In the salvage mode, a preformed base can be converted to a nucleoside if ribose 1-phosphate or deoxyribose 1-phosphate is present, or when a sugar phosphate or base is taken from one available nucleoside to make another. For example, deoxyadenosine, by providing deoxyribose 1-phosphate, can enable a thymidylate synthase mutant to use thymine (Fig. 2-17 and Section 11).

Because of the general absence of an inosine kinase, the inosine ribo- and deoxynucleosides are converted to inosinate by an indirect route (Fig. 2-18). Cells defective in de novo inosinate synthesis utilize hypoxanthine derived from inosine. They do so by a sequence of phosphorylase, mutase, synthase, and transferase reactions. Thus, hypoxanthine and ribose 1-phosphate are generated in the phosphorylase reaction; ribose 5-phosphate is produced by phosphoribomutase and is the substrate for PRPP synthase; inosinate is then generated by HGPRTase.

Base Exchange into Deoxynucleosides

Bases, including a large variety of chemically synthesized purines and pyrimidines, can be exchanged in nucleosides by enzymes (glycosyltransferases) found in some bacteria. The general reaction is shown in Table 2-5. The direct exchange avoids phosphate or a phosphorylated intermediate.[111]

Interconversion by Base Alterations

Deaminations of adenine and guanine (as free bases or in nucleosides and nucleotides) to the corresponding hypoxanthine and xanthine forms, and of cyto-

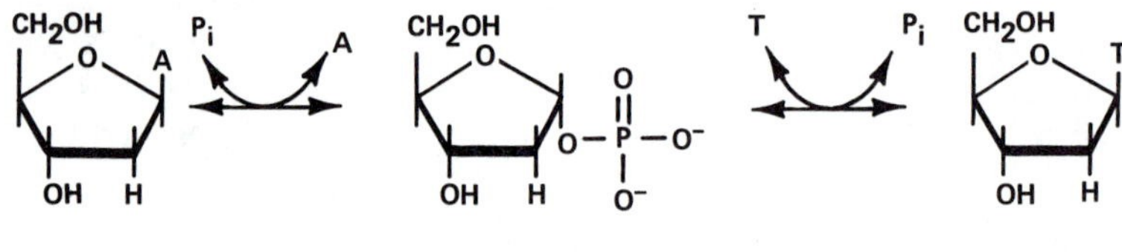

Figure 2-17
Interconversion of deoxyadenosine and thymidine by nucleoside phosphorylase.

108. Yamada EW (1976) *Meth Enz 51*:423.
109. Scocca JJ (1978) *Meth Enz 51*:432; Hoffee PA, Blank J (1978) *Meth Enz 51*:437; Schwartz M (1978) *Meth Enz 51*:442.
110. Hoffee PA, May R, Robertson BC (1978) *Meth Enz 51*:517; Glantz MD, Lewis AS (1978) *Meth Enz 51*:524; Stoeckler JD, Agarwal RP, Agarwal KC, Parks RE Jr (1978) *Meth Enz 51*:530; Milman G (1978) *Meth Enz 51*:538.
111. Beck WS, Levin M (1963) *JBC 238*:702.

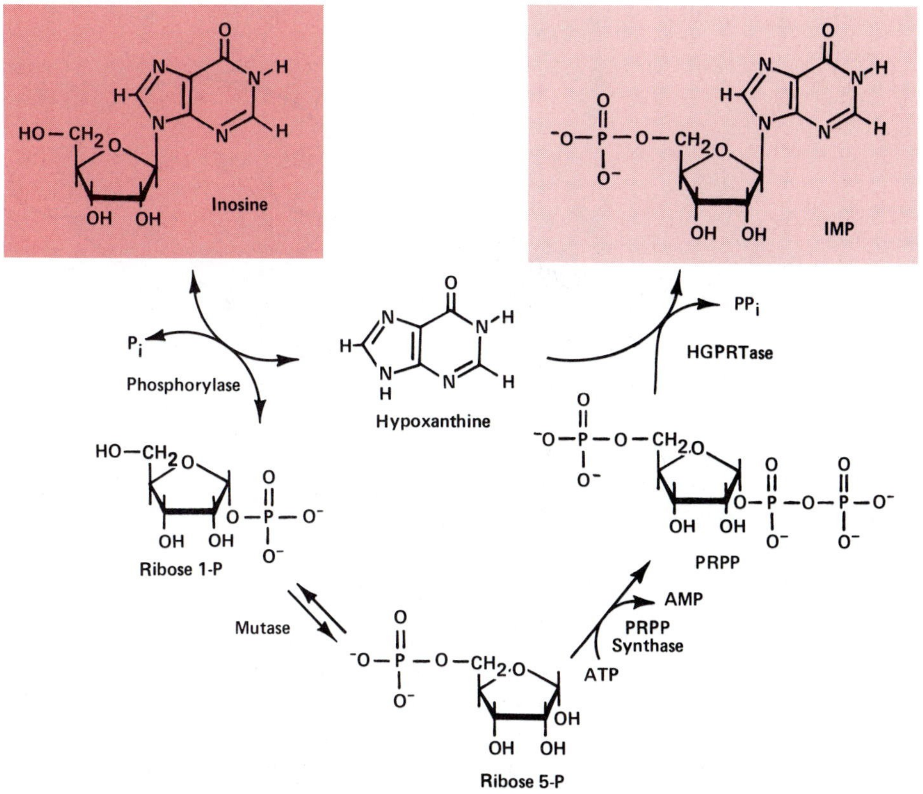

Figure 2-18
Synthesis of inosinate (IMP) by salvage of inosine via hypoxanthine and PRPP.

sine to the uracil forms, by specific enzymes are relatively common. Deamination (dCMP → dUMP) and aminations of the keto forms of ribonucleotides (IMP → AMP, XMP → GMP, UTP → CTP) occur in the de novo pathways.

Reutilization of Nucleotides

Perhaps the most important and least recognized form of salvage involves the nucleotides generated from nucleic acid degradation by nucleases. With rare exceptions, nucleotides do not traverse cell membranes and must therefore be dephosphorylated to be released from or taken into cells. The cycling of nucleotides out of and back into cells employs nucleotidase and kinase actions (see Substrate Cycles in Section 15). Nucleotides retained in intact cells and reused for nucleic acid synthesis suffer relatively little dilution from external nucleotides, nucleosides, or free bases.

Recovery from stressful exercise or other dangerous depletions of ATP depends on nucleotide salvage. Deamination of adenylate to inosinate and ammonia (Fig. 2-19) in response to vigorous muscular or neuronal activity[112] is a short-term measure for restoring "energy charge," the important signal expressed by the amount of ATP relative to ADP and AMP. The interconversions, catalyzed by adenylate kinase (Section 5), are displaced toward ATP synthesis,

112. Schultz V, Lowenstein JM (1978) *JBC 253*:1938.

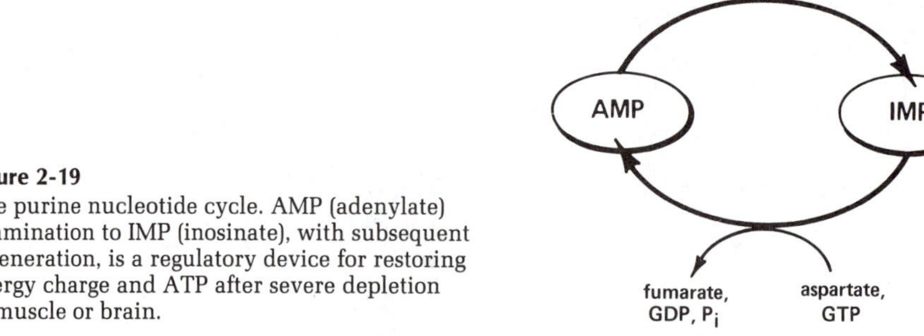

Figure 2-19
The purine nucleotide cycle. AMP (adenylate) deamination to IMP (inosinate), with subsequent regeneration, is a regulatory device for restoring energy charge and ATP after severe depletion in muscle or brain.

and allosteric effects are imposed on glycolytic enzymes that lead to ATP production. Restoration of AMP by salvage of IMP to complete the "purine nucleotide cycle"[113] employs the terminal reactions of the de novo biosynthetic pathway (Fig. 2-5).

Human Enzyme Deficiency Diseases

The importance of purine salvage is illustrated in humans by the *Lesch-Nyhan syndrome*,[114] a severe disease in infants who lack HGPRTase. Hypoxanthine and guanine arise as breakdown products of nucleic acids. Inability to utilize PRPP in this crucial salvage reaction results in its accumulation. Overproduction of purine nucleotides ensues because glutamine–PRPP amidotransferase is stimulated by the elevated PRPP levels. The overproduced purines are diverted by oxidation to uric acid in excessive amounts, and gout-like tissue damage results. The severe central nervous system disturbance seen in this syndrome may be due to inability of the brain to salvage the purines. Remarkably, transgenic mice lacking HGPRTase show no symptoms of disease.[115] Salvage of guanine by HGPRTase is essential for *Schistosoma mansoni*, a pathogenic parasite in humans, which cannot synthesize purines de novo; the enzyme[116] affords a chemotherapeutic target. The purine salvage and disposal enzymes (i.e., adenosine deaminase and purine nucleoside phosphorylase) have special significance for lymphocyte differentiation and proper immune function.[117]

Adenosine deaminase deficiency leads to a severe combined immunodeficiency disease resulting from defects in both T cell and B cell development. A 100-fold elevation of the dATP level is observed in T cells and results from the accumulation of deoxyadenosine, which is a powerful negative regulator of ribonucleotide reductase (Section 7) and depresses the level of the other dNTPs. The basis for B cell toxicity is less clear. Increased adenosine levels may inhibit

---

113. Tornheim K, Lowenstein JM (1975) *JBC 250*:6304.
114. Stout JT, Caskey CT (1989) in *The Metabolic Basis of Inherited Disease* (Scriver CR, Beaudet AL, Sly WS, Valle D, eds.) 6th ed. McGraw-Hill, New York, p. 1007; Davidson BL, Pashmforoush M, Kelley WN, Palella TD (1989) *JBC 264*:520.
115. Hooper ML, Hardy K, Handyside A, Hunter S, Monk M (1987) *Nature 326*:292; Kuehn MR, Bradley A, Robertson EJ, Evans MJ (1987) *Nature 326*:295; Thompson S, Clarke AR, Pow AM, Hooper ML, Melton DW (1989) *Cell 56*:313.
116. Yuan L, Craig SP, McKerrow JH, Wang CC (1990) *JBC 265*:13528.
117. Barton RW, Goldschneider I (1979) *Mol Cell Biochem 28*:135; Hershfield MS, Kredich NM (1978) *Science 202*:757; Martin DW Jr, Gelfand EW (1981) *ARB 50*:845; Kredich NM, Hershfield MS (1989) in *The Metabolic Basis of Inherited Disease* (Scriver CR, Beaudet AL, Sly WS, Valle D, eds.) 6th ed. McGraw-Hill, New York, p. 1045.

S-adenosylhomocysteine hydrolase. The resulting increased levels of S-adenosylhomocysteine would inhibit a number of essential methylation reactions. The absence of dNTP accumulation in B cells may in part be explained by a higher level of deoxynucleotidase activity.[118]

Purine nucleoside phosphorylase deficiency results in a huge accumulation of dGTP in T cells, affecting their development and function. B cells do not appear to be affected. A suggested mechanism[119] is based on the observed differences in metabolism of deoxyguanosine in B cells and T cells and on the toxicity of deoxyguanosine only for T lymphoblastoid cells. The major route of deoxyguanosine metabolism in B cells is degradation to guanine and then conversion to GMP, whereas in T cells deoxyguanosine is converted directly by a nonspecific deoxycytidine kinase to dGMP and then phosphorylated to dGTP to become a negative effector for ribonucleotide reductase. Supporting this mechanism is the differential toxicity of 8-aminoguanosine, an inhibitor of purine nucleoside phosphorylase:[120] T cells but not B cells are affected, with a fourfold increase in dGTP even at low levels of deoxyguanosine.

2-11 Thymine and Thymidine Conversion to Thymidylate[121]

Salvage Pathway of Thymidylate Synthesis

The salvage route by which thymidylate is synthesized from thymine and thymidine (Fig. 2-20) appears to be common to all cells except fungi. These path-

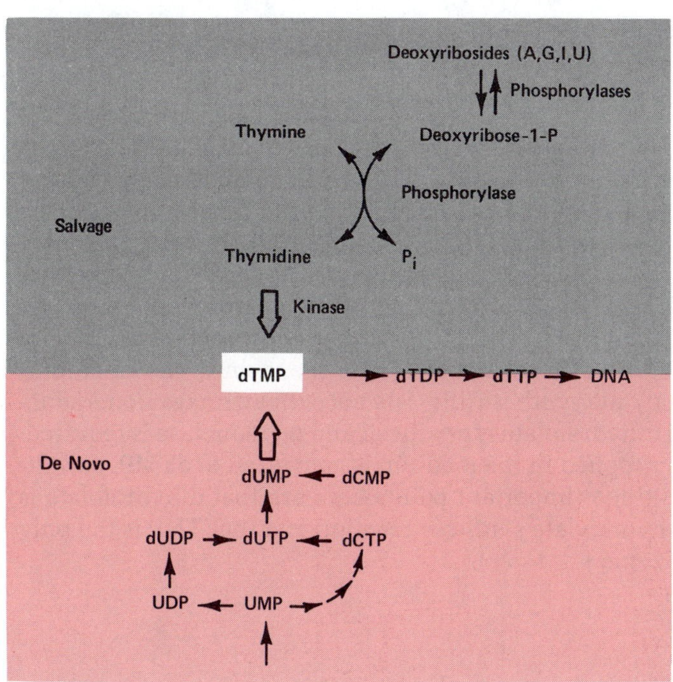

Figure 2-20
Salvage and de novo pathways of thymidylate (dTMP) biosynthesis.

118. Carson DA, Kaye J, Wasson DB (1981) *J Immunol 126*:348.
119. Osborne WRA (1981) *TIBS 6*:80.
120. Kazmers IS, Mitchell BS, Dadonna PE, Wotring LL, Townsend LB, Kelley WN (1981) *Science 214*:1137.
121. O'Donovan GA (1978) *DNA Synthesis, NATO A17*:219.

ways are also frequently exploited experimentally for labeling DNA (Section 12) and for influencing the rate of DNA synthesis.

Thymine is converted to the nucleotide in two steps. The first is the conversion to thymidine by thymidine phosphorylase:

$$\text{thymine} + \text{deoxyribose 1-P} \longleftrightarrow \text{thymidine} + P_i$$

The second step is the phosphorylation of thymidine by thymidine kinase:

$$\text{thymidine} + \text{ATP} \longrightarrow \text{dTMP} + \text{ADP}$$

Deoxyribose 1-phosphate is usually made available through the phosphorolysis of another deoxynucleoside (e.g., deoxyguanosine):

$$\text{deoxyguanosine} + P_i \longleftrightarrow \text{deoxyribose 1-P} + \text{guanine}$$

Thymine is also salvaged in some cells by a nonphosphorolytic reaction that exchanges purine and pyrimidine bases directly into a deoxynucleoside (e.g., deoxyadenosine) by a *trans*-N-deoxyribosylase:

$$\text{thymidine} + \text{adenine} \longleftrightarrow \text{thymine} + \text{deoxyadenosine}$$

Mutants defective in thymidylate synthetase (*thy*⁻) require thymine or thymidine to make DNA. Mutants of *E. coli* that lack thymidine phosphorylase (*tpp*⁻) fail to utilize thymine; those that lack thymidine kinase (*tdk*⁻) fail to utilize either thymine or thymidine.

Selection of Thymidylate Synthase–Defective Mutants

A technique for selecting mutants defective in thymidylate synthase exploits the salvage pathway and the pivotal role of dihydrofolate reductase. Exposure of normal cells to aminopterin or to amethopterin (methotrexate) inhibits the reductase and thus the regeneration of tetrahydrofolate (THF) coenzymes (Section 8). Even incomplete inhibition of the reductase piles up the dihydrofolate form of the coenzyme as a result of repeated cycling through thymidylate synthase. The cell dies for lack of purines and certain amino acids, whose biosynthesis depends on the THF state of the folate coenzymes. However, mutants defective in thymidylate synthase (*thy*⁻) do not suffer damage from reductase inhibition since no dihydrofolate is produced and no reductase is required. Thymine or thymidine supplied in the medium is converted to dTMP, permitting the mutants to multiply. (Important points here are that dihydrofolate is generated only in the thymidylate synthase reaction and that THF is the only active form of the coenzyme.)

Thymidine Kinase

All growing cells[122] except fungi possess thymidine kinase. The enzyme is induced, or is raised to much higher levels, during certain virus infections or

122. Bresnick E (1978) *Meth Enz 51*:360; Cheng Y-C (1978) *Meth Enz 51*:365; Sherley JL, Kelley TJ (1988) *JBC 263*:8350; Seiser C, Knöfler M, Rudelstorfer I, Haas R, Wintersberger E (1989) *NAR 17*:185.

under conditions of increased growth rate, as in regenerating liver. A remarkably nonspecific thymidine kinase[123] encoded by herpesvirus has been exploited therapeutically by its action on a synthetic guanosine analog, acyclovir (Section 14-3): the synthetic nucleotide, upon its subsequent conversion to a triphosphate, is a potent inhibitor of viral DNA replication in infected cells.

The kinase in *E. coli* was purified and was shown to phosphorylate deoxyuridine and its 5-fluoro-, 5-chloro-, 5-bromo-, and 5-iodo- derivatives.[124] When the various halogen analogs are administered as free pyrimidines or nucleosides, they are readily converted in the cell to nucleotides. Thymidylate synthase is thus inhibited by fluorodeoxyuridylate (Fig. 2-14), produced from fluorodeoxyuridine by thymidine kinase. Bromo analogs follow the same route, but they do not inhibit thymidylate synthase and are incorporated into DNA with great facility. Such analogs are useful in the laboratory as density markers in separating DNAs, as mutagens, and as radiation-sensitizing agents.

Regulation of Salvage Pathways

Although allosteric controls are generally found only in de novo biosynthetic pathways, the strict regulation of thymidine kinase,[125] deoxycytidine kinase (Section 10), and uridine kinase[126] indicates the importance of salvage. A variety of deoxynucleoside diphosphates serve as stimulatory effectors, whereas dTTP is a powerful inhibitor. Nucleotides may stimulate or inhibit, depending on the multimeric conformations of the enzyme they induce.[127]

2-12 Labeling of DNA in Cells[128]

A cogent reason to stress the salvage pathways of nucleotide biosynthesis is the wide use of radioactive labeling of DNA by exogenous radioactive precursors to monitor in vivo events. The design and interpretation of labeling experiments must consider the details of salvage metabolism. Short-term (pulse) labeling requires that the precursors be efficiently and rapidly taken up by the cell, be stable during uptake, be converted rapidly to the nucleotide level, and be specifically incorporated into DNA—essentially without dilution by intracellular precursor pools.

Inorganic phosphate, despite its ready and efficient uptake, enters large and diverse phosphate pools and equilibrates slowly. It is therefore suitable only for long-term labeling. Nucleotides as a rule do not pass through plasma membranes, but the bases and nucleosides often meet the criteria for pulse labeling.

Use of Thymine and Thymidine in DNA Labeling

Although thymine and thymidine are most commonly used and the ensuing discussion deals mainly with *E. coli*, these and other bases and nucleosides have also been widely used to label DNA in eukaryotic cells.

123. Sanderson MR, Freemont PS, Murthy HMK, Krane JF, Summers WC, Steitz TA (1988) *JMB 202*:917.
124. Okazaki R, Kornberg A (1964) *JBC 239*:269, 275; Chen MS, Prusoff WH (1978) *Meth Enz 51*:354.
125. Okazaki R, Kornberg A (1964) *JBC 239*:269, 275; Chen MS, Prusoff WH (1978) *Meth Enz 51*:354.
126. Cheng N, Payne RC, Traut TW (1986) *JBC 261*:13006.
127. Iwatsuki N, Okazaki R (1967) *JMB 29*:155.
128. Schultz V, Lowenstein JM (1978) *JBC 253*:1938.

Thymine and thymidine, which generally have no fate other than as DNA building blocks, have the advantage of rapid conversion to the nucleotide level. Furthermore, mutants are available that lack thymidylate synthase and have very small pools of thymidylate as a consequence. Thymidine is accumulated in the cell and converted to dTTP more rapidly and efficiently than thymine. However, several pitfalls (in some instances avoidable) are inherent in thymidine labeling.

Problems in DNA Labeling

Absence of Thymidine Kinase. This seemingly ubiquitous enzyme, essential for thymidine labeling, is lacking in fungi. A mutant strain that permits the entry of deoxythymidylate is available in yeast.

Dilution. Dilution with a large intracellular pool of thymidylate occurs in wild-type cells in which synthesis de novo dilutes thymidylate as much as threefold. Thymidylate synthase mutants, available for most species, can be used to avoid this dilution. In *B. subtilis*, which has two unlinked thymidylate synthases,[129] the double mutant *thyA*$^-$ *thyB*$^-$ must be used.

Incorporation of Uracil into DNA. In *thy*$^-$ cells, uracil may be incorporated into DNA in increased amounts because of lowered dTTP levels. An increase in the intracellular dUTP : dTTP ratio would increase both uracil incorporation into DNA and, consequently, the number of small nascent fragments upon uracil excision (Sections 9 and 15-10). Thus, mutants must also be deficient in uracil N-glycosylase (*thy*$^-$ *ung*$^-$) so as not to accumulate these short fragments.

Destruction of Thymidine by Phosphorolysis. Thymidine phosphorylase, induced by thymidine, within a few minutes converts thymidine to thymine. In wild-type *E. coli* (but not mammalian cells), the facility for incorporating thymine into thymidylate is rather limited. The problem is dealt with in several ways:

(1) A source of deoxyribose 1-phosphate is supplied by phosphorylase action to restore thymidine from thymine. Effective exogenous sources of deoxyribose 1-phosphate are deoxyadenosine and deoxyguanosine, through cleavage by purine nucleoside phosphorylase. Deoxyguanosine is more generally useful because deoxyadenosine is toxic for some strains.

(2) In *thy*$^-$ mutants, deoxyuridylate accumulation results in an endogenous supply of deoxyribose 1-phosphate by successive nucleotidase and phosphorylase actions.

(3) Deoxyribose 1-phosphate may be conserved by preventing its conversion by deoxyribomutase to deoxyribose 5-phosphate, which is further catabolized by deoxyriboaldolase to acetaldehyde and glyceraldehyde 3-phosphate. Mutants in the genes for deoxyribomutase (*drm*$^-$) or deoxyriboaldolase (*dra*$^-$) achieve this block.

(4) Thymidylate can be an excellent source of thymidine through surface nucleotidase (phosphatase) action. By achieving a *gradual* release and uptake of thymidine, its loss by cleavage to thymine is reduced.

---

129. Neuhard J, Price AR, Schack L, Thomassen E (1978) *PNAS* 75:1194.

(5) Mutants in thymidine phosphorylase (*tpp*$^-$) prevent the loss of thymidine. Partial activity in these mutants can be offset by supplying uridine, a potent inhibitor of the enzyme.

(6) Uridine inhibition of thymidine phosphorylase in *tpp*$^+$ cells makes it possible to switch from long-term labeling of DNA with [^{14}C]thymine to pulse-labeling, simply by adding [^{3}H]thymidine and uridine to the medium.[130] Multidrug treatments carry risks for *E. coli* as they do for humans. For example, uridine may be cleaved by uridine phosphorylase to uracil to an extent sufficient to inhibit uracil N-glycosylase, thus interfering with removal of uracil residues from DNA (Section 9).

Deficient Uptake of Thymidine from the Medium. Various *thy*$^-$ strains differ strikingly in their transport of nucleosides. A common defect in nucleoside uptake (*nup*$^-$) appears to be identical with a deficiency in the receptor for T6 phage and for colicin K (*tsx*$^-$); the defect is in the action of an outer membrane pore. Strains vary in thymidine requirements, ranging from 0.2 to 100 μg/ml. Inadequate levels can be rate-limiting for DNA chain elongation; excessive levels may be toxic to the cell and costly as a reagent.

Toxicity of Thymidine and Radioactivity. Most *thy*$^-$ strains sustained with thymidine attain only 20% to 50% of the dTTP levels of otherwise identical *thy*$^+$ strains. Exogenous thymidine inhibits some still unidentified step in dTTP synthesis. Radioactivity of [^{3}H]thymidine at levels commonly used in labeling procedures (e.g., 10 μCi per ml for 15 minutes) can perturb replication and cell cycle progression in Chinese hamster ovary cells.[131]

The above examples illustrate the complexities in genetic makeup and physiologic state that influence the metabolism of nucleoside precursors. These factors become very important when pulse-labeling techniques are used to analyze the kinetics and mechanisms of DNA replication in animal cells as well as in *E. coli*.

Dynamics of Biosynthesis and Turnover[132]

Measurements of the specific radioactivity of a precursor [in counts per minute per mol (cpm/mol)] and the extent of its incorporation into a product provide definitive information about the rate of synthesis of the product. Application of this basic approach to DNA biosynthesis in vivo has been pioneered with mammalian cells in culture, and avoids both the need for mutants and the numerous pitfalls cited above.

A trace of a labeled nucleoside, [^{3}H]thymidine in this instance (Fig. 2-21), is administered, and the attainment of a steady state of the immediate product (i.e., dTTP) is monitored. At this point, knowing the value of the constant specific radioactivity of the dTTP pool, one can determine from its incorporation into DNA the precise rate of DNA synthesis during this interval. The rate of synthesis can be verified by determining the turnover of the precursor, indicated by the rate at which the specific radioactivity of the dTTP pool decays as unlabeled thymidine replaces the labeled thymidine.

130. Womack JE (1977) *MGG 158*:11.
131. Hoy CA, Lewis ED, Schimke RT (1990) *Mol Cell Biol 10*:1584.
132. Reichard P, Nicander B (1985) Curr Top Cell Regul 26:403; Reichard P (1987) *Biochemistry 26*:3245.

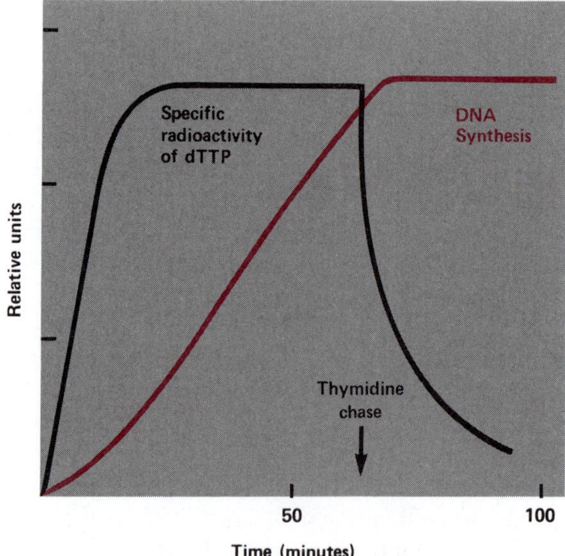

Figure 2-21
Dynamics of biosynthesis and turnover of dTTP, a DNA precursor. [After Reichard P, Nicander B (1985) *Curr Top Cell Regul* 26:403]

Similar assays of pool size and specific radioactivity can be applied to any stage in the DNA biosynthetic pathway, allowing absolute rates of in situ enzyme reactions (e.g., ribonucleotide reductase, dCMP deaminase, or thymidylate synthase) to be determined.

2-13 Uncommon Nucleotides

In addition to the four common purine and pyrimidine nucleotides, DNA normally contains methylated bases in very small amounts. Certain phage DNAs, however, include an uncommon nucleotide to the exclusion of a common one. Analogs of the natural nucleotides may also be incorporated into DNA experimentally. The most varied assortment of uncommon nucleotides is found in transfer RNA. However, their nature and their origin through modification of the common nucleotides in the RNA precursor is beyond the scope of this discussion. Although the functions served by the uncommon nucleotides are in most instances still unknown, it is clear that the exclusion of a common base or its modification by glucosylation or methylation can protect the DNA against destruction by nucleases in the cell (see below and Section 21-9).

Routes of Entry into DNA

The uncommon nucleotides (Fig. 2-22) enter DNA by three routes.

Incorporation of a Base Analog. Synthetic analogs are converted by salvage pathways to the dNTPs and can compete with the natural nucleotide for base pairing with the template; 5-bromouracil competes effectively enough to replace a large fraction of thymidylate residues in DNA, whereas incorporation of 8-azaguanine deoxynucleotide is very small. Uracil may enter DNA from dUTP in competition with dTTP (Section 9).

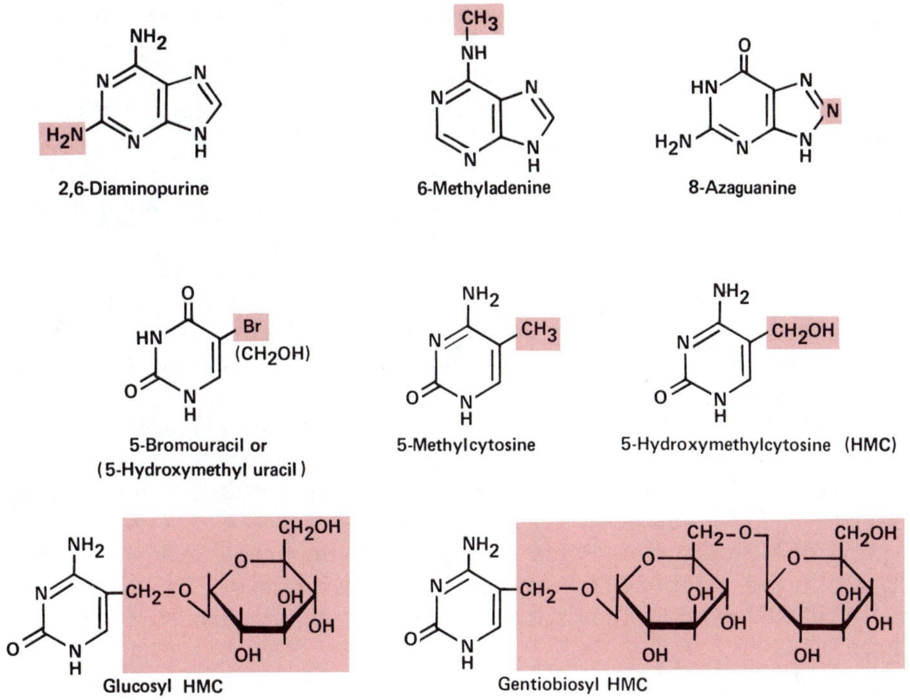

2,6-Diaminopurine

6-Methyladenine

8-Azaguanine

5-Bromouracil or
(5-Hydroxymethyl uracil)

5-Methylcytosine

5-Hydroxymethylcytosine (HMC)

Glucosyl HMC

Gentiobiosyl HMC

Figure 2-22
Bases infrequently found or incorporated in DNA.

Incorporation of an Enzymatically Altered Nucleotide. Examples of incorporation of altered nucleotides are found in certain phage DNAs. Hydroxymethylation of deoxycytidylate or of deoxyuridylate by phage-encoded enzymes leads to complete replacement of the cytosine or thymine residues, respectively, with the 5-hydroxymethyl analogs in the DNA of the viral progeny. Uracil may replace thymine by phage-encoded inhibitors and a nucleotidase (see below).

Modification of DNA. The natural occurrence of 6-methyladenine in bacterial DNAs and 5-methylcytosine in animal and higher plant DNAs is due to specific methyltransferases that modify certain of the adenine or cytosine residues of an assembled DNA chain. Varied and distinctive patterns of glycosyl hydroxymethylcytosine residues observed in the DNAs of phages T2, T4, and T6 are generated by specific glycosyl transferase enzymes that act on the polymerized DNA chain (Section 17-6).

Virus-Induced Novelties in Nucleotide Biosynthesis

Phages and eukaryotic viruses often encode proteins that interrupt certain host functions, nucleases that degrade host DNA to supply deoxynucleotides, and biosynthetic enzymes that augment host levels of kinases, reductases, thioredoxin, etc. Beyond these devices, phages that possess novel DNAs contrive ways of replacing a common nucleotide with the uncommon one.

Replacement of Cytosine by Hydroxymethylcytosine in T-Even Phage DNAs. For all the cytosine residues to be replaced by hydroxymethylcytosine (HMC) in the synthesis of T-even phage DNAs (Section 17-6), the dCTP produced in the host cell must be excluded. The ingenious solution by the infecting phage lies in a set of novel encoded enzymes (Fig. 2-23).[133]

(1) A hydroxymethylase[134] (gp42) converts dCMP to hydroxymethyl dCMP, a likely site for inhibition by 5-azacytidine.

(2) A nucleoside monophosphate kinase[135] (gp1) phosphorylates hydroxymethyl dCMP to the diphosphate, the precursor of hydroxymethyl dCTP.

(3) A family of induced glycosyl transferases[136] glucosylates the hydroxymethylcytosine groups in the DNA in patterns characteristic for the different T-even phage DNAs (Section 21-9).

(4) A dCDPase – dCTPase (gp56)[137] removes these nucleotides as DNA precursors; by converting them back to dCMP, the substrate is provided for hydroxymethylase (step 1, above) and deaminase (step 5). This multifunctional protein also cleaves dUTP and dUDP.

(5) A dCMP deaminase[138] (*cd* gene product) removes dCMP, to generate dUMP, the precursor of dTMP, and thus achieves a ratio of dTMP to hydroxymethyl dCMP near 2 : 1, the thymine to hydroxymethylcytosine ratio found in phage DNA.[139]

(6) Nucleases (e.g., endonucleases II and IV) are specific for cytosine-containing DNA. The host DNA, as well as any phage DNA with cytosine residues, is degraded to generate dCMP and the other deoxynucleotide precursors needed for phage DNA synthesis.

Aggregates of T-even phage – encoded enzymes — dCMP hydroxymethylase,

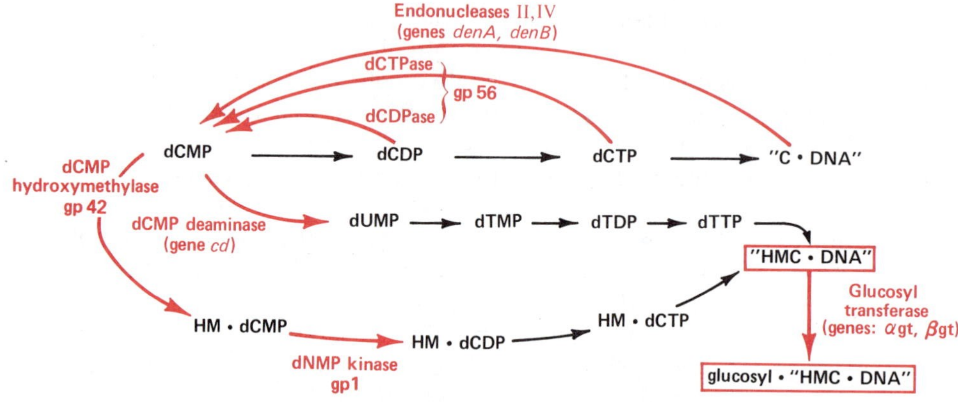

Figure 2-23
Pathway of biosynthesis of glucosyl hydroxymethylcytosine (HMC) DNA induced by T-even phage infection.

---

133. Cohen SS (1968) *Virus-Induced Enzymes.* Columbia University Press, New York.
134. Mathews CK, Brown F, Cohen SS (1964) *JBC 239*:2957.
135. Duckworth DH, Bessman MJ (1967) *JBC 242*:2877.
136. Josse J, Kornberg A (1962) *JBC 237*:1968.
137. Zimmerman SB, Kornberg A (1961) *JBC 236*:1480.
138. Maley GF, Duceman BW, Wang A-M, Martinez J, Maley F (1990) *JBC 265*:47.
139. Chiu C-S, Ruettinger T, Flanegan JB, Greenberg GR (1977) *JBC 252*:8603.

nucleotide kinase, dCMP deaminase, thymidine kinase, ribonucleotide reductase, dTMP synthase, dihydrofolate reductase, and dUTPase—and the host nucleoside diphosphate kinase have been observed.[140] Such complexes might channel precursors to facilitate their incorporation into phage DNA.

Replacement of Thymine by Hydroxymethyluracil and Uracil in *B. subtilis* Phage DNAs. Mechanisms comparable to those described for replacement of cytosine by hydroxymethylcytosine are employed by several *B. subtilis* phages (e.g., SP01; Section 17-8) to replace thymine by hydroxymethyluracil (Fig. 2-24, top). Four phage-encoded enzymes analogous to those observed in T-even phage infections are dUMP hydroxymethylase, hydroxymethyl dUMP kinase, dTTPase-dUTPase, and dCMP deaminase. A novel protein inhibitor of host cell thymidylate synthase blocks the formation of dTMP.

B. subtilis phages with uracil-containing DNA (e.g., PBS1;[141] Section 17-8) introduce novel proteins to exclude thymine and replace it with uracil (Fig. 2-24, bottom): dCTP deaminase, dUMP kinase, and dTMP nucleotidase.[142] An inhibitor of host cell dUTPase[143] preserves dUTP for incorporation into DNA; inhibitors of nuclease and uracil N-glycosylase prevent the loss of uracil from DNA.

Replacement of Adenine by 2-Aminoadenine (2, 6-Diaminopurine). The adenine analog 2-aminoadenine (Section 14-3) replaces adenine in the DNA of the cyanophage S-2L. The phage-encoded mechanisms for producing 2-amino dATP and excluding dATP have not been explained.

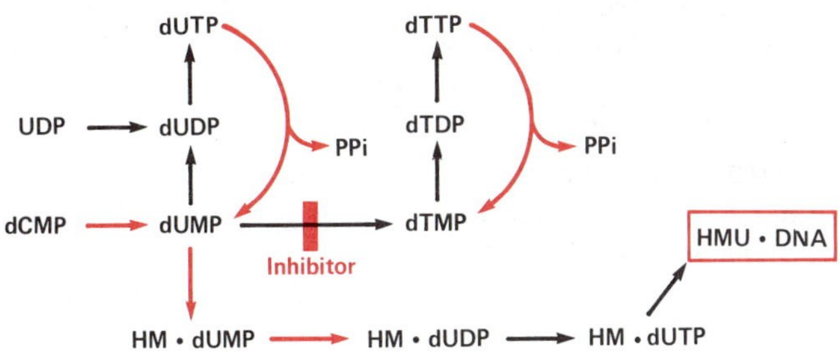

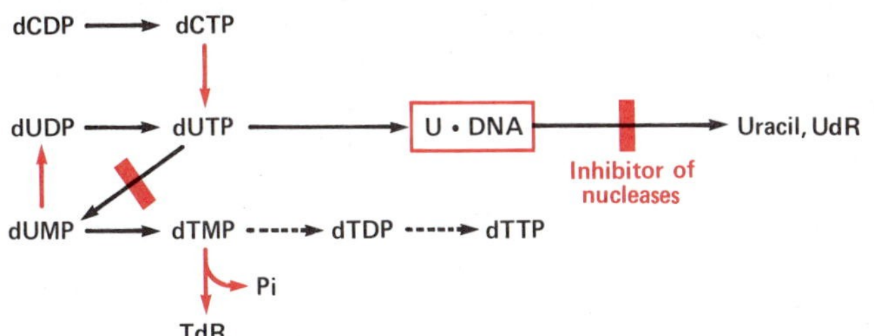

Figure 2-24
Pathways of biosynthesis of hydroxymethyluracil (HMU) and uracil DNAs induced by phages in *B. subtilis*. TdR = thymidine; UdR = uridine.

140. Chiu C-S, Cook KS, Greenberg GR (1982) *JBC 257*:15087; Moen LK, Howell LM, Lasser GW, Mathews CK (1988) *J Mol Recognition 1*:48; Mathews CK, Moen LK, Wang Y, Sargent RG (1988) *TIBS 13*:394.
141. Hitzeman RA, Price AR (1978) *DNA Synthesis, NATO A17*:255.
142. Price AR (1978) *Meth Enz 51*:285.
143. Price AR, Frato J (1975) *JBC 250*:8804.

Escape from Regulatory Controls. Rapid, free-running synthesis of viral DNA is facilitated not only by augmented levels of phage-encoded enzymes, but also by their insensitivity to regulators that modulate the host biosynthetic enzymes. Phage T4-encoded ribonucleotide reductase resists inhibition by dATP (Section 7), the principal inhibitor of host reductase. Herpesvirus-induced reductase[144] is similarly insensitive to regulation by dTTP and dATP, as is the deoxypyrimidine kinase[145] to dCTP and dTTP.

2-14 One-Carbon Metabolism

Attention to two-carbon compounds had for decades overshadowed interest in one-carbon biochemistry. After all, the acetyl unit is the key intermediate in oxidative and fermentative metabolism and in the biosynthesis of fatty acids, amino acids, and steroids. Nevertheless, with regard to DNA biosynthesis, concern with the biochemistry of one-carbon compounds is more immediate. They are crucial, as folate derivatives, in the formation of purines and thymine (Figs. 2-4 and 2-14) and, as methionine derivatives, in the methylation of DNA.

Formation of the One-Carbon Donor

Folic acid (pteroylglutamic acid) was recognized first as a vitamin and bacterial growth factor and later as the coenzyme of one-carbon metabolism. It is active only in the tetrahydro form [tetrahydropteroylglutamate, or *tetrahydrofolate (THF)*—Fig. 2-25], as a carrier of three of the redox states of one-carbon compounds (Table 2-6). Nutritional studies of bacteria demonstrated that the vitamin requirement for folate could be replaced by supplying four cellular metabolites: a purine (adenine), thymine, methionine, and serine.

Serine, which is derived from the glycolytic breakdown of glucose via 3-phosphoglycerate, is the primary source of one-carbon units (Fig. 2-26). Transfer of a hydroxymethyl group from serine to THF yields the intermediate 5,10-methyl-

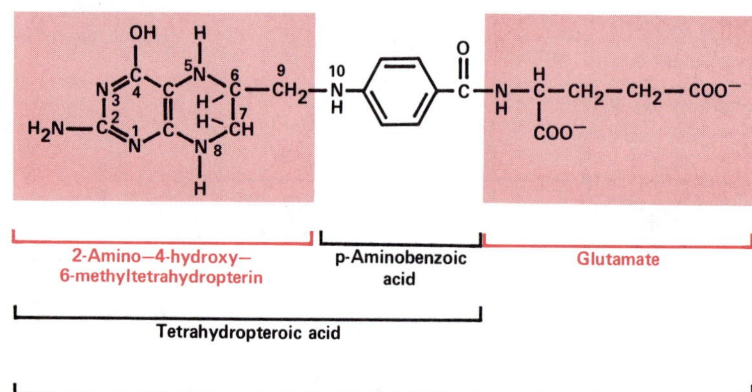

Figure 2-25
Tetrahydrofolic acid (tetrahydrofolate, THF, H_4-folate).

144. Langelier Y, De'Champs M, Buttin G (1978) *J Virol 26*:547.
145. Ogino T, Otsuka T, Takahashi M (1977) *J Virol 21*:1232.

Table 2-6
Chemical, biological, and coenzyme forms of the three redox states of one-carbon metabolic intermediates

Chemical	methanol H_3COH	formaldehyde H_2CO	formic acid HCOOH
Biological	methyl $-CH_3$	hydroxymethyl $-CH_2OH$ or methylene $-CH_2-$	formyl $-CHO$ or methenyl $=CH-$
Coenzyme (THF)	5-methyl	5,10-methylene	5-formyl or 10-formyl or 5,10-methenyl

ene THF, the one-carbon donor for thymidylate synthase (Fig. 2-14). A specific methylene THF reductase produces 5-methyl THF, the donor for homocysteine methyl transferase, a cobalamin-containing enzyme that forms methionine in mammalian tissues. Methylene THF dehydrogenase forms 5,10-methenyl THF, which, upon hydrolysis by a cyclohydrolase to 10-formyl THF, serves as the one-carbon donor for purine biosynthesis.

A Trifunctional Enzyme

Three catalytic activities in the interconversion of the THF coenzymes are on a single protein[146] (enzymes 1, 2, and 3 in Figs. 2-27 and 2-28); two of these catalytic centers are able to channel 5,10-methylene THF to 10-formyl THF.[147] This

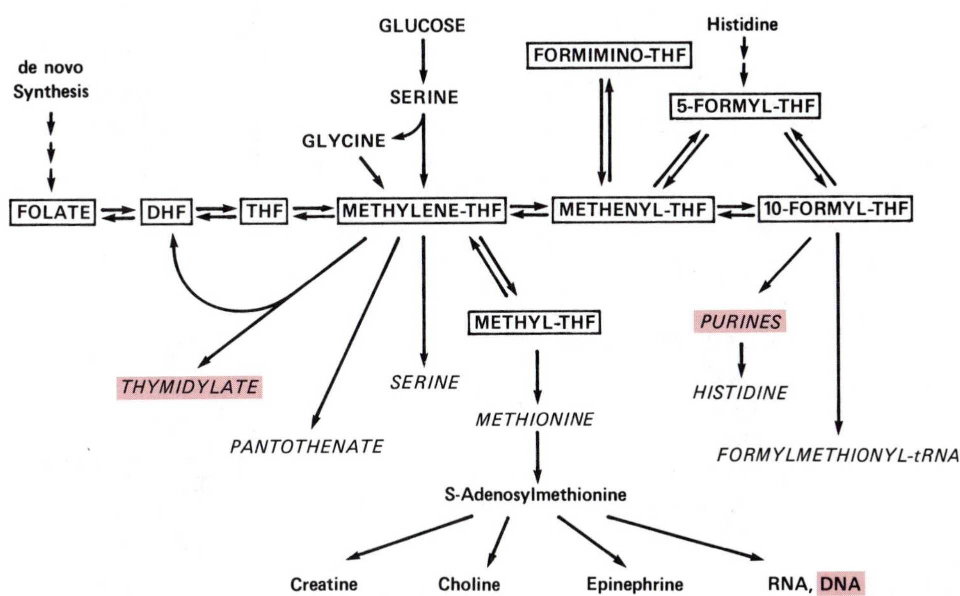

Figure 2-26
Pathways of one-carbon metabolism showing folate derivatives as intermediary metabolites in biosynthetic reactions. (Modified scheme of Professor J Rabinowitz)

146. Mackenzie RE (1984) in *Folates and Pterins: Chemistry and Biochemistry of Folates* (Blakley RL, Benkovic SJ, eds.). Wiley, New York, Vol 1, p. 225; Hum DW, Bell AW, Rozen R, MacKenzie RE (1988) *JBC 263*:15946.
147. Cohen L, MacKenzie RE (1978) *BBA 522*:311.

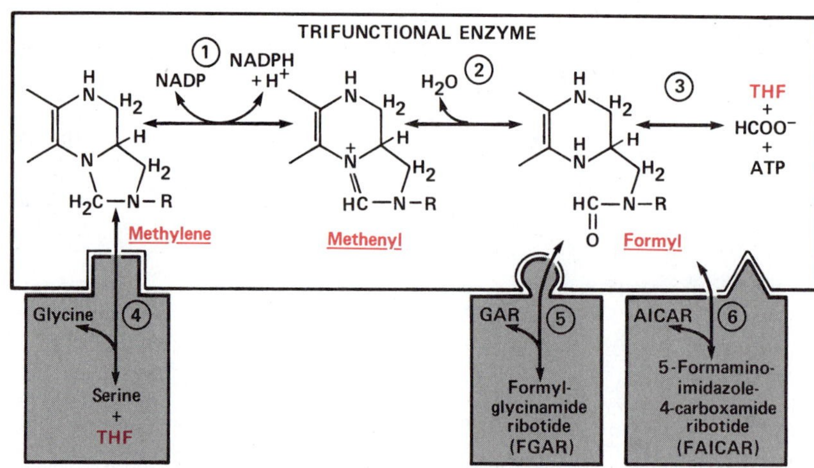

Figure 2-27
Trifunctional enzyme that interconverts methylene, methenyl, and formyl tetrahydrofolate coenzymes and complexes with transferases to glycine, glycinamide ribotide (GAR), and 5-aminoimidazole-4-carboxamide ribotide (AICAR). Numbered reactions correspond to those in Figure 2-28. (Courtesy of Professor M. E. Jones)

trifunctional protein exists in a multienzyme aggregate that also contains serine hydroxymethyl transferase (enzyme 4 in Figs. 2-27 and 2-28), which regenerates, from THF, the 5,10-methylene THF required for thymidylate synthesis.[148] The complex also contains the two transformylases (enzymes 5 and 6 in the figures) of de novo purine biosynthesis. Some of the folate cofactors are quite labile in an aqueous solvent, and the multienzyme complex, with its ability to channel the THF intermediate, helps to solve this problem. A similar purpose may be served in the protozoan *Crithidia fasciculata* by linkage of the dihydrofolate reductase and thymidylate synthase activities in a single polypeptide.[149]

Polyglutamate Forms of Folate

THF commonly contains additional glutamate residues attached to the coenzyme. This polymeric state may serve in maintaining effective intracellular

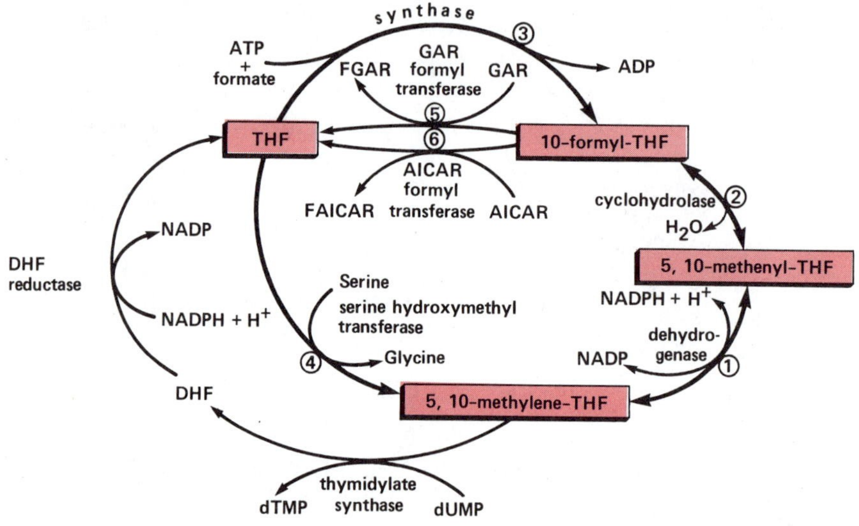

Figure 2-28
Tetrahydrofolate coenzyme cycle linked to transferases, thymidylate synthase, and dihydrofolate reductase. Numbered reactions correspond to those in Figure 2-27. (Courtesy of Professor M. E. Jones)

148. Caperelli CA, Benkovic PA, Chettur G, Benkovic SJ (1980) *JBC 255*:1885.
149. Ferone R, Roland S (1980) *PNAS 77*:5802.

concentrations of the coenzyme and may be favored by some enzymes. The pentaglutamyl form of 5,10-methylene THF has a K_m of 7 μM for the yeast thymidylate synthase compared to a value of 70 μM for the monoglutamyl form.[150]

Clinical Use of Methotrexate

The strikingly successful treatment of childhood leukemia is based on the knowledge of one-carbon metabolism. Methotrexate (amethopterin), as an inhibitor of dihydrofolate reductase, blocks the regeneration of THF from dihydrofolate and thus shuts off the synthesis of purines, thymine, histidine, and methionine. Salvage pathways are unable to make up the demands for precursors of these essential building blocks, and rapidly growing cells, normal as well as malignant, are thereby destroyed. Massive doses of methotrexate are required and would be lethal were it not that subsequent counterdoses of 5-methyl THF (leucovorin) can restore residual normal cells and rescue the patient.

2-15 Traffic Patterns on the Pathways[151]

The pathways of nucleotide metabolism can be likened to a city's traffic patterns. Salvage routes can be used instead of de novo pathways either as energetically attractive alternatives or out of necessity. Problems arise when the traffic gets heavy (e.g., from excess adenine or thymidine) and when some routes are blocked. Then, the capacities of available pathways are exceeded and pile-ups occur. These may be aggravated by the absence or improper intervention of repressor and allosteric controls, causing the system to fail at crucial places.

Traffic Control Schemes

In addition to providing a supply of dNTPs adequate to sustain the desired rate of DNA replication, biosynthetic pathways must maintain the dNTPs in nice balance. Distortion of the ratios of dNTP concentrations results in significant increases in spontaneous mutagenesis[152] and chromosomal abnormalities. The controls, finely tuned to the stage in the cell cycle, operate at several levels: supply of ribonucleoside diphosphates, reduction of ribonucleotides, deamination of dCMP (a branch point between biosynthesis of dCTP and dTTP), and in "substrate cycles" (see below).

Eukaryotic cells have two strategies to facilitate their pathways of nucleotide metabolism (see Sections 3, 4, 14, and 15): (1) Enzymes with sequential functions that are discrete polypeptides in prokaryotes may, in eukaryotes, be connected in a multifunctional polypeptide; (2) channeling and preservation of labile intermediates may be achieved by complexes containing many distinct enzymes, multifunctional ones included.

---

150. Bisson LF, Thorner J (1981) *JBC 256*:12456.
151. Elion GB (1989) *Science 244*:41; Reichard P (1988) *ARB 57*:349; Elliott K, Fitzsimons DW (eds.) (1977) *Purine and Pyrimidine Metabolism*. Ciba Foundation Symposium No. 48. Elsevier, Amsterdam; Mathews CK (1989) in *Molecular Biology of Chromosomal Proteins* (Adolph KW, ed.). Springer-Verlag, New York, p. 3.
152. Meuth M (1984) *Mutat Res 126*:107; Phear G, Nalbantoglu J, Meuth M (1987) *PNAS 84*:4450; Sargent RG, Mathews CK (1987) *JBC 262*:5546; Caras IW, Martin DW Jr (1988) *Mol Cell Biol 8*:2698.

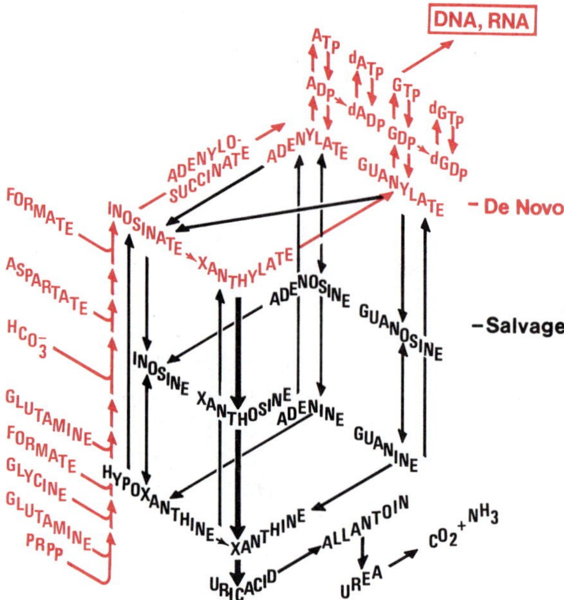

Figure 2-29
De novo and salvage pathways of purine nucleotide biosynthesis.

Mapping the Pathways

The major pathways of biosynthesis of purine and pyrimidine nucleotides (Figs. 2-29 and 2-30) and their breakdown have been mapped largely by explorations with cell extracts and purified enzymes from a variety of microorganisms and animal cells. To what extent these reaction patterns apply to a given cell or organism in vivo is often difficult to determine. Thus, there are significant gaps in our knowledge about the metabolic pathways in most cells and tissues, especially on how these pathways fit together in the economy of a complex organism.

Not depicted in these metabolic schemes, and only sketchily known, are the controls that regulate the traffic on these pathways. The mechanisms for switching among alternative routings are sometimes disclosed by the use of an inhibitory drug or by an enzyme deficiency related to a disease, such as gout.[153] Discoveries of multiple sites for metabolic control of purine metabolism and the regulation of these sites in humans have come from studies of this complex disease. For lack of a comparable disease of pyrimidine metabolism, less is known about human pyrimidine economy. However, much has been learned from studies of substrate cycles in cell cultures (see below).

Insights Gained from Gout and Lesch-Nyhan Syndrome

Several examples can be given of insights into purine metabolism gained from studies of gouty patients and cell cultures derived from them.

(1) *Lack of purine salvage, PRPP accumulation, and purine overproduction.* A severe lack of HGPRTase, as in Lesch-Nyhan syndrome in humans, sharply reduces the consumption of PRPP. The deficiency in transferase may be due to

153. Palella TD, Fox IH (1989) in *The Metabolic Basis of Inherited Disease* (Scriver CR, Beaudet AL, Sly WS, Valle D, eds.) 6th ed. McGraw-Hill, New York, p. 965.

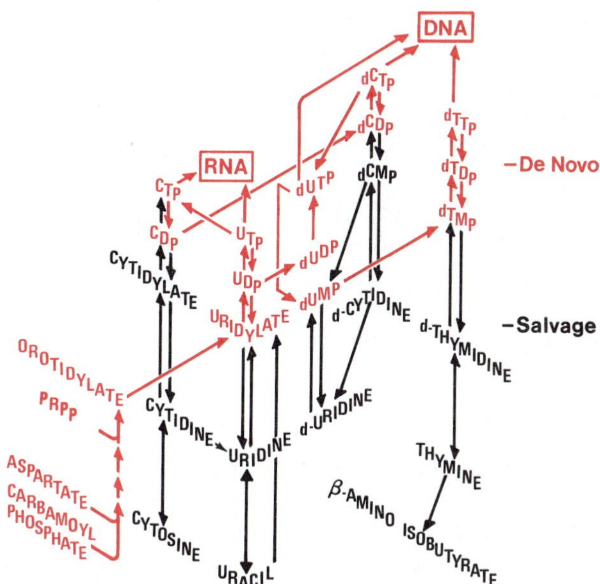

Figure 2-30
De novo and salvage pathways of pyrimidine nucleotide biosynthesis.

an intrinsically lower catalytic efficiency, an elevated K_m for one of the substrates, or a reduced amount of enzyme. The elevated PRPP level, resulting from its underutilization, triggers increased activity of PRPP amidotransferase. This first committed step in de novo purine biosynthesis (Fig. 2-4) responds to PRPP concentration in a sharp, sigmoidal fashion. The resulting increased synthesis of phosphoribosylamine leads to the overproduction of purines and the elevated blood and tissue levels of uric acid associated with gout.

(2) *Feedback-resistant PRPP synthetase and overproduction of PRPP and purines.* Fibroblasts from certain patients with gout have a PRPP synthetase that is less inhibited by feedback regulators (i.e., ADP and GDP) than is the enzyme from normal cells. The mutant cells also demonstrate increases in PRPP synthetase activity and in purine nucleotide production de novo. Presumably, the purine overproduction is due to an increased activity of PRPP amidotransferase resulting from the elevated PRPP levels. Potent activation of PRPP synthetase by inorganic orthophosphate and complexities in the allosteric responses of both the synthetase and the amidotransferase provide additional loci for control.

(3) *Prevention of purine overproduction by inhibition of xanthine oxidase.* The drug allopurinol is an analog of hypoxanthine in which the positions of N7 and C8 are reversed. Allopurinol inhibits xanthine oxidase and prevents the successive oxidations of hypoxanthine and xanthine, thus preventing uric acid formation. Allopurinol also reduces the rate of de novo purine biosynthesis. This effect may be attributed to increased removal of PRPP by salvage of free purines and to allosteric feedback inhibition of PRPP amidotransferase. The drug, if converted to its ribonucleotide, might inhibit orotidylate decarboxylation and pyrimidine biosynthesis.

(4) *Excretion: alternative response to purine overproduction.* Emphasis on the choice between salvage and de novo biosynthesis pathways neglects the importance of excretion in disposing of an oversupply of purines. The human is unique in the genetically supplied means for excreting purines. Only primates and birds do not degrade uric acid to soluble products, and only humans among the primates lack the renal capacity to remove uric acid effectively from the

blood stream. Drugs that improve the renal clearance of uric acid help to relieve the symptoms of gout. The human trait of maintaining high blood uric acid levels, which may lead to gout, has been regarded as an evolutionary misfortune, but it may instead turn out to be an advantage should the potent antioxidant property of uric acid prove to be an important defense against aging and cancer caused by oxidants and radicals.[154]

The intensive study of gout has illuminated certain regulatory pathways in purine metabolism. The immunodeficiency diseases due to lack of adenosine deaminase or purine nucleoside phosphorylase (Section 10) promise further insights into the regulation of purine metabolism, and into basic features of ribonucleotide reduction to deoxyribonucleotides as well.

Compartments, Pools, and Metabolons[155]

The intracellular pools of dNTPs are tiny (<0.2 mM; see Table 2-1). They are at the limit of measurement in quiescent mammalian cells and are sufficient for only a few minutes of DNA synthesis even in S-phase cells. Except for mitochondria and a separate dCTP pool for deoxyliponucleotides,[156] there is only scanty evidence for compartmentation of these pools and for the channeling of synthetic pathways to optimize their function. Early claims of a separate pool of dTTP, derived by thymidine salvage, have not been supported. Whereas compartmentation of dCTP derived from deoxycytidine salvage does exist, this applies only to the biosynthesis of membrane lipids.

Catalysis of sequential metabolic reactions by large enzyme complexes—dubbed "metabolons"[157]—might achieve the efficient channeling of intermediates. Biosynthesis of the deoxyribonucleotides is organized into a multienzyme complex in T4 phage–infected cells,[158] but channeling of the dNTPs to the polymerase has not been established. Moreover, claims for a "replitase,"[159] a miniorganelle in mammalian cells that both synthesizes the building blocks and assembles them into DNA, are in serious doubt[160]. These claims must accommodate the fact that synthesis of dNTPs is in the cytosol and replication is in the nucleus.[161] At this time, considerable progress has been made in casting of the actors in the drama of supplying the building blocks of replication, but an adequate script for their associations and actions has not yet been written.

Substrate ("Futile") Cycles[162]

For fine tuning of the dNTP pool sizes for optimal, high-fidelity replication, precursor deoxynucleotides may be shunted (as deoxynucleosides) to the outside of the cell and reabsorbed later when needed. Their flux is mediated by deoxynucleotidase action to convert them to deoxynucleosides for passage to the outside and by subsequent kinase action on the deoxynucleosides after they have reentered by means of permease action or facilitated diffusion.

154. Ames BN, Cathcart R, Schwiers E, Hochstein P (1981) *PNAS 78*:6858.
155. Reichard P (1988) *ARB 57*:349; Mathews CK, Moen LK, Wang Y, Sargent RG (1988) *TIBS 13*:394.
156. Spyrou G, Reichard P (1987) *JBC 262*:16425.
157. Srere PA (1987) *ARB 56*:21.
158. Mathews CK, Moen LK, Wang Y, Sargent RG (1988) *TIBS 13*:394; Thylen C, Mathews CK (1989) *JBC 264*:15169.
159. veer Reddy GP, Pardee AB (1982) *JBC 257*:12526.
160. Wawra E (1988) *JBC 263*:9908.
161. Leeds JM, Slabaugh MB, Mathews CK (1985) *Mol Cell Biol 5*:3443.
162. Reichard P (1988) *ARB 57*:349; Bianchi V, Pontis E, Reichard P (1987) *Mol Cell Biol 7*:4218; Höglund L, Pontis E, Reichard P (1988) *Can Res 48*:3681.

CHAPTER 3

DNA Synthesis

3-1 Early Attempts at Enzymatic Synthesis of Nucleic Acids[1]

The double helix model of DNA[2] could account for the replication of DNA by the complementarity of A to T and of G to C. The simplicity of the model dispelled a variety of speculations on how the DNA molecule might be reproduced and assembled, including the possibility of its direct origin from RNA by reduction.

The suggestion was made that nucleotides paired to the DNA template might be "zippered" together without any enzyme action.[3] However, to the biochemist it is implicit that all biosynthetic and degradative events are catalyzed by enzymes to achieve control, specificity, and rapid rates of reaction.

The first attempts at understanding the mechanism of nucleic acid synthesis were guided by what had been learned from studies of nucleotide and coenzyme biosynthesis. The primary products of nucleotide biosynthesis were found to be nucleoside 5′-monophosphates, and not the familiar 2′ and 3′ nucleotide isomers obtained from the breakdown of RNA by alkali or ribonucleases. Synthesis of the coenzyme nicotinamide adenine dinucleotide (NAD), a relatively simple nucleotide condensation product, illustrated the basic mechanism of nucleotidyl transfer: a nucleophilic attack by nicotinamide mononucleotide (NMN) on the α-phosphate of ATP, with the elimination of PP_i (Fig. 3-1). This nucleotidyl transfer mechanism, in which the nucleotide is activated as a nu-

1. Kornberg A (1957) in *The Chemical Basis of Heredity* (McElroy WD, Glass B, eds.). Johns Hopkins Press, Baltimore, p. 579; Kornberg A (1959) *Rev Mod Phys 31*:200; Kornberg A (1960) *Science 131*:1503; Kornberg A (1962) *Enzymatic Synthesis of DNA*. Wiley, New York; Kornberg A (1989) *For the Love of Enzymes*. Harvard University Press, Cambridge; Kornberg A (1989) *ARB 58*:1.
2. Watson JD, Crick FHC (1953) *Nature 171*:737; Watson JD, Crick FHC (1953) *CSHS 18*:123.
3. Watson JD, Crick FHC (1953) *Nature 171*:737; Watson JD, Crick FHC (1953) *CSHS 18*:123.

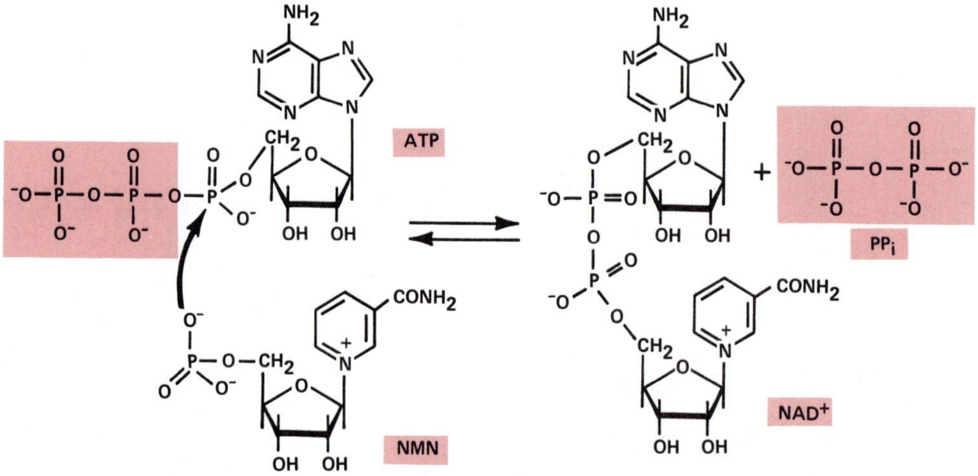

Figure 3-1
Coenzyme biosynthesis illustrated by the condensation of NMN and ATP to yield NAD and PP_i.

cleoside 5′-triphosphate, is the prototype for transfers of nucleotides to form all the other nucleotide coenzymes and also the mixed acid anhydrides, the form in which activated fatty acids, amino acids, and sulfate are found (Fig. 3-2).[4]

It might seem unreasonable today to have considered any other than the 5′ isomers as nucleic acid precursors. Yet, as mentioned previously (Section 2-4), biochemistry before 1950 emphasized energy production, and relatively little work had been done on biosynthetic pathways. It was assumed that since individual enzymatic reactions could be pulled and pushed back and forth, so could entire pathways. Glycogen could be synthesized by glycogen phosphorylase. Why not protein synthesis by reversal of peptidases? Lipids by lipases? And RNA by ribonuclease? Inasmuch as the 2′,3′ cyclic nucleotide produced by ribonuclease degradation of RNA was also utilized by it to reverse the reaction, these nucleotides were commonly regarded as the likely building blocks in nucleic acid synthesis. During the 1950s, however, it became clear that pathways of energy production and biosynthesis are distinct. They are dual and divided highways. How rational it is that the physical integration and control of a sequence of reactions should be sharply separated from another sequence whose purpose is unrelated or even conflicting. Now it seems obvious why there are separate routes of biosynthesis and degradation.

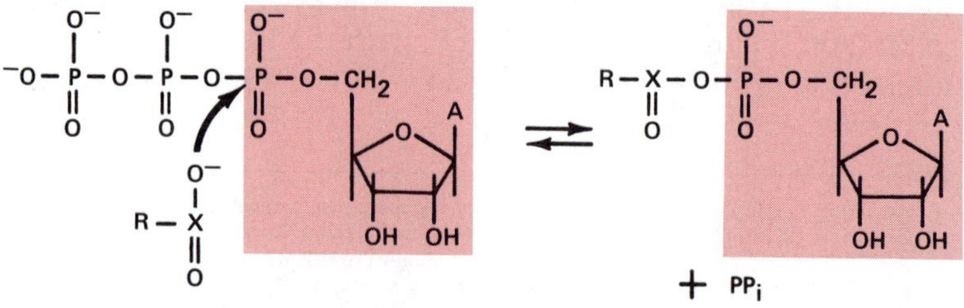

Figure 3-2
Nucleotidyl transfer as a general mechanism for synthesis of coenzymes and activated intermediates in the biosynthesis of fatty acids and proteins.

4. Kornberg A (1962) in *Horizons in Biochemistry* (Kasha M, Pullman B, eds.). Academic Press, New York, p. 251.

Efforts to reconstruct in cell-free systems what we now regard as the biosynthetic route of nucleic acids started with the observation that ^{14}C-labeled ATP was incorporated by *E. coli* extracts into an acid-insoluble product presumed to be RNA.[5] Similarly, [^{14}C]thymidine was incorporated by such extracts into an acid-insoluble compound that was sensitive to deoxyribonuclease (DNase) and thus was presumed to be DNA.[6]

Attempts to characterize the RNA-synthesizing action were overshadowed by the simultaneous discovery of polynucleotide phosphorylase. In extracts of the bacterium *Azotobacter vinelandii*, this enzyme polymerizes ADP and other nucleoside diphosphates into a polyribonucleotide polymer.[7] At physiologic concentrations of phosphate, however, this enzyme rapidly phosphorolyzes RNA. With the discovery of RNA polymerase a few years later, the function of polynucleotide phosphorylase was assigned to the salvage of nucleotides from RNA rather than to their polymerization to RNA.

The incorporation of [^{14}C]thymidine was found to occur by way of the 5′-triphosphate, dTTP. With the purification of the enzyme responsible for conversion of dTTP to DNA,[8] it became apparent that the other three common deoxyribonucleotides and preformed DNA were required for the reaction. Since the purified enzyme, called DNA polymerase, used the preformed DNA as a template for directing the assembly of a new DNA chain, it was clear at once why all four nucleotides were absolutely required.

The chronology of the publication of the Watson-Crick model in 1953, followed two years later by the discovery of DNA polymerase, has led to an assumption that DNA was included to serve as a *template* in our earliest attempts to incorporate thymidine. Not so. We added DNA expecting that it would serve as a *primer* for growth of a DNA chain, by analogy with the work of Cori and Cori, in which the growth of a carbohydrate chain, catalyzed by glycogen phosphorylase, depended on preformed glycogen as a primer. There was no presumption of discovering a phenomenon so utterly unprecedented in biochemistry: the absolute dependence of an enzyme upon instruction from one of its substrates, serving as a template.

3-2 Basic Features of Polymerase Action: The Template-Primer[9]

Types of Polymerases

DNA-directed enzymes (*DNA polymerases*), as catalysts of DNA chain growth, are found in extracts of all cells—bacterial, plant, and animal. Enzymes that transcribe DNA chains into RNA are called *RNA polymerases* (Chapter 7), and those polymerases that transcribe RNA into DNA are RNA-directed DNA polymerases, called *reverse transcriptases* (Section 6-9).

The DNA and RNA polymerases are unique among enzymes in that the choice of substrate is determined by a template as well as by the enzyme. Bacterial DNA polymerases can make animal DNA, and an animal polymerase can make bacte-

5. Kornberg A (1989) *ARB 58*:1.
6. Kornberg A (1989) *ARB 58*:1.
7. Grunberg-Manago M, Ochoa S (1955) *JACS 77*:3165; Ochoa S, Heppel LA (1957) in *The Chemical Basis of Heredity* (McElroy WD, Glass B, eds.). Johns Hopkins Press, Baltimore, p. 615.
8. Lehman IR, Bessman MJ, Simms ES, Kornberg A (1958) *JBC 233*:163.
9. Kornberg A (1969) *Science 163*:1410.

rial DNA, depending on the DNA template. Thus, there is no recognition of a genetic sequence of bases in the template chain by a DNA polymerase.

Primer versus Template Functions

Polymerization of DNA has two basic features (Fig. 3-3). First, a phosphodiester bridge forms between the 5′-phosphate group of an incoming deoxynucleotide and the 3′-hydroxyl group at the growing end of a DNA strand (the DNA *primer strand*). Chain growth is exclusively 5′→3′. Second, each deoxynucleotide added to the primer strand is selected by base-pair matching to the DNA *template strand*. No known DNA polymerase is able to *initiate* chains; that function and others related to chromosome replication require additional proteins.

Distinguishing between the terms *primer* and *template* is important in discussing polymerase action. The term *primer* refers strictly to the DNA chain that is growing at its 3′-OH terminus; the *primer terminus* is its terminal nucleotide that bears this 3′-OH group (see Fig. 3-3). The term *template* refers to the companion DNA chain, which furnishes the instructions for the sequence of nucleotides to be added to the primer strand. Thus the term *template-primer* embraces two different functions and meanings.

Template-Primer Structures

The enormous variety of secondary structures of DNA can be divided into roughly five categories of template-primer (Fig. 3-4). These include synthetic polymers as well as natural DNA. Polymerases are often distinguished from one another on the basis of their capacity to utilize the different kinds of template-primer structures. This will be considered in detail for each of the enzymes discussed; the properties of *E. coli* DNA polymerase I (Chapter 4), illustrated in Figure 3-4, are given as an example.

Intact Duplex. Intact double-stranded DNA, such as linear phage T4 DNA and circular duplex DNA, is inert as a substrate for all known DNA polymerases. The circular duplex has no ends. The linear duplex, in which the strands are the

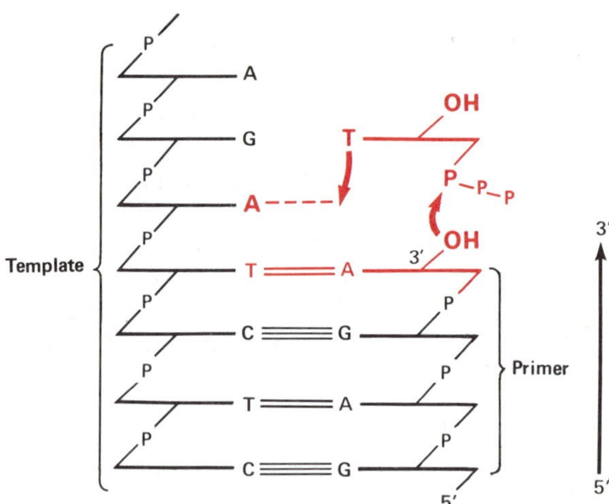

Figure 3-3
Template-primer in action.

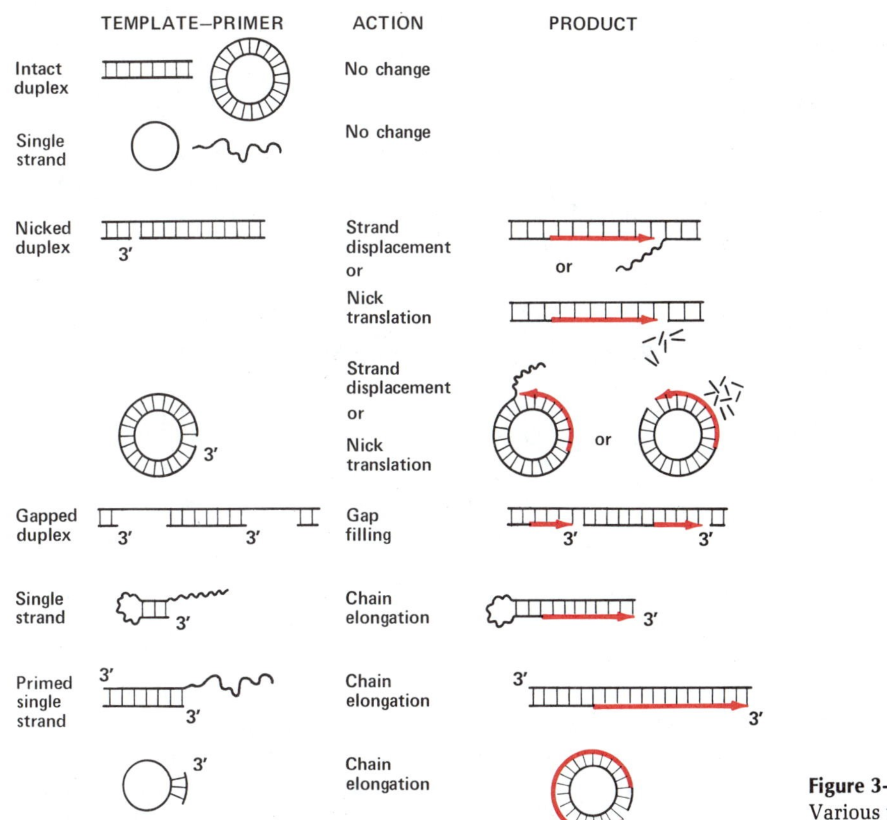

Figure 3-4
Various template-primers and their use by *E. coli* DNA polymerase I.

same length, cannot be extended by DNA polymerase, nor can either strand be replicated. RNA polymerase (Chapter 7), however, can initiate transcription on intact duplexes. Part of its multisubunit structure is devoted to recognizing signals for the initiation of an RNA chain.

Nicked Duplex. Duplex DNA may be "activated" as a template-primer for certain DNA polymerases by single-strand scissions introduced by an endonuclease such as pancreatic DNase I (Section 13-5). A *nick* is defined as a break introduced in a DNA strand that can be sealed by the enzyme DNA ligase, which links contiguous 3′-OH and 5′-P groups. Extensive nicking may cause the duplex to fragment when scissions on opposite strands are within a few nucleotides of each other. Such fragments may have protruding 5′ ends which can serve as templates for all DNA polymerases.

Gapped Duplex. DNA is made more active as a template when short gaps are introduced at nicks by the action of an enzyme, such as exonuclease III, that degrades one strand of a duplex structure in the 3′→5′ direction at a nick.

Single Strand. Single-stranded DNA molecules are not active as template-primers unless they have duplex regions. They may self-anneal to produce such regions, or they may anneal partially with other molecules. Duplex DNA may become effectively ssDNA when large sections are removed from its ends or when gaps in the duplex are greatly enlarged. A synthetic hook-shaped DNA with a particular length and sequence in which the unmatched 5′ ended stem is the template can be an abundant source of a well-defined template-primer.

Primed Single Strand. Annealing of an oligonucleotide to a synthetic homopolymer chain or to a natural circular single strand such as some viral DNAs (Section 4-16) yields primed ssDNA. This priming renders the single strands active as templates for DNA polymerases.

3-3 Basic Features of Polymerase Action: Polymerization Step[10]

The overall reaction catalyzed by DNA polymerase is as follows:

$$\underset{\text{DNA}}{(\text{dNMP})_n} + \text{dNTP} \longleftrightarrow \underset{\text{DNA}}{(\text{dNMP})_{n+1}} + \text{PP}_i$$

The enzyme has two substrates. One is the base-paired primer terminus of a template-primer containing a free 3′-OH group; the other substrate is a deoxynucleoside 5′-*tri*phosphate (dNTP).

The reaction consists of a nucleophilic attack by the 3′-OH group of the primer terminus on the α-phosphate of the incoming dNTP. The products are a new primer terminus—that is, a nucleotide linked to the primer chain by a 3′,5′-phosphodiester bond—and inorganic pyrophosphate. Subsequent hydrolysis of the pyrophosphate by inorganic pyrophosphatase (Section 2-6) can provide an additional driving force for polymerization and renders the reaction highly irreversible.

Mechanisms of Chain Growth

The overall direction of nucleic acid chain growth is 5′→3′. The nucleotide is added to the 3′-OH end of the chain. General mechanisms of polymer biosynthesis may be grouped in two classes, called tail growth and head growth (Fig. 3-5). Chain growth of RNA and DNA are examples of tail growth.

In *tail growth*, the monomer, activated by an energy-rich moiety with high group-transfer potential (e.g., ATP, glucose 1-P), is attacked by the nonactivated, growing tail of the polymer. This growth mechanism of RNA and DNA synthesis is also used in the polymerization of many polysaccharides, in which the nonactivated hydroxyl end of the sugar polymer grows by attacking the phosphoryl-activated end of a monomer.

In *head growth*, the nonactivated substituent of a monomer attacks the activated, growing end of the polymer. Since the monomer also contains an activated substituent not altered in the reaction, the lengthened chain again contains this activated substituent at its growing head.

The head-growth mechanism is exemplified in three processes: protein synthesis on a ribosome, fatty acid synthesis in a synthetase complex, and coupling

---

10. Kornberg A (1969) *Science* *163*:1410.

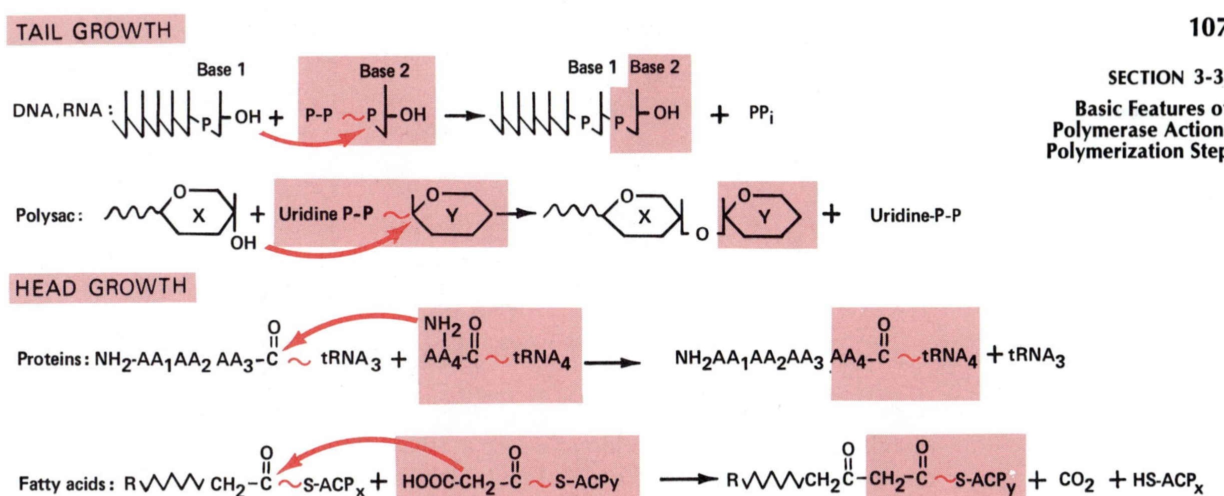

Figure 3-5
Mechanisms of tail growth and head growth in the biosynthesis of macromolecules. X and Y = sugar moieties; AA = amino acids; tRNA = transfer RNA of the indicated amino acid; ACP = acyl carrier protein.

of trisaccharide units on the bacterial inner cell membrane in O antigen synthesis. These processes have one feature in common. The activated monomer, as well as the activated head, is fixed and oriented on the surface of a cellular particle. Tail-growth synthesis seems, on the other hand, to be used when the geometry of assembly is not so rigidly defined, and when either component, usually the monomer, is brought in from the soluble phase.

Chain Growth in Replication

The opposite directions of the strands in a DNA double helix posed a problem to biochemists trying to understand how DNA is replicated. How could both strands of the helix be replicated at the same time if growth by the DNA polymerase mechanism is strictly 5′→3′? An adequate answer to this question is now in hand and will be presented in Chapter 15. However, for a number of years this dilemma led to a search for a mechanism involving 3′→5′ chain growth.

A possible tail-growth mechanism for simultaneous replication of the antiparallel strands of the DNA duplex invoked the 3′ isomers of the dNTPs. It postulated attack by the 5′-OH tail end of the chain on the activated monomer, with a consequent 3′→5′ growth (Fig. 3-6). Thus far there is no evidence for the biosynthesis of nucleoside 3′-mono- or triphosphates nor any indication, despite an intensive search, of an enzyme that can polymerize chemically synthesized 3′-dNTPs (Fig. 3-6, top).

A possible head-growth mechanism for achieving 3′→5′ growth invokes an attack by the 3′-OH group of the incoming dNTP on the 5′-α-phosphate of a triphosphate-activated primer terminus (Fig. 3-6, bottom). This would entail an

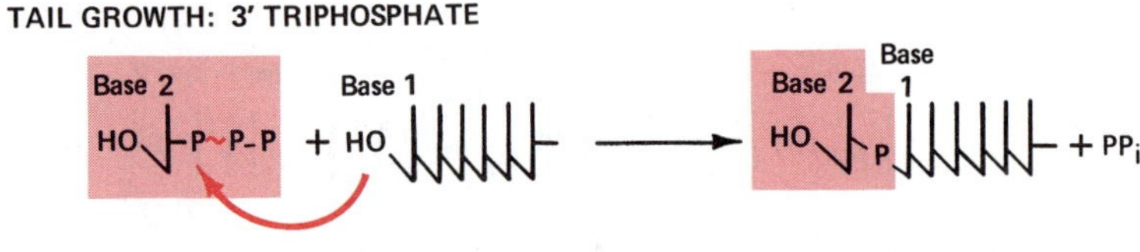

Figure 3-6
Hypothetical mechanisms for chain growth in a 3′→5′ direction.

attack by an activated monomer on an activated polymer. Once again, no evidence exists for this kind of DNA polymerization. Furthermore, the requirement for an activated growing polymer end would be inconsistent with "proofreading." In this operation, DNA polymerase checks the fidelity of base pairing at the primer terminus and, in the rare event of a mispairing, removes a mismatched nucleotide (see below); in a head-growth mechanism, this would remove the activated primer terminus and terminate replication.

Proofreading for Base-Pair Matching

Accurate base pairing by DNA polymerase ensures the fidelity of replication (Sections 4-11 and 15-9). Before the polymerization step can occur, the incoming nucleoside triphosphate must be paired with its template partner. If the correct base pair has not been formed, the enzyme rejects the mispaired nucleotide (Section 4-7).

Even before matching the incoming nucleotide to its template partner, the enzyme can check the accuracy of base pairing between the primer terminus and the template. If these bases are unpaired, subsequent polymerization will not occur. Rather, the mispaired primer terminus will be excised by an associated function of polymerases, their 3′→5′ exonuclease action. This activity degrades unpaired nucleotides from the 3′ end of the chain. When a correctly paired duplex structure is reached, polymerization can proceed (Section 4-9).

This proofreading activity, a feature of many DNA polymerases, provides a rationale for the tail-growth, rather than the head-growth, mechanism of polymerization. As mentioned already, were a mismatch present at the primer terminus, in the head-growth mechanism, removal of the activated triphosphate end of the polymer would prevent further polymerization.

Generalizations about DNA Polymerase Action

What we know thus far of the DNA polymerases allows these ten generalizations:

(1) The nucleotide added is a monomer activated as a 5′-triphosphate. Suggestions that the diphosphate,[11] or another activated form of the monophos-

---

11. Rubinow SI, Yen A (1972) *NNB* *239*:73.

phate,[12] is the DNA precursor in the cell are probably in error for technical reasons or because of compartmentalization (metabolic channeling) of the dNTP in different cellular pools.[13]

(2) The phosphodiester bridge is made by a nucleophilic attack by the 3′-OH group of a primer terminus on the α-phosphate atom of the activated monomer and elimination of the terminal pyrophosphate.

(3) A primer with a 3′-OH group is required; new chains cannot be initiated.

(4) Chain elongation takes place by a tail-growth mechanism and is 5′→3′.

(5) Correct base pairing is required at the primer terminus, as well as in matching the incoming nucleotide to the template. In the case of *E. coli* DNA polymerase I, if a base at the primer terminus is unpaired, because of mismatching or fraying, it is removed by the 3′→5′ exonuclease activity of the polymerase. This proofreading ensures high fidelity of replication.

(6) Polymerization may be *processive* (Sections 4-11 and 15-8); that is, a succession of several polymerization steps may occur without release of the enzyme from the template. A *nonprocessive* or *distributive* polymerase dissociates from the template after the addition of each nucleotide, and may depend on accessory proteins to confer processivity.

(7) Polarity of the newly synthesized chain is opposite to that of the template; the growing chain and template are antiparallel.

(8) Any genetic sequence in the template will be copied, with little particular preference or specificity.

(9) Chains are generally started (Section 15-7) by covalent extension of a short RNA transcript laid down by a primase. Starts can also be made by two other mechanisms: by attachment of the initial nucleotide to a protein oriented at the 3′ end of the template chain and by the nicking of one strand of a duplex template to generate a primer terminus.

(10) The polymerase is commonly associated (in a holoenzyme assembly) with accessory proteins that supply DNA melting, priming, proofreading, and processivity functions.

3-4 Measurement of Polymerase Action[14]

Techniques of Measurement

The activity of a polymerase is quantitatively determined by measuring conversion of the monomer to polymeric form. There are at least four ways in which this change has been measured.

(1) Conversion of a radioactive mononucleotide (using ^{32}P in the α-phosphate, ^{14}C or ^{3}H in the nucleoside) into an acid-insoluble form. Polynucleotides of ten or more residues are rendered insoluble by adding acid (perchloric or trichloracetic acid). The precipitate is collected by centrifugation or on a glass filter and is washed. The control values, resulting from residual substrate, may be reduced

12. Werner R (1971) *NNB 233*:99.
13. Fridland A (1973) *NNB 243*:105.
14. Lehman IR, Bessman MJ, Simms ES, Kornberg A (1958) *JBC 233*:163; Jovin TM, Englund PT, Bertsch LL (1969) *JBC 244*:2996.

to less than 0.01% of the starting radioactivity. More discriminating techniques for characterizing the macromolecular product and separating it from substrate depend on methods such as gel filtration, gel electrophoresis, and sedimentation gradients.

(2) Conversion of a radioactive mononucleotide to polyelectrolyte form, which is then tightly adsorbed to an ion exchanger[15] (e.g., 2,-diethylaminoethanol–cellulose), which binds highly negatively charged compounds. A small circle of DEAE–cellulose paper can be washed free of the substrate while retaining polynucleotides down to the size of a trinucleotide. Advantages of this method are that many samples can be processed at once and that acid-soluble polynucleotides (too small to be precipitated) are detected.

(3) Identification of hypochromic shift as mononucleotides are polymerized.[16] The decrease in absorbance at 260 nm, caused by stacking interactions of the parallel bases in the helix, provides a direct measure of the extent of reaction. Most suitable for kinetic analysis, the method is, however, relatively insensitive. It is useful only when a substantial fraction of the substrate is converted to polymer and the template concentration is not high.

(4) Measurement of increase in viscosity when mononucleotides are polymerized.[17] This method has much the same advantages and limitations as the optical method.

Polymerase activity can be measured, in some special instances, by observing a change in the macromolecular properties of the template-primer. One example is the conversion of a radioactive ssDNA template, such as labeled phage ϕX174 or M13 DNA, to a particular duplex form separable by sedimentation or gel electrophoresis and resistant to single-strand–specific nucleases. A change or increase in the biological activity of such DNA molecules may also be observed.

Substrates

Template-Primers. The template-primers used by DNA polymerases vary in secondary structure, as discussed in Section 2. Since the polymerases differ in their capacity to use various template-primers, the nature of this substrate is an important parameter of the assay.

Simple homopolymer pairs, such as poly dA · oligo dT or poly dG · oligo dC, are attractive template-primers; only a single nucleotide substrate, dTTP for the former or dCTP for the latter, is required for polymerization. These homopolymers are variable and complex in structure and often fail to substitute for natural DNA, as in the initiation of DNA chains in the reconstituted replication systems of the coliphages (Sections 17-3 and 17-4).

Concentrations of template-primer used in the assays are generally between 10^{-5} and 10^{-4} M in nucleotide residues. Depending on the size of the DNA, molecular concentrations are three to six orders of magnitude lower.

Nucleotides. The dNTPs, or precursors convertible to them, are generally required in concentrations near 10^{-5} M. The purine and pyrimidine nucleotides required are those needed for pairing with the template.

15. Brutlag D, Kornberg A (1972) *JBC 247*:241.
16. Radding CM, Kornberg A (1962) *JBC 237*:2877.
17. Schachman HK, Adler J, Radding CM, Lehman IR, Kornberg A (1960) *JBC 235*:3242.

Commercial dNTPs, though readily available, have some major drawbacks. Nonradioactive preparations may contain significant amounts of other contaminating nucleotides (deoxy- or ribo-); highly radioactive samples are often contaminated with inhibitors, and these may accumulate as the preparation ages.

Other Factors Affecting Polymerase Studies

Requirements for various factors and responses to them are important in the assay and characterization of the various polymerases (Chapters 4 to 6). Certain factors are particularly noteworthy.

Nucleases. The effects of nucleases (Chapter 13) on the measurement[18] and analysis of polymerase action cannot be overemphasized. A trace of endonuclease may render an inert duplex active for a polymerase. Alternatively, an endonuclease that creates 3′-P termini may convert an active template-primer into an inhibitor that binds the enzyme in an unproductive complex. Finally, an exonuclease such as exonuclease III, which acts as a phosphomonoesterase on 3′-P termini, enlarging nicks into gaps, can easily increase the reaction rates of certain polymerases by 100- or 1000-fold. Obviously, an excess of nuclease would lead to net loss of DNA, and without adequate DNA pools to trap and protect incorporated radioactive substrates, high levels of DNAse would obscure polymerase activity.

Metal. A divalent metal ion, preferably Mg^{2+}, is invariably required. With Mn^{2+} as a substitute, polymerases discriminate less against analogs (e.g., dideoxyNTPs) and mismatches.[19] Sharp optima in metal concentration, especially the inhibitory levels of Mn^{2+}, can be avoided by the use of a weak chelator ("metal buffer"), such as citrate or isocitrate.

Ionic Strength. Profound inhibition by salt, generally chlorides, can be circumvented by the use of other anions. In the case of *E. coli* enzymes, the physiologic anion is glutamate.[20]

Macromolecular Concentration. Dissociation of multimacromolecular assemblies and diminished macromolecular interactions can result from a severe dilution in cell extracts and from fractionation procedures. This can be minimized by adding hydrophilic polymers (e.g., 10% polyethylene glycol), which help to restore the molecular crowding of the cellular milieu.[21]

Agents Affecting DNA Secondary Structure. Temperature, pH, or the presence of spermidine or certain proteins can influence the melting and annealing of template-primers, with crucial consequences for polymerase actions.

Adjuvants. Polymerases, like many enzymes, may require adjuvants such as sulfhydryl protective agents, chelators for heavy metals, detergents, and coatings for glass surfaces.

---

18. Lehman IR (1967) *ARB 36*:645.
19. Tabor S, Richardson CC (1989) *PNAS 86*:4076.
20. Leirmo S, Harrison C, Cayley DS, Burgess RR, Record MT Jr (1987) *Biochemistry 26*:2095; Griep MA, McHenry CS (1989) *JBC 264*:11294.
21. Minton A (1983) *Mol Cell Biochem 55*:119; Zimmerman SB, Harrison B (1985) *NAR 13*:2241; Jarvis TC, Ring DM, Daube SS, von Hippel PH (1990) *JBC 265*:15160.

CHAPTER 4

DNA Polymerase I of *E. coli*

4-1 Isolation and Physicochemical Properties

DNA polymerase I (pol I) of *E. coli* is so designated because it was the first polymerase recognized.[1] We will consider this enzyme in detail because it has been studied more extensively than the others and appears to embody the salient features of DNA polymerase action. Further, structural and genetic studies have revealed impressive homologies among prokaryotic and eukaryotic polymerases, and pol I and closely related polymerases have been used extensively in recombinant DNA technology, attracting additional interest in the properties of this prototypical enzyme.

DNA polymerase I has been isolated by conventional steps of protein fractionation. From 100 kg of *E. coli* paste, about 1.0 g of pure enzyme was obtained, with an overall yield of about 20% from the extract.[2] The number of enzyme molecules per cell [assuming that 1 g of *E. coli* cell paste ($A_{595} = 400$) contains 6×10^{11} cells] is estimated to be 400, based on the formula $10^9 (X)/(Y)(Z)$, where X is the number of enzyme units per gram of cell paste, Y the number of units per milligram of pure enzyme, and Z the molecular weight of the enzyme.

1. Kornberg A, Lehman IR, Bessman MJ, Simms ES (1956) *BBA 21*:197; Lehman IR, Bessman MJ, Simms ES, Kornberg A (1958) *JBC 233*:163.
2. Kornberg A (1960) *Science 131*:1503; Jovin TM, Englund PT, Bertsch LL (1969) *JBC 244*:2996; Kornberg A (1981) *Enzymes 14*:3; Lehman IR (1981) *Enzymes 14*:15.

E. coli strains infected with a lytic λ phage carrying the gene for pol I (*polA*) overproduce the enzyme nearly 100-fold.[3] At this level of amplification, pol I represents about 4% of the weight of cellular protein. Such strains provide a source for rapid and efficient purification, as do strains with plasmids bearing the whole or part of the *polA* gene.[4]

As deduced from the sequence of the *polA* gene, the enzyme is a single chain of 928 residues (103 kDa).[5] A remarkable feature of the amino acid composition is the presence of only two cysteines, one of which is in the C-terminal two-thirds of the polypeptides (large fragment; Section 5). Neither cysteine is judged to be part of the active site because the enzyme modified with iodoacetate or mercuric ion retains full activity.[6] The reaction of mercuric ion with a molar excess of enzyme produces a dimer of polymerases linked by a mercury atom. The use of ^{203}Hg provides a convenient way of incorporating a radioactive label of about 10,000 cpm per microgram of enzyme, without affecting enzymatic activity. This label has served as a marker in DNA binding studies (Section 5). Mercury has also served on occasion to block the action of a sulfhydryl-sensitive enzyme (e.g., dGTPase[7]) that may contaminate pol I preparations.[8]

Metal binding[9] by both the enzyme and the substrates is essential in the pol I reaction; there is (1) binding of a divalent cation (Mg^{2+}, Mn^{2+}, or Zn^{2+}) at two separate sites in the $3' \rightarrow 5'$ exonuclease proofreading domain (Section 8), (2) binding of one or more divalent cations (e.g., Mg^{2+}) in an unidentified locus in the large domain of the large (polymerization) fragment, (3) chelation of dNTP and Mg^{2+} (Section 4), and (4) possibly, binding to other enzyme sites (e.g., the small fragment) and to the DNA. Substitution of Mn^{2+} for Mg^{2+} sharply reduces the ability of pol I to discriminate against dNTP analogs (e.g., dideoxy-NTPs)[10] and mismatches, the latter being responsible for a decrease in fidelity. Pol I contains no firmly bound zinc,[11] yet it is inhibited by o-phenanthroline, a zinc chelator: with a thiol present, contaminating traces of cupric ion are reduced and form phenanthroline–cuprous ion complexes that cause DNA strand breaks due to peroxides.[12]

Fluorescent derivatives of pol I have been prepared with the acylating agent N-carboxymethyl isatoic anhydride.[13] Labeling and modification of the DNA and dNTP binding sites will be discussed in Sections 6 and 7.

4-2 Preview of Polymerase Functions

Pol I performs many functions, which are manifested and measured in a variety of ways. Without reference to the amino acid sequence, the folded structure of

3. Kelley WS, Chalmers K, Murray NE (1977) *PNAS 74*:5632; Murray NE, Kelley WS (1979) *MGG 175*:77.
4. Joyce CM, Grindley NDF (1983) *PNAS 80*:1830; Minkley EG Jr, Leney AT, Bodner JB, Panicker MM, Brown WE (1984) *JBC 259*:10386.
5. Joyce CM, Kelley WS, Grindley NDF (1982) *JBC 257*:1958.
6. Jovin TM, Englund PT, Kornberg A (1969) *JBC 244*:3009.
7. Kornberg SR, Lehman IR, Bessman MJ, Simms ES, Kornberg A (1958) *JBC 233*:159; Seto D, Bhatnagar SK, Bessman MJ (1988) *JBC 263*:1494.
8. Englund PT, Huberman JA, Jovin TM, Kornberg A (1969) *JBC 244*:3038.
9. Mullen GP, Serpersu EH, Ferrin LJ, Loeb LA, Mildvan AS (1990) *JBC 265*:14327.
10. Tabor S, Richardson CC (1989) *PNAS 86*:4076.
11. Walton KE, Fitzgerald PC, Herrmann MS, Behnke WD (1982) *BBRC 108*:1353.
12. Marshall LE, Graham DR, Reich KA, Sigman DS (1981) *Biochemistry 20*:244.
13. Jovin TM, Englund PT, Kornberg A (1969) *JBC 244*:3009.

part of the enzyme (Section 5), and kinetic mechanisms (Section 11), some general facts about its actions should be stated at this point.

The principal, discrete activities of the enzyme are polymerization, pyrophosphorolysis, pyrophosphate exchange, and two independent exonucleolytic hydrolyses, one of which degrades in the 3′→5′ and the other in the 5′→3′ direction. Pyrophosphorolysis and pyrophosphate exchange are simply demonstrations of the reversal of the polymerization reaction. In view of the large concentrations of pyrophosphate required and the potency of inorganic pyrophosphatase in all cells, the physiologic significance of these reactions is probably only minor.

DNA chain growth in the 5′→3′ direction is dictated by the specificity for two substrates: the 3′-hydroxyl primer terminus and the deoxynucleoside 5′-triphosphate. The chain is synthesized antiparallel to the template; a synthetic template-primer with parallel strands is not extended by the enzyme, but is a substrate for the 3′→5′ exonuclease.[14]

The 3′→5′ exonuclease is a component of the polymerase machinery that monitors base pairing at the primer terminus, supplementing the capacity of polymerase to match the nucleotide substrate to the template. A mismatched terminal nucleotide on the primer chain binds to a site on the enzyme that results in its hydrolysis and removal.

The 5′→3′ exonuclease activity, to date found only in pol I, degrades duplex DNA from a 5′ end, releasing mono- or oligonucleotides. Its capacity to excise distorted sections of DNA that lie in the path of the chain extension by polymerase seems well suited for removing a DNA lesion such as a thymine dimer or for removing RNA from a DNA–RNA hybrid duplex (Section 18).

Pol I contains its several enzyme activities in one polypeptide chain. Proteolytic cleavage divides the polypeptide into a large and a small fragment (Section 4). The polymerization, pyrophosphorolysis, pyrophosphate exchange, and 3′→5′ exonuclease activities, all features of the polymerization operation, are in the large, C-terminal fragment. The 5′→3′ exonuclease activity is in the small, N-terminal fragment, distinct from the other activities.

A striking feature of pol I, observed in few other polymerases, is its capacity to promote replication at a nick, unaided by other proteins. This entails melting of the duplex beyond the nick and progressive *strand displacement* (Section 4) of the 5′ chain. Other polymerases similarly known to advance a replicating fork require a DNA-binding protein with a strong affinity for the displaced single strand or the assistance of a helicase (Chapter 11). Strand displacement coupled with 5′→3′ exonuclease action leads to progression of the nick *(nick translation)* along the template (Section 13). Uncoupled from this nuclease, strand displacement followed by use of the displaced strand as a template *(template-switching)* leads to net synthesis of DNA (Section 16).

In the cell, pol I may be associated with proteins of related function. However, it is difficult to determine whether the persistence of a particular activity in an enzyme preparation represents a significant association or merely an adventitious one. A trace of nucleoside diphosphate kinase supports the utilization of deoxynucleoside diphosphates as substrates.[15] But pure polymerase preparations have little or none of this kinase activity and are specific for triphosphates.

---

14. Germann MW, Kalisch BW, van de Sande JH (1988) *Biochemistry 27*:8302; Rippe K, Jovin TM (1989) *Biochemistry 28*:9542.
15. Miller LK, Wells RD (1971) *PNAS 68*:2298.

4-3 Multiple Sites in an Active Center[16]

DNA polymerase is globular, with a diameter near 65 Å. It can therefore contact nearly two DNA helical turns (20 bp, ~70 Å) and a helical width of about 20 Å. The enzyme is readily cleaved by proteases (Section 4) to a small N-terminal fragment of 35 kDa which contains the 5′→3′ exonuclease activity, and a large C-terminal fragment of 68 kDa (popularly known as the *Klenow fragment*), which contains the polymerase and 3′→5′ exonuclease activities. High-resolution crystal-structure analysis shows the large fragment to be further divided into two domains (Section 5), the larger of which includes the polymerase and the smaller the 3′→5′ exonuclease[17] (Fig. 4-1). Until the three-dimensional structure of the *intact* polymerase bound to DNA is known, its disposition on a template-primer, gapped or nicked, must remain uncertain.

Even prior to structural studies, evidence clearly indicated that the active center includes at least eight sites that recognize and bind the nucleotides, DNA, and other reactants.[18] The enzyme contains a site for the

(1) *Template* chain, including the point where the base pair is to be formed and several nucleotides on either side.
(2) Growing chain, the *primer*, with a polarity opposite to that of the template.
(3) *Primer terminus*, the 3′-OH nucleotide of the primer at the growing end of the chain.
(4) Activated monomeric substrate, *deoxynucleoside 5′-triphosphate* (dNTP).
(5) Product, inorganic pyrophosphate (PP_i).
(6) Product of cleavage of the primer terminus by the 3′→5′ exonuclease, a nucleoside monophosphate (dNMP).
(7) Divalent metal cation.
(8) Displaced *5′-terminal nucleotide* or *oligonucleotide*, positioned in the path of the growing chain (beyond the triphosphate site), for 5′→3′ exonucleolytic cleavage.

4-4 Proteolytic Cleavage: Three Enzymes in One Polypeptide[19]

Proteolytic cleavage of pol I separates the polypeptide chain into two active fragments[20] and provides an unusual opportunity to examine the two parts of this complex enzyme: (1) a large C-terminal fragment[21] contains the polymerase and 3′→5′ exonuclease activities and the binding sites for dNTPs; (2) a smaller N-terminal fragment[22] contains only the 5′→3′ exonuclease activity (Fig. 4-1).

The large fragment, obtained by proteolytic cleavage of intact enzyme or by cloning this segment of the *polA* gene, carries out DNA synthesis on the 3′-OH side of a nick in dsDNA. It catalyzes 3′→5′ exonuclease action on both ssDNA

16. Kornberg A (1969) *Science 163*:410; Joyce CM, Steitz TA (1987) *TIBS 12*:288.
17. Freemont PS, Ollis DL, Steitz TA, Joyce CM (1986) *Proteins 1*:66; Hall JD (1988) *Trends Genet 4*:42; Freemont PS, Friedman JM, Beese LS, Sanderson MR, Steitz TA (1988) *PNAS 85*:8924.
18. Joyce CM, Steitz TA (1987) *TIBS 12*:288.
19. Klenow H, Henningsen I (1970) *PNAS 65*:168; Brutlag D, Atkinson MR, Setlow P, Kornberg A (1969) *BBRC 37*:982; Setlow P, Brutlag D, Kornberg A (1972) *JBC 247*:224; Setlow P, Kornberg A (1972) *JBC 247*:232.
20. Brutlag D, Atkinson MR, Setlow P, Kornberg A (1969) *BBRC 37*:982.
21. Brutlag D, Atkinson MR, Setlow P, Kornberg A (1969) *BBRC 37*:982.
22. Setlow P, Brutlag D, Kornberg A (1972) *JBC 247*:224.

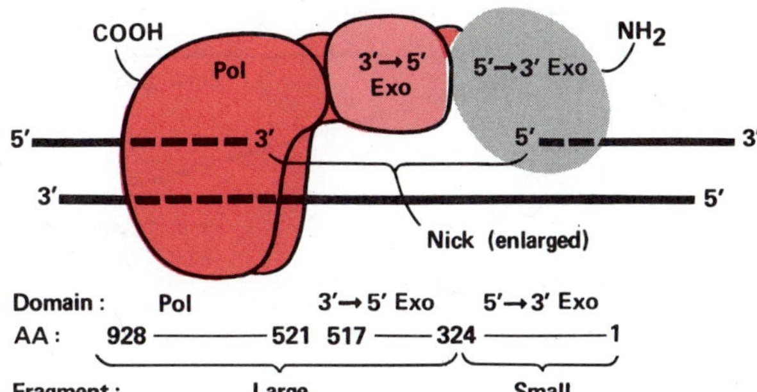

Figure 4-1
Proposed domain structure of pol I. The colored areas represent the experimentally determined large fragment structure; the shaded area indicates a possible location for the small fragment produced by proteolytic cleavage of pol I. The 5′→3′ exonuclease domain interacts with the DNA 5′ end downstream of the primer terminus. Me^{2+} = divalent metal cation; AA = amino acid residues; Exo = exonuclease. [Adapted from Joyce CM, Steitz TA (1987) *TIBS 12*:288]

and unpaired regions in dsDNA, with production of deoxynucleoside monophosphates (Section 9). The large fragment also catalyzes primed poly d(AT) synthesis at a relatively linear rate, generating exceptionally long poly d(AT) molecules, but it is inactive in unprimed (de novo) poly d(AT) synthesis (Section 16). Complete utilization of the dATP and dTTP substrates, as well as the kinetics of synthesis and the size of the poly d(AT), show that 5′→3′ exonuclease activity is absent. During synthesis at the 3′-OH side of a nick, the 5′-P end of the nick is not degraded.

The small fragment retains only the 5′→3′ exonuclease activity. This fragment resembles the 5′→3′ exonuclease of the intact enzyme in degrading DNA to mono- and oligonucleotides and in its capacity to excise mismatched regions such as thymine dimers. But it differs from the intact enzyme in that dNTPs fail to stimulate the exonuclease or increase the proportion of oligonucleotides among the products.

No evidence exists of a physical interaction between the separated large and small fragments when they are brought together.[23] However, when the two fragments are mixed with nicked DNA and suitable dNTPs, evidence of interaction is striking. The same strong stimulation of polymerization by 5′→3′ exonuclease activity is seen with this mixture as with the intact enzyme (Section 13). Excision of UV-induced thymine dimers illustrates the contribution of DNA synthesis to the repair process (Fig. 4-2). These findings suggest that the two fragments, when properly juxtaposed at a nick, effectively coordinate polymerization and 5′→3′ exonuclease action.

Cleavage of pol I into two active fragments that show no measurable affinity for one another and are similar to the intact pol I in catalytic activity indicates that the single polypeptide chain possesses three distinct enzymes: one fragment contains the polymerase and 3′→5′ exonuclease functions; the other 5′→3′ excision activity. The two fragments are held together by a polypeptide link susceptible to proteases such as subtilisin or trypsin. This polypeptide hinge ensures that both the polymerase and excision functions act simultaneously at the same nick in a DNA molecule.

Separation of the enzyme activities has also been achieved by inhibitors or mutations which inactivate the polymerase or the 3′→5′ exonuclease selectively (see Table 4-1, in Section 5), and others which labilize the 5′→3′ exonuclease without affecting the large fragment activities (see Table 4-9, in Section 18). The sharp physical and functional distinctions of the several domains of pol

23. Setlow P, Kornberg A (1972) *JBC 247*:232.

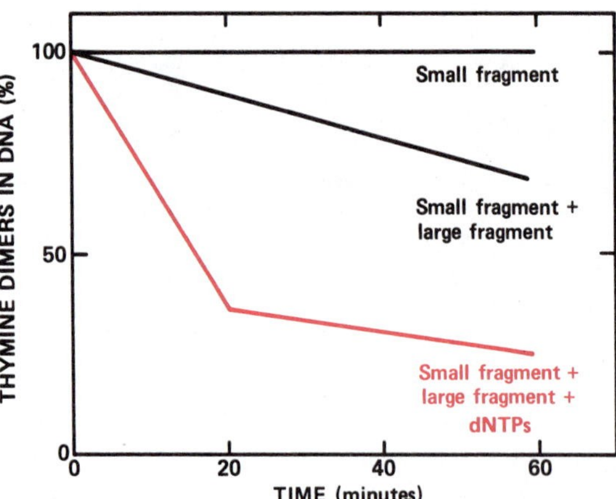

Figure 4-2
Thymine dimer excision. The small fragment (5′→3′ exonuclease) is active only in the presence of the large fragment (polymerase) and is most active during concurrent polymerization. [After Friedberg EC, Lehman IR (1974) *BBRC* *58*:132]

I suggest that they have evolved as separate genes, subsequently fused to encode multifunctional proteins or subunits.

Pol I shares with a few other polymerases (e.g., those of phages T5 and ϕ29) the capacity to displace a strand without the assistance of an energy-driven helicase or other factors. The isolated large fragment when polymerizing from a nick can displace the 5′-ended strand without hydrolysis, a finding initially obscured by the 5′→3′ exonuclease activity in the small fragment. Displaced single-stranded "overhangs" are transiently generated (see Fig. 4-20) and then cleaved by the exonuclease.[24] After each polymerization, the displacement or exonuclease action occurs with about equal probability. Generation of an overhang of about 12 ntd is expected for every 1300 nucleotides polymerized and may offer an opportunity for recombination. Discontinuities in DNA, which include nicks and gaps, may also include such overhangs.

4-5 Structure of the Large Fragment[25]

Analysis of crystals of the large fragment, prepared directly by cloning[26] rather than by proteolysis of pol I, has provided a three-dimensional structure at 2.75 Å resolution (Fig. 4-3). The C-terminal chain of 605 amino acids is folded into a large domain (residues 521 to 928) and a smaller one (residues 324 to 517).

The large domain contains a cleft, about 20 to 24 Å wide and 25 to 35 Å deep, with sides made up of many α helices and a base of β sheets (Fig. 4-3). A model of duplex DNA, 8 bp long and in the B form, can fit in the cleft (Plate 4); the α helices may even provide threads that direct a groove of DNA along a spiral path. Electrostatic field calculations[27] of the large fragment show the most positive potential to be appropriately located within the cleft.

Mutations in certain residues have furnished important insights into the separate functions of the large and small domains. The mutation *polA6*, an

24. Lundquist RC, Olivera BM (1982) *Cell 31*:53.
25. Joyce CM, Steitz TA (1987) *TIBS 12*:288; Freemont PS, Ollis DL, Steitz TA, Joyce CM (1986) *Proteins 1*:66; Hall JD (1988) *Trends Genet 4*:42; Freemont PS, Friedman JM, Beese LS, Sanderson MR, Steitz TA (1988) *PNAS 85*:8924.
26. Joyce CM, Grindley NDF (1983) *PNAS 80*:1830.
27. Warwicker J, Ollis DL, Richards FM, Steitz TA (1985) *JMB 186*:645.

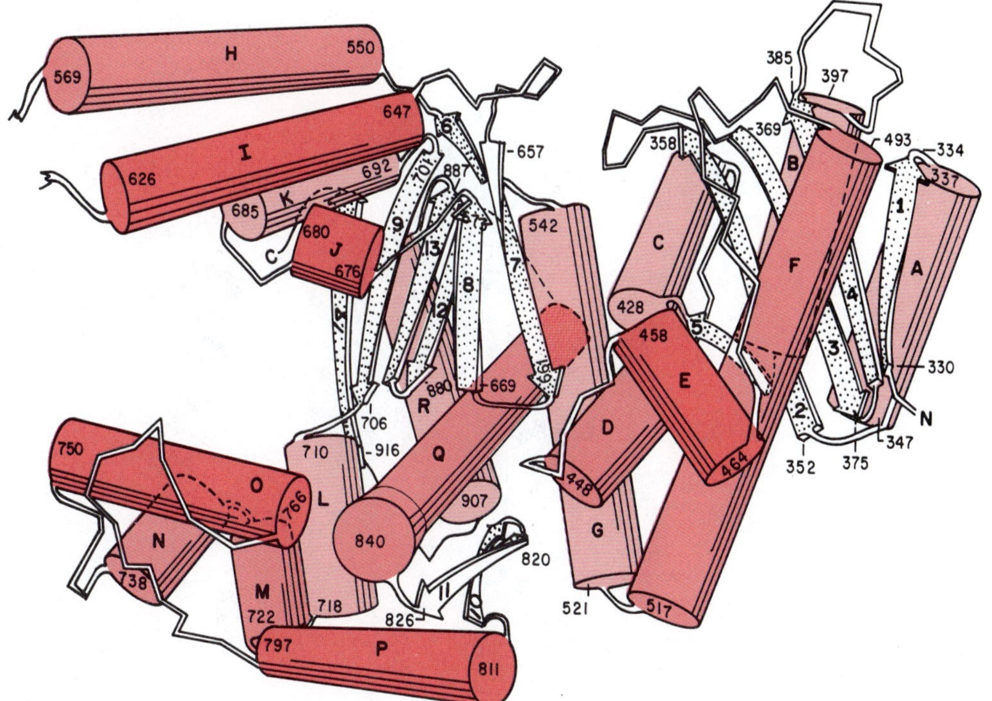

Figure 4-3
A schematic three-dimensional structure of the large fragment of pol I. Regions of polypeptide that form α helices are represented by cylinders and those that form β sheets by arrows. The open region between helices I and O is the cleft for DNA. The polypeptide discontinuity between helices H and I is a disordered subdomain forming a moveable flap (Fig. 4-16). [After Joyce CM, Steitz TA (1987) *TIBS 12*:288]

arginine to histidine (arg→his) change in helix K[28], weakens the interaction with DNA. Site-directed mutagenesis has identified a number of residues in or near the cleft that may be responsible for DNA binding (arg668, arg841, asn845, gln849, and his881) and others that form or influence the dNTP binding site (asn845, arg841, and tyr766).[29]

The small domain contains the site for binding NMP and two sites, A and B, for divalent metal ions (Fig. 4-4; see also Fig. 4-1). The NMP site is occupied by both dNMP and riboNMP; NMP is used here in the generic sense to apply to both forms. Inasmuch as an NMP profoundly inhibits 3′→5′ exonuclease action,[30] the position which the NMP occupies in the crystal marks the locus of this enzymatic function. Site A may be occupied by Mg^{2+}, Mn^{2+}, or Zn^{2+}; the metal ion is coordinated with the carboxylates of asp355, asp501, and glu357 and the phosphate of NMP.[31]

Effects of mutations in the small domain are summarized in Table 4-1. A double mutant in which asp355 and glu357 are each replaced with ala removes metal site A, thereby preventing binding to metal site B and the NMP site and

28. Joyce CM, Fujii DM, Laks HS, Hughes CM, Grindley NDF (1985) *JMB 186*:283.
29. Polesky AH, Steitz TA, Grindley NDF, Joyce CM (1990) *JBC 265*:14579.
30. Que BG, Downey KM, So AG (1978) *Biochemistry 17*:1603; Que BG, Downey KM, So AG (1979) *Biochemistry 18*:2064.
31. Ollis DL, Brick P, Hamlin R, Xuong NG, Steitz TA (1985) *Nature 313*:762.

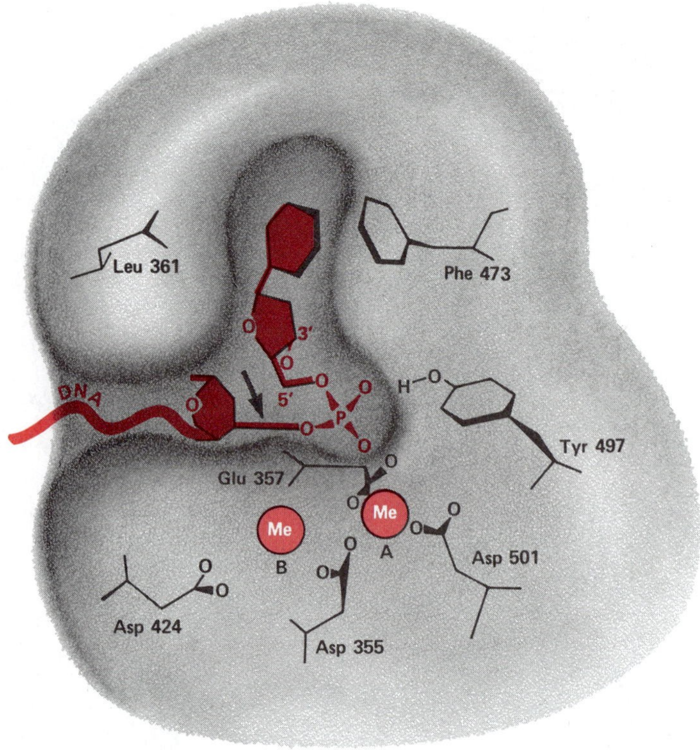

Figure 4-4
The 3′→5′ exonuclease active site with the bound 3′ end (dCMP) of a DNA chain. The position of chain cleavage (arrow), the side chains of residues interacting with the metal ions and dCMP, and the two binding sites for divalent metal ions (Me) are shown.

Table 4-1
Mutations in the 3′→5′ exonuclease domain[a]

Property	Residues with mutations to ala	
	asp355 glu357	asp424
Crystal structure	unchanged	unchanged
Polymerase activity	present	present
3′→5′ Exonuclease activity	absent[b]	absent[b]
NMP binding	absent	present
Metal site A binding	absent	present
Metal site B binding	absent	absent

[a] From Derbyshire V, Freemont PS, Sanderson MR, Beese L, Friedman J, Joyce CM, Steitz TA (1988) *Science* *240*:199.
[b] The ratio of exonuclease to polymerase activity was reduced by 10^5.

destroying exonuclease activity.[32] Removal of metal site B by asp424→ala also eliminates exonuclease activity but does not affect binding at the NMP site or at metal site A. Thus, metal binding is needed at site A for NMP binding and at site B for exonuclease action. Remarkably, the removal of binding and exonuclease functions does not perturb polymerase activity or the overall enzyme structure. Structural studies confirm the two-metal-ion mechanism.[33]

The distinctiveness of the polymerase and nuclease sites is also demonstrated by the addition of the 2′,3′-epoxynucleotide of adenosine 2′,3′-epoxide 5′-triphosphate (epoxy-ATP) to the primer terminus. The polymerase remains tightly bound to the modified DNA duplex and is suicidally inactivated as a polymerase.[34] (The epoxy terminus is also resistant to 3′→5′ exonuclease action.) Remarkably, the 3′→5′ exonuclease activity of the inactivated enzyme remains intact and can bind a second (unmodified) DNA duplex and carry out exonucleolytic cleavage at an undiminished rate.[35]

A striking feature of the structure of the large fragment, and totally unexpected from enzymologic studies, is the enormous distance—25 to 30 Å—between the primer terminus at the site of polymerization, in the large domain, and proofreading at the 3′→5′ exonuclease site, in the small domain.

Cocrystals of duplex or single-stranded oligomers with the large fragment[36] have revealed important interactions that also were not anticipated from earlier structural, as well as functional, studies. One such cocrystal contains a duplex DNA of 8 bp with a 3-bp protrusion at one 5′ end:

```
CCGGCGG
GGCCGCCAGA-5′
```

The duplex crystallizes in a 1:1 complex with the large fragment (EDTA was present to inhibit exonuclease activity). At the high ionic strength needed for crystallization, the DNA is observed to be mainly bound to the small domain rather than in the cleft and is largely separated into single strands. One 3′ end of the DNA is located in the NMP site, and from it the single-stranded DNA extends back three to four residues toward the cleft in the large domain.

When a single-stranded tetranucleotide $d(pT)_4$ is bound to crystals of either of the two mutants lacking exonuclease activity (Table 4-1), it is observed to be bound to the small domain, with the phosphate of its 5′-terminal phosphodiester in the same position as the phosphate in the NMP binding site (Fig. 4-4).

With this first fine-structure analysis of the polymerase in hand, the functional features—kinetics of polymerization, proofreading, reversibility, and processivity (Sections 9 to 11)—need to be accommodated to the physical facts. Clearly, more information is needed about the large fragment of pol I and its interactions with the small fragment in the complete enzyme. Fine-structure analyses of other polymerases, particularly the closely related polymerase encoded by phage T7 (Sections 5-9 and 5-11) and the core of polymerase III holoenzyme (Section 5-3), would be of great value in providing insights into the evolution and functions of these remarkable enzymes.

---

32. Derbyshire V, Freemont PS, Sanderson MR, Beese L, Friedman J, Joyce CM, Steitz TA (1988) *Science* *240*:199; Derbyshire V, Grindley NDF, Joyce CM (1991) *EMBO J* *10*:17.
33. Beese LS, Steitz TA (1991) *EMBO J* *10*:25.
34. Abboud MM, Sim WJ, Loeb LA, Mildvan AS (1978) *JBC* *253*:3415.
35. Catalano CE, Benkovic SJ (1989) *Biochemistry* *28*:4374.
36. Freemont PS, Friedman JM, Beese LS, Sanderson MR, Steitz TA (1988) *PNAS* *85*:8924.

4-6 DNA Binding Site[37]

Fine-structure analysis of the large fragment (Section 5) shows that two-thirds of the C-terminal portion is folded into the deep, wide cleft that can accommodate duplex B-form DNA (see Plate 4). The surface features of the cleft,[38] ferrate oxidation[39] of residue met512, pyridoxal 5′-P modification[40] of lys635, phenylglyoxal labeling[41] of arg841, several point mutations[42] (Sections 5 and 18), oligopeptide (728 to 777) binding of DNA,[43] and footprinting all argue that this cleft is the binding site for the template-primer.

Binding of DNA to pol I had been studied earlier by measuring the sedimentation velocity, in sucrose density gradients, of mixtures containing a variety of DNA structures (Fig. 4-5) labeled with ^{3}H or ^{32}P and pol I labeled with ^{203}Hg. The mixtures were layered on top of the gradients, and the enzyme–DNA complexes were identified after sedimentation. Binding occurs along single-stranded chains and especially at nicks and ends of duplexes, but not to any measurable extent at covalently closed duplex regions.

Single-stranded DNA produced by denaturing duplex DNA is bound by pol I in proportion to its length and to the same extent as ϕX174 ssDNA—about one molecule per 300 nucleotide residues. The introduction of a nick containing 3′-OH and 5′-P termini into a circular duplex DNA by DNase I activates the molecule for replication. On the other hand, nicks with 3′-P and 5′-OH termini introduced by micrococcal nuclease (Section 13-5) inhibit replication by binding the enzyme nonproductively. Thus, regardless of the kind of nick, pol I binds each nick efficiently (Table 4-2 and Fig. 4-6).

An alternating copolymer of dA and dT of 23 residues [$d(AT)_{12}$] assumes a hairpin or nicked conformation upon melting and quick cooling at low ionic strength and low oligomer concentration (Fig. 4-7). With excess enzyme, $d(AT)_{12}$ sediments almost quantitatively with the enzyme, indicating very high binding affinity. With the oligomer present in excess, all the enzyme sediments as a

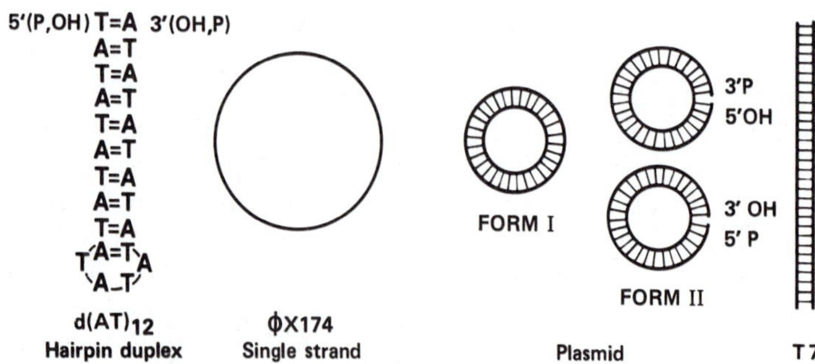

Figure 4-5
DNA structures tested for binding to pol I.

37. Englund PT, Kelly RB, Kornberg A (1969) *JBC 244*:3045; Griffith J, Huberman JA, Kornberg A (1971) *JMB* 55:209; Freemont PS, Ollis DL, Steitz TA, Joyce CM (1986) *Proteins 1*:66; Hall JD (1988) *Trends Genet* 4:42; Freemont PS, Friedman JM, Beese LS, Sanderson MR, Steitz TA (1988) *PNAS 85*:8924.
38. Joyce CM, Fujii DM, Laks HS, Hughes CM, Grindley NDF (1985) *JMB 186*:283.
39. Basu A, Williams KR, Modak MJ (1987) *JBC 262*:9601.
40. Basu S, Basu A, Modak MJ (1988) *Biochemistry 27*:6710.
41. Mohan PM, Basu A, Basu S, Abraham KI, Modak MJ (1988) *Biochemistry 27*:226.
42. Polesky AH, Steitz TA, Grindley NDF, Joyce CM (1990) *JBC 265*:14579.
43. Mullen GP, Shenbagamurthi P, Mildvan AS (1989) *JBC 264*:19637.

Table 4-2
Influence of DNA structure on binding of DNA to pol I

Conformation	Per DNA molecule: Nicks or ends	Per DNA molecule: Polymerase molecules bound
$d(AT)_{12}$ oligomer		
Hairpin	1	1
ϕX174 DNA		
Circular, single-stranded	0	20
Circular, duplex	0	<0.1
Plasmid DNA		
Irreversibly denatured	0	21
3′-Hydroxyl nick	1	1
3′-Phosphate nick	5	6
T7 DNA		
Linear, duplex	2	2
Single-stranded	2	240

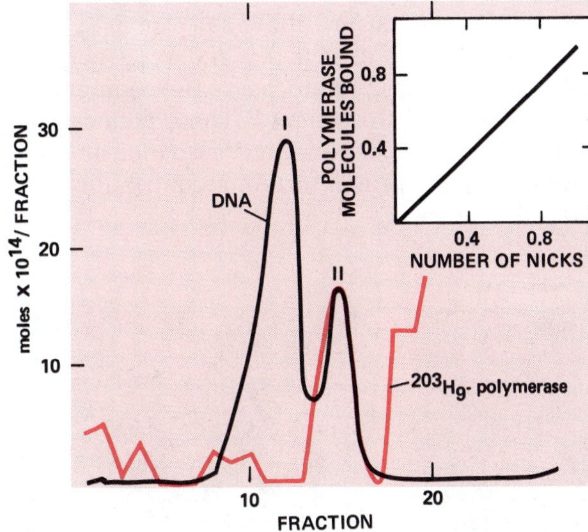

Figure 4-6
Influence of DNA structure on binding of DNA to pol I. The enzyme fails to bind to circular duplex Form I but binds to Form II.

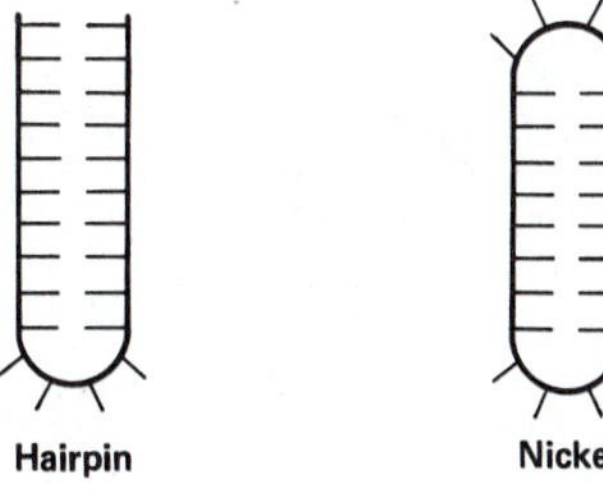

Figure 4-7
Conformations of the copolymer $d(AT)_{12}$.

complex containing an equimolar amount of oligomer. The same results obtain whether the oligomers terminate with a 3′-OH or a 3′-P. With linear duplex DNA of phage T7, one pol I molecule binds at each end of the duplex.

Electron microscopy[44] provides an independent means of visualizing DNA (Fig. 4-8) and its binding to pol I (Fig. 4-9). When a DNA duplex of dA_{4000} and dT_{200} is spread with pol I as the only protein present, the pol I molecules are seen (Fig. 4-9, left) as a string of beads on the poly dA·poly dT strand spaced at distances of approximately 600 Å, at the nicks where the dT_{200} chains abut each other. Enzyme molecules are never seen bound to fully helical, intact DNA. When enzyme dimers, linked through their sulfhydryl groups by Hg, are bound to the nicked polymer (Fig. 4-9, right), the binding involves only one enzyme molecule; the other enzyme molecule in the dimer assumes a variety of positions on the grid, owing to the flexibility of the Hg bridge.

Binding of DNA duplexes by pol I at ends or nicks depends not only on fraying and single-strandedness in these regions, but probably includes binding specifically directed to the 3′ and 5′ nucleotide termini at these points. The binding to this single site is reversible, is independent of Mg^{2+}, and is strongly inhibited by 0.1 M phosphate.

An interesting exception to the failure of pol I to bind intact duplex DNA is seen in the complexes formed with the palindromic units present in about 1000 copies as clusters in the *E. coli* genome.[45]

With regard to primer binding, functional studies with poly dA and poly dT templates reveal that even a mononucleotide can be utilized as a primer by pol I and several eukaryotic polymerases.[46] Values of both K_m and V_{max} increase linearly with primer chain length until it is near 10 ntd, a size in keeping with that of the cleft in modeling of the large pol I fragment.[47] These data can be interpreted in terms of free binding energies; other findings[48] also indicate that the nucleoside and phosphate moieties of the primer terminus contribute decisively in binding by a particular polymerase.

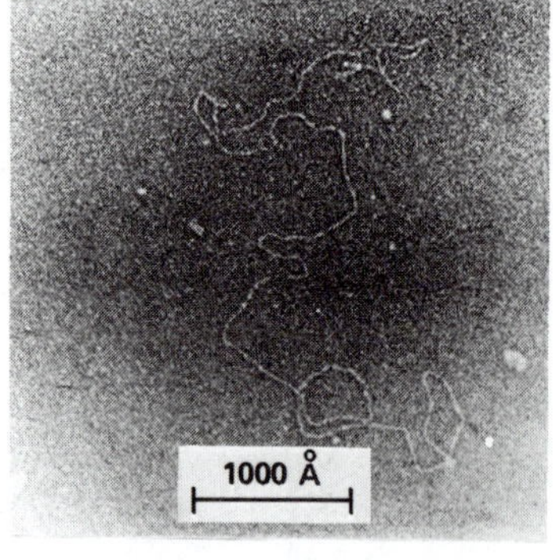

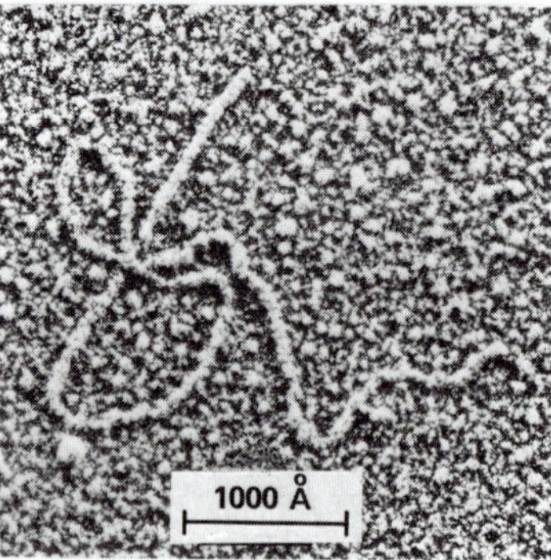

Figure 4-8
Electron micrographs of DNA ($dA_{4000} \cdot dT_{200}$) visualized (left) by a high-resolution procedure (Griffith) and (right) by the protein film method (Kleinschmidt). (Courtesy of Professor J Griffith)

44. Griffith J, Huberman JA, Kornberg A (1971) *JMB 55*:209.
45. Gilson E, Clément JM, Perrin D, Hofnung M (1987) *Trends Genet 3*:226; Higgins CF, McLaren RS, Newbury SF (1988) *Gene 72*:3; Yang Y, Ames GF-L (1988) *PNAS 85*:8850; Gilson E, Perrin D, Hofnung M, (1990) *NAR 18*:3941.
46. Nevinsky GA, Veniaminova AG, Levina AS, Podust VN, Lavrik OI, Holler E (1990) *Biochemistry 29*:1200.
47. Joyce CM, Steitz TA (1987) *TIBS 12*:288.
48. Veniaminova AG, Levina AS, Nevinsky GA, Podust VN (1987) *Mol Biol USSR 21*:1378; Nevinsky GA, Levina AS, Doronin SV, Podust VN, Lavrik OI (1987) in *Biophosphates and Their Analogues; Synthesis, Structure, Metabolism and Activity* (Brusik KS, Stec ED, eds.). Elsevier, Amsterdam, p. 391.

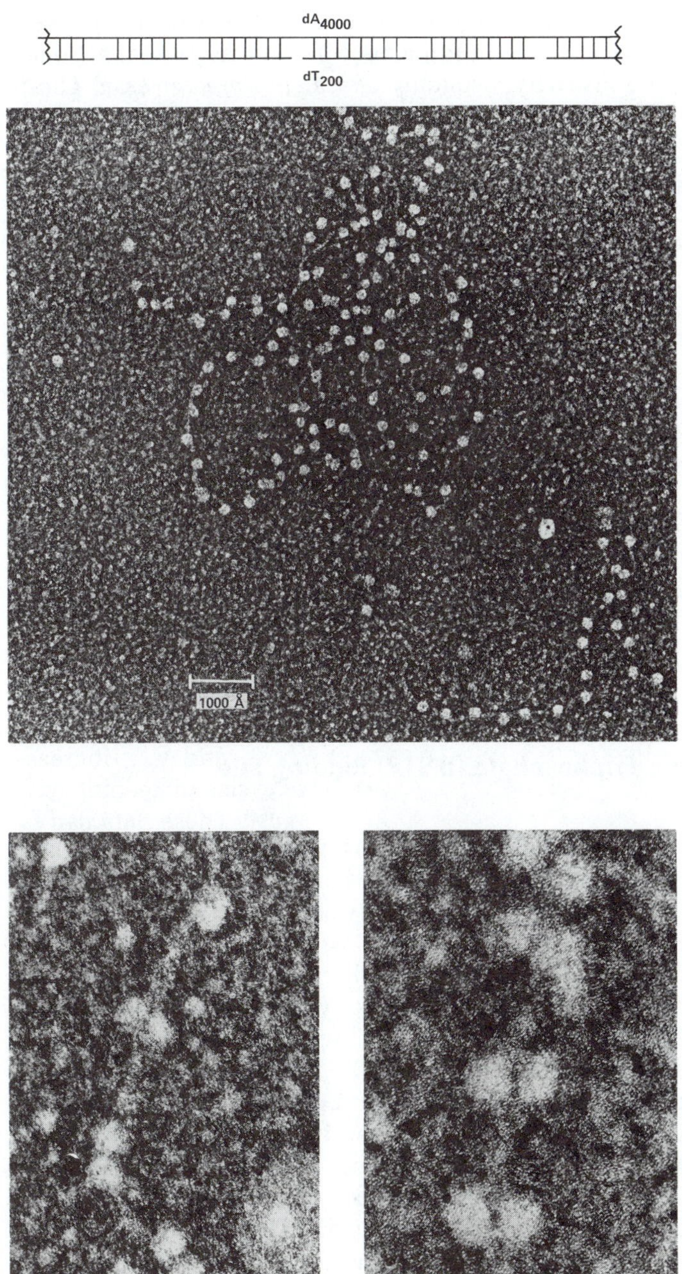

Figure 4-9
(Top) DNA polymerase bound to $dA_{4000} \cdot dT_{200}$. Molecules are bound where dT_{200} oligomers aligned on dA_{4000} abut each other, as diagrammed above. These nicks are at regular intervals of about 600 Å. (Bottom) DNA polymerase dimers bound to $dA_{4000} \cdot dT_{200}$. (Courtesy of Professor J Griffith)

Beyond the hydrodynamic and electron microscopic methods cited here, other techniques for measuring DNA binding have become available that offer novel information, more sensitivity, economy of scale, or greater ease. These methods include:

(1) X-ray crystallography: fine-structure analysis of synthetic oligonucleotides cocrystallized with the protein (Section 5).
(2) Nitrocellulose-filter binding: DNA (labeled) is retained when bound to protein.
(3) Footprinting: DNA is protected against cleavages in regions covered by the protein, detected by accessibility to endonucleases, exonucleases, and chemical agents.
(4) Gel electrophoresis: when bound to the protein, DNA shows retarded migration.
(5) UV-mediated cross-linking: DNA (labeled) is covalently attached to protein at the binding site.
(6) Fluorescent labeling of specific residues in the DNA: progress through the binding site is monitored by emission intensities.[49]
(7) Pre–steady-state kinetic methods: these provide the best estimate of the K_d for the *productive* enzyme–DNA complex;[50] values are 5 nM for pol I and 18 nM for T7 polymerase.

4-7 Deoxynucleoside Triphosphate (dNTP) Binding Site

The dNTPs bind to a single site on pol I, as determined by equilibrium[51] and nonequilibrium dialysis.[52] With the six possible combinations of two dNTPs labeled distinctively, there is in every instance only one site for which all dNTPs compete. The purines (dGTP, dATP) are bound with K_d values near 10 μM, while the values for the pyrimidines (dTTP, dCTP) are near 50 μM. The triphosphate moiety is of primary importance for specificity of binding; the sugar and base components contribute far less (Fig. 4-10). Thus, ATP is bound as well as dATP, dideoxy-TTP is bound even better than TTP, a dNTP with an acetyl or nucleotidyl substituent attached to the 3′-OH group also binds firmly, and a variety of base analogs are also bound, as inferred from their effectiveness as substrates (Section 11). Even 3′-dNTPs, deoxyribose 5′-triphosphate, and inorganic tripolyphosphate are bound, but more weakly.

Although Mg^{2+} is not required for binding in the dNTP site, the β-γ bidentate chelate is probably the active dNTP substrate (β and γ are the two outer phosphates; Section 11). The strength of dNTP binding declines with increasing temperature and ionic strength.

Nucleoside diphosphates, nucleoside monophosphates, and inorganic pyrophosphate show no detectable binding to or competition for the dNTP binding site. However, firm binding of a nucleoside monophosphate at a site in another domain of the large fragment (Fig. 4-1) will be described in Section 8.

---

49. Allen DJ, Darke PL, Benkovic SJ (1989) *Biochemistry 28*:4601; Catalano CE, Allen DJ, Benkovic SJ (1990) *Biochemistry 29*:3612.
50. Patel SS, Wong I, Johnson KA (1991) *Biochemistry 30*:511.
51. Englund PT, Huberman JA, Jovin TM, Kornberg A (1969) *JBC 244*:3038.
52. Muise O, Holler E (1985) *Biochemistry 24*:3618.

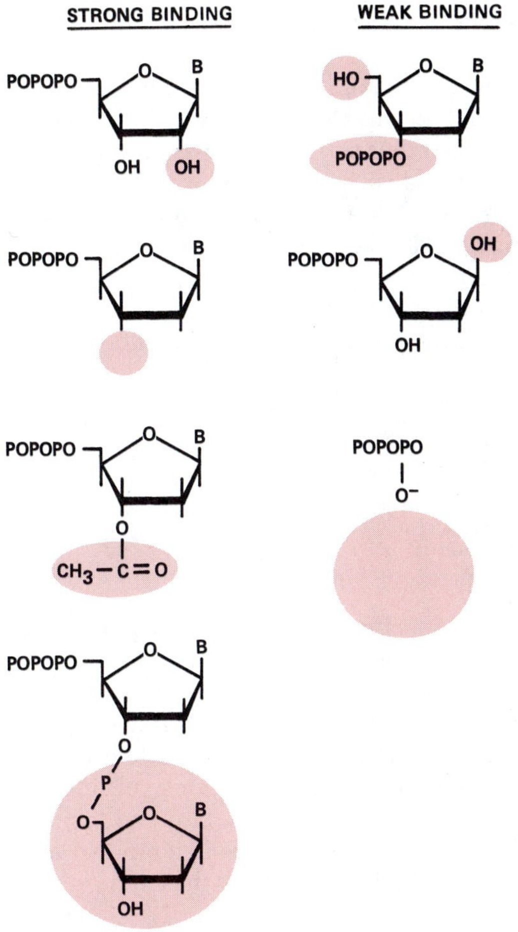

STRONG BINDING

WEAK BINDING

POPOPO B OH OH

HO B POPOPO

POPOPO B

POPOPO B OH OH

POPOPO B O $CH_3-C=O$

POPOPO O^-

POPOPO B O P O B OH

Figure 4-10
Structural requirements for binding to the triphosphate site, determined by analogs of a dNTP. The deviant features of the analogs that bind strongly (left column) or weakly (right column) are shown in color.

Inasmuch as binding of pol I to DNA may precede its dNTP interactions (Section 11), the influence of the template-primer on dNTP binding can be of paramount importance. The apparent K_m values for incorporation of a dNTP vary from little or no influence of a noncomplementary template[53] up to increases of 6- to 20-fold[54] and even greater than 100-fold;[55] binding of a dNTP to the enzyme is not affected by a complementary homopolymer.[56] Thus, occupancy of the DNA binding site inhibits the binding of a noncomplementary dNTP in the dNTP site to an extent that depends on the particular kind of misincorporation, the DNA context (i.e., the surrounding DNA sequence) and other parameters.

The incorporation of a large variety of analogs (Section 11), including some with bulky base adducts, at reasonable rates implies that these analogs have

53. Kuchta RD, Mizrahi V, Benkovic PA, Johnson KA, Benkovic SJ (1987) *Biochemistry 26*:8410; Kuchta RD, Benkovic P, Benkovic SJ (1988) *Biochemistry 27*:6716; Cowart M, Gibson KJ, Allen DJ, Benkovic SJ (1989) *Biochemistry 28*:1975.
54. El-Deiry WS, Downey KM, So AG (1984) *PNAS 81*:7378; El-Deiry WS, So AG, Downey KM (1988) *Biochemistry 27*:546.
55. Griffith J, Huberman JA, Kornberg A (1971) *JMB 55*:209; Fersht AR, Shi J-P, Tsui W-C (1983) *JMB 165*:655.
56. Muise O, Holler E (1985) *Biochemistry 24*:3618.

been bound with dissociation constants comparable to those of the natural dNTPs.

Thus far, the dNTP binding site has not been identified in the crystal structure of the large fragment (Section 5), although localization in or near the cleft (the presumed template-primer binding site) is rather certain (Fig. 4-1). In support of this inference is the finding that polymerase activity in the large fragment is retained even when the entire 3′→5′ exonuclease domain is removed (Section 5). Photoaffinity labeling[57] of residues his881, arg682, and tyr766, pyridoxal phosphate labeling[58] of lys758, mutagenesis at sites such as asn845, arg841, and tyr766,[59] and dNTP binding to an oligopeptide (728 to 777),[60] all implicate dNTP binding sites in the cleft of the large fragment of the enzyme. Hydrophobic interactions between the dNTP base moiety and residues ile765 and tyr766, and ionic forces between the triphosphate and lys758, may be crucial to the binding.

4-8 Nucleoside Monophosphate Binding Site[61]

The properties of the single and highly specific nucleoside monophosphate (NMP) binding site link it to the 3′→5′ exonucleolytic cleavage activity. Crystal structure analysis (Section 5) and site-directed mutagenesis (Section 5) have located the NMP and associated metal-binding sites in the large-fragment domain responsible for the 3′→5′ exonuclease function (Fig. 4-1). As determined initially by equilibrium dialysis measurements of a large variety of nucleotides,[62] polymerase binds an NMP that bears a *free 3′-OH* group and is *not* a triphosphate (Fig. 4-11). Each of the four common dNMPs is bound to and competes for a single site on the enzyme, entirely distinct from the triphosphate site (Section 5).

Nucleotides with non-ionized base analogs or with substitutions of sulfur or carbon for oxygen in the furanose ring are bound. However, such molecules become inactive for binding if the 5′-P is absent, if it is replaced by a phosphinate or phosphonate, or if other phosphates or another nucleotide are added to it. Nucleotides are also unable to bind in the monophosphate site if the 3′-OH group is replaced by a 3′-deoxy, 3-O-methyl, or 3′-P substitution, or if an arabinosyl, lyxosyl, or xylosyl configuration replaces the ribosyl (Fig. 4-11).

In earlier studies[63] of the binding properties of the NMP site, correlations with primer terminus functions seemed possible, but current structural evidence of the relationship of the site to 3′→5′ exonuclease activity is compelling. All indications from studies of binding, crystal structure, and exonuclease activity define the need for a free 3′-OH group and a 5′-P group. Chains terminated by a 2′,3′-dideoxynucleotide are relatively inert to 3′→5′ exonuclease action, presumably because the terminal nucleotide fails to bind at the NMP site. A series

57. Pandey VN, Williams KR, Stone KL, Modak MJ (1987) *Biochemistry 26*:7744; Pandey VN, Modak MJ (1988) *JBC 263*:6068.
58. Basu A, Modak MJ (1987) *Biochemistry 26*:1704.
59. Polesky AH, Steitz TA, Grindley NDF, Joyce CM (1990) *JBC 265*:14579.
60. Mullen GP, Shenbagamurthi P, Mildvan AS (1989) *JBC 264*:19637.
61. Huberman JA, Kornberg A (1970) *JBC 245*:5326; Joyce CM, Grindley NDF (1983) *PNAS 80*:1830.
62. Huberman JA, Kornberg A (1970) *JBC 245*:5326.
63. Huberman JA, Kornberg A (1970) *JBC 245*:5326.

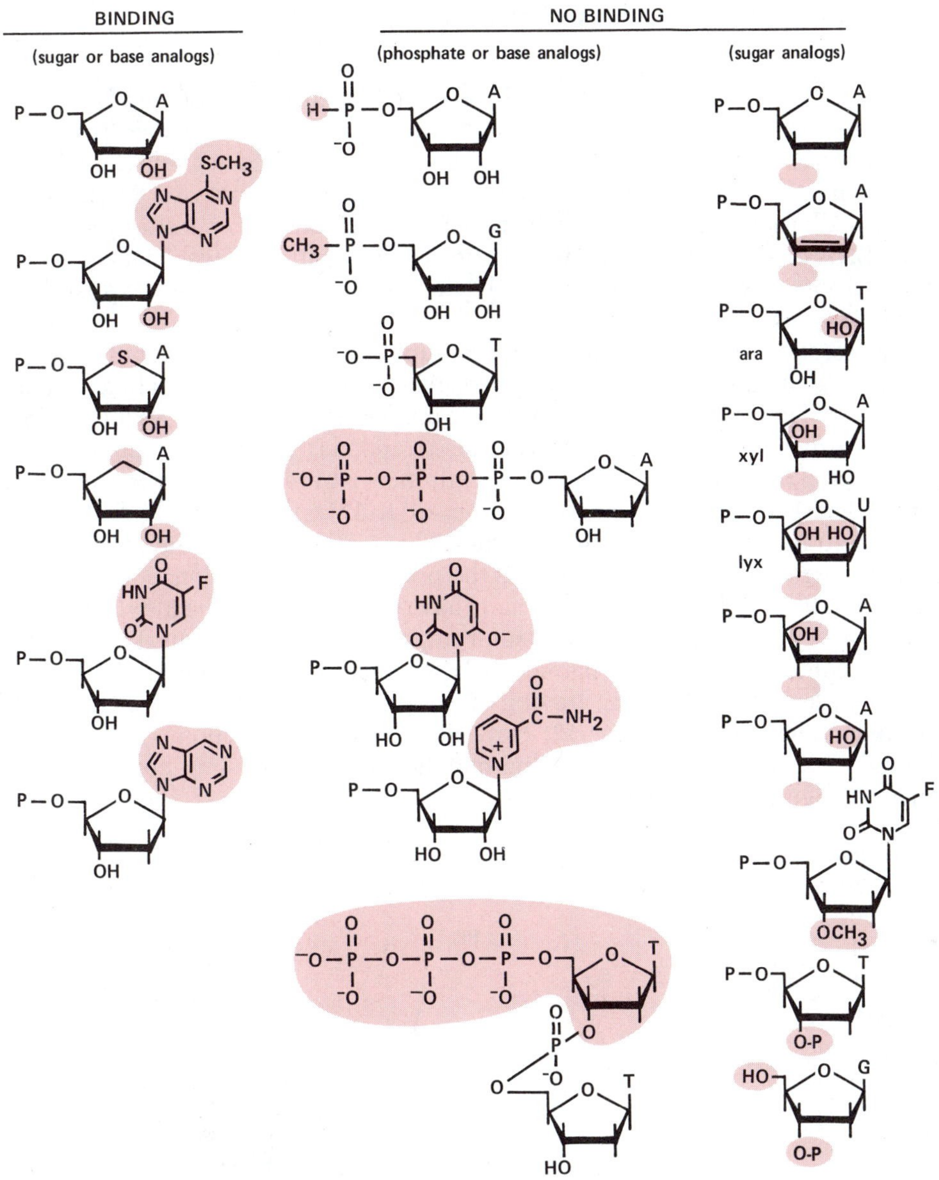

Figure 4-11
Structural requirements for binding to the NMP binding site, determined by nucleotide analogs that bind (left column) and those that do not (other columns). The deviant features of the analogs are shown in color. The sugar analogs arabinose, xylose, and lyxose are designated ara, xyl, and lyx.

of NMPs, including a number whose binding to the enzyme has been measured by equilibrium dialysis (as in Fig. 4-11), have been tested for their capacity to inhibit the 3′→5′ exonuclease function on dT_{50}.[64] A close correlation exists between the binding (K_d) and the inhibition (K_i); the inhibition is competitive with the dT_{50} substrate. Yet polymerization is not inhibited!

64. Que BG, Downey KM, So AG (1978) *Biochemistry 17*:1603; Que BG, Downey KM, So AG (1979) *Biochemistry 18*:2064.

4-9 The 3′→5′ Exonuclease: Proofreading[65]

A crucial feature in the fidelity of pol I replication is the ability of the enzyme to degrade DNA from the primer terminus in a 3′→5′ direction. This 3′→5′ exonuclease function is entirely distinct from the ability of pol I to degrade DNA from the 5′ end of the chain in the 5′→3′ direction (5′→3′ exonuclease; Sections 12 and 13). The nature of the two nucleolytic functions acting in opposite directions on a DNA chain and their significance for DNA metabolism was made clearer when it was possible to separate them by proteolytic cleavage of the polypeptide chain (Section 4).

The essential function of the 3′→5′ exonuclease, located in the small domain of the large fragment (Fig. 4-1), is to recognize and cleave a mismatched primer terminus. This conclusion first emerged from observing that although the exonuclease degrades both single- and double-stranded DNA, the influence of temperature on single-strand degradation is minimal, whereas the effect is profound on duplex DNA: a rise in temperature from 30° to 37° enhances the hydrolysis of double-stranded DNA more than tenfold. This indicates that a frayed end created by partial melting of the helix is the preferred site of action.

Studies with synthetic DNA polymers have confirmed that the 3′→5′ exonuclease acts preferentially on a frayed or unpaired 3′-OH terminus. In polymer chains containing correctly paired 3′-OH ends, and in the absence of dNTPs required for polymerization, fraying of the primer terminus exposes it to repeated nuclease action (Fig. 4-12). However, this nuclease action is suppressed by a dNTP, but only by the one that matches the template and hence can be incorporated into the polymer.

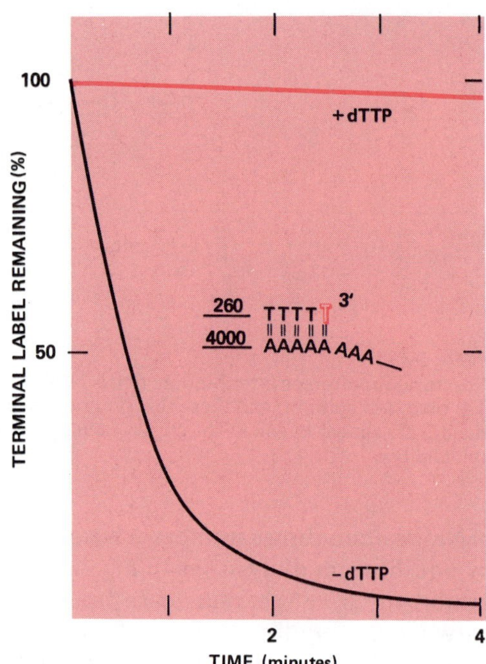

Figure 4-12
Retention of base-paired primer terminus. The terminus is preserved under polymerization conditions (+dTTP) and removed by repeated 3′→5′ exonuclease action in the absence of polymerization (−dTTP). The open T is labeled, and numbers are residues in each polymer.

65. Brutlag D, Kornberg A (1972) *JBC 247*:241; Kunkel TA (1988) Cell 53:837.

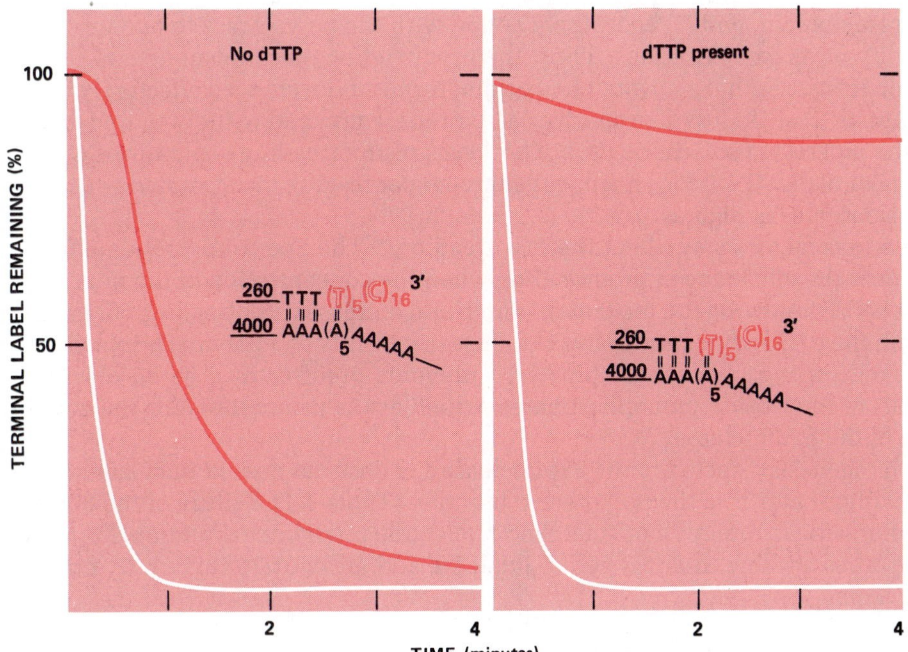

Figure 4-13
Removal of a mismatched primer terminus. The 3′→5′ exonuclease activity removes the mismatched C residues whether or not dTTP is present to support polymerization. When the T residues are reached, they are retained in the presence of dTTP but are removed, after a brief lag period, in the absence of dTTP. The open letters represent labeled residues; numbers are residues in each polymer.

Polymers with mismatched ends lose the nonmatching termini before polymerization is initiated (Fig. 4-13). The mispaired termini tested included purine–purine, pyrimidine–pyrimidine, and purine–pyrimidine mismatches. The dNTPs that can base-pair with the template but cannot be incorporated into polymer fail to suppress nuclease action. For example, phosphonate analogs of dTTP, which bind to the triphosphate site and base-pair with a poly dA template but cannot serve as substrates, are correspondingly ineffective in suppressing nuclease.

Of special interest is the dideoxynucleotide analog ddTTP, which is incorporated at a rate 1000-fold slower than dTTP and to the extent of only one nucleotide per chain.[66] Nevertheless, ddTTP inhibits 3′→5′ exonuclease activity as much as does dTTP. These results suggest a stage in which the primer terminus, in a firm complex with template and triphosphate and physically separated from the exonuclease site, is relatively invulnerable to nuclease action even when the rate-limiting step in polymerization is exceedingly slow.

The characteristics expected of the 3′→5′ nuclease, acting as a proofreading or editing mechanism to correct rare errors in the polymerization process, are as follows: (1) preference for a single-stranded (i.e., mismatched) 3′-OH chain terminus over that of a base-paired duplex, (2) intimate association with the poly-

66. Atkinson MR, Deutscher MP, Kornberg A, Russell AF, Moffatt JG (1969) *Biochemistry* 8:4897.

merase domain, and (3) coordinate action with the polymerase to enhance the fidelity of its condensations. Thus, the proofreading mechanism operates best when it ensures that a mismatched primer terminus is removed effectively with minimal "cost" in loss of properly base-paired termini and without undue delay in the progress of polymerization. The "cost" of proofreading—measured as the number of dNMP (or PP_i) residues generated per residue incorporated—can be considerable, as high as 10%.[67]

Various factors may affect the proofreading.[68] The "next-nucleotide effect" reduces the efficiency of proofreading when the concentration of the next correct dNTP following the mismatch is high and increases proofreading efficiency when the dNTP concentration is low. Insusceptibility of a polymerizable dNTP to excision (e.g., phosphorothioates)[69] prevents proofreading, as do elevated levels of nucleoside monophosphates, which act as competitors for the active site of the exonuclease.

Mutagenesis directed to the exonuclease site destroys the exonuclease activity without affecting the polymerase function (Table 4-1). Failure of the 3′→5′ exonuclease to remove a mismatched nucleotide produces a mutation. Mutator properties are due to a less active nuclease and antimutator effects to a more active one.

4-10 Pyrophosphorolysis and Pyrophosphate Exchange[70]

Before considering the refined kinetic mechanisms of pol I (Section 11), the gross features of the reversal of polymerization should be in mind. The pyrophosphorolysis reaction catalyzed by pol I is the degradation of a DNA chain by inorganic pyrophosphate:

$$\underset{\text{DNA}}{(\text{dNMP})_n} + \text{x(PP)} \longleftrightarrow \underset{\text{DNA}}{(\text{dNMP})_{n-x}} + \text{x(dNPPP)}$$

Pyrophosphorolysis reaches a steady state when the accumulation of triphosphates supports synthesis at a rate that balances their removal. This reverse reaction, as with polymerization, requires a proper primer terminus. With a chain terminating in a dideoxynucleotide, the reaction rate with pol I is reduced at least 1000-fold;[71] for T7 polymerase the factor is only four.[72] Unlike 3′→5′ exonuclease hydrolysis, pyrophosphorolysis absolutely requires that the primer be annealed to a complementary strand. The reaction is therefore truly a reversal of polymerization.

The specificity of pyrophosphorolysis for the product of polymerization is illustrated with pairs of synthetic polymer chains (Table 4-3). This specificity clearly distinguishes pyrophosphorolysis, which attacks only base-paired ter-

---

67. Fersht AR, Knill-Jones JW, Tsui W-C (1982) *JMB 156*:37.
68. Kunkel TA (1988) *Cell 53*:837.
69. Kunkel TA, Eckstein F, Mildvan AS, Koplitz RM, Loeb LA (1981) *PNAS 78*:6734.
70. Deutscher MP, Kornberg A (1969) *JBC 244*:3019.
71. Atkinson MR, Deutscher MP, Kornberg A, Russell AF, Moffatt JG (1969) *Biochemistry 8*:4897.
72. Tabor S, Richardson CC (1990) *JBC 265*:8322.

Table 4-3
Nature of cleavage of the primer terminus in the presence of inorganic pyrophosphate[a]

Structure at 3′ primer terminus	Products released (percentage of total)	
	Monophosphate (hydrolysis)	Triphosphate (pyrophosphorolysis)
Single-stranded chain: $dT_{260}[^3H]dT_1$	100	<0.5
Double-stranded, base-paired: $dT_{260}[^3H]dT_1 \cdot dA_{4000}$	53	47
Single-stranded chain: $dT_{260}[^3H]dC_1$	100	<0.5
Double-stranded, mispaired: $dT_{260}[^3H]dC_1 \cdot dA_{4000}$	100	<0.5

[a] From Brutlag D, Kornberg A (1972) *JBC 247*:241.

mini, from hydrolysis by the exonuclease, in which an attack on unpaired termini is strongly favored. Thus, an enzyme–DNA complex with a base-paired primer terminus allows the incorporation of a triphosphate or attack by pyrophosphate, while another, in which the 3′ terminus is frayed, allows only hydrolysis or, rarely, extension that may seal in a mismatch.

The enzyme also supports exchange of pyrophosphate into the β-γ groups of a dNTP:

$$^{32}P^{32}P_i + dNPPP \xrightleftharpoons{DNA} PP_i + dNP^{32}P^{32}P$$

This reaction occurs when only a single triphosphate is present, but otherwise requires the primer and template conditions demanded of DNA synthesis. There is no evidence for a nucleotidyl–enzyme intermediate. Instead, inorganic pyrophosphate exchange appears to result from a sequence of a polymerization step and a pyrophosphorolytic step repeated many times over.

4-11 Kinetic Mechanisms[73]

For two triose phosphates to be condensed by aldolase to form a hexose diphosphate in a relatively simple, reversible reaction, at least nine intermediate steps are needed.[74] How numerous must be the stages in achieving the rapid, high-fidelity, and processive addition of a nucleotide to a DNA chain. Whatever the complexities of the kinetic mechanisms, they must fit with extant observations on the many functions of the enzyme and how these functions are organized in discrete structural domains.

Kinetic Scheme

Whereas pol I and T7 DNA polymerase are structurally related (Section 5-9) and conform to similar reaction pathways, they exhibit significant kinetic and

73. Kuchta RD, Mizrahi V, Benkovic PA, Johnson KA, Benkovic SJ (1987) *Biochemistry 26*:8410; Kuchta RD, Benkovic P, Benkovic SJ (1988) *Biochemistry 27*:6716; Cowart M, Gibson KJ, Allen DJ, Benkovic SJ (1989) *Biochemistry 28*:1975; Carroll SS, Benkovic SJ (1990) *Chem Rev 90*:1291.
74. Rose IA, Warms JVB (1985) *Biochemistry 24*:3952; Rose IA, personal communication.

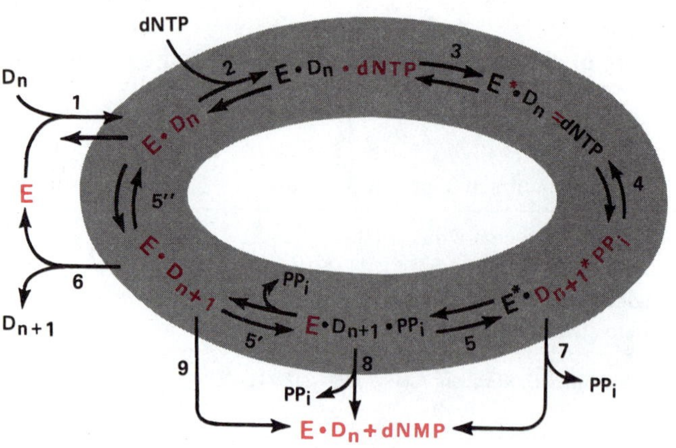

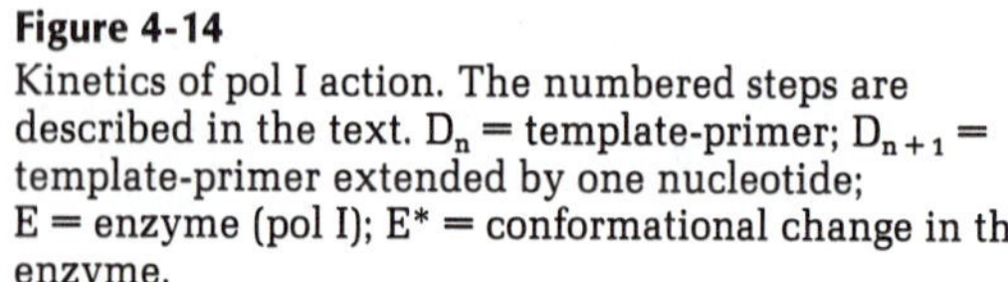

Figure 4-14
Kinetics of pol I action. The numbered steps are described in the text. D_n = template-primer; D_{n+1} = template-primer extended by one nucleotide;
E = enzyme (pol I); E* = conformational change in the enzyme.

mechanistic differences.[75] Compared to pol I, T7 polymerase (1) selects against an incorrect dNTP at the initial binding step, whereas pol I binds correct and incorrect dNTPs with comparable affinities; (2) incorporates nucleotides at 300 ntd per second, a sixfold faster rate than that of pol I; (3) is highly processive; (4) has a more active 3′→5′ exonuclease proofreading function; and (5) remains bound to the template and corrects an error, whereas pol I dissociates following a mismatching, which is proofread upon rebinding to the enzyme. These mechanistic differences appear to reflect the principal roles of T7 as a replicative polymerase and pol I as devoted to patching and repair.

A scheme (Fig. 4-14) which attempts to embody kinetic features of pol I action on a simple oligomeric template-primer[76] can serve as a basis for discussion.

Step 1. Binding of the template-primer (D_n in Fig. 4-14) orients it in the cleft of the large fragment of the enzyme (E in Fig. 4-14), with the primer terminus in the proper general location (Plate 4).

Step 2. Available dNTPs compete for binding in the single dNTP site to form a ternary complex with E and D_n.

Step 3. Complementary base pairing of the dNTP with the template is accompanied by a conformational change in the enzyme (E*). [Mg^{2+}-dNTP is bound as a β-γ bidentate[77] in a unique configuration (Fig. 4-15);[78] β and γ are the two outer phosphates.] This and the succeeding Step 4 are the crucial, rate-limiting events and generally proceed at rates near 1000 ntd per second.

Step 4. Nucleophilic attack by the 3′-OH of the primer terminus on the α-phosphate of the dNTP (Fig. 4-15) forms the diester bond and the products D_{n+1} and PP_i (Fig. 4-14). Inversion of the phosphate configuration is compatible with an in-line mechanism.[79]

---

75. Patel SS, Wong I, Johnson KA (1991) *Biochemistry 30*:511.
76. Kuchta RD, Mizrahi V, Benkovic PA, Johnson KA, Benkovic SJ (1987) *Biochemistry 26*:8410; Kuchta RD, Benkovic P, Benkovic SJ (1988) *Biochemistry 27*:6716; Cowart M, Gibson KJ, Allen DJ, Benkovic SJ (1989) *Biochemistry 28*:1975; Carroll SS, Benkovic SJ (1990), a review, in press.
77. Burgers PMJ, Eckstein F (1979) *JBC 254*:6889.
78. Cornelius RD, Cleland WW (1978) *Biochemistry 17*:3279; Brody RS, Frey PA (1981) *Biochemistry 20*:1245.
79. Burgers PMJ, Eckstein F (1979) *JBC 254*:6889; Cornelius RD, Cleland WW (1978) *Biochemistry 17*:3279.

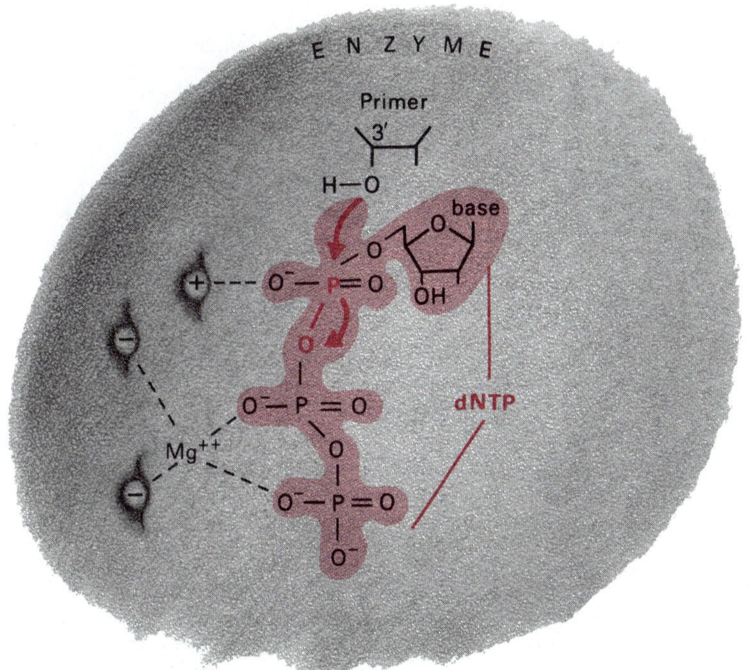

Figure 4-15
Binding of Mg^{2+}·dATP as a β-γ bidentate chelate in the dNTP binding site of pol I, and nucleophilic attack by the 3′-OH of the primer terminus on the α-phosphate of the dATP. Plus and minus charges are presumed to reside in the enzyme surface.

Step 5. The enzyme reverts to its original conformation ($E^* \rightarrow E$), one of the products, PP_i, is released, and the template-primer translocates without dissociation to restore the primer terminus to the position it occupied at the start of the reaction. Reversal by pyrophosphorolysis (see below) is feeble because of the low affinity of PP_i ($K_d = 0.14$ mM for pol I; 2 mM for T7).

Cycles of polymerization continue (processively) until all the available template is filled in.

Step 6. Alternatively, the template-primer dissociates from the enzyme—a distributive (nonprocessive) event.

Steps 7, 8, and 9. The newly formed primer terminus of the produce (D_{n+1}) translocates to the 3′→5′ exonuclease site to be hydrolyzed to the original $E \cdot D_n$, with release of dNMP and PP_i.

Fidelity of Nucleotide Incorporation. The fidelity of incorporation (Section 15-9) in this scheme is determined by binding of the complementary dNTP (Steps 2, 3, and 4) and by 3′→5′ exonuclease proofreading (Steps 7, 8, and 9). For T7 polymerase, the fidelity due to dNTP binding, expressed as the selectivity of a correct versus an incorrect dNTP (i.e., the ratio of k_{cat}/K_m values), is $>10^5$; for exonuclease proofreading, kinetic constants show some selectivity for removal of the incorrect dNTP (Table 4-4).[80] Stalling of the polymerase caused by a mismatched primer terminus increases the window of time during which exonuclease action can remove a mismatched nucleotide, increasing the selectivity for exonucleolytic action to 250 for T7. However, for pol I, the most rapid

80. Wong I, Patel SS, Johnson KA (1991) *Biochemistry 30*:526; Donlin MJ, Patel SS, Johnson KA (1991) *Biochemistry 30*:538; Bebenek K, Joyce CM, Fitzgerald MP, Kunkel TA (1990) *JBC 265*:13878.

Table 4-4
Kinetic values for polymerase and exonuclease activities of T7 polymerase and pol I[a]

Reaction	T7 dNTP selected		Pol I dNTP selected	
	Correct	Incorrect	Correct	Incorrect
Polymerization	300	0.002	50	0.005
Pyrophosphorolysis	1	$<10^{-5}$	0.5	ND[b]
Exonuclease (3′→5′)	0.2	3	0.003	0.006
Dissociation	0.2	0.4	0.2	3
dNTP binding (K_d)	18 μM	>10 μM	5 μM	5–10 μM

[a] The values (per second) were kindly furnished by Professor KA Johnson.
[b] ND = not determined.

reaction following a misincorporation is the dissociation of the enzyme from the DNA, with repair of the mismatch taking place upon subsequent rebinding of the enzyme to the DNA or by the action of other enzymes, in vivo.

Incorporation of Analogs. Selection of the correct dNTP depends on distinguishing between the nearly identical dimensions and geometry of all four base pairs: AT, TA, GC, and CG. An early, and still impressive, demonstration of the crucial importance of base pairing in polymerization is found in the substitutions of natural bases with analogs.[81] As shown in Table 4-5, the utilization of analogs by polymerase can be described simply by their base-pairing potentialities: uracil and 5-substituted uracils can take the place of thymine, 5-substituted cytosines can serve for cytosine, and hypoxanthine can serve for guanine. Whereas hypoxanthine, capable of forming two hydrogen bonds with cytosine, is effective at a reduced rate of incorporation, xanthine, which can form only one hydrogen bond with cytosine, is inert.

Table 4-5
Replacement of natural bases by analogs in polymerase action

Analog (as the deoxynucleoside triphosphate)	dNTP replaced by analog (percentage of value with natural dNTP)			
	dTTP	dATP	dCTP	dGTP
Uracil	54	0	0	0
5-Bromouracil	97	0	0	0
5-Fluorouracil	32	0	0	0
5-Hydroxymethylcytosine[a]	0	0	98	0
5-Methylcytosine	0	0	185	0
5-Bromocytosine	0	0	118	0
5-Fluorocytosine	0	0	63	0
N-Methyl-5-fluorocytosine	0	0	0	0
Hypoxanthine	0	0	0	25
Xanthine	0	0	0	0

[a] Values were obtained with phage T2 polymerase, the others with pol I.

81. Bessman MJ, Lehman IR, Adler J, Zimmerman SB, Simms ES, Kornberg A (1958) *PNAS 44*:633; Kornberg A (1962) *Enzymatic Synthesis of DNA*. Wiley, New York.

Although the analogs behave essentially the same when tested with DNA polymerases from various sources, there are notable exceptions. For example, 5,6-dihydrothymidine triphosphate is a substrate for pol I, but not for phage T4 polymerase or the reverse transcriptase of avian myeloblastosis virus.[82] Incorporation of this analog, which lacks the planar pyrimidine ring and has a higher K_m and lower V_{max}, is sharply reduced when a cluster of adenines is encountered in the template; the incorporation of multiple analogs generates disorder at or near the primer terminus site and enfeebles replication. The pH, the sequence of the template, and other parameters may each have a significant influence on the incorporation of an analog.[83]

As for dNTP base analogs, the alkyl[84] and bulky adducts are accepted as substrates. Examples are a 5- or 6-membered ring nitroxide tethered by an alkane or alkene to the 5 position of the pyrimidine ring[85] and a biotinyl-ϵ-amino caproyl-3-amino alkyl moiety attached to the same position. On the other hand, the thymine glycol analog (5,6-dihydroxydihydrothymine) is not a substrate. Phosphorothioate analogs in which sulfur replaces one of the oxygens in the α-phosphate [(2′-deoxynucleoside 5′-*O*-(1-thiotriphosphate)] are incorporated at the same rate, but the phosphorothio-NMP in the primer terminus perturbs the structure enough to decrease the overall rate by 20% to 65%, depending on the template.[86]

Incorporation at Abnormal Bases or Specific Sequences. Pyrimidine dimers, the most extensively studied photochemical lesions, are mutagenic or lethal and are thought to terminate replication. Yet the thymine cyclobutane (*cis–syn*) dimer in a synthetic oligonucleotide template can be matched with A residues by pol I.[87] This infrequent bypass requires high levels of enzyme and dNTP.

Pausing at specific sequences, commonly at regions of secondary structure, is observed with T4 DNA polymerase[88] and other polymerases and can be countered by SSBs (e.g., gp32) and by helicases (e.g., gp41). The nonviability of phages with long palindromes is attributable to the retarded rate of their replication.[89]

The addition of a nucleotide to a blunt-ended duplex, without apparent template direction, has been observed for pol I and for several eukaryotic polymerases.[90] The rate is very feeble; dATP is favored over the other dNTPs, and the process may be similar to the incorporation of a nucleotide at a point where the template has been damaged and lacks a base[91] (Section 21-2).

Alternative Kinetic Schemes. These have been suggested for stages in the polymerization cycle. "Kinetic proofreading"[92] proposes that hydrolysis of the dNTP precedes diester bond formation, but the experimental evidence is con-

---

82. Ide H, Wallace SS (1988) *NAR 16*:11339.
83. Ferrin LJ, Mildvan AS (1986) *Biochemistry 25*:5131; Lai M-D, Beattie KL (1988) *Biochemistry 27*:1722; Driggers PH, Beattie KL (1988) *Biochemistry 27*:1729.
84. Singer B, Chavez F, Spengler SJ, Kusmierek JT, Mendelman L, Goodman MF (1989) *Biochemistry 28*:1478.
85. Pauly GT, Thomas IE, Bobst AM (1987) *Biochemistry 26*:7304
86. Vosberg H-P, Eckstein F (1977) *Biochemistry 16*:3633.
87. Taylor J-S, O'Day CL (1990) *Biochemistry 29*:1624.
88. Bedinger P, Munn M, Alberts BM (1989) *JBC 264*:16880.
89. Lindsey JC, Leach DRF (1989) *JMB 206*:779.
90. Clark JM, Joyce CM, Beardsley GP (1987) *JMB 198*:123; Clark JM (1988) *NAR 16*:9677.
91. Sagher D, Strauss B (1983) *Biochemistry 22*:4518; Randall SK, Eritja R, Kaplan BE, Petruska J, Goodman MF (1987) *JBC 262*:6864.
92. Hopfield JJ (1974) *PNAS 71*:4135.

tradictory.[93] For example, polymerization results in an inversion of dNMP configuration; were hydrolysis to dNMP to occur first, the configuration would be retained or randomized. Other schemes propose that with successive polymerization cycles, discrimination increases in the matching of a dNTP ("energy relay")[94] and in the proofreading by exonucleolytic hydrolysis ("mnemonic warm-up");[95] here, as well, the experimental evidence is conflicting.[96]

Reversal of Polymerization. Polymerization can be reversed by PP_i (Section 10). Reversal can be measured by exchange of PP_i into a dNTP (in a polymerization cycle) or by pyrophosphorolysis. Observed discrepancies between the two measurements are probably due to factors extrinsic to the basic mechanism outlined in Figure 4-14. For example, pyrophosphorolysis is more dependent on the sequence at the terminus; the exchange, weighted by the polymerization activity, is higher than the net rate of removal of the terminus. The effect of PP_i in decreasing fidelity can be explained by the failure of pyrophosphorolysis to remove a mismatch.

Processivity

Processivity,[97] the uninterrupted association of an enzyme with its substrate (i.e., template) for many cycles of reaction, is a property shared by many polymerases (Section 15-8). Pol I is processive for about 20 residues on an activated (nicked and gapped) thymus DNA template and on the nicked circular duplex of plasmid ColE1 (Table 4-6). Processivity is at least twice as great with a gapped ColE1 template, implying added instability at a locus of nick translation. With a poly d(AT), processivity is near 200 residues, but this number is drastically reduced at low temperature and high ionic strength, presumably as the result of effects on both the template-primer and the enzyme.

Table 4-6
Processivity of pol I as influenced by template-primer, temperature, and ionic strength[a]

DNA template-primer	Conditions[a]	Processivity (residues)
Nicked ColE1	37°, low salt	8
Gapped ColE1	"	47
Nicked–gapped thymus	"	24
Poly d(AT)	37°, low salt	188
"	5°, low salt	14
"	5°, high salt	3

[a] Ionic strengths at low and high salt levels were about 0.1 and 0.3, respectively.

93. Kuchta RD, Mizrahi V, Benkovic PA, Johnson KA, Benkovic SJ (1987) *Biochemistry 26*:8410; Kuchta RD, Benkovic P, Benkovic SJ (1988) *Biochemistry 27*:6716; Cowart M, Gibson KJ, Allen DJ, Benkovic SJ (1989) *Biochemistry 28*:1975; Fersht AR, Shi J-P, Tsui W-C (1983) *JMB 165*:655.
94. Hopfield JJ (1980) *PNAS 77*:5248.
95. Papanicolaou C, Lecomte P, Ninio J (1986) *JMB 189*:435; Lecomte PJ, Ninio J (1987) *FEBS Lett 221*:194.
96. Kuchta RD, Mizrahi V, Benkovic PA, Johnson KA, Benkovic SJ (1987) *Biochemistry 26*:8410; Kuchta RD, Benkovic P, Benkovic SJ (1988) *Biochemistry 27*:6716; Cowart M, Gibson KJ, Allen DJ, Benkovic SJ (1989) *Biochemistry 28*:1975.
97. Fay PJ, Johanson KO, McHenry CS, Bambara RA (1982) *JBC 257*:5692; Huang C-C, Hearst JE, Alberts BM (1981) *JBC 256*:4087; Matson SW, Bambara RA (1981) *J Bact 146*:275.

Polymerization Cycles[98]

The evidence for separation of the polymerization and exonuclease sites, based on x-ray crystallography, mutations, and physical behavior in solution,[99] is conclusive. However, integrating the kinetic mechanisms of polymerization, proofreading, and processivity (Fig. 4-14) with the structure of the large fragment (Section 5 and Fig. 4-3) presents several problems: (1) this extraordinarily complex series of reactions depends on an enormous number of kinetic constants, (2) structural information about the binding sites for the dNTP and the primer terminus is still inadequate, and (3) the vast distance (nearly 30 Å) separating the sites for polymerization and for proofreading needs to be accommodated.

Movement of the Primer between Polymerization and 3′→5′ Exonuclease Sites

In a scheme that reconciles some of the structural and kinetic data, the primer separates from the template for a 4-bp length and may slide rapidly back and forth the required distance between the polymerization and exonuclease sites (Fig. 4-16) on a time scale (10 to 100 milliseconds) compatible with the catalytic rates of the enzyme. Examples of similar rapid linear diffusion (sliding) of a protein on DNA include the numerous RNA polymerases, nucleases, and repressors.

A matched primer terminus is held in the polymerization site for attack on a juxtaposed dNTP that is base-paired to the template; the terminus then melts and traverses the path to the exonuclease site where, as a single-stranded, 4-ntd frayed end, it may be hydrolyzed. A mismatched primer terminus, already partially melted, more readily takes the route to the exonuclease site or occupies it longer because of the very slow rate of polymerization of the next dNMP on a

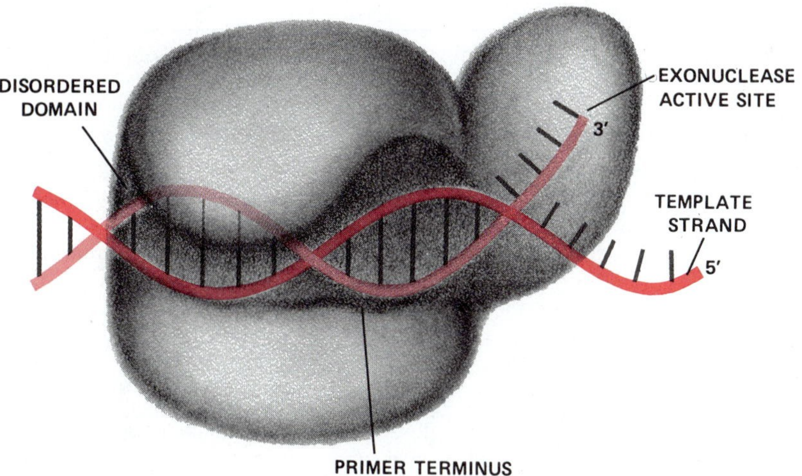

Figure 4-16
The way in which the large fragment of pol I may bind to a DNA molecule when the 3′ terminus of the primer strand is in the 3′→5′ exonuclease active site.

98. Kuchta RD, Mizrahi V, Benkovic PA, Johnson KA, Benkovic SJ (1987) *Biochemistry 26*:8410; Kuchta RD, Benkovic P, Benkovic SJ (1988) *Biochemistry 27*:6716; Cowart M, Gibson KJ, Allen DJ, Benkovic SJ (1989) *Biochemistry 28*:1975.
99. Cowart M, Gibson KJ, Allen DJ, Benkovic SJ (1989) *Biochemistry 28*:1975.

mismatched primer terminus, and so becomes a target for hydrolysis more often than the matched terminus.

With each polymerization event, the newly formed primer terminus slides back to the polymerization site or forward to the proofreading site. Direct evidence for such sliding between sites has been obtained with T7 polymerase in a single turnover in the presence of a trapping agent.[100] An alternative to this intramolecular shuttling has also been observed.[101] With a certain frequency, the template-primer dissociates enough from its numerous attachments to interrupt processivity and reassociate with the exonuclease site of another polymerase molecule.

4-12 The 5′→3′ Exonuclease: Excision–Repair[102]

The 5′→3′ exonuclease differs from the 3′→5′ exonuclease (Table 4-7) not only in the terminus selected and the direction of cleavage, but in another fundamental way: the 5′→3′ nuclease cleaves a diester bond only at a base-paired region, whereas the 3′→5′ nuclease cleaves a bond only at a non–base-paired region. Furthermore, the 5′→3′ nuclease can excise oligonucleotides up to ten residues long from the 5′ end, whereas the 3′→5′ nuclease removes only a single nucleotide at a time. Phosphorothioate linkages are not hydrolyzed by the 5′→3′ activity; when placed near the end of an oligonucleotide with a mismatch at its center, a phosphorothioate group will protect the chain from degradation.[103]

Table 4-7
Differences between 3′→5′ and 5′→3′ exonuclease activities of pol I

Property	3′→5′	5′→3′
Terminus of substrate	very specific: 3′-OH in d-ribo configuration; lyxo-, xylo-, arabino-, dideoxy- inert	nonspecific: 5′-OH, mono-, di-, or triphosphate; DNA or RNA: matched or mismatched base pairs
Secondary structure	single-stranded chain; frayed or non–base-paired end of duplex	base-paired site in duplex; duplex region near a D-loop
Product	5′-mononucleotide exclusively	5′-mononucleotides ~80%; oligonucleotides ~20%
Endonuclease action (excision of oligomers)	none	cleaves diester bonds as distant as 8 from the end or even further
Influence of concurrent polymerization	inhibited completely	increases rate tenfold; increases relative proportion of oligonucleotides
Proposed roles	proofreading in replication	excision–repair in nick translation; removal of RNA primer from 5′ end of DNA chain

100. Donlin MJ, Patel SS, Johnson KA (1991) *Biochemistry 30*:538.
101. Joyce CM (1989) *JBC 264*:10858.
102. Klett RP, Cerami A, Reich E (1968) *PNAS 60*:943; Deutscher MP, Kornberg A (1969) *JBC 244*:3029; Kelly RB, Atkinson MR, Huberman JA, Kornberg A (1969) *Nature 224*:495; Cozzarelli NR, Kelly RB, Kornberg A (1969) *JMB 45*:513.
103. Ott J, Eckstein F (1987) *Biochemistry 26*:8237.

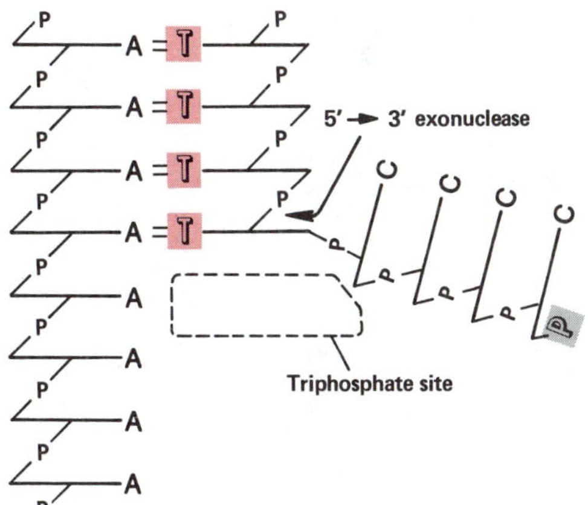

Figure 4-17
Excision of a mismatched 5′ terminal region by the 5′→3′ exonuclease activity of pol I. Analysis of the products was facilitated by ^{32}P labeling (open letter P) of the 5′ terminus and ^{3}H labeling (open letters) of the T residues.

A convenient way to assay each nuclease activity in the presence of the other is to use ssDNA as substrate for the 3′→5′ exonuclease and duplex DNA with 3′ termini blocked by dideoxyribonucleotides as substrate for the 5′→3′ exonuclease.

The nature of the 5′→3′ nuclease action is illustrated in experiments with a synthetic substrate, the polymer $d(pC)_4(pT)_{200}$.[104] This polymer, though inert as a substrate, is cleaved when annealed to poly dA (Fig. 4-17). Cleavage occurs at the diester bond beyond the first or second base pairing in the chain, releasing $d(pC)_4pT$ and $d(pC)_4(pT)_2$. Small amounts of longer oligonucleotides, containing additional T residues, are also produced, possibly due to the lack of complete T pairings to poly dA by $d(pC)_4(pT)_{200}$ or due to occasional failures to cleave the polymer as it is drawn through the active site on the enzyme.

In UV-irradiated DNA and in duplexes of poly dA with irradiated poly dT, thymine dimers fail to base-pair with adenine and so distort the helix (Fig. 4-18). These DNAs, after a 5′ endonucleolytic incision near the lesion (Section 13-9), are cleaved in the same manner as the mispaired homopolymers.[105] Under certain circumstances, the 5′→3′ exonuclease can make an endonucleolytic scission even in the absence of a 5′ terminus. When negatively supertwisted circular DNA is used as a template with an interpolated ssDNA or RNA fragment to serve as primer (forming a D-loop structure, as in Fig. 16-1), polymerization leads to nicking of the circular DNA.[106] Most probably this is because the 5′→3′ exonuclease activity responds to a distortion near the "replicating fork" much as it does in excising an oligonucleotide from a mismatched 5′ terminal region.

This ability to excise mismatches and distortions resulting from UV irradiation suggests that pol I removes and repairs faulty regions in DNA in vivo. The physiologic role of the 5′→3′ exonuclease and the in vivo evidence for excision and repair of UV, x-ray, and related DNA lesions will be discussed later (Section 18), as will the crucial role of pol I in the *uvr* ABCD nuclease excision of damaged regions in vitro (Sections 13-9 and 21-2).

The coupling of polymerizing and excising functions in a single enzyme may be of significant advantage in repairing damage in vivo. As a thymine dimer is

---

104. Kelly RB, Atkinson MR, Huberman JA, Kornberg A (1969) *Nature 224*:495.
105. Kelly RB, Atkinson MR, Huberman JA, Kornberg A (1969) *Nature 224*:495.
106. Champoux J, McConaughy BL (1975) *Biochemistry 14*:307; Liu LF, Wang JC (1975) in *DNA Synthesis and Its Regulation* (Goulian M, Hanawalt P, eds.). Benjamin-Cummings, Menlo Park, Calif., p. 38.

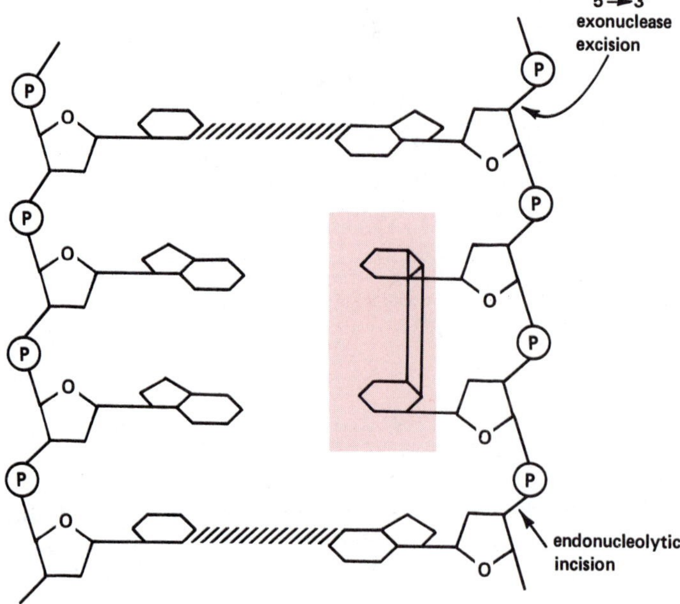

Figure 4-18
Disruption of base pairing in the region of a cyclobutyl thymine dimer.

being removed by the 5'→3' exonuclease, the polymerase could concurrently fill in the gap. Therefore, excision–repair by pol I would at no time leave a single-stranded gap exposed to attack at the 3' end by exonucleases and endonucleases. Rather, only a nick with a 3'-OH and a 5'-P would be left, from which the polymerase might be displaced by DNA ligase. Perhaps the extensive DNA degradation that is observed after UV irradiation of *E. coli polA* mutants results from production of single-stranded gaps during the excision of thymine dimers by secondary repair systems. These gaps may last long enough for enzymes such as exonuclease III to enlarge them and thereby cause extensive degradation of the DNA.

An important, and even vital, physiologic repair function probably served by the 5'→3' exonuclease is removal of the RNA fragment that primes the initiation of DNA synthesis. The RNA in model RNA–DNA hybrid polymers is removed by the exonuclease, as is the RNA priming fragment of enzymatically synthesized DNA (Section 18).

4-13 The 5'→3' Exonuclease: Nick Translation

DNA polymerase binds at a nick and, in the presence of the required dNTPs, extends the primer terminus. Before the 5'→3' exonuclease activity of pol I was recognized, studies showed that polymerization is accompanied by a burst of hydrolysis of the template-primer that matches the extent of polymerization (Fig. 4-19). With dA_{4000} as template and sufficient $[^3H]dT_{300}$ to cover it, synthesis of a poly dT chain proceeds to the extent of matching the input dA_{4000} with a concurrent hydrolysis of the dT_{300}. How could the known 3'→5' exonuclease and polymerase activities operate simultaneously and antagonistically on the same primer terminus? Even more puzzling then was the observation that

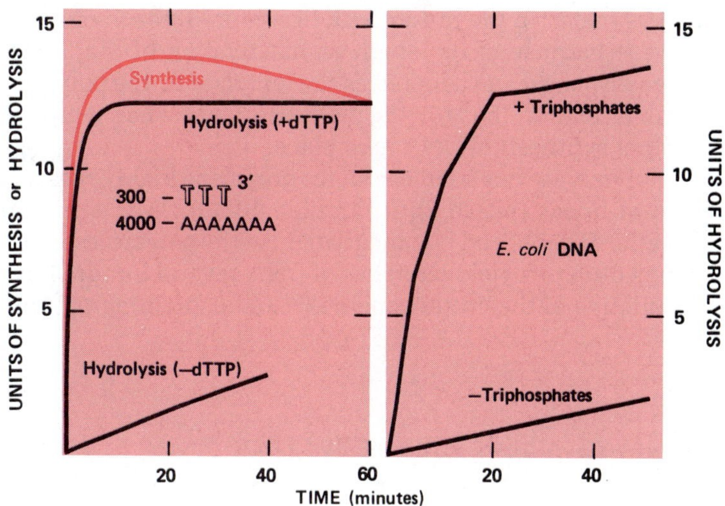

Figure 4-19
Nick translation by pol I. When concurrent polymerization and 5'→3' exonuclease actions balance each other (as on the left), a nick in the DNA is linearly advanced (translated) along the chain. Hydrolysis is stimulated many-fold by concurrent synthesis (right).

hydrolysis of DNA is enhanced tenfold upon the addition of the dNTPs that make DNA synthesis possible (Fig. 4-19, right).[107]

With the discovery of the 5'→3' exonuclease activity,[108] it became clear that polymerization at a nick is coordinated with nuclease action, so as to move *(translate)* the nick linearly along the helix without a change in mass of the DNA. The nuclease action at the nick is exclusively 5'→3' and occurs at a rate ten times faster in advance of a growing chain than in the absence of polymerization.

This concurrent polymerization and hydrolysis proceeds for a brief period and stops, presumably for one of three reasons (Fig. 4-20): (1) replication reaches the end of the template or a nick in it and therefore terminates; (2) polymerase is

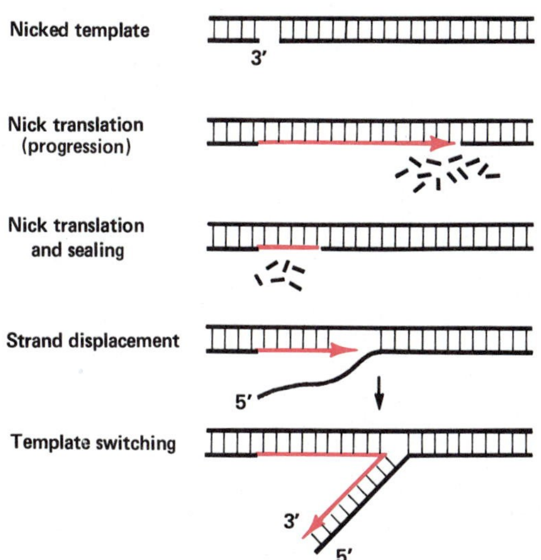

Figure 4-20
Polymerase action at a nick. Concurrent polymerization and nuclease action may lead to nick translation (to an extent determined by ligase sealing), strand displacement, or template-switching.

107. Lehman IR (1967) *ARB 36*:645.
108. Klett RP, Cerami A, Reich E (1968) *PNAS 60*:943; Deutscher MP, Kornberg A (1969) *JBC 244*:3029.

dislodged and the nick is sealed in the presence of ligase; or (3) the 5′ oligonucleotide escapes cleavage because of its length or inactivation of the 5′→3′ exonuclease—it is then displaced and adopted as the template by the polymerase, producing a branch or fork on the growing molecule. Only when template switching occurs does net synthesis of DNA take place.

Nick translation by polymerase has been useful for preparing highly radioactive DNA. A template of linear (phage λ) or circular (SV40) DNA is nicked randomly by pancreatic DNase and is incubated with polymerase and [α-^{32}P]dNTPs. This procedure can yield stretches of DNA several hundred nucleotides long, representative of the entire genome,[109] and containing 10^8 cpm per μg.

4-14 Apparent de Novo Synthesis of Repetitive DNA[110]

The synthesis, in the absence of any template-primer, of a huge polymer of deoxyadenylate and thymidylate in which the residues are arranged in perfectly ordered alternation was one of the most remarkable and unanticipated events in the history of DNA polymerase studies.[111] The event was remarkable because this synthesis of DNA was observed in the total absence of *added* template to direct the reaction or of primer to start it. Since the initial discovery, the apparent de novo synthesis of a large variety of homopolymers and copolymers by pol I has been observed and studied.[112]

The first DNA-like polymer discovered was the alternating copolymer of A and T (Fig. 4-21), now designated poly d(AT).[113] Since then, other examples of de novo synthesis have also been observed, including homopolymer pairs such as poly dG·poly dC, poly dI·poly dC, and poly dA·poly dT, and the alternating copolymers poly d(GC) and poly d(IC).

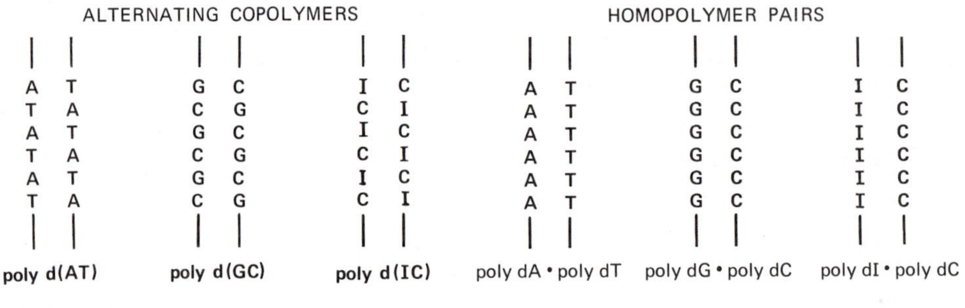

ALTERNATING COPOLYMERS			HOMOPOLYMER PAIRS		
A T	G C	I C	A T	G C	I C
T A	C G	C I	A T	G C	I C
A T	G C	I C	A T	G C	I C
T A	C G	C I	A T	G C	I C
A T	G C	I C	A T	G C	I C
T A	C G	C I	A T	G C	I C
poly d(AT)	poly d(GC)	poly d(IC)	poly dA • poly dT	poly dG • poly dC	poly dI • poly dC

Figure 4-21
Alternating copolymers and homopolymers synthesized by pol I.

109. Rigby PWJ, Dieckmann M, Rhodes C, Berg P (1977) *JMB 113*:237.
110. Kornberg A (1965) in *Evolving Genes and Proteins* (Bryson V, Vogel HJ, eds.). Academic Press, New York, p. 403; Burd JF, Wells RD (1970) *JMB 53*:435; Schachman HK, Adler J, Radding CM, Lehman IR, Kornberg A (1960) *JBC 235*:3242; Radding CM, Josse J, Kornberg A (1962) *JBC 237*:2869.
111. Schachman HK, Adler J, Radding CM, Lehman IR, Kornberg A (1960) *JBC 235*:3242.
112. Burd JF, Wells RD (1970) *JMB 53*:435; Radding CM, Josse J, Kornberg A (1962) *JBC 237*:2869.
113. Burd JF, Wells RD (1970) *JMB 53*:435.

There are important reasons for examining this subject in some detail:

(1) The apparent violation of basic requirements by polymerase for template and primer demands explanation.

(2) Studies of de novo synthesis have demonstrated the template-primer capacities of molecules containing as few as six base pairs.

(3) The growth of large polymers directed by a small template-primer can be explained by a mechanism called reiterative replication.

(4) The kind of polymer produced—whether homopolymer or copolymer and whether composed of A and T or G and C—can be influenced by a variety of agents, including intercalating dyes, pH, ionic strength, and temperature, in addition to the particular polymerase employed and the nuclease activities present.

(5) Repetitive tracts that closely resemble these polymers in their simplicity (e.g., satellite DNAs) constitute significant fractions of eukaryotic chromosomes, especially in the region of the centromere. The evolutionary origin of these unusual DNAs and even, perhaps, of all DNA may be reflected in some ways in the reiterative replication observed in polymerase-catalyzed de novo synthesis.

(6) Finally, having an assortment of repetitive DNAs with sharply defined composition and sequence is invaluable for the isolation and characterization of polymerases and nucleases and for physicochemical analyses of DNA structure, as well as for a variety of related studies.

Stages of de Novo Synthesis

Kinetic studies of de novo polymer development have been assisted by the use of chemically prepared, *small* template-primers. De novo synthesis occurs in four stages: (1) initiation, (2) reiterative replication, (3) polymer-directed replication, and (4) degradation of the polymer when the dNTP substrates have been exhausted. The first two stages and part of the third do not reach the level of detection by spectrophotometry, viscometry, or tracer incorporation into the polymer. However, once the polymer accumulates in significant amounts, template-directed polymer synthesis ensues exponentially.

Initiation

Polymer synthesis de novo is detectable only after a lag period of many hours (Fig. 4-22) and then proceeds at an exponential rate.[114] The longer the lag period, the lower the exponential rate constant, suggesting that observable polymer synthesis depends on a few macromolecules serving as template-primers. The fewer template-primers present, the longer the lag and the slower the subsequent rate of synthesis. Sampling and examining the reaction mixture during the lag period establishes that lag-reducing macromolecules are produced by the time 20% of the lag period has elapsed and that these are then reproduced exponentially.

Two principal possibilities for the initiation of the de novo reaction can be suggested. One is that a chain of alternating A and T residues really starts de

114. Radding CM, Josse J, Kornberg A (1962) *JBC 237*:2869.

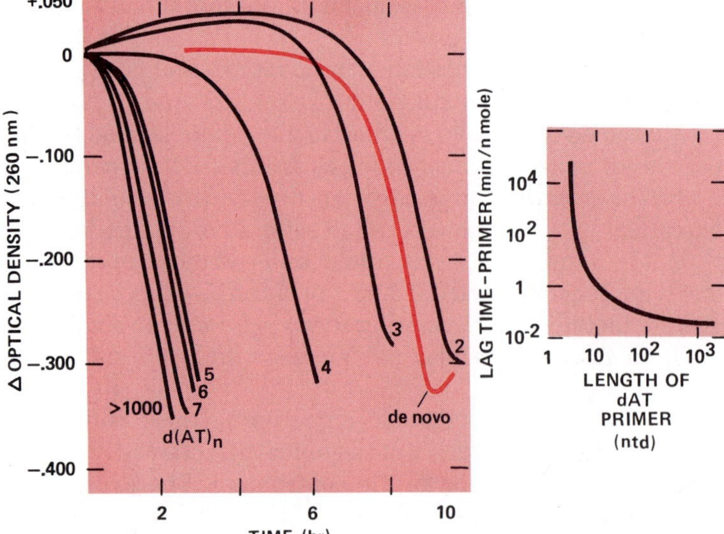

Figure 4-22
Kinetics of poly d(AT) synthesis primed by the oligomers $d(AT)_{2-7}$ and poly d(AT).

novo and, upon reaching a minimal length of six nucleotides, two such chains undergo reiterative replication, as described below. However, the probability is exceedingly small that DNA polymerase initiates a chain and then either adds a residue to a terminus not already base-paired or extends a chain without forming a base pair with a template. On the contrary, the 3′→5′ exonuclease action of polymerase rapidly degrades oligonucleotides and also degrades dinucleotides. Even though de novo initiation can be accepted as an exceedingly rare event, this combination of improbable and successive reactions building up a polymer with a specific, alternating sequence is highly unlikely.

The second alternative is that DNA fragments persist as impurities in the enzyme and serve as template-primers to produce oligonucleotides, which then undergo reiterative replication. Analyses of the best available enzyme preparations have shown less than one phosphate residue per molecule of enzyme.[115] Nevertheless, the affinity of polymerases for DNA is strong, and it seems plausible that among the 10^{14} enzyme molecules in a reaction mixture a few still have some tightly bound *E. coli* DNA fragments. Assuming that a minimal sequence of ATATAT is required somewhere in a DNA fragment, it would occur on a random basis once in approximately every 4000 residues. Thus a DNA contaminant in a pol I preparation at a level as low as 0.1% (0.3 nucleotide residue per enzyme molecule) would still allow 4000 nucleotide residues (which might include one ATATAT sequence) per 15,000 enzyme molecules. The suggestion[116] that such fragments are made by a contaminating transferase that can polymerize dNDPs without primer or template direction has yet to be confirmed.

---

115. Jovin TM, Englund PT, Bertsch LL (1969) *JBC 244*:2996.
116. Nazarenko IA, Potapov VA, Romashchenko AG, Salganik RI (1978) *FEBS Lett 86*:201; Nazarenko IA, Bobko LE, Romaschenko AG, Khripin YL, Salganik RI (1979) *NAR 6*:2545.

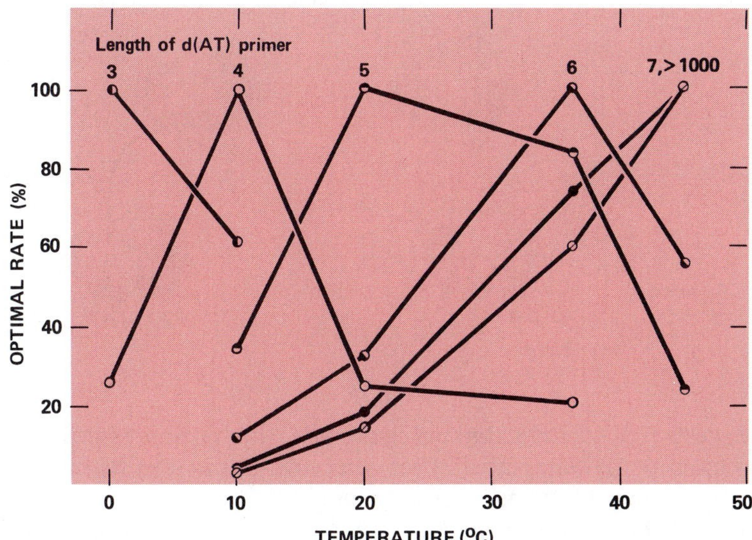

Figure 4-23
Influence of temperature and the length of d(AT) oligomers on their priming action.

Reiterative Replication

A novel capacity of pol I to reiterate a short sequence of template may prove to be at the crux of de novo synthesis.[117] Indications of this property come largely from studies of the priming action of chemically synthesized poly d(AT) oligomers. A reiterative mechanism is also suggested by de novo poly d(AT) synthesis stimulated by natural DNAs with a relatively high content of A and T residues.

A reduction in lag time for poly d(AT) synthesis is observed in the presence of d(AT) oligomers of 4 to 14 residues in length (Fig. 4-23; see also Fig. 4-22, right). Although the optimal temperature for replication of a polymer is near 40°, the optima for priming by oligomers are strikingly different: near 0° for the hexamer $d(AT)_3$, near 10° for $d(AT)_4$, and near 20° for $d(AT)_5$ (Fig. 4-23), reflecting the temperature at which the oligomer exists as a stable duplex in solution and can function as a template-primer.

Assuming that one strand of the oligomeric duplex is bound and held fixed by the polymerase in the template site, the other strand may then serve as the primer for extensive chain growth. Slippage of the primer strand on the template (Fig. 4-24) exposes additional template for elongation. Successive episodes of slippage and polymerization reiterate sequences of AT residues until the primer chain has grown to a size sufficient to fold back on itself into a hairpin-like duplex.

An important feature of the influence of temperature on priming by the oligomers is that, as shown in Figure 4-23, *each size has and maintains a distinctive temperature optimum.* Were the duplex of $d(AT)_3$ hexamers that serves as a nucleating element to grow immediately to the size of a $d(AT)_6$ duplex or larger, the optimal temperature for polymer synthesis for this oligomer would change to near 40° rather than the 0° optimum for $d(AT)_3$. This remarkable behavior of

117. Kornberg A, Bertsch LL, Jackson JF, Khorana HG (1964) *PNAS 51*:315.

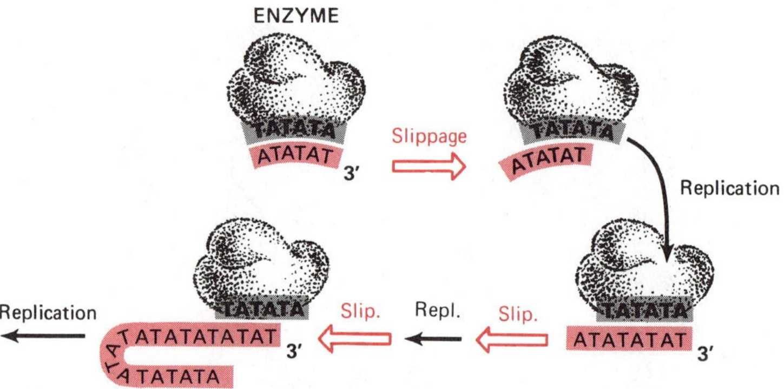

Figure 4-24
Reiteration and slippage mechanism to account for priming by d(AT) oligomers. Note that the template strand of the oligomer is held fixed by the enzyme and remains constant in size after repeated stages of slippage and replication.

the oligomers suggests that, as indicated in Figure 4-24, the hexamer serving as template is held fixed by the enzyme and remains constant in size while supporting the growth of the primer that can slip along it.

The suggestion that the reiteration of AT sequences in DNA accounts for de novo synthesis is supported by the reduction in lag time when AT-rich DNAs are added to a reaction mixture of dATP, dTTP, and pol I. A positive correlation is observed with the AT content of the DNA, the most profound effect being with a very AT-rich fraction or crab DNA.

Polymer-Directed Replication

A linear rate of synthesis by reiterative replication can become exponential only when new polymer molecules are generated to serve as template-primers for the huge excess of idle polymerase molecules. Clear indications that nuclease action is required for rapid and exponential rates of polymer synthesis come from two directions. One is the failure to obtain de novo poly dG·poly dC synthesis when pol I preparations are freed of traces of endonuclease I, and the restoration of polymer synthesis when small amounts of endonuclease are added back to the pure polymerase reaction mixture. Another line of evidence is the behavior of the large fragment of pol I in poly d(AT) synthesis.[118] This enzyme preparation, lacking the 5′→3′ exonuclease, is unable to carry out de novo synthesis. When primed with poly d(AT), the large fragment sustains a linear rate of polymerization rather than the exponential rate seen with the intact enzyme. It seems likely that the 5′→3′ exonuclease of the intact enzyme generates fragments that possess priming activity.

Degradation

A peak level of polymer synthesis is passed when the rate of degradation of the polymer by nuclease exceeds the rate of synthesis. The rate of degradation becomes overwhelming when synthesis ceases entirely, upon exhaustion of one of the required dNTPs. In primed poly d(AT) synthesis, catalyzed by the large fragments, there is 100% conversion of dATP and dTTP to polymer. This is in contrast to the situation observed with the intact enzyme, whose consumption of the substrates at the peak extent of synthesis never exceeds 70%. Presumably,

---

118. Setlow P, Brutlag D, Kornberg A (1972) *JBC 247*:224.

degradation by 5′→3′ exonucleolytic excisions proceeds throughout the period of synthesis as well as after exhaustion of the dNTPs.

Factors Affecting Polymer Synthesis

In view of the complex kinetics of de novo polymer synthesis and the nice coordination of nuclease and polymerase action required, one might expect that a number of reaction parameters would influence the rate and extent of reaction. Nevertheless, there are striking effects which remain largely unexplained. Proflavin[119] or nogalamycin,[120] compounds that intercalate between bases in the DNA helix, bring about the synthesis of homopolymer chains of poly dA·poly dT rather than the copolymer, poly d(AT). Polymerization of dITP and dCTP in 50 mM phosphate buffer yields the homopolymer pair exclusively at pH 7.3, whereas only the copolymer is produced at pH 7.9. Simply changing to an amine buffer from phosphate, while maintaining the same pH (7.3) and ionic strength, also causes a complete shift from homopolymer to copolymer synthesis. These seemingly slight perturbations in reaction conditions produce profound changes in the kinetics and specificity of de novo polymer synthesis, presumably as the consequence of conformational changes in the polymerizing and nucleolytic activities and in the DNA molecules upon which they act.

Natural Occurrence of Repetitive DNA

Satellite DNA in Crabs. The DNA of several species of crab, when analyzed by equilibrium sedimentation in a buoyant-density gradient, reveals, in addition to the major component characteristic of animal DNA, a satellite band of very light density.[121] In some species, the satellite band is as much as one-fourth of the total DNA. The crab satellite is startlingly similar in composition and sequence to the alternating copolymer poly d(AT).[122] It is composed of approximately 97% A and T residues, of which all but a few are alternating; the 3% of G and C residues are peppered throughout the DNA. Similar DNAs are also found in other animals (Section 1-5).

The function of this satellite DNA remains obscure. The satellite DNA is confined to the heterochromatin fraction in the centromere region of the chromosome. Protein coded by a DNA rich in an -ATAT- sequence would contain copolymeric runs of isoleucine and tyrosine, but no such proteins have been found, nor does this DNA appear to be transcribed. Furthermore, the amount of AT satellite varies widely from one crab species to another.

Repetitive Sequences in Yeast. High levels of A and T residues, organized into both alternating copolymers and homopolymeric regions with some G and C residues interspersed, also occur in the mitochondrial DNA of several yeast species.[123] Here, too, the origins and significance of these sequences are uncertain. Perhaps interruptions in the normal course of DNA replication have permitted these sequences, so susceptible to slippage, to be reiterated many times over. Other repetitive tracts of DNA, as well as a novel transposable genetic

---

119. McCarter JA, Kadohama N, Tsiapalis C (1969) *Can J B 47*:391.
120. Olson K, Luk D, Harvey CL (1972) *BBA 227*:269.
121. Sueoka N (1961) *JMB 3*:31.
122. Swartz MN, Trautner TA, Kornberg A (1962) *JBC 237*:1961.
123. Borst P (1972) *ARB 41*:333.

element arising in *E. coli* from an insertion sequence (IS2) (Section 18-1), are likely to have been formed by a slippage-reiterative form of replication.[124]

4-15 Ribonucleotides as Substrate, Primer Terminus, and Template[125]

Another deviation from the basic rules of DNA polymerase behavior, besides the apparent synthesis of DNA-like polymers without added template or primer, is the capacity of pol I to use ribonucleotides. Under certain conditions, pol I substitutes an rNTP, a ribonucleotide primer terminus, or an RNA template for the appropriate deoxyribo counterparts.

Ribonucleotide Substrates[126]

Pol I can use an rNTP in place of a dNTP when Mn^{2+} replaces Mg^{2+}. Although the reaction is not extensive, due to the feeble incorporation of UTP, the capacity of the enzyme to use rNTPs is of interest because it provides an additional measure of the specificity of the polymerization event, a possible clue to the function of Mg^{2+}, and a suggestion for the in vivo mutagenic role of Mn^{2+}.

The *rate* of rCTP incorporation in place of dCTP, in the presence of the other three dNTPs and Mn^{2+}, approaches that of dCTP. Incorporation of rGTP as substitute for dGTP is also effective, but that of rATP for dATP is much less so. With rUTP in place of dTTP, incorporation is barely detected, a behavior also observed with dUTP in place of dTTP under standard assay conditions. The poor performance of uracil as a substitute for thymine may be due to some structural feature of the AU base pair or the excision of uracil from DNA chains by a trace contaminant of uracil N-glycosylase (Section 13-9).

Mn^{2+} distorts polymerization not only by permitting ribonucleotide incorporation, but also by permitting misincorporation of deoxynucleotides themselves.[127] When the repair of short regions of defined DNA sequences is observed with only dTTP present, Mn^{2+} permits T to form a base pair with G in the template (Fig. 1-18). Limited replication in which GTP or CTP is incorporated also leads to a significant percentage of misincorporations. However, under certain experimental conditions good fidelity of incorporation is observed with CTP replacing dCTP.[128] Whether Mn^{2+} exerts its permissive effect on ribonucleotide substitutions and misincorporations at the stage of base pairing, proofreading, or binding of the template-primer is not yet clear.

Ribonucleotide Primer Terminus and Template

The incorporation of ribonucleotides into a DNA chain synthesized by pol I indicates not only that the enzyme uses them as substrates in place of deoxynu-

---

124. Ghosal D, Saedler H (1978) *Nature 275*:611; Ghosal D, Gross J, Saedler H (1978) *CSHS 43*:1193.
125. Berg P, Fancher H, Chamberlin M (1963) in *Informational Macromolecules* (Vogel HJ, Bryson B, Lampen JO, eds.). Academic Press, New York, p. 467; Van de Sande JH, Loewen PC, Khorana HG (1972) *JBC 247*:6140; Wells RD, Flügel RM, Larson JE, Schendel PF, Sweet RW (1972) *Biochemistry 11*:621; Karkas JD, Stavrianopoulos JG, Chargaff E (1972) *PNAS 69*:398.
126. Berg P, Fancher H, Chamberlin M (1963) in *Informational Macromolecules* (Vogel HJ, Bryson B, Lampen JO, eds.). Academic Press, New York, p. 467; Van de Sande JH, Loewen PC, Khorana HG (1972) *JBC 247*:6140.
127. Van de Sande JH, Loewen PC, Khorana HG (1972) *JBC 247*:6140; Klenow H, Henningsen I (1969) *EJB 9*:133; Hall ZW, Lehman IR (1968) *JMB 36*:321.
128. Klenow H, Henningsen I (1969) *EJB 9*:133.

cleotides, but also that the added ribonucleotide serves effectively as a primer terminus. The ribohomopolymer pair, poly rA · poly rU, synthesized by polynucleotide phosphorylase, also serves as a template-primer. RNA-directed DNA polymerases (reverse transcriptases) of RNA tumor viruses use an RNA template-primer in the normal replication cycle (Section 6-9). The reason that some animal cell DNA polymerases prefer RNA (Section 6-5) is less clear.

When pol I utilizes poly rA · poly rU as template-primer, the corresponding poly dA · poly dT is produced. The stages in this replication may be represented by these equations:[129]

Formation of hybrids	poly rA · poly rU + dATP + dTTP → poly rA · poly dT + poly dA · poly rU
Replication of hybrids	poly rA · poly dT + dATP → poly dA · poly dT + poly rA
	poly dA · poly rU + dTTP → poly dA · poly dT + poly rU
Reannealing	poly rA + poly rU → poly rA · poly rU
Sum	poly rA · poly rU + dATP + dTTP → poly rA · poly rU + poly dA · poly dT

What overall might appear to be conservative replication of ribohomopolymers can be accounted for by ribo–deoxyribo hybrid pairs as intermediates. Such hybrid pairs are excellent template-primers.

These results make it clear that ribohomopolymers do support deoxypolymer synthesis. As members of hybrid duplexes, such as poly rA · poly dT or poly rG · poly dC, the ribohomopolymers are as effective as templates as are the unmixed deoxyhomopolymer pairs.

The hybrid intermediates would be expected to form by covalent extension of a ribo primer terminus. Although the deoxyribo product appears to be cleanly separated from the template-primer by equilibrium sedimentation in a density gradient, a result that seems to rule out covalent linkage, the use of [α-^{32}P]dNTPs in the reaction mixture would probably demonstrate covalent attachment of the DNA homopolymer to a trace amount of RNA primer.[130]

The relative rates with which the enzyme uses ribopolymers as templates and primers vary considerably, often under apparently identical conditions.[131] In the case of poly rA · poly rU, rates for its template-priming action range from 1% to 25% of those observed with poly dA · poly dT. Single-stranded homopolymers, either ribo or deoxy, are generally inert, but in some instances appear to support significant synthesis. Factors that profoundly alter the effectiveness of a template-primer, whether ribo or deoxyribo, can be classified as follows:

(1) State of the template-primer: the number of 3′-OH primer termini with a sufficient length of template to support extension; the presence of annealed oligonucleotide primer fragments, even in homopolymer samples isolated by isopycnic sedimentation; the presence of inhibitory 3′-P termini.

(2) Purity of the enzyme: the presence of nucleases that stimulate extension by creating additional primer termini, by removing inhibitory termini, and by producing oligonucleotide primer fragments; the presence of nucleases that inhibit extension by destroying the template-primer or the product itself.

(3) Incubation conditions: changes, however small, in temperature, pH, ionic strength, buffer, and metal ions.

---

129. Chamberlin M (1965) *Fed Proc 24*:1446; Modak MJ, Marcus SL, Cavalieri LF (1974) *JBC 249*:7373.
130. Wells RD, Flügel RM, Larson JE, Schendel PF, Sweet RW (1972) *Biochemistry 11*:621.
131. Karkas JD (1973) *PNAS 70*:3834.

Natural RNA template-primers have been used in vitro to direct DNA synthesis by pol I.[132] Ribosomal (28S) RNA from *Drosophila* or from mammalian cells, globin mRNA, and RNAs from tobacco mosaic virus and avian myeloblastosis virus are active; the addition of oligo dT_{6-9} is essential for replication of some templates. The rates of these DNA syntheses are only a small percentage of the rates with natural DNA as template-primer. In evaluating these observations, the influence of the state of the template-primer, the purity of the enzymes, and incubation conditions should be borne in mind. Nevertheless, the capacity of pol I to be directed by RNA is irrefutable, although its physiologic significance is still not apparent.

4-16 Products of Pol I Synthesis[133]

The DNA produced by pol I after many rounds of replication resembles the original template-primer in physical and chemical properties. The product directed by a long DNA has the sedimentation and viscoelastic characteristics of a DNA fiber several microns in length. But it is unusual in two ways.[134] One is the readiness with which the DNA renatures after thermal or alkaline melting: despite rapid cooling or neutralization, the product, unlike natural DNA, regains its original optical density and native properties. The other unusual feature is the branched appearance of the DNA in electron micrographs.[135]

Both the rapid renaturability and the branching can be attributed to the presence of hairpin-like structures. Such a structure may come about by *template-switching*, some time during replication, at a nick in a duplex. At this nick point, first one strand of the duplex could be used as a template and then later the complementary strand. Such template-switching can be more easily visualized in a space-filling DNA model than from a diagram on paper. Free rotation of the phosphodiester bridge in the DNA backbone permits a DNA strand to displace its complement without any looping or distortion of the helix, and would enable a polymerase molecule to continue replication of the new template without "jumping" or changing direction. The factors that might prompt template-switching are still unknown. Subsequently, by a process called *branch migration*,[136] the displaced strand of the template may be restored to the duplex state, permitting the product to anneal into a hairpin structure (Fig. 4-25). Endonucleolytic cleavage is required to release the covalent attachment of the product to the template.

The chemical composition of the product reflects the template;[137] with an ssDNA as template the synthetic complementary strand has exactly the composition predicted by base pairing. After extensive replication of duplex DNA, the AT and GC contents of the product are the same as in the template. Inasmuch as the accuracy of analysis of base composition is only about 1%, a more sensitive measure of the fidelity of replication may be gained by the use of a template

---

132. Loeb LA, Tartoff KD, Travaglini EC (1973) *NNB 242*:66; Modak MJ, Marcus SL, Cavalieri LF (1973) *BBRC* 55:1; Modak MJ, Marcus SL, Cavalieri LF (1974) *BBRC 56*:247.
133. Schildkraut CL, Richardson CC, Kornberg A (1964) *JMB 9*:24; Inman RB, Schildkraut CL, Kornberg A (1965) *JMB 11*:285; Goulian M, Kornberg A (1967) *PNAS 58*:1723; Goulian M, Kornberg A, Sinsheimer RL (1967) *PNAS 58*:2321.
134. Schildkraut CL, Richardson CC, Kornberg A (1964) *JMB 9*:24.
135. Inman RB, Schildkraut CL, Kornberg A (1965) *JMB 11*:285.
136. Lee CS, Davis RW, Davidson N (1970) *JMB 48*:1.
137. Lehman IR, Zimmerman SB, Adler J, Bessman MJ, Simms ES, Kornberg A (1958) *PNAS 44*:1191.

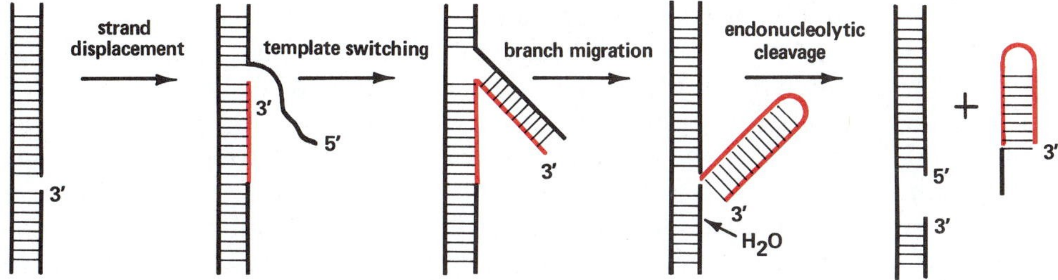

Figure 4-25
Synthesis of hairpin and branched structures by pol I.

lacking one of the bases. Thus, the misincorporation of guanine nucleotides with poly d(AT) as template can be looked for, and is ordinarily not found even at a level of one nucleotide per 10^5 A and T nucleotides polymerized.[138]

Nearest-Neighbor Analysis[139]

Two decades before the techniques of sequence analysis were available, the base analysis of the product confirmed that the overall composition matched the template-primer. However, it did not measure the fidelity of copying the correct sequence of linkages of one nucleotide to another.

The frequency of each of the 16 possible dinucleotide arrangements in the product can be determined by *nearest-neighbor analysis.*[140] This procedure is based on the incorporation of [α-^{32}P]dNTPs, followed by enzymatic hydrolysis of the product to 3′-deoxynucleotides (Fig. 4-26). Thus, from a product in

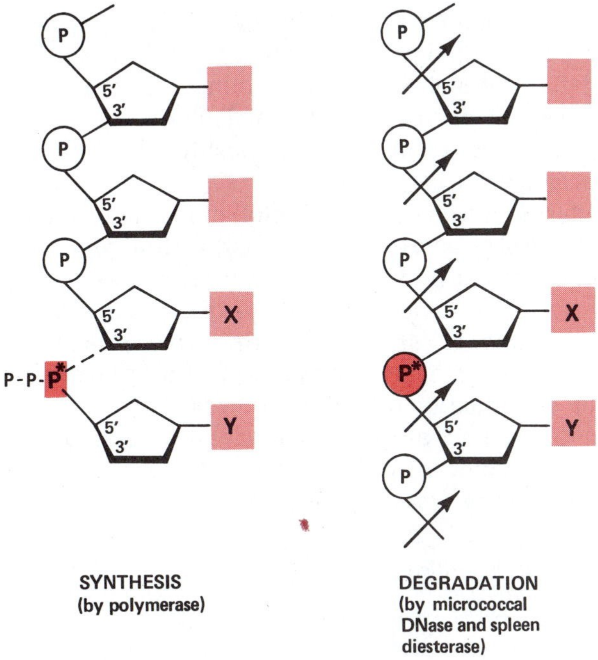

Figure 4-26
Steps in the nearest-neighbor analysis of DNA. The ^{32}P label (indicated by P*) is originally in the α position of dYTP, the nucleotide that becomes the nearest neighbor of the terminal X residue. The ^{32}P label remains attached to the X residue upon 3′-nucleolytic cleavage.

138. Trautner TA, Swartz MN, Kornberg A (1962) *PNAS 48*:449.
139. Josse J, Kaiser AD, Kornberg A (1961) *JBC 236*:864.
140. Josse J, Kaiser AD, Kornberg A (1961) *JBC 236*:864

which [α-^{32}P]dATP has been incorporated, one obtains the labeled 3′-deoxynucleotides (i.e., Ap*, Gp*, Cp*, Tp*) in relative amounts that indicate the frequency with which A was linked to these nucleotides (i.e., Ap*A, Gp*A, Cp*A, Tp*A). By repeating the analysis three more times with [α-^{32}P]-labeled dGTP, dCTP, and dTTP, the remaining 12 nearest-neighbor frequencies can be determined.

The simplest application of nearest-neighbor frequency analysis is to the synthetic copolymers and homopolymers. With the copolymer poly d(AT) as template-primer and [α-^{32}P]dATP incorporated, only Tp* is obtained upon hydrolysis; with [α-^{32}P]dTTP, it is only Ap*. Thus, the sequences are exclusively Tp*A and Ap*T. Since the amounts of A and T in poly d(AT) are equal, it follows that the product has the alternating sequence . . . p*Tp*Ap*Tp*Ap*Tp*A. . . . When the homopolymer poly dA·poly dT serves as template-primer, the results indicate that the sequences are exclusively Ap*A and Tp*T. With sufficient care, this method can detect a dinucleotide sequence as infrequent as 1 in 1000. Within this range, there is no detectable failure in the fidelity of pol I in copying copolymers, homopolymers, and DNAs of defined sequence.

Products primed by natural DNAs from diverse sources show dinucleotide frequencies that are unique for each DNA.[141] Although these frequencies usually reflect the abundance of the bases in the DNA, there are some interesting exceptions. The frequency of GpC in animal and bacterial DNAs is nearly random, but the CpG sequence, particularly in vertebrate DNAs, is extraordinarily low (Table 4-8).[142] (The diagrams at the top of the table are included as a reminder that the CpG sequence in one strand matches a CpG sequence in the complementary strand. The other dinucleotide in low abundance in virtually all DNAs is TpA.[143])

Among the DNA viruses that infect animal cells, in polyoma and Shope papilloma, two very small viruses with limited genetic content (Section 19-2), the CpG frequencies reflect the pattern of their host. On the other hand, the CpG frequencies of three large viruses are distinctly different. These observations have suggested to some investigators a more intimate relationship to host cells in the evolutionary origin of the smaller viruses.

The CpG sequence is the preferred site for cytosine methylation in animal and plant cells and is linked to the control of gene expression; methylation generally reduces transcription. The frequency of CpG is lowest in transcriptionally inactive DNA (e.g., introns), while that of TpA is lowest in DNA expressed in cytosolic mRNA (i.e., cDNA).[144] The evolutionary selection of TpA may be due to the relative instability of its UpA transcript; the depletion of CpG[145] may be due to the mutational transition upon deamination of methylated CpG.

Although the two chains in the initial Watson-Crick model were oriented in opposite directions (polarity), firm evidence for this was lacking until provided by nearest-neighbor frequency analysis.[146] As Figure 4-27 shows, a given tem-

---

141. Swartz MN, Trautner TA, Kornberg A (1962) *JBC 237*:1961.
142. Subak-Sharpe H, Bürk RR, Crawford LV, Morrison JM, Hay J, Keir HM (1966) *CSHS 31*:737; Reddy VB, Thimmappaya B, Dhar R, Subramanian KN, Zain BS, Pan J, Ghosh PK, Celma ML, Weissman SM (1978) *Science 200*:494.
143. Subak-Sharpe H, Bürk RR, Crawford LV, Morrison JM, Hay J, Keir HM (1966) *CSHS 31*:737; Reddy VB, Thimmappaya B, Dhar R, Subramanian KN, Zain BS, Pan J, Ghosh PK, Celma ML, Weissman SM (1978) *Science 200*:494.
144. Beutler E, Gelbart T, Han J, Koziol JA, Beutler B (1989) *PNAS 86*:192.
145. Sved J, Bird A (1990) *PNAS 87*:4692.
146. Josse J, Kaiser AD, Kornberg A (1961) *JBC 236*:864.

Table 4-8
CpG and GpC frequencies in DNA from animal and bacterial sources[a]

5′ —C≡G— 3′ / 3′ —G≡C— 5′ (P) 5′ —G≡C— 3′ / 3′ —C≡G— 5′ (P)

Source of template	CpG	GpC	Ratio: GpC/CpG
Animal cells			
Human spleen	.010	.043	4.3
Rabbit liver	.013	.048	3.7
Chicken red cell	.011	.052	4.7
Crab			
Main band	.019	.030	1.6
Satellite	.0007	.0015	2.2
Animal viruses			
Shope papilloma	.024	.055	2.3
Polyoma	.018	.052	2.9
Herpes	.109	.106	0.99
Pseudorabies	.146	.132	0.91
Vaccinia	.035	.026	0.74
Bacteria and phage			
Mycobacterium phlei	.076	.067	0.88
Micrococcus luteus	.069	.060	0.87
Escherichia coli	.072	.079	1.10
Phage T1	.070	.076	1.08

[a] From Subak-Sharpe H, Bürk RR, Crawford LV, Morrison JM, Hay J, Keir HM (1966) *CSHS 31*:737.

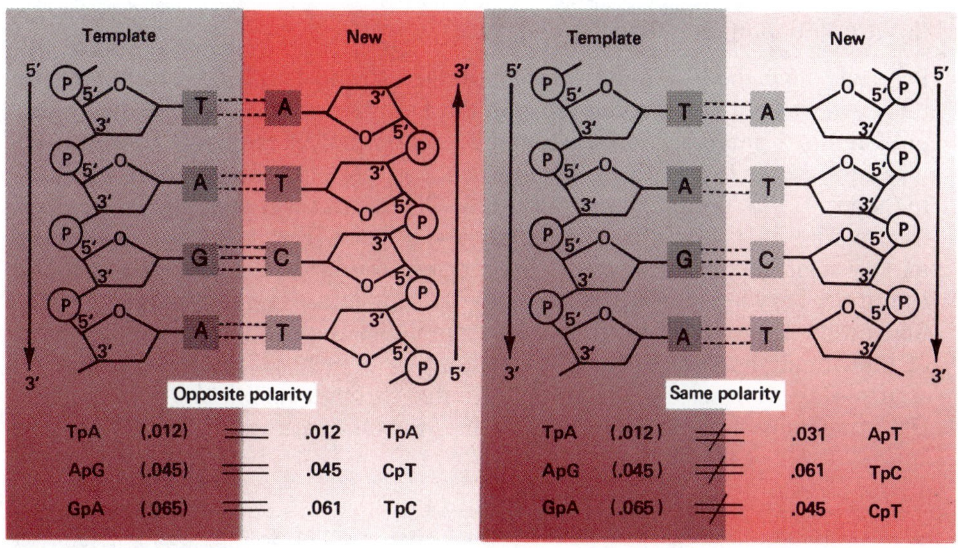

Figure 4-27
Nearest-neighbor sequences expected when replication of the template produces a chain of the opposite polarity (left) or the same polarity (right). [After Josse J, Kaiser AD, Kornberg A (1961) *JBC 236*:864.]

plate sequence could be copied in either the opposite or the same polarity. The frequencies measured in the enzymatic product showed that the nearest-neighbor sequences matched those predicted for strands with opposite polarity but not for those with the same polarity.

This property of the enzymatically synthesized strands implies that the natural DNA template has the same characteristic. It must be understood, however, that all the information obtained from nearest-neighbor analysis is derived from product strands in which labeled dNTPs are incorporated and only by inference applies to the natural DNA template. Inherent in this method is the assumption that the product strands, as well as the original template, have been replicated extensively and unselectively.

Biological Activity

Despite the accuracy with which the base sequences of template-primers have been replicated, biological activity in the enzymatically synthesized DNA was not readily reproduced. For example, the transforming activity of *B. subtilis* DNA was not increased despite a severalfold net synthesis of the DNA;[147] the branched nature of the product has been offered as a reason for the lack of biological activity. With the demonstration in 1967 (12 years after discovery of the enzyme) that pol I can produce fully infectious copies of circular ϕX174 DNA,[148] several things were made clear. No amino acids or unusual nucleotides were needed for the in vitro synthesis; the A, T, G, and C nucleotides, in a high grade of purity, sufficed. The 5386-ntd stretch of ϕX174 DNA was copied error-free, as inferred from the infectivity of the product. The polymerase action, when assisted by ligase, produced covalently closed circles, rather than nicked products. The success in replicating this rather simple viral chromosome of ten genes opened up possibilities for achieving the synthesis, in vitro, of altered genomes and other important and more complicated DNAs.

In Vitro Initiation on a Circular Template

Precisely how the synthetic complementary strand is initiated on a circular template was not resolved by these in vitro studies of ϕX174 replication using pol I alone. A small amount of boiled *E. coli* extract,[149] required for initiation, was later shown to provide fragments of *E. coli* DNA that fortuitously annealed to the viral DNA template; unmatched 3′-OH termini were tailored to a correct base-paired fit by the 3′→5′ exonuclease of pol I (Fig. 4-28). Replication by extension of the properly base-paired primer terminus then proceeded around the circle. Very few nucleotide residues of the priming fragment persisted in the final product (about one residue in five completed circles),[150] due to the excision of the unmatched 5′ end of the fragment by 5′→3′ exonuclease action. Of course, DNA synthesis in vivo is not initiated through an existing population of DNA fragments, but, rather, generally involves the synthesis of an RNA priming

---

147. Richardson CC, Schildkraut CL, Aposhian HV, Kornberg A, Bodmer W, Lederberg J (1963) in *Informational Macromolecules* (Vogel HJ, Bryson V, Lampen JO, eds.). Academic Press, New York, p. 13.
148. Goulian M, Kornberg A (1967) *PNAS 58*:1723; Goulian M, Kornberg A, Sinsheimer RL (1967) *PNAS 58*:2321.
149. Goulian M, Kornberg A (1967) *PNAS 58*:1723; Goulian M, Kornberg A, Sinsheimer RL (1967) *PNAS 58*:2321.
150. Goulian M (1968) *CSHS 33*:11; Goulian M, Goulian SH, Codd EE, Blumenfeld AZ (1973) *Biochemistry 12*:2893.

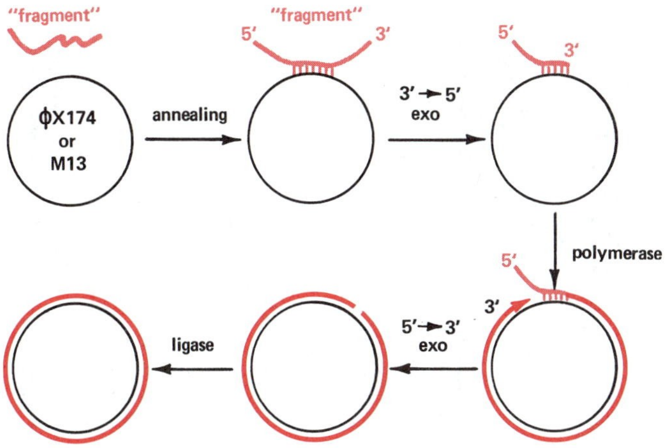

"fragment"
"fragment"
5′ 3′
φX174 or M13
annealing
3′→5′ exo
5′ 3′
polymerase
5′ 3′
ligase
5′→3′ exo

Figure 4-28
Replication of a single-stranded circle to the replicative form by pol I. An oligonucleotide fragment is needed to prime the process. The sequence of operations illustrates the successive tailoring actions of the 3′→5′ exonuclease, polymerase, and 5′→3′ exonuclease functions of the enzyme.

fragment that is extended by polymerases and later excised through the action of 5′→3′ exonuclease (Section 12).

4-17 The Polymerase Chain Reaction (PCR)[151]

A 50-bp DNA fragment from a prehistoric creature, one virus particle in 10^6 cells, or DNA in the bloodstain from a crime can be amplified a millionfold or more for an unequivocal identification. This astonishing and highly useful technique, known as the *polymerase chain reaction* (PCR), exploits pol I (or a related polymerase) and its now familiar capacities for sequencing and synthesizing DNA. The journal *Science* selected PCR as the major scientific development of 1989 and chose for its first "Molecule of the Year" the DNA polymerase molecule that drives the reaction.[152] The capacity of the polymerase to amplify a DNA segment had long been appreciated,[153] but for lack of techniques to sequence and synthesize primers, this capacity could not be widely used.

The method (Fig. 4-29) requires knowing the sequence of the regions that flank the gene to be amplified. Then, specific synthetic oligomers, annealed to the flanking regions on each of the strands, can serve as primers. After the primers are extended by polymerase for the length of the unmatched template, a second cycle is initiated by heat-denaturing the duplex, thereby melting it and separating the strands, then annealing fresh primers and extending them again. The primers delimit the ends of the DNA that will be amplified. An automated device is used for the many repetitions of the polymerase chain reaction, and a heat-stable polymerase[154] avoids the need for fresh additions of enzyme. After 20 such cycles—completed in an hour—the target sequence has been amplified a millionfold, and after 30 cycles, a billionfold.

Amplification of two sequences (e.g., for detection of two related viruses) can

151. Mullis K, Faloona F, Scharf S, Saiki R, Horn G, Ehrlich H (1986) *CSHS 51*:263; Mullis KB, Faloona FA (1987) *Meth Enz 155*:335; Gibbs RA, Chamberlin JS (1989) *Genes Dev 3*:1095.
152. Guyer RL, Koshland DE Jr (1989) *Science 246*:1543
153. Kleppe K, Ohtsuka E, Kleppe R, Molineux I, Khorana HG (1971) *JMB 56*:341.
154. Saiki RK, Gelfand DH, Stoffel S, Scharf SJ, Higuchi R, Horn GT, Mullis KB, Ehrlich HA (1988) *Science 239*:487, Eckert KA, Kunkel TA (1990) *NAR 18*:3739.

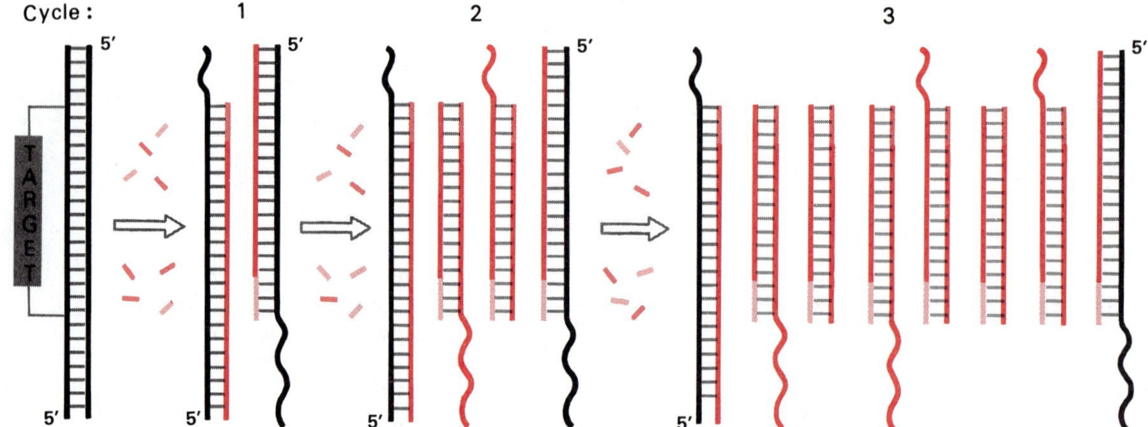

Figure 4-29
Polymerase chain reaction (PCR) for amplifying specific gene sequences. The double-stranded DNA is heated and the separated chains are allowed to bind the primers (dark bars) that define the ends of the sequence to be amplified. The primers then initiate the synthesis of two new chains complementary to the originals. This series of events is repeated 20 to 30 times, with each cycle giving a doubling of the DNA.

be accomplished simultaneously. The technique can be applied in a variety of ways: to amplify an RNA sequence, with the initial replication carried out by reverse transcriptase; to amplify only one strand, by limiting the primer available for the other; to induce mutagenesis at a site defined by a primer with a mismatched, deleted, or inserted base pair; or to manage with only one known sequence, using an anchor-modified restriction site to define the other limit of the amplified segment.[155]

To avoid the repeated additions of polymerase at each cycle, a thermostable enzyme such as that from the thermophile *Thermus aquaticus* (Taq; Section 5-7) has been used. Replication by the Taq polymerase at an elevated temperature (e.g., 55° rather than 37°) imposes more stringent hybridization of the primers, which improves specificity, homogeneity of product size, and yield of product, as well as simplifying the procedure. A typical incubation mixture contains from 10 pg to 10 μg of the target DNA, 1000-fold or greater excess of primers, 0.2 to 1 μM of dNTPs, and levels of $MgCl_2$, salt, and enzyme empirically adjusted for optimal results.

Because the Taq polymerase lacks a proofreading exonuclease, base-substitution errors are more common than with pol I, and frameshifts may also occur with greater frequency.[156] With the added possibility of thermal damage to the target DNA, mutations, if introduced in an early cycle, will appear among the final products. For that reason, caution or, better still, duplication of the procedure is warranted in obtaining the authentic sequence of a clone derived directly from an amplified sequence.

---

155. Roux KH, Dhanarajan P (1990) *BioTechniques 8*:48.
156. Tindall KR, Kunkel TA (1988) *Biochemistry 27*:6008; Dunning AM, Talmud P, Humphries SE (1988) *NAR 16*:10393; Krawczak M, Reiss J, Schmidtke J, Rösler U (1989) *NAR 17*:2197.

4-18 Mutants and the Physiologic Role of DNA Polymerase in Replication and Repair

Identification of *polA* Mutants

For more than a decade (1956–1969), only one DNA polymerase was known in *E. coli*, and it was assumed to be the enzyme solely responsible for chromosome replication. Intimations that DNA synthesis in vivo was more complex came from the enzyme's low turnover number, its inability to start the chains required in discontinuous replication,[157] and the large number of genes identified as essential for the replication of T4 phage and the *E. coli* chromosome.[158] The most telling advance was the isolation of a mutant that was viable even though severely deficient in DNA polymerase.[159] A brute-force search through several thousand strains from a heavily mutagenized population had yielded one such mutant.

The mutant gene, *polA1*, was located at 87 minutes on the genetic map.[160] The amino terminal end of the polymerase molecule is coded by the end of the *polA* gene near the *metE* locus (86 min), and transcription and translation proceed clockwise on the map toward the *rha* locus (88 min). Several mutant alleles at *polA* have since been identified (see below). This first mutant allele, *polA1*, is an amber mutation suppressed by the amber suppressors *suI*, *suII*, or *suIII*, and is recessive to the wild-type gene.

Because the *polA1* mutant grows at a normal rate, extracts were searched for other DNA polymerase activities. Drastic reduction of polymerase activity by the *polA* defect and further reduction by the use of *polA* antibody provided the low background against which a residual polymerase activity was eventually detected. This activity, later recognized as a mixture of two distinct enzymes, named *DNA polymerases II and III* in the order of their discovery, will be discussed in the next chapter. The *polA* gene product, the prototypic enzyme, was designated *DNA polymerase I* (pol I) and was thought to be entirely absent in *polA1* cells.

The most apparent defect of the *polA1* mutant is its incapacity to repair DNA damage from UV irradiation and radiomimetic drugs like methyl methanesulfonate. Sensitivity of the mutant to these agents is many times greater than that of wild-type cells.

Based on a selection of strains defective in DNA repair capacity, a number of additional *polA* mutants were later isolated, among them a nonsuppressible *polA5* and one defective in repair only at elevated temperature, *polA12*. The temperature sensitivity of the pol I activity of the *polA12* mutant provided evidence that the *polA* locus is the structural gene for pol I. Studies of these and other *polA* mutants furnished insights into the structure of the enzyme (see below). Although they grow at a normal rate and support the growth of lysogenic and virulent phages, *polA* mutant cells display a number of defects that are generally consistent with a deficiency in filling DNA gaps in a variety of situations (Table 4-9).

---

157. Kornberg A (1969) *Science 163*:1410
158. Gross JD (1972) *Curr Top Microbiol Immunol 57*:39.
159. De Lucia P, Cairns J (1969) *Nature 224*:1164.
160. Kelley WS, Grindley NDF (1976) *MGG 147*:307.

Table 4-9
Phenotypic defects of *polA* mutants[a]

Sensitivity to thymine starvation[b]
Sensitivity to irradiation by UV[c] and x-rays[d]
Sensitivity to methyl methanesulfonate[e]
Extensive degradation of DNA during repair of UV damage[e]
Increased frequency of deletions[f] and recombination[g]
Low viability (5%) on nutritional shift up[h]
Instability of certain plasmids;[i] lack of replication[e,j]
Poor joining of 10S fragments[k]
Single-strand gaps in phage ϕX174 RF[l]
Poor growth of phage λ; defective in exonuclease (β or γ protein)[m]
Deficiency in phage T4 replication and recombination[n]
Nonviability as *lig*, *recA*, *recB*, or *uvrB* double-mutants
Inadequate suppression of *dnaQ* mutant[o]

[a] Lehman IR, Uyemura DG (1976) *Science 193*:963; Gross JD (1972) *Curr Top Microbiol Immunol 57*:39.
[b] Nakayama H, Hanawalt PC, personal communication.
[c] Gross J, Gross M (1969) *Nature 224*:1166.
[d] Town CD, Smith KC, Kaplan HS (1971) *Science 172*:851.
[e] Cooper PK, Hanawalt PC (1972) *PNAS 69*:1156.
[f] Coukell MB, Yanofsky C (1970) *Nature 228*:633.
[g] Konrad EB, Lehman IR, personal communication.
[h] Rosenkranz HS, Carr HS, Morgan C (1971) *BBRC 44*:546.
[i] Kingsbury DT, Helinski DR (1970) *BBRC 41*:1538; Goebel W (1972) *NNB 237*:67.
[j] Goebel W, Schrempf H (1972) *BBRC 49*:591.
[k] Okazaki R, Arisawa M, Sugino A (1971) *PNAS 68*:2954; Kuempel PL, Veomett GE (1970) *BBRC 41*:973.
[l] Schekman RW, Iwaya M, Bromstrup K, Denhardt DT (1971) *JMB 57*:177.
[m] Zissler J, Signer E, Schaefer F (1971) in *The Bacteriophage Lambda* (Hershey AD, ed.). CSHL, p. 469.
[n] Mosig G, Bowden BW, Bock S (1972) *NNB 240*:12.
[o] Lancy ED, Lifsics MR, Kehres DG, Maurer R (1989) *J Bact 171*:5572.

Inasmuch as the *polA1* mutant synthesizes DNA despite the apparent lack of pol I, the enzyme was first assumed to have no part in replicating the *E. coli* chromosome and was regarded simply as a component of one of the many DNA repair systems. But assigning physiologic functions on the basis of such data is inherently risky. Failure to detect an enzyme activity in a cell-free extract is not conclusive. Among numerous reasons for such failure may be instability of the enzyme upon extraction, vulnerability to proteolytic enzymes, or tight binding to a component that masks its activity. Because special assay conditions may be required, quantitative estimations of enzyme levels are uncertain.

With appropriate attention to enzyme lability, extracts of these *polA* mutants were shown to contain at least measurable levels of pol I activity, with functions not only in the excision and replicative repair of DNA lesions but also in a stage of chromosome replication. The properties of the isolated mutant enzymes have also provided insights into structural relationships of the several activities of this multifunctional enzyme. For these reasons, the properties of some of these mutants deserve special attention.

Polymerizing Defect:[161] *polA1* (trp342 → am)

Extracts of this mutant have about 1% of the polymerizing activity of wild-type cells, but near-normal levels of the 5′→3′ exonuclease (measured uncoupled to

161. Lehman IR, Chien JR (1973) *JBC 248*:7717.

polymerization). However, the exonuclease is not part of an intact polymerase molecule. Instead, it is associated with a smaller polypeptide that corresponds to the small fragment generated by proteolytic cleavage of intact polymerase (Section 13). The low residual level of polymerizing activity in *polA1* mutants is probably due to infrequent "read-through" of the amber codon; the small polypeptide encoded at the 5′ end of the gene with its 5′→3′ exonuclease activity is either the amber fragment or derived from it by proteolysis.

Were the low level of pol I activity in extracts of *polA1* mutants a valid reflection of the actual intracellular level, the number of pol I molecules might still be sufficient to perform an essential function in replication. Wild-type cells are estimated to have about 300 molecules each of pol I and of DNA ligase. Studies with ligase mutants are instructive: with only 1% of the normal level of ligase, mutants show no impairment in growth, whereas with levels depressed tenfold further (less than one molecule per cell), mutants accumulate short pieces of nascent DNA and die (Section 9-5). Probably, then, the reactions at the replication fork require only a small fraction of the ligase and polymerase molecules, and the rest are deployed in repair and recombination functions along the great length of the chromosome. Thus, the polymerase depletion observed in *polA1* mutants, despite profoundly affecting repair of lesions, still permits an adequate rate of replication.

Nick-Translation Defect:[162] *polA12*

This defect produces a temperature-sensitive mutant that is viable at 43° but is far more sensitive than wild-type cells at this elevated temperature to DNA damage by UV irradiation and methyl methanesulfonate. The pol I isolated from the mutant exhibits reduced electrophoretic mobility and slower sedimentation, indicating a misfolded, less compact tertiary structure, possibly the source of the thermolability and the sensitivity of the enzyme to low ionic strength.

Although the polymerase and nuclease activities of the mutant enzyme are all unstable at 43°, the most striking defect is the lack of coordination between polymerase and 5′→3′ exonuclease required for effective polymerization at a nick. When provided with a nicked circular duplex DNA, the wild-type enzyme incorporates nucleotides at one end of the nick, matched by equimolar release of nucleotides from the other—that is, *nick translation* (Fig. 4-30). The mutant, by contrast, shows virtually no synthesis or degradation. With nicks in the template enlarged to gaps by exonuclease III, gap-filling by both wild-type and mutant enzymes is found, but the mutant enzyme fails to carry the process on to nick translation. Even at 30°, where levels of polymerizing and exonuclease activities of the mutant enzyme are near normal, nick translation is clearly deficient. Possibly, a critical spatial arrangement of these functional sites in the active center of the enzyme required for nick translation is perturbed by the imperfect folding caused by the *polA12* mutation.

The discrepancy in the *polA12* mutant between maintenance of viability and failure to repeat DNA lesions may be explained as in the case of *polA1* mutant: the number of mutant molecules or their level of function, while sufficient for supporting replication, may be inadequate to repair a large number of lesions.

162. Monk M, Kinross J (1972) *J Bact 109*:971; Uyemura D, Lehman IR (1976) *JBC 251*:4078.

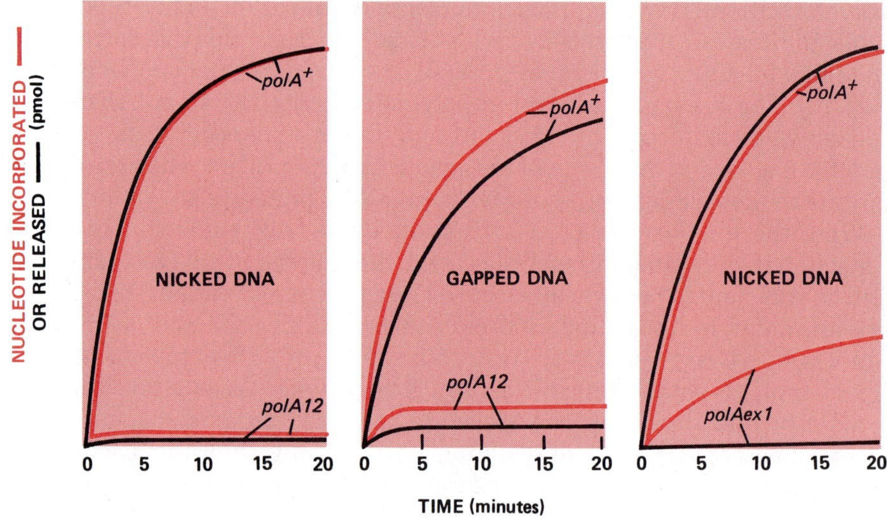

Figure 4-30
Comparison of mutant *polA12* or *polAex1* and wild-type (*polA*+) DNA polymerase at 43°. Lack of nick translation characterizes *polA12*; *polAex1* is deficient in 5′→3′ exonuclease activity.

Conditionally Lethal 5′→3′ Exonuclease Defect:[163] *polAex1 (polA480ex)* (gly184 → asp)

This temperature-sensitive mutant with a lethal 5′→3′ defect had been selected for its hyperactivity in recombination. Like the other *polA* mutants, *polAex1* is sensitive to several agents that damage DNA and is defective in enabling the joining of nascent DNA fragments. Its polymerase activity remains at the normal level. The singular defect in the *polAex1* mutant is in the 5′→3′ exonuclease, and this appears to be its undoing.

The 5′→3′ exonuclease activity of the isolated mutant polymerase is depressed even at 30°, but at 43° it is further inhibited by concurrent polymerization, rather than stimulated by it. With a nicked DNA as a template-primer (Fig. 4-30), polymerization by strand displacement takes place but without hydrolysis. The lower polymerization rate of *polAex1* relative to wild-type enzyme is probably due to the constraints of strand displacement, and was not observed with a gapped template.

Another conditionally lethal *polA* mutant, *BT4113 (polA4113)* (gly103 → glu),[164] was selected on the basis of direct assay of mutagenized colonies for their ability to synthesize and degrade DNA. This temperature-sensitive mutant was defective in both polymerase and 5′→3′ exonuclease activities at 45°, although the primary lesion appears to be in the polypeptide region containing the nuclease function. When the mutant polymerase was cleaved by proteolysis before being tested at 45°, the polymerase activity remained normal, implying that the distortion of the small-fragment region is responsible for a secondary collapse of the entire polymerase molecule.

The message conveyed from these studies of *polA12*, *polAex1*, and *BT4113* is that severe deficiency in the 5′→3′ exonuclease is not tolerated (Table 4-10).

---

163. Konrad EB, Lehman IR (1974) *PNAS 71*:2048; Uyemura D, Eichler DC, Lehman IR (1976) *JBC 251*:4085.
164. Olivera BM, Bonhoeffer F (1974) *Nature 250*:513.

Table 4-10
Viability of *polA* mutants with defects in 5′→3′ exonuclease

Mutant	Mutation	Defective in repair and joining of nascent fragments	Enzyme defect in vitro	Amino acid affected	Conditionally lethal
polA1	amber	yes	polymerization	trp342→am	no
polA12	ts[a]	yes	nick translation	unknown	no
polAex1	ts	yes	5′ → 3′ exo	gly184→asp	yes
BT4113	ts	yes	polymerization, 5′ → 3′ exo	gly103→glu	yes

[a] ts = temperature-sensitive.

Conditionally Lethal *polA lig* Double Mutant

In either of the temperature-sensitive *polA* or *lig* mutants, a shift to the restrictive temperature leads to a massive accumulation of short DNA fragments. Nevertheless, DNA synthesis continues for hours and the cells remain viable. The *polA lig* double mutant, when shifted to restrictive temperature, promptly ceases DNA synthesis and dies. The indispensability of the *polA* gene is further confirmed by the failure of attempts to introduce *polA* mutation into cells deficient in recombination and repair (e.g., *recA, recBC, uvrB*).

Null (Deletion) *polA* Mutants[165]

Mutants lacking both the large fragment (polymerase and 3′→5′ exonuclease domains) and the small fragment (5′→3′ exonuclease), designated Δ*polA*, have proved to be viable in minimal media but not in rich media. Remarkably, mutants which retained either domain could grow on the rich medium. Thus, auxiliary pathways can replace the pol I functions when growth is slow but not when accelerated rates of replication (and repair) demand higher levels of activity. These mutants may help to explain the way in which one pol I domain complements the other: the polymerase function, when present, may displace the 5′ chain end, rendering it more susceptible to other nucleases in the cell, while the retained 5′→3′ exonuclease, by creating gaps, provides templates for other polymerases [e.g., pol II or the pol III core (Section 5-3)]. The conditional lethality of several 5′→3′ exonuclease mutants (Table 4-10), generally demonstrated in rich media, may be reconciled by quantitative aspects of growth rate, leakiness, and the contributions of auxiliary enzymes and factors.

Mutants in the Active Sites of the Large Fragment

Site-directed mutagenesis has identified key residues in the 3′→5′ exonuclease site in the small domain at which mutations destroy that function without any effect on polymerase activity (Section 5). In a similar way, numerous mutations in the cleft region of the large domain have indicated those that reduce binding of the template-primer and others that raise the K_m for dNTPs (Section 5). The decrease in k_{cat} can be profound, up to 10^4-fold, without impairment of the 3′→5′ exonuclease activity.

165. Joyce CM, Grindley NDF (1984) *J Bact* 158:636.

Other Mutants Relevant to Pol I Functions in Replication and Repair

Principal Replicative Role. A mutant, *pcbA1*, identified as an allele of *gyrB*,[166] enables pol I to suppress the conditionally lethal mutations (temperature-sensitive or nonsense) in *dnaE*, the gene that encodes the α (polymerase) subunit of pol III holoenzyme (Section 5-3). By replacing the polymerase function and by interacting with accessory subunits (Section 5-4) in some significant way, an effective alternative is created, with pol I serving a key role in the principal replicative pathway of this strain. How *gyrB* deficiency enables pol I to substitute in the holoenzyme is still unclear.

Error-Prone Repair. Upon induction by SOS conditions, pol I acquires a new form (pol I*)[167] with a demonstrably lower fidelity that may contribute to error-prone (mutagenic) repair. The pol I modification does not entail any diminution in size, but instead shows an increase in complexity by the binding of another polypeptide, possibly one of the accessory subunits (e.g., τ, β, ϵ) in pol III holoenzyme.[168] This interaction may even be related to that in which pol I substitutes for the pol III α subunit in a *gyrB* mutant (see above).

Postreplication Repair.[169] The role of pol I in the *recA*-dependent repair of UV lesions—daughter-strand gaps and double-strand breaks—was made clearer by examining the Δ*polA* strain. The pol I deletion made a *uvrA* strain very sensitive to UV, about as much as the severely deficient *uvrA,recF* double mutant. The 5′→3′ exonuclease activity of pol I appears to be the major factor, inasmuch as *uvrA*, *polA546* (defective in 5′→3′ exonuclease) was almost as deficient as the *uvrA*, Δ*polA* strain.

Summary

The main points establishing the role of pol I in replication and repair may be summarized in the behavior of mutants (Tables 4-9 and 4-10) and enumerated as follows:

(1) Because the two mutant *E. coli* strains *polAex1* and *BT4113* are conditionally lethal, the enzyme must perform some essential functions.

(2) Retarded joining of nascent DNA fragments and defective repair of DNA damage in all *polA* mutants can be explained by the impaired gap-filling and excision functions of the enzyme.

(3) The failure of certain mutants to maintain viability, carry on repair, and join fragments is best explained in terms of their loss of a singular property of the enzyme, namely, the refined coordination between polymerization and 5′→3′ exonuclease action.

(4) Retarded joining of nascent DNA fragments and loss of viability are in turn explained by failure during semiconservative chromosome replication to excise and replace RNA at the 5′ termini of the nascent fragments.

---

166. Bryan S, Chen H, Sun Y, Moses RE (1988) *BBA 951*:249; Maki H, Bryan SK, Horiuchi T, Moses RE (1989) *J Bact 171*:3139.
167. Lackey D, Krauss SW, Linn S (1985) *JBC 260*:3178.
168. Ruscitti TM, Polayes DA, Martin SA, Linn S (1988) *J Cell Biol 107*:228a.
169. Sharma RC, Smith KC (1987) *J Bact 169*:4559.

CHAPTER 5

Prokaryotic DNA Polymerases Other Than *E. coli* Pol I

5-1 Discovery of *E. coli*[1] DNA Polymerases II and III

The discovery of mutants of *E. coli* that had little or no detectable DNA polymerase activity in extracts but nevertheless synthesized DNA at a normal rate (Section 4-18) spurred a search for an unknown DNA-synthesizing activity in these strains. When conditions for the preparation of active extracts and for adequate assays were eventually found, two polymerases distinct from DNA polymerase I were isolated. Designated DNA polymerase II[2] and III[3] in the order of their discovery, these enzymes are commonly referred to as *pol II* and *pol III*. Once their properties were known, pol II and pol III were demonstrated in extracts of wild-type cells as well. Under assay conditions standardized for pol I, the combined pol II and pol III activity in extracts is usually less than 5% of pol I activity. This explains why the specific antibody neutralization of pol I activity in extracts of wild-type cells had failed to uncover the residual activity of pol II and pol III. Later studies revealed that pol III, made up of three subunits, is the core of a pol III holoenzyme with multiple subunits (Sections 3 and 4).

1. Gefter ML, Molineux IJ, Kornberg T, Khorana HG (1972) *JBC 247*:3321.
2. Kornberg T, Gefter ML (1970) *BBRC 40*:1348; Moses RE, Richardson CC (1970) *BBRC 41*:1567; Moses RE, Richardson CC (1970) *BBRC 41*:1565; Knippers R (1970) *Nature 228*:1050.
3. Kornberg T, Gefter ML (1971) *PNAS 68*:761.

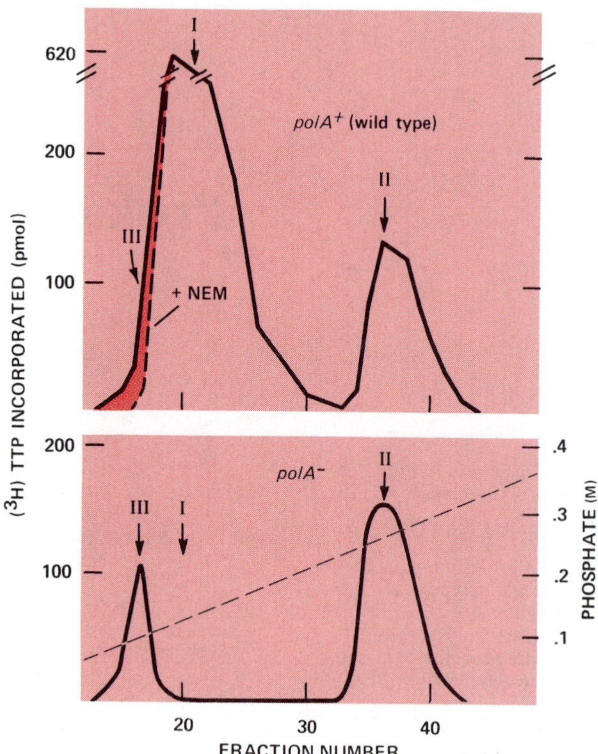

Figure 5-1
Separation of DNA polymerases I, II, and III by phosphocellulose chromatography. Cell-free extracts were analyzed. Assays of the pol I peak in the presence of the pol III inhibitor N-ethylmaleimide (NEM) are shown by the heavy dashed line. The relative absence of polymerase I in extracts of *polA*− cells made it possible to detect polymerases II and III more readily. [From Gefter ML, Hirota Y, Kornberg T, Wechsler JA, Barnoux C (1971) *PNAS* *68*:3150]

5-2 DNA Polymerase II of *E. coli*

Pol II has been purified to homogeneity by procedures similar to those used for pol I. Separation from pol I and pol III was achieved by phosphocellulose chromatography (Fig. 5-1).

Comparison with Pol I and Pol III

The properties of pol I, II, and III are compared in Table 5-1. Pol II, unlike pol I, is unable to use a template-primer that is simply nicked; it is also less effective than pol I with a template that is extensively single-stranded. The optimal template-primer for both pol I and pol II is a duplex with short gaps. There is no accurate determination of the optimal gap size; estimates place it at under 100 ntd. Pol II lacks the 5′→3′ exonuclease activity possessed by pol I.

Pol II resembles pol III (Table 5-1) in its sensitivity to sulfhydryl-blocking agents, but differs in its stimulation by single-strand binding protein (SSB) on primed single strands (see below), greater affinity for phosphocellulose, lesser heat-lability, greater resistance to salt, indifference to ethanol, and effectiveness at lower concentrations of dNTPs.

Pol II and pol III have a greater specificity than pol I for the 2′-deoxyribo configuration of the dNTP.[4] Neither ribosyl nor arabinosyl analogs can be sub-

4. Helfman WB, Hendler SS, Shannahoff DH, Smith DW (1978) *Biochemistry* *17*:1607.

Table 5-1
Properties of polymerases I, II, and III of *E. coli*

	pol I[a]	pol II[a]	pol III[a] (core)
Functions			
Polymerization: 5′→3′	+	+	+
Exonuclease: 3′→5′	+	+	+
Exonuclease: 5′→3′	+	−	−
Pyrophosphorolysis and PP_i exchange	+		+
Template primer			
Intact duplex	−	−	−
Primed single strands,[b]	+	−	−
stimulation by SSB	−	+	−
Nicked duplex [poly d(AT)]	+	−	−
Duplex with gaps or protruding single-strand 5′ ends of: < 100 nucleotides	+	+	+
> 100 nucleotides	+	−	−
Polymer synthesis de novo	+	−	−
Activity			
Effect of KCl (percent of optimal activity) 20 mM	60	60	100
50 mM	80	100	50
100 mM	100	70	10
150 mM	80	50	0
K_m for dNTPs	low	low	high
Stimulation by β subunit	−	+	+
Inhibition by 2′-deoxy analogs	−	+	+
Inhibition by arabinosyl CTP	−	+	−
Inhibition by sulfhydryl (—SH)-blocking agents	−	+	+
Inhibition by pol I antiserum	+	−	−
General			
Size (kDa)	103	90	130, 27.5, 10[d]
Affinity to phosphocellulose: molarity of phosphate required for elution	0.15	0.25	0.10
Molecules/cell, estimated	400		10–20
Turnover number,[c] estimated	600	30	9000
Structural genes	*polA*	*polB*	*polC*[e]
Conditional lethal mutant	yes	no	yes

[a] + and − represent the presence and absence, respectively, of the property listed.
[b] A primed single strand is a long single strand with a short length of complementary strand annealed to it. SSB = single-strand binding protein.
[c] Nucleotides polymerized at 37°/min/molecule of enzyme.
[d] Sizes of the α, ϵ, and θ subunits.
[e] Also known as *dnaE*, the gene for the large (α) subunit.

stituted to a significant extent, even in the presence of Mn^{2+}. Pol II is the most susceptible to inhibition by arabinosyl analogs (Section 14-3).

Effects of SSB

Use of a single-stranded template by pol II, is stimulated 50- to 100-fold by *E. coli* SSB (Section 10-5), bringing the activity to a level near that observed with gapped DNA. SSB does not stimulate the use of gapped DNA by pol II, nor does it influence pol I action on either template. The effect of SSB is thought to be in promoting the processivity of pol II by melting out regions of duplex secondary structure that would cause the enzyme to dissociate. SSB may also serve in preventing nonproductive association of the polymerase with single-stranded regions, thus favoring its association with a primer terminus. The β subunit and other pol III holoenzyme subunits[5] stimulate pol II replication of single-stranded templates, suggesting a relationship to a subassembly of the holoenzyme.[6]

Role in DNA Repair

A mutant, *polB1*, with no detectable (<0.5%) pol II activity in extracts, has been obtained by assaying extracts from many surviving colonies of mutagenized *polA1* cells.[7] (The *polB* locus is at 2 minutes on the genetic map.[8]) This *polA1 polB1* mutant grows normally at 25° and 42° and supports the growth of a variety of phages. The *polB1* mutation by itself does not make the cells sensitive to UV irradiation (Fig. 5-2) nor does it affect the frequency of recombination in transductions and mating transfers. Yet, in mutants deficient in pol I and III, pol II serves in the repair of UV lesions.[9]

Even more strikingly, pol II may have a role in repair in the DNA-damage-induced mutagenic (SOS) response[10] (Section 21-1). When a sequence with an abasic (apurinic or apyrimidinic) lesion (Δ) is used as a template, for example:

5′-TCCCACTCACGACGT
   AGGGTGAGTGCTGCAAGAΔTTTTTTTTT

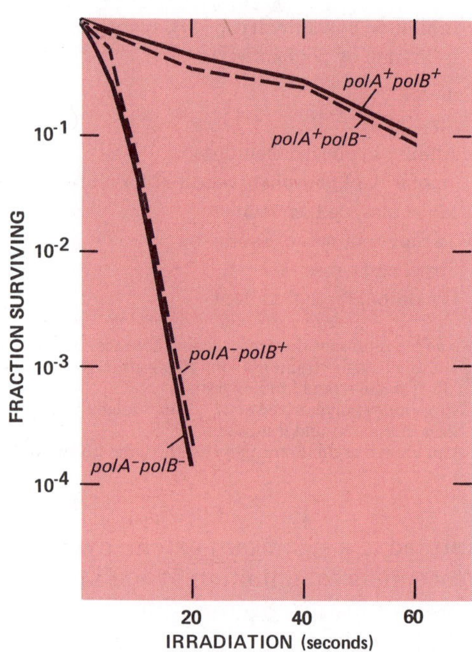

Figure 5-2
Sensitivity to ultraviolet irradiation of *polA* but not *polB* mutants. [From Campbell JL, Soll L, Richardson CC (1972) *PNAS 69*:2090]

5. Wickner S (1977) *PNAS 74*:2815.
6. McHenry CS (1988) *ARB 57*:519.
7. Campbell JL, Soll L, Richardson CC (1972) *PNAS 69*:2090.
8. Campbell JL, Shizuya H, Richardson CC (1974) *J Bact 119*:494.
9. Masker W, Hanawalt PC, Shizuya H (1973) *NNB 244*:242; Tait RC, Harris AL, Smith DW (1974) *PNAS 71*:675.
10. Bonner CA, Randall SK, Rayssiguier C, Radman M, Eritja R, Kaplan BE, McEntee K, Goodman MF (1988) *JBC 263*:18946.

a polymerase activity can be identified in extracts of *E. coli* which incorporates a nucleotide opposite the deleted base and extends the chain beyond it. This activity is increased sevenfold by SOS induction; when purified to near homogeneity, it appears to be pol II activity, as judged by a mass of 84 kDa, stimulation by the β subunit of pol III holoenzyme and SSB, and dependence on a functional *polB* gene. *PolB* has been identified as an allele of *dinA*,[11] induced in the SOS response (Section 21-3) but not shown to be essential for mutagenic repair.

Upon isolation from overproducing cells bearing the cloned *polB* gene, pol II is 99 kDa, but smaller polypeptides of 82 and 55 kDa, presumably degradation products, retain activity.[12] The size of pol II deduced from the *polB* gene is 88 kDa and it appears structurally homologous to the eukaryotic pol α polymerases.[13] Strains lacking the gene show no growth or replication defect.

5-3 DNA Polymerase III Holoenzyme of *E. coli*[14]: The Core

The Holoenzyme and Its Components

The most complex of the DNA polymerases, pol III holoenzyme is the principal replicative enzyme of *E. coli*. Only about ten molecules are present per cell. Although isolable, after a 7500-fold purification,[15] as an entity of 20, or more polypeptides with a collective mass near 900 kDa, it dissociates readily upon dilution into a variety of subassemblies[16] (Table 5-2). The first recognized of these was pol III, the catalytic core made up of three tightly associated polypeptides. The instability and low activity of pol III when assayed at moderate ionic strengths obscured its presence in extracts, even in those of *polA1* cells. Only upon phosphocellulose chromatography was its existence, distinct from pol I

Table 5-2
Major subunits and subassemblies of pol III holoenzyme

Subunit	Mass (kDa)	Gene	Subassembly			
α	130[a]	*dnaE*	pol III (core)			
ϵ	27.5[a]	*dnaQ* (*mutD*)				
θ	10			pol III′		
τ	71[a]	*dnaX*				
γ	47.5[a]	*dnaX*			pol III*	holoenzyme
δ	35					
δ'	33		γ complex			
χ	15					
ψ	12					
β	40.6[a]	*dnaN*				

[a] Based on DNA sequence; others are based on electrophoresis.

11. Bonner CA, Hays S, McEntee K, Goodman MF (1990) *PNAS 87*: 7663; Iwasaki H, Nakata A, Walker GC, Shinagawa H (1990) *J Bact 172*: 6268.
12. Chen H, Bryan SK, Moses RE, (1989) *JBC 264*:20591.
13. Chen H, Sun Y, Stark T, Beattie W, Moses RE (1990) *DNA and Cell Biol 9*:631; Iwasaki H, Ishino Y, Toh H, Nakata A, Shinagawa H, *Mol Gen Genet*, in press.
14. McHenry CS (1988) *ARB 57*:519; Maki H, Maki S, Kornberg A (1988) *JBC 263*:6570.
15. McHenry CS, Kornberg A (1977) *JBC 252*:6478.
16. Maki H, Maki S, Kornberg A (1988) *JBC 263*:6570.

and pol II, established.[17] An essential locus, called *dnaE* (*polC*), was identified as the structural gene for the polymerase subunit of pol III,[18] later identified as the α subunit.[19] The subunit composition of pol III holoenzyme and some of its subassemblies are listed in Table 5-2.

Discovery of the holoenzyme form of pol III came with the recognition that pol I, pol II, and pol III, while effective in filling gaps in DNA, are all incapable of rapidly replicating single-stranded viral circles that are several thousand residues long. The low *processivity* of these polymerases (i.e., their frequent dissociations from the template) resulted in feeble rates of synthesis on templates with infrequently spaced primers, such as singly primed phage M13. However, a novel activity was observed in *E. coli* extracts which was highly efficient in replicating long stretches of template.[20] This proved upon purification to be pol III complexed with many auxiliary subunits that clamp the core to the template and endow it with high processivity (i.e., synthesis of long stretches of DNA without dissociating from the template) (Fig. 5-3).

Pol III holoenzyme action is distinguished from the behavior of its subassemblies—pol III, pol III′, and pol III*—by its absolute dependence on activation by ATP (or dATP) and its high processivity even at high ionic strengths. The rate of synthesis by the holoenzyme on ssDNA coated with SSB is about 700 ntd per second at 37°, a value close to the in vivo rate of fork movement of 1000 ntd per second. The basic mechanism of polymerization is presumably similar, if not identical, to that of pol I (Section 4-11).

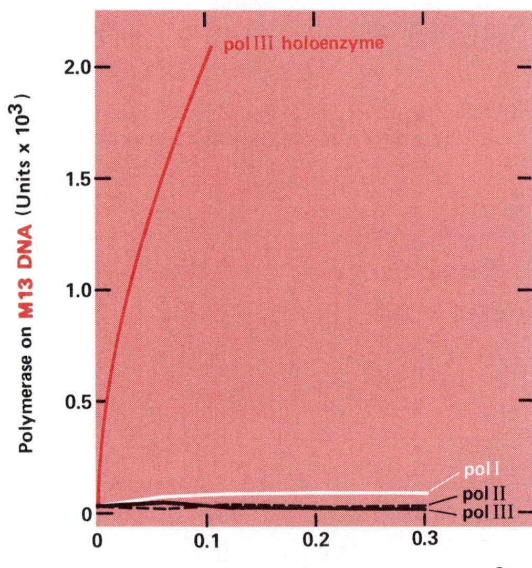

Figure 5-3
Requirement for a complex form of DNA polymerase III (pol III holoenzyme) for DNA synthesis on a primed single strand.

17. Kornberg T, Gefter ML (1971) *PNAS 68*:761; Kornberg T, Gefter ML (1972) *JBC 247*:5369.
18. Gefter ML, Hirota Y, Kornberg T, Wechsler JA, Barnoux C (1971) *PNAS 68*:3150.
19. Livingston DM, Hinkle DC, Richardson CC (1975) *JBC 250*:461; Livingston DM, Richardson CC (1975) *JBC 250*:470; McHenry C, Crow W (1979) *JBC 254*:1748; Maki H, Kornberg A (1985) *JBC 260*:12987.
20. Wickner W, Schekman R, Geider K, Kornberg A (1973) *PNAS 70*:1764; Wickner W, Kornberg A (1974) *JBC 249*:6244.

Pol III: The Core (Subunits α, ϵ, and θ)[21]

Pol III, the catalytic core of the holoenzyme, when purified nearly 30,000-fold, is a 167-kDa complex, resolvable into its constituents only under denaturing conditions. The subunits are α, the polymerase; ϵ, the $3' \rightarrow 5'$ exonuclease; and θ, with no known function. The core complex has been isolated as a dimer[22] as well as a monomer, but interconversion between these two forms has not been observed. Even though the principal substrate of pol III is a nucleic acid, the protein, like pol I, is acidic (pI near 5) rather than basic.

Initiation of DNA synthesis by binding of the core polymerase to the template-primer (to form the *initiation complex*; Section 6) may be 1000-fold slower than the subsequent polymerizations. Thus, pol III, with a processivity of only about ten residues, is effective only in filling short gaps. In addition to those which are part of the holoenzyme, some 40 free core molecules are also present in a cell and perhaps are used for their gap-filling activity.[23]

The α Subunit. The α subunit (130 kDa), the polymerase and product of the *dnaE (polC)* gene, is found only as part of the core unless artificially overproduced.[24] Purified to homogeneity, the isolated α subunit differs from the intact core in being more thermolabile and devoid of any nuclease activity. Remarkably, a defective α, resulting from conditionally lethal temperature-sensitive or nonsense mutations in the *dnaE* gene, can be functionally replaced by pol I, provided that the cell also has a *gyrB* mutation (Section 4-18).

A sharp limitation in the overproduction of α (to 2% of that of an adjacent gene on the same plasmid) suggests that expression of the *dnaE* gene is tightly regulated and cannot be bypassed by use of an alternate promoter. Even so, a 40-fold overproduction[25] of α results in only a doubling of the number of cores and no change in the level of holoenzyme,[26] clearly indicating that factors other than α determine the levels of these assemblies.

Among the conditionally lethal *dnaE* mutants, some exhibit a modest increase in the rate of spontaneous mutations—a mutator phenotype. The *dnaE* gene is located in an operon (Fig. 5-4) downstream from a gene (*lpxB*) responsible for the synthesis of the disaccharide of lipid A, a key outer membrane component.[27] Between *lpxB* and *dnaE* are a consensus sequence (dnaA box) for binding

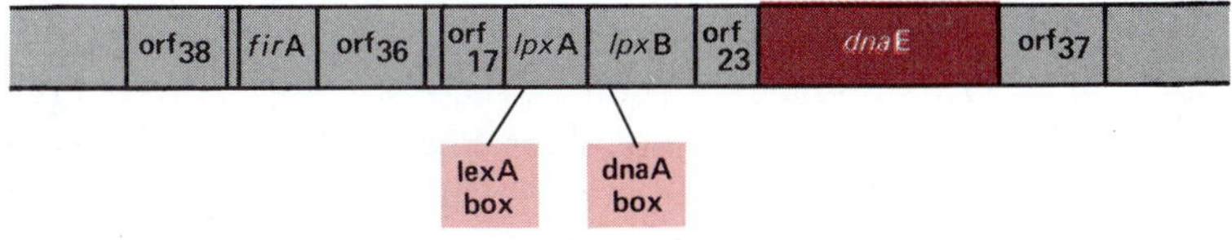

Figure 5-4
Location of the *dnaE* gene, which encodes the α subunit of pol III holoenzyme, and surrounding features of potential regulatory significance.

21. McHenry C, Crow W (1979) *JBC 254*:1748; Maki H, Kornberg A (1985) *JBC 260*:12987; Maki H, Kornberg A (1987) *PNAS 84*:4389.
22. Maki H, Maki S, Kornberg A (1988) *JBC 263*:6570.
23. Maki H, Kornberg A (1985) *JBC 260*:12987.
24. Maki H, Horiuchi T, Kornberg A (1985) *JBC 260*:12982.
25. Maki H, Horiuchi T, Kornberg A (1985) *JBC 260*:12982.
26. Maki H, Kornberg A (1985) *JBC 260*:12987.
27. Crowell DN, Anderson MS, Raetz CRH (1986) *J Bact 168*:152; Crowell DN, Reznikoff WS, Raetz CRH (1987) *J Bact 169*:5727.

dnaA protein (the chromosome initiator protein), an internal promoter, and an open reading frame (orf) for a 23-kDa protein.[28] This operon structure suggests that expression of *dnaE* may be coordinately regulated with cell growth and division and with factors related to membrane biosynthesis.[29]

The ϵ Subunit. The ϵ subunit (27.5 kDa), encoded by the *dnaQ* gene, is the 3′→5′ exonuclease.[30] An allele of *dnaQ*,[31] *mutD*,[32] increases the rate of spontaneous (presumably replicative) mutations by as much as 10^5-fold. This *mutD* mutator activity depends on high levels of thymidine in the medium, presumably to attain high levels of dTTP in the cell; UV cross-linkage of dTTP uniquely to ϵ[33] had implicated this subunit as the mutator. A mutant form of ϵ, discovered later, was demonstrably defective in the exonuclease activity.[34] Overproduction of the ϵ subunit offsets the increased mutation rate that occurs during induction of the SOS response to DNA damage[35] (Section 21-3), suggesting a normal exonuclease role for ϵ in this response.

A null mutant in *dnaQ* that occurs in the nearly identical pol III of *Salmonella typhimurium* (Section 7) is defective in growth as well as being mutagenic. The growth defect is suppressed by a mutation in the α subunit and an increased use of pol I.[36] Thus, the ϵ subunit is crucial for the α function in replication beyond its contribution to fidelity.

The α · ϵ Complex. Mixing purified α and ϵ subunits generates a 1 : 1 complex with the polymerase activity increased two-fold and the 3′→5′ exonuclease increased 50- to 100-fold (Fig. 5-5), approaching the activities in the core.[37] Clearly, the proofreading capacity of the core resides in the α–ϵ interaction. The basis of this remarkable exonuclease enhancement is indicated by the increased affinity of ϵ for DNA (>100-fold decrease in apparent K_m) as a result of its binding to α (Fig. 5-6).

In the reconstitution of a highly processive holoenzyme from its subunits (Section 6), the α·ϵ complex can substitute for the core, but the α subunit, without ϵ, cannot.[38] In the α·ϵ complex, the proofreading subunit may contribute structurally to the polymerase site, as in polymerases such as pol I, in which these activities reside in the linked domains of a single polypeptide (Fig. 4-1).

The θ Subunit. The gene encoding this 10-kDa subunit is still unknown. Inasmuch as the polymerase and exonuclease activities of the α·ϵ complex approximate those of the core, and the holoenzyme can be reconstituted with α and ϵ (Section 4), the function of the θ subunit is not clear. Possibly, θ contributes to the linkage of α to ϵ and of one core to another, or to the binding of other subunits in the build-up of large subassemblies of the holoenzyme.

---

28. McHenry CS (1988) *ARB 57*:519.
29. Sakka K, Watanabe T, Beers R, Wu HC (1987) *J Bact 169*:3400.
30. Scheuermann RH, Echols H (1984) *PNAS 81*:7747.
31. Horiuchi T, Maki H, Sekiguchi M (1978) *MGG 163*:277.
32. Cox EC (1976) *ARG 10*:135; Ehrlich HA, Cox EC (1980) *MGG 178*:703.
33. Biswas SB, Kornberg A (1984) *JBC 259*:7990.
34. Echols H, Lu C, Burgers PMJ (1983) *PNAS 80*:2189; Scheuermann R, Tam S, Burgers PMJ, Lu C, Echols H (1983) *PNAS 80*:7085; DiFrancesco R, Bhatnagar SK, Brown A, Bessman MJ (1984) *JBC 259*:5567.
35. Jonczyk P, Fijalkowska I, Ciesla Z (1988) *PNAS 85*:9124; Foster PL, Sullivan AD (1988) *MGG 214*:467.
36. Lancy ED, Lifsics MR, Kehres DG, Maurer R (1989) *J Bact 171*:5572.
37. Maki H, Kornberg A (1987) *PNAS 84*:4389.
38. Studwell PS, O'Donnell M (1990) *JBC 265*:1171.

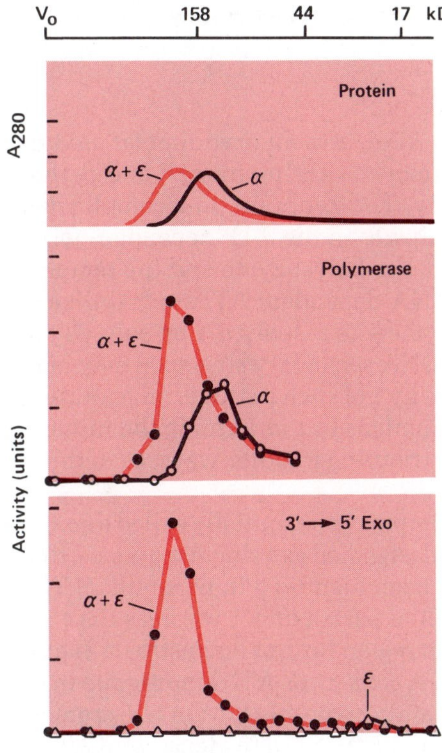

Figure 5-5
Isolation of the $\alpha\cdot\epsilon$ complex. A mixture of the α and ϵ subunits ($\alpha\cdot\epsilon$ complex) was analyzed by high-pressure liquid chromatography (HPLC) on a gel filtration column (V_o = void volume). In separate experiments, free α (open circles) or free ϵ (open triangles) was also filtered. Proteins were detected in the column fractions by UV absorbance (A_{280}); ϵ subunit was not detected, presumably because of its low content of tryptophan and tyrosine. Polymerase and 3′→5′ exonuclease (exo) activities were determined with a template-primer of hook DNA (a synthetic ssDNA with a self-complementary sequence at the 3′ end).

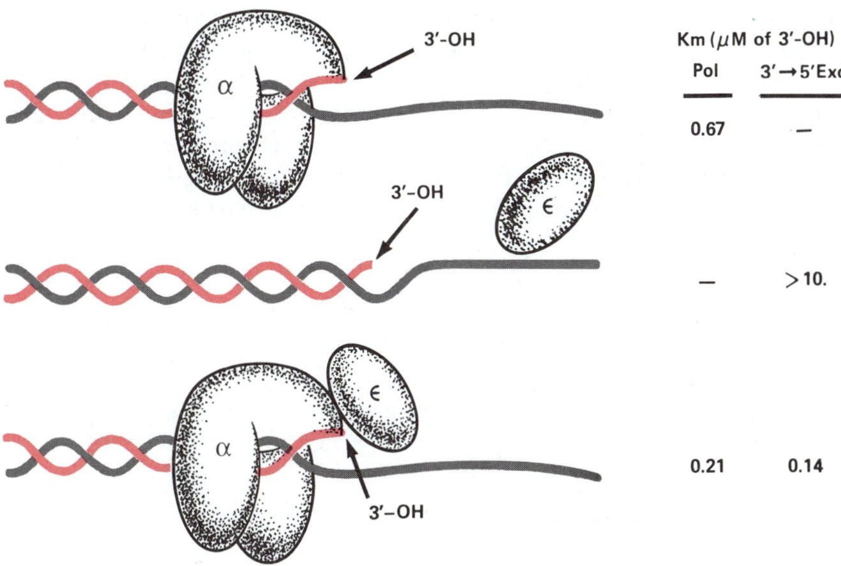

Km (μM of 3′-OH)	
Pol	3′→5′Exo
0.67	—
—	>10.
0.21	0.14

Figure 5-6
Proofreading by the pol III α and ϵ subunits independently and as the $\alpha\cdot\epsilon$ complex.

Subunits γ and τ

The auxiliary subunits γ (47 kDa) and τ (71 kDa), present in equimolar amounts in pol III holoenzyme, are encoded by one open reading frame[39] and perform discrete functions essential for processivity. The γ subunit, in a complex with four other polypeptides (see below), participates in the ATP-dependent formation of the initiation complex consisting of the holoenzyme and the template-primer (Section 6). The τ subunit is an ssDNA-dependent ATPase.[40] Unlike the case with other such ATPases, homopolymers (e.g., poly dA or poly dT) are inactive effectors,[41] indicating that some DNA secondary structure may be essential. In the holoenzyme, the recognition by τ of such a structural feature in a template may contribute to the several functions of τ in forming the initiation complex, increasing processivity, and contributing in other ways, as in the pol III′ subassembly (see below).

Both subunits are encoded by a single complex gene, initially called *dnaX* (for τ) and *dnaZ* (for γ) and later more properly designated as a single sequence, *dnaX* (Fig. 5-7).[42] An homologous sequence has been identified in *B. subtilis*.[43] The γ subunit complements mutations in the initial part of *dnaX*, whereas the τ subunit can also complement mutations in the region further downstream. Thus, γ constitutes the N-terminal two-thirds of τ. A consensus ATP binding site in this N-terminal sequence is probably the domain responsible for the UV-promoted cross-linkage of ATP observed with both γ and τ.[44] The DNA-dependent ATPase activity expressed by τ is lacking in γ, presumably due to the need for a DNA binding domain in the C-terminal region possessed only by τ.

Translational Frameshifting to Generate the γ Subunit.[45] Although protease action in vitro[46] can convert τ to a stable product with the functions and approximate size of γ, this processing mechanism is not responsible for the generation of γ in vivo. Instead, a (−1) frameshift during translation allows the use of a UGA stop codon to terminate translation of the γ polypeptide (Fig. 5-7). The frameshift occurs in the mRNA in a stretch of six adenines preceding a stable hairpin structure which might form an obstructive pseudoknot.[47] This frameshifting is remarkable for its high frequency (about 80% in an overproducing cell). The sequence has a striking structural similarity to the frameshifting signal responsible for expression of the *pol* and *pro* genes in many retroviruses.[48]

---

39. Kodaira M, Biswas SB, Kornberg A (1983) *MGG 192*:80; Mullin DA, Woldringh CL, Henson JM, Walker JR (1983) *MGG 192*:73; Flower AM, McHenry CS (1986) *NAR 14*:8091; Yin K-C, Blinkowa A, Walker JR (1986) *NAR 14*:8091; Maki S, Kornberg A (1988) *JBC 263*:6547.
40. Lee S-H, Walker JR (1987) *PNAS 84*:2713.
41. Tsuchihashi Z, Kornberg A (1989) *JBC 264*:17790.
42. Kodaira M, Biswas SB, Kornberg A (1983) *MGG 192*:80; Mullin DA, Woldringh CL, Henson JM, Walker JR (1983) *MGG 192*:73; Flower AM, McHenry CS (1986) *NAR 14*:8091; Yin K-C, Blinkowa A, Walker JR (1986) *NAR 14*:8091; Maki S, Kornberg A (1988) *JBC 263*:6547.
43. Alonso JC, Shirahige K, Ogasawara N (1990), in press.
44. Biswas SB, Kornberg A (1984) *JBC 259*:7990.
45. Tsuchihashi Z, Kornberg A (1990) *PNAS 87*:2516; Blinkowa AL, Walker JR (1990) *NAR 18*:1725.
46. Garcia GM, Mar PK, Mullin DA, Walker JR, Prather NE (1986) *Cell 45*:453; Lee S-H, Kanda P, Kennedy RC, Walker JR (1987) *NAR 15*:7663.
47. Schimmel P (1989) *Cell 58*:9.
48. Jacks T, Madhami HD, Masiarz FR, Varmus HE (1988) *Cell 55*:447; Weiss RB, Dunn DM, Shuh M, Atkins JF, Gesteland RF (1989) *New Biologist 1*:159.

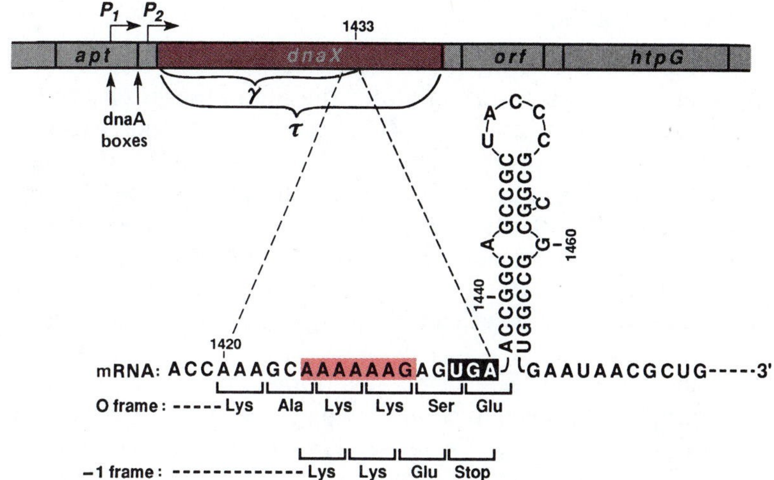

Figure 5-7
Location of the *dnaX* gene, which encodes subunits τ and γ of pol III holoenzyme, and surrounding features of potential regulatory significance. P = transcription promoter.

Pol III′[49]

Among the several subassemblies generated by dissociation of the holoenzyme is pol III′, in which τ is bound to pol III and improves its processivity about sixfold (Table 5-3). The core in pol III′ is stimulated by spermidine and appears to be dimeric ($core_2\tau_2$) and nonspherical. This basic pol III′ unit, with two cores and two τ subunits, is also present in larger oligomeric assemblies.[50] An activity resembling pol III′ can be produced by reconstituting core with τ in the presence of a template-primer preloaded with the β subunit (Section 6). Irregularity in the shape of pol III′ leads to anomalous rates of sedimentation and migration upon gel filtration. The elongated shape of pol III′ may be important in functions proposed for τ in the synthesis of the leading strand at the replication fork (Section 6). A polypeptide slightly larger than τ with a potent DNA-dependent ATPase activity,[51] possibly helicase II, is generally found in holoenzyme preparations but can be removed without affecting polymerase function.

Table 5-3
Processivity of Pol III subassemblies

Subassembly	Processivity	Stimulatory factors
Pol III	10	None
Pol III′	60	Spermidine
Pol III*	200	SSB
Holoenzyme	$> 10^5$	SSB

49. McHenry CS (1982) *JBC 257*:2657.
50. Maki H, Maki S, Kornberg A (1988) *JBC 263*:6570.
51. Meyer RR, Brown CL, Rein DC (1984) *JBC 259*:5093.

The γ Complex (γ, δ, δ', χ, and ψ)[52]

Originally, this complex was recognized as an entity resolved from holoenzyme and needed, together with core and β, to attain the high processivity of the holoenzyme.[53] The complex was found to contain two each of two polypeptides — γ and δ (about 30 kDa) — and, surprisingly, was ten times as active (on a molar basis) as the core and β subunit in reconstituting polymerase activity. The presumed catalytic action of the $\gamma \cdot \delta$ complex was confirmed when an activity needed to reconstitute holoenzyme activity from core and β (and not supplied by τ or γ or both) was purified from cell extracts. This activity, called the γ complex, contained polypeptides in addition to γ and δ and was catalytic in its function.[54]

With a total mass exceeding 200 kDa, calculated from hydrodynamic properties, the purified γ complex is judged to contain two γ subunits and two each of four other polypeptides: δ, δ', χ, and ψ (Table 5-2). For lack of genetic identification of any of these four polypeptides, their role as subunits remains uncertain. However, the capacity of a δ subunit to complex with γ or τ in reconstituting the holoenzyme clearly indicates a function (Section 6).

Pol III*[55]

An early indication of the structural complexity of pol III holoenzyme was its resolution, upon phosphocellulose chromatography, into two components: (1) an adsorbed entity, pol III*,[56] with greatly diminished processivity (Table 5-3), and (2) an unadsorbed component, originally named copol III* and later identified as the β subunit (see below), that could restore pol III* to the holoenzyme level of processivity. Pol III* is a complex of nine polypeptides (Fig. 5-8).

The holoenzyme dissociates readily:[57]

$$\text{pol III}^* \cdot \beta_2 \longleftrightarrow \text{pol III}^* + \beta_2$$

The K_d is very high but is reduced to near 1 nM in the presence of ATP. When the holoenzyme is isolated by the standard procedure, β is present in substoichiometric amounts and generally must be added to elicit the processivity and replication activity of the holoenzyme preparation.

Subunit β[58]

The β subunit is the most important processivity factor for subassemblies of pol III holoenzyme and appears to interact with pol II as well (Section 2). A dimer of 40.6-kDa polypeptides, β dissociates readily from the holoenzyme to yield pol

---

52. McHenry CS, Kornberg A (1977) *JBC 252*:6478; McHenry CS, Oberfelder R, Johanson K, Tomasiewicz H, Franden MA (1987) in *DNA Replication and Recombination* (McMacken R, Kelly TJ, eds.). UCLA, Vol. 47. AR Liss, New York, p. 47; Maki S, Kornberg A (1988) *JBC 263*:6555.
53. McHenry CS, Kornberg A (1977) *JBC 252*:6478.
54. Maki S, Kornberg A (1988) *JBC 263*:6555.
55. Maki H, Maki S, Kornberg A (1988) *JBC 263*:6570.
56. Wickner W, Schekman R, Geider K, Kornberg A (1973) *PNAS 70*:1764; Wickner W, Kornberg A (1974) *JBC 249*:6244.
57. Lasken R, Kornberg A (1988) *JBC 262*:1720.
58. Burgers PMJ, Kornberg A, Sakakibara Y (1981) *PNAS 78*:5391; Johanson KO, Haynes TE, McHenry CS (1986) *JBC 261*:11460.

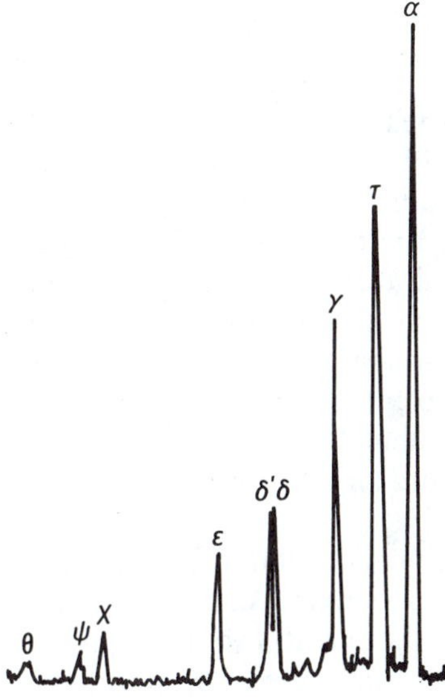

Figure 5-8
Subunits of pol III* as revealed by sodium dodecyl sulfate–polyacrylamide gel electrophoresis (SDS-PAGE).

III*, a complex of vastly reduced processivity. At a cellular abundance of about 300 dimers, the β subunit exceeds the level of pol III* by some ten- to twentyfold. The excess of β serves to maintain pol III* in the holoenzyme form[59] and may contribute processivity to free core molecules or to other subassemblies of holoenzyme by clamping to the template-primer.

Although it is not demonstrably a DNA-binding protein, β can become tightly attached to a template-primer to form a *preinitiation complex* (Section 6) by the ATP-driven action of the γ complex. But even without the γ complex, the β subunit can, at very high concentrations and independent of ATP, form a highly processive complex with just the core and template-primer.[60] As part of firm core·DNA complexes, β is inaccessible to antibodies that neutralize its action in the free form.[61] The β subunit dissociates from the holoenzyme and reassociates in the cycling of the holoenzyme from one template to another.

The *dnaN* gene,[62] which encodes the β subunit, is located between *dnaA*, the structural gene for the chromosome initiator protein, and *recF*, the gene for RecF protein, active in RecA-mediated recombination (Fig. 5-9). The *dnaA* gene is autogenously regulated and is largely independent of *dnaN* expression, which, with *recF*, depends on promoters within the *dnaA*-translated region.[63]

Figure 5-9
Location of the *dnaN* gene, which encodes the β subunit of pol III holoenzyme.

59. Lasken R, Kornberg A (1988) *JBC 262*:1720.
60. LaDuca RJ, Crute JJ, McHenry CS, Bambara RA (1986) *JBC 261*:7550; Kwon-Shin O, Bodner JB, McHenry CS, Bambara RA (1987) *JBC 262*:2121.
61. Johnson KO, McHenry CS (1982) *JBC 257*:12310.
62. Burgers PMJ, Kornberg A, Sakakibara Y (1981) *PNAS 78*:5391.
63. Armengod ME, Garcia-Sogo M, Lambies E (1988) *JBC 263*:12109.

5-5 DNA Polymerase III Holoenzyme of *E. coli:* ATP Activation

The high processivity of pol III holoenzyme, the feature that distinguishes it from its subassemblies (pol III, pol III′, and pol III*; Table 5-3), depends on activation by ATP (or dATP).[64] Other NTPs, including nonhydrolyzable ATP analogs, fail to substitute for ATP. ATP binds tightly to holoenzyme, but only feebly to pol III*. Binding of ATP to the free β subunit is not detectable, but reconstitution of the holoenzyme from pol III* and β subunit restores tight binding. Approximately two or three ATPs are bound per holoenzyme engaged in the formation of the initiation complex with a primed template, but ATP is not needed for subsequent DNA synthesis. The nucleotide analog dAMPPNP (2′-deoxyadenyl-5′-imidodiphosphate) can substitute for dATP in chain elongation but not in holoenzyme activation.

The ATPase activity of the holoenzyme is specifically elicited by template-primer DNA.[65] Reaction of a preformed ATP–holoenzyme complex with a primed template results in hydrolysis of the ATP bound to the holoenzyme, release of ADP and P_i, and formation of the initiation complex. Approximately two ATPs are hydrolyzed for each initiation complex formed, a value in keeping with the number of ATPs bound in the ATP–holoenzyme complex. In the absence of ATP or dATP, holoenzyme behaves like pol III or pol III* (with or without ATP) in its sensitivity to salt and low processivity. The initiation complex formed by ATP-activated holoenzyme is resistant to a level of KCl (150 mM) that completely inhibits unactivated holoenzyme and its subassemblies. Upon completing replication of the available template, holoenzyme can dissociate and form an initiation complex with another primed template, provided ATP for reactivation is available.

The subunits involved in the binding of ATP to the holoenzyme, first identified by UV-induced cross-linking,[66] are most clearly τ and γ. Both, as isolated subunits, were subsequently shown to bind ATP with a K_d of 2 μM. The τ subunit is a DNA-dependent ATPase (Section 4), whereas the ATPase activity of γ is observed only as part of the γ complex, presumably a key element in its action in forming the preinitiation complex (Section 6).

5-6 DNA Polymerase III Holoenzyme of *E. coli:* Structure and Dynamics

Reconstitution of the Holoenzyme[67]

Beyond the association of the β subunit with pol III* to form the holoenzyme (Section 4), reconstitution of a highly processive polymerase has been achieved with an isolated core, τ, γ complex, and β (Fig. 5-10).

The primary event in the reconstitution is an action by the γ complex to convert β from a state that fails to bind DNA to one that binds a template-primer, forming a firm *preinitiation complex*. Hydrolysis of ATP bound to γ (as part of the γ complex) appears to be required. In this relatively slow reaction (1 to 2 min-

64. Burgers PMJ, Kornberg A (1982) *JBC 257*:11468; Burgers PMJ, Kornberg A (1982) *JBC 257*:11474.
65. Burgers PMJ, Kornberg A (1982) *JBC 257*:11468; Burgers PMJ, Kornberg A (1982) *JBC 257*:11474.
66. Biswas SB, Kornberg A (1984) *JBC 259*:7990.
67. Maki S, Kornberg A (1988) *JBC 263*:6561; O'Donnell ME (1987) *JBC 262*:16558.

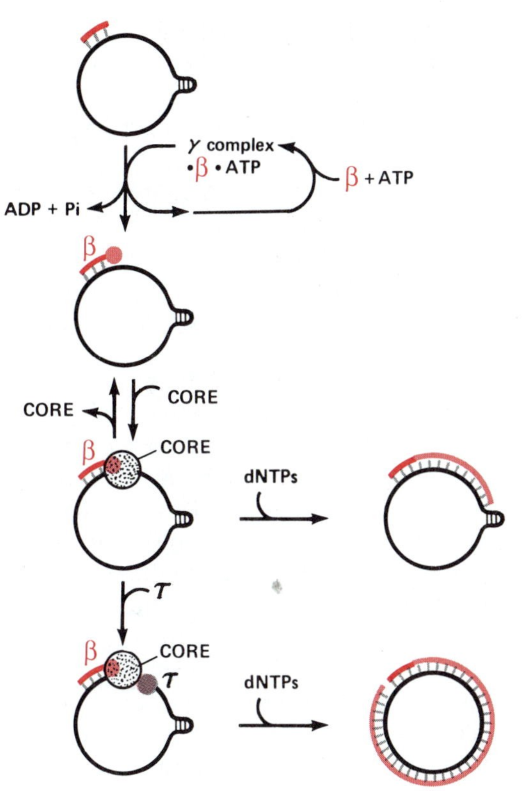

Figure 5-10
Scheme for reconstitution of a processive polymerase.

utes), the γ complex participates catalytically. In place of the γ complex, the isolated δ subunit combined with γ ($\gamma \cdot \delta$) or the isolated δ' subunit combined with τ ($\tau \cdot \delta'$) can each form an effective preinitiation complex with β and template-primer.[68]

In the next event, completed in a few seconds, the preinitiation complex accepts a core, with or without τ, to form an *initiation complex,* in which the γ complex is present in substoichiometric amounts relative to core, β, and template-primer. ATP is not required at this stage. Elongation complexes, when formed by adding all the dNTPs but one, proved to have compositions similar to those of the initiation complexes. With τ available, formation of the initiation complex is more rapid, the complex is more stable, and the processivity attained by the elongation complex is increased considerably to a level comparable to that of the holoenzyme. The enzyme when reconstituted with τ retains this subunit in stoichiometric amounts and pauses neither at secondary structures in the single-stranded template (Fig. 5-10) nor at duplex regions formed by strands annealed to the template.[69]

Pol III Holoenzyme: An Asymmetric Dimer[70]

Whether, in vivo and in vitro, the discontinuous synthesis of the often-called "lagging" strand (involving repeated primary starts) is nicely coordinated with

68. O'Donnell M, Studwell PS (1990) *JBC 265:*1179.
69. Maki S, Kornberg A (1988) *JBC 263:*6561.
70. Maki H, Maki S, Kornberg A (1988) *JBC 263:*6570; McHenry CS (1988) *ARB 57:*519.

the continuous synthesis of the "leading" strand, so as to achieve essentially concurrent synthesis of both strands, is still uncertain. That such coupling of the synthesis of both strands might occur in the progress of replication at a fork seems plausible and is supported by several structural and functional features of the holoenzyme and its subassemblies.

The core subassembly and the holoenzyme have been isolated as dimeric units when care was taken to minimize dissociation by dilution or salts. With an estimated molecular mass near 900 kDa, pol III* should contain two of each of its subunits, and the holoenzyme should have two θ dimers in addition.[71] The dimeric forms of the core, free or complexed with τ (i.e., pol III′), are consistent with such an organization.

As an asymmetric dimer, the holoenzyme can be imagined to have one core complexed with τ and the other with the γ complex (Fig. 5-11). However, the locations of the subunits within the core, the associations between core and both τ and β, and other details of the holoenzyme organization are unknown.

When the ATP γ-thioate analog (ATPγS) is substituted for ATP in the formation of the initiation complex and in cycling from one template-primer to another, the behavior of the holoenzyme suggests two functionally distinct components, a further indication of an asymmetric dimer.[72] Reconstitution of holoenzyme subassemblies yields structures with functions also in accord with this view. The subassembly containing τ, core, and β, with only small amounts of γ complex, is highly processive and clearly competent to sustain the continuous synthesis of the leading strand. By contrast, the subassembly reconstituted without τ dissociates much more readily, consistent with the interrupted synthesis of the lagging strand.

Dynamics of Pol III Holoenzyme[73]

The holoenzyme rapidly and processively replicates a primed ssDNA virtually to completion.[74] On a circular template, the duplex product contains only a nick

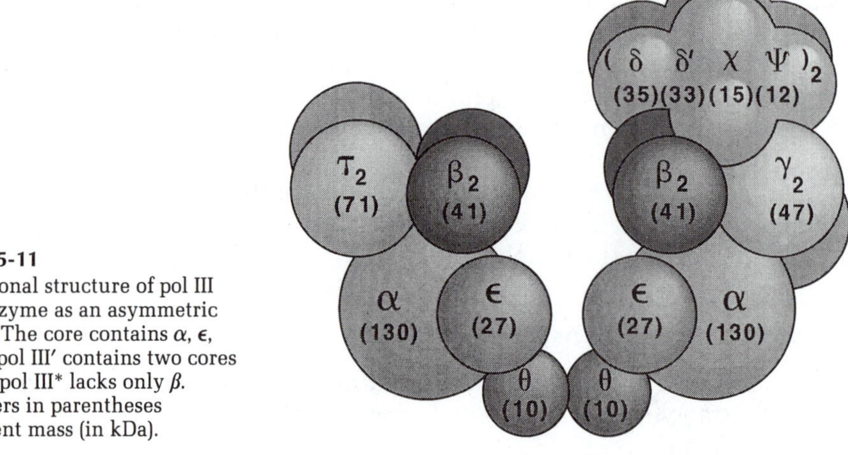

Figure 5-11
Provisional structure of pol III holoenzyme as an asymmetric dimer. The core contains α, ϵ, and θ; pol III′ contains two cores plus τ; pol III* lacks only β. Numbers in parentheses represent mass (in kDa).

71. Maki H, Maki S, Kornberg A (1988) *JBC 263*:6570.
72. Johanson KO, McHenry CS (1984) *JBC 259*:4589.
73. O'Donnell ME, Kornberg A (1985) *JBC 260*:12875; O'Donnell ME, Kornberg A (1985) *JBC 260*:12884.
74. O'Donnell ME, Kornberg A (1985) *JBC 260*:12884.

at the termination of the newly synthesized strand. When primed by DNA, the nicked structures are sealable by ligase, except for about 10% in which replication proceeds and displaces one nucleotide of the 5′ end of the primer; with an RNA primer, replication goes a little farther, displacing one to five nucleotides. On a linear duplex, synthesis generally terminates one base shy of the end of the template. Thus, the holoenzyme utilizes essentially all of the available template and shows no 5′→3′exonuclease activity and little or none of the strand displacement and nick-translation activities characteristic of pol I (Section 4-13).

Studies have been made of the movements of the holoenzyme in replicating a template primed with several synthetic 15-mers annealed at known positions on a single-stranded circular or linear template[75] (Fig. 5-12). After extension of one 15-mer to the full length of the available template, the holoenzyme moves in the elongation direction, traverses a duplex region 15 bp or even 400 bp long, and exploits a 3′-OH end as the next primer. This downstream polarity may be due in part to an inability of the enzyme to move (upstream) along *ssDNA.* These holoenzyme actions, unlike formation of the initiation complex, do not require ATP. The time elapsed between completion of a segment (Fig. 5-12) and initiation at the next primer downstream is rapid (1 second or less). Thus, holoenzyme associates with a new primer on the template without dissociating from the template. By diffusing rapidly on *duplex DNA,* probably in both directions, the holoenzyme forms a complex with the first primer terminus it encounters.

Whereas the dissociation from an initiation complex or a completed synthetic strand is very slow (several minutes), the enzyme can move intermolecularly from one template-primer to another, activated as a preinitiation complex within 10 seconds.[76] Unlike the intramolecular diffusion, this efficient cycling implies the existence of a subassembly that can dissociate from some of its accessory proteins and then reassociate with a fresh preinitiation complex, strongly suggesting subunit movements which have considerable significance for holoenzyme dynamics at the replication fork.

Overview of Pol III Holoenzyme Dynamics

Knowledge of the composition, organization, and functions of the pol III holoenzyme is still grossly incomplete and suffers from an inadequate recognition of interactions with other entities linked in replication. In view of the tight complexes that primases form with eukaryotic α polymerases, some association of

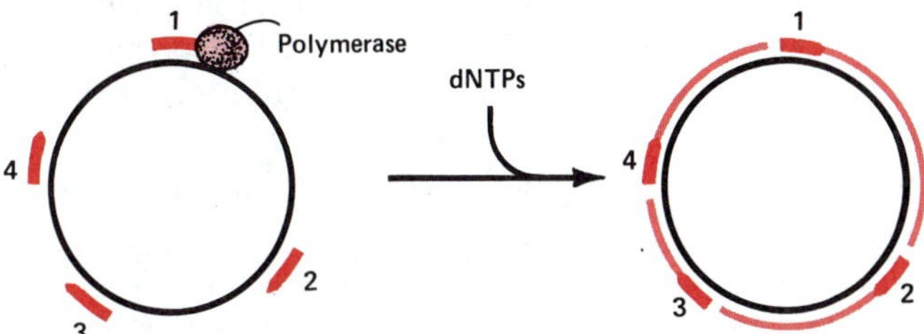

Figure 5-12
Scheme for replication by pol III holoenzyme on a template primed by four synthetic polymers.

75. O'Donnell ME, Kornberg A (1985) *JBC 260*:12875.
76. O'Donnell ME (1987) *JBC 262*:16558.

primase with holoenzyme, particularly with its γ-complex arm, seems likely. The helicase components in the primosome (i.e., dnaB and PriA proteins; Section 8-4) may coordinate their actions with holoenzyme and primosomal operations at the replication fork, in effect behaving as a superassembly of all these components, a "*replisome.*" Mention should also be made of possible interactions between the holoenzyme and SSB coating the template, and the displacement of SSB from the path of synthesis; the association of the SSB of phage T4 (i.e., gp32) with its polymerase may provide an example of such a "treadmilling" interaction.

In addition to serving as the principal replicative enzyme for the host chromosome, the holoenzyme is also responsible for the replication of extrachromosomal elements—certain phages and plasmids—which may reach rather high copy numbers. Further, the holoenzyme is clearly involved in DNA synthesis during recombination and repair of DNA damage caused by UV, methyl methanesulfonate, and hydrogen peroxide;[77] the ϵ subunit has also been implicated in the mutagenic (SOS) response to DNA damage.[78]

Given that virtually all of the ten or so copies of holoenzyme in a rapidly growing cell are deployed at the chromosomal replication forks, the question arises as to how enough holoenzyme molecules are furnished to support the replication of numerous extrachromosomal elements and the synthesis associated with repair and recombination. A reasonable supposition is that the gap-filling activities needed for these nonreplicative processes may be provided by the core, pol III′, and other subassemblies of the holoenzyme. The capacity of pol I to replace a defective α subunit in a *gyrB* mutant[79] raises further possibilities for novel replicative units.

Finally, little is known about the regulation of the biosynthesis and assembly of the holoenzyme subunits.[80] Perhaps, as with ribosome synthesis, there are feedback-regulated pathways that allow the level of holoenzyme to respond to the changing needs for DNA synthesis in the cell.

5-7 Other Bacterial DNA Polymerases

Bacillus subtilis Polymerases[81]

Understanding the developmental programs of *B. subtilis,* in which a complete chromosome is formed for inclusion in a highly resistant, dormant spore and is then replicated upon germination, is a strong reason for a closer study of the *B. subtilis* DNA polymerases.[82] As the genetics and biochemistry of sporulation and germination are further advanced, the operation and regulation of these replicative structures and events will attract increasing interest.

Investigations of these polymerases have yielded several significant findings:

77. Hagensee ME, Bryan SK, Moses RE (1987) *J Bact 169*:4608.
78. Jonczyk P, Fijalkowska I, Ciesla Z (1988) *PNAS 85*:9124; Foster PL, Sullivan AD (1988) *MGG 214*:467.
79. Maki H, Bryan SK, Horiuchi T, Moses RE (1989) *J Bact 171*:3139; Moses RE, personal communication.
80. McHenry CS (1988) *ARB 57*:519.
81. Low RL, Rashbaum SA, Cozzarelli NR (1976) *JBC 251*:1311; Ott RW, Barnes MH, Brown NC, Ganesan AT (1986) *J Bact 165*:951; Ott RW, Goodman LE, Ganesan AT (1987) *MGG 207*:335; Ganesan AT, Townsend T (1988) in *Genetics and Biotechnology of Bacilli* (Ganesan AT, Hoch JA, eds.). Academic Press, New York, Vol. 2, p. 317; Sanjanwala B, Ganesan AT (1989) *PNAS 86*:4421.
82. Okazaki T, Kornberg A (1964) *JBC 239*:259; Falaschi A, Kornberg A (1966) *JBC 241*:1478.

(1) a method for preparing a nucleic acid–free cell extract, (2) the remarkable similarity of these distinctive polymerases to those of the phylogenetically distant *E. coli*, (3) the only example to date of an antimicrobial agent directed against a microbial DNA polymerase, and (4) a striking correlation, in spore germination, between the appearance of one of the DNA polymerases and the onset of replication.

Partition-Phase Removal of Nucleic Acids. This procedure for separating a polymerase from nucleic acids, an essential early step in purification, deserves attention. Whereas the procedure for *E. coli* pol I depends in part on a nucleolytic digestion, the low nuclease content of *B. subtilis* extracts makes such an operation unfeasible. Partition between polyethylene glycol and polydextran phases has been effective in the first step of polymerase purification from *B. subtilis*[83] extracts and has been adopted in other instances as well.

Polymerases. Pol I and pol II, the products of *polA* and *polB*, respectively, are remarkably similar in *B. subtilis* to pol I and pol II of *E. coli* in structural and functional properties. *B. subtilis* pol III, also very much like *E. coli* pol III, is the replicative polymerase in a holoenzyme form, but differs in two ways: (1) it combines within a single polypeptide the α subunit polymerase and the ϵ subunit exonuclease activities of the *E. coli* core, and (2) the enzyme is specifically inhibited by arylazopyrimidines (see below).

The cloned *B. subtilis* α subunit,[84] the product of *polC*, has a mass of 163 kDa as inferred from the genetic sequence. The 3′→5′ proofreading activity forms an integral part of the enzyme, presumably in the N-terminal domain. The amino acid residues in this region show a 26% homology to those of the ϵ subunit of *E. coli*.[85] The increment in size of the *B. subtilis* α subunit compared to that of *E. coli* is nearly equivalent to the mass of the *E. coli* ϵ subunit. Mutants analogous to temperature-sensitive *dnaE* mutants of *E. coli* possess a thermolabile pol III.[86] Not only is the α subunit similar in function to that of *E. coli* pol III, but also the primary sequence shows significant homology.[87]

As in the case of *E. coli* pol II, the analog arabinosyl CTP preferentially inhibits (in vitro) pol II among the *B. subtilis* polymerases. Inasmuch as arabinosyl CTP can be covalently fixed to the primer, it behaves as a chain terminator for further growth unless removed by 3′→5′ exonuclease. The action of this inhibitor may therefore depend on its persistence on the primer ends of chains, as well as on its capacity to be incorporated. The inhibitory effect in vivo appears to be targeted to pol III, as this polymerase shows increased resistance to arabinosyl analogs in mutants selected for resistance against them.

Arylazopyrimidine Inhibition of Pol III. The antimicrobial agent 6-(p-hydroxyphenylazo)-uracil (HP uracil) in its reduced form (Fig. 5-13) inhibits DNA synthesis specifically and most strikingly in certain Gram-positive bacteria.[88]

---

83. Sanjanwala B, Ganesan AT (1989) *PNAS 86*:4421; Sanjanwala B, Ganesan AT, personal communication.
84. Ott RW, Barnes MH, Brown NC, Ganesan AT (1986) *J Bact 165*:951; Ott RW, Goodman LE, Ganesan AT (1987) *MGG 207*:335; Ganesan AT, Townsend T (1988) in *Genetics and Biotechnology of Bacilli* (Ganesan AT, Hoch JA, eds.). Academic Press, New York, Vol. 2 p. 317.
85. Sanjanwala B, Ganesan AT (1989) *PNAS 86*:4421.
86. Cozzarelli NR, Low RL (1973) *BBRC 51*:151.
87. Barnes MH, Brown NC, unpublished.
88. Brown NC (1971) *JMB 59*:1.

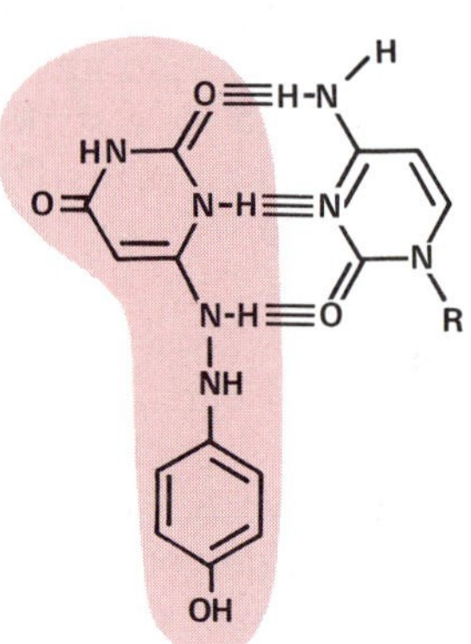

Figure 5-13
Models suggested for inhibition of pol III by arylazopyrimidines through base pairing. [From Mackenzie JM, Neville MM, Wright GE, Brown NC (1973) *PNAS 70*:512; Coulter CL, Cozzarelli NR (1975) *JMB 91*:329]

Inhibition in *B. subtilis* is limited to replicative synthesis. The DNA synthesis associated with repair and the replication of some *B. subtilis* phages is resistant to effects of the drug. Synthesis in permeable cells, considered to be chromosomal replication (Section 15-4), is inhibited by HP uracil. Of the three *B. subtilis* polymerases, only pol III is affected by the drug in very low concentration.[89] None of the DNA polymerases of *E. coli* or the T4 phage–induced enzyme are inhibited by reduced HP uracil. A mutant of *B. subtilis* (*azp12*), isolated for resistance to HP uracil, was also found to have a resistant pol III activity. The mutation is a single base change near the C terminus at nucleotide 3523 (TCA → GCA), producing a change from serine 1175 to alanine.[90]

The analog 6-(*p*-hydroxyphenylazo)-isocytosine (HP isocytosine) in reduced form (Fig. 5-13) also inhibits pol III. The greater sensitivity of certain templates and the antagonism by dATP of HP isocytosine (and by dGTP of HP uracil) suggest a model[91] (Fig. 5-14) in which HP uracil and HP isocytosine form reversible ternary complexes with the DNA template-primer and the polymerase. The drugs bind in the active center of the enzyme in a way that overlaps the dNTP binding site. Binding to the template is by three hydrogen bonds between a C residue in the template and HP uracil or between a T residue in the template and HP isocytosine. The base-pairing scheme in Figure 5-13 is the only one consistent with NMR[92] and x-ray data.[93]

The mechanism of inhibition by the arylazopyrimidines is not limited to blocking the access of a particular dNTP to the active center of the enzyme since the drug–enzyme–DNA complex effectively blocks the incorporation of all four dNTPs. Inhibition of the enzyme may depend on an interaction between the phenol ring of the drug and a hydrophobic region unique to the pol III of *B. subtilis*.

---

89. Mackenzie JM, Neville MM, Wright GE, Brown NC (1973) *PNAS 70*:512; Gass KB, Low RL, Cozzarelli NR (1973) *PNAS 70*:103.
90. Sanjanwala B, Ganesan AT (1989) *PNAS 86*:4421.
91. Coulter CL, Cozzarelli NR (1975) *JMB 91*:329; Cozzarelli NR (1977) *ARB 46*:641; Brown NC, Wright GE (1978) *DNA Synthesis, NATO*, 467.
92. Mackenzie JM, Neville MM, Wright GE, Brown NC (1973) *PNAS 70*:512.
93. Mackenzie JM, Neville MM, Wright GE, Brown NC (1973) *PNAS 70*:512; Gass KB, Low RL, Cozzarelli NR (1973) *PNAS 70*:103.

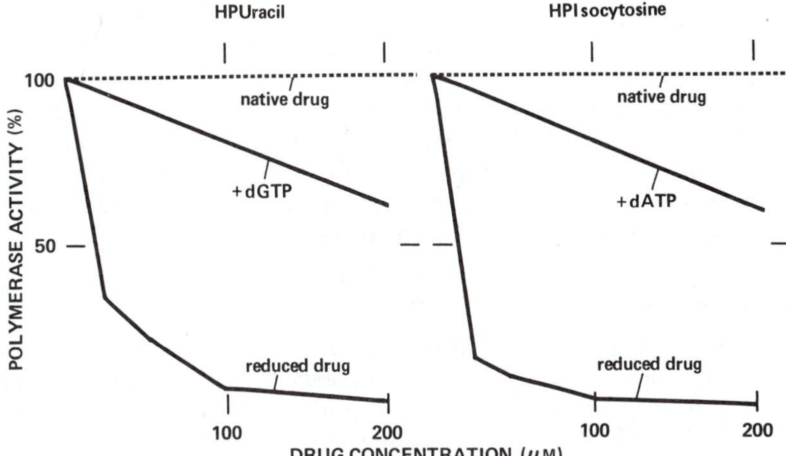

Figure 5-14
Effect of arylazopyrimidines on DNA polymerase activity of a partially purified fraction of *B. subtilis*. HP uracil inhibition is antagonized by dGTP; HP isocytosine, by dATP. [From Mackenzie JM, Neville MM, Wright GE, Brown NC (1973) *PNAS 70*:512]

Spore Germination. RNA synthesis and protein synthesis are under way within a few minutes of initiation of germination, but DNA synthesis is delayed for about an hour. Nevertheless, both pol I and dNTPs are present in the spore at levels near those in the vegetative cell. The total absence of pol III from the dormant spore and its appearance about 30 minutes after the start of germination clearly suggest that its presence is crucial for chromosomal DNA replication.[94]

Salmonella typhimurium,[95] *Streptococcus pneumoniae*,[96] and *Micrococcus luteus*[97] DNA Polymerases

Pol I from *S. typhimurium* (a close cousin of *E. coli*) and from the unrelated Gram-positive species *S. pneumoniae* (pneumococcus) and *M. luteus* (known earlier as *M. lysodeikticus* because the cells are easily lysed) have enzymatic properties much like those of pol I from *E. coli* and *B. subtilis*.

The homogeneous wild-type *S. typhimurium* enzyme, near 100 kDa, is completely inhibited by antibody against *E. coli* pol I, and grossly resembles the *E. coli* enzyme in the parameters examined. A mutant pol I purified from a mutator strain of *S. typhimurium* has poor fidelity (high error frequency) due to a reduced nucleotide specificity during polymerization rather than a defect in proofreading.

Although *S. pneumoniae* is separated from the Gram-negative bacilli *E. coli* and *S. typhimurium* by more than one billion years of evolution, the pol I of *S. pneumoniae* and the pol I of *E. coli* are remarkably similar in size and multiple functions, and share a 40% amino acid identity. The *polA* gene of *S. pneumo-*

94. Ciarrocchi G, Attolini D, Cobianchi F, Riva S, Falaschi A (1977) *J Bact 131*:776.
95. Harwood SJ, Schendel PF, Wells RD (1970) *JBC 245*:5614; Hamilton L, Mahler I, Grossman L (1974) *Biochemistry 13*:1885.
96. Lopez P, Martinez S, Diaz A, Espinosa M, Lacks SA (1989) *JBC 264*:4255.
97. Engler MJ, Bessman MJ (1979) *CSHS 43*:929.

niae has been cloned in multicopy plasmids in both *S. pneumoniae* and *E. coli*, and the polymerase has been overproduced to levels of 5% of cellular protein.

Special attention has been given to de novo synthesis of polymers by the *M. luteus* enzyme, particularly the alternating GC copolymer which is not readily obtained with the *E. coli* enzyme. In view of the presence of pol II and III in both *E. coli* and *B. subtilis*, it seems likely that these enzymes are present in other bacteria as well. In fact, the pol III α subunits of *S. typhimurium* and *E. coli* are identical in size and in 97% of their amino acid residues.[98]

Thermophilic and Halophilic DNA Polymerases

The archaebacterial kingdom,[99] derived from an ancestral prokaryote that relates the archaebacteria to the eubacteria and eukaryotes, includes most of the species that grow in unusual environments: *thermoacidophiles* in hot sulfur springs at 80° and pH 2, and *halophiles* in high salt concentrations. Their genomes, unremarkable in size and composition, and replicated at these extremes of temperature, pH, and salt, imply remarkable structural adaptations in their polymerases that enable them to function under these extraordinary conditions. Studies of these polymerases should prove to be of general interest and have already, in the case of the thermophiles, provided highly useful reagents.

Thermophilic DNA Polymerases. These polymerases are astonishing in their ability to exploit a template-primer for replication at temperatures well above the T_m of the duplex DNA. The archaebacterial thermophiles may also reveal important links in the evolutionary development of polymerases. Reliance on thermophilic polymerases in the burgeoning use of the polymerase chain reaction (PCR; Section 4-17) demands a better understanding of their capacities and limitations.

The polymerase purified from the archaebacterium *Sulfolobus acidocaldarius* (growing in hot sulfur springs at up to 85° at pH 2) is a single polypeptide of 100 kDa.[100] The enzyme, while optimal in activity at 70° and stable at 80°, can synthesize chains 150 ntd long even at 100°. In defiance of all experience with nonthermophilic polymerases, the isolated *S. acidocaldarius* enzyme can use an M13 circular template primed with a 20-mer (at 1 : 1 stoichiometry!) and extend this primer by 100 ntd or more at 100°! The evident capacity of the enzyme to clamp the primer so efficiently to the template at a temperature 40° above the T_m and proceed to extend it is most impressive. Because the absence of any proofreading exonuclease activity makes a high error rate likely, an auxiliary enzyme with such an activity can be anticipated to account for the viability of the species.

From *Thermus aquaticus*, a eubacterium, the gene for the DNA polymerase has been cloned and overexpressed, and the 94-kDa enzyme, stable at 95° (half-life, 45 minutes), has been isolated in good yield.[101] Truncated 63-kDa forms of the enzyme are also thermostable and active. The *T. aquaticus* (Taq) enzyme, lacking the proofreading 3′→5′ exonuclease activity present in most prokaryotic polymerases, might be more error-prone than *E. coli* pol I,[102] and this

98. Lancy ED, Lifsics MR, Munson P, Maurer R (1989) *J Bact 171*:5581.
99. Woese CR (1987) *Microbiol Rev 51*:221; Woese CR, Kandler O, Wheelis ML (1990) *PNAS 87*:4576.
100. Elie C, Salhi S, Rossignol J-M, Forterre P, De Recondo A-M (1988) *BBA 951*:261; Salhi S, Elie C, Forterre P, De Recondo A-M, Rossignol J-M (1989) *JMB 209*:635.
101. Lawyer FC, Stoffel S, Saiki RK, Myambo K, Drummond R, Gelfand DH (1989) *JBC 264*:6427.
102. Tindall KR, Kunkel TA (1988) *Biochemistry 27*:6008; Dunning AM, Talmud P, Humphries SE (1988) *NAR 16*:10393; Eckert KA, Kunkel TA (1990) *NAR 18*:3739.

may prove to be a matter of concern when used under PCR conditions (Section 4-17). The Taq enzyme does possess a 5′→3′ exonuclease activity that is strongly coupled to chain growth. The rate of synthesis, about 100 ntd per second per enzyme molecule at 70°, is what might be anticipated from *E. coli* pol I were it stable at this elevated temperature.

While clearly different from the polymerases of *E. coli*, the Taq enzyme shows considerable homology to *E. coli* pol I in the N-terminal domain responsible for the 5′→3′ exonuclease activity and to pol I and phage T7 polymerase in the cleft regions (Section 4-5) responsible for polymerization, but shows no homology to the 3′→5′ exonuclease domain of pol I. The absence of cysteine and the increased ratios of arginine to lysine, glutamate to aspartate, and serine to threonine are characteristic of the composition of other thermophilic enzymes but do not clarify their extraordinary heat resistance.

Halophilic DNA Polymerases. The archaebacterium *Halobacterium halobium* has several DNA polymerase activities, two of which resemble the eukaryotic polymerases.[103] One multisubunit, α-like polymerase, is sensitive to aphidicolin and N-ethylmaleimide, is resistant to dideoxynucleotides, and has associated primase and 3′→5′ exonuclease activities. Another small, β-like polymerase is resistant to aphidicolin and N-ethylmaleimide, is sensitive to dideoxynucleotides, and appears to possess both 3′→5′ and 5′→3′ exonuclease activities.

Reverse Transcriptases

RNA-directed DNA polymerases, for many years identified only in retroviruses and in eukaryotic cells (Section 6-9), have also been discovered in myxobacteria[104] and in certain strains of *E. coli*.[105] The reverse transcriptases, identified largely by homologies to the polymerase, spacer, and RNase H domains encoded in the genetic sequences of retroviral reverse transcriptases, are remarkable for the primer and template employed and the DNA produced. An internal guanylate residue of an oligoribonucleotide is the primer. When extended by dNTPs from the 2′-OH of the guanylate, using the RNA (of like length) as template, the product is an ssDNA of about 100 ntd emerging as a branch from the RNA. The function in the cell of the several hundred copies of this highly conserved, unique, multicopy RNA–DNA single strand (ms DNA) is still unknown.

5-8 Phage T4 DNA Polymerase[106]

How does a T-even phage, within a few minutes of infecting *E. coli*, replicate its own unique DNA at a rate ten times faster than the synthesis rate of host DNA in

103. Sorokine I, Ben-Mahrez K, Nakayama M, Kohiyama M (1990) *EMBO J*, in press.
104. Lampson BC, Inouye M, Inouye S (1989) *Cell 56*:701; Inouye S, Hsu M-Y, Eagle S, Inouye M (1989) *Cell 56*:709; Inouye S, Herzer PJ, Inouye M (1990) *PNAS 87*:942.
105. Lim D, Maas WK (1989) *Cell 56*:891; Lampson BC, Sun J, Hsu M-Y, Vallejo-Ramirez J, Inouye S, Inouye M (1989) *Science 243*:1033; Lampson BC, Viswanathan M, Inouye M, Inouye S (1990) *JBC 265*:8490; Eckert KA, Kunkel TA (1990) *NAR 18*:3739.
106. Cha T-A, Alberts BM (1988) in *Cancer Cells 6: Eukaryotic DNA Replication* (Kelly TJ, Stillman B, eds.). CHSL, p. 1; Mace DC, Alberts BM (1984) *JMB 177*:313; Selick HE, Barry J, Cha T-A, Munn M, Nakanishi M, Wong ML, Alberts BM (1987) in *DNA Replication and Recombination* (McMacken R, Kelly TJ, eds.). UCLA, Vol. 47. A R Liss, New York, p. 183; Richardson RW, Ellis RL, Nossal NG (1990) in *Molecular Mechanisms in DNA Replication and Recombination* (Richardson CC, Lehman IR, eds.). UCLA, Vol. 127. A R Liss, New York, p. 247.

the uninfected cell? This question motivated the earliest work on the enzymatic synthesis of DNA.[107] Yet the discovery of DNA polymerase was made in uninfected cells, because at that time no such activity was detectable in extracts of infected cells.

Demonstrating an increased rate of DNA synthesis in extracts of phage-infected cells required the use of hydroxymethylated dCTP as a substrate, because the phage-encoded dCTPase in the crude extracts rapidly destroys dCTP (Section 2-13). The increased rate of DNA synthesis was explained by the appearance of a novel polymerase encoded by the phage and not by an alteration in the level or efficiency of the host enzyme.[108] An unusual requirement for ssDNA as template-primer and a resistance to antibody directed against the host enzyme indicated that the heightened polymerase activity after infection was new and different. Some years later, the T4 phage–encoded enzyme was isolated in pure form[109] and identified as the product of a well-analyzed genetic locus (gene *43*).[110] Thus, the T4 enzyme (gp43) was the first DNA polymerase shown to be essential for growth, and presumably for replication in vivo.

T4-encoded polymerase, rather than the T2-encoded enzyme, was selected for purification because infection with an available T4 gene *44* mutant did not shut off synthesis of polymerase and other phage proteins produced during the early phase of a wild-type infection. Intensive genetic investigation[111] of T4, the availability of a large number of carefully mapped mutants, the cloning of these genes, and the overproduction of their products have all been persuasive in maintaining the focus on T4 for biochemical studies of DNA synthesis and the relation of polymerase structure to function.[112] Furthermore, there are no indications that the DNA polymerases encoded by T2 or T6 differ significantly from that of T4.

As with other T4 replication proteins, the regulation of gene *43*, encoding DNA polymerase, is at the translational level. Whereas many of the genes are controlled by *regA* (the translational repressor), gene *43* (like gene *32*, which encodes T4 SSB; Section 10-3), is also autogenously regulated by binding of the DNA polymerase to the 5′ end of its mRNA.[113]

T4 DNA polymerase (gp43) is a single polypeptide similar in size to pol I of *E. coli*, but differing in having at least one essential sulfhydryl group among its 15 cysteine residues. The sequence is surprisingly homologous to a family of polymerases that includes those of certain animal viruses, eukaryotic α polymerases, and some phage polymerases, but does not include *E. coli* pol I or pol III (Section 11).

The striking difference between T4 polymerase and pol I is in the template-primer requirement. Unlike pol I, T4 polymerase, unaided by accessory proteins, cannot use a nicked duplex but requires a primed single-stranded template. Such templates may have short or long gaps. In its ability to replicate DNA with extensive single-stranded regions, the T4 enzyme differs from *E. coli* pol II and pol III (the core), which operate only on short gaps. The T4 enzyme displays the same requirements as pol I for a primer terminus, template, and dNTPs in

---

107. Kornberg A, Lehman IR, Simms ES (1956) *Fed Proc 15*:950.
108. Aposhian HV, Kornberg A (1962) *JBC 237*:519.
109. Goulian M, Lucas ZJ, Kornberg A (1968) *JBC 243*:627.
110. deWaard A, Paul AV, Lehman IR (1965) *PNAS 54*:1241.
111. Edgar RS (1969) *Harvey Lectures 63*:263.
112. Reha-Krantz LJ (1988) *JMB 202*:711, Reha-Krantz LJ (1990) *Genetics 124*:213; Reha-Krantz LJ, Liesner S, Parmaksizoglu S, Stocki S (1989) *J Virol 63*:4762.
113. Andrake M, Guild N, Hsu T, Gold L, Tuerk C, Karam J (1988) *PNAS 85*:7942.

the assembly of a chain in the 5′→3′ direction. Its turnover number, about 400 ntd per second under optimal conditions, is near the rate of replication fork movement.

Accessory Proteins[114]

T4 polymerase functions as a holoenzyme composed of a gp43 polymerase core and three distinctive accessory phage-encoded proteins (gp44, gp62, and gp45) akin to the host pol III holoenzyme subunits in facilitating replication by the core polymerase. The four-protein assembly replicates a single-stranded template at an optimal rate and processivity and, together with gp41 helicase, can efficiently utilize an otherwise inaccessible nicked duplex DNA. The functions of each of these accessory proteins have been clarified by model systems.[115]

The rate of replicating a primed circular DNA by gp43 is stimulated fourfold by the accessory proteins gp44, 62, and 45 or sixfold by gp32; with the combined action of all five proteins (gp43, 44, 62, 45, and 32), the rate is more than 40-fold that of gp43 alone. The accessory proteins clamp the polymerase to the template, thereby enabling the enzyme to traverse and replicate the strongest hairpin helices in a highly processive manner (>20,000 nucleotides synthesized without dissociation). While ATP hydrolysis is needed to form the holoenzyme-like complex of polymerase and accessory proteins, the usually nonhydrolyzable analog ATPγS is thereafter able to sustain elongation.

The polymerase complex can start replication at a nick in an otherwise intact duplex DNA. The synthesis rate of the five-protein system is only 20% of that estimated for fork movement in vivo. Increasing the gp32 concentration to a great excess over that needed to bind the displaced single strand increases fork movement by promoting melting of the helix. The rate of fork movement is increased threefold by the addition of gp41 helicase without the large excess of gp32 but is not affected by the presence or priming action of the primase gp61 alone (Section 17-6).

3′→5′ Exonuclease Activity[116]

T4 polymerase has an extraordinarily active 3′→5′ exonuclease, similar to that of pol I (Section 4-9) but 200 times more active. The base-pairing fidelity of the T4 polymerase multienzyme complex is high; mutator and antimutator features reside in the nuclease,[117] as does the error-prone mutagenic repair observed after UV damage.[118]

The 3′→5′ exonuclease degrades ssDNA to the terminal dinucleotide and has been a useful reagent (as has exonuclease I) for identifying the dinucleotide at the 5′ end of a chain or the sequence at the 3′-OH ends of a duplex.[119] Nuclease action at the end of a frayed duplex, coupled to polymerization, leads to nucleotide turnover at the terminus, allowing it to be identified or labeled.

---

114. Jarvis TC, Dolejsi MK, Hockensmith JW, Kubasek WL, Paul LS, Reddy MK, von Hippel PH (1990) in *Molecular Mechanisms in DNA Replication and Recombination* (Richardson CC, Lehman IR, eds.). UCLA, Vol. 127. A R Liss, New York, p. 261.
115. Selick HE, Barry J, Cha T-A, Munn M, Nakanishi M, Wong ML, Alberts BM (1987) in *DNA Replication and Recombination* (McMacken R, Kelly TJ, eds.). UCLA, Vol. 47. A R Liss, New York, p. 183.
116. Bessman MJ, Muzyczka N, Goodman MF, Schnaar RL (1974) *JMB 88*:409; Bedinger P, Alberts BM (1983) *JBC 258*:9649; Sinha NK (1987) *PNAS 84*:915.
117. Bessman MJ, Muzyczka N, Goodman MF, Schnaar RL (1974) *JMB 88*:409.
118. Drake JW (1988) *MGG 214*:547.
119. Englund PT, Price SS, Weigel PH (1974) *Meth Enz 29*:273.

5′→3′ Exonuclease[120]

T4 polymerase lacks the overt 5′→3′ exonuclease activity observed in pol I. However, protein sequences near mutation sites in the N-terminal region are similar to sequences that appear in the 5′→3′ exonuclease domain of pol I and the RNase H of *E. coli*. Thus, the apparently dissimilar pol I and T4 polymerase (Section 11) seem to share the same structural–functional organization: an N-terminal domain that may encode a 5′→3′ exonuclease (or RNase H), an adjacent polymerase domain, and a C-terminal 3′→5′ exonuclease domain.

5-9 Phage T7 DNA Polymerase[121]

The remarkably streamlined polymerase employed by phage T7 for its replication is composed of T7 gp5 (79.7 kDa) in a tight 1:1 complex with an accessory subunit, the 11.7-kDa thioredoxin of *E. coli* (Section 2-7). Such appropriating of a host protein to convert a phage protein into an effective enzyme has been observed in another instance, the association of the replicase of an RNA phage (e.g., f2, Qβ, MS2) with the *E. coli* elongation factor, Tu, which normally functions in protein synthesis. The T7 polymerase, functionally comparable and structurally homologous to the large fragment of *E. coli* pol I (Section 11), gains the high processivity of a holoenzyme (e.g., *E. coli* pol III or phage T4 polymerase) by this union with thioredoxin.

Gene *5* Protein (Gp5)

The initial difficulty in isolating gp5 from the T7 polymerase holoenzyme was circumvented by the use of the *E. coli* mutant *trxA* (formerly *tsnC*), which is unable to support T7 growth because of a defective thioredoxin. Since then, gp5 has been cloned and isolated from overproducing cells. Whereas gp5 resembles the holoenzyme in 3′→5′ exonuclease activity on ssDNA, it has only about 2% of the polymerase and 3′→5′ exonuclease activities on dsDNA. The addition of thioredoxin to isolated gp5 in vitro restores all functions to the level of the native complex.

Although the polymerase lacks the 5′→3′ exonuclease activity of *E. coli* pol I, such an exonuclease is encoded by gene *6*, just downstream of gene *5*, and is essential for phage replication.[122] The combined masses of gp5 and gp6 approximate that of pol I and may cooperate in the excision of primers during replication. Also, the helicase action of gp4 (see below) is essential for T4 polymerase action at the replication fork.

Thioredoxin–Gp5 Interactions

As an accessory protein to gp5, thioredoxin (1) stimulates both the polymerase and dsDNA exonuclease activities by extending the lifetime of a gp5 complex

120. Reha-Krantz L (1990) *Genetics 124*:213.
121. Beauchamp BB, Huber HE, Ikeda RA, Myers JA, Nakai H, Rabkin SD, Tabor S, Richardson CC (1983) *Cell 33*:315; Richardson CC (1987) in *DNA Replication and Recombination* (McMacken R, Kelly TJ, eds.). UCLA, Vol. 47. A R Liss, New York, p. 151.
122. Frenkel GD, Richardson CC (1971) *JBC 246*:4738; Frenkel GD, Richardson CC (1971) *JBC 246*:4848; Kerr C, Sadowski PD (1972) *JBC 247*:305; Shinozaki K, Okazaki T (1978) *NAR 5*:4245.

with template-primer from less than a second to about 5 minutes,[123] and (2) increases the processivity of polymerization some 1000-fold.[124] Just how thioredoxin makes gp5 a processive polymerase is not clear, but the sulfhydryl function that is essential for the redox coenzyme role of thioredoxin in ribonucleotide reductase (Section 2-7) is not required.[125] Whereas alkylation of either of the two active-site cysteines (cys-gly-pro-cys) in thioredoxin destroys its crucial contribution to processivity activity, replacement of both cysteines by serines (ser-gly-pro-ser) does not. Evidently, the occupation of these sites by either cysteine or serine residues (but not by an alkylated cysteine) generates a conformation acceptable for the interaction with gp5, although the serine substitutions do severely reduce polymerase activity.

Holoenzyme Interactions with Gp4 and Gp6[126]

The product of gene *4* possesses both helicase and primase activities. The helicase activity processively opens the duplex at a replication fork, and the primase activity distributively generates tetranucleotide primers; the extension of these primers by the polymerase depends on sustained helicase activity. Coordination of gp4 and polymerase actions is based on both a firm interaction between these proteins and the affinity of the polymerase for ssDNA, allowing the formation of a stable complex which includes both gp4 and the DNA. The binding to DNA of both gp4 and the polymerase in the complex is important for efficient operation in the synthesis of DNA at a replication fork. An interaction with gp6 would provide the polymerase with the 5′→3′ exonuclease activity that it lacks. Gene *6*, which maps next to gene *5*, has strong homologies to the N-terminal 5′→3′ exonuclease domains of pol I of *E. coli* and *S. pneumoniae*.[127]

Modifications of T7 Polymerase

Two Forms of the Polymerase. The puzzling difference[128] between the T7 polymerase form isolated in the absence of EDTA and the form prepared in its presence has been resolved. Without EDTA, the exonuclease activities are destroyed in the course of the purification procedure, by a selective oxidation (requiring iron, molecular O_2, and a reducing agent):[129] reactive oxygen species modify amino acid residues in the vicinity of the iron bound to essential exonuclease sites. The chemically modified enzyme has become a desirable reagent (Sequenase) for DNA sequencing by the dideoxy-NTP (ddNTP), chain-termination method.[130] Unlike the large fragment of *E. coli* pol I, the altered T7 enzyme does not discriminate against ddNTPs and tolerates nucleotide analogs (e.g., dITP in place of dGTP) that improve the electrophoretic resolution of bands in gels. Mutants of gp5 that lack the 3′→5′ exonuclease activity but retain full polymerase activity are especially useful for mechanistic and analytic studies.[131]

---

123. Huber HE, Tabor S, Richardson CC (1987) *JBC 262*:16224.
124. Tabor S, Huber HE, Richardson CC (1987) *JBC 262*:16212.
125. Huber HE, Russel M, Model P, Richardson CC (1986) *JBC 261*:15006.
126. Nakai H, Richardson CC (1986) *JBC 261*:15208; Nakai H, Richardson CC (1986) *JBC 261*:15217.
127. Lopez P, Martinez S, Diaz A, Espinosa M, Lacks SA (1989) *JBC 264*:4255.
128. Engler MJ, Lechner RL, Richardson CC (1983) *JBC 258*:11165.
129. Tabor S, Richardson CC (1987) *JBC 262*:15330.
130. Tabor S, Richardson CC (1987) *PNAS 84*:4767.
131. Wong I, Patel SS, Johnson KA (1991) *Biochemistry 30*:526.

Mn^{2+} Substitution for Mg^{2+}. This substitution[132] maintains the rate of T7 polymerase activity, provided that a weak chelator (e.g., citrate or isocitrate) buffers the inhibitory effect of Mn^{2+} concentrations beyond 1 mM. The remarkable feature of Mn^{2+} in replacing Mg^{2+} is that it completely removes the discrimination between dNTPs and ddNTPs and substantially reduces the selection against rNTPs and other analogs. With Mg^{2+}, the incorporation of dideoxynucleotides varies greatly as a function of the sequence context of the template; with Mn^{2+}, dideoxynucleotide incorporation becomes virtually uniform. How metals interact with the furanose ring of the mono- or dideoxynucleotide to influence its behavior as a substrate needs to be clarified.

5-10 DNA Polymerases of Other Phages: ϕ29, M2, PRD1, N4, T5

Polymerases of Phages Dependent on Protein-Primed Initiation

With a linear duplex template, the problems of replacing the role of a primer RNA at the initiating 5′ end of a DNA strand are avoided by using a protein primer (Section 17-8). Phages with linear duplex genomes that use this mechanism have polymerases with dual nucleotidyl transferase functions: (1) attachment of the initial (5′) deoxynucleotide to a certain residue of the terminal (primer) protein (the serine residue in ϕ29; the tyrosine residue in PRD1), and (2) as with other polymerases, elongation of the chain from that first residue along the length of the genome. The most thoroughly studied of these polymerases is that of phage ϕ29 of *B. subtilis*.

Phage ϕ29 Polymerase.[133] The gene for the ϕ29 DNA polymerase (gene *2*) has been cloned and the overproduced protein (gp2) has been isolated.[134] The 68-kDa enzyme interacts with the terminal protein (gp3)[135] and catalyzes the nucleotidyl transfer from dATP to serine 232 at each end of the duplex genome. In this initiation complex, the 5′ dAMP covalently linked to gp3 matches the 3′-terminal T residue at the end of each template chain. Formation of the initiation complex is strongly stimulated by 10 mM ammonium sulfate and by gp6 (a dimer of 12-kDa subunits), an effect accompanied by a reduction in K_m for dATP from 6 to 0.4 μM.[136] The 3′-OH group of dAMP in the initiation complex is used as a primer by the polymerase for replication of the remainder of the template strands.

In its unique binding of duplex DNA at the terminal regions of the genome, gp6 may also stimulate the elongation reaction.[137] Elongation by the enzyme's strand displacement activity and high processivity account for the efficient, symmetrical replication of the entire phage genome.[138] The polymerase also possesses a 3′→5′ exonuclease activity on ssDNA and a 5′→3′ exonuclease.

132. Tabor S, Richardson CC (1989) *PNAS 86*:4076.
133. Salas M (1988) *Curr Top Microbiol Immunol 136*:71; Salas M, Bernad A, Zaballos A, Martin G, Otero MJ, Garmendia C, Serrano M, Blasco MA, Lázaro JM, Parés E, Hermoso JM, Blanco L (1990) in *Molecular Mechanisms in DNA Replication and Recombination* (Richardson CC, Lehman IR, eds.). UCLA, Vol. 127. AR Liss, New York, p. 277.
134. Blanco L, Salas M (1984) *PNAS 81*:5325; Blanco L, Garcia JA, Salas M (1984) *Gene 29*:33.
135. Zaballos A, Salas M (1989) *NAR 17*:10353.
136. Blanco L, Salas M (1985) *PNAS 82*:6404.
137. Blanco L, Gutierrez J, Lázaro JM, Bernad A, Salas M (1986) *NAR 14*:4923.
138. Blanco L, Bernad A, Lázaro JM, Martin G, Garmendia C, Salas M (1989) *JBC 264*:8935.

Inasmuch as the dATP concentration for initiation complex formation can be as low as 0.1 μM, and for elongation must be more than ten times higher, the polymerase may have separate domains for initiation and for elongation, an inference supported by the different actions of inhibitors:[139] aphidicolin and phosphonacetic acid act mainly on elongation, whereas nucleotide analogs inhibit both initiation and elongation. The 3′→5′ exonuclease activity in a specific domain of the enzyme is affected by all the inhibitors. Yet there is clear evidence that the initiation and elongation activities share the dNTP-binding site that includes the highly conserved YGDTDS motif (Section 11) in the C-terminal domain.[140]

Conserved domains throughout the protein are homologous with the polymerase and exonuclease domains of a wide assortment of prokaryotic and eukaryotic polymerases (Section 11). Some of these homologous regions are probably the sites exploited by the above inhibitors, which act similarly on many polymerases.

Phage M2[141] and PRD1[142] Polymerases. Phage M2 polymerase has an 82% overall homology to the ϕ29 polymerase and resembles it very closely in structural and functional features.[143] PRD1 is a small, lipid-containing phage that infects a wide variety of Gram-negative bacteria. It also uses a terminal protein as a primer for DNA synthesis, and its 63.3-kDa polymerase is possibly the very smallest replicative polymerase. The initial phosphodiester bond made by the polymerase is between a terminal-protein tyrosine residue and dGMP. Chain elongation and other features resemble those described for ϕ29 polymerase.

Phage N4 Polymerase[144]

The polymerase encoded by N4, a 72-kb linear coliphage, is a polypeptide of 87 kDa with a strong 3′→5′ exonuclease activity. Other phage and host proteins are relied upon for strand-displacing and 5′→3′ exonuclease functions.

Phage T5 Polymerase[145]

The T5 polymerase, a single polypeptide of 94.3 kDa (deduced from its DNA sequence),[146] has several remarkable properties: (1) high processivity for both polymerization and 3′→5′ exonuclease action;[147] (2) use of a nicked template and strand displacement without the aid of helicases or other proteins (a capacity also shown by *E. coli* pol I and its large fragment, by ϕ29 polymerase, and by T7 polymerase modified chemically or genetically); and (3) participation in a complex which contains the host RNA polymerase, transcriptional factors, a DNA-binding protein, and gpD15 nuclease.[148]

---

139. Blanco L, Salas M (1986) *Virology 153*:179.
140. Bernad A, Lázaro J, Salas M, Blanco L (1990) *PNAS 87*:4610.
141. Matsumoto K, Takano H, Kim CI, Hirokawa H (1989) *Gene 84*:247.
142. Jung G, Leavitt MC, Hsieh J-C, Ito J (1987) *PNAS 84*:8287.
143. Matsumoto K, Takano H, Kim CI, Hirokawa H (1989) *Gene 84*:247.
144. Lindberg GK, Rist JK, Kunkel TA, Sugino A, Rothman-Denes LB (1988) *JBC 263*:11319.
145. Leavitt MC, Ito J (1989) *PNAS 86*:4465.
146. Leavitt MC, Ito J (1989) *PNAS 86*:4465; Chatterjee DK, Fukimura RK, Campbell JH, Gerard GF (1990) *Gene*, in press.
147. Das SK, Fujimura RK (1980) *JBC 255*:7149; Fujimura RK, Tavtigian SV, Choy TL, Roop BC (1985) *J Virol 53*:495; Schneider SS, Roop BC, Fujimura RK (1985) *J Virol 56*:245.
148. Ficht TA, Moyer RW (1980) *JBC 255*:7040.

5-11 Motifs and Homologies among the DNA Polymerases

The amino acid sequences of the many DNA polymerases, including the RNA-directed reverse transcriptases (Section 6-9) and those of the DNA- and RNA-directed RNA polymerases, reveal impressive structural similarities that point to functional and evolutionary relationships. Especially striking are major familial homologies: (1) between *E. coli* pol I and the DNA polymerases of *S. pneumoniae*, phages T7 and T5, and *Thermus aquaticus*; (2) between other prokaryotic (e.g., T4) and most α-like eukaryotic DNA polymerases in their polymerase domains; (3) between a wide assortment of polymerases in their $3'\rightarrow5'$ exonuclease domains; and (4) between eukaryotic polymerase β and terminal deoxynucleotidyl transferase. Further, (5) among the various kinds of polymerases, a sequence motif may apply to them all.

(1) The familial relationships of *E. coli* pol I and some other bacterial polymerases (e.g., that of *S. pneumoniae*) suggest an accretion of at least six separate genetic regions.[149] In the comparison of *E. coli* pol I and T7 polymerase (gp5), there is a strong similarity between gp5 and the nine polypeptide segments that form the large DNA-binding cleft of the C-terminal domain of pol I.[150] In matching gp5 to these segments, 38% to 58% of the residues are identical, with only a single amino acid deletion in two of the T7 peptides. Furthermore, the dissimilarities in these T7 peptides can be reconciled to fit the structural and functional features of the pol I active site. As for phage T5 polymerase, its high degree of homology with pol I is largely in the $3'\rightarrow5'$ exonuclease domain.[151] The DNA polymerase from *T. aquaticus*, a eubacterial thermophile, shows similar homologies.[152]

Short segments of T7 RNA polymerase also bear a weak homology to the pol I and T7 DNA polymerases; by contrast, T7 RNA polymerase showed no clear homologies with some other prokaryotic or the eukaryotic DNA polymerases. The homologous α subunits of pol III of *E. coli*, *S. typhimurium*, and *B. subtilis* seem to form a class of their own.

(2) A second family of homologies is based on six conserved regions with a similar linear spatial arrangement in the C-terminal half of DNA polymerases, encoded in rather diverse genomes (Fig. 5-15).[153] These genomes include those of phages T4, ϕ29, and PRD1; adenovirus, vaccinia, and assorted herpes viruses; yeast DNA killer plasmids and maize mitochondrial elements; and the whole gamut of α-like eukaryotic polymerases. One or more of the six conserved regions embraces the active site of this class of polymerases, as judged by their common sensitivity to the drugs aphidicolin and phosphonacetic acid, the location of some mutations conferring resistance to these drugs, and resistance to the nucleotide analogs butylaniline dATP and butylphenyl dGTP. A widely shared YGDTDS consensus motif is most likely associated with metal binding in the dNTP binding site.[154]

---

149. Lopez P, Martinez S, Diaz A, Espinosa M, Lacks SA (1989) *JBC 264*:4255.
150. Ollis DL, Kline C, Steitz TA (1985) *Nature 313*:818.
151. Leavitt MC, Ito J (1989) *PNAS 86*:4465.
152. Lawyer FC, Stoffel S, Saiki RK, Myambo K, Drummond R, Gelfand D (1989) *JBC 264*:6427.
153. Larder BA, Kemp SD, Darby G (1987) *EMBO J 6*:169; Bernad A, Zaballos A, Salas M, Blanco L (1987) *EMBO J 6*:4219; Spicer EK, Rush J, Fung C, Reha-Krantz LJ, Karam JD, Konigsberg WH (1988) *JBC 263*:7478; Wong SW, Wahl AF, Yuan P-M, Arai N, Pearson BE, Arai K, Korn D, Hunkapiller M, Wang TSF (1988) *EMBO J 7*:37; Wang TS-F, Wong SW, Korn D (1989) *FASEB J 3*:14.
154. Bernad A, Lázaro J, Salas M, Blanco L (1990) *PNAS 87*:4610.

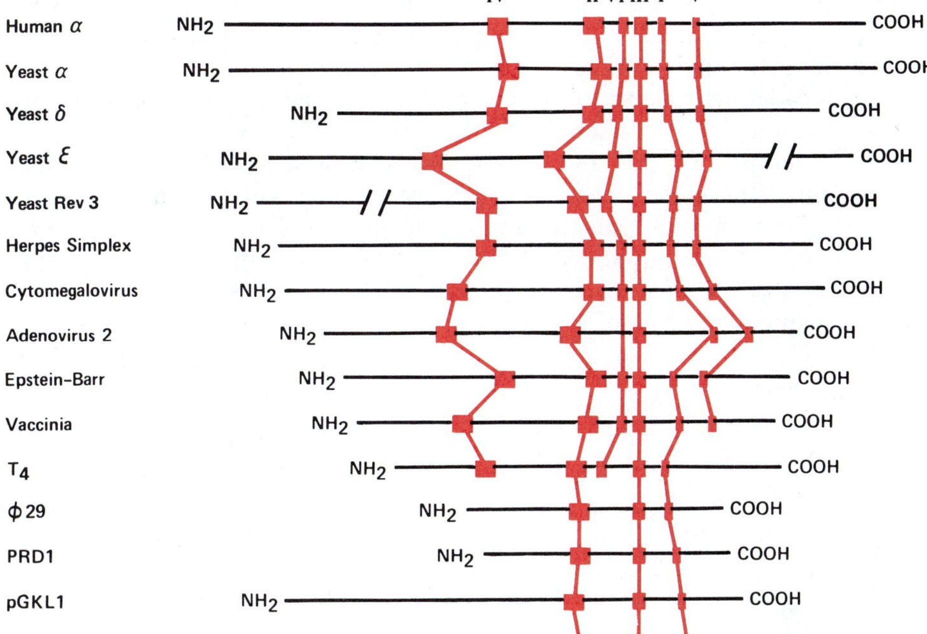

Figure 5-15
Conserved regions (I–VI) of DNA polymerases. (Courtesy of Dr Teresa Wang)

The foregoing findings constitute evidence that similar active sites have been encoded by one of the most phylogenetically stable genes in evolution. Within this family further subdivisions are apparent. T4 polymerase is more closely related to the polymerases of eukaryotic viruses (e.g., vaccinia) than to the polymerases of prokaryotic viruses. The polymerases (inferred from sequences) of linear plasmids (pGKL1 and maize S1 mitochondrial DNA) show the greatest homology with the protein-primed polymerases, suggesting a closer phylogenetic connection between these plasmid and viral polymerases.

(3) Three discrete regions associated with an overt or potential 3′→5′ exonuclease activity are observed in the N-terminal half of the polymerases of phages T4 and φ29,[155] and are also identifiable in *E. coli* pol I, in phage T7 and SPO2 polymerases, in the ε subunit of *E. coli* pol III core and holoenzyme, and, surprisingly, in the conserved region IV of the eukaryotic α-like polymerases. The last-named polymerases almost always lack this exonuclease activity, probably due to the loss of a crucial residue.

(4) The homology between eukaryotic polymerase β (Section 6-4) and terminal deoxynucleotidyl transferase (Section 6-10) is intriguing in view of the small size and lack of processivity of both enzymes.[156] Their relationship may be clarified when their physiologic functions are better understood.

---

155. Spicer EK, Rush J, Fung C, Reha-Krantz LJ, Karam JD, Konigsberg WH (1988) *JBC 263*:7478; Bernad A, Blanco L, Lazaro JM, Martin G, Salas M (1989) *Cell 59*:219.
156. Matsukage A, Nishikawa K, Ooi T, Seto Y, Yamaguchi M (1987) *JBC 262*:8960; Anderson RS, Lawrence CB, Wilson SH, Beattie KL (1987) *Gene 60*:163.

(5) Upon alignment of the sequences of 11 DNA-directed and 20 RNA-directed DNA polymerases (including four inferred from transposon segments) and 26 RNA-directed RNA polymerases, a consensus sequence of 15 amino acids was deduced in which a central pair of aspartates is flanked by certain residues, some of which are hydrophobic.[157] In addition, the asp–asp span was observed in the β subunits of E. *coli* and eukaryotic (pol II) RNA polymerases. The consensus sequence can be pictured as a β hairpin (strand–loop–strand) structure in which the aspartates bind an essential divalent metal and also act directly in the active site. The uniqueness of this motif among polymerases is attested to by its virtual absence in a search of the 6800 primary structures available in a protein-sequence database.

---

157. Argos P (1988) *NAR* *16*:9909.

CHAPTER 6

Eukaryotic DNA Polymerases

6-1 Introduction[1]

The Evolution of Assay Procedures

Early attempts to purify and characterize animal cell DNA polymerases used the calf thymus gland as the source.[2] Rich in DNA and obtainable in large quantities, this tissue was an obvious choice. However, thymus gland proved to be unusual in that it possesses terminal deoxynucleotidyl transferase (Section 10); since this enzyme polymerizes nucleotides without template direction, it complicated the assays of template-directed DNA polymerases.

Studies of replication enzymes from thymus and other eukaryotic tissues were confused and delayed by several factors: (1) heterogeneity of cells within a tissue, (2) heterogeneity in stages of cell growth and development, (3) low enzyme levels even in growing cells, (4) multiple organelles within a cell, (5) severe proteolysis, and (6) fractionation procedures that had to optimize the clean separation of organelles without damaging them. (Detergents, lytic enzymes, and salts used in preparing cell extracts may reduce contamination, but at the same time they cause distortion and leakiness of organelles.) Vigorous efforts with cultured cells, *Drosophila* embryos, and yeast, using refined purification procedures (including those using immunochemical reagents), protease inhibitors, and genetic analysis, have resolved most of the conflicting observations.

1. Weissbach A (1977) *ARB 46*:25; Fry M, Loeb LA (1986) *Animal Cell DNA Polymerases*. CRC Press, Boca Raton, Florida, p. 221; Huberman JA (1987) *Cell 48*:7; Stillman B, Kelly TJ (eds.) (1990) *Eukaryotic DNA Replication*. CSHL.
2. Bollum FJ, Potter VR (1958) *JBC 233*:478.

In assessing the nature and level of the DNA polymerase activities in a crude preparation, the most critical variable is the nature of the template-primer DNA—whether native or denatured, nicked or gapped, or a particular homopolymer pair. Nucleases may convert an inactive template to active or the reverse. Endogenous DNA and RNA, polyanions, salts, metal ions, and other components in crude extracts significantly influence both the template-primer and the enzymes. Effects of antibodies and inhibitors may be strikingly different, depending on whether polymerases are native, truncated, or complexed. In short, one must regard with great caution the characterization of a polymerase in a crude fraction that depends largely on a preference for a particular template-primer and sensitivity to a certain inhibitor.

The Varieties of Eukaryotic Polymerases

The properties of an assortment of eukaryotic polymerases provide fascinating variations on the basic themes of structure and function characteristic of prokaryotic polymerases. Eukaryotic cells have at least four nuclear DNA polymerases: α, β, δ, and ϵ (Table 6-1). Polymerases α and δ are clearly responsible for chromosomal replication; ϵ resembles δ and appears to be essential in yeast and to be linked to repair functions in mammalian cells; β, the smallest of the polymerases, is strongly associated with repair. An additional polymerase, γ, is located in mitochondria and presumably is responsible for the replication of mitochondrial DNA.

Intensified interest in yeast has propelled genetic and biochemical studies of its replicative enzymes: in yeast, polymerase I is analogous to α, pol II to ϵ, pol III to δ, and the mitochondrial polymerase is akin to γ. Still another yeast polymerase, yet to be isolated, is inferred from the sequence of gene *REV3*; the structural domains are unequivocally those of a polymerase, and loss of *REV3* interferes with mutagenic repair.[3] Similarities between the several yeast polymerases and animal cell polymerases bring the power of yeast genetics and molecular analysis and an abundance of source material to bear on studies of eukaryotic polymerases.

Based on the size, subunit composition, and properties of the purified polymerases (Table 6-1), several summarizing statements can be made about them and related polymerases.

(1) Polymerase α (Section 2), essential for replication, contains a catalytic core subunit (approximately 170 kDa), two associated primase subunits (50 and 60 kDa), and an additional polypeptide (70 to 75 kDa) of uncertain function. A 3′→5′ exonuclease, if present, is cryptic. Polymerase α is especially sensitive to aphidicolin and butylphenyl dGTP, but not to dideoxy-NTPs. Yeast pol I, encoded by the essential *POL1* gene, is highly similar in structure and properties.

(2) Polymerase δ (Section 3) is clearly identified with replication, whereas pol ϵ (Section 3) may be employed in repair. Pol δ is dependent on proliferating cell nuclear antigen (PCNA, also called *cyclin*), but pol ϵ is not affected by PCNA. Yeast pol III, encoded by the essential *CDC2* gene, is similar to pol δ; pol II, entirely distinct from pol III, corresponds to ϵ. However, a relationship between

3. Morrison A, Christensen RB, Alley J, Beck AK, Bernstine EG, Lemontt JF, Lawrence CW (1989) *J Bact* *171*:5659.

Table 6-1
Eukaryotic DNA Polymerases[a]

	α	δ	ϵ	β	γ
Previous mammalian designation	α	δ,δ_1	δ,δ_2,δII$,\delta^*$	β	γ
Yeast designation	pol I	pol III	pol II		mitoch.[b]
Yeast gene	*POL1 (CDC17)*	*POL3 (CDC2)*	*POL2*[c]		*MIP1*
Mass (kDa)					
Native	>250	170	256	36–38	160–300
Catalytic core	165–180	125	215	36–38	125
Other subunits	70, 50, 60	48	55	none	35, 47
Location	nucleus	nucleus	nucleus	nucleus	mitoch.
Associated functions					
3′→5′ exonuclease	no[d]	yes	yes	no	yes
Primase	yes	no	no	no	no
Properties					
Response to α auxiliary factors	yes	no	no	no	no
Response to PCNA	no	yes	no	no	no
Preferred template	gap	poly dA · oligo dT	poly dA · oligo dT	gap	poly rA · oligo dT
Divalent cation	Mg	Mg	Mg	Mg/Mn	Mn/Mg
Processivity	low	high[e]	high	low	high
Fidelity	high	high	high	low	high
Inhibitors					
NaCl (0.15 M)	strong	strong	strong	none	none
Aphidicolin	strong	strong	strong	none	none
N-Ethylmaleimide (NEM)	strong	strong	strong	none	strong
Butylphenyl dGTP (I_{50})	1 μM	100 μM	100 μM	100 μM	none
Dideoxy-NTPs	none	weak	weak	strong	strong
Replication	yes	yes	yes	no	yes
Repair	no			yes	no

[a] Burgers PMJ, Bambara RA, Campbell JL, Chang LMS, Downey KM, Hübscher U, Lee MYWT, Linn SM, So AG, Spadari S (1990) *EJB 191*:617.
[b] mitoch. = mitochondria.
[c] Morrison A, Araki H, Clark AB, Hamatake RK, Sugino A (1990) *Cell 62*:1143.
[d] Cryptic in *Drosophila*.
[e] In the presence of PCNA (proliferating cell nuclear antigen).

polymerases δ and ϵ in mammalian cells has yet to be excluded. In both δ and ϵ, the polypeptide of 120 to 170 kDa contains a 3′→5′ exonuclease, is highly processive (with PCNA available to δ), lacks primase activity, and is less sensitive than pol α to butylphenyl dGTP.

(3) Uncertainty about the functions of pol α, pol δ, and pol ϵ may be resolved by the yeast polymerases and the replication of SV40 virus in extracts from human cells. It seems likely that among these polymerases, at least pol α and pol δ work together as part of a larger entity (i.e., a "replisome"). The high processivity of a δ polymerase equipped with a 3′→5′ exonuclease would be suited for the continuous synthesis of the leading strand, while the α polymerase, with its associated primase and lesser processivity, could fulfill the functions needed for the discontinuous synthesis of the lagging strand.

(4) Polymerase β (Section 4), which probably is involved in repair but not in

replication, is only 36 to 38 kDa, has no 3′→5′ exonuclease activity, and is not affected by the inhibitors of pol α but is sensitive to dideoxy-NTPs.

(5) Polymerase γ (Section 5), the mitochondrial enzyme, has a catalytic subunit of 125 kDa with considerable activity on RNA-like templates, and an additional polypeptide of unknown function. The 3′→5′ exonuclease function is remarkably mismatch-specific. Responses to inhibitors are unlike those of pol α.

(6) Among a variety of unicellular organisms (e.g., *Ustilago, Neurospora, Tetrahymena*), the bewildering array of polymerase activities may be due more to proteolysis and other limitations in the assays of crude mixtures and partially purified fractions (see above) than to truly unique polymerases. Intensive studies, as with yeast, will probably disclose a high degree of similarity and conservation among the replicative enzymes of these species.

(7) Among the DNA polymerases encoded by animal viruses (Section 8), those of herpes and vaccinia strongly resemble pol δ, and to some extent pol α, in structure and in many properties. No functional counterparts of the RNA-directed DNA polymerases (reverse transcriptases) of the retroviruses (Section 9) have been prepared from uninfected cells.

(8) Telomerase (Section 9), the most recently discovered DNA polymerase, takes directions from an oligoribonucleotide complexed with the enzyme to furnish the characteristic termini of the linear chromosomes of *Tetrahymena* and presumably of yeast and animal chromosomes.

(9) Terminal deoxynucleotidyl transferase (Section 10) does not respond to template direction, and therefore by strict definition is not a DNA polymerase. The striking sequence homology with mammalian pol β (Section 5-11) suggests that both enzymes have a common ancestry.[4] Since the enzyme is found only in T lymphocytes at an early stage of development, a function for it in the genetic rearrangements of immunoglobulin genes seems probable. As a reagent in recombinant DNA manipulations, its role has been clear and important.

6-2 DNA Polymerase α[5]

Controversies that swirled over the subcellular location and structure of pol α have been settled by refinements in technique, such as avoidance of proteolysis, and by genetic analysis. The enzyme from mouse lymphoblasts,[6] from *Drosophila*, and from yeast is demonstrably essential for replication and resides in the nucleus. It has a four-subunit structure: a large polypeptide core (about 170 kDa) and three accessory subunits of 70 to 75, 55 to 60, and 48 to 50 kDa. Because the two smaller subunits possess primase activity (Section 8-6), an appropriate designation for the holoenzyme structure is *DNA polymerase–primase*. The 70- to 75-kDa subunit, also shown to be essential (in yeast), may have some regulatory role (see below).

In some cell extracts, the polymerase and primase subunits are dissociated, a striking example being mouse cells infected with a parvovirus.[7] Polypeptide

4. Matsukage A, Nishikawa K, Ooi T, Seto Y, Yamaguchi M (1987) *JBC 262:*8960.
5. Kaguni LS, Lehman IR (1988) *BBA 950:*87; Lehman IR, Kaguni LS (1989) *JBC 264:*4265; So AG, Downey KM (1988) *Biochemistry 27:*4591; Burgers PMJ (1989) *Prog N A Res 37:*235; Wang TS-F, Wong SW, Korn D (1988) *FASEB J 3:*14.
6. Goulian M, Heard CJ (1989) *JBC 264:*19407.
7. Ho TFL, Gupta JS, Faust EA (1989) *Biochemistry 28:*4622.

fragments, some with enzymatic activities, have been reported, but are now known to be the result of proteolysis during the course of purification.

Pol α achieves moderate processivity and surprisingly high fidelity, even without an apparently active 3′→5′ exonuclease.

Function[8]

An essential role for pol α in chromosomal replication is based on many lines of evidence. (1) Mouse cells with a temperature-sensitive mutant pol α fail to replicate DNA at the restrictive temperature. (2) Antibodies against pol α inhibit DNA replication when introduced into permeabilized nuclei or cultured mammalian cells; depletion of pol α in HeLa cell extracts by antibody precipitation prevents the replication of SV40 DNA. (3) The levels of pol α are high in rapidly growing mammalian cells in culture and in proliferating lymphocytes, but are low in quiescent, differentiated cells; by contrast, pol β and pol γ levels remain unchanged upon induction of rapid cell growth.[9] (4) *POL1*, the gene for the yeast enzyme homologous to pol α, is essential and is allelic to *CDC17;* conditional mutants in this single-copy gene are blocked in DNA synthesis and in the cell cycle at elevated temperatures.[10] Such *POL1* mutants are not diminished in repair of DNA damage (single- and double-strand breaks by x-rays); thus, by implication, pol α is not responsible for DNA repair.

Gene and Structure of the Large α Subunit, the Polymerase

The *polα* cDNA in humans[11] predicts a size of 165 kDa for the large subunit, the catalytic core of pol α; the comparable *POL1* gene *(CDC17)* in yeast[12] encodes a subunit of 167 kDa. Reports of larger sizes (around 200 kDa) are probably due to posttranslational modifications such as phosphorylation and glycosylation. The human and yeast amino acid sequences show extensive homology in six ordered regions found in the carboxyl half of the protein; a strikingly similar structural organization characterizes the DNA polymerases from diverse animal viruses and prokaryotic sources (Section 5-11). Based on the locations of mutations, two of the regions contain residues involved in dNTP binding. Potential metal-binding domains ("zinc fingers"), also in the C-terminal region, suggest sites involved in binding DNA.

Although the N-terminal portion is the least conserved, it nevertheless possesses three regions identified with the 3′→5′ exonuclease activity in the polymerases of phages T4 and ϕ29 and recognizable in the sequences of a wide range of polymerases[13] (Section 5-11). Thus, the presence of 3′→5′ exonuclease activity in *Drosophila* pol α[14] (see below), albeit cryptic, no longer seems anomalous. Absence of the activity from the pol α of other species (e.g., yeasts and mammals) may be due to the substitution of one or a few critical residues. The N-terminal regions may also be responsible for the binding of primase and other accessory polypeptides.

---

8. Kaguni LS, Lehman IR (1988) *BBA 950*:87; Lehman IR, Kaguni LS (1989) *JBC 264*:4265.
9. Schneider E, Müller B, Schindler R (1985) *BBA 825*:375.
10. Johnson LM, Snyder M, Chang LMS, Davis RW, Campbell JL (1985) *Cell 43*:369; Campbell JL (1986) *ARB 55*:733.
11. Wang TS-F, Wong SW, Korn D (1988) *FASEB J 3*:14.
12. Pizzagalli A, Valsasnini P, Plevani P, Lucchini G (1988) *PNAS 85*:3772.
13. Spicer EK, Rush J, Fung C, Reha-Krantz LJ, Kram JD, Konigsberg WH (1988) *JBC 263*:7478; Bernad A, Blanco L, Lázaro JM, Martín G, Salas M (1989) *Cell 59*:219.
14. Cotterill SM, Reyland ME, Loeb LA, Lehman IR (1987) *PNAS 84*:5635.

Assays of polymerase levels as a function of the cell cycle depend critically on whether the cells are induced from a quiescent state or fractionated from an asynchronously growing culture (Section 7).[15] Both pol α[16] and yeast pol I[17] are strongly induced during the transition from G_0 to the G_1–S phase, the level of the pol I transcript rising about 100-fold. In cycling cells, the yeast enzyme shows a tenfold increase in the transcript at the G_1–S boundary, but human cells show no change and express pol α constitutively throughout the cell cycle.

Regulation of posttranslational modification is suggested by the presence of phosphate in the 180 and 77 kDa subunits and in the 55 kDa primase subunit of pol α in mammalian cells.[18] The large subunit in rat embryonic fibroblasts may be a phosphoprotein responding to dephosphorylation and kinase treatments.[19] Other instances of activity responses to protein kinases indicate the probable importance of phosphorylation in regulation.[20]

Mechanism

Pol α, like other DNA polymerases, operates by an ordered sequential mechanism in which strong binding to a single-stranded template is followed by binding to the primer and then to the complementary dNTP. The original kinetic studies on pol α, performed with a proteolyzed heterodimer of 66 and 76 kDa,[21] have been confirmed with an intact four-subunit enzyme.[22] However, the K_m values for template-primer and dNTPs were about tenfold higher for the proteolyzed form than for the intact enzyme; the sensitivities to aphidicolin and salt were also considerably less in the degraded forms. The catalytic rate depends on pH and Mg concentration and is optimal at pH 7.0 and 5 mM Mg.

The recognition, binding, and template functions of DNA observed with the *Drosophila* holoenzyme were profoundly decreased by abasic (AP) sites, even when these were as far as four residues from the primer terminus in the base-paired region of the template-primer.[23] Specificity for dNTP and template interactions have been revealed by affinity labeling.[24]

Processivity values of pol α vary from 2 to 2000 (nucleotides polymerized per binding event), rising with a decrease in temperature, pH, or Mg concentration.[25] A variety of accessory proteins and effector molecules may enhance processivity, depending on the source of pol α.

Fidelity of pol α, despite the absence of a measurable 3′→5′ exonuclease activity, can be high in some instances, with as few as one mismatch remaining

15. Wong SW, Wahl AF, Yuan PM, Arai N, Pearson BE, Arai K, Korn D, Hunkapiller MW, Wang TSF (1988) *EMBO J 7*:37; Wahl AF, Geis AM, Spain BH, Wong SW, Korn D, Wang TS-F (1988) *Mol Cell Biol 8*:5016.
16. Wong SW, Wahl AF, Yuan PM, Arai N, Pearson BE, Arai K, Korn D, Hunkapiller MW, Wang TSF (1988) *EMBO J 7*:37; Wahl AF, Geis AM, Spain BH, Wong SW, Korn D, Wang TS-F (1988) *Mol Cell Biol 8*:5016.
17. Johnston LM, White JHM, Johnson AL, Lucchini G, Plevani P (1987) *NAR 15*:5017; Nasheuer H-P, Moore A, Wahl AF, Wang TS-F, personal communication.
18. Wong SW, Paborsky LR, Fisher PA, Wang TS-F, Korn D (1986) *JBC 261*:7958.
19. Donaldson RW, Gerner EW (1987) *PNAS 84*:759.
20. Krauss SW, Mochly-Rosen D, Koshland DE Jr, Linn S (1987) *JBC 262*:3432; Krauss SW, Linn S (1989) *J Cell Biochem [Suppl] 13D*:162; Cripps-Wolfman J, Henshaw EC, Bambara RA (1989) *JBC 264*:19478.
21. Fisher PA, Korn D (1981) *Biochemistry 20*:4560; Korn D, Fisher PA, Wang TS-F (1983) in *New Approaches in Eukaryotic DNA Replication* (de Recondo AM, ed.). Plenum Press, New York, p. 17.
22. Wong SW, Paborsky LR, Fisher PA, Wang TS-F, Korn D (1986) *JBC 261*:7958.
23. Ng L, Weiss SJ, Fisher PA (1989) *JBC 264*:13018.
24. Doronin SV, Lavrik OI, Nevinsky GA, Podust VN (1987) *FEBS Lett 216*:221; Lavrik OI, Levina S, Nevinsky GA, Podust VN (1987) *FEBS Lett 216*:225.
25. Tan CK, So MJ, Downey KM, So AG (1987) *NAR 15*:2269; Hohn KT, Grosse F (1987) *Biochemistry 26*:2870; Sabatino RD, Myers TW, Bambara RA, Kwan-Shin O, Marraccino RL, Frickey PH (1988) *Biochemistry 27*:2998.

per 10^6 nucleotides incorporated.[26] Inasmuch as noncatalyzed base pairing predicts a far higher error frequency, the enzyme must be responsible for the improved matching. In addition to amplifying the free energy difference between matched and unmatched base pairs by water exclusion on its hydrophobic surface,[27] the polymerase also dissociates rapidly from a mismatched primer terminus and rebinds to it very feebly.

An especially interesting, and possibly singular, case is the cryptic 3′→5′ exonuclease activity of the large subunit of the *Drosophila* enzyme, unmasked by removal of the accessory 73-kDa polypeptide:[28] not only is the fidelity of the enzyme improved, but its processivity (in the presence of SSB) is also enhanced[29] and sensitivities to inhibitors are reduced.[30] Proteolyzed forms of the *Drosophila* and calf thymus enzymes display reduced fidelity, as do preparations of the native enzyme after storage for prolonged periods at −70°.

Accessory Factors

In view of the low catalytic rate of the isolated enzyme, assistance by accessory proteins can be anticipated and, indeed, a number of accessory factors have been identified. Considering the diversity of enzyme sources, assays, and authors, some of these factors may prove to be the same.

Stimulation of the enzyme by an SSB has been observed in the replication of SV40 DNA,[31] especially by protein A (also called RF-A or RP-A). This SSB is made up of three subunits (70, 32, and 13 kDa),[32] the largest of which binds ssDNA and may assist T antigen (Section 19-2) in its helicase function. RF-A from both human and yeast cells is phosphorylated in a cell-cycle dependent manner.[33] A yeast SSB, which may stimulate pol I by increasing processivity,[34] has proved to be a cytosolic RNA-binding protein.[35]

RF-C (RP-C), one of the multisubunit factors needed for SV40 DNA replication, behaves as a pol α accessory protein similar to the T4 gp44–gp62 pair (Section 17-6) that forms a complex with gp43, the T4 DNA polymerase (Section 5-8). RF-C possesses a DNA-dependent ATPase activity, and its principal function is to facilitate binding of the polymerase to the primer terminus.[36]

Another distinctive α-accessory factor (AAF),[37] isolated from cultured mouse cells, contains polypeptides of 132 and 45 kDa and is notable for stimulating the self-primed synthesis on unprimed single strands. By increasing affinity to the template, this AAF raises processivity about tenfold to yield products of proper size, mammalian Okazaki fragments.[38]

Pol α is stimulated by RNase H (Section 13-10) when priming and chain extension are coupled, presumably by an increase in the frequency of initia-

---

26. Cotterill SM, Reyland ME, Loeb LA, Lehman IR (1987) *PNAS 84*:5635; Loeb LA, Kunkel TA (1982) *ARB 51*:429; Loeb LA, Liu PK, Fry M (1986) *Prog N A Res 33*:57; Reyland ME, Loeb LA (1987) *JBC 262*:10824.
27. Petruska J, Sowers LC, Goodman MF (1986) *PNAS 83*:1559.
28. Cotterill SM, Reyland ME, Loeb LA, Lehman IR (1987) *PNAS 84*:5635.
29. Cotterill SW, Chui GSJ, Lehman IR (1987) *JBC 262*:16100.
30. Reyland ME, Lehman IR, Loeb RA (1988) *JBC 263*:6518.
31. Li JJ, Kelly TJ (1984) *PNAS 81*:6973; Stillman B, Gluzman Y (1985) *Mol Cell Biol 5*:2051; Wobbe CR, Dean F, Weissbach L, Hurwitz J (1985) *PNAS 82*:5710.
32. Wold MS, Kelly TJ (1988) *PNAS 85*:2523; Fairman MP, Stillman B (1988) *EMBO J 7*:1211; Ishimi Y, Claude A, Bullock P, Hurwitz J (1988) *JBC 263*:19723; Kenny MK, Lee S-H, Hurwitz J (1990) *PNAS 86*:9757.
33. Din S-D, Brill SJ, Fairman MP, Stillman B (1990) *Genes and Develop* 4:968.
34. Jong AY-S, Campbell JL (1986) *PNAS 83*:877.
35. Burgers PMJ (1989) *Prog N A Res 37*:235.
36. Tsurimoto T, Stillman B (1990) *PNAS 87*:1023.
37. Goulian M, Heard CJ, Grimm SL (1990) *JBC, 265*:13221.
38. Goulian M, Heard CJ (1990) *JBC, 265*:13231.

tions.[39] Stimulation of yeast pol I by a yeast RNase H may operate by increasing the affinity for primer termini.[40] Protein–protein interactions may account for these effects.

Large complexes with additional activities were isolated when pol α was purified, with ATP present, by methods that minimize dissociation. A 640-kDa polymerase–primase complex[41] isolated from synchronized HeLa cells contained a diadenosine tetraphosphate [$A(p)_4A$] binding protein, a 3′→5′ exonuclease, and rate-enhancing primer-recognition proteins (C_1 and C_2).[42] In an early stage of purification from calf thymus, fractions with the polymerase–primase complex also contained topoisomerase II, helicase, methyl transferase, 3′→5′ exonuclease, RNase H, and a calmodulin-binding protein.[43] Whether these diverse activities are physically associated with the polymerase in replicative functions is still uncertain.

A system reconstituted with purified mammalian proteins[44] that is effective in the discontinuous synthesis of the lagging strand consists of five entities: the α polymerase–primase, α-accessory factor, RNase H-1, DNA ligase, and a polypeptide with 5′ → 3′ exonuclease activity. The exonuclease, probably identical to DNase IV[45] and DNase VIII,[46] is required in this in vitro system to remove the residual one or two primer ribonucleotides resistant to the RNase H.

Whereas the functional importance of the linkage of primase to pol α is demonstrable,[47] the role of the 70 to 75-kDa subunit is unclear. Separation of this polypeptide from the *Drosophila* polymerase unmasks the 3′→5′ exonuclease activity and raises the processivity; disruption of the gene for the corresponding yeast 84-kDa subunit is lethal.[48]

6-3 DNA Polymerases δ and ϵ[49]

Overview

Recognition of the multiple forms and replication function of pol δ is one of the most significant advances in the eukaryotic polymerase field in the past decade.

---

39. DiFrancesco RA, Lehman IR (1985) *JBC 260*:14764; Masutani C, Enomoto T, Suzuki M, Hanaoka F, Ui M (1990) *JBC 265*:10210.
40. Karwan R, Blutsch H, Wintersberger U (1983) *Biochemistry 22*:5500.
41. Skarnes W, Bonin P, Baril EF (1986) *JBC 261*:6629.
42. Pritchard CG, Weaver DT, Baril EF, De Pamphilis ML (1983) *JBC 258*:9810; Kaguni LS, Rossignol J-M, Conaway RC, Lehman IR (1983) in *Mechanisms of DNA Replication and Recombination* (Cozzarelli NR, ed.). AR Liss, New York, p. 495; Kawasaki K, Enomoto T, Suzuki M, Seki M, Haraoka F, Yamada M (1986) *Biochemistry 25*:3044; Malkas LH, Hickey RJ, Li C, Pedersen N, Baril EF (1990) *Biochemistry 29*:6362.
43. Ottiger H-P, Hübscher U (1984) *PNAS 81*:3993; Hübscher U, Gassmann M, Spadari S, Brown NC, Ferrari E, Buhk HJ (1987) *Philos Trans R S Lond [Biol] 317*:421.
44. Goulian M, Richards SH, Heard CJ, Bigsby BM (1990) *JBC 265*:18461.
45. Lindahl T, Gally JA, Edelman GM (1969) *PNAS 62*:597; Ishimi Y, Claude A, Bullock P, Hurwitz J (1988) *JBC 263*:19723; Goulian M, Richards SH, Heard CJ, Bigsby BM (1990) *JBC 265*:18461.
46. Pedrini AM, Grossman L (1983) *JBC 258*:1536.
47. Suzuki M, Enomoto T, Masutani C, Hanaoka F, Yamada M, Ui M (1989) *JBC 264*:10065.
48. Hinkle D, personal communication.
49. So AG, Downey KM (1988) *Biochemistry 27*:4591; Burgers PMJ (1989) *Prog N A Res 37*:235; Wahl AF, Crute JJ, Sabatino RD, Bodner JB, Maraccino RL, Harwell LW, Lord EM, Bambara RA (1986) *Biochemistry 25*:7821; Bambara RA, Myers TW, Sabatino RD (1990) in *The Eukaryotic Nucleus: Molecular Biochemistry and Macromolecular Assemblies* (Strauss P, Wilson SH, eds.). Telford Press, Caldwell, NJ, p. 69; Randahl H, Elliott GC, Linn S (1988) *JBC 263*:12228; Syvaoja J, Linn S (1989) *JBC 264*:2489; Wong SW, Syvaoja J, Tan C-K, Downey KM, So AG, Linn S, Wang TS-F (1989) *JBC 264*:5924; Focher F, Gassmann M, Hafkemeyer P, Ferrari E, Spadari S, Hübscher U (1989) *NAR 17*:1805; Syväoja J, Susmensaari S, Nishida C, Goldsmith JS, Chui GSJ, Jain S, Linn S (1990) *PNAS 87*:6664.

Discovered in calf thymus and hyperplastic rabbit bone marrow[50] as unique in having an intrinsic 3′→5′ exonuclease proofreading activity, pol δ was found to resemble pol α in its sensitivity to aphidicolin and N-ethylmaleimide (NEM) and its resistance to dideoxy-NTPs. Any lingering doubts about pol δ as a structurally and immunochemically distinct entity essential in replication were dispelled with two revelations: one was the finding that SV40 DNA replication in vitro[51] requires PCNA, which is a processivity factor for δ (see below) and is the proliferating cell nuclear antigen, needed for initiation of DNA replication in the cell cycle.[52] The other was the recognition that pol III, the yeast counterpart of pol δ,[53] is encoded by *CDC2*, an essential cell-cycle gene.

The similarities of pol δ and pol ϵ warrant that they still be considered together. They share many features (Table 6-1), some of which distinguish them from pol α: an overt 3′→5′ exonuclease activity, lack of primase activity, template-primer preference for poly dA·oligo dT, high processivity, relative sensitivities to certain inhibitors, a simple subunit structure, distinctive responses to accessory factors, and immunologic specificity. The processivity of pol δ on the one hand and the lack of this feature in pol α, which on the other hand does possess primase activity, have made it seem that pol δ is specialized for the continuous replication of the leading strand and pol α for the discontinuous replication of the lagging strand.[54] (Why pol α lacks the proofreading activity present in pol δ needs to be explained.) The *E. coli* pol III holoenzyme, organized as an asymmetric dimeric polymerase at a replication fork, may show a similar distribution of functions for continuous synthesis of one strand and discontinuous synthesis of the other (Section 5-6).

Pol δ: PCNA-Dependence

Isolated from calf thymus, pol δ is made up of two subunits, 125 and 48 kDa;[55] the yeast equivalent is also composed of two subunits (125 and 55 kDa).[56] The larger yeast subunit, with a calculated weight of 124.5 kDa, is encoded by the yeast cell division cycle gene *CDC2*;[57] mutations in this gene cause arrest in the S phase of the cell cycle at an elevated (nonpermissive) temperature. Another form of the mammalian pol δ, isolated from human placenta, contains a 170-kDa polypeptide and some smaller ones.[58]

A key feature of pol δ, its dependence on PCNA, was revealed by the isolation, from calf thymus, of a 36-kDa protein which increased the pol δ processivity by

---

50. Byrnes JJ, Downey KM, Black VL, So AG (1976) *Biochemistry 15*:2817; Lee MYWT, Tan C-K, So AG, Downey KM (1980) *Biochemistry 19*:2096; Goscin LP, Byrnes JJ (1982) *Biochemistry 21*:2513.
51. Wold MS, Kelly TJ (1988) *PNAS 85*:2523; Fairman MP, Stillman B (1988) *EMBO J 7*:1211; Ishimi Y, Claude A, Bullock P, Hurwitz J (1988) *JBC 263*:19723; Weinberg DH, Kelly TJ (1989) *JBC 264*:9742.
52. Tan C-K, Castillo C, So AG, Downey KM (1986) *JBC 261*:12310; Bravo R, Frank R, Blundell PA, MacDonald-Bravo H (1987) *Nature 326*:515; Liu Y-C, Marraccino RL, Keng PC, Bambara RA, Lord EM, Chow W-G, Zain SB (1989) *Biochemistry 28*:2967; Suzuka I, Daidoji H, Matsuoka M, Kadowaki K-I, Takasaki Y, Nakane PK, Moriuchi T (1989) *PNAS 86*:3189.
53. Bauer GA, Heller HM, Burgers PMJ (1988) *JBC 263*:917; Burgers PMJ, Bauer GA (1988) *JBC 263*:925; Boulet A, Simon M, Faye G, Bauer GA, Burgers PMJ (1989) *EMBO J 8*:1849; Sitney KC, Budd ME, Campbell JL (1989) *Cell 56*:599.
54. So AG, Downey KM (1988) *Biochemistry 27*:4591; Focher F, Ferrari E, Spadari S, Hübscher U (1988) *FEBS Lett 229*:6; Tsurimoto T, Stillman B (1989) *EMBO J 8*:3883; Tsurimoto T, Stillman B (1991) *JBC 266*:1961.
55. Lee MYWT, Tan C-K, Downey KM, So AG (1984) *Biochemistry 23*:1906.
56. Burgers PMJ (1989) *Prog N A Res 37*:235; Bauer GA, Heller HM, Burgers PMJ (1988) *JBC 263*:917; Burgers PMJ, Bauer GA (1988) *JBC 263*:925.
57. Boulet A, Simon M, Faye G, Bauer GA, Burgers PMJ (1989) *EMBO J 8*:1849; Sitney KC, Budd ME, Campbell JL (1989) *Cell 56*:599.
58. Lee MYWT, Toomey NL (1987) *Biochemistry 26*:1076.

40-fold, making it able to replicate a singly primed phage DNA circle.[59] Reminiscent of the β subunit of *E. coli* pol III holoenzyme in its size and function as an essential protein, the 36-kDa auxiliary protein shows sequence homology to the T4 polymerase accessory protein gp45; its physiologic importance was established by the demonstration of its identity with PCNA (also called cyclin)[60] and the absolute requirement for it in the replication of SV40 DNA in HeLa cell extracts.[61] The stimulation of yeast pol III by yeast PCNA is in chain elongation rather than in binding of the polymerase to the template-primer;[62] PCNA had no effect on the distributive nature of the 3′→5′ exonuclease activity.

Along with PCNA, two other proteins are needed in SV40 replication for pol δ action: RF-A (RP-A) for binding ssDNA[63] and RF-C (RP-C) for binding the primer terminus. Pol δ, like pol α, is sensitive to aphidicolin and NEM, and resistant to dideoxy-NTPs; unlike pol α, it is only slightly inhibited by butylphenyl dGTP (Table 6-1). In further support of the pol δ role in replication are its elevated levels in tissues, paralleling those of pol α, during chromosomal replication.[64] Synthesis and maturation of nascent DNA in permeabilized cells appear to require both pol α and pol δ.[65]

Assays of pol δ (and also of pol ϵ; see below) generally employ poly dA · oligo dT as template-primer because of the relatively feeble activity observed with primed ssDNA. However, reconstitution of a complex of pol δ with PCNA, RF-A (RP-A), RF-C (RP-C, A1), and ATP,[66] reminiscent of *E. coli* pol III holoenzyme and T4 polymerase, does support vigorous replication of primed natural DNA at physiologic rates in excess of 100 ntd per second.

Pol ϵ: PCNA-Independence

Despite its immunologic relatedness[67] and numerous other similarities to pol δ, pol ϵ is distinguished from pol δ principally by its high processivity independent of PCNA (Table 6-1). Pol ϵ has been identified in a variety of cells; its physical features are less uniform than those of pol δ. The yeast gene is designated *POL2*, but the mammalian genetic origins are still undefined. A pol ϵ association with DNA repair or replication in vivo has not been proved as yet.

Mammalian pol ϵ has been isolated in several ways. In one, it emerges as a factor needed to restore DNA repair synthesis to UV-irradiated human fibroblasts (HeLa cells) that have been permeabilized and depleted of soluble material. In this instance, pol ϵ appears as a large polypeptide of 215 kDa and possibly one of 55 kDa.[68] In low salt concentration, the molecular mass is approximately

---

59. Tan C-K, Castillo C, So AG, Downey KM (1986) *JBC 261*:12310; Bravo R, Frank R, Blundell PA, MacDonald-Bravo H (1987) *Nature 326*:515.
60. Liu Y-C, Marraccino RL, Keng PC, Bambara RA, Lord EM, Chow W-G, Zain SB (1989) *Biochemistry 28*:2967; Suzuka I, Daidoji H, Matsuoka M, Kadowaki K-I, Takasaki Y, Nakane PK, Moriuchi T (1989) *PNAS 86*:3189.
61. Prelich G, Tan C-K, Kostura M, Mathews MB, So AG, Downey KM, Stillman B (1987) *Nature 326*:517.
62. Bauer GA, Burgers PMJ (1988) *PNAS 85*:7506; Bauer GA, Burgers PMJ (1988) *BBA 951*:274; Bauer GA, Burgers PMJ (1990) *NAR 18*:261.
63. Kenny MK, Lee S-H, Hurwitz J (1989) *PNAS 86*:9757.
64. Zhang SJ, Lee MYWT (1987) *Arch B B 252*:24.
65. Hammond RA, Byrnes JJ, Miller MR (1987) *Biochemistry 26*:6817.
66. Tsurimoto T, Stillman B (1990) *PNAS 87*:1023; Tsurimoto T, Melendy T, Stillman B (1990) *Nature 346*:534; Lee S-H, Hurwitz J (1990) *PNAS 87*:5672.
67. Wong SW, Syvaoja J, Tan C-K, Downey KM, So AG, Linn S, Wang TS-F (1989) *JBC 264*:5924.
68. Nishida C, Reinhard P, Linn S (1988) *JBC 263*:501; Randahl H, Elliott GC, Linn S (1988) *JBC 263*:12228; Syvaoja J, Linn S (1989) *JBC 264*:2489; Wong SW, Syvaoja J, Tan C-K, Downey KM, So AG, Linn S, Wang TS-F (1989) *JBC 264*:5924; Syväoja J, Susmensaari S, Nishida C, Goldsmith JS, Chui GSJ, Jain S, Linn S (1990) *PNAS 87*:6664.

doubled. The pol ϵ activity resembles pol δ and is not replaceable by pol α. Coupled with phage T4 endonuclease V, pol ϵ also supports DNA synthesis in UV-treated, repair-defective xeroderma pigmentosum cells.

A second example of pol ϵ is the polymerase isolated from calf thymus.[69] As with the HeLa enzyme, no evidence exists for a cryptic PCNA or for physical identity with pol δ. Still another instance of a pol δ-type but PCNA-independent enzyme is the polymerase purified from rabbit bone marrow, a single polypeptide of only 122 kDa.[70] A similar PCNA-independent polymerase isolated from cultured mouse cells[71] and closely resembling pol δ in several of its enzymatic properties may well prove to be the 125-kDa polypeptide of pol δ.

Finally, the pol II identified in yeast,[72] possessing a proofreading 3′→5′ exonuclease activity, is assigned to the pol ϵ category; the principal polypeptide of 256 kDa (deduced from pol 2) is highly processive. Isolated active polypeptides of 145 to 170 kDa appear to be proteolytic products. Yeast pol II is stimulated by a factor containing three polypeptides (66, 37, and 13.5 kDa), the largest of which is associated with ssDNA-binding activity.[73]

6-4 DNA Polymerase β[74]

This nuclear enzyme, found in a wide variety of animal tissues and chick embryos, is a 36- to 38-kDa polypeptide,[75] highly conserved in sequence.[76] In cultured mammalian cells, the levels of the enzyme and its transcript are low and remain independent of growth and cell-cycle stages.[77] In *Drosophila* embryos, a polymerase of 110 kDa has the features of pol β that distinguish it from the other polymerases.[78] The lack of a replicative function implies participation in other DNA metabolic processes, notably repair.[79]

Properties and Mechanism

This smallest of eukaryotic polymerases, with no other physically connected functions, has many distinctive properties. Pol β, by virtue of its size and basicity, is readily separated from the other polymerases. (Associations in large

---

69. Wahl AF, Crute JJ, Sabatino RD, Bodner JB, Maraccino RL, Harwell LW, Lord EM, Bambara RA (1986) *Biochemistry 25*:7821; Bambara RA, Myers TW, Sabatino RD (1990) in *The Eukaryotic Nucleus: Molecular Biochemistry and Macromolecular Assemblies* (Strauss P, Wilson SH, eds.). Telford Press, Caldwell, NJ, p. 69; Focher F, Gassmann M, Hafkemeyer P, Ferrari E, Spadari S, Hübscher U (1989) *NAR 17*:1805.
70. Goscin LP, Byrnes JJ (1982) *Biochemistry 21*:2513.
71. Goulian M, Herrmann SM, Sackett JW, Grimm SL (1990) *JBC 265*:16402.
72. Burgers PMJ (1989) *Prog N A Res 37*:235; Chang LMS (1977) *JBC 252*:1873; Wintersberger E (1978) *EJB 84*:167; Budd ME, Sitney KC, Campbell JL (1989) *JBC 264*:6557; Hamatake RK, Hasegawa H, Clark AB, Bebenek K, Kunkel TA, Sugino A (1990) *JBC 265*:4072; Morrison A, Araki H, Clark AB, Hamatake RK, Sugino A (1990) *Cell 62*:1143.
73. Brown WC, Smiley JK, Campbell JL (1990) *PNAS 87*:677.
74. Wilson SH, Abbotts J, Widen S (1988) *BBA 949*:149; Rein DC, Recupero AJ, Reed MP, Meyer RR (1990) in *The Eukaryotic Nucleus: Molecular Biochemistry and Macromolecular Assemblies* (Strauss P, Wilson SH, eds.). Telford Press, Caldwell, NJ, p. 95; Kumar A, Abbots, J, Karawya EM, Wilson SH (1990) *Biochemistry 29*:7156.
75. Matsukage A, Nishikawa K, Ooi T, Seto Y, Yamaguchi M (1987) *JBC 262*:8960; Sakaguchi K, Boyd JB (1985) *JBC 260*:10406.
76. SenGupta DN, Zmudzka BZ, Kumar P, Cobianchi F, Skowronski J, Wilson SH (1986) *BBRC 136*:341; Widen SG, Kedar P, Wilson SH (1988) *JBC 263*:16992.
77. Chang LMS, Brown M, Bollum FJ (1983) *JMB 74*:1; Mitchell J, Karawya E, Kinsella T, Wilson SH (1985) *Mutat Res 146*:295.
78. Sakaguchi K, Boyd JB (1985) *JBC 260*:10406.
79. Smith CA, Okumoto DS (1984) *Biochemistry 23*:1383; Yamada K, Hanaoka F, Yamada M (1985) *JBC 260*:10412.

complexes have proven neither reproducible nor functionally significant.) Pol β, unlike pol α, is not inhibited by aphidicolin, sulfhydryl-blocking reagents, or salt, is inhibited only weakly by butylphenyl dGTP, and is especially sensitive to dideoxy-NTPs (Table 6-1).

Kinetic studies of pol β from several sources[80] confirm that polymerization is almost exclusively distributive (i.e., pol β has a processivity value of one). The reaction mechanism is ordered and sequential,[81] with DNA binding first, followed by a dNTP. An initial burst of incorporation which decays to a lower steady-state velocity is a kinetic complexity identified as enzyme hysteresis.[82] Pyrophosphorolysis and exchange of inorganic pyrophosphate are not demonstrable.[83] There is limited strand displacement synthesis on nicked duplex DNA substrates. Reaction velocities are maximal on DNA primer-templates containing short gaps of 10 to 20 ntd.[84] While the affinity of the enzyme for nicked duplex DNA molecules is high, that for single-stranded or intact duplex DNA is feeble or absent. Thus, a base-paired primer adjacent to a very short single-stranded region (actual or potential) is the preferred template-primer.

Stimulatory proteins[85] have been found that reduce the K_m values for template-primer and dNTPs and increase the rate and extent of synthesis on nicked DNA and on templates with long single-stranded stretches. Among six such factors, one behaves like the C_1 and C_2 primer-recognition proteins[86] specific for pol α; others are DNA-binding proteins, and still others include an exonuclease (DNase V)[87] and an ATPase (III).[88] The specifity and significance of the interactions of these proteins with pol β are still uncertain.

Mn substitution[89] lowers the K_m for template-primer and can replace the need for one of the stimulatory proteins at low DNA concentrations. With Mg, the minimum template length is 5 ntd, whereas with Mn a single template nucleotide suffices. Mn also increases processivity to four to six nucleotides inserted per binding cycle.[90]

Physiologic Role

Many studies conclude that pol β levels in tissues as well as in cultured cells show no correlation with chromosome replication,[91] in contrast to those of pol α. In regenerating rat liver and in hepatomas, pol α levels increase enormously while those of pol β remain the same.[92] Levels of the pol β transcript respond to doses of DNA-damaging agents.[93] Neuronal nuclei of adult rats that lack pol α respond to damage by UV irradiation[94] with a sevenfold to tenfold elevation of

80. Tanabe K, Bohn EW, Wilson SH (1979) *Biochemistry 18*:3401; Wang TS-F, Korn D (1980) *Biochemistry 19*:1782; Wang TS-F, Korn D (1982) *Biochemistry 21*:1597.
81. Wang TS-F, Korn D (1982) *Biochemistry 21*: 1597.
82. Di Giuseppe JA, Wright SE, Dresler SL (1989) *NAR 17*:3079.
83. Wang TS-F, Eichler DC, Korn D (1977) *Biochemistry 16*:4917; Wang TS-F, Korn D (1982) *Biochemistry 21*:1597.
84. Wang TS-F, Korn D (1980) *Biochemistry 19*:1782.
85. Rein DC, Recupero AJ, Reed MP, Meyer RR (1990) in *The Eukaryotic Nucleus: Molecular Biochemistry and Macromolecular Assemblies* (Strauss P, Wilson SH, eds.). Telford Press, Caldwell, NJ, p. 95.
86. Skarnes W, Bonin P, Baril EF (1986) *JBC 261*:6629.
87. Mosbaugh DW, Meyer RR (1980) *JBC 255*:108.
88. Thomas DC, Rein DC, Meyer RR (1988) *NAR 16*:6447.
89. Wang TS-F, Korn D (1982) *Biochemistry 21*:1597.
90. Wang TS-F, Korn D (1982) *Biochemistry 21*:1597.
91. Zmudska BZ, SenGupta DN, Matsukage A, Cobianchi R, Kumar P, Wilson SH (1986) *PNAS 83*:5106.
92. Zmudzka BZ, Fornace A, Collins J, Wilson SH (1988) *NAR 16*:9587.
93. Fornace AJ, Zmudzka B, Hollander C, Wilson SH (1988) *Mol Cell Biol 9*:851.
94. Waser JU, Hübscher U, Kuenzle CC, Spadari S (1979) *EJB 97*:361; Hübscher U, Kuenzle CC, Spadari S (1979) *PNAS 76*:2316.

pol β, but no change is detected in the level of pol γ. In correction of a GT mispair by a thymine DNA glycosylase in extracts of human cells, pol β fills the single nucleotide gap in the role of short-patch repair.[95] Pol β may also function in recombination late in spermatocyte meiosis.[96]

6-5 DNA Polymerase γ[97]

Properties

Pol γ is the enzyme in mitochondria[98] responsible for replication of mitochondrial DNA. Its presence in nuclear preparations is attributable to contamination. When an RNA-directed DNA polymerase resembling the reverse transcriptase of tumor viruses was discovered in normal tissues,[99] it was at first regarded as a fourth type of cellular DNA polymerase. However, recognition that pol γ strongly prefers ribohomopolymer templates indicated that this mitochondrial enzyme could account for cellular "reverse transcriptase-like" activity.

Pol γ differs from viral reverse transcriptases in its inability to use natural RNA templates and its lack of associated RNase H activity (Section 9). The cellular and viral enzymes are also antigenically unrelated. Pol γ is antigenically distinct from pol α and β as well, and is readily separated from them by chromatography on DEAE-cellulose or DNA-cellulose. While pol β activity with a poly rA · oligo dT template is significant, it can be completely inhibited by phosphate; pol γ is fully active on this template in phosphate. Among other distinguishing features (Table 6-1) are its inhibition by both NEM and a dideoxy-NTP, resistance to aphidicolin, and stimulation by salt.

Pol γ purified from *Drosophila* embryos possesses two polypeptides of 125 and 35 kDa in equimolar amounts.[100] An enzyme of 140 kDa with similar properties has been isolated from *Xenopus laevis* ovaries.[101] Catalytically active polypeptides of smaller sizes[102] are probably proteolytic products, as observed with pol α. The polymerase function, remarkable for its ability to use a wide variety of template-primers—gapped DNA and a singly primed phage circle, as well as poly rA · $(dT)_{12-18}$—resides in the large polypeptide. The enzyme displays a very low K_m (1 μM) for dNTPs (at least tenfold lower than that of pol α or pol β) and shows a preference for dTTP over dUTP, not seen in other polymerases.[103] Processivity of the *Drosophila* enzyme is unusual in its synthesis of multiples of a 30-ntd unit.[104]

The limited number of DNA repair systems in mitochondria[105] may be

95. Wilbauer K, Kirieny J (1990) *PNAS 87*:5842.
96. Grippo P, Geremia R, Locorotondo G, Monesi V (1978) *Cell Differ 7*:237.
97. Wernette CM, Kaguni LS (1986) *JBC 261*:14764; Yamaguchi M, Matsukage A, Takahashi T (1980) *JBC 255*:7002; Yamaguchi M, Matsukage A, Takahashi T (1980) *Nature 285*:45.
98. Bolden A, Pedrali-Noy G, Weissbach A (1977) *JBC 252*:3351; Bertazzoni U, Scovassi AI, Brun GM (1977) *EJB 81*:237; Hübscher U, Kuenzle CC, Spadari S (1977) *EJB 81*:249; Tanaka S, Koike K (1978) *BBRC 81*:791; Dube DK, Kunkel TA, Seal G, Loeb LA (1979) *BBA 561*:369.
99. Fridlender B, Fry M, Bolden A, Weissbach A (1972) *PNAS 69*:452; Scolnick EM, Aaronson SA, Todaro GJ, Parks WP (1971) *Nature 229*:318.
100. Wernette CM, Kaguni LS (1986) *JBC 261*:14764.
101. Insdorf NF, Bogenhagen DF (1989) *JBC 264*:21491; Insdorf NF, Bogenhagen DF (1989) *JBC 264*:21498.
102. Yamaguchi M, Matsukage A, Takahashi T (1980) *JBC 255*:7002; Yamaguchi M, Matsukage A, Takahashi T (1980) *Nature 285*:45.
103. Mosbaugh DW (1988) *NAR 16*:5645.
104. Wernette CM, Conway MC, Kaguni LS (1988) *Biochemistry 27*:6046.
105. Tomkinson AE, Bonk T, Bartfeld N, Linn S (1988) *J Cell Biochem [Suppl] 12A*:278.

explained by the numerous copies of mitochondrial DNA—several thousand organelles per cell—and may also account for the fivefold more rapid rate of change of this genome compared to the nuclear genome. The low error rate of pol γ (near 1 per 10^6 nucleotides incorporated) can be attributed to a potent intrinsic $3' \rightarrow 5'$ exonuclease.[106] Because the activity is 15 to 30 times more selective in removing mismatched primer termini than matched ones, the rate of dNMP turnover is very low compared to that of other polymerases.

A unique anatomical feature of brain tissue makes it simple to separate mitochondria (in synaptosomes) from nuclei.[107] In the particulate synaptic end knobs, the presence of one species of mitochondrial DNA polymerase was established by fractionation, sedimentation, and electrophoretic properties. Its similarity to pol γ was evident from these criteria, in addition to its optimal metal and pH requirements and template preferences.[108] These findings agree with similar conclusions reached in studies of pol γ from HeLa cells, rat liver, and chick embryo.

Although pol γ is found in mitochondria of all mammalian cells, it is far less abundant than pol α and β. When each enzyme in a total cell lysate is assayed under optimal conditions, pol γ represents only about 2% as much activity as α and 10% to 25% as much as β. Levels of pol γ are generally not enhanced by factors that stimulate pol α.

Plant Chloroplast Polymerase

A polymerase purified from spinach chloroplasts[109] shares many properties with the mammalian mitochondrial enzyme: preference for a poly rA·$(dT)_{12-18}$ template, Mn^{2+} optimum of 0.1 to 1 mM, KCl optimum of 100 mM, pH optimum of 8 to 9, large size (105 kDa), resistance to aphidicolin, and sensitivity to NEM. In view of the resemblance of chloroplast DNA to mitochondrial DNA in its lack of associated nucleosomes and the presence of displacement loops in its replicative forms, the physical and functional similarities of the plant γ-like polymerase to the mammalian enzyme are not surprising.

6-6 DNA Polymerases of Fungi, Slime Molds, Protozoa, and Sea Urchins[110]

A phylogenetic survey[111] of polymerases from vertebrates to fungi is necessarily limited by the paucity of characterizations and the confusion introduced by proteolysis. In view of the remarkable similarity between yeast and mammalian polymerases, the different polymerase activities observed in the large assortment of lower eukaryotes will most likely, upon further study, fall into the relatively well defined types described in the preceding sections. The attractive-

106. Kunkel TA, Soni A (1988) *JBC 263*:4450; Kaguni LS, Wernette CM, Conway MC, Yang-Cashman P (1988) in *Cancer Cells 6: Eukaryotic DNA Replication* (Kelly TJ, Stillman B, eds.). CSHL, p. 425; Kaguni LS, Olson MW (1989) *PNAS 86*:6469.
107. Hübscher U, Kuenzle CC, Limacher W, Scherrer P, Spadari S (1978) *CSHS 43*:625; Hübscher U, Kuenzle CC, Spadari S (1977) *EJB 81*:249.
108. Hübscher U, Kuenzle CC, Limacher W, Scherrer P, Spadari S (1978) *CSHS 43*:625; Hübscher U, Kuenzle CC, Spadari S (1977) *EJB 81*:249.
109. Sala F, Amileni AR, Parisi B, Spadari S (1980) *EJB 112*:211.
110. Burgers PMJ (1989) *Prog N A Res 37*:235; Scovassi AI, Plevani P, Bertazzoni U (1980) *TIBS 5*:335.
111. Scovassi AI, Plevani P, Bertazzoni U (1980) *TIBS 5*:335.

ness of future studies of these lower eukaryotes rests on access to kilogram quantities of genetically and nutritionally defined cells subjectable to a variety of developmental controls and evolutionary patterns that shape the structure and functions of their polymerases.

Basidiomycete Polymerases[112]

The major DNA polymerase of the smut fungus, *Ustilago maydis*, has been purified to homogeneity. Although the major polymerase activity in crude extracts has a subunit size of 125 kDa, the purified enzyme contains a pair of subunits of 50 and 55 kDa, presumably the result of proteolysis. Aggregation of the 6.3S dimer to an 8.4S tetramer is favored by low salt or Mg^{2+}. The enzyme is inhibited by sulfhydryl-blocking agents.

In functional features, the enzyme resembles the prokaryotic more than the animal polymerases. Reversibility of polymerization by inorganic pyrophosphate is readily demonstrable. There is an associated 3′→5′ exonuclease with proofreading properties resembling those of *E. coli* pol I or phage T4 polymerase. A noncomplementary primer terminus is not extended and is rapidly removed by hydrolysis; a base-paired terminus is conserved. In its ability to hydrolyze DNA with a single-stranded 3′-terminal region more than 3 or 4 residues long, the *Ustilago* exonuclease resembles *E. coli* exonuclease I more than it does any of the DNA polymerases. An interesting dependence on monovalent cations for polymerization but not for exonuclease activity warrants a broader examination of the prokaryotic polymerases for this property.

The *Ustilago* enzyme can use a poly rA template as well as deoxyhomopolymers and activated thymus DNA. Although oligoribohomopolymers fail to serve as primers, transcribed RNA fragments used by prokaryotic polymerases and by DNA polymerase α have yet to be tested.

The major DNA polymerase of *Ustilago* is regarded as essential for nuclear DNA replication in the cell. Polymerase activity from mutants thermosensitive for nuclear DNA synthesis in vivo is also thermosensitive in cell extracts. Discrepancies between the thermosensitivity of nuclear and mitochondrial DNA synthesis in certain of these mutants suggest the existence of one or more additional polymerases or auxiliary replication proteins. A less active, high-molecular-weight DNA polymerase activity with a preference for poly dA·oligo dT over gapped DNA and an ability to use an RNA primer has been identified but not extensively purified.

Yeast Polymerases

Commercial baker's yeast or wild-type *Saccharomyces cerevisiae*, when grown aerobically to near stationary phase, each yields comparable amounts of polymerase activity upon extraction. In addition to the mitochondrial enzyme (pol γ), at least three nuclear polymerases (pol I, II, and III) have been isolated (Sections 2 to 5). The existence of a fifth DNA polymerase has been inferred from the typical polymerase sequence homology observed in *REV3*, a nonessential gene required for the mutagenic repair of UV damage,[113] but the polymerase activity has yet to be detected in a cell extract. Little attention has yet been given to the

---

112. Banks GR, Yarronton GT (1976) *EJB 62*:143; Spanos A, Sedgwick SG, Yarranton GF, Hübscher U, Banks GR (1981) *NAR 9*:1825; Hübscher U, Spanos A, Albert W, Grummt F, Banks GR (1981) *PNAS 78*:6771.
113. Matsukage A, Nishikawa K, Ooi T, Seto Y, Yamaguchi M (1987) *JBC 262*:8960.

fission yeast *Schizosaccharomyces pombé*, in which the linkage of mitotic processes to replication can be studied.

Polymerases of Slime Molds (Myxomycetes)[114]

Synchronously growing *Physarum polycephalum* is a widely used model of a differentiating unicellular system. Pol α purified from its exponentially growing microplasmodia is represented in two forms, 204 kDa and 116 kDa, with similar properties: sensitivity to sulfhydryl inhibitors, stimulation by 30 mM NaCl, template preference for activated DNA, and no template action of poly rA · dT_{12}.

The polymerase is strikingly inhibited by an anionic polyester of L-malic acid isolated from the microplasmodia.[115] The polymalate competes at low levels with the DNA binding site and can be antagonized by polyamines and histone H1. The nuclear location of the inhibitor, its specificity for pol α, and its cyclical production suggest some role in regulating DNA synthesis.

Protozoan Polymerases

Two DNA polymerases have been partially purified from *Euglena gracilis*,[116] a flagellate capable of photosynthetic or heterotrophic growth: polymerase A, 190 kDa, and polymerase B, 240 kDa. Polymerase B differs from polymerase A but resembles pol II of yeast in using poly rA as template and in having an associated exonuclease activity. Further purification is needed to establish the nature of associated nucleases and to compare the polymerases with those from mitochondria and chloroplasts in the same cell.

From *Tetrahymena pyriformis*,[117] a ciliated protozoan, two highly purified DNA polymerases have also been obtained. One was induced by DNA damage and has been regarded as a highly amplified form of a mitochondrial enzyme. The other was derived from a nonmitochondrial fraction of untreated, exponentially growing cultures. Each source yielded an 80-kDa enzyme. Their generally similar physical and functional features make it uncertain whether the "induced mitochondrial" enzyme is distinct from the "nuclear" one.

Replication of DNA in the macronucleus of paramecia[118] is concerned with gene expression and vegetative existence, while in the micronucleus it is responsible for maintaining the germ line. Three polymerases were separated by chromatographic procedures from macronuclei that had been freed of micronuclei and contaminating bacteria. The polymerases had similar properties and molecular weights near 100,000, but varied in abundance with the growth cycle of the cultures.

Sea Urchin Polymerases[119]

The DNA polymerase isolated from nuclei of blastula-stage embryos of *Strongylocentrotus purpuratus* was one of the first characterizations of an α-type en-

---

114. Baer A, Schiebel W (1978) *EJB 86*:77; Weber C, Fischer H, Holler E (1988) *EJB 176*:199.
115. Holler E, Fischer H, Weber C, Stopper H, Steger H, Simek H (1987) *EJB 163*:297; Fischer H, Erdmann S, Holler E (1989) *Biochemistry 28*:5219.
116. McLennan AG, Keir HM (1975) *BJ 151*:227; McLennan AG, Keir HM (1975) *BJ 151*:239.
117. Sakai A, Watanabe Y (1982) *J Biochem 91*:845; Sakai A, Watanabe Y (1982) *J Biochem 91*:855; Ostergaard E, Brams P, Westergaard O, Nielsen OF (1987) *BBA 908*:150.
118. Tait A, Cummings DJ (1975) *BBA 378*:282.
119. Loomis LW, Rossomando EF, Chang LMS (1976) *BBA 425*:469; Morris PW, Racine PM (1978) *NAR 5*:3959; Baril EF, Scheiner C, Pederson T (1980) *PNAS 77*:3317.

zyme with an associated primase activity. It was considered the principal polymerase in the developing embryo. With improved separation techniques, additional polymerases have been found in eggs and embryos. One of these has the physical and functional properties of the β polymerase from mammalian sources. The enzyme sediments at 3.3S, is insensitive to NEM, uses poly dA · oligo dT as a template 200 times more effectively than activated DNA, and is sensitive to phosphate with a homopolymer template. The striking similarity of the sea urchin β-type polymerases to the human enzyme suggests a conservation of structure over a long evolutionary span.

Polymerases from Other Unicellular Eukaryotes

A survey[120] of polymerases from *Neurospora*, *Chlorella*, *Chlamydomonas*, and *Trypanosoma* makes it clear that the information is too fragmentary to reach conclusions about the nature of the replication system in any of these organisms or their relationship to the characterized animal and yeast enzymes.

6-7 DNA Polymerases in Growth and Development

Studies of growing, differentiating, and aging animal cells and tissues can provide both information about the functions of polymerases and insights into controls of replication. Among many tissues examined, there is a clear association between DNA polymerase activity, generally identified as α, and mitotic activity. With recognition of the critical role of pol δ in replication, earlier measurements need to be reexamined. As assays and separation methods become more refined, the relative abundances of polymerases and changes in their structures and properties can be detected and evaluated.

An important distinction has been observed between two techniques for cell-cycle studies. When a cell cycle is initiated from a G_0 quiescent state (e.g., serum starvation), the entry into S phase is accompanied by a sharp rise in pol α activity. However, when ultracentrifugal elutriation is used to separate asynchronously growing cells into their several cell-cycle-stage groups, the level of pol α activity, the level of its mRNA, and the nascent protein synthesis rate are each virtually the same in all stages.[121]

Numerous systems are attractive for study (Table 6-2). The sea urchin, often used for embryological studies, yielded one of the first homogeneous DNA polymerases. Each system has distinctive features to recommend it: (1) a proliferative response to mitogenic stimuli by cells in culture, (2) extraordinarily rapid DNA synthesis in fertilized eggs and developing embryos, (3) striking tissue changes during embryonic and perinatal development and in senescence, and (4) finely programmed regeneration of a tissue after its partial extirpation.

Cellular Proliferation in Postnatal Spleen Development[122]

Late in the rat's gestation period, two days before birth, when the rate of cell division in the embryonic spleen is very low, the percentage distributions of

120. Burgers PMJ (1989) *Prog N A Res 37*:235.
121. Wong SW, Wahl AF, Yuan PM, Arai N, Pearson BE, Arai K, Korn D, Hunkapiller MW, Wang TSF (1988) *EMBO J 7*:37; Wahl AF, Geis AM, Spain BH, Wong SW, Korn D, Wang TS-F (1988) *Mol Cell Biol 8*:5016.
122. Knopf KW, Yamada M, Weissbach A (1976) *Biochemistry 15*:4540; Hübscher U, Kuenzle CC, Spadari S (1977) *NAR 4*:2917.

Table 6-2
Growth and development systems for studies of DNA replication

Cell or tissue	Distinctive features
Lymphocytes, fibroblasts in culture[a]	mitogenesis by lectins and other agents
Eggs and embryos (*Drosophila*)[b]	DNA doubling in 3 minutes; plentiful source
Eggs and embryos (*Xenopus*, sea urchin)[c]	active, abundant source
Testes (mouse)[d]	spermatogenesis
Spleen and heart (mouse, rat)[e]	perinatal development and aging
Brain (rat)[f]	neuronal development
Liver (rat)[g]	regeneration after partial removal

[a] Bertazzoni U, Stefanini M, Pedrali-Noy G, Giulotto E, Nuzzo F, Falaschi A, Spadari S (1976) *PNAS 73*:758; Coleman MS, Hutton IJ, Bollum FJ (1974) *Nature 248*:407; Jazwinski SM, Wang JL, Edelman GM (1976) *PNAS 73*:2231; Mayer R, Smith GR, Gallo RC (1975) *Blood 46*:509; Wahl HF, Geis AM, Spain BH, Wong SW, Korn D, Wang TS-F (1988) *Mol Cell Biol 8*:5016.
[b] Brakel CL, Blumenthal AB (1977) *Biochemistry 16*:3137; Brakel CL, Blumenthal AB (1978) *NAR 5*:2565.
[c] Fansler B, Loeb LA (1972) *Exp Cell Res 75*:433; Hobart PM, Infante AA (1978) *JBC 253*:8229.
[d] Chevallier P, Phillipe M (1976) *Exp Cell Res 99*:237; Hecht NB, Farrel D, Davidson D (1976) *Dev Biol 48*:56.
[e] Hübscher U, Kuenzle CC, Spadari S (1977) *NAR 4*:2917; Claycomb WC (1975) *JBC 250*:3229.
[f] Hübscher U, Kuenzle CC, Spadari S (1977) *NAR 4*:2917.
[g] William LE, Surrey S, Lieberman I (1975) *JBC 250*:8179; La Brecque DR, Feigenbaum A, Bachur NR (1978) *Science 199*:1082.

DNA polymerases α, β, and γ are 1%, 85%, and 14%, respectively. By the second day after birth, dividing and differentiating lymphocytes in the germinal centers have nearly doubled in DNA content, and the distribution of DNA polymerases has changed dramatically: their relative activities are now 74%, 21%, and 5%. In the adult animal, a persistently high level of pol α activity (about 45% of the total) reflects the sustained growth and turnover of the lymphocyte population.

Postnatal Differentiation of Rat Brain and Cardiac Muscle

DNA pol α activity in forebrain neurons is highest several days before birth.[123] With the decline in mitotic activity during terminal differentiation of the neuron, pol α activity declines precipitously and vanishes by the 23rd postnatal day (Fig. 6-1). On the other hand, the activities of pol β and pol γ do not alter appreciably during perinatal development and remain at measurable levels in the adult tissue. The relative prenatal levels of polymerases α, β, and γ of about 70%, 30%, and 2% change to adult levels of 0, 95%, and 5%. More recent assays of pol δ levels during the course of rat neuronal development[124] show them to be coincident with pol α levels; presumably, the two enzymes are coordinated in function.

The DNA polymerase patterns in ventricular cardiac muscle[125] during perinatal development are qualitatively similar to those in neurons. They correlate with the virtually complete cessation of ventricular cellular proliferation before birth. All three DNA polymerase activities are low by the second prenatal day. Pol α can no longer be assayed by the seventh postnatal day, but the activities of β and γ persist into adulthood. The levels of pol α and pol δ largely coincide.

123. Knopf KW, Yamada M, Weissbach A (1976) *Biochemistry 15*:4540; Hübscher U, Kuenzle CC, Spadari S (1977) *NAR 4*:2917.
124. Spadari S, Focher F, Hübscher U (1988) *In Vivo 2*:317.
125. Zhang SJ, Lee MYWT (1987) *Arch B B 252*:24.

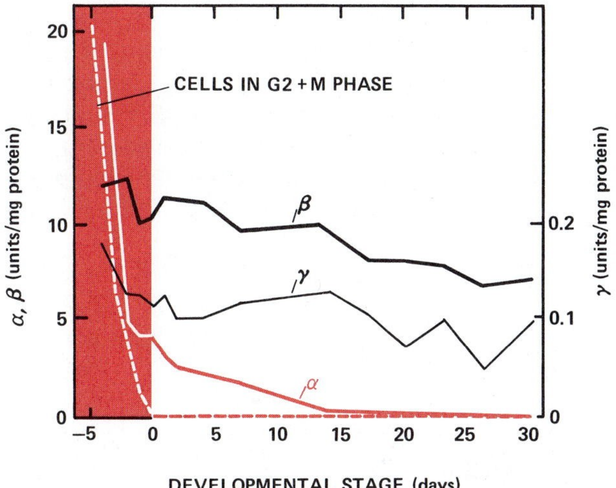

Figure 6-1
DNA polymerase levels during postnatal differentiation of rat brain. Crude extracts of neuronal (forebrain) tissue were assayed for polymerases α, β, and γ. The decline in percentage of cells (left ordinate) in G_2 and M phases indicates the absence of proliferation in neuronal tissue postnatally.

6-8 DNA Virus–Induced DNA Polymerases

The animal DNA viruses (Chapter 19) can be placed in three groups with respect to their dependence on DNA polymerases. (1) The large herpes and vaccinia viruses and the intermediate size adenovirus, analogous to the T phages, encode specific DNA polymerases as well as other replication proteins. (2) The small oncogenic viruses (polyoma, SV40), analogous to the small DNA phages ϕX174 and M13, rely on host cellular DNA polymerases and induce them to increase their level of activity. (3) The small hepatitis B virus carries within its virion a DNA polymerase with unique properties that resembles the reverse transcriptases (Section 9).

Herpesvirus Polymerase[126]

Herpesviruses (Section 19-5) attract great interest because of their associations with cancer, with latent infections causing recurrent disease, and with severe outbreaks in immunodeficient states (e.g., AIDS). The many herpesviruses include herpes simplex virus (HSV-1, HSV-2), which infect human mucous membranes; varicella–zoster virus, which causes chicken pox and shingles; human cytomegalovirus (CMV); Epstein-Barr virus (EBV) of infectious mononucleosis and Burkitt's lymphoma; equine herpes virus; and Marek's disease virus, which causes malignant lymphoma in fowl.

The HSV-encoded polymerase is a nuclear enzyme. Upon infection, the activity exceeds that of the host polymerases several-fold. The purified enzyme of 136 kDa is complexed with the product of herpes gene *UL42* (unique long 42), a duplex DNA–binding protein of about 60 kDa.[127] The enzyme is inhibited by aphidicolin and dideoxy-NTPs, and also shares the characteristic six domains of the eukaryotic polymerases (Section 5-11). However, it is distinguishable from pol α by several features: (1) it possesses a $3' \rightarrow 5'$ exonuclease, (2) it lacks a tightly associated primase, (3) it is inhibited by phosphonoacetate and phos-

126. Challberg MD, Kelly TJ (1989) *ARB 58*:671; Marcy AI, Olivo PD, Challberg MD, Goen DM (1990) *NAR 18*:1207.
127. Crute JJ, Lehman IR (1989) *JBC 264*:19266.

phonoformate, (4) it is stimulated, rather than severely inhibited, by high salt, (5) it is more sensitive to certain nucleoside analogs (e.g., acyclovir triphosphate), (6) it possesses an endonucleolytic RNase H activity which resembles the 5′→3′ exonuclease of *E. coli* pol I in attacking DNA as well as RNA,[128] and (7) it displays distinctive chromatographic properties. The enzyme is also immunologically distinct from all host polymerases and relies on six other herpes-encoded polypeptides (including a helicase–primase and SSB) for replication from a unique herpes origin.

Although the enzyme is highly processive on an ssDNA template, the rate is very slow, about 5 ntd per second. The rate is stimulated 20-fold by *E. coli* SSB and to a lesser extent by ICP8, the herpes SSB. Presumably, other factors account for the efficiency of replication in vivo.

Potent inhibitors of the enzyme act at the PP_i binding site. Phosphonoacetate binds with a K_i near 1 μM, 1000-fold tighter than PP_i. Whereas the toxicity of phosphonoacetate precludes its clinical use, phosphonoformate,[129] a comparable inhibitor, has significant therapeutic value. The most effective agent in inhibiting herpes simplex and varicella–zoster (but not CMV) replication is acyclovir (acycloguanosine),[130] which owes its effectiveness to its being a substrate for the herpes-encoded (but not the host) thymidine kinase. (Because CMV, unlike herpes, encodes no thymidine kinase, acyclovir is not converted to the triphosphate, and CMV is therefore unaffected by the drug.) After successive conversions of acyclovir to the mono-, di-, and triphosphate forms, the analog is incorporated by the herpes polymerase in place of dGTP and, like a dideoxy-NTP, terminates the elongation of a chain. Furthermore, the K_m of the acyclovir NTP (0.1 μM) for the herpes polymerase is 30-fold lower than for the host pol α. Analog-terminated chains, while very tightly bound by the polymerase, are not substrates for its 3′→5′ exonuclease activity. The acyclovir triphosphate not only aborts chain elongation, but also inactivates the herpes polymerase by forming a tight dead-end complex when the next nucleotide encoded by the template is bound.[131]

Mutations in several positions in the viral *pol* gene alter sensitivity to drugs that are relatively specific for the viral polymerase.[132] These mutations identify regions of the protein that participate in substrate recognition (e.g., PP_i and dNTPs). Inasmuch as the loci are in several of the conserved regions of the polymerase[133] and confer simultaneous resistance to analogs of PP_i (e.g., phosphonoformate) and of nucleosides (e.g., acyclovir triphosphate or aphidicolin), it seems likely that substrate binding depends on interactions between these regions.

Vaccinia Virus Polymerase

DNA polymerase encoded by vaccinia, a cytoplasmic poxvirus[134] (Section 19-6), is localized in the cytoplasm. Purified to homogeneity, the single polypeptide is

---

128. Crute JJ, Lehman IR (1989) *JBC 264*:19266.
129. Eriksson B, Larsson A, Helgstrand E, Johansson N-G, Oberg B (1980) *BBA 607*:53.
130. Furman PA, St Clair MH, Fufe JA, Rideout JL, Keller PM, Elion GB (1979) *J Virol 32*:72; Miller WH, Miller RL (1980) *JBC 255*:7204; Darby G, Field HJ, Salisbury SA (1981) *Nature 289*:81; Dorsky DJ, Crumpacker CS (1987) *Ann Int Med 107*:859.
131. Reardon JE, Spector T (1989) *JBC 264*:7405.
132. Gibbs JS, Chiou HC, Bastow KF, Cheng Y-C, Coen DM (1988) *PNAS 85*:6672; Hall JD, Wang Y, Pierpont J, Berlin MS, Rundlett SE, Woodward S (1989) *NAR 17*:9231.
133. Larder BA, Kemp SD, Darby G (1987) *EMBO J 6*:169.
134. Challberg MD, Englund PT (1979) *JBC 254*:7812; Challberg MD, Englund PT (1979) *JBC 254*:7820.

about 110 kDa. Its level of activity in cytoplasmic extracts of HeLa cells 6 hours after infection is at least ten times that of DNA polymerase α. Unlike the host pol α, the enzyme has an active 3′→5′ exonuclease. The polymerase is ineffective in using an ssDNA template, due to pausing at secondary structures. Like pol α and the herpes polymerase, the vaccinia enzyme is strongly inhibited by aphidicolin.

Adenovirus Polymerase[135]

This viral DNA polymerase is a 140-kDa polypeptide responsible for replicating the entire 35-kb linear duplex of adenoviruses. The enzyme is remarkable for catalyzing two types of synthetic events: (1) the start of a DNA chain by covalent extension from the β-hydroxyl of a particular serine residue in a virus-encoded, 80-kDa terminal protein, and (2) the conventional elongation of a chain from a primer terminus. The start, by a template-directed insertion of a cytidyl deoxynucleotide at each of the identical ends of the genome, is greatly facilitated by host proteins which bind near these ends. These proteins, nuclear factor I (NFI, CTF) and nuclear factor III (NFIII, OTF-1, ORP-C), also play important roles in transcription.

Once started, highly processive replication by the polymerase relies on a specific interaction in a 1 : 1 complex with a virus-encoded 59-kDa phosphoprotein which binds ssDNA; SSBs from other sources are inactive. In its dual capacity to exploit protein priming and to achieve continuous and complete replication of a genome, the adenoviral polymerase is unique among animal virus enzymes, but shares these properties with a number of phage polymerases, of which that of ϕ29 in *B. subtilis* (Section 5-10) is the best-studied example.

Beyond these special features, the adenoviral polymerase is distinguished from any one of the host polymerases by (1) sensitivity to dideoxy-NTPs and resistance to aphidicolin, (2) a 3′→5′ exonuclease, and (3) use of a variety of deoxyhomopolymers as template-primers, but failure to use an RNA primer.

Baculovirus Polymerase[136]

The nuclear polyhedrosis viruses (e.g., *Autographa californica* of lepidoptera and the virus of the silkworm, *Bombyx mori*) encode a polymerase whose domains strongly resemble those of herpesviruses, poxviruses, adenoviruses, and eukaryotic cells. The 114-kDa polypeptide includes a distributive 3′→5′ exonuclease activity.

6-9 RNA-Directed DNA Polymerases: Reverse Transcriptases[137] (RTs) and Telomerase[138]

DNA polymerases that are directed by ssRNA perform transcription in reverse. These polymerases were first found among the RNA tumor viruses (retrovir-

---

135. Challberg MD, Kelly TJ (1989) *ARB 58*:671; Stillman B (1989) *Ann Rev Cell Biol 5*:197; Watson CJ, Hay RT (1990) *NAR 18*:1167.
136. Miller LK (1988) *Ann Rev Microbiol 42*:177; Maeda S (1989) *Ann Rev Entomol 34*:351.
137. Varmus H, Swanstrom R (1985) in *Molecular Biology of RNA Tumor Viruses* (Weiss R, Teich N, Varmus H, Coffin J, eds.). CSHL, p. 75; Mason W, Taylor J, Hull R (1987) *Adv Virus Res 17*:129.
138. Greider CW, Blackburn EH (1989) *Nature 337*:331; Morin GB (1989) *Cell 59*:521; Shippen-Lentz D, Blackburn EH (1990) *Science 247*:546.

uses), where they convert the RNA genome into duplex DNA which is then integrated and maintained in the host chromosome as a provirus. Since this initial discovery, information flow from RNA to DNA (reverse transcription) has been observed in eukaryotes from yeast to humans, as well as in bacteria.[139]

In addition to the retroviral reverse transcriptases (RTs), another such enzyme has been identified in certain DNA viruses (e.g., hepatitis B; Section 19-8). The existence of an RT has also been inferred in various species from a genetic sequence. Retrovirus-like transposable elements, called *retrotransposons*, have been identified in yeast *(Ty)*, in *Drosophila (copia)*, and in rodents *(IAP)*. A candidate RT activity has been observed in yeast, but has not been fully characterized.[140] The *telomerases*, enzymes that synthesize the unique termini of chromosomes under direction by RNA, are in some ways functionally analogous to the RTs.

Besides these clear and putative examples of reverse transcriptase, there is the presumption of such an activity in the origin of certain sequences in eukaryotic genomes. In some species, up to 10% of the DNA, including the *Alu* repeat elements, is made up of pseudogenes called "processed retropseudogenes."[141] These appear to be DNA copies of a huge variety of RNA polymerase II and III processed transcripts, resembling cDNAs. Whether the presumed RT action responsible for these sequences is due to an infectious agent or to a cellular enzyme with a biological function remains to be determined in each instance.

Retroviral Reverse Transcriptases

Despite the extremely small amounts of source material available, RTs were purified from several RNA tumor viruses; they can now be obtained in substantial quantities by cloning and amplification.[142]

Functions of Retroviral RTs. These efficient polymerases, located in the virion core particle, perform a most remarkable variety of functions. First, the bifunctional DNA polymerase domain synthesizes a DNA–RNA duplex using a tRNA primer and the viral 70S RNA genome as a template. Next, the endonucleolytic RNase H domain specifically and extensively cleaves the RNA, preserving a primed DNA template.[143] In the elaborately choreographed transfer of the growing DNA chain from one template to another, the RNase H plays the essential role of removing the RNA template from the newly synthesized DNA, thereby allowing the primer to engage the new template. Finally, the DNA polymerase domain of the RT uses the DNA as template to produce a DNA duplex suitable for integration into the host genome. In each case, the RT in vivo is complexed with its template—in effect, it is committed to that particular template.

Avian, Murine, and Feline RT Structures. The structures isolated from avian tumor viruses (avian myeloblastosis and chicken Rous sarcoma) differ from those of murine and feline leukemia viruses. The avian enzyme, near 160 kDa, is made up of two structurally related subunits of 63 kDa (α) and 95 kDa (β); the murine and feline molecules are a single subunit of about 80 kDa.

139. Lim D, Maas WK (1989) *Cell 56*:891; Lampson BC, Sun J, Hsu MY, Vallejo-Ramirez J, Inouye S, Inouye M (1989) *Science 243*:1033.
140. Garfinkel DJ, Boeke JD, Fink GR (1985) *Cell 42*:507.
141. Baltimore D (1985) *Cell 40*:481; Weiner AM, Deininger PL, Efstratiadis A (1986) *ARB 55*:631.
142. Roth MJ, Tanese N, Goff SP (1985) *JBC 260*:9326; Soltis DA, Skalka AM (1988) *PNAS 85*:3372.
143. Krug MS, Berger SL (1989) *PNAS 86*:3539.

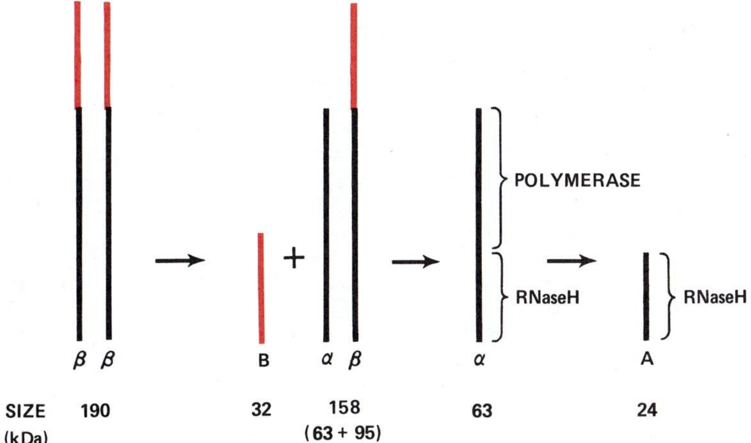

Figure 6-2
Avian reverse transcriptase: $\beta\beta$ precursor, $\alpha\beta$ holoenzyme, subunits, and derivatives.

The avian holoenzyme ($\alpha\beta$), the most abundant form of the enzyme, appears to be derived by proteolytic cleavage from a minor, less active, $\beta\beta$ precursor (Fig. 6-2). The β subunit is phosphorylated, but the α subunit, derived from it by cleavage, is not.[144] Thus the phosphorylated peptide is removed by proteolysis or is dephosphorylated during proteolysis. The released 32-kDa B fragment is a DNA endonuclease (integrase) and is found in virions.[145] The α subunit, which may also be isolated from virions, has both polymerase and RNase H activities, but the holoenzyme may be required for optimal template-primer binding[146] and processive polymerase action. Sustained proteolysis in vitro yields a 24-kDa fragment A that manifests only RNase H activity; its amino acid sequence is found in the α and β subunits but not in fragment B.[147] The avian α subunit and the murine enzyme appear to be distributive. No evidence has been found in murine virions of an $\alpha\beta$ holoenzyme form of RT.

The domain structures of the murine enzyme have been defined by refined mutational analysis and expression from bacterial plasmids: both DNA polymerase activities are confined to the N-terminal two-thirds of the gene, and the RNase H is in the remaining one-third; the activities encoded by each of the two domains can be expressed from plasmids independently.[148]

Properties of Retroviral RTs. The viral enzymes resemble most known DNA polymerases in the $5' \rightarrow 3'$ direction of synthesis, catalysis of PP_i exchange, and pyrophosphorolysis. They also require a template, a 3′-OH primer terminus, the four dNTPs, and Mg^{2+} (or Mn^{2+}). There is no detectable exonuclease activity. The RTs display a variety of template-primer preferences. One case[149] (Table 6-3) illustrates the need for template and primer and the effectiveness of homopolymers, especially the ribopolymers. There is a sharp preference for poly rA·oligo dT over poly dA·oligo dT. Surprisingly, the ribopolymer poly rU is ineffective whereas the deoxypolymer poly dC is utilized. The enzyme can use natural RNAs as templates provided an oligomer of dT is present to serve as

144. Hizi A, Joklik WK (1977) *Virology 78*:571.
145. Golomb M, Grandgenett DP (1979) *JBC 254*:1606.
146. Modak MJ, Marcus SL (1977) *J Virol 22*:243; Hizi A, Leis JP, Joklik WK (1977) *JBC 252*:6878.
147. Lai MT, Verma IM (1978) *J Virol 25*:652.
148. Tanese N, Goff SP (1988) *PNAS 85*:1777; Jean-Jean O, Moyret C, Bernard D, de Recondo A-M, Rossignol J-M (1989) *BBRC 158*:595; Repaske R, Hartley JW, Kavlick MF, O'Neill RR, Austin JB (1989) *J Virol 63*:1460.
149. Baltimore D, Smoler D (1971) *PNAS 68*:1507.

Table 6-3
Use of homopolymer template-primers by retroviral RT[a]

	Templates and synthesis rates[b]			
Primer	Ribopolymer		Deoxyribopolymer	
dT_{10}	rA	100	dA	1
dA_{14}	rU	1	dT	1
dG_{12}	rC	50	dC	25
dC_{14}	rI	25	dI	1

[a] From Baltimore D, Smoler D (1971) *PNAS 68*:1507.
[b] Relative rates of DNA synthesis; no activity was observed with any primer alone or with any template alone.

primer (Table 6-4). In the presence of the oligomer, even the excellent 70S RNA template-primer is more effective, and various mRNAs prove to be very efficient templates. Because of the location of the polyadenylate tail at the 3′ end of mRNA and of the primer tRNAs in the 70S RNA complex, the annealed oligo dT stimulates replication greatly because it can prime synthesis for the full length of the template.

The weakly processive viral enzyme bears some resemblance to pols II and III of *E. coli* in failing to use nicked DNA duplexes and in preferring a DNA template-primer with short single-strand gaps. The enzyme fills such gaps completely, as judged by the capacity of ligase to seal the chains.

The viral polymerase is often distinguished from the host DNA pol α by its capacity to use ribopolymers, particularly poly $rC \cdot dG_{12}$, under standard conditions.[150] It is unique among all DNA polymerases in two respects: (1) the capacity to use the 70S viral RNA complex, thereby producing a DNA product that hybridizes specifically and completely with this RNA, and (2) the possession of an inherent RNase H activity. This RNase H can act as a processive exonuclease that specifically degrades the RNA of a RNA–DNA hybrid starting from either the 5′ or 3′ end. The oligonucleotide products are two to eight residues long, each terminated in a 3′-OH and a 5′-P. The RNase H of RT also possesses endonucle-

Table 6-4
Use of natural template-primers by retroviral RT

	dT_{12-18}	
Template	Absent	Present
Poly rA	<1[a]	100[a]
70S RNA	6	25
70S RNA, denatured	<1	17
HeLa rRNA	<1	<1
Globin mRNA	<1	108
Ovalbumin mRNA	<1	60

[a] Rates of DNA synthesis relative to poly rA · dT_{12-18} set at 100.

150. Kiessling AA, Goulian M (1976) *BBRC 71*:1069.

ase activity[151] that is difficult to detect because of its substrate specificity, implied in its capacity to generate a correctly positioned primer for synthesis of the viral DNA (+) strand.[152] *E. coli* RNase H displays a comparable specificity in its action on the ColE1 primer (Section 18-2).

Even though the RT replicates 70S viral RNA and shows a general template preference for a ribohomopolymer over a deoxy chain, the viral enzyme does copy DNA as well as RNA templates and in vivo produces a duplex DNA product. It should be recalled that the classic DNA-directed DNA polymerases can also use ribohomopolymer templates, sometimes even preferentially over deoxypolymers (Section 4-15).

Preferences among secondary structures of DNA templates vary among different forms of the avian RT. Certain preparations are more active than others in utilizing nicked DNA and in repairing lengthy gaps in DNA.[153] Conceivably, once the behavior of RNA-directed viral polymerases under a variety of conditions is better understood, the distinctions between them and the DNA-directed enzymes will appear to be even less sharp than they seem now.

Polymerase activity can be measured in virus particles rendered permeable to substrates by damaging the lipid envelope with nonionic detergents or diethyl ether. The activities of the enzyme in the virus particle and those of the purified enzyme are remarkably similar. Yet the replication of related RNA viruses and the progress of infection may differ markedly due to the conditions in the various cells they infect.

HIV (Human Immunodeficiency Virus) RT[154]

HIV-1, the agent responsible for acquired immunodeficiency syndrome (AIDS), possesses an RT that resembles other retroviral RTs in most respects of organization, structure, and function, but with several unique characteristics: (1) the enzyme is made up of two subunits of 66 and 51 kDa, the latter most likely derived from cleavage of one of the polypeptides in a homodimer of 66-kDa chains; (2) it contains a flexible, nonessential region—detected by insertion mutagenesis—in the middle, with the polymerase domain, localized as with other RTs, in the N-terminal portion; (3) it has the C-terminal RNase H of the 66-kDa subunit in a separable 15-kDa polypeptide not found in the purified enzyme, although it is present in the virion in a 1:1 ratio with the 51-kDa subunit;[155] and (4) it responds to a small insertion in this C-terminal region with a loss of polymerase activity, presumably due to interference with proper folding of the enzyme.[156]

The kinetic features[157] examined with poly rA·oligo dT reveal an ordered mechanism similar to that of *E. coli* pol I and mammalian pol α, with a strong affinity for binding the primer portion of the template-primer. Processivity varies with the template, in the following order: poly rA > ssRNA > ssDNA >

---

151. Krug MS, Berger SL (1989) *PNAS 86*:3539.
152. Rattray AJ, Champoux JJ (1989) *JMB 208*:445.
153. Leis J, Hurwitz J (1972) *PNAS 69*:2331.
154. Majumdar C, Abbotts J, Broder S, Wilson SW (1988) *JBC 263*:15657; Prasad VR, Goff SP (1989) *PNAS 86*:3104; Wong-Staal F (1988) in *HIV and Other Highly Pathogenic Viruses* (Smith RA, ed.), p. 33. Academic Press, San Diego.
155. Hansen J, Schulze T, Mellert W, Moelling K (1988) *EMBO J 7*:239.
156. Hizi A, Barber A, Hughes SH (1989) *Virology 170*:326.
157. Huber HE, McCoy JM, Seehra JS, Richardson CC (1989) *JBC 264*:4669; Kedar PS, Abbotts J, Kovacs T, Lesiak K, Torrence P, Wilson SH (1990) *Biochemistry 29*:3603.

dsDNA > poly dA. Noteworthy features are strand displacement synthesis on dsDNA and sliding on or switching between poly rA templates to generate products longer than the template chain.

The great diversity of the HIV-1 genome may be the consequence of hypermutability in its replication, due at least in part to the very high error rate of the RT of HIV-1. As observed in vitro, a misincorporation rate of 1 per 2000 to 4000 nucleotides polymerized ranks this enzyme as the least accurate of the RTs examined.[158]

Hepatitis B [HBV, Hepadna (Hepatitis DNA)] RT[159]

The HBV genome (Section 19-8), a partially duplex, 3-kb DNA (Fig. 19-18), is replicated via an mRNA intermediate which is then transcribed back into viral DNA. The presumed RT, a 90-kDA polypeptide, has been identified in the core particle of a human HBV line,[160] but has not been purified and characterized. Several enzymatic activities are found in particles isolated from the serum of HBV-infected ducks; among these are DNA polymerase actions on the endogenous HBV template and on poly dA·$(dT)_{12-18}$, and a RT action on poly rA·dT_{12}. Of several inhibitors tested,[161] aphidicolin is inactive, but all three HBV RT activities are inhibited by chloroquine, an intercalator (Section 14-4) also active against retroviral enzymes, and by phosphonoformate, also active against herpes polymerase.

Most of the information about the HBV RT is inferred from the sequence of the *pol (P)* gene, one of the three open reading frames of the virus.[162] The *P* gene encodes a polyprotein containing the terminal protein in the N-terminal part, followed by a nonessential region and then the DNA polymerase and RNase H domains. Protease action separates the terminal protein from the rest some time before or after the terminal protein acts in priming. The domains encoding the enzyme activities show significant homologies in organization and sequence to those of the retroviral RTs.[163]

Telomerase[164]

Short sequences tandemly repeated up to 100 times *(telomeres)* terminate the duplex linear chromosomes of organisms from yeast to mammals. The highly conserved sequences [e.g., $(TTGGGG)_n$ at the 3′ ends in *Tetrahymena;* $(TTAGGG)_n$ in humans] are synthesized by a partially purified polymerase in the absence of a DNA template. Instead, a sequence contained within an intrinsic RNA (about 160 ntd long) of the enzyme very likely provides the template. An analogy to ribozyme catalysis in splicing[165] has been suggested;[166] alterna-

158. Preston BD, Poiesz BJ, Loeb LA (1988) *Science 242*:1168; Roberts JD, Bebenek K, Kunkel TA (1988) *Science 242*:1171.
159. Ganem D, Varmus HE (1987) *ARB 56*:651.
160. Bavand MR, Laub O (1988) *J Virol 62*:626.
161. Offensperger W-B, Walter E, Offensperger S, Zeschnigk C, Blum HE, Gerok W (1988) *Virology 164*:48.
162. Bartenschlager R, Schaller H (1988) *EMBO J 7*:4185; Chang L-J, Pryciak P, Ganem D, Varmus HE (1989) *Nature 337*:364.
163. Miller RH (1988) *Virology 164*:147.
164. Greider CW, Blackburn EH (1989) *Nature 337*:331; Morin GB (1989) *Cell 59*:521; Shippen-Lentz D, Blackburn EH (1990) *Science 247*:546.
165. Cech TR, Bass BL (1986) *ARB 55*:599; Been MD, Cech TR (1988) *Science 239*:1412.
166. Greider CW, Blackburn EH (1989) *Nature 337*:331.

tively, the RNA-templated action by telomerase can be viewed as akin to that of RTs[167] and some DNA polymerases in the optional use of an RNA template.

6-10 Terminal Deoxynucleotidyl Transferase (TdT)[168]

Found in extracts of the thymus gland and bone marrow, TdT behaves like a DNA polymerase in synthesizing a DNA chain by $5' \rightarrow 3'$ polymerization of dNTPs. Unlike a DNA polymerase, the enzyme neither requires nor copies a template.

Functions and Properties

How TdT serves the cell is still unclear. The role of thymus cells in antibody synthesis suggests some novel genetic chemistry in which the enzyme contributes to antibody diversity. Overexpression of TdT in B cells leads to increased nucleotide insertions at the junctions of the V, D, and J regions of immunoglobulins. Although the physiologic role of TdT may still be vague, its usefulness in the synthesis of homopolymers and in recombinant DNA technology has been unequivocal. As a reagent, it ranks with polynucleotide phosphorylase, whose random assembly of ribonucleotides was so important for making the homopolymers and mixed RNA polymers that helped break the genetic code.

TdT is also found in neoplastic cells from patients with acute lymphoblastic leukemia and has been used diagnostically as a marker for the presence of immature (blast) T cells.[169] The enzyme is found in T cell leukemias and lymphomas but not in normal B cells or in neoplastic cells with B cell characteristics.

TdT resembles all other polymerases in the mechanism of chain elongation, but differs in not needing a template nor being directed by one. The enzyme, as prepared from thymus glands, is a 58-kDa polypeptide[170] with a high degree of homology from birds to humans. Active forms of the enzyme, as small as 32 kDa, are known to be proteolytic cleavage products. TdT is similar to pol β in its small size (Section 4) and is structurally related to pol β;[171] further, TdT is nonprocessive and exhibits no PP_i exchange, pyrophosphorolysis, or exonuclease activities.

The major conservation of size and structure of TdT, its occurrence only in certain classes of thymic prelymphocytes, and the nature of its in vitro actions all point to an important, though still undefined, function for the enzyme. A plausible role is in the generation of somatic diversity at the junctions of V, D, and J regions of immunoglobulins and T-cell receptor genes[172] (Section 21-6). The isolation from young thymic nuclei of a large TdT complex (1600 kDa) containing DNA ligase and $3' \rightarrow 5'$ exonuclease activities suggests its involvement in a recombinase system responsible for the gene arrangements.[173]

---

167. Boeke JD (1990) *Cell 61*:193.
168. Bollum FJ (1978) *Adv Enz 47*:347; Ratliff RL (1981) Enzymes *14*:105.
169. Harrison TA, Barr RD, McCaffrey RP, Sarna G, Silverstone AE, Perry S, Baltimore D (1976) *BBRC 69*:63.
170. Nakamura H, Tanabe K, Yoshida S, Morita T (1981) *JBC 256*:8745; Bollum FJ, Chang LMS (1981) *JBC 256*:8767.
171. Matsukage A, Nishikawa K, Ooi T, Seto Y, Yamaguchi M (1987) *JBC 262*:8960.
172. Yancopoulos GD, Blackwell TK, Suh H, Hood L, Alt FW (1986) *Cell 44*:251; Landau NR, Schatz DG, Rosa M, Baltimore D (1987) *Mol Cell Biol 7*:3237; Alt FW, Blackwell TK, Yancopoulos GD (1987) *Science 238*:1079.
173. Pandey VN, Dave VP, Patil MS, Pradhan DS, Amrute SB, Modak MJ (1990) *Biochemistry 29*:4037.

Table 6-5
Effect of metals on TdT discrimination between purines and pyrimidines[a]

dNTP	Polymerization rate[b]	
	Mg^{2+}	Co^{2+}
A	(1.0)	1.1
G	1.6	2.3
C	0.13	14
T	0.10	16

[a] From Johnson D, Morgan AR (1976) *BBRC 72*:840.
[b] DNA polymerization rates are relative to dATP (245 nmol/mg protein per hour).

Structural analyses of the human TdT gene and its 5′ flanking region may also contribute to an understanding of TdT functions.[174]

Mechanisms

The enzyme, in phosphate buffer, needs a primer at least as large as a trinucleotide, a free 3′-OH group for extension, and a free 5′-P as well. Active primers are d(pApApA) and d(pppTpTpT), but not d(ApApA). In cacodylate buffer [$(CH_3)_2AsO_2^-$], polymerization takes place without a primer but only after some delay. The rate and extent of incorporation of nucleotides are greatly influenced by buffer, metal ion, and enzyme concentration.[175] Striking effects of metal ions are observed in differences between polymerizations of purines and pyrimidines. Purine incorporation occurs to about the same extent with Co^{2+} and Mg^{2+}, while pyrimidine incorporation is over 100 times greater with Co^{2+} (Table 6-5). Zn^{2+} increases the incorporation of dATP more than tenfold in assays with Mg^{2+} present, but has no effect on that of dCTP with Co^{2+} present.

Inhibition by ATP

Profound inhibition of TdT action by ATP distinguishes TdT polymerizing activity from that of all known DNA polymerases.[176] ATP as low as 50 μM inhibits it almost completely, even when dNTP substrate levels are 20 times greater. The effect can be attributed to the incorporation of an adenylate residue that renders the chain virtually inert as a primer terminus. The same enzyme site, containing two cysteine residues, is responsible for binding ATP and dNTPs.[177]

All dNTPs, including a variety of unnatural bases, are polymerized, with varying V_{max} and K_m values. With a primer present, the polymer lengths are 50 to 500 residues.

TdT as a Reagent

The enzyme has been used in many ways. For example, copolymers have been made in which a homopolymer chain of one nucleotide is followed by one or

174. Riley LK, Morrow JK, Danton MJ, Coleman MS (1988) *PNAS 85*:2489; Chang LMS, Rafter E, Rusquet-Valerius R, Peterson RC, White ST, Bollum FJ (1988) *JBC 263*:12509.
175. Chirpich TP (1977) *BBRC 78*:1219.
176. Bhalla RB, Schwartz MK, Modak MJ (1977) *BBRC 76*:1056; Modak MJ (1978) *Biochemistry 17*:3116; Modak MJ (1979) *Biochemistry 18*:2679.
177. Panday VN, Modak MJ (1989) *JBC 264*:867.

more stretches of another kind of nucleotide. The polymer $d(pT)_6(pG)_{10}(pA)_{313}$ was obtained in this way.[178] In other studies, cellulose to which a short oligomer is attached in ester linkage through its 5′-P was extended by TdT and then used successfully for assay of ligase and studies of its mechanism.[179]

It has also been possible by careful adjustment of conditions to add one or a very few nucleotide residues, in random distribution, to the end of a synthetic polymer or natural DNA. At low ratios of dNTP to primer, the distribution of chain lengths is determined by the ratio. In this way, a duplex homopolymer with one of the 3′ primer termini containing a mismatched base, namely $dA_{4000} \cdot dT_{200}dC_1$, has been made and used to demonstrate the proofreading function of the 3′→5′ exonuclease of *E. coli* pol I (Section 4-9).

Important use has been made of TdT in adding one or a few residues of highly radioactive nucleotides for sequence analysis or for oligonucleotide "tailing" in preparing recombinant DNAs (Section 22-4). The optimal primer is a 3′-OH terminus of a single-stranded chain at least 3 residues long. To make such a terminus from duplex DNA with flush ends or with a protruding 5′ end, the 5′ → 3′ exonuclease action of λ exonuclease has been employed. However, under appropriate incubation conditions, fraying at the 3′ end can be sufficient for adequate and controlled addition of nucleotides. Blunt-end extensions with purines take place in the presence of Mg^{2+} and at low ionic strength; pyrimidines are added in the presence of Co^{2+} at higher ionic strength.[180] These methods are effective, for example, in adding short tails of dG or dC to DNA. For the synthesis of poly dG chains, which tend to form tight duplexes, N-acetyl dGTP is preferable; subsequent alkaline hydrolysis can be used to remove the protecting acetyl groups.[181] TdT has also proved useful for adding biotinylated or fluoresceinated probes to synthetic oligonucleotides.[182]

---

178. Ratliff RL, Hayes FN (1967) *BBA 134*:203.
179. Cozzarelli NR, Melechen NE, Jovin TM, Kornberg A (1967) *BBRC 28*:578.
180. Brutlag D, Fry K, Nelson T, Hung P (1977) *Cell 10*:509; Roychoudhury R, Jay E, Wu R (1976) *NAR 3*:863.
181. Lefler CF, Bollum FJ (1969) *JBC 244*:594.
182. Kumar A, Tchen P, Roullet F, Cohen J (1988) *Anal Biochem 169*:376.

CHAPTER 7

RNA Polymerases

7-1 Comparison of RNA and DNA Polymerases

RNA synthesis directed by a DNA template is too intimately related to DNA replication to be omitted entirely, even from a book that dwells on replication rather than transcription. Aside from the formal similarities of the two processes, there are several persuasive reasons for paying close attention to RNA synthesis when one studies DNA synthesis.

To begin with, RNA polymerase (RNAP) is a protein that carries out coordinated, DNA-directed synthesis of a chain from a particular starting point, and a great deal is known about the process. Also, RNA transcripts serve as specific priming devices for initiating DNA synthesis, and transcription activates the initiation of replication on some templates. Eukaryotic promoters and enhancers are often inseparable from replication origins and intimately involved in the initiation process. Transcription and replication are interdependent at several levels of metabolic regulation. Finally, RNA itself may have a structural role in the organization of the chromosome.

Fortunately for studies of transcription, RNAP is abundant in the cell and remains nearly intact despite vigorous isolation procedures. By contrast, there are many fewer copies of the replication enzymes per cell, and an intact physiologic replication complex has not been obtainable. As a result, more is known about the biochemistry of transcription from studies of RNA polymerases than about replication by DNA polymerases. However, the functions of factors held less tightly by RNAP (such as termination factors in *E. coli*) are as elusive as the missing pieces of the replisome puzzle.

Before considering RNAPs in detail it would be helpful to cite the similarities and distinctions between the RNA and DNA polymerases of *E. coli* (Table 7-1). Until the discovery of DNA polymerases II and III and the holoenzyme form of DNA polymerase III, the contrast between the multisubunit structure of RNAP and the single polypeptide of DNA polymerase I was striking. Now the similarities are more evident (Section 5-3).

RNA and DNA polymerases have basically identical catalytic properties in the growth of polynucleotide chains. RNAPs require all four rNTPs for complementary base pairing with a DNA template, just as DNA polymerases require all four dNTPs. Nucleophilic attack by the 3′-hydroxyl terminus of the growing chain on the α-phosphorus of the monomeric 5′-triphosphate propagates the chain in a 5′→3′ direction. Both polymerases at times show the apparent lack of template direction and dependence that has been described for DNA polymerase in the reiterative and seemingly de novo synthesis of homopolymers and copolymers (Section 4-14).

Table 7-1
Comparison of *E. coli* RNA polymerase with DNA polymerases I and III

	RNA polymerase	DNA polymerases I and III
SIMILARITIES		
DNA template-directed	yes	yes
Requires 4 triphosphates	yes (rNTPs)	yes (dNTPs)
Requires divalent cation	yes	yes
Synthesis: 5′→3′, antiparallel	yes	yes
Anomalous use of NTPs	rare for dNTPs	rare for rNTPs
Uses RNA primer	yes	yes
Uses RNA template	yes	yes[a]
DISTINCTIONS		
Functions	transcription, linked to translation	replication, repair
Mass (kDa)	448	109[a], 900[b]
Subunits	5	1[a], 20[b]
Starts chains	yes	no
NTP binding sites	2	1[a]
K_m for triphosphates:		
Initiation (apparent K_m)	high (> 100 μM)	low (5 μM)
Elongation	low (15 μM)	low (5 μM)
Terminates chains	yes	no
Recognizes sequences	yes	no
Intact duplex as template	yes	no
Product	single-strand	part of duplex
Ultrahigh fidelity	uncertain	yes
Proofreading	uncertain	yes
Turnover number (ntd/sec)	~ 50	~ 10[a], > 500[b]
Inhibition by rifampicin, actinomycin	yes	no

[a] Applies to DNA polymerase I.
[b] Applies to DNA polymerase III holoenzyme.

RNAP is sharply distinguished from DNA polymerases by its capacity to both start and terminate a new chain when copying a duplex template. This property is essential for the selective transcription of a certain gene among the many available on the chromosome. Selective initiation is accomplished by recognition of a nucleotide sequence that identifies the precise location and strand where the start is to be made. Such sequences are called *promoters*, and they direct the initiation of transcription of a given gene or group of genes. DNA polymerases, by contrast, cannot start chains and generally rely on transcription by an RNA polymerase or primase to do it for them. RNAPs terminate chains at the end of a gene or operon, either unaided or with the assistance of a termination factor. Once in motion, the very processive replicative DNA polymerases normally copy a template until it is exhausted. Specific replication termination signals have been identified, but they probably exert their effect by impeding helicase action in separating the strands of a duplex, rather than by affecting a DNA polymerase directly (Section 15-12).

The true template for RNAP is a duplex that undergoes localized and transient melting during transcription. By contrast, DNA polymerases are inert on such intact duplexes and require nicked or frayed regions with auxiliary proteins to melt the helical duplex and expose relatively long single-stranded segments for replication.

DNA polymerases are exquisitely designed for ultrahigh fidelity in copying any template and do not discriminate among sequences being copied. In the interest of error-free copying, certain DNA polymerases have proofreading capacities built into their operation; DNA polymerase I can even excise lesions from the 5′ end of one strand of a template duplex as it replicates (Sections 4-12 and 4-13). The proofreading function, a 3′→5′ exonuclease that recognizes and excises a mismatched primer terminus, depends on a domain distinct from the one that polymerizes.

As more is learned about where the polymerases are located in the cell and how they are regulated, further similarities and distinctions will become evident.

7-2 Structure of *E. coli* RNA Polymerase[1]

RNA polymerase of *E. coli*, the most extensively studied bacterial RNAP, is representative of the enzyme isolated from a number of bacterial genera, including *Salmonella*, *Serratia*, *Proteus*, *Aerobacter*, and *Bacillus*.[2] All these RNAPs are complex in organization and similar in properties. By contrast, the RNAPs encoded by phages T7 and N4 are distinctive (Section 13).

In considering the structure of an RNAP, the methods for purifying and assaying the enzyme are especially important.[3] Problems with persistent contaminants, dissociated subunits and factors, proteolytic alteration, modification,

1. Reznikoff W, Burgess R, Dahlberg J, Gross C, Record MT, Wickens M (eds.) (1987) *RNA Polymerase and Regulation of Transcription: A Steenbock Symposium*. Elsevier, New York; Chamberlin MJ (1976) in *RNA Polymerase* (Losick R, Chamberlin M, eds.). CSHL, p. 17; Burgess RR (1976) in *RNA Polymerase* (Losick R, Chamberlin M, eds.). CSHL, p. 69; Zillig W, Palm P, Heil A (1976) in *RNA Polymerase* (Losick R, Chamberlin M, eds.). CSHL, p. 101.
2. Fukuda R, Ishihama A, Saitoh T, Taketo M (1977) *MGG* 154:135.
3. Chamberlin MJ, Kingston R, Gilman M, Wiggs J, deVera A (1983) *Meth Enz* 101:540.

denaturation, and aggregation plague studies of all complex enzymes. They have been especially troublesome in studies of RNAP, and are often responsible for inconsistent results obtained in different laboratories or even in successive preparations of the enzyme in the same laboratory. Quantitative assays of active enzyme should include measurements of rates of promoter site selection, initiation and chain elongation, and the specificity of termination.

Three-Dimensional Structure

E. coli RNAP has been crystallized by association with a lipid monolayer. These two-dimensional crystals diffract to about 25 Å resolution and reveal an irregularly shaped protein with dimensions of about 100 × 100 × 160 Å.[4] The structure of the enzyme[5] contains a channel (Plate 5) similar to the active-site cleft of DNA polymerase I (Section 4-5).[6] This channel, about 25 Å in diameter and 55 Å in length, could accommodate a segment of double-helical DNA. However, this length would encompass only about 16 bp, significantly shorter than the 80 bp protected by the holoenzyme during initiation and the 30 bp bound during elongation. The DNA not bound in the channel may be bent around the surface of the enzyme. The structural similarity to DNA pol I is supported by a small amount of amino acid sequence homology between pol I and the β' subunit of RNAP.[7]

Subunit Composition

RNAP exists in two distinct active forms. The *core enzyme* is composed of α, β, and β' subunits in the ratio 2 : 1 : 1. The masses of the *E. coli* enzyme subunits are 36.5 kDa for α, 151 kDa for β, and 156 kDa for β'. Association of a σ subunit with the core constitutes the *holoenzyme*. The composite mass of the *E. coli* holoenzyme is 448 kDa. The σ subunit is present at only about one-third the abundance of the core.[8] Required only during initiation of transcription at a specific site, σ dissociates from the RNAP–template complex shortly thereafter and cycles to a new core for each round of transcription. Both *E. coli* and *B. subtilis* have multiple forms of σ (see Table 7-3), each responsible for recognizing a particular class of promoters (Section 4). The predominant σ is 70 kDa (σ^{70}) in *E. coli* and 43 kDa (σ^{43}) in *B. subtilis*.

Upon chromatography of RNAP on phosphocellulose, the major σ subunit fails to bind to the adsorbent and can be separated from the $\alpha_2\beta\beta'$ core. Unlike the holoenzyme, the core cannot use an intact duplex template such as phage T2 DNA (Table 7-2), but it can act as well as the holoenzyme on a nicked duplex or on ssDNA.[9] Poly d(AT), if frayed enough to be susceptible to the single-strand–specific exonuclease I, can serve as a template for the core polymerase (Table 7-2).

The genes for the *E. coli* RNAP subunits lie within operons encoding ribosomal proteins, which facilitates the coregulation of their synthesis with that of the translation machinery.[10] The gene for α (*rpoA*) is located at 72 minutes on the

---

4. Darst SA, Ribi HO, Pierce DW, Kornberg RD (1988) *JMB 203*:269.
5. Darst SA, Kubalek EW, Kornberg RD (1989) *Nature 340*:730.
6. Ollis DL, Brick P, Hamlin R, Xuong NG, Steitz T (1985) *Nature 313*:762.
7. Allison LA, Moyle M, Shales M, Ingles CJ (1985) *Cell 42*:599.
8. Harris JD, Heilig JS, Martinez II, Calendar R, Isaksson LA (1978) *PNAS 75*:6177; Burgess RR, Travers AA, Dunn JJ, Bautz EKF (1969) *Nature 221*:43; Reznikoff WS, Siegele DA, Cowing DW, Gross CA (1985) *ARG 19*:355.
9. Burgess RR, Travers AA, Dunn JJ, Bautz EKF (1969) *Nature 221*:43.
10. Jaskunas SR, Burgess RR, Nomura M (1975) *PNAS 62*:5036; Yasamoto M, Nomura M (1978) *PNAS 75*:3891; Linn T, Scaife J (1978) *Nature 276*:33.

Table 7-2
Relative activities of RNAP holoenzyme and core

	Activity on template (units/mg)	
Polymerase	Poly d(AT)	T2 DNA
Holoenzyme	20,000	8,000
Core	16,000	< 300

E. coli map. The protein has been sequenced; it is 329 amino acids and acidic.[11] The structural gene for σ (*rpoD*) is at 66 minutes, immediately downstream of the *dnaG* gene, which encodes DNA primase. The σ operon also encodes the ribosomal protein S21, and thus contains genes involved in the initiation of translation, replication, and transcription.[12]

The genes for β (*rpoB*) and β′ (*rpoC*) are located within an operon at 88 minutes and are transcribed in the order β β′. Although the operon encodes several ribosomal subunits, the transcription of *rpoB* and *rpoC* is partially uncoupled from that of these more abundant proteins by an intercistronic transcriptional terminator. Both the genes and their proteins have been sequenced:[13] β, with 1342 amino acids, and β′, with 1407, are among the largest of *E. coli* proteins. Although similar in size, sequence analysis confirms that β and β′ are nevertheless unique, unrelated proteins. Mutations that confer a cellular resistance to several antibiotics [e.g., rifampicin (rifampin)] are within the *rpoB* gene, once designated *rif*.

A 10-kDa ω subunit is consistently present in preparations of *E. coli* RNAP. The ω subunit can be dissociated from the intact enzyme (i.e., the holoenzyme) without apparent loss of activity, and it is not required in the reconstitution of an active holoenzyme after disaggregation of native enzyme by urea treatment. The gene for ω has been cloned, sequenced, mapped at 82 minutes, and named *rpoZ*.[14] Disruption of the *rpoZ* gene imparts a slow-growth phenotype to cells.[15] The poor growth is attributed to interruption of expression of the downstream *spoT* (a gene involved in regulation of the stringent response to amino acid starvation), rather than to the absence of ω. The presence of ω in RNAP is required, however, for transcription to be inhibited by guanosine tetraphosphate (ppGpp) in vitro.[16] Guanosine tetraphosphate is thought to be a signalling molecule involved in stringent control. Thus the ω subunit is implicated in the regulation of transcription, rather than directly in RNA synthesis.

Other proteins, Rho and NusA, which influence the termination of transcription (Section 7), interact with RNAP but have not as yet been regarded as part of a holoenzyme.

The holoenzyme can be dissociated and resolved into its component subunits, and then reconstituted at a 50% to 60% yield.[17] The separated polypeptides,

11. Ovchinnikov YA, Lipkin VM, Modyanov NN, Chertov OY, Smirnov YV (1977) *FEBS Lett 76*:108.
12. Burton ZF, Gross CA, Watanabe KK, Burgess RR (1983) *Cell 32*:335.
13. Ovchinnikov YA, Monastyrskaya GS, Gubanov VV, Guryev SO, Salomatina IS, Shuvaeva TM, Lipkin VM, Sverdlov ED (1982) *NAR 10*:4035; Ovchinnikov YA, Monastyrskaya GS, Gubanov VV, Guryev SO, Chertov OY, Modyanov NN, Grinkevich VA, Makarova IA, Marchenko TV, Polovnikova IN, Lipkin VM, Sverdlov ED (1981) *EJB 116*:621.
14. Gentry DR, Burgess RR (1986) *Gene 48*:33.
15. Gentry DR, Burgess RR (1989) *J Bact 171*:1271.
16. Igarashi K, Fujita N, Ishihama A (1989) *NAR 17*:8755.
17. Ishihama A (1981) *Adv Biophys 14*:1.

individually inert, become active when reassembled in the following order:

$$\alpha + \alpha \rightarrow \alpha_2 \rightarrow \alpha_2\beta \rightarrow \alpha_2\beta\beta' \rightarrow \alpha_2\beta\beta'\sigma$$

The presence of σ appears to stimulate assembly at an intermediate step before it is itself incorporated into the holoenzyme. The requirement for all of the subunits in reconstituting enzyme activity indicates that each is an essential component of RNAP. Analysis of subunit mutants defective in enzyme assembly suggests that this pathway is also used in vivo.[18] In an α mutant that fails to form the dimer, the β and β' subunits are rapidly degraded. In a β' mutant, the $\alpha_2\beta$ complex accumulates and the β' polypeptide is rapidly degraded. Assembly of the enzyme is required for subunit stability; overproduced normal polypeptides are also rapidly degraded in vivo for lack of subunits with which to assemble.[19]

Structure and Function of the β Subunit

Antibiotics that bind RNAP and inhibit its action have provided important insights into its structure and subunit functions. Rifampicin (Fig. 14-12), for example, which binds firmly to the β subunit, completely blocks productive initiation of RNA chains by the enzyme in vitro and in vivo.[20] The polymerase–rifampicin complex apparently fails in performing the translocation step that follows formation of the first or second phosphodiester bond.[21] The inactive enzyme complex, bound at the promoter site, becomes a barrier to transcription through this region by an active RNAP molecule.[22] Cells gain resistance to rifampicin by virtue of an altered β subunit that fails to bind the drug. Used this way, inhibition by rifampicin has shown RNAP to be the enzyme responsible for transcription in the cell, and has provided a means for exploring the mechanism of the enzyme's action. Streptolydigin (Fig. 14-12), like rifampicin,[23] binds the β subunit, but it blocks chain growth rather than initiation.

Numerous rifampicin-resistant mutant enzymes show altered transcription properties. In addition, the incomplete polypeptide products of nearly 100 amber mutations and the "missense" proteins generated by suppression with different tRNAs have provided insights into the functional organization of the β subunit.[24] Mutations to rifampicin resistance map in three separate regions within a 200-amino-acid stretch in the center of the 1342-residue β subunit and define the rifampicin-binding pocket.[25] The isolated mutant β subunit does not bind rifampicin, and the $\alpha_2\beta$ complex does so only weakly. These observations and results of cross-linking studies suggest that rifampicin contacts subunits in addition to β.

The rifampicin-binding domain defines a region of the protein involved in transcription termination. Of 17 alleles imparting resistance, 14 cause altered termination behavior.[26] Mutations in the N-terminal and the furthermost C-ter-

18. Ishihama A, Fujita N, Glass RE (1987) *Proteins 2*:42.
19. Ishihama A, Rukuda R (1980) *Mol Cell Biochem 31*:177.
20. Wehrli W, Staehelin M (1971) *Bact Rev 35*:290; Wehrli W, Handschin J, Winderli W (1976) in *RNA Polymerase* (Losick R, Chamberlin M, eds.). CSHL, p. 397.
21. McClure WR, Cech CL (1978) *JBC 253*:8949.
22. Kassavetis GA, Kaya KM, Chamberlin MJ (1978) *Biochemistry 17*:5798.
23. Halling SM, Burtis KC, Doi RH (1978) *Nature 272*:837.
24. Nene V, Glass RE (1984) *MGG 194*:166; Glass RE, Honda A, Ishihama A (1986) *MGG 203*:492.
25. Jin DJ, Gross CA (1988) *JMB 202*:45.
26. Jin DJ, Walter WA, Gross CA (1988) *JMB 202*:245.

minal portions of the rifampicin-binding region increase the sensitivity to terminators, whereas centrally located mutations decrease termination efficiency. Half the alleles affect both Rho-independent and Rho-dependent termination (Section 7), suggesting action in a step common to both pathways.

Up to 20% of the C-terminal region is dispensable in the assembly of β with the other core subunits.[27] The σ subunit, however, fails to associate with these complexes, implying that the C terminus of β is involved in σ binding. Production of β fragments that are competent for assembly into core polymerase but are nonfunctional in transcription probably explains the dominant lethal phenotype of certain amber alleles.

In addition to functioning in rifampicin binding and in transcription termination, and interacting with the σ subunit, β also binds rNTPs[28] and possesses an atom of Zn^{2+}.

Compared to β, little is known about the functional organization of the other core subunits. The β' polypeptide also has a tightly bound Zn^{2+}, is involved in DNA binding, and is the site of action of polyanionic inhibitors, such as heparin. No specific functions have been ascribed to the α subunits. Several important functional domains within the σ polypeptides have been identified by homologies between the various σ factors, by specific mutants with altered properties, and by protein chemistry (Section 4).

7-3 Overview of Functions and Regulation of *E. coli* RNA Polymerase[29]

RNAP action can be simplified to the following equation:

$$(\mathrm{ATP})_n + (\mathrm{GTP})_n + (\mathrm{UTP})_n + (\mathrm{CTP})_n \longrightarrow (\mathrm{AMP{-}GMP{-}UMP{-}CMP})_n + 4n\mathrm{PP_i}$$

This equation resembles the one for DNA polymerases, except that the DNA template is neither extended as a primer nor generally incorporated into a stable duplex with the product. RNAP both initiates and terminates the chain. Formation of the phosphodiester bonds for chain growth is thus only one stage in a complicated sequence of reactions (Fig. 7-1): *template binding and site selection* (Section 4), *initiation* (Section 5), *elongation* (Section 6), and *termination* (Section 7).

Several factors regulate and modify *E. coli* RNAP in all stages of transcription but particularly in site selection, initiation, and termination (Section 8). Various σ factors and transcription activators affect site selection by the polymerase, thereby enabling the transcription pattern to change in response to environmental cues. In addition, repressor proteins inhibit certain stages of initiation by binding to the promoter. Modifications in the subunits may affect transcription and are especially evident after phage T4 infection (Section 13). At the termination stage, attenuators, which may be in the leader regions upstream of the coding sequence or between genes in an operon, are of particular regulatory

27. Glass RE, Honda A, Ishihama A (1986) *MGG 203*:492.
28. Grachev MA, Kolocheva TI, Lukhtanov EA, Mustaev AA (1987) *EJB 163*:113; Grachev MA, Lukhtanov EA, Mustaev AA, Zaychikov EF, Abdukayumov MN, Rabinov IV, Richter VI, Skoblov YS, Chistyakov PG (1989) *EJB 180*:577.
29. McClure WR (1985) *ARB 54*:171.

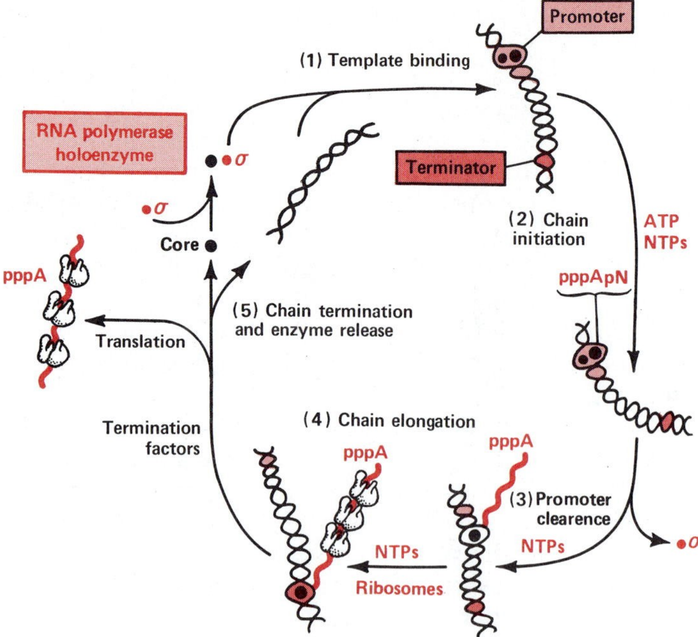

Figure 7-1
Transcription cycle of *E. coli* RNA polymerase, leading to synthesis of a single, specific RNA transcript. [After Chamberlin MJ (1976) *RNA Polymerase*. CSHL, p. 17]

importance (Sections 7 and 8). The Rho protein catalyzes termination by interacting with both the RNA chain and the polymerase, setting the stage for release of the transcript and the enzyme (Section 7); the Nus proteins (Section 7) affect termination as well. Still other proteins bind RNAP and enable it to go through "stop signs;" the actions of such *antitermination factors* (Section 7) are especially significant in programmed gene expression during the phage λ life cycle.

7-4 Template Binding and Site Selection by *E. coli* RNA Polymerase[30]

Productive binding of RNAP to a promoter can be separated from the subsequent steps in transcription because it occurs in the absence of the rNTP substrates. Under appropriate ionic conditions and at physiologic termperature (37°), the enzyme locates a promoter on the template DNA and forms a tight complex resistant to dissociation. The ability of the enzyme to read promoters as specific start signals for transcription is due to a selective interaction with these DNA sequences during the initial phase of the transcription cycle.

Promoter Sequences

Certain promoter sequences are recognized by *E. coli* RNAP containing the σ^{70} form of the σ subunit. These sequences have been extensively defined genetically as the primary location of mutations affecting gene expression, and biochemically as sites of RNAP binding and initiation of transcription. Based on an

30. McClure WR (1985) *ARB 54*:171; von Hippel PH, Bear DG, Morgan WD, McSwiggen JA (1984) *ARB 53*:389; Travers AA (1987) *Crit Rev B 22*:181.

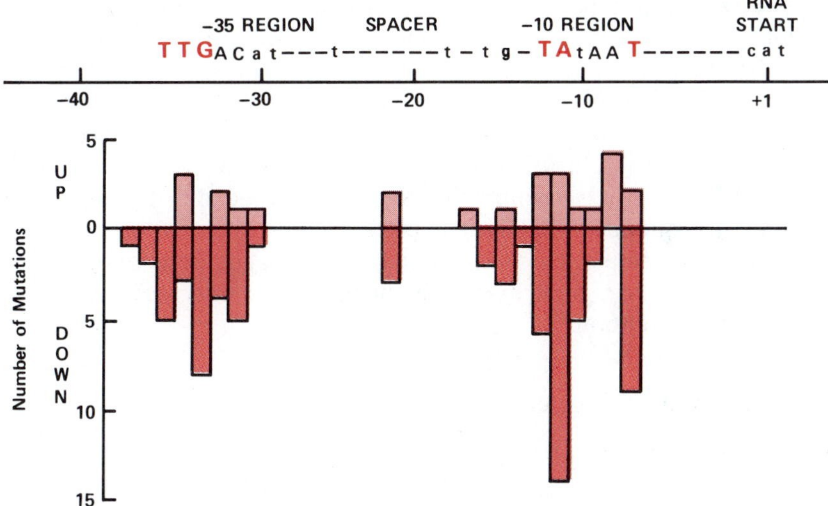

Figure 7-2
E. coli promoter consensus sequence and distribution of mutations within known promoters. Sequence homologies for bases within the promoter are distinguished by typeface: large capitals in color, highly conserved (>75%); small capitals, conserved (>50%); lower case, weakly conserved (<40%). The distance between the −35 and −10 regions (the spacer) is ordinarily 17 ± 1 bp. Below each position, including the spacer, the number and the nature (up = increase in promoter activity, down = decrease in promoter activity) of known promoter mutations are shown as bars. [Adapted from Hawley DK, McClure WR (1983) *NAR 11*:2237]

examination of nearly 200 examples, a consensus promoter sequence has been established[31] (Fig. 7-2). The most important features are two hexameric sequence blocks, located approximately 35 and 10 bp upstream of the transcription start site and separated by 17 ± 1 bp. The preferred sequence of the −35 region is TTGACA; that of the −10 region is TATAAT. Although the consensus sequence is a strong promoter, and the sequences defined by homology are clearly important, it is not necessarily the best promoter. The promoters used to establish the consensus vary in strength and in their response to positive activators, repressors, and chemical modification. Different classes of promoters also vary in the choice made for optimizing one or another of the multiple stages of initiation (Section 5). Promoters therefore diverge considerably from the consensus nucleotide sequence. Additional nucleotides, undetected by homology studies, may also be involved to compensate for the lack of certain consensus residues in some promoters.

Locating a Promoter Sequence[32]

In order to locate a specific promoter sequence, RNAP holoenzyme must conduct a random hunt through the vast excess of nonspecific sites on the template. The very rapid rate of initiation observed with some promoters suggests that periods of exceedingly rapid linear diffusion of the enzyme along the DNA are involved.

31. Hawley DK, McClure WR (1983) *NAR 11*:2237.
32. von Hippel PH, Bear DG, Morgan WD, McSwiggen JA (1984) *ARB 53*:389; von Hippel PH, Berg OG (1989) *JBC 264*:675; Berg OG, von Hippel PH (1985) *Ann Rev Biophys Biophys Chem 14*:131.

RNAP holoenzyme may have one conformation for binding to nonspecific sites and another that allows for the additional contacts required for the specific interaction with the promoter. Affinity for a strong promoter is 10^3 to 10^4 times higher than that for random sequences. Nonspecific binding is electrostatic, while specific interaction with promoters involves both electrostatic contacts and specific hydrogen-bond formation between the enzyme and the base pairs in the promoter. One way by which the "electrostatic" conformation might locate a promoter sequence is by recognizing a specific structure in the DNA helix. Since nucleotide sequence is expressed in the local geometry of the helix, the enzyme may exploit a change in groove width or helix curvature for assistance in locating the promoter.[33]

Formation of the RNAP–Promoter Open Complex

Binding of the RNAP to the promoter is a multi-step process. Several stages precede formation of the stable, transcriptionally competent open complex.[34] The pathway for interaction between RNAP and the promoter DNA can be summarized as follows:

$$\text{RNAP + DNA (nonspecific)} \longrightarrow \text{RNAP + promoter} \longrightarrow \text{closed complex} \longrightarrow \text{intermediate complex} \longrightarrow \text{open complex}$$

The *closed complex* initially formed at the promoter is short-lived and is in rapid equilibrium with enzyme molecules both free in solution and bound nonspecifically to the template. RNAP bound to the promoter undergoes two sequential isomerization reactions to give first an *intermediate complex*[35] and then the *open complex*. The first isomerization is primarily a conformational change in the enzyme, with few changes in the protein–DNA contacts. In the second conversion, about 12 base pairs (10 to 17) of the DNA duplex at the transcription start site are melted; the resulting open complex is extremely stable, having a half-life of hours or even days. Which of these stages is rate-limiting for initiation of transcription depends on the promoter sequence and the reaction conditions, with incubation temperature being especially significant.

RNAP bound in the open complex can initiate an RNA chain in less than 0.2 second when presented with rNTPs. The melted region is necessary for base pairing between rNTPs and the template DNA strand. Thus, the open-promoter complex is the critical mediator for initiating a specific RNA chain.

Promoter Recognition by the σ Subunit[36]

The σ subunit, as part of the holoenzyme, recognizes both the -35 and -10 sequences in the promoter and probably contributes to the melting of the duplex to form the open complex. The multiple σ forms that are normally present in different bacteria, those induced during spore formation in *B. subtilis*, and those appearing during infection by certain phages have provided insights into their structure and mechanism.

---

33. Travers AA (1989) *ARB 58*:427; Travers AA (1987) *Crit Rev B 22*:181.
34. Chamberlin MJ (1974) *ARB 43*:5019; McClure WR (1980) *PNAS 77*:5634; McClure WR (1985) *ARB 54*:171.
35. Buc H, McClure WR (1985) *Biochemistry 24*:2712; Cowing DW, Mecsas J, Record MT Jr, Gross CA (1989) *JMB 210*:521.
36. Helmann JD, Chamberlin MJ (1988) *ARB 57*:839.

At least 17 σ factors have been identified, along with the promoter sequences which they recognize (Table 7-3); most factors are designated by a superscript denoting the molecular weight of the protein. The major σ forms from *E. coli* (σ^{70}) and *B. subtilis* (σ^{43}) recognize the same sequence and direct the initiation of most genes. The minor σ factors, commonly around 30 kDa, redirect the core polymerase to different classes of promoters, such as those for the heat-shock genes in *E. coli* and the sporulation genes in *B. subtilis*. Promoters recognized by these σ factors vary from the canonical sequence in either the −35 region, the −10 region, or both. Inasmuch as most promoters contain two regularly spaced sequence elements, even though their specificities differ, the σ factors apparently share a common mode of binding.

As inferred from their amino acid sequences, the σ factors are a family of related proteins.[37] Four regions of homology and presumably of functional relatedness are especially evident (Fig. 7-3). Of these, regions 2 and 4 are most conserved and are responsible, respectively, for recognition of the −10 and −35 promoter elements.

Table 7-3
Representative bacterial and phage σ factors[a]

Factor	Encoding gene	Genes transcribed	Consensus sequence recognized[b]	
			−35	−10
E. coli:				
σ^{70}	*rpoD*	most genes	TTGACA	TATAAT
σ^{54}	*rpoN, ntrA, glnF*	nitrogen-regulated	CTGGCACN$_5$	TTGCA
σ^{32}	*rpoH, htpR*	heat-shock regulated	CTTGAA	CCCATNTA
$\sigma^{28}(\sigma^F)$	*flaI* ?	flagellin and chemotaxis	TAAA	GCCGATAA[c]
$\sigma^{24}(\sigma^F)$	unknown	*rpoH* and *htrA*	GAACTT	TCTGA[d]
σ^{gp55}	T4 gene 55	phage late genes	none	TATAAATA
B. subtilis:				
σ^{43}	*rpoD, sigA*	most genes	TTGACA	TATAAT
σ^{28}	*sigD*	flagellin	CTAAA	CCGATAT[e]
σ^{29}	*sigE, spoIIGB*	sporulation	TTNAAA	CATATT
σ^H	*sigH, spoOH*	sporulation	unknown	
σ^{32}	*sigC*	unknown	AAATC	TANTGNTTNTA
σ^{37}	*sigB*	unknown	AGGNTT	GGNATTGNT
σ^G	*spoIIIG*	*ssp, gdh*, forespore-specific	PyGNATPu	CAN(T/A)NTA[f]
σ^F	*spoIIAC*	sporulation	similar to that for σ^G	
σ^K	*spoIVCB, spoIIIC*	sporulation, mother cell–specific	unknown[g]	
σ^{gp28}	*SPO1, 28*	phage middle genes	TNAGGAGAN$_2$	TTTNTTT
$\sigma^{gp33/34}$	*SPO1, 33, 34*	phage late genes	CGTTAGA	GATATT

[a] Helmann JD, Chamberlin MJ (1988) *ARB 57*:839.
[b] N = nonconsensus nucleotides, P$_y$ = C or T, Pu = A or G.
[c] Arnosti DN, Chamberlin MJ (1989) *PNAS 86*:830.
[d] Erickson JW, Gross CA (1989) *Genes Dev 3*:1462; Wang QP, Kaguni JM (1989) *J Bact 171*:4248.
[e] Mirel DB, Chamberlin MJ (1989) *J Bact 171*:3095.
[f] Sun DX, Stragier P, Setlow P (1989) *Genes Dev 3*:141.
[g] Kross L, Kunkel B, Losick R (1989) *Science 243*:526.

37. Gribskov M, Burgess RR (1986) *NAR 14*:6745; Helmann JD, Chamberlin MJ (1988) *ARB 57*:839; Stragier P, Parsot C, Bouvier J (1985) *FEBS Lett 187*:11.

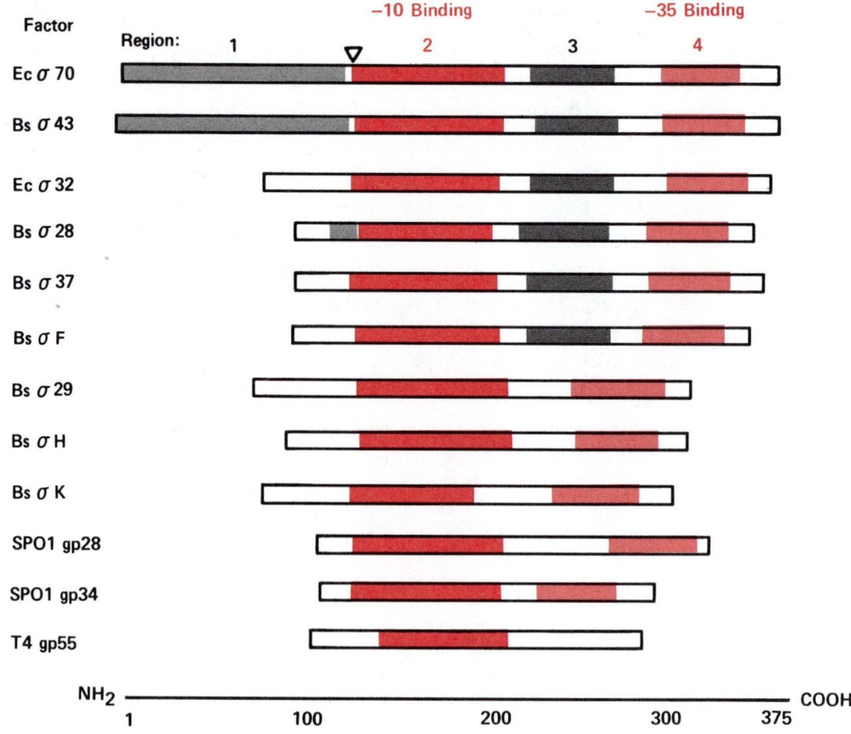

Figure 7-3
Alignment of the protein sequences of some bacterial and phage σ factors. Conserved regions are numbered and a deletion of 245 amino acids in Ecσ70 is indicated by a triangle. Bs = *B. subtilis*, Ec = *E. coli*. Also see Table 7-3. [Adapted from Helmann JD, Chamberlin MJ (1988) *ARB* 57:839]

The C-terminal region (region 4) contains a helix–turn–helix motif, common among site-specific DNA-binding proteins (Section 10-9). The evidence for recognition of the −35 promoter sequence by the C-terminal region is threefold: (1) Two mutations in this region of the *E. coli* σ^{70} protein specifically suppress base-pair changes in the −35 promoter sequence.[38] Specific amino acid changes in the "recognition helix" of the helix–turn–helix motif alter the interaction of the proteins with the fifth and third bases of the −35 region, but fail to alter expression of other types of promoter mutations. This suppression pattern most strongly indicates that these amino acids form specific hydrogen bonds with the −35 base pairs. (2) The σ factors which normally recognize the same −35 sequences have very similar amino acid sequences in region 4. *E. coli* σ^{70} and *B. subtilis* σ^{43} have identical amino acid sequences in the recognition helix; this correlation can be seen for other pairs of σ factors as well. (3) Finally, the σ^{gp55} of phage T4, unique in lacking the region 4 protein sequence, recognizes promoters also unique in lacking a −35 region.

Region 2 of the σ proteins is responsible for recognition of the −10 promoter sequence. In *E. coli* σ^{70} and in *B. subtilis* σ^{30}, mutations which alter the recogni-

38. Siegele DA, Hu JC, Walter WA, Gross CA (1989) *JMB 206*:591; Gardella T, Moyle H, Susskind MM (1989) *JMB 206*:579.

tion of promoters with specific changes in the −10 sequence probably identify contacts between the promoter and the protein.[39] Furthermore, σ factors that recognize the same −10 sequences have strongly related amino acid sequences within region 2.

The structural motif in the protein responsible for DNA binding in the −10 region is not a helix–turn–helix. The region contains clusters of aromatic amino acids which may promote binding to single-stranded nucleic acids by intercalation of their side chains between the bases.[40] A region similar to the ribonucleoprotein consensus sequence, established from comparison of eukaryotic RNA and DNA SSBs, is evident.[41] Were region 2 of the σ subunit responsible for melting the DNA during formation of the open complex, then its recognition of the −10 region might be based on an ssDNA structure.

Immediately N-terminal to the area of region 2 homology, and extending into the conserved area, is a part of the protein implicated by deletion analysis as important in the interaction between σ^{70} and core polymerase.[42] Furthermore, a synthetic peptide containing this sequence also binds the core enzyme. The specificity of the peptide–polymerase interaction, is, however, uncertain.

E. coli σ^{70}, in addition to specific contacts with the −10 and −35 regions in the promoter, also contributes significantly by reducing the nonspecific DNA binding by core polymerase.[43] This latter activity is not conserved among σ factors: *B. subtilis* σ^{43} lacks this capacity to influence the affinity of the core polymerase, which then relies on an additional core-binding protein, called δ^{21}, to fulfill this function.[44] Presumably, the 245-amino-acid domain of σ^{70}, absent from σ^{43}, is responsible.

Single-Stranded Templates

Single-stranded DNA and RNA are bound by RNAP and transcribed by it. Even though it is clear that duplex DNA is the physiologic template for the enzyme, the behavior of single-stranded polynucleotides and of damaged DNA is important to the understanding of its actions in vivo and in vitro. For lack of such knowledge, it is puzzling how the replication of M13 viral ssDNA is initiated by RNAP, with the synthesis of a short RNA at a specific sequence that bears no resemblance to a promoter (Section 8-2).

Many more starts are made on ssDNA than on duplex DNA, and the chains produced are shorter. Masking by a DNA-binding protein, or by a viral capsid protein, may prevent initiation of RNA chains at biologically incorrect sites. In the transcription of single-stranded polynucleotides such as poly dT and poly rU, base-pairing errors are rather common. Since single-stranded polynucleotides can displace σ from the holoenzyme, it is not clear which form of the enzyme is involved in initiation or whether the site occupied by the polynucleotide is the template site, the product site, or even both.

---

39. Siegele DA, Hu JC, Walter WA, Gross CA (1989) *JMB 206*:591; Zuber P, Healy J, Carter HL, Cutting S, Moran CP, Losick R (1989) *JMB 206*:605.
40. Helmann JD, Chamberlin MJ (1988) *ARB 57*:839.
41. Swanson MS, Nakagawa TY, LeVan K, Dreyfuss G (1987) *Mol Cell Biol 7*:1731; Adam SA, Nakagawa T, Swanson MS, Woodruff TK, Dreyfuss G (1986) *Mol Cell Biol 6*:2932.
42. Lesley SA, Burgess RR (1989) *Biochemistry 28*:7728.
43. Helmann JD, Chamberlin MJ (1988) *ARB 57*:839.
44. Hyde EI, Hilton MD, Whiteley HR (1986) *JBC 261*:16565; Pero J, Nelson J, Fox TD (1975) *PNAS 72*:1589; Tijan R, Losick R, Pero J, Hinnebusch A (1977) *EJB 74*:149; Hilton MD, Whiteley HR (1985) *JBC 260*:8121.

7-5 Chain Initiation by *E. coli* RNA Polymerase[45]

In the initiation of RNA synthesis, promoter recognition is the first of several steps. Detailed knowledge of the initiation pathway is essential for an understanding of the regulatory processes that enable different genes to be transcribed by the same enzyme over a 10^4 range in rate and enable a particular gene to be either repressed or induced 10^3-fold.

Formation of the closed complex, conformational change in the enzyme to give the intermediate complex, and melting of the DNA to generate the open complex have been introduced in Section 4. The addition of rNTPs to the preformed open complex generates the *initial transcribing complex* (Fig. 7-4), similar in structure to the open complex but containing short RNA transcripts. Once synthesis has progressed far enough (about 10 ntd) and the enzyme clears the promoter, the polymerase adopts the characteristic *elongation complex* conformation.

Commonly, the steps governing open complex formation (Section 4) determine the rate of RNA synthesis. Dominant factors are the equilibrium constant for binding of RNAP to the promoter (K_B) and the rate of the two isomerization steps (which can be considered together in one term, k_f) leading to the open complex. However, since a second RNAP cannot bind the promoter until it is vacated, clearance of RNAP from the promoter may be significant in the rate of initiation of some genes.

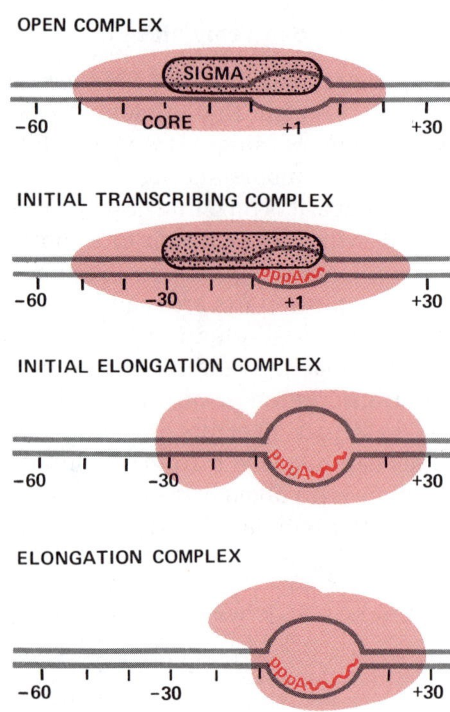

Figure 7-4
E. coli RNAP–promoter complexes during the various stages of initiation. The approximate position of the protein on the DNA as revealed by protection studies is shown, as is the region of the template melted by the polymerase. [Adapted from Krummel B, Chamberlin MJ (1989) *Biochemistry 28*:7829]

45. McClure WR (1985) *ARB 54*:171; von Hippel PH, Bear DG, Morgan WD, McSwiggen JA (1984) *ARB 53*:389.

Measuring the Initiation Stages

Several assay methods have been applied to the transcription initiation reactions to provide a detailed picture of the events occurring at the promoter region.

Filter binding[46] is based on the retention of protein–DNA complexes, but not free DNA, by nitrocellulose filters. Retention of radiolabeled DNA by RNAP is thus evidence of complex formation. This technique can measure the rate and extent of DNA binding by RNAP; salt and temperature can be varied to favor isolation of closed or open complexes. Filter retention does not, however, provide any information about the location of the enzyme on the template or the transcription competence of the isolated complex. Thus tight nonspecific complexes, such as those formed between core polymerase and the ends of a fragment, although inactive as transcription intermediates, may be scored in this assay.

In *footprinting*, the presence of bound protein protects the DNA from chemical or enzymatic attack. Protection leaves a blank space, or "footprint," when digested DNA is examined by gel electrophoresis. The method has the distinct advantage of revealing both the locality and the extent of binding by the protein. RNAP occupancy of more than one site on the same fragment can be analyzed, as can the presence of additional regulatory proteins. Conformational changes in RNAP, reflected in its DNA contacts, have also been identified (see below).

Neutral-gel electrophoresis separates the different RNAP–promoter complexes (closed, open, initial transcribing, and elongation) from one another, so that the RNA and protein components can be determined.[47] This technique is especially useful in conjunction with footprinting: the purification of complexes prior to determining the DNA–protein contacts greatly increases the accuracy of analysis. Transcription intermediates that contain identical or nearly identical components have been refractory to purification by other techniques.

Abortive initiation assays identify open complexes by their ability to synthesize di- and trinucleotide products when the rNTPs are limited to the first two or three complementary rNTPs required for transcription.[48] (Abortive initiation in the presence of all four rNTPs also occurs, although less often, and may be part of the normal initiation pathway.) The steady-state rate of synthesis of the "abortive" transcripts is proportional to the concentration of open complexes; RNAP synthesizes and releases these short products, generating them in greater than stoichiometric amounts. Transcription is restarted without dissociation of the enzyme from the template because the RNAP–promoter contacts are not changed during synthesis of the short RNAs. From the kinetics of abortive initiation, both the K_B and k_f values can be determined. The assay can also measure the relative strengths of two competitive promoters present in the same reaction mixture.

Transcription measures initiation directly by analysis of the RNA product. Open complexes or initial transcribing complexes can be formed and then allowed to finish synthesis under conditions that block initiation (Table 7-4), such as the addition of rifampicin or heparin. This single round of transcription measures the number of completed initiations; product analysis directly reveals

---

46. Melancon P, Burgess RR, Record MT Jr (1982) *Biochemistry 21*:4318; Melancon P, Burgess RR, Record MT Jr (1983) *Biochemistry 22*:5169; Roe J-H, Burgess RR, Record MT Jr (1984) *JMB 176*:495.
47. Straney DC, Crothers DM (1985) *Cell 43*:449.
48. McClure WR (1980) *PNAS 77*:5634; Hawley DK, McClure WR (1980) *PNAS 77*:6381.

Table 7-4
Effects of various agents on RNA polymerase

Agent	RNAP action inhibited		
	Formation of open promoter complex	Initiation of synthesis	Chain elongation
Rifampicin	no	yes[a]	no
Low temperature (< 5°)	yes	yes	no
Heparin	yes	no	no
High salt concentration (> 0.1 M NaCl)	yes[b]	no	no

[a] Can synthesize a dinucleotide in the presence of rifampicin.
[b] Some promoters can function at higher salt concentrations (0.3–0.4 M).

the start site and extent. However, pausing and premature terminations are common and complicate the quantitation of initiation frequency based on the yield of RNA.

RNAP–Promoter Contacts and Conformational Changes During Initiation[49]

The initiation of transcription has a sharp temperature transition, characteristic for each promoter. The higher temperature is required specifically for isomerization from the closed to the intermediate or the open complex.[50] RNAP–promoter contacts observed below the transition temperature are those of the closed complex. As judged by DNase I footprinting, RNAP in the closed complex protects the nucleotides −60 to −5 upstream of the RNA start site; periodic cleavage enhancements in the −60 to −35 region are indicative of wrapping of the DNA around the enzyme surface. The effect of superhelicity on enzyme binding also suggests that wrapping is involved.

Measurement of the energetics of initiation suggests that in the isomerization from the closed complex to the intermediate complex, the major change that occurs involves the structure of the protein. Apparently, hydrophobic surfaces of the enzyme become shielded from the aqueous solvent at this stage. This conformational change is, however, accompanied by little alteration in the RNAP–DNA contacts.

Because of its stability, the open complex has been the most studied RNAP–promoter complex generated during initiation of transcription. In the case of the open complex, protection of the DNA by RNAP extends to 20 bp downstream of the transcription start site (Fig. 7-4). Rather minor changes in the upstream contacts accompany extension of the protected region; essentially no contacts are sacrificed by the additional interactions. A large change in RNAP conformation or in the path of the DNA around the enzyme, enabling it to cover more than 250 Å (~75 bp) of DNA, presumably accompanies open complex formation.

About 10 to 17 base pairs of the DNA duplex near the transcription start site are melted in the open complex, bases in the −9 to +3 region showing the sensitivity to chemical reagents[51] characteristic of ssDNA. Topological experi-

49. Krummel B, Chamberlin MJ (1989) *Biochemistry 28*:7829; Metzger W, Schickor P, Heumann H (1989) *EMBO J 8*:2745.
50. Buc H, McClure WR (1985) *Biochemistry 24*:2712; Spassky A, Kirkegaard K, Buc H (1985) *Biochemistry 24*:2723.
51. Kirkegaard D, Buc H, Spassky A, Wang JC (1983) *PNAS 80*:2544.

ments on DNA structure are consistent with the helix being unwound to this extent.[52] Mutations in the −10 promoter region primarily affect the isomerization rate, further implicating this region in the melting reaction.

Each promoter has a characteristic strength with which it binds RNAP and a rate at which isomerization occurs. The product of these two parameters ($K_B \times k_f$) measures overall promoter strength, with the strongest promoters possessing optimal values for K_B (10^7 to 10^9 M^{-1}) and k_f (10^{-1} to $10^{-3}/sec^{-1}$).[53] Some promoters bind very tightly but isomerize slowly, while others with a fast k_f may be limited in binding. The kinetic characteristics of a promoter are important in its susceptibility to different modes of regulation (Section 8).[54]

RNAP bound in an open complex can initiate RNA synthesis. The chain is nearly always started with a purine (51% ATP and 42% GTP) and initiates 6 or 7 bp downstream from the 3′ end of the −10 promoter element. The apparent K_m* for the initiating nucleotide (~150 μM) is tenfold higher than that for elongation, suggesting either two distinct rNTP binding sites or, alternatively, a change in RNAP conformation driven by the free energy of binding the first rNTP. AMP, ADP, and dinucleotides, in addition to ATP and GTP, can bind to the initiation site (Table 7-5).[55] Whereas AMP and ADP have much lower affinities than ATP and are not efficiently extended in chain growth, dinucleotides, particularly those with A or G at the 3′ end, have a relatively strong affinity and are extended. Mg^{2+} is not required for binding the initial nucleotide. By contrast, binding of an elongation nucleotide depends on Mg^{2+} and occurs only with the triphosphates, all four of which are accepted equally, with the choice among them determined by the template. A variety of analogs that base-pair with the template are bound in this site. These include the triphosphate forms of the potent formycin and tubercidin antibiotics which compete with ATP (Section 14-5).

During synthesis of a 6- to 9-ntd chain, the contacts between RNAP and the promoter remain essentially unchanged from those of the open complex[56] (Fig. 7-4), and short abortive transcripts are continually synthesized without dissoci-

Table 7-5
NTP binding sites of RNA polymerase

Property	Binding sites: Initiation	Binding sites: Elongation
Specificity	ATP, GTP	ATP, GTP, CTP, UTP
Analogs	ITP only	all that base-pair
Mono- and diphosphates	AMP, ADP	no
Dinucleotides	yes	no
NTP concentration	high: 150 μM	low: 15μM
Mg^{2+} required	no	yes
Half-life of rNTP–enzyme complex	0.2 sec	0.02 sec

* The apparent K_m is the nucleotide concentration required for half-maximal rate of initiation.

52. Gamper HB, Hearst JE (1982) *Cell 29*:81.
53. McClure WR (1983) in *Biochemistry of Metabolic Processes* (Lennon DLF, Stratman FW, Zahltern RN, eds.). Elsevier, New York, p. 207.
54. Lanzer M, Bujard H (1988) *PNAS 85*:8973.
55. Minkley EG, Pribnow D (1973) *JMB 77*:255.
56. Krummel B, Chamberlin MJ (1989) *Biochemistry 28*:7829.

ation of RNAP. Abortive initiation is especially pronounced in the absence of all four of the rNTPs needed for elongation. On some promoters, the ternary complex, containing RNAP holoenzyme, promoter DNA, and a short RNA transcript, is stable enough to be isolated and characterized. The enzyme does not progressively move down the DNA during elongation, but, rather, synthesizes the short transcripts without translocation on the DNA.[57] How the promoter contacts are preserved is a fascinating question.

The initial transcribing complex is converted into the fully stabilized elongation complex at a position on the template which varies with different DNA molecules; conversion at chain lengths of 2 to 3 and 9 to 16 have been reported.[58] Change of the enzyme to the elongation mode is characterized by (1) an increased resistance to dissociation by high salt levels ($>$300 mM), (2) resistance of the enzyme to trypsin,[59] (3) reduction of the number of base pairs protected by the enzyme from nearly 80 to about 30, (4) loss of the σ subunit,[60] and (5) an end to abortive initiations.

7-6 Chain Elongation and Pausing by *E. coli* RNA Polymerase[61]

The core enzyme carries out elongation, assisted by factors that affect pausing, the use of termination sites, and coupling with translation. Each nucleotide addition involves successive cycles of selection, polymerization, and translocation along the template. Because transcription takes place on duplex DNA, in contrast to DNA synthesis, which uses an ssDNA template, RNAP must generate and maintain a transcription bubble. This involves unwinding the DNA in the enzyme's path, rewinding it behind, and displacing the RNA–DNA hybrid. Transcription is very processive; dissociation of RNAP occurs only at a termination site, and reassociation with a terminated chain is precluded.

Transcription Fidelity

Base selection depends on recognition of the proper size and shape of the nucleotide. It is difficult to assess the fidelity of chain initiations, but correctness of base pairing in chain elongation can be gauged from the transcription of defined homopolymer and copolymer templates. With alternating copolymers, errors have been observed with a frequency of 0.002% to 0.04%.[62] With the homopolymers poly rU and poly dT, incorporation of the noncomplementary GTP in place of ATP is as common as 1 in 30. However, much lower error frequencies of about 10^{-5} have been observed,[63] a level of accuracy suggesting that some type of proofreading may accompany transcription.

57. Krummel B, Chamberlin MJ (1989) *Biochemistry 28*:7829.
58. Carpousis AJ, Gralla JD (1980) *Biochemistry 19*:3245; Munson LM, Reznikoff WS (1981) *Biochemistry 20*:2081; Levin JR, Krummel B, Chamberlin MJ (1987) *JMB 196*:85.
59. Novak RL, Doty P (1968) *JBC 243*:6068.
60. Travers AA, Burgess RR (1969) *Nature 222*:537; Hansen UM, McClure WR (1980) *JBC 255*:9564; Stackhouse TM, Telesnitsky AP, Meares CF (1989) *Biochemistry 28*:7781.
61. Yager TD, von Hippel PH (1987) in *Escherichia coli and Salmonella typhimurium* (Neidhardt FC, ed.). Am Soc Microbiol, Washington, DC, p. 1241.
62. Springgate DR, Loeb LA (1975) *JMB 97*:577; Strniste GF, Smith DA, Hayes FN (1973) *Biochemistry 12*:603.
63. Blank A, Gallant JA, Burgess RR, Loeb LA (1986) *Biochemistry 25*:5920; Rosenberger RF, Hilton J (1983) *MGG 191*:207.

RNAP lacks a 3′→5′ proofreading exonuclease, common in DNA polymerases. However, the template-dependent hydrolysis of noncomplementary rNTPs into diphosphates has been detected.[64] An *rpoB* mutation, which causes an increased frequency of transcription errors in vivo and an increased rate of misincorporation in vitro, is specifically defective in this hydrolysis. Thus, the hydrolysis of noncognate rNTPs may be an intrinsic activity of RNAP, improving the fidelity of RNA synthesis. In addition, pyrophosphorolysis by RNAP can remove the nucleotide at the growing chain end by reversal of the incorporation reaction.[65] Although elimination of pyrophosphate from a transcription reaction increases the rate of misincorporation, the significance of pyrophosphorolysis as a proofreading mechanism is uncertain. As with DNA synthesis, maintaining the fidelity of RNA synthesis may depend not only on base-pair selection at the stage of inserting the nucleotide, but also on slowing the rate at which a mismatched chain end is utilized during the next steps.

The Transcription Bubble[66]

About 18 base pairs of the DNA duplex are unwound as part of the transcription bubble, 12 of which are involved in an RNA–DNA hybrid with the newly synthesized chain. Bubble length is determined by the effects of transcription on the topology of the template, and appears to be quite uniform. However, many aspects of template composition and topology have not been systematically investigated, and these may affect the architecture of the transcribing complex.

Among the functions of RNAP are the displacement of the nascent RNA and the limiting of the length of the hybrid.[67] Although hybrids are generally short and transient, there are instances of more extensive, biologically active RNA–DNA hybrids generated during transcription. The ColE1 primer RNA (Section 18-2) is the most thoroughly studied example. Hybrid formation depends critically on the secondary structure of the 5′ end of the transcript displaced from the template before the RNAP reaches the site of the origin of replication. The unique folded structure of the transcript may interact with the template DNA to stabilize a melted conformation of the DNA in the region of hybridization.[68]

Details of how RNAP translocates on the template are still unknown. If it moves in "corkscrew" fashion, rotating once per helical turn of the DNA, there should be no topological problems associated with elongation. Free rotation of the transcription complex relative to the template is apparently not so simple. The moving transcription complex has a significant impact on the topology of the template, transiently overwinding the DNA ahead of the complex and underwinding it behind.[69] This effect is readily demonstrable for divergent transcription units on a circular plasmid.[70] The RNAP–RNA complex, perhaps with translating ribosomes bound to it, must be bulky enough to interfere with rotation around the DNA. By unwinding and rewinding the template, topoisomerases may play a critical role in elongation during transcription (Section 12-1).

---

64. Libby RT, Nelson JL, Calvo JM, Gallant JA (1989) *EMBO J 8*:3153.
65. Kahn JD, Hearst JE (1989) *JMB 205*:291.
66. Yager TD, von Hippel PH (1987) in *Escherichia coli and Salmonella typhimurium* (Neidhardt FC, ed.). Am Soc Microbiol, Washington, DC, p. 1241.
67. Richardson JP (1975) *JMB 98*:565.
68. Masukata H, Tomizawa J (1990) *Cell 62*:331.
69. Pruss GJ, Drlica K (1989) *Cell 56*:521; Tsao YP, Wu HY, Liu LF (1989) *Cell 56*:111; Wu HY, Shyy SH, Wang JC, Liu LF (1988) *Cell 53*:433.
70. Wu HY, Shyy SH, Wang JC, Liu LF (1988) *Cell 53*:433.

Activities of the Individual Subunits

The specific functions of the core subunits in elongation have been probed with antibodies and inhibitors. Antibiotics specific for the α subunit are not available, and its role is still unclear. The β subunit has a tightly bound Zn^{2+}, which may be involved in catalysis. Binding and cross-linking studies suggest that β contains the nucleotide binding site (or sites). Streptolydigin inhibits elongation by binding to β competitively with an rNTP in the elongation site, but not with the rNTP involved in initiation. By contrast, a β-specific monoclonal antibody inhibits transcription competitively with an initiating dinucleotide, but not with rNTPs during elongation. Photoaffinity labeling of the nascent RNA to the enzyme reveals that β contacts the growing chain. The β' subunit, which is also labeled by the nascent RNA, is the target of inhibition by heparin, suggesting that β' is responsible for template binding. A monoclonal antibody to β actually increases the enzyme's affinity for DNA. Antibodies which inhibit initiation and elongation, but not DNA binding, suggest an additional role of β' in catalysis, as does the atom of bound Zn^{2+}.

Elongation and Pausing

The maximum rate of chain elongation in vitro, about 30 to 60 ntd per second at 37° under optimal conditions, is comparable to that observed in vivo. The overall rate on natural DNA templates is somewhat slower, 12 to 19 ntd per second, due to the contribution of sequence-directed pausing. Even with long pauses, elongation is processive; RNAP does not dissociate from the template until it reaches the termination sequence directing transcript release.

Pausing, or stalling, which is quite common in vitro, very likely occurs in vivo in the coupling of transcription with translation and as a preliminary to termination. Pause times vary from 10 seconds to several minutes. Concentrations of rNTPs below 20 μM cause pausing in vitro predominantly in regions containing runs of the sequence requiring incorporation of the limiting rNTP. Certain sequences in the template also cause stalling of the transcribing RNAP. One type of pause site depends on formation of a stem–loop structure in the nascent RNA and thus appears 16 to 20 ntd downstream of a dyad symmetry element in the DNA. Both the use of nucleotide analogs with altered base-pairing properties and mutagenesis of the dyad symmetry element directly implicate a hairpin structure in the nascent RNA in pausing. A second type of pause site, lacking such hairpin structures, is characterized by GC-rich sequences, an indication that the stability of the RNA–DNA hybrid within the transcription bubble may be involved.

NusA protein (Section 7) binds to both core polymerase and RNA, increasing the frequency and duration of pauses. The effect on pausing occurs at approximately stoichiometric ratios of RNAP and NusA, indicating that polymerase binding is involved in its action. The NusA complex decreases the affinity of rNTPs for the polymerase in the elongation complex. The affinity of NusA for polymerase, as well as its behavior in the transcription reactions, suggests that NusA associates and dissociates rapidly from the transcription complex. NusA protein also plays an important role in termination and antitermination by assisting other protein factors in their association with RNAP (Section 7).

7-7 Termination of Transcription by *E. coli* RNA Polymerase[71]

Termination, like initiation, proceeds in an ordered series of steps, including the pausing of elongation, transcript release, and dissociation of the enzyme. Termination requires not only the cessation of elongation, but also the release of the nascent RNA and enzyme from the template. Two types of termination signals are utilized by *E. coli* RNAP: (1) Rho-independent signals are recognized by core RNAP and cause termination of up to 90% of the transcripts at a distinct site. The Rho-independent pathway has been misleadingly called "factor-independent." Although core RNAP by itself responds to these signals, other factors such as the Nus proteins have substantial effects on their efficient use. (2) Rho-dependent termination depends on a series of weak signals spread over about 100 bp and requires the Rho protein.

Rho-Independent Termination

A Rho-independent terminator consists of a region of dyad symmetry followed by a run of uridine residues immediately preceding the 3′ end of the RNA. This structure is sufficient for termination. The dyad symmetry element functions as a stem-and-loop structure in the RNA, while the run of U residues is thought to destabilize the RNA–DNA hybrid, thereby promoting dissociation of the transcript. Some terminators have a run of A residues just upstream of the hairpin structure and thus have T residues on the complementary strand downstream, enabling them to function in both directions.

The Rho-independent terminators are influenced by the sequences that surround them. Classic examples of the effects of upstream sequences are found in the attenuators in the leader regions of amino acid biosynthetic operons (Section 8). The hairpin sequence of the Rho-independent terminator can fold into alternative conformations by interacting with complementary regions in the upstream RNA sequence. Some folded structures preclude formation of the terminator hairpin, causing transcription to read through the region. Adoption of one or another of the available RNA secondary structures depends on translation of the leader region, enabling transcription of the downstream genes to respond to the supply of the key amino acids.

The efficiency of Rho-independent termination is also influenced by the sequence of the region transcribed early, about 30 ntd downstream of the promoter. These sequences can affect termination up to 20-fold and can operate several hundred nucleotides upstream of the terminator.[72] While an effect on RNA secondary structure is possible, a more likely explanation is that a conformational change occurs in RNAP, giving it a higher probability of reading through termination sequences. Mutations in the β subunit can both increase and decrease transcription termination (Section 2), further indicating that the conformation of core RNAP can critically affect the efficiency of termination.

Although Rho-independent terminators can be recognized by core RNAP

---

71. Platt T (1986) *ARB* 55:339; Yager TD, von Hippel PH (1987) in *Escherichia coli and Salmonella typhimurium* (Neidhardt FC, ed.). Am Soc Microbiol, Washington, DC, p. 1241.
72. Goliger JA, Yang XJ, Guo H-C, Roberts JW (1989) *JMB* 205:331; Telesnitsky APW, Chamberlin MJ (1989) *JMB* 205:315.

alone, the NusA protein (see below) clearly affects termination at some of these signals. Transcription termination is not a simple localized occurrence, but, rather, is modulated by changes in the structures of RNAP and RNA, as well as by the availability of accessory proteins, the translation machinery, and the small molecules that affect the rates of transcription and translation.

Rho-Dependent Termination

The structure of Rho-dependent terminators is less well defined than that of the Rho-independent ones. The target of Rho action is the nascent RNA 80 to 100 ntd upstream of the stop site; the sequence at the 3′ end of the RNA has relatively little effect. The upstream region that allows Rho-dependent termination includes an RNA sequence that is relatively devoid of secondary structure and that contains C residues, preferably spaced every 12 ntd.

Rho protein, as purified from *E. coli,* is a hexamer of 50-kDa subunits. It does not complex with RNAP nor does it bind tightly to DNA. It does have a strong affinity for RNA,[73] and in the presence of RNA it hydrolyzes rNTPs to the rNDPs and P_i. Optimal NTPase activity is expressed with ATP and poly rC. Rho ATPase activity is essential for termination of transcription at specific sites on various phage and bacterial templates in vitro. Rho is at least a two-domain protein, the N-terminal domain being involved in RNA binding and the C-terminal being required for the NTPase activity. The protein apparently has two distinct RNA binding sites for activation of the NTPase.

Purified Rho protein is a helicase, displacing the RNA portion of an RNA–DNA hybrid (Section 11-3). This activity occurs in the absence of RNAP on appropriate substrates and most likely reflects the involvement of Rho protein in transcript release during termination. The helicase activity has a 5′→3′ polarity with respect to the RNA chain to which the protein is bound and is active on RNA–DNA hybrid substrates known to be targets of Rho-dependent termination. RNAs that lack extensive secondary structure and contain C residues are most readily displaced. The nascent RNA from a stalled ternary complex becomes a target for release by Rho, provided the transcript binds the protein and activates the NTPase required for the helicase action.

Mutations in the *rho* gene[74] can suppress (relieve) the *genetic polarity* often observed as a result of nonsense or insertion mutations: Not only is the function of a specific gene destroyed by a nonsense mutation, but the expression of the genes downstream from it in the operon is also abolished. When the RNA is left bare by the absence of translating ribosomes, Rho protein can recognize cryptic signals as Rho-dependent termination sites, and cause the premature termination of transcription. For lack of Rho protein, continuation of transcription beyond the stop codon becomes more extensive.

Antitermination

Antitermination mechanisms dominate the control of gene expression during the lytic cycle of phage λ (Section 11-7). Two such systems involve host proteins and the phage-encoded *N* and *Q* gene products. Delayed early transcription

73. McSwiggen JA, Bear DG, von Hippel PH (1988) *JMB 199*:609.
74. Ratner D (1976) in *RNA Polymerase* (Losick R, Chamberlin M, eds.). CSHL, p. 645; Inoko H, Shigesada K, Imai M (1977) *PNAS 74*:1162.

requires the N antitermination system, while expression of the late genes responds to the Q protein.

N-Dependent Antitermination.[75] Although more complex than the Q-dependent pathway (see below), N-dependent antitermination has been studied in more detail. The *N* gene product binds to RNAP, causing it to behave like a juggernaut, rumbling along the DNA template and ignoring both Rho-dependent and -independent terminators. Several host factors and a specific sequence in the RNA are required. The *cis*-acting sequence, called the *nut* (*N* utilization) site, located between the promoter and terminator, includes the sequence to which NusA protein may bind (boxA) and a region of dyad symmetry (boxB). Host mutants that block antitermination, called *nus* (*N* utilization substance), reveal five essential loci (Table 7-6). The *nusA* and *nusB* gene products are involved in a variety of reactions in transcription elongation and termination. Mutations with the *nus* phenotype also identify special alleles within the following genes: *nusC* is in *rpoB* (the gene for the β subunit of RNAP), *nusD* is within *rho*, and *nusE* is within the gene for the ribosomal protein S10. The β subunit of RNAP, Rho protein, and S10 are thus implicated in antitermination. Numerous mutations in *rpoB* affecting Rho-dependent and -independent termination also argue for the involvement of the β subunit in these processes; genetic evidence suggests that both NusA protein and N protein bind to the β subunit. The involvement of translating ribosomes in both termination pathways is also clear (see discussion of polarity, above).

NusA protein mediates the binding of N protein to RNAP; the *nut* site may provide a signal for NusA binding or cause a conformational change in RNAP. In cell extracts, the NusB and NusE proteins are required for correct antitermination. The S10 protein apparently can function independently from the ribosome, but the physiologic relevance of this isolated action is not known. The ability of the N-dependent system to overcome termination at both Rho-dependent (e.g., the signal t_{L1}, in the leftward operon) and Rho-independent (e.g., t_{L2}) terminators (Section 17-7) emphasizes that both these pathways have stages in common.

Table 7-6
Termination and antitermination factors

Gene	Protein	Mass (kDa)	Activities and comments
rho (*nusD*)	Rho	50	termination factor; RNA–DNA helicase promoting transcript release; RNA-dependent rNTPase
nusA	NusA	54	core RNAP-binding protein; may mediate binding of other factors, especially N; increases pausing; required for antitermination by N
nusB	NusB	14.5	required for N-dependent and rRNA operon antitermination
nusE	S10	11.7	S10 ribosomal protein; required for N-dependent antitermination
λ *N*[a]	N	13.5	antitermination at rho-dependent (t_{L1}) and rho-independent (t_{L2}) terminators
λ *Q*[a]	Q	23	antitermination at rho-independent terminator ($t_{R'}$)

[a] Phage λ genes.

75. Friedman DI, Granston AE, Thompson D, Schauer AT, Olson ER (1989) *Genome* 31:491.

Q-Dependent Antitermination.[76] The Q-dependent antitermination at $t_{R'}$, one of the termination signals in the rightward transcript (Section 17-1), is brought about in vitro with only RNAP and the Q protein, although NusA increases the efficiency. A sequence in the RNA is involved, perhaps to cause pausing of RNAP for loading of the Q protein. NusA binds core RNAP and may function in assisting with the loading of accessory factors onto the polymerase.

Avoidance of Premature Termination.[77] Coupling of RNA and protein synthesis is revealed by the polar effect on downstream transcription of mutations that cause premature termination of translation. In the absence of normal coupling, Rho-dependent termination sites uncovered in the nascent RNA cause premature termination. The *E. coli* rRNA operons, which synthesize structural RNAs that are never translated, have a special mechanism to avoid premature termination of transcription. The early transcribed region imparts antitermination properties to the RNAP. The presence of sequences similar to the *nut* site for N-dependent antitermination in λ indicates that a similar mechanism is involved. NusB protein is also required. Rho-independent terminators present at the ends of some of these operons suggest that the antitermination conformation of RNAP does not ignore all terminators.

7-8 Regulation of Transcription by *E. coli* RNA Polymerase[78]

Transcriptional regulation is the primary means by which gene expression adjusts in response to environmental stimuli. RNA synthesis can be regulated at several stages: closed complex formation, open complex formation (Section 4), promoter clearance (Section 5), or termination (Section 7). Although positive and negative regulatory mechanisms exist for most of these stages, regulation at promoter recognition and open complex formation are especially prominent.

Different regulatory mechanisms may better serve different types of responses; numerous examples can be given of types of responses and the mechanisms by which they are achieved: (1) A permanent temporal change in gene expression, as occurs during endospore formation in *B. subtilis* or lytic infection by a phage, can be achieved by modifying RNAP with new σ factors. (2) In response to DNA damage, many unlinked genes are co-induced for the repair interval, by proteolytic inactivation of a common repressor protein. (3) More localized changes in expression, such as modulation of genes involved in the biosynthesis of a specific metabolite, are mediated by factors that directly sense the intracellular status of the product; the *lactose repressor*, which dissociates from the promoter upon binding of a metabolite of lactose, allowing transcription of the *lac* operon, is the classic example. (4) Coupling the production of topoisomerases (which change DNA structure) to the requirement for their use is achieved by directly relating their expression to the topology of the DNA.

Some actual and possible regulatory motifs that function at various stages in the transcription pathway are outlined in the following paragraphs.

---

76. Yang XJ, Goliger JA, Robert JW (1989) *JMB 210*:453.
77. Gourse RL, de Boer HA, Nomura M (1986) *Cell 44*:197; Theissen G, Eberle J, Zacharias M, Tobias L, Wagner R (1990) *NAR 18*:3893.
78. Hoopes BC, McClure WR (1987) in *Escherichia coli and Salmonella typhimurium* (Neidhardt FC, ed.). Am Soc Microbiol, Washington, DC, p. 1231.

Positive activators of initiation are of two general types: σ subunits and site-specific DNA-binding proteins. By binding to RNAP, a σ factor changes the DNA recognition properties of the enzyme, enabling it to initiate at a different set of promoters. The DNA-binding activator stimulates the action of RNAP at certain promoters by virtue of being bound close by.

The various σ factors are distinguished by their size and the sequences they recognize (Table 7-3). Genes transcribed by RNAP that contains an alternate form of σ have promoter sequences that are significantly different from the standard σ^{70} consensus sequence. How these alternate σ factors perform their regulatory roles is not completely understood. A "σ cascade" may temporally control gene expression.[79] In the proposed model, the cascade is started by induction of a new σ, which replaces the normal subunit and directs expression of a new class of genes, including, perhaps, one for still another form of σ. Each new σ, by replacing the existing form on core RNAP can, in turn, activate another class of genes. Although simple in principle, the actual situation is probably more complex. The multiple coexistent forms of σ compete for the RNAP core, influenced by many factors that are largely unknown. In some instances, such as in phage T4 development, additional RNAP-binding proteins are known to influence which σ can interact with core.

A different mechanism pertains in the use of alternate σ factors for regulation of the heat-shock response, nitrogen metabolism genes, and flagellar synthesis. By this device, the "housekeeping" genes, under σ^{70} control, can be expressed at the same time as the genes under control of the alternate σ factor. The regulatory loops involved in heat-shock gene synthesis are the most thoroughly understood. The level of σ^{32} directly determines synthesis of the mRNAs for the heat-shock proteins after a shift in growth temperature. The σ^{32} form is normally expressed at only low levels and in an unstable form; transient changes in both the synthesis and the stability of σ^{32} allow the level to be adjusted in response to the ambient temperature and other environmental stimuli.[80]

The regulation of genes by alternate σ factors is a general mechanism that operates in both Gram-positive and Gram-negative bacteria and their phages. The pathway used for initiation of transcription by RNAP containing these alternate σ forms is probably similar to that used by the predominant *E. coli* holoenzyme, but this assumption needs to be tested. As for the regulatory implications of alternate σ factors, the problem becomes focused on how the alternate holoenzymes are generated and maintained. The repertoire of regulatory mechanisms for the standard RNAP that includes repressors, activators, and attenuators is probably applicable to the transcription initiated by the alternate holoenzyme forms as well.

Promoters requiring positive activator proteins for efficient expression are characterized by a poor fit to the consensus sequence in the −35 region. The protein activators (Section 10-9), by binding to the region near −35, stimulate initiation either by improving the affinity of polymerase for the sequence or by increasing the rate of isomerization. The complex of cAMP with catabolite activator protein (the CAP·cAMP complex; Section 10-9) enhances the affinity

79. Losick R, Pero J (1981) *Cell 25*:582.
80. Straus DB, Walter WA, Gross CA (1987) *Nature 329*:348.

of RNAP to the lactose promoter about 20-fold. The phage λ cI protein stimulates transcription from one specific promoter, P_{RM}, 11-fold by increasing the rate of open complex formation, while the λ cII protein activates both of these initiation stages at a second promoter, P_{RE}.

Protein–protein interactions between the activator and RNAP are involved in some instances. Mutants defective specifically in positive activation of P_{RM} by λ cI but not affected in DNA binding may define a surface on cI that contacts RNAP. The CAP·cAMP complex is less likely to interact directly with RNAP because the CAP binding sites are located in different places among the various promoters that the complex activates. The CAP·cAMP complex also introduces a dramatic bend (>140°) into the DNA upon binding (Section 1-10). Inasmuch as the DNA is bent around RNAP during initiation, CAP may stimulate promoter recognition by facilitating this bending. However, further indications of the role of bending in the activation of transcription are needed.

Negative Regulation of Initiation

Proteins that bind to DNA and interfere with any of the essential template–RNAP interactions can negatively regulate transcription. Examples of protein inhibitors of both RNAP binding and isomerization to the open complex are known. The phage λ cI repressor inhibits transcription by binding to the −35 region of the promoter (Section 4), blocking the binding of RNAP. By contrast, the *arc* repressor protein of phage P22 inhibits progression of RNAP from the closed to the open complex by binding DNA between the −35 and −10 promoter elements. The LexA protein, which represses expression of the various genes induced by DNA damage, binds to different regions, from upstream of the −35 sequence to within the transcribed part of the gene. Thus, LexA protein very likely inhibits transcription at several different stages, which may range from RNAP binding to promoter clearance. Repressor proteins can also act by occluding access to the promoter of a required activator protein (see below), rather than affecting RNAP–DNA interactions directly.

Inhibition of initiation by a change in the DNA structure is another mechanism of negative regulation. Several repressor proteins form intramolecular DNA loops which might inhibit initiation. Whether loop formation simply increases the occupancy of the DNA binding site by the repressor or affects the initiation reactions directly is uncertain.

RNAP itself can act as a transcription repressor. When two promoters overlap, RNAP binding to one promoter can repress initiation of transcription at the other. By the same device, repression of one promoter may simultaneously activate a second promoter. The incidence of promoters occurring in close proximity makes the potential for this type of regulation quite common.

Regulation by Modification of the DNA

The responsiveness of transcription to DNA structure is demonstrated by the effects of DNA methylation and superhelicity. These signals can be used for either activation or repression of gene expression. Some promoters contain the GATC sequence, which is recognized and methylated by DNA adenine methyltransferase, the *dam* gene product of *E. coli* (Section 21-9). Regulation by methylation has been clearly demonstrated for the transposase promoter of the prokaryotic transposon, Tn10 (Section 21-7), which is activated tenfold by the absence of methylation. The transposase gene is expressed primarily during the brief period of the cell cycle when the promoter is transiently hemimethylated after passage of the replication fork. In contrast, transcription of the *dnaA* gene, which contains several methylation sites in its promoter, is inhibited in the unmethylated state;[81] thus *dnaA* expression may be transiently repressed by the passage of the replication fork.

Negative supercoiling, which has a large and variable effect on promoter activity in vitro, regulates the gyrase and topoisomerase genes in vivo. The *topA* gene, encoding topoisomerase I, the enzyme which relaxes negative supercoils (Section 12-2), is activated when the DNA is highly supercoiled. Alternatively, the *gyrA* and *gyrB* genes, encoding DNA gyrase (Section 12-3), are induced by relaxation of the template. No special sequences or protein factors are known to be involved in this topological regulation.

Regulation by Termination

Attenuation is a device to regulate transcription after an RNAP molecule has started transcribing but before it has reached the structural gene sequence.[82] The device is used for early termination of transcription of an amino acid biosynthetic operon when an abundance of the charged cognate tRNA signals a sufficiency of the amino acid. For example, in the presence of high levels of charged tryptophan tRNA, over 90% of the transcriptional starts of the *trp* operon are aborted.[83] The attenuator site is at the end of a leader sequence of 166 nucleotides, which contains a ribosome binding site and an in-phase translational terminator, as well as the sequence for transcriptional termination.

The complex mechanism for attenuation includes the action of Rho protein and the translation of the leader transcript. In vitro, 95% of the initiated RNAP molecules terminate elongation at the attenuator site but remain stably complexed until dissociated by Rho action.[84] The short peptides encoded in the leader sequences of the *trp, his,* and *phe* operons contain two or more codons in tandem for the corresponding amino acid. This remarkable coincidence, together with additional facts, suggests that in vivo a translated leader transcript, in the presence of an excess of the amino acid, makes it possible for Rho to dissociate a stalled termination complex.[85]

Transcriptional attenuation is also used to couple expression of the *pyrBI* operon to the availability of UTP. Transcription read-through occurs after polymerase pauses at multiple uridines in the transcript. Pausing allows the ribosomes to catch up to the polymerase and prevent formation of the attenuator hairpin required for termination.

Regulation by antitermination is most clearly demonstrated in the programmed gene expression during lytic development of the lambdoid phages[86] (Section 7). A surprising turn is the capacity of the λ-related phage HKO22 to

---

81. Braun RE, Wright A (1986) *MGG 202*:246.
82. Landick R, Yanofsky C (1987) in *Escherichia coli and Salmonella typhimurium* (Neidhardt FC, ed.). Am Soc Microbiol, Washington, DC, p. 1276.
83. Bertrand K, Korn LJ, Lee F, Yanofsky C (1977) *JMB 117*:227; Lee F, Yanofsky C (1977) *PNAS 74*:4365.
84. Fuller RS, Platt T (1978) *NAR* 5:4613.
85. Zurawski G, Elseviers D, Stauffer GV, Yanofsky C (1978) *PNAS 75*:5988.
86. Roberts JW (1988) *Cell 52*:5.

exclude λ by virtue of encoding a protein which converts *nut* sites from antitermination signals to terminators![87] The ability of specific factors to convert RNAP into a form that ignores a terminator or recognizes a new sequence for termination opens many regulatory possibilities.

7-9 Eukaryotic RNA Polymerases[88]

The first instance of DNA-directed RNA synthesis with rNTPs was observed with a preparation of rat liver nuclei. Yet much of the subsequent work on the regulation and operation of transcription is based on the isolated and characterized RNAPs from bacterial systems. The mechanism and control of eukaryotic gene expression at the transcriptional level is properly regarded as one of the most important problems in biology today. An impediment to its solution has been the inadequate knowledge about eukaryotic RNAPs.

Before embarking on a more detailed discussion of transcription in eukaryotes, we should note several general differences between the transcription strategies of prokaryotic and eukaryotic organisms:

(1) In contrast to the single enzyme of prokaryotic cells, eukaryotes have three nuclear RNAPs, specialized for transcription of different classes of genes.
(2) Eukaryotic RNAPs are more complex and contain twice as many subunits as do their prokaryotic counterparts. Yet they lack subunits analogous to the σ subunit of the prokaryotic enzymes. Promoters are first recognized by transcription factors, some of which are site-specific DNA-binding proteins. Binding of these factors (and additional transcription factors that recognize the bound proteins rather than the DNA sequence) recruits RNAP to the promoter, presumably through protein–protein interactions.
(3) Promoter sequences are not defined regions localized immediately upstream of the transcription start site but may be made up of sequences internal to the transcribed region and located as far as several thousand base pairs upstream or downstream of the start site.
(4) Transcription and translation are uncoupled in eukaryotes. Processing of the primary transcript precedes its packaging and transport from the nucleus to the cytoplasm for translation.

Nuclear RNAPs have been purified extensively from a large variety of animal tissues and cultured cells, plants, and yeast. Each is a complex, multisubunit enzyme with a mass between 400 and 700 kDa. Class I (or A) synthesizes rRNA; II (or B), pre-mRNAs; and III (or C), tRNAs and 5S RNA. An additional small polymerase, about 145 kDa, is found in mitochondria (Section 12).

In addition to their specialized functions, the RNAPs are distinguished by their degree of sensitivity to α-amanitin, by chromatographic properties, by subunit composition, and to a lesser extent by preferences for templates and cations in assays in vitro (Table 7-7). All show the same basic requirements for a

87. Robert J, Sloan SB, Weisberg RA, Gottesman ME, Robledo R, Harbrecht D (1987) *Cell 51*:483.
88. Sentenac A (1985) *Crit Rev B 18*:31; Lewis MK, Burgess RR (1982) Enzymes *15*:109.

Table 7-7
Comparison of classes of nuclear RNA polymerases

Property	Polymerase class I(A)	II(B)	III(C)
Location	nucleolus	nucleoplasm	nucleoplasm
Products	rRNA	pre-mRNA, snRNA	tRNA, 5S RNA
α-Amanitin inhibition (μg/ml for 50%)			
Animal cells	> 1000	0.01–0.05	10–25
Insects	> 1000	0.03–0.06	> 1000
Yeast	~ 500	~ 1	> 1000
Preferred DNA template	nicked or gapped	nicked or gapped	duplex poly d(AT)
Mn^{2+}/Mg^{2+} activity ratio[a]	1–2	5–10	2

[a] Values are for nonspecific transcription. Promoter-dependent initiation requires Mg^{2+} and is inhibitied by Mn^{2+}.

DNA template, the four complementary rNTPs, and divalent metal (Mg^{2+} or Mn^{2+}).

Inhibition of class II polymerases by very low levels of α-amanitin (Fig. 14-12), which inhibits transcription after synthesis of the first phosphodiester bond, sets them apart from the others. Class III polymerases from animal cells are 100 times less sensitive, and those of yeast and the silkworm, *Bombyx mori*, are resistant even to the highest levels. The class I enzymes are inert to the action of the drug. Class II polymerases were judged to be essential for cell growth because of the in vitro resistance to α-amanitin of the enzyme obtained from cells that had become resistant to the drug (Section 10).

The polypeptide components of each class of polymerases are remarkably constant in eukaryotic cells throughout nature, from yeast to silkworms to humans. Each polymerase has two large polypeptides (>100 kDa) and a collection of 4 to 12 smaller ones present in molar ratios of one or two. The two large subunits are unique to each class of enzyme. The largest and second largest subunits share regions of amino acid sequence homology with their cognates from the different forms of eukaryotic RNAP (e.g., the largest subunits of yeast RNA pol I, pol II, and pol III share amino acid homology) and with either the β′ or the β subunits of *E. coli* RNAP. Thus, the large subunits of all these RNAPs most likely arose from a common progenitor and provide conserved functions, critical to the mechanism of RNA synthesis. Some of the small subunits are common to the different enzyme classes. In yeast, for example, all three classes have 27-, 23-, and 14.5-kDa subunits in common, and in addition, the class I and III enzymes share 40- and 19-kDa subunits.

Reconstitution of enzyme activity from the purified polypeptides has not yet been achieved, thus limiting an unequivocal establishment of their subunit functions in transcription. In addition to the copurification with enzyme activity and presence in stoichiometric ratios, the appearance of the same polypeptides in different classes of polymerase and in enzymes purified from different sources argues for their legitimacy as subunits. Cloning of the yeast genes for the three nuclear RNAPs identifies at least 23 that encode the polypeptides.

7-10 Eukaryotic RNA Polymerase II

Structure and Function[89]

RNA pol II synthesizes all the pre-mRNAs in eukaryotic cells. It thus transcribes the largest fraction of the genome and is presumably subject to the most variable regulation of the three classes of polymerases. For these reasons, and also because of its sensitivity to α-amanitin and the consequent genetic selection, RNA pol II has been studied in the most detail.

The enzyme from all sources contains two large subunits, 215 and 139 kDa, and a collection of smaller polypeptides, each less than 50 kDa. The enzyme purified from vertebrates usually has six small subunits while those from plants and lower eukaryotes have eight. The small subunits from the yeast enzyme have molecular masses of 44.5, 32, 27, 23, 16, 14.5, 12.6, and 10 kDa; the 27-, 23-, and 14.5-kDa polypeptides are also present in RNA pol I and pol III.

The genes for the two largest subunits have been cloned and sequenced, and encode 215- and 139-kDa proteins essential for viability. The largest shares amino acid sequence homology with the β subunit of *E. coli* RNAP, while the second largest is related to β'. The two large RNA pol II subunits are also related to those of eukaryotic RNA pol I and pol III. Thus, these large subunits probably have similar functions during transcription by the various enzymes (for functions of *E. coli* subunits, see Sections 2, 6, and 7). Potential Zn^{2+} binding sites are evident in both proteins. The 139-kDa subunit also contains a purine nucleotide–binding sequence and can be labeled with a purine nucleoside analog, suggestive of an rNTP binding site. Mutations imparting resistance to α-amanitin lie within the gene for the largest subunit.

RNA pol II has been isolated in three forms, II0, IIA, and IIB, that differ in the apparent molecular weights of their largest subunits. The large subunits of the three forms are denoted by subscripts of the same letter: II_0, II_a, and II_b. The primary translation product, II_a, is 215 kDa and contains an unusual C-terminal domain composed of a seven-amino-acid sequence, tyr-ser-pro-thr-ser-pro-ser, repeated (with some degeneracy) 26 times in yeast, 42 times in *Drosophila*, and 52 times in mammalian II_a. The smallest form of the subunit, II_b (180 kDa), is derived from II_a by proteolytic removal of this domain, while in the largest form, II_0 (~240 kDa), this region is multiply phosphorylated (Fig. 7-5).

After purification (depending on the source and method), IIA and IIB are the major forms of the enzyme. However, in freshly lysed nuclei, II0 predominates; form IIB is probably a nonphysiologic proteolytic product. The II0 form is ten times as active as IIA in transcription from the adenovirus major late promoter, and II0 is preferentially labeled by a nucleotide analog probe incorporated into RNA during synthesis. Thus, II0 is the most transcriptionally active form of the enzyme in a promoter-dependent reaction.

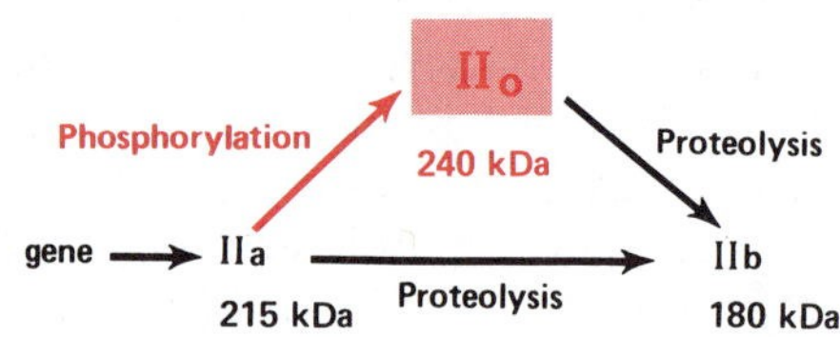

Figure 7-5
Pathway for generation of the three forms of the large subunit of RNA pol II.

89. Sawadogo M, Sentenac A (1990) *ARB* 59:711.

Deletion of the repeat units from the yeast gene shows that at least ten copies of the sequence are required for cell viability. This C-terminal domain is, however, not essential for template binding or RNA synthesis, as judged by the enzymatic activity of the IIB form. Inhibition by a monoclonal antibody against this domain indicates that it functions in initiation of transcription at promoter sites. However, all three forms can accurately initiate transcription in vitro.[90] Protein kinases which specifically phosphorylate II_a have been purified, and phosphorylation is correlated with the transition from initiation to the elongation mode during promoter-dependent transcription.[91] Further work should soon clarify the mechanistic and regulatory importance of this unusual protein domain.

Promoters and Their Organization[92]

Most promoters of mRNA genes are composed of three different elements (Fig. 7-6 and Table 7-8): (1) a selector region, including the TATA box and start sequence, (2) an upstream regulatory (promoter) element, and (3) an enhancer [or, in yeast, an upstream activating sequence (UAS)].

The *TATA box* [consensus sequence: TATA(A/T)A(A/T)], located about 25 bp upstream (40 to 120 bp in yeast), determines the transcription start site. Often, deletion of the TATA box changes only the specificity of the start site rather than decreasing the efficiency of initiation. However, in some promoters the TATA box is essential and clearly does more than just "select" the exact position of initiation. The sequence CA (GT in the template) is often found just at the transcription start site, with the A being the first residue incorporated into RNA.

Upstream regulatory elements and enhancers affect the efficiency of transcription. Upstream promoter elements are found approximately between −40 bp and −110 bp with respect to the start site. In contrast, enhancer elements increase expression from promoters, seemingly independent of their orientation and position (upstream or downstream) with respect to the start site

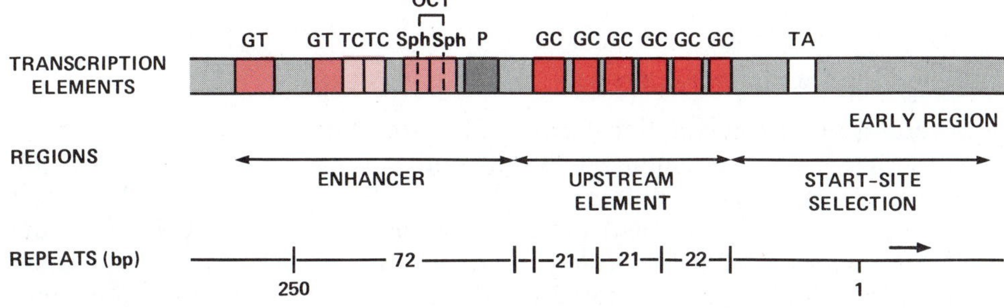

Figure 7-6
The SV40 early promoter region, showing the multiple types of sequence motifs that bind the cellular transcription factors (GT, OCT, P, Sph, and TC boxes) and the viral large T antigen (21 bp repeats). Selector region (TA = TATA box), upstream sequence, and enhancer are all present in this promoter. Arrow = start and direction of transcription. [Adapted from Sentenac A (1985) *Crit Rev B 18*:31]

90. Kim WY, Dahmus ME (1989) *JBC 264*:3169.
91. Payne JM, Laybourn PJ, Dahmus ME (1989) *JBC 264*:19621.
92. Dynan WS, Tjian R (1985) *Nature 316*:774; Maniatis T, Goodbourn S, Fischer JA (1987) *Science 236*:1237; Serfling E, Jasin M, Schaffner W (1985) *Trends Genet 1*:224; Struhl K (1987) *Cell 49*:295.

Table 7-8
Promoter organization

Property	Eukaryotic promoters: TATA box	Eukaryotic promoters: Upstream element	Eukaryotic promoters: Enhancer (UAS)[a]	Prokaryotic promoter
Selection of initiation	yes	no	no	yes
Efficiency	no	yes	yes	yes
Iterons	no	yes	yes	no
Location	−25[b]	−40 to −110	distant	−10, −35
Orientation specificity	yes	yes	no	yes
Promoter specificity	yes	yes	no	yes
Cell type and developmental specificity	no	no	yes	no
Recognized by RNAP	no	no	no	yes
Recognized by transcription factors	yes	yes	yes	no

[a] UAS = upstream activating sequence.
[b] −40 to −120 in yeast.

and over large distances (many kb). Furthermore, enhancers can activate virtually any promoter, while upstream promoter elements are usually more promoter-specific. Enhancer activation appears often to be cell-type–specific or developmentally specific and responsive to environmental signals. Both regulatory and enhancer elements are composed of multiple, short (<20 bp) sequence regions to which transcription factors bind,[93] and the distinction between an upstream promoter element and an enhancer is not always clear: multiple copies of an upstream element often impart "enhancer activity," and some transcription factors interact with both types of sequences.

The activity of all three types of promoter elements depends on their association with transcription factors. In contrast to *E. coli* RNAP, eukaryotic RNA pol II does not recognize a promoter sequence. Transcription factors which bind specific DNA sequences first form a complex with the templates, allowing the association of other transcription factors and RNA pol II, initiation by the enzyme, and subsequent RNA synthesis.

General Transcription Factors

Several general transcription factors (TFs) are required for the initiation of transcription at all promoters, including simple promoters that contain no upstream activating regions. Interestingly, only the TATA box binding factor (TFIID) is a site-specific DNA-binding protein. At least two of the TFs bind RNAP, emphasizing the importance of protein–protein as well as DNA–protein interactions in the specific assembly of the transcription machinery at the promoter.

General transcription factors (Table 7-9) have been identified as components required for promoter-dependent transcription in vitro. Fractionation has been primarily from human cells,[94] although yeast extracts,[95] rat liver,[96] and *Drosoph-*

93. Dynan WS (1989) *Cell 58*:1; Maniatis T, Goodbourn S, Fischer JA (1987) *Science 236*:1237.
94. Matsui T, Segall J, Weil PA, Roeder RG (1980) *JBC 255*:11992; Moncollin V, Miyamoto NG, Zheng X-M, Egly J-M (1986) *EMBO J 5*:2577; Saltzman AG, Weinmann R (1989) *FASEB J 3*:1723.
95. Flanagan PM, Kelleher RJ, Feaver WJ, Lue NF, LaPointe JW, Kornberg RD (1990) *JBC 265*:11105.
96. Conaway JW, Conaway RC (1989) *JBC 264*:2357; Conaway JW, Reines D, Conaway RC (1990) *JBC 265*:7552; Conaway RC, Conaway JW (1990) *JBC 265*:7559; Conaway JW, Travis E, Conaway RC (1990) *JBC 265*:7564.

Table 7-9
General transcription factors for RNA polymerase II

Factor	Source	Mass (kDa)	Activities and comments
TFIID	human cells	37	TATA box binding factor; acts first in initiation; conserved between yeast and human; essential gene in yeast
TFIIB	human cells		
TFIIA	human cells		Stabilizes TFIID bound to the promoter; binds nonspecifically to DNA.
TFIIE	human cells	34, 57	Interacts with RNAP; may be the same as RAP74[a]
TFIIF	human cells	30, 72	Binds RNAP; the same as RAP30, 74[a]
α	rat liver	35	
$\beta-\gamma$	rat liver	67 and 31	Native mass of complex ~ 250 kDa; the same as human RAP30/74[a] complex
δ	rat liver		ATPase (dATPase) activity stimulated by DNA containing the TATA box
B	*Drosophila* cells		TATA box binding factor
Factor 5	*Drosophila* cells		Required for initiation and reduces pausing during elongation; probably analogous to human TFIIE
a	yeast		
b	yeast	50, 75, 85	DNA-dependent ATPase
TFIID	yeast	27	
e	yeast		

[a] RAP = RNAP-associated protein; the number following signifies the mass (kDa) of the factor.

ila cells[97] have proved to be useful sources as well. TFIID[98] (the TATA box binding factor) has been purified from yeast subsequent to its identification in the human transcription system. The yeast protein substitutes for the mammalian factor in the human in vitro transcription system, indicating that the eukaryotic general transcription apparatus is well conserved throughout evolution. This functional conservation is especially impressive considering the multiple protein–protein interactions apparently involved in initiation.

A picture of how these factors and RNA pol II associate with one another and with the template during initiation is beginning to be resolved.[99] TFIID binds to the DNA in the promoter region first, as assayed by the protection of a region around the TATA box from nuclease digestion. The TFIID–promoter complex is apparently stabilized by addition of TFIIA; TFIID and TFIIA together form the preinitiation complex, which is resistant to challenge with other templates and is rapidly recognized by RNA pol II in the presence of the other initiation factors. Some of these factors bind RNA pol II, and assembly may involve their association with the enzyme in solution prior to binding the promoter complex. ATP (or dATP) hydrolysis is required prior to RNA synthesis.[100] The δ factor from rat liver also has an ATPase (dATPase) activity, stimulated by DNA that contains the TATA box.[101] Upon addition of the rNTPs, RNA synthesis can begin.

97. Price DH, Sluder AE, Greenleaf AL (1987) *JBC 262*:3244; Parker CS, Topol J (1984) *Cell 36*:357.
98. Cavallini B, Huet J, Plassat J-L, Sentenac A, Egly J-M, Chambon P (1988) *Nature 334*:77; Buratowski S, Hahn S, Sharp PA, Guarente L (1988) *Nature 334*:37; Horikoshi M, Wang CK, Fujii H, Cromlish JA, Weil PA, Roeder RG (1989) *PNAS 86*:4843.
99. Davison BL, Egly J-M, Mulvihill ER, Chambon P (1983) *Nature 301*:680; Hawley DK, Roeder RG (1985) *JBC 260*:8163; Hawley DK, Roeder RG (1987) *JBC 262*:3452.
100. Bunick D, Zadomeni R, Ackerman S, Weinmann R (1982) *Cell 29*:877; Ernst H, Filipowicz W, Shatkin AJ (1983) *Mol Cell Biol 3*:2172; Sawadogo M, Roeder RG (1984) *JBC 259*:5321; Conaway RC, Conaway JW (1988) *JBC 263*:2962.
101. Conaway RC, Conaway JW (1989) *PNAS 86*:7356.

Activation of Promoters by Enhancers and Upstream Sequences

Some of the best-characterized sequence elements in RNA pol II promoters and the transcription factors which bind them are listed in Table 7-10. Of the enormous number of transcription factors already identified,[102] only a sampling of the specific DNA-binding factors are included. As more is learned about the reactions carried out by the general transcription factors, mechanisms for stimulation of these reactions by the enhancer sequences and the proteins that bind them will be clarified.

Four general models have been proposed to explain how upstream elements and enhancers, with their bound factors, stimulate the initiation reactions:[103] (1) the upstream elements and enhancer sequences serve as entry sites for transcription factors, providing access to the neighboring promoters; or (2) they alter the chromatin structure in the region of the promoter, facilitating its use; or (3) they stabilize factors at the initiation site by protein–protein interactions and looping out of the intervening DNA; or, alternatively, (4) transcription of DNA attached to the nuclear matrix requires enhancer or activator sequences to provide anchoring sites for localization.

None of these mechanisms has either been established or ruled out. Furthermore, different mechanisms may pertain in different instances. The transcription factors, and thus probably their mechanisms of action, are highly conserved among eukaryotic species. Several factors originally identified in mammalian cells have been found in yeast, and heterologous transcription systems, involving interactions between factors from different sources, are functional. Genetic analysis (most readily done in yeast) and biochemical fractionations[104] promise to illuminate these important and apparently complex interactions.

Table 7-10
Examples of RNA polymerase II transcription factors and their promoter motifs

Factor	Promoter region	Sequence element	Source
TFIID	selector	TATA box	human, yeast
B	selector	TATA box	*Drosophila*
Sp1	upstream and enhancer	GC box	human
CTF (NFI)	upstream and enhancer	CAAT box	human, yeast
NF-κB	enhancer	GGGACTTTCC	mammalian cells
GAL4	UAS[a]	several related motifs	yeast
GCN4	UAS[a]	TGACTC	yeast
HSTF	upstream	heat-shock element	*Drosophila*, human, yeast
NF-A1	upstream and enhancer	AAG(T/C)ATGCA	human
AP-1	enhancer	TTAGTCAG	human
GR	enhancer	glucocorticoid response element	human, rat, chicken

[a] UAS = upstream activating sequence

102. Wingender E (1988) *NAR 16*:1879.
103. Atchison ML (1988) *Ann Rev Cell Biol 4*:127; Guarente L (1988) *Cell 52*:303; Wasytyk B (1988) *Crit Rev B 23*:77.
104. Lue NF, Buchman AR, Kornberg RD (1989) *PNAS 86*:486; Lue NF, Kornberg RD (1987) *PNAS 84*:8839.

Elongation and Termination

Although initiation at promoters clearly requires transcription factors as well as RNA pol II, this pathway can be bypassed by using a linear template containing single-stranded poly dC extensions on the 3′ ends. Efficient, specific initiation is achieved on such templates using only the purified polymerase; these reactions have been used to characterize the elongation and termination phases of transcription by RNA pol II.

During transcription on the poly dC–tailed templates, the newly synthesized RNA forms an extensive RNA–DNA hybrid, while the nontemplate strand of the DNA duplex is continually displaced. This behavior is in marked contrast to that of *E. coli* RNAP, which actively peels off the RNA from the template during transcription.[105] Addition of RNase H, which specifically degrades RNA in an RNA–DNA hybrid, digests the 5′ portion of the nascent RNA, allowing displacement of the transcript and renaturation of the template.[106] An RNase H role in a physiologic transcription complex is, however, not yet clear.

Transcription in the purified system on tailed template is 10 to 20 times slower than the in vivo rate, due to pausing of the polymerase. This pausing occurs at specific sequences, but the characteristics defining such pause sites are uncertain. One factor, SII, purified from mammalian cells, stimulates elongation by RNA pol II by reducing this pausing.[107] A second protein, factor 5, purified from *Drosophila* cells, is required both for specific initiation at promoter sequences and for reduced pausing at sites on poly dC–tailed templates.[108] Factor 5 appears to be analogous to the human initiation factor TFIIE, which binds tightly to the polymerase and is required for its association with the preinitiation complex. The effect of factor 5 on elongation does not require its stable association with the transcription complex. Rather, the protein continually associates and dissociates, to minimize the time spent at pause sites.

In vivo, the transcribing RNA pol II must encounter histones bound to and condensing the DNA template. The inhibitory effect of nucleosomes on transcription in vitro has long been known. While histones bound to the promoter region inhibit transcription initiation, nucleosomes bound elsewhere to the template do not block elongation.[109] The transcribing enzyme can effectively displace the bound proteins.

The rapid 3′-end processing and polyadenylation of mRNAs makes it difficult to identify the termination signals used by RNA pol II. Because termination occurs downstream of the processing sites, the sequences utilized are not reflected in the mature RNA. Identification of termination sites has been attempted using two systems: (1) the transcription by purified RNA pol II of poly dC–tailed templates, and (2) nuclear runoffs, in which transcripts initiated in vivo are radioactively labeled during elongation in isolated nuclei.

Use of the poly dC–tailed template system has identified intrinsic termination sites, which stop transcription and cause transcript release. Some of these sites are in regions which block transcription in vivo, attesting to their potential

105. Dedrick RL, Chamberlin MJ (1985) *Biochemistry 24*:2245.
106. Kane CM (1988) *Biochemistry 27*:3187.
107. Rappaport J, Reinberg D, Zandomeni R, Weinmann R (1987) *JBC 262*:5227; Reinberg D, Roeder RG (1987) *JBC 262*:3310.
108. Price DH, Sluder AE, Greenleaf AL (1989) *Mol Cell Biol 9*:1465; Price DH, Sluder AE, Greenleaf AL (1987) *JBC 262*:3244.
109. Lorch Y, LaPointe JW, Kornberg RD (1987) *Cell 49*:203.

importance for gene expression. These sequences, rich in T residues, usually contain a run of T residues at the termination site. However, no direct correlation is evident between the length of the T stretch and the termination efficiency, nor do all T runs cause termination. Detailed analysis of several intrinsic termination sites suggests that a bend in the DNA, caused by phased runs of A residues (and hence T residues; Section 1-10) may signal the enzyme to halt synthesis.

Nuclear runoff experiments clearly indicate that transcription termination in pre-mRNA genes is sometimes coupled to mRNA processing[110] (see below). Termination of transcription occurs at heterogeneous locations anywhere from a hundred to several thousand nucleotides downstream of the polyadenylation site. In the human α_2-globin and mouse β-globin genes, the 3′ processing signal is required for efficient termination. Mutations in the sequence required for poly A addition also cause extensive transcription past the normal termination site. The generality of such coupling of processing with termination remains unknown.

In genes encoding the snRNAs (which lack introns and are not polyadenylated), termination occurs at or very near the 3′ end of the mature RNA and involves the recognition of a specific sequence. Utilization of this 3′ termination signal requires that transcription initiates from an snRNA promoter, indicating that a transcription complex different from that involved in synthesis of pre-mRNAs is essential.

Premature termination of transcription results in truncated mRNAs, suggesting that control of intragenic termination is a mechanism of controlling gene expression. The human immunodeficiency virus (HIV) apparently encodes an antitermination factor (the *tat* gene product) which functions to block premature termination of transcription from the LTR promoter (Section 19-7). Induction of the heat-shock response in *Drosophila* causes a shift from premature termination (or continual abortive initiation) within the *hsp70* gene to full-length transcription.

Processing of mRNA

In contrast with prokaryotic mRNA, several remarkable processing steps are required to convert the nuclear RNA transcript to an active translatable form. First, a nucleotidyl cap is affixed to the 5′ end, and methylation of this structure aids in ribosomal recognition and translation. Second, intervening sequences (introns), up to 20 or even more, must be spliced out precisely. Third, the 3′ end is cleaved and a poly A tail of about 200 residues is appended to the newly generated end. The mechanics and control of these operations have become fruitful areas for studies of nucleic acid biochemistry.

7-11 Eukaryotic RNA Polymerase III[111]

Transcription of a class III gene in vitro by RNA pol III was first observed with the 5S RNA of *Xenopus*. In contrast to prokaryotic promoters and those used by eukaryotic RNA pol II, the initiation signals recognized by the RNA pol III–

110. Proudfoot NJ (1989) *TIBS* 14:105.
111. Geiduschek EP, Tocchini-Valentini GP (1988) *ARB* 57:873.

specific transcription factors are primarily within the transcribed region of the gene. One such factor, TFIIIA, is currently the best characterized eukaryotic transcription factor and is the prototype of the "zinc finger" class of DNA-binding proteins (Section 10-9).

Polymerase Structure

RNA pol III is the most complex of the three eukaryotic polymerases. The yeast enzyme, with a molecular weight of 700 kDa, consists of 14 distinct polypeptides. Although in vitro reconstitution of enzyme activity from the isolated components has not been achieved, there is evidence, beyond their copurification, that at least ten of these serve as subunits in RNA synthesis. Five of the small subunits are shared with class I enzymes, and three with both class I and class II.

The two large subunits, 160 and 128 kDa, have regions of amino acid sequence homologous to those of RNA pol II, indicative of their functioning in template, substrate, and Zn^{2+} binding (Section 10). The 128-kDa protein also cross-links specifically to rNTPs during transcription. Antibody directed against this subunit inhibits template binding. RNA pol III, in contrast to pol I and pol II, also possesses an intermediate-sized subunit of 82 kDa. Antibodies against this polypeptide and those against the 53-kDa component each inhibit transcription in vitro; one against the 23-kDa subunit (as with the 128-kDa subunit) inhibits binding of the enzyme to the template.

Several of the genes for the RNA pol III subunits have been cloned and shown to be essential for cell viability. A temperature-sensitive allele of the yeast *RPO31* gene, which encodes the largest subunit, is specifically defective in tRNA and 5S RNA synthesis in vivo.

Promoter Structure and Initiation of Transcription[112]

In addition to the genes for 5S and tRNA, RNA pol III transcribes genes for several other cellular RNAs (7SK, 7SL, M, U6, and 4.5S) and for RNAs encoded by adenoviruses and Epstein-Barr viruses. Typically, promoters for RNA pol III lie internal to the transcribed region—at sites called *internal control regions* (ICRs)—and serve as recognition sites for binding of specific transcription factors (Fig. 7-7). An additional class of RNA pol III promoters, located entirely upstream of the start site, similar to those used by RNA pol II, has also been described (see below).

The tRNA-type promoters consist of two sequence elements, boxA, located between +8 and +19, and box B, between +52 and +62. Both are highly conserved among tRNA genes; however, the spacing between the two sequences varies from 31 to 93 bp. Two transcription factors, TFIIIC and TFIIIB, direct the polymerase to the promoter. TFIIIC specifically binds to the boxB element, and probably also interacts with boxA. TFIIIC most likely contains two components, one which binds boxB and the other which extends the interaction upstream. Addition of TFIIIB, not itself a specific DNA-binding protein, stabilizes the TFIIIC·DNA complex on some genes and is required for initiation of transcription. RNA pol III associates with this preinitiation complex and starts transcription a measured distance upstream of boxA.

The 5S promoter, situated between +50 and +97, contains, in addition to

112. Kassavetis GA, Braun BR, Nguyen LH, Geiduschek EP (1990) *Cell 60*:235.

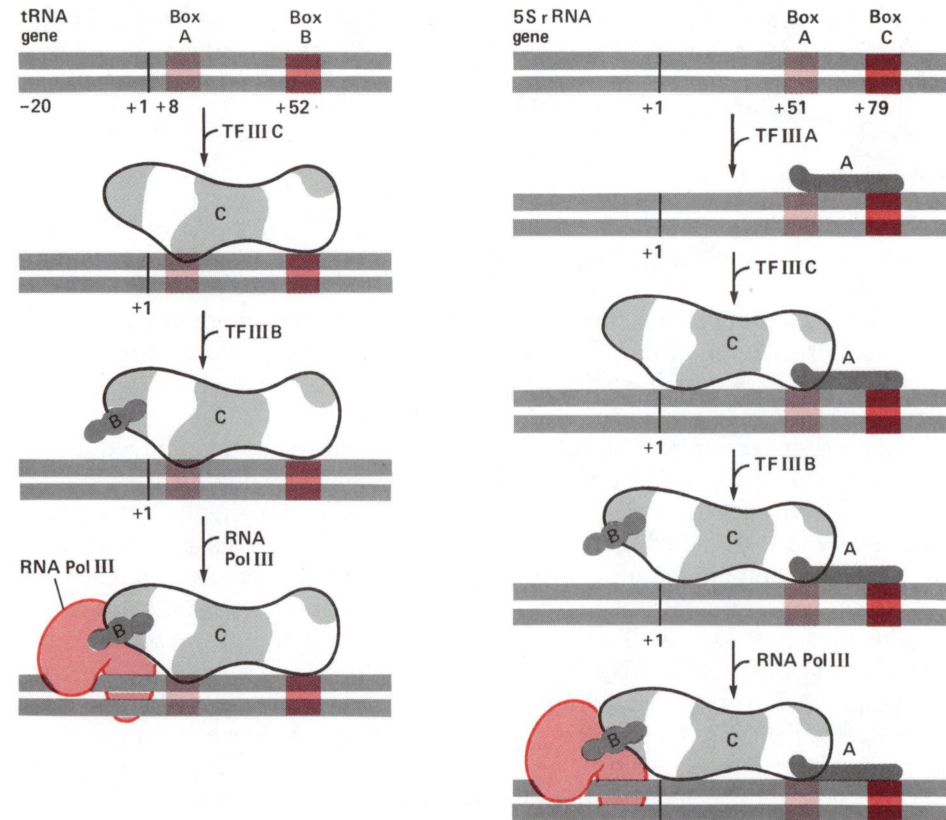

Figure 7-7
Assembly of RNA polymerase III on the two classes of RNAP III promoters (*left*, tRNA gene promoter; *right*, rRNA gene promoter) by three transcription factors, TFIIIA, B, and C, that bind the internal control regions. [Adapted from Rawn JD (1989) *Biochemistry*, Neil Patterson Publishers, Burlington NC]

boxA and in place of boxB, a boxC sequence which is recognized by the 5S-specific transcription factor TFIIIA. TFIIIA of the *Xenopus* oocyte has been purified to homogeneity. It is 38 kDa and contains nine repeats of a zinc finger domain responsible for its DNA binding (Section 10-9); its special structure allows it to bind over an extended region of DNA. Some 50 base pairs are protected by the binding of TFIIIA, an extremely large region for a protein of its size. TFIIIA also binds to the product 5S RNA. Expression of 5S RNA may be regulated by a competition between the product and the promoter for binding of the transcription factor.

Binding of TFIIIA is the initiating event in assembly of the transcription complex at the 5S promoter. Its presence enhances the association of TFIIIC, which by itself binds poorly to this promoter, presumably due to the lack of boxB sequences. Interaction with the boxA sequence is apparently involved. After the association of TFIIIA and TFIIIC, the TFIIIB factor can join the promoter complex, where it solidifies the interactions. This preinitiation complex promotes multiple rounds of transcription by transient associations with RNA pol III. Since it lies within the transcribed region, the ability of this complex to remain stably bound throughout multiple rounds is intriguing. The complex, however, is destroyed by the passage of a replication fork. Histones, which block activity of the promoter, are excluded from the promoter region by this complex. Whether the preinitiation complex or nucleosomes assemble first at the promoter may contribute to the regulation of 5S expression.

The third type of promoter recognized by RNA pol III differs from those for the tRNA and 5S RNA genes, resembling more the promoters for RNA pol II (Fig. 7-8).[113] The gene for the snRNA U6 (all other snRNAs are transcribed by RNA pol II) and the 7SK RNA gene lack sequences homologous to boxB or boxC. Although

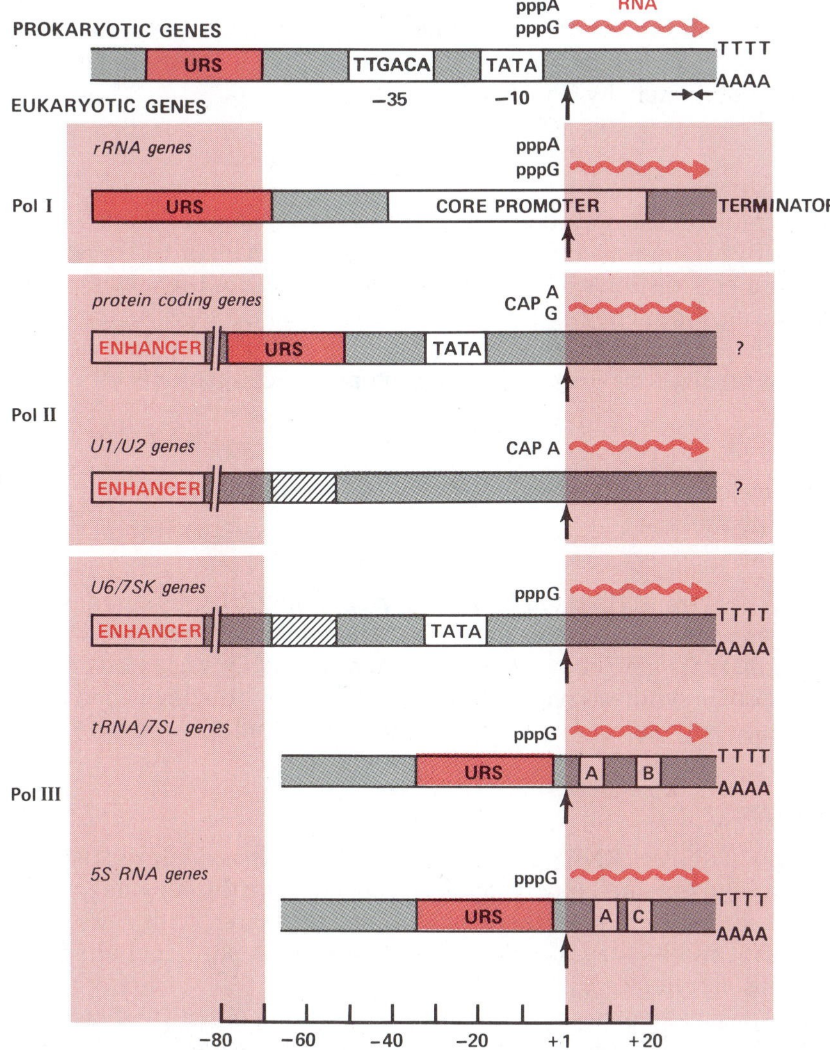

Figure 7-8
The organization of eukaryotic promoters recognized by the three types of nuclear RNA polymerases. URS = upstream regulatory sequence, vertical arrows = start site of transcription, intersecting arrows under the prokaryotic gene = direct repeat, ? = sequences required for terminating transcription are unknown. [Adapted from Murphy S, Moorefield B, Pieler T (1989) *Trends Genet* 5:122]

113. Moenne A, Camier S, Anderson G, Margottin F, Beggs J, Sentenac A (1990) *EMBO J* 9:271.

they contain a boxA sequence, the entire transcribed region, including this sequence, is not essential for promoter function; the primary signals required for transcription must lie upstream.

The sequence required for initiation of transcription includes an AT-rich element at −27, similar to a TATA box, and an upstream activating element similar to those in the genes for the other snRNAs. Octamer elements common in RNA pol II mRNA promoters and enhancers are present in this sequence, strongly suggesting that the two enzymes can use a common transcription factor. The *Xenopus* gene for U6 is in fact transcribed at a low level by RNA pol II. In the human gene for U6, the AT-rich element is responsible for determining which polymerase will transcribe the gene. Mutation of this element induces RNA pol II transcription from the U6 promoter, whereas insertion of the 9-bp sequence into the U2 promoter converts it from a RNA pol II–transcribed gene to one used predominately by RNA pol III.

Termination

RNA pol III terminates transcription at defined sequences: four or more T residues, flanked by GC-rich sequences. Inability of the enzyme to transcribe though a cluster of T residues limits the variety of sequences that can be synthesized by the enzyme. The reason that the U1 through U5 snRNAs are transcribed by RNA pol II instead of pol III may be the run of T residues within their sequence required for packaging into protein particles (snRNPs).[114]

7-12 Eukaryotic RNA Polymerase I[115] and Mitochondrial RNA Polymerase

Although RNA pol I transcribes only the precursor of the rRNA, its synthesis of the multiple gene copies represents nearly half the total RNA produced by the cell. In addition to its efficiency, this polymerase is highly regulated in the coupling of ribosome synthesis with cell growth. RNA pol I is distinguished from the other cellular polymerases by its resistance to α-amanitin.

Enzyme Subunits

Like class II and III RNAPs, RNA pol I is composed of two large subunits (190 and 130 kDa) and four to ten smaller polypeptides, depending on the enzyme source. As mentioned earlier, several of these small units are shared with class III or with both class II and III enzymes. Subunits of yeast RNA pol I, including the largest, are phosphorylated.

A temperature-sensitive mutation in the largest subunit of the yeast enzyme reveals that rDNA transcription (synthesis of rRNA) is defective, while 5S RNA synthesis, which is carried out by RNA pol III, is relatively insensitive to the mutation.[116] Mutations in the putative Zn^{2+} binding domain, shared by the large subunits of all the eukaryotic polymerases and the *E. coli* enzyme, cause tem-

114. Lobo SM, Hernandez N (1989) *Cell 58*:55.
115. Sollner-Webb B, Tower J (1986) *ARB 55*:801.
116. Wittekind M, Dodd J, Vu L, Kolb JM, Buhler J, Sentenac A, Nomura M (1988) *Mol Cell Biol 8*:3997.

perature-sensitive enzyme activity, consistent with the importance of this domain in catalysis.

Promoter Structure and Initiation of Transcription

The rRNA genes occur in tandem head-to-tail repeats of 100 to 5000 copies per cell. Between the genes is a region of variable length, called the *nontranscribed spacer*, which contains signals that both stimulate initiation and cause transcription termination.[117] Transcription, at least in some species, continues into this spacer region in spite of its name.

The human rDNA promoter is made up of two domains: an essential core element located between −45 and +20, and an upstream element between −107 and −186 which stimulates initiation. This two-part organization is conserved in the rDNA promoters of other organisms, even though little homology exists at the nucleotide-sequence level.

Two transcription factors essential for initiation by RNA pol I have been purified and characterized from the human cell-free transcription system.[118] UBFI is a DNA-binding protein that recognizes sequences in both the upstream element and the core promoter. The other factor, SL1, lacks DNA-binding activity, interacting instead with the UBFI protein to enhance its association with both DNA sites. Both factors are essential for the association of RNA pol I with the promoter and for initiation of transcription. The SL1 protein, present in much lower amounts than is the UBFI protein, may act catalytically in assembling the transcription complex. As is true for RNA pol III preinitiation complexes, those of RNA pol I are stable and activate more than one round of transcription without dissociation.

In contrast to the other eukaryotic transcription systems, heterologous (cross-species) rRNA synthesis reactions are not usually functional: extracts from human cells transcribe only the human rDNA genes or those from a closely related primate species, and are nonfunctional on templates from other organisms. This specificity is at the level of transcription factors, rather than the highly conserved polymerase. The SL1 protein appears to be responsible for the species-specific activity of RNA pol I transcription. Human SL1 protein, when added to extracts from mouse cells, is sufficient to redirect the endogenous polymerase to a human rDNA promoter.

Mitochondrial RNA Polymerase

This fourth RNAP in eukaryotic cells transcribes the mitochondrial genome to produce rRNAs, tRNAs, and a few mRNAs, as well as the primer for DNA replication (Section 18-11). Most proteins needed for mitochondrial function, including the RNAP and its transcription factor, are encoded by the nuclear genome and imported into the organelle.

The mitochondrial RNAPs from mammalian cells, *Xenopus*, and yeast all appear to be single-subunit enzymes of about 140 kDa. The gene for the yeast enzyme has been cloned and sequenced, revealing considerable homology to the phage T7 and T3 RNAPs[119] but no relationship to the nuclear or *E. coli* enzymes. The evolutionary basis for the relationship between the mitochondrial and phage RNAPs is obscure.

---

117. Baker SM, Platt T (1986) *Cell 47*:839.
118. Bell SP, Learned RM, Jantzen H-M, Tjian R (1988) *Science 241*:1192.
119. Masters BS, Stohl LL, Clayton DA (1987) *Cell 51*:89.

Transcription of mitochondrial promoters by each of the purified polymerases requires the action of at least one transcription factor. Transcription is species-specific: while promoters from the mitochondrial genomes of different organisms share a rather simple organizational structure, there are few similarities at the nucleotide sequence level. The specificity of promoter recognition most likely lies within the transcription factors, which are more variable (with molecular weights between 43 and 70 kDa) than the polymerases.[120]

7-13 RNA Polymerases in Viral Infections[121]

Studies of transcription, as well as replication, require attention to the DNA template as much as to the polymerase. Because viral chromosomes are available in relatively intact form, are well characterized biochemically and genetically, and can be isolated in sufficient quantity, they serve as ideal templates. Furthermore, the lessons learned apply generally to the operations of the more complex and less accessible host chromosomes.

The very small phages and animal viruses rely on transcriptional machinery of the host cell. The larger viruses (T7, T4, λ, and vaccinia virus) introduce novel mechanisms to ensure preferential selection of their promoters and to arrange the switching from early to late gene expression. Recognition of very early genes by host cell polymerases depends on strong viral promoters and the absence of repressors. Switching at appropriate times to transcription of the late viral genes may be managed in several different ways: introduction of a new polymerase, modification of the host polymerase, production of regulatory proteins functioning as repressors and antiterminators, and combinations of these devices.

Encoding of a New Polymerase: T7, T3, and PBS2 Infections[122]

Initially the host RNAP is used to transcribe the "left" 20% of the T7 chromosome, which includes gene *1*, the gene that encodes the viral RNAP. Translation of this early message produces the T7-specific RNAP, which then transcribes the rest of the T7 chromosome. Certain T7-induced proteins shut off the host RNAP (Section 17-5).

The most remarkable and distinctive feature of T7 RNAP, compared to the host enzyme, is its simplicity (Table 7-11). Within a single polypeptide of 110 kDa it contains the capacity to start and terminate chains and to transcribe a DNA template faithfully. Its exceedingly rapid chain growth and efficient termination may account, in part at least, for the short growth cycle of the phage. Rifampicin insensitivity of the T7 polymerase also distinguishes it from the host enzyme.[123]

How, despite its severe streamlining, does the T7 enzyme manage the job as well as the very complex host enzyme? Part of the answer may be in the relatively simpler demands for initiation and termination. The T7 enzyme recognizes only six or so major promoters, all with an identical sequence of 16 bp

---

120. Fisher RP, Topper JN, Clayton DA (1987) *Cell 50*:247.
121. Losick R, Pero J (1976) in *RNA Polymerase* (Losick R, Chamberlin M, eds.). CSHL, p. 227.
122. Chamberlin M, Ring J (1973) *JBC 248*:2245; Bautz EKF (1976) in *RNA Polymerase* (Losick R, Chamberlin M, eds.). CSHL, p. 273.
123. Chamberlin M, Ring J (1973) *JBC 248*:2245; Chamberlin M, Ryan T (1982) *Enzymes 15*:87.

Table 7-11
Comparison of host and T7-encoded RNA polymerases

Property	RNA polymerase	
	E. coli	T7
Mass (kDa)	480	110
Subunits	5	1
Initiating rNTP	ATP, GTP	GTP
Templates		
All phage DNAs	yes	no
All bacterial DNAs	yes	no
T7 DNA	yes	yes
Poly dG, poly dA, poly dT	yes	no
Poly dC	yes	yes
Anomalous de novo synthesis	yes	no
Mn^{2+} substitutes for Mg^{2+}	yes	no
Inhibition by rifampicin, streptolydigin, streptovaricin	yes	no
Inhibition of chain growth by heparin, high salt, poly rU	no	yes
Efficient termination	variable	yes
Turnover number (ntd/sec)	~ 50	~ 250

(compared to the thousands of promoters recognized by the host enzyme), and responds to a single, well-defined termination signal. The enzyme seems to be specially and narrowly designed for transcribing the T7 template, and is indifferent to the diverse group of regulatory factors and transcriptional signals which affect the host enzyme.

The analogous phage T3 RNAP[124] is almost inert on any genome except T3. Despite the similarity of the T3 and T7 genomes, their RNAPs show no cross-recognition of promoters. The T3 polymerase is designed for a unique starting signal, different from that of the T7 enzyme, and may start each RNA with the same nucleotide sequence.

Another instance of a phage-encoded polymerase[125] is the rifampicin-resistant, multisubunit enzyme responsible for the late transcription of *B. subtilis* phage PBS2 (Section 17-8). Early transcription, which is also rifampicin-resistant, is thought to be carried out by the host RNAP modified by a phage-encoded polypeptide that enters the host cell with the phage DNA during infection.

Modification of the Host Polymerase: T4 and SP01 Infections[126]

The T4 phage, unlike T7, does not encode a novel RNAP. Instead, it exploits the host enzyme by interaction with phage-encoded proteins and by direct alteration of the polymerase subunits (Section 17-6). The T4 regulatory proteins, by modifying the transcriptional specificity of the host RNAP, switch it away from expression of host genes to expression of the late T4 genes. The 10-kDa regulatory protein strongly inhibits the σ subunit and destabilizes its attachment to the core enzyme. Additional properties are imparted by other regulatory polypeptides, and transcription of late genes requires a specific structure of the template DNA provided by simultaneous replication (Section 17-6).

124. Chakraborty PR, Salvo RA, Majumder HK, Maitra J (1977) *JBC 252*:6485.
125. Clark S, Losick R, Pero J (1974) *Nature 252*:21; Clark S (1978) *J Virol 25*:224.
126. Losick R, Pero J (1976) in *RNA Polymerase* (Losick R, Chamberlin M, eds.). CSHL, p. 227.

Other alterations observed in the host enzyme following T4 infection include ADP-ribosylation of the α subunits, phosphorylation of the β and β' subunits, and still other changes observed in the peptide map of a tryptic digest of all four subunits.

SP01 and SP82, related *B. subtilis* phages (Section 17-8), also encode several proteins that bind to the host RNAP[127], including alternate σ factors (Section 4) that endow it with the clear-cut specificities needed to transcribe the early, middle, and late viral genes while shutting off the transcription of host genes.

Repressors and Antiterminators: λ Infection[128]

The choice between temperate and lytic modes of λ infection, and the switch during lytic infection from transcription of early genes to late ones, are determined by highly sensitive modulation of transcription by regulatory proteins (see Fig. 17-3). The characterization of these proteins and the DNA sequences they affect represents the most advanced molecular model available of how chromosomal information is used in a living cell. In the prophage state, the λ repressor (the *cI* gene product) stimulates its own synthesis, at the same time blocking transcription of genes N and *cro*, which are required for normal vegetative growth. Inactivation of the λ repressor enables the initiation of transcription of the early genes by the host RNAP. Antiterminator proteins encoded by genes N and Q (Section 7) ensure that transcription will not be aborted by early or late terminators, respectively. The Cro protein has dual functions: the turning off of early gene functions and turning on of late ones.

A Viral Polymerase in the Virion: Vaccinia and Coliphage N4 Infections

Vaccinia (Section 19-6) was the first virion observed to contain a nucleic acid polymerase. The reverse transcriptases of retroviruses (Section 6-9) were recognized later. The vaccinia virion contains the RNAP needed to produce correct mRNA, and the enzymes needed for capping and methylating the 5′ end and attaching a poly A tail to the 3′ end. The viral polymerse,[129] which also appears in the cytoplasm of infected cells, is unlike the host cell polymerase in its requirement for Mn^{2+} and its insensitivity to α-amanitin, but is similar in complexity. The viral enzyme contains nine subunits ranging in size from 13 to 135 kDa. The extensive sequence similarities between the largest subunits of the vaccinia virus and cellular RNAPs indicate an evolutionary relationship.[130] Studies of vaccinia virus polymerase afford opportunities to examine the specificity and efficiency of transcription and how it is linked to the unique eukaryotic features of splicing, capping, methylation, and polyadenylation.

Coliphage N4 (Section 17-8) encodes two rifampicin-resistant RNAPs.[131] A polymerase which is packaged within the virion capsid, and which enters the host cell with the infecting DNA, transcribes the early genes,[132] including those

---

127. Duffy JJ, Geiduschek EP (1977) *Nature 270*:28.
128. Friedman DI (1987) in *The Bacteriophages* (Calendar R, ed.). Plenum Press, New York, Vol. 2 p. 263.
129. Nevins JR, Joklik WK (1977) *JBC 252*:6930; Whitely HR, Spiegelman GB, Lawie JM, Hiatt WR (1976) in *RNA Polymerase* (Losick R, Chamberlin M, eds.). CSHL, p. 587.
130. Broyles SS, Moss B (1986) *PNAS 83*:3141.
131. Zivin R, Zehring W, Rothman-Denes LB (1978) *JMB 152*:335.
132. Haynes LL, Rothman-Denes LB (1985) *Cell 41*:597.

for the N4 RNAP II, which synthesizes the middle transcripts. In contrast to most other phages, N4 relies on the host RNAP only *late* in infection. The virion polymerase is a single-subunit enzyme of 320 kDa, one of the largest proteins known. This enzyme requires a denatured N4 DNA template, which it transcribes asymmetrically.[133] N4 RNAP II is a heterodimer of 30 and 40 kDa subunits, which are capable of RNA synthesis on ssDNA templates but lack specificity for N4 DNA. An inner membrane protein, P17, is required in vivo for accurate polymerase initiation of the middle genes.[134] Whether the activation of specific transcription involves interactions with the template, the polymerase, or both components is not yet known.

7-14 Relationships of RNA Synthesis to DNA Replication

Beyond the obvious similarities between RNA and DNA synthesis, many interrelationships between the processes deserve attention. Three important ones will be considered here.

(1) *RNA synthesis is required for initiation at most prokaryotic replication origins.* Transcription by RNAP directly primes leading-strand replication on the filamentous and T-odd phages (Sections 17-3 and 17-5), the plasmid ColE1 (Section 18-2), and mitochondrial DNA (Section 18-11). Transcription, or the action of RNAP, is required for initiation on several other templates, where it serves a role other than priming. Phage λ provides the classic example (Section 17-7), and the contribution of transcription to initiation at the *E. coli* chromosomal origin is similar (Section 16-3). While the molecular details of the "transcriptional activation" mechanism are not fully known, generation of an R-loop during transcription in the vicinity of the origin very likely stimulates the assembly of the initiation complex by affecting the structure of the template. The impact of transcription on DNA structure has only recently been appreciated; it is certain to have wide-ranging effects on a variety of DNA transactions.

The involvement of transcription in the initiation of replication of highly divergent replicons, including the genome of *B. subtilis* (Sections 16-4 and 20-3), coliphages T4 (Section 17-6) and N4 (Section 17-8), and the plasmids F (Section 18-6) and R6K (Section 18-5), attests to its wide-ranging importance. The flexibility of RNAP responses to changes in the cellular environment, and its early role in replication, point to the transcription step as a potential point for regulation.

(2) *Enhancers, promoters, and transcription factors are commonly part of eukaryotic origins.*[135] Studies of eukaryotic DNA viruses establish a role for enhancer and promoter elements, and the factors that bind them, in replication origins (Section 16-5). The origins of SV40 and polyoma virus overlap with the transcription control region, which contains the early and late promoters and their associated enhancer. Polyoma replication requires the enhancer, and

---

133. Falco SC, Zehring W, Rothman-Denes LB (1978) *PNAS 75*:3220.
134. Abravaya K, Rothman-Denes LB (1989) *JBC 264*:12695.
135. DePamphilis ML (1988) *Cell 52*:635.

SV40 replication is stimulated by it. Exchanging the endogenous enhancer with one of a different cell specificity changes the cell type supportive of replication of the virus. Two transcription factors, CTF and NFIII, stimulate formation of the initiation complexes at the ends (replication origins) of the adenovirus genome by binding to recognition sequences located nearby (Section 19-4). The sequences required for extrachromosomal maintenance of "plasmid-like" eukaryotic viruses (Epstein-Barr and bovine papilloma viruses) consist of origins and enhancer sequences which function independently of their orientation or location with respect to one another.

In these instances, the transcription factors, rather than the act of transcription, stimulate initiation of replication. Quite likely, these control regions promote replication and transcription simultaneously. The consequences of enhancer action—organization of chromatin structure, stabilization of the binding of protein factors, creation of entry sites for factors or attachment sites for the DNA to the nuclear matrix—affect the machinery of replication, as well as that of transcription. Furthermore, the regulatory features conferred on promoters by enhancers may also be conveyed to the replication origins.

Identification of the replication origins of cellular chromosomes has just begun, but the influence of transcription is already evident. Sequences required for amplification (Section 16-6) overlap promoter and enhancer elements and activate initiation of replication over considerable distances. Transcriptionally active regions of DNA are characteristically replicated earlier in S phase than are silent genes, suggesting a potential coupling of the processes (Section 20-7).

(3) *Intiation of transcription and replication involve similar stages.* In both processes, assembly of the "sequence-blind" synthesis machinery is directed to a particular site to start synthesis at a specific time. Thus, origins can be regarded as promoter sequences for replication initiation, and the component reactions can be divided into stages of assembly and strand melting.

Assembly establishes the polymerase at the initiation sequence through a combination of specific protein–DNA and protein–protein interactions, as observed in initiation at eukaryotic promoters. Sequence specificity resides in the DNA-binding transcription factors. Association of the polymerase requires sequence-independent factors, which bind to both the polymerase and the protein–DNA complex at the promoter. The parallels to the assembly of the replication complex at origins, as seen for phage λ, *oriC*, and SV40 replication, are striking.

Strand melting—that is, opening the duplex DNA strands—the intensively studied stage in initiation by *E. coli* RNAP, must also occur at replication origins. The activities of the prokaryotic σ factors and the dnaA (replication initiator) protein have remarkable similarities: both proteins recognize the initiation region in the template sequence and help melt the DNA duplex to provide a single-stranded region for entry of the synthesis machinery. The σ factor binds two sequences in the promoter, interacting with the -35 sequence as duplex DNA while apparently melting the -10 sequence; the dnaA protein binds the duplex *oriC* sequence at the four dnaA boxes and subsequently melts the DNA within the three 13-mer repeats (Section 16-6).

Separating the sequence-specific recognition properties of the σ subunit required for initiation from the RNAP core may allow the enzyme to be optimized for translocating along DNA. The slow rate of promoter clearance and the synthesis of abortive products prior to the loss of σ by *E. coli* RNAP may reflect the disadvantages of having a strong sequence-specific affinity for an enzyme that needs to translocate along DNA. Thus the assignment of sequence recognition to separate proteins may be a feature generally employed by both the eukaryotic RNAPs and DNA polymerases. Initiation can occur at specific locations without sacrificing the nonspecific binding characteristics needed for rapid movement of the polymerase along the template.

Inasmuch as the mechanisms of initiation of transcription and replication resemble each other, the repertoire of regulatory strategies may be similar as well. Studies of regulation of replication should profit from lessons learned from transcription. Activators and repressors that act at both the assembly and strand-opening stages of initiation, modification of the template and the polymerase, changes in DNA topology, the use of multiple "promoters," and control by both termination and antitermination are all potential means of regulating replication, as they are of transcription.

CHAPTER 8

Primases, Primosomes, and Priming

8-1 The Need for Priming in DNA Synthesis

One of the basic rules of replication (Section 15-1) is that a DNA polymerase cannot start a chain and must rely on a priming device. Every initiation event requires a primer.

The opposite polarities of the two strands of the DNA duplex and the exclusively 5′→3′ polarity of synthesis by DNA polymerases necessitates a semidiscontinuous mechanism of replication (Section 15-3). Two different modes of priming operate for the continuous (leading) and discontinuous (lagging) strands (Fig. 8-1). The continuous strand needs to be primed only once, usually at or near a chromosomal origin, the special sequence where replication begins (Chapter 16); uninterrupted extension from this primer replicates the entire

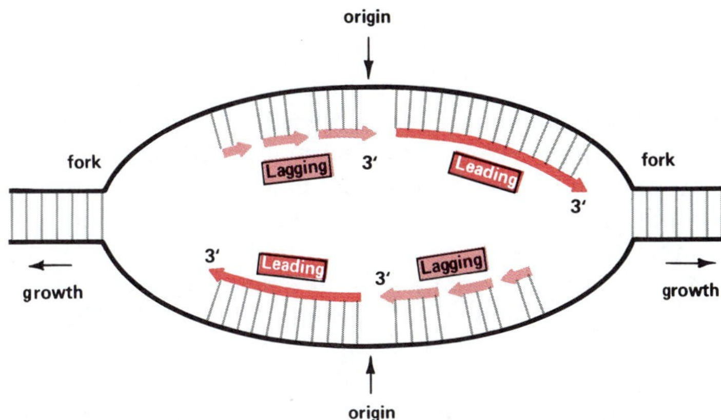

Figure 8-1
Bidirectional fork movement from the origin. Replication is semidiscontinuous: continuous on the "leading" strand and discontinuous on the "lagging" strand.

strand. In contrast, the discontinuous strand must be primed repeatedly to generate the short nascent strands (Okazaki fragments), which are 1000 to 2000 ntd long in prokaryotes and 100 to 200 ntd in eukaryotes. Thus, devices used for priming the synthesis of the two strands may differ.

A preview of the means used for priming the continuous strand reveals a diverse set of mechanisms:

(1) An RNA primer is synthesized by either an RNA polymerase or a primase (Sections 3 to 6).

(2) A DNA primer terminus is generated by one of several mechanisms: (a) A nick may be made at a specific place in duplex DNA by an endonuclease (Section 7). (b) A 3′ end of DNA may be introduced into a duplex by recombination to make a D-loop, which can then be extended (as in T4 late replication, Section 17-6). The 3′ termini generated by the integration of viral DNA into a host chromosome may also fall into this class (Section 17-8). (c) A hairpin structure at the end of a genome may provide the 3′ end, as in parvoviruses (Section 15-7).

(3) A preformed RNA might serve as a primer, as in the tRNA initiation of the replication of retroviral genomes (Section 19-7).

(4) A primer moiety can be provided by a protein rather than by a nucleic acid, a mechanism used by certain linear duplex viruses of *B. subtilis* (φ29), *E. coli* (PRD1), and eukaryotic cells (adenovirus) (Section 8).

In contrast to the many ways used to prime continuous-strand replication, the multiple primers on the discontinuous strand are invariably synthesized by primases specialized for this function. In addition to their capacity in priming, these primases are likely to be integral parts of the replication machinery (replisome). They often function as part of a multiprotein complex, a *primosome*, containing a helicase activity as well as a primase. Primosomes may further interact with DNA polymerases and DNA-binding proteins to maintain coordinated, concurrent synthesis of both strands at the advancing replication fork.

8-2 Discovery of RNA-Primed DNA Synthesis

The first evidence for the role of RNA in priming DNA replication came from studies of the initiation mechanisms used by the small single-stranded phages of bacteria. M13, an Ff phage (Section 17-3), revealed that DNA replication could be primed at a unique position by the host RNA polymerase. Through studies of initiation of φX174 and the related G4 phage (Section 17-4), the *E. coli* primase was discovered, along with the proteins that associate with primase to generate the mobile multiprotein complex called the primosome. Clarification of the mechanisms of RNA priming of strand initiation of these three small phages helped reveal how host chromosome strands are initiated first at the origin and subsequently at the replication forks.

Synthesis of the M13 strand complementary to the single-stranded circular viral DNA in its conversion to the duplex replicative form (SS → RF; Fig. 8-2, and Section 17-3) does not require the expression of viral genes. Rather, this synthesis depends on host cell enzymes. If one of these were RNA polymerase (RNAP), then rifampicin, an antibiotic that specifically inhibits initiation of

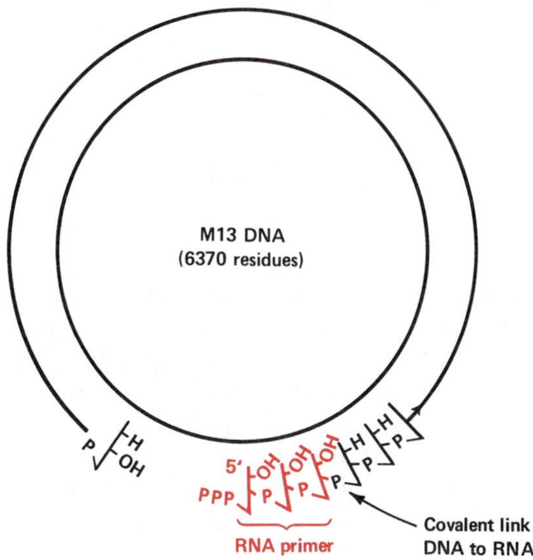

Figure 8-2
A scheme for RNA-primed initiation of synthesis of a strand complementary to the viral DNA of M13. A short chain of RNA is extended covalently by DNA. A short gap remains between the 3′ DNA terminus and RNA at the 5′ end.

transcription by RNAP (Section 7-2), should prevent this first step in M13 DNA synthesis. This is in fact the case.[1]

Rifampicin completely inhibits conversion of the viral ssDNA to the duplex form. Dependence on RNAP (or, at least, its β subunit) is further attested to by the failure of rifampicin to be inhibitory in an *E. coli* mutant with a rifampicin-resistant RNAP. Finally, a soluble enzyme extract prepared by gentle lysis converts the single-stranded M13 circle to the duplex form in a rifampicin-sensitive reaction. The product contains a small gap in the synthetic complementary strand at the point of the origin (Fig. 8-2) and is called RFII to distinguish it from the covalently closed circular RFI duplex. Evidence that RNAP synthesizes the RNA priming fragment in this system can be summarized as follows:

(1) Inhibition by rifampicin, streptolydigin, and actinomycin—all inhibitors of RNAP—and lack of inhibition by rifampicin in extracts of mutants with a rifampicin-resistant RNAP.

(2) Requirement for all four rNTPs.

(3) Conversion of SS to RF in two stages, an initial stage (A) of RNA synthesis to produce a primed single strand, which can be isolated and converted to RF in a second stage (B) in the absence of the rNTPs (except for ATP) and in the presence of rifampicin:

$$\text{SS} \underset{\text{RNA synthesis}}{\text{—(A)}\longrightarrow} \text{Primed SS} \underset{\text{DNA synthesis}}{\text{—(B)}\longrightarrow} \text{RFII}$$

(4) Demonstration of a phosphodiester linkage of a deoxynucleotide to a ribonucleotide in the isolated RF, in stoichiometric equivalence to the number of RF molecules formed. The radiolabel ^{32}P is transferred from [α-^{32}P]dNTPs to a ribonucleotide (2′- and 3′-AMP) upon alkaline cleavage of the produce (Fig. 8-3).

1. Brutlag D, Schekman R, Kornberg A (1971) *PNAS* *68*:2826.

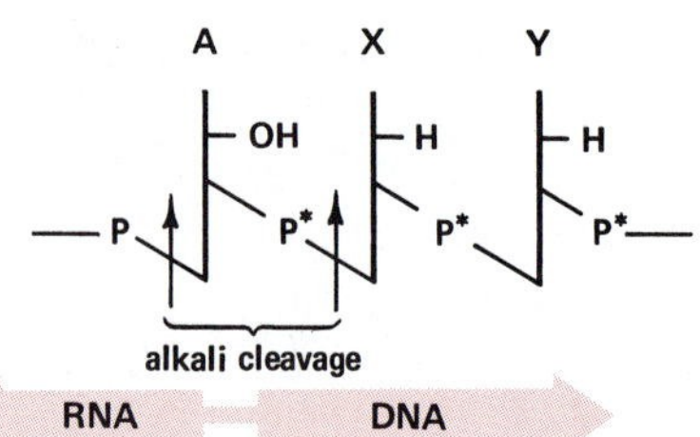

Figure 8-3
Procedure for demonstrating covalent linkage between the RNA primer and the DNA chain. Transfer of ^{32}P from an α-labeled deoxynucleoside triphosphate to a ribonucleotide, in this case to adenylate, is observed after cleavage by alkali. A = adenine; X and Y = any of the four bases.

(5) Persistence of an RNA segment at the 5′ end of the complementary strand, as inferred from the requirements for closure of the RFII product to RFI (Fig. 8-4). To produce an alkali-stable RFI, implying an absence of ribonucleotides, the 5′→3′ exonuclease of *E. coli* pol I is required to excise the primer and fill the gap. Either T4 or *E. coli* ligase can then seal the circle. To produce an alkali-labile RFI, a DNA polymerase is needed which fills gaps but lacks the excising function of a 5′→3′ exonuclease. T4 DNA polymerase fills this role. T4 ligase, which can join both DNA and RNA chains, can then seal the circle; *E. coli* ligase, which joins only DNA chains, cannot be substituted.[2]

These exploratory studies of RNA priming of M13 replication have been extended with purified enzymes and analysis of the sequence of the primer. The approaches and mechanisms apply to the priming of other phage and animal viruses and to the replication of bacterial and eukaryotic chromosomes.

Phage M13 proved to be a lucky choice for testing the RNA-priming hypothesis: the initial assay, which depended on rifampicin inhibition of the initiation of DNA synthesis, would not have applied to ϕX. This seemingly similar chromosome does not depend on the host RNAP in conversion of its viral strand to RF, and so is not inhibited by rifampicin. Nevertheless, an RNA fragment does prime ϕX DNA synthesis and is generated by an entirely different RNA-synthesizing system, previously unrecognized (Section 4).

Initially, the small phages M13, G4, and ϕX174 and polyoma virus provided the clearest examples of RNA priming of DNA chains.[3] Because these tiny viruses rely on the host cell for virtually all their synthetic needs, the proteins used in viral replication proved to be the ones the cell uses to replicate its own

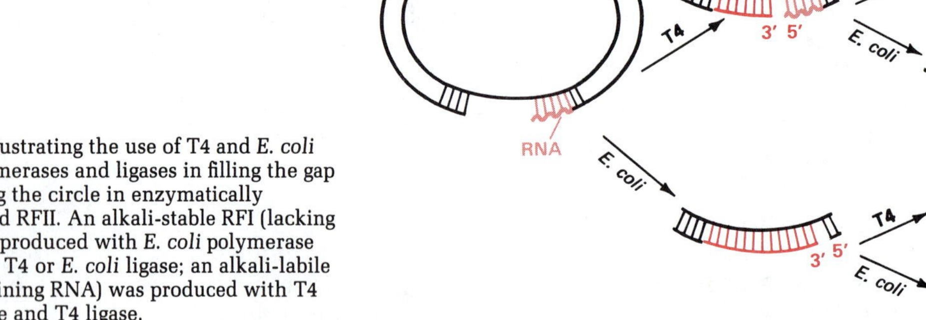

Figure 8-4
Scheme illustrating the use of T4 and *E. coli* DNA polymerases and ligases in filling the gap and closing the circle in enzymatically synthesized RFII. An alkali-stable RFI (lacking RNA) was produced with *E. coli* polymerase and either T4 or *E. coli* ligase; an alkali-labile RFI (containing RNA) was produced with T4 polymerase and T4 ligase.

2. Westergaard O, Brutlag D, Kornberg A (1973) *JBC 248*:1361.
3. Reichard P, Eliasson R, Söderman G (1974) *PNAS 71*:4901.

DNA. The difficult demonstration that RNA primes the nascent fragments in discontinuous replication in vivo was achieved in studies of *E. coli*,[4] T7 phage,[5] *B. subtilis*,[6] SV40,[7] and lymphocytes. RNA priming is the most generally used mechanism for initiation of DNA chains.

The basis for using RNA rather than DNA to start DNA chains may be to ensure the ultrahigh fidelity of the DNA record. At or near the start of a DNA chain, the proofreading and error-correcting designs built into DNA replicative mechanisms may not function as well in removing base-pairing errors as during chain growth. RNA messages are designed to be transient, whereas DNA is essentially indelible. Thus, the start of a DNA chain marked by a "foreign" RNA–DNA hybrid can be readily recognized and excised, followed by nearly error-free DNA replication to fill the resulting gap.

8-3 *E. coli* Primase (*dnaG* Gene Product)

Resolution and reconstitution of the factors needed for replication of the ϕX174 viral strand produced a list of more than ten proteins whose functions, except for SSB and DNA polymerase, were unknown.[8] Among them was dnaG protein, which had also been purified by complementation of dnaG-deficient lysates; the larger size of nascent fragments in these lysates suggested they had a deficiency in initiation.[9]

The fortuitous discovery that phage G4, closely related to ϕX174, has strikingly simpler requirements for conversion of its single-stranded chromosome to the duplex RF[10] (Section 17-4) helped to clarify the function of dnaG protein. Inasmuch as only SSB, dnaG protein, and DNA polymerase III holoenzyme are needed to assay G4 replication, purification of the dnaG protein was greatly facilitated, and its primase function was revealed. Improved procedures for isolating SSB[11] and DNA pol III holoenzyme[12] were welcome byproducts.

Primase is a single polypeptide of 60 kDa;[13] there are about 50 to 100 copies per cell. When primase acts on G4 DNA (coated with SSB) in the presence of rNTPs, a 29-ntd complementary copy of a unique region is formed.[14] Upon the subsequent addition of DNA polymerase and dNTPs, this RNA chain becomes covalently extended by DNA synthesis (Fig. 8-5). The primer, starting with ATP, contains a potential hairpin structure with seven GC base pairs in its sequence (Fig. 8-6). The RNA synthesized by primase is at the origin of replication and matches precisely the sequence in the origin assigned earlier from in vitro studies. Furthermore, initiation occurs at the same sequence in vivo.[15]

---

4. Ogawa T, Hirose S, Okazaki T, Okazaki R (1977) *JMB 112*:121; Miyamoto C, Denhardt DT (1977) *JMB 116*:681; Okazaki T, Kurosawa Y, Ogawa T, Seki T, Shinozaki S, Hirose S, Fujiyama A, Kohara Y, Machida T, Tamanoi F, Hozumi T (1976) *CSHS 43*:203.
5. Okazaki T, Kurosawa Y, Ogawa T, Seki T, Shinozaki S, Hirose S, Fujiyama A, Kohara Y, Machida Y, Tamanoi F, Hozumi T (1978) *CSHS 43*:203.
6. Tamanoi F, Okazaki T, Okazaki R (1977) *BBRC 77*:290.
7. DePamphilis ML, Anderson S, Bar-Shavit R, Collins E, Edenberg H, Herman T, Karas B, Kaufmann G, Krokan H, Shelton E, Su R, Tapper D, Wassarman PM (1978) *CSHS 43*:679.
8. Schekman R, Weiner A, Kornberg A (1974) *Science 186*:987; McMacken R, Kornberg A (1978) *JBC 253*:3313; Wickner S, Hurwitz J (1974) *PNAS 71*:4120.
9. Lark KG (1972) *NNB 240*:237.
10. Zechel K, Bouché J-P, Kornberg A (1975) *JBC 250*:4684; Bouché J-P, Zechel K, Kornberg A (1975) *JBC 250*:5995.
11. Weiner JH, Bertsch LL, Kornberg A (1975) *JBC 250*:1972.
12. McHenry C, Kornberg A (1977) *JBC 252*:6478.
13. Rowen L, Kornberg A (1978) *JBC 253*:758.
14. Bouché J-P, Rowen L, Kornberg A (1978) *JBC 253*:765.
15. Hourcade D, Dressler D (1978) *PNAS 75*:1652.

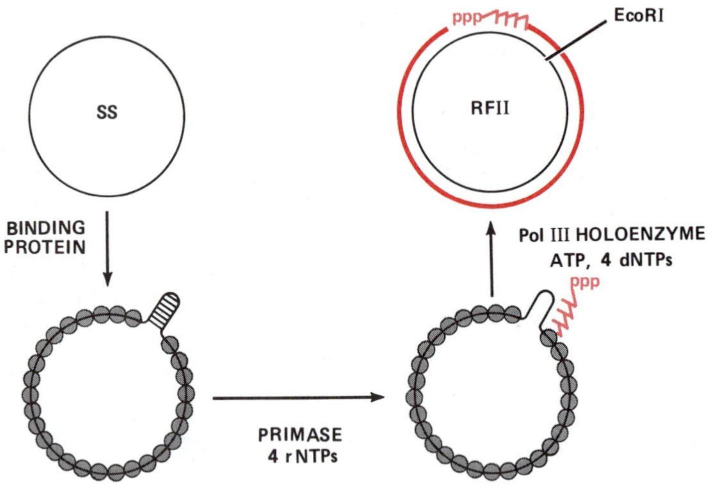

Figure 8-5
Scheme showing the stages and proteins involved in the conversion of G4 viral DNA into RFII DNA (SS → RF). The unique cleavage site of EcoRI restriction nuclease is shown in the upper right.

The hairpin region in G4 DNA, which resists coating and melting by SSB, serves as the recognition signal for primase. A complex of two molecules of primase bound to the SSB-coated G4 DNA circle most likely represents the active complex.[16] When isolated in the absence of rNTPs this complex is virtually inactive in primer formation. However, the complex formed in the presence of ATP, GTP, and UTP (which support synthesis of a 9-ntd transcript) is fully active in replication.[17]

When primase action is coupled directly to replication, the primer size is abbreviated to a few nucleotides by its early extension by DNA polymerase. Furthermore, primase can substitute dNTPs for rNTPs in all but the first, and possibly second, position to produce hybrid primers interspersed with ribo- and deoxyribonucleotides.[18] Under physiologic conditions, the relative concentrations of rNTPs and dNTPs and their affinities for primase, among other factors, determine the composition and size of the primer. The energy barrier in opening the template hairpin (Fig. 8-6) may terminate the primer at six or fewer residues. Priming by very short oligonucleotides implies an additional function of primase in stabilizing the short primer on the template.

In the replication of most templates, primase acts in concert with other factors; interaction with the dnaB protein is essential for priming of most DNAs. For primase to act on ϕX DNA, still more proteins are required: dnaC, dnaT, PriA, PriB, and PriC proteins (Section 4). These factors and ATP assemble a multiprotein complex, the primosome, on the template that allows subsequent priming by primase; which factors become stably bound and function during priming is not clearly established. Detailed analysis of the G4 origin and other priming sites may provide insights into how primase recognizes DNA and an understanding of how, in the absence of such special sequences, dnaB protein activates primase.

Like G4, the closely related α3, St-1, and ϕK phages are also primed by primase acting alone.[19] Comparison of the origin sequences of these three phages, in addition to deletion analysis, shows that a region of about 139 ntd containing

16. Stayton M, Kornberg A (1983) *JBC 258*:13205.
17. Bates DL, Kornberg A, unpublished.
18. Rowen L, Kornberg A, unpublished.
19. Sims J, Capon D, Dressler D (1979) *JBC 254*:12615.

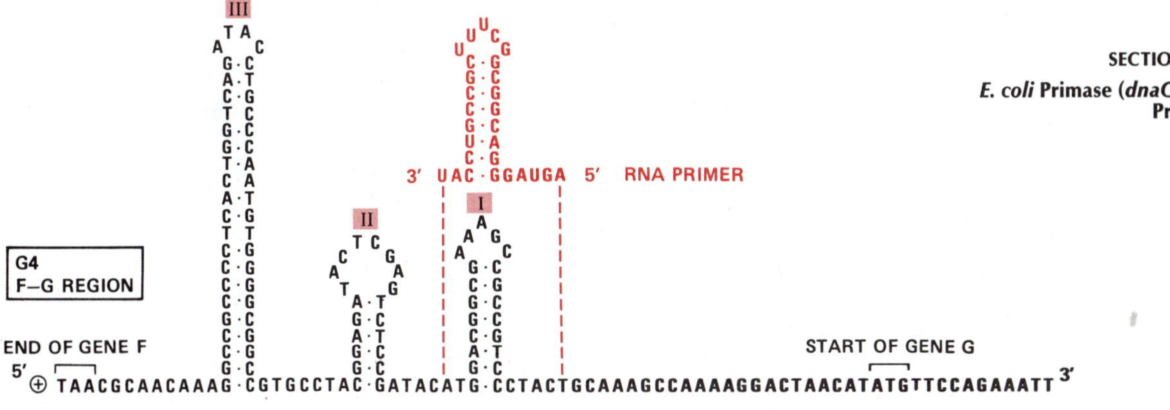

Figure 8-6
Nucleotide sequences and proposed secondary structures at the origin for synthesis of the complementary strand of phage G4 and for the oligoribonucleotide primer synthesized by primase. Three stem–loop structures (I, II, and III) are required for priming phages α3, St-1, and ϕK, as well as phage G4. The phage G4 origin lies in the region between genes F and G. Much shorter primers are produced when primase action is coupled to replication.

three stem–loop (hairpin) structures is required for priming[20] (Fig. 8-6). The secondary structure of stem–loops I and II (Fig. 8-6) is important rather than their nucleotide sequence; insertions and substitutions that preserve this structure are compatible with origin function.[21] Stem–loop III is not as tolerant of changes, indicating that its exact sequence is required. Footprinting analysis of primase bound to the ϕK origin in the presence of SSB suggests that the protein interacts directly with stem–loop III and protects the stem from nuclease digestion.[22]

The final essential determinant in the G4 origin is the sequence immediately at the start of RNA synthesis. The nucleotides 3′-GTC are required in the template at the start site, with initiation of the RNA primer (5′-AG . . .), complementary to TC.[23] The distance between this trinucleotide sequence and stem–loop I is important. The consensus spacing is 5 ntd; a spacing of 8 ntd is tolerated but one of 12 ntd is not. The sequence 3′-GTC is the preferred recognition sequence for initiation by primase, both when acting alone and when assisted by the primosomal proteins (see below). This preference for a short sequence in the template is common to other primases (e.g., T4, T7; see Section 5).

Replication of the continuous strand of the plasmid R1 is primed solely by *E. coli* primase.[24] The nascent continuous strands isolated from replication intermediates have a unique 5′ end, and initiate at a single site about 380 bp downstream of the origin. (The origin of replication in this case is the *cis*-acting sequence required for initiation; Sections 16-1 and 16-4.) Cloning of this region in a single-stranded phage confirms that it directs primer synthesis by primase. Synthesis of the complementary strand on this ssDNA requires only SSB, primase, and DNA pol III holoenzyme and is initiated at the same site as is used

20. Lambert PF, Waring DA, Wells RD, Reznikoff WS (1986) *J Virol 58*:450.
21. Hiasa H, Tanaka K, Sakai H, Yoshida K, Honda Y, Komano T, Godson GN (1989) *Gene 84*:17.
22. Sims J, Benz EW (1980) *PNAS 77*:900.
23. Hiasa H, Sakai H, Tanaka K, Honda Y, Komano T, Godson GN (1989) *Gene 84*:9.
24. Masai H, Arai K (1989) *JBC 264*:8082.

during plasmid replication. This R1 sequence is thus functionally analogous to the G4 complementary-strand origin.

In the closely related plasmid R100, a sequence equivalent to the R1 priming site also functions as a priming site in vivo for complementary-strand synthesis of single-stranded phage. Similar sequences have been detected in plasmids R6K and F, indicating that recognition of specific sequences by primase is more common than believed from earlier studies. Yet secondary sites can operate in the absence of the R1 sequence to support priming of the lagging strand efficiently enough to sustain plasmid replication in vivo.

The extensive regions of secondary structure that dominate the G4 origin are not in evidence around the R1 priming site. Homology between the single-stranded phage and plasmid-derived sequences is restricted to a region immediately surrounding the start of RNA synthesis (Table 8-1). The 3′-GTC sequence is present at the start site and several A and C residues are similarly located upstream. Although the stem–loop structures of G4 are absent from the plasmid-priming regions, suggesting that they are not uniquely essential for priming, extensive mutational analysis is needed to delineate the essential features of these priming sites.

The preference of primase for the 3′-GTC sequence in the template is also apparent when primase is activated by dnaB protein (Section 4). Primers synthesized during replication of the *E. coli* chromosome in vivo (where primase probably interacts with dnaB protein and perhaps with other primosomal proteins) are predominantly 11 ± 1 ntd long.[25] They start with a purine rNTP, A being used five times more often than G; the second nucleotide is also most commonly a purine. Of the isolated primers, 50% start with pppApG at their 5′ end, the complement of TC in the 3′-GTC signal. The fixed-lengthed primers (11 ± 1 ntd) resemble the size of those synthesized by eukaryotic primases (Section 6). Primers are heterogeneous in length when synthesized in vitro by *E. coli* primase alone or together with dnaB protein in the ϕX-type primosome.

The RNA–DNA junctions near the replication origins of phage λ[26] and the *E.*

Table 8-1
Sequences of template strands around the start site and the RNA–DNA transition point of primers synthesized by primase acting alone

Template	Sequence																			
						|— RNA										—| DNA →				
						pppA	G	.	.	.										
						−11										−1	+1			
G4	A	A	A	C	G	T	C	A	T	C	C	C	T	G	C	C	G	C	C	G
St-1,ϕK,α3	A	A	T	C	G	T	C	C	T	C	C	C	T	T	C	C	G	C	C	G
R100	G	A	C	C	G	T	C	C	C	G	A	C	C	C	C	C	T	A	C	C
R1	G	A	C	C	G	T	C	C	C	C	G	A	C	C	C	C	T	A	C	C
Ff6	G	A	C	C	G	T	C	C	T	C	C	G	A	C	G	C	G	T	C	A
R6K *ori*α	G	A	C	C	G	T	C	C	C	C	C	G	C	G	T	T	C	G	C	G

25. Kitani T, Yoda K, Ogawa T, Okazaki T (1985) *JMB 184*:45.
26. Yoda K, Yasuda H, Jiang X-W, Okazaki T (1988) *NAR 16*:6531.

coli chromosome[27] provide further evidence for a specific primase recognition sequence[28]. Of 48 junctions mapped in the λ phage origin region, 92% had a 3′ purine–pyrimidine sequence on the template strand at a position −12 and −11 ntd upstream of the start of DNA synthesis. Similarly, 43% of the junctions around the *E. coli* origin (*oriC*) contain the 3′-GTC trinucleotide centered at −11 ntd upstream.

Based on analysis of priming sites, primers, and RNA–DNA junctions, 3′-GTCpy (where py = pyrimidine) is clearly the sequence preferred by *E. coli* primase for the initiation of RNA primers. The initial residue (e.g., G) is used for recognition but not as a template, as is also done by the phage T7 and T4 primases (Section 5). Thus, T7 and T4 need an initial nucleotide in the template that does not appear in the primer; however, it is not a G but rather a C for T7 and a T for T4.

8-4 *E. coli* Primosomes

E. coli primase seldom acts alone; most commonly it teams up with the multifunctional dnaB protein in the synthesis of primers to start DNA chains. Assembling the dnaB protein–primase complex on a template, whether it be ssDNA (e.g., ϕX174) coated with SSB or duplex DNA (e.g., pBR322, *E. coli* chromosome), requires additional "prepriming" proteins. The resulting mobile protein complexes, containing dnaB protein and primase and sometimes additional components track processively along the DNA template, and are called *primosomes.*[29]

General (Nonspecific) Priming: Activation of Primase by dnaB Protein

Primase and dnaB protein are needed throughout replication of the *E. coli* chromosome. The functional interaction between these two essential enzymes can be demonstrated on ssDNAs upon which primase alone is inert. The dnaB protein and then primase associate with a single-stranded template (Fig. 8-7) in the presence of ATP.[30] On any ssDNA template, including poly dT, the complex synthesizes numerous oligoribonucleotides that are 10 to 60 residues long; coating of the template by SSB completely inhibits this synthesis.

The association of dnaB protein and primase with DNA during general priming is distributive, with dnaB protein continually dissociating from the template and reassociating.[31] When binding of dnaB to the template limits the rate of priming, dnaC protein, complexed with dnaB, stimulates priming by increasing the affinity of dnaB for the template about fivefold.[32]

The mechanism whereby dnaB protein enables single-stranded templates to be primed by primase is not clear. The dnaB protein, a hexameric complex of 50-kDa monomers, is a potent DNA-dependent ATPase and helicase that tracks

---

27. Kohara Y, Tohdoh N, Jiang X-W, Okazaki T (1985) *NAR 13*:6847; Hirose S, Hiraga S, Okazaki T (1983) *MGG 189*:422.
28. Yoda K, Okazaki T (1990), submitted.
29. McMacken R, Uedo K, Kornberg A (1977) *PNAS 74*:4190.
30. Arai K, Kornberg A (1979) *PNAS 76*:4308.
31. Arai K, Kornberg A (1981) *JBC 256*:5260.
32. Wahle E, Lasken RS, Kornberg A (1989) *JBC 264*:2469.

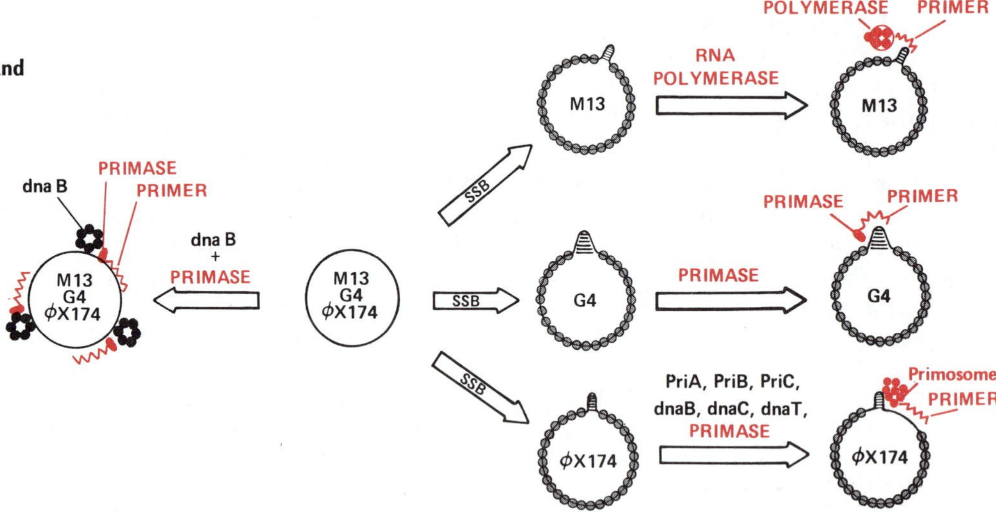

Figure 8-7
General (nonspecific) priming by dnaB protein and primase on uncoated DNA and the specific priming systems for M13, G4, and ϕX174 DNAs coated with SSB.

along DNA in the 5′→3′ direction (Section 11-2). General priming is supported by ATPγS, which is not hydrolyzed by the dnaB protein (but is utilized in primer synthesis by primase); ATPγS actually stimulates priming by increasing the stability of the dnaB·DNA complex. Thus, the ATP hydrolysis and translocation activities of dnaB are not required for activation of primase by dnaB.

Although a *functional interplay* between dnaB protein and primase is apparent, a *physical interaction*, either in solution or on the template, has not been demonstrated. If the role of dnaB protein were the generation of DNA secondary structures to serve as templates for primase,[33] then a physical interaction between the two proteins would not be required.

Specific Priming Mechanisms: The ϕX-Type Primosome

The dnaB protein and primase are unable to assemble on and prime most ssDNAs coated with SSB, nor can they associate unaided with a duplex template. Thus, with general priming blocked on SSB-coated single-stranded templates, specific priming mechanisms are needed (Fig. 8-7). One such mechanism is the specific recognition of secondary structures, as in the recognition of the M13 template by RNAP (Section 2) and of the G4 template by primase (Section 3).

The ϕX phage reveals still another mechanism of priming, one that depends on the assembly of a multiprotein, mobile primosome.[34] The primosome supports multiple priming events and is insensitive to the inhibitory effect that SSB normally exerts on the activities of dnaB protein and primase. The ϕX mechanism affords an attractive model for how the priming machinery for discontinuous-strand synthesis might be assembled and might operate in the replication of duplex DNA.

The proteins required for primosome assembly, in addition to dnaB protein

33. Arai K, Kornberg A (1981) *JBC 256*:5260.
34. Arai K, Low R, Kobori J, Shlomai J, Kornberg A (1981) *JBC 256*:5273.

and primase, are PriA, PriB, and PriC (formerly named n′, n, and n″, respectively), dnaC protein, and dnaT protein (formerly called protein i). These *prepriming proteins* are responsible for the ordered assembly of the primosome on the template. Although all seven are involved in assembly, the identity of all the proteins stably incorporated into the mobile complex has not been established, nor are the roles of all the components clear.

Primosome Assembly[35]

Assembly of the primosome (Fig. 8-8) starts by recognition of *pas* (primosome assembly site) by the PriA protein (see below). The PriB and PriC proteins also act early. In solution, the dnaB and dnaC proteins form a $B_6 \cdot C_6$ complex, stabilized by binding ATP to the dnaC protein. This B·C complex, with the help of dnaT protein, transfers dnaB protein to the template; the dnaT protein may also bind to the DNA, while dnaC protein is released. The association of primase with the assembled prepriming proteins (the *preprimosome*) generates the ϕX-type *primosome*.

The mobile primosome tracks along the ϕX template, synthesizing primers at numerous places, in contrast to the unique location of the primer synthesized by

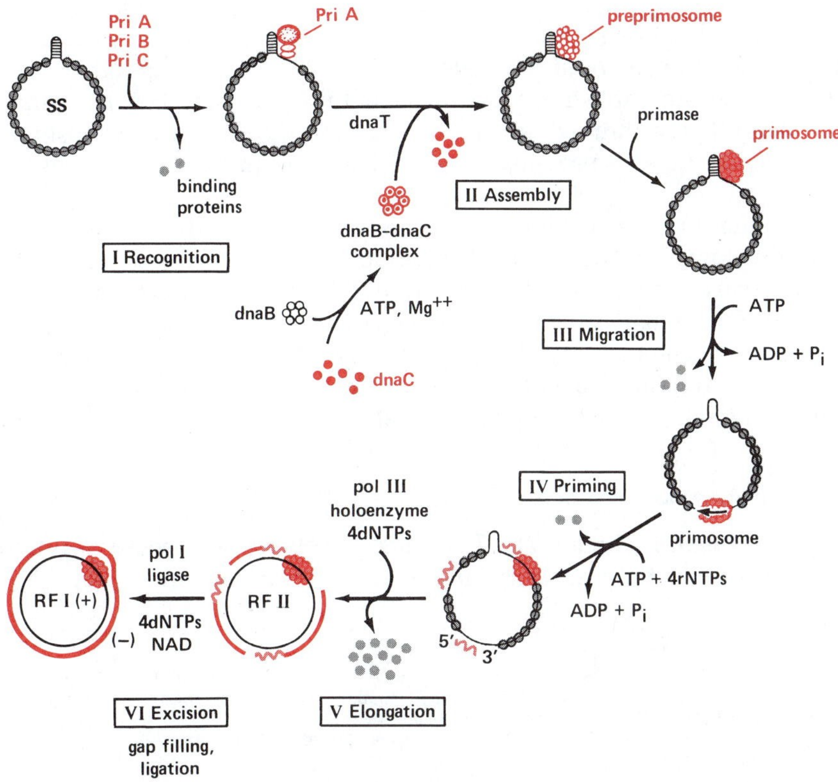

Figure 8-8
Scheme for assembly and migration of the primosome and the stepwise displacement of SSB in the ϕX SS → RF reaction.

35. Arai K, Low R, Kobori J, Shlomai J, Kornberg A (1981) *JBC* 256:5273.

primase acting alone on the G4 template. The dnaB protein and the PriA protein are each able to fuel primosome movement on a ssDNA template (Section 11-2).[36] The action of primase is not essential for either the stability or mobility of the complex; in fact, the preprimosome complex may be the real mobile and processive entity, with primase continually associating and dissociating to carry out the actual RNA synthesis.

Primosome Assembly Site

Assembly of the seven-protein primosome is initiated by interaction of the PriA protein with a specific sequence called a *primosome assembly site (pas)*. The *pas* of ϕX174 is contained on a 70-ntd fragment between genes F and G, a location analogous to that of the unique origin of complementary-strand synthesis of phage G4 (Fig. 8-6). *Pas* sequences have been isolated from several plasmids: ColE1, pBR322, pACYC177, CloDF13, F, R100, and ColE2.[37] Similar sequences have been isolated from the *E. coli* chromosome,[38] but whether they function in an analogous manner remains to be established.

The *pas* sequences, when present on SSB-coated ssDNA, have these features: (1) they serve as effectors of the DNA-dependent ATPase activity of PriA protein,[39] (2) they bind a single molecule of PriA protein specifically, as revealed by DNase I footprinting,[40] and (3) they support complementary-strand synthesis dependent on the seven primosomal proteins. Comparison of nine *pas* sequences reveals at least two structural classes, represented by the ϕX site and the ColE1 discontinuous-strand template site.[41] All but one of the other *pas* sequences show considerable homology to one of these two sites. However, the two groups, as well as the unrelated sequence, have no discernible relationship. The recognition by the PriA protein of the various *pas* sequences thus needs to be explained.

The *pas* sequences have considerable secondary structure that is undoubtedly responsible for their recognition. The resistance of some *pas* sequences to digestion by *E. coli* exonuclease VII is indicative of dsDNA, as is their sensitivity to DNase I while the neighboring SSB-coated ssDNA remains resistant. The *pas* sequences contain regions of dyad symmetry and can be folded into several alternative structures; physical characteristics and mutational analysis favor a conformation involving the interaction of two stem–loop structures (Fig. 8-9).[42]

The Primosome-Assembly Proteins

PriA Protein[43] (Protein n′, Factor Y). This 76-kDa polypeptide recognizes *pas* sequences as the first step in assembly of the active primosome (Fig. 8-8). The

36. Lee MS, Marians KJ (1989) *JBC 264*:14531.
37. Zipursky SL, Marians KJ (1980) *PNAS 77*:6521; Nomura N, Ray DS (1980) *PNAS 77*:6566; Imber R, Low RL, Ray DS (1983) *PNAS 80*:7132; van der Ende A, Teertstra R, van der Avoort HGAM, Weisbeek PJ (1983) *NAR 11*:4957; Hiasa H, Sakai H, Komano T (1989) *FEBS Lett 246*:21.
38. van der Ende A, Teertstra R, van der Avoort HGAM, Weisbeek PJ (1983) *NAR 11*:4957; Stuitje AR, Weisbeek PJ, Meijer M (1984) *NAR 12*:3321.
39. Shlomai J, Kornberg A (1980) *JBC 255*:6789; Shlomai J, Kornberg A (1980) *JBC 255*:6794; Shlomai J, Kornberg A (1980) *PNAS 77*:799; Wickner S, Hurwitz J (1975) *PNAS 72*:3342.
40. Greenbaum JH, Marians KJ (1984) *JBC 259*:2594.
41. Masai H, Nomura N, Kubota Y, Arai K (1990) *JBC 265*:15124.
42. Greenbaum JH, Marians KJ (1985) *JBC 260*:12266; Abarzúa P, Soeller W, Marians KJ (1984) *JBC 259*:14286; Soeller W, Abarzúa P, Marians KJ (1984) *JBC 259*:14293; Marians KJ, Soeller W, Zipursky SL (1982) *JBC 257*:5656.
43. Shlomai J, Kornberg A (1980) *JBC 255*:6789; Shlomai J, Kornberg A (1980) *JBC 255*:6794.

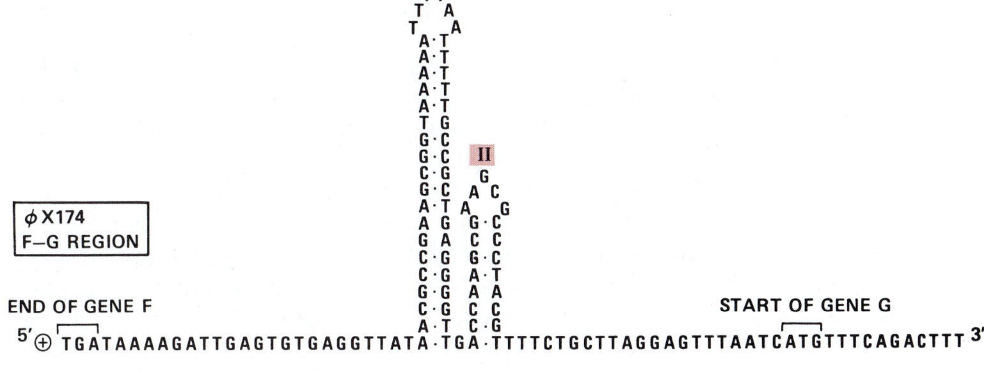

Figure 8-9
*φ*X *Pas* sequence folded into the two-hairpin conformation.

PriA protein can translocate on ssDNA, using the energy of ATP (or dATP) hydrolysis. Translocation displaces bound SSB, overcoming its inhibitory effects on dnaB protein and primase. On partially duplex DNA molecules, PriA protein is a DNA helicase that translocates on DNA in the 3′→5′ direction (Section 11-2).[44] The primosome thus contains two enzymes, the PriA and dnaB proteins, capable of translocating in opposite directions on ssDNA and displacing a strand from dsDNA in the primosome's path (Section 11-2).

The *priA* gene has been cloned and sequenced, and is located at 88.5 minutes on the *E. coli* map.[45] Strains with mutations in this gene may be defective in chromosomal and episomal replication in vivo.[46]

PriB Protein[47] (Protein n). This 11.5-kDa polypeptide is a dimer in solution. It is heat- and acid-resistant, and sensitive to N-ethylmaleimide (NEM). Binding to DNA depends on a direct interaction with SSB. Although 30 monomers can be bound to the *φ*X174 circle of 5386 ntd, only one monomer is incorporated in the primosome. The *priB gene* lies between *rpsF* and *rpsR* at 96 minutes on the *E. coli* map, within an operon of ribosomal proteins.

PriC Protein[48] (Protein n″). This polypeptide of 23 kDa shares the heat- and acid-resistant properties of the PriB protein, but is also resistant to NEM. The low abundance of PriC protein confounded its isolation, and its presence as a trace contaminant in preparations of replication proteins made its assay difficult. An early report of a molecular weight of 17 kDa was due to a confusion with the most prominent component in partially purified PriC protein fractions, namely, the 17-kDa ribosomal protein L13. When ribosomes are removed early in purification, fractions retain their PriC activity, and further fractionation separates the PriC activity from L13. Thus, PriC is clearly unique from the rRNA-binding proteins artifactually released from the ribosomes during

44. Lee MS, Marians KJ (1987) *PNAS 84*:8345; Lasken RS, Kornberg A (1988) *JBC 263*:5512.
45. Lee EH, Masai H, Allen GC Jr, Kornberg A (1990) *PNAS 87*:4620; Nurse P, DiGate RJ, Zavitz, KH, Marians KJ (1990) *PNAS 87*:4615.
46. Lee EH, Kornberg A (1991) *PNAS 88*, in press.
47. Low RL, Shlomai J, Kornberg A (1982) *JBC 257*:6242; Lee EH, Masai H, Allen GC Jr, Kornberg A (1990) *PNAS 87*:4620; Allen GC Jr, Kornberg A (1991) *JBC*, in press.
48. Lee EH, Kornberg A (1991), unpublished.

lysozyme–EDTA cell lysis. The PriC protein acts early in primosome assembly; about one molecule per single-stranded template is required.

DnaT Protein[49] (Protein i). The dnaT protein is a trimer of 22-kDa polypeptides. Action of this protein in primosome assembly is required at a stage involving the dnaB·dnaC complex (see below) and after the function of the PriA, B, and C proteins. The 22-kDa protein was originally purified as an essential factor for ϕX174 SS → RF replication, and was called protein i. N-terminal sequencing of the purified protein led to the identification of its gene,[50] located immediately upstream of *dnaC*, in a position occupied by *dnaT*. Complementation of *dnaT* mutants by the cloned protein i gene, overproduction of protein i activity in such cells, and the lack of protein i activity in extracts from the *dnaT* mutant confirmed that *dnaT* is the gene for the primosomal protein i.

Cells with a *dnaT* mutation are temperature-sensitive for growth and DNA synthesis, indicating that the protein, and thus perhaps the primosome, is essential for chromosomal replication. The "slow-stop" phenotype (Section 15-4) is indicative of a defect in initiation, although the behavior of *dnaT* mutants also suggests a role for this protein in termination and in stable DNA replication (Section 20-3).

DnaB Protein. The dnaB protein is made up of subunits near 50 kDa in size which form a hexamer of about 300 kDa. Its abundance is estimated to be about 20 molecules per cell. The level of dnaB protein is amplified over 200-fold in cells with "runaway" plasmids bearing the *dnaB* gene. Large quantities of crystalline protein are easily obtained (320 mg from 1.5 kg of cell paste),[51] inviting studies of dnaB protein structure and function. As a DNA-dependent ATPase (or rNTPase),[52] the protein had been regarded as responsible for the mobility of the primosome; it is now clear that either dnaB protein or PriA protein can fuel primosomal translocation (Section 11-2).

The basic function of dnaB protein in priming is illustrated in the general (nonspecific) priming on uncoated DNA by primase when dnaB is present.[53] ATP (or a nonhydrolyzable analog) induces a high-affinity complex with dnaB protein that forms or stabilizes a DNA secondary structure which is apparently exploitable by primase. Upon ATP hydrolysis, the low-affinity ADP·dnaB complex dissociates.[54] For specific priming of replication on SSB-coated DNA, dnaB protein must be part of the stable, mobile, and processive primosome complex, but the mechanism by which it activates primase is probably unchanged.

The multifaceted structure of dnaB protein[55] has specific sites required for (1) activation of primase, (2) interaction with dnaC protein (see below), (3) binding to ssDNA, (4) binding to dsDNA, and (5) bringing about the hydrolysis of ATP. Sites for allosteric effects of ATP and dATP on the protein are also inferred.[56] The binding of ATP and ADP is also manifested by the stabilizing effect of these nucleotides, which protect dnaB protein against inactivation by heat.[57] ATP also

49. Arai K, McMacken R, Yasuda S, Kornberg A (1981) *JBC 256*:5281.
50. Masai H, Bond MW, Arai K (1986) *PNAS 83*:1256; Masai H, Arai K (1988) *JBC 263*:15083.
51. Arai K, Yasuda S, Kornberg A (1981) *JBC 256*:5247.
52. Reha-Krantz LJ, Hurwitz J (1978) *JBC 253*:4043; Arai K, Kornberg A (1981) *JBC 256*:5253; Arai K, Kornberg A (1981) *JBC 256*:5260.
53. Arai K, Kornberg A (1981) *JBC 256*:5267.
54. Arai K, Kornberg A (1981) *JBC 256*:5260.
55. Arai K, Kornberg A (1981) *JBC 256*:5253.
56. Arai K, Kornberg A (1981) *JBC 256*:5260.
57. Arai K, Yasuda S, Kornberg A (1981) *JBC 256*:5247.

dramatically activates mutant forms of the protein that otherwise appear inert.[58] Limited proteolytic cleavage reveals the domain structure of dnaB protein.[59] With removal of a 12-kDa fragment from the N terminus, replicative interactions are lost; ATPase activity and binding to ssDNA are preserved in a 30-kDa C-terminal polypeptide.

DnaC Protein and the DnaB·DnaC Complex. Cloning of the *dnaC* gene and a 50-fold overproduction of the 29-kDa dnaC protein have facilitated its isolation and characterization.[60] The formation of a complex between six dnaC protein monomers and a dnaB protein hexamer can be monitored by the resulting protection of the dnaC protein from NEM.[61] ATP (or dATP) binds the dnaC protein in the complex and stabilizes the B–C interaction.[62]

$$B_6 + 6\,C \rightleftharpoons B_6 \cdot C_6 \xrightarrow[6\,ATP]{} (B \cdot C \cdot ATP)_6 \xrightarrow[DNA]{} \boxed{DNA \cdot B_6} + 6\,C + 6\,ADP + 6\,P_i \longrightarrow$$

Prepriming on ssDNA
Prepriming at *ori*C
Helicase activity

DNA:
ssDNA
ϕX preprimosome
*ori*C • dnaA

The B·C complex, with the participation of dnaT protein, transfers one hexamer of dnaB to the developing primosome. In the assembly of the prepriming complex at the site where dnaA protein binds to the *E. coli oriC* sequence (see below and Section 16-3), dnaB protein is also donated from a B·C complex.

Although serving a critical role in assembly, the dnaC protein is apparently not maintained in the developing primosome.[63] The amount of dnaC protein recovered in the isolated primosome assemblies at either the ϕX *pas* or the *oriC* sequence is variable. The presence of dnaC protein apparently holds dnaB protein in an inactive state. Thus, after serving in delivery of dnaB to the template, dnaC protein must leave the complex to allow dnaB protein to function. How dissolution of the tight dnaB·dnaC protein complex is achieved at the site of the developing primosome, and the role of ATP hydrolysis by dnaC protein in this process, are not yet clear.

The phage λ encodes an analog of the cellular dnaC protein, the product of gene *P* (Section 17-7). The dnaB·λP protein complex is the active form of dnaB protein for assembly at *ori*λ. However, λ P protein is a potent inhibitor of the functions of dnaB protein and must be removed by the action of several cellular heat-shock proteins for the functions of dnaB protein to be revealed and replication to proceed.[64]

The ϕX-Type Primosome in Duplex DNA Replication

Several features of the ϕX-type primosome make it an attractive model for how priming of the discontinuous strand may take place in the replication of duplex DNA:

---

58. Günther E, Mikolajczyk M, Schuster H (1981) *JBC 256*:11970.
59. Nakayama N, Arai N, Kaziro Y, Arai K (1984) *JBC 259*:88.
60. Kobori JA, Kornberg A (1982) *JBC 257*:13763.
61. Wickner S, Hurwitz J (1975) *PNAS 72*:921; Kobori JA, Kornberg A (1982) *JBC 257*:13770.
62. Wahle E, Lasken RS, Kornberg A (1989) *JBC 264*:2469.
63. Wahle E, Lasken RS, Kornberg A (1989) *JBC 264*:2469; Wahle E, Lasken RS, Kornberg A (1989) *JBC 264*:2463.
64. Mallory JB, Alfano C, McMacken R (1990) *JBC 265*:13297.

(1) It is made up entirely of *E. coli* replication proteins, five of which (dnaB, dnaC, dnaT, PriA, and primase) are implicated genetically in chromosomal replication, and three of which (dnaB, dnaC, and primase) are essential for *oriC* plasmid replication in vitro.

(2) The complex moves in the 5′→3′ direction along the DNA strand to which it is bound (Section 11-2). On duplex DNA this polarity would place it on the template for the discontinuous strand, moving in the same direction as the fork.

(3) The preprimosome is processive, with the dnaB protein, at least, remaining stably associated with the template for multiple rounds of priming.

Involvement of the ϕX-type primosome in discontinuous-strand synthesis is indicated in pBR322 plasmid replication[65] (Section 18-2). Although both strands are synthesized essentially simultaneously, replication of the continuous and discontinuous strands involves different priming mechanisms. Thus, studies of this plasmid are helpful in dissecting priming processes.

Discontinuous-strand synthesis on the pBR322 template (Fig. 8-10) initiates at a *pas* located about 150 bp downstream of the continuous-strand origin and depends on the same seven proteins essential for priming of ϕX SS → RF replication.[66] The primosome, probably due to the dnaB protein within it, functions as the helicase at the replication fork (Section 11-2), as well as activating the priming by primase.

The ϕX-type primosome, an attractive model for the discontinuous-strand replication complex, can clearly function in this capacity. On pBR322, it seems to be the major complex responsible for priming the discontinuous strand. Isolation of *pas* sequences from other plasmids, and potentially analogous sequences from the *E. coli* chromosome, argue for involvement of the ϕX-type primosome in the replication of duplex DNA. The essential nature of the dnaT protein also favors the requirement of this type of primosome in chromosomal replication. Yet several components of the ϕX-type primosome (i.e., PriA, PriB, PriC, and dnaT proteins) are not essential for duplex DNA replication initiated at *oriC* or *ori*λ in vitro. Protein complexes similar to the ϕX-type primosome but simpler in architecture suffice in these systems.

The *oriC*-Type Primosome

The *E. coli* dnaA protein and the phage λ O and P proteins can substitute for dnaT and the Pri proteins in assembling a functional primosome.[67] The protein complex generated at *oriC* and *ori*λ qualifies as a primosome by the following criteria: (1) mobility of the protein complex, (2) multiple priming on the template, and (3) stable binding of dnaB protein, acting processively; the dnaB protein presumably activates primer synthesis by the same mechanism as is used during general priming.

The details of assembly of dnaB protein and primase at *oriC* and *ori*λ are presented elsewhere (*oriC* in Section 16-3; *ori*λ in Section 17-7). Briefly, dnaA protein (or λ O protein) binds specifically to the origin sequence and causes a

65. Minden JS, Marians KJ (1985) *JBC 260*:9316.
66. Minden JS, Marians KJ (1985) *JBC 260*:9316; Masai H, Arai K (1988) *JBC 263*:15016.
67. Baker TA, Funnell BE, Kornberg A (1987) *JBC 262*:6877; Baker TA, Sekimizu K, Funnell BE, Kornberg A (1986) *Cell 45*:53; Dodson M, Echols H, Wickner S, Alfano C, Mensa-Wilmot K, Gomes B, LeBowitz J, Roberts JD, McMacken R (1986) *PNAS 83*:7638.

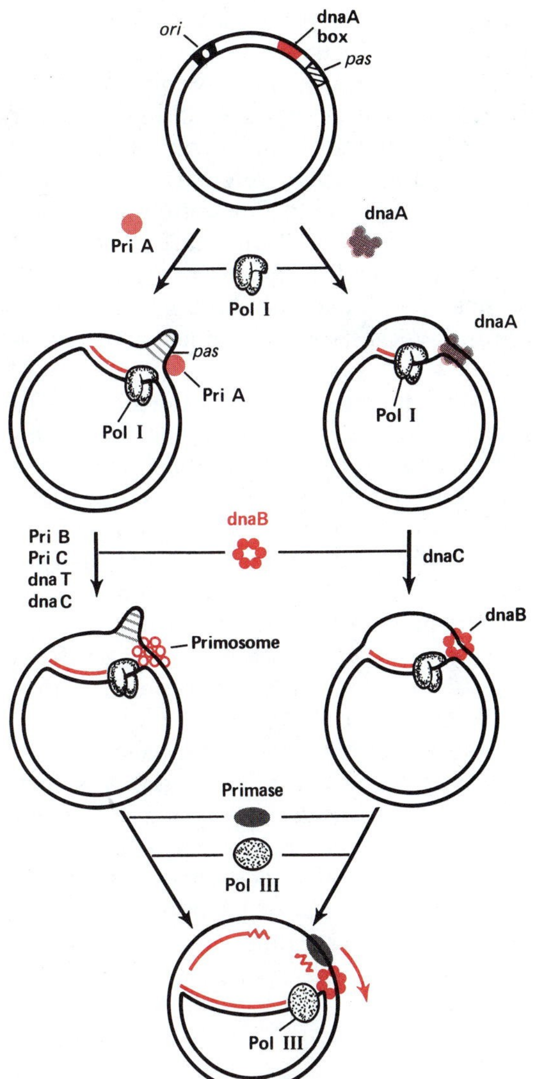

Figure 8-10
Scheme for the initiation of discontinuous-strand synthesis on pBR322 DNA, showing equivalence of the alternative PriA and dnaA protein-dependent pathways.

localized melting of the DNA duplex.[68] The dnaB protein is transferred to this complex from either a dnaB·dnaC complex for *oriC* or a dnaB·λ P complex for *ori*λ. The dnaB protein becomes stably bound to the template, where it unwinds the DNA as a helicase (Section 11-2) and activates priming by primase.[69]

Assembly of the *ori*C-type primosome is similar to assembly of the φX-type that occurs as *pas* sequences. Recognition of the template by a specific binding protein occurs first; either PriA, dnaA, or λ O qualifies. Through subsequent protein–protein interactions, the dnaB protein becomes stably established on the template, and with the final association of primase with dnaB protein, the primosome comes into being.

68. Bramhill D, Kornberg A (1988) *Cell 52*:743; Bramhill D, Kornberg A (1988) *Cell 54*:915; Schnos M, Zahn K, Inman RB, Blattner FR, (1988) *Cell 52*:385.
69. LeBowitz JH, McMacken R (1986) *JBC 261*:4738; Baker TA, Funnell BE, Kornberg A (1987) *JBC 262*:6877; Baker TA, Sekimizu K, Funnell BE, Kornberg A (1986) *Cell 45*:53; Dodson M, Echols H, Wickner S, Alfano C, Mensa-Wilmot K, Gomes B, LeBowitz J, Roberts JD, McMacken R (1986) *PNAS 83*:7638.

The alternative replication pathways used during discontinuous-strand synthesis on the plasmid pBR322 (Fig. 8-10) provide further evidence for the analogous roles of the dnaA and PriA proteins in primosome assembly. Although dnaA protein is not essential for pBR322 replication, a dnaA binding site near the discontinuous-strand-template *pas*[70] can function in primosome assembly.[71] In the presence of excess dnaA protein, the primosomal protein dnaT, normally essential for pBR322 replication,[72] becomes dispensable. Furthermore, deletion of the *pas* sequence makes the replication dependent on dnaA protein and independent of dnaT protein. Either pathway leads to extensive replication of the discontinuous strand; in the absence of both pathways, replication is very inefficient. Thus, *for discontinuous-strand synthesis, the* pas *and the dnaA binding site are functionally equivalent.*

A novel sequence from the plasmid R6K functions as a *pas* for the initiation of complementary-strand synthesis when present on a single-stranded phage template.[73] This sequence provides further evidence for the similar functions of dnaA and PriA proteins. Initiation of DNA synthesis at this site requires, in addition to dnaB and dnaC proteins, the dnaA protein, while the PriA, PriB, PriC, and dnaT proteins are not essential.[74] The sequence forms a duplex hairpin which contains a dnaA box; dnaA protein, instead of PriA protein, binds to this region and loads dnaB protein onto the template to generate an *oriC*-type primosome capable of DNA helicase activity as well as activation of primase. This dnaA-dependent loading site can also substitute for the discontinuous-strand-template *pas* during replication of pBR322 in vitro.

The λ O and P proteins can assemble the dnaB protein and primase onto SSB-coated ssDNA.[75] The strong affinity of the λ O protein for ssDNA is very likely responsible, inasmuch as *ori*λ is not required. This system, although doubtful in physiologic importance, is useful because it separates two stages: (1) the protein–protein interactions involved in transferring the replication enzymes onto the template, and (2) the complex protein–DNA interactions that occur at the origin. An analogous reaction in which the dnaA and dnaC proteins transfer dnaB protein and primase to ssDNA has been reported but not characterized in detail.[76]

8-5 Phage and Plasmid Primases

Gene *4* Protein of Phage T7[77]

Gene 4 protein (gp4), both a primase and a helicase (Section 11-2), binds specifically and stably to ssDNA and translocates in a 5′→3′ direction[78] in a reaction

70. Fuller RS, Funnell BE, Kornberg A (1984) *Cell 38*:889.
71. Seufert W, Messer W (1987) *Cell 48*:73.
72. Masai H, Arai K (1988) *JBC 263*:15016.
73. Nomura N, Masai H, Inuzuka M, Miyazaki C, Ohtsubo E, Itoh T, Sasamoto S, Matsai M, Ishizaki R, Arai K (1990), submitted.
74. Masai H, Nomura N, Arai K (1990) *JBC 265*:15134.
75. LeBowitz J, Zylicz M, Georgopoulos C, McMacken R (1985) *PNAS 82*:3988.
76. Wahle E, Lasken RS, Kornberg A (1989) *JBC 264*:2469.
77. Richardson CC, Beauchamp BB, Huber HE, Ikeda RA, Myers JA, Nakai H, Rabkin SD, Tabor S, White J (1987) in *DNA Replication and Recombination* (McMacken R, Kelly TJ, eds.). UCLA, Vol. 47. AR Liss, New York, p. 151.
78. Tabor S, Richardson CC (1981) *PNAS 78*:205.

fueled by NTP hydrolysis. The K_m for the preferred substrate, dTTP, is 4×10^{-4} M. Nonhydrolyzable NTPs, such as β,γ-methylene-dTTP, induce tighter binding but inhibit both the primase and helicase activities. NTP hydrolysis appears to be essential for unidirectional movement of the gp4 to unwind DNA and localize primase recognition sites.

When gp4 reaches a specific recognition site, it catalyzes the synthesis of a tetraribonucleotide primer. On DNA templates of known sequence, the predominant recognition sequences are 3′-CTGG(G/T) and 3′-CTGTG. Synthesis starts opposite T, and RNA primers with the sequence pppACC(C/A) or pppACAC are generated.[79]

Primer syntheses observed in vivo and in vitro are similar. The 5′-terminal sequences of the tetra- and pentanucleotides linked to nascent DNA strands isolated from T7-infected cells[80] are mainly pApCp(A/C)p(A/C). The complementary initiation site most frequently used to signal the synthesis of a primer RNA must therefore be 3′-CTG(G/T).[81] Several sites of transition from primer RNA to DNA can be observed within a short stretch of DNA, indicating that although primer synthesis starts at a precisely defined nucleotide, the point of transition to DNA can vary within a few nucleotides.

Although both the helicase and primase of the T7 replication fork are encoded by a single gene,[82] more than one molecule of gp4 is required at the replication fork (Fig. 8-11). During coupled continuous- and discontinuous-strand synthesis, the helicase activity is highly *processive,* with the protein remaining stably bound to the template, while the primase is *distributive,* dissociating and reassociating throughout the reaction.[83] Thus, the same molecule of gp4 cannot be performing both functions at the replication fork.

This requirement for at least two molecules of gp4 during replication may be related to the two distinct forms of the protein made during infection. The 63- and 56-kDa proteins are translated from the gene 4 mRNA in the same reading frame from start sites separated by 189 ntd.[84] The larger protein contains a 7-kDa region at the N terminus, absent from the smaller form. Both proteins are DNA-dependent dTTPases and helicases, and both proteins stimulate the T7 DNA polymerase in the elongation of primers, but only the 63-kDa protein has primase activity.[85]

Although it lacks primase activity, the 56-kDa form retains the capacity to synthesize dinucleotides from rNTPs. The polymerization active center is apparently wanting in the ability to synthesize primers in a DNA-dependent fashion.[86] The N-terminal domain of the 63-kDa form may serve in the recognition of the specific priming sequences in the template. An amino acid sequence homologous to a zinc finger motif (Section 10-9) is found in this region and could be responsible for this DNA recognition.[87] The very absence of this sequence-

---

79. Scherzinger E, Lanka E, Morelli G, Seiffert D, Yuki A (1977) *EJB 72*:543; Romano LJ, Richardson CC (1979) *JBC 254*:10476; Romano LJ, Richardson CC (1979) *JBC 254*:10483; Tabor S, Richardson CC (1981) *PNAS 78*:205.
80. Sugimoto K, Miyasaka T, Fujiyama A, Kohara Y, Okazaki T (1988) *MGG 211*:400; Seki T, Okazaki T (1979) *NAR 7*:1603; Ogawa T, Okazaki T (1979) *NAR 7*:1621.
81. Fujiyama A, Kohara Y, Okazaki T (1981) *PNAS 78*:903.
82. Dunn JJ, Studier FW (1983) *JMB 166*:477.
83. Nakai H. Richardson CC (1988) *JBC 263*:9818.
84. Dunn JJ, Studier FW (1983) *JMB 166*:477.
85. Bernstein JA, Richardson CC (1989) *JBC 264*:13066.
86. Nakai H, Richardson CC (1988) *JBC 263*:9818; Bernstein JA, Richardson CC (1988) *PNAS 85*:396; Bernstein JA, Richardson CC (1988) *JBC 263*:14891.
87. Bernstein JA, Richardson CC (1988) *PNAS 85*:396.

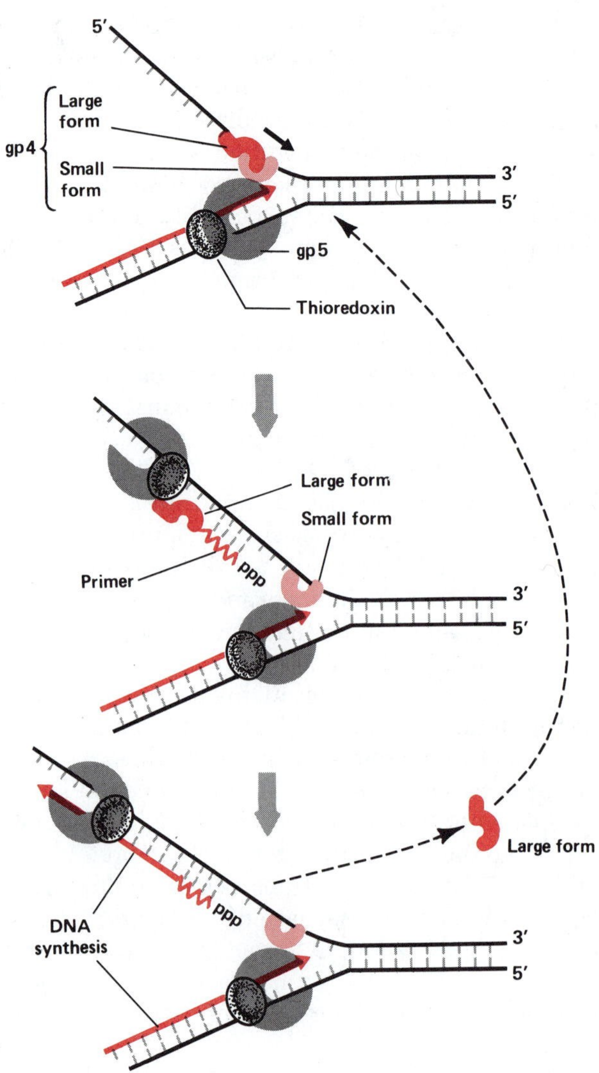

Figure 8-11
Roles of the 63- and 56-kDa forms of T7 gp4 at the replication fork.

specific binding domain in the small form perhaps favors helicase activity by preventing "stalling" at primase recognition sites during translocation.

Priming by the large protein is stimulated fourfold by the small protein. Thus, as with the dnaB protein–primase pair in *E. coli* and the gp41/61 pair of phage T4 (see below), a helicase–primase interaction at the replication fork is employed for efficient priming. The mechanism of this helicase stimulation of primase is not clear. Inasmuch as the gp4 NTPase activity is most active in protein multimers,[88] heteromultimers between the large and small forms may constitute the most efficient complexes. A specific interaction of gp4 with the T7 DNA polymerase is notable: gp4 fails to stimulate DNA synthesis by *E. coli* pol I or by T4 DNA polymerase, and the primers synthesized by gp4 are elongated only by the T7 DNA polymerase. The ability of T7 DNA polymerase to facilitate binding of gp4 to DNA suggests that this specificity reflects a physical interaction.

88. Bernstein JA, Richardson CC (1989) *JBC 264*:13066.

Gene *41* and *61* Proteins of Phage T4

The protein products of genes *41* (54 kDa) and *61* (40 kDa) synthesize pentaribonucleotides on any natural ssDNA; these can serve as primers for replication by T4 DNA polymerase (Section 17-6). Gp41 alone exhibits DNA-dependent GTPase (or ATPase)[89] and helicase[90] activities and stimulates primer synthesis by gp61.[91] Although gp61 alone can utilize rNTP substrates to synthesize dimers, production of the biologically active tetramer and pentamer primers requires both proteins.[92] This coupled effort of gp41 and gp61 resembles that of the dnaB protein–primase team and the T7 gp4 pair.

The pentaribonucleotide primers synthesized in vitro on a T4 DNA template start with a 5′-pppApC dinucleotide, followed by many different sequences. Similarly, the nascent DNA fragments isolated from T4-infected cells[93] have one to five ribonucleotide residues at their 5′ ends with the sequence $pApC(pN)_3$. The gp41/61 complex recognizes the template sequence 3′-TTG, in which the first T is not copied. On DNA containing cytosine rather than the hydroxymethylcytosine of T-even phage DNA, 3′-TCG sequences are also used; the few primers synthesized by high levels of gp61 in the absence of gp41 are initiated exclusively at these TCG sites. Such sites are absent in natural T4 DNA which contains hydroxymethylcytosine in place of cytosine; therefore gp61, which undoubtedly contains the polymerization active center essential for primer synthesis, could never prime T4 DNA replication operating alone.

Although the functional interaction is clear, a physical interaction between gp41 and gp61 has not been demonstrated. Optimal activity occurs at a ratio of five molecules of gp41 to one of gp61.[94] An increase in the apparent molecular weight of gp41 in the presence of GTP or GTPγS indicates that a multimer may be the active form.[95] Although gp41 has ssDNA-stimulated GTPase activity, it fails to form stable complexes with the template; gp61 stabilizes the association between gp41 and ssDNA, indicative of a physical complex between the proteins bound to DNA.

The P4 α Protein[96]

Replication of the small linear phage P4 (Section 17-7) is resistant to rifampicin and independent of the host *dnaG* gene; thus P4 is thought to encode its own priming system. The gene α protein, the only phage product required for replication, is most likely a primase. This 85-kDa protein synthesizes poly rG on a poly dI·poly dC or poly dG·poly dC template.

Replication in vitro that depends on the α protein, *E. coli* SSB, and DNA pol III holoenzyme is specific for templates containing the P4 origin and a *cis*-acting replication region, suggesting that the α protein interacts specifically with these sequences. The protein that provides the helicase function, presumably needed during P4 replication, is unknown. Perhaps, analogous to the phage T7 system, the α protein provides both the helicase and primase activities.

---

89. Nossal NG (1979) *JBC 254*:6026; Morris CF, Moran LA, Alberts BM (1979) *JBC 254*:6797.
90. Venkatesan M, Silver LL, Nossal NG (1982) *JBC 257*:12426.
91. Nossal NG (1980) *JBC 255*:2176; Liu CC, Alberts BM (1981) *JBC 256*:2821.
92. Liu CC, Alberts BM (1980) *PNAS 77*:5698; Cha TA, Alberts BM (1986) *JBC 261*:7001; Cha TA, Alberts BM (1989) *JBC 264*:12220; Hinton DM, Nossal NG (1987) *JBC 262*:10873; Cha TA, Alberts BM (1990) *Biochemistry 29*:1791.
93. Kurosawa Y, Okazaki T (1979) *JMB 135*:841.
94. Richardson RW, Nossal NG (1989) *JBC 264*:4725.
95. Venkatesan M, Silver LL, Nossal NG (1982) *JBC 257*:12426.
96. Krevolin MD, Calendar R (1985) *JMB 182*:509; Flensburg J, Calendar R (1987) *JMB 195*:439.

Plasmid-Encoded Primases[97]

Certain conjugative-type plasmids in *E. coli* specify the synthesis of primases that can substitute for the host primase and also serve the plasmid in conjugative transfer. A primase of 118 kDa encoded by RP4 is needed for priming the transferred strand, despite the presence of the host priming systems. An enzyme has been purified from cells harboring the plasmid R64 (Sections 18-4 and 18-7); the enzyme primes DNA synthesis on single-stranded viral circles (M13, ϕX174, and G4) by its synthesis of oligoribonucleotides. The plasmid-encoded primase appears to be responsible for priming the synthesis of the strand complementary to the R64 strand transferred during conjugation.

8-6 Eukaryotic Primases[98]

Primase activity is commonly a component of DNA polymerase α, a nuclear enzyme responsible for chromosomal replication (Section 6-2). The relatively low processivity of pol α, in addition to its associated primase, makes it a plausible candidate for synthesis of the discontinuous strand during duplex DNA replication. The more processive pol δ, which lacks primase activity, may be responsible for synthesis of the continuous strand.

Pol α purified from numerous sources, including yeast and human cells, is composed of four subunits:[99] a large DNA polymerase subunit (~180 kDa), two small subunits (~60 and ~50 kDa) associated with primase activity, and a ~70-kDa protein with no known catalytic activity.

The primase activity of the small subunits is evident when they are part of the polymerase–primase complex, and also when they are separated from the 180- and 70-kDa polypeptides.[100] The 60- and 50-kDa proteins seem to function as a unit and have not been separated from each other under conditions that permit primase activity to be maintained. Their sedimentation behavior is consistent with a heterodimer of ~110 kDa and they are coprecipitated by an antibody against the smaller polypeptide. While the rNTP binding site resides specifically in the 50-kDa protein, adenosine analogs that probe the active center of primases will label both proteins with similar efficiency.[101] Thus, the proteins appear to act together in primer synthesis. The genes encoding both of the small subunits cloned from yeast are present as single copies, and are essential.[102]

Oligonucleotides of a distinct length are synthesized by primase acting both as part of the polymerase–primase complex and in the separated form. The characteristic length is 12 to 14 ntd[103] for the *Drosophila* enzyme, while the mouse[104]

---

97. Lanka E, Scherzinger E, Günther E, Schuster H (1979) *PNAS 76*:3632.
98. Roth Y-F (1987) *EJB 165*:473; Lehman IR, Kaguni LS (1989) *JBC 264*:4265; Kaguni LS, Lehman IR (1988) *BBA 950*:87.
99. Lehman IR, Kaguni LS (1989) *JBC 264*:4265.
100. Cotterill SM, Reyland ME, Loeb LA, Lehman IR (1987) *PNAS 84*:5635; Kaguni LS, Rossignol J-M, Conaway RC, Banks GR, Lehman IR (1983) *JBC 258*:9037; Suzuki M, Enomoto T, Hanaoka F, Yamada M (1985) *J Biochem 98*:581; Plevani P, Foiani M, Valsasnini P, Badaracco G, Cheriathundam E, Chang LMS (1985) *JBC 260*:7102.
101. Foiani M, Lindner AJ, Hartmann GR, Lucchini G, Plevani P (1989) *JBC 264*:2189.
102. Lucchini G, Francesconi S, Foiani M, Badaracco G, Plevani P (1987) *EMBO J 6*:737; Lehman IR, Kaguni LS (1989) *JBC 264*:4265.
103. Conaway RC, Lehman IR (1982) *PNAS 79*:4585.
104. Tseng BY, Ahlem CN (1983) *JBC 258*:9845.

and yeast[105] primases synthesize chains of 8 to 12 ntd. When DNA synthesis is blocked, for lack of either the dNTPs or the polymerase, multimers of this characteristic length (e.g., 24-mers and 36-mers) are generated.[106] Multimer synthesis, at least for the yeast enzyme, requires dissociation and rebinding of the protein to the template, followed by a second round of "monomer" synthesis.[107] Thus eukaryotic primases, unlike prokaryotic primases, *count out* a length of primer, rather than *read* a particular recognition sequence.

Primers are initiated with a purine; the K_m for these substrates is high, ranging from 1 to 5 mM for ATP and about tenfold lower for GTP.[108] ATP is used for initiation about four times more commonly than GTP, but this preference is influenced by the nucleotide-pool size and the template sequence.[109] Most primases can utilize dNTPs in place of rNTPs during primer synthesis for all but the initial nucleotide;[110] efficient primer synthesis and subsequent DNA synthesis can take place in the presence of only ATP, GTP, or the four dNTPs. Sometimes primase incorporation of dNTPs is feeble, as in the case of the yeast enzyme.[111] Fidelity, assessed for calf thymus primase, is very low.[111a]

Whether primase exerts any sequence preference on natural DNAs is uncertain. Some specificity is inferred from the nonrandom placement of primers.[112] On defined polymer templates, poly dC is highly favored; poly dT is also active, while poly dA and poly dG are essentially inert.[113] Thus, pyrimidines in the template sequence are apparently important for recognition by primase, as would be anticipated from the necessity of a purine rNTP for initiation.

Drosophila primase seems to pay little attention to the sequence of the template; once primer synthesis has been initiated, complementarity between the rNTP substrates and the template is not essential.[114] Lack of the appropriate rNTPs has little inhibitory effect on the rate or extent of primer synthesis. These properties contribute to the characterization of eukaryotic primases as "polymerases that can count but cannot read." Whether this lack of template direction is uniform among eukaryotic primases has not been determined.

In view of their coexistence in a single, multifunctional enzyme, coupling of the primase and DNA polymerase activities of pol α seems likely. In the absence of dNTPs, primer synthesis by the intact polymerase–primase is blocked following an initial burst; when DNA replication is impeded, apparently only a single round of primer synthesis can occur.[115] Furthermore, the presence of polymerase prevents the synthesis of the multimer-length primers (e.g., 24-mers, 36-mers) by primase.[116] The means whereby the primer is moved from the active site on the small primase subunits to the DNA polymerase site on the 180-kDa polypeptide is not known.

105. Singh HJ, Brooke RG, Pausch MH, Williams GT, Trainor C, Dumas LB (1986) *JBC 261*:8564.
106. Cotterill S, Chui G, Lehman IR (1987) *PNAS 79*:4585.
107. Brooks M, Dumas LB (1989) *JBC 264*:3602.
108. Kaguni LS, Lehman IR (1988) *BBA 950*:87; Gronostajski RM, Field J, Hurwitz J (1984) *JBC 259*:9479.
109. Yamaguchi M, Hendrickson EA, DePamphilis ML (1985) *Mol Cell Biol 5*:1170; Yamaguchi M, Hendrickson EA, DePamphilis ML (1985) *JBC 260*:6254; Eliasson R, Reichard P (1979) *JBC 253*:7469.
110. Gronostajski RM, Field J, Hurwitz J (1984) *JBC 259*:9479; Conaway RC, Lehman IR (1982) *PNAS 79*:4585.
111. Brooks M, Dumas LB (1989) *JBC 264*:3602.
111a. Zhang S, Grosse F (1990) *JMB 216*:475
112. Yamaguchi M, Hendrickson EA, DePamphilis ML (1985) *Mol Cell Biol 5*:1170.
113. Roth Y-F (1987) *EJB 165*:473.
114. Cotterill S, Chui G, Lehman IR (1987) *JBC 262*:16105.
115. Cotterill S, Chui G, Lehman IR (1987) *JBC 262*:16105.
116. Cotterill S, Chui G, Lehman IR (1987) *JBC 262*:16105; Singh HJ, Brooke RG, Pausch MH, Williams GT, Trainor C, Dumas LB (1986) *JBC 261*:8564; Badaracco G, Bianchi M, Valsasnini P, Magni G, Plevani P (1985) *EMBO J 4*:1313.

Among a few examples of eukaryotic primases not associated with pol α are two activities from yeast. These are distinct from pol I (the yeast equivalent of pol α) on physical and immunologic grounds.[117] The role of these additional primases is as yet unclear.

Herpes Simplex Virus–Encoded Primase[118]

A helicase–primase complex has been isolated from cells infected with HSV-1. The complex contains three proteins with molecular weights of 120, 97, and 70 kDa. The proteins are products of the *UL52*, *UL5*, and *UL8* genes, three of the seven genes essential for viral DNA replication. Although the function of each of these proteins has not been established, amino acid sequence homologies suggest that *UL5* encodes the helicase (Section 11-5). The combination of primase with a helicase in herpes virus replication resembles the pairing of these activities in prokaryotes, rather than the alignment with DNA polymerase common in eukaryotes. The 40-kDa primase activity retrieved from herpes-infected cells[119] has been shown to be the host mitochondrial RNA polymerase, found in the cytosol as a consequence of herpes-induced disruption of the mitochondrial membrane.[120]

8-7 Endonucleolytic Priming

Covalent starts by extension of a 3′-OH DNA terminus serve in several schemes for initiating the synthesis of a continuous strand. This type of primer is generated by nicking one strand of a duplex DNA molecule. Several viruses and plasmids encode proteins that cleave a specific phosphodiester bond to introduce the nick that generates the replication origin. These endonucleases function in place of primases in that they generate a primer terminus. While suited for priming the continuous strand, these enzymes are not appropriate for priming discontinuous replication; genomes that depend on endonucleolytic priming of the continuous strand invariably employ another mechanism for initiating synthesis of the discontinuous strand.

Other examples of covalent extension of a 3′-OH DNA terminus include *terminal protein priming* (Section 8-8), the *rolling-hairpin mechanism* for a linear duplex with palindromic sequences at its ends (Section 15-4), and elongation of DNA ends introduced into the duplex by recombination (Sections 17-6, 21-5, and 21-6).

The *rolling-circle* pathway (Fig. 8-12) is used in the replication of circular genomes initiated by endonucleolytic priming; it entails the following steps:

(1) Site-specific nicking within the origin generates a free 3′-OH group for extension by a DNA polymerase.

---

117. Jazwinski SM, Edelman GM (1985) *JBC 260*:4995; Biswas EE, Joseph PE, Biswas SB (1987) *Biochemistry 26*:5377.
118. Crute JJ, Tsurumi T, Zhu L, Weller SK, Olivo PD, Challberg MD, Mocarski ES, Lehman IR (1989) *PNAS 86*:2186; Dodson MS,Crute JJ, Bruckner RC, Lehman IR (1989) *JBC 264*:20835; Crute JJ, Lehman IR (1991) *JBC 266*, in press.
119. Holmes AM, Weitstock SM, Ruyechan WT (1988) *J Virol 62*:1038.
120. Tsurumi T, Lehman IR (1990) *J. Virol 64*:450.

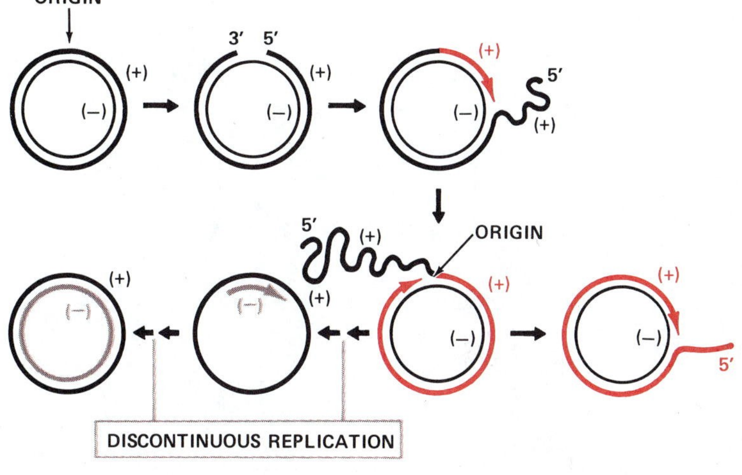

Figure 8-12
Rolling-circle mechanism of replication exemplified by replication of ϕX RFI. Continuous synthesis directed by the circular complementary (−) strand produces unit-length viral (+) strands that then direct discontinuous synthesis of (−) strands to form RFI. Synthesis of another (+) strand immediately follows completion of the first.

(2) Unwinding by a DNA helicase displaces the cleaved strand from the template; the displaced ssDNA is coated by SSB.

(3) Covalent extension of the 3′-OH end of the nick generates a new strand by continuous synthesis.

(4) The round of synthesis terminates by cleavage of the regenerated duplex replication origin, releasing the displaced strand.

(5) Ligation of the 3′ and 5′ ends circularizes the single strand.

Endonucleolytic cleavage is involved in both the initiation and the termination of rolling-circle replication; the sequence-specific cleavage that starts the reaction is also needed to terminate the round of synthesis. By virtue of their nicking and closing activities, the endonucleases specifically relax the superhelicity of plasmids containing their cognate origins.

ϕX174 Gene *A* Protein

The best characterized of these replicative endonucleases is the 60-kDa gene *A* protein (gpA), the sole replication protein encoded by ϕX174. The enzyme is exquisitely designed to cleave and religate the replication intermediates (Fig. 8-13) without the expenditure of energy. GpA introduces a specific nick into the (+) strand of the supercoiled ϕX174 RFI. The 3′-OH end of the nick site, a G residue at position 4305, forms the primer terminus,[121] and the protein covalently attaches, via a tyrosine residue, to the 5′-P group[122] of the A at position 4306. After a round of (+) strand synthesis, gpA, bound to the 5′ end of the strand that has been displaced during replication, cleaves the origin sequence that has been regenerated during replication. The second cleavage releases the displaced single strand from the replicating complex while generating a new 3′-OH group for covalent extension in a second round of rolling-circle replication. Simulta-

121. Eisenberg S, Scott JF, Kornberg A (1976) *PNAS 73*:1594; Ikeda J-E, Yudelevich A, Hurwitz J (1977) *PNAS 73*:2669; Langeveld SA, van Mansfeld ADM, Baas PD, Jansz HS, van Arkel GA, Weisbeek PJ (1978) *Nature, 271*:417.

122. Eisenberg S, Griffith J, Kornberg A (1977) *PNAS 74*:3198, Ikeda J-E, Yudelevich A, Shimamoto N, Hurwitz J (1979) *JBC 254*:9416.

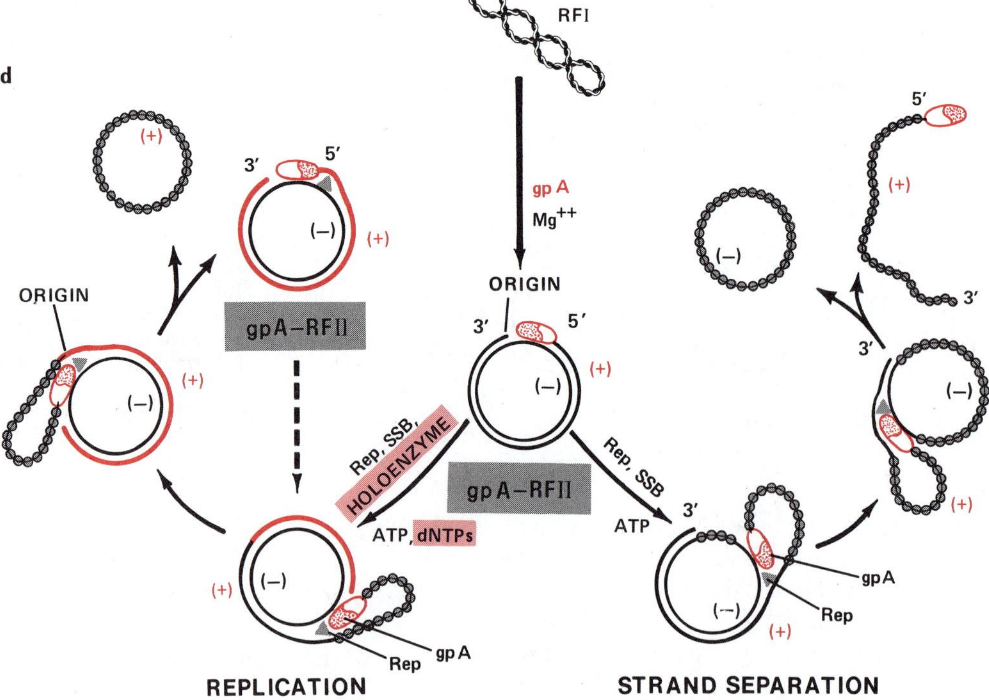

Figure 8-13
Scheme for gpA action, illustrating its multiple functions. The looped rolling-circle intermediate form is used in strand separation, uncoupled from replication, as well as in the synthesis of viral (+) strands. Rep = the Rep helicase.

neously with the second cleavage, the gpA ligates the 3′ and 5′ ends of the displaced strand to generate a (+) strand circle.

Thus, two cleavages and a ligation of the displaced strand are carried out by the gpA. The reactions require only Mg^{2+} and a negatively supercoiled template for initiation; no ATP or other nucleotide cofactor is involved. The energy liberated by the cleavage and stored in the protein–DNA covalent intermediate is used to sustain the subsequent ligation.

The enzyme has dual active sites that participate in the covalent protein–DNA intermediate.[123] The reactive groups are two tyrosyl residues (amino acids 343 and 347) separated by three amino acids in the sequence *tyr*-val-ala-lys-*tyr*-val-asn-lys. The two tyrosines are covalently linked to the DNA with equal efficiencies, and probably act in alternation during the multiple rounds of cleavage and ligation. If, as predicted, these tyrosines are present in an α-helical portion of the protein, they would be closely spaced on the same side of the helix and symmetrically positioned with respect to the DNA phosphodiester bond. Simultaneous cleavage of the origin and ligation of the displaced strand switches the covalently associated 5′ end of the (+) strand of the DNA from one tyrosine to the other (Fig. 8-14).

During the production of multiple copies of the RF (RF → RF replication; Section 17-4) gpA is responsible for priming the continuous strand, while the ϕX-type primosome (Section 4) primes the discontinuous strand. Conservation

123. van Mansfeld ADM, van Teeffelen HAAM, Baas PD, Jansz HS (1986) NAR 14:4229.

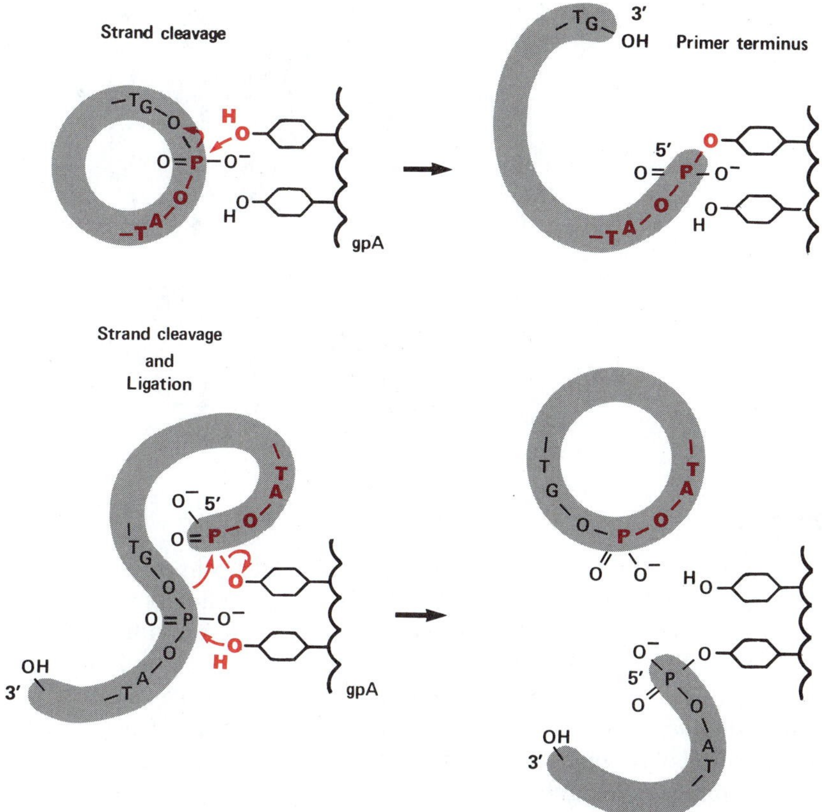

Figure 8-14
Model for the gpA-catalyzed cleavage and cleavage–ligation reactions which occur during initiation and termination of ϕX rolling-circle replication. The origin sequence, shown schematically as -TG-O-P-O-AT-, is in the viral (+) strand of the duplex rolling-circle intermediate shown in Fig. 8-13.

of the primosome, initially used for SS → RF replication, during the RF → RF stage is indicated. RF molecules generated by the action of the primosome in conversion of the viral single strand are superior substrates for gpA cleavage despite their relatively low superhelix density (Fig. 8-15). The lag in the reaction is eliminated and viral strands are produced rapidly and efficiently. The parental RF bearing the primosome may be the sole template for replication, while the numerous supercoiled progeny RF are produced for transcription.

Some curious features of ϕX174 gpA deserve notice. In vivo, the protein acts preferentially *in cis* (on the same template molecule from which it is transcribed) in spite of an abundance of 3000 gpA molecules per infected cell. A second product of gene *A*, the A* protein, lacks the N-terminal third of gpA but shares its endonuclease and ligase properties. The A* protein is responsible for the shutoff of host cell replication, but the mechanism of this action is unknown. GpA is also involved in replication fork movement, interacting with the *E. coli* Rep helicase (Section 11-2) and traveling with it around the template (Fig. 8-13). Thus, like the primases of *E. coli* and phages T7 and T4, gpA is functionally associated with a helicase. Similarities between gpA and other origin-binding proteins in their recognition of the origin sequence and assembly of the replication complex are given their deserved attention in Section 16-4.

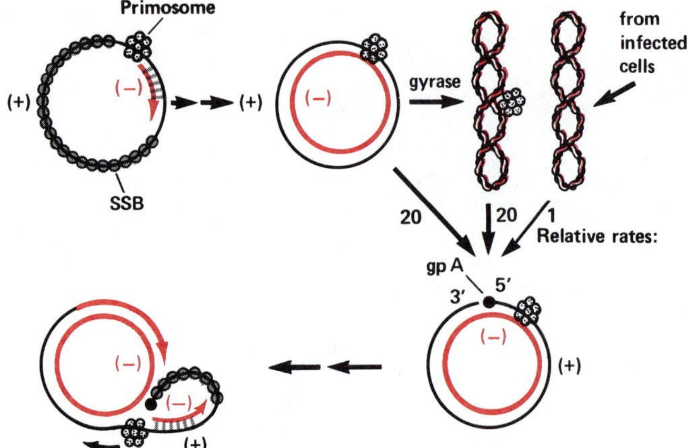

Figure 8-15
Conservation of the primosome throughout the first two stages of the ϕX DNA replicative cycle.

Ff Gene *2* Protein

The gene *2* protein of Ff (F plasmid–specific, filamentous) phages carries out reactions analogous to those of ϕX gpA. The 46-kDa gp2 specifically binds to the (+) strand origin,[124] perhaps as a dimer, and introduces a nick, providing a 3′-OH DNA primer for rolling-circle replication.[125] In contrast to gpA, gp2 does not form a stable covalent complex with the 5′ end of the cleavage junction but simply holds this end very tightly, maintaining it in the correct position for ligation after a round of replication. Ligation also depends on a hairpin structure in the DNA template. Inasmuch as DNA joining by gp2 does not require the input of energy, it follows that the second cleavage of the origin must somehow donate its energy to the ligation (Fig. 8-16). RFI molecules precleaved by gp2 are active in the subsequent replication reactions—providing further evidence that the ligation does not depend on the first cleavage. Despite the prior cleavage, addition of gp2 is still required, indicative of its participation in the replication stage.[126]

When Mn^{2+} replaces the Mg^{2+} required for gp2 activity, the protein introduces a double-stranded break in the DNA instead of a nick.[127] (This also occurs with ϕX174 gpA*,[128] but has been studied in more detail with gp2.) Double-strand cleavage occurs in two distinct stages: nicking at the normal cleavage site occurs first, followed by cleavage of the opposite strand, one nucleotide 3′ to the first break. Ligation of the initial 3′-OH to the 5′-P produced by the second cleavage results in a double-strand break with an interstrand cross-link (hairpin structure) on one side and a one-base 3′ overhang on the other:

$$\begin{array}{l} {}^{5'}\text{-T-T-T}\overset{\downarrow}{}\text{-A-A-} \\ {}_{3'}\text{-A-A-A-T-T-} \end{array} \longrightarrow \begin{array}{l} \quad\ \text{HO}\ \ \text{OPO}_3\text{H}_2 \\ \text{-T-T-T}\ \ \text{A-A-} \\ \text{-A-A-A-T-T-} \end{array} \longrightarrow \begin{array}{l} {}^{5'}\text{-T-T}\diagdown_{\text{T}} \\ {}_{3'}\text{-A-A}\diagup \end{array} \begin{array}{l} {}^{5'}\text{OPO}_3\text{H}_2 \\ \ \text{A-A-} \\ \text{A-T-T-} \\ {}_{3'}\text{OH} \end{array}$$

124. Geider K, Baumel I, Meyer TF (1982) *JBC 257*:6488. Horiuchi K (1986) *JMB 188*:215; Greenstein D, Horiuchi K (1987) *JMB 197*:157.
125. Meyer TF, Geider K, Kurz C, Schaller H (1979) *Nature 278*:365.
126. Geider K, Baumel I, Meyer TF (1982) *JBC 257*:6488.
127. Greenstein D, Horiuchi K (1989) *JBC 264*:12627.
128. van der Ende A, Langeveld SA, Teertstra R, van Arkel GA, Weisbeek PJ (1981) *NAR 9*:2037.

Gp2 clearly has the capacity to carry out two successive cleavages, with the second cleavage coupled to ligation. This unnatural reaction probably reflects the steps that occur during termination of a replication cycle. The presence of Mn^{2+} relaxes the cleavage requirements for a negatively supercoiled template and for certain sequences within the origin. The mechanism by which Mn^{2+} changes the reaction pathway of gp2 cleavage may be through an effect on DNA structure, such as stabilizing a bend.

In contrast to the many rounds of gpA-dependent rolling-circle replication of ϕX174, gp2 apparently carries out only one round of (+) strand synthesis at a time.[129] After displacement of a circular (+) strand, the gp2 dissociates from the template; the RFII molecule is ligated and supercoiled by gyrase and thus is readied for a second round of (+) strand synthesis. This difference between the rolling-circle pathways of ϕX and Ff is evident in the kinetics of appearance of (+) strands during replication of the two phages.

pT181 Gene *repC* Protein[130]

The small plasmids of Gram-positive bacteria share the rolling-circle replication pathway of the small ssDNA phages of *E. coli*. Plasmid pT181 is a prototypical plasmid of Gram-positive bacteria; its 38-kDa RepC initiator protein is a replicative endonuclease. The RepC protein cleaves the replication origin at a specific sequence, generating a 3′-OH group for replication, and apparently becomes covalently attached to the 5′-P. In contrast to the phage proteins, RepC does not require a supercoiled template for activity, and recognizes a rather GC-rich sequence. The mechanism of action[131] of RepC protein is otherwise presumed to be similar to that of ϕX gpA.

8-8 Terminal Protein Priming[132]

The primer terminus required for DNA replication can be provided by the 3′-OH group of an amino acid within a protein, rather than by an RNA or DNA molecule. Serine, threonine, and tyrosine residues have each been selected to serve this function of covalent linkage to the nucleic acid. Protein primers are invariably covalently attached to the ends of duplex linear genomes, and hence are commonly called *terminal proteins*. Numerous phages and some linear plasmids are known to or thought to use the terminal protein priming mechanism. The adenoviruses of animal cells and certain *B. subtilis* phages are the best-studied examples. Terminal protein priming is suitable only for the synthesis of a continuous strand. However, a single priming event for each of the two strands of a linear genome should suffice for replication of the templates, and no discontinuous replication need be involved.

Even though the protein priming mechanism is similar for eukaryotic and prokaryotic viruses, the terminal proteins show no relationship, suggesting that

---

129. Horiuchi K, Ravetch JV, Zinder ND (1979) *CSHS 43*:389.
130. Koepsel RR, Murray RW, Rosenblum WD, Khan SA (1985) *PNAS 82*:6845; Koepsel RR, Murray RW, Khan SA (1986) *PNAS 83*:5484.
131. Murray RW, Koepsel RR, Khan SA (1989) *JBC 264*:1051.
132. Salas M (1988) in *The Bacteriophages* (Calendar R, ed.). Plenum Press, New York, Vol. 1, p. 169; Challberg MD, Kelly TJ (1989) *ARB 58*:671.

this useful mechanism has been discovered independently several times during evolution.

Adenovirus Terminal Protein and Mechanism of Priming[133]

The 55-kDa terminal protein of adenovirus is bound to the 5′ end of each of the strands of the duplex linear genome.[134] The covalent linkage is via an ester bond between a particular serine OH group and the phosphate of dCMP, the 5′-terminal nucleotide of the genome. The 55-kDa protein is a proteolytic product of the 80-kDa preterminal protein that primes DNA synthesis;[135] proteolytic processing occurs during packaging of the protein–DNA complex into virions.[136]

The preterminal protein, the adenovirus DNA polymerase, and two cellular DNA-binding proteins are required for the formation of the preterminal protein–dCMP primer[137] (Fig. 8-16). The preferred DNA template contains the duplex ends of the adenovirus genome with the bound terminal protein that acted as a priming protein during the previous round of replication. However, if the template is partially single-stranded at the ends, the viral 55-kDa terminal protein is dispensable, suggesting that its role is to open the duplex template in order to facilitate the initial association of the preterminal protein. ATP, but not its hydrolysis, is involved in this step, and is essential for primer formation only on the fully duplex template.[138]

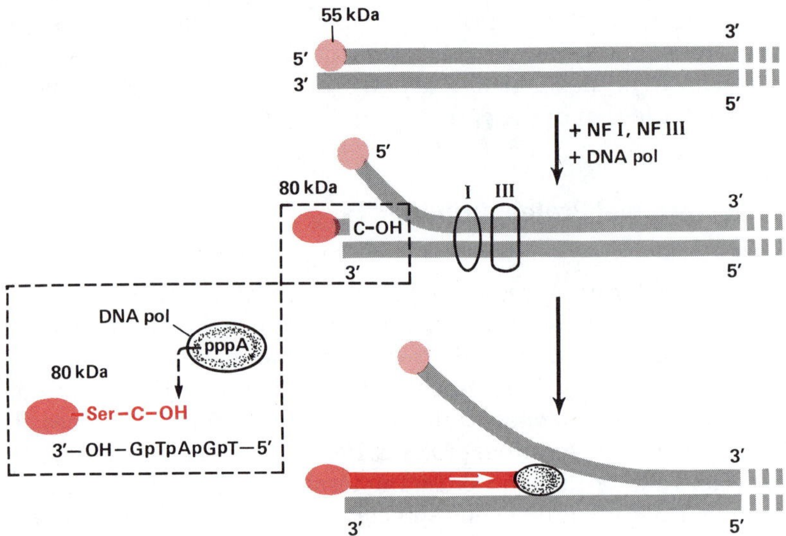

Figure 8-16
Scheme for the initiation of adenovirus DNA replication. I, III = nuclear factors I and III; DNA pol = adenovirus DNA polymerase.

133. Challberg MD, Kelly TJ (1989) *ARB 58*:671.
134. Robinson AJ, Younghusband HB, Bellett AJD (1973) *J Virol 56*:54; Rekosh DMK, Russell WC, Bellett AJD, Robinson AJ (1977) *Cell 11*:283.
135. Challberg MS, Desiderio SV, Kelly TJ (1980) *PNAS 77*:5105; Stillman BW, Lewis JB, Chow LT, Mathews MB, Smart JE (1981) *Cell 23*:497; Challberg MD, Kelly TJ (1981) *J Virol 38*:272; Smart JE, Stillman BW (1982) *JBC 257*:13499.
136. Challberg MD, Kelly TJ (1981) *J Virol 38*:272.
137. Nagata K, Guggenheimer RA, Hurwitz J (1983) *PNAS 80*:6177; Rosenfeld PJ, Kelly TJ (1986) *JBC 261*:1398; Pruijn GJM, van Driel W, van der Vliet PC (1986) *Nature 322*:656; Rosenfeld PJ, O'Neill EA, Wides RJ, Kelly TJ (1987) *JCB 7*:875.
138. Kenny MK, Hurwitz J (1988) *JBC 263*:9809.

The preterminal protein forms a 1:1 complex with the adenovirus polymerase. Association of this complex with the template requires that the cellular transcription factors NFI and NFIII (Section 19-4) be bound near the ends of the genome.[139] How these factors stimulate initiation is not known, nor is the mechanism by which the covalent bond forms between dCTP and the preterminal protein. Although the adenovirus DNA polymerase is probably responsible for this step, a role for the preterminal protein has not been eliminated. Interestingly, the adenovirus polymerase, unlike other polymerases, is relatively inactive on substrates which have an RNA primer.[140]

The preterminal protein primes replication of both strands of adenovirus DNA.[141] Yet the replicative intermediates in these events need to be known. The products of continuous elongation of the first primer are (1) a duplex genome with one parental and one new strand, and (2) a displaced parental strand with the terminal protein bound to its 5′ end. Priming of complementary-strand synthesis on the displaced single strand may occur by the annealing of the self-complementary ends of the genome to generate a duplex "panhandle" structure, the "handles" being identical to the natural ends of the duplex genome. Priming by a fresh preterminal protein–adenovirus polymerase complex could then take place by the mechanism used to prime synthesis of the first strand.

The ϕ29 Terminal Protein and Priming[142]

The linear genome of the *B. subtilis* phage ϕ29 has a 31-kDa terminal protein covalently attached to each of its 5′ ends. This terminal protein, the gene *3* product, primes replication of both DNA strands. An OH group of a serine residue near the C terminus of the protein is linked by a phosphodiester bond to the first nucleotide of the chromosome, a 5′-dAMP.[143]

Gp3 has been cloned and overproduced in *E. coli* and subjected to mutagenesis in vitro. Serine 232, normally the site of covalent linkage within the 266–amino acid protein,[144] cannot be replaced by threonine. In the C-terminal region of the protein, residues 240 to 262 are required, but deletion of residues 17 to 47 from the N terminus results in a protein that retains some priming activity, and removal of residues 5 to 13 actually activates priming. Complementation studies with gene *3* mutants and terminal proteins from related phages indicate the importance of protein–protein interactions between the parental terminal protein, bound to the ends of the genome, and the terminal protein which will serve as the primer for the next round of DNA synthesis.

As in adenovirus, the ϕ29 terminal protein forms a complex with the viral

139. Pruijn GJM, van Driel W, van der Vliet PC (1986) *Nature 322*:656; Rosenfeld PJ, O'Neill EA, Wides RJ, Kelly TJ (1987) *JCB 7*:875; O'Neill EA, Kelly TJ (1988) *JBC 263*:931.
140. Field J, Gronostajski RM, Hurwitz J (1984) *JBC 259*:9487.
141. Stillman BW (1983) *Cell 35*:7; Kelly TJ (1984) in *The Adenoviruses* (Ginsberg HS, ed.). Plenum Press, New York, p. 271; Kelly TJ, Wold MS, Li J (1988) *Adv Virus Res 34*:1.
142. Salas M (1988) in *The Bacteriophages* (Calendar R, ed.). Plenum Press, New York, Vol. 1, p. 169; Salas M (1988) *Curr Top Microbiol Immunol 136*:71; Blanco L, Prieto I, Gutiérrez J, Bernad A, Lázaro JM, Hermoso JM, Salas M (1987) *J Virol 61*:3983; Gutiérrez J, Vinós J, Prieto I, Méndez E, Hermoso JM, Salas M (1986) *Virology 155*:474; Blanco L, Salas M (1985) *PNAS 82*:6404; Blanco L, Bernad A, Salas M (1988) in *DNA Replication and Mutagenesis* (Moese RE, Summers WC, eds.). Am Soc Microbiol, Washington, DC, p. 122; Prieto I, Serrano M, Lázaro JM, Salas M, Hermoso JM (1988) *PNAS 85*:314.
143. Hermoso JM, Salas M (1980) *PNAS 77*:6425; Hermoso JM, Méndez E, Soriano F, Salas M (1985) *NAR 13*:7715.
144. Hermoso JM, Mendez E, Soriano F, Salas M (1985) *NAR 13*:7715.

DNA polymerase prior to the priming reaction.[145] The preferred template is a fragment with the normal ϕ29 genome ends; the reaction is five to ten times more efficient when the parental terminal protein is in place. The terminal genome sequences, in the absence of bound terminal protein, are superior templates for the priming reaction compared to fragments derived from internal regions of the genome, indicating the importance of nucleotide sequence. The ϕ29 DNA polymerase catalyzes the formation of the covalent bond between dAMP and the protein.[146] The K_m for dATP in this initiation reaction is lower than that for elongation of DNA chains, suggestive of two active sites on the DNA polymerase, one specific for the protein and a second for adding nucleotides to an existing DNA chain. The gene 6 protein, a dimeric DNA-binding protein, stimulates primer formation by lowering the K_m for dATP about fivefold.[147]

Discontinuous DNA synthesis is not required for replication of the ϕ29 genome. Both strands are initiated by protein priming, although not necessarily synchronously. Replication from each end toward the center of the chromosome results in two daughter duplexes, each with one parental and one newly synthesized strand.

Other Terminal Proteins

The terminal protein of *Streptococcus pneumoniae* phage Cp-1 is attached via a threonine–5′-dAMP linkage and primes replication by a mechanism similar to that used by ϕ29. PRD1, a phage of *E. coli* and *Salmonella typhimurium*, has a tyrosine–5′-dGMP–linked terminal protein which very likely serves an analogous function. Other genomes with proteins covalently attached to their 5′ ends include mitochondrial DNA from maize, a linear *Streptomyces* plasmid, and the linear yeast plasmids pGKL1 and pGKL2. Although not firmly established, these genomes most likely represent further examples of protein primers.

---

145. Blanco L, Prieto I, Gutiérrez J, Bernad A, Lázaro JM, Hermoso JM, Salas M (1987) *J Virol 61*:3983.
146. Blanco L, Salas M (1985) *PNAS 82*:6404; Blanco L, Bernad A, Sales M (1988) in *DNA Replication and Mutagenesis* (Moese RE, Summers WC, eds.). Am Soc Microbiol, Washington, DC, p. 122.
147. Blanco L, Gutiérrez J, Lázaro JM, Bernad A, Salas M (1986) *NAR 14*:4923; Prieto I, Serrano M, Lázaro JM, Salas M, Hermoso JM (1988) *PNAS 85*:314.

CHAPTER 9

Ligases and Polynucleotide Kinases

9-1 Assays and Discovery of DNA Ligases[1]

By 1967, models for recombination, repair, and replication predicted an enzyme that could reseal the broken backbone of a DNA chain. A joining enzyme would be needed for the breakage and reunion events postulated at that time: excision–repair of lesions, the opening and closing of chromosomes in recombination, and the joining of nascent pieces in discontinuous replication. It is nonetheless remarkable that in that single year, five independent reports, based on five different assays, described the discovery of DNA ligase.[2]

Since 1967, DNA ligase activity has been found in a variety of sources, and an RNA ligase was discovered a few years later. DNA ligases catalyze the formation of a phosphodiester bond between adjacent 3′-OH and 5′-P termini of polynucleotides hydrogen-bonded to a complementary strand (Fig. 9-1), whereas RNA ligases join single-stranded polynucleotides.

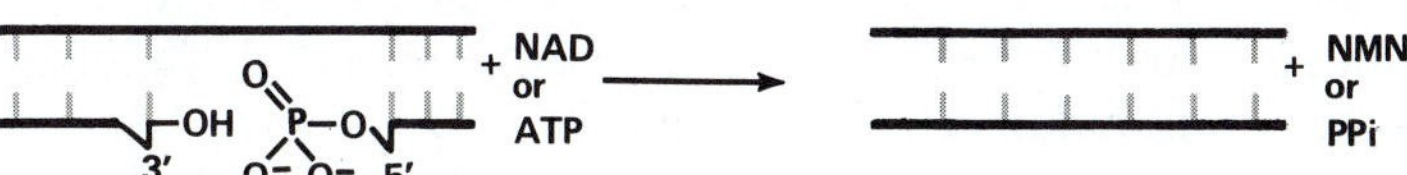

Figure 9-1
Overall reaction of DNA ligases.

1. Lehman IR (1974) Science *186*:790; Engler MJ, Richardson CC (1982) Enzymes *15*:3.
2. Lehman IR (1974) Sciences *186*:790; Engler MJ, Richardson CC (1982) Enzymes *15*:3.

DNA ligases join DNA polymers, and RNA ligases (Section 6) join RNA polymers, by the same mechanism. DNA ligase is ubiquitous and essential to all cells, whereas the distribution of RNA ligase is more restricted and its function is less certain. DNA ligases are less efficient in joining single-stranded than double-stranded molecules, are more active with DNA than with RNA, and have a well-defined role in nucleic acid metabolism.

In the several assays used in the nearly simultaneous discoveries of DNA ligase, and in some assays introduced since, the substrates are enlightening with respect to the understanding and use of the enzyme (Fig. 9-2):

(1) *Phage λ DNA, a linear duplex with cohesive, or "sticky," ends,* which join and form a doubly-nicked circle (Section 17-7). The assay[3] [Fig. 9-2 (1)] measures conversion of the nicked circle into a sealed covalent one. Joining the two nicks causes the DNA to sediment more rapidly in alkaline sucrose gradients. Also, at alkaline pH, the covalent circle adheres to hydroxyapatite; the linear duplex does not.

(2) *A long synthetic polymer, poly dA_{4000}, to which are base-paired contiguous stretches of oligo dT_{150}, each labeled with ^{32}P at the 5′ end*[4] [Fig. 9-2 (2)]. The labeled ends of the short chains are susceptible to phosphomonoesterase action unless they are joined by DNA ligase. The extent of ligase action on oligo dT_{150} molecules can be measured by loss of ^{32}P-phosphate that is susceptible to phosphomonoesterase.

(3) *Oligo dC attached to cellulose and free ^{32}P-labeled oligo dC, annealed together to poly dI* to form a sedimentable complex[5] [Fig. 9-2 (3)] at neutral pH. In alkali, the poly dI·oligo dC complex is not stable, and the label does not sediment with the cellulose derivatized with oligo dC. When joined by ligase to the 3′-OH of the cellulose-bound oligo dC, the labeled oligo dC remains sedimentable after denaturation with alkali.

(4) *A nicked ^3H-labeled poly d(AT)* which is readily degraded by exonuclease III[6] [Fig. 9-2 (4)]. After ligase seals the nick, no ends remain, so that the nuclease cannot release acid-soluble, labeled mononucleotides.

(5) *Transforming DNA inactivated by DNase.* The transforming activity of nicked DNA can be reactivated by ligase.

Two further assays, which depend on the initial reaction of ligase with its coenzyme (ATP or NAD^+), will be referred to in Section 3.

9-2 Properties and Abundance of DNA Ligases

DNA ligases from *E. coli* and *B. subtilis* use NAD^+ as coenzyme and as the energy source for synthesis of the phosphodiester bond. DNA ligases encoded by phages T4 and T7 and those found in eukaryotic cells use ATP.

---

3. Gellert M (1967) *PNAS 57*:148; Gefter ML, Becker A, Hurwitz J (1967) *PNAS 58*:240.
4. Olivera BM, Lehman IR (1967) *PNAS 57*:1426; Olivera BM, Lehman IR (1967) *PNAS 57*:1700; Weiss B, Richardson CC (1967) *PNAS 57*:1021.
5. Cozzarelli NR, Melechen NE, Jovin TM, Kornberg A (1967) *BBRC 28*:578.
6. Modrich P, Lehman IR (1970) *JBC 245*:3626.

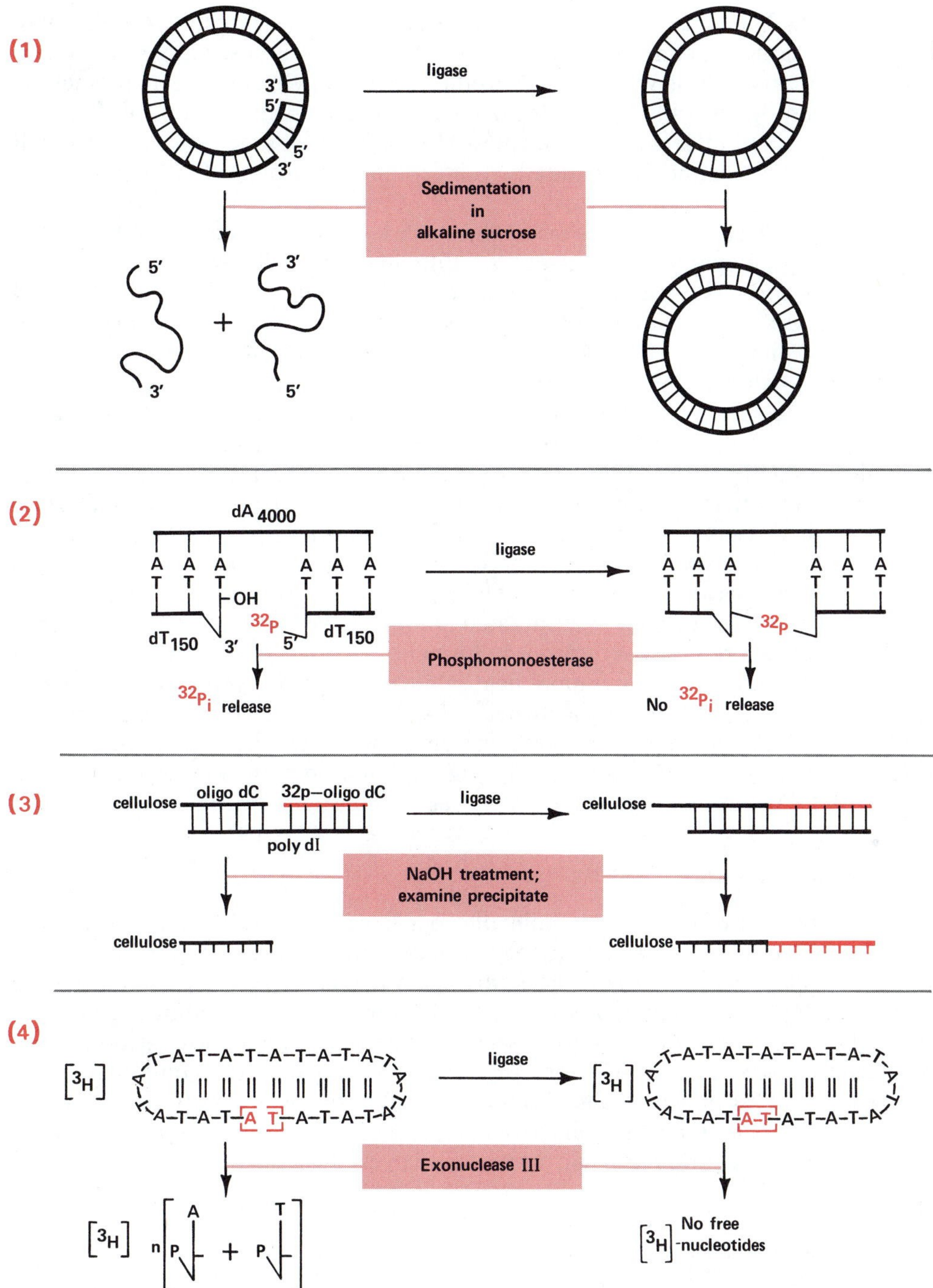

Figure 9-2
Some procedures used in the assay of ligase.

Properties

The *E. coli* enzyme is a 75-kDa polypeptide with a sedimentation coefficient of 3.9S, suggestive of an elongated shape. The enzyme is susceptible to adventitious proteolysis as well as to controlled cleavage by trypsin.[7] The proteolytic product, significantly smaller in size than the intact enzyme, can still form the first covalent intermediate, enzyme–AMP, at an unaltered rate and to the full extent, but it cannot subsequently transfer the AMP group to DNA for phosphodiester bond formation.

Phage T4–encoded ligase resembles the *E. coli* enzyme in size and shape.[8] It is a 60-kDa polypeptide and, given its sedimentation coefficient of 3.5S, may also be elongated. The enzyme shows marked inhibition by salt, being virtually inert at 0.2 M KCl. Spermine inhibits T4 ligase by raising the K_m for the DNA substrate.[9] The phage T7 ligase is 41 kDa, as determined from its DNA sequence.[10]

Mammalian cells[11] and *Drosophila* embryos,[12] unlike prokaryotes (and yeast, thus far), possess two forms of DNA ligase. *Ligase I* is a 125-kDa monomer of asymmetric shape;[13] it has also been identified in multiple smaller, active forms, the result of proteolysis in vitro. In particular, an 85-kDa domain, representing the C-terminal two-thirds of the enzyme, is produced. This fragment with undiminished ligase activity is similar in size and shows significant sequence homology to yeast DNA ligases; the function of the N-terminal part of the enzyme is unknown. Ligase I is the dominant form in proliferating cells (e.g., regenerating rat liver and mitogenically stimulated lymphocytes) and is defective in Bloom's syndrome (a replicative disorder in humans; Sections 5 and 21-4). *Ligase II* is a 68-kDa polypeptide and is more heat-labile than ligase I. As the principal activity in resting cells, it may be associated with repair rather than replication.

The two mammalian ligases are serologically unrelated: neutralizing antibodies against ligase I do not recognize ligase II, nor do antibodies against II react with I. Further distinguishing properties are considered in Section 4.

Abundance

The number of DNA ligase molecules per *E. coli* cell (estimated from the extracted activity divided by the specific activity per molecule of pure enzyme) is about 300, a value close to that estimated for DNA polymerase I (Section 4-1). In view of their closely related functions in filling gaps and sealing segments of DNA, it seems reasonable to expect that physical interactions between them would improve their efficiency in replication, repair, and recombination. An extract of T4-infected cells is 500 times higher in specific activity than is the most active mammalian cell extract, largely because the T4 ligase has a much higher turnover number than do mammalian cell ligases. Amplified levels of the

---

7. Panasenko SM, Modrich P, Lehman IR (1976) *JBC 251*:3432.
8. Knopf K-W (1977) *EJB 73*:33; Murray NE, Bruce SA, Murray K (1979) *JMB 132*:493; Tait RC, Rodriguez RL, West RW Jr (1980) *JBC 255*:813; Davis RW, Botstein D, Roth JR (1980) *Advances in Bacterial Genetics*, p. 196. CSHL.
9. Raae AJ, Kleppe RK, Kleppe K (1975) *EJB 60*:437.
10. Dunn JJ, Studier FW (1981) *JMB 148*:303.
11. Arrand JE, Willis AE, Goldsmith I, Lindahl T (1986) *JBC 261*:9079; Teraoka H, Sumikawa T, Tsukada K (1986) *JBC 261*:6888; Elder RH, Rossignol J-M (1990) *Biochemistry 29*:6009.
12. Takahashi M, Senshu M (1987) *FEBS Lett 213*:345; Rabin BA, Chase JW (1987) *JBC 262*:14105.
13. Tomkinson AE, Lasko DD, Daly G, Lindahl T (1990) *JBC 265*:12611; Lasko DD, Tomkinson AE, Lindahl T (1990) *JBC 265*:12618.

E. coli and T4 ligases[14] induced by cloning the genes have provided ample sources for studies of the enzymes and for their widespread use (particularly T4) in recombinant DNA technology.

9-3 Enzymatic Mechanism of DNA Ligases[15]

Ligases utilize the group transfer potential of the phosphoanhydride bonds of NAD^+ or ATP to form a phosphodiester bond between nucleic acid chains. The reaction (Fig. 9-3) occurs in three discrete steps: (1) formation of an enzyme–nucleotide intermediate by transfer of the adenylyl group of NAD or ATP to the ϵ-NH_2 of a lysine residue in the enzyme; when the cofactor is NAD^+, the other product is NMN; when ATP, the other product is PP_i; (2) adenylyl activation of the 5′-P terminus of DNA by transfer of the adenylyl group from the enzyme; (3) phosphodiester bond formation by attack of the 3′-OH terminus of the DNA on the activated 5′-P group, with release of AMP. This formulation is based on the isolated covalent intermediates, reversal of each step, and steady-state kinetic analysis.

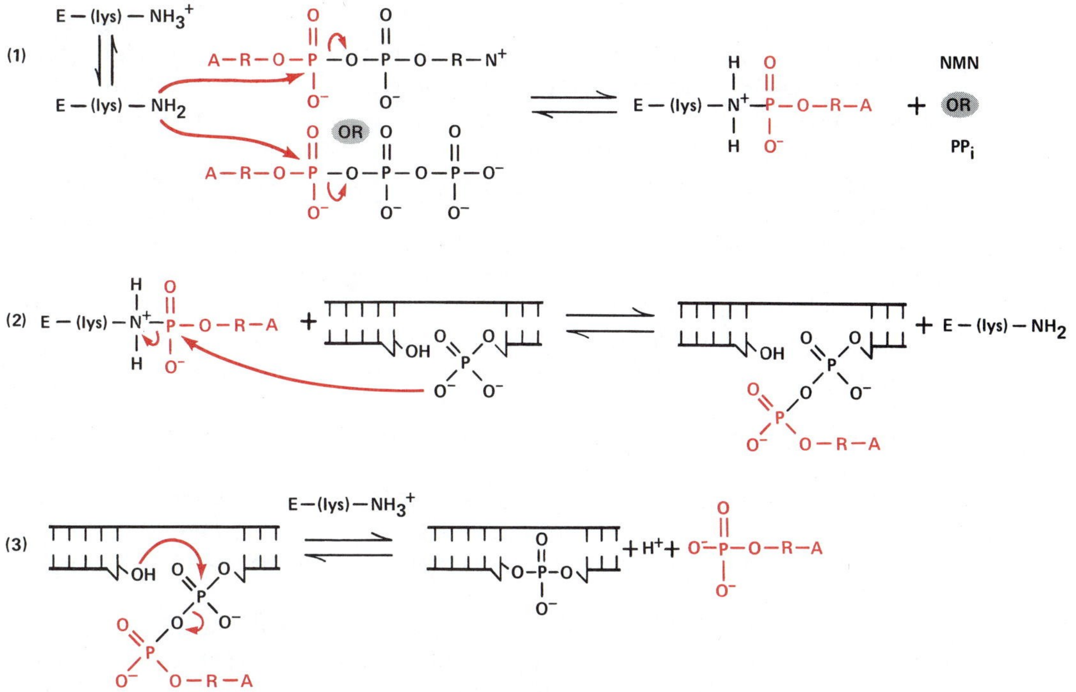

Figure 9-3
The three steps in the overall ligase reaction. E = enzyme; lys = lysine residue.

14. Murray NE, Bruce SA, Murray K (1979) *JMB 132*:493; Tait RC, Rodriguez RL, West RW Jr (1980) *JBC 255*:813; Davis RW, Botstein D, Roth JR (1980) *Advances in Bacterial Genetics*, p. 196. CSHL.
15. Lehman IR (1974) *Science 186*:790; Engler MJ, Richardson CC (1982) *Enzymes 15*:3.

Ligase-adenylate, the first covalent intermediate, is readily isolated. Starting with AMP-labeled NAD^+ or ATP, measurement of the acid-insoluble, labeled fraction gives a direct assay of the enzyme. So does the exchange of labeled NMN or PP_i in the reversal of the reaction when nucleic acid substrate is absent. In crude enzyme fractions, other reactions may interfere with these assays. Another drawback is that the small ligases created by proteolysis form the ligase-adenylate intermediate at an unaltered rate and to the full extent, but cannot transfer the AMP group to DNA for phosphodiester bond formation.

In forming enzyme–AMP intermediate, the enzyme makes use of an ϵ-amino group as a nucleophile in nucleotidyl transfer. The phosphoamide linkage in the intermediate behaves like authentic ϵ-adenylyl lysine in its relative acid lability and alkaline stability and in cleavage by acidic but not neutral hydroxylamine. These criteria distinguish a phosphoamide from a phosphodiester or a mixed anhydride.

The isolated adenylylated ligase can react directly with nicked DNA to transfer the adenylyl group to the 5′-P terminus and thus generate *DNA-adenylate*, the second covalent intermediate. Even under conditions of low pH and temperature, only small amounts of DNA-adenylate accumulate. This intermediate, separated from the enzyme that generated it, still reacts rapidly with the unadenylylated form of the enzyme to form a phosphodiester bond and release AMP.

Since nicked polynucleotides with 2′,3′-dideoxy instead of 3′-OH ends cannot form phosphodiester bonds, these unreactive nicks might presumably trap a DNA-adenylate intermediate formed by ligase-adenylate. However, the rate of DNA-adenylate synthesis was found to be at least 1000 times slower on a dT_{200} chain with a 2′,3′-dideoxy end substituted for a 3′-OH. Recognition of a 3′-OH group may be critical for orienting ligase-adenylate at a nick to effect the adenylate transfer to the adjacent 5′-P group. RNA ligase, which has the same basic mechanism as DNA ligase (Section 6), also requires a 3′-OH–terminated polynucleotide as an effector for RNA-adenylate formation.

DNA-adenylate has been identified in two ways. Denatured DNA-[^{32}P]adenylate, sequentially hydrolyzed by exonuclease I from the 3′-OH end, does not release the usual 5′ terminal dinucleotide but, instead, a trinucleotide containing labeled AMP. It may be inferred that the AMP is linked to the 5′ end by a pyrophosphate rather than a phosphodiester bond. The second line of evidence is from synthetic dT_{200}-adenylate in which the 5′ terminus of dT_{200} is linked by a phosphoanhydride bond to 5′-AMP. This compound behaves like the enzymatically generated DNA-adenylate. It is readily joined by ligase to an adjacent dT_{200}-adenylate unit aligned on poly dA_{5000}, releasing AMP in stoichiometric amounts.

The joining reaction is inhibited by the addition of coenzyme (NAD^+ or ATP), apparently because the enzyme accepts the adenylyl group of the coenzyme. In the adenylylated form the enzyme cannot catalyze the final joining reaction. Adenylyl transfer and activation by ligase is rather specific for AMP. The T4 enzyme is less than 10% as active when dAMP is substituted for AMP, and it is inert with GMP. Neither will *E. coli* ligase join a dT_{200} activated with a 5′-triphosphate to an adjacent 3′-OH group, the essence of a DNA polymerase–catalyzed reaction.

Reversal of the Reactions

As already indicated, each of the covalent intermediates participates in a reverse reaction. Ligase-adenylate generates NAD^+ (or ATP) from NMN (or PP_i); DNA-adenylate reacts with free ligase and NMN to form NAD^+. Reversal of the entire sequence of reactions can be demonstrated in the AMP-dependent conversion of a covalently closed, superhelical circle (via a nicked, adenylylated form) to a relaxed circle which may be covalently closed (resealed) or nicked (Fig. 9-4).

The ligase-catalyzed relaxation of phage superhelical circles is, as with topoisomerase I, stepwise and suggests the existence of an intermediate — a complex of the DNA with ligase and AMP — that is supportive of this distributive behavior.[16] Unlike topoisomerase I, ligase uses positive as well as negative supercoils, and employs coenzyme activation in nucleotidyl transfer rather than direct covalent association between DNA and the enzyme.

The *nicking–closing activity* of ligases is displayed in the reversal of its reactions, and bears some resemblance to the enzymes that perform the reaction without the intervention of a cofactor. These include the gene *A* protein of phage ϕX174 that initiates and terminates a rolling circle in replication (Section 8-7), the Int protein of phage λ and other specific recombinases (Section 21-6), and the variety of topoisomerases (Chapter 12) that provide the relaxing, supercoiling, swiveling, knotting, catenating, and decatenating functions for replication, recombination, and transcription of duplex DNA. In addition to requiring a cofactor, the reverse ligation reaction is kinetically sluggish, by a factor of 1000 or more, compared to the topoisomerase reactions.

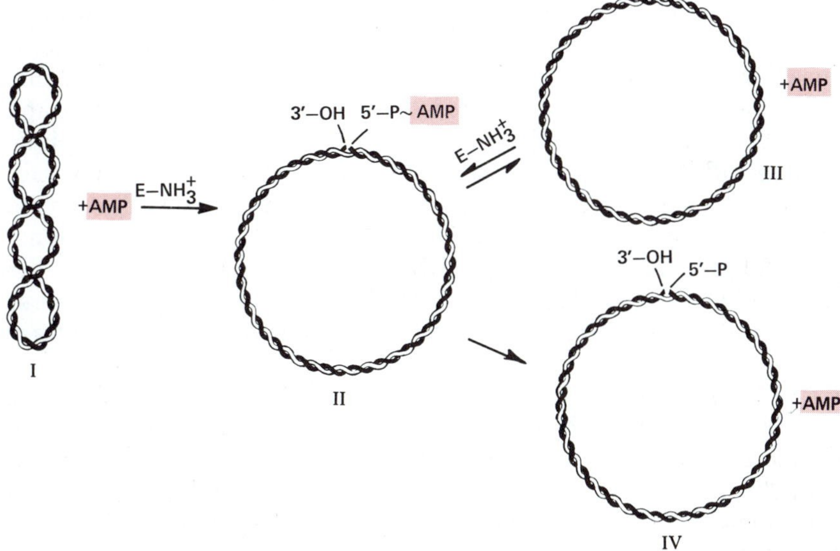

Figure 9-4
Reversal of the ligase reaction. A superhelical circle (I) is relaxed or nicked in the presence of AMP to produce an adenylylated intermediate (II), which may be either rejoined after relief of the superhelical twist (III) or hydrolyzed (IV).

16. Montecucco A, Ciarrocchi G (1988) *NAR 16*:7369; Montecucco A, Pedrali-Noy G, Spadari S, Ciarrocchi G (1988) *BBA 951*:330.

Steady-State Kinetic Analysis

If the reactive covalent intermediates isolated are authentic participants in the overall reaction, the rate constants for their formation and use must be at least equal to that of the overall reaction. This was readily demonstrated in the first step, the formation of ligase-adenylate, but was technically difficult to prove in the next two steps, the formation and consumption of DNA-adenylate.

An interesting kinetic feature is the activation of *E. coli* ligase by NH_4^+. As little as 5 mM NH_4Cl raises the V_{max} of the overall joining reaction 20-fold. Since NH_4^+ has no demonstrable influence on the first and third reactions in the sequence (Fig. 9-3), it would seem that its effect is on the second step, the adenylyl transfer from ligase to DNA. NH_4^+ has no effect on any aspect of T4 ligase action.

9-4 Substrate Specificity of DNA Ligases[17]

In addition to their differences in cofactor and NH_4^+ activations, *E. coli* and T4 ligases have distinguishable substrate specificities. T4 ligase, but not *E. coli* ligase, can link homopolymer RNA chains aligned on complementary DNA, DNA chains aligned on complementary RNA, and the 3′-OH end of RNA (heteroduplexed with DNA) to DNA. T4 ligase can use a duplex length as short as 3 bp on either side of a nick, but does so less efficiently when a single-bp mismatch exists at either side of the ligation junction.[18]

Most remarkably, T4 ligase can link base-paired (blunt) ends of two duplex natural DNA molecules.[19] Neither *E. coli*, T4, nor other DNA ligases can link free single strands, but RNA ligase can do so (Section 6). The joining of two blunt-ended linear DNA complexes is noteworthy because it once was assumed that an intact complementary strand is needed to orient the ends of a broken strand for joining. T4 DNA ligase is aided in this reaction by RNA ligase or hydrophilic polymers. While RNA ligase by itself is unable to catalyze the reaction, it stimulates low concentrations of T4 DNA ligase in this activity (but not others) up to 20-fold. The rather low affinity between DNA ligase and blunt-ended DNA is not raised by RNA ligase, as K_m plots show.[20] RNA ligase probably adenylylates the 5′-P of the DNA to make it a substrate for T4 DNA ligase, which can thus serve as a reagent for synthesizing chimeric (recombinant) nucleic acids and varieties of copolymers.

Ligase joining of blunt ends or the short cohesive ends of phage λ can be stimulated 1000-fold or more by the presence of high concentrations of hydrophilic polymers.[21] Polyethylene glycol (molecular weight 8000) at 15% facilitates intermolecular joining or the interaction between the distant ends of a DNA molecule by virtue of the excluded volume effect, aptly termed "macromolecular crowding."[22]

---

17. Lehman IR (1974) *Science 186*:790; Engler MJ, Richardson CC (1982) *Enzymes 15*:3.
18. Wu DY, Wallace RB (1989) *Gene 76*:245.
19. Sgaramella V, Ehrlich SD (1978) *EJB 86*:531; Deugau KV, van de Sande JH (1978) *Biochemistry 17*:723.
20. Sugino A, Goodman HM, Heyneker HL, Shin J, Boyer HW, Cozzarelli NR (1977) *JBC 252*:3987.
21. Zimmerman SB, Pheiffer BH (1983) *PNAS 80*:5852; Zimmerman SB, Harrison B (1985) *NAR 13*:2241; Teraoka H, Tsukada K (1987) *J Biochem 101*:225; Sobczak J, Duguet M (1988) *EJB 175*:379.
22. Minton A (1983) *Mol Cell Biochem 55*:119.

Table 9-1
Comparison of *E. coli* and T4 ligases

	E. coli	T4
Cofactor	NAD^+	ATP
NH_4^+ activation	yes	no
Substrates		
Poly dA · oligo dT	yes	yes (100)[a]
Poly rA · oligo dT	no	yes (2)[a]
Poly dT · oligo rA	no	yes (2)[a]
Poly dA · oligo rU	no	no
Base pairs at nick	>10	≥6
Blunt ends	no[b]	yes

[a] Relative to poly dA · oligo dT set at 100.
[b] Some activity in the presence of hydrophilic polymers.

Much of the work on substrate specificity of ligases was carried out with synthetic homopolymer pairs. Quite different results have been obtained with apparently similar polymers. For example, poly dT·oligo rA is active with T4 ligase, but poly dA·oligo rU is not (Table 9-1). *E. coli* ligase was discovered with $dA_{4000} \cdot dT_{150}$, but it would have been missed had $dT_{4000} \cdot dA_{150}$ been used instead, because the enzyme fails to act on that substrate. The capacity of homopolymers to form triple helices (e.g., A·U·U· from poly A and poly U)[23] may be responsible.

What the enzymes do with natural RNA, DNA, and hybrids cannot be predicted from their actions on particular homopolymer pairs. The T4 DNA ligase can join RNA to DNA, as well as DNA to DNA; on the other hand, the *E. coli* enzyme that joins DNA to DNA can link a 5′-P terminus of a DNA to the 3′-OH terminus of RNA but not the 3′-OH end of DNA to a 5′-P of RNA.

An abasic residue (apurinic or apyrimidinic) at either the 3′ or 5′ end of oligonucleotides (in the presence of a complementary strand) can be ligated by T4 ligase, although the rate is diminished.[24] Remarkably, a gap between the polynucleotides to be joined also does not prevent their being ligated. With these suboptimal substrates, the step affected was the attack by the 3′-OH end on the 5′-adenylylated DNA rather than the formation of the DNA intermediate.

In contrast to the use of ATP by all phage and eukaryotic ligases, all bacterial ligases choose NAD^+ as the energy source for synthesis of the diester bond. The basis for this choice is not evident. When the NAD^+ ligase in *Salmonella typhimurium* is inactivated and the ATP enzyme of phage T4 is substituted instead, there is no loss of function even under the stress of DNA damage and repair.[25] Thus, functional interactions between ligase and replication and repair enzymes are not essential, nor are any metabolic consequences observed in the nicotinamide nucleotide cycle when ATP is substituted for NAD^+ in DNA ligations.

The distinctive features and substrate specificities of the two ligases in eukaryotes are given in Table 9-2.

23. Fresco J (1966) *JMB 19*:145; Fresco J (1972) Prog N A Res *12*:2.
24. Goffin C, Bailly V, Verly WG (1987) *NAR 15*:8755; Wiaderkiewicz R, Ruiz-Carillo A (1987) *NAR 15*:7831.
25. Park UE, Olivera BM, Hughes KT, Roth JR, Hillyard DR (1989) *J Bact 171*:2173.

Table 9-2
Comparison of eukaryotic ligases I and II

	Ligase I	Ligase II
Mass (kDa)	85–125	68
ATP, K_M (μM)	2	40
Substrates		
Blunt end	yes	no
Poly rA · oligo dT	no	yes
5′-P deoxy end abutting a 3′-OH ribo end	yes	nd[a]
Function	replication	repair
Location	nucleus	nucelus
Fraction of cellular activity (%)	90	≤10

[a] nd = not determined.

9-5 Functions in Vivo of DNA Ligases

Studies in Prokaryotes[26]

Ligase is demonstrably the sole agent for sealing DNA chains in replication, repair, and recombination. Apportioning of the ligase molecules among these functions and balancing their use between the needs of the host cell and those of an infecting phage can be correlated with the levels of the enzymes supplied by the host and introduced as a consequence of the infection. The numerous physiologic roles of ligase, as well as the influence of other metabolic processes upon them, have been disclosed through the use of cellular and phage mutants.

Two *E. coli* mutants with changes in the structural gene for ligase are *lig4* and *lig*^{ts}*7*. Ligase activities in extracts of both mutants are heat sensitive (Table 9-3), *lig*^{ts}*7* at a temperature as low as 25°. The *lig4* mutant, even at 42°, is only partially deficient in DNA repair and grows at a nearly normal rate. The *lig*^{ts}*7* mutation is lethal at 42°; the rapid loss of cell viability can be correlated with a large accumulation of short nascent DNA fragments. Even at 25°, *lig*^{ts}*7* is deficient in sealing replication fragments and repairing DNA damage.

The growth of *lig4* mutants at 42° and *lig*^{ts}*7* mutants at 25° (Table 9-3) can be explained by the persistence of ligase function, even with levels depressed to

Table 9-3
Ligase levels and functions of two temperature-sensitive *E. coli* mutants

	Ligase level[a]		Growth		DNA repair[b]	
Mutant strain	25°	42°	25°	42°	25°	42°
lig4	100	1	+	+	+	±
lig^{ts}*7*	1	<1	+	−	−	−

[a] Assayed at 25° or 42° and expressed as percent of wild-type activity at these temperatures.
[b] Measured by sensitivity to UV irradiation or an alkylating agent.

26. Engler MJ, Richardson CC (1982) *Enzymes* 15:3.

only about 1% of those in wild-type cells. As discussed in connection with DNA polymerase I deficiency (Section 4-18), relatively few molecules may be needed to maintain the reactions at a replication fork. Neglect of repair and recombination functions, to which most of the ligase molecules are devoted, is evidently tolerated.

Assuming 300 ligase molecules per cell, a turnover number of 25 joining events per minute per molecule, and 200 joining events per replicative fork per minute at 30°, only about 2% of the ligase molecules would be required to sustain replication. Below this critical number of molecules, deficiencies in replication as well as in repair become evident and, if severe enough, viability is lost. For a particular function, such as recombination, the level of enzyme can have different effects at different stages. Depending on the rate-limiting step, decreased ligase may be stimulatory: *lig*⁻ mutants are found among strains isolated for hyperrecombinational activity, possibly because the lifetime of DNA recombination intermediates is longer in such cells.

A fivefold increase in ligase level due to T4 infection is abolished by a mutation in the T4 ligase gene *30*. Although DNA synthesis in the mutant starts at a normal rate, it is almost entirely arrested in a few minutes. Short pieces of DNA accumulate and very few phage particles are produced. Recombination between phage DNA molecules leads to the accumulation of intermediates with gaps and branches. The normal level of host-cell ligase is clearly insufficient.

Increased as well as decreased levels of cellular ligase are reflected in the support of growth of various phages (Table 9-4). A tenfold increase in ligase by an *E. coli* ligase overproducer (*lop8*) can overcome the deficiency in T4 *lig*⁻. An additional phage mutation in the T4 *rII* gene illustrates the complexity of the in vivo economy of ligase. Mutants in this gene reproduce in *E. coli* B but not in *E. coli* K12 lysogenic for phage λ. (No other phage mutants make this distinction.) A mutation in the *rII* gene suppresses T4 *lig*⁻ (Table 9-5),[27] provided the cells are not themselves deficient in ligase. Reduced levels of nuclease activity in the *rII* mutant may diminish the need for the cellular ligase and thus permit the phage to grow in wild-type cells. However, reduction of ligase to 5% of normal in the cell, by introduction of the *lig4* mutation into the *lop8* strain, no longer allows growth of T4 *lig*⁻, *rII*⁻.

Table 9-4
Dependence of phage growth and DNA replication on ligase levels in *E. coli*[a]

E. coli mutant	Temp.	Ligase level (%)	Growth of T4 phage (*lig*⁻)	Growth of T4 phage (*lig*⁻ *rII*⁻)	Growth of T7 phage (*lig*⁻)	Sealing nascent DNA	Growth of λ phage
lop8	37	1000	+	+	+	+	+
Wild type	37	100	−	+	+	+	+
lop8 lig4	42	5	−	−	+	+	+
lig4	42	1	−	−	−	+	+
*lig*ᵗˢ*7*	42	<1	−	−	−	−	+

[a] Tabulated by Professor N Cozzarelli from data in Gottesman MM, Hicks ML, Gellert M (1973) *JMB 77*:531, and in Konrad EB, Modrich P, Lehman IR (1973) *JMB 77*:519.

27. Ebisuzaki K, Campbell L (1969) *Virology 38*:701; Karam JD (1969) *BBRC 37*:416.

Table 9-5
Ligase mutation in T4 phage: suppression in a wild-type host cell by a second mutation in the phage *rII* gene

	T4 phage growth on *E. coli* B strains	
T4 phage mutant	Wild-type	*lig*⁻
rII⁻	+	+
lig⁻[a]	−	−
rII⁻ *lig*⁻	+	−

[a] Mutation in T4 ligase gene *30*.

The *lig4* mutation reduces ligase levels (in the absence of *lop8*) to only about 1% of normal; this is too drastic to sustain phage T7 replication (which is less demanding than that of T4) when the T7 ligase gene is also defective. As indicated above, the more severe $lig^{ts}7$ mutation interferes with cellular DNA replication (Table 9-3) but is adequate for phage λ growth (Table 9-4).

Studies in Eukaryotes[28]

Eukaryotic cells deficient in ligase may exhibit profound phenotypic changes. *Bloom's syndrome* (Section 21-4) is an inherited disease with sun-sensitive lesions and an increased risk of cancer. Cells from affected people show frequent chromosomal rearrangements, a tenfold elevation in mutation rate, defective replicative intermediates, and an abnormality in the postincision step of DNA excision repair. The basic defect is a structural alteration in ligase I due to point mutations[29] that reduce the activity and increase the heat-sensitivity of the enzyme. The levels of ligase II remain normal in all the affected patients.

A conditionally lethal mutant of the yeast *Schizosaccharomyces pombé* with DNA ligase activity *(cdc17)* has been identified.[30] At the permissive temperature (25°), *cdc17* has less than 20% of the normal activity, while at the restrictive temperature (35°), no activity has been detected. At the restrictive temperature, mutant cells can replicate DNA during S phase, but mitosis does not occur. Most of the new DNA synthesized at the restrictive temperature remains as very small pieces, suggesting that DNA ligase is needed for joining of the fragments. The mutant cells are also highly sensitive to UV light even at the nonpermissive temperature. A mutant of *Saccharomyces cerevisiae* shows similar properties but has no detectable DNA ligase activity at either the permissive or nonpermissive temperature.[31] A 63-kDa DNA ligase encoded by vaccinia virus[32] shares 30% amino acid identity with the *S. pombé* and *S. cerevisiae* ligases, which posses masses near 85 kDa.

28. Lindahl T, Willis AE, Lasko DD, Tomkinson A (1989) in *DNA Repair Mechanisms and Their Biological Implications in Mammalian Cells* (Lambert M, Laval J, eds.). Plenum Press, New York.
29. Willis AE, Weksberg R, Tomlinson S, Lindahl T (1987) *PNAS 84*:8016; Chan J Y-H, Becker FF (1988) *JBC 263*:18231; Rünger TM, Kraemer KH (1989) *EMBO J 8*:1419.
30. Barker DG, White JHM, Johnston LH (1987) *EJB 162*:659; Smith GL, Chan YS, Kerr SM (1989) *NAR 17*:9051.
31. Johnston LH, Nasmyth KA (1978) *Nature 274*:891.
32. Smith GL, Chan YS, Kerr SM (1989) *NAR 17*:9051.

T4 Phage RNA Ligase

The best characterized of RNA ligases is gp63 *(rli)*, encoded by T4 phage. Its only known essential function is the noncovalent attachment of the phage's tail fibers to its base plates, a feat that bears little mark of the capacities of gp63 in synthetic nucleic acid chemistry (see below). Tail-fiber attachment by gp63 appears to be completely independent of its RNA ligase activity. Mutants (rli^-) have been isolated that lack the RNA ligase activity but are normal in tail-fiber attachment.[34] Such mutants grow normally in wild-type *E. coli* but are defective in T4 DNA replication and particularly in T4 late gene expression in certain restrictive *E. coli* strains. This phenotype is identical to that of *pseT (pnk)* mutations in the T4 polynucleotide kinase gene (Section 7), suggesting that the ligase and the kinase take part in the same pathway.

The gene *63* protein, a 43-kDa monomer, has been purified to homogeneity. The enzymatic activities of RNA ligase on nucleic acids have yet to be assigned a definitive function in T4 phage development. The enzyme resembles T4 phage DNA ligase in its basic reaction mechanism[35] but also differs from it in the following ways.

(1) Gp63 prefers single-stranded to double-stranded nucleic acids. Intramolecular circles form easily, but intermolecular joining can be directed by blocking the 3′ end of the donor molecule, the 5′-P end of which forms the phosphodiester bond. Gp63 is astonishingly permissive in accepting a wide variety of 5′-P donors, 3′-OH acceptors,[36] base analogs,[37] and ssDNA molecules up to 40 ntd long.[38]

(2) The 5′-P donors can be RNA or DNA, including oligomers as short as a 3′,5′-nucleoside diphosphate. Joining an oligoribonucleotide to the 5′ terminus of a short or long DNA chain provides a means of extending the DNA with a defined sequence.[39] The 5′-P donors also include a large number of compounds with the general structure AMP–P–X. The P–X moiety can be in any nucleotide; it can also be nicotinamide ribose P, ribose 5-P, or even cyanoethyl P. Prior adenyl activation of the 5′-P–X donor makes the ligation more efficient.[40]

(3) Among the 3′-OH acceptors, RNA is much preferred over DNA, and oligoribonucleotides as short as trimers function well. RNA ligase can be used to extend the 3′ end of virtually any RNA molecule, even an oligodeoxyribonucleotide.[41]

---

33. Romaniuk PJ, Uhlenbeck OC (1983) *Meth Enz 100*:52; Brennan CA, Manthey AE, Gumport RI (1983) *Meth Enz 100*:38.
34. Runnels JM, Soltis D, Hey T, Snyder L (1982) *JMB 154*:273.
35. Harnett SP, Lowe G, Tansley G (1985) *Biochemistry 24*:7446.
36. England TE, Gumport RI, Uhlenbeck OC (1977) *PNAS 74*:4839; England TE, Uhlenbeck OC (1978) *Biochemistry 17*:2069; Kikuchi Y, Hishinuma F, Sakaguchi K (1978) *PNAS 75*:1270.
37. Brennan CA, Gumport RI (1985) *NAR 13*:8665.
38. Tessier DC, Brousseau R, Vernet T (1986) *Anal Biochem 158*:171.
39. Higgins NP, Geballe AP, Cozzarelli NR (1979) *NAR 6*:1013.
40. Hoffmann PU, McLaughlin LW (1987) *NAR 15*:5289.
41. Hinton DM, Baez JA, Gumport RI (1978) *Biochemistry 17*:5091.

(4) The 3′-OH polynucleotide acceptor is required for a second role as catalyst (effector) in the second stage of the reaction. An acceptor may serve as effector for itself or another acceptor; this would imply ready exchange between acceptors on an enzyme with only one site.

(5) The intermediate DNA- or RNA-adenylate often accumulates in large amounts under standard reaction conditions.

(6) Reversal of the reaction, in the presence of AMP, occurs readily and can cause translocation of a segment from one end of an RNA chain to the other.

(7) RNA ligase does not catalyze the joining of blunt-ended duplexes, but at low concentrations stimulates this action by T4 DNA ligase.[42]

(8) RNA ligase has no role in replication or recombination, but a repair role in tRNA metabolism has been proposed (Section 7).

(9) RNA ligase has an unrelated activity in tail-fiber attachment.[43]

The unique occurrence in prokaryotes of introns in a few phage T4 genes (Section 17-6) and the coexistence of RNA ligase suggest that the enzyme might be involved in removal of the introns. Yet splicing proceeds normally upon infection with T4 RNA ligase mutants.

Yeast[44] and Wheat Germ[45] RNA Ligases

The RNA ligases from yeast and wheat germ have multiple activities that qualify them for splicing of RNA. In addition to the characteristic actions and mechanism resembling those of T4 RNA ligase, there are two additional activities: (1) a kinase for a 5′-OH RNA terminus, and (2) a phosphodiesterase that cleaves a 2′,3′-cyclic P RNA terminus, leaving the phosphate attached in the 2′-position. In a concerted reaction on RNA, the 2′,3′-cyclic P terminus is converted to 2′-P,3′-OH, and the 5′-OH terminus to a 5′-P. These termini can then be circularized to generate an RNA with a 2′-P marking the site of covalent closure. Two molecules of ATP are consumed, one for the kinase reaction that yields ADP, and the other for the ligase activation that yields AMP.

A ligase activity involved in splicing intron-containing pre-tRNA in HeLa cell extracts has been separated from the splicing endonuclease but not further characterized.[46]

9-7 Polynucleotide Kinases

Polynucleotide kinase, first recognized as a T4 phage-encoded enzyme, has become an important biochemical reagent. The properties of this enzyme, and of similar kinases that are widely distributed in animal cells, argue for a role associated with DNA and RNA ligases in repair reactions.

---

42. Sugino A, Goodman HM, Heyneker HL, Shin J, Boyer HW, Cozzarelli NR (1977) *JBC 252*:3987.
43. Runnels JM, Soltis D, Hey T, Snyder L (1982) *JMB 154*:273.
44. Phizicky EM, Schwartz RC, Abelson J (1986) *JBC 261*:2978.
45. Konarska M, Filipowicz W, Gross HJ (1982) *PNAS 79*:1474; Pick L, Hurwitz J (1986) *JBC 261*:6684; Pick L, Furneaux H, Hurwitz J (1986) *JBC 261*:6694.
46. Herrenknecht K, van Tol H, Gross HJ (1988) *NAR 16*:7713.

T4 Phage Polynucleotide Kinase–Phosphatase[47]

The enzyme, encoded by the T-even phages, has not been detected in uninfected bacteria. (Presumably, the 5′→3′ exonuclease activity of DNA polymerase I can be relied on to generate 5′-P termini for DNA ligase.) Purified to homogeneity from T4-infected cells, the polynucleotide kinase (Table 9-6) is a dimer or tetramer of identical 34-kDa polypeptides.[48] The polypeptide is a product of the *pseT* gene, located between genes *63* and *31* in phage T4[49] (Section 17-6).

The enzyme uses virtually all NTPs to phosphorylate the 5′-OH termini of DNA, RNA, and oligonucleotides. While a nucleoside 3′-monophosphate and a 5′-terminal nucleotide with the base protected by a bulky adduct[50] are excellent substrates, the cyclobutane pyrimidine dimer is not.[51] A protruding chain end of a duplex is preferred over flush ends or a recessed terminus; nicked DNA is a very poor substrate. Unlike most polynucleotide kinases, this enzyme catalyzes two other reactions: (1) reversal of the phosphorylation reaction and thereby the promotion of an exchange between the γ-P of an NTP donor and the 5′-P group of a polynucleotide, and (2) phosphatase action on a 3′-P group on DNA or on oligodeoxynucleotides. Mutations in *pseT* generally affect both the kinase and phosphatase activities, but one mutant (*pseT1*) produces an altered enzyme deficient only in the phosphatase.[52]

Table 9-6
Properties of phage and eukaryotic polynucleotide kinases

	Phage T4 polynucleotide kinase	Eukaryote	
		DNA kinase	RNA kinase
Mass (kDa)	68	80	nd[a]
pH optimum	7.7	5.5	8.4
Sulfate sensitivity	low	high	low
Acceptor specificity			
RNA vs. DNA	RNA and DNA	DNA	RNA
Nicks in DNA	no	yes	
Size	mono to poly	oligo (>10) to poly	poly
Donor NTP			
Specificity	broad	broad	nd
ATP, K_m (μM)	14	2	500
3′-Phosphatase activity	yes	yes	nd
ATP–P_i exchange activity	yes	yes	nd

[a] nd = not determined.

47. Richardson CC (1981) *Enzymes 14*:299; Kleppe K, Lillehaug JR (1979) *Adv Enz 48*:245.
48. Lillehaug JR (1977) *EJB 73*:499.
49. Sirotkin K, Cooley W, Runnels J, Snyder LR (1978) *JMB 123*:221; Mileham AJ, Revel HR, Murray NE (1980) *MGG 179*:227.
50. van de Sande JH, Bilsker M (1973) *Biochemistry 12*:5056; Reddy MV, Randerath K (1986) *Carcinogenesis 7*:1543.
51. Weinfeld M, Liuzzi M, Paterson MC (1989) *JBC 264*:6364.
52. Cameron V, Soltis D, Uhlenbeck OC (1978) *NAR 5*:825.

The enzyme is an indispensable reagent for labeling DNA and RNA chains for many uses, among which are sequence analysis, monitoring the progress of DNA and RNA synthesis, identifying the ends of chains and determining their lengths, fingerprinting, and the physical mapping of restriction enzyme fragments. The capacity of the kinase to phosphorylate the depurinated dinucleotide dApX but not dXpA (where X represents the apurinic deoxyribose group) forms the basis for a ^{32}P-postlabeling assay for apurinic sites in DNA.[53]

Yet the function of the enzyme in phage development remains obscure. The combination of 5′-kinase and 3′-phosphatase activities in one oligomeric enzyme suggests a role allied to DNA ligase when the DNA break bears a 3′-P. However, phenotypic defects in DNA repair, recombination, or replication have not been demonstrated with *pseT* mutants in wild-type cells. On the other hand, the enzyme may cooperate with RNA ligase to restore the host tRNAs cleaved by a phage anticodon nuclease[54] that generates 2′,3′-cyclic P and 5′-OH termini. The phosphatase action to remove the cyclic phosphate combined with the kinase action would produce the 5′-P needed by RNA ligase to seal the break and restore the tRNA.[55]

Eukaryotic DNA Kinase – Phosphatase[56]

Unlike the phage-encoded enzyme, eukaryotic polynucleotide kinase–3′-phosphatase is specific for DNA, works at nicks in duplex DNA, and requires an oligonucleotide at least 10 residues long (Table 9-6). When the enzyme is used as a reagent in conjunction with the phage enzyme, RNA and DNA chain ends can be distinguished, as can nicks and free ends.

DNA kinase purified from rat liver chromatin[57] or calf thymus nuclei is a polypeptide of 80 kDa. When freed of an RNA kinase,[58] the strict specificity for DNA becomes apparent.[59] These mammalian kinases are distinguishable from the kinase activity intrinsic to the yeast and wheat germ RNA ligases (Section 6).

53. Weinfeld M, Liuzzi M, Paterson MC (1990) *Biochemistry 29:*1737.
54. Chapman D, Morad I, Kaufmann G, Gait MJ, Jorissen L, Snyder L (1988) *JMB 199:*373.
55. Amitsur M, Levitz R, Kaufmann G (1987) *EMBO J 6:*2499; Amitsur M, Morad I, Kaufmann G (1989) *EMBO J 8:*2411.
56. Zimmerman SB, Pheiffer BH (1981) *Enzymes 14:*315; Pheiffer BH, Zimmerman SB (1982) *BBRC 109:*1297.
57. Habraken Y, Verley WG (1983) *FEBS Lett 160:*46; Habraken Y, Verly WG (1988) *NAR 14:*8103.
58. Shuman S, Hurwitz J (1979) *JBC 254:*10396.
59. Tamura S, Teraoka H, Tsukada K (1981) *EJB 115:*449.

CHAPTER 10

DNA-Binding Proteins

10-1 Introduction

The many proteins that recognize and bind DNA embrace every aspect of cellular DNA function and chemistry (Table 10-1). They are discovered by their presence in isolated DNA complexes, by their binding to DNA in vitro, and by the need for them in DNA-dependent functions.

Proteins may affect the structure of DNA either with or without catalyzing the breakage of covalent bonds. Considered in this chapter are (1) the single-strand binding proteins (SSBs), which impart a regular structure to DNA single strands, a structure recognized and exploited by a variety of enzymes in replication, repair, and recombination; (2) the duplex-binding proteins that regulate the functions of the genomes and organize them for packaging; and (3) the covalent complexes that participate in a variety of DNA transactions. Other chapters deal with the energy-driven helicases (translocases and strand-displacing proteins) and some other DNA-dependent ATPases (Chapter 11); the topoisomerases that control DNA conformation (Chapter 12); the replication initiator proteins, which bind and activate the origins of chromosomes (Chapter 16); and the strand-exchange proteins responsible for homologous recombination (Chapter 21).

Domains are a recurring theme in the organization and operations of the DNA-binding proteins. Certain domains recognize DNA; others bind ATP, promote oligomerization, or interact specifically with other proteins. Sometimes, separated by other regions, discrete domains carry out the several actions of a multifaceted protein.

Table 10-1
DNA-binding proteins

Class	Function	Examples	Chapter-section reference
I. DNA structure, packaging	single-strand binding	T4 gp32	10-3
	DNA condensation	histones, HU, IHF	10-7,8
	unwinding (helicase)	Rep protein, dnaB protein	11-2
	untwisting	topoisomerase I	12-2
	twisting	gyrase	12-3
	chromosome organization	scaffolding protein	10-7
	sperm DNA packing	protamine	10-6
	virus DNA packing	core and capsid proteins	10-8
II. Replication, recombination, repair	chromosome initiation	dnaA protein, T antigen	16-3; 19-4
	DNA chain initiation	dnaB protein	11-2
	RNA primer formation	primase	8-3
	DNA polymerization	DNA polymerase III holoenzyme	5-3 to 6
		λ integrase	21-6
	site-specific recombination	RecA protein	21-5
	strand exchange	repair nucleases	13-9
	nucleotide excision	uracil N-glycosylase	13-9
	base excision	photolyase	21-2
	base repair	DNA ligase	9-1 to 5
	DNA joining		
III. Transcription	mRNA synthesis	RNA polymerase	7-3,10
	positive regulation	cAMP receptor	10-9
	negative regulation	*lac* repressor	10-9
IV. Degradation	digestion	pancreatic DNase	13-1,5
	restriction	restriction nucleases	13-6 to 8
	nucleotide salvage	endo-, exonucleases	13-2,3,5
V. Modification	DNA methylation	adenine methyl transferase	21-10
	DNA glycosylation	T4 β-glycosyl transferase	17-6; 21-10
VI. Transport	conjugal transfer	F plasmid TraI	18-6
	viral DNA entry	M13 gp3	17-3
	transformation	*H influenzae* receptor protein	21-7

DNA Site Selection[1]

It is unlikely that a site-specific binding protein generally finds its cognate DNA sequence by tediously testing all possible positions along the DNA. In many cases, the interaction involves three stages in which the binding becomes increasingly specific. (1) Territorial binding results from the interaction of cationic amino acids with the negatively charged cloud surrounding DNA.[2] This limits the protein to its substrate, where it moves freely and rapidly along the length of the DNA and may even switch strands without an expenditure of energy. (2) The rapid scan is slowed when the protein drops into a permissible site or interacts

1. Berg OG, Winger RB, von Hippel PH (1981) *Biochemistry 20*:6929; Steitz TA (1990) *Quart Rev Biophys 23*:205.
2. Manning GS (1978) *Q Rev Biophys 11*:179; Manning GS (1980) *Biopolymers 19*:37; Wilson RW, Bloomfield VA (1979) *Biochemistry 18*:2192.

preferentially with some conformational feature of DNA: a supercoiled domain, a single-stranded region, or an altered backbone structure, such as a bend. AT-richness, GC-richness, inverted repeats of DNA sequences, and looped-out or mismatched residues may form the basis of these conformational cues for binding. (3) With region-specific binding achieved, fewer interactions need to be tried to achieve site specificity. Direct interactions in the major or minor grooves may suffice, or binding to groups on a single strand may follow localized melting of the duplex.

RNA-Binding Proteins

An exception to the arbitrary exclusion of RNA subjects from *DNA Replication* is made for RNA polymerases (Chapter 7) and could also be made for RNA-binding proteins. Important principles can be learned about the proteins that form discrete particles[3] in addition to those that constitute ribosomes and viruses. One such protein from *Xenopus* binds both RNA and DNA.[4] It associates with 5S RNA to form a stable 7S particle that accumulates in young oocytes in anticipation of accelerated ribosome synthesis. The protein is also an essential positive transcription factor that binds specifically to the center of the gene encoding the 5S RNA (Section 7-11).

10-2 Single-Strand Binding Proteins (SSBs)[5]

SSBs are defined as binding proteins with a strong preference for DNA over RNA, and for ssDNA over duplex DNA. They bind tightly and cooperatively, and do not catalyze associated activities, such as the DNA-dependent ATPase activities of helicases and topoisomerases.

Studies of T4 phage mutants deficient in genetic recombination and replication suggested that the product of T4 gene *32* (T4 gp32) might be a key factor in these processes. The approach taken to search for gp32 in extracts of infected cells was to screen first for proteins with a great affinity for DNA. Chromatography on ssDNA–cellulose separates proteins with varying affinities for ssDNA.[6] Among some 20 such proteins, the most tenaciously bound is gp32, which is resistant to elution by the polyanionic dextran sulfate; a high salt concentration is required to detach gp32 from the DNA–cellulose column.

The Nature of SSB Binding to DNA

Gp32 is one of a widely distributed group of proteins (Sections 3 to 6) that convert duplex DNA to single strands at a temperature far below the normal midpoint of the melting transition (T_m) for the DNA. This is achieved by their tight, cooperative binding to ssDNA and relatively weak binding to dsDNA. These proteins are distinguished from helicases by not requiring the energy of ATP hydrolysis and

3. Picard B, Wegnez M (1979) *PNAS 76*:241; Lerner MR, Steitz JA (1981) *Cell 25*:298.
4. Pelham HRB, Brown DD (1980) *PNAS 77*:4170; Sakonju S, Brown DD, Engelke D, Ng S-Y, Shastry BS, Roeder RG (1981) *Cell 23*:665.
5. Chase JW, Williams KR (1986) *ARB 55*:103.
6. Alberts B, Herrick G (1971) *Meth Enz 22*:198.

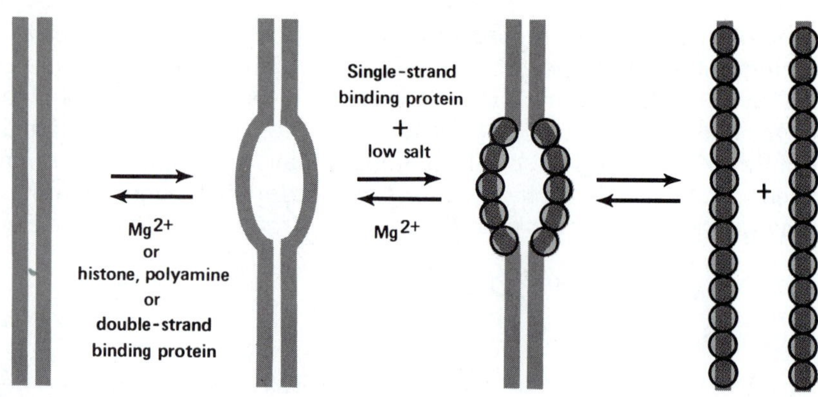

Figure 10-1
Single-strand binding protein displaces the equilibrium toward the single-stranded form; histones, polyamines, or double-strand binding proteins favor the duplex form.

by extensively coating the ssDNA, helicase II (UvrD) being an exception to the latter property.

Binding of an SSB in a region of duplex DNA depends on transient melting of the duplex. The energy of binding drives the melting to completion even at temperatures 40° below the normal T_m (Fig. 10-1). A denaturation map of phage λ similar to that produced with alkali or heating can be made with this type of protein.[7] On single strands, the binding involves the DNA backbone and shows clear base preferences but no sequence specificity. Thus, binding to easily melted (i.e., AT-rich) regions of the DNA duplex is preferred simply because there the energy cost of releasing single strands from the helix is least.

Proteins such as gp32 bind cooperatively to DNA strands. In other words, the individual protein molecules prefer to line up successively in close juxtaposition along a DNA strand (Fig. 10-2). Each protein molecule thus binds adjacent to another protein molecule, rather than to an unoccupied stretch of DNA. This occurs because of favorable protein–protein interactions and perhaps because

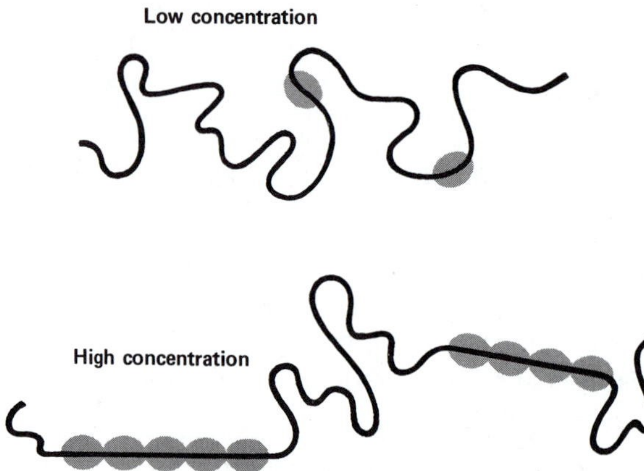

Figure 10-2
Cooperative binding of the single-strand binding protein to DNA. (Courtesy of Professor B Alberts)

7. Sigal N, Delius H, Kornberg T, Gefter ML, Alberts B (1971) *PNAS* 69:3537.

binding is easier to a polynucleotide extended by an attached protein. As a result, when single-stranded circles are mixed with sufficient gp32 to bind one-third of the DNA, then one-third of the circles, as seen in the electron microscope, are covered nearly completely with protein while the rest are nearly bare.[8]

Effects of SSB Binding to DNA

Preferential binding of the single-stranded backbone by a gp32-type protein, besides destabilizing a duplex, can also have the reverse effect of facilitating strand renaturation. By melting out persistent regions of secondary structure within the single strands, the SSB enables two complementary strands colliding at random to align their base pairs more readily. Thus, extended conformation of single strands induced by the protein increases the rate of reannealing of homologous strands by as much as 1000-fold under ionic conditions that favor stability of the duplex.[9]

Renaturation of DNA at 37° may take days to achieve because of the short, hairpin-like base-paired loops that form at many sites within a single strand of natural DNA. Strand recombination can occur within minutes when these duplex regions are melted at an elevated temperature or at 37°, either by the action of a base-pair destabilizing agent, such as formamide, or by an SSB.

The rate of replication by T4 polymerase can also be increased many-fold by the binding of gp32 to single-stranded template regions.[10] In this instance, the SSB probably functions in an analogous way to melt out secondary structures. Additional interactions with the DNA polymerase may be inferred from the specificity of an SSB for the polymerase of the same organism.

Nomenclature

These proteins were initially designated "unwinding" proteins, and, later, "melting" or "helix-destabilizing" or "single-strand binding" proteins. As a catalyst of renaturation, a gp32-type protein might be more appropriately designated a "winding" protein or "annealing" protein. Selecting the most suitable name for these proteins is troublesome, because it is not clear whether the cell is using the unwinding or the winding function, nor whether all the proteins perform one or both functions. Other types of single-strand binding proteins include the helicases and some topoisomerases. To designate the group of single-strand binding proteins represented by gp32 as "helix-destabilizing" proteins[11] might suggest binding to duplex DNA and may in some cases obscure a key role in renaturation. The name "single-strand binding protein" (SSB) for this group has the virtue of being most general,[12] and conforms to the *ssb* designation for the genetic locus of the *E. coli* SSB.[13]

8. Delius H, Mantell NJ, Alberts B (1972) *JMB 67*:341.
9. Christansen C, Baldwin RL (1977) *JMB 115*:441.
10. Huberman JA, Kornberg A, Alberts BM (1971) *JMB 62*:39.
11. Alberts B, Sternglanz R (1977) *Nature 269*:655.
12. Geider K (1978) *EJB 87*:617.
13. Meyer RR, Glassberg J, Kornberg A (1979) *PNAS 76*:1702; Glassberg J, Meyer RR, Kornberg A (1979) *J Bact 140*:14.

Properties of SSBs

The properties of an SSB can be summarized as follows:

(1) Specificity for binding ssDNA, with an indifference generally to base sequence.

(2) Cooperative binding. This cooperativity may be "unlimited," as in the formation of long protein clusters (e.g., by T4 gp32), or "limited," as in the pairing of *E. coli* tetramers to form octamers.

(3) Stoichiometric action. However, an SSB can be cycled in a "treadmill" fashion, going on and off the ends of an aggregate, as tubulin does in solution.[14] The limited supplies of the *E. coli* and T4 phage proteins must be used in this "catalytic" fashion, as opposed to the relatively stoichiometric use of the large depots of M13 phage and adenoviral SSBs that coat entire genomes.

(4) Coating with protein at a weight ratio to nucleic acid that depends on the SSB. The chain may have an extended or condensed conformation when visualized in the electron microscope, influenced by salt (type and concentration) and by other factors. The coated backbone resists nucleases, leaving the bases unpaired and interactive. Thus, certain SSBs may facilitate renaturation and stimulate replication by a homologous polymerase.

(5) Distinctive properties. For example, T4 gp32, M13 gp5, and *E. coli* SSB each have unique aspects to the way in which they bind DNA and interact with catalytically active proteins. These interactions, in turn, influence the nature and kinetics of DNA binding. (These three well-characterized prokaryotic SSBs are compared in Table 10-2 and will be considered in the sections that follow.)

Table 10-2
Properties of three prokaryotic SSBs

Property	T4 gp32	M13 gp5	*E. coli* SSB
Binding			
DNA > RNA	yes	yes	yes
ss >> double strand	yes	yes	yes
DNA strands in protein complex	one	two	one
Nucleotides bound per monomer	10	4	8–16
Weight ratio, protein/DNA	12	8	4–8
Cooperativity	unlimited	unlimited	unlimited and limited
Biological functions	stimulates replication and combination	inhibits complementary strand synthesis	stimulates replication and repair
Copies per cell	10,000	75,000	800
Copies per replication fork	170	—	270
Isoelectric pH	5.5	—	6.0
Mass, native (kDa)	33.5	19.4	75.6
Native form	monomer	dimer	tetramer

14. Hill TL, Tsuchiya T (1981) *PNAS 78*:4796.

(6) Kinetics of dissociation from DNA. This can be as important in the overall function of an SSB as the kinetics of binding to the DNA, and may rely even more on the auxiliary actions of other proteins and competitive DNAs.[15]

(7) Fine regulation of cellular levels for optimal function.

10-3 T4 Phage Gene *32* Protein (Gp32)[16]

Gp32, the first discovered and one of the most studied of the SSBs (Table 10-2), is a single polypeptide of 301 residues with a mass of 33.5 kDa, as deduced from its gene sequence.

Binding[17]

The affinity of gp32 for ssDNA can be as much as 10^4 or 10^5 times greater than for native, duplex DNA. The basic unit of the single-strand structure that can associate with a single binding site of the protein is a dinucleoside monophosphate.[18] Affinity increases with oligonucleotide size to a length of five to seven residues, which span the binding site. Binding is independent of base sequence; RNA polymers are bound much less effectively.

The crux of binding by gp32, in addition to strong selectivity for single strands, is its striking cooperativity.[19] Whereas a single gp32 molecule can bind to duplex DNA and to RNA, binding to ssDNA is stronger and, with cooperativity, attains the high specificity for ssDNA. The affinity for binding to a second protein molecule next to a first is increased 10^3-fold over binding to a bare region of DNA. The free energy gain specifically associated with one such neighbor–neighbor binding is about −4 kcal. Gp32 molecules will therefore preferentially bind to and fill all adjacent single-stranded sequences in their path, forming a continuous tract of protein. The wave will advance until it encounters a region of secondary structure, such as a large hairpin loop, with net stability exceeding the protein binding energies. At this point, binding will begin on a new bare region of DNA.

As the total energy involved in gp32 binding inside a cell is about −1 to −2 kcal per base pair, hairpin loops, long and nearly perfectly matched, are left unmelted in ssDNA unless other factors supervene to affect the nature of the DNA or the properties of gp32. The tendency of gp32 to aggregate at high concentration[20] is thought to influence its cooperative behavior, but the significance of observations made in the absence of polynucleotides remains uncertain.

When gp32 binds to a single-stranded region in a duplex, the polynucleotide backbone in the duplex is appreciably distorted. The bases appear to be stacked, according to some circular dichroism measurements,[21] and unstacked, according to UV hyperchromicity.[22] Presumably, some regularly arranged structure is

---

15. Lohman TM (1984) *Biochemistry 23*:4665.
16. Chase JW, Williams KR (1986) *ARB 55*:103.
17. Lohman TM (1984) *Biochemistry 23*:4656; Lohman TM (1984) *Biochemistry 23*:4665.
18. Kelly RC, Jensen DE, von Hippel PH (1976) *JBC 251*:7240.
19. Jensen DE, Kelly RC, von Hippel PH (1976) *JBC 251*:7215.
20. Carroll RB, Neet K, Goldthwait DA (1975) *JMB 91*:275.
21. Greve J, Maestre MF, Moise H, Hosoda J (1978) *Biochemistry 17*:887.
22. Jensen DE, Kelly RC, von Hippel PH (1976) *JBC 251*:7215.

formed that responds differently to the two types of measurement. In the electron microscope, a DNA circle coated with the protein appears elongated. The DNA chain is extended to 4.6 Å per nucleotide, compared to 3.4 Å for a strand in the B form of a duplex in solution.[23]

Structure

Functionally discrete domains in gp32 are responsible for binding to DNA, cooperative interaction between bound molecules, aggregation of unbound protein, and interactions with functionally related proteins. Mutations at various sites in gene *32* and proteolytic cleavages of gp32 have made it possible to map these regions in the molecule.[24] The basic N-terminal domain (up to residue 21) contains essential sites for self-aggregation and cooperative binding to DNA; the acidic C-terminal domain (starting from residues 253 to 275) is needed to catalyze the annealing of homologous strands and for interactions that generate the replication complex; gp32 that lacks both terminal domains still binds DNA.[25] Susceptibility to tryptic hydrolysis of these regions when free or bound to DNA, as well as physical studies, suggests a model in which monomers line up on DNA with their N-terminal regions overlapping, leaving other regions free to destabilize the helix and interact with recombination, repair, and replication protein complexes.

Gp32 within its DNA-binding core domain contains 1 mol of Zn(II) tightly complexed with three cysteines and one histidine as ligands,[26] resembling the zinc fingers of regulatory proteins (Section 9). The zinc is an important conformational element that affects the cooperative binding to DNA.

Replication

Gp32 is required throughout replication. It is not as catalytic as the phage replication enzymes, being needed in amounts stoichiometric with the nascent DNA. Unlike an enzyme, the binding protein is a reactant that changes both the rate and equilibrium of the reaction. Thus the protein has a structural function, though it is not incorporated into the mature phage. By binding to template strands at the replicating fork, the protein may effect local unwinding, extend the template into a conformation optimal for base pairing and replication, and protect the single strands against nucleases (Fig. 10-3).

About 60 replication forks maintain the very rapid rate of T4 DNA replication in the infected cell. Since an infected cell contains about 10,000 gp32 molecules, there would be 170 molecules per replicating fork were all of them engaged in replication. An assembly of binding-protein molecules at a replicating fork could serve as a moving scaffold and be used continuously until each DNA molecule is completed. The amount of gp32 would thus limit the rate of DNA production by determining the number of forks in use at one time.

The salt concentration and levels of Mg^{2+} in the cell would favor renaturation by gp32 rather than melting of the double helix in advance of the replicating

---

23. Delius H, Mantell NJ, Alberts B (1972) *JMB 67*:341.
24. Breschkin AM, Mosig G (1977) *JMB 112*:279; Breschkin AM, Mosig G (1977) *JMB 112*:295; Mosig G, Luder A, Garcia G, Dannenberg R, Bock S (1978) *CSHS 43*:501.
25. Pan T, King GC, Coleman JE (1989) *Biochemistry 28*:8833.
26. Giedroc DP, Keating KM, Williams KR, Konigsberg WH, Coleman JE (1986) *PNAS 83*:8452; Giedroc DP, Johnson BA, Armitage IM, Coleman JE (1989) *Biochemistry 28*:2410.

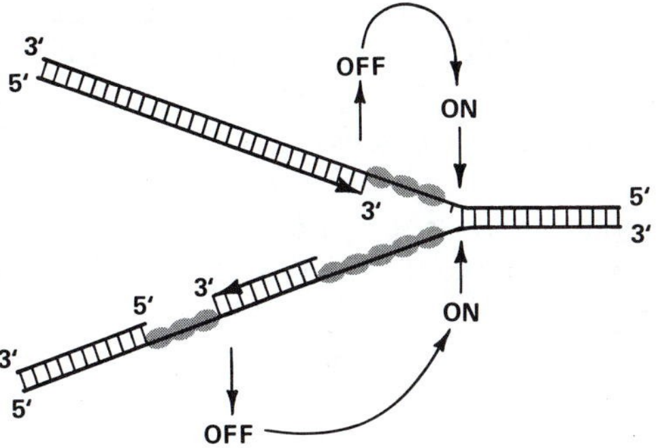

Figure 10-3
Hypothetical scheme for single-strand binding protein action at a replicating fork. The protein is recycled after binding single-stranded regions of the template and facilitating replication. (Courtesy of Professor B Alberts)

fork. To facilitate winding and unwinding, other replication proteins and factors are clearly needed (e.g., the helicase activity of the gp41/61 complex). Direct interaction between gp32 and T4 polymerase is known; gp32 specifically stimulates this polymerase, but not others (such as *E. coli* pol I and phage T7 polymerase), in replication of a single-stranded template.[27] It also enables T4 DNA polymerase to operate at a nick in a duplex structure (Fig. 10-4 and Section 5-8), presumably by a strand-displacing action analogous to that of *E. coli* pol I (Section 4-13).

In a reconstituted replication mixture which includes gene products 41 (helicase), 43 (polymerase), and 44, 45, and 62 (polymerase accessory factors), gp32 is essential for any synthesis on duplex templates (Section 17-6).

Recombination and Protection against Nucleases

Little can be said about the molecular details of how the binding properties of gp32 account for its indispensability in recombination. Aside from lowering the melting temperature of DNA, the major function may be the tight and cooperative binding to single-stranded recombination intermediates, including specific interactions with phage proteins (e.g., UvsX and Dda). It may stabilize and

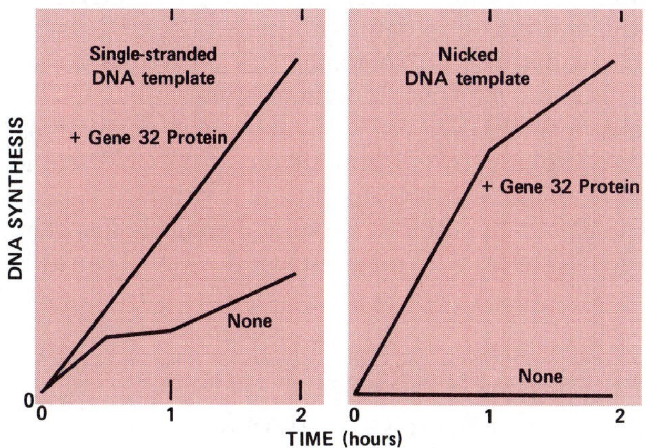

Figure 10-4
T4 phage gene 32 protein stimulates T4 DNA polymerase (left) in replication of a single-stranded template and (right) enables the polymerase to utilize a nicked duplex template.

27. Huberman JA, Kornberg A, Alberts BM (1971) *JMB 62*:39.

protect these active regions of recombination and replication against nucleases that specifically attack them. Gp32 protects ssDNA against many nucleases in vitro. Also, it is inferred from infections with mutant phages that gp32 moderates the degradation of T4 DNA by host RecBCD nuclease and by nucleases[28] controlled by T4 genes *46* and *47*. T4 DNA extracted from cells infected with mutants in gene *32* is more degraded, and recombination intermediates lack the single-stranded regions, termini, and branched structures associated with gp32 functions.

Autoregulation of Supply[29]

The quantity of gp32 determines the rate and amount of replication and recombination. This vital supply is regulated at the level of translation. When the quantity of gp32 exceeds what is needed for the ssDNA, excess protein then covers the mRNA for gp32 and shuts off synthesis of the protein.

The key feature in this model of self-regulation is the titration by binding a series of single-stranded nucleic acid regions by gp32: ssDNA first, followed by the mRNA for gp32, which specifically inhibits its translation.[30] Recognition of gp32 mRNA most likely depends on a unique secondary structure rather than on its base sequence.

10-4 M13 Phage Gene *5* Protein; Other Phage and Plasmid SSBs[31]

M13 Gp5

The protein (Table 10-2) encoded by gene *5* of M13 phage (and by closely related Ff phages) is essential for the production of viral single strands from the duplex replicative form, the last stage in DNA replication (Section 17-3). Unlike the stimulatory functions of T4 gp32 and *E. coli* SSB in replication, recombination, and repair, the role of gp5 is to interrupt the synthesis of the duplex replicative forms and to prepare the DNA for packaging into the virus particle in the membrane.

Gp5 is the SSB for which the most detailed structural information is available. The protein in solution is a dimer of 9.69-kDa subunits. It binds cooperatively to single strands and reduces the melting temperature of duplex DNA.[32] It is not apparent from the amino acid composition and sequence,[33] as it is with histones, for example, that the protein is a nucleic acid–binding protein.

Gp5 molecules coat the entire viral DNA circle of about 6400 nucleotides, each dimer spanning 8 ntd[34] and holding two distant regions of the DNA strand in juxtaposition (Plates 6 and 7), so that the DNA folds back, hairpin-like, in an antiparallel manner. The structure of the dimeric protein of phage fd has been solved to 2.3-Å resolution, and cocrystals with oligonucleotides have been ana-

---

28. Mosig G, Bock S (1976) *J Virol 17*:756.
29. Gold L, O'Farrell PZ, Russel M (1976) *JBC 251*:7251; Russel M, Gold L, Morrissett H, O'Farrell PZ (1976) *JBC 251*:7263; Krisch HM, van Houwe G, Belin D, Gibbs W, Epstein RH (1977) *Virology 78*:87.
30. Gold L (1988) *ARB 57*:199.
31. Chase JW, Williams KR (1986) *ARB 55*:103; Coleman JE, Oakley JL (1980) *Crit Rev B 7*:247; Kowalczykowski SC, Bear DG, von Hippel PH (1981) *Enzymes 14*:373.
32. Sang B-C, Gray DM (1987) *Biochemistry 26*:7210.
33. Nakashima Y, Dunker AK, Marvin DA, Konigsberg W (1974) *FEBS Lett 40*:290.
34. Pretorius HT, Klein M, Day LA (1975) *JBC 250*:9262; Cavalieri SJ, Neet KE, Goldthwait DA (1976) *JMB 102*:697.

lyzed at even higher resolution.[35] The unliganded protein has a cleft 30 Å long that qualifies in size, shape, and amino acid composition as the DNA binding site. Arrayed along the external edges of the groove are the aromatic residues that can stack on the nucleotide bases; binding of a tyrosine residue has also been observed by NMR.[36] Basic amino acids, including specific lysine residues,[37] in the interior of the cleft may attract and bind the DNA backbone. The protein secondary structure consists solely of antiparallel β sheets, one of which may participate in the interactions for cooperative binding of gp5 to DNA. The nucleotide spacing when bound to the protein is 4.5 Å, nearly the same as that for the T4 gp32·DNA complex. Binding to dsDNA [poly dA·poly dT and poly d(AT)] has also been detected.[38]

About 100 viral single strands can be isolated from the infected cell as nucleoprotein complexes,[39] yielding about 75,000 dimeric gp5 molecules per cell. Essentially all the viral DNA and well over half the gp5 molecules are contained in these complexes. M13 mutants with a temperature-sensitive defect in the protein do not produce viral strands, but accumulate the duplex replicative form instead.

Based on its properties, gp5 can serve in several ways: in packaging the viral DNA by inhibiting the use of viral strands as templates that would lead to more replicative duplexes; by protecting the viral strands against nucleases and nonproductive binding by RNA polymerase or other DNA-binding proteins; and by conferring on the viral strands the conformation required for insertion into the cell membrane for encapsidation by coat protein.

Phage T7, T5, ϕ29, N4, and Plasmid SSBs

SSBs are employed in the replication of the linear duplex genomes of phages T7, T5, ϕ29, and N4. The gene for the T7 SSB, located between genes *2* and *3*, is named *2.5*. An amber mutant, whose binding protein is 28 kDa instead of 32 kDa, is defective in replication, especially in a host with a mutant SSB. Host SSB stimulates replication by increasing the affinity of T7 DNA polymerase for the coated template. Gp2.5 increases the frequency of initiating discontinuous-strand synthesis tenfold but has little effect on the continuous strand.[40] A deficiency in DNA repair and recombination, more striking than that in replication, is observed with the T7 mutant even when the host is overproducing wild-type SSB. Thus, the host SSB appears to be inadequate as a sole substitute for the phage protein, especially in repair and recombination.

A binding protein encoded by phage T5 gene *D5* has a mass of 29 kDa and is produced in quantities near 2% of the cell protein. Cooperative binding to duplex DNA as well as noncooperative binding to ssDNA[41] suggests a requirement for this protein in viral DNA synthesis and transcription.

Replication of the *B. subtilis* phage ϕ29 relies on its gp5 to bind single strands[42]

35. McPherson A, Jurnak FA, Wang AHJ, Molineux I, Rich A (1979) *JMB 134*:379; McPherson A, Jurnak F, Wang A, Kolpak F, Rich A, Molineux I, Fitzgerald P (1980) *Biophys J 32*:155.
36. Garssen GJ, Tesser GI, Schoenmakers JGG, Hilbers CW (1980) *BBA 607*:361.
37. Dick LR, Sherry AD, Newkirk MM, Gray DM (1988) *JBC 263*:18864.
38. Sang BC, Gray DM (1987) *Biochemistry 26*:7210.
39. Weiner JH, Bertsch LL, Kornberg A (1975) *JBC 250*:1972.
40. Nakai H, Richardson CC (1988) *JBC 263*:9831.
41. McCorquodale DJ, Gossling J, Benzinger R, Chesney R, Lawhorne L, Moyer RW (1979) *J Virol 29*:322; Rice AC, Ficht TA, Holladay LA, Moyer RW (1979) *JBC 254*:8042.
42. Martin G, Lázara JM, Méndez E, Salas M (1989) *NAR 17*:3663.

in order to sustain elongation but not noticeably to assist either the protein-priming or the initial propagation of strands.

An SSB specific for stimulating the DNA polymerase of the coliphage N4 genome (72 kb) has the functional and physical characteristics of other SSBs.[43] Among the residues of the 31-kDa N4 protein are ten tyrosine residues that have their fluorescence quenched by binding to ssDNA; the binding site size is 11 ntd per monomer of SSB.

Proteins with SSB features are encoded in transmissible plasmids. An SSB associated with the F plasmid has considerable homology to that of *E. coli*.[44]

10-5 *E. coli* SSB[45]

Discovery of the T4 gp32 and M13 gp5 SSBs stimulated the successful search for an SSB in extracts of uninfected *E. coli*. These proteins (Table 10-2) were isolated by DNA–cellulose chromatography and characterized by their stimulation of the homologous DNA polymerases. For replicative conversion of phage ssDNA circles to the duplex form in vitro, *E. coli* SSB is essential. Fractionation of the replication proteins in a crude extract led to the fortuitous discovery of a factor, identified as *E. coli* SSB, that retains activity after boiling for several minutes and thus can be isolated easily in high yield.[46]

Structure and Properties[47]

The protein, a tetramer of 18.9-kDa subunits, product of the *ssb* gene, is extraordinarily stable. However, aggregation at high concentrations (>2 mg/ml) may destroy its capacity to support DNA polymerase activity on single strands; in addition, NaCl at 0.2 M inhibits binding. The protein binds single strands selectively and cooperatively. The cooperativity can be "limited," as in the pairing of two tetramers to form the octamer, or "unlimited" in forming long clusters. A polynucleotide at least four to six residues long is required for binding.[48]

The binding constant increases 1000-fold when $d(pT)_{16}$, which interacts with only one subunit of the tetramer, is replaced by $d(pT)_{35}$, a size sufficient to interact with two subunits; a stretch of 65 ntd is needed to fully wrap around the tetramer. Binding to natural DNA is far slower than to poly dT due to the presence of secondary structures.[49]

Largely indifferent to base sequence, SSB binds even the apurinic analog of $d(pCpA)_3$, yet the complex with oligo dT is far stronger than that with oligo dA. RNA is bound but less tightly than DNA. A complex between a ϕX174 ssDNA

43. Lindberg G, Kowalczykowski SC, Rist JK, Sugino A, Rothman-Denes LB (1989) *JBC 264*:12700.
44. Chase JW, Merrill BM, Williams KR (1983) *PNAS 80*:5480; Kolodkin AL, Capage MA, Golub EI, Low KB (1983) *PNAS 80*:4422; Golub EI, Low KB (1985) *J Bact 162*:235.
45. Chase JW, Williams KR (1986) *ARB 55*:103; Lohman TM, Bujalowski W, Overman LB (1988) *TIBS 13*:250; Bujalowski W, Lohman TM (1989) *JMB 207*:249; Bujalowski W, Lohman TM (1989) *JMB 207*:269; Meyer RR, Laine PS (1990) *Microbiol Rev 54*:342.
46. Weiner JH, Bertsch LL, Kornberg A (1975) *JBC 250*:1972.
47. Meyer RR, Glassberg J, Kornberg A (1979) *PNAS 76*:1702; Glassberg J, Meyer RR, Kornberg A (1979) *J Bact 140*:14; Meyer RR, Glassberg J, Scott JV, Kornberg A (1980) *JBC 255*:2897.
48. Ruyechan WT, Wetmur JG (1976) *Biochemistry 15*:5057; Bandyopadhyay PK, Wu C-W (1978) *Biochemistry 17*:4078.
49. Urbanke C, Schaper A (1990) *Biochemistry 29*:1744.

circle and about 150 SSB molecules appears stable in the presence of unbound ssDNA, an indication of cooperative binding; nevertheless, free and bound SSB molecules exchange rapidly.

The extent of DNA bound by SSB depends strikingly on the ionic environment.[50] The site size rises from 35 to 65 ntd per tetramer as salt, Mg^{2+}, or polyamine concentrations are increased, reflecting the increase from two to four in the number of subunits that interact with the DNA. At low ratios of SSB to DNA, a beaded structure is observed in the electron microscope; with the DNA wrapped around the tetramer (or octamer), the contour length is reduced by 75%.[51] With higher ratios of SSB to DNA, the morphology of the coated strand is smoother and more extended, and the compaction of the contour length is less extreme.[52] As with nucleosomes, uncoated regions can be revealed by digestion with a small endonuclease (e.g., DNase I).

Single strands can be distinguished from double strands in the electron microscope only with difficulty, but when coated with SSB, the single strands stand out in micrographs as thick cords. Contrast can be enhanced further by exposing the protein–DNA complex to antibody directed against SSB.

The organization of domains within the protein resembles that of T4 gp32. The N-terminal region (residues 1 to 105) is responsible for binding DNA. The remainder of the molecule (residues 106 to 165) contributes to modulating the DNA binding, the helix destabilization, and interactions with other proteins in DNA transactions.

The SSB regains its tetrameric form and function after being heated at 100°. The *ssb-1* mutation, a temperature-sensitive lesion in which tyrosine is substituted for histidine 55, favors dissociation of the active tetrameric form of SSB into monomers. After exposure to 100° for 10 minutes, the mutant protein retains its replicative functions in assays at 30° but is virtually inactive at 42°. The essential tetrameric conformation sustained at 30° is evidently lacking in the mutant protein at 42°; despite its structural resistance to recovery after heating at 100°, it is functionally thermosensitive within a narrow temperature range.

Functions in Vitro

SSB performs a number of replicative, renaturing, and protective functions that are readily demonstrable in vitro. The protein:

(1) Contributes to the opening and unwinding of a duplex at an origin of replication in a supercoiled DNA, as with *oriC* of the *E. coli* chromosome, or at the nicked origin of rolling-circle replication, as with ϕX174 phage DNA.

(2) Directs priming of DNA synthesis to specific loci (origins) by covering single-stranded regions (e.g., on phage M13, G4, and ϕX174 DNA) which might otherwise be used.

(3) Sustains unwinding of duplex DNA by helicase actions at replication forks and loci of repair.

50. Lohman TM, Overman LB (1985) *JBC 260*:3594; Bujalowski W, Lohman TM (1986) *Biochemistry 25*:7799; Bujalowski W, Overman LB, Lohman TM (1988) *JBC 263*:4629.
51. Chrysogeolos S, Griffith J (1982) *PNAS 79*:5803.
52. Griffith JD, Harris LD, Register J (1984) *CSHS 49*:553.

(4) Improves fidelity of and elongation by *E. coli* pol II and pol III holoenzyme, but not pol I. SSB stimulates the T7 polymerase by enhancing the affinity for the template-primer of the enzyme at limiting concentrations;[53] gp32 stimulation of T4 DNA polymerase cannot be replaced by SSB.

(5) Renatures denatured DNA. With spermidine at the intracellular concentration of 2 mM, the renaturation rate is increased 5000-fold. The rate diminishes with decreasing chain length to the point at which no catalysis is observed in the joining of the 12-ntd cohesive ends of phage λ.

(6) Inhibits both the endo- and exonucleolytic activities of nucleases, as in contributing to the conversion of RecBCD nuclease into an ATP-driven helicase that generates the long single-stranded intermediates in recombination.

(7) Participates in recombination and recombinational repair by RecA protein. By coating the single strands, SSB spares RecA protein and effects its efficient use,[54] perhaps interacting directly with RecA protein as well; the mutant SSB (*ssb-1*) is inactive.

In addition, a complex of SSB with the primosomal PriB protein[55] (formerly called n protein) and interactions with other proteins, like that observed between T4 gp32 and DNA polymerase, may be important in the contributions made by *E. coli* SSB to a variety of functions in *E. coli*.

Functions in Vivo

Replicative functions anticipated from in vitro studies have been confirmed with thermosensitive (*ts*) *ssb* mutants, particularly *ssb-1* and *ssb-113* (formerly *lexC-113*).[56] DNA synthesis stops abruptly when the temperature is raised to 42°, and the cells soon die. The mutants are also UV-sensitive, fail in the SOS response to DNA damage, and are defective in recombination. Strangely, the *ssb-113* mutant, with an SSB that seems to behave normally in vitro, displays repair and recombination deficiencies in vivo, even at a permissive growth temperature.

There are only about 300 (tetrameric) molecules of SSB in the cell, compared to 10,000 of gp32 in the T4-infected cell. However, when the number of replication forks is taken into account, both proteins are probably present in equal amounts relative to the ssDNA at a fork. Each protein is present at a level sufficient to cover about 1400 ntd of nascent ssDNA at each replication fork.

10-6 Eukaryotic SSBs[57]

In view of the essential functions performed by the prokaryotic SSBs in replication, repair, and recombination, one might anticipate that comparable proteins play similar roles in these processes in eukaryotic cells. Yet discoveries of such SSBs have been limited, perhaps due to the abundance and prominence of the

53. Myers TW, Romano LJ (1988) *JBC 263*:17006.
54. Cox MM, Lehman IR (1987) *ARB 56*:229; Chow SA, Rao BJ, Radding CM (1988) *JBC 263*:200.
55. Low RL, Shlomai J, Kornberg A (1982) *JBC 257*:6242.
56. Johnson BF (1977) *MGG 157*:91.
57. Chase JW, Williams KR (1986) *ARB 55*:103.

histones in the distinctive nucleosomal organization of eukaryotic genomes. The most convincing examples of SSBs are those disclosed by viral infections.

Adenovirus SSB (Ad DBP)[58]

Inasmuch as replication of the linear duplex genome of adenoviruses proceeds by way of single-stranded intermediates, the essential role of an SSB in this process seemed likely. The 59-kDa polypeptide encoded by the virus (initially considered to be 72 kDa because of its anomalous electrophoretic mobility) is essential for replication in vivo, as demonstrated with thermosensitive mutants, and for reconstitution of the initiation and elongation of adenoviral DNA in vitro.[59]

Ad SSB (also known as Ad DBP, for *a*denovirus *D*NA-*b*inding *p*rotein) is the most abundant of the early adenoviral proteins, with 5×10^6 molecules per cell. It binds tightly, cooperatively, and stoichiometrically to ssDNA and resembles T4 gp32 in its structure and functions. The binding site covers about 7 ntd and extends the chain. The C-terminal domain, defined by proteolytic cleavage and including about two-thirds of the protein, binds DNA[60] and is sufficient for DNA replication. (In prokaryotic SSBs, the N-terminal portion is the DNA-binding domain.) A region in which temperature-sensitive mutations that affect replication and ssDNA binding have been mapped contains a zinc finger motif (Section 9).[61] The function of the N-terminal domain is uncertain, as is the significance of phosphorylation of some 9 to 11 serine and threonine residues.

Unlike the prokaryotic SSBs, AdSSB binds to the termini of duplex DNA and does not lower the thermal melting transition of poly d(AT), properties which may account for a role in the initiation of replication of the adenoviral genome (Section 19-4). Ad SSB may also affect gene expression, regulating its own synthesis, perhaps like T4 gp32.

Herpesvirus SSB (ICP8)[62]

The herpes-encoded SSB (originally designated infected-cell protein 8, or ICP8, and also known as DBP; Section 19-5) is essential for viral DNA replication.[63] Thermosensitive mutants stop DNA synthesis promptly at a restrictive temperature. The abundant 128-kDa protein binds ssDNA tightly and cooperatively, much like *E. coli* SSB.[64] Surprisingly, the herpesvirus SSB is not as effective as *E. coli* SSB in enabling the DNA polymerase to replicate an extensively single-stranded template.[65] Presumably, additional nuclear factors are needed to supplement its action.

Cellular SSBs

One of the elusive cellular SSBs has been revealed by fractionation of host proteins needed for replication of SV40 and related small DNA tumor viruses in

---

58. Klessig DF, Quinlan MP (1982) *J Mol Appl Genet 1*:263; van Amerongen H, van Grondelle R, van der Vliet PC (1987) *Biochemistry 26*:4646.
59. Stillman B (1989) *Ann Rev Cell Biol 5*:197; Challberg MD, Kelly TJ (1989) *ARB 58*:671.
60. Neale GAM, Kitchingman GR (1989) *JBC 264*:3153.
61. Vos HL, van der Lee FM, Reemst AMCB, van Loon AE, Sussenbach JS (1988) *Virology 163*:1.
62. Quinn JP, McGeoch DJ (1985) *NAR 13*:8143.
63. Weller SK, Lee KJ, Sabourin DJ, Schaffer PA (1983) *J Virol 45*:354.
64. Ruyechan WT (1983) *J Virol 46*:661.
65. O'Donnell ME, Elias P, Funnell BE, Lehman IR (1987) *JBC 262*:4260.

extracts of human cells.[66] One such protein is RF-A (human SSB; also called RP-A and protein A), made up of three tightly associated polypeptides of 70–76, 32–34, and 11–14 kDa. The purified protein binds tightly to ssDNA and is required for the helicase action of the virus-encoded T antigen (Section 19-4) in opening the SV40 origin. Although *E. coli* SSB can substitute partially in unwinding parental DNA, the more complex RF-A contributes essential activities by stimulation of DNA polymerases in the progress of the replication forks.

Several SSB-like proteins have been isolated from calf thymus and various tissue culture cells by fractionation on ssDNA–cellulose columns. The predominant species, and the one that most resembles T4 gp32 and *E. coli* SSB, is the 24-kDa protein UP1.[67] Like the prokaryotic SSBs, UP1 binds ssDNA 1000 times more tightly than dsDNA, shows some preference for DNA over RNA, destabilizes poly d(AT), and stimulates a replication protein, the activity of DNA polymerase α. However, UP1 differs in the heterogeneity of its size and charge, and shows no tendency to aggregate or bind cooperatively, nor does it stimulate the renaturation of duplex DNA.

With the later recognition that UP1 is most probably a proteolytic product, corresponding to the N-terminal 195 amino acids of the 34-kDa A_1 heterogeneous nuclear ribonucleoprotein (hnRNP) and lacking the glycine-rich C termini, its status as an SSB is dubious.[68] Despite the probably artifactual nature of UP1, its structure and properties are instructive in comparing nucleic acid–binding proteins. For both prokaryotic SSBs and eukaryotic hnRNPs, the binding of single-stranded nucleic acids transiently stabilizes certain extended structures that are essential to the nucleic acid functions. To achieve this, UP1 and A_1, like gp32, employ repeated domains rich in aromatic amino acid sequences for binding DNA. Unlike UP1, the A_1 hnRNP does bind cooperatively to nucleic acid[69] and has a powerful annealing activity.[70]

Among other SSB candidates derived from calf thymus are two ssDNA-binding proteins of 48 and 61 kDa.[71] They stimulate the activity of DNA polymerase α 100-fold, apparently by blocking nonproductive binding sites on ssDNA and thereby directing the polymerase to associate with the primer terminus.

A yeast SSB of 34 kDa[72] stimulates a strand-exchange protein, much as *E. coli* SSB does its cognate RecA protein. Still other SSBs identified in yeast are less well defined in function.[73] One, which stimulates yeast polymerase I, has proved to be a cytosolic RNA-binding protein.[74] An SSB of 20 kDa is mitochondrial in location.

A DNA-binding protein from the lily plant (*Lilium*) is found only in the germ cells and is presumed to be active in meiotic recombination.[75] A similar protein is also present in rat spermatocytes.[76] The lily protein complexes with ssDNA

---

66. Wobbe CR, Weissbach L, Borowiec JA, Dean FB, Murakami Y, Bullock P, Hurwitz J (1987) *PNAS 84*:1834; Fairman MP, Stillman B (1988) *EMBO J 7*:1211; Wold MS, Kelly T (1988) *PNAS 85*:2523; Wold MS, Weinberg DH, Virshup DM, Li JJ, Kelly TJ (1989) *JBC 264*:2801; Kenny MK, Lee S-H, Hurwitz J (1989) *PNAS 86*:9757.
67. Herrick G, Alberts B (1976) *JBC 251*:2124; Herrick G, Alberts B (1976) *JBC 251*:2133.
68. Valentini O, Biamonti G, Pandolfo M, Morandi C, Riva S (1985) *NAR 13*:337; Merrill BM, LoPresti MB, Stone KL, Williams KR (1986) *JBC 261*:878.
69. Cobianchi F, Karpel RL, Williams KR, Notario V, Wilson SH (1988) *JBC 263*:1063.
70. Pontius B, Berg P, (1990) *PNAS 87*:8403.
71. Sapp M, Konig H, Riedel HD, Richter A, Knippers R (1985) *JBC 260*:1550.
72. Heyer W-D, Kolodner RD (1989) *Biochemistry 28*:2856; Hamatake RK, Dykstra CC, Sugino A (1989) *JBC 264*:13336.
73. Jong AY-S, Campbell JL (1986) *PNAS 83*:877.
74. Burgers PMJ (1989) *Prog N A Res 37*:235.
75. Hotta Y, Stern H (1979) *EJB 95*:31.
76. Mather J, Hotta Y (1977) *Exp Cell Res 109*:181.

even at 2 M NaCl and requires Mg^{2+} or Ca^{2+} for binding. Upon phosphorylation by a cAMP-dependent protein kinase, the lily protein will catalyze the melting of duplex DNA and the reannealing of ssDNA.

10-7 Histones and Chromatin[77]

Histones

Histone proteins[78] bind and stabilize DNA in the duplex state but also bind ssDNA. They are found firmly associated with DNA in the eukaryotic cell. Because histones contain a high proportion of the basic amino acids lysine and arginine, they and polyamines like spermine and spermidine can neutralize the phosphate groups in DNA and contribute to the stability and folding of the double helix. Whatever other functions histones may prove to have in the regulation of replication and transcription, they are certainly fundamental to the structure of chromosomes in nuclei and to the organization of replicative forms of infecting viruses.

Five kinds of histone (H1, H2A, H2B, H3, and H4) are found almost invariably in the chromosomes of eukaryotes (Table 10-3). H3 and H4, relatively arginine-rich histones, are strikingly similar in different tissues and in diverse species. A most remarkable conservation of amino acid sequence is found in H4, which is identical in the cow, rat, and pig, and differs in pea seedlings in only two residues: a valine for an isoleucine and a lysine for an arginine.

This unusual conservation of sequence through widely diverging lines of evolution most likely developed from a structural requirement for multiple precise contacts between H3 and H4 and DNA and other histones, as well as with the replication and transcription proteins. Comparable evolutionary stability is found in the amino acid sequences of the structural proteins actin[79] and tubulin.[80]

Posttranslational modifications introduced by methylation, acetylation, phosphorylation, and ADP ribosylation result in diversities in histones, but the significance of these modifications for the function of histones is still unknown.

Table 10-3
Properties of calf thymus histones

Name	Ratio of lysine to arginine	Total residues	Mass (kDa)
H1	20	213	21
H2A	1.25	129	14.0
H2B	2.5	125	13.8
H3	0.72	135	15.3
H4	0.79	102	11.3

77. Kornberg RD (1977) *ARB 46*:931; Kornberg RD, Klug A (1981) *Sci Amer 244*:52; Wassarman PM, Kornberg RD (eds.) (1989) *Meth Enz 170*. Academic Press, New York.
78. Huberman JA (1973) *ARB 42*:355; Isenberg I (1979) *ARB 48*:159.
79. Elzinga M, Collins JH, Kuehl WM, Adelstein RS (1973) *PNAS 70*:2687.
80. Luduena RF, Woodward DO (1973) *PNAS 70*:3594.

DNA Renaturation by Histones[81]

Either singly or in combination, histones from several sources promote the renaturation of single strands of denatured DNA. The reaction can be completed in a minute and involves the histones in a stoichiometric rather than catalytic way. Since the reaction can be observed even with crude extracts of *Drosophila* embryos, it should provide a functional and sensitive assay for histones and other proteins that bind preferentially to duplex DNA.

Chromatin[82]

The organization of DNA by histones in chromatin is crucial in the duplication of DNA, its segregation, and its mode of expression in the cell cycle and during development. Inasmuch as histones appear to be absent from prokaryotes, the distinctive features of eukaryotic DNA replication (such as slower movement of the replication fork, smaller replication fragments, and limitation to only part of the cell cycle) may be associated with the histones. The discovery of the remarkable nucleosome structure of chromatin[83] provides a starting point from which these differences may be better understood.

A chromatin fiber is now known in fine detail to be a string of nucleosome beads (Fig. 10-5). A *nucleosome* consists of a set of eight histones about which 200 base pairs of DNA are wrapped. The set of histones comprises two each of proteins H2A, H2B, H3, and H4. The DNA content of nucleosomes varies, ranging from about 160 base pairs in fungi to about 240 in sea urchin sperm. Marked differences in histone packing are found even within a given tissue of a species: in the rabbit brain cortex, neuronal cells have 162 base pairs; glial cells, 197. Yet the same amount of DNA is organized in nucleosomes assembled in vitro, regardless of the origin of the histones.[84]

Digestion of chromatin with certain nucleases results in a *core particle* containing a length of protected DNA that is constant, about 140 bp, in the nucleosomes of all tissues and species. The remainder of the DNA in a nucleosome (20 to 100 bp), which is comparatively accessible to digestion, serves as a *linker*

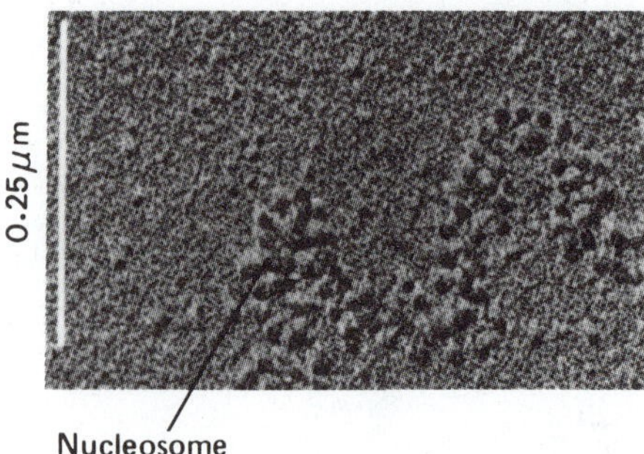

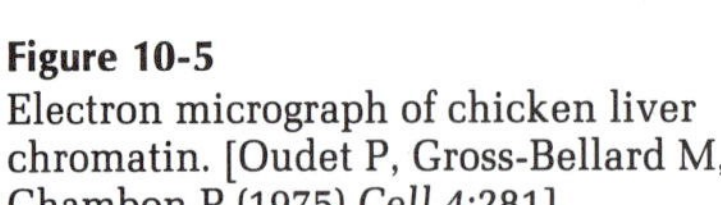

Figure 10-5
Electron micrograph of chicken liver chromatin. [Oudet P, Gross-Bellard M, Chambon P (1975) *Cell* 4:281]

81. Cox MM, Lehman IR (1981) *NAR* 9:389.
82. Kornberg RD (1977) *ARB* 46:931; Kornberg RD (1977) *Chromatin* CSHS, p. 42; Kornberg RD, Klug A (1981) *Sci Amer* 244:52; McGhee JD, Felsenfeld G (1980) *ARB* 49:1115; van Holde KE (1989) *Chromatin*, Springer-Verlag, New York; Wassarman PM, Kornberg RD (eds.) (1989) *Meth Enz* 170. Academic Press, New York.
83. Kornberg RD (1974) *Science* 184:868.
84. Spadafora C, Oudet P, Chambon P (1978) *NAR* 5:3479.

between cores. Another histone, H1, is associated with this linker region but its location is uncertain. The H1 attachment is apparently flexible enough to allow the higher orders of coiling and folding that must occur in the condensed states of chromosomes.[85]

Three-Dimensional Arrangement of Histones in the Nucleosome[86]

Crystals of a nucleosome core particle containing a defined 146-bp DNA sequence have been examined by x-ray diffraction to a resolution of 3 to 5 Å. Histone–DNA cross-linking, neutron scattering (Fig. 10-6), and electron microscopy have added further information. The DNA core length of 140 bp is coiled in a shallow helix with a pitch (axial repeat) of 28 Å and a diameter of about 90 Å, making 1.75 turns around the histones (Fig. 10-7). The DNA remains in the B conformation, but the periodicity of the double helix is reduced from 10.4 bp per turn (the value for DNA in solution; Section 1-3) to about 10.0. This change in

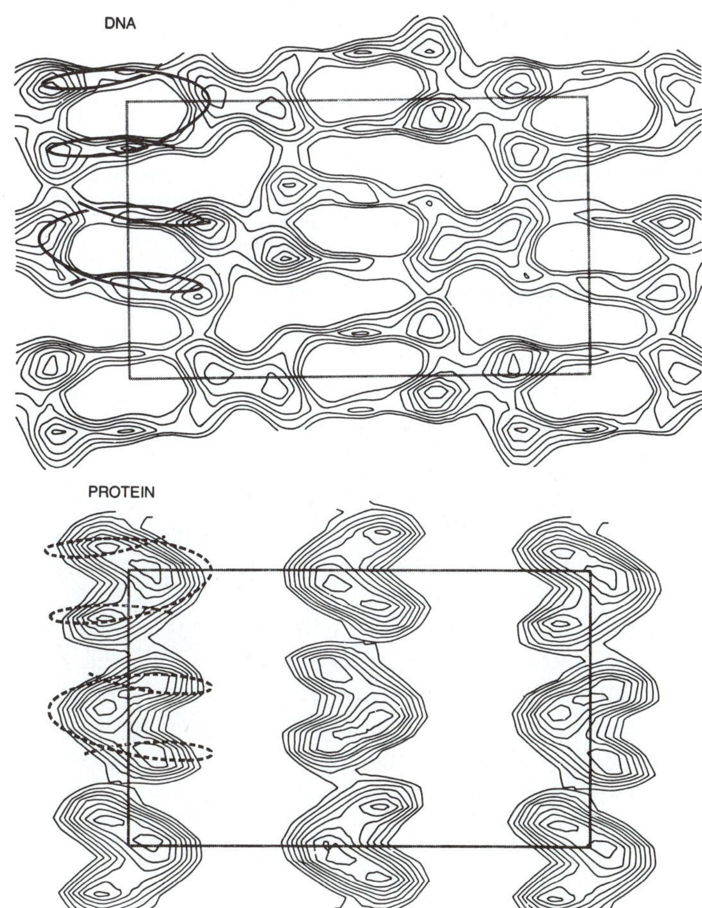

Figure 10-6
Neutron-scattering maps of nucleosome core crystals. These maps show separately the density of the DNA (top) and of the protein (bottom). The DNA density correlates well with the projection of about $1\frac{3}{4}$ turns of the DNA superhelix (solid line). The DNA superhelix (broken line) also appears to fit well around the protein, the histone octamer. [From Kornberg RD, Klug A (1981) *Sci Amer 244*:52]

85. Worcel A, Benyajati C (1977) *Cell 12*:83.
86. Richmond TJ, Searles MA, Simpson RT (1988) *JMB 199*:161.

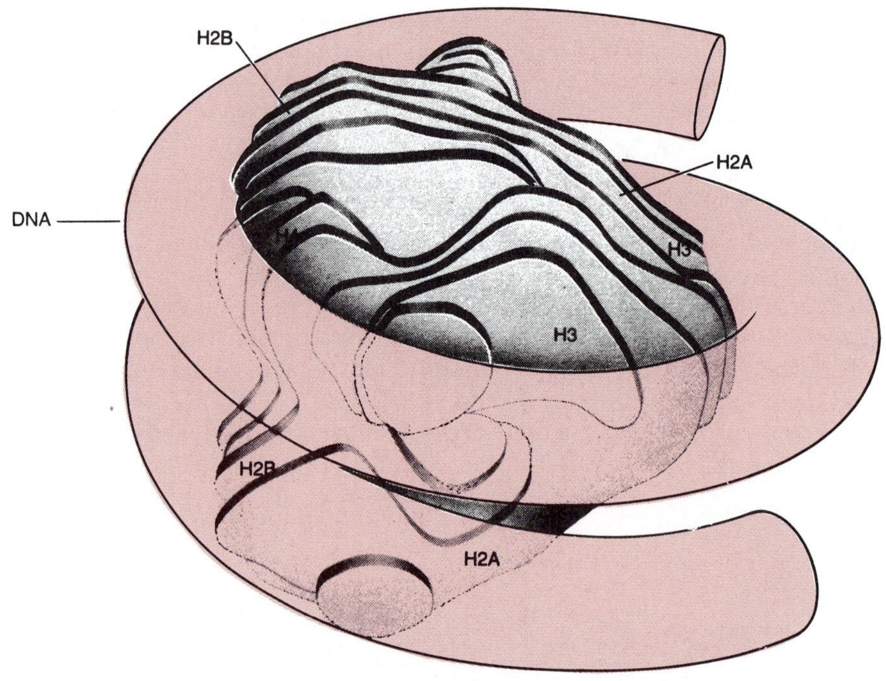

Figure 10-7
Model of the nucleosome core. The model was made by winding a tube simulating the DNA superhelix on a model of the histone octamer, which was built from a three-dimensional map derived from electron micrographs of the histone octamer. The ridges on the periphery of the octamer form a more or less continuous helical ramp on which a 146-bp length of DNA can be wound. The locations of individual histone molecules (whose boundaries are not defined at this resolution) are proposed here on the basis of chemical cross-linking data. [From Kornberg RD, Klug A (1981) *Sci Amer 244*:52]

periodicity is required to correlate the 1.75-DNA-turn core-particle model with the degree of supercoiling found in viral chromatin. Formation of an SV40 nucleosome introduces the equivalent of approximately one negative superhelical turn (Section 1-9). The importance of such negative supertwisting for many prokaryotic DNA functions (Section 1-9) may prove equally pertinent to chromatin.

A prominent nucleosome feature is a ridge that winds in a left-handed sense around the octamer and corresponds to the path followed by the DNA (Fig. 10-7). The locations of the individual histones along this path may be assigned on the basis of the map derived from chemical cross-linking.[87] The three-dimensional data permit some clarification and simplification of the map, leading to the following linear arrangement of the histones along the DNA:

H2A—H2B—H4—H3—H3—H4—H2B—H2A

In three dimensions, this arrangement appears as a central disk-shaped $(H3)_2(H4)_2$ tetramer with an H2A–H2B dimer stacked on each face. The pro-

87. Mirzabekov AD, Shick VV, Belyavsky AV, Bavykin SG (1978) *PNAS 75*:4184.

posed central role of the tetramer in the nucleosome[88] is thus confirmed, and the dissociation of the octamer in solution by the loss of H2A–H2B dimers[89] is readily understood.

The location of nucleosomes along the DNA appears to be related not to the specific sequence but rather to the bendability of the DNA[90] imparted by certain sequences. In its path around the histone octamer, the DNA does not curve uniformly but shows several sharp bends. An underlying sequence periodicity in the nucleosomal DNA can be related to bendability that may favor or exclude interactions with histones (Section 1-10).

Nucleosome Disassembly and Assembly

Nucleosomes may be disassembled during DNA replication by the separation of strands at the fork. Disassembly as the rate-limiting step for movement of the replication fork could explain why the unreplicated regions ("ss gaps") and Okazaki fragments at replication forks in eukaryotes are only about 200 bases, as compared with about 1000 in prokaryotes.[91] The similarity of the ss gap size to the DNA content of a nucleosome suggests that a nucleosome unfolds for replication as a unit.

Nucleosome disassembly and reassembly in replication[92] may involve a separation of $(H3)_2(H4)_2$ tetramers from H2A–H2B dimers, which may then reassociate with newly synthesized histones. New and old H3 and H4 also intermix, as do H2A and H2B. The kinetics of disassembly are further complicated by nucleosome dissolution due to transcription and other nonreplicative events.

The assembly of nucleosomes from newly synthesized histones may be catalyzed by a protein such as the nicking–closing enzyme,[93] a chromatin assembly factor (CAF-1),[94] or an acidic protein found in *Xenopus* oocytes.[95] CAF-1 is a multisubunit protein which promotes chromatin assembly of SV40 DNA that is being replicated by a reconstituted system; the minichromosomes contain the four histones in correctly spaced nucleosomes. The *Xenopus* acidic protein forms a complex with all four types of histone and mediates their association with DNA. Intermediates in the folding of newly synthesized DNA can be detected by virtue of altered sensitivity to digestion by DNases.[96]

Histones appear to be synthesized throughout the cell cycle,[97] suggesting that they are continually synthesized but assemble into chromatin only when new DNA appears during S phase. At other times during the cell cycle the newly synthesized histones form a pool, possibly associated with the abundant acidic protein in *Xenopus* oocyte nuclei. Factors for chromatin assembly in *Drosophila* embryos that modulate the transfer of histones to DNA include RNA or other polyanions.[98]

---

88. Cobianchi F, Karpel RL, Williams KR, Notario V, Wilson SH (1988) *JBC 263*:1063.
89. Thomas JO, Kornberg RD (1975) *PNAS 72*:2626.
90. Calladine CR, Drew HR (1986) *JMB 192*:907; Travers AA (1987) *TIBS 12*:108; Travers AA (1989) *ARB 58*:427.
91. Kriegstein HJ, Hogness DS (1974) *PNAS 71*:135.
92. Jackson V, Chalkley R (1985) *Biochemistry 24*:6921; Jackson V, Chalkley R (1985) *Biochemistry 24*:6930; Senshu T, Yamasu K, Ohsawa T (1985) *EJB 150*:575; Jackson V (1990) *Biochemistry 29*:719.
93. Germond JE, Brutlag D (1979) *PNAS 76*:3775.
94. Smith S, Stillman B (1989) *Cell 58*:15.
95. Laskey RA, Honda BM, Mills AD, Finch JT (1978) *Nature 275*:416.
96. Hancock R (1978) *PNAS 75*:2130; Worcel A, Han S, Wong ML (1978) *Cell 15*:969; Crèmisi C, Chestier A, Yaniv M (1977) *Cell 12*:947; Jackson V, Granner D, Chalkley R (1976) *PNAS 73*:2266.
97. Groppi VE Jr, Coffino P (1980) *Cell 21*:195; Searle RL (1981) *Biochemistry 20*:6432.
98. Stein A (1979) *JMB 130*:103; Stein A, Whitlock JP Jr, Bina M (1979) *PNAS 76*:5000.

Nonhistone DNA-Binding Proteins[99]

Nonhistone chromosomal proteins must also participate as enzymes and regulators in replication and transcription in chromatin. This complex group contains, in the rat, 12 to 18 major components and many minor ones. Some apparently serve as structural elements.

Metaphase chromosomes, depleted of histones and most of the nonhistone proteins, still retain the highly organized structure that gives the chromosome its familiar morphology.[100] After the extraction of chromatin and soluble components from isolated nuclei, a proteinaceous structure remains, the *nuclear matrix*, or *scaffold*. The residual nonhistone proteins with small amounts of tightly bound DNA and RNA retain many of the structural features of the intact nucleus, including a surrounding nuclear lamina with nuclear pore complexes and the internal nuclear matrix. Similar nuclear structures have been obtained from all cells, from yeast to human. Functional proteins released by mild sonication from the matrix of regenerating rat liver nuclei include the DNA polymerase α-primase complex, other replicative proteins,[101] and topoisomerases.

Bound to the nonhistone proteins of the nuclear matrix are specific DNA sequences called MARs or SARs (*m*atrix- or *s*caffold-*a*ssociated *r*egions). They form a halo of DNA organized as 30- to 90-kb loops, each anchored to the matrix at its base. In this model of chromosome organization, the nucleoprotein fibers form radially oriented loops that converge to the central scaffolding. Still another protein is known to bind certain satellite DNA sequences in the highly condensed heterochromatin region of *Drosophila* chromosomes.[102]

The nuclear matrix of rat brain nuclei may be involved in the incorporation of circular DNA into massive catenated networks.[103] Oocyte nuclei, when microinjected with circular DNA, also convert it to large catenated clusters.[104] The presumed action of topoisomerases associated with the nuclear scaffold (Section 12-6) may be assisted by agents that may crowd DNA in vivo [e.g., spermidine,[105] HMG17 (see below),[106] and a yeast protein[107]], much as hydrophilic polymers (e.g., polyvinyl alcohol)[108] do in vitro.

High-mobility-group (HMG) proteins, derived from mammalian cells, are so designated because of their electrophoretic mobility at low pH (the result of their small size and high charge). HMG proteins fall into four groups:[109] HMG1 and HMG2, the most prevalent nonhistone chromatin proteins, are closely related and are distinct from HMG14 and HMG17. Despite much effort to relate their abundance to differentiation and cellular proliferation, their physiologic functions are still undefined. A possible role of HMG1 in the regulation of transcription and replication may reside in the relaxation of negatively supercoiled domains in chromatin, thereby conserving some structural features and enabling the binding of sequence-specific regulatory factors.[110]

---

99. Berezney R (1984) in *Chromosomal Nonhistone Proteins* (Hnilica LS, ed.). CRC Press, Boca Raton, Fla, Vol. 4, p. 119.
100. Adolph KW, Cheng SM, Laemmli UK (1977) *Cell 12*:805; Paulson JR, Laemmli UK (1977) *Cell 12*:817.
101. Tubo RA, Berezney R (1987) *JBC 262*:1148; Tubo RA, Berezney R (1987) *JBC 262*:5857.
102. Hsieh T, Brutlag DL (1979) *PNAS 76*:726.
103. Tsutsui K, Tsutsui K, Oda T (1989) *JBC 264*:7644.
104. Mertz JE, Miller TJ (1983) *Mol Cell Biol 3*:126.
105. Shlomai, J, Zadok A (1983) *NAR 11*:4019.
106. Tse Y, Javaherian K, Wang JC (1984) *Arch B B 231*:169.
107. Goto T, Wang JC (1982) *JBC 257*:5866.
108. Holden JA, Low RL (1985) *JBC 260*:14491.
109. Walker JM, Goodwin GH, Johns EW (1979) *FEBS Lett 100*:394; Walker JM, Gooderham K, Hastings JRB, Mayes E, Johns EW (1980) *FEBS Lett 122*:264.
110. Sheflin LG, Spaulding SW (1989) *Biochemistry 28*:5658.

The bacterial genome is condensed several thousandfold but is not organized into the nucleosomal units found in eukaryotic chromatin. Lacking a nuclear membrane, the *bacterial nucleoid* is confined to a ribosome-free, vacuole-like space in which the DNA is highly concentrated. At a density of 20 to 40 mg/ml, its packing is similar to that of vegetative phage T4 DNA and the DNA of the interphase nucleus of hepatocytes. The relatively loose DNA meshwork can still accommodate the operations of replication, transcription, recombination, and repair.

The denser DNA packing in phage heads (up to 800 mg/ml in T4) is accomplished either by a scaffold of capsid and core proteins that guide the spooling of DNA into crystalline-like arrays (e.g., in T4, P2, λ, P22, T5, and T7)[112] or by a tight association with core proteins to form a nucleocapsid (e.g., in PM2). In this regard, the packed phage DNA differs from that of the highly condensed metaphase chromosomes, in which the nucleosomes are compacted by further levels of helical organization.

Although *E. coli* DNA can be visualized in the electron microscope as a condensed, beaded fiber resembling chromatin,[113] histone-like proteins are not abundant enough to account for such a structure. Instead, polyamines (spermidine and putrescine) and Mg ions are the more likely source of counterions to achieve DNA compaction. Nucleoid studies should take into account that glutamate is the anion preferred by *E. coli* (Section 15-5).

The only protein candidate associated with the nucleoid in significant amounts is HU protein (see below), but even so its level is less than 20% of that needed to organize the DNA, and its distribution is around the periphery of the nucleoid, near the ribosomes. The several duplex DNA-binding proteins, including HU, may be more adapted to functional features (e.g., transcription) of the nucleoid than to its packaging.

Despite some similarities between bacterial protein–DNA complexes and chromatin, the differences are fundamental. The retention of DNA superhelicity seen after γ-irradiation of eukaryotic cells, measured by psoralen binding, is not observed in *E. coli*.[114] Instead, such treatment or a DNA gyrase inhibitor relaxes the supercoiling of *E. coli* DNA, implying that the torsional tension is not restrained, as in eukaryotic cells, by nucleosomal structures.

HU Protein[115]

HU is a 19-kDa heterodimer of two polypeptides: HU1 (also called HUβ and NS1) and HU2 (HUα, NS2). The 90-amino-acid sequences of these subunits are 70% homologous. HU is present in *E. coli* at a level of 20,000 dimers per genome equivalent, the most abundant of the DNA-binding proteins. It resembles the histones in being a small, basic, heat-stable protein, rich in lysine and alanine

---

111. Drlica K, Rouvière-Yaniv J (1987) *Microbiol Rev 51*:301; Kellenberger E (1989) in *The Bacterial Chromosome* (Drlica K, Riley M, eds.). Am Soc Microbiol, Washington DC, p. 173; Gualerzi CO, Pon CL (eds.) (1986) *Bacterial Chromatin*. Springer-Verlag, New York.
112. Earnshaw WC, Casjens SR (1980) *Cell 21*:319.
113. Griffith JD (1976) *PNAS 73*:563.
114. Sinden RR, Carlson JO, Pettijohn DE (1980) *Cell 21*:773.
115. Drlica K, Rouvière-Yaniv J (1987) *Microbiol Rev 51*:301; Pettijohn D (1988) *JBC 263*:12793; Rouvière-Yaniv J, Bonnefoy E, Huisman O, Almeida A (1989) in *The Bacterial Chromosome* (Drlica K, Riley M, eds.). Am Soc Microbiol, Washington DC, p. 247.

and lacking tryptophan and cysteine. Similar proteins have been isolated from a wide variety of species. The one from cyanobacteria is similar in size, composition, and properties to *E. coli* HU and cross-reacts immunologically.[116] This high degree of evolutionary conservation is reminiscent of the histones in eukaryotes.

HU binds single- and double-stranded DNA and RNA. Binding to dsDNA is less affected by helix-destabilizing conditions (e.g., low salt, high temperature) than is the binding of SSBs to ssDNA. HU causes dsDNA to bend and fold and, in sufficient amounts, condenses the DNA into a wreath-like structure resembling beaded chromatin. Structural studies[117] suggest that two β-ribbon arms of the dimer encircle and cover one turn of the DNA helix. Neither the cooperativity of SSBs nor the sequence specificity of regulatory proteins is observed in the binding. The rate of HU dissociation from DNA is rapid ($t_{1/2}$, 0.6 minute at 50 mM NaCl) and becomes even faster at higher ionic strengths.

HU makes possible the ring closure of a short linear DNA fragment (~100 bp), which is otherwise prevented by the natural rigidity of a duplex of this length.[118] By mediating very tight DNA curvatures and by its rapid association–dissociation cycling in binding DNA, the protein may facilitate the generation of numerous alternate structural distortions of the DNA helix axis.[119]

HU is required in vitro for an early stage of initiation of replication at the unique *E. coli* chromosomal origin *(oriC)*;[120] it also plays a role in transpositions of Tn10[121] (Section 21-7) and of phage Mu[122] (Section 17-7), and in phase variation (genetic inversion) of *Salmonella typhimurium*.[123] At higher levels, HU inhibits the replication of *oriC*[124] and phage λ[125] plasmids, presumably by reducing superhelicity; this effect can be counteracted by transcriptional activation.[126] HU also influences transcription, stimulating the actions of the *lac* repressor and catabolite activator protein, while inhibiting *aroH* DNA recognition by the *trp* repressor.[127]

The contributions in vivo of HU can be inferred from mutants in either of the HU genes (*hupA* or *hupB*) or in both.[128] The *hupB* mutant (lacking HU1) shows no phenotypic change, but the *hupA* mutant (lacking HU2), like the double mutant, grows poorly, is defective in cell division, and fails to support the growth of phage Mu. The severity of the *hupA* mutation can be attributed to instability of HU1 in the absence of HU2, resulting in a low level of total HU; apparently the reverse, instability of HU2 in the absence of HU1, is not true. The homodimers are functional and are found in wild-type cells. Overproduction of either subunit also creates problems: filamentation, induction of the SOS system, and reduction in the copy number of plasmids.

---

116. Haselkorn R, Rouvière-Yaniv J (1976) *PNAS 73*:1917.
117. Tanaka I, Appelt K, Dijk J, White S, Wilson K (1984) *Nature 310*:376.
118. Hodges-Garcia Y, Hagerman PJ, Pettijohn DE (1989) *JBC 264*:14621.
119. White SW, Appelt K, Wilson KS, Tanaka I (1989) *Proteins 5*:281.
120. Dixon NE, Kornberg A (1984) *PNAS 81*:424; Funnell BE, Baker TA, Kornberg A (1986) *JBC 261*:5616.
121. Morisato D, Kleckner N (1987) *Cell 51*:101.
122. Craigie R, Mizuuchi K (1987) *Cell 51*:493; Surette MG, Buch SJ, Chaconas G (1987) *Cell 49*:253.
123. Johnson R, Bruist M, Simon M (1986) *Cell 46*:531.
124. Ogawa T, Baker TA, van der Ende A, Kornberg A (1985) *PNAS 82*:3562.
125. Mensa-Wilmot K, Carroll K, McMacken R (1989) *EMBO J 8*:2393.
126. Ogawa T, Baker TA, van der Ende A, Kornberg A (1985) *PNAS 82*:3562; Baker TA, Kornberg A (1988) *Cell 55*:113.
127. Flashner Y, Gralla JD (1988) *Cell 54*:713.
128. Wada M, Kano Y, Ogawa T, Okazaki T, Imamoto F (1988) *JMB 204*:581; Huisman O, Faelen M, Girard D, Jaffé A, Toussaint A, Rouvière-Yaniv J (1989) *J Bact 171*:3704; Bonnefoy E, Almeida A, Rouvière-Yaniv J (1989) *PNAS 86*:7691.

Whereas chromosomal replication appears unaffected in the double HU mutant, mini-F and mini-P plasmids cannot be introduced.[129] The stability of an *oriC* plasmid (minichromosome) is reduced when the *mioC* promoter near the origin is also deleted.[130] Along with observations in vitro that HU and transcription near the origin (Section 16-3) stimulate initiation at *oriC*, dependence on HU and the *mioC* promoter in vivo suggest that activation by these factors can strongly influence replication.

Integration Host Factor (IHF)[131]

This histone-like protein was discovered as a participant in the site-specific recombination of phage λ[132] and has been rediscovered time and again in other site-specific recombinations, in the regulation of a number of phage and bacterial genes,[133] and in plasmid partition.[134] Thus, IHF has been humorously called "a protein for all reasons."[135] Despite many structural and functional similarities to HU protein, there are also striking differences.

Like HU, IHF is a dimer with similar-sized subunits: IHF-α of 11.2 kDa, encoded by *himA*, and IHF-β of 10.6 kDa, encoded by *hip (himD)*; whether the dimer is heterotypic (α,β) or homotypic (α_2 or β_2) is still unproven. The protein is heat-stable and about 35% identical in sequence to HU. Like HU and histones, it wraps duplex DNA around its surface, generating a bend (Section 1-10). IHF is far less abundant than HU, is not essential for the growth and survival of *E. coli*, and recognizes a 13-bp sequence, which it binds and bends in a unique way.[136] By virtue of this bending, IHF enables the two autonomous DNA-binding domains of λ integrase to create a DNA loop for the cleavage and rejoining reactions.

Other Histone-Like Proteins[137]

Of the various other histone-like proteins that have been described, four examples are given below. Other histone-like proteins of great interest include those that confer UV resistance on the genomes of bacterial spores.[138]

Archaebacterial HTa. This histone-like protein[139] is associated with the DNA of *Thermoplasma acidophilum*. HTa remains stably bound to the DNA even at high salt concentrations, condenses the DNA into globular nucleosome-like particles, and protects it from thermal denaturation.[140] HTa bears considerable homology to the HU protein of *E. coli* and the HB_S protein of *Bacillus stearothermophilus*.[141]

---

129. Ogura T, Niki H, Kano Y, Imamoto F, Hiraga S (1990) *MGG 220*:197.
130. Ogawa T, Wada M, Kano Y, Imamoto F, Okazaki T (1989) *J Bact 171*:5672.
131. Drlica K, Rouvière-Yaniv J (1987) *Microbiol Rev 51*:301; Nash HA (1981) *ARG 15*:143; Friedman DI (1988) *Cell 55*:545.
132. Johnson R, Bruist M, Simon M (1986) *Cell 46*:531.
133. Tsui P, Freundlich M (1988) *JMB 203*:817; Flashner Y, Gralla JD (1988) *Cell 54*:713.
134. Funnell BE (1988) *PNAS 85*:6657.
135. Ogawa T, Baker TA, van der Ende A, Kornberg A (1985) *PNAS 82*:3562.
136. Yang C-C, Nash HA (1989) *Cell 57*:869; Moitoso de Vargas L, Kim S, Landy A (1989) *Science 244*:1457.
137. Hsieh T, Brutlag D (1980) *Cell 21*:115.
138. Setlow P (1988) *Ann Rev Microbiol 42*:319.
139. Searcy D, Stein D (1980) *BBA 609*:180; Searcy D, Montenay-Garastier T, Hélène C (1989) *Biochemistry 28*:9058.
140. Searcy DG (1986) in *Bacterial Chromatin* (Gualerzi CO, Pons CL, eds.). Springer-Verlag, Heidelberg, p. 175; Reddy TR, Suryanarayana T (1989) *JBC 264*:17298.
141. Drlica K, Rouvière-Yaniv J (1987) *Microbiol Rev 51*:301.

H Protein.[142] Remarkably similar in amino acid composition to histone H2A, H protein was recognized in fractionated extracts of *E. coli* because its binding to DNA inhibited replication, transcription, and other DNA-dependent reactions. The protein is a dimer of 28-kDa polypeptides and binds single-stranded or duplex DNA. H protein is neutralized by rabbit antibodies against calf thymus H2A, and resembles that histone not only in amino acid composition but also in DNA-binding features and in its support of the reannealing of denatured DNA. However, the identity of H protein with the S3 ribosomal protein[143] indicates that it is naturally an RNA-binding protein, and its presence in cell extracts is most likely an artifact of the dissolution of ribosomes.

FirA Protein.[144] FirA is a heat-stable, acid-soluble, 17-kDa protein that behaves as a tetramer, binds DNA, and is present at about 4000 monomers per cell. The *firA* gene is allelic with *skp;* its product counteracts the rifampicin resistance in *E. coli* observed with mutations in the β subunit of RNA polymerase. The role of firA protein, possibly in transcription, is uncertain.

Fis Protein. Purified as a participant in several related site-specific recombination reactions[145] (the Hin, Gin, and Cin inversion systems; Section 21-6), Fis (*f*actor for *i*nversion *s*timulation) is an additional small, heat-stable dsDNA-binding protein. The active form is a homodimer of 98 amino acid subunits. Nucleotide sequence shows the protein to be structurally unrelated to HU and IHF. Rather, a region near the C terminus is similar to the helix–turn–helix motif of bacterial repressors (Section 9).[146] The protein has been crystallized for x-ray diffraction.[147]

During recombination, binding of Fis protein to an enhancer sequence, which can be located at many places on the DNA substrate, stimulates inversion more than 100-fold.[148] Two 20-bp sequences within the 60-bp enhancer are protected from chemical cleavage by Fis. Presumably, each monomer of the Fis dimer interacts with a DNA site. Activation of recombination may involve a physical interaction between Fis protein and the recombinase, and a changing of a DNA segment to the inverted conformation prior to ligation of the cleaved strands.[149]

10-9 Regulatory Proteins[150]

Binding proteins that alter DNA structure and affect both replication and transcription have already been treated in this chapter and will be considered in

142. Hübscher U, Lutz H, Kornberg A (1980) *PNAS 77*:5097.
143. Bruckner RC, Cox MM (1989) *NAR 17*:3145.
144. Lathe R, Buc H, Lecocq J, Bautz E (1980) *PNAS 77*:3548; Aasland R, Coleman J, Holck AL, Smith CL, Raetz CRH, Kleppe K (1988) *J Bact 170*:5916.
145. Koch C, Kahmann R (1986) *JBC 261*:15673; Johnson RC, Bruist MF, Simon MI (1986) *Cell 46*:531.
146. Koch C, Vandekerckhove J, Kahmann R (1988) *PNAS 85*:4237.
147. Choe HW, Labahn J, Itoh S, Koch C, Kahman R, Saenger W (1989) *JMB 208*:209.
148. Johnson RC, Glasgow AC, Simon MI (1987) *Nature 329*:462.
149. Johnson RC, Bruist MF (1989) *EMBO J 8*: 1581.
150. Pabo CO, Sauer RT (1984) *ARB 53*:293; Struhl K (1989) *TIBS 14*:137; Brennan RG, Matthews BW (1989) *TIBS 14*:286.

other chapters. However, a thorough treatment of the numerous DNA-binding proteins identified as regulators of gene expression is beyond the scope of this book, as is a detailed consideration of transcription itself. Yet the RNA polymerases (Chapter 7) and the activator and repressor proteins that modulate their actions have so much in common with the replicative proteins that some notice, however brief, should be taken of the motifs that guide the functions and structures of these regulatory proteins. The actions of these proteins in a variety of combinations may explain how each of the enormous number of genes can be finely regulated in cells of diverse types displaying unique developmental patterns and responding to fluctuating physiologic conditions. Among the basic structural patterns of the regulatory DNA-binding proteins recognized thus far, four are well characterized: the helix–turn–helix, two types of zinc fingers, and the leucine zipper.

Helix–Turn–Helix[151] Domains and Homeodomains[152]

The first studied and clearest examples of the helix–turn–helix motif are the prokaryotic repressors and activators. Two of these, the *cro* repressor and the λ repressor (cI protein), bind to six phage λ operator sequences in a manner that determines whether the phage will follow a lytic or a lysogenic pathway. The *trp* (tryptophan) repressor blocks transcription by binding the operator of the *trp* operon. CAP [catabolite *a*ctivator *p*rotein; also called CRP (cyclic AMP receptor protein)] regulates several catabolite-sensitive operons in *E. coli*. The overall manner by which these proteins bind to operator DNA is similar, as deduced from analysis of protein–DNA cocrystals. Despite variations in their size, domains, and tertiary structure, these proteins, binding as dimers, use an α helix (in a helix–turn–helix) to contact major grooves along one face of the double helix.

Cro.[153] With a monomer size of 66 amino acids, Cro is made up of three helices and three extended segments that form an antiparallel β sheet. A C-terminal region of β structure interacts with that of another monomer to form a dimer with twofold rotational symmetry (Fig. 10-8). In its association with each of the six 17-bp sites in the operator DNA, the pair of symmetrically related DNA-binding (recognition) helices of the dimer are separated by 34 Å, the same distance separating successive major grooves in the palindromic sequence of the operator.

The second (recognition) helix of the helix–turn–helix contains most of the side chains that interact with the site-specific base pairs. The first helix of the motif mediates contacts with the DNA backbone and often provides side-chain–side-chain interactions that stabilize the base-specific interactions of the second helix. The helix–turn–helix domain of Cro and that of the other proteins of known structure are homologous with the sequences in a large number of other DNA-binding proteins, including *E. coli lac* repressor.

---

151. Brennan RG, Matthews BW (1989) *JBC 264*:1903; Schleif R (1988) *Science 241*:1182.
152. Wright CVE, Cho KWY, Oliver G, De Robertis EM (1989) *TIBS 14*:52.
153. Ohlendorf DH, Anderson WF, Matthews BW (1983) *J Mol Evol 19*:109; Wolberger C, Dong Y, Ptashne M, Harrison SC (1988) *Nature 335*:789.

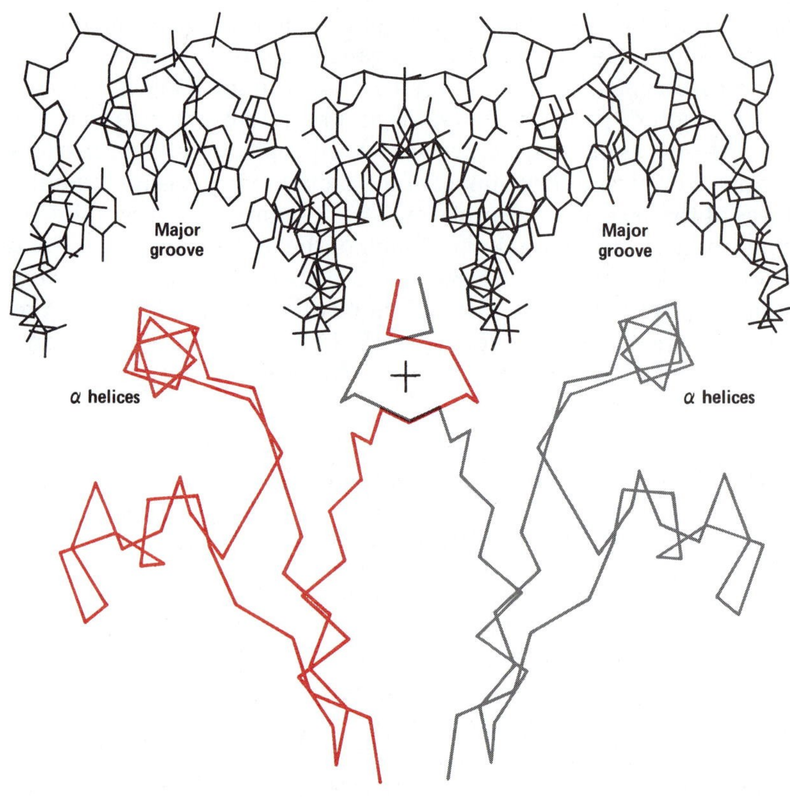

Figure 10-8
Presumed interaction of cro (the *cro* repressor, bottom) with DNA (top). Two cro monomers, related by a twofold symmetry axis (+ symbol), interact with DNA in such a way that a pair of symmetry-related α helices occupy successive major grooves of the DNA and two extended polypeptide strands run parallel to the backbone of the DNA. (Courtesy of Professor BW Matthews)

The λ Repressor.[154] Having a mass of 26 kDa as a monomer, the λ repressor forms dimers and higher oligomers which bind to the same six operator sites as does Cro. The helix–turn–helix motif is part of an N-terminal DNA-binding domain, whereas crucial protein–protein interactions are governed by a C-terminal domain. A protease-sensitive "connector" joins these two independent domains (which denature at very different temperatures).[155] This 40-amino-acid connector region contains the site for autolytic cleavage facilitated by RecA protein upon induction of a λ lysogen. Removal of repression upon cleavage results both from a loss of cooperative binding to adjacent operator sites and from destabilization of the active dimeric form of λ repressor.

The *trp* Repressor.[156] This 58-kDa dimer of extensively interlocked subunits depends on tryptophan as a corepressor. When bound to the repressor, tryptophan alters the conformation of the helix–turn–helix entity and increases the distance between the subunits by 8 Å, thereby enabling the recognition helices

154. Jordan SR, Pabo CO (1988) *Science 242*:893.
155. Pabo CO, Sauer RT, Sturtevant JM, Ptashne M (1979) *PNAS 76*:1608.
156. Otwinowski Z, Schevitz RW, Zhang R-G, Lawson CL, Joachimiak A, Marmorstein RQ, Luisi BF, Sigler PB (1988) *Nature 335*:321.

to fit into adjacent major grooves of the operator (Plates 8 and 9). Many of the base-specific interactions of the *trp* repressor are mediated via water molecules.

CAP.[157] A monomer of 209 amino acids, CAP is active only as a dimer. Each subunit is folded into a C-terminal domain which contains the helices responsible for binding DNA, and an N-terminal domain which binds cyclic AMP and contains most of the dimer contacts (Plate 10).

Homeodomain.[158] A homeodomain is a 60-amino-acid stretch first discovered in genes responsible for regulating segmental development in *Drosophila*. Nearly 100 genes encoding such proteins have been identified, each gene possessing a *homeobox* of 180 ntd that is highly conserved from yeast to humans. The homeodomains of the yeast mating-type proteins MATa1 and MATα2 not only resemble those of flies and frogs, but also possess the helix–turn–helix motif of the prokaryotic regulatory proteins responsible for binding specific DNA sequences. The CAATTAAA sequence is the common target for binding by homeodomains of the proteins that regulate transcription of unspecified genes in *Drosophila* and transcription of the prolactin and growth hormone genes in mammals. The crystal structure[159] of a complex of the *Drosophila* engrailed homeodomain and a duplex DNA site provides insights into the homeodomain-DNA interaction.

Sigma Subunits of RNA Polymerases. Among the regulatory proteins that employ a helix–turn–helix for DNA binding are the sigma (σ) subunits responsible in virtually all RNA polymerases for recognition of the −35 promoter sequence (Section 7-4); mutation of this helix–turn–helix region of a sporulation-specific σ diminishes by more than 10^6-fold the incidence of sporulation in *B. subtilis*.[160]

Zinc Fingers[161]

This motif, identified in over 100 DNA-binding proteins and found in virtually all species, was first recognized in a factor for the transcription of 5S RNA genes by RNA polymerase III. This 40-kDa protein, TF IIIA, contains nine zinc fingers, each of which possesses an invariant pair of cysteines on one side and a pair of histidines on the other, tetrahedrally coordinated with a zinc ion; the finger-like projection of intervening amino acids includes three invariant large hydrophobic residues: phenylalanine, leucine, and tyrosine. NMR studies of a single zinc finger[162] indicate a compact globular structure with a well-defined helix possessing a high density of exposed basic and polar amino acid side chains. The finger binds DNA (Fig. 10-9), probably in the large groove, with five base pairs footprinted periodically by each finger.

---

157. Matthews BW, Ohlendorf DH, Anderson WF, Fisher RG, Takeda Y (1983) *CSHS 47*:427; Schultz SC, Shields GC, Steitz TA (1990) *JMB 213*:159.
158. Gehring WJ (1987) *Science 236*:1245; Coleman KG, Poole SJ, Weir MP, Soeller WC, Kornberg T (1987) *Genes Dev 1*:19; Scott MP, Tamkun JW, Hartzell GW (1989) *BBA 989*:25; Qian YQ, Billeter M, Otting G, Müller M, Gehring WJ, Wüthrich K (1989) *Cell 59*:573.
159. Kissinger CR, Liu B, Martin-Blanco E, Kornberg TB, Pabo CO (1990) *Cell 63*:579.
160. Yudkin MD, Millonig JH, Appleby L (1989) *Mol Microbiol 3*:257.
161. Klug A, Rhodes D (1987) *TIBS 12*:464; Berg J (1988) *PNAS 85*:99; Parraga G, Young L, Klevit RE (1989) *TIBS 14*:398; Parraga G, Horvath S, Hood L, Young ET, Klevit RE (1990) *PNAS 87*:137.
162. Lee MS, Gippert GP, Soman KV, Case DA, Wright PE (1989) *Science 245*:635.

Figure 10-9
Model for the interaction between DNA and a protein consisting of three tandemly repeated zinc finger domains, based on the structure predicted for the individual domains. The protein (colored) lies in the DNA major groove, with the α helices (coils) making contact with the base pairs and the β sheets (arrows) making contact with the DNA backbone. Stippled ball = zinc ion.

From a minimum of two such fingers up to 37 (six "hands" of six fingers, plus one), these proteins are even more represented in the regulation of development than in the control of transcription. Over 40 zinc finger proteins have been inferred from the sequences of the Krüffel genes of *Xenopus*. The diversity required for the intricate patterns of differentiation may be achieved by exploiting the combinational potential of the multiple fingers.

Another type of zinc finger[163] is prominent in the yeast GAL4 transcriptional activator and in mammalian steroid receptors. Here, the zinc ion is coordinated with two pairs of cysteines. In this class, just two fingers with an essential intervening region confers the specificity. A single GAL4 zinc finger can have all but one of its 28 amino acids replaced with the analogous zinc finger region of another yeast activator protein (PPR1) without changing the sequence-specificity of GAL4 binding. However, the region of 14 amino acids adjacent to the zinc finger is essential for sequence-specific recognition.[164] Whether a zinc finger of this type engages DNA directly, contributes to function via the overall protein structure, does both, or even does neither, most likely depends on the particular protein.

The Leucine Zipper, a Coiled Coil[165]

Some DNA-binding proteins, including a yeast transcriptional activator (GCN4) and several oncoproteins (Jun, Fos, and Myc), use a leucine-zipper motif (Plate 11) to form homo- or heterodimers. The dimerization depends on hydrophobic packing between parallel α helices (coiled coils) that contain four to five leucines spaced exactly seven residues apart—in effect, every two turns of the α helix. As with the second class of zinc fingers, the crucial DNA interactions lie outside these 30-residue regions. Leucine zipper domains in two tobacco DNA-

---

163. Evans RM, Hollenberg SM (1988) *Cell 52*:1.
164. Corton JC, Johnston SA (1989) *Nature 340*:724.
165. O'Shea EK, Rutkowski R, Staford WF III, Kim PS (1989) *Science 245*:646; Turner R, Tjian R (1989) *Science 243*:1689; Abel T, Maniatis T (1989) *Nature 341*:24; Vinson CR, Sigler PB, McNight SL (1989) *Science 246*:911; Abate C, Luk D, Gentz R, Rauscher FJ III, Curran T (1990) *PNAS 87*:1032.

binding proteins[166] are homologous to those in animal proteins (e.g., nuclear factor CREB), indicating the ancient origin of this motif. The presence of the leucine zipper in proteins that also possess a homeodomain or zinc fingers is indicative of discrete functions for these different motifs.

Other Motifs

Still other structures are known to be the interactive elements in DNA-binding proteins. For example, the phage P22 *arc* and *mnt* repressors depend, in part, on N-terminal arms in β conformation for binding functionally important base positions in the major groove.[167] These proteins may be structurally similar to the *metJ* repressor, whose crystal structure has been solved.[168]

Of related interest, enfolding arms of each subunit of EcoRI endonuclease wrap around the DNA, reminiscent of *E. coli* HU and IHF proteins. Sequence-specific recognition is mediated by α helices which approach the DNA end-on. The DNA cleavage site is located in a cleft that binds the DNA backbone in the vicinity of the scissile bond.[169]

A helix–loop–helix family of regulatory proteins (binding to immunoglobulin enhancers and to genes for myogenesis, Myc, and *Drosophila* development) contains two amphipathic helices with hydrophobic residues clustered on one side.[170] Still another family of proteins, with no identified motif, binds a CCAAT box and is important in the regulation of transcription and replication.[171]

10-10 Covalent Protein–DNA Complexes

Among the most remarkable of the proteins acting on DNA are those that, through covalent linkage, alter its shape or start and direct its replication. These proteins are treated separately elsewhere in this and other chapters, but it is useful here to survey some of these complexes and review briefly the functions they perform.

Topoisomerases (Chapter 12)

The topoisomerase–DNA linkage conserves the energy of the nicked phosphodiester bond and orients the freed ends for resealing. In the covalent intermediate, a tyrosine in the enzyme is linked to the 5′-P end of the nicked DNA (except for eukaryotic topoisomerase I, in which the linkage is to the 3′-P); the intermediate may be observed when the sequence of nicking and ligation is interrupted. After the DNA has been twisted or untwisted, the covalent bond is attacked in an energy-independent transesterification by the adjacent free end, to reseal the chain.

166. Katagiri F, Lam E, Chua N-H (1989) *Nature 340*:727.
167. Knight KL, Sauer RT (1988) *Biochemistry 27*:2088; Knight KL, Sauer RT (1989) *PNAS 86*:797; Knight KL, Sauer RT (1989) *JBC 264*:13706.
168. Rafferty JB, Somers WS, Saint-Girons I, Phillips SEV (1989) *Nature 341*:705.
169. McClarin JA, Frederick CA, Wang B-C, Greene P, Boyer HW, Grable J, Rosenberg JM (1986) Science *234*:1526; Jen-Jacobson L, Lesser D, Kurpiewski M (1986) *Cell 45*:619.
170. Murre C, McCaw PS, Baltimore D (1989) *Cell 56*:777.
171. Santoro C, Mermod N, Andrews PC, Tjian R (1988) *Nature 334*:218.

ϕX174 Gene *A* Protein (Sections 8-7 and 17-4)

The ϕX gpA complex with DNA is an example of a readily isolable complex in which the covalently linked protein remains functional. The protein–DNA linkage defines the origin of replication of duplex circular ϕX174, retains the energy of the broken phosphodiester bond, keeps the 5′-P end oriented at the progressing replicating fork (looped rolling circle), and closes the viral strand circle when the linkage is attacked by a 3′-OH group (a second nicking) (Fig. 8-13). Covalent linkage of protein and DNA is essential for the specificity and efficiency of this complex sequence.

Plasmid Transfer Complexes (Section 18-7)

The protein–DNA linkage may facilitate the transfer of plasmid DNA from one cell to another and then the recircularization of the DNA in the recipient cell. Virtually all isolated plasmids contain supercoiled DNA with one or more proteins tightly bound, but not covalently linked, to the DNA in a "relaxation complex." Nicking of the DNA and covalent linkage of the protein may be demonstrated in vitro by treatment with sodium dodecyl sulfate, ethidium, or certain proteases. (Presumably, nicking is induced in some specific way in vivo.) The nicking and covalent attachment are at a specific location on one of the two strands of the plasmid. Among the plasmid complexes studied are those carrying colicinogenic (Col), resistance (R), and fertility (F) factors. In the case of F plasmids, the protein linkage is only to the DNA strand that transfers the sex factor and the *E. coli* chromosome. The 5′ chain end, known to lead this transfer, is the one to which the protein is linked.

Viral Priming Proteins (Sections 5-10 and 6-8)

Some phages (e.g., ϕ29) and adenovirus initiate the replication of their linear genomes by a virus-encoded terminal protein which is linked by the viral DNA polymerase to each 5′ end of the viral genome (Sections 8-8, 17-8, and 19-4). The covalent protein–DNA linkage is between a serine residue and either dAMP (in ϕ29) or dCMP (in adenovirus). Similar protein–nucleic acid linkages are found as intermediates in the replication of RNA of poliovirus (linked to tyrosine in VPg) and foot-and-mouth disease virus.

Other Complexes

Phosphotyrosine is the link between a uridylyl residue and glutamine synthase in the allosteric regulation of the enzyme. It is also the modified amino acid in cell-surface receptors and cellular proteins that mediate the signals for gene expression and for neoplastic transformations by oncogenic viruses.

CHAPTER 11

DNA Helicases

11-1 Helicases: Actions and Polarity[1]

Helicases use the energy of nucleoside triphosphate hydrolysis to break the hydrogen bonds that hold the strands together in duplex DNA. Providing single strands as templates for DNA polymerase is essential for progress of the replication fork. Thus, almost all DNA polymerases acting on duplex DNA depend on the action of a helicase.

Assays and Characteristics

Helicase activity can be detected both coupled to and apart from DNA synthesis. The most common substrates (Fig. 11-1) are partially duplex DNA molecules with single-stranded extensions (tails).[2] The helicase, fueled by NTP hydrolysis, displaces one of the strands of the duplex. The single strand can be detected by its altered mobility in electrophoresis on a nondenaturing gel or by its sensitivity to single-strand–specific nucleases.

Helicases may be classified by (1) structural preference for the DNA substrate, (2) polarity of translocation, (3) processivity, and (4) associated functions.

(1) *Substrate structure.* Some helicases can operate at a blunt end, whereas others require a 3′ or 5′ single-stranded tail, or even a forked molecule. When one encounters a DNA-dependent ATPase in vitro, a helicase function is a strong possibility. Exploration should include a variety of substrates, inasmuch

1. Matson SW, Kaiser-Rogers KA (1990) *ARB 59*:289.
2. LeBowitz JH, McMacken R (1986) *JBC 261*:4738; Matson SW, Tabor S, Richardson CC (1983) *JBC 258*:14017; Matson SW, George JW (1987) *JBC 262*:2066.

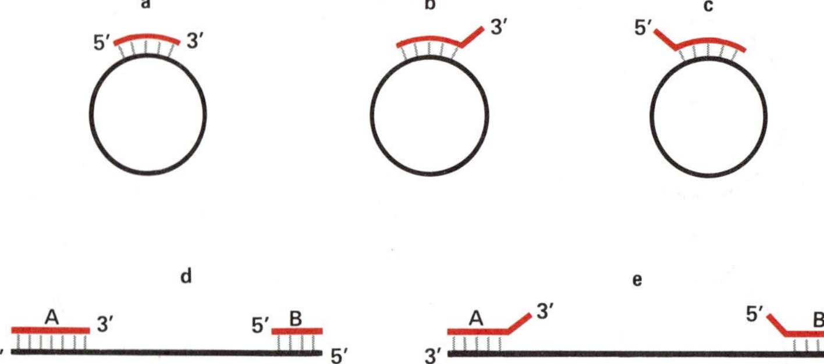

Figure 11-1
Substrates commonly used to assay helicase activity.

as helicases requiring tails or forks would be missed if tested on substrates with fully annealed 3′ and 5′ ends.

(2) *Polarity.* Helicases translocate along DNA in the 5′→3′ or in the 3′→5′ direction; one helicase is known to move in either direction. A single-stranded molecule with a duplex portion at each of its two ends (Fig. 11-1d) can be used to determine the polarity of helicase movement.[3] Assuming that the enzyme initially binds the single-stranded region, displacement of the fragment annealed to the 3′ end of the single strand (fragment A in Fig. 11-1d) indicates a 5′→3′ translocation, while displacement of the 5′-annealed fragment (fragment B) indicates a 3′→5′ polarity. The polarity of a helicase that operates at a forked structure can appear ambiguous. When the central single-stranded region is longer than the tails and provides a larger target for binding (as in Figure 11-1e), the preferential release of fragment B is indicative of 3′→5′ movement.

(3) *Processivity.* Analogous to polymerase processivity (Section 15-8), helicase processivity is measured by the number of nucleotides separated during each association event. A highly processive helicase will remain bound and will continue to melt an entire duplex once started; such enzymes only rarely cycle to a "challenge template," a second DNA added after the reaction has been initiated. The DNA-dependent NTPase activity of a helicase can provide information about processivity. A helicase prefers a circular or long linear strand if it tracks along DNA but is slow in binding to or dissociating from the DNA molecule. A distributive helicase shows a lesser dependence on the length of the DNA effector (for NTPase activity), once the minimal length for binding the enzyme is satisfied. The NTP preference of a helicase can be determined simply by ssDNA-dependent hydrolysis, but differences are observed when these affinities and rates are compared with those of hydrolysis coupled to strand displacement.

(4) *Associated functions.* Although the strand-displacement assay itself detects the action of a helicase, DNA melting is normally coupled to the action of other replication proteins, such as initiator proteins, primases, and polymerases. SSBs are often essential for displacement of long fragments and for blocking their reannealing. However, SSBs can also inhibit helicase action by competing for binding to the single-stranded portion of the substrate, blocking access of the helicase.

Duplex DNA circles with a nick, gap, or single-stranded tail are used to analyze helicase action coupled to DNA synthesis[4] (Fig. 11-2). Concerted action of a

3. LeBowitz JH, McMacken R (1986) *JBC 261*:4738; Matson SW (1986) *JBC 261*:10169.
4. Mok M, Marians KJ (1987) *JBC 262*:16644; Lechner RL, Richardson CC (1983) *JBC 258*:11185; Cha TA, Alberts BM (1989) *JBC 264*:12220.

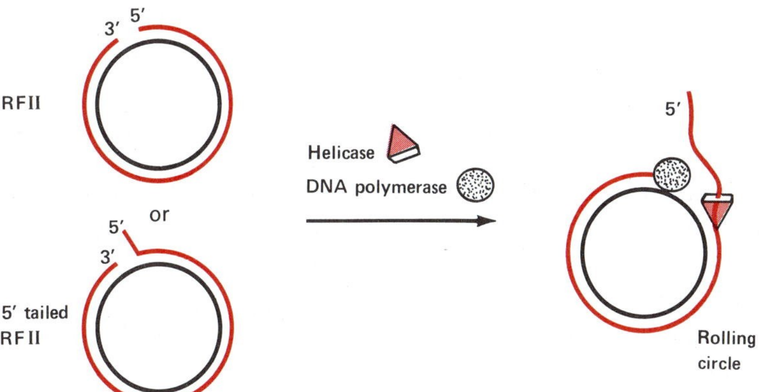

Figure 11-2
Substrates used to assay helicase activity coupled to DNA synthesis.

helicase and polymerase can generate a rolling circle to produce a long stretch of ssDNA. The rate of DNA synthesis is usually limited by the rate of unwinding; the length of products produced (under conditions that prevent cycling of the helicase on and off the template) is a measure of processivity.

On topologically constrained DNA, a topoisomerase is required for extensive helicase action. For every ten or so base pairs melted by a helicase, the two strands must be untwisted by one turn. If the ends of the DNA are not free to rotate, as is the case on a circular molecule or a linear DNA with tethered ends, positive superhelicity accumulates ahead of the helicase-generated fork. This tension must be removed by a topoisomerase (Section 12-1) for melting to proceed. The combined action of a helicase, an SSB, and a topoisomerase on covalently closed, circular DNA generates a highly underwound DNA species (Fig. 11-3) that has been called form I* or form U. These underwound structures, when generated by a replicative helicase (e.g., dnaB protein) and topoisomerase "swivel" (e.g., gyrase) in the absence of DNA synthesis, are useful in assaying the helicase activity of replication proteins.

In interactions with other replication enzymes, the helicase occupies a central position at the replication fork. An association between helicase and primase is common. Kinetic evidence for an interaction between helicase and both SSB and DNA polymerase is also frequently observed. In addition to their critical role in replication, helicases are implicated in recombination, in repair, in strand transfer during conjugation, and in transcription termination. Their importance in so many aspects of DNA metabolism argues for more attention to their structure and to actions which are not yet well understood.

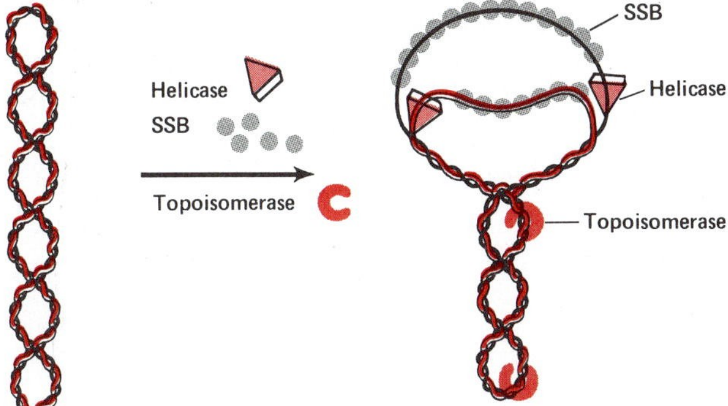

Figure 11-3
Concerted action of a helicase, SSB, and topoisomerase to generate a highly underwound DNA circle.

11-2 Replicative Helicases of *E. coli* and Phages[5]

The *E. coli* and coliphage helicases responsible for opening duplex DNA during replication are among the best characterized. The properties of these helicases, acting alone and in combination with other replication proteins, provide insights into the mechanism of melting and the dynamic nature of the replisome.

E. coli dnaB Protein

The helicase activity of dnaB protein is central to replication in *E. coli*. The dnaB protein, a hexamer of 50-kDa subunits (Section 8-4), activates priming by DNA primase (Section 8-4) and melts the DNA strands at the replication fork. Displacement of a single-stranded fragment from a partially duplex molecule reveals the intrinsic helicase activity.[6] Melting requires ssDNA on both strands at the border of the duplex region (a "preformed fork"). Although the minimal length of these single-stranded extensions is not known precisely, 100 ntd at the 3′ end is sufficient but 40 is inadequate; considerably longer 5′ single-stranded tails have been used. Despite the ambiguities of a forked substrate, there is a tenfold preference for displacing a fragment hybridized to the 3′ end of a doubly forked molecule (Fig. 11-1e), over a fragment on the 5′ end. Thus, the dnaB protein translocates with a 5′→3′ polarity along the strand to which it is bound.

As fuel for DNA melting, ATP is preferred;[7] GTP and CTP are also used and to a similar extent, but UTP is a poor substrate. Unlike the case with some helicases, dATP is a very feeble substitute for ATP. Analogs of ATP, such as AMPPNP, which are not hydrolyzed by dnaB protein but allow it to bind ssDNA, do not sustain the helicase. High levels of rNTPs (>2 mM), well above the K_m for the ssDNA-dependent ATPase,[8] are required for optimal helicase activity.

E. coli SSB and primase together stimulate the dnaB helicase approximately sixfold.[9] Association of dnaB with the template (supported by a nonhydrolyzable NTP analog) can be separated from helicase activity. SSB inhibits binding by competing with dnaB protein for the ssDNA, while stimulating melting, probably by preserving the separation of the melted strands. Stimulation by primase is not seen in the absence of SSB, and this may reflect an increased stability of the dnaB protein–template association. A 100-fold molar excess of dnaB protein over DNA is required for binding, thereby limiting its efficiency and making it difficult to determine its processivity and stoichiometry in the active complex.

On a nicked duplex circle with a protruding 5′ tail to promote binding of the protein (e.g., 5′-tailed RFII; see Fig. 11-2), dnaB helicase action can be coupled to DNA synthesis by DNA polymerase III holoenzyme.[10] Long single-stranded linear molecules are generated from a rolling-circle intermediate at a rate of approximately 700 bp per second. This rate, 20 times faster than that estimated on a linear forked substrate, may indicate a functional interaction between the polymerase and dnaB protein. Neither primase nor SSB is required to achieve this high velocity. Were the space between the two enzymes on the DNA exten-

---

5. Matson SW, Kaiser-Rogers KA (1990) *ARB 59*:289.
6. LeBowitz JH, McMacken R (1986) *JBC 261*:4738.
7. LeBowitz JH, McMacken R (1986) *JBC 261*:4738.
8. Reha-Krantz LJ, Hurwitz J (1978) *JBC 253*:4051; Arai K, Kornberg A (1981) *JBC 256*:5253.
9. LeBowitz JH, McMacken R (1986) *JBC 261*:4738.
10. Mok M, Marians KJ (1987) *JBC 262*:16644.

sive, SSB coating of the single-stranded gap would very likely be required for rapid synthesis by the polymerase. The lack of such a requirement for SSB is evidence for the coupling of melting and DNA synthesis.

On the tailed RFII template, strand separation by dnaB protein is highly processive.[11] The addition of poly dA, to which dnaB protein binds efficiently, does not inhibit the helicase, indicating that dnaB protein can melt more than 50 kb of duplex DNA without dissociating. In contrast, the ssDNA-dependent rNTPase action and general priming activities of dnaB protein are quite distributive.[12] The parameters that determine whether dnaB protein is processive or distributive have not been identified. Possibly, interaction of dnaB protein with both DNA strands at the replication fork confers a processive helicase conformation on the protein.

The dnaB helicase is also active within the initial replication forks set up at the *E. coli* chromosomal origin (*oriC*),[13] at the origin of phage λ (*ori*λ),[14] and at primosome assembly sites.[15] Several features about unwinding by dnaB protein at a chromosomal origin have been revealed by these reactions:

(1) The use of dnaB protein is efficient. Poor template binding by dnaB protein, which dictates a requirement for large excesses of protein, is overcome by the origin-initiator proteins. At *oriC*, the dnaA and dnaC proteins efficiently load dnaB protein at a ratio of only two to four molecules per template (i.e., one to two molecules per fork);[16] the λ O and λ P proteins[17] perform the same role at *ori*λ, and the preprimosomal proteins[18] do the same at a primosome assembly site (*pas*).

(2) Melting is origin-dependent and fueled by dnaB protein. The dnaB protein action can be supported by GTP or CTP, as well as by ATP; it is the only enzyme present that can utilize all these nucleotides. GTPγS blocks melting and specifically inhibits rNTP hydrolysis by dnaB protein.[19]

(3) Melting can be independent of most of the replication machinery. Melting is not inhibited by SSB, is processive, and does not require the presence of primase or DNA pol III holoenzyme, although when present these proteins may stimulate melting or stabilize the fork.[20]

The dnaB and dnaC proteins form a 6 : 6 complex that activates dnaB protein for entry into the *oriC*·dnaA protein complex or the *pas*·PriA protein complex (Section 8-4). An analogous λ P·dnaB protein complex functions during initiation at an *ori*λ complexed with λ O protein. The dnaB protein complexed either with dnaC or with λ P protein appears to be inactive as a helicase.[21] Only when

11. Mok M, Marians KJ (1987) *JBC 262*:16644.
12. Arai K, Kornberg A (1981) *JBC 256*:5253; Arai K, Kornberg A (1981) *JBC 256*:5260.
13. Baker TA, Funnell BE, Kornberg A (1987) *JBC 262*:6877; Baker TA, Sekimizu K, Funnell BE, Kornberg A (1986) *Cell 45*:53.
14. Dodson M, Echols H, Wickner S, Alfano C, Mensa-Wilmot K, Gomes B, LeBowitz J, Roberts JD, McMacken R (1986) *PNAS 83*:7638.
15. Mok M, Marians KJ (1987) *JBC 262*:16644.
16. Baker TA, Funnell BE, Kornberg A (1987) *JBC 262*:6877; Baker TA, Sekimizu K, Funnell BE, Kornberg A (1986) *Cell 45*:53.
17. Dodson M, Echols H, Wickner S, Alfano C, Mensa-Wilmot K, Gomes B, LeBowitz J, Roberts JD, McMacken R (1986) *PNAS 83*:7638.
18. Mok M, Marians KJ (1987) *JBC 262*:16644.
19. Baker TA, Sekimizu K, Funnell BE, Kornberg A (1986) *Cell 45*:53.
20. Baker TA, Funnell BE, Kornberg A (1987) *JBC 262*:6877.
21. Wahle E, Lasken RS, Kornberg A (1989) *JBC 264*:2469; Liberek K, Georgopoulos C, Zylicz M (1988) *PNAS 85*:6632; Alfano C, McMacken R (1989) *JBC 264*:10709.

dnaC protein or λ P protein is ejected from its respective complex, after delivery of dnaB protein, does the dnaB protein become active. Unlike the spontaneous release of dnaC protein, the ejection of λ P protein requires the action of heat-shock proteins[22] (Section 17-7).

Evidence that dnaB protein is the principal helicase for *E. coli* chromosomal replication is as follows. (1) Conditionally lethal mutations with a fast-stop replication defect indicate that dnaB protein is essential for fork movement in vivo.[23] (2) In purified in vitro systems, the dnaB protein is the only intrinsic helicase required for replication of plasmids that depend on *oriC* or *oriλ*. (3) When coupled to synthesis by DNA pol III holoenzyme, dnaB protein can melt the DNA at the rapid rate and high processivity expected of an enzyme participating in chromosomal replication.

The PriA Protein[24] (Protein n′, Factor Y)

The 76-kDa PriA protein (also called protein n′ and factor Y) is a component of the φX-type primosome, a mobile protein complex which synthesizes primers on DNAs that contain a site for its assembly (*pas*, primosome assembly site; Section 8-4). Recognition of *pas* by PriA protein is the first step in assembly and activates its DNA-dependent ATPase (or dATPase) activity. The PriA protein can provide mobility to the primosome[25] and displaces SSB in its path, allowing dnaB protein and primase to function.[26]

The ATP- (or dATP-) dependent helicase activity of the PriA protein displaces a single-stranded fragment from a partially duplex substrate with one molecule per template, does not require a forked duplex, and moves in a 3′→5′ direction. The presence of a *pas* sequence on the ssDNA portion of the template stimulates the helicase activity 10- to 15-fold. SSB is required for melting. Binding to the *pas* sequence, which forms a secondary structure resistant to melting by SSB, most likely explains why binding of PriA, unlike binding of other helicases, is not inhibited by SSB. Processive melting by PriA displaces fragments over 1 kb long, at about 50 bp per second, and is not diverted by challenge templates.

The Preprimosome and the Primosome[27]

The φX-type preprimosome is a mobile multiprotein complex whose assembly requires six *E. coli* replication proteins (PriA, PriB, PriC, dnaB, dnaC, and dnaT) and a *pas* sequence on the DNA template. Upon addition of primase to form the primosome, the complex synthesizes multiple primers while translocating along a template (Section 8-4).

The preprimosome contains the PriA and dnaB proteins and displays the helicase activities of both, as does the primosome (Fig. 11-4). The 3′→5′ helicase activity, which is supported only by ATP or dATP and acts at a flush end, is attributable to PriA protein, while the 5′→3′ activity, which uses ATP, GTP, or CTP but not dATP and requires a preformed fork, is characteristic of dnaB

22. Dodson M, Echols H, Wickner S, Alfano C, Mensa-Wilmot K, Gomes B, LeBowitz J, Roberts JD, McMacken R (1986) *PNAS 83*:7638; Alfano C, McMacken R (1989) *JBC 264*:10709.
23. Wechsler JA, Gross JD (1971) *MGG 113*:273.
24. Lee MS, Marians KJ (1987) *PNAS 84*:8345; Lasken RS, Kornberg A (1988) *JBC 263*:5512.
25. Lee MS, Marians KJ (1989) *JBC 264*:14531.
26. Shlomai J, Kornberg A (1980) *JBC 255*:6794.
27. Lee MS, Marians KJ (1989) *JBC 264*:14531; Mok M, Marians KJ (1987) *JBC 262*:2304.

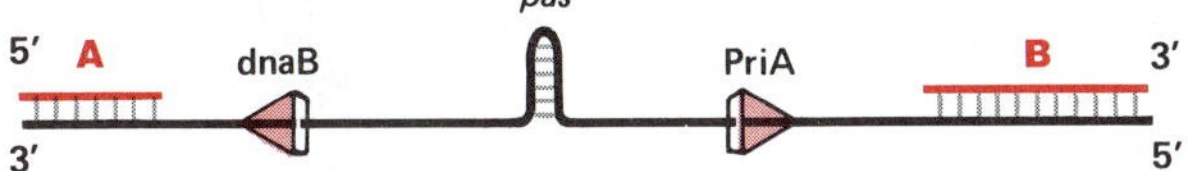

Enzyme	Fragment displaced	
	− SSB	+ SSB
dnaB*	A	—
Pri A	B	B
Primosome:		
+ GTP	nd	A
+ dATP	nd	B
+ ATP	nd	A, B

Figure 11-4
Conditions that reveal the bidirectional movement of the ϕX-type primosome containing two proteins, PriA and dnaB, with intrinsic helicase activity. Since 10 μM ATP is required for primosome assembly, it is also present along with GTP or dATP; nd = not determined, since primosome assembly requires SSB.

protein. An absolute requirement for the other preprimosomal proteins and the *pas* sequence, and the lack of inhibition by SSB, indicate that the helicase activities reflect an intact complex rather than loosely associated PriA and dnaB proteins.

The preprimosome thus contains two DNA binding sites and translocation activities that would appear to pull the complex in opposite directions. On single-stranded templates, both proteins are bound to the same strand. Possibly, other proteins, low-molecular-weight effectors, or template structures may influence the direction of primosome movement.[28] Alternatively, the two activities might work simultaneously if a loop of template were pulled through the protein complex (Section 15-6).

The response of the primosome to the concentration of NTPs and the ability of dnaB protein to support fork movement on its own suggest that this protein is the principal helicase. The activity of PriA protein may result from its capacity to translocate processively along a strand, rather than to melt a duplex. In a model that employs the two translocating activities of the primosome to couple continuous- and discontinuous-strand synthesis[29] (Section 15-6), the dnaB protein separates the DNA strands to generate the fork while the PriA protein, translocating behind it, causes the DNA to loop. A further conjecture is that one half of the dimeric DNA pol III holoenzyme acts in discontinuous-strand synthesis on the looped region, while the other half proceeds with continuous synthesis of the other strand.

Rep Protein

This 68-kDa helicase was identified by its absolute requirement during the rolling-circle replication phase in the life cycles of ϕX174 and Ff phages (Sections 17-3, 17-4); it is also essential for the growth of phage P2. However, Rep

28. Lee MS, Marians KJ (1989) *JBC 264*:14531.
29. Lee MS, Marians KJ (1989) *JBC 264*:14531; Lasken RS, Kornberg A (1988) *JBC 263*:5512.

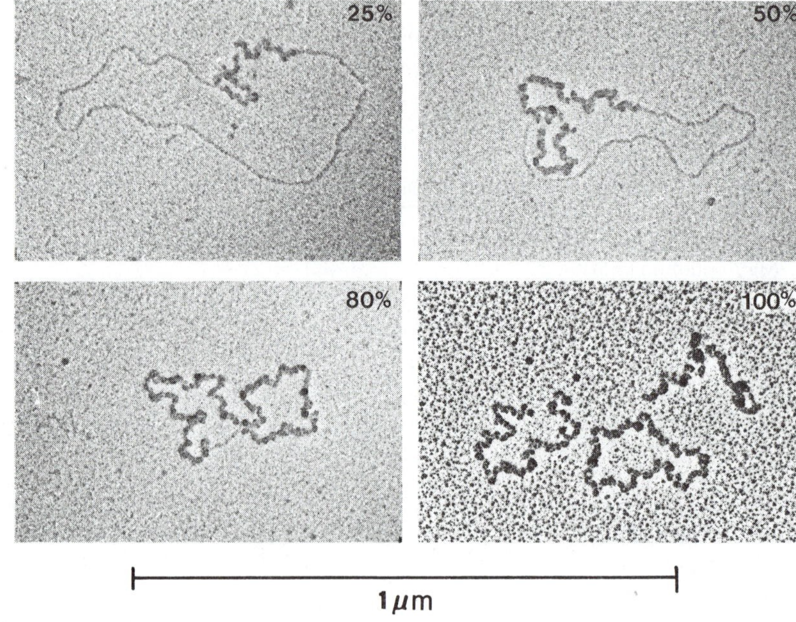

Figure 11-5
Looped rolling-circle intermediates in strand separation of the ϕX gpA–RFII complex catalyzed by Rep protein. The ssDNA generated is coated by SSB, giving it a thick, beaded appearance and reducing the contour length by a factor of 2.56. The intermediates shown here, obtained after 30 seconds of incubation, represent strand separations of 25%, 50%, and 80%. The final product was made up of linear and circular single strands in equal number.

protein is not essential for cell growth.[30] Mutations that block phage replication cause only an accumulation of replication forks and slower fork movement,[31] but not the dramatic cessation of DNA synthesis seen for genes intimately involved in replication *(dnaB, dnaG, dnaN)*. The homology of Rep protein to helicase II[32] (gene *uvrD* protein; Section 3) and the nonviability of *uvrD, rep* double mutants[33] suggest a role in repair synthesis. The ability of Rep to melt RNA–DNA hybrids[34] and the incompatibility of *rep* and *rho* mutations[35] suggest a role also in transcription (Section 3).

In concert with a phage initiator endonuclease (gpA for ϕX and gp2 for M13; see Section 8-7), Rep melts the duplex replicative form of DNA (RFII) after cleavage at the viral strand origin[36] (Fig. 11-5). A stable, isolable complex containing gpA and one molecule of Rep is competent to unwind the entire duplex, indicating the processivity of this helicase activity. Translocating in the 3′→5′ direction,[37] Rep separates the duplex strands, consuming two molecules of ATP (or dATP) for each base pair melted.[38] SSB is needed for this action, presumably to prevent the strands from reannealing.

30. Colasanti J, Denhardt DT (1987) *MGG 209*:382.
31. Lane HED, Denhardt DT (1975) *JMB 97*:99.
32. Gilchrist CA, Denhardt DT (1987) *NAR 15*:465.
33. Kumura K, Sekiguchi M (1984) *JBC 259*:1560; Taucher-Scholz G, Abdel-Monem M, Hoffmann-Berling H (1983) in *Mechanisms of DNA Replication and Recombination* (Cozzarelli NR, ed.). ICN-UCLA Symposia on Molecular and Cellular Biology, Vol. 10. Academic Press, p. 65.
34. Sheaff RJ, Matson SW, personal communication.
35. Fassler JS, Tessman I, Tessman ES (1985) *J Bact 161*:609.
36. Scott JF, Eisenberg S, Bertsch LL, Kornberg A (1977) *PNAS 74*:193; Eisenberg S, Griffith J, Kornberg A (1977) *PNAS 74*:3198.
37. Duguet M, Yarranton G, Gefter M (1978) *CSHS 43*:63; Yarranton GT, Gefter ML (1979) *PNAS 76*:1658.
38. Kornberg A, Scott JF, Bertsch LL (1978) *JBC 253*:3298.

Table 11-1
Characteristics of Rep protein interactions with single-stranded DNA and the replication fork[a]

Property	ssDNA	Replication fork
ATPase activity		
K_m for ATP or dATP (μM)	high (200)	low (25) and high (200)[b]
Action on other rNTPs, dNTPs	yes	none or little
ATP analogs (relative K_i)	high	low
Mg^{2+} optimum (mM)	broad (0.1–2.0)	sharp (5–10)
Effect of SSB	inhibits	activates
Effect of ϕX gpA	no effect	required
Binding		
Stoichiometry: one Rep molecule per	20 residues	fork
Salt inhibition (100 mM NaCl)	yes	no
Effect of ATP or ADP	destabilizes	no effect
Effect of SSB	inhibits	no effect
Effect of ϕX gpA	no effect	required
Kinetic mechanism	distributive	processive

[a] Arai N, Arai K, Kornberg A (1981) *JBC 256*:5287; Kornberg A, Scott JF, Bertsch LL (1978) *JBC 253*:3298.
[b] Enzyme has two apparent K_m values under these conditions.

The helicase activity of Rep protein can be observed on partially duplex molecules in the absence of gpA. In contrast to its behavior on the ϕX RFII molecules, Rep has only limited processivity on these partially duplex substrates. A duplex region of 70 bp is melted but not one of 350 bp,[39] even in the presence of SSB. The high processivity of Rep on the ϕX RF template is probably a result of its interaction with gpA. In Ff phage replication, the gp2 protein is required for replication fork movement, even when the template has been cleaved at the origin, suggesting that Rep and gp2 also interact throughout the unwinding.

Clear differences exist between the behavior of Rep when it acts alone on ssDNA and when it operates at the replication fork generated by gpA. As observed by ATPase and binding activities (Table 11-1), the character of Rep is fundamentally altered by interactions at the replication fork.

The importance of the phage proteins for Rep activity at a replication fork suggests the existence of a cellular equivalent of gpA to assist Rep. A small protein (15 kDa) has been purified[40] which stimulates the helicase but not the ATPase activity of Rep and apparently functions by binding to the DNA. Optimal stimulation requires 350 to 600 molecules of this 15-kDa protein per template, indicating a mechanism of action distinct from that of the phage proteins.

Phage T7 Gene *4* Protein (Gp4)

T7 gene 4 encodes both the helicase and primase required for phage DNA replication.[41] Two proteins are produced; the larger (63 kDa) has both activities, whereas the smaller (56 kDa), which lacks an N-terminal region of 63 amino

39. Smith KR, Yancey JE, Matson SW (1989) *JBC 264*:6119.
40. Smith KR, Yancey JE, Matson SW (1989) *JBC 264*:6119.
41. Bernstein JA, Richardson CC (1989) *JBC 264*:13066.

acids, also lacks the capacity to synthesize primers[42] (Section 8-5). Both forms also possess ssDNA-dependent NTPase activity, with dTTP as the preferred substrate. Hydrolysis drives the helicase in the 5′→3′ direction.[43] ATP, dATP, and the other ribotriphosphates and deoxytriphosphates, with the exception of CTP, also support duplex melting.

As with dnaB helicase, gp4 requires a preformed fork structure in order to melt the duplex portion of a hybrid molecule. In addition to the 5′ tail, required for gp4 binding, a 3′ tail of 6 to 7 ntd is essential for helicase action. Helicase activity has also been analyzed coupled to DNA synthesis on a nicked circular template with a 5′ tail.[44] The helicase moves 5′→3′ into the duplex, while elongation at the 3′ junction by T7 polymerase generates a rolling-circle replication complex. In this configuration, gp4 melts the template processively, moving at about 300 bp per second in a manner appropriate for a replicative helicase.

The central role of gp4 at the replication fork is reflected in its interactions with other replication proteins. A single molecule of gp4 is not sufficient for both melting the duplex and synthesizing primers. Although both activities are intrinsic to the large form of the protein, the helicase acts processively, while the primase does not; priming is inhibited by dilution of the reaction mixture or by the addition of a challenge DNA template.[45] Thus, the primase continually cycles on and off the template during replication (Section 8-5), much as the host primase appears to do (Section 8-4). Possibly, the large and small forms move together, functioning as a helicase on the template for the discontinuous strand. When a priming site is encountered, the large form stops to synthesize a primer while the small form continues melting the fork. Assuming that the N-terminal region of gp4 (found only in the large form) contains the specific sequence-recognition domain for primer synthesis, the absence of this region in the small form might make the small form more efficient as a helicase because it would not be diverted by primase recognition sites.

The ability of gp4 to stimulate displacement synthesis by T7 DNA polymerase but not by the *E. coli* or T4 enzymes suggests a specific interaction between T7 polymerase and gp4. In fact, these proteins appear to form a complex on the DNA template.[46] The presence of the polymerase relieves the requirement for a nucleotide in binding gp4 to DNA. *E. coli* SSB and its T7 equivalent (gene *2.5* protein) have different effects on the T7 primase activity,[47] another indication that the T7 replication proteins — the polymerase and gp2.5 with the gp4 helicase–primase as the central element — form a discrete interactive complex.

Phage T4 Gene *41* Protein (Gp41)

Gp41 is the major helicase in T4 DNA replication (Fig. 11-6).[48] Combined with gp61, it forms the T4 primosome responsible for melting the duplex during continuous-strand synthesis and for the multiple priming events during discontinuous-strand synthesis. Gp41 (59 kDa) is a monomer in solution but probably functions as a multimer at the replication fork.[49] Incubation of the protein with

42. Bernstein JA, Richardson CC (1988) *PNAS 85*:396.
43. Matson SW, Tabor S, Richardson CC (1983) *JBC 258*:14017.
44. Lechner RL, Richardson CC (1983) *JBC 258*:11185.
45. Nakai H, Richardson CC (1988) *JBC 263*:9818.
46. Nakai H, Richardson CC (1986) *JBC 261*:15208.
47. Nakai H, Richardson CC (1988) *JBC 263*:9831.
48. Venkatesan M, Silver LL, Nossal NG (1982) *JBC 257*:12426.
49. Venkatesan M, Silver LL, Nossal NG (1982) *JBC 257*:12426; Richardson RW, Nossal NG (1989) *JBC 264*:4725.

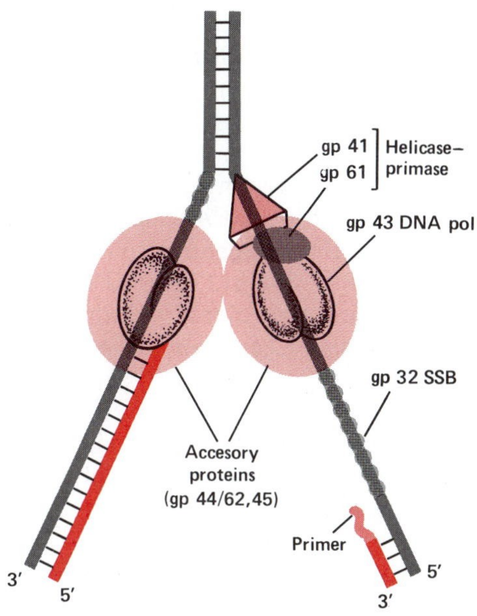

Figure 11-6
Model of the proposed interaction of T4 polymerase (gp43) and its accessory proteins (gp 44/62 and 45) with the helicase–primase complex (gp 41/61) at the replication fork; T4 SSB (gp32) binds to any ssDNA.

GTP or GTPγS stimulates multimer formation. Maximal activity of the gp41/61 complex occurs at a 5 : 1 ratio of the two proteins. Single-stranded DNAs activate hydrolysis by gp41: GTP is the optimal NTP; ATP, dATP, and dGTP are also used but at slower rates. Circular or long linear molecules are the best NTPase effectors, indicating that gp41 tracks processively along the DNA.[50]

On a partially duplex DNA substrate, gp41 moves 5′→3′ in displacing a strand. A 3′ tail of 29 ntd or longer is needed on the strand to be displaced.[51] Gp61 stimulates gp41 activity, most notably when the 3′ tail is lacking at the duplex junction; presumably gp61 influences the initial fork formation. Even without forming a primer, gp61 stimulates gp41 in melting long duplexes by stabilizing the gp41 · DNA complex. Phage T4 DNA polymerase accessory proteins (Section 5-8) also stimulate helicase activity.[52]

The helicase activity of gp41 is evident when coupled to rolling-circle DNA synthesis.[53] On a nicked duplex (RFII) with a 5′ tail, gp41 bound to the tail combines with T4 polymerase holoenzyme (gp43, 44/62, and 45; Section 5-8) to promote extensive rolling-circle replication. Strand separation proceeds at 250 bp per second and is highly processive. With this template, neither gp32 (the T4 SSB) nor gp61 stimulates helix melting, in contrast to their effects on the partially duplex substrates. When the template for displacement synthesis is simply a nicked circle (lacking the 5′ tail), gp32 is essential, suggesting a role in fraying the DNA at the nick to allow the loading of gp41.

Proteolytic cleavage of gp41 to remove the C-terminal 20 amino acids identifies a region of the protein involved in coupling helicase and polymerase actions.[54] The truncated protein has normal ssDNA-dependent NTPase activity and can still activate primer synthesis by gp61 (Section 8-5); however, it fails to catalyze strand displacement synthesis with T4 polymerase and is inhibited by

50. Morris CF, Moran LA, Alberts BM (1979) *JBC 254*:6797; Liu CC, Alberts BM (1981) *JBC 256*:2813.
51. Richardson RW, Nossal NG (1989) *JBC 264*:4725.
52. Venkatesan M, Silver LL, Nossal NG (1982) *JBC 257*:12426.
53. Cha TA, Alberts BM (1989) *JBC 264*:12220.
54. Richardson RW, Nossal NG (1989) *JBC 264*:4732.

gp32. This inhibition is not relieved by the T4 polymerase accessory proteins, in contrast to their effect on full-length gp41 when inhibited by gp32.

Phage Analogs of dnaB Protein

P22 Gp12. Gene *12* of phage P22 of *Salmonella typhimurium* encodes a protein that substitutes for the host *dnaB* function during phage replication. The gp12 and dnaB proteins are homologous;[55] both have ssDNA-dependent ATPase activity and the ability to activate priming by primase.[56] Gp12 most likely possesses helicase activity as well, although this is not shown directly. P22 replication is independent of both *dnaB* and *dnaC*. Thus, gp12 appears to be a dnaB-like protein that does not require dnaC protein.

P1 Ban Protein. The P1 *ban* (*b* *an*alog) product is a dnaB-like protein made during the phage lytic cycle which renders replication independent of *dnaB*. Constitutive expression of the *ban* gene from a prophage suppresses the ts phenotype of *dnaB* mutants for chromosomal replication as well.[57] The Ban protein probably has all the properties of the dnaB protein that are required for host cell replication, including helicase activity. Action of Ban protein requires *dnaC* function,[58] and the dnaB and Ban proteins form active heteromultimers.[59]

Summary of the Properties of Replicative Helicases

The dnaB protein, T7 gp4, and T4 gp41 (Table 11-2) form a family of replicative helicases with similar properties in unwinding duplex DNA and in complex

Table 11-2
Comparison of *E. coli* and phage replicative helicases

Protein	Mass (kDa), form	Polarity	Rate (bp/sec)	Processivity	Effect of primases	Effect of DNA polymerase
dnaB	50 hexamer	5′→3′	35–60 750[b]	high[b] > 50 kb	stimulated by *E. coli* dnaG	stimulates rolling-circle synthesis
T7 gp4	63, 56[a] multimer	5′→3′	300[b]	high[b]	same protein	specific for T7 pol
T4 gp41	59 multimer	5′→3′	250[b]	high[b]	stimulated by T4 gp61	accessory proteins stimulate
PriA	76 monomer	3′→5′	50	< 500 bp	none[c]	none
Rep	68 monomer	3′→5′	150[d]	high[d] < 500 bp alone	none	none

[a] Two forms, 63 and 56 kDa; the large form also has primase activity.
[b] When DNA polymerase is present during rolling-circle synthesis.
[c] Part of primosome.
[d] In the presence of gpA endonuclease.

55. Backhaus H, Petri JB (1984) *Gene 32*:289.
56. Wickner S (1984) *JBC 259*:14038; Wickner S (1984) *JBC 259*:14044.
57. D'Ari R, Jaffe-Brachet A, Touati-Schwartz D, Yarmolinsky M (1975) *JMB 94*:341.
58. Hay N, Cohen G (1983) *Virology 131*:193.
59. Lanka E, Mikolajczyk M, Schlicht M, Schuster, H (1978) *JBC 253*:5847.

formation at a replication fork. These three proteins processively unwind the duplex, moving with a 5′→3′ polarity. At a replication fork, this would place them on the discontinuous-strand template, moving in the same direction as the continuous-strand polymerase but on the opposite DNA strand. Melting requires, or is most efficient with, a duplex containing a preformed fork structure, indicating that these proteins cannot initiate melting and that they may bind both strands at the replication fork.

These helicases all interact with a DNA primase, promoting the initiation of primer synthesis. In coupled synthesis of the continuous and discontinuous strands, they melt the duplex for continuous-strand synthesis while activating primer synthesis by primase for the discontinuous strand. Interactions of helicases with DNA polymerases are less clear, yet the effects of DNA polymerase on the rate of melting and the stability of the helicase–DNA complex are evident. The distance that separates the helicase from the polymerase on DNA appears small, indicating that melting is what limits the rate of DNA synthesis.

The properties of PriA protein and Rep protein (Table 11-2) are distinct from those of the dnaB helicase class. PriA and Rep move in the 3′→5′ direction and interact with other replication proteins less directly. Although involved in the priming of chain starts, PriA protein has not been shown to drive replication fork movement. Rather than functioning in the melting of the duplex at the fork, its helicase activity may reflect the capacity to move on ssDNA. On the other hand, Rep protein, the first helicase discovered to have a replication role, is identified clearly only in certain life-cycle stages of single-stranded phages; the contributions of Rep protein to the host cell are less clear.

11-3 Prokaryotic Helicases That Function in Repair, Recombination, Conjugal Transfer, or Transcription

Transactions on DNA often involve at least transient melting of the duplex. The repair of DNA damage, like replication, requires a single-stranded template. Pairing of homologous regions of DNA during genetic recombination may require both the melting of duplex DNA for strand invasion and the generation of extended single-stranded segments. The displacement of a single strand from a duplex genome for transfer to a new cell during conjugation clearly requires DNA melting. Finally, transcription termination calls for melting of the R-loop within the transcription bubble. The properties of the various nonreplicative prokaryotic helicases that serve in these functions are summarized in Table 11-3.

Repair and Recombination[60]

***E. coli* Helicase II.** Present at much higher levels than the replicative helicases, helicase II appears to contribute to both DNA repair and recombination. Cells deficient in helicase II (*uvrD* mutants) are sensitive to DNA-damaging agents and exhibit an increased rate of spontaneous mutations, manifestations of a

60. Matson SW, Kaiser-Rogers KA (1990) *ARB* 59:289.

Table 11-3
Nonreplicative prokaryotic helicases

Protein	Mass (kDa)	Polarity	Comments
Helicase II (*uvrD* gene product)	82[a]	3′→5′	stoichiometric mechanism; DNA repair function
UvrAB complex	180[b]	5′→3′	DNA repair function
Helicase IV	75[a]	3′→5′	possible function in repair; low intracellular abundance
RecBCD complex	330[b]	both	involved in recombination
T4 Dda	49[a]	5′→3′	stoichiometric mechanism; displaces proteins bound to DNA
Helicase I	180[a]	5′→3′	encoded by F plasmid; required for DNA transfer during conjugation
Rho	50[a]	5′→3′	specific for RNA–DNA hybrids; functions in transcription termination at specific sites
Helicase III	20[a]	5′→3′	unknown function

[a] Monomer.
[b] Complex.

defect in DNA repair. Furthermore, *polA uvrD* double mutants are not viable, indicating that this repair role is essential under some circumstances.[61]

Helicase II, in contrast to many helicases, does not require a DNA tail to bind and melt a duplex fragment; fragments with blunt (flush) ends or with 3′ or 5′ tails, forked molecules, and nicked circles are all substrates.[62] Thus, DNA melting may be initiated in a duplex region rather than at an end; a nick on circular DNA might serve only to relieve the topological strain that accompanies melting. Yet ssDNA appears to be the preferred effector of the DNA-dependent ATPase (or dATPase) activity. Helicase II unwinds DNA–RNA hybrids more efficiently than it does duplex DNA,[63] a capacity without a defined physiologic role.

Melting of fragments with flush ends makes it difficult to determine the polarity of a helicase. Measurement depends on the preference of the enzyme for loading onto partially duplex molecules via the single-stranded regions, as with substrates used to determine polarity (Fig. 11-1d and e). Although helicase II appears to displace fragments from both directions, the 3′→5′ polarity dominates at low enzyme concentrations.[64] The apparent 5′→3′ movement can be explained by a 3′→5′ attack on the duplex from the flush (external) end.

Unlike the replicative helicases, helicase II must be present at a high level for initiation of unwinding. Half-maximal activity is achieved at a ratio of 1 monomer per 1.6 nucleotides of substrate DNA. This ratio is independent of the length of the DNA and is similar for blunt-end molecules and nicked circles. Although the number of protein monomers bound during melting is not clear, the levels

61. Kushner SR, Shepherd J, Edwards G, Maples VR (1978) in *DNA Repair Mechanisms* (Hanawalt PC, Friedberg EC, Fox CF, eds.). ICN-UCLA Symposia on Molecular and Cellular Biology, Vol. 9. Academic Press, p. 251; Arthur HM, Lloyd RG (1980) *MGG 180*:185; Siegel EC (1973) *J Bact 113*:161.
62. Runyon GT, Lohman TM (1989) *JBC 264*:17502.
63. Matson SW (1989) *PNAS 86*:4430.
64. Matson SW (1986) *JBC 261*:10169.

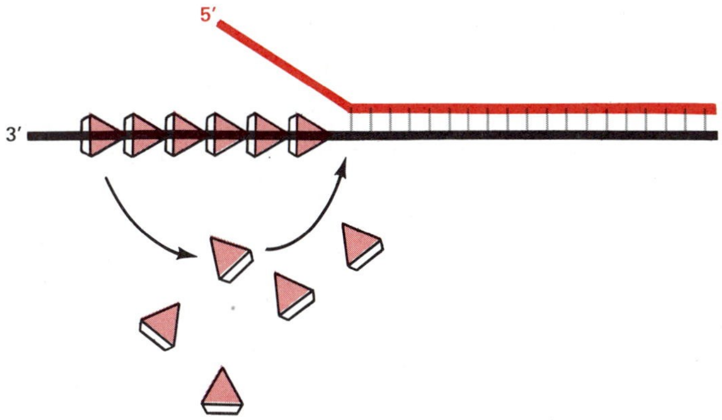

Figure 11-7
Melting of duplex DNA by stoichiometric levels of helicase II; this may proceed by unidirectional loading of the protein along the DNA.

needed are sufficient to coat the single-stranded products (Fig. 11-7). Consistent with a stoichiometric mechanism for helicase II, its abundance is between 3000 and 50,000 molecules per cell, compared to 10 to 100 molecules for the catalytic dnaB, PriA, and Rep proteins.

Both the excision–repair pathway and the methyl-directed mismatch repair pathway (Section 21-2) depend on helicase II. In the excision of adducts caused by DNA-damaging agents, helicase II acts with DNA polymerase I to release a single-stranded fragment of 12 to 13 ntd generated by the UvrABC excinuclease (excision-nuclease) complex[65] (Sections 13-9 and 21-2). This activity seems to depend on the special capacity of helicase II to initiate melting at a nick. Helicase II and pol I are also required for turnover of the UvrABC excinuclease; in their absence, the UvrABC complex participates in only a single repair event.[66] In methyl-directed mismatch repair (Section 21-2), helicase II probably displaces a fragment as long as several kilobases between the hemimethylated site and the mismatched base pair cleaved by the MutSLH complex[67] (Section 21-2). After removal of the damaged fragment, the gap can be filled by DNA pol III holoenzyme.

The increased frequency of recombination observed in some *uvrD* mutants suggests that helicase II is an "antirecombinase." By unwinding fortuitously paired regions, it may inhibit recombination between duplicated regions of DNA.

***E. coli* UvrAB Complex.**[68] The A and B components of the UvrABC excinuclease form a complex that has a helicase activity capable of melting D-loops and short stretches of duplex DNA. Movement is in the 5′→3′ direction and does not require a preformed fork in the substrate; 10 ntd of ssDNA at the 5′ end are required for enzyme binding. Melting is supported by hydrolysis of ATP or dATP but not by hydrolysis of other nucleotides. How this helicase activity is

65. Sancar A, Sancar GB (1988) *ARB 57*:29.
66. Caron PR, Kushner SR, Grossman L (1985) *PNAS 82*:4925; Husain I, Van Houten B, Thomas DC, Abdel-Monem M, Sancar A (1985) *PNAS 82*:6774; Selby CP, Sancar A (1988) *Biochemistry 27*:7184.
67. Modrich P (1989) *JBC 264*:6597.
68. Oh EY, Grossman L (1987) *PNAS 84*:3638; Oh EY, Grossman L (1989) *JBC 264*:1336.

combined with that of helicase II in the action of UvrABC excinuclease is unknown.

***E. coli* Helicase IV.**[69] Purified on the basis of its DNA melting activity, helicase IV has properties similar to those of helicase II and Rep. It is an ssDNA-dependent ATPase (or dATPase) and melts DNA with a 3′→5′ polarity. A limited processivity is indicated by an efficient displacement of a 70-bp fragment but a much poorer activity against one that is 350 bp long; SSB increases the rate of unwinding of these longer regions about fourfold. The nucleotide sequence of *helD*, the gene for helicase IV, reveals substantial homology to the sequences for helicase II and Rep, although helicase IV is not as closely related to these two proteins as they are to one another (Section 5). Helicase IV was also identified as the protein responsible for suppressing the lethality of *uvrD*^{ts} *rep3* double mutants, indicating that it can substitute for an essential function provided by either helicase II or Rep protein.[70] Helicase IV is present at only about 10 molecules per cell, and its role in normal cellular metabolism remains obscure.

***E. coli* RecBCD Complex.**[71] The helicase activity of this 330-kDa protein complex is revealed under conditions in which the endo- and exonuclease activities are minimized. Melting starts at the end of a duplex molecule. In contrast to most other helicases, RecBCD binds both strands of DNA during unwinding. Tails (3′ or 5′) on a duplex inhibit rather than promote the helicase action. The DNA-dependent ATPase is activated by dsDNA more effectively than by ssDNA, a further indication that binding to the duplex form is preferred. Because the complex binds to both strands, an assignment of polarity to helicase movement is not meaningful.

Hydrolysis of ATP (or dATP), but not that of the other NTPs, fuels unwinding. Two to three ATPs are hydrolyzed per base pair unwound. Movement is processive and rapid, to the extent that 25 kb of DNA can be melted in one association event at a rate of 250 to 470 bp per second.

Translocation on the DNA duplex occurs at different rates on the two strands, so that a loop of ssDNA is formed behind the moving complex (Fig. 11-8). Both the RecB and RecD polypeptides bind ATP.[72] The B subunit, acting alone, also binds DNA and hydrolyzes ATP, making it the likely helicase candidate.[73] Furthermore, the *recB* nucleotide sequence is homologous to those of the *rep* and *uvrD* family (Section 5). The D subunit also shows a weak amino acid sequence homology to the protein products of this family. Possibly, both the B and D subunits translocate along DNA, one on each strand, and thus account for the duplex DNA binding feature of the intact complex.

Mutations in *recB* and *recC* create a sensitivity to DNA-damaging agents, a lowered cell viability, and a deficiency in recombination. Alleles that affect only helicase activity are specifically impaired in recombination. During DNA-pairing reactions in vitro, the RecBCD helicase generates ssDNA that can be utilized by RecA protein in homologous strand exchanges[74] (Section 21-5).

---

69. Wood ER, Matson SW (1987) *JBC 262*:15269; Wood ER, Matson SW (1989) *JBC 264*:8297.
70. Matson SW, Kaiser-Rogers KA (1990) *ARB 59*:289.
71. Taylor A, Smith GR (1980) *Cell 22*:447; Telander-Muskavitch KM, Linn S (1982) *JBC 257*:2641; Roman LJ, Kowalczykowski SC (1989) *Biochemistry 28*:2863.
72. Julin DA, Lehman IR (1987) *JBC 262*:9044.
73. Lieberman RP, Oishi M (1974) *PNAS 71*:4816; Hickson ID, Robson CN, Atkinson KE, Hutton L, Emmerson PT (1985) *JBC 260*:1224.
74. Roman LJ, Kowalczykowski SC (1989) *JBC 264*:18340.

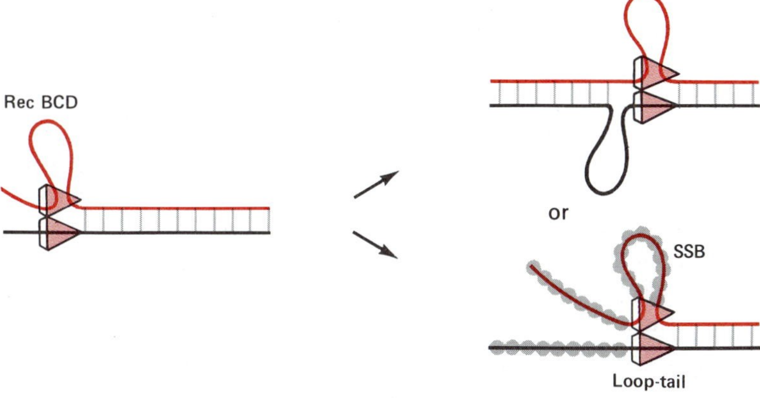

Figure 11-8
Loop formation during melting and translocation by RecBCD enzyme, based on structures seen in electronmicrographs. [After Taylor A, Smith G (1980) *Cell 22*:447]

T4 Dda Protein.[75] This 49-kDa protein, first recognized as a *D*NA-*d*ependent *A*TPase, is a T4-encoded helicase, as is gp41 (Section 2). The Dda protein melts duplex DNA, using the energy of ATP (or dATP) hydrolysis and moving in the 5′→3′ direction. A 5′ tail of at least 12 ntd is required for enzyme binding; an additional 3′ tail stimulates unwinding but is not essential.

As with *E. coli* helicase II, a high ratio of Dda protein to DNA (1 monomer per 1 to 3 nucleotides) is required for efficient unwinding. Competitive inhibition of melting by ssDNA is in keeping with a stoichiometric binding mechanism.

Although the role of Dda protein in T4 DNA metabolism is unknown, it contributes to replication in vitro in several ways. (1) It stimulates strand displacement synthesis by T4 DNA polymerase and T4 SSB (gp32).[76] (2) It removes proteins bound to the template in the path of the replication fork and thus overcomes the inhibition of replication by concurrent transcription.[77] (3) It contributes to recombination-dependent DNA replication, catalyzed by T4 DNA polymerase, gp32, and UvsX protein[78] (Section 17-6), by increasing the rate of strand exchange by UvsX protein.[79]

Conjugal Transfer:[80] Helicase I (F Plasmid Gene *traI* Protein)

Helicase I, the 180-kDa product of the F plasmid *traI* gene, promotes the transfer of DNA between *E. coli* cells in conjugation.[81] A single strand is displaced and transferred after the introduction of a nick into the plasmid genome at the *oriT* sequence (Section 18-7). Replication in both the donor and recipient cells, occurring along with or after the transfer, regenerates the duplex state. To effect the F transfer, the entire F plasmid, 100 kb long, must be melted by helicase I; the F-transfer functions can also mobilize the bacterial chromosome.

75. Jongeneel CV, Formosa T, Alberts BM (1984) *JBC 259*:12925; Kuhn B, Abdel-Monem M, Krell H, Hoffmann-Berling H (1979) *JBC 254*:11343.
76. Jongeneel CV, Bedinger P, Alberts BM (1984) *JBC 259*:12933.
77. Formosa T, Alberts BM (1986) *Cell 47*:793.
78. Kodadek T, Alberts BM (1987) *Nature 326*:312.
79. Kodadek T, Alberts BM (1987) *Nature 326*:312.
80. Ippen-Ihler KA, Minkley EG Jr (1986) *ARG 20*:593; Willets N, Shurray R (1987) in *Escherichia coli and Salmonella typhimurium* (Neidhardt FC, ed.). Am Soc Microbiol, Washington DC, Vol. 2, p. 1110.
81. Abdel-Monem M, Taucher-Scholz G, Klinkert MQ (1983) *PNAS 80*:4659.

Helicase I[82] unwinds DNA with a 5′→3′ polarity[83] after binding a 5′ tail on a duplex fragment. Hydrolysis of any of the NTPs supports unwinding. Although a large multimer of helicase I was thought to be the active form, efficient melting can take place at a low ratio of protein to template.

Melting by helicase I is very processive.[84] The same level of enzyme is required to melt duplexes of different lengths; challenges by ssDNAs are not inhibitory. As a result of this high processivity, circular ssDNAs are superior to linear molecules as effectors of the DNA-dependent ATPase. Circles support a strong ATPase, whereas the activity with linearized DNA is very feeble; yet a duplex region is melted on either form. Presumably, the end of a linear molecule is reached rapidly by the processive helicase and the protein is slow to dissociate or to reassociate with another DNA; on a circular molecule, the helicase can traverse the DNA endlessly. A slow rate of dissociation or reassociation can also explain the lack of enzyme turnover in a helicase assay.

During conjugal DNA transfer, helicase I most likely initiates melting from the nick at *oriT* even though a single-stranded region longer than 200 ntd is required for enzyme binding in vitro. In vivo, the *traI* product appears to be responsible for nicking as well as unwinding.[85] Recognition and cleavage of *oriT* may account for the plasmid-specific nature of helicase I function in vivo (Section 18-7).

Transcription: *E. coli* Rho Protein, an RNA–DNA Helicase[86]

The transcription termination factor, Rho (Section 7-7), catalyzes ATP-dependent melting of RNA–DNA hybrids. Strand separation proceeds 5′→3′ with respect to the RNA, to which the enzyme is bound. Effective substrates are RNA molecules containing sequences similar to Rho-dependent terminators. Lack of secondary structure in the RNA and the presence of C residues appear to be important (Section 7-7). The Rho protein, a hexamer of 50-kDa subunits, catalyzes unwinding at a molecular ratio of about one Rho to one RNA transcript.

During transcription termination, Rho may bind to the nascent RNA transcript at a region with appropriate sequence and structure. Pausing of transcription occurs, probably due to a direct interaction between Rho and RNA polymerase. Melting of the RNA–DNA duplex then ensues, leading to release of the transcript.

Unknown Function: *E. coli* Helicase III[87]

This 20-kDa enzyme, purified from cells deficient in Rep protein, possesses 5′→3′ DNA melting and ssDNA-dependent ATPase (or dATPase) activity. No role for helicase III has been found in any in vitro system of replication, recombination, or repair, and the responsible gene has not yet been identified. A covalent intermediate of helicase III and phosphate suggests a mechanism for coupling ATP hydrolysis to helicase action (Section 5).

---

82. Abdel-Monem M, Hoffmann-Berling H (1976) *EJB 65*:431.
83. Lahue EE, Matson SW (1988) *JBC 263*:3208.
84. Lahue EE, Matson SW (1988) *JBC 263*:3208.
85. Traxler BA, Minkley EG Jr (1988) *JMB 204*:205.
86. Brennan CA, Dombroski AJ, Platt T (1987) *Cell 48*:945.
87. Yaranton GT, Das RH, Gefter ML (1979) *JBC 254*:11997; Yaranton GT, Das RH, Gefter ML (1979) *JBC 254*:12002; Das RH, Yarranton GT, Gefter ML (1980) *JBC 255*:8069.

11-4 Eukaryotic Helicases[88]

Proteins from many sources have the capacity to unwind DNA duplexes, RNA–DNA hybrids, and duplex RNA. For lack of adequate genetic foundations and purified systems for replication, recombination, and repair, the functions of most of these helicases are still unknown. The exceptions are certain viral systems in which the in vitro replication is advanced and genetic analysis is available.

Viral Replicative Helicases

SV40 Large T Antigen.[89] The only viral protein required for SV40 replication (Section 19-2), the large T antigen (92 kDa as a monomer) binds specifically to the origin sequence, unwinds it during initiation, and has NTP-dependent DNA and RNA helicase activities. By genetic analysis, T antigen is essential for replication.

The intrinsic helicase activity can be detected by the displacement of a single-stranded fragment from a partially duplex DNA. Translocation is in the 3′→5′ direction, and depends on NTP hydrolysis, with ATP preferred. Despite preferential binding to the SV40 origin, T antigen can also bind nonspecifically to initiate duplex melting. A 3′ tail of 5 to 10 ntd stimulates binding, possibly because T antigen associates preferentially with the junction between single- and double-stranded DNA. A 5′ tail or preformed fork does not stimulate melting.

The rate of duplex melting is only 75 to 100 bp per minute, compared to that of 300 to 800 bp per second achieved by the prokaryotic replicative helicases. This slow rate is similar to that of fork movement on an SV40 chromosome (about 145 bp per minute) both in vivo and in vitro. The processivity of T antigen is high; short and long regions of duplex are melted with like efficiency, and ssDNA, when added after the reaction is under way, offers no challenge.

On both linear and circular SV40 templates, melting starts and proceeds bidirectionally from the origin. With the host replication machinery in place, T antigen continues to open the duplex, even to completion of replication. The active form of T antigen is two hexamers. An SSB from either HeLa cells or *E. coli* enables T antigen to open the SV40 origin, but replication specifically requires the mammalian SSB.

Although T antigen can unwind the entire SV40 template in vitro, it may be replaced or augmented by a cellular helicase during replication. T antigen differs from the prokaryotic replicative helicases (Table 11-2), the 5′→3′ polarity of which keeps them on the template for the discontinuous strand at the replication fork. The 3′→5′ movement of T antigen would imply translocation on the continuous strand template ahead of the polymerase. Also unlike the prokaryotic helicases, T antigen has no associated primase, although it appears to interact directly with DNA polymerase α (Section 6-2), the intrinsic primase of which makes it a candidate for synthesizing the discontinuous strand. Thus

88. Matson SW, Kaiser-Rogers KA (1990) *ARB 59*:289.
89. Stahl H, Dröge P, Knippers R (1986) *EMBO J 5*:1939; Goetz GS, Dean FB, Hurwitz J, Matson SW (1988) *JBC 263*:383; Wiekowski M, Schwarz MW, Stahl H (1988) *JBC 263*:436; Scheffner M, Wessel R, Stahl H (1989) *NAR 17*:93.

the organization of polymerases, primase, helicase, and SSB at the eukaryotic replication fork remains an important question.

T antigen melts duplex RNA as well as DNA.[90] In a reaction similar to that on DNA, melting proceeds in the 3′→5′ direction and is independent of the sequence of the substrate. The NTP cofactors optimal for RNA melting differ from those for DNA. TTP and UTP are preferred, whereas ATP and dATP are more efficient for DNA; CTP, dCTP, GTP, and dGTP also work better for helicase action on RNA than on DNA. These nucleotide preferences are also manifested by T antigen in its RNA- and DNA-dependent NTPase activities. The localization of some T antigen in the cytoplasm, bound to ribonucleoproteins, suggests that the efficient RNA melting observed in vitro may have a biological function.

Herpesvirus Helicases. Four of the seven proteins required for HSV-1 DNA replication[91] are associated with helicase activity.

A helicase–primase complex is composed of the products of genes *UL51* (~120 kDa), *UL5* (~97 kDa), and *UL8* (~70 kDa).[92] This complex possesses DNA-dependent ATPase and GTPase activities in addition to melting and primase activities. The roles of the individual subunits have not been established; however, the amino acid sequence encoded by the *UL5* gene shows homologies to the Rep helicase class[93] (Section 5), suggesting that this polypeptide is a helicase.

The preferred substrate for DNA melting by the helicase–primase complex is a partially duplex DNA molecule with a 3′ tail on the fragment to be displaced (a preformed fork), indicating a 5′→3′ translocation. This polarity and the association with primase resemble the properties of helicase–primase complexes in prokaryotic replication.

The fourth protein, the product of the *UL9* gene (~83 kDa), binds specifically to the viral origin sequences,[94] is essential for replication in vivo,[95] and has an intrinsic 3′→5′ helicase activity.[96] The protein sequence contains an ATP-binding consensus sequence similar to that found in the SV40 large T antigen. Thus, the *UL9* gene protein and the T antigen, with similar features, may have analogous roles in the initiation of genome replication.

Cellular Helicases

Fractionation procedures that remove nuclease activities which interfere with helicase assays have uncovered several DNA-melting enzymes (Table 11-4). Aside from the viral enzymes and the yeast RAD3 protein, associated with excision repair, their functions have not been defined.

11-5 Homologies and Mechanisms

Recurring motifs in the amino acid sequences of helicases offer clues to the structural and functional organization of members of a family. Such motifs,

90. Scheffner M, Knippers R, Stahl H (1989) *Cell 57*:955.
91. Wu CA, Nelson NJ, McGeoch DJ, Challberg MD (1988) *J Virol 62*:435.
92. Crute JJ, Mocarski ES, Lehman IR (1988) *NAR 16*:6585; Crute JJ, Tsurumi T, Zhu LA, Weller SK, Olivo PD, Challberg MD, Mocarski ES, Lehman IR (1989) *PNAS 86*:2186.
93. McGeoch DJ, Dalrymple MA, Dolan A, McNab D, Perry LJ, Taylor P, Challberg MD (1988) *J Virol 62*:444.
94. Elias P, O'Donnell ME, Mocarski ES, Lehman IR (1986) *PNAS 83*:6322; Elias P, Lehman IR (1988) *PNAS 85*:2959.
95. Wu CA, Nelson NJ, McGeoch DJ, Challberg MD (1988) *J Virol 62*:435.
96. Challberg M, personal communication; Bruckner RC, Crute JJ, Dodson MS, Lehman IR (1991) *JBC 266*:2669.

Table 11-4
Eukaryotic helicases

Protein; source	Mass[a] (kDa)	Polarity	NTPs	Comments
T antigen; SV40 infected cells	92	3′→5′	ATP, dATP, CTP dCTP, GTP, dGTP	melts duplex RNA in addition to DNA
Helicase–primase; HSV-1 infected cells	120 97 70	5′→3′	ATP, GTP	amino acid sequence suggests the 97-kDa subunit (*UL5* gene product) as the helicase.
ATPase III; yeast[b]	63	nd[i]	ATP, dATP	stimulates yeast pol I
RAD3; yeast[c]	90	5′→3′	ATP, dATP	essential gene; excision repair
ATPase B; mouse[d]	58	5′→3′	ATP, dATP, UTP, CTP, GTP	ss- and dsDNA-dependent NTPase; melts > 140bp
Helicase; *Xenopus*[e]	140	nd	ATP, dATP	not homogeneous; two subunits: 75 and 62 kDa
pol α associated[f]; calf thymus	47	nd	ATP, dATP, CTP, dCTP	low K_m for ATP hydrolysis (< 200 μM)
pol δ associated[g]; calf thymus	nd	nd	ATP, dATP	confers strand displacement activity to pol δ
Helicase; lily[h]	nd	none	ATP	unwinds duplex from either end; binds to nicks; recombination

[a] Subunit masses are listed except for the *Xenopus* enzyme, which is the apparent native mass.
[b] Sugino A, Ryu BH, Sugino T, Naumovski L, Friedberg E (1986) *JBC 261*:11744.
[c] Sung P, Prakkash L, Matson SW, Prakash S (1987) *PNAS 84*:8951.
[d] Seki M, Enomoto T, Yanagisawa J, Hanaoka F, Ui M (1988) *Biochemistry 27*:1766.
[e] Poll EHA, Benbow RM (1988) *Biochemistry 27*:8701.
[f] Hübscher U, Stalder HP (1985) *NAR 13*:5471; Thömmas P, Hübscher U (1990) *JBC 265*: 14347.
[g] Downey KM, Andrews DM, Li X, Castillo C, Tan CK, So AG (1990) in *Molecular Mechanisms in DNA Replication and Recombination* (Richardson CC, Lehman IR, eds). UCLA, Vol. 127. AR Liss, New York, p. 141.
[h] Hotta Y, Stern H (1978) *Biochemistry 17*:1872.
[i] nd = not determined.

when uncovered in proteins of unknown function, may in turn suggest a role in melting of duplex DNA or RNA.

ATP-Binding Motifs

Helicases invariably are ATPases dependent on binding to DNA. Most contain a characteristic purine nucleotide binding motif (Table 11-5). The binding pocket, composed of two elements (A and B sequences),[97] is found in proteins that bind and hydrolyze purine nucleotides as well as in helicases. The A sequence forms a loop which binds one of the phosphates of the nucleotide; the B sequence, containing a conserved aspartate residing within a stretch of hydrophobic residues, binds a second phosphate via magnesium.[98]

The Rep Protein Family[99]

Amino acid sequence homologies among several helicases extend beyond the nucleotide binding region. The protein sequences of Rep and helicase II (*uvrD* gene product) are nearly 40% identical; this relationship speaks for a common

97. Walker JE, Saraste M, Runswick MJ, Gay NJ (1982) *EMBO J 1*:945; Higgins CF, Hiles ID, Salmond GPC, Gill DR, Downie JA, Evans IJ, Holland IB, Gray L, Buckel SD, Bell AW, Hermodson MA (1986) *Nature 323*:448.
98. Hodgman TC (1988) *Nature 333*:22.
99. Gorbalenya AE, Koonin EV, Donchenko AP, Blinov VM (1989) *NAR 17*:4713; Gorbalenya AE, Koonin EV, Donchenko AP, Blinov VM (1988) *FEBS Lett 235*:16; Hodgman TC (1988) *Nature 333*:22; Hodgman TC (1988) *Nature 333*:578.

Table 11-5
Amino acid sequences of proposed purine nucleotide binding sites (A sequence) of several prokaryotic and eukaryotic helicases

Helicase	A sequence							
dnaB	A	R	P	S	M	G	K	T
PriA	G	V	T	G	S	G	K	T
Rep	A	G	A	G	S	G	K	T
Helicase II	A	G	A	G	S	G	K	T
RecB	A	S	A	G	T	G	K	T
RecD	G	G	P	G	T	G	K	T
Helicase IV	A	G	A	G	S	G	K	T
T7 gp4	S	G	S	G	M	G	K	S
Rho	A	R	P	K	A	G	K	T
Yeast RAD3	M	P	S	G	T	G	K	T
SV40 T antigen	G	P	I	D	S	G	K	T
HSV-1 UL5	G	N	A	G	S	G	K	S
HSV-1 UL9	A	P	M	G	S	G	K	T
Consensus	A or G	—	—	G	—	G	K	T or S

progenitor gene.[100] Helicase IV bears significant homology to both Rep and helicase II beyond its N-terminal 200 amino acids, which are absent from the other two proteins.[101] The B and D subunits of the RecBCD nuclease–helicase complex also belong in the Rep family. Both subunits bind ATP by virtue of purine nucleotide binding motifs. However, the homology between RecB protein and Rep or helicase II is only about 20%; RecD protein is even more divergent. The PriA protein is also a distantly related member of this protein family.[102]

Clustered regions of conserved amino acid sequence most likely form the structures responsible for DNA binding, hydrolysis, and translocation, as well as nucleotide binding. A short stretch of sequence homology to viral DNA polymerases is also evident.

Sequence homology to the Rep protein family, suggestive of a helicase, is found in the product of the *UL5* gene of HSV-1.[103] Essential for herpesvirus replication, the UL5 protein is part of the helicase–primase enzyme complex purified from infected cells. Strong homologies to the Rep helicase family are also present along much of the length of the 134-kDa protein encoded by the *RADH* gene of *Saccharomyces cerevisiae*. Mutations in *RADH* sensitize cells to the lethal effect of UV radiation specifically during the G_1 phase.[104] The protein, thought to belong to the error-prone repair machinery, is significantly longer than the 68-kDa Rep protein and has a completely unrelated C terminus.[105] The product of the yeast *PIF* gene, required for recombination and repair of the

100. Gilchrist CA, Denhardt DT (1987) *NAR 15*:465.
101. Wood ER, Matson SW (1989) *JBC 264*:8297.
102. Lee EH, Masai H, Allen GC Jr, Kornberg A (1990) *PNAS 87*: 4620; Nurse P, DiGate RJ, Zavitz KH, Marians KJ (1990) *PNAS 87*: 4615.
103. Hodgman TC (1988) *Nature 333*:22.
104. Aboussekhra A, Cahnet R, Zgaga Z, Cassier-Chauvat C, Heude M, Fabre F (1989) *NAR 17*:7211.
105. Aboussekhra A, Cahnet R, Zgaga Z, Cassier-Chauvat C, Heude M, Fabre F (1989) *NAR 17*:7211.

mitochondrial genome, is also homologous to the Rep helicase family and may function as an organelle-specific helicase.[106] These known and putative helicases of the Rep family are also structurally related to a conserved class of proteins involved in viral RNA replication[107] and to the eukaryotic translation initiation factor, eIF-4A, an RNA-unwinding enzyme.[108]

Several helicases bear no discernible homology to the Rep protein sequence beyond the nucleotide-binding motif. These include dnaB protein, the T7 and T4 replicative helicases, Rho protein, and SV40 T antigen. The dnaB protein is homologous to the phage P22 gp12 and, to a limited extent, to T7 gp4, suggesting that they belong to a second helicase protein family.[109]

Mechanisms of Melting

Helicases couple the energy of NTP hydrolysis to melting the hydrogen bonds and dissipating the other forces that hold together the strands of the DNA duplex. Although the mechanism of enzymatic coupling is not known in any detail and doubtless varies among the helicases, some facts are noteworthy.

Although NTP hydrolysis is required for unwinding, the two activities can be uncoupled, as shown by the NTPase activity effected simply by the binding of the helicase to ssDNA. A base-paired secondary structure may not be essential, as indicated by the effectiveness of homopolymers, in which standard base pairing is precluded. Differences in enzymatic activity observed with ssDNA and with a proper helicase substrate are evident, however (see Table 11-1 for examples with Rep protein), and provide clues to the coupling of hydrolysis to DNA melting. Because of the obligatory linkage of DNA binding with NTP hydrolysis, helicases are invariably DNA-dependent NTPases. The NTP hydrolysis is associated with translocation of the protein along the DNA strand. Movement stops when hydrolysis is blocked; hydrolysis is inhibited when translocation is blocked. Hydrolysis of TTP by T7 gp4 is halted when the enzyme encounters a bulky DNA adduct in its path;[110] ATP hydrolysis by helicase I nearly ceases when the end of a linear DNA is reached.[111]

With most helicases, specific binding of ssDNA is expressed by the enzyme's need for a single-stranded tail on a partially duplex molecule. The requirement for a preformed fork by some suggest that this structure is their principal locus of action. However, this preference could also reflect an inability of a helicase to initiate melting at a duplex end rather than a high affinity for a junction between single- and double-stranded DNA. The binding of helicase to ssDNA but not to dsDNA suggests that during melting the protein separates the strands of the duplex before or along with translocation, rather than binding the encountered duplex region and then opening it.

Unlike most helicases, the RecBCD complex binds both strands of a DNA duplex, and dsDNA is ten times more efficient as an effector of NTPase than ssDNA. In further contrast to other helicases, melting by RecBCD is inhibited,

106. Foury F, Lahaye A (1987) *EMBO J 6*:1441.
107. Gorbalenya AE, Koonin EV, Donchenko AP, Blinov VM (1989) *NAR 17*:4713; Gorbalenya AE, Koonin EV, Donchenko AP, Blinov VM (1988) *FEBS Lett 235*:16; Hodgman TC (1988) *Nature 333*:22; Hodgman TC (1988) *Nature 333*:578.
108. Rhoads RE (1988) *TIBS 13*:52; Ray BK, Lawson TG, Kramer JC, Cladaras MH, Grifo JA, Abramson RD, Merrick WC, Thach RE (1985) *JBC 260*:7651.
109. Wong A, Kean L, Maurer R (1988) *J Bact 170*:2668.
110. Brown WC, Romano LJ (1989) *JBC 264*:6748.
111. Lahue EE, Matson SW (1988) *JBC 263*:3208.

rather than stimulated, by single-stranded tails on the duplex molecule. Possibly, the B and D subunits of this enzyme represent two translocases, one for each strand.

Two to three NTP molecules are hydrolyzed per base pair melted by several of the helicases. Based on crude estimates of the free-energy release upon ATP hydrolysis (~10 kcal) and the melting of an average base pair (~4 kcal), the use of two ATP molecules per base pair represents an efficiency of about 20%. Included in the energy expenditure are the conformational changes imparted on the helicase to endow it with polarity of translocation. Not only is this a reasonably efficient operation, but the added cost to replication is trivial: the investment of two ATPs to provide the templates for assembly of a pair of nucleotides should be compared with the investment of about 140 ATP equivalents to synthesize the pair of nucleotides.

A Model for Translocation and Unwinding

In coupling the unidirectional translocation of T7 gp4 to TTP hydrolysis, as an example of helicase movement on DNA, a DNA-bound enzyme is complexed to TTP. Hydrolysis of the TTP to TDP and P_i is accompanied by a conformational change in the enzyme. Binding of a fresh TTP powers the translocation in the $5' \rightarrow 3'$ direction. Dissociation of the helicase–DNA complex is infrequent.

Migration of Rho protein along RNA may be driven by a similar mechanism.[112] The hexameric structure of Rho (and of dnaB protein and T antigen as well) provides six potential DNA binding sites which might alternately bind the old and new segments of DNA as the enzyme moves (rolls) along the template.

In the coupling of ATP hydrolysis to translocation by *E. coli* helicase III, two alternative conformations of the enzyme have been proposed,[113] one binding ssDNA, the other dsDNA. A phosphorylated form of the enzyme may represent one of these. The existence of such a form is inferred from the ability of the enzyme to generate ATP from ADP. The following scheme is suggested: (1) Before binding ATP, the protein binds to the duplex DNA just ahead of the junction between the single- and double-stranded regions. (2) Upon the binding of ATP, DNA-dependent hydrolysis generates an enzyme–phosphate complex. (3) This complex binds single- but not double-stranded DNA, and thus melting of the duplex accompanies complex formation.

Other Melting Mechanisms

In addition to the dual-conformation tracking model just described, other mechanisms need to be considered. One scheme employs many molecules of protein in translocating along a DNA strand to propagate the fork, as in the case of *E. coli* helicase II and T4 Dda protein (Section 3). Another mechanism is radically different. The enzyme from *Xenopus*[114] that melts double-stranded RNA does so by deaminating many of the adenosine residues to inosine and thus destroying their pairing with thymine.

---

112. Bear DG, Hicks PS, Escudero KW, Andrews CL, McSwiggen JA, von Hippel PH (1988) *JMB 199*:623.
113. Das RH, Yarranton GT, Gefter ML (1980) *JBC 255*:8069.
114. Bass BL, Weintraub H (1988) *Cell 55*:1089.

CHAPTER 12

Topoisomerases

12-1 Topoisomerases: Assays and Functions[1]

Topoisomerases catalyze the interconversion of topological isomers (topoisomers) of a DNA molecule. By introducing a transient break in the phosphodiester backbone through formation of a covalent protein–DNA intermediate, the enzyme allows the DNA strands to pass through one another, thereby changing the topological state of the molecule. Other DNA characteristics, such as the sequence, are not affected. The linking number, the total number of times the two chains in a covalently closed duplex cross one another (Section 1-9), may be changed by the introduction or the relaxation of superhelical turns. Topoisomerases can also catenate (link as rings in a chain) or decatenate DNA molecules and knot or unknot them.

Topoisomerases are classified as type I or type II (Table 12-1) on the following basis.[2] Type I enzymes (e.g., *E. coli* topo I and topo III) introduce a break into one strand of DNA to allow strand passage, while type II enzymes (e.g., *E. coli* gyrase) generate a break in both strands. As a result, type I enzymes change the linking number of DNA in steps as small as one; they catenate (or decatenate) single-stranded circular DNAs or duplexes providing that at least one of the molecules contains a nick or gap. Type II enzymes introduce a linking-number change in steps of two, and can link (or unlink) two covalently closed duplex circles and knot (or unknot) a completely duplex DNA.

1. Wang JC (1985) *ARB 54*:665; Wang JC (1987) *BBA 909*:1; Gellert M (1981) *ARB 50*:879; Cozzarelli NR (1980) *Science 207*:953.
2. Liu LF, Liu C-C Alberts BM (1980) *Cell 19*:697.

Table 12-1
Properties of type I and type II topoisomerases

Property	Type I		Type II	
	E. coli[a]	Eukaryotic[b]	Gyrase	Eukaryotic
DNA strands cleaved	one	one	two	two
Subunit mass (kDa)	~100	~95	97, 90	~150
Subunits	monomer	monomer	A_2B_2	homodimer
ATP requirement	no	no	yes	yes
Mg^{2+} requirement	yes	no	yes	yes
DNA-dependent ATPase	no	no	yes	yes
Makes (−) supercoils	no	no	yes	no
Relaxes (−) supercoils	yes	yes	no[c]	yes
Relaxes (+) supercoils	no	yes	yes[d]	yes
Catenation, knotting	yes[e]	yes[e]	yes	yes

[a] *E. coli* topo I and topo III.
[b] Yeast *TOP3* most likely encodes a type I enzyme with characteristics similar to those of the *E. coli* rather than the eukaryotic enzymes.
[c] Yes in the absence of ATP.
[d] By introduction of negative supercoils.
[e] Requires a nick or gap in one strand of duplex.

Replication

The strands of the parental duplex must be completely untwisted (bringing the linking number to zero) during the replication process. For every approximately 10.5 base pairs of DNA melted by a helicase, one link between the two strands must be removed. If the ends of the DNA molecule are not free to rotate, as in a circular template or in a linear molecule that either is very long or has tethered ends, a *swivel* (relaxing) function of a topoisomerase is needed to prevent overwinding (positive supercoiling) of the DNA ahead of the fork[3] (Fig. 12-1).

When replication of the duplex is nearly complete, the final interlinks be-

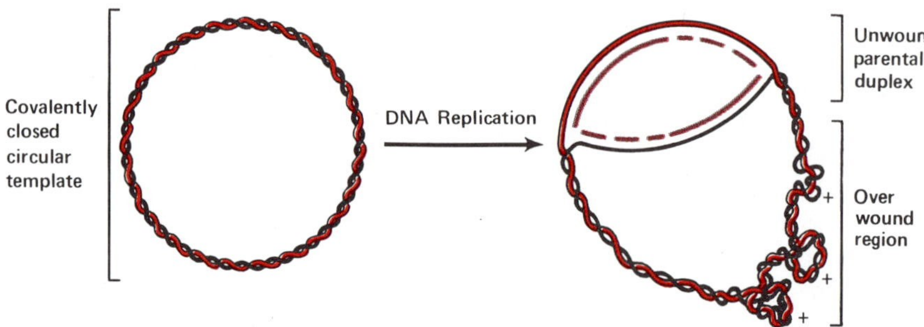

Figure 12-1
Swivel function of topoisomerase during replication. Overwound (positively supercoiled) regions accumulate ahead of the fork as the parental strands separate for replication. A topoisomerase to remove these twists is therefore required as a swivel, to allow extensive replication.

3. Delbruck M, Stent G (1957) in *The Chemical Basis of Heredity* (McElroy WD, Glass B, eds.). John Hopkins Press, Baltimore, p. 699; Cairns J (1963) *JMB 6*:208; Cairns J (1963) *CSHS 28*:43.

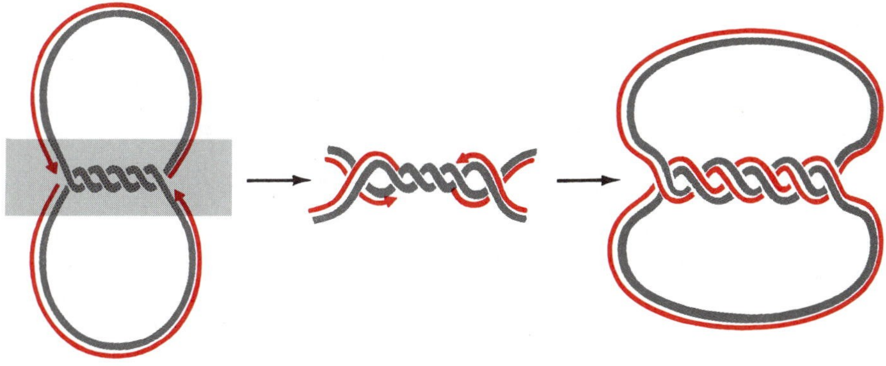

Figure 12-2
Decatenation by topoisomerase. Replication of the final region of the DNA template without prior unwinding generates multiply interlinked catenated daughter molecules which must be separated by a topoisomerase. [After Weaver DT, Fields-Berry SC, DePamphilis ML (1985) *Cell* *41*:565]

tween the two strands may be converted into links between the duplex portions of the daughter molecules[4] (Fig. 12-2). Decatenation by a type II topoisomerase can remove these remaining intertwines to allow the separation of the daughter molecules. Nicks and gaps in replication intermediates may enable a type I enzyme to perform this step as well.[5]

In bacteria, a negatively supercoiled template is essential for initiation of replication.[6] The free energy of this DNA conformation is used for assembly of the replication complex at the origin and the initial melting of the DNA strands (Section 16-3). Maintenance of this high-energy state depends on topoisomerase action.

Transcription[7]

As with replication, initiation of transcription is, in some cases, favored by a negatively supercoiled template. During RNA synthesis, topoisomerases relieve the strain caused by translocation of the polymerase along the DNA. Movement of the enzyme along the DNA causes the superhelicity of the template ahead of the moving enzyme to become more positive, while negative supercoils accumulate behind (Fig. 12-3). When other ways to dissipate these superhelical domains, such as rotating either the RNA polymerase around the DNA template or fusing the oppositely supercoiled regions, are not available, the actions of one or more topoisomerases may be required.

The impact of transcription on the template structure can be observed in *E. coli* topoisomerase mutants or by the use of topoisomerase inhibitors. Because topoisomerase I relaxes only negative supercoils, and DNA gyrase most likely relaxes primarily positive ones[8] (Table 12-1; Sections 2 and 3), inactivation of

4. Sundin O, Varshavsky A (1981) *Cell 25*:659.
5. Minden JS, Marians KJ (1986) *JBC 261*:11906.
6. Funnell BE, Baker TA, Kornberg A (1987) *JBC 262*:10327.
7. Liu LF, Wang JC (1987) *PNAS 84*:7024; Wu H-Y, Shyy S, Wang JC, Liu LF (1988) *Cell 53*:433; Tsao Y-P, Wu H-Y, Liu LF (1989) *Cell 56*:111; Giaever GN, Wang JC (1988) *Cell 55*:849.
8. Wu H-Y, Shyy S, Wang JC, Liu LF (1988) *Cell 53*:433.

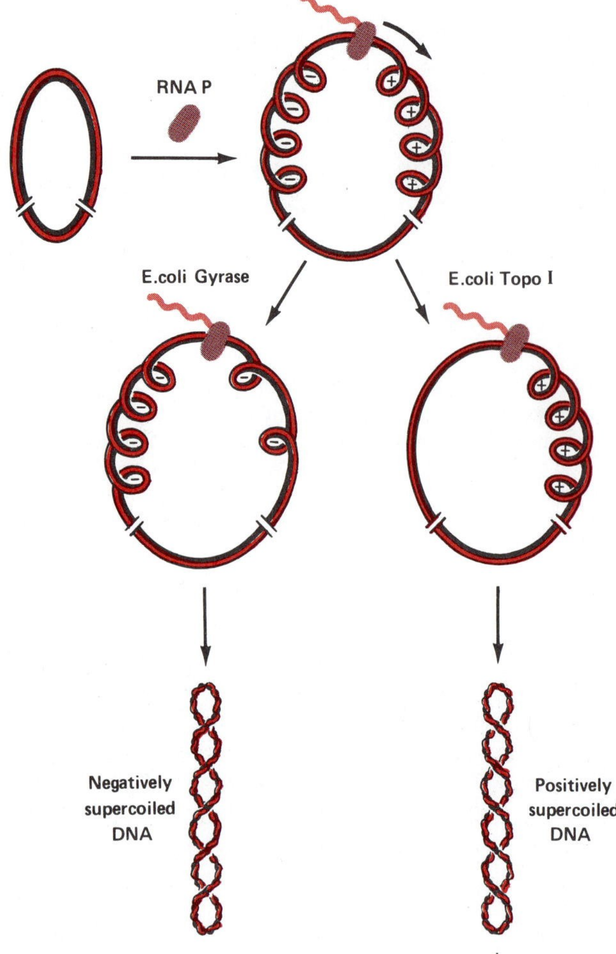

Figure 12-3
Scheme showing how the specificities of the *E. coli* topoisomerases can be used to demonstrate the positively and negatively supercoiled domains generated during transcription. RNAP = RNA polymerase.

topo I causes highly negatively supercoiled DNA to accumulate in response to transcription, while inactivation of gyrase generates positively supercoiled DNA. When the activities of the enzymes are in balance, the positive and negative turns are both dissipated, leaving the structure of the DNA template relatively unchanged from its initial conformation rather than completely unlinked as occurs during DNA replication.

Recombination and Other Functions

Topoisomerases are involved in several aspects of recombination:

(1) Initial pairing between molecules is favored by negative superhelicity, which increases their single-stranded character and makes it energetically feasible for a third DNA strand to invade the duplex (Section 1-13). Assembly of the proper synaptic structures in some site-specific recombination also depends on negatively supercoiled DNA[9] (Section 21-6).

9. Nash HA (1981) *ARG 15*:143; Craig NL (1988) *ARG 22*:77; Landy A (1989) *ARB 58*:913.

(2) In homologous recombination, intertwining of the DNA strands of two molecules can be achieved through the concerted action of a strand exchange enzyme (Section 21-5) and a topoisomerase.[10]

(3) Unknotting and decatenating the recombinant products depends on topoisomerase action.

(4) Nonhomologous (illegitimate) recombination may result from strand breakage by topoisomerases[11] (Section 21-5).

(5) In yeast, increased recombination between repeated sequences can result from a topoisomerase deficiency.[12]

Topoisomerases may contribute both structurally and functionally, as part of the proteinaceous nuclear scaffold, to the organization of the genome of eukaryotic cells into domains.[13] Inasmuch as topoisomerases are the prime targets of many potent antibacterial and anticancer drugs (Section 14-5), the mechanism by which these agents interrupt cellular functions may furnish additional insights into the physiological contributions of topoisomerases.[14]

12-2 *E. coli* Topoisomerases I and III

The first topoisomerase discovered was the ω protein of *E. coli*.[15] Since then, relaxing enzymes have been found in other bacteria and in eukaryotes.[16] The ω protein has been renamed *E. coli* DNA topoisomerase I (topo I); gyrase, which introduces supertwists into relaxed DNA (Section 3), is occasionally called *E. coli* DNA topoisomerase II; a second type I enzyme is known as *E. coli* DNA topoisomerase III (topo III).

E. coli Topoisomerase I

Topo I, a single polypeptide of about 100 kDa, contains three to four atoms of tightly bound zinc, apparently required for activity, and three repeats of a putative zinc-binding domain.[17] The amino acid sequence[18] is related to that of *E. coli* topo III (see below) and that of the initiator proteins of plasmids R1 and R100 (Sections 7 and 18-4), but not to those of yeast TOP1 protein or of vaccinia virus and calf thymus type I enzymes (Section 4). This divergence is in keeping with the mechanistic differences between these two groups of enzymes (see below). However, *E. coli* topo I is 40% homologous to TOP3 protein (most likely a type I enzyme encoded by the *TOP3* gene in yeast[19]; see Section 4). Structurally distinct from the TOP1 protein, TOP3 has properties similar to those of the *E. coli* enzymes.

---

10. Cunningham RP, Wu AM, Shibata T, Das Gupta C, Radding CM (1981) *Cell 24*:213; Wu AM, Bianchi M, Das Gupta C, Radding CM (1983) *PNAS 80*:1256; Kmiec EB, Kroeger PE, Brougham MJ, Holloman WK (1983) *Cell 34*:919.
11. Ikeda H, Aoki K, Naito A (1982) *PNAS 79*:3724.
12. Wallis JW, Chrebet G, Brodsky G, Rolfe M, Rothstein R (1989) *Cell 58*:409; Kim RA, Wang JC (1989) *Cell 57*:975; Fink GR (1989) *Cell 58*:225; Christman MF, Bietrich FS, Fink GR (1988) *Cell 55*:413.
13. Gasser SM, Laroche T, Falquet J, Boy de la Tour E, Laemmli UK (1986) *JMB 188*:613.
14. Liu LF (1989) *ARB 58*:351.
15. Wang JC (1971) *JMB 55*:523.
16. Wang JC (1985) *ARB 54*:665.
17. Tse-Dinh Y-C, Beran-Steed RK (1988) *JCB 263*:15857.
18. Tse-Dinh Y-C, Wang JC (1986) *JMB 191*:321.
19. Wallis JW, Chrebet G, Brodsky G, Rolfe M, Rothstein R (1989) *Cell 58*:409.

The relaxing activity of topo I has several features:[20] (1) Negative superhelical turns are removed but no other change in the DNA occurs. (2) Relaxation of the DNA occurs gradually; intermediates between supercoiled and completely relaxed DNA are evident throughout the reaction. (3) A covalent protein–DNA intermediate conserves the energy of the nicked phosphodiester bond for concerted resealing. (4) Unlike DNA ligase, topo I does not employ a cofactor (e.g., ATP or NAD).

The type I topoisomerases of *E. coli*[21] and *Micrococcus luteus*[22] differ significantly from the eukaryotic enzymes (Table 12-1) in that they depend on Mg^{2+} and form a transient nick by covalent linkage at the 5′-P group;[23] the eukaryotic topoisomerase I linkage[24] is to the 3′-P. Furthermore, bacterial topoisomerases act only on negative superhelical turns and lose catalytic efficiency as the degree of superhelicity decreases.

Reactions Catalyzed by *E. coli* Topoisomerase I. In addition to relaxing negatively supercoiled DNA, topo I carries out three additional reactions (Fig. 12-4). Topo I (1) interacts with single-stranded circles to introduce topological knots,[25] (2) forms covalently closed duplex circles from complementary single-stranded ones,[26] and (3) catenates and decatenates duplex DNA circles, provided that at least one of the molecules contains a nick or a gap in one strand.[27]

The complex of topoisomerase with the DNA strand has been examined directly[28] (Fig. 12-5). Mg^{2+} permits the various catalytic events to ensue, while high salt concentrations cause the complex to dissociate. Alkaline treatment leads to a scission of the DNA, leaving the enzyme covalently linked to the 5′ end. Treatment of the covalent complex with proteases also produces a break within the DNA, while digesting the topoisomerase.

Reaction Mechanism of *E. coli* Topoisomerase I.[29] In the relaxation of DNA by this enzyme, one strand of the helix is passed through its complementary strand. The mechanism is likely to involve the following steps:

(1) Topo I binds to and disrupts a segment of duplex DNA, associating with a single-stranded region to form the enzyme–DNA complex.

(2) Topo I cleaves one strand of this bubble and becomes covalently linked to the 5′-P at the cleavage site by a phosphotyrosine bond.[30] Noncovalent contacts are made with DNA both 5′ and 3′ to the cleavage site, the complementary strand, and the duplex regions flanking the bubble.

(3) A strand of DNA is passed through the cleavage site.

(4) The broken strand is religated and the enzyme is released.

This mechanism is analogous to that for type II enzymes (Section 3) except that a transient enzyme-bridged break involves only one strand rather than two and the change in linking number is one.[31]

20. Wang JC (1971) *JMB 55*:523.
21. Liu LF, Depew RE, Wang JC (1976) *JMB 106*:439.
22. Kung VT, Wang JC (1977) *JBC 252*:5398.
23. Tse YC, Kirkegaard K, Wang JC (1980) *JBC 255*:5560.
24. Champoux JJ (1977) *PNAS 74*:3800.
25. Liu LF, Depew RE, Wang JC (1976) *JMB 106*:439.
26. Kirkegaard K, Wang JC (1978) *NAR 5*:3811.
27. Tse Y-C, Wang JC (1980) *Cell 22*:269; Brown PO, Cozzarelli NR (1981) *PNAS 78*:843.
28. Depew RE, Liu LF, Wang JC (1978) *JBC 253*:511.
29. Kirkegaard K, Wang JC (1985) *JMB 185*:625.
30. Tse YC, Kirkegaard K, Wang JC (1980) *JBC 255*:5560.
31. Lovett MA, Katz L, Helinski DR (1974) *Nature 251*:337.

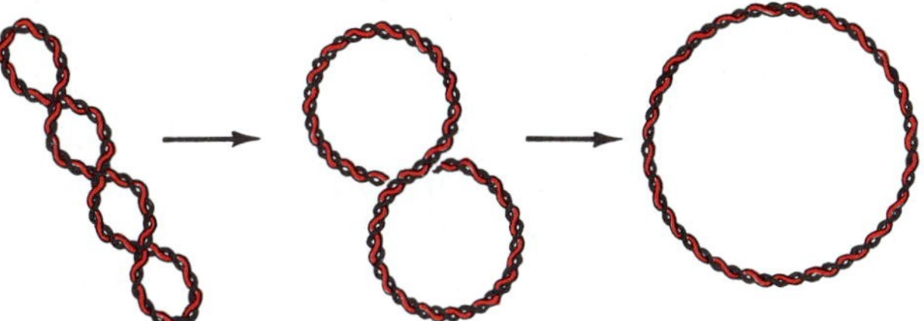

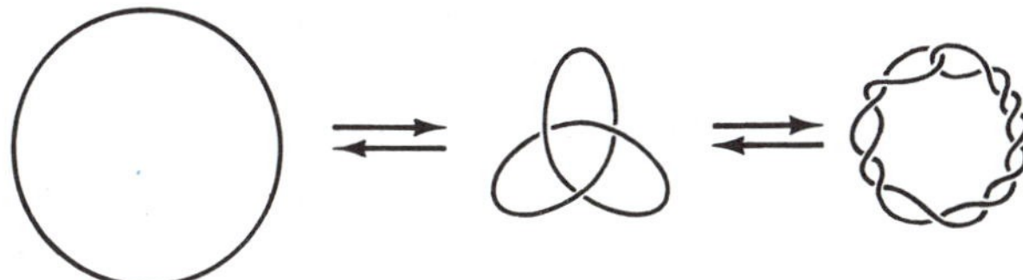

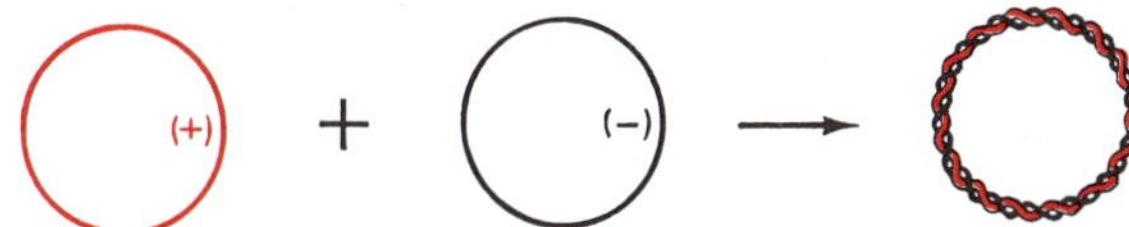

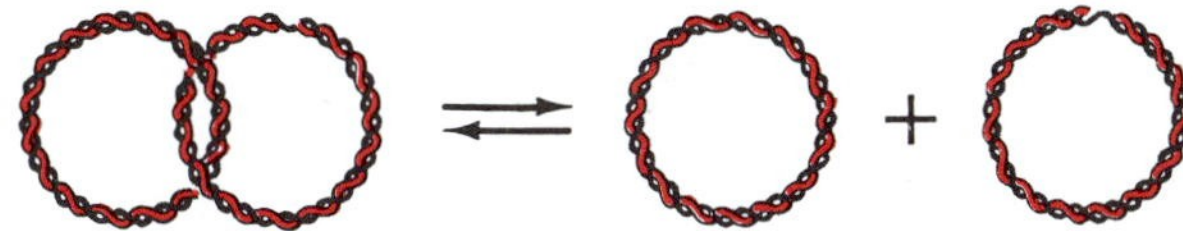

Figure 12-4
The four types of topological conversions catalyzed by *E. coli* topoisomerase I. (Courtesy of Professor JC Wang)

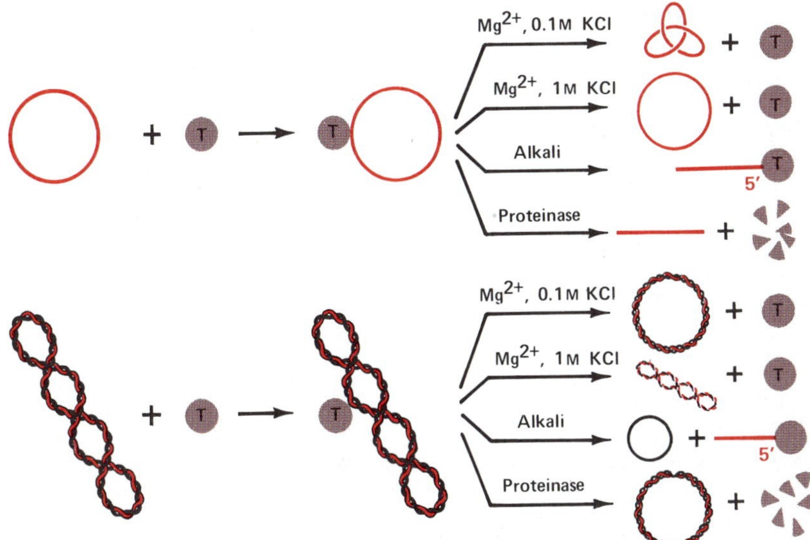

Figure 12-5
E. coli topoisomerase I complexes formed with single-stranded or supercoiled DNA. The complexes are catalytically active in Mg^{2+} + 0.01 M KCl, dissociate in 1 M KCl, are nicked in alkali, and are nicked and proteolyzed by proteinase. (Courtesy of Professor JC Wang)

The interaction of topo I with ssDNA during strand passage was suspected from the inhibition of relaxation by single-stranded fragments[32] and the ability of the protein to bind and cleave ssDNA.[33] Its specificity for relaxing negative but not positive superhelical turns can be attributed to the increased single-stranded character of the negatively supercoiled form (Section 1-9). The enzyme can also relax a positively supercoiled circle containing a single-stranded loop to which the enzyme can bind and upon which it can act.[34]

Functions of Topoisomerase I. Mutants of the gene for topo I (*topA*) provide clues to its function. Deletion, nonsense, and insertion *topA* mutations cause only a modest decrease in growth rate.[35] However, viability requires extracistronic suppressors, some of which are in the genes for the subunits of DNA gyrase.[36] Gyrase, by introducing negative supercoils into DNA (Section 3), antagonizes topo I. The improved growth of *topA* mutants that also contain gyrase mutations with decreased activity argues that maintenance of the chromosome within a rather narrow range of negative superhelicity is important for cell viability. The lethality of *topA* deletions in the absence of suppressor mutations indicates that topo I is in fact essential.[37]

The consequences of *topA* mutations include suppression of the phenotype of *dnaA* mutations,[38] enhanced sensitivity to UV,[39] and changes in the rate of transcription of certain genes. The specificity of topo I for relaxing negative but not positive supercoils has been used to show the topological effects of transcription on the template DNA structure, indicating that topoisomerases normally have a role in transcription elongation[40] (Section 1). Some of these effects of *topA* mutations may be modulated through global changes in chromosomal superhelicity, whereas other effects result from a more direct role of the enzyme.

Topo I is important in several in vitro replication reactions. It functions as a specificity factor during initiation from the *E. coli* chromosomal origin (*oriC*) by aborting starts elsewhere than at *oriC*[41] (Section 16-3). This effect appears to result from relaxation of the template, inasmuch as the effect is opposed by gyrase. A similar role in maintaining origin specificity is seen in replicaton of the plasmid pBR322.[42] Topo I can also decatenate the nicked and gapped daughter molecules that accumulate late in pBR322 replications,[43] a function usually attributed to gyrase, though direct evidence for this is sparse.

E. coli Topoisomerase III

A second type I enzyme, encoded by *topB*, was identified in *E. coli*, first as a relaxing activity present in *topA* mutant cells[44] and subsequently as a protein

32. Wang JC (1971) *JMB 55*:523.
33. Tse Y-C, Kirkegaard K, Wang JC (1980) JBC 255:5560.
34. Kirkegaard K, Wang JC (1985) *JMB 185*:625.
35. Sternglanz R, DiNardo S, Voelkel KA, Nishimura Y, Hirota Y, Becherer K, Zumstein L, Wang JC (1981) *PNAS 78*:2747.
36. DiNardo S, Voelkel KA, Sternglanz R, Reynolds AE, Wright A (1982) *Cell 31*:43.
37. Wang JC (1984) *J Cell Sci [Suppl] 1*:21.
38. Louarn J, Bouché J-P, Patte J, Louarn J-M (1984) *MGG 195*:170.
39. Overbye KM, Basu SK, Margolin P (1983) *CSHS 47*:785.
40. Wu H-Y, Shyy S, Wang JC, Liu LF (1988) *Cell 53*:433.
41. Kaguni JM, Kornberg A (1984) *JBC 259*:9578.
42. Minden MS, Marians KJ (1985) *JBC 260*:9316.
43. Minden JS, Marians KJ (1986) *JBC 261*:11906.
44. Dean FB, Krasnow MA, Otter R, Matzuk MM, Spengler S, Cozzarelli NR (1982) CSHS *47*:769; Srivenugopal KS, Lockshon D, Morris DR (1984) *Biochemistry 23*:1899.

efficient in decatenating pBR322 replication intermediates (Section 18-2).[45] Topo III, a monomer of 74 kDa, resembles topo I in relaxing negative supercoils, but is totally inactive against positive supercoils.[46] The ability of ssDNA to compete with supercoiled DNA for the enzyme implies a requirement for melted regions in the duplex for activity.

Topo III efficiently separates nicked or gapped interlocked circles with a specific activity 10-fold higher than that of topo I and 70-fold higher than that of gyrase.[47] Decatenation is also more efficient than relaxation and proceeds readily at 30°, whereas 52° is optimal for relaxation. This property may be due to the inefficiency of topo III in melting DNA to generate the single-stranded region required for strand passage, a stage not required in unlinking nicked or gapped molecules.

At the amino acid level,[48] topo I and topo III show an identity of 24% and a similarity of 46%. Homology is concentrated in the center of the proteins; their N- and C-terminal regions are not related. Topo III is present in only one to ten copies per cell[49] and is unlikely to be essential, as *topB* lies within a region of the chromosome that can be deleted. Little role in regulating DNA superhelicity is indicated; in a *topA*-deficient strain, upon inactivation of gyrase, the superhelical density of the intracellular DNA was unchanged in spite of the presence of topo III. The preference for decatenating molecules in vitro suggests a similar role in vivo, although gyrase has been implicated in this process.[50]

12-3 *E. coli* DNA Gyrase[51] (*E. coli* Topoisomerase II)

DNA gyrase was discovered as an activity needed for the integrative recombination of phage λ DNA in vitro[52] (Section 21-6). Gyrase converts relaxed, closed-circular duplex DNA to the negatively superhelical form, a more favorable substrate for recombination both in vitro and in vivo. ATP is required and is hydrolyzed in the reaction. The activity of gyrase has since been associated with events in DNA replication, transcription, repair, and recombination (see below). The double-strand-break mechanism of gyrase classifies it as a type II enzyme. A similar topoisomerase has been identified and purified from *B. subtilis*, a Gram-positive bacterium; the genes for its two subunits have been sequenced[53] and are related to those of the *E. coli* enzyme.[54]

Subunit Composition

Early studies of two DNA replications inhibitors, nalidixic acid[55] and novobiocin,[56] aided in purification of the enzyme and in characterization of its subunit

45. DiGate RJ, Marians KJ (1988) *JBC 263*:13366.
46. Srivenugopal KS, Lockshon D, Morris DR (1984) *Biochemistry 23*:1899.
47. DiGate RJ, Marians KJ (1988) *JBC 263*:13366.
48. DiGate RJ, Marians KJ (1989) *JBC 264*:17924.
49. DiGate RJ, Marians KJ (1988) *JBC 263*:13366.
50. Bliska JB, Cozzarelli NR (1987) *JMB 194*:205; Steck TR, Drlica K (1984) *Cell 36*:1081.
51. Maxwell A, Gellert M (1986) *Adv Protein Chem 38*:69.
52. Gellert M, Mizuuchi K, O'Dea MH, Nash HA (1976) *PNAS 73*:3872.
53. Moriya S, Ogasawara N, Yoshikawa H (1985) *NAR 13*:2251.
54. Adachi T, Mizuuchi K, Menzel R, Gellert M (1984) *NAR 12*:6389; Swanberg SL, Wang JC (1987) *JMB 197*:729.
55. Gellert M, Mizuuchi K, O'Dea MH, Itoh T, Tomizawa J-I (1977) *PNAS 74*:4772; Sugino A, Peebles CL, Kreuzer KN, Cozzarelli NR (1977) *PNAS 74*:4767.
56. Gellert M, Mizuuchi K, O'Dea MH, Itoh T, Tomizawa J-I (1977) *PNAS 74*:4772.

Table 12-2
Subunits of *E. coli* gyrase

Property	A	B
Mass (kDa)	97	90
Gene	*gyrA* (*nalA*)	*gyrB* (*cou*)
Inhibitors	nalidixic acid oxolinic acid norfloxacin	coumermycin A1 novobiocin
DNA binding, cleavage	yes	no
ATP binding, energy transduction	no	yes

structure as a 400-kDa A_2B_2 tetramer (Table 12-2). The gene for the A subunit, *gyrA* (formerly *nalA*), controls resistance to nalidixic acid and the more potent synthetic quinolones (Section 14-5). The gene for the B subunit, *gyrB* (formerly *cou*), specifies the protein target of novobiocin and coumermycin A1. The isolated tetramer[57] is readily resolved and reconstituted from its subunits. Better yields of the enzyme are obtained by purification of the subunits separately from overproducing strains, guided by complementation assays.[58]

Reactions Catalyzed by Gyrase[59]

The several activities of the *E. coli* and *M. luteus* enzymes (Fig. 12-6) include:

(1) *Negative supercoiling of circular DNA* to yield a superhelical density ($\sigma \sim -0.11$) up to about 1½ times that found in native DNA.[60]

(2) *Relaxation of negatively supercoiled DNA* in the absence of ATP.

(3) *Double-strand breakage of DNA;* this takes place even in the absence of ATP and is stimulated by a quinolone drug (Section 14-5), Mg^{2+}, and subsequent treatment with sodium dodecyl sulfate or alkali to denature the protein. Staggered cuts are made with a moderate site preference;[61] the 5′-P extension in the four-base overlap has the denatured A subunit covalently attached.

(4) *ATP hydrolysis* to ADP and P_1 in the presence of duplex DNA.

(5) *Positive wrapping of DNA around gyrase;* formation of this complex, which does not require ATP and is not affected by drug actions, provides a clue to the gyrase mechanism (see below).

(6) *Catenation and decatenation of duplex circles*[62] and the related activities of *knotting and unknotting DNA.* Catenation, unlike all other reactions, requires a polyvalent cation (e.g., spermidine) or DNA-binding proteins (e.g., histones). By neutralizing the negative charge, polycations compact the DNA into aggregates in which the high local DNA concentration drives intermolecular linking.

57. Higgins NP, Peebles CL, Sugino A, Cozzarelli NR (1978) *PNAS 75*:1773; Mizuuchi K, O'Dea MH, Gellert M (1978) *PNAS 75*:5960.
58. Mizuuchi K, Mizuuchi M, O'Dea MH, Gellert M (1984) *JBC 259*:9199.
59. Peebles CL, Higgins MP, Kreuzer KN, Morrison A, Brown PO, Sugino A, Cozzarelli NR (1978) *CSHS 43*:41; Maxwell A, Gellert M (1986) *Adv Protein Chem 38*:69.
60. Bates AD, Maxwell A (1989) *EMBO J 8*:1861.
61. Morrison A, Cozzarelli NR (1979) *Cell 17*:175.
62. Kreuzer KN, Cozzarelli NR (1980) *Cell 20*:245.

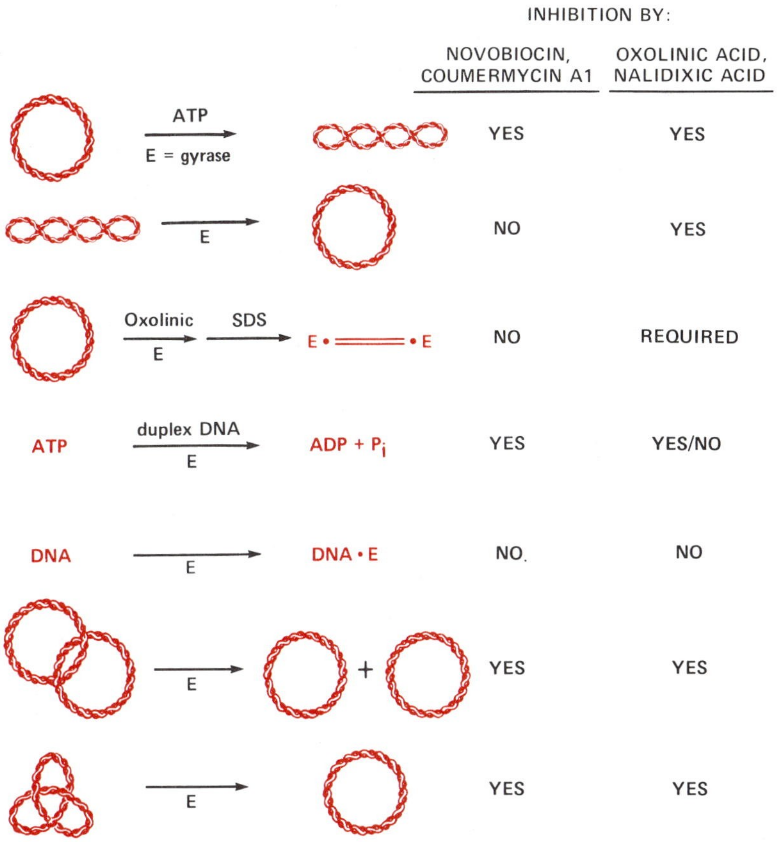

		INHIBITION BY:	
		NOVOBIOCIN, COUMERMYCIN A1	OXOLINIC ACID, NALIDIXIC ACID
	ATP → (E = gyrase)	YES	YES
	→ E	NO	YES
	Oxolinic → E, SDS → E•═══•E	NO	REQUIRED
ATP	duplex DNA → E, ADP + P_i	YES	YES/NO
DNA	→ E, DNA • E	NO	NO
	→ E, +	YES	YES
	→ E	YES	YES

Figure 12-6
Multiple activities of gyrase. E (enzyme) = gyrase. (Adapted from a scheme by Dr NR Cozzarelli)

DNA Binding

A segment of approximately 140 bp of duplex DNA is bound by gyrase, with a central core of 20 bp strongly protected against digestion by DNase I. Flanking regions are more extensively digested, with enhanced sensitivity at every 10 to 11 base pairs, indicating that the DNA is wrapped around the surface of the protein.[63] The preferred sequence of the core region was determined from 19 cleavage sites generated by gyrase after quinolone and detergent treatment.[64] While this 20-bp sequence is important for protein binding, the flanking regions are required in order to generate a strong binding site.[65]

Gyrase has a 40-fold higher affinity for a 100-bp fragment from the plasmid pSC101 than it does for most DNA.[66] AT-rich sequences in this fragment occur with a pronounced 10-bp periodicity, shifted by 5 bp from a similar periodicity of GC-rich sequences. This pattern conforms to the rules established for the preferred positioning of nucleosomes along DNA (Section 1-10) and allows the DNA to bend due to the relative ease of narrowing the DNA grooves on the inside of the bend and of expanding them on the outside. In fact, as judged by sensitivity to DNase I, the GC sequences are exposed on the outside of the DNA wrapped

63. Liu LF, Wang JC (1978) *PNAS 75*:2098; Liu LF, Wang JC (1978) *Cell 15*:979.
64. Lockshon D, Morris DR (1985) *JMB 181*:63.
65. Fisher LM, Barot HA, Cullen ME (1986) *EMBO J 5*:1411.
66. Wahle E, Kornberg A (1988) *EMBO J 7*:1889.

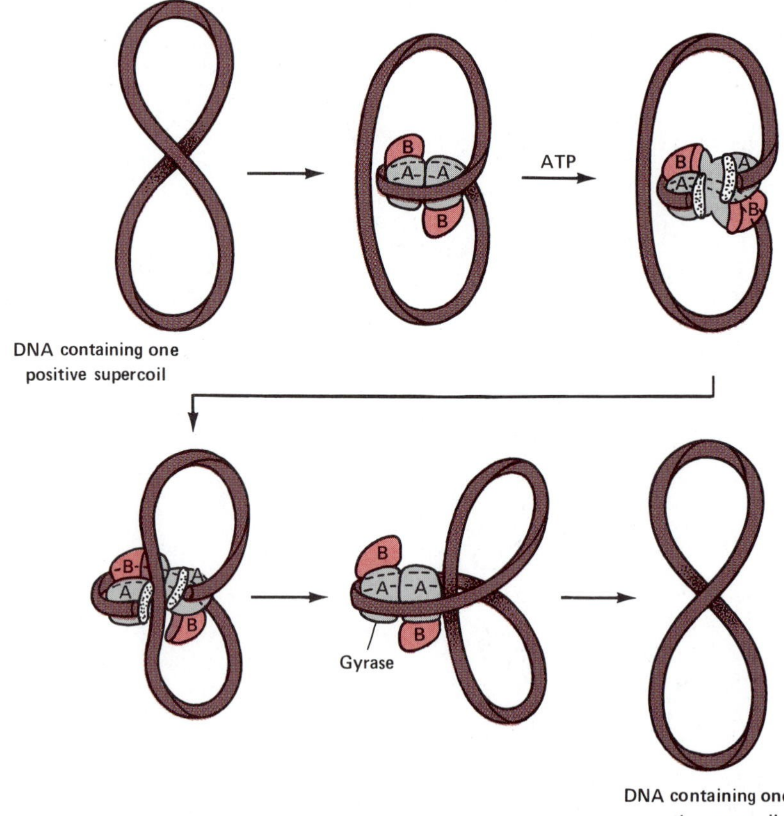

Figure 12-7
Schematic of strand passage that must occur during supercoiling by DNA gyrase. (Courtesy of Professor NR Cozzarelli.)

around gyrase. Thus, bendable arms flanking the 20-bp core of conserved nucleotide sequence apparently contribute to an effective gyrase binding site.

Reaction Mechanism[67]

Gyrase binds DNA in a way that enables two DNA segments to cross and form two supercoiled loops, one positive and the other negative. Gyrase also enables one of the DNA segments to pass through a reversible break in the other (Fig. 12-7), so that the positive supercoil is converted to a negative one. In the process, two negative supercoils are introduced in each catalytic step. Demonstration of the "step-of-two" change in supercoiling by DNA gyrase provided initial evidence for this model.[68]

The steps in supercoiling can be explained as follows:

(1) Gyrase binds DNA, wrapping a segment of about 105 to 140 bp in a positive sense around itself.

(2) The DNA is then cleaved in both strands, with a 4-bp stagger; cleavage is near the center of the wrapped segment. A covalent bond is formed between the 5′ terminus at the break in each strand and tyrosine 122 in each A subunit.

---

67. Maxwell A, Gellert M (1984) *JBC 259*:14472; Maxwell A, Gellert M (1986) *Adv Protein Chem 38*:69; Rau DC, Gellert M, Thoma F, Maxwell A (1987) *JMB 193*:555.
68. Brown PO, Cozzarelli NR (1979) *Science 206*:1081; Liu LF, Liu, C-C, Alberts BM (1980) *Cell 19*:697; Mizuuchi K, Fisher LM, O'Dea MH, Gellert M (1980) *PNAS 77*:1847.

(3) A second region of DNA, lying within or near the wrapped segment, is translocated through the break. This conformational change is facilitated by the binding of ATP to the B subunit.

(4) The cleaved strands are resealed using the energy stored in the protein–DNA covalent complex.

(5) These events are supported by a nonhydroyzable ATP analog, but enzyme turnover, for a second cycle of supercoiling, requires ATP hydrolysis.

The B subunit is the ATPase. A low level of hydrolysis can be detected without DNA but in the presence of heat or urea, which probably induce certain conformational changes.[69] However, the A subunit and linear DNA greatly stimulate the ATPase. DNA effectors of 100 bp or longer are best, as anticipated by the length of DNA bound by the enzyme. Shorter DNA fragments are also active, but much higher concentrations are required, indicating that DNA binding to more than one site on the enzyme may be involved.

Gyrase Functions in DNA Replication[70]

The inhibition of DNA replication by gyrase-targeted drugs provided early indications for direct involvement of this enzyme in this process. However, inferences from the action of the quinolones (Section 14-5) pointed as well to poisoning of the template by the drug–enzyme complex.

That gyrase is required for *E. coli* chromosomal replication is proved by *gyrA* and *gyrB* mutations[71] and the dependence on the enzyme for replication in vitro.[72] Mutations fall into both the fast- and slow-stop classes of replication inhibition (Section 15-4), indicating that different severities of enzyme defects may preferentially affect stages of initiation, elongation, or termination.

The functions of gyrase in the several stages of replication from the *E. coli* chromosomal origin (*oriC*) in vitro are as follows (Section 16-3):

(1) To maintain a negatively supercoiled template for assembly of the initiation complex and opening of the duplex at the origin.[73]

(2) To provide the swivel for progress of the replication fork.[74] Because of the slow turnover rate of gyrase (about two supercoils per second), coordinate action by 50 gyrase molecules is needed to support melting of the parental DNA associated with a synthesis of 1000 bp per second.

(3) To decatenate and supercoil the daughter molecules during the terminal stages.[75]

The important role of gyrase in all stages of *oriC* plasmid replication—initiation, elongation, and termination—should not obscure the share that topo I and topo III may also have in these several functions (Section 2).

---

69. Maxwell A, Gellert M (1984) *JBC 259*:14472.
70. Wang JC (1985) *ARB 54*:665; Gellert M (1981) *ARB 50*:879.
71. Filutowicz M (1980) *MGG 177*:301; Orr E, Fairweather NF, Holland IB, Pritchard RH (1979) *MGG 177*:103.
72. Baker TA, Sekimizu K, Funnell BE, Kornberg A (1986) *Cell 45*:53; Baker TA, Funnell BE, Kornberg A (1987) *JBC 262*:6877.
73. Funnell BE, Baker TA, Kornberg A (1987) *JBC 262*:10327; Baker TA, Kornberg A (1988) *Cell 55*:113.
74. Baker TA, Sekimizu K, Funnell BE, Kornberg A (1986) *Cell 45*:53; Baker TA, Funnell BE, Kornberg A (1987) *JBC 262*:6877.
75. Funnell BE, Baker TA, Kornberg A (1987) *JBC 262*:10327.

Other Functions of DNA Gyrase

Cells maintain the superhelicity of the chromosome within rather narrow limits. This homeostatic control is achieved by the antagonistic actions of gyrase and topo I.[76] Some of the mutations that compensate for a lethal deficiency in topo I are within *gyrA* and *gyrB*, serving, it would seem, to keep the level of negative superhelicity down to a tolerable level (Section 2).

Transcription elongation involves the participation of gyrase to remove the supercoils that accumulate ahead of the transcription apparatus[77] (Section 1). The ability of mutations in *rpoB* (encoding the β subunit of RNA polymerase) to suppress the temperature-sensitive growth of gyrase mutants[78] is a further indication of the interdependence of gyrase, transcription, and template superhelicity. Mutations in both the β subunit of RNA polymerase and Rho, the transcription termination factor (Section 7-7), affect the topological state of intracellular DNA.[79] Transcription, its impact on the template structure, and resolution of these changes by the topoisomerases are clearly important in the establishment and maintenance of the conformation of the chromosomal DNA.

Additional instances in which gyrase action is implied, but not well understood, include effects on partitioning of plasmid and chromosomal DNA into daughter cells at division,[80] illegitimate (nonhomologous) recombination,[81] and genetic suppression of mutations in the *dnaE* gene.[82] Two new genes involved in partitioning of chromosomal DNA (Section 20-5) encode proteins homologous to the A and B subunits of gyrase, indicating that *E. coli* has an additional gyrase-like enzyme.[83]

Control of Gyrase Gene Expression

The synthesis of both gyrase subunits is induced by coumermycin[84] and occurs more rapidly in *topA*$^+$ than in *topA* mutant cells. These findings suggest that the basis of induction is the relaxation of the DNA; relaxed templates are also better for transcription of the gyrase genes in vitro. By this device, production of gyrase is feedback-regulated by the cell's need for supercoiling activity.

12-4 Eukaryotic Type I Topoisomerases

Type I topoisomerases isolated from yeasts, amphibians, insects, mammals, and plants have similar characteristics, and the yeast and human enzymes are related in amino acid sequence as well. However, they are distinct from *E. coli* topo I and topo III (Section 2) in both mechanism and physical characteristics (Table 12-1).

Eukaryotic topo I is a monomeric protein of approximately 95 kDa (between

---

76. DiNardo S, Voelkel KA, Sternglanz R, Reynolds AE, Wright A (1982) *Cell 31*:43.
77. Wu H-Y, Shyy S, Wang JC, Liu LF (1988) *Cell 53*:433; Pruss GJ, Drlica K (1989) *Cell 56*:521.
78. Filutowicz M, Jonczyk P (1983) *MGG 191*:282.
79. Arnold GF, Tessman I (1988) *J Bact 170*:4266; Drlica K, Franco RJ, Steck TR (1988) *J Bact 170*:4983.
80. Wahle E, Kornberg A (1988) *EMBO J 7*:1889; Norris V, Alliotte T, Jaffe A, D'Ari R (1986) *J Bact 168*:494; Miller CA, Beacage SL, Cohen SN (1990) *Cell 62*:127.
81. Ikeda H, Aoki K, Naito A (1982) *PNAS 79*:3724.
82. Maki H, Bryan SK, Horiuchi T, Moses RE (1989) *J Bact 171*:3139.
83. Kato J, Nishimura Y, Imamura R, Niki H, Hiraga S, Suzuki H (1990) *Cell 63*:393.
84. Menzel R, Gellert M (1987) *J. Bact 169*:1272.

90 and 135 kDa). Although the reactions catalyzed are generally the same as those of *E. coli* topo I (Section 2), positively and negatively supercoiled DNAs are relaxed equally well. The lack of inhibition of relaxation by ssDNA fragments implies that the enzyme has a preference for duplex DNA. Relaxation is independent of ATP and can occur in the presence of EDTA, yet Mg^{2+} stimulates enzyme activity. The covalent protein–DNA intermediate is formed with a specific tyrosine residue via a 3′-P group rather than the 5′-P used by the *E. coli* type I enzymes.

Protein Structure

Topo I enzymes from yeast (*Saccharomyces cerevisiae, Schizosaccharomyces pombé*) and from human cells are related proteins. An internal 400-amino-acid stretch and about 50 residues at the C terminus are conserved, but the N-terminal portions are different. The active site for the yeast enzymes has been mapped to a single tyrosine residue that lies near the C terminus[85] and within the region shared by the human enzyme. Homology with the vaccinia virus type I topoisomerase, a distinctive smaller protein,[86] is also apparent (Section 7).

Substrate Preferences

Topo I cleavage of DNA, induced by the addition of protein-denaturing agents, occurs at preferred locations. Camptothecin, an inhibitor (Section 14-5), traps the enzyme–DNA intermediate and increases the yield of cleaved product. The correlation between the strength of topo I binding to a site and its cleavage of a site is only fair: good binding sites may not be strong cleavage sites. Yet there is a reasonable consensus for the sequence of cleavage sites. The most important residues are the four base pairs 5′ to the nick (−4 to −1) and the base pair just 3′ (+1). Similar sequences are recognized by topo I from yeast, *Drosophila*, wheat, and humans.

Strongly favored binding sites have also been described. A 16-bp sequence recognized by the *Tetrahymena* enzyme occurs in the spacer regions in the ribosomal DNA,[87] suggesting a topo I role in transcription or recombination within these repeats (Section 6). In addition to acting as a strong cleavage site, this sequence imparts a ten-fold preference for relaxation by the enzyme in vitro.[88]

Topo I has structural preferences in addition to favored nucleotide sequences. The enzyme is active on both positively and negatively supercoiled DNA. In contrast to the *E. coli* enzymes, eukaryotic topo I favors duplex regions over ssDNA. Preferred cleavage sites are also found at the junctions of DNA bends.[89]

Yeast TOP3, a Distinctive Type I Enzyme[90]

An *S. cerevisiae* gene (*TOP3*) is closely related in sequence to *E. coli topA* and *topB* and probably encodes a type I topoisomerase similar to the prokaryotic

---

85. Eng W, Pandit SD, Sternglanz R (1989) *JBC 264*:13373; Lynn RM, Bjornsti MA, Caron PR, Wang JC (1989) *PNAS 86*:3559.
86. Lynn RM, Bjornsti MA, Caron PR, Wang JC (1989) *PNAS 86*:3559.
87. Stevnsner T, Mortensen UH, Westergaard O, Bonven BJ (1989) *JBC 264*:10110.
88. Busk H, Thomsen B, Bonven BJ, Kjeldsen E, Nielsen OF, Westergaard O (1987) *Nature 327*:638.
89. Caserta M, Amadei A, Di Mauro E, Camilloni G (1989) *NAR 17*:8463.
90. Wallis JW, Chrebet G, Brodsky G, Rolfe M, Rothstein R (1989) *Cell 58*:409.

enzymes. Mutants of *TOP3* were selected on the basis of an increased frequency of recombination between repeated elements (Section 6); these mutations also caused a reduced growth rate. Complementation of this phenotype enabled the wild-type gene to be cloned and sequenced, but the enzyme has not yet been identified. Expression of the *E. coli* topo I in yeast complements the slow-growth phenotype, an argument that these two enzymes have related functions. The synergistic effects on reducing the growth rate shown by *top3 top1* and *top3 top2* double mutants suggest that the three yeast topoisomerases have overlapping roles in cellular DNA metabolism (Section 6).

12-5 Eukaryotic Type II Topoisomerases

The type II topoisomerases from yeast, *Drosophila*, *Xenopus*, calf thymus, and human cells have similar structures and biochemical properties. These homodimers of 150- to 180-kDa subunits relax negative and positive supercoils at similar rates, but in contrast to the bacterial DNA gyrase do not introduce negative supercoils; relaxation is accompanied by changes in linking number in steps of two. Duplex molecules without nicks, gaps, or supercoils can be catenated, decatenated, knotted, and unknotted. Enzymatic activity requires ATP hydrolysis and Mg^{2+}.

Protein Structure

The sequences of topo II from *S. cerevisiae*,[91] *S. pombae*,[92] *Drosophila*,[93] and humans[94] reveal colinear homology over most of the protein, with the greatest divergence in the C-terminal region. Overall, the homology for 1000 amino acids at the N terminus is 46% between *Drosophila* and each of the yeast enzymes, 49% between the two yeast proteins, and 63% between *Drosophila* and human proteins.

Based on homology to the prokaryotic type II enzymes,[95] two regions of the eukaryotic proteins appear to be functionally distinct (Fig. 12-8). The eukaryotic N-terminal region is related to the B subunit of gyrase and probably corresponds to the ATPase domain. A central portion is similar in sequence to the gyrase A subunit and is involved in the nicking and closing activity. Homology to the phage T4 topoisomerase subunits (Section 7) also confirms the functions of these regions. Location of the tyrosine that forms the covalent protein–DNA linkage within the middle region of the *S. cerevisiae* enzyme is consistent with the homology to the gyrase A subunit.[96] Although the reactions catalyzed by eukaryotic topo II and gyrase are clearly different, the structural similarity of the enzymes indicates that some steps in their catalytic mechanisms are very likely shared.

---

91. Giaever G, Lynn R, Goto T, Wang JC (1986) *JBC 261*:12448.
92. Uemura T, Morikawa K, Yanagida M (1986) *EMBO J 5*:2355.
93. Wyckoff E, Natalie D, Nolan JM, Lee M, Hsieh T-S (1989) *JMB 205*:1.
94. Tsai-Pflugfelder M, Liu LF, Liu AA, Tewey KM, Whang-Peng J, Knutsen T, Heubner K, Croce DM, Wang JC (1988) *PNAS 85*:7177.
95. Wyckoff E, Natalie D, Nolan JM, Lee M, Hsieh T-S (1989) *JMB 205*:1.
96. Worland ST, Wang JC (1989) *JBC 264*:4412.

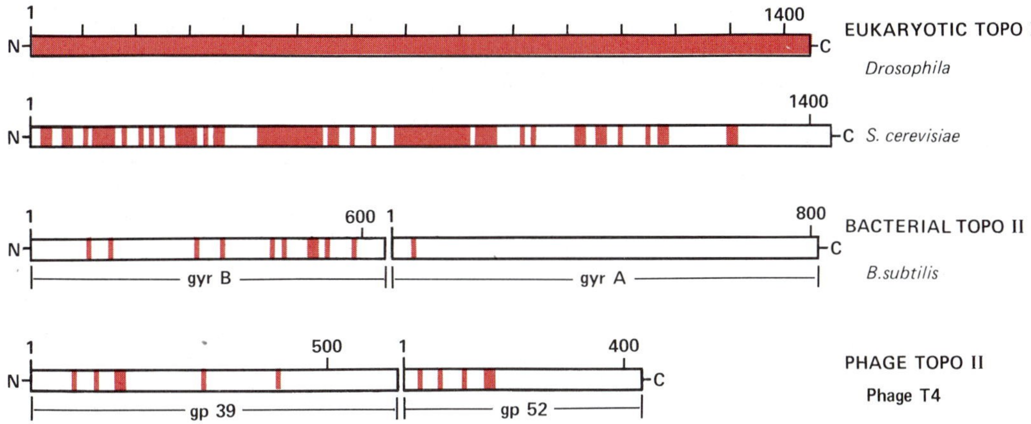

Figure 12-8
Amino acid sequence relationships between type II topoisomerases. Regions in which five out of ten amino acids are identical to the *Drosophila* protein are shown in color. [After Wyckoff E, Natalie DM, Nolan JM, Lee M, Hsieh T-S (1989) *JMB 205*:1]

DNA Binding and Cleavage and the Reaction Mechanism

DNA cleavage by topo II is stimulated by drugs (Section 14-5) that interfere with the subsequent ligation step. As with bacterial gyrase (Section 3), a ternary complex of drug, enzyme, and DNA is formed in which both strands of DNA are cleaved to generate a 4-bp staggered cut. The 5′ overhanging strands are covalently attached by the terminal phosphoryl group to a tyrosine in each subunit of the enzyme. A modest degree of sequence specificity is seen at the break sites; the consensus sequence for cleavage by the *Drosophila* enzyme[97] is

5′ GTN(A/T)AYATTNATNNG

Cleavage of the two DNA strands can be uncoupled[98] in Mg^{2+}-EDTA, at low pH, or when Mg^{2+} is replaced by Ca^{2+}, Mn^{2+}, or Co^{2+}. Divalent cations may act to coordinate cleavage by the two subunits in the dimeric enzyme. The kinetics suggests that cleavage of the two strands is sequential, as is religation, with reunion of the second strand being slower than that of the first[99] (Fig. 12-8). The asymmetry of the preferred cleavage sites also indicates a strand specificity by the enzyme in its interactions with DNA.[100]

12-6 Functions of the Eukaryotic Topoisomerases

In contrast to the topological specificities of the prokaryotic topoisomerases, eukaryotic topo I and topo II relax negative as well as positive supercoils. Their

97. Sander M, Hsieh T-S (1985) *NAR 13*:1057.
98. Zechiedrich EL, Christiansen K, Andersen AH, Westergaard O, Osheroff N (1989) *Biochemistry 28*:6229; Lee MP, Sander M, Hsieh T-S (1989) *JBC 264*:13510.
99. Zechiedrich EL, Christiansen K, Andersen AH, Westergaard O, Osheroff N (1989) *Biochemistry 28*:6229; Lee MP, Sander M, Hsieh T-S (1989) *JBC 264*:13510.
100. Andersen AH, Christiansen K, Zechiedrich EL, Jensen PS, Osheroff N, Westergaard O (1989) *Biochemistry 28*:6237.

functions appear redundant compared to those of their prokaryotic counterparts, but some distinctive roles are suggested.

Roles in DNA Replication

Among the genes for the yeast enzymes, *TOP1* encodes the classic eukaryotic type I enzyme, *TOP2* encodes topo II, and *TOP3* is responsible for a second type I enzyme which resembles the prokaryotic topoisomerases.

The viability of *top1* deletion mutants (Δ*top1*) argues that topo I is not essential for DNA replication, cell growth, or cell division. In contrast, *TOP2* is essential and is required specifically during mitosis.[101] The S phase proceeds normally in the absence of either enzyme, but DNA synthesis shuts off rapidly after a shift to the nonpermissive temperature in a Δ*top1*, *top2*^{ts} strain,[102] indicating that one of the two proteins is required. Apparently either topo I or topo II (but not the *TOP3* gene product) can provide the swivel during DNA synthesis. Extension of short nascent chains into long replicated strands is delayed in *top1* mutants, however, suggesting that topo I is more effective.[103] Topo II function is required uniquely during cell division for separation of the daughter chromosomes. The synthesis of short chains in the absence of both topo I and topo II suggests that neither enzyme is required for initiation.[104]

Similar conclusions can be drawn from the roles of topoisomerases in the SV40 in vitro replication system[105] (Section 19-4). During DNA synthesis on this small circular viral genome, topo I or topo II can support fork movement, but decatenation of the late intermediates depends specifically on topo II. Initiation of SV40 replication requires neither negatively supercoiled DNA nor the action of a topoisomerase.

Roles in Transcription[106]

Because eukaryotic topo I and topo II can remove both positive and negative supercoils, it was more difficult to establish in eukaryotic than in prokaryotic cells that transcription generates supercoiled domains. However, in a yeast strain that lacks both enzymes but expresses the *E. coli topA* gene, positively supercoiled DNA accumulates during transcription. Thus, as in *E. coli*, supercoiled domains are normally generated in eukaryotic cells, and either topo I or topo II can adequately remove them.

Several lines of evidence suggest that topo I, rather than topo II, normally functions in transcription elongation: (1) Topo I is found associated with transcribed regions in *Drosophila* DNA.[107] (2) Exceptionally strong binding sites flank highly transcribed regions, such as the ribosomal RNA genes.[108] (3) Topo I is localized in the nucleolus.[109] (4) Treatment of cells with the topo I inhibitor

---

101. Holm C, Goto T, Wang JC, Botstein D (1985) *Cell 41*:553; Uemura T, Yanagida M (1986) *EMBO J 5*:1003; Uemura T, Morikawa K, Yanagida M (1986) *EMBO J 5*:2355.
102. Brill SJ, DiNardo S, Voelkel-Meiman K, Sternglanz R (1987) *Nature 326*:414.
103. Kim RA, Wang JC (1989) *JMB 208*:257.
104. Kim RA, Wang JC (1989) *JMB 208*:257.
105. Yang L, Wold MS, Li J-J, Kelly TJ, Liu LF (1987) *PNAS 84*:950.
106. Giaever GN, Wang JC (1988) *Cell 55*:849.
107. Gilmour DS, Pflugfelder G, Wang JC, Lis JT (1986) *Cell 44*:401.
108. Bonven BJ, Gocke E, Westergaard O (1985) *Cell 41*:541; Christiansen K, Bonven BJ, Westergaard O (1987) *JMB 193*:517.
109. Higashinakagawa T, Wahn H, Reeder RH (1977) *Dev Biol 55*:375; Fleischmann G, Pflugfelder G, Steiner EK, Javaherian K, Howard GC, Wang JC, Elgin SCR (1984) *PNAS 81*:6958; Muller MT, Pfund WP, Mehta VB, Trask DK (1985) *EMBO J 4*:1237.

camptothecin inhibits rRNA synthesis.[110] (5) Vaccinia virus core particles[111] (Section 7) and immature *Xenopus* oocytes[112] contain only topo I and are transcriptionally active but do not replicate DNA. (6) Topo I is present throughout the cell cycle and in quiescent cells, in contrast to topo II, which is present principally in dividing cells[113] just before mitosis.[114] Thus, it seems that topo II functions in cell division, while topo I is used for most other processes, including transcription.

Roles in Recombination

That topo I and topo II directly participate in illegitimate (nonhomologous) and homologous recombination (Section 21-5) is based on the following: (1) Topo I binding sequences are found near the junctions generated by excision of integrated viral sequences.[115] (2) Matrix-associated regions of chromosomal DNA which include topo II cleavage sites are associated with the recombination break points of several chromosome rearrangements.[116] (3) Anti–topo II drugs increase the frequency of sister chromatid exchange.[117] These roles can be reconciled with the mechanistic similarity between topoisomerases and the proteins that mediate site-specific recombination and transposition (Sections 21-6 and 21-7).

In addition to their stimulatory effect on recombination, yeast mutants deficient in topoisomerases indicate that these enzymes normally repress genetic exchange between repeated elements.[118] Inactivation of topo I results in a 100-fold increase in recombination between the copies of the rRNA genes;[119] if, in addition, topo II is partially inactivated, the rRNA genes are excised from the chromosome at high frequency.[120] The increased recombination frequency resulting from the reduced levels of topoisomerases may result from the accumulation of superhelicity generated during transcription. Considering the high transcriptional activity of the rRNA genes and the involvement of topoisomerases, this explanation is attractive. Furthermore, the rRNA transcription control region promotes recombination when it is moved to other regions of the chromosome.[121]

Repeated sequences arising from the ends of the yeast retrotransposon Ty, δ elements (Section 21-7), are dispersed throughout the chromosome. A mutation that increases recombination between δ sequences was traced to *TOP3*, indicating that the TOP3 topoisomerase also represses recombination.[122] However, efficient removal of transcription-induced supercoiling is unlikely to be involved in the repression of recombination by the TOP3 protein, as most δ elements are transcriptionally silent,[123] and TOP3 protein appears not to have a major influence on DNA relaxation in vivo.

---

110. Zhang H, Wang JC, Liu LF (1988) *PNAS 85*:1060.
111. Bauer WR, Ressner EC, Kates J, Patzke JV (1977) *PNAS 74*:1841.
112. Higashinakagawa T, Wahn H, Reeder RH (1977) *Dev Biol 55*:375.
113. Heck MMS, Earnshaw WC (1986) *JCB 103*:2569; Heck MMS, Hittelman WN, Earnshaw WC (1988) *PNAS 85*:1086; Fairman R, Brutlag DL (1988) *Biochemistry 27*:560.
114. Heck MMS, Hittelman WN, Earnshaw WC (1988) *PNAS 85*:1086.
115. Bullock P, Champoux JJ, Botchan M (1985) *Science 230*:954; Bullock P, Forrester W, Botchan M (1984) *JMB 174*:55.
116. Sperry AO, Blasquez VC, Garrard WT (1989) *PNAS 86*:5497.
117. Marshall B, Ralph RK, Hancock R (1983) *NAR 11*:4251.
118. Fink GR (1989) *Cell 58*:225.
119. Christman MF, Dietrich FS, Fink GR (1988) *Cell 55*:413.
120. Kim RA, Wang JC (1989) *Cell 57*:975.
121. Stewart SE, Roeder GS (1989) *Mol Cell Biol 9*:3464; Voelkel-Meiman K, Keil RL, Roeder GS (1987) *Cell 48*:1071.
122. Wallis JW, Chrebet G, Brodsky G, Rolfe M, Rothstein R (1989) *Cell 58*:409.
123. Fink GR (1989) *Cell 58*:225.

On the basis of their opposing actions, the topoisomerases, like "double-edged swords," can cut both ways in recombination, forming recombinational intermediates or disentangling paired DNA strands.[124]

Chromatin Assembly, Chromosome Condensation, and the Chromosome Scaffold

The topoisomerases help to organize eukaryotic chromosomes. During the assembly of nucleosomes, a topoisomerase may relieve the strain associated with the tight wrapping of the DNA around the histones. In *Xenopus* egg or oocyte extracts, which assemble chromatin efficiently, topo I appears to be the dominant activity.[125] The enzyme relaxes the strain in the DNA generated during assembly, rather than introducing negative supercoils to drive histone binding.

Topo II as a component of the mitotic chromosome scaffold and the interphase nuclear matrix[126] may contribute structurally as well as functionally to chromosomal organization. These proteinaceous structures maintain looped domains within the chromosome. Topo II might define the end points of these topologically and functionally distinct regions. Attachment at the bases of these loops may be important for topo II function in chromosome condensation[127] and in separation of the daughter DNAs at cell division.[128] However, topo II is present principally in dividing cells, whereas the chromosomal loops persist in quiescent cells.

12-7 Other Topoisomerases

Topoisomerases are found nearly everywhere that DNA is found. In addition to the four main classes (Sections 2, 3, 4, and 5), different types have been isolated from viral, plasmid, and organelle sources.

Phage T4 Topoisomerase[129]

Functionally and structurally, this enzyme resembles the eukaryotic type II topoisomerases more than it resembles gyrase. The protein catalyzes ATP-dependent relaxation but is not capable of introducing superhelical turns. Furthermore, in contrast to gyrase, T4 topoisomerase is sensitive to amsacrine (M-AMSA; Section 14-5), an inhibitor of eukaryotic topo II.[130]

The enzyme is a hexamer, containing two each of the products of genes *39* (57 kDa), *52* (48 kDa), and *60* (18 kDa). Gp39 contains the active site for ATP hydrolysis and is structurally related to both the bacterial gyrase B subunit [131] and the N-terminal region of eukaryotic topo II (Fig. 12-8); gp52 becomes cova-

---

124. Wang JC, Caron PR, Kim RA (1990) *Cell 62*:403.
125. Almouzni G, Méchali M (1988) *EMBO J 7*:4355.
126. Earnshaw WC, Halligan B, Cooke CA, Heck MMS, Liu LF (1985) *JBC 100*:1706; Gasser SM, Laroche T, Falquet J, Boy de la Tour E, Laemmli UK (1986) *JMB 188*:613; Berrios M, Osheroff N, Fisher PA (1985) *PNAS 82*:4142.
127. Newport J, Spann T (1987) *Cell 48*:219; Uemura T, Ohkura H, Adashi Y, Morino K, Shiozaki K, Yanagida M (1987) *Cell 50*:917.
128. Holm C, Goto T, Wang JC, Botstein D (1985) *Cell 41*:553; Uemura T, Yanagida M (1986) *EMBO J 5*:1003; Uemura T, Morikawa K, Yanagida M (1986) *EMBO J 5*:2355.
129. Kreuzer KN, Jongeneel CV (1983) *Meth Enz 100*:144; Seasholtz AF, Greenberg GR (1983) *JBC 258*:1221.
130. Huff AC, Leatherwood JK, Kreuzer KN (1989) *PNAS 86*:1307.
131. Huang WM (1986) *NAR 14*:7751.

lently attached to the DNA during cleavage[132] and is related in sequence to the A subunit of gyrase and the internal portion of topo II, including the region containing the active-site tyrosine.[133] The homology of the T4 protein with the eukaryotic enzymes is greater than with the *E. coli* proteins.

The T4 topoisomerase is required for normal phage DNA replication specifically during initiation (Section 17-6), although host gyrase can partially substitute. Analysis of cleavage sites indicates that the enzyme interacts differently with T4 DNA (containing glycosylated hydroxymethylcytosine residues) than with normal, unmodified cytosine DNA.[134] Roles for T4 topoisomerase in mutagenesis, illegitimate recombination,[135] and regulation of gene expression[136] have also been indicated. Some of these functions may involve the membrane association of the topoisomerase subunits.[137]

Reverse Gyrase[138]

This type I enzyme, isolated from the thermoacidophilic archaebacterium *Sulfolobus acidocaldarius*, appears to have the unique ability to introduce positive supercoils. The enzyme, purified to apparent homogeneity, is a monomer of 128 kDa and can utilize ATP to relax and also to positively supercoil DNA. Low ionic strength and high temperature (e.g., 75°) favor supercoiling. No role for this enzyme has been established, although positively supercoiled DNA has been isolated from a virus-like particle of *Sulfolobus*.[139] Possibly, at the extremely high temperature and low pH in the environments of these organisms, a high GC content and positive superhelicity may help to maintain the template in a duplex conformation.

Agrobacterium Ti Plasmid VirD1 Protein[140]

The gene for this protein lies within an operon encoding the factors required for transfer of the T-DNA region of Ti plasmid from *Agrobacterium* into plants. The product of the *virD1* gene has been overproduced in *E. coli* and partially purified; it appears to have DNA-relaxing activity. Negatively supercoiled DNA is relaxed in a reaction which requires Mg^{2+}, but is independent of ATP or other nucleotide cofactors. VirD1, at about 21 kDa, is the smallest topoisomerase known. Along with a second protein, VirD2, VirD1 appears to catalyze the site-specific cleavage of the T-DNA prior to DNA transfer.

Vaccinia Virus Topoisomerase[141]

The cytoplasmically replicating vaccinia virus encodes a small type I topoisomerase which is encapsidated in the phage particle. Similar to the standard

---

132. Rowe TC, Tewey KM, Liu LF (1984) *JBC 259*:9177.
133. Huang WM (1986) *NAR 14*:7379.
134. Kreuzer KN, Alberts BM (1984) *JBC 259*:5339.
135. Chiba M, Shimizu H, Fujimoto A, Nashimoto H, Ikeda H (1989) *JBC 264*:12785.
136. Wirak DO, Cook KS, Greenberg GR (1988) *JBC 263*:6193; Cook KS, Wirak DO, Seasholtz AF, Greenberg GR (1988) *JBC 263*:6202.
137. Huang WM (1975) *Virology 66*:508; Takacs BJ, Rosenbusch JP (1975) *JBC 250*:2339.
138. Nadal M, Jaxel C, Portemer C, Forterre P, Mirambeau G, Duguet M (1988) *Biochemistry 27*:9102; Mirambeau G, Forterre P, Duguet M (1984) *JMB 179*:559; Kikuchi A, Asai K (1984) *Nature 309*:677.
139. Nadal M, Mirambeau G, Forterre P, Reiter W-D, Duguet M (1986) *Nature 321*:256.
140. Ghai J, Das A (1989) *PNAS 86*:3109.
141. Shuman S, Moss B (1987) *PNAS 84*:7478.

eukaryotic topo I, it relaxes both positive and negative supercoils, is stimulated by Mg^{2+}, and attaches to the 3′-P of the DNA strand upon cleavage; some amino acid homology with the eukaryotic topo I can be recognized around the active-site tyrosine. At 32 kDa, it is much smaller than the host enzymes and is also resistant to inhibition by camptothecin. Furthermore, enzyme activity is stimulated by ATP and inhibited by coumermycins and nonhydrolyzable ATP analogs, properties normally associated with type II rather than type I enzymes; ATP does not, however, appear to be hydrolyzed.[142]

Although essential for viral growth,[143] the function of this topoisomerase is not known. Encapsidation suggests that it plays a role in the transcription of early genes, which takes place in the virus particle. Other possible functions include participation in DNA packaging, replication, or recombination. When overproduced in *E. coli*, vaccinia topo I stimulates illegitimate recombination, a fact of interest in view of the high rate of recombination during vaccinia virus infection.[144]

Chloroplast, Mitochondrial, and Kinetoplast Topoisomerases

A type I enzyme isolated from chloroplasts is similar to the prokaryotic enzymes.[145] Relaxation occurs only on negatively supercoiled DNA and requires Mg^{2+}. A type II enzyme from rat liver and human lymphoblastic leukemic cell mitochondria[146] has been reported but has not been studied in detail.

The single large mitochondrion (kinetoplast) of trypanosomes contains a network of several thousand small circular DNA molecules in addition to the 20 to 30 copies of the mitochondrial genome (Section 18-11). Replication of this kDNA network involves removal of each minicircle for replication as a free molecule, after which it is returned to the periphery of the network. The properties of two topoisomerases isolated from trypanosomes suggest their involvement in this process. An enzyme with characteristics typical of eukaryotic type II topoisomerases both catenates and decatenates minicircles, in addition to relaxing DNA.[147] The protein is a homodimer of 132-kDa subunits and requires ATP and Mg^{2+} for activity. The presence of the enzyme in the kinetoplast has been confirmed by subcellular fractionation. Immunofluorescence localizes it to two regions on opposite sides of the kDNA network, suggesting its involvement in rejoining minicircles to the structure following replication.[148]

A 60-kDa type II enzyme efficiently removes covalently closed minicircles from the network but is inactive on those that are nicked.[149] The specificity appears to involve recognition of the altered structure of the kDNA bend (Section 1-10) on the two plasmid forms. This property is attractive, for it may make it possible for nicked molecules that have already been replicated to remain attached to the network while unreplicated, covalently closed circles are specifically removed. It remains to be shown that this enzyme resides specifically within the organelle.

142. Tsubota Y, Waqar MA, Burke JF, Milavetz BI, Evans MJ, Kowalski D, Huberman JA (1979) *CSHS 43*:693.
143. Shuman S, Golder M, Moss B (1989) *Virology 170*:302.
144. Shuman S (1989) *PNAS 86*:3489.
145. Siedlecki J, Zimmerman W, Weissbach A (1983) *NAR 11*:1523.
146. Castora FJ, Sternglanz R, Simpson MV (1982) in *Mitochondrial Genes* (Attardi G, Slonimski P, Borst P, eds.). CSHL, p. 143; Castora FJ, Lazarus GM, Kunes D (1985) *BBRC 130*:854.
147. Melendy T, Ray DS (1989) *JBC 264*:1870.
148. Melendy T, Sheline C, Ray DS (1988) *Cell 55*:1083.
149. Shlomai J, Zadok A (1983) *NAR 11*:4019; Shlomai J, Linial M (1986) *JBC 261*:16219.

Replication and Recombination Enzymes with Topoisomerase Activity

The ϕX174 gene *A* and Ff gene *II* proteins, endonucleases which initiate rolling-circle replication (Section 8-7), are sequence-specific type I topoisomerases by virture of being able to nick and reseal DNA at the origin of replication. Phage λ Int protein and Tn3 resolvase, site-specific recombination enzymes (Section 21-6), also show type I relaxing activity associated with the cleavage and ligation reactions they perform during recombination. The plasmid R1 initiator protein, RepA, is homologous in amino acid sequence to *E. coli* topo I. RepA relaxes DNA; it is not clear, however, that the relaxation activity has a role in plasmid replication (Section 18-4).

CHAPTER 13

Deoxyribonucleases

13-1 Deoxyribonucleases in Vitro and in Vivo[1]

Deoxyribonucleases (DNases) hydrolyze the phosphodiester linkage between deoxyribonucleosides. Exonucleases act from the end of a DNA chain, whereas endonucleases attack interior linkages. The great value of DNases as laboratory reagents[2] has tended to overshadow an appreciation of their importance in vivo. DNases contribute significantly to the biosynthesis of nucleic acids and have a central role in repair, recombination, and restriction of DNA.

The utility of DNases as reagents is shown by several examples. (1) Pancreatic DNase and snake venom phosphodiesterase were used to establish the 3′,5′-phosphodiester backbone structure of DNA. (2) These enzymes were also used to prepare 5′-deoxynucleotides as substrates for DNA polymerase. (3) Micrococcal nuclease and spleen phosphodiesterase were used for nearest-neighbor analyses of DNA (Section 4-16). (4) Single-strand–specific endonucleases have served for analysis of nucleic acid homologies, heteroduplex formation, mutations, and structural features. (5) Restriction endonucleases are used to dissect, map, sequence, clone, and reconstitute genes and chromosomes.

The discoveries of biosynthetic pathways distinct from degradative routes, in all branches of biochemistry, have now displaced the old notion that biosynthesis occurs by reversal of breakdown reactions. The tide of these modern views tends to obscure the key importance of hydrolytic steps in biosynthesis. To cite some instances: in glycogen synthesis, fructose-1,6-diphosphatase; in triglyceride and phospholipid synthesis, phosphatidic acid phosphatase; in protein

1. Linn S (1982) in *Nucleases* (Linn SM, Roberts RJ, eds.). CSHL, p. 291.
2. Laskowski M (1982) in *Nucleases* (Linn SM, Roberts RJ, eds.). CSHL, p. 1.

synthesis, proteolytic processing of enzymes, hormones, secreted proteins, and viral proteins; and in the manufacture of ribosomal, messenger, and transfer RNAs, nucleolytic scission of precursors.

Nucleolytic cleavages contribute to DNA biosynthesis in several ways:

(1) Nicking in certain instances creates an origin for replication (Section 8-7).

(2) Concerted nicking and ligating relieves obstructive positive supertwisting and introduces essential negative supertwisting (Section 12-3).

(3) Excision of the RNA fragment priming the initiation of a DNA chain (Section 4-12) allows it to be replaced by DNA.

(4) Proofreading of mismatched primer ends by the 3′→5′ exonuclease of pol I (Section 4-9) enhances the fidelity of replication.

(5) Specific scissions enable a viral chromosome to be assembled and packaged (Section 17-7).

(6) Excision of DNA defects prevents mutations and interruptions of replication (Section 13-9).

(7) Breakdown of DNA generates precursors for DNA synthesis by salvage pathways, as in the utilization of host DNA by invading viruses (Sections 2-13 and 17-6).

Ribonucleases, except for RNase H and nucleases that act on both RNA and DNA, are not covered in this volume. Yet RNases deserve attention, for the great variety of their mechanisms, structures, and functions contributes to the understanding of the DNases.

13-2 Exonucleases: 3′→5′[3]

Exonucleases do not act on circular DNA (single- or double-stranded) but require the terminus of a chain. Some utilize a 3′ end, others a 5′ end; still others can act at either end. Some are sensitive to the presence or absence of a phosphomonoester at the terminus, and others are not. Classification of these nucleases is based on their action on substrates radioactively labeled with different markers at the two ends. A 5′ terminus is readily labeled by [γ-^{32}P]ATP and polynucleotide kinase (after the unlabeled 5′-P is first removed with phosphatase). A 3′ terminus can be labeled by a radioactive dNTP and terminal deoxynucleotidyl transferase or, in duplexes, by DNA polymerase in the presence of a single radiolabeled dNTP.

The 3′→5′ exonucleases abound in all cells (Tables 13-1 and 13-2). They may be further distinguished by a preference for single-stranded or duplex DNA and by production of mono- or oligonucleotides. How their actions are used to shape and modify a primer terminus for DNA synthesis (following repair and recombination) is illustrated by the reactions of pol I, exo I, and exo III on a variety of 3′ termini (Table 13-3).

Binding of an exonuclease may not be limited to the chain terminus. From the action of the 3′→5′ exonuclease of T4 phage DNA polymerase on single strands of different chain length, it may be inferred that the enzyme binds strongly to

3. Linn S (1981) *Enzymes 14*:121; Linn S (1982) in *Nucleases* (Linn SM, Roberts RJ, eds.). CSHL, p. 291; Weiss B (1981) *Enzymes 14*:203; Phillips GJ, Kushner SR (1987) *JBC 262*:455.

the interior of a DNA chain as well as to the 3′ end. For example, when poly dT_{300} and poly dT_{3000} are compared as substrates at the same concentration of 3′-OH termini, the exonuclease rate is nearly 100 times greater with the shorter chain. When the longer DNA chain was covered with T4 gp32, the rate increased sharply; sequestering of enzyme by unproductive binding to internal regions of the DNA is also seen with *E. coli* pol I and *B. subtilis* pol III and may be a widely shared property among the exonucleases.

Exonuclease I[4]

E. coli exo I, a polypeptide of 55 kDa, is a product of the gene that suppresses *recBC* mutations [*sbcB;* also called *xonA* (exo *one* A)]. Exo I is absolutely specific for ssDNA, degrading from the 3′ end.[5] (Activity observed from the 5′ end, in the

Table 13-1
Prokaryotic exonucleases:[a] 3′→5′

Enzyme	Source	DNA strand[b]	Comments[c]
Exo I	*E. coli (xonA, sbcB)*	ss	degrades to terminal dinucleotide; degrades glycosylated DNA
DNA pol I (exo II)	*E. coli (polA)*	ss	acts on frayed or mismatched terminus of duplex; proofreading function; formerly called exo II; complete degradation; see also Table 13-4 and Section 4-9
Exo IV	*E. coli*	ss	complete degradation of oligonucleotides
Exo V	*E. coli*	ss	see RecBCD endonuclease, below
Exo VII	*E. coli (xse)*	ss	produces oligonucleotides from 3′ and 5′ ends; Mg^{2+} not required; see also Table 13-4
Bal 31	*Alteromonas espejiana*	ss	exo- and endonucleases; nibbles away 3′ and 5′ ends of duplex DNA; see also Tables 13-4 and 13-6
ϵ subunit of DNA pol III	*E. coli (dnaQ)*	ss	resembles 3′→5′ exo activity of pol I; see also Section 5-3
T4 exo IV	Phage T4 encoded *(dexA)*	ss	complete degradation of oligonucleotides
T4 DNA polymerase	Phage T4 encoded (gene *43*)	ss	resembles pol I; degrades up to terminal dinucleotide; see also Section 5-8
RecBCD endonuclease (exo V)	*E. coli*	ss, ds	dependent on ATP; acts as a DNA-dependent ATPase; no action at a nick; also a 5′→3′ exonuclease and an endonuclease; oligonucleotide products; functions in recombination; see also Section 9 and Tables 13-4 and 13-6
Exo III (endo II,VI)	*E. coli (xth)*	ds	phosphomonoesterase on a 3′-P terminus; exonuclease on dsDNA at end or nick; endo II and VI; *xthA* mutant is sensitive to H_2O_2; see also Tables 13-6 and 13-12
Exonuclease	*B. subtilis*	ds	extracellular; produces 3′ nucleotides; also acts 5′→3′ on single strand (see also Table 13-4); acts on RNA

[a] Linn S (1981) Enzymes *14*:121; Linn S (1982) in *Nucleases* (Linn SM, Roberts RJ, eds). CSHL, p. 291.
[b] ss = single-stranded; ds = double-stranded.
[c] Products are 5′-mononucleotides except for the *B. subtilis* exonuclease.

4. Weiss B (1981) Enzymes *14*:203; Phillips GJ, Kushner SR (1987) *JBC 262*:455.
5. Lehman IR (1971) Enzymes *4*:251.

Table 13-2
Eukaryotic exonucleases: 3′→5′

Enzyme	Source	DNA strand[a]	Comments[b]
Exo I[c]	Yeast	ss	associated with pol II; proofreading
Exo III[d]	Yeast	ss	associated with pol III; proofreading
DNA pol α	*Drosophila*	ss	cryptic activity in the DNA pol–primase
DNA pol γ	*Drosophila*, mammals	ss	proofreading
DNA pol δ	Mammals	ss	proofreading
DNA pol ϵ	Mammals	ss	proofreading
Snake venom phosphodiesterase[e]	*Crotalus adamanteus*	ss	complete degradation; also acts on RNA and nucleoside di- and triphosphates
DNase V[f]	HeLa cells (human)	ds	bidirectional (see also Table 13-5); associates with DNA pol β

[a] ss = single-stranded; ds = double-stranded.
[b] Products are 5′-mononucleotides.
[c] Arendes J, Kim KC, Sugino A (1983) *PNAS 80*:673; Tsuchiya E, Kimura K, Miyakawa T, Fukui S (1984) *NAR 12*:3143.
[d] Bauer GA, Burgers PMJ (1988) *JBC 263*:917.
[e] Khorana HG (1961) *Enzymes 5*:79.
[f] Randahl H, Elliott GC, Linn S (1988) *JBC 263*:12228.

Table 13-3
Actions of several 3′→5′ exonucleases on various 3′ termini

	Action on 3′ primer terminus			
	Chain extension	3′→5′ hydrolysis		
Structure	pol I	pol I	exo I	exo III[a]
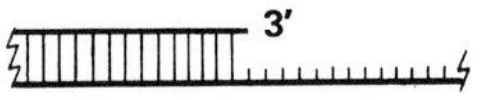 3′	yes	no[b]	no	yes
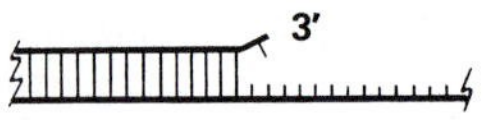 3′	no	yes	no[c]	yes
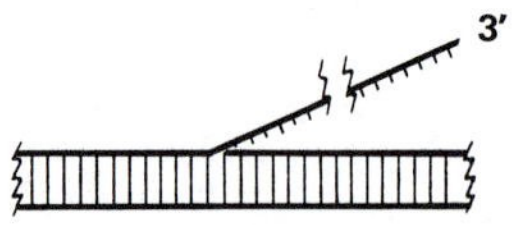 3′	no	yes	yes	no
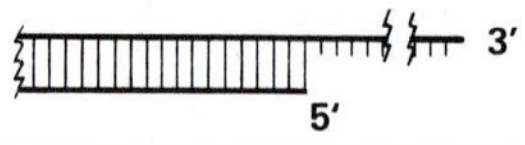 3′ 5′	no	yes	yes	no

[a] Exo III is unique among exonucleases in acting as a phosphomonoesterase on a 3′-phosphate terminus.
[b] No, in the presence of triphosphates; yes, in their absence.
[c] Hydrolysis is very slow.

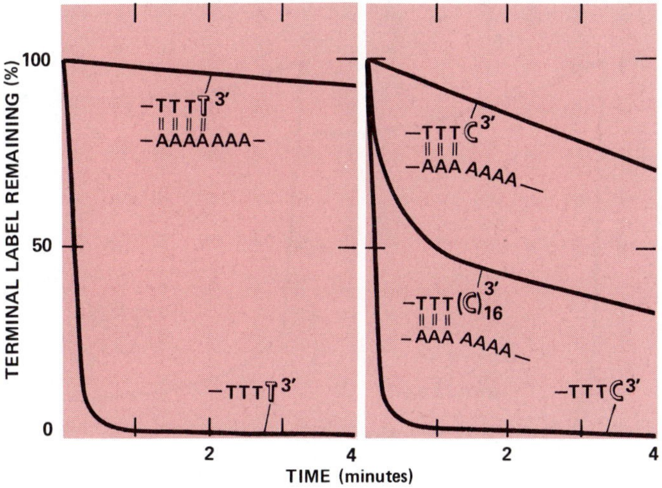

Figure 13-1
Exonuclease I action on termini of base-paired and single-stranded chains (left) and on mismatched (C) terminal regions, 1 and 16 residues long, in duplex chains (right). The polymer substrates were like those in Figures 4-12 and 4-13.

ratio of 1 : 40,000 relative to the 3′ end, apparently arose from a trace of exo VII.) The products of exo I action are mononucleotides except for the residual terminal 5′-dinucleotide.

Exo I rapidly removes the terminal nucleotide from the synthetic single strand $dT_{260}[^3H]dT_1$; it acts slowly, if at all, when this polymer is completely annealed to dA_{4000} (Fig. 13-1, left). Similarly, the enzyme removes $[^3H]dC$ from $dT_{260}[^3H]dC_1$, but only slowly from the polymer annealed to dA_{4000} (Fig. 13-1, right). Thus, removal of the labeled dCMP residue from the single strand is rapid, but removal of the mismatched end from the same substrate annealed to dA_{4000} is very slow. When the number of terminal dCMP residues on the annealed chain is increased to 16, exo I removes up to 60% of these residues rapidly; the remaining dCMP nucleotides are released very slowly (Fig. 13-1, right). Thus, removal of unpaired nucleotides is very rapid until exo I reaches within 6 to 8 residues of a base-paired region, at which point the rate is slowed.

Mutants deficient in the enzyme may appear normal, yet the *sbcB* mutation is manifested by suppression of the UV and mitomycin sensitivities of a *recBCD* mutant (Section 9).

An interaction of exo I with *E. coli* SSB stimulates the hydrolysis of ssDNA,[6] resembling the exonuclease action of T4 DNA polymerase with gp32.

Exonuclease III[7]

This 28-kDa enzyme, despite its modest size, has several distinctive catalytic features.

(1) It is a 3′→5′ exonuclease relatively specific for dsDNA. The terminal residue of $dT_{260}[^3H]dT_1$ is hydrolyzed very rapidly only when the polymer is annealed to dA_{4000} (Fig. 13-2, left);[8] the same is true for $dT_{260}[^3H]dC_1$ (Fig. 13-2, right). Polymers with a length of 3′-terminal dCMP residues resist hydrolysis in

6. Molineux IJ, Gefter ML (1975) *JMB 98*:811.
7. Weiss B (1981) *Enzymes 14*:203; Phillips GJ, Kushner SR (1987) *JBC 262*:455.
8. Brutlag D, Kornberg A (1972) *JBC 247*:241.

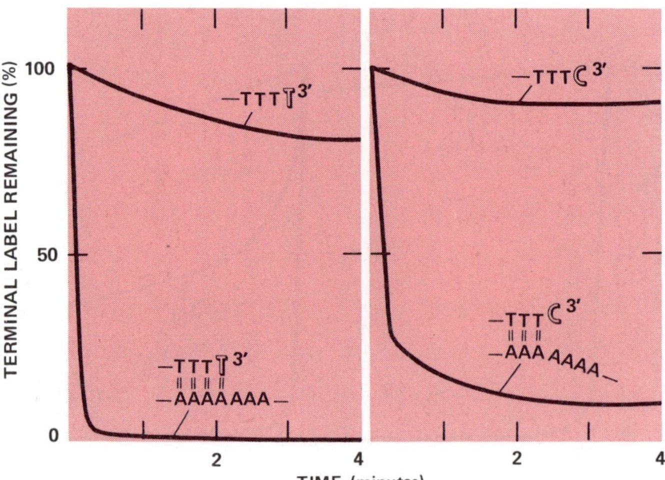

Figure 13-2
Exonuclease III action on termini of single-stranded chains (upper curves) and on base-paired or mismatched termini of duplex chains (lower curves). The polymer substrates were like those in Figures 4-12 and 4-13.

proportion to the length of sequence not annealed to dA_{4000}. From these and related experiments it appears that exo III hydrolyzes duplex polymers containing up to three mispaired terminal nucleotides. Acting at a nick in a duplex, the enzyme can enlarge it to a gap; at a 3′ duplex end, it can create a protruding 5′-ended chain.

(2) The enzyme is a phosphomonoesterase on a chain terminated by a 3′-P group. Such a terminus, generated by enzymatic or chemical cleavage, is not only inert as a primer for pol I, but through unproductive binding is a potent inhibitor. Removal of the phosphate group restores activity to the primer terminus.

(3) The enzyme is endonuclease II (also called endonuclease VI),[9] an activity discovered in studies of incision at apurinic or apyrimidinic (AP) sites induced chemically, or by spontaneous depurinations, or by DNA glycosylases acting on the products of spontaneous deaminations of cytosine, adenine, and guanine or on other damaged bases (Section 9). The enzyme also cleaves DNA that possesses a urea residue of a fragmented base at the glycosylic bond. Cleavage at the phosphodiester bond next to the AP site places the lesion at the 5′ chain end at the break.

(4) The enzyme removes sugar fragments (e.g., glycolate) from the 3′ end.

These apparently unrelated hydrolytic activities of exo III suggest that the enzyme has at least three discrete binding sites in its active center (Fig. 13-3). One site recognizes a base-paired locus in a duplex, the second recognizes the 3′-P linkage at that locus, and the third requires the absence of a base pair in the region adjacent to the 3′-P group. Thus, the exonuclease function is suited to a 3′ terminus, naturally frayed or slightly mismatched (Fig. 13-2). The enzyme is a 3′-phosphatase when the 3′-P bond is on a terminal base pair; it is an endonuclease when the space created by the loss of a base mimics the end of a chain. Exo III prefers to cleave the RNA chain of an RNA–DNA hybrid duplex.

Hydrolysis of a duplex by exo III proceeds to a 50% limit because at this point

9. Little JW, Lehman IR, Kaiser AD (1967) *JBC 242*:672; Little JW (1967) *JBC 242*:679.

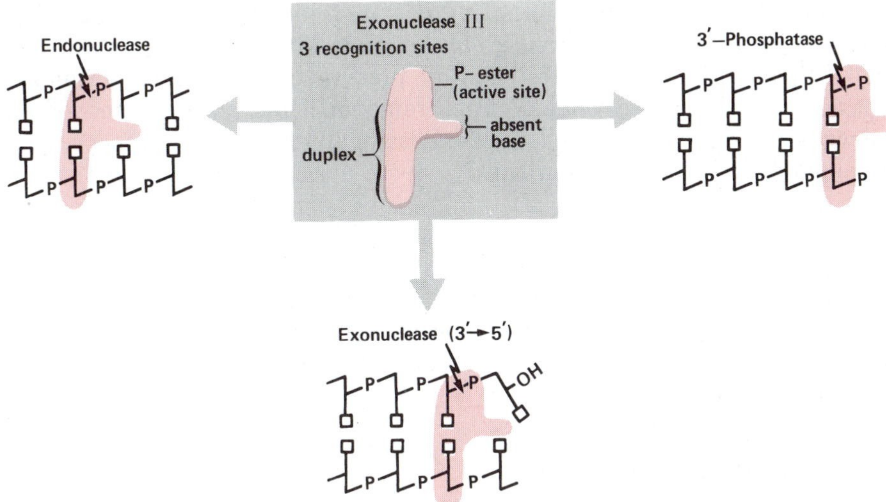

Figure 13-3
Model for active site of exonuclease III to reconcile its three distinctive activities: endonuclease II, exonuclease III, and 3′-phosphatase. [After Weiss B (1976) *JBC 251*:1896]

no duplex DNA remains. Should one 3′ chain be blocked to exo III action, then the other chain can be degraded completely, leaving the chain with the blocked end intact. Coupled with S1 nuclease action (Section 5), exo III can degrade both strands from the unblocked end of the duplex, a method that has proved useful for the generation of a nested set of deletions for spanning a cloned segment of DNA. Blockage can be achieved by nucleotides with sulfur substituted in the α-phosphate (α-phosphorothionates).[10] Such sulfur-substituted dNTPs can be polymerized by pol I and sealed by DNA ligase but are resistant to the 3′→5′ exonuclease action of pol I, as well as that of exo III.

Exo III mutants *(xthA)*, like exo I mutants, show no significant physiologic deficiency in a wild-type genetic background, aside from a heightened sensitivity to H_2O_2.[11] The properties of these and other 3′→5′ exonucleases suggest that their actions may be complementary in some instances, redundant in others (Table 13-3). For example, the rate of exo I degradation of a protruding single strand decreases as the enzyme approaches a duplex region, a point at which exo III may act. The switch-over from exo I to exo III would occur about 3 residues from the base-paired region. On the other hand, the 3′→5′ exo activity of pol I on displaced single strands appears to combine the activities of exo I and III. The activities of these several enzymes indicate that, while each may be involved in more than one physiologic process, none is absolutely indispensable.

The importance of the endonucleolytic function of exo III is seen, however, in mutants deficient in dUTPase *(dut)*, in which many abasic sites accumulate because of excessive uracil incorporation and its subsequent removal (Section 2-8). Although a deletion *xth* mutant with only 10% to 15% of wild-type endonucleolytic activity for AP sites appears normal, the combination of both *dut* and

10. Spitzer S, Eckstein F (1988) *NAR 16*:11691.
11. Imlay JA, Linn S (1986) *J Bact 166*:519; Imlay JA, Linn S (1988) *Science 240*:1302.

xth mutations is conditionally lethal. At 37° these double mutants form filaments, and fewer than 0.1% survive. On the other hand, a third mutation, *ung* (uracil-DNA glycosylase), restores viability, presumably by avoiding formation of abasic sites as a consequence of uracil removal. Thus, exo III is needed for repair when a large number of bases are missing in the DNA and especially to overcome the damage by oxyradicals.[12]

Exonuclease VII[13]

This enzyme is readily distinguished from exo I and exo III in that it acts from either end of a single-stranded chain, forms oligonucleotides, and retains full activity in EDTA, which chelates divalent cations. The enzyme is discussed in more detail in Section 3.

Eukaryotic 3′→5′ Exonucleases

Unlike the DNases found in *E. coli*, comparable enzymes from mammalian, fungal, and other eukaryotic sources (Table 13-2) are less well known due to fewer studies of their involvement in replication, repair, and recombination and the meager genetic information about them. Among the 3′→5′ exonuclease activities, those associated with the proofreading function of some DNA polymerases (e.g., pol δ) are best known (Section 6-3). Snake venom phosphodiesterase, celebrated as a reagent, acts on both DNA and RNA and, as a phosphatase, on NDPs and NTPs.

13-3 Exonucleases: 5′→3′[14]

The nucleases in this group all act from the 5′ end; some may also act from the 3′ end. They differ among themselves in producing mono- or oligonucleotides, and in preferring a single-stranded or a duplex substrate (Tables 13-4 and 13-5). Despite the variety in this group, the products are mostly oligo- rather than mononucleotides, perhaps indicating that these enzymes generally excise 5′ chain ends that are aberrant because of an RNA primer or a DNA lesion. Exemplifying these actions are the 5′→3′ exonuclease domain of pol I, the homologous polypeptide encoded by gene *6* of phage T7, and the T4 repair exonucleases. Exo VII and RecBCD nuclease may also perform these functions, inasmuch as mutants defective in xse (exo seven) and *recBCD* are deficient in excision repair. The HeLa cell nuclease, DNase V, by complexing with DNA polymerase β, provides an effective means for generating and filling small gaps; the limited nick translation (strand displacement synthesis) functions in short-patch DNA repair.[15]

12. Imlay JA, Linn S (1986) *J Bact 166*:519; Imlay JA, Linn S (1988) *Science 240*:1302.
13. Weiss B (1981) *Enzymes 14*:203; Phillips GJ, Kushner SR (1987) *JBC 262*:455.
14. Linn S (1981) *Enzymes 14*:121; Linn S (1982) in *Nucleases* (Linn SM, Roberts RJ, eds.). CSHL, p. 291; Weiss B (1981) *Enzymes 14*:203; Phillips GJ, Kushner SR (1987) *JBC 262*:455.
15. Randahl H, Elliott GC, Linn S (1988) *JBC 263*:12228.

Table 13-4
Prokaryotic exonucleases:[a] 5′→3′

Enzyme	Source	DNA strand[b]	Comments[c]
Exonuclease	*B. subtilis*	ss	extracellular; produces 3′ nucleotides; also acts 3′→5′ on double strand (see also Table 13-1); acts on RNA
Exo VII[c]	*E. coli (xse); M. luteus*	ss	produces oligonucleotides from 3′ and 5′ ends; Mg^{2+} not required; see also Table 13-1
Bal 31[d]	*Alteromonas espejiana*	ss	exo- and endonucleases; removes frayed 3′ and 5′ ends of duplex DNA; see also Table 13-1
T5 exonuclease[e]	phage T5-encoded	ss, ds	produces mono- and oligonucleotides; mutant-infected cells show arrested DNA synthesis
RecBCD endonuclease (exo V)	*E. coli (recB,recC,recD)*	ss, ds	dependent on ATP; acts as DNA-dependent ATPase; no action at a nick; also a 3′→5′ exonuclease; helicase; oligonucleotide products; functions in recombination; see also Section 9 and Table 13-1
λ exonuclease[c]	phage λ-encoded	ss, ds	strong preference for 5′-P; no action at a nick; preference for duplex over denatured DNA, but acts on oligonucleotides; functions in recombination
DNA pol I (exo VI)	*E. coli (polA)*	ds	produces mononucleotides and oligonucleotides of 2–10 residues; acts at 5′-OH, mono-P, tri-P; may function in excision repair; see also Table 13-1 and Section 4-12
T7 exonuclease	phage T7-encoded (gene *6*)	ds	essential for replication; see also Section 17-5
SP3 exonuclease[f]	*B. subtilis* phage SP3	ds	produces dinucleotides only
T4 exo B	phage T4-encoded	ds	resembles 5′→3′ exonuclease of DNA polymerase I; see also Section 9
T4 exo C	phage T4-encoded	ds	both 5′→3′ and 3′→5′ exonucleases; see also Section 9
Exo VIII (RecF)	*E. coli (recF)*	ds	expressed only in *sbcA* mutants; can substitute for RecBCD
RecJ exo[c]	*E. coli*	ss	Mg^{2+} required

[a] ss = single-stranded; ds = double-stranded.
[b] Linn S (1981) *Enzymes* 14:121; Linn S (1982) in *Nucleases* (Linn SM, Roberts RJ, eds.). CSHL, p. 291.
[c] See Section 3 for references.
[d] Legerski RJ, Hodnett JL, Gray HB Jr (1978) *NAR* 5:1445.
[e] Frenkel GC, Richardson CC (1971) *JBC* 246:4839.
[f] Aposhian HV, Friedman N, Nishikara M, Heimer EP, Nussbaum AL (1970) *JMB* 49:367.

Phage λ Exonuclease[16]

One of two λ *red* genes, called *exo*, encodes the phage λ exonuclease, the counterpart of the host Rec system associated with recombination. The other gene in the *red* locus is *β*, whose function is essential for recombination in vivo. The *β* protein forms a complex with the exonuclease in vitro but has no known enzymatic function. The λ exo has a very strong preference for a 5′-P–terminated duplex chain. Unlike exo III, it does not act at a nick, but it does remove a single-stranded tail from a partially duplex DNA. The use of these properties will be illustrated in a discussion of recombination (Section 17-7).

16. Little JW, Lehman IR, Kaiser AD (1967) *JBC* 242:672; Little JW (1967) *JBC* 242:679.

Table 13-5
Eukaryotic exonucleases: 5′→3′

Enzyme	Source	DNA strand[a]	Comments[b]
Exo II[b]	yeast	ss	
Exo V[c]	yeast	ss	produces dinucleotides processively; nuclear
Phosphodiesterase[d]	spleen	ss	produces 3′ nucleotides; requires 5′-OH terminus; also acts on RNA
Exonuclease[e]	*Neurospora crassa*	ss	also acts on RNA
Exo IV[f]	yeast	ss, ds	also acts on RNA
Mammalian "exo VII"	human placenta; KB cells	ss, ds	resembles *E. coli* exo VII; see also Section 9
Mammalian DNase IV	rabbit bone marrow and lung; human KB cells	ds	resembles 5′→3′ exo of pol I; see also Section 9
DNase V	HeLa cells (human)	ds	bidirectional (see also Table 13-2)

[a] ss = single stranded; ds = double stranded.
[b] Villadsen IS, Bjorn SE, Vrang A (1982) *JBC 257*:8177.
[c] Burgers PMJ, Bauer GA, Tam L (1988) *JBC 263*:8099.
[d] Bernardi G (1971) *Enzymes* 4:271.
[e] Rabin EZ, Tenenhouse H, Fraser MJ (1972) *BBA 259*:50.
[f] Bauer GA, Burgers PMJ (1988) *JBC 263*:917.

Exonuclease VII[17]

This *E. coli* nuclease was mentioned in the preceding section; a similar enzyme is present in *Micrococcus luteus*. Except for the 5′→3′ exonuclease activities of pol I and the RecBCD enzymes, it is the only other prominent exonuclease in *E. coli* acting from the 5′ end of a chain. Its specificity for ssDNA and its capacity to attack from either chain end can be employed to degrade long single strands protruding from dsDNA. By processive excision of oligonucleotides from either end of a chain, it can excise pyrimidine dimers and other lesions. Mutants deficient in exo VII (*xse*) show a slightly increased sensitivity to UV but are most notable for their hyperrecombinational character, which approaches that of *polAex1* (Section 4-18). The persistence of single-stranded tails on DNA in cells deficient in exo VII may explain the mutants' facility in recombination.

RecJ Exonuclease[18]

The *recJ* locus, which is required for proficiency in conjugational recombination and for UV repair in cells harboring a *recD* mutation, encodes a 60-kDa exonuclease. This activity, dependent on Mg^{2+}, is specific for ssDNA and is presumed to be 5′→3′ in direction, based on its ability to hydrolyze a 5′ but not a 3′ strand tail extending from a duplex region.

13-4 Processivity of the Exonucleases[19]

Enzymes that synthesize and degrade polymers may dissociate after each catalytic event (a *distributive action*), or they may remain bound to the polymer until

17. Weiss B (1981) *Enzymes 14*:203; Phillips GJ, Kushner SR (1987) *JBC 262*:455.
18. Lovett ST, Kolodner RD (1989) *PNAS 86*:2627.
19. Thomas KR, Olivera BM (1978) *JBC 253*:424.

many cycles of reaction are completed (a *processive action*). Determining the polarity of action of a highly processive enzyme in limiting amounts can on occasion be difficult, inasmuch as the label at either end of a chain appears early in the digestion. The mechanism employed by a nuclease is significant for metabolism in vivo and important in the use of purified enzymes as reagents.

Processivity of exonuclease action can be determined from the kinetics of release of nucleotides from a polymer differentially labeled at the two ends. The polymer $[^3H]dT_{125}[^{32}P]dT_{500}$ can be used either as a single strand or, when annealed to dA_{4000}, as a duplex with unlabeled protruding single-stranded tails; for measuring processivity, the substrate must always be in excess over the enzyme.

Among exonucleases with a single-strand preference, exo I is clearly processive (Fig. 13-4); the rate of nucleotide release from the 5′ end almost matches that from the 3′ end, even though attack from the 3′ end is by far preferred. After 60% digestion of the substrate, the remaining polymer is the original length, showing that many molecules remain untouched while others are completely digested. When dA_4 oligonucleotides are annealed to the poly dT substrate, exo I becomes much less processive, indicating that approach to a duplex region causes the enzyme to dissociate. Three other enzymes with a single-strand preference—spleen exonuclease, pol I, and T4 polymerase—are all distributive. For example, release of nucleotides from the 5′ end by the spleen enzyme is nearly complete before degradation of the 3′ end begins (Fig. 13-4).

Among exonucleases preferring a duplex substrate, the only one found to be processive, of four examined, is λ exonuclease (Fig. 13-4). Exo III is distributive (Fig. 13-4), releasing all nucleotides from the 3′ end before the 5′ end is degraded. Pol I and T7 exo also remove nucleotides distributively, but from the 5′ end.

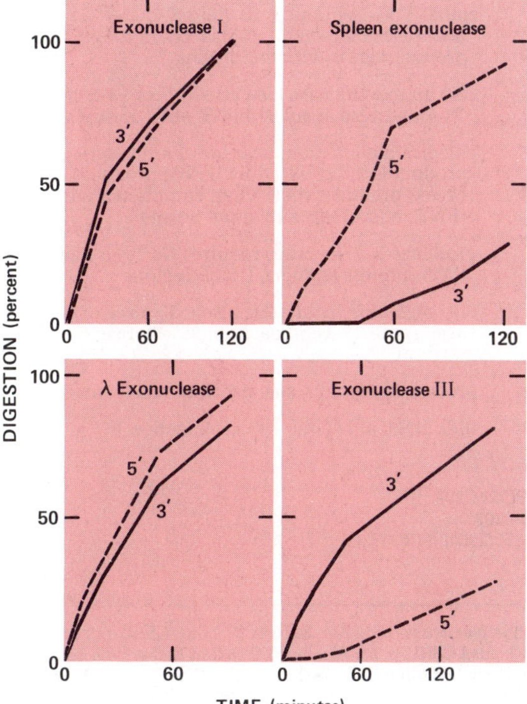

Figure 13-4
Processivity of exonucleases as judged by release of nucleotides from a DNA homopolymer with a ^{3}H-labeled sector (about 50 ntd long) at one end of the chain and a ^{32}P-labeled sector at the other end. Exonuclease I (3′→5′) and phage λ exonuclease (5′→3′) are processive; once started, a molecule is degraded rapidly from one end to the other. Spleen exonuclease (5′→3′) and exonuclease III (3′→5′) dissociate frequently and thus display their polar specificities repeatedly.

Processivity implies that an enzyme retains a site for polymer binding after the terminal nucleotide has been released. The view that exo I must bind the penultimate phosphodiester bond of ssDNA in order to cleave the terminal bond explains two features of the enzyme: its inability to hydrolyze the 5′-terminal dinucleotide of a chain, and its dissociation when it reaches a base-paired region. For the same reason, processive action by λ exo may terminate when it reaches a nick, where the absence of a phosphodiester bond beyond the one to be cleaved may prevent proper binding.

13-5 Endonucleases[20]

Endonucleases do not require a chain end and commonly show a strong preference for ssDNA or for dsDNA. Prominent among the endonucleases are the prokaryotic restriction endonucleases (Sections 6 to 8) and the nucleases for repair of DNA damage (Section 9) (Table 13-6). Another class of endonucleases, considered in this section, are the single-strand–specific enzymes acting on both RNA and DNA. Known chiefly as reagents, their metabolic significance is generally neglected.

Table 13-6
Prokaryotic endonucleases

Enzyme	Source	DNA strand[a]	Comments
RecBCD endonuclease	*E. coli (recB, recC, recD)*	ss	partially ATP-dependent; also an exonuclease (see Tables 13-1 and 13-4); functions in recombination and repair (see Section 9)
T7 endonuclease[b]	phage T7 encoded (gene *3*)	ss	essential for replication; preference for ss over ds about 150:1
T4 endonuclease IV[c]	phage T4 encoded *(denA)*	ss	splits -TpC- sequences to yield 5′-dCMP–terminated oligonucleotides; chain length of product varies with conditions
Bal 31 nuclease	*Alteromonas espejiana*	ss	also an exonuclease (see also Tables 13-1 and 13-4); nibbles away 3′ and 5′ ends of duplex DNA
Endonuclease I[d] (endo I)	*E. coli* (*endA*)	ss, ds	periplasmic location; average chain length of product is 7; inhibited by tRNA; produces ds break; produces nick when complexed with tRNA; *endo I* mutants grow normally
Micrococcal nuclease[e]	*Staphylococcus*	ss, ds	produces 3′-P termini; requires Ca^{2+}; also acts on RNA; prefers ss and AT-rich regions
Endonuclease II (endo VI, exo III)	*E. coli* (*xth*)	ds	cleavage next to AP site; also a 3′→5′ exo (see also Table 13-1); phosphomonoesterase on 3′-P termini
Restriction endonucleases		unmodified ds	produce ds breaks; see also Sections 7 and 8
Repair endonucleases		ds with lesions	nick DNA at lesions; see also Section 9

[a] ss = single-stranded; ds = double-stranded.
[b] Center MS, Studier FW, Richardson CC (1970) *PNAS 65*:242; Sadowski PD (1971) *JBC 246*:209.
[c] Sadowski PD, Bakyta I (1972) *JBC 247*:405; Vetter D, Sadowski PD (1974) *Virology 14*:207.
[d] Lehman IR, Roussos GG, Pratt EA (1962) *JBC 237*:819; Dürwald H, Hoffman-Berling H (1968) *JMB 34*:331.
[e] Anfinsen CB, Cuatrecasas P, Taniuchi H (1971) Enzymes 4:177.

20. Linn S (1981) *Enzymes 14*:121; Linn S (1982) in *Nucleases* (Linn SM, Roberts RJ, eds.). CSHL, p. 291; Bickle TA (1982) in *Nucleases* (Linn SM, Roberts RJ, eds.) CSHL, p. 85; Bickle TA (1987) in *Escherichia coli and Salmonella typhimurium* (Neidhardt FC, ed.). Am Soc Microbiol, Washington DC, Vol. 1 p. 692; Modrich P, Roberts RJ (1982) in *Nucleases* (Linn SM, Roberts RJ, eds.). CSHL, p. 109; Roberts RJ (1982) in *Nucleases* (Linn SM, Roberts RJ, eds.). CSHL, p. 311; Shishido K, Ando T (1982) in *Nucleases* (Linn SM, Roberts RJ, eds.). CSHL, p. 155.

An important characteristic of most endonucleases is their recognition of base sequences. This was once assumed to be a property only of restriction endonucleases, which recognize a sequence of four to eight nucleotides of a duplex in making a double-stranded break (Sections 7 and 8). Yet sequence preference or specificity appears to characterize all endonucleases. A striking example is cytosine-specific cleavage by T4 endonuclease IV. Further, when pancreatic DNase I, *E. coli* endo I, or spleen DNase II are freed of all contaminant activities, these nucleases produce digests with characteristic frequencies of oligonucleotides. The sequences of the 3′ and 5′ termini of the products form a pattern distinctive for each enzyme throughout the course of digestion. For these endonucleases, the recognition is limited to sequences of three to four nucleotides. The relative susceptibility of the many available sequences is reflected in both K_m and V_{max}, and can be affected by cations, ionic strength, or other reaction conditions.

Single-Strand – Specific Endonucleases[21]

These endonucleases, isolated mostly from fungi, yeast, plants, and animals, include (Table 13-7): S1 nuclease from *Aspergillus oryzae,* P1 nuclease from *Penicillium citrinum,* Bal 31 nuclease from *Alteromonas espejiana, Neurospora* nuclease from the mycelia and conidia of *Neurospora crassa,* mung bean nuclease I from sprouts, *Ustilago maydis* nuclease, and a number of others. The enzymes generally show the following properties: (1) they are remarkably specific (10,000-fold or more) for single-stranded over duplex DNA, while also recognizing distorted regions in dsDNA; (2) they are active on both RNA and DNA; (3) they produce 5′-P–terminated mono- and oligonucleotides; (4) in

Table 13-7
Eukaryotic endonucleases

Enzyme	Source	DNA strand[a]	Comments
Neurospora endonuclease[b]	*Neurospora crassa*	ss	also acts on RNA
S1-nuclease[c]	*Aspergillus oryzae*	ss	"
P1-nuclease	*Penicillium citrinum*	ss	"
Mung bean nuclease I	mung bean sprouts	ss	"
Ustilago nuclease (DNase I)	*Ustilago maydis*	ss	"
DNase I[d]	bovine pancreas	ss, ds	average chain length of product is 4; produces ds break in presence of Mn^{2+}
DNase II	thymus,[e] spleen[f]	ss, ds	produces 3′-P termini; no Mg^{2+} requirement or pH optimum
AP endonucleases	nucleus, mitochondria	ds	see also Section 9
Endo R	HeLa cells	ds	specific for GC sites[g]

[a] ss = single-stranded; ds = double-stranded.
[b] Rabin EZ, Tenenhouse H, Fraser MJ (1972) *BBA 259*:50.
[c] Sutton WD (1971) *BBA 240*:522; Beard P, Morrow JF, Berg P (1973) *Virology 12*:1303.
[d] Matsuda M, Ogoshi H (1966) *J Biochem 59*:230; Laskowski M Sr (1971) *Enzymes* 4:289; Zimmerman SB, Coleman NF (1971) *JBC 246*:309.
[e] Bernardi G (1961) *BBA 53*:216.
[f] Bernardi G, Griffe M (1964) *Biochemistry 3*:1419.
[g] Gottlieb J, Muzyczka N (1990) *JBC 265*:10836, 10842.

21. Bickle TA (1982) in *Nucleases* (Linn SM, Roberts RJ, eds.). CSHL, Bickle TA, p. 85; (1987) in *Escherichia coli and Salmonella typhimurium* (Neidhardt FC, ed.). Am Soc Microbiol, Washington DC, Vol. 1 p. 692.

many instances, they are thermostable; (5) they are zinc or cobalt metalloproteins, often with acid pH optima; and (6) they possess a tolerance for denaturants and high salt concentrations.

The uses to which these nucleases have been put are legion; as examples: detection of locally altered structures in dsDNA, mapping of mutations, mapping mRNA termini, preparation of region-specific deletion mutants, cleavage of loops in cDNA cloning, recovery of 5′ caps of eukaryotic mRNAs, industrial production of 5′-mono- and 5′-dinucleotides, estimation of duplex content, isolation of rapidly annealing regions, and elimination of short single-stranded ends.

Among these nucleases, each provides some special advantage. The Bal 31 nuclease[22] removes the frayed ends of a linear DNA distributively. With both strands of the duplex chewed away without the introduction of strand breaks elsewhere, such DNA fragments are suitable for blunt-end ligation to other DNA duplexes. Small terminal deletions can be introduced by measured exposure to Bal 31.

The S1 nuclease often introduces a single break into a supercoiled DNA where the duplex is melted or contains a cruciform structure, and then proceeds to nick the opposite strand to generate a linear duplex. The P1 nuclease, used judiciously at neutral pH, can identify the regions of localized melting at the unique origin of chromosome replication *(oriC)* in *E. coli*. The mung bean enzyme, much like P1, is also a 3′-nucleotidase, a phosphomonoesterase activity that prefers 3′-mononucleotides.

What appeared to be a motley group of *Neurospora* endonucleases[23] has proved to be a related set of polypeptides with novel features that may have wide metabolic significance. An inactive precursor polypeptide (95 kDa) is processed proteolytically into several other forms. These include a single-strand–specific endonuclease, a single-strand–specific exonuclease, a mitochondrial DNase, and a periplasmic DNase. All forms of the enzyme are antigenically related. A mutant strain appears to lack the protease responsible for two processing events: (1) conversion of an integral mitochondrial inner membrane form of the enzyme to one loosely associated with the mitochondrial membrane, and (2) conversion of a cytosolic form to one secreted by lysosome-like vesicles. The mutant is characterized by defective meiosis, abnormal mitotic recombination, and a high spontaneous mutation rate, but is not associated with sensitivity to UV– or x-ray–induced mutagenesis. Details of the various enzyme forms, functions, and interconversions should help to illuminate how proteases regulate the activities, physiologic actions, and localization of these nucleases.

13-6 Restriction Endonucleases: General Considerations[24]

Detection of DNA foreign to the cell and its degradation by restriction endonucleases occur in most prokaryotic cells. This immunity mechanism depends on

---

22. Legerski RJ, Hodnett JL, Gray HB Jr (1978) *NAR 5*:1445.
23. Lehman IR (1981) *Enzymes 14*:193.
24. Little JW, Lehman IR, Kaiser AD (1967) *JBC 242*:672; Little JW (1967) *JBC 242*:679; Yuan R (1981) *ARB 50*:285; Endlich B, Linn S (1981) *Enzymes 14*:137; Wells RD, Klein RD, Singleton CK (1981) *Enzymes 14*:157; Kannan P, Cowan GM, Daniel AS, Gann AAF, Murray NE (1989) *JMB 209*:335.

protecting the cell's own DNA by specific modifications, such as methylation of one or both strands at sites of recognition (Section 21-9). Thus far, restriction endonucleases have not been detected in any eukaryote.

Distinctions between Type I and Type II Enzymes

Two classes of restriction endonucleases are known, type I and type II (Table 13-8). Type I enzymes are found mainly in enteric bacteria; they recognize a rather large sequence and cleave the DNA at a nonspecific place 1 kb or more from the recognition sequence. Type II enzymes are ubiquitous in prokaryotes; they recognize and cleave at or very close to a specific 4- to 8-ntd sequence.

The type I class includes the ATP-dependent, complex type, of which EcoK (from *E. coli* K12) and EcoB (from *E. coli* B) are examples. These enzymes are multisubunit complexes that possess both nuclease and methylase activity. They recognize a specific and unmodified site for binding and either methylate it or translocate on the DNA to cleave it at some distant, nonspecific site. The products are heterogeneous in size; the precise structures for the 5′ termini have not been determined, but the 3′ termini are random, normal nucleotides. ATP and S-adenosylmethionine (AdoMet) induce essential conformational changes, and ATP is hydrolyzed extensively.

A class related to type I, sometimes designated type III, includes EcoP1 and like endonucleases (Section 7). These enzymes differ from type I in their specific cleavages at a point 25 to 27 bp from the recognition sequence; they require ATP, but do not hydrolyze it.

Table 13-8
Restriction endonucleases

Characteristics	Type I	Type II
Occurrence	Enterobacteriaceae	ubiquitous in bacteria
Examples	EcoK	EcoRI
Restriction, methylation[a]	single enzyme	separate enzymes
Subunits	distinctive (3)	often a simple dimer
Essential restriction factors	AdoMet,[b] ATP, Mg^{2+}	Mg^{2+}
Host-specificity site	3-mer and 4-mer, hyphenated by 6–8 ntd	symmetry (two fold); 4, 6, 8 bp
Cleavage site	random: > 1 kb from host-specificity site	at host-specificity site
Enzymatic turnover	no	yes
DNA translocation	yes	no
Methylation factors		
Essential	AdoMet[b]	AdoMet[b]
Stimulatory	ATP, Mg^{2+}	
ATPase	extensive	none
Methylation at host-specificity site	yes	yes

[a] A type I enzyme performs one or the other; a type II endonuclease is distinct from the methylase.
[b] AdoMet = S-adenosylmethionine.

Type II enzymes (Section 8) are simpler in some ways. The specific sites of binding and cleavage are nearly identical. Only Mg^{2+} is required; no ATPase or methylase activities are observed. A second enzyme, which recognizes the same sequence, provides the methylation function. The value of these nucleases as reagents in genetic research has prompted the most massive hunt in the history of enzymology. At least 400 are now known.

13-7 Restriction Endonucleases: Nonspecific Cleavage (Type I)

EcoK, EcoB, and Related Type I Restriction Enzymes: ATP-Dependent Restriction Nucleases[25]

As members of closely related families, these type I restriction enzymes recognize different but specific sequences. The enzymes are made up of three types of subunits in several oligomeric forms. Relatedness within the EcoK family (including EcoB and some *Salmonella* members) is demonstrable by interchange of the subunits between enzymes, cross-hybridization between their genes, and cross-reactivity of antibodies raised against their subunits. Members of the EcoA and EcoE families are distinct from those of EcoK by the same criteria.

The subunits have clearly distinguishable functions: α (130 kDa; the product of *hsdR*) is the endonuclease, β (60 kDa; *hsdM*) is the methyl transferase, and γ (50 kDa; *hsdS*) is responsible for DNA site recognition. (The gene names stand for *h*ost *s*pecificity for *D*NA; *r*estriction, *m*odification, and *s*equence.)

The reactions in vitro can be divided into several stages.

Site Recognition and Complex Formation. The enzyme, activated by binding of AdoMet, recognizes a specific site in the DNA, where it forms a stable complex. The recognition site on several DNAs has the 13- or 15-nucleotide sequence shown in Table 13-9.[26] The recognition sequence consists of three domains: (1) a trimer (AAC for EcoK, TGA for EcoB, GAG for EcoE) and (2) a tetramer, separated by (3) a 6- or 8-ntd region of variable sequence. The recognition sequence

Table 13-9
Type I restriction endonuclease recognition site[a]

Enzyme	Recognition site
EcoK	5′-A A* C N N N N N N G T G C T T G N N N N N N C A* C G-5′
EcoB	5′-T G A* N N N N N N N N T G C T A C T N N N N N N N N A C G A-5′

[a] N = nonspecific nucleotides; * = methylation site.

25. Modrich P, Roberts RJ (1982) in *Nucleases* (Linn SM, Roberts RJ, eds.). CSHL, p. 109; Roberts RJ (1982) in *Nucleases* (Linn SM, Roberts RJ, eds.). CSHL, p. 311.
26. Kan, NC, Lautenberger JA, Edgell MH, Hutchinson CA III (1979) *JMB 130*:191; Brooks JE, Roberts RJ (1982) *NAR 10*:913.

lacks the twofold rotational symmetry that characterizes the recognition sites of most type II restriction endonucleases. A single base change by mutation in the trimer or tetramer or a change in size of the intervening sequence is sufficient to abolish sensitivity to both restriction and modification. The domain in the sequence-specificity S (γ) subunit that recognizes the trimer component of the DNA target is located in the N-terminal region of the polypeptide and shows a 50% amino acid identity among related lines.[27]

Interference with restriction is observed when T7 phage prepared on an *E. coli* strain without a modification system grows efficiently on *E. coli* strains B and K despite having five sites susceptible in vitro both to cleavage by EcoB endonuclease and to modification by EcoB methylase. This seeming paradox is explained by the action of a protein encoded by gene *0.3*, the first gene expressed after infection, which overcomes both the modification and the restriction systems of the host cell.[28] The protein presumably inhibits by interacting with the site-recognition subunit of the endonucleases (Section 17-5).

A Three-Way Choice. After recognition of the DNA sequence by the enzyme and formation of a complex, a choice is made among three potential reactions, the effect of which is to degrade unmethylated foreign DNA and keep the resident DNA intact. Unmethylated DNA, invariably cleaved in vitro, is almost always restricted in vivo as well.

(1) Fully methylated DNA is neither cleaved nor further modified, and the enzyme is released.

(2) The complex with hemimethylated DNA (one strand methylated and one unmethylated) is not a substrate for restriction but is the best substrate for methylation. ATP stimulates this methylation; whether the bound AdoMet that activates the enzyme for DNA binding serves as a methyl donor or whether additional AdoMet is needed is uncertain.[29]

(3) Within the complex on unmethylated DNA, ATP or a nonhydrolyzable analog induces conformational changes in the enzyme.

In the third reaction, translocation of DNA relative to the complex[30] is 200 bp per second. DNA cleavage takes place at a random site 1000 bp or more away from the recognition site and requires ATP hydrolysis. Electron micrographs[31] show the enzyme associated with a loop of duplex DNA, as would result from the simultaneous binding of both the recognition and cleavage sites. The enzyme is consumed in the reaction.

ATP Hydrolysis. After sequential cleavage of the two strands, extensive ATP hydrolysis ensues, but there is no evidence for further endonucleolytic activity. Over 10,000 molecules of ATP are hydrolyzed per event, with an ATPase turnover number of over 1000 per minute, even though the enzyme as a nuclease may not turn over at all.

27. Tomkinson AE, Bonk RT, Linn S (1988) *JBC 263*:8066.
28. Studier FW (1975) *JMB 94*:283; Spoerel N, Herrlich P, Bickle TA (1979) *Nature 278*:30.
29. Studier FW (1975) *JMB 94*:283; Spoerel N, Herrlich P, Bickle TA (1979) *Nature 278*:30.
30. Studier FW, Bandyopadhyay PK (1988) *PNAS 86*:4677.
31. Endlich B, Linn S (1985) *JBC 260*:5720; Endlich B, Linn S (1985) *JBC 260*:5729.

The ATPase behavior is intriguing.[32] It depends on Mg^{2+} and AdoMet and also requires DNA with the correct and unmodified sequence. It continues for hours, long after the restrictive cleavages have been made in the DNA. Pancreatic DNase degradation of the DNA aborts the ATPase action, presumably by dissociating the enzyme from the DNA. ATP in the cell may perhaps serve to maintain the oligomeric endonuclease in an active form; in vitro, in its isolated and damaged state, the protein may dissipate the ATP needlessly.

EcoP1, EcoP15, HinfIII: Type III Endonucleases[33]

Resembling the EcoK and EcoA families in complexity, these enzymes are characterized by genes encoding two subunits (approximately 100 and 75 kDa) that restrict and modify a specific sequence. The smaller subunit can carry out the modification reaction without the larger subunit.

The recognition sequences of these enzymes[34] are as follows:

EcoP1	EcoP15	HinfIII
5′-AGACC TCTGG	5′-CAGCAG GTCGTC	5′-CGAAT GCTTA

Because methylation is confined to adenine, the EcoP1 and EcoP15 sequences can be modified on only one strand. The HinfIII recognition sequence is also methylated on only one of its strands. It is still unclear how the daughter duplex resulting from replication of the unmethylated parental strand can be protected from restriction directly following replication.

Each of these endonucleases cleaves DNA at a point 25 to 27 bp to the right of the host-specificity sequence shown above, and each leaves a 5′ single-stranded tail 2 or 3 ntd long. Type III restriction enzymes require ATP but, unlike the type I enzymes, do not cleave it; AdoMet stimulates nuclease activity but is not essential. In the absence of ATP, each endonuclease can act specifically as the appropriate modification methylase. When ATP and AdoMet are both present, methylation and cleavage are competing reactions.

13-8 Restriction Endonucleases: Specific Cleavage (Type II)[35]

The first characterization of a type II restriction enzyme and of a specific, symmetrical cleavage sequence came from studies of HindII.[36] Two other nucleases that occur in this *Hemophilus* strain were at first overlooked because they do not restrict phage T7 DNA, the substrate used for HindII.

Type II nucleases usually recognize specific, symmetrical sequences of 4, 5, 6, or 8 nucleotides, and generally cleave within the recognition sequence (see below). More than 100 distinct recognition sequences are known, a few exam-

32. Eskin B, Linn S (1972) *JBC 247*:6183; Eskin B, Linn S (1972) *JBC 247*:6192; Linn S, Lautenberger JA, Eskin B, Lackey D (1974) *Fed Proc 33*:1128.
33. Lovett ST, Kolodner RD (1989) *PNAS 86*:2627.
34. Bickle TA (1982) in *Nucleases* (Linn SM, Roberts RJ, eds.). CSHL, p. 85; Bickle TA (1987) in *Escherichia coli and Salmonella typhimurium* (Neidhardt FC, ed.). Am Soc Microbiol, Washington DC, Vol. 1 p. 692.
35. Thomas KR, Olivera BM (1978) *JBC 253*:424; Roberts RJ (1988) *NAR 16* (*Suppl*):r271.
36. Kelly TJ, Smith HO (1970) *JMB 51*:393; Smith HO, Wilcox KW (1970) *JMB 51*:379.

ples of which are shown in Table 13-10. Virtually every group of prokaryotes—Gram-positive and Gram-negative, aerobes and anaerobes, rods and cocci, *Myxobacteria* and *Actinomycetes*—possess such enzymes. The *Hemophilus* genus leads the list with 22 examples isolated from 29 strains examined.

Agarose gel electrophoresis in the presence of ethidium permits the discrete fragments obtained from digestion to be resolved and visualized by fluorescence. Unique banding patterns, determined by the frequency and location of particular recognition sites in the DNA, are obtained with each enzyme. In a list of 146 nucleases tested on phage γ DNA, some produce five or fewer cleavages, and others produce 20 or more.

The type II restriction nucleases are relatively stable enzymes and require only Mg^{2+} for activity. The double-stranded breaks introduced are either flush-ended or staggered by a few residues. In the latter case, the protruding cohesive ("sticky") ends are readily joined by ligase action (recombined) in vitro to those of other DNA molecules generated by cleavage by the same enzyme. Some

Table 13-10
Recognition and cleavage sites for several of the restriction nucleases

Restriction nuclease	Recognition sequence[a] (Axis of Symmetry)	Number of cleavage sites in DNA: ΦX174	λ	Ad2[b]	SV40
EcoRI (*E. coli RTFI*)	5′pG↓pApA*|pTpTpC / CpTpT|pA*pApG↑p	0	5	5	1
EcoRII (*E. coli RTFII*)	↓CC*|TGG	2	>35	>35	16
HindII (*Hemophilus influenzae Rd*)	GTPy↓|PuA*C	13	34	>20	7
HindIII (*Hemophilus influenzae Rd*)	*↓AAG|CTT	0	6	11	6
HaeIII (*Hemophilus aegyptius*)	GG↓|CC	11	50	>50	19
HpaII (*Hemophilus parainfluenzae*)	C↓C|GG	5	50	>50	1
PstI (*Providencia stuartii 164*)	CTG|CA↓G	1	18	25	2
SmaI (*Serratia marcescens Sb_b*)	CCC↓|GGG	0	3	12	0
Bam I (*Bacillus amyloliquefaciens H*)	G↓GA|TCC	0	5	3	1
Bgl II (*Bacillus globiggi*)	A↓GA|TCT	0	5	12	0

[a] Asterisks = methylation; arrows = cleavage sites.
[b] Ad2 = adenovirus 2.

nucleases produce 3′ tails that cannot be ligated or repaired by polymerase action but are suited for RecA-mediated recombination.

Modification by methylation of a residue within the recognition sequence (at the amino group of adenine or at the C5 of cytosine) blocks nuclease action (Section 21-9). Modification methylases, in the few instances examined, are entirely distinct structurally and functionally from the corresponding type II restriction endonucleases.

Some enzymes derived from different sources recognize the same nucleotide sequence; they are called "*isoschizomers*."[37] Among them, one of a pair may be inhibited by methylation and the other not; such isoschizomers can serve as reagents to probe a genome for the presence or absence of a methylated residue. For example, Sau3A, MboI, and DpnI, all of which cleave GATC, can be used to distinguish the state of methylation of adenine (N6, principally in prokaryotic DNA) and cytosine (C5, principally in eukaryotic DNA) (Table 13-11).

Whether the restriction endonucleases are designed to be an immunity barrier to foreign DNA, to promote recombination, or to serve some other physiologic role (Section 21-8) is uncertain. In any case, the type II nucleases are clearly one of nature's greatest gifts to science. By their use, a large eukaryotic chromosome, a chain 10^8 bp long, can be reduced to specific fragments of manipulable size for mapping, sequence analysis, and amplification by cloning.

Of special interest in the mapping and sequencing of large genomes are two nucleases with an 8-ntd specificity. Sf1 from *Streptomyces fimbriatus*[38] cleaves the sequence

```
             ↓
5'GGCCNNNNNGGCC
  CCGGNNNNNCCGG
       ↑
```

and Not1 from *Nocardia otitidis-caviarum*[39] cleaves the sequence

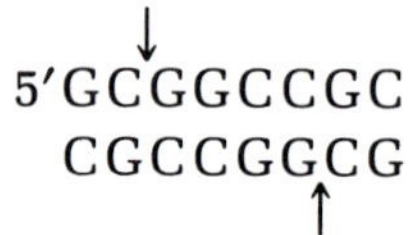

Table 13-11
Isoschizomeric restriction (type II) nucleases that distinguish the methylation states of GATC in DNA[a]

	Sequence		
Enzyme	GATC	G$\overset{*}{\text{A}}$TC	GAT$\overset{*}{\text{C}}$
MboI	+	−	+
Sau3A	+	+	−
DpnI	−	+	−

[a] * = N6-methyladenine or C5-methylcytosine; + = cleavage; − = no cleavage.

37. Roberts RJ (1976) *Crit Rev B* *4*:123; Smith HO (1979) *Science* *205*:455.
38. Qiang B-Q, Schildkraut I (1984) *NAR* *12*:4507; 1988–89 Catalog, New England Biolabs Inc., Beverly, MA.
39. Roberts RJ (1988) *NAR* *16* (*Suppl*):r271.

Digests of the *E. coli* chromosome appear as a discrete number of bands in a pulsed-field gradient gel; the DNA fragments can be ligated and recut.

Despite their abundance, variety, and the attention shown them, only a few of the nucleases have been purified to homogeneity and studied seriously as enzymes (see below). Because of a remarkably effective method for detecting restriction endonuclease activity and the low degree of contamination by nonspecific nucleases, preparations of even limited or undefined purity are generally reliable reagents.

EcoRI Endonuclease[40]

The best understood of the type II restriction systems is the restriction of DNA by EcoRI endonuclease and its prevention by methylation (at the adenine residue adjacent to the axis of symmetry in the recognition sequence; Table 13-10). The methylase functions independently of the nuclease and recognizes and acts on the sequence in a different way (see below). Cocrystallization of EcoRI and its recognition sequence reveals the features of the protein in its binding and action on a DNA substrate.[41]

The restriction nucleases offer a most attractive opportunity for studying the structure–function relationships of proteins that interact with DNA, a central question in biology. EcoRI, for example, modest in size, crystallized and analyzed at high resolution, can be altered by mutagenesis in order to examine every detail of its structure, while its hexanucleotide substrate can likewise be altered at every residue. The interactions can be examined by a variety of techniques at the levels of both binding and cleavage of the scissile bond, and can be compared with those of the methylation enzyme and isoschizomers that act on the same substrate.

EcoRI endonuclease is a dimer of 31-kDa subunits. The minimal recognition sequence in DNA or in a synthetic octanucleotide duplex is a hexanucleotide with twofold rotational symmetry perpendicular to the helix axis of the DNA (Table 13-10). The sites of cleavage are symmetrically located about the same twofold axis. Such cleavage may depend on a similar symmetrical orientation of the two subunits of the enzyme. The products of the staggered cleavage have 3′-OH and 5′-P ends.

At 37° with ColE1 DNA as substrate, the turnover number is 4 double-strand scissions per minute; the K_m for DNA is 10^{-8} M. One strand is cleaved to produce a transient nicked intermediate that is enzyme-bound, after which the opposite strand is cleaved. The K_A values[42] for pBR322 are: 1.9×10^{11} M for one site, and 1×10^{5} M for nonspecific sites. The rate-limiting step is the very slow dissociation of the enzyme from its product.

The endonuclease can act processively over distances of several hundred base pairs, as judged by the manner in which it cleaves circular and linear molecules

40. Shishido K, Ando T (1982) in *Nucleases* (Linn SM, Roberts RJ, eds.). CSHL, p. 155; Newman AK, Rubin RA, Kim SH, Modrich P (1981) *JBC 256*:2131; Greene PJ, Gupta M, Boyer HW, Brown WE, Rosenberg JM (1981) *JBC 256*:2143; McClarin JA, Frederick CA, Wang B-C, Greene P, Boyer HW, Grable J, Rosenberg JM (1986) *Science 234*:1526.
41. Newman AK, Rubin RA, Kim SH, Modrich P (1981) *JBC 256*:2131; McClarin JA, Frederick CA, Wang B-C, Greene P, Boyer HW, Grable J, Rosenberg JM (1986) *Science 234*:1526; Kim Y, Grable JC, Love R, Greene PJ, Rosenberg JM (1990) *Science 249*:1307.
42. Terry BJ, Jack WE, Rubin RA, Modrich P (1983) *JBC 258*:9820.

containing two EcoRI sites.[43] A circular DNA substrate is better than a linear one because the enzyme can move around the circle and is not sequestered at an end. Similarly, the enzyme has more leeway to find the cleavage sites if the sites are centrally located on a linear DNA rather than near an end. Diffusion along the helix contour enables the enzyme to search efficiently for recognition and cleavage sites. Although type II restriction nucleases generally recognize and cleave only duplex DNA efficiently, EcoRI also cleaves single strands.

Analysis to 2.8 Å resolution of crystals of the endonuclease bound (in the absence of Mg^{2+}) to the duplex 13-mer substrate (5′-TCGCGAATTCGCG) and additional studies provide insights into the enzyme structure.[44] Each subunit of the symmetrical dimer is organized into domains composed of a β sheet sandwiched by α helices. The enzyme binds symmetrically to the DNA, filling the major groove over the length of the recognition site, with extended arms of the β sheet holding the DNA in a scissors grip. A striking feature of the complex is a kink in the DNA at the center of the palindromic twofold axis, unwinding the DNA by 25°, widening the major groove by about 3.5 Å, and everting the two central AT base pairs into the major groove.

Two α helices from each subunit that project into the major groove bear residues that form 12 hydrogen bonds with the purines (GAA), two between arg200 and the guanine and four between glu144–arg145 and the two adenines. Although originally proposed to mediate the specificity of the enzyme, mutation of these residues results in reduction of catalytic activity and binding affinity, but does not alter the basic specificity of the enzyme. For DNA scission,[45] binding of the catalytic cleft in the globular domain of one subunit to the backbone at the GpA target bond is facilitated by the enfolding arm of the other subunit. Catalysis proceeds with inversion of configuration of the phosphate, and may require an essential glutamate residue at position 111 which is positioned near the cleaved bond in the crystal structure.

Certain changes in buffer conditions (low salt, basic pH, glycerol, Mn^{2+}) stimulate cleavage of sites that differ from the canonical site by one base pair, a phenomenon known as EcoRI* activity.[46] Large conformational changes, perhaps in both the protein and the DNA, that are induced only by the correct target site and not by nonspecific sequences may be essential both for specific binding and for catalysis. Evidence for such conformational changes includes: (1) the distorted conformation of the DNA substrate in the crystal structure, (2) a larger than expected change in the water-exposed nonpolar surface area on formation of the complex, (3) a large change in the free energy of interaction of the protein and DNA in formation of the transition state, and (4) the isolation of mutations that result in relaxed, EcoRI*-like specificity.

The EcoRI restriction and modification enzymes are rather dissimilar in their contacts with the DNA substrate, even though they recognize the same se-

---

43. Terry BJ, Jack WE, Modrich P, (1985) *JBC 260*:13130.
44. Alves J, Rüter T, Geiger R, Fliess A, Maass G, Pingoud A (1989) *Biochemistry 28*:2678; Needels MC, Fried SR, Love R, Rosenberg JM, Boyer HW, Greene PJ (1989) *PNAS 86*:3579; Heitman J, Model P (1990) *Proteins: Struc Funct Genet 7*:185.
45. Connolly BA, Eckstein F, Pingoud A (1984) *JBC 259*:10760; King K, Benkovic SJ, Modrich P (1989) *JBC 264*:11807; Kim Y, Grable JC, Love R, Greene PJ, Rosenberg JM (1990) *Science 249*:1307.
46. Heitman J, Model P (1990) *EMBO J 9*:3369; Ha J-H, Spolar RS, Record MT Jr (1989) *J Mol Biol 209*:801; Lesser DR, Kurpiewski MR, Jen-Jacobson L (1990) *Science 250*:776; Thielking V, Alves J, Fliess A, Maass G, Pingoud A (1990) *Biochemistry 29*:4682.

quence. Whereas the endonuclease, a dimer, interacts with the 6-bp sequence to form a complex with twofold symmetry appropriate for cleavage of both strands, the methylase binds as a monomer, asymmetrically transferring methyl groups one at a time and dissociating from the DNA between transfers.

Overlapping (Cohesive) Ends

The staggered cleavages produced by some type II endonucleases provide overlapping ends that facilitate the joining of any two DNA duplexes generated by enzymes with the same specificity. This kind of endonucleolytic cleavage may be seen as an invitation to recombination of the cleaved DNA, as well as to degradation, and could conceivably serve both functions in vivo (Section 21-5).

In a few instances, a type II restriction endonuclease recognizes an asymmetrical pentanucleotide sequence. The enzyme from *Hemophilus gallinarum* (HgaI) produces staggered cuts 5 bp from the recognition sequence to yield overlapping pentanucleotide extensions (Fig. 13-5). Each of these extensions is different in sequence and unique for the DNA from which the fragment is derived. These fragments therefore cannot anneal in a random manner, as do EcoRI fragments, and may be used profitably in the exchange of regions within a gene or in the cloning of a specific fragment from very complex mixtures of HgaI fragments.

13-9 Nucleases in Repair and Recombination[47]

The same enzymes often participate in repair, recombination, and replication. As examples, DNA polymerase I and ligase are not limited to any one of these functions. The versatility of a nuclease to perform in diverse repair and recombinational systems in vitro is also apparent in the physiologic features of a mutant in vivo. Whereas a defect in any one of several nuclease genes (e.g., *polA*, *xse*, *recBCD*, *uvrABC*) may be tolerated, defects in two or more are more likely to be lethal. Although enzymologic studies have concentrated on DNA repair, it seems best to regard nucleases as cellular reagents with the potential for many uses.

Repair processes (Sections 21-2 to 21-4) include those involving photoreacti-

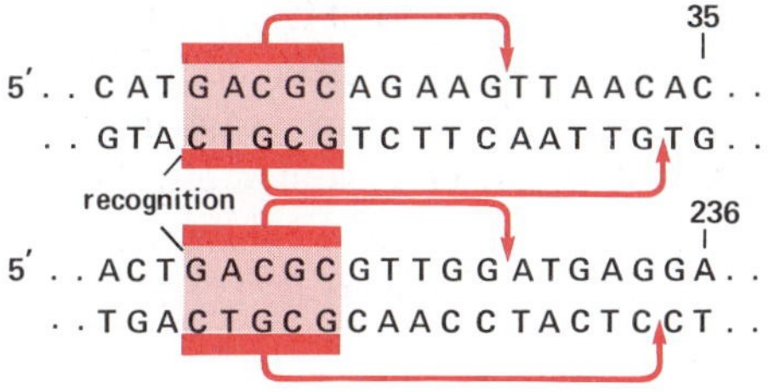

Figure 13-5
Sequences around two of the six HgaI cleavage sites in ϕX174 RF DNA. Recognition sites are boxed and the cleavage sites are shown by arrows. Numbers refer to the distance of the 3′ end of the viral strand sequence from the *Providentia stuartii* (PstI) restriction cleavage site.

47. Friedberg EC (1985) *DNA Repair*. WH Freeman, New York; Linn S (1982) in *Nucleases* (Linn SM, Roberts RJ, eds.). CSHL, p. 59; Sancar A, Sancar GB (1988) *ARB* 57:29; Sadowski PD (1982) in *Nucleases* (Linn SM, Roberts RJ, eds.). CSHL, p. 23.

vation (Section 21-2), postreplicational mismatch repair (Section 21-2), direct excision of modified, unnatural, or mismatched bases by N-glycosylases (see below), and two kinds of phosphodiesterase actions: *incision* of the backbone at or near the point of a lesion, and then *excision* of an oligonucleotide that includes the defective segment. Some stages in recombination may also employ these nucleases.

N-Glycosylases[48]

A large variety of enzymes recognize an altered or improper base and excise it by hydrolysis of the N-glycosyl bond that links the base to the deoxyribose of the DNA backbone. The N-glycosylases recognize specific lesions and fall into four groups, based on their removal of (1) deaminated bases (i.e., uracil, thymine,[49] hypoxanthine,[50] xanthine), (2) alkylated purines (e.g., N3- and N7-adducts and imidazole ring–opened derivatives[51]), (3) oxidized pyrimidines, or (4) oxidized purines. The AP (apurinic or apyrimidinic) sugar formed by the N-glycosylases of groups 1 and 2 is removed by a class II AP endonuclease (see below). Glycosylases of groups 3 and 4 possess both N-glycosylase and endonuclease activities (see Class I AP Endonucleases, below); the N-glycosylase of *E. coli* that removes the formamidopyrimidine produced by breakdown of N7-alkylguanine, the major product of DNA alkylation, is such a class I endonuclease.[52]

Uracil N-glycosylase of *E. coli* is a 35-kDa protein that acts on single-stranded as well as double-stranded DNA containing uracil either incorporated in place of thymine (Sections 2-9 and 15-10) or derived from cytosine by spontaneous deamination. This glycosylase activity has been identified in other bacteria, calf thymus, and human cells; it is present in mitochondria as well. The *E. coli* and human enzymes are highly homologous.[53]

The ubiquity of the enzyme[54] may be a response to the mutagenic consequences of the significant amount of cytosine deamination in DNA. The UG base pair is relatively undistorted in conformation compared to the CG it replaces and would not arouse most repair mechanisms; mutations would then be induced in the next round of replication of the uracil-containing chain.[55] It seems unlikely that the enzyme evolved to fill the need to remove the not uncommon substitutions of uracil for thymine, which are well tolerated phenotypically and do not induce mutations. Indeed, the highly mutable sites in *E. coli*, recognized as "hot spots" for spontaneous base substitutions,[56] appear to result from the failure to remove the thymine produced by the spontaneous deamination of 5-methylcytosine. A mismatch-specific thymine glycosylase cooperates with DNA polymerase β in nuclear extracts of human cells to correct G-T mispairs.[57]

In explorations for eukaryotic repair systems, genes have been sought that complement defects in these systems. One such gene cloned from yeast encodes an N-glycosylase for 3-methyladenine and protects *E. coli* from methylating

48. Friedberg EC (1985) *DNA Repair*. WH Freeman, New York; Lindahl T (1982) *ARB 51*:61.
49. Wiebauer K, Jiricny J (1990) *PNAS 87*:5842.
50. Harosh I, Sperling J (1988) *JBC 263*:3328.
51. O'Connor TR, Laval J (1989) *PNAS 86*:5222.
52. O'Connor TR, Laval J (1989) *PNAS 86*:5222.
53. Olsen LC, Aasland R, Wittwer CU, Krokan HE, Helland DE (1989) *EMBO J 8*:3121.
54. Sancar A, Sancar GB (1988) *ARB 57*:29; Duncan BK (1981) *Enzymes 14*:565.
55. Hayakawa H, Sekiguchi M (1978) *BBRC 83*:1312.
56. Coulondre C, Miller JH, Farabaugh PJ, Gilbert W (1978) *Nature 274*:775.
57. Wiebauer K, Jiricny J (1990) *PNAS 87*:5842.

agents.[58] Yeast mutants with defects in this gene are lacking in this N-glycosylase activity in cell extracts.

Repair Endonucleases[59]

Cells have endonucleases (Table 13-12) that patrol the chromosome for uracil, missing bases, bulky adducts to bases, UV damage, and other lesions in the DNA. The endonucleases respond to a defect by making phosphodiester bond cleavages (nicks) that generally leave a 5′-P group next to the defect, to be removed subsequently by an excision exonuclease. UvrABC nuclease, by making cuts on both sides of a lesion, is an excision as well as an incision nuclease. What seems a redundancy of five or more endonucleases in *E. coli* and other cells is probably the arsenal needed to cope with the great variety of lesions that need to be excised (Section 21-1).

Assuming that N-glycosylases remove the majority of lesions (in numbers rather than variety), then it falls to the endonucleases to repair most of the bulky lesions or the helix distortions they cause. These lesions include (1) a base modified by alkylation or by another addition or transformation, such as the damage caused by ionizing radiation ("monoadduct damage"); (2) a base improperly paired (mismatch); (3) more than one base modified, as in pyrimidine

Table 13-12
Repair endonucleases

Enzyme	Source	DNA damage[a]	Comments
AP endonucleases (Class I), N-glycosylases			concerted N-glycosylase and AP endonuclease; produce 5′-P termini
Endo III	*E. coli*	UV, x-rays	
Formamidopyrimidine N-glycosylase	*E. coli*	alkylation	may complement endo III
T4 UV endo (endo V)	T4 phage (*denV*)	UV	complements UvrABC and human mutants for UV sensitivity
M. luteus γ-endo (UV endo)	*M. luteus*	redox damage	resembles endo III
Eukaryotic endonucleases	yeast, mammals		
AP endonucleases (Class II)			AP endonuclease only; produce 5′-P termini
Endo II (endo VI)	E. coli (*xthA*)	AP site	same as exo III (see also Tables 13-1 and 13-6); major activity for AP sites
Endo IV	*E. coli, Bacillus*	AP site	no metal requirement; like endo II
Endo V	*E. coli*	AP site, uracil, UV, x-rays	
Eukaryotic endo II	human cells		
AP endonucleases (unclassified)	*Drosophila*, human	UV	produce 3′-P termini
UvrABC endo (excinuclease)	*E. coli (uvrA, uvrB, uvrC)*	bulky adducts	ATP-dependent; helicase
Mammalian endonucleases	various tissues	UV	no action on AP site

[a]AP = apyrimidinic or apurinic; UV = ultraviolet radiation.

58. O'Connor TR, Laval J (1989) *PNAS 86*:5222.
59. Jen-Jacobsen L, Lessor D, Kurpiewski M (1986) *Cell 45*:619; Lu A-L, Jack WE, Modrich P (1981) *JBC 256*:13200.

dimers produced by UV irradiation ("diadduct damage"); (4) absence of a base due to the action of an N-glycosylase, heat, or acid; and (5) cross-linkage of strands.

In the evolution of T4 phage, even the imposing array of repair systems of *E. coli* must have proved inadequate. The phage carries information for a highly active endonuclease of its own (endo V; see below) to ensure that certain lesions, specifically pyrimidine dimers, in its DNA will be efficiently excised and repaired.

Class I AP Endonucleases[60]

AP (apurinic or apyrimidinic) sites, generated in DNA by spontaneous acid-catalyzed hydrolysis, by radiation, or by the action of an N-glycosylase, are cleaved by specific endonucleases. The phosphodiester bond is broken on either side of the AP site to initiate the removal of the abasic deoxyribose for its eventual replacement by the proper nucleotide. On the basis of the initial incision, most of the AP endonucleases can be placed in one of the two classes (Fig. 13-6).

***E. coli* Endo III.** All class I AP enzymes possess both N-glycosylase and AP endonuclease activities. As an example, *E. coli* endo III[61] concertedly hydrolyzes the N-glycosyl bond of a thymine glycol or one of its breakdown products to deoxyribose and then cleaves the phosphodiester bond on the 3′ side of the AP site. The nuclease cleavage is remarkable in that the mechanism does not involve hydrolysis but rather a lyase reaction in the course of which the 2-deoxyribose is converted to a 2,3-dideoxy-2,3-didehydro sugar. Subsequent release of this novel abasic sugar phosphate can be achieved by a class II nuclease hydrolysis of the phosphodiester bond on the 5′ side (see below).

Thus, class I AP endonucleases act by a β-elimination mechanism to generate a 3′-unsaturated sugar residue which is then removed by a class II AP endonuclease. The latter endonucleases act by hydrolysis and depend on separate diesterases to remove the products of their action.[62]

Although this endonuclease activity appears to be ubiquitous in pro- and eukaryotes, mutants in the endo III gene (*nth*, for *endo three*) do not exhibit enhanced sensitivity to DNA-damaging agents unless the genes for other DNA

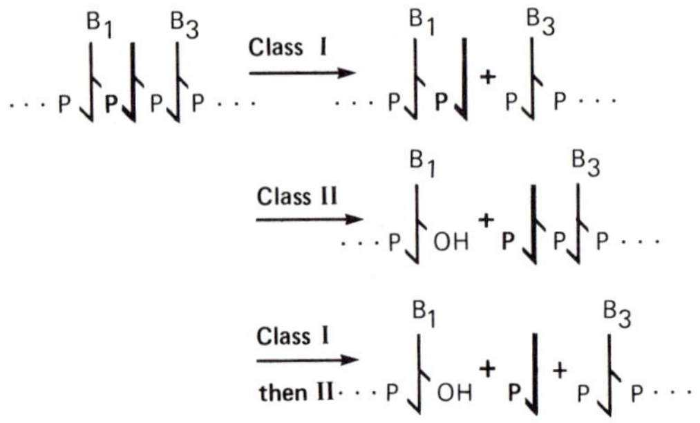

Figure 13-6
Cleavage mechanisms of class I and class II AP endonucleases. (Courtesy of Professor S Linn)

60. Kim J, Linn S (1988) *NAR 16*:1135.
61. Kow YW, Wallace SS (1987) *Biochemistry 26*:8200; Kim J, Linn S (1988) *NAR 16*:1135; Weiss RB, Gallagher PE, Brent TP, Duker NJ (1989) *Biochemistry 28*:1488.
62. Franklin WA, Lindahl T (1988) *EMBO J 7*:3617; Bernelot-Moens C, Demple B (1989) *NAR 17*:587.

repair enzymes are also affected by mutations. This observation is indicative of the extensive back-up potential for the enzymatic repair of DNA lesions. Possibly, the activity in *E. coli*[63] that removes the oxyradical-damaged imidazole ring derived from N7-alkylguanine, and also acts as an AP endonuclease, may complement endo III in its disposal of ring-damaged pyrimidines. These two related activities could recognize a broad spectrum of base alterations.

Phage T4 UV Endo (Endo V).[64] This 18-kDa protein, encoded by the *denV* gene, possesses both DNA glycosylase and AP endonuclease activities. Separate domains of the protein are responsible for each of these activities. T4 endo V is highly specific for incising the chain next to pyrimidine dimers, preferably in duplex DNA,[65] and also acts at AP sites (Fig. 13-7). It locates a dimer by processive scanning of DNA in vitro and in vivo.[66] The enzyme is indifferent to whether the DNA is glycosylated, linear, or supercoiled. The purified enzyme complements an incision-enzyme deficiency for DNA repair of UV damage when administered to mutant cell extracts as diverse as *E. coli uvrA*$^-$ and the fibroblasts from a patient with xeroderma pigmentosum (Section 21-2).[67] The *denV* gene, cloned into *E. coli recA uvrA* or *recA uvrB* mutants, complements the deficiency in incision at pyrimidine dimers in vivo, as well.[68]

***M. luteus* γ-Endo (UV Endonuclease).**[69] This enzyme is strikingly similar in size and actions to the T4 endo V enzyme (Fig. 13-7), with a preference for thymine over cytosine dimers. Mutants deficient in the enzyme display no lack of UV resistance, presumably because the nucleotide excision repair system is adequate for removing pyrimidine dimers.[70]

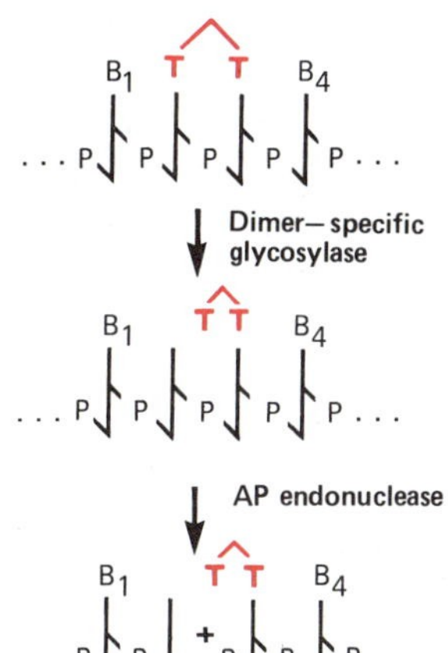

Figure 13-7
Cleavage mechanism of phage T4 (T4 endo V) and *Micrococcus luteus* (γ-endo) UV endonucleases. Following the cyclobutane-dimer–specific N-glycosylase cleavage, incision by an AP (apurinic or apyrimidinic) endonuclease takes place.

63. Chen J, Derfler B, Maskati A, Samson L (1989) *PNAS 86*:7961.
64. Nakabeppu Y, Yamashita K, Sekiguchi M (1982) *JBC 257*:2556.
65. Inaoka T, Ishida M, Ohtsuka E (1989) *JBC 264*:2609; Ishida M, Kanamori Y, Hori N, Inaoka T, Ohtsuka E (1990) *Biochemistry 29*:3817.
66. Gruskin EA, Lloyd RS (1988) *JBC 263*:12728.
67. Ciarrocchi G, Linn S (1978) *PNAS 75*:1887; Smith CA, Hanawalt PC (1978) *PNAS 75*:2598.
68. Lloyd RS, Hanawalt PC (1981) *PNAS 78*:2796.
69. Grafstrom RH, Park L, Grossman L (1982) *JBC 257*:13465; Jorgensen TJ, Kow YW, Wallace SS, Henner WD (1987) *Biochemistry 26*:6436.
70. Tao K, Noda A, Yonei S (1987) *Mutat Res 183*:231.

Eukaryotic AP Endonucleases. Endonucleases from yeast[71] and at least three from mammalian cells[72] act on DNA subjected to reduction or oxidation reactions (initiated by ionizing radiation or oxidizing agents); they have been called "redoxyendonucleases." (This is a confusing term inasmuch as no oxidoreductions are involved in the enzyme actions.) The N-glycosylase–AP endonuclease mechanism of these enzymes resembles that of *E. coli* endo III.

Other AP eukaryotic endonucleases have also been identified. In general, all cells have at least one major class II AP endonuclease and perhaps a mitochondrial one (see below) and several class I AP endonucleases associated with functions related to repair of oxidative damage. Of the several AP endonucleases in yeast, one purified to near homogeneity[73] is distinct from the other yeast AP endonucleases described above and is unrelated to any of the products of the numerous *RAD* (*ra*diation-sensitive) loci. Several mammalian AP endonucleases, including those from mitochondria (see below), have yet to be sorted out for their unique mechanistic and physiologic features. The repair deficiency of xeroderma pigmentosum, which involves eight or more distinct complementation groups, is due at least in part to a lack of incision activity; group D lacks one of the class I AP endonucleases.[74] T4 endo V, when introduced into cultured cells from patients with this disease, stimulates the repair of UV lesions.

Class II AP Endonucleases

The major volume of endonuclease action at AP sites is attributable to the class II enzymes, which possess no N-glycosylase activity. (However, the largest number of distinctive endonucleases are found in class I due to the great variety of glycosylase specificities.) The class II enzymes cleave the DNA 5′ to the lesion, generating an intermediate from which an abasic residue can be released by an alkali-catalyzed β-elimination reaction as a 2,3-dideoxy-2,3-didehydro sugar phosphate (Fig. 13-8). The same result can be achieved by combining the activities of class I and class II AP endonucleases (Fig. 13-6). In this combination, the nonhydrolytic cleavage of the phosphodiester bond 3′ to the lesion via a β-elimination mechanism (see above) by a class I enzyme is followed by the class II enzymatic hydrolysis of the bond 5′ to the lesion. For the removal of abasic residues from 5′ termini of class II products, exonucleases or specific deoxyribophosphodiesterases appear to be required.[75]

***E. coli* Endonucleases.** *E. coli endonuclease II* (alias endo VI and exonuclease III) accounts for the bulk of the AP endonuclease activity in *E. coli*.[76] In addition to its functions as a 3′→5′ exonuclease (Section 2), a 3′ phosphatase, a 3′-sugar fragment hydrolase, and an AP endonuclease, endo II also incises the phosphodiester bond 5′ to a urea residue of a fragmented base in DNA.

E. coli endonuclease IV[77] acts chiefly at AP sites. However, endo IV has only a

71. Gossett J, Lee K, Cunningham RP, Doetsch PW (1988) *Biochemistry 27*:2629.
72. Doetsch PW, Henner WD, Cunningham RP, Toney JH, Helland DE (1987) *Mol Cell Biol 7*:26; Kim J, Linn S (1989) *JBC 264*:2739.
73. Grafstrom RH, Park L, Grossman L (1982) *JBC 257*:13465; Jorgensen TJ, Kow YW, Wallace SS, Henner WD (1987) *Biochemistry 26*:6436.
74. Tomkinson AE, Bonk RT, Kim J, Bartfeld N, Linn S (1990) *NAR 18*:929.
75. Franklin WA, Lindahl T (1988) *EMBO J 7*:3617.
76. Lindahl T (1982) *ARB 51*:61; Weiss B, Grossman L (1987) *Adv Enz 60*:1; Siwek B, Bricteux-Grégoire S, Bailly V, Verly WG (1988) *NAR 16*:5031.
77. Levin JD, Johnson AW, Demple B (1988) *JBC 263*:8066.

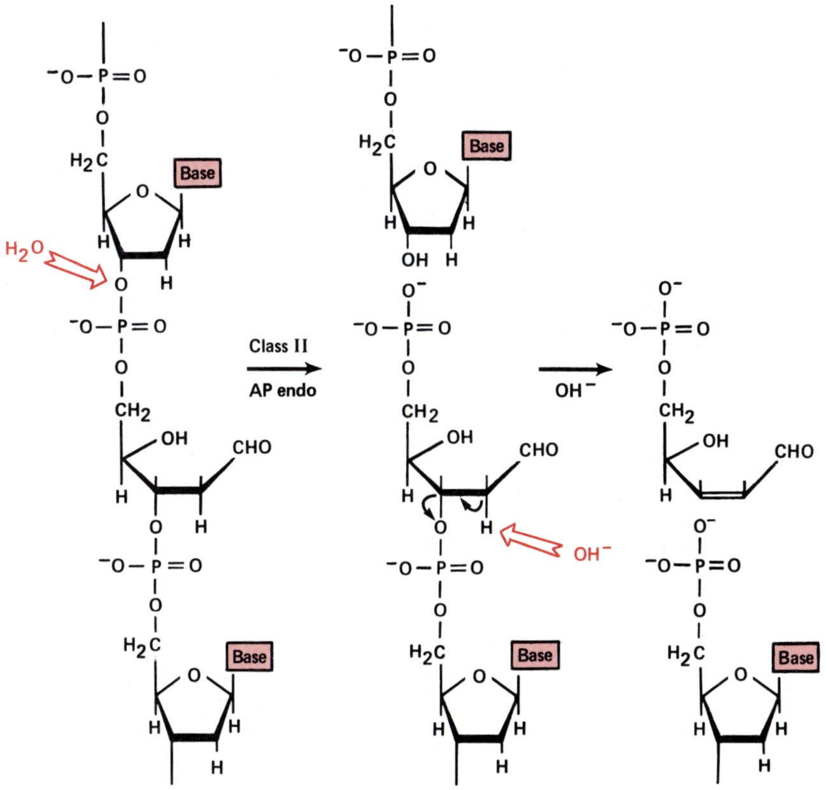

Figure 13-8
Mechanisms of cleavage by class II AP endonucleases and β-elimination by alkali.

minor role in repair compared to endo II, but resembles AP endo II, the major eukaryotic repair enzyme (Table 13-12). Like *E. coli* endo II, endo IV removes a 3′-terminal phosphoglycolaldehyde and similar fragments generated by oxidants and x-rays. The enzyme can release a variety of fragments, including deoxyribose 5-P, phosphoglycolaldehyde, and even phosphate, from the 3′ end of a chain. This unusual capacity to hydrolyze both phosphomonoesters and -diesters is shared with *E. coli* endo II (exo III) and a yeast endonuclease, as well as the secreted scavenging enzymes (e.g., P1 and mung bean nucleases) that act preferentially on ssDNA (Table 13-7). Mutants (*nfo*, for endo *four*) show sensitivities to several damaging agents, especially when combined with mutations in other repair endonuclease genes (e.g., *nth, xth*).

E. coli endonuclease V acts not only on a variety of DNA lesions, but also cleaves at uracil residues and attacks nondamaged DNA. Neither the enzyme nor the gene has yet been characterized.

Eukaryotic Endonucleases. All eukaryotic cells most probably possess an array of nuclear class II enzymes comparable to that of *E. coli*. An AP endonuclease identified in HeLa cells seems to be altered[78] in a cell line from a patient with ataxia telangiectasia, a repair-deficiency disease.

78. Kuhnlein U (1985) *JBC 260*:14918.

Unclassified AP Endonucleases

Instead of generating a 5′-P terminus, as do the class I and II enzymes, a *Drosophila* enzyme,[79] like a class I enzyme, incises on the 3′ side of the abasic lesion, but produces a 3′-P terminus (Fig. 13-9). Conceivably, other AP endonucleases will be found which also produce a 3′-P terminus but with an incision on the 5′ side of the lesion. An AP endonuclease from human placenta[80] incises on either the 3′ or 5′ side of the AP site and may be considered as a member of either class I or class II.

Mitochondrial Endonucleases.[81] The role of these enzymes is unclear. Despite uncertainty about whether mitochondria require or possess a nucleotide excision repair system for their multicopy genomes, AP endonuclease activities have been isolated from a mouse cell line with properties resembling those of extramitochondrial class I and class II enzymes, but clearly distinguishable from them. Inasmuch as mitochondrial nucleases can generate a primer terminus active for extension by a DNA polymerase in vitro, these enzymes may be designed for base-excision DNA repair. An alternative role for the nucleases might be to degrade and eliminate a defective genome from the multicopy pool.

UvrABC Endonuclease (Excinuclease)[82]

The major system in *E. coli* for the removal of pyrimidine dimers, bulky carcinogenic adducts,[83] and other lesions[84] depends on a complex of the products of the unlinked genes *uvrA*, *uvrB*, and *uvrC*. Whereas mutants deficient in this enzyme system are not unduly sensitive to oxidizing agents or ionizing radiation, the isolated excision repair enzyme can also remove the products of oxidative damage, such as thymine glycol and AP sites.[85]

The three proteins of the complex are UvrA (104 kDa), B (76 kDa), and C (66 kDa). UvrA, the damage-recognition subunit, has two identified zinc fingers;[86] its DNA-binding and ATPase activities are enhanced by UvrB, which also possesses a cryptic ATPase.[87] The UvrAB complex has helicase activity, melting supercoiled DNA and displacing the third strand in D-loops[88] with a 5′→3′ polarity. The translocation to a lesion and the unwinding prepares the

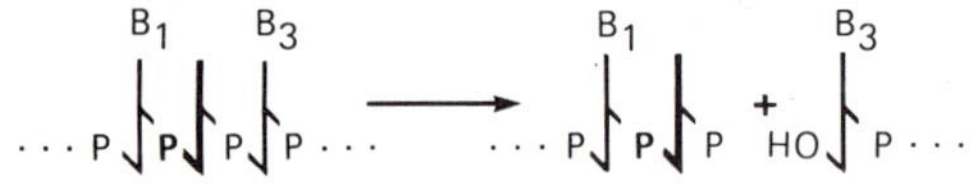

Figure 13-9
Cleavage mechanism of *Drosophila* AP endonuclease.

79. Spiering AL, Deutsch WA (1986) *JBC 261*:3222.
80. Grafstrom RH, Shaper NL, Grossman L (1982) *JBC 257*:13459.
81. Tomkinson AE, Bonk RT, Linn S (1988) *JBC 263*:12532; Tomkinson AE, Bonk RT, Kim J, Bartfeld N, Linn S (1990) *NAR 18*:929.
82. Seeberg E, Steinum AL (1982) *PNAS 79*:988; Caron PR, Kushner SR, Grossman L (1985) *PNAS 82*:4925; Weiss B, Grossman L (1987) *Adv Enz 60*:1; Sancar A, Sancar GB (1988) *ARB 57*:29; Van Houten B (1990) *Microbiol Rev 54*:18.
83. Selby CP, Sancar A (1988) *Biochemistry 27*:7184; Jones BK, Yeung AT (1988) *PNAS 85*:8410; Pierce JR, Case R, Tang M-S (1989) *Biochemistry 28*:5821.
84. Voigt JM, Van Houten B, Sancar A, Topal MD (1989) *JBC 264*:5172; Snowden A, Kow YW, Van Houten B (1990) *Biochemistry 29*:7251.
85. Lin J-J, Sancar A (1989) *Biochemistry 28*:7979.
86. Navaratnam S, Myles GM, Strange RW, Sancar A (1989) *JBC 264*:16067.
87. Caron PR, Grossman L (1988) *NAR 16*:9651.
88. Oh EY, Grossman L (1987) *PNAS 84*:3638; Oh EY, Grossman L (1989) *JBC 264*:1336.

DNA for incision by the excinuclease complex, which in vitro contains only the UvrB and UvrC subunits.[89] Mutations in an ATP-binding motif in UvrB protein impair the excision event, implying a second nucleotide-dependent step in incision following initial complex formation with UvrA.[90]

The incision entails two endonucleolytic breaks, one at the eighth phosphodiester bond 5′ to the lesion and then a second at the fourth or fifth bond 3′ to the same lesion (Fig. 13-10). However, for excision of the damaged fragment (12 to 13 ntd long), turnover of the UvrABC complex, and filling of the gap, helicase II (the product of *uvrD*) and the coordinated polymerase and 5′→3′ exonuclease actions of pol I are also required.[91] Expression of the *uvrA, B, C,* and *D* genes is enhanced by the *recA–lexA* regulatory circuit (SOS system; Section 21-3), but evidence is lacking that these higher induced levels of the Uvr proteins are needed for optimal rates of excision repair of damaged DNA.

Excision Exonucleases

For the removal of damaged DNA, three routes are prominent: (1) excision of the oligonucleotide containing the lesion (e.g., by UvrABC endonuclease), (2) successive incisions by AP endonucleases, and (3) excision by exonucleases at a break in the phosphodiester backbone made by an AP endonuclease. With the lesion marked by the break, in the third case, the aberrant residues at the 3′- or 5′-end of the nicked strand are removed by exonucleases, most commonly those acting in the 5′→3′ direction (Tables 13-4 and 13-5).

Pol I displays its 5′→3′ exonuclease activity by binding at a nick where nuclease activity coupled to polymerization advances the nick relative to the template strand (nick translation; Section 4-13). The importance of this activity for the removal of DNA lesions in vivo has been proved in a variety of ways (Section 4-18). Studies of mutants lacking pol I, II, or III show that pol I has the

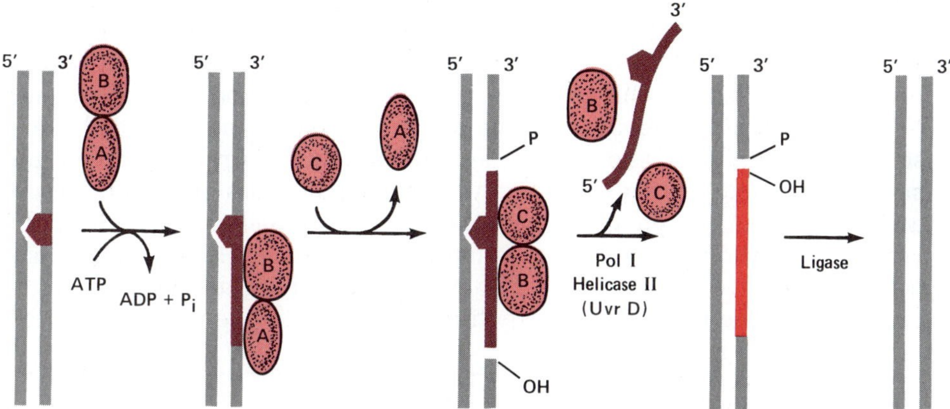

Figure 13-10
Nucleotide excision by the uvr system of *E. coli.* A, B, and C are the respective proteins of the UvrABC excision nuclease. Helicase II is the *uvrD* gene product.

89. Orren DK, Sancar A (1989) *PNAS 86*:5237.
90. Seeley TW, Grossman L (1989) *PNAS 86*:6577.
91. Caron PR, Kushner SR, Grossman L (1985) *PNAS 82*:4925; Weiss B, Grossman L (1987) *Adv Enz 60*:1.

predominant role in repair. As for AP sites at a 3′ chain end, these can be excised by the proofreading function of the 3′→5′ exonuclease of pol I[92] or by exo III (i.e., endo II).

Exonuclease VII differs from the pol I exonuclease activity in two ways: (1) it has high specificity for ssDNA and (2) it acts in both the 5′→3′ and 3′→5′ directions. Thus, exo VII and pol I can complement each other in responding to the conformation of a chain in need of excision repair.

T4 exonucleases B and C are responsible for excision following the incision by T4 endo V. The 35-kDa exo B resembles the 5′→3′ exonuclease of pol I, whereas the 20-kDa exo C appears to act in both directions.

Mammalian exonucleases in rodent and human cells are 5′→3′ exonucleases of both major classes, as found in *E. coli;* one, like pol I, has a preference for duplex DNA and the other, like exo VII, has a specificity for ssDNA. Evidence for additional types of mammalian excision exonucleases also exists.

Phage λ exonuclease, which exhibits its 5′→3′ activity chiefly on dsDNA, generates the single-stranded regions essential for recombination (Section 17-7).

RecBCD DNase[93]

Also designated exonuclease V and previously thought to contain only the subunits encoded by the *recB* and *recC* genes, this enzyme is now known to require the product of *recD* as well.[94] The 330-kDa RecBCD has five discrete activities: DNA-dependent ATPase, ssDNA exonuclease, dsDNA exonuclease, ssDNA endonuclease, and DNA helicase (Section 11-3). The RecBCD enzyme, together with the RecA protein, constitutes the major recombination pathway in *E. coli.* The enzyme nicks at *chi* sites (recombinational "hot spots")[95] and also contributes to DNA repair synthesis.

A plausible scheme for RecBCD action embodies many of the nuclease and ATPase functions and explains the generation of long single strands for participation in recombination. The RecBCD subunits bind both strands at the end of a duplex (Fig. 13-11). The enzyme unwinds the duplex, forming a loop on the strand with a 3′ end at the entry point.[96] One of the separated strands is split into fragments of 100 to 500 ntd, while the other strand remains intact, with a subunit bound to its end. This reaction can then proceed in both the 3′→5′ and 5′→3′ directions. Processive enzyme action generates a nondegraded single strand of several thousand nucleotides, which may in turn be degraded into smaller fragments, perhaps because the enzyme switches orientation. In its action on ssDNA, the enzyme is regulated by RecA protein and by SSB.[97] The regulation of the enzyme's degradation of linear duplex DNA is uncertain, especially in vivo where susceptible duplex termini are normally absent or inaccessible. The novel looped structures generated by the helicase activity,

---

92. Mosbaugh DW, Linn SJ (1982) *JBC 257*:575.
93. Telander-Muskavitch K, Linn S (1981) *Enzymes 14*:233; Telander-Muskavitch K, Linn S (1982) *JBC 257*:2641; Taylor AF (1988) in *Genetic Recombination* (Kucherlapati R, Smith GR, eds.). Am Soc Microbiol, Washington DC, p. 231.
94. Amundsen SK, Taylor AF, Chaudhury AM, Smith GR (1986) *PNAS 83*:5558.
95. Stahl FW (1986) *Prog N A Res 33*:169; Smith GR (1987) *ARG 21*:179.
96. Braedt G, Smith GR (1989) *PNAS 86*:871.
97. Williams JGK, Shibata T, Radding CM (1981) *JBC 256*:7573; Rosamond J, Telander KM, Linn SJ (1979) *JBC 254*:8646.

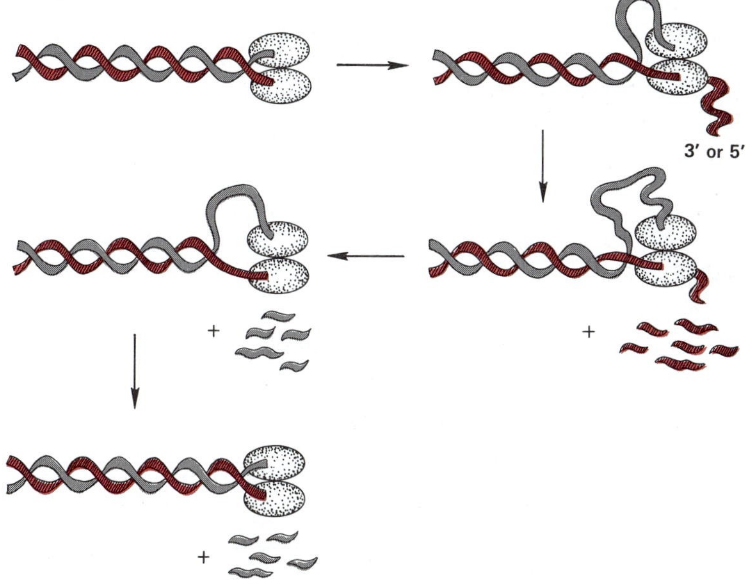

Figure 13-11
Scheme for digestion of duplex DNA by RecBCD DNase. The two subunits of the enzyme are arbitrarily placed at each chain end of the terminus of the duplex. The small fragments represent single strands several hundred nucleotides long. (Courtesy of Professor S Linn)

observed microscopically among RecBCD reaction intermediates,[98] may help in defining the enzyme's role in recombination (Section 11-3).[99]

Exonuclease action from either end further hydrolyzes the oligonucleotide fragments to a limit size of about 5 ntd. However, in the presence of enough SSB to cover the liberated single strands, no DNA hydrolysis occurs. Here the RecBCD DNase acts as a helicase,[100] resembling Rep protein in its displacement of long DNA strands (Section 11-3). In hydrolyzing ATP during unwinding, the enzyme consumes two to three ATP molecules per base pair of DNA unwound.[101]

ATPase and helicase activities of RecBCD persist even when nuclease action is eliminated because the DNA has been either coated with SSB, cross-linked with psoralen, or replaced with an RNA–DNA hybrid. Even in the absence of these measures, the extent of hydrolysis of duplex DNA becomes limited at a concentration of ATP near the physiologic level of 3 mM. This level of ATP thus favors the formation of long single-stranded regions within the duplex, a suitable intermediate for recombination.

The capacity of RecA protein to catalyze the formation of D-loops (Section 21-5) suggests that RecBCD protein may have a collaborative role in generating the single strand for the key heteroduplex intermediate in recombination. A contribution to replication by RecBCD protein is indicated in the failure of some *recBCD* mutants to synthesize DNA, presumably because of their inability to repair or replicate damaged DNA.[102]

---

98. Selby CP, Sancar A (1988) *Biochemistry 27*:7184; Jones BK, Yeung AT (1988) *PNAS 85*:8410; Pierce JR, Case R, Tang M-S (1989) *Biochemistry 28*:5821.
99. Kowalczykowski SC, Roman LJ (1990) in *Molecular Mechanisms in DNA Replication and Recombination* (Richardson CC, Lehman IR, eds.). UCLA, Vol. 127. AR Liss, New York, p. 357.
100. Roman LJ, Kowalczykowski SC (1989) *Biochemistry 28*:2873.
101. Roman LJ, Kowalczykowski SC (1989) *Biochemistry 28*:2863, 2873.
102. Capaldo FN, Barbour SD (1975) *JMB 91*:53; Miller JE, Barbour SD (1977) *J Bact 130*:160.

The importance of controlling RecBCD DNase activity is evident from the fact that a number of phages, including λ, T7, T4, and Mu, introduce inhibitors of the enzyme. Interaction with γ protein, encoded by phage λ (Section 17-7), is thought to prevent the RecBCD DNase from destroying a crucial intermediate in λ DNA replication at the stage of concatemer formation. Success in the transformation of linear DNA into *E. coli* also relies on the use of a *recB, C,* or *D* mutant.

13-10 Ribonuclease H (RNase H)[103]

An endonuclease specific for the RNA strand of an RNA–DNA duplex (i.e., a *h*ybridase acting only on the RNA, hence RNase H) is found in all cells. The activity is also a key element of the viral polypeptide responsible for reverse transcription (Section 6-9).

RNases H produce oligonucleotides with 5′-P and 3′-OH termini; the products are from 2 to 9 ntd in length except for yeast and ascites cell enzymes, which produce mono- and dinucleotides. Homopolymers and heteropolymers are accepted as substrates.

Remarkably, plasmids (e.g., ColE1) and mitochondrial DNAs with only a peppering of ribonucleotides are cleaved, as is an RNA at the region to which only a tetramer of deoxyribonucleotides is annealed.[104] A 14-mer RNA duplex containing only two deoxy residues on one of the strands directs cleavage to that site in the RNA by *E. coli* RNase H.[105] With an RNA strand covalently linked to DNA (as in the case of an RNA-primed nascent DNA), various RNases H differ in their capacity to cleave the RNA–DNA junction and thereby generate a 5′-P–terminated DNA strand. The enzymes of yeast[106] and of retroviruses[107] [e.g., avian myeloblastosis virus (AMV)] can do so, whereas the *E. coli* RNase H cannot remove the ribonucleotide esterified to DNA and so *E. coli* must rely on the 5′→3′ exonuclease activity of pol I for this function.

In calf thymus, where the activity was first discovered,[108] two distinctive size classes (30 and 70 kDa) are found. In yeast, also, two enzymes (55 and 70 kDa) can be identified, with distinguishable properties.[109] The 70-kDa polypeptide of yeast possesses a cryptic reverse transcriptase activity resembling the products of the retroviral *pol* genes.[110] Whereas in reverse transcription the RNase H functions in the removal of viral RNA from the RNA–DNA intermediate, a related role of a cellular RNase H for eukaryotic retrotransposons (Section 6-9) has not been established.

The replicative functions of RNase H (Section 19-7) have been most clearly demonstrated with the RNase HI of *E. coli*, the product of the *rnhA* gene. With plasmids bearing the unique chromosomal origin (*oriC*), RNase HI ensures specificity of initiation at *oriC*, presumably by removing RNA primers laid down

---

103. Crouch RJ, Dirksen M-L (1982) in *Nucleases* (Linn SM, Roberts RJ, eds.). CSHL, p. 211; Kogoma T (1986) *J Bact 166*:361; Wintersberger U (1990) *Pharmacol Therap 48*:259.
104. Donis-Keller H (1979) *NAR 7*:179.
105. Wyatt JR, Walker GT (1989) *NAR 17*:7833.
106. Karwan R, Wintersberger U (1986) *FEBS Lett 206*:189.
107. Champoux JJ, Gilboa E, Baltimore D (1984) *J Virol 49*:686.
108. Stein H, Hausen P (1969) *Science 166*:393.
109. Karwan R, Wintersberger U (1988) *JBC 263*:14970.
110. Karwan R, Kühne C, Wintersberger U (1986) *PNAS 83*:5919; Wintersberger U, Kühne C, Karwan R (1988) *BBA 951*:322.

elsewhere on the DNA[111] (Section 16-3). By allowing replication to start at other places on the chromosome ("stable DNA replication"; Section 20-3), mutants in *rnhA* can suppress the deletion of *oriC* or the mutations in *dnaA* responsible for the dnaA initiator protein acting at *oriC*. In replication of the ColE1 plasmid (Section 18-2), RNase HI is responsible for processing of the RNA primer as well as for its subsequent removal.[112] Although the 5′→3′ exonuclease of pol I is the principal activity in removing the RNA primers that initiate DNA strands, RNase HI makes a significant auxiliary contribution.[113]

As with eukaryotes, *E. coli* possesses a second, but very feeble RNase H activity[114] (RNase HII), the gene (*rnhB*) for which is next to *dnaE* that encodes the α subunit of DNA polymerase III holoenzyme. Curiously, *rnhA* is located next to *dnaQ*, the gene responsible for the ϵ subunit of the holoenzyme.

111. Ogawa T, Pickett GG, Kogoma T, Kornberg A (1984) *PNAS 81*:1040.
112. Itoh T, Tomizawa J (1980) *PNAS 77*:2450.
113. Funnell BE, Baker TA, Kornberg A (1986) *JBC 261*:5616.
114. Itaya M (1990) PNAS 87:8587.

CHAPTER 14

Inhibitors of Replication

14-1 Inhibitors as Drugs and Reagents[1]

Inhibitors of DNA replication continue to serve the clinician as prime drugs for suppressing proliferative, viral, bacterial, and autoimmune diseases, and to provide the laboratory investigator with the means to analyze biochemical pathways in vivo and in vitro. To the well-known classes of inhibitors that affect nucleotide biosynthesis, replication, or DNA itself are now added agents that act on regulatory processes, including postreplicational methylation.

With more refined biochemical and structural analysis of inhibitor–oligonucleotide and inhibitor–enzyme complexes, chemical rationale is supplementing the systematic screening of natural products for the discovery of superior inhibitors—for example, the bifunctional intercalators, which rival polymerases and regulatory proteins in their affinity for DNA.

While studies with mutants have been especially useful in elucidating complex biochemical events, they are applicable for analyzing DNA replication only when the mutant proteins are conditionally defective, such as being temperature-sensitive in production, stability, or function. Even so, the protein defect of these mutants may be only partly manifested under the stressful conditions or may be slow in developing. Thus, inhibitors, particularly antibiotics, offer promising alternatives to mutants.

Inhibitors are especially attractive for studies of eukaryotic systems in which mutant selection and genetic analysis are difficult. The large variety of antibiotics known to bind to and inhibit ribosomal proteins suggests that selective

1. Elion GB (1989) *Science* *244*:41.

pressures in nature should have led to the evolutionary development of antibiotics for virtually every essential bacterial protein. A concerted search for natural products and the chemical modification of promising ones might uncover antibiotic agents directed against most proteins and reactions essential to cells.[2] Antibiotic inhibitors, having already permitted dissection of the intricate mechanisms of protein biosynthesis, may advance nucleic acid research as well. Some examples are already at hand.

Chemical inhibitors of nucleic acid synthesis have various effects (Fig. 14-1). Some block the synthesis of nucleotide precursors or their polymerization; some are incorporated as nucleotide analogs into DNA; others interfere with template functions by binding, modifying, or degrading DNA; and still others bind and inactivate the replication proteins. In addition, physical agents such as UV light, x-rays, and gamma rays inhibit replication by damaging DNA. Inhibitors of RNA synthesis, as well as inhibitors of DNA replication, must be considered in this volume, not only because of the direct role of RNA transcription in initiating DNA chains, but also because of the analogies between DNA-dependent RNA synthesis and DNA synthesis. Some inhibitors such as colchicine and other microtubule disrupters (e.g., vincristine and vinblastine) act only indirectly on DNA synthesis. However, by inhibiting mitosis and prereplicative events, these agents profoundly affect DNA synthesis by preventing its initiation.

The rationale for employing the action of inhibitors in the treatment of bacterial diseases differs from that for viral, neoplastic, and autoimmune diseases.

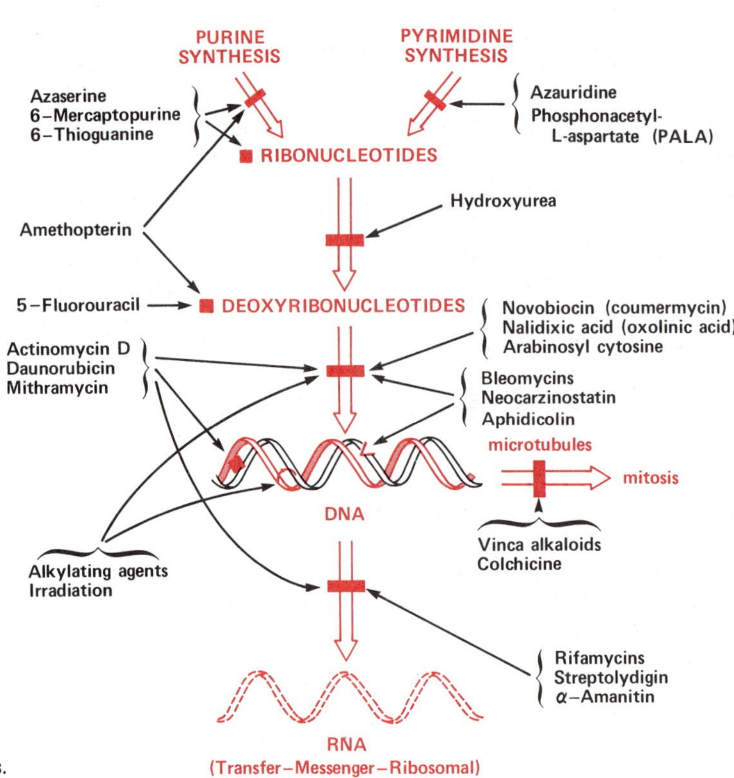

Figure 14-1
Agents used to block pathways of DNA biosynthesis.

2. Umezawa H (1972) *Enzyme Inhibitors of Microbial Origin*. Univ. Tokyo Press, Tokyo.

Antibacterial action generally depends on a uniquely vulnerable protein target in the bacterium. However, for curbing the growth of a virus, a tumor cell, or a lymphocyte, success depends less often on exploiting a virus- or tumor-specific enzyme and may rely instead on the differences between growth rates, drug destruction, and repair in the affected and the normal cells. Pharmacologic factors, including permeation and oxygenation, are also crucial in achieving therapeutic effectiveness with tolerable toxicity. These metabolic factors may be more skillfully managed when the affinities and kinetics of interaction of a drug with its target are known. Choosing the best drug and devising combinations of drugs (or drugs and physical agents) with complementary effectiveness and minimal toxicity depend on this knowledge.

The inhibitors considered in this chapter were selected from a large number of synthetic compounds and natural products used experimentally and clinically. The choice is intended to illustrate the use of inhibitors in analyzing biological and biochemical mechanisms of nucleic acid synthesis and in treating disease. Aside from their value in studying DNA replication and in treatment, some of the compounds are worth study as notable mutagens and carcinogens.

14-2 Inhibitors of Nucleotide Biosynthesis[3]

Reducing the supply of precursors may eventually limit the rate of nucleic acid biosynthesis and lead to mutations in the DNA template. In cells with ample supplies of precursors from de novo or salvage pathways, an inhibitor can function (1) as an antimetabolite blocking the enzymatic use of a substrate, (2) as an alternative substrate incorporated into DNA or RNA, or (3) as an analog of a feedback inhibitor ("pseudofeedback inhibitor") of an enzyme responsible for allosteric regulation of a biosynthetic pathway. Effective inhibitors are available for blocking the synthesis of purine and pyrimidine nucleotides, folate and folate coenzymes, and the deoxyribonucleotides (Table 14-1; Figs. 14-2 and 14-3).

Inhibitors of Purine Synthesis

Azaserine and *6-diazo-5-oxo-L-norleucine* (DON),[4] analogs of glutamine, were first recognized in extracts of *Streptomyces*. They inhibit three reactions in purine biosynthesis (Figs. 2-4 and 2-5), but their activities in inhibiting cell growth or division arise primarily from inhibition of formylglycinamide ribonucleotide amidotransferase (Fig. 2-4). As diazo keto compounds, they resemble diazomethane in chemical reactivity. They form a covalent bond with a cysteine residue in the active site, DON being 50 times more potent than azaserine. They also inhibit cytidine triphosphate synthase (Fig. 2-8).

Analogs of the purine and pyrimidine bases, as nucleotides, may inhibit reactions of nucleotide biosynthesis (Table 14-1). When converted to dNTPs, they also may be incorporated into DNA and interfere with its functions (see Tables 14-3 and 14-4). *6-Mercaptopurine* (6-MP) and *6-thioguanine*[5] (6-TG) are analogs

3. Langen P (1975) *Antimetabolites of Nucleic Acid Metabolism*. Gordon and Breach, New York.
4. Langen P (1975) *Antimetabolites of Nucleic Acid Metabolism*. Gordon and Breach, New York.
5. Langen P (1975) *Antimetabolites of Nucleic Acid Metabolism*. Gordon and Breach, New York.

Table 14-1
Inhibitors of nucleotide biosynthesis

Inhibitor	Metabolite antagonized	Enzyme or function inhibited
PURINE SYNTHESIS		
Azaserine 6-Diazo-5-oxo-L-norleucine (DON)	glutamine	formylglycinamide ribotide amidotransferase
6-Mercaptopurine, thioIMP, methylthioIMP 6-Thioguanine, thioGMP	IMP GMP	purine nucleotide synthesis and interconversions; feedback of PRPP amidotransferase
8-Azaguanine, azaGMP	GMP	purine nucleotide synthesis
Ribavirin MP	IMP	IMP dehydrogenase
Mycophenolic acid	IMP	IMP dehydrogenase
Thiadiazole	IMP	IMP dehydrogenase
Hadacidin	aspartic acid	adenylosuccinate synthase
PYRIMIDINE SYNTHESIS		
N-Phosphonoacetyl-L-asparatate (PALA)	carbamoylaspartate	aspartate carbamoyl transferase
5-Azaorotate	orotate	OMP synthase
6-Azauridine, azaUMP	OMP	OMP decarboxylase
5-Azacytidine	OMP	OMP decarboxylase
6-Azacytidine, azaCMP	dCMP	dCMP hydroxymethylase
FOLATE SYNTHESIS		
Sulfonamides	*p*-aminobenzoate	folate biosynthesis
Trimethoprin Methotrexate (amethopterin) Aminopterin	folate, dihydrofolate	folate (dihydrofolate) reductase
DEOXYNUCLEOTIDE SYNTHESIS		
Hydroxyurea	enzyme free radical	ribonucleotide reductase
5-Fluorouracil (FU), FU deoxynucleoside, FdUMP	dUMP	thymidylate synthase
NUCLEOSIDE TRIPHOSPHATE SYNTHESIS		
Cyclamidomycin (desdanine)	NDPs	nucleoside diphosphate kinase

of hypoxanthine and guanine in which sulfur replaces oxygen at position 6. The incorporation of 6-MP or 6-TG into DNA is an important feature in their use in cancer treatment (Section 3). In immunosuppressive treatment, *azathioprine* (Imuran), a prodrug which liberates 6-MP, has made kidney transplantation a more effective therapy. *ThioIMP* (thioinosinate) inhibits the conversion of IMP to adenylosuccinate in the synthesis of AMP, and inhibits the oxidation of IMP by IMP dehydrogenase in the synthesis of GMP (Fig. 2-5). *ThioGMP* (thioguanylate), *thioIMP*, and *8-azaguanosineMP* are negative allosteric effectors of 5-phosphoribosyl pyrophosphate (PRPP) amidotransferase, the first committed step in purine biosynthesis. Among the most potent is *6-methylthioIMP*, formed readily from thioIMP.

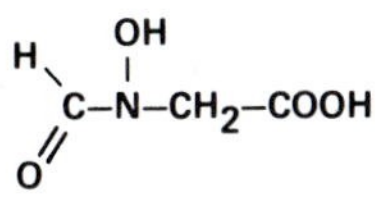

$N \equiv \overset{+}{N} - CH_2 - CO - O - CH_2 - CH(NH_2) - COOH$

Azaserine

$N \equiv \overset{+}{N} - CH_2 - CO - CH_2 - CH_2 - CH(NH_2) - COOH$

Hadacidin

6-Diazo-5-oxo-norleucine (DON)

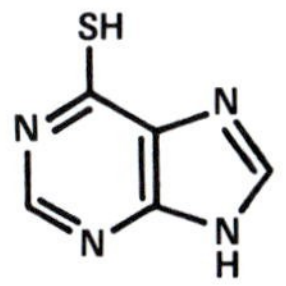

6-Mercaptopurine

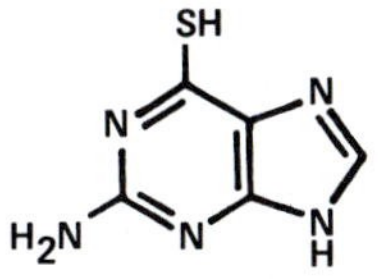

6-Thioguanine

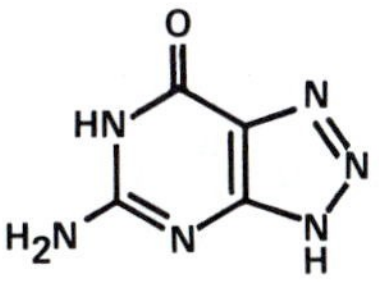

8-Azaguanine

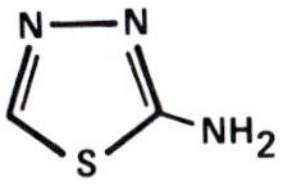

Thiadiazole

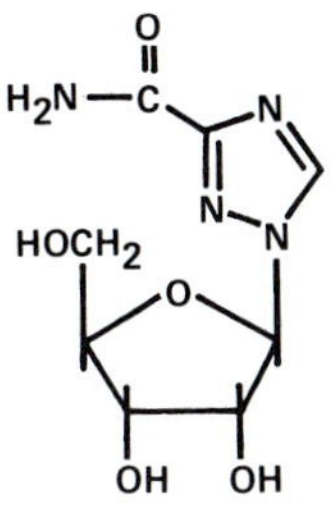

Ribavirin

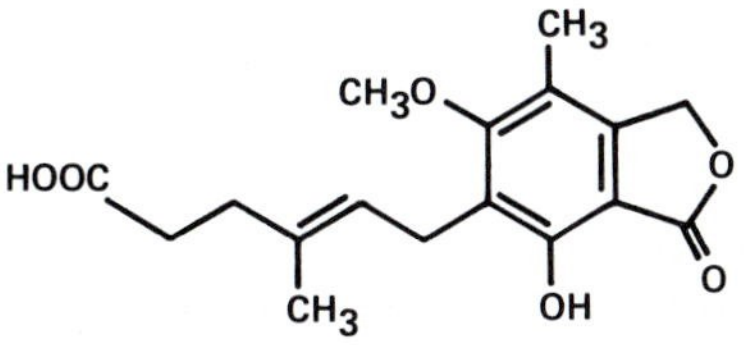

Mycophenolic Acid

Figure 14-2
Inhibitors of purine synthesis.

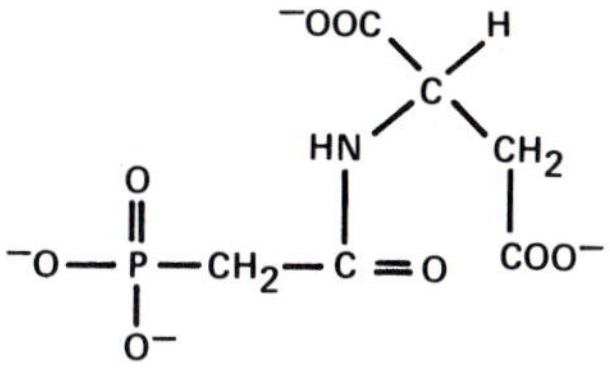

Phosphonacetyl-L-aspartate (PALA)

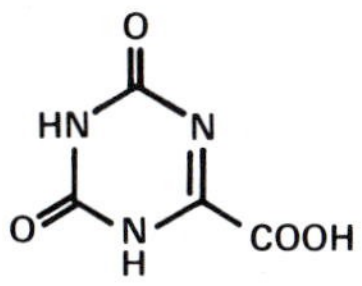

5-Azaorotate

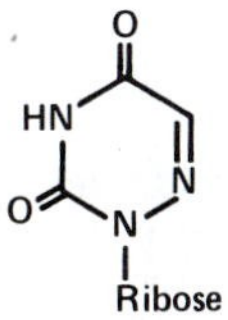

6-Azauridine
(Azaribine)

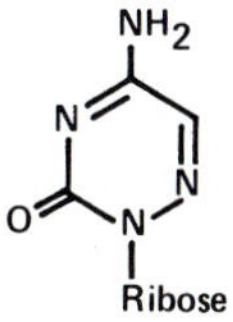

6-Azacytidine

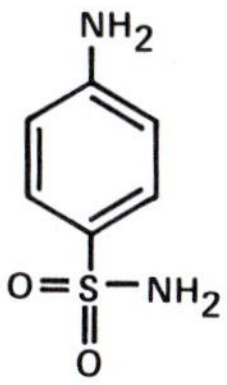

Sulfanilamide

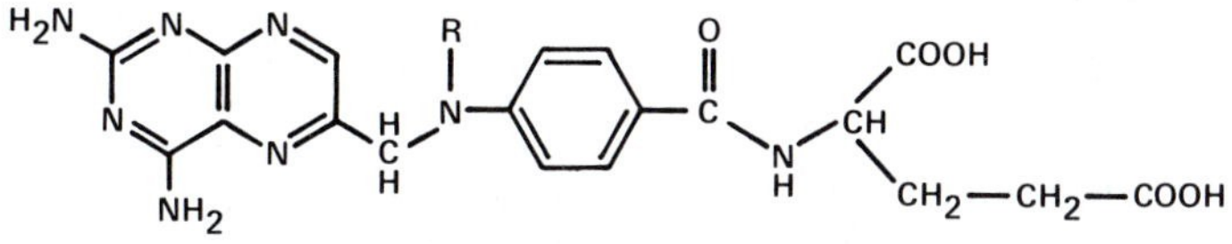

Aminopterin: R = H
Amethopterin (Methotrexate): R = CH_3

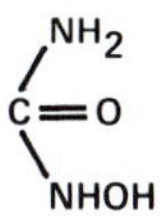

Hydroxyurea

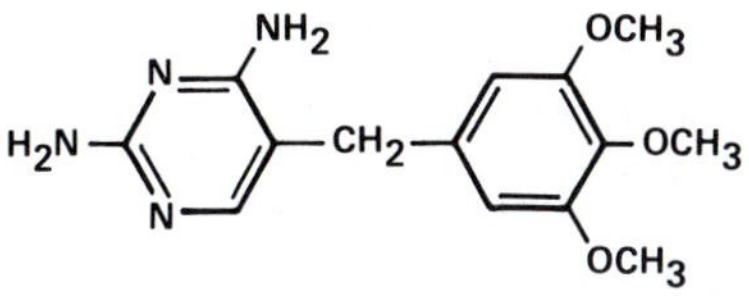

Trimethoprim

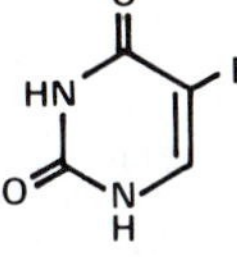

5-Fluorouracil

Figure 14-3
Inhibitors of nucleotide biosynthesis.

Mycophenolic acid[6] (Fig. 14-2 and Table 14-2), *ribavirin MP*[7] (Fig. 14-2), and *thiadiazole*[8] (2-amino-1,3,4-thiadiazole; Fig. 14-2) inhibit IMP dehydrogenase, thus interrupting guanylate biosynthesis specifically. *Hadacidin*[9] (N-formyl-N-hydroxyglycine; Fig. 14-2) is an aspartic acid analog isolated from fungi and inhibitory to the growth of bacteria, tumor cells, and plant tissues. By competing with aspartic acid, it inhibits adenylosuccinate synthetase and thereby the de novo synthesis of AMP.

Inhibitors of Pyrimidine Synthesis

N-phosphonoacetyl-L-aspartate (PALA; Fig. 14-3) was synthesized to serve as an inhibitor of aspartate carbamoyl transferase (ATCase). Designed to mimic both substrates of the enzyme, aspartate and carbamoyl phosphate,[10] PALA has more than 100 times greater affinity than carbamoyl phosphate for ATCase, and thus performs as a potent inhibitor of this key enzyme in pyrimidine nucleotide synthesis (Fig. 2-7). PALA inhibits certain kinds of tumor cells selectively because they have lower concentrations of ATCase.[11] For similar reasons, PALA may not be immunosuppressive.

5-Azaorotate inhibits the synthesis of orotidine 5′-phosphate (OMP); *6-azauridine* and *5-azacytidine*, as nucleotides, inhibit OMP decarboxylation. 5-Azacytidine is more prominent in the inhibition of macromolecular biosynthesis, acting primarily via incorporation into RNA and DNA and activation of gene expression through interference with DNA methylation.

6-Azacytidine inhibits a reaction unique to T-even phages, the hydroxymethylation of dCMP (Section 2-13). The inhibition prevents the synthesis of hydroxymethylcytosine nucleotide, and dCMP or 6-aza-dCMP residues are incorporated into replicating phage DNA. As a result, this DNA is susceptible to a phage-induced nuclease that attacks DNA containing cytosine residues.

Table 14-2
Inhibition of nucleotide biosynthesis by methotrexate and mycophenolic acid[a]

	Cellular nucleotide level (% of normal)			
	Methotrexate		Mycophenolate	
Base	NTP	dNTP	NTP	dNTP
A	26	56	89	110
G	11	31	22	38
U or T	230	32	171	250
C		250	166	74

[a] From Henderson JF, Lowe JK, Barankiewicz J (1977) in *Purine and Pyrimidine Metabolism*. Ciba Foundation Symposium 48, Elsevier, Amsterdam, p. 3.

6. Nery R, Nice E (1971) *J Pharm Pharmacol 23*:842.
7. Miles DL, Miles DW, Redington P, Eyring H (1976) *PNAS 73*:4257.
8. Nelson JA, Rose LM, Bennett LL Jr (1977) *Canc Res 37*:182.
9. Shigeura HT, Gordon CN (1962) *JBC 237*:1932, 1937; Rossomando EF, Maldonado B, Crean EV (1978) *Antimicrob Agents Chemother 14*:476.
10. Swyryd EA, Seaver SS, Stark GR (1974) *JBC 249*:6945; Yoshida T, Stark GR, Hoogenraad NJ (1974) *JBC 249*:6951; Johnson RK, Inouye T, Goldin A, Stark GR (1976) *Canc Res 36*:2720.
11. Johnson RK, Swyryd EA, Stark GR (1978) *Canc Res 38*:315.

Inhibitors of Folate Synthesis

The antibacterial action of *sulfa drugs* is due to their competitive antagonism of p-aminobenzoic acid in the microbial biosynthesis of folate. By contrast, in animals, folate must be supplied as an essential dietary element. Conversion of folate to the tetrahydro form is required for its coenzyme function in all cells. The inhibition of dihydrofolate (also folate) reductase (DHFR)—by *trimethoprim* for the bacterial enzyme, *pyrimethamine* for the plasmodial enzyme, and *methotrexate* (amethopterin)[12] and *aminopterin* for animal cell enzymes—depletes the coenzyme, thereby blocking the synthesis of purine nucleotides, thymidylate, and methionine (Table 14-2). These "magic bullets" of microbial disease and cancer chemotherapy emerged from the discoveries of the sulfonamides and folates and their modes of action.[13]

The affinity of an analog for DHFR (e.g., that of methotrexate for the mammalian cell enzyme) may be 10^4 to 10^5 times that of folate or dihydrofolate—great enough to make the binding virtually stoichiometric. Binding varies widely, depending on the analog and the source of DHFR. For example, trimethoprim has a very high affinity for the bacterial enzyme and pyrimethamine for the malarial enzyme, but they bind poorly to the DHFR of mammalian cells.

The folate analogs effective against mammalian DHFR are used to treat neoplastic diseases (especially childhood leukemias) and psoriasis, a proliferative skin disease, and to suppress the immune system. The methotrexate treatment regimen for neoplastic disease may use doses massive enough to kill both rapidly growing normal cells and the neoplastic cells. The normal cells are then rescued by large doses of leucovorin, the 5-formyl-tetrahydrofolate form of the coenzyme, which bypasses the blockade of tetrahydrofolate regeneration.

Cell cultures resistant to 3000 times the methotrexate concentration that kills sensitive cells can be obtained by selection. These strains produce 200 times the normal level of DHFR because the DHFR gene has been amplified to thousands of copies per cell.[14]

Inhibitors of Deoxynucleotide Synthesis

Hydroxyurea[15] destroys the free radical of *E. coli* ribonucleotide reductase and is a potent reversible inhibitor of the mammalian reductase. These actions block the production of all deoxyribonucleotides and, hence, block DNA synthesis (Section 2-7).

5-Fluorouracil,[16] in the form of a deoxyribonucleotide, covalently binds and inactivates thymidylate synthase (Section 2-8) and shuts down de novo biosynthesis of thymidylate. Fluoro-dUTP (FdUTP) degradation by dUTPase (Section 2-9) may prevent its incorporation into DNA, whereas persistence of fluoro-UTP may permit it to be incorporated into RNA, leading to inhibited growth and cell death.

Methotrexate and 5-fluorouracil used together in the treatment of cancer may have antagonistic as well as synergistic effects.[17] Inhibition of thymidylate syn-

12. Hryniuk WM, Brox LW, Henderson JF, Tamaoki T (1975) *Canc Res 35*:1427; Williams JW, Morrison JF, Duggleby RG (1979) *Biochemistry 18*:2567.
13. Jukes TH (1987) *Canc Res 47*:5528.
14. Schimke RT, Kaufman RJ, Alt FW, Kellems RF (1978) *Science 202*:1051.
15. Timson J (1975) *Mutat Res 32*:115.
16. Bellisario RL, Maley GF, Galivan JH, Maley F (1976) *PNAS 73*:1848.
17. Cadman E, Heimer R, Benz C (1981) *JBC 256*:1695.

thesis by 5-fluorodeoxyuridylate (FdUMP) requires methylene tetrahydrofolate for covalent binding to thymidylate synthase. Because the inhibition of DHFR by methotrexate prevents regeneration of the tetrahydrofolate, the supply of methylene tetrahydrofolate is depleted and FdUMP is unable to form the ternary complex necessary for prolonged inhibition. With the level of dTTP and its ratio to dUTP diminished, the incorporation of uracil into DNA is increased (Section 15-10) and breakage of DNA is more frequent.

An Inhibitor of Nucleoside Triphosphate Synthesis

Cyclamidomycin (pyracrimycin A, desdanine)[18] is a specific inhibitor of nucleoside diphosphate kinase in *E. coli*. This ubiquitous enzyme produces NTPs, including ATP, and is thus crucial to all macromolecular syntheses.

Catabolite Analogs

The effectiveness of biosynthetic analogs often depends on their being converted to the nucleotide form and avoiding removal by catabolic enzymes. Potentiation of the antitumor and toxic activity of 6-mercaptopurine by *allopurinol* (Fig. 14-4) may be due to inhibition of xanthine oxidase by the latter (Section 3). Inhibitors of adenosine deaminase not only potentiate the antitumor activity of adenosine analogs, but one of these, *2′-deoxycoformycin* (pentostatin; Fig. 14-4), even when administered alone, produces remissions in patients with acute leukemia derived from T lymphocytes.[19]

14-3 Nucleotide Analogs Incorporated into DNA or RNA[20]

Certain analogs of the NTPs, modified in the sugar or base, are accepted by polymerases for pairing with the DNA template and are incorporated into nucleic acid, but subsequently block further chain growth or interfere with nucleic acid functions (Table 14-3 and Fig. 14-5).

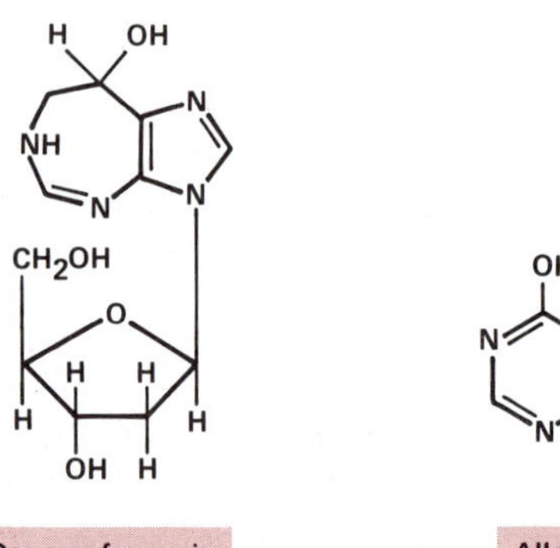

Figure 14-4
Catabolite analogs.

18. Saeki T, Hori M, Umezawa H (1975) *J Antibiot 28*:974.
19. Cass CE (1979) in *Antibiotics V* (Hahn FE, ed.). Springer-Verlag, New York, p. 85.
20. Elion GB (1989) *Science 244*:41.

Table 14-3
Nucleotide analogs incorporated into DNA or RNA

Analog	Incorporated into DNA or RNA	Inhibition
CHAIN TERMINATORS		
2′,3′-Dideoxy-NTPs AZT (azidothymidine) Arabinosyl NTPs (araC, araA)	DNA	chain growth, 3′→5′ exonuclease
Acyclovir NTP	DNA (analog of G)	chain growth (herpes DNA polymerase)
Cordycepin TP (3′-deoxyATP) 3′-Amino ATP	DNA, RNA	chain growth
DEFECTIVE NUCLEIC ACID		
Uracil dNTP (dUTP)	DNA (analog of T)	DNA integrity: excision leads to chain breakage
5-Hydroxyuridine TP 5-Aminouridine TP	RNA	syntheses and functions of DNA and RNA
5-Bromouracil dNTP 5-Iodouracil dNTP	DNA (analog of T)	fidelity of replication (mutation); differentiation
5-Azacytidine TP	RNA, DNA (analog of C)	processing of rRNA (defective)
Allopurinol (NTP)[a]	RNA (analog of A)	xanthine oxidase; antiprotozoal
Tubercidin TP Toyocamycin TP Formycin 7-Deazanebularin	DNA, RNA	syntheses and functions of DNA and RNA
2-Aminopurine dNTP 2-Aminoadenine dNTP (2,6-diaminopurine)	DNA (analog of A)	fidelity of replication (mutation)
6-Thioguanine dNTP	DNA (analog of G)	fidelity of replication (mutation)
UNCLASSIFIED		
2′-Deoxy-2′-azidocytidine NTP		initiation of polyoma DNA synthesis; *E. coli* primase

[a] 4-Hydroxypyrazolo(3,4-d)pyrimidine NTP.

Chain Termination

The *2′,3′-dideoxyribonucleosides* (Fig. 14-5), if converted to the triphosphates (ddNTPs), are incorporated into DNA at a very slow rate. In studies with *E. coli* pol I, the discrimination does not appear to be in the binding or base pairing of the analog but occurs at a subsequent stage in the reaction when the analog proves inadequate as a primer for the next polymerization event (Section 4-7). Because the analog lacks a 3′-OH group, proofreading excision of the analog is also exceedingly slow, and thus the block of chain growth is maintained. The strikingly specific inhibition of eukaryotic DNA polymerases β and γ (Section 6-1) suggests that strongly competitive binding by ddNTPs blocks the actions of these enzymes.

Acyclovir [9-(2-hydroxy-ethoxymethyl) guanine, acycloguanosine; Fig. 14-5] is one of several synthetic nucleoside analogs [9-(2,3-dihydroxypropyl) adenine is another][21] that are potent inhibitors of herpes simplex viruses. Acyclovir is

21. De Clercq E, Holy A (1979) *J Med Chem 22*:510; King GSD, Sengier L (1981) *J Chem Res (M)*:1501.

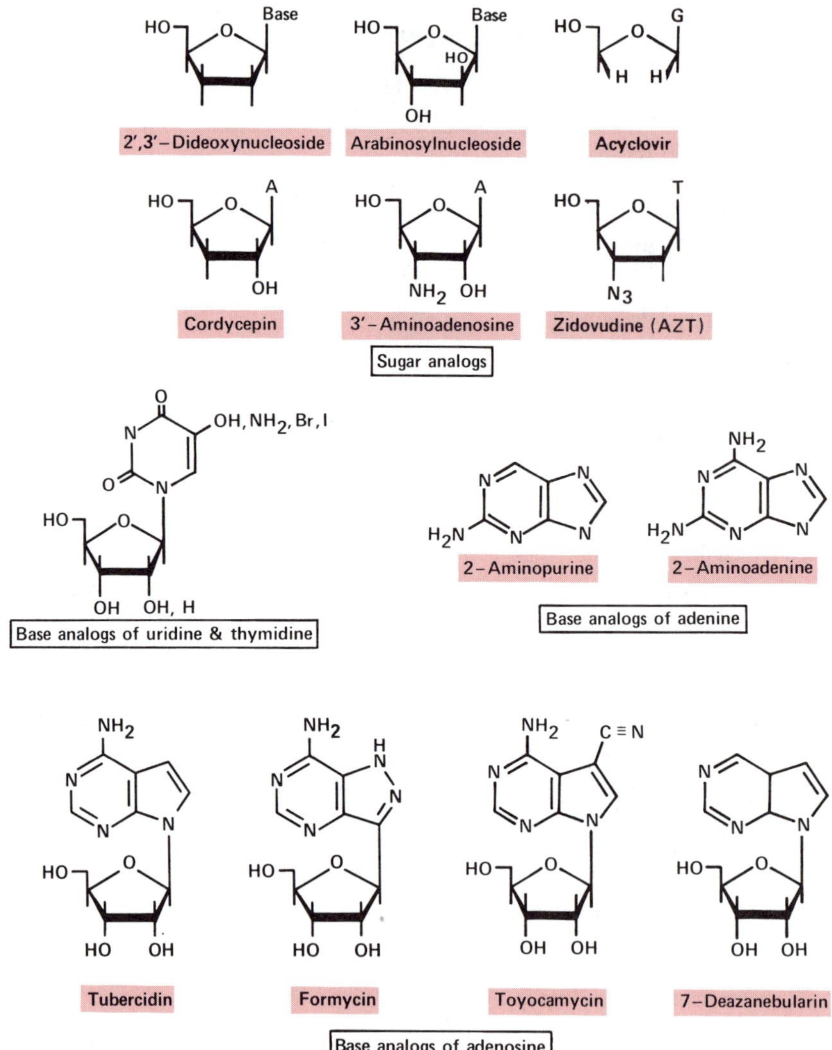

Figure 14-5
Nucleotide analogs incorporated into DNA or RNA.

effective as a drug because the analog is phosphorylated to acycloGMP by the herpes-encoded thymidine kinase, but not by the host cell enzyme. AcycloGMP, when further phosphorylated by host cell kinases to acycloGTP, inhibits the herpes-encoded DNA polymerase by a novel mechanism[22] more profoundly than it does the host DNA polymerase. An x-ray diffraction analysis of the acycloguanosine structure[23] attempts to rationalize the remarkable affinities of the nucleoside analog and its phosphorylated derivatives for these otherwise highly specific enzymes.

Azidothymidine (3'-deoxy-3'-azidothymidine, AZT, zidovudine; Fig. 14-5) and *2',3'-dideoxycytidine* (ddC) are among a group of nucleoside analogs[24] that

22. Reardon JE, Spector T (1989) *JBC 264*:7405.
23. Birnbaum GI, Cygler M, Kusmierek JT, Shugar D (1981) *BBRC 103*:968.
24. Mitsuya H, Broder S (1986) *PNAS 83*:1911; Halmos T, Montserret R, Antonakis K (1989) *NAR 17*:7663; Yarchoan R, Broder S (1989) *Pharmacol Ther 40*:329; Huang P, Farquhar D, Plunkett W (1990) *JBC 265*:11914.

are potent inhibitors of human immunodeficiency virus (HIV) replication in cell culture; several have been used in the treatment of the disease AIDS. The nucleosides, converted to the active triphosphate form, inhibit the HIV reverse transcriptase activity, as well as terminating chains once incorporated. Adverse effects of the active forms of AZT and ddC on the host polymerases include the inhibition of both mitochondrial[25] and nuclear replication.

Arabinosides[26] (arabinosyl nucleosides; Fig. 14-5) are notable drugs; *cytarabine* (cytosine arabinoside, araC) is used to treat cancer, and *vidarabine* (adenine arabinoside, araA)[27] is an antiviral agent. Among many possibilities and claims for how the drugs act, the most tenable are based on incorporation of the arabinosides into DNA, where they distort the primer-template and block further DNA synthesis by chain termination. Still, the inhibitory action of araC at the stage of chain growth may be marked with certain of the prokaryotic (Section 5-2) and eukaryotic (Section 6-1) polymerases.

Because araC must be in the nucleoside triphosphate form to be active, circumstances that favor conversion of the nucleoside by deoxycytidine kinase enhance its clinical value. Competing with this kinase in some cells and tissues is a potent cytidine deaminase, which destroys the usefulness of araC as a drug Curiously, the related arabinosyl analogs araT and araU occur naturally in sponges.[28]

Cordycepin triphosphate[29] (3′-deoxyATP; Fig. 14-5) inhibits chain elongation by RNA and DNA polymerases by generating an inactive primer terminus. The nucleoside does not inhibit bacterial growth, probably because it is phosphorylated very poorly. The triphosphate strongly inhibits RNA synthesis in ascites tumor and HeLa cell lines and affects the addition of polyA segments to the 3′ end of mRNAs. Like 3′-deoxyATP, *3′-amino ATP*[30] inhibits RNA synthesis in isolated nuclei and extracts of ascites tumor cells by terminating elongation after addition to the primer. Although the nucleoside inhibits DNA synthesis in these cells, extracts show no inhibition of DNA synthesis by the triphosphate. Ribonucleotide reduction or RNA-synthesis–dependent DNA synthesis may be the targets of the inhibition in vivo.

Defective Nucleic Acid

Uracil incorporation into DNA by way of dUTP probably occurs to a significant extent under normal circumstances and can be extensive when the ratio of dUTP to dTTP is elevated (Sections 2-9 and 15-10). Uracil in DNA is recognized as foreign and is excised by an N-glycosylase. Mutants defective in the glycosylase can accumulate uracil to levels nearly equimolar with thymine without obvious malfunction of the DNA. Extreme examples are phages in which thymine is completely replaced by uracil or hydroxymethyluracil (Section 17-8).

Uridine and *deoxyuridine analogs* with *5-hydroxy* or *5-amino substituents* inhibit the synthesis of DNA, RNA, and protein in *E. coli* and interfere, in undetermined ways, with the functions of the RNA and DNA molecules into which they are incorporated.

---

25. Chen C-H, Cheng Y-C (1989) *JBC 264*:11934.
26. Matsushita T, Kubitschek HE (1975) *Adv Microb Physiol 12*:247; Cozzarelli NR (1977) *ARB 46*:641.
27. Fridland A (1977) *Biochemistry 16*:5308; Dicioccio RA, Srivastava BIS (1977) *EJB 79*:411.
28. Bergmann W, Feeney RJ (1950) *JACS 72*:2809; Bergmann W, Feeney RJ (1950) *J Org Chem 16*:981.
29. Suhadolnik RJ (1970) *Nucleoside Antibiotics*. Wiley, New York; Roy-Burman P (1970) *Analogues of Nucleic Acid Components*. Springer-Verlag, New York.
30. Langen P (1975) *Antimetabolites of Nucleic Acid Metabolism*. Gordon and Breach, New York.

5-Bromouracil (BU) and *5-iodouracil* are deoxyuridine nucleotides containing bromine or iodine, respectively, in the 5 position. Readily incorporated into DNA, they cause replication errors and are mutagenic in UV-irradiated DNA when converted to the cyclobutane dimers. The *lac* operator in BU-containing phage λ DNA is an order of magnitude more effective in binding the *lac* repressor than is normal DNA. Thus, BU in DNA may also alter the recognition of specific replication signals.

5-Azacytidine (Section 6) incorporation into RNA and DNA may account for defective processing of ribosomal RNA,[31] but is not demonstrably mutagenic. This analog also inhibits pyrimidine nucleotide synthesis and the methylation of DNA (Section 2).

Allopurinol (Fig. 14-4) is the synthetic analog of hypoxanthine in which the nitrogen and carbon in positions 7 and 8 are reversed. It is in widespread clinical use to inhibit xanthine (and hypoxanthine) oxidase and thereby suppress the accumulation of uric acid in gout and related disorders. The drug is also an antiprotozoal agent by virtue of being incorporated into RNA, and it potentiates 6-mercaptopurine (Section 2). Allopurinol is converted first to the ribonucleotide, poorly in mammals and more so in protozoa; only in protozoa is the nucleotidyl form accepted by adenylosuccinate synthase and lyase to become the AMP analog, which, as a triphosphate, is incorporated into RNA.[32]

Tubercidin[33] (7-deazaadenosine), *formycin*,[34] *toyocamycin*,[35] and *7-deazanebularin*[36] (Fig. 14-5) are cytotoxic analogs of adenosine. In vitro, they serve in the form of NTPs as substitutes for ATP in RNA synthesis. They may inhibit by forming nonfunctional RNA or by acting as antagonists in other ATP-dependent reactions. The 2′-deoxy derivative can be formed from the NTP by ribonucleotide reductase (Section 2-7).[37] Deazanebularin substitutes for both ATP and GTP in RNA synthesis, although it pairs significantly only with uracil.

2-Aminopurine[38] is incorporated into DNA as an analog of adenine. It is mutagenic, presumably because of mispairing with cytosine, which can explain its capacity to reverse mutations induced by 5-bromouracil. Powerful inhibition of adenosine deaminase (Section 2-10) explains its immunosuppressive effects.

2-Aminoadenine (2,6-diaminopurine) as a dNTP is an analog of dATP and effectively substitutes for dATP with *E. coli* pol I.[39] It forms three hydrogen bonds with thymine and may give rise to mispairing with cytosine. Remarkably, 2-aminoadenine has been found in place of adenine in the duplex DNA of a phage that infects blue-green algae;[40] an elevated DNA-melting temperature indicates the presence of triply hydrogen-bonded base pairings with thymine. In this unique example of the complete substitution of a DNA purine with a novel base, it would be of interest to know how dATP is excluded from DNA and how 2-aminoadenine influences the genetic stability of these phages.

6-Thioguanine, beyond interfering with purine nucleotide synthesis (Section

---

31. Weiss JW, Pitot HC (1975) *Biochemistry 14*:316.
32. Nelson DJ, Bugge CJL, Elion GB, Berens RL, Marr JJ (1979) *JBC 254*:3959; Spector T, Jones TE, Elion GB (1979) *JBC 254*:8422.
33. Nishimura S, Harada F, Ikehara M (1966) *BBA 129*:301.
34. Ikehara M, Murao K, Harada F, Nishimura S (1968) *BBA 155*:82; Ward DC, Cerami A, Reich E, Acs G, Altwerger L (1969) *JBC 244*:3243.
35. Corcoran JW, Hahn FE, eds. (1975) *Antibiotics*. Springer-Verlag, New York, Vol. 3.
36. Ward DC, Reich E (1972) *JBC 247*:705.
37. Brinkley SA, Lewis A, Critz WJ, Witt LL, Townsend LB, Blakeley RL (1978) *Biochemistry 17*:2350.
38. Langen P (1975) *Antimetabolites of Nucleic Acid Metabolism*. Gordon and Breach, New York.
39. Cerrami A, Reich E, Ward DC, Goldberg IH (1967) *PNAS 57*:1036.
40. Kirnos MD, Khudyakov IY, Alexandrushkina NI, Vanyushin BF (1977) *Nature 270*:369.

2), exerts an important part of its antitumor activity by being incorporated into DNA.[41] *6-Mercaptopurine* can also be incorporated via thioGMP into DNA, but is a less efficient precursor than thioguanine.[42]

An Unclassified Analog

2′-Deoxy-2′-azidocytidine[43] appears to be an analog of either ribo- or deoxyribonucleosides. It inhibits primase[44] (Section 8-3) and also inhibits an early step in the synthesis of polyoma virus DNA in hamster ovary cells, possibly at the initiation of each replication cycle. Although nucleoside diphosphates bearing a 2′-azido group specifically inactivate ribonucleotide reductase in vitro (by destroying the free radical in the enzyme), the in vivo effect cannot be attributed to this action. It is not yet known whether the analog is incorporated into DNA or RNA.

14-4 Inhibitors That Modify DNA

Although DNA is relatively unreactive chemically, the need to preserve its conformation and continuity for extraordinary lengths makes it vulnerable to agents that bind to it noncovalently or introduce occasional covalent modifications. Clustering of AT and GC pairs in regions that serve as origins of replication, promoters of transcription, and other vital signals may offer especially sensitive targets for certain of these agents. Inhibitors of topoisomerases exert reverberating effects by alterations in the crucially important topology of DNA (Section 5).

Noncovalent DNA Binders (Table 14-4)[45]

Actinomycin D[46] (deactinomycin; Fig. 14-6) is one of the most extensively used and studied inhibitors of nucleic acid synthesis, particularly RNA synthesis. It has a lesser effect on DNA synthesis. In an attractive model, based on x-ray diffraction patterns of nucleotide–drug crystals, the planar phenoxazone ring system is intercalated between alternating GpC base pairs of poly d(GC), and the cyclic peptide portion is hydrogen-bonded with the 2-amino group of guanine in the minor groove. However, actinomycin D also binds with high affinity to a duplex that lacks this classic GpC site.[47]

Netropsin and *distamycin A*[48] (Fig. 14-6) are basic oligopeptide antibiotics which inhibit the growth of plant and animal cells as well as that of bacteria and viruses. They form extremely stable complexes with duplex DNA (Plate 12),

41. Maybaum J, Morgans CW, Hink LA (1987) *Canc Res 47*:3083.
42. Maybaum J, Hink LA, Roethel WM, Mandel HG (1985) *Biochem Pharmacol 34*:3677; Liliemark J, Petterson B, Engberg B, Lafolie P, Masquelier M, Peterson C (1990) *Canc Res 50*:108.
43. Skoog L, Bjursell G, Thelander L, Hägerström T, Hobbs J, Eckstein F (1977) *EJB 72*:371; Bjursell G, Skoog L, Thelander L, Söderman G (1977) *PNAS 74*:5310.
44. Reichard P, Rowen L, Eliasson R, Hobbs J, Eckstein F (1978) *JBC 253*:7011.
45. Wang AH-J (1987) in *Nucleic Acids and Molecular Biology* (Eckstein F, Lilley DM, eds.). Springer-Verlag, New York, Vol. 1, p. 32.
46. Sobell HM (1973) *Prog N A Res 13*:153; snyder JG, Hartman NG, D'Estantoit BL, Kennard O, Remeta DP, Breslauer KJ (1989) *PNAS 86*:3968.
47. Snyder JG, Hartman NG, D'Estantoit BL, Kennard O, Remeta DP, Breslauer KJ (1989) *PNAS 86*:3968.
48. Zimmer C, Wahnert U (1986) *Prog Biophys Mol Biol 47*:31

Table 14-4
Inhibitors that bind or modify DNA noncovalently

Inhibitor	Mechanism	Inhibition
Actinomycin D	intercalation (GC base pairs)	RNA, DNA chain growth
Netropsin Distamycin A	AT regions in minor groove	DNA, RNA chain growth
Anthracyclines Daunorubicin (daunomycin) Doxorubicin (Adriamycin) Mithramycin Olivomycin Chromomycin A_3	intercalation	DNA, RNA chain growth
Nogalamycin	alternating AT base pairs; also GC base pairs	
Acridine dyes Acriflavine Quinacrine (Atabrine) Proflavine	intercalation	fidelity of replication (frameshift mutation); RNA chain initiation; plasmid replication
Ethidium Propidium	intercalation	DNA replication; fidelity of replication (mutation)
Echinomycin (quinomycin A)	intercalation	inhibits growth of bacteria, viruses, and tumors
Kanchanomycin	Mg^{2+}–DNA ternary complex	DNA chain growth
8-Aminoquinolines		DNA chain growth
Chloroquine	intercalation	
Benzo[*a*]pyrene 4,5-epoxide	intercalation	fidelity of synthesis

especially with the base pairs of poly d(AT) and poly d(IC). Unlike actinomycin D, they do not bind by intercalation but by hydrogen-bonding and electrostatic interactions. Crystal and NMR structural studies have elucidated the precise nature of these drug–DNA complexes and have demonstrated a bifurcated hydrogen-bonded conformation in the d(AT) base pair at the site of drug binding.[49] Despite blocking the activity of RNA and DNA polymerases on DNA templates,[50] distamycin surprisingly stimulates the replication of oligo dA · poly dT, presumably because its binding stabilizes the short primer-template duplex and promotes initiations.[51]

Anthracycline glycosides include hundreds of antibiotics, among them the *daunorubicin*[52] family [*daunorubicin* (daunomycin, rubidomycin) and *doxorubicin* (Adriamycin)], *mithramycin*[53] (plicamycin), and *nogalamycin*[54] (Fig. 14-6

49. Coll M, Frederick CA, Wang AH-J, Rich A (1987) *PNAS 84*:8385; Pelton JG, Wemmer DE (1988) *Biochemistry 27*:8088.
50. Grehn L, Ragnarsson U, Eriksson B, Oberg B (1983) *J Med Chem 26*:1042.
51. Levy A, Weisman-Shomer P, Fry M (1989) *Biochemistry 28*:7262.
52. Goodman MF, Lee GM (1977) *JBC 252*:2670; Phillips DR, DeMarco A, Zunnio F (1978) *EJB 85*:487; Patel DJ, Canuel LL (1978) *EJB 90*:247.
53. Kersten H, Kersten W (1974) *Inhibitors of Nucleic Acid Synthesis*. Springer-Verlag, New York; Corcoran JW, Hahn FE, eds. (1975) *Antibiotics*. Springer-Verlag, New York, Vol. 3.
54. Liaw YC, Gao Y-G, Robinson H, van der Marel GA, van Boom JH, Wang AH-J (1989) *Biochemistry 28*:9913; Williams LD, Egli M, Gao Q, Bash P, van der Marel GA, van Boom JH, Rich A, Frederick CA (1990) *PNAS 87*:2225.

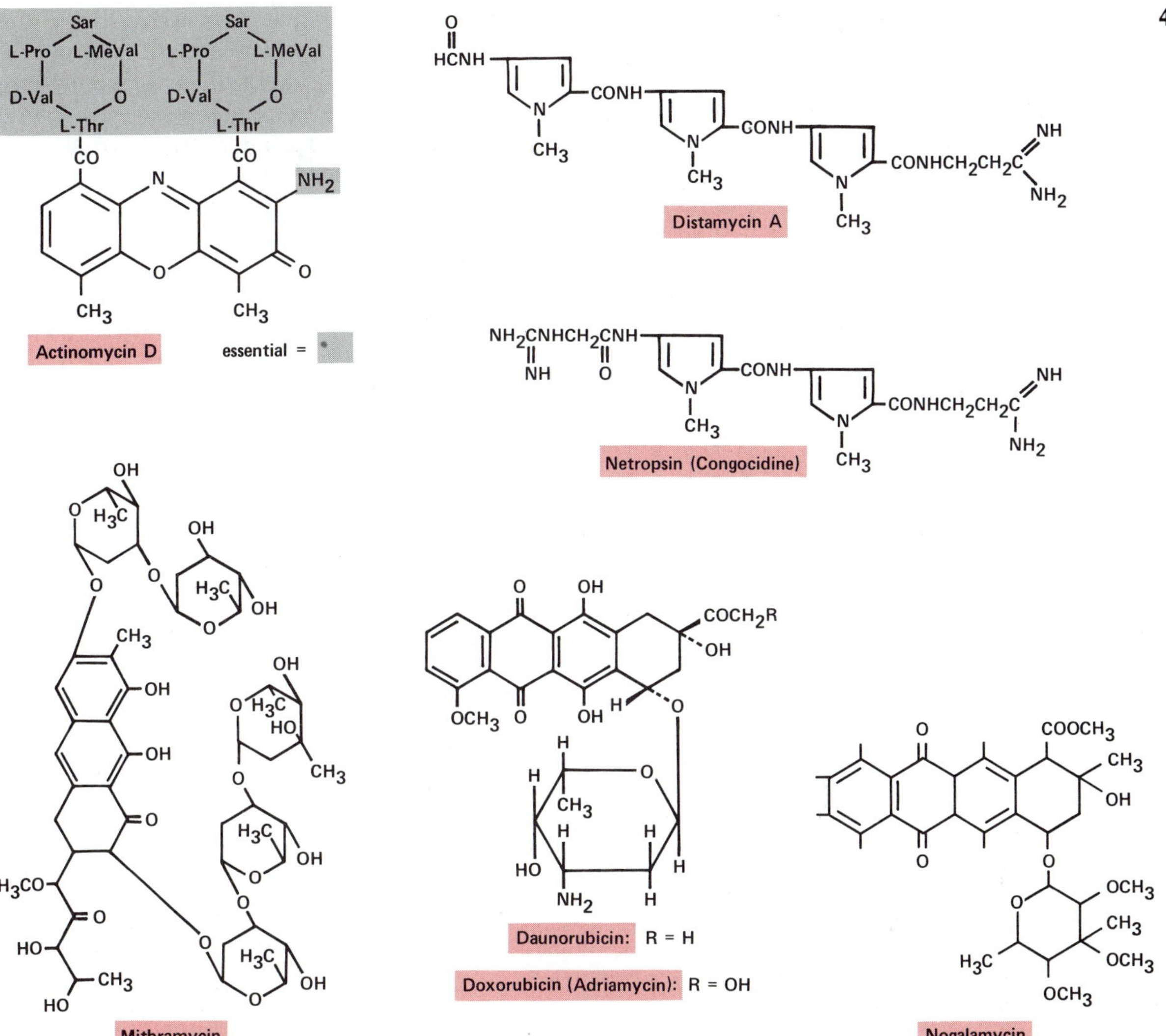

Figure 14-6
Inhibitors that bind DNA noncovalently.

and Plate 13). Intercalation of the tetracycline ring structure between base pairs is influenced by the hydrogen-bonding of the amino group of the distinctive sugar moiety to the DNA backbone. These inferences are drawn from studies with substituents by means of x-ray diffraction,[55] spectroscopy,[56] molecular model building, and the capacity of analogs to inhibit phage T4 DNA polymerase in vitro. Nogalamycin prevents the de novo synthesis by *E. coli* pol I of the alternating copolymer poly d(AT), thus enabling the homopolymer pair poly

55. Liaw YC, Gao Y-G, Robinson H, van der Marel GA, van Boom JH, Wang AH-J (1989) *Biochemistry 28*:9913.
56. Eriksson M, Nordén B, Eriksson S (1988) *Biochemistry 27*:8144.

dA·poly dT to be formed. Strong binding of AT-rich regions is striking for the alternating dA·dT base pairs and also for GC sequences embedded in a stretch of AT sequences. The exact mechanisms for the remarkable effectiveness of these compounds against a variety of cancers remain to be elucidated.

Bifunctional intercalators (Fig. 14-7), which intercalate at two DNA sites simultaneously, include several classes of compounds:[57] the synthetic[58] *acridines, chloroquines, quinaldines, phenanthridines,* and *pyridocarbazoles,*[59] and the

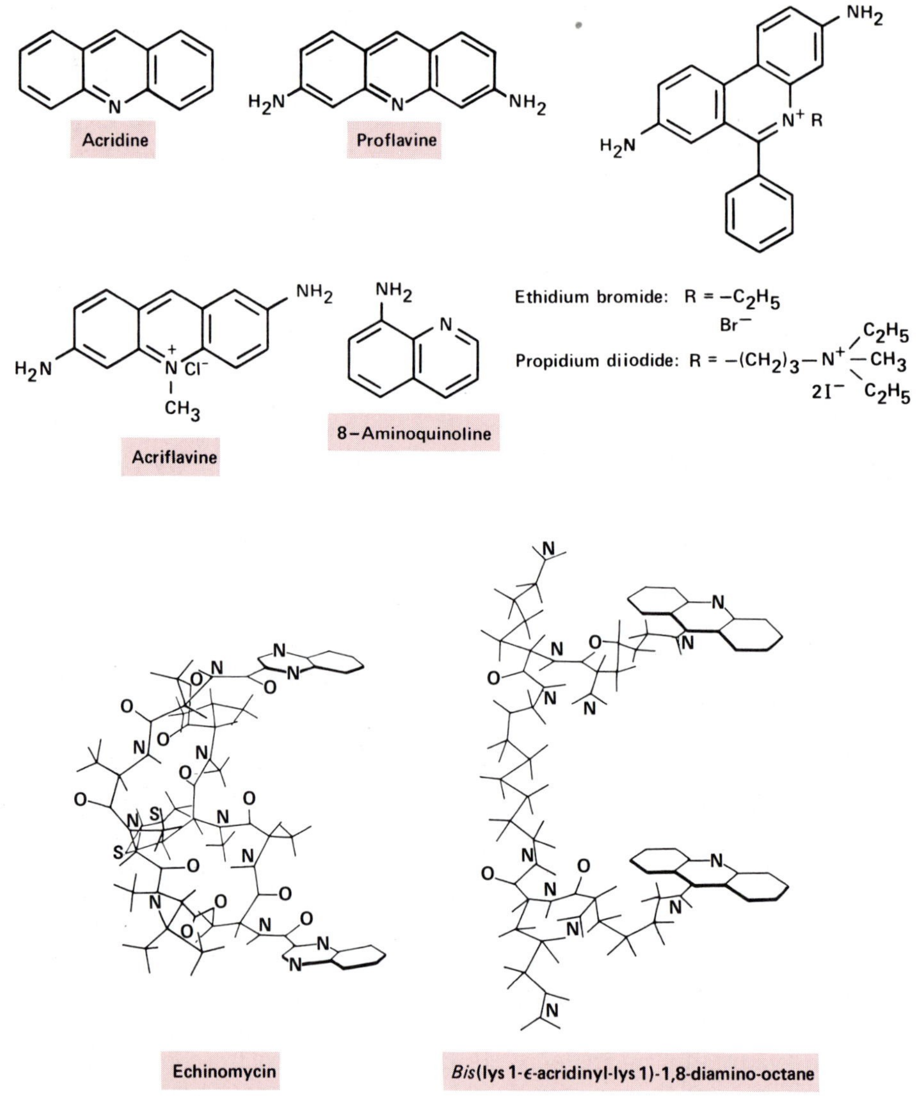

Acridine

Proflavine

Ethidium bromide: R = $-C_2H_5$ Br^-

Propidium diiodide: R = $-(CH_2)_3-N^+(C_2H_5)_2-CH_3$ $2I^-$

Acriflavine

8-Aminoquinoline

Echinomycin

Bis(lys 1-ε-acridinyl-lys 1)-1,8-diamino-octane

Figure 14-7
Inhibitors that bind DNA noncovalently.

57. Le Pecq J-B, Le Bret M, Barbet J, Roques BP (1975) *PNAS 72*:2915.
58. Kuhlmann KF, Mosher CW (1981) *J Med Chem 24*:1333.
59. Pelaprat D, Delbarre A, Le Guen I, Roques BP, Le Pecq J-B (1980) *J Med Chem 23*:1336.

natural antibiotics, such as *echinomycin*,[60] a quinoxaline. The very high affinity of these compounds (K_d near 10^{-11} M^{-1}) approaches that of polymerases and potent regulatory proteins.[61] A curious biological effect, not observed with antitumor monointercalators (e.g., actinomycin D or doxorubicin), is illustrated by a *7H-pyridocarbazole* dimer. Exposed cells grow for several generations, without apparent effect on macromolecular biosynthesis, and then permanently cease dividing. This delayed, drug-induced mortality may be related to a cellular program of controlled suicide called "apoptosis"[62] and may be useful as a model for similar events in stages of embryonic development.

Acridine, ethidium, and *propidium* dyes (Fig. 14-7) have received much attention because of their use in buoyant-density separation of supercoiled from other forms of DNA.[63] The buoyant density of the DNA decreases in proportion to the amount of compound intercalated; more dye molecules bind to the open than to the constrained, closed duplexes. Intercalation of these agents into DNA inhibits its cleavage by restriction endonucleases and can be applied in the analysis of DNA.[64] Acridine dimers may have DNA affinities comparable to those of repressors and RNA polymerases.[65]

Acridines[66] (*acriflavine, quinacrine, proflavine;* Fig. 14-7) cause bacterial cells to lose (be "cured" of) their episomes, an effect ascribed to the failure of these closed circles to replicate, causing them to disappear by dilution in the multiplying cell population. Intercalating dyes produce frameshift mutations by causing the deletion or addition of a base upon replication. Mutagenic effects on the twisted, circular mitochondrial genome of yeast are greater than on nuclear genes.[67] *Proflavine* inhibits RNA synthesis at initiation rather than in elongation, perhaps by blocking formation of the DNA–RNA polymerase complex.[68] Acridine drugs are useful in treating malignant and parasitic diseases.

Echinomycin (quinomycin A; Fig. 14-7), a quinoxaline antibiotic produced by *Streptomyces,* is highly active against Gram-positive bacteria, viruses, and tumors. It binds different DNA sites with different affinities, showing some preference for GC-rich regions.[69] The echinomycin structure is the basis for the design of peptidic bifunctional intercalating agents such as a lysyl-lysine bifunctional derivative of 9-aminoacridine (Fig. 14-7).[70]

Kanchanomycin[71] is a potent antibiotic of still undefined structure. It inhibits pol I action on duplex DNA by blocking template function; thus, the inhibition can be overcome by increasing the template concentration. RNA polymerase inhibition, however, may involve enzyme inactivation, because inhibition is not antagonized by increased template concentration but by increased enzyme levels.[72] Likewise, the antimalarial *8-aminoquinolines* (Fig. 14-7) bind to DNA and inhibit DNA polymerase activity.[73] As with kanchanomycin, there is evidence for enzyme inactivation as well as template binding.

---

60. Waring M (1981) *ARB 50*:159; Fox KR, Kentebe E (1990) *NAR 18*:1957.
61. Capelle N, Barbet J, Dessen P, Blanquet S, Roques BP, Le Pecq J-B (1979) *Biochemistry 18*:3354.
62. Wyllie AH (1980) *Nature 284*:555.
63. Radloff R, Bauer W, Vinograd J (1967) *PNAS 57*:1514.
64. Parker RC, Watson RM, Vinograd J (1977) *PNAS 74*:851.
65. Capelle N, Barbet J, Dessen P, Blanquet S, Roques BP, Le Pecq J-B (1979) *Biochemistry 18*:3354.
66. Corcoran JW, Hahn FE, eds. (1975) *Antibiotics.* Springer-Verlag, New York, Vol. 3.
67. Slonimski PP, Perrodin G, Croft JH (1968) *BBRC 30*:232.
68. Richardson JP (1966) *JMB 21*:83.
69. Fox KR, Wakelin LPG, Waring MJ (1981) *Biochemistry 20*:5768.
70. Bernier J-L, Henichart J-P, Catteau J-P (1981) *Biochem J 199*:479.
71. Kersten H, Kersten W (1974) *Inhibitors of Nucleic Acid Synthesis.* Springer-Verlag, New York.
72. Goldberg IH, Friedman PA (1971) *ARB 40*:775.
73. Whichard LP, Washington ME, Holbrook DJ Jr (1972) *BBA 287*:52.

Benzo[a]pyrene 4,5-epoxide[74] is the best known of the polycyclic aromatic hydrocarbon mutagens and carcinogens. A major target is the exocyclic amino group of guanine. Covalent binding of the diol epoxide, which takes minutes, is preceded by more extensive noncovalent DNA interactions on a time scale of milliseconds; these short-lived complexes may be the more significant in the ultimate biological activities of these compounds.

Covalent DNA Binders (Modifiers) (Table 14-5)

Alkylating agents[75] (Fig. 14-8) or their metabolic products form carbonium ion or other reactive intermediates that link them to a variety of nucleophiles, including phosphate, amino, sulfhydryl, hydroxyl, carboxyl, and imidazole groups. The prime target in DNA is N7 of guanine; less reactive are N1 and N3 of adenine, N3 of cytosine, and O6 of guanine. The resulting 7-alkylguanine (1) may form an improper base pair with thymine, (2) may be excised directly or by enzymatic repair, or (3) may become cross-linked by a polyfunctional alkylating agent to a second guanine residue or to a nucleophilic moiety of a protein. Among the classes of agents effective as DNA modifiers in the chemotherapy of neoplasms and in immunosuppression are *nitrogen mustards, ethyleneimines, alkyl sulfonates, nitrosoureas*, a *triazine* [*dacarbazine*, DTIC; 5-(3,3-dimethyl-1-triazenyl)-1H-imidazole-4-carboxamide] and a *hydrazine* [*procarbazine*; 1-methyl-2-p-(isopropylcarbamo-benzyl-hydrazine)] (Fig. 14-8). Some agents,

Table 14-5
Inhibitors that bind or modify DNA covalently

Inhibitor	Mechanism	Inhibition
Alkylating agents Nitrogen mustards Ethyleneimines Alkyl sulfonates Nitrosoureas Triazine dacarbazine (DTIC) Hydrazine procarbazine	monoalkylation and cross-linking of DNA strands; also cross-linking of DNA to protein	DNA synthesis and functions
Bleomycins, phleomycins	chain breakage	DNA synthesis and functions
Mitomycin C Porfiromycin (semiquinone form) Anthramycin	cross-linking	DNA synthesis and functions
Psoralens	intercalation, cross-linking	DNA synthesis and functions
Neocarzinostatin, auromomycin Esperamicin A_1, daunomycin, A447-C, calichemicin $\gamma 1$	chain breakage	DNA synthesis and functions
cis-Diamminedichloroplatinum (II) (cisplatin)	intercalation, cross-linking	DNA synthesis and functions
Benzo[*a*]pyrene 4,5-epoxide	intercalation	fidelity of synthesis

74. Geacintov NE, Shahbaz M, Ibanez V, Moussaoui K, Harvey RG (1988) *Biochemistry 27*:8380.
75. Calabresi P, Parks RE Jr (1985) in *Goodman and Gilman's The Pharmacological Basis of Therapeutics* (Gilman AG, Goodman LS, Rall TW, Murad F, eds.) 7th ed. Macmillan, New York, p. 1247.

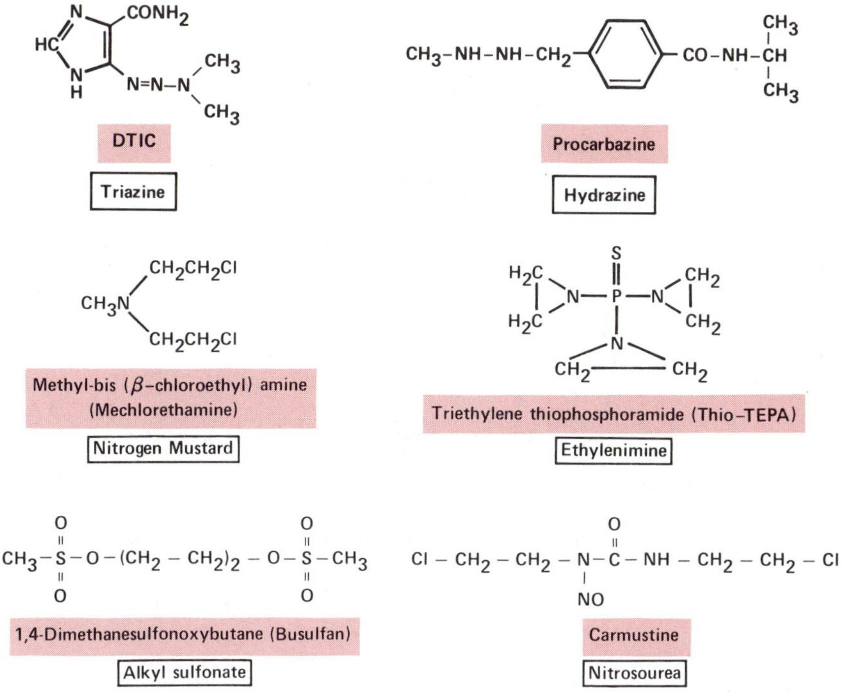

DTIC
Triazine

Procarbazine
Hydrazine

Methyl-bis (β-chloroethyl) amine (Mechlorethamine)
Nitrogen Mustard

Triethylene thiophosphoramide (Thio-TEPA)
Ethylenimine

1,4-Dimethanesulfonoxybutane (Busulfan)
Alkyl sulfonate

Carmustine
Nitrosourea

Figure 14-8
Inhibitors that bind and modify DNA.

such as DTIC and procarbazine, must be metabolically altered to become alkylating agents.

Haloethylnitrosoureas[76] include carmustine (BCNU, bis-chloroethylnitrosourea; Fig. 14-8), which acts by transfer of a chloroethyl group to O6 of guanine, which in turn provides an electrophilic center for cross-linking duplex DNA.[77] Mutants unable to excise the O6-alkylguanine are far more sensitive to the cytotoxic action of the drug. The frequent occurrence in human cancer cells and SV40-transformed cells of phenotypes similar to these mutants may explain the preferential chemotherapeutic efficiency of the nitrosoureas.

Bleomycins[78] (Fig. 14-9), isolated from *Streptomyces*, are copper chelates of a complex mixture of basic glycopeptides. Their remarkable antitumor activity has spurred interest in exploring their properties and enlarging their variety. Differences from one another at the terminal amine groups can be multipled by incorporating specific amines biologically from fermentation mixtures. The bleomycin family has been further expanded synthetically, to more than a hundred compounds, by the addition of amines to bleomycinic acid. The *phleomycins*[79] differ only in having the partially reduced (dihydro) form of the inner thiazole moiety.

The activated drug is a transient complex of iron and oxygen; Fe(II) with O_2 is effective, as is Fe(III) and H_2O_2 (or other hydroperoxides).[80] The activated forms

76. Ludlum DB (1990) *Mutat Res 233*:117.
77. Erickson LC, Laurent G, Sharkey NA, Kohn KW (1980) *Nature 288*:727.
78. Hecht SM (1979) *Bleomycin: Chemical, Biochemical, and Biological Aspects.* Springer-Verlag, New York.
79. Sleigh MJ, Grigg GW (1976) *BJ 155*:87; Sleigh MJ (1976) *NAR 3*:891.
80. Burger RM, Peisach J, Horwitz SB (1981) *JBC 256*:11636; Sakai TT, Riordan JM, Booth TE, Glickson JD (1981) *J Med Chem 24*:279; Kuo MT (1981) *Canc Res 41*:2439.

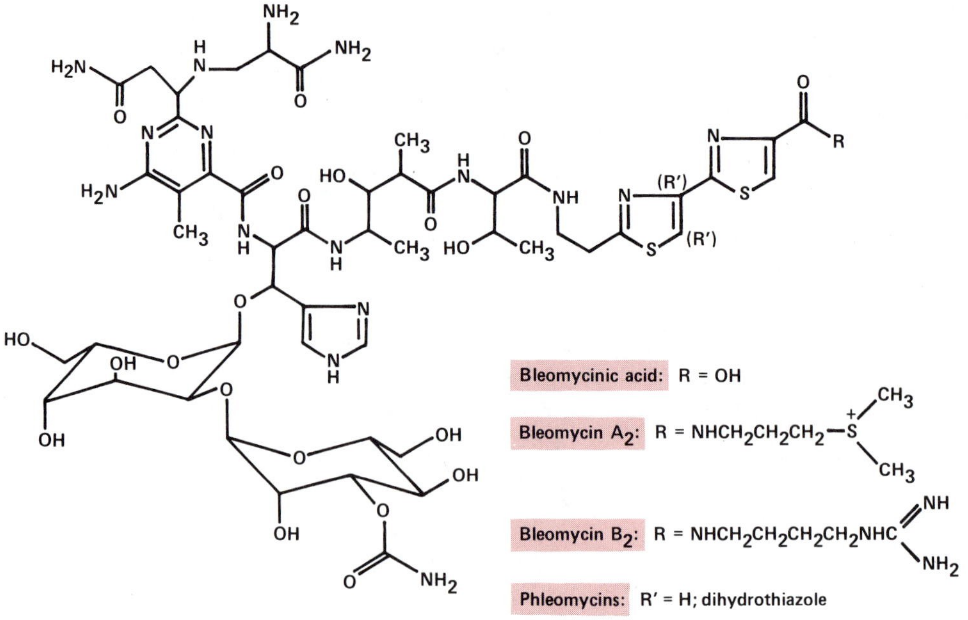

Bleomycinic acid: R = OH

Bleomycin A_2: R = $NHCH_2CH_2CH_2-\overset{+}{S}(CH_3)_2$

Bleomycin B_2: R = $NHCH_2CH_2CH_2CH_2NHC(=NH)NH_2$

Phleomycins: R′ = H; dihydrothiazole

Figure 14-9
Bleomycins: antitumor agents that bind and degrade DNA.

of the bleomycin family exert their cytotoxic activity by cleaving DNA at purine–pyrimidine (GpT and GpC) sequences and pyrimidine–pyrimidine sequences.[81] Interaction with DNA (in vitro) is a two-step process. First, the agent intercalates through its tripeptide cationic terminal region by binding along a potential Cu(II) chelation site localized around the bleomycin pyrimidine moiety. Next, bases, principally pyrimidines, are released by rupture of the N-glycosylic bond; fragmentation of the labilized AP site follows. There are one-tenth as many double-strand breaks as single-strand breaks. Excision–repair at AP sites is induced in *E. coli* and yeast.

Bleomycins, through their nuclease-like actions, can preferentially break DNA sequences in chromatin with a transcriptionally active configuration. Some drug actions are amplified by sulfhydryl and other reducing agents and by a wide range of aromatic compounds, such as acridines, anthracenes, and triphenylmethane dyes. *Phleomycin* increases the melting temperature of DNA; other bleomycins lower it. DNA synthesis is inhibited more strongly than RNA synthesis.

Mitomycin C[82] and *porfiromycin* (Fig. 14-10) inhibit DNA synthesis in cells by interstrand cross-linking of DNA. Either drug is active only when reduced to the semiquinone,[83] and then it acts as a bifunctional alkylating agent between guanine residues. In *E. coli* mutants defective in DNA repair and infected with small DNA phages, synthesis of host DNA but not viral DNA is blocked. Inasmuch as a single lesion inactivates the chromosome, the host DNA offers a target 1000 times greater than that of the phage. Also the host genome is represented by only one or two, rather than many, copies.

81. Takeshita M, Kappen LS, Grollman AP, Eisenberg M, Goldberg IH (1981) *Biochemistry 20*:7599.
82. Borowy-Borowski H, Lipman R, Chowdary D, Tomasz M (1990) *Biochemistry 29*:2992.
83. Borowy-Borowski H, Lipman R, Chowdary D, Tomasz M (1990) *Biochemistry 29*:2992.

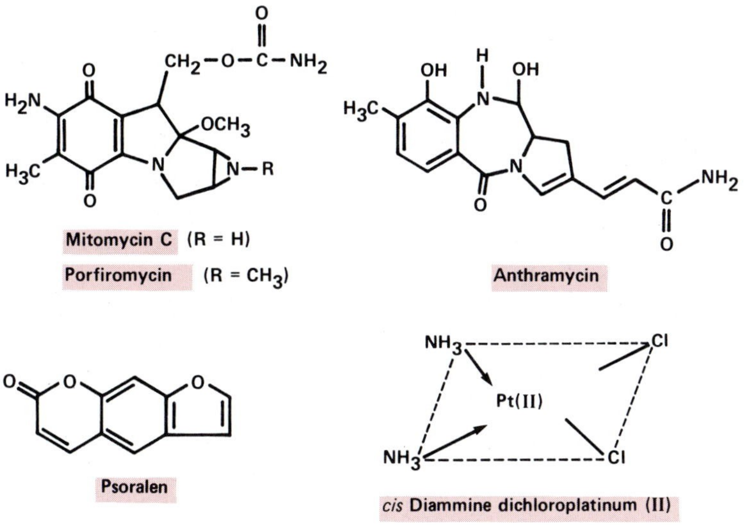

Mitomycin C (R = H)
Porfiromycin (R = CH_3)
Anthramycin
Psoralen
cis Diammine dichloroplatinum (II)

Figure 14-10
Inhibitors that bind and cross-link DNA.

Anthramycin[84] (Fig. 14-10) is an effective antitumor antibiotic that reacts covalently with guanosine in duplex DNA, spanning a 4-bp region; it remains associated with the denatured strands at alkaline pH. It binds the homopolymer pair poly dG·poly dC but not poly dA·poly dT. The drug inhibits both RNA and DNA synthesis, although inhibition of RNA synthesis requires a helical template. The parameters of DNA structure and composition for binding anthramycin have been examined by NMR.[85]

Psoralens[86] (furocoumarins; Fig. 14-10) are linear tricyclic heterocyclics that intercalate in duplex DNA and, when they absorb long-wave UV light, form covalent linkages to pyrimidine bases and cross-link pyrimidines on opposite strands. These compounds have become useful probes in examining the dynamics of DNA and RNA secondary structures as well as promising drugs for controlling proliferative cell growth in psoriasis and mycosis fungoides.[87]

Neocarzinostatin[88] (zinostatin), an antibiotic and antitumor agent from *Streptomyces*, is an acidic protein made up of a single chain of 109 residues. It is a specific inhibitor of DNA synthesis in bacteria and animal cells at concentrations near 10^{-9} M. Its actions in cleaving DNA and releasing free thymine, strongly amplified by sulfhydryl agents, resemble those of the bleomycins. As with bleomycin damage, the DNA with multiple AP sites is susceptible to further fragmentation or to repair.

The apoprotein (acidic polypeptide) of neocarzinostatin has no biological activity, whereas the nonprotein chromophore possesses full activity; the apopro-

84. Hurley LH, Needham-Van Devanter DR (1986) *Accts Chem Res 19*:230; Krugh TR, Graves DE, Stone MP (1989) *Biochemistry 28*:9988.
85. Hurley LH, Needham-Van Devanter DR (1986) *Accts Chem Res 19*:230; Krugh TR, Graves DE, Stone MP (1989) *Biochemistry 28*:9988.
86. Hyde JE, Hearst JE (1978) *Biochemistry 17*:1251.
87. Povirk LF, Goldberg IH (1980) *Biochemistry 19*:4773; Tatsumi K, Bose KK, Ayres K, Strauss BS (1980) *Biochemistry 19*:4767; Goldberg IH, Hatayama T, Kappen LS, Napier MA, Povirk LF (1981) in *Molecular Actions and Targets for Cancer Chemotherapeutic Agents* (Sartorelli AC, Lazo TJ, Bertino JR, eds.). Academic Press, New York, p. 163.
88. Chin D-H, Zeng C, Costello CE, Goldberg IH (1988) *Biochemistry 27*:8106.

tein stabilizes the labile chromophore and may control its release for interaction with DNA. The chromophore ($C_{35}H_{35}NO_{12}$) contains 2,6-dideoxy-2-methylaminogalactose and 2-hydroxy-5-methoxy-7-methylnaphthoate covalently linked to a highly saturated $C_{15}H_{10}O_4$ unit that contains a five-membered cyclic carbonate ring (1,3-dioxolan-2-one).[89]

Neocarzinostatin and the closely related *auromomycin* introduce single-strand breaks in linear duplex or superhelical DNA by damaging the deoxyribose to release free bases, in an oxygen-dependent reaction greatly stimulated by mercaptans. Neocarzinostatin primarily attacks at thymine residues (75% of the total), but also at adenine (19%) and cytosine (6%); auromomycin attacks at guanine (67%), thymine (24%), and adenine (9%).[90] Oxidation of the 5′ carbon of DNA nucleosides to the aldehyde results in a strand break and a DNA fragment bearing a nucleoside-5′-aldehyde at its 5′ end. The main drug action entails the abstraction of hydrogen from the deoxyribose, with the consequent strand breakage and related damages.

Analogous to the polypeptide that stabilizes the active chromophore of neocarzinostatin are the carbohydrate residues of other potent antitumor natural products. Binding of the carbohydrate in the minor groove of DNA directs the aglycone chromophore in its double-strand cleavage activity. Included in this class are *esperamicin* A_1,[91] *daunomycin, A447-C*, and *calichemicin* $\gamma 1$.[92]

cis-*Diamminedichloroplatinum (II)*[93] (cisplatin, *cis*-DDP; Fig. 14-10) is an effective antitumor agent; the *trans* isomer is not. The inhibition of cell division in *E. coli* by neutral platinum coordination compounds (e.g., cis-$Pt(NH_3)_2Cl_2$) led to trials in cancer chemotherapy. For antitumor activity, the compounds must have at least two labile ligands in a *cis* configuration. Covalent binding to DNA is responsible for the cytotoxicity. For an equal number of platinum atoms bound to DNA, *cis*-$Pt(NH_3)_2Cl_2$ inhibits DNA synthesis more efficiently than does the *trans* isomer;[94] cell killing is 5 to 10 times greater, and mutagenesis is at least several hundredfold higher.

In vitro, *cis* and *trans* platinum coordination complexes bind guanine primarily, most likely at N7. Two major *intrastrand* cross-links join two adjacent guanine residues and an adenine adjacent to a guanine. This bifunctional adduct blocks synthesis by *E. coli* pol I and eukaryotic pol α, whereas the monofunctional lesion does not.[95] *Interstrand* cross-links, probably the most lethal lesions, also favor guanine residues and distort the DNA.[96] However, the platination lesion most crucial for antitumor activity is still uncertain.

Inasmuch as the *cis* and *trans* platinum isomers appear to react equally with DNA, the failure of *trans* adducts to accumulate in certain cells is most likely due in part to the greater efficiency with which the *trans* isomer lesion is

---

89. Napier MA, Holmquist B, Strydom DJ, Goldberg IH (1981) *Biochemistry 20*:5602.
90. Takeshita M, Kappen LS, Grollman AP, Eisenberg M, Goldberg IH (1981) *Biochemistry 20*:7599; Povirk LF, Goldberg IH (1982) *PNAS 79*:369; Kappen LS, Goldberg IH, Liesch JM (1982) *PNAS 79*:744.
91. Hawley RC, Kiessling LL, Schreiber SL (1989) *PNAS 86*:1105; Sugiura Y, Uesawa Y, Takahashi Y, Kuwahara J, Golik J, Doyle TW (1989) *PNAS 86*:7672.
92. Sherman SE, Lippard SJ (1987) *Chem Rev 87*:1153; Johnson NP, Lapetoule P, Razaka H, Butour JL (1988) in *Platinum and Other Metal Coordination Compounds in Cancer Chemotherapy* (Nicolini M, ed.). Martinus Nijhoff, Boston, p. 3.
93. Povirk LF, Goldberg IH (1980) *Biochemistry 19*:4773; Tatsumi K, Bose KK, Ayres K, Strauss BS (1980) *Biochemistry 19*:4767; Goldberg IH, Hatayama T, Kappen LS, Napier MA, Povirk LF (1981) in *Molecular Actions and Targets for Cancer Chemotherapeutic Agents* (Sartorelli AC, Lazo TJ, Bertino JR, eds.). Academic Press, New York, p. 163.
94. Bernges F, Holler E (1988) *Biochemistry 27*:6398.
95. Hoffman J-S, Johnson NP, Villani G (1989) *JBC 264*:15130.
96. Schwartz A, Marrot L, Leng M (1989) *Biochemistry 28*:7975.

repaired.[97] The sensitivity of repair-deficient *E. coli* mutants to *cis*-$Pt(NH_3)_2Cl_2$ suggests that the DNA lesions are removed by both UV-repair and SOS mechanisms[98] (Section 21-3). In studies with pol I, the $5'\rightarrow3'$ exonuclease (repair) activity is the only enzyme function that shows a clear distinction between the platinum isomers: the *cis* inhibits the exonuclease while the *trans* does not.

14-5 Inhibitors of Topoisomerases[99]

DNA replication at many stages, as well as other DNA transactions (e.g., recombination, transcription), depends on the topological state of the DNA: superhelical tension, catenation, and knotting (Sections 1-9 and 12-1). Type I and type II topoisomerases (Chapter 12), found in all cells, relax supercoils; the bacterial type II enzyme, gyrase, is unique in its capacity to introduce negative supercoils. Inhibitors of the topoisomerases have proved to be outstanding antibacterial drugs and highly promising as antitumor agents.

The *coumarins*[100] (*novobiocin, coumermycin* A_1, and *chlorobiocin*; Table 14-6 and Fig. 14-11) are related *Streptomyces*-derived antibiotics containing coumarin and sugar moieties. Coumermycin is more effective than novobiocin.

Table 14-6
Inhibitors of topoisomerases

Class of drug and examples	Topoisomerases inhibitied
Coumarins	bacterial gyrase (B subunit), eukaryotic topo II, reverse gyrase, vaccinia type I topo
Novobiocin	
Coumermycin A_1	
Chlorobiocin	
Quinolones	bacterial gyrase (A subunit), phage T4 topo
Nalidixic acid	
Oxolinic acid	
Norfloxacin	
Acridines	eukaryotic topo II, phage T4 topo
Amsacrine (m-AMSA)	
Anthracyclines	eukaryotic topo II
5-Iminodaunorubicin	
Ellipticine	eukaryotic topo II
Epipodophyllotoxins	eukaryotic topo II
VP-16-213 (etoposide)	
VM-26 (teniposide)	
Alkaloids	eukaryotic topo I
Camptothecin	
Amilorides	eukaryotic topo II
Amiloride	
Peptides	bacterial gyrase (B subunit)
Microcin B17	

97. Ciccarelli RB, Solomon MJ, Varshavsky A, Lippard SJ (1985) *Biochemistry 24*:7533; Eastman A, Schulte N (1988) *Biochemistry 27*:4730.
98. Brouwer J, Van de Putte P, Fichtinger-Schepman AMJ, Reedijk J (1981) *PNAS 78*:7010.
99. Drlica K, Franco RJ (1988) *Biochemistry 27*:2253.
100. Gellert M, O'Dea MH, Itoh T, Tomizawa J (1977) *PNAS 74*:4772; Gellert M (1981) *ARB 50*:879; Foglesong PD, Bauer WR (1984) *J Virol 49*:1; Nakasu S, Kikuchi A (1985) *EMBO J 4*:2705.

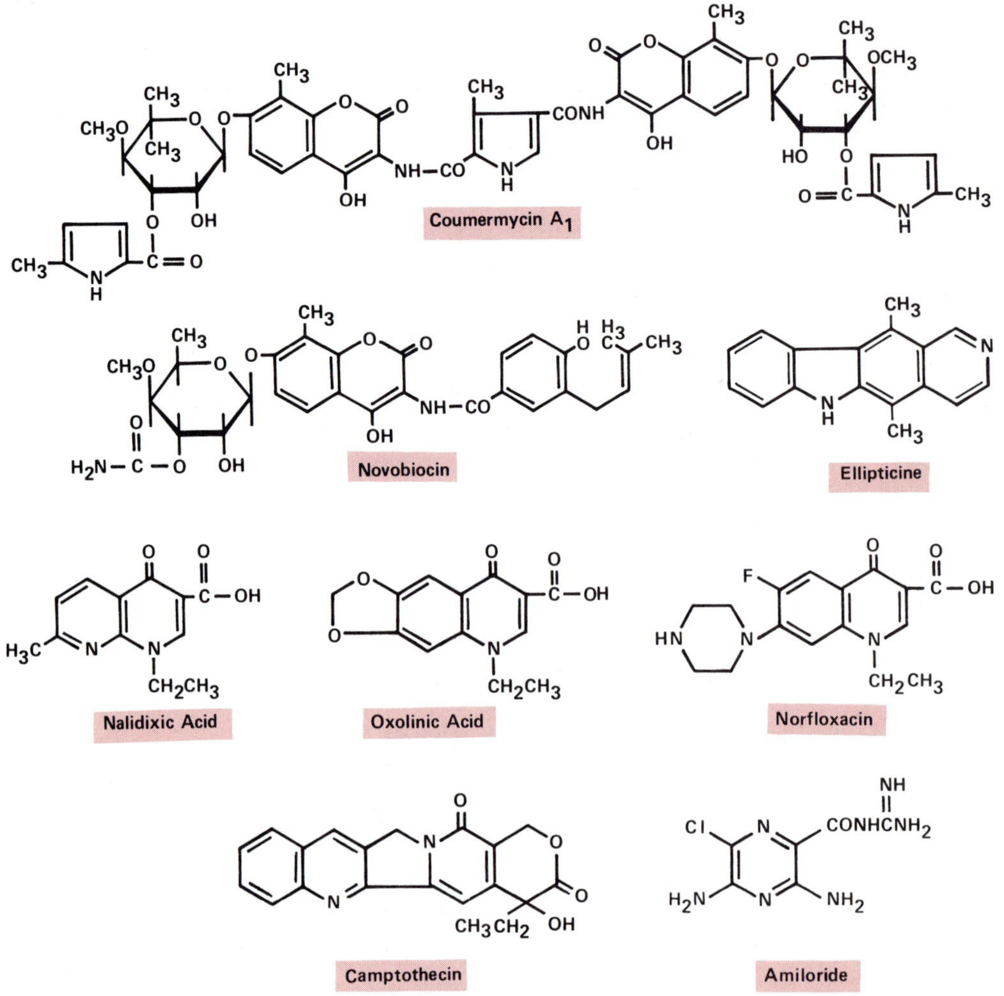

Coumermycin A_1

Novobiocin

Ellipticine

Nalidixic Acid

Oxolinic Acid

Norfloxacin

Camptothecin

Amiloride

Figure 14-11
Inhibitors that bind DNA topoisomerases.

Their inhibitory effects on bacterial growth and multiple DNA functions are due to binding and inhibition of the gyrase B subunit (the *cou* gene product), probably via competition with its ATPase activity. Their action on type II topoisomerases in eukaryotes, where the bulk of the DNA is not torsionally strained, may be at the level of chromatin assembly or chromosome segregation at cell division.

Microcins constitute a family of polypeptide antibiotics secreted by diverse enteric bacteria that inhibit DNA replication[101] in a large number of Gram-negative enteric bacteria. *Microcin B17*, a processed product containing 43 amino acids (3.2 kDa), 26 of which are glycines,[102] interacts with a number of different proteins. The primary effect in *E. coli* is a prompt cessation of DNA synthesis, most likely by the binding and inhibition of the gyrase B subunit.[103]

101. Herrero M, Moreno F (1986) *J Gen Microbiol 132*:393.
102. Davagnino J, Herrero M, Furlong D, Moreno F, Kolter R (1986) *Proteins: Struct Funct Genet 1*:230; Genilloud O, Moreno F, Kolter R (1989) *J Bact 171*:1126.
103. Moreno F, personal communication.

The *quinolones*[104] *nalidixic acid, oxolinic acid,* and *norfloxacin* (Fig. 14-11), listed in the order of their increasing clinical potency, are highly effective synthetic antibacterial agents. They block the gyrase A subunit [the *gyrA* (formerly *nalA*) gene product], the nicking–ligating component of the enzyme. Preferential binding of these drugs to ssDNA may indicate a reaction with a single-stranded pocket of DNA generated during gyrase-induced strand passage[105] (Section 12-3). Lesions in bacterial DNA result from the trapping of the gyrase–DNA reaction intermediate by the drug; addition of a protein denaturant (e.g., sodium dodecyl sulfate) leads to chain scissions, with the gyrase A subunit bound covalently to the 5′ end of each DNA strand. Because they complex the topoisomerase with DNA, rather than simply removing the enzyme activity, these drugs can be regarded as poisons, comparable to the antitumor drugs active against the eukaryotic topoisomerases (see below).

Four categories of eukaryotic type II topoisomerase inhibitors (Table 14-6)—the *acridines* (Fig. 14-7), *anthracyclines, ellipticines* (Fig. 14-11), and *epipodophyllotoxins*—are active as antitumor agents, producing drug–DNA–enzyme "cleavable complexes." These DNA lesions occur preferentially in nascent DNA[106] and are analogous to those observed with the quinolones on bacterial DNA (see above). During SV40 DNA replication in vivo, the type II topoisomerase inhibitors slow replication of the last 5% of the genome and prevent termination of the replicative cycle.[107]

Amilorides[108] (Fig. 14-11) include a class of substituted pyrazines with a guanidino carbonyl moiety that can form planar structures which intercalate into DNA to inhibit topoisomerase II in vitro and in vivo.

Camptothecin (Fig. 14-11), an alkaloid inhibitor of eukaryotic type I topoisomerase, may trap the enzyme in cleavable DNA complexes resembling those produced by the antitumor drugs directed against the type II enzyme.[109] Detergent treatment of the drug–DNA–enzyme complex reveals the protein bound covalently to the 3′ end of the nicked strand. Although the type I topoisomerase is nonessential in yeast, binding of camptothecin turns it into a poison that kills the cell. Strains lacking the enzyme are resistant to the drug, whereas those with elevated enzyme levels are hypersensitive.[110]

14-6 Inhibitors of Polymerases and Replication Proteins

Compared to the large number of antibiotics that bind and inhibit the transcription and translation proteins or the topoisomerases, relatively few are known that bind and inhibit DNA polymerases and other replication proteins (Table 14-7). Because of the key role of transcription in some stages of replication, inhibitors of RNA polymerases are included in this chapter.

---

104. Snapka RM (1986) *Mol Cell Biol 6*:4221; Snapka RM, Powelson MA, Strayer JM (1988) *Mol Cell Biol 8*:515.
105. Shen LL, Kohlbrenner WE, Weigl D, Baranowski J (1989) *JBC 264*:2973; Shen LL, Baranowski J, Pernet AG (1989) *Biochemistry 28*:3879; Shen LL, Mitscher LA, Sharma PN, O'Donnell TJ, Chu DWT, Cooper CS, Rosen T, Pernet AG (1989) *Biochemistry 28*:3886.
106. Woynarowski JM, Sigmund RD, Beerman TA (1988) *BBA 950*:21.
107. Richter A, Strausfeld U, Knippers R (1987) *NAR 15*:3455.
108. Besterman JM, Elwell LP, Cragoe EJ Jr, Andrews CW, Cory M (1989) *JBC 264*:2324.
109. Kjeldsen E, Thomsen B, Mollerap S, Bonven B, Bolund L, Westergaard O (1988) *JMB 202*:333; Hertzberg RP, Caranfa MJ, Hecht SM (1989) *Biochemistry 28*:4629; Kjeldsen E, Bonven B, Andoh T, Ishii K, Okada K, Bolund L, Westergaard O (1988) *JBC 263*:3912.
110. Nitiss J, Wang JC (1988) *PNAS 85*:7501.

Table 14-7
Inhibitors of replication proteins

Inhibitor	Protein bound	Inhibition
Rifampicin Streptovaricin	β-subunit of bacterial RNA polymerase	RNA initiation, DNA initiation
Streptolydigin	β-subunit of bacterial RNA polymerase	RNA chain growth
α-Amanitin	eukaryotic RNA polymerase II	mRNA synthesis
Arylhydrazinopyrimidines HP uracil HP isocytosine	*B. subtilis* pol III	DNA replication
Phosphonoacetic acid Phosphonoformic acid	Herpes and vaccinia DNA polymerase	DNA replication
Aphidicolin	Eukaryotic pol α, pol δ	DNA replication
Butylphenyl dGTP Butylanilino dNTP	Eukaryotic pol α	DNA replication
Edeine	?	DNA synthesis

Inhibitors of RNA Polymerase (Fig. 14-12)

The *ansamycin* antibiotics[111] (which contain a ring that bridges parts of the chromophore; Fig. 14-13) include the rifamycins, the streptovaricins, streptolydigin, and numerous other natural compounds of related structure.

Rifamycins[112] are a group of *Streptomyces* antibiotics, of which one of the derivatives, *rifampicin* (rifampin; Fig. 14-12), specifically inhibits RNA synthesis in prokaryotes. The drug binds to the β subunit of *E. coli* RNA polymerase; mutations that produce resistance to the drug are within the *rpoB* gene, which encodes this subunit. Some replication systems (M13, ColE1) are inhibited because RNA polymerase provides the primer for DNA replication. Rifampicin blocks the formation of the first or second phosphodiester bond, but prevents neither the binding of RNA polymerase to the DNA template nor the propagation of RNA chains (Section 7-4).

Mitochondrial RNA synthesis in yeast and in virtually all eukaryotes is insensitive to rifampicin but not to a hydrophobic octyloxime derivative, *rifampicin AF-013*. This and other hydrophobic derivatives of rifampicin (at the position indicated by color in Fig. 14-12) also inhibit RNA-directed DNA synthesis by tumor viral polymerases. At high concentrations, these derivatives inhibit not only certain mammalian RNA polymerases, but also several other related enzymes, and their value as specific reagents is therefore questionable.

Streptovaricin,[113] a drug similar in origin to rifampicin, shows the same specificity and mode of action. *Streptolydigin*[114] (Fig. 14-12) also binds the β subunit of

111. Wehrli W (1977) *Top Curr Chem* 72:21.
112. Wehrli W (1977) *Top Curr Chem* 72:21.
113. Wehrli W (1977) *Top Curr Chem* 72:21.
114. Wehrli W (1977) *Top Curr Chem* 72:21.

Rifampicin

Streptolydigin

α, R = NH_2; β, R = OH

Amanitin

Figure 14-12
Inhibitors that bind RNA polymerases.

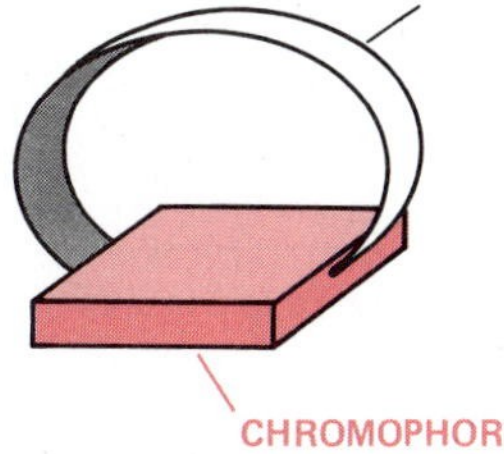

Figure 14-13
The flat aromatic nucleus and the aliphatic bridge, shaped like a handle (Latin, ansa), that joins two nonadjacent parts of the chromophore in the ansamycin antibiotics (which include the rifamycins shown in Fig. 14-12).

RNA polymerase but, unlike rifampicin, preferentially blocks RNA chain elongation.

The sulfur-containing bicyclic polypeptide *α-amanitin*[115] (Fig. 14-12), a potent poison[116] from the mushroom *Amanita phalloides*, specifically inhibits mammalian mRNA synthesis by binding RNA polymerase II. The semisymmetrical structure of α-amanatin may fit into a symmetrical site in pol II or in a pol II–DNA complex. RNA pol III, responsible for the synthesis of tRNA and 5S RNA, is relatively insensitive, and pol I and mitochondrial RNA polymerase are virtually unaffected even by very high levels.

Inhibitors of DNA Polymerase (Fig. 14-14)

Arylhydrazinopyrimidines[117] (arylazopyrimidines; Fig. 5-13) are purine dNTP analogs that produce unusually strong and specific inhibition of *B. subtilis* DNA polymerase III in vivo and in vitro. They bind the enzyme and DNA in a ternary complex despite their lack of sugar or phosphate groups. As discussed earlier (Section 5-7), the mechanism apparently involves base pairing with template pyrimidines. *6-Anilinopyrimidines, 6-(benzylamino)pyrimidines*, and specific *N^2-(arylamino)purines*[118] designed on the basis of the arylhydrazinopyrimidine prototype[119] are also potent inhibitors of DNA pol III and function by a similar mechanism. These analogs have been most useful in identifying the replication patterns and DNA polymerases of *B. subtilis*, and encourage a search for inhibitors of other polymerases and replication proteins.

Phosphonoacetic acid and *phosphonoformic acid*[120] (Fig. 14-14), despite their simple structures, are effective antiviral drugs. They selectively inhibit the DNA polymerases encoded by herpes simplex and vaccinia viruses. The viral polymerases are generally over 100 times more sensitive than the host cell enzymes, but the margin may be less in some cells and tissues. Evidence for specificity against viral polymerase is the 200-fold increase in the concentration

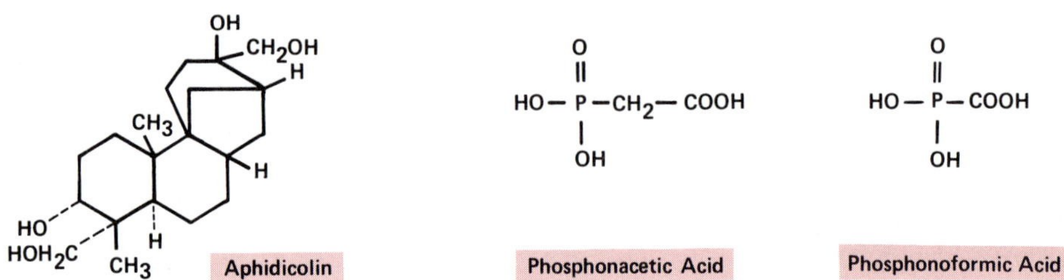

Figure 14-14
Inhibitors that bind DNA polymerases.

115. Kersten H, Kersten W (1974) *Inhibitors of Nucleic Acid Synthesis*. Springer-Verlag, New York; Corcoran JW, Hahn FE, eds. (1975) *Antibiotics*. Springer-Verlag, New York, Vol. 3.
116. Wieland T, Faulstich H (1978) *Crit Rev B* 5:185.
117. Oberg B (1989) *Pharmacol Ther* 40:213; Brown NC, Dudycz LW, Wright GE (1986) *Drugs Exp Clin Res* 12:555; Wright GE, Brown NC (1990) *Pharmacol Ther*, 47:447.
118. Brown NC, Dudycz LW, Wright GE (1986) *Drugs Exp Clin Res* 12:555; Wright GE, Brown NC (1990) *Pharmacol Ther*, 47:447.
119. Brown NC, Dudycz LW, Wright GE (1986) *Drugs Exp Clin Res* 12:555; Wright GE, Brown NC (1990) *Pharmacol Ther*, 47:447.
120. Oberg B (1989) *Pharmacol Ther* 40:213.

of drug needed to inhibit drug-resistant herpes mutants. These phosphonic acids operate as analogs of inorganic pyrophosphate, occupying an essential site on the enzyme (Sections 6-8 and 19-5).

Aphidicolin (Fig. 14-14), a tetracyclic diterpenoid antibiotic, inhibits replicative eukaryotic DNA polymerases[121] (α and δ polymerases of yeast and animals, viral-encoded polymerases, and α-like polymerases of plants); it does not affect the β and γ polymerases. Using aphidicolin as a selective reagent, aphidicolin-resistant DNA pol α mutants have been isolated from *Drosophila*[122] and other cultured cells.[123] Aphidicolin is a strong competitor of dCMP incorporation, less so of dTMP, and shows little if any competition with dAMP and dGMP.[124] Aphidicolin inhibits DNA replication of phage ϕ29 but not of the *B. subtilis* host or other *B. subtilis* phages; its action is on the ϕ29-encoded phage DNA polymerase.[125]

N(2)-(Butlyphenyl)dGTP and *N(2)-butylanilino)dATP*, designated *BuPdGTP* and *BuAdATP*, respectively, are analogs specific for DNA pol α. They were designed on the basis of the structure of the *B. subtilis* pol III–specific arylhydrazinopyrimidines.[126] With a potency 100 times that of aphidicolin, these inhibitors are the most highly selective of the pol α–dNTP analogs.

Edeines[127] (Fig. 14-15) are basic, linear, oligopeptide antibiotics from *Bacillus brevis*. They contain spermidine in addition to some novel amino acids. Their rather selective inhibition of DNA synthesis, without evidence of DNA binding, suggests a specific interaction with DNA polymerases or other essential replica-

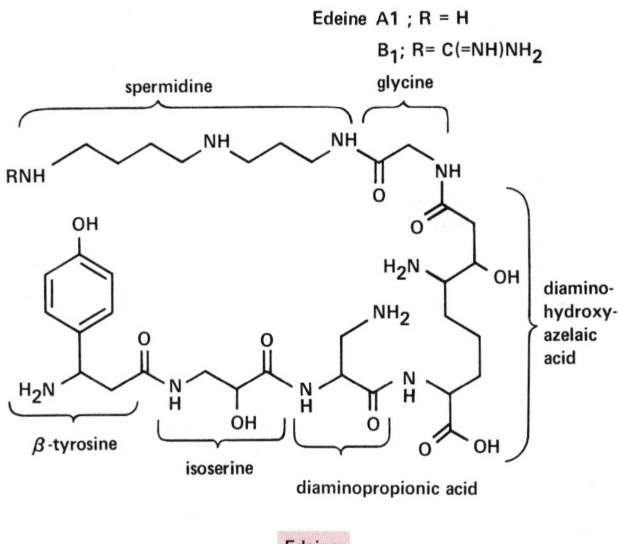

Figure 14-15
An inhibitor of DNA synthesis with unknown locus of action.

121. Liu PK, Chang C-C, Trosko JE, Dube DK, Martin GM, Loeb LA (1983) *PNAS 80*:797.
122. Sugino A, Nakayama K (1980) *PNAS 77*:7049.
123. Nishimura M, Yasuda H, Ikegami S, Ohashi M, Yamada M (1979) *BBRC 91*:939.
124. Chang C-C, Boezi JA, Warren ST, Sabourin CLK, Liu PK, Glatzer L, Trosko JE (1981) *Somatic Cell Genet 7*:235; Krokan H, Wist E, Krokan RH (1981) *NAR 9*:4709.
125. Blanco L, Salas M (1986) *Virology 153*:179; Bernad A, Zaballos A, Salas M, Blanco L (1987) *EMBO J 6*:4219.
126. Brown NC, Dudycz LW, Wright GE (1986) *Drugs Exp Clin Res 12*:555; Wright GE, Brown NC (1990) *Pharmacol Ther 47*:447.
127. Oberg B (1989) *Pharmacol Ther 40*:213.

tion proteins. Further investigations have been held up by the limited availability of purified edeines.

Other antibiotics, such as *ficellomycin*,[128] *feldamycin*,[129] and *sarubicin*,[130] are known to inhibit DNA replication, but because their toxicity to animals has discouraged their use as drugs they have become unavailable for experimental studies. These agents, and presumably many others abandoned by pharmaceutical companies for lack of clinical promise, might prove as useful in replication research as have α-amanitin and aphidicolin.

Postreplicational Modifications

Except in some phage systems which synthesize alternative precursors, modifications of DNA and RNA are made after the chain has been assembled. As the functional significance of these modifications becomes clearer, there is greater interest in the development of agents that affect these processes. For example, the methylation of certain cytosine residues in mammalian DNA is likely to have an important influence on gene activation and cell differentiation (Section 21-9).[131] The methyl donor is invariably S-adenosylmethionine (AdoMet; Fig. 14-16), and compounds that affect either (1) methyl transferases, (2) hydrolytic enzymes that remove inhibitory products, or (3) the enzymes of AdoMet regeneration can have major consequences. AdoMet is also the source of the propyl-

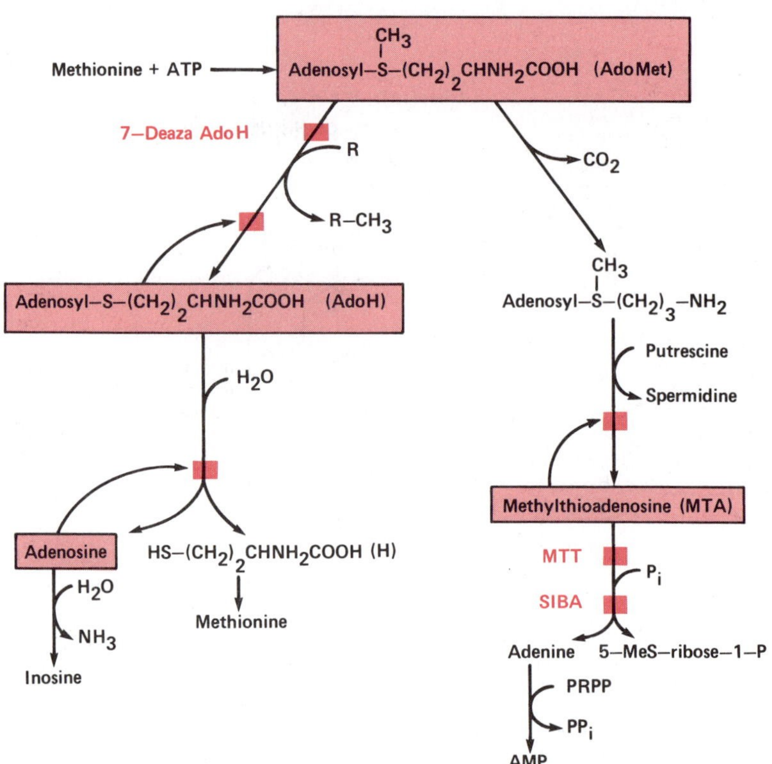

Figure 14-16
Pathways of S-adenosylmethionine utilization and the inhibitors that affect them. MTT = 5′-deoxy-5-methylthio-tubercidin; SIBA = S-isobutyladenosine.

128. Reusser F (1977) *Biochemistry 16*:3406.
129. Reusser F (1977) *Biochemistry 16*:3406.
130. Reinhardt G, Bradler G, Eckardt K, Tresselt D, Ihn W (1980) *J Antibiot 33*:787; Slechta L, Chidester CG, Reusser F (1980) *J Antibiot 33*:919.
131. Jones PA, Taylor SM (1980) *Cell 20*:85; Jones PA, Taylor SM (1981) *NAR 9*:2933.

amine moiety in the biosynthesis of spermidine and spermine, the polyamines with pervasive influence on nucleic acid functions and cellular metabolism.[132]

The cytidine analog *5-azacytidine* (Sections 2 and 3) and other analogs modified in the methyl acceptor position of the pyrimidine ring induce undermethylation of mammalian DNA, presumably by inhibiting the methyl transferase. Three related analogs, *5-aza-2′-deoxycytidine, pseudoisocytidine,* and *5-fluoro-2′-deoxycytidine,*[133] are thought to exert similar effects, as do *3-deazaadenosine* and *ethionine* (Fig. 14-17).[134] Ethionine, long known to be the cause of liver cancer and an antagonist of methionine, may exert its carcinogenic effect through the alteration of DNA methylation patterns.

The products of methyl and propylamine transfers are, respectively, S-adenosylhomocysteine (AdoH) and 5′-deoxy-5-methyl-thioadenosine (MTA). These products inhibit the respective group transfer reactions but they are actively

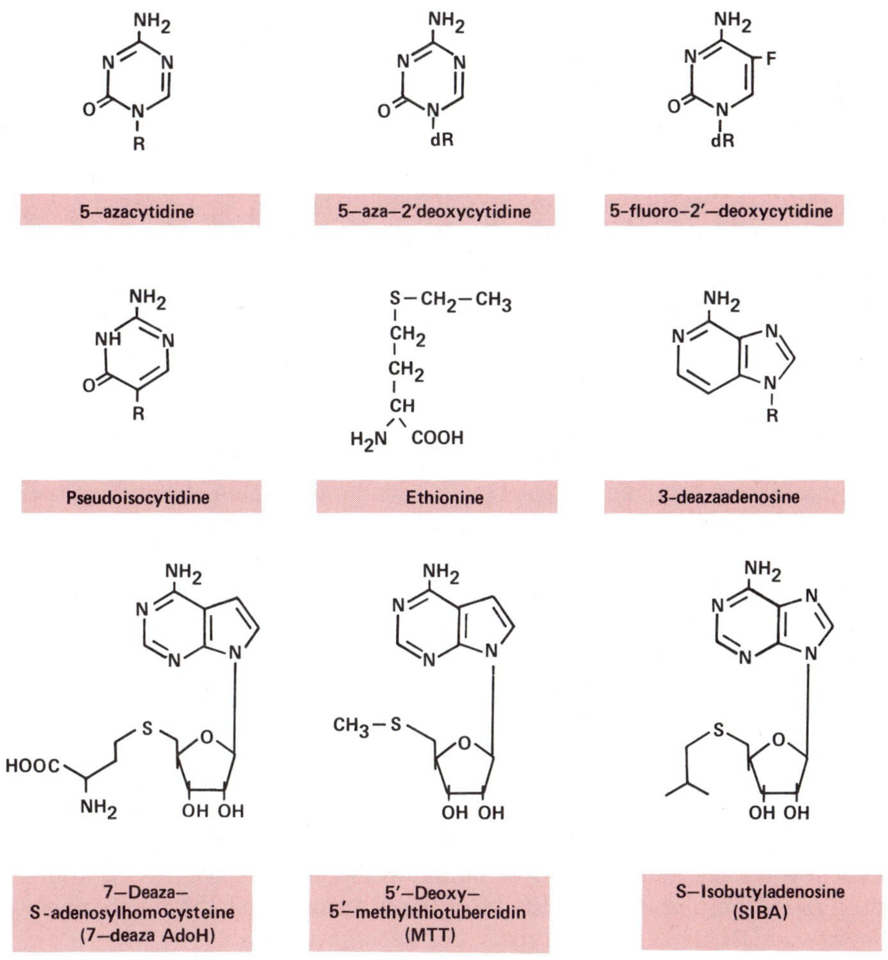

Figure 14-17
Inhibitors that affect nucleic acid methylation.

132. Heby O (1981) *Differentiation 19*:1.
133. Jones PA, Taylor SM (1980) *Cell 20*:85; Jones PA, Taylor SM (1981) *NAR 9*:2933.
134. Southern EM (1975) *JMB 98*:503; Maniatis T, Jeffrey A, Kleid DG (1975) *PNAS 72*:1184.

catabolized (Fig. 14-16). Some metabolically stable synthetic analogs of AdoH and MTA have been prepared.[135] *7-Deaza-S-adenosylhomocysteine* (7-deaza AdoH) is a potent inhibitor of AdoMet-dependent methylases and decreases the level of methylation of tRNAs and mRNAs in animal cells. *5′-Deoxy-5-methylthiotubercidin* (MTT) and *5′-deoxy-5′-S-isobutylthioadenosine* (SIBA) (Fig. 14-17) are inhibitors of MTA phosphorylase.

14-7 Radiation Damage of DNA[136]

UV light and the ionizing radiations of x-rays and gamma rays affect rapidly growing tissues more than adult tissues. Replication and cell division are profoundly inhibited (Section 21-1) because DNA is the most sensitive of the biological components affected by radiation. Despite enormous effort devoted to this important class of inhibitors of replication, chemical mechanisms for their actions on DNA in vivo are often ill-defined.

Dimerization of adjacent pyrimidines, particularly thymine, is commonly regarded as the major effect of UV irradiation. Where such dimers exist in DNA, base pairing is impossible, and unless the DNA is repaired (Chapter 21), all of its functions are interrupted or distorted. Substitution of bromouracil for thymine enhances the sensitivity to damage by UV light of 313 nm. Intercalated psoralens (Section 4) become covalently linked to strands and form cross-links between strands upon absorption of long-wave (347-nm) UV light. UV damage to purines has received less attention because of the greater ease of monitoring pyrimidine lesions, yet this focus on pyrimidines may obscure highly significant lesions sustained in vivo by the chemically vulnerable purines.

The gross effects of ionizing radiations are (1) arrested or abnormal mitoses, (2) chromatid breaks, and (3) increased mutation rates. The damage to DNA occurs in many ways (Section 21-1). Adducts of hydroxyl and peroxide radicals alter the pyrimidines and purines, cause single-strand and double-strand breaks in the phosphodiester backbone, and cross-link DNA strands to each other and to adjacent proteins. It is difficult to determine the concentrations and lifetimes of the radicals responsible for damage and how they are affected by levels of oxygen and radical scavengers in the cell. It is difficult also to assess what fraction of strand breakage is the direct result of irradiation rather than the secondary result of enzymatic excision of damaged bases and sections of the phosphodiester backbone to which they were attached.

Because of the numerous variables in the direct and indirect effects of ionizing radiations on DNA, extrapolations from in vitro and in vivo responses remain tenuous, and so must inferences drawn from studies with a given cell under certain conditions when applied to another cell, even under similar circumstances. Beyond the complexities of radiation chemistry, progress in understanding radiation biology depends heavily on advances in the biochemistry and biology of DNA replication, repair, and recombination.

---

135. Parks RE, Stoeckler JD, Cambor C, Savarese TM, Crabtree GW, Chu SH (1981) in *Molecular Actions and Targets for Cancer Chemotherapeutic Agents* (Sartorelli AC, Lazo JS, Bertino JR, eds.). Academic Press, New York, p. 229; Kaehler M, Coward J, Rottman F (1979) *NAR* 6:1161.
136. Ward JF (1975) *Adv Rad Biol* 5:181; Elkind MM, Redpath JL (1977) in *Cancer*. Plenum Press, New York, Vol. 6, p. 51; Friedberg EC (1985) in *DNA Repair*. Freeman, New York, p. 40.

CHAPTER 15

Replication Mechanisms and Operations

15-1 Basic Rules of Replication[1]

One of the major revelations of this century is the universality of biochemistry. First observed in the near-identity of the pathways and mechanisms of alcohol fermentation in yeast and glycolysis in human muscle, this insight was reinforced by the discoveries of the intricate patterns of biosynthesis of amino acids, fatty acids, and nucleotides. It could have been expected that the devices for polymerizing these precursors would also be preserved throughout evolution. In fact, the replication of *E. coli* and its phages and plasmids has proven prototypical. The variations on these basic themes in other prokaryotes and in eukaryotes are endlessly fascinating.

The basic rules:

(1) *Replication is a semiconservative process.*[2] During replication the parental DNA duplex is not conserved as such, but its strands are maintained (with infrequent scissions)[3] and copied by base pairing with complementary nucleotides. The result of replication is two progeny duplexes identical to each other and to the parental duplex. When cells, whether bacterial, plant, or animal, containing DNA labeled with ^{15}N are shifted to a medium containing $^{14}NH_4Cl$,

1. Moses RE, Summers WC (eds.) (1988) *DNA Replication and Mutagenesis.* Am Soc Microbiol, Washington DC; McMacken R, Kelly TJ (eds.) (1987) *DNA Replication and Recombination.* UCLA, Vol. 47. AR Liss, New York; Richardson CC, Lehman IR (eds.) (1990) *Molecular Mechanisms in DNA Replication and Recombination.* UCLA, Vol. 127. AR Liss, New York.
2. Meselson M, Stahl FW (1958) *PNAS 44*:671.
3. Strayer DR, Boyer PD (1978) *JMB 120*:281.

the progeny duplexes after one generation are hybrid in buoyant density; that is, one strand contains the heavy isotope, the other the light (Fig. 15-1). Progeny duplexes after a second generation are hybrid in density or fully light, in equal numbers.

(2) *Initiation of replication occurs at specific sequences, called origins* (Chapter 16). In a rapidly growing bacterium, reinitiation at the same origin can occur before the first round of replication is complete. Thus the rate of replication is increased by the device of a multiforked chromosome. In eukaryotic chromosomes, replication begins at many origins located at different positions along the chromosomal DNA; reinitiations do not occur during a single cell cycle. Because of the large number of simultaneously active forks, the overall rate of replication in eukaryotes may be faster than in bacteria, even though the rate of movement of each replicating fork is much slower.

(3) *Control is generally at initiation of replication* (Section 16-3). Rates of chain elongation and intervals of chromosome segregation are relatively constant.

(4) *Fork movement is uni- or bidirectional.* Replication proceeds from the origin site either in one or in both directions sequentially to the terminus. Bidirectional movement, involving two growing points or forks emanating from a common origin, appears to be the more common (Fig. 8-1). The movement of one or two forks away from the origin creates a loop which appears as a "bubble" or "eye" or theta-like ("Cairns") structure in electron micrographs. The speed or extent of movement of the two forks is not necessarily the same in both directions.

(5) *Strands are elongated in the 5′→3′ direction by the addition of nucleotide monomers.* Replication by DNA polymerases (Section 3-4) does not proceed simultaneously along both strands; near the fork, a transient single-stranded region is created on one side of the duplex while the other (antiparallel) strand is being replicated. Although replication appears to be simultaneous on both strands when measured by autoradiography or genetic techniques, these methods resolve the process only at the micron (10^{-6} m) level, whereas the polymerization of nucleotide monomers proceeds at the angstrom (10^{-10} m) level. Electron microscopy, with a resolving power of about 100 nucleotides

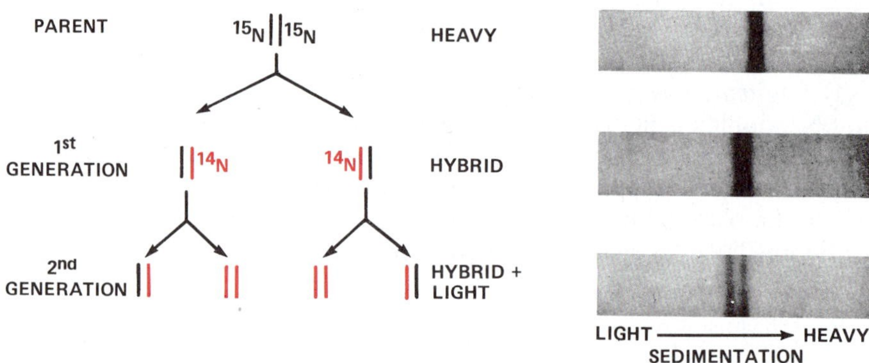

Figure 15-1
Semiconservative replication of *E. coli* DNA. ^{15}N (heavy) parental, ^{14}N (light) progeny, and hybrid first-generation DNAs are separated by sedimentation in a cesium chloride equilibrium density gradient.

(3×10^{-8} m), has clearly shown single-stranded regions at one side of the growing fork.

(6) *Replication is semidiscontinuous for the most part: relatively continuous on one (the leading) strand and discontinuous on the other (lagging) strand* (Fig. 8-1). The short fragments of the nascent discontinuously synthesized strand are in the range of 100 to 200 ntd long (4S to 5S) in animal cells and 1000 to 2000 ntd (8S to 10S) in prokaryotes. These fragments are later joined, so that *both* newly synthesized strands appear uninterrupted. Either strand of the chromosome may support continuous or discontinuous replication.

(7) *Short fragments are started by a short segment of RNA to serve as a primer for DNA polymerase.* This initiator RNA is later excised, and the resulting gap is filled with DNA. The nascent pieces (called *Okazaki fragments*) are then joined to the growing chromosome.

(8) *Various mechanisms exist for starting a DNA strand.* In addition to RNA priming, other devices include covalent attachment of the DNA strand to a terminal protein and covalent extension of a nick or at a looped end of a parental strand.

(9) *Termination may be at a fixed point in replication.* In *E. coli* and *B. subtilis* chromosomes and in some plasmids, the termination point is defined by a *ter* (terminus) region that inhibits replication forks (Section 12). However, a unique termination site is not essential. For the circular chromosomes of phage λ and SV40, which lack a *ter* region, replication terminates simply at the meeting place of two forks moving bidirectionally from the origin at about equal rates.

(10) *Replication mechanisms depend on genome structure and conformation to ensure production of complete chromosomes.* Redundancy of DNA sequence, a keynote of prokaryotic chromosomes, takes several forms: a terminal duplex, as in phage T7 (Section 17-5), a terminal single strand, as in phage λ (Section 17-7), concatemeric duplication in tandem of the entire genome, as in several viral replicative forms (Sections 17-5 and 19-5), and circularity of the genome. These structural features ensure that replication at terminal, as well as interior, regions produces a complete, high-fidelity copy. *The conformation of the chromosome, particularly the degree of superhelicity and susceptibility to melting and bending,*[4] *influenced by many parameters, may be crucial in every stage of replication.*

(11) *Multiple replication mechanisms may operate even within a single cell.* Choices among alternative replication pathways may affect where a cycle of replication starts; the mechanism of initiation; aspects of chain elongation that influence continuity, velocity, processivity, and fidelity; and the means for completing and segregating the daughter chromosomes. Circumstances such as relative abundance of enzymes, temperature, and type of nutrients determine the choices made.

15-2 Replication Fork: Origin, Direction, Structure[5]

The replication fork marks the advance of semiconservative replication of a duplex chromosome. It is part of either a replication bubble generated by a

---

4. Travers AA (1990) *Cell 60*:177.
5. Schnös M, Inman RB (1970) *JMB 51*:61; Schnös M, Inman RB (1971) *JMB 55*:31; Gyurasits EB, Wake RG (1973) *JMB 73*:55; Wake RG (1973) *JMB 77*:569; Wolfson J, Dressler D, Magazin M (1972) *PNAS 69*:499; Wolfson J, Dressler D, Magazin M (1972) *PNAS 69*:998.

primed start or a rolling circle generated by a covalent start. Examples of such structures (Fig. 16-1) are theta structures, displacement loops, bubble (eye) and fork structures, and sigma or lariat forms.

Origins

Although origins are considered in detail in Chapter 16, they are mentioned briefly here because the replication forks emanate from them. Among the techniques employed to identify and analyze origins and forks are autoradiography, electron microscopy, denaturation mapping, gel electrophoresis, and genetic analysis. Advances in pulsed-field gel electrophoresis[6] and amplification of origins and forks by the polymerase chain reaction (PCR; Section 4-17) have added major dimensions to these techniques.

These techniques identify replication origins as the starting sites of DNA synthesis. Another definition of a replication origin is that region of the genome near which the replication fork develops, as discussed in Chapter 16.

Conditionally lethal *E. coli* mutants defective in chromosome initiation (Section 16-3) can be rescued if the origin of another genome (e.g., that of phage P2 or F plasmid) is integrated somewhere in the host chromosome.[7] This phenomenon is called *integrative suppression.*

Structure and Movement of the Fork[8]

In eukaryotic chromosomes,[9] numerous starts generate loops which are seen as bubbles along the DNA fiber. These replicating units *(replicons)* are 20 to 300 kb in size, as judged by the number of starts in a length of chromosome or the spacing between origins, and are generally closer to the size of viral than bacterial genomes. Whereas the rate of fork movement is slow (0.5 to 5 kb per minute) compared to that in prokaryotes (near 100 kb per minute), simultaneous functioning of many replicating units compensates, allowing the entire genome to be replicated in an hour or less. The *Drosophila* cleavage embryo, for example, duplicates its DNA in 3 minutes, whereas *E. coli* requires 40 minutes for a genome only one-fortieth as long.

Replication in the $5'\rightarrow3'$ direction on both strands of a duplex requires the transient presence in the discontinuously synthesized strand of a single-stranded region near the fork. By an electron microscopy technique that clearly distinguishes ssDNA from dsDNA, single-stranded template regions have been seen at replicating forks in *Drosophila*, λ phage, and T4 phage. These regions are almost invariably on only one strand of the fork. In λ, the regions range in length up to 0.4 μm, suggesting that this single-stranded unit is approximately 1000 base pairs in size. The size in T7 is somewhat larger. The signals for initiation of DNA replication along these single-stranded stretches are determined in part by the sequence specificity of the primase, but other factors are commonly involved.

Complexities may be introduced into the structure of the growing fork during

---

6. Ohki M, Smith CL (1989) *NAR 17*:3479.
7. Lindahl G, Hirota Y, Jacob F (1971) *PNAS 68*:2407.
8. Inman RB, Schnös M (1971) *JMB 56*:319; Wolfson J, Dressler D (1972) *PNAS 69*:2682; Kriegstein HJ, Hogness DS (1974) *PNAS 71*:135.
9. Edenberg HJ, Huberman JA (1975) *ARG 9*:245.

isolation of the DNA. For example, in the electron microscope, replicating molecules of phages P2 and T4 and *Drosophila* show an extra single strand (a "whisker") protruding at the fork. This single strand is very likely the nascent leading strand displaced by the reannealing of the parental helix strands *(branch migration).* Rather than its being an obligatory stage in replication, branch migration to an extent that would displace nascent strands and interfere with replication is probably prevented by some mechanisms in vivo. Replicating DNA of sea urchin embryos has numerous microbubbles[10] less than one-tenth the size of the bubbles (eyes) in *Drosophila.* These may reflect frequent starts that have not yet coalesced.

The DNA growing point, identified in the electron microscope or by pulse-labeling, is consistent with replication being *semidiscontinuous;* that is, synthesis is continuous in one strand, generally called the *leading strand,* and discontinuous in the other, generally called the *lagging strand.* The leading and lagging designations assume that synthesis of the leading strand opens the duplex, thereby providing the template upon which subsequent synthesis of the lagging strand depends. However, opening of the duplex is usually effected by a helicase, which may be associated with the discontinuous-strand functions. Thus, synthesis of the lagging strand may even *precede* that of the leading strand.

15-3 Semidiscontinuous Replication[11]

Nascent Short Fragments as Replication Intermediates

The most recently synthesized, or nascent, DNA molecules in preparations of replicating phage T4 are short pieces *(Okazaki fragments).* They sediment at about 8S to 10S in a gradient of alkaline sucrose, representing chain lengths of 1000 to 2000 residues. Nascent DNA in a large variety of bacterial viruses and prokaryotic forms consists of short fragments of this general size; in eukaryotic cells[12] it consists of smaller 4S fragments (100 to 200 ntd).

To capture the T4 replication pieces before they join the main body of growing chains, it is necessary to (1) reduce the replication rate by lowering the growth temperature (e.g., to 8°), (2) use a brief pulse of a DNA precursor of high specific radioactivity that enters DNA rapidly and directly (e.g., [^{3}H]thymidine, which enters DNA in 5 to 60 seconds), and (3) quench the pulse efficiently. Under these conditions, much of the precursor label is captured as small pieces. After a prolonged pulse, however, or after subsequent exposure to large concentrations of unlabeled precursors (a "chase"), the radioactive precursor is found exclusively in high-molecular-weight DNA. Mutant *E. coli* cells deficient in ligase or in pol I, or in both, accumulate large amounts of *E. coli* replication fragments.

From these early experiments and from others with different phages and cells, it had been inferred that both strands of the duplex are synthesized as short

---

10. Baldari CT, Amaldi F, Buongiorno-Nardelli M (1978) *Cell 15*:1095.
11. Okazaki R, Okazaki T, Hirose S, Sugino A, Ogawa T, Kurosawa Y, Shinozaki K, Tamanoi F, Seki T, Machida Y, Fujiyama A, Kohara Y (1975) in *DNA Synthesis and Its Regulation* (Goulian M, Hanawalt P, Fox FC, eds.). Benjamin, Menlo Park, New Jersey, p. 832; Kurosawa Y, Ogawa T, Hirose S, Okazaki T, Okazaki R (1975) *JMB 96*:653.
12. Blumenthal AB, Clark EJ (1977) *Cell 12*:183.

fragments, that is, discontinuously. Sequential labeling by two different isotopes, followed by nuclease digestion, showed the direction of synthesis of the fragments to be 5′→3′. Moreover, finding RNA linked to DNA at the 5′ ends implied that at least some fragments are products of de novo initiation.

However, evidence that short fragments are truly nascent replication intermediates is difficult to establish and can also be misleading (see below). The fragment may arise from the breakdown of a continuously synthesized strand, as well as being an Okazaki fragment. Correction for the loss of the 5′ RNA end in vivo or during isolation is a serious problem. The use of appropriate mutants, in vitro systems, and refinements of technical procedures have helped to solve most of these difficulties. It seems likely that, in all systems, initiations are limited to one of the two daughter strands at the fork and that the semidiscontinuous model best accounts for kinetic observations as well as for the DNA growing point seen in the electron microscope.

Problems in Demonstrating Semidiscontinuous Replication

Excision of uracil incorporated into DNA is a source of "pseudo"-Okazaki fragments. Incorporation of dUMP into nascent DNA, in place of dTMP, can give rise to pseudo-Okazaki fragments. Because the cell recognizes uracil as foreign in DNA and cleaves the polynucleotide chain to remove it, the growing strand may exist in a fragmented state in the interval between the excision step and the completion of repair synthesis (Section 4). Uracil excision is largely absent in *E. coli* mutants defective in uracil N-glycosylase (*ung*$^-$) (Section 13-9). Conversely, exaggerated production of small Okazaki fragments (100 to 200 ntd) occurs in *dut* mutant cells: an elevated dUTP concentration due to a reduced level of dUTPase results in the overincorporation of uracil into DNA during replication (Section 10). Introduction of the *ung* mutation into *dut* mutants suppresses the excision and repair of uracil-containing sites. The *ung* mutation does not, however, eliminate replication fragments completely; about half the label still appears in 8S to 10S fragments. This result argues that about 50% of the replication fragments normally observed in wild-type cells are true intermediates in discontinuous replication (unless an excision system other than uracil N-glycosylase is responsible for part or all of them).

Whereas in vitro systems may suffer from the lack of an essential component, they may gain from the absence of an interfering one. Such is the case with uracil incorporation and repair. In the cellophane-disk lysate system (Section 4), excision–repair reactions that fragment DNA are less active, and the analysis of replication intermediates is facilitated. In pulse experiments with this system, 50% of the nascent DNA appears as 10S fragments and the remainder as larger (30S) DNA. This result can be taken to mean that replication is continuous on one strand and discontinuous on the other. Fragment size and number can be manipulated by varying the ratios of dUTP to dTTP, the levels of dUTPase (by using the *dut* mutation), or the capacity for uracil excision (by using the *ung* mutation).

The influence of physiologic states and drug treatments, as well as mutations, on uracil incorporation into DNA in vivo can be profound. Likewise, the content and balance of enzymes and dNTPs in complex systems in vitro affect uracil incorporation with significant consequences. These factors will be considered further in Section 10.

RNA linked to DNA is rapidly removed in normal cells. Quantitative determination of the initiating nucleotides (RNA primers) at the 5′ end of nascent DNA chains proves that these chains indeed arise from de novo initiation (discontinuous replication) and not from strand breakage. However, this is difficult to demonstrate because initiating RNA primers are removed quickly in vivo. *E. coli* mutants (*polAex1*) deficient in the 5′→3′ exonuclease activity of DNA polymerase I (Section 4-18) can partially prevent this removal. With a spleen exonuclease digestion technique (see below), designed to assay Okazaki fragments tipped with RNA, such fragments are barely detectable in wild-type cells,[13] but a substantial percentage is observed in the *E. coli* mutant.

In preparations of animal cell nuclei, the initiating primers in nascent fragments are relatively easy to detect. The nuclei are accessible to labeled NTPs as substrates for the synthesis of RNA and DNA and allow the survival of RNA linked to DNA. In replicating polyoma and SV40 DNAs, RNA clearly initiates nascent 4S fragments (Section 19-2), but it is not certain that *each* fragment has a primer. Approximately 90% of the 4S nascent chains synthesized in broken animal cell systems have a nonanucleotide RNA primer;[14] most of the primers are intact, as judged by the presence of a purine NTP at the 5′ terminus of a nonanucleotide. Only about 10% of the nascent chains isolated from intact growing cells[15] have a full primer; 20% have a shorter RNA moiety, apparently the result of partial degradation.

Isolation of RNA linked to DNA is not achieved by standard buoyant-density separations. Isopycnic sedimentation after denaturation of a DNA sample does not reliably separate the minute amounts of RNA-tipped replicative intermediate from the large quantities of other fragments that contain RNAs adhering noncovalently to DNA. The problem has been solved for polyoma and SV40 by preliminary sedimentation to enrich for the replicative intermediate and by subsequent enzymatic digestion with DNase to permit electrophoretic separation of discrete initiator oligoribonucleotides.[16]

Analysis of 5′ fragment ends by labeling can be imprecise. Polynucleotide kinase and [γ-^{32}P]ATP are used to assay 5′-OH ends on DNA chains presumably generated by alkaline hydrolysis of covalently linked RNA. Since the kinase also catalyzes an exchange reaction between ATP and 5′-P chain ends (Section 9-7), the incubation conditions must reduce exchange to negligible levels. Alternatively, spleen exonuclease digestion, which starts only at 5′-OH ends, may be used.[17] These methods are valid with well-defined substrates, but some uncertainties remain when they are applied to complex mixtures with a variety of structures.

Nascent fragments are extracted under neutral conditions.[18] Pulse-labeled DNA analyzed directly on a sucrose gradient, *without denaturation*, yields some 8S to 10S fragments. This unexpected subclass of *free* ssDNA fragments may represent a considerable fraction of the total nascent fragments, suggesting that the growing point is often disrupted during the isolation procedures.

Validity of the semidiscontinuous replication mechanism should be judged by its generality rather than universality. Many factors may operate to interrupt

13. Kurosawa Y, Ogawa T, Hirose S, Okazaki T, Okazaki R (1975) *JMB 96*:653.
14. Tseng BY, Goulian M (1977) *Cell 12*:483.
15. Tseng BY, Erickson JM, Goulian M (1979) *JMB 129*:531.
16. Reichard P, Eliasson R, Söderman G (1974) *PNAS 71*:4901.
17. Okazaki R, Hirose S, Okazaki T, Ogawa T, Kurosawa Y (1975) *BBRC 62*:1018.
18. Raggenbaas M, Caro L (1978) *DNA Synthesis, NATO*:299.

continuous synthesis and favor fresh starts. Low temperatures imposed experimentally to trap nascent fragments may favor discontinuous synthesis, as may an imbalance of replication proteins or dNTPs, or unreplicatable lesions and stable secondary structures in the template. In such cases, failure of continuous synthesis to keep pace with helicase melting of the parental duplex might create opportunities for primed starts of a discontinuous synthesis.

15-4 Replication Genes and Crude Systems

In every replication system, the multiple stages and complexity of the replication process imply that numerous genes and proteins must be involved. Phage T4, for example, despite the modest 160-kb size of its genome, has at least seven genes encoding proteins essential for DNA synthesis and more than 20 other genes that serve related functions.

Inasmuch as replication proteins are essential for viability, defining the genes encoding these proteins requires the isolation of alleles that are *conditionally defective*. Phage mutations in essential replication functions are generally obtained as nonsense mutants that can be propagated only in host cells that possess the appropriate suppressor tRNA. Mutations in proteins for essential cellular processes are more readily isolated through the use of mutants that grow at an intermediate temperature but are temperature-sensitive to heat or cold. The protein defects are tolerated under the *permissive* conditions but can be elicited under the *restrictive* ones at which the alteration in the protein makes it deficient. A number of *E. coli* mutants are known that grow and synthesize DNA at 30° but fail to do so at 42°. They maintain RNA and protein synthesis at nearly normal rates upon being shifted to the restrictive temperature, until growth and cell division fail for lack of DNA synthesis (Section 20-3). These temperature-sensitive (*dna*^{ts}) mutants[19] have helped to define many replication proteins, as well as to guide their assay and isolation. Bacterial strains with ts nonsense suppressors are also available for analyzing essential bacterial functions.

The several *dna*^{ts} mutants fall into two general groups with respect to responses in DNA synthesis when the temperature is switched from permissive to restrictive: the *quick-stop* mutations encode proteins defective in chain growth or extension; the *slow-stop* mutants, which sustain synthesis until completion of the ongoing round of replication, have proteins defective in origin initiation (or in termination of the preceding round). About a dozen *dna*^{ts} and other replication mutants have been isolated and mapped to distinctive loci in *E. coli* (Table 15-1), in *B. subtilis* (Table 15-2), and in yeast (Table 15-3). In some instances, the defects proved to be in nucleotide synthesis rather than in replication (i.e., *nrd*, *dut*) and are mentioned only for historical reference. Although a gene product may be identified as quick-stop, it may prove essential for initiation of a round of replication as well; dnaB and dnaC proteins and gyrase are examples. Loci for the mutator genes that affect the fidelity of replication are collected in Table 15-6 in Section 9.

To identify gene products and their functions in metabolism it is essential to resolve each system into its parts and then reconstitute it. In practical terms, this

---

19. Gross JD (1971) *Curr Top Microbiol Immunol* 57:39; Wechsler JA (1978) *DNA Synthesis NATO*:49; Sevastopoulos CG, Wehr CT, Glaser DA (1977) *PNAS* 74:3485.

Table 15-1
Replication genes and proteins of *E. coli*[a]

Gene	Map location (minutes)	In vivo phenotype of mutant	Protein and in vitro function
dnaA	83	slow-stop; defective origin initiation	dnaA; initiation at the origin
dnaB	92	quick-stop	dnaB; helicase
dnaC	99	slow- or quick-stop	dnaC; complex with dnaB
dnaE (polC)	4	quick-stop	α subunit of pol III holoenzyme
dnaG	67	quick-stop; defective initiation of fragments	primase
dnaJ	0	slow-stop	dnaJ; phage λ initiation; heat-shock response
dnaK	0	slow-stop	dnaK; phage λ initiation; heat-shock response
dnaN	83		β subunit of pol III holoenzyme
dnaQ	5	mutator	ϵ subunit of pol III holoenzyme
dnaT	99	slow-stop; stable replication	dnaT (protein i); primosome assembly
dnaX	11	quick-stop	γ and τ subunits of pol III holoenzyme
dnaY		quick-stop	arginine tRNA for rare codon
dnaZ	11	quick-stop	γ subunit of pol III holoenzyme; renamed *dnaX*
dut	82	very short nascent fragments	dUTPase
grpE[b]			initiation of phages λ and P1; heat-shock response
gyrA (nalA)	48	quick- and slow-stop; nalidixate-(oxolinate-)sensitive	DNA gyrase subunit α; nicking–closing
gyrB (cou)	83	quick- and slow-stop; coumermycin-(novobiocin-)sensitive	DNA gyrase subunit β; ATPase
lig	52	accumulation of replication fragments	DNA ligase; covalently seals DNA nicks
nrdA (dnaF)	49	quick-stop	R1 subunit of ribonucleotide reductase
nrdB	49	quick-stop	R2 subunit of ribonucleotide reductase
ori	84	nonviability	origin of chromosomal replication
polA	87	defective in DNA repair	pol I; gap filling, RNA excision
polB	2	defective in DNA repair	DNA polymerase II
priA[c]	88		protein PriA (n′); primosome
priB[c]	96		protein PriB (n); primosome
priC[c]			protein PriC (n″); primosome
rep	85	slowed fork movement	helicase
rnhA	5	"stable" replication; mutants suppress *dnaA*$^-$ or *oriC*$^-$	RNase H1; removal of RNA primers
rpoA	73		α subunit of RNA polymerase
rpoB	90	defective in transcription; chromosome initiation	β subunit of RNA polymerase
rpoC	90		β' subunit of RNA polymerase
rpoD	(67)[c]		σ subunit of RNA polymerase
ssb	92	quick-stop; defective in repair, recombination	SSB (single-strand binding protein)
ter	27–36		terminus of chromosomal replication
topA	28	mutants suppress *dnaA*$^-$	topoisomerase I
trxA	86	phage T7 negative	thioredoxin: coenzyme of ribonucleotide reductase, subunit of phage T7 DNA polymerase
tus (tau)	36	none	*ter*- binding protein; termination

[a] Unless noted, references are in Bachman BJ (1990) *Microbiol Rev 54*:130.
[b] Zylicz M, Ang D, Georgeopoulos C (1987) *JBC 262*:17437.
[c] Lee EH, Masai H, Allen GC Jr, Kornberg A (1990) *PNAS 87*:4620.

Table 15-2
Replication genes and proteins of *Bacillus subtilis*[a]

Gene	Analogous *E. coli* gene	Phenotype	Protein and in vitro function
dnaA (formerly *nrdA,B*)	*dnaA*	slow stop of DNA synthesis	initiator protein
dnaB		slow stop of DNA synthesis of chromosome and plasmids	polypeptide of 472 amino acids; membrane binding
dnaC		initiation of DNA synthesis	
dnaD		slow stop of DNA synthesis	
dnaE	*dnaG*	inhibition of DNA synthesis; linked to *rpo* gene as in *E. coli*	primase; primer formation
dnaF	*dnaE*+*dnaQ*	slow- and fast-stop DNA synthesis alleles; 3′→5′ exo-defective allele	polymerase includes the 3′→5′ exonuclease; analogous to *E. coli* pol III holoenzyme α and ϵ subunits combined
dnaG	*dnaN*	fast stop of DNA synthesis; gene location resembles that of *E. coli*	analogous to β subunit of *E. coli* pol III holoenzyme
dnaH		fast stop of DNA synthesis	
dnaI		inhibition of DNA synthesis	
dnax[b]	*dnaX*		
gyrA	*gyrA*	mutants resistant to nalidixate	A subunit of DNA gyrase
gyrB	*gyrB*	mutants resistant to novobiocin	B subunit of DNA gyrase
nrdA, B (formerly *dnaA*)	*nrdA, nrdB*	inhibition of DNA synthesis	nucleoside diphosphate reductase

[a] Prepared with the advice of Professor A. Ganesan.
[b] Struck JCR, Alonso JC, Toschka HY, Erdmann VA (1990) *Mol Gen Genet 222*:470.

Table 15-3
Replication genes and proteins of yeast[a]

Gene	In vivo phenotype	Protein[b]
POL1 (cdc17)	essential	DNA polymerase I (167) (pol α-like)[c]
POL2	essential	DNA polymerase II (170) (pol ϵ-like)[c]
POL3 (cdc2)	essential	DNA polymerase III (125) (pol δ-like)[c]
MIP1		DNA polymerase M (147) (pol γ-like)[c]
POL 30	essential	PCNA[d]
PR11	essential	primase subunit
PR12	essential	primase subunit
CDC9	cell-cycle defect	DNA ligase
TOP1	essential in *TOP3*⁻	topoisomerase I
TOP3	chromosome separation	topoisomerase II

[a] Burgers PMJ, Bambara RA, Campbell JL, Chang LMS, Downey KM, Hübscher U, Lee MYWT, Linn SM, So AG, Spadari S (1990) *EJB 191*:617.
[b] Mass (kDa) in parentheses.
[c] Mammalian polymerases.
[d] PCNA = proliferating cell nuclear antigen.

means breaking the cell barrier and isolating the molecular components. This has been much more difficult for the replication proteins than for the transcription and translation enzymes. Replication proteins are relatively scarce because they are needed for a process that occurs only once in a cell generation. Also, unlike the compact components of RNA polymerase and ribosomes, replication proteins are dispersed in cell lysates. The forces holding the putative "replisome" together are dissipated by the gentlest procedures for dispersing the molecular contents of the cell.

Experimental systems of varying intactness have been used to probe mechanisms and resolve the enzymes of chromosome replication. The crude systems are (1) permeable cells that are accessible to low-molecular-weight compounds such as nucleotide precursors but not to large proteins or nucleic acids, and (2) partial lysates and membrane complexes that retain many structural elements of the cell but are accessible to both precursors and proteins.

Permeable Cells[20]

Rendering *E. coli* cells permeable to nucleotide precursors by treatment with EDTA[21] destroys their capacity for semiconservative DNA replication. A better method uses a lipid solvent[22] (ether or toluene) that affects the membrane and appears to leave cells intact but unable to multiply and form colonies. Fortified with ATP, the four dNTPs, and Mg^{2+}, these cells sustain semiconservative DNA replication at nearly physiologic rates (2000 ntd per cell per second). ATP is required to maintain dNTP levels and for many additional functions.

With permeable cells, typical nascent replication fragments are the initial DNA products. These are later joined to yield an intact chromosome in the presence of added NAD, the cofactor of *E. coli* ligase. Nicotinamide mononucleotide (NMN), which interrupts ligation by promoting the reverse reaction, prevents this joining.

DNA synthesis is interrupted in permeable *dna*^{ts} mutants when the temperature is shifted to 43° or upon addition of drugs such as nalidixic acid or mitomycin, which are known to interrupt DNA synthesis in vivo. Sulfhydryl-blocking agents such as N-ethylmaleimide also inhibit, whereas cyanide and azide, which block DNA synthesis in the intact cell, have no influence in permeable cells, since precursor nucleotides and ATP are supplied exogenously.

Among the procedures for preparing permeable animal cells,[23] lysolecithin treatment works well with suspensions from the Chinese hamster ovary. The cells retain their morphology, and they synthesize DNA, RNA, and protein at near-normal rates when supplied with appropriate precursors and cofactors. DNA synthesis is limited to the S period in the permeable cells, as in a synchronized intact culture (Sections 20-6 and 20-7), showing that synchrony is not triggered by nucleotide levels. A reversible permeabilizing procedure is the hypertonic salt treatment of kidney cell monolayers; after the salt solution is removed, the cells behave as if intact.

The use of permeable cells has yielded information not obtainable from studies of intact cells. In *E. coli*, for example, DNA synthesis attributable to repair has

---

20. Vosberg H-P, Hoffmann-Berling H (1971) *JMB 58*:739; Billen D, Olson AC (1976) *J Cell Biol 69*:732; Miller MR, Castellot JJ Jr, Pardee AB (1978) *Biochemistry 17*:1073.
21. Buttin G, Kornberg A (1966) *JBC 241*:5419.
22. Vosberg H-P, Hoffmann-Berling H (1971) *JMB 58*:739.
23. Billen D, Olson AC (1976) *J Cell Biol 69*:732; Miller MR, Castellot JJ Jr, Pardee AB (1978) *Biochemistry 17*:1073.

been distinguished from chromosome replication. With yeast cells,[24] a number of replication mutants become accessible to biochemical analysis. However, permeable cells have inherent risks. Disrupting the cell membrane and the organization of protein complexes that rely on its integrity may distort in vivo events without gaining the advantage of resolving the components of those events. With permeable cells, only a limited range of *small* proteins can be added exogenously. The next step in analysis of replication requires more extensive disruption of the permeable cell.

Partial Lysates[25]

Extensive lysis of cells by lysozyme, detergents, or mechanical breakage destroys most DNA-replicating activity. However, when lysis is gentle and measures are taken to avoid dilution of the macromolecules in the cell, DNA replication resembling that of permeable cells in rate and character is observed. Three such systems will be discussed: porous cells prepared by toluene treatment and rendered permeable to macromolecules by a nonionic detergent; washed nuclei; and cell lysates on a cellophane disk.

Porous Cells. *E. coli* cells treated with toluene and 1% Triton X-100 maintain DNA replication even though enzymes the size of pol I can diffuse into and out of the cell.[26] Replication blocked at an elevated temperature in certain dna^{ts} mutants can be reestablished in the presence of Triton X-100 by extracts of wild-type cells.[27] Both the toluene and the detergent appear to have multiple effects on membrane and hydrophobic proteins, altering the normal patterns of replication processes.[28]

A system comparable to the toluene–Triton preparation of *E. coli* is the Brij 58 treatment of azide-poisoned *B. subtilis*.[29] Such cells are permeable to proteins and lose DNA-synthesizing activity attributable to pol I, but they still can initiate new rounds of replication at the chromosome origin.

Plasmolysis of cells by exposure to 2 M sucrose[30] introduces a porosity approaching that of total lysates (see below), with retention of the relative intactness of the porous cell.

Washed Nuclei. Preparations of nuclei from a variety of animal cells retain the cellular chromosomal DNA and a limited capacity to replicate it. Such nuclei are useful for assaying precursor requirements and the need for large molecules supplied by the cytoplasm or lost from the nuclei during preparation. Nuclei from cells infected with polyoma virus (Section 19-2), which can complete but not initiate rounds of viral replication, have been very instructive nevertheless. The long survival of RNA linked to nascent fragments in isolated nuclei afforded the first clear insight into RNA priming of DNA synthesis in animal cells.[31]

Cell Lysates on a Cellophane Disk. Cells may be lysed on a cellophane disk held on an agar surface and then transferred with the disk onto a droplet of incuba-

24. Oertel W, Goulian M (1978) *DNA Synthesis NATO*:1033.
25. Schaller H, Otto B, Nüsslein V, Huf J, Herrmann R, Bonhoeffer F (1972) *JMB 63*:183.
26. Moses RE (1972) *JBC 247*:6031; Moses RE, Moody EM (1975) *JBC 250*:8055.
27. Moses RE (1972) *JBC 247*:6031; Moses RE, Moody EM (1975) *JBC 250*:8055.
28. Hanson MA, Moses RE (1978) *Arch B B 190*:671.
29. Ganesan AT (1971) *PNAS 68*:1296.
30. Wickner RB, Hurwitz J (1972) *BBRC 47*:202.
31. Reichard P, Eliasson R, Söderman G (1974) *PNAS 71*:4901.

tion mixture. Small molecules can diffuse through the cellophane; macromolecular components of the replication machinery are preserved at high concentration in the lysate on top of the disk and macromolecules can be added to it.[32] DNA synthesis in the lysate is prevented if the cells are pretreated with agents that block DNA replication.

Properties of Lysate Systems. Lysed cells have the same precursor requirements for DNA synthesis and the same response to inhibitors as do permeable cells. DNA synthesis by the lysates is also temperature-sensitive when they are prepared from *dna*^{ts} mutants; the slow-stop and quick-stop properties of initiation and extension mutants are also demonstrable in the lysate. Although this system was used to assay the purification of replication proteins from wild-type cells that were missing or defective in mutant cell lysates, such assays are severely limited because it is difficult to define the function of a purified protein introduced into a crude and complex system. That the *dnaE* gene product is pol III was observed with the gentle lysate assay,[33] but a separate polymerase assay and prior knowledge of the properties of pol III were necessary. Although dnaG protein could be partially purified by the gentle lysate assay, its primase function could not be resolved with this system,[34] nor could RNA priming be demonstrated.

At a certain stage of resolving the replication proteins and explaining their functions, advances with permeable cells and partial lysates become limited. Such "in vivtro" preparations may suffer the combined deficiencies of both in vivo and in vitro systems. Biochemical studies with a soluble enzyme system then become essential.

15-5 Replication Proteins and Reconstituted Systems[35]

DNA replication with all the characteristics obtained in permeable-cell preparations has not yet been attained with soluble enzyme fractions. However, lysozyme lysis of *E. coli* at low temperature without detergents yields, after sedimentation, extracts virtually free of host DNA and containing the complex array of soluble components that replicate simple phage chromosomes (Section 17-1). Since some phages rely almost entirely on the host replication machinery, enzymatic components of the host can be probed and isolated and their functions defined. Eukaryotic viruses have served a similar role in illuminating the replication pathway of the cells that play host to them. Purification of each component by standard methods has provided key insights into mechanisms of replication of the more complex host chromosome.

Methods of Purification

The creation of strains with multiple copies of a gene and amplified levels of its product is of prime importance for purification. With or without that advantage,

32. Schaller H, Otto B, Nüsslein V, Huf J, Herrmann R, Bonhoeffer F (1972) *JMB 63*:183.
33. Nüsslein V, Otto B, Bonhoeffer F, Schaller H (1971) *NNB 234*:285; Otto B, Bonhoeffer F, Schaller H (1973) *EJB 34*:440.
34. Lark KG (1972) *NNB 240*:237.
35. Schekman R, Weiner A, Kornberg A (1974) *Science 186*:987; McMacken R, Kornberg A (1978) *JBC 253*:3313; Kornberg A (1988) *JBC 263*:1.

it is important to select a strain that lyses easily, harvest the cells in log phase, freeze the cells rapidly, and store them at low temperature. Lysis and extraction conditions optimal for one replication protein may be inappropriate for another. Patient attention to all these details can raise the yield of, say, dnaC protein 100-fold.

In practice, few if any of the replication proteins at unamplified levels can be assayed in crude lysates, despite the use of optimal conditions for preparing lysates. The reason is the low concentration of replication proteins in the lysate, or the presence of inhibitors, or both. Precipitating the replication proteins with low levels of ammonium sulfate may eliminate most of the inhibitors, and the pellet can be dissolved in a small volume to concentrate the proteins. Only at this stage can a reasonably valid assay be made. A concentrated lysate prepared by direct evaporation on a cellophane disk[36] may be a useful guide to some features of a system.

Fraction II. A long and arduous search for a cell-free system to support the replication of an *oriC* plasmid (Section 16-3) culminated successfully when functionally inactive *E. coli* lysates were subjected to a refined ammonium sulfate fractionation.[37] A precipitate (called *fraction II*) that contained the numerous required proteins and eliminated most of the inhibitory factors was effective, provided that it was dissolved in a small volume to concentrate the proteins (in excess of 10%) and that a large hydrophilic polymer (e.g., polyethylene glycol 8000) was present in the assay at a high concentration (near 10%). The availability of fraction II has made it possible to explore other replication systems and attempt their resolution; these include other plasmids (e.g., ColE1, R1, RK2) and phages λ and Mu.

Assays of Soluble Enzymes

A prime requirement in enzyme purification is a simple and quick assay. With the standard technique for measuring the incorporation of labeled deoxyribonucleotides into DNA, enzyme activity in some 20 fractions of a separation procedure can be assayed in an hour's time. Of course, it is crucial in these assays to know that the nucleotide incorporation being measured depends on the template-primer used and that the nucleotide is in the DNA product expected. With such assays in hand, purification of viral and host replication enzymes has been successful.

In the *complementation assay*, an extract from a dna^{ts} mutant that contains a particular gene product, inactive or inactivated at elevated temperature, is "complemented" by extracts of wild-type cells that contain this protein in a form that functions at high temperature. Assays for purification of dnaB, dnaC, dnaE, dnaG, and dnaZ proteins were provided in this manner.

The alternative classic *resolution and reconstitution assay* relies on applying standard enzyme fractionation techniques to extracts of wild-type cells in order to separate and sort out the proteins. When activity is lost upon fractionation, it may be restored by combining some apparently inactive fractions. In effect, "deletions" are generated that can be "complemented."

---

36. Schneck PK, van Dorp B, Staudenbauer WL, Hofschneider PH (1978) *NAR* 5:1689.
37. Fuller RS, Kaguni JM, Kornberg A (1981) *PNAS* 78:7370.

If the proteins of a replication system are naturally bound in a replisome complex, both the complementation and the resolution–reconstitution assays must depend on reassembling the complex correctly as a prerequisite for its catalytic function. There are limitations to each approach and a combination of both is most desirable.

Complementation assays have these drawbacks: (1) Inasmuch as the thermosensitive protein is still present, fractionation may select compounds that stabilize the mutant protein instead of identifying the wild-type protein. (2) The mutant protein in a complex must be replaced (or dominated) by the wild-type protein. (3) Multiple deficiencies, common in extracts of mutant cells, may lead to purifying the wrong component; thus, rate-limiting levels of pol III in extracts of *dnaG* mutants may lead to separating the polymerase instead of dnaG protein. (4) When a protein has finally been purified, it is unlikely that the mechanism of its action can be deduced simply by an assay of net DNA synthesis in a crude extract. The purified protein may bypass the deficiency rather than correct it. (5) Finally, when all the enzymes for which mutants are available have been purified, essential enzymes may still be lacking because mutants deficient in them have not yet been recognized.

Resolution–reconstitution assays of wild-type extracts have their limitations as well. (1) Identifying an essential factor as a valid component of the system often requires a touchstone of in vivo behavior, such as drug sensitivity, mutant thermosensitivity, or exact knowledge of the reaction product. (2) Many components are needed, each sufficiently free of the factor under assay. (3) Inhibitors are likely to be present, and factors may be purified simply because of their anti-inhibitory activity; their dispensability becomes apparent only when the inhibitor is removed. (4) Instability and scarcity make it difficult to maintain supplies of numerous components; stabilizing agents (e.g., salt, glycerol) may interfere with assays. (5) The order of adding components to the assay mixture and their relative amounts may be crucial. (6) Even when all the essential purified components appear to be in hand, the list may lack some component essential for optimal function in a crude system or in vivo.

Macromolecular Crowding and the Physiologic Anion[38]

Any attempt to reconstitute a fragile multimacromolecular assembly must recreate the friendly milieu of the particular cell or organelle from which the components were derived. Of the drastic consequences of cellular lysis, one of the most severe is the huge dilution of macromolecules: from a normally crowded gelatinous habitat in the cell with a protein concentration of 10% or greater to the thin transparent soup of a crude lysate with 0.1% or less protein. Fortunately, this disastrous dilution can often be corrected by the addition of a hydrophilic polymer (e.g., polyethylene glycol) of high molecular weight (e.g., 10,000 kDa) and in high concentration (near 10% by volume). These polymers generally do not affect the conformations and functions of proteins and nucleic acids except to crowd them into a small volume in which their reactivity may increase a thousandfold.

An astonishing innocence about the anionic preference of a particular cell has

---

38. Minton AP (1983) *Mol Cell Biochem 55*:119; Zimmerman SB, Trach SO (1988) *NAR 16*:6309; Jarvis TC, Ring DM, Daube SS, von Hippel PH (1990) *JBC 265*:15160.

commonly led to the arbitrary use of chlorides, inherited from classic studies of cells in seawater. However, the anion which a cell uses for buffering and osmolality may be strikingly different, as in the choice of glutamate by *E. coli*.[39] Numerous replication proteins, known to be completely inactive at relatively low salt concentrations (e.g., 0.1 M), tolerate glutamate at ten times that level. Akin to the innocuousness of the hydrophilic polymer, glutamate appears not to interfere with the ionic environment and hydration of a protein, as chloride does.

Reverse Genetics

Beyond high-resolution gel filtration and chromatography, beyond immunoaffinity adsorption and other advanced techniques for purifying a protein to homogeneity, modern technology provides a novel approach. By identifying the gene that encodes the purified protein, one can enhance its abundance, isolate it in generous quantities, and clarify its physiologic and biochemical properties. The procedure, sometimes referred to as "reverse genetics," entails these steps: (1) determine the amino acid sequence of one or several polypeptide fragments of the purified protein (a picomolar quantity of a protein band separated on a polyacrylamide gel often suffices), (2) synthesize oligonucleotides that could encode these protein sequences, (3) probe cDNA or genomic libraries with these sequences to find a gene that hybridizes with one or more of them, (4) isolate the gene by molecular cloning, and (5) highly overproduce the protein by increasing the gene dosage, transcription rate, and translational efficiency. Not only does this procedure facilitate the easy isolation of large quantities of the pure protein for biochemical studies, but it provides, when the gene sequence is determined, insights into the structure of the protein and, by directed mutagenesis and gene disruption, a means for determining its physiologic function.

Folded Bacterial Chromosome[40]

The successful isolation of replication proteins has depended heavily on the use of the intact viral chromosomes and plasmids as templates. Standard DNA preparations from salmon sperm, calf thymus, and *E. coli*, and synthetic DNA polymers as well, are valuable reagents but they may fail as guides to the isolation of all the replication machinery. The fragmentation of these large chromosomes by shear forces destroys essential structures and makes the DNA a substrate for nucleases and repair systems. For these reasons, some attempts to reconstitute replication in *E. coli* have tried to employ its intact chromosome as template.

The isolated folded and supercoiled *E. coli* chromosome can support semiconservative replication. It requires a crude soluble enzyme fraction in addition to pol I, ligase, and ATP, and displays many features predicted by chromosome behavior in vivo.[41] Unfortunately, such DNA preparations contain significant amounts of many different proteins which confuse the picture and have been difficult to resolve.

---

39. Richey B, Cayley DS, Mossing MC, Kolka C, Anderson CF, Farrar TC, Record MT Jr (1987) *JBC 262*:7157; Levino S, Harrison C, Cayley DS, Burgess RR, Record MT Jr (1987) *Biochemistry 26*:2095.
40. Kornberg T, Lockwood A, Worcel A (1974) *PNAS 71*:3189.
41. Kornberg T, Lockwood A, Worcel A (1974) *PNAS 71*:3189.

Reconstruction of the Replisome[42]

The isolation of proteins and chromosomes needed to sustain replication in vitro is an important stage in resolving the multimacromolecular assembly called a replisome. However, as with all enzymologic studies, the isolated enzymes and substrate (i.e., template) must be evaluated physiologically. Parts derived from other cellular machinery may be substituting in the reconstituted reaction for those used in vivo. Ultimately, the identity and behavior of isolated components need to be verified by in vivo genetic and physiologic studies and with crude in vitro systems, such as porous cells and lysates. With the enzymes in pure form to complement a system, and with specific antibodies to neutralize their activities, their actual roles in semiconservative replication can be evaluated in a more complex and natural setting.

15-6 Enzymology of the Replication Fork[43]

The many proteins that move the replication fork and the means they use to do it have come into clearer view. If organized into a single entity, a *replisome*, comparable in complexity to the ribosome, these proteins might achieve the rapid and coordinated continuous and discontinuous synthesis of both strands concurrently (Plate 14).

Our current understanding of the enzymology of the replication fork is based largely on insights gained from the replication of phages, *E. coli oriC* plasmids, and animal viruses. Homogeneous preparations of intact chromosomes and plasmids free of gaps, nicks, or structural disorder provide suitable and abundant substrates for in vitro studies. The small viruses that rely almost entirely on host equipment for replication have proved to be effective probes for identifying the replication enzymes and illuminating how they function in replication of the host chromosome.

Components of the machinery to initiate the chromosome and advance the replication forks may vary structurally, but generally they display remarkable functional similarities (Table 15-4). Opening the duplex so as to advance replication relies on the action of helicases (Chapter 11). Topological impediments, such as the positive supertwisting that develops during the unwinding of a physically constrained chromosome, can be overcome by topoisomerases (Chapter 12). The bared single strands are immediately covered by binding proteins (Chapter 10) that protect the DNA and put the strands into a favorable conformation to function as templates. Synthesis by a polymerase holoenzyme is highly processive and continuous. The repeated initiations needed in discontinuous strand synthesis require RNA primers, generated by a primase or a more complex primosome (Chapter 8), and when extended by a polymerase holoenzyme (Chapters 5 and 6), yield the nascent fragments of discontinuous strand synthesis.

---

42. Kornberg A (1977) *Biochem Soc Trans* 5:359; Kornberg A (1978) *CSHS 43*:1; Kornberg A (1988) *JBC 263*:1.
43. Moses RE, Summers WC (eds.) (1988) *DNA Replication and Mutagenesis*. Am Soc Microbiol, Washington DC; McMacken R, Kelly TJ (eds.) (1987) *DNA Replication and Recombination*. UCLA, Vol. 47. AR Liss, New York; Richardson CC, Lehman IR (eds.) (1990) *Molecular Mechanisms in DNA Replication and Recombination*. UCLA Vol. 127. AR Liss, New York; Kornberg A (1988) *JBC 263*:1.

Table 15-4
Components of well-characterized replication systems[a]

Function	*E. coli*[b]	ColE1	T7	T4	SV40	Herpesvirus
Helicase	dnaB	dnaB, PriA	gp4	gp41	T antigen	UL5, UL8, UL52 (helicase/primase)
Primase	primase (dnaG)	primase (primosome)	gp4	gp61	pol α primase	UL5, UL8, UL52 (helicase/primase)
Polymerase						
Core	pol III (α,ϵ,θ)	pol III (α,ϵ,θ)	gp5	gp43	pol α, pol δ	UL30 or pol
Accessory subunits	β,γ complex, τ	β,γ complex, τ	trx	gp44,62,45	PCNA, RF-C	UL42
Processivity factors	β,τ	β,τ	trx	gp44,62,45	PCNA, RF-C, AAF	
SSB	SSB	SSB	gp2.5	gp32	RF-A	ICP8
Swivel	gyrase	gyrase		T4 topo	topo I or topo II	
Decatenation	gyrase	topo III (?)			topo II	
Initiation at the origin	dnaA, RNAP	RNAP, pol I, RNase H	T7 RNAP		T antigen transcription factors	UL9

[a] AAF = α accessory factor; PCNA = proliferating cell nuclear antigen; RF = replication factor; RNAP = RNA polymerase; SSB = single-stranded binding protein; topo = topoisomerase; trx = thioredoxin; UL = unique long (gene fragment).
[b] Phage λ replication forks have the same components. For initiation at *ori*λ, gpO substitutes for dnaA protein; gpP is analogous to dnaC protein.

Helicase and Primase at the Fork

Melting of the duplex by helicases is nearly always necessary for progress of a replication fork; coordination of this strand separation with the multiple priming events in discontinuous strand synthesis is common and may be required for achieving essentially concurrent synthesis of both strands at the replication fork. Whereas the structural and kinetic features to integrate these two key replication functions into a replisome superassembly (see below) are still unclear, some component interactions are well defined.

Helicase–Primase Interactions. At least nine helicases are encoded by the *E. coli* genome (Sections 11-2 and 11-3). Among them, dnaB protein, PriA protein, and Rep protein all have roles in replication: dnaB protein as probably the principal helicase in chromosomal replication (Section 11-2), PriA protein as a key element of the primosome (Section 8-4), and Rep protein as essential for the replication fork that rolls around the circular template of single-stranded phages ϕX174 and Ff (Section 11). All three helicases cooperate with other components of the replication machinery: dnaB and PriA with primase and the primosome, and Rep protein with phage initiator endonucleases.

The dnaB and PriA proteins are part of a mobile preprimosome which prepares the template for primase to lay down a primer at or near the replication fork (Section 8-4). The opposing translocation polarities of these two helicases on DNA may be synergistic in replication (Section 8-4), as may those of Rep and dnaB proteins. The relatively simple complex of dnaB protein with primase constitutes a miniprimosome that is sufficient in the reconstituted enzyme systems for complete replication of *oriC* plasmids and λ phage initiated at their origins. Similar couplings of helicase and primase activities are notable: between the T4 gp41 helicase and gp61 primase; between the polypeptides encoded by gene *4* of phage T7; and among the three polypeptides encoded by herpes simplex virus.

Polymerase Interactions. Primase is almost invariably a subunit component of the eukaryotic α polymerases; in effect, part of the holoenzyme assembly (Section 6-2). The integration of helicase–primase activities with polymerase is evident in the fork movements of the phage T4 and T7 systems. Formation of a complex between a polymerase holoenzyme, helicases, and primase (Table 11-2) seems probable from their synergistic effects, but in most cases this has not been demonstrated physically.

Polymerase Holoenzyme at the Fork

The replicative polymerases of *E. coli* and phage T4 and those of yeast and animals all share a number of structural and functional features that qualify them for achieving the concurrent continuous and discontinuous synthesis of both strands at the replication fork. Pol III holoenzyme of *E. coli*, composed of ten or more pairs of subunits, includes twin polymerase cores, each with a distinctive arm (Section 5-3). One arm, with high processivity, seems designed for continuous synthesis; the other, with lesser processivity, for discontinuous synthesis.

This model of a dimeric, asymmetric organization, as suggested in pol III holoenzyme, could also be achieved in eukaryotic cells by a combination of the two multisubunit replicative polymerases: the highly processive pol δ being responsible for the continuous synthesis of one strand, and the less processive, primase-containing pol α for the discontinuous synthesis of the other.

Phage T4 polymerase (Section 17-6), less complex in organization than the host *E. coli* enzyme, is demonstrably effective as a single molecular assembly in the synthesis of both strands. The holoenzyme, made up of gp43 (the polymerase) and gp44, 62, and 45 (accessory proteins), requires only the gp41 helicase for optimal continuous-strand synthesis; the T4 Dda helicase (Section 11-2) does not substitute for gp41. With the addition of gp61 (the primase), discontinuous-strand synthesis accompanies continuous-strand synthesis, and the two become properly coupled in the presence of gp32 (the SSB).

The DNA polymerase of phage T7 (Section 17-5), like the RNA polymerase of this phage, is remarkably streamlined and efficient. With the host thioredoxin as the only accessory subunit, this polymerase barely qualifies as a holoenzyme, yet it attains processivity and a high rate of fork movement. The T7 mechanism responsible for a possibly concurrent synthesis of both strands needs to be clarified further.

The Replisome

The notion of a supramolecular assembly, a *replisome*—with helicases, primosome, and polymerase holoenzyme operating coordinately at a replication fork—is highly plausible. Whereas a replisome may exist in crude enzyme fractions or in association with an active replication fork, the isolation of such an intact structure has yet to be achieved. Possibly, membranous or other cytoskeletal attachments are needed to preserve its existence. Such attachments may also be important in the dynamics of a replisome at a replication fork.

With the awareness of the mobility of a primosome on the SSB-coated, single-stranded circle of ϕX174 came the recognition that components of the primosome could displace the SSB and that its polar movement on DNA directed it to keep up with the progress of the replication fork. Thus, regarded as a "*locomotive*," the primosome had several key components: the PriA protein as the "engine," equipped with an SSB "cowcatcher," and the dnaB protein as the "engineer" that may generate or recognize secondary structures in the template to enable the primase to lay down primers on the DNA track.

However, this locomotive analogy had to cope with the knowledge that during primer synthesis, the primase must be translocated in the direction opposite to primosomal and fork movements, and that subsequent findings showed the translocation of PriA protein on DNA, like that of primase, to be counter to the polarities of dnaB protein and the primosome. In fact, the primosome was observed to move in both directions, drawn by the opposite movements of PriA and dnaB proteins (Section 8-4).

As an alternative to the image of a locomotive moving to and fro, the primosome and, by implication, the replisome, might be regarded instead as a stationary element in the cell, a "*sewing machine*" through which the DNA is drawn, rather than being a mobile unit that moves along a relatively fixed DNA track (Fig. 15-2). Pulling a loop of DNA through the primosome would be an attractive device for the template strand in discontinuous replication and could at the same time explain how integral components of the primosome can be translocated, relative to DNA, in opposite directions.

15-7 Start of DNA Chains

The DNA chain starts in replication are limited largely to the nascent short fragments in the discontinuous growth of the "lagging" strand and to the rare

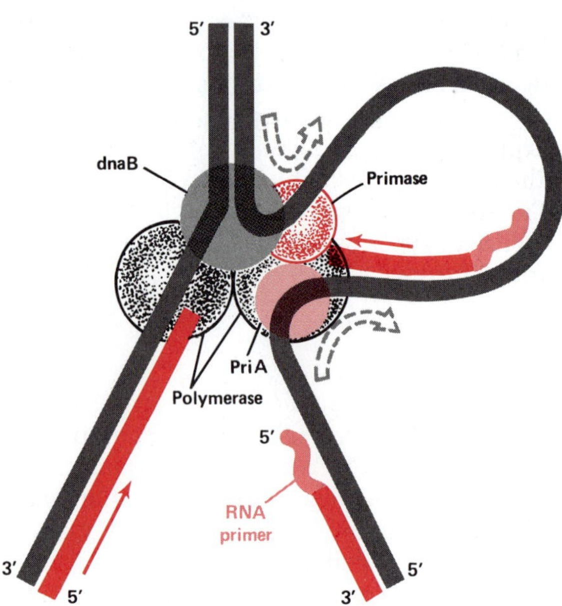

Figure 15-2
Hypothetical scheme for concurrent replication of leading and lagging strands by DNA polymerase associated with a primosome to form a "replisome." Note that the dnaB and PriA proteins in the primosome are translocated on the DNA template in opposite directions.

initiation event at the origin of a chromosome. Unlike RNA polymerases, DNA polymerases cannot start a chain, but can only extend chains that have a 3′-OH terminus paired to a template strand extended beyond it. The several ways in which termini are used as primers and in which primer termini are generated by primases and terminal proteins (Section 8-8) will be reviewed briefly in this section.

RNA Primer

A very brief RNA transcript is the most general device for priming the covalent start by a DNA polymerase. The discovery of RNA priming was prompted by the knowledge that RNA polymerase can start a chain, that *E. coli* pol I can extend a ribonucleotide-ended chain, and that the foreign RNA at the 5′ end of a chain can be excised and replaced by the polymerase and exonuclease activities of pol I. Fortunately, the first test of the validity of RNA priming depended on the rifampicin sensitivity of the conversion of a single-stranded, circular M13 viral DNA to its duplex replicative form. The implied action of RNA polymerase, verified by enzymatic studies, proved that a brief transcript synthesized by the enzyme primed the start of the complementary DNA chain. (The sequence selected by RNA polymerase for transcription, remarkably, lacks any of the characteristic features of a promoter.)

Rather few RNA priming events depend on the host RNA polymerase. Aside from filamentous phages and some plasmids (e.g., ColE1), RNA priming in *E. coli* is rifampicin-insensitive; in eukaryotic cells it is α-amanitin–insensitive. The action of a primase is required and each species has a distinctive form. The variety of primases among plasmids, phages, bacteria, yeast, and animal viruses and cells is enormous. Prokaryotic primases generally "read" a specific tetranucleotide sequence in the template strand to produce a transcript of like or slightly greater length. Eukaryotic primases generally "count." For lack of a capacity to read a recognition sequence, they produce primers of about 10 ntd (or multiples of this unit length) at virtually any template sequence.

Primases usually can bind and transcribe a template unaided, but accessory proteins may play a crucial role. For example, dnaB protein of *E. coli*, in addition to its helicase activity, is required by primase for its function at most template sites. Eukaryotic primase activity commonly resides in a heterodimeric protein, possibly in one of the subunits serving an accessory factor function.

The extent to which the various DNA polymerases discriminate an RNA from a DNA terminus during polymerization and proofreading is not well known. The T7 polymerase is notable for its capacity to exploit a tetranucleotide primer, even at temperatures at which the small primer should be melted away from the template. Effective channeling of the primase product to the primer terminus binding site of the polymerase may also be achieved in vivo by an intimate physical association of a prokaryotic primase with a polymerase, an association resembling the interaction observed between the eukaryotic enzymes in vitro.

In one instance, an RNA primer is preformed rather than made in situ. A tRNA annealed to the 3′ end of a retroviral RNA primes its replication by reverse transcriptase (Section 19-7). Aside from the ColE1 RNA I, which is complementary and antagonistic to the primer RNA, no other examples are known of a *trans*-acting RNA that serves or regulates priming.

The need to remove and replace the RNA primer by DNA polymerase poses a problem for linear genomes (Section 12) that is solved by concatemer formation, as with T7 phage, or by resorting to other priming devices, such as the use of a terminal protein.

Other Priming Devices

Terminal Protein Primer. Some animal viruses (e.g., adenovirus) and phages (e.g., ϕ29) with linear genomes use a virus-encoded terminal protein that recognizes and binds a sequence at each 3′ end of the viral template. The viral polymerase covalently links the initiating base-paired dNTP to an amino acid side chain of the terminal protein and then uses this nucleotide to prime the extension of the chain. The terminal protein, or a processed form of it, remains linked to the 5′ chain ends of the packaged genome. This priming device ensures the high fidelity of the DNA polymerase operation for all but the 5′-terminal nucleotide and obviates the need for a device to replace RNA primers.

Parental Strand Primer. As distinct from the de novo starts by RNA or protein primers, covalent extension of a parental DNA strand can be achieved by a *rolling-circle* or *rolling-hairpin* mechanism (Section 11). Cleavage of one strand of a circular duplex DNA provides the 3′-OH primer terminus from which the complementary circular strand can be copied. Repeated excursions around the template—in effect a rolling circle of replication—generates multigenome-length progeny strands. Palindromic (self-complementary) sequences at the ends of a linear genome (single- or double-stranded) may form hairpins and thereby provide a base-paired 3′-OH primer terminus. Progress along the genome generates palindromic termini in the progeny and the means to maintain a rolling-hairpin mechanism of replication.

Hairpin (Loopback) Model for Template-Primer Function of a Single Strand. This mechanism, though hypothetical, can be demonstrated experimentally by using a variety of intrinsic and exogenous exonuclease activities and T4 polymerase functions.[44] A new chain can be started at the 3′-OH terminus of the single-stranded template (Fig. 15-3a) if it loops back by intrastrand base pairing (Fig. 15-3b). Any unpaired region at the 3′ end is hydrolyzed by a 3′→5′ exonuclease back to a base-paired primer terminus (Fig. 15-3c); chain growth then proceeds (Fig. 15-3d). Exonuclease III, by attacking the 3′ terminus of the duplex product, removes the newly synthesized chain (Figs. 15-3e and 15-4a), whereas phage λ exonuclease (Section 13-3), acting on the 5′ terminus, removes only template nucleotides (Figs. 15-3f and 15-4). Inasmuch as exo III and λ exo act only on duplex DNA, a portion of the template or product is spared from digestion. Upon conversion to a single strand, the DNA is again susceptible to a single-strand–specific exonuclease (e.g., exo I).

Further hairpin formation at this stage (Fig. 15-3g), followed by trimming of an unpaired 3′ end by a 3′→5′ exonuclease and a paired end by λ exo (Fig. 15-3h), enables the polymerase to start synthesis again (Fig. 15-3i). Still another round of λ exo digestion, annealing, 3′→5′ exo digestion, and polymerase action (Fig. 15-3j) can ultimately digest and replace the original template.

---

44. Goulian M, Lucas ZJ, Kornberg A (1968) *JBC 243*:627.

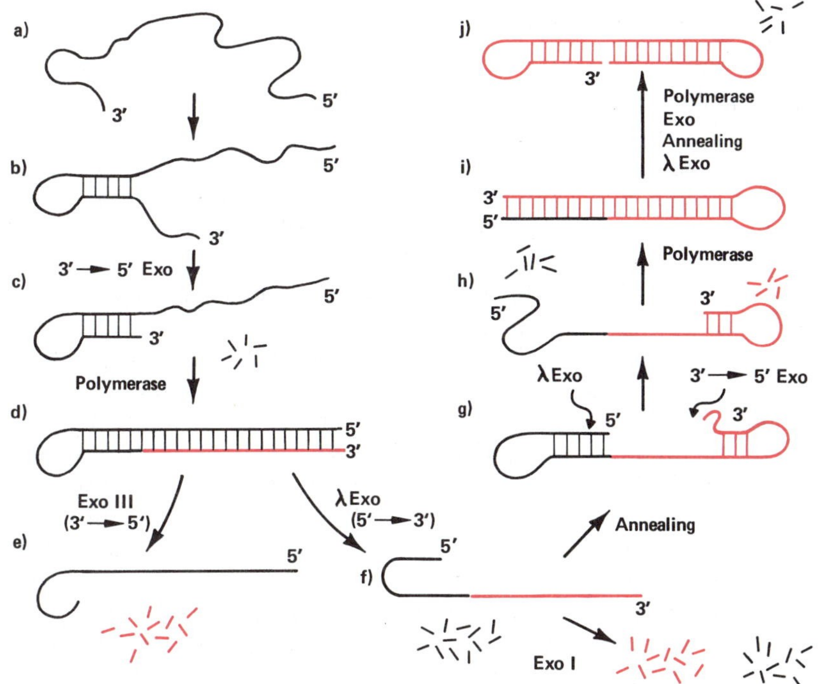

Figure 15-3
Scheme for single-stranded DNA functioning as a "hairpin" template-primer for T4 polymerase, as indicated by the action of various exonucleases on the polymerase product. [After Goulian M, Lucas ZJ, Kornberg A (1968) *JBC 243*:627]

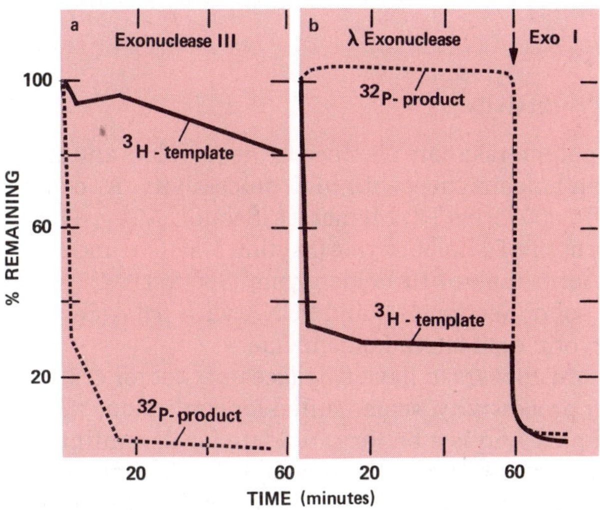

Figure 15-4
Susceptibility of the template-primer and the T4-polymerase products (at stage d in Fig. 15-3) to exonuclease III (a) and λ exonuclease (b). [From Goulian M, Lucas ZJ, Kornberg A (1968) *JBC 243*:627]

15-8 Processivity of Replication[45]

A crucial feature in polymerase design is the frequency with which the enzyme dissociates from the template-primer after the addition of a nucleotide to the growing chain end, compared to translocating directly to the newly generated primer terminus. When dissociation follows each nucleotide addition (i.e., when one elongation step occurs per binding event), the *processivity value* is one, and the replication mode is regarded as *distributive* (dispersive, nonprocessive). When the polymerase remains clamped to the template-primer for repeated additions to the chain, the replication mode is *processive* (nondistributive, nondispersive); processivity values can range beyond the many thousands.

Processivity is influenced by many factors—the structure of the template-primer and the local DNA sequence, the temperature and ionic strength—all superimposed on the inherent design of the polymerase for its specialized function. Clearly, a replicative polymerase—as with RNA polymerase in its movement along DNA to transcribe a long message, or the ribosome in its translocation along that message—needs to avoid time-consuming dissociations and reassociations if it is to copy an extensive genome in a reasonable length of time. By contrast, a repair polymerase that needs only to fill a short gap does not operate under the same time constraint and hence can dispense with the domains and accessory polypeptides needed to achieve high processivity. In the repair of UV lesions in *E. coli* by the various DNA polymerases, the patch lengths are proportional to their processivity,[46] a nice correlation of in vivo and in vitro activities.

Whichever the processive enzyme, whether polymerase, helicase, or nuclease, the detailed mechanisms of processivity are largely obscure. Inasmuch as the numerous forces through which the enzyme interacts with a long-chain substrate are still undefined, it is no wonder that the way in which these many contacts are serially released and reformed with each cycle of polymerization must also remain a mystery.

Role of Accessory Factors in Processivity

Accessory factors that keep a polymerase core clamped to the template and slide with it during replication form holoenzymes with high processivity. As examples: thioredoxin with phage T7 gp5 forms T7 polymerase (Section 5-9); gp44, 62, and 45 with phage T4 gp43 form the T4 holoenzyme (Section 5-8); β, τ, and the γ complex with *E. coli* pol III form the pol III holoenzyme (Section 5-4); RF-C (RP-C) and AAF (AAP) with pol α (Section 6-2) and PCNA with pol δ (Section 6-3) probably form the eukaryotic replicative holoenzyme.

The dimeric *E. coli* pol III core appears to have two accessory-factor arms, of which the τ arm, with greater processivity, seems suited for continuous strand synthesis, whereas the γ-complex arm is a better candidate for discontinuous strand synthesis. In the same vein, features of the replicative eukaryotic polymerases suggest that the highly processive pol δ, combined with PCNA, carries

---

45. McClure WR, Chow Y (1980) *Meth Enz 64*:277; Fay PJ, Johanson KO, McHenry CS, Bambara RA (1981) *JBC 256*:976; Fay PJ, Johanson KO, McHenry CS, Bambara RA (1982) *JBC 257*:5692; Huang C-C, Hearst JE, Alberts BM (1981) *JBC 256*:4087.
46. Matson SW, Bambara RA (1981) *J Bact 146*:275.

out continuous synthesis, but the less processive pol α, complexed with primase, is better qualified for discontinuous synthesis. A highly processive helicase (Section 11-1), by melting the parental duplex and ensuring template availability, may be regarded as an accessory factor for polymerase processivity.

Measures of Processivity

The *processivity value* is readily measured by the chain lengths of the DNA products synthesized under conditions of template excess to minimize the reassociation of an enzyme with the template it has abandoned. Both the product size and the template sequence that favor dissociation can be determined by gel electrophoretic examination of the products made from a defined template-primer. Polymerases differ widely. Eukaryotic polymerase β, a repair enzyme, is nearly fully distributive, whereas the processivities of *E. coli* pol III holoenzyme and phage T7 polymerase seem virtually unlimited. The processivity values of *E. coli* pol I, employed chiefly in gap repair, are intermediate (Table 15-5).[47]

In the absence of one of the four dNTPs, a sharp reduction in the rate of DNA synthesis is observed with pol I but not with eukaryotic pol β. In view of the long time consumed in the association and dissociation stages, it follows that with the template-primer in large excess, it makes little difference to a distributive polymerase if one of the dNTPs is missing. Synthesis is aborted after each polymeri-

Table 15-5
Processivity of various DNA polymerases[a]

Source	Polymerase	Template	Processivity value[b]
E. coli	pol I	nicked ColE1	15–20
		gapped ColE1	45–55
		oligo dT·poly dA	11–13
		poly d(AT)	170–200
	polA5	nicked ColE1	3–5
	pol III core	primed ss circle[c]	10–15
	pol III′	primed ss circle[c]	40
	pol III*	primed ss circle[c]	190
	pol III holoenzyme	primed ss circle[c]	> 5000
Phage T4	T4 polymerase[d]	oligo dT·poly dA	11–13
Phage T5	T5 polymerase	oligo dT·poly dA	155–170
Calf thymus	polymerase α	oligo dT·poly dA	7–9
	polymerase β	oligo dT·poly dA	9–11
KB cell	polymerase α	nicked–gapped calf thymus	6–16
	polymerase β	nicked–gapped calf thymus	1–2
AMV (virus)	viral polymerase	nicked ColE1	22–30

[a] Adapted from data supplied by Professor R. A. Bambara.
[b] Nucleotide residues; measured at 37°, ionic strength of 0.1 μ.
[c] Primed phage single-stranded circle.
[d] With accessory proteins on a primed ss circle, the processivity value is > 20,000.

47. Matson SW, Bambara RA (1981) *J Bact* 146:275.

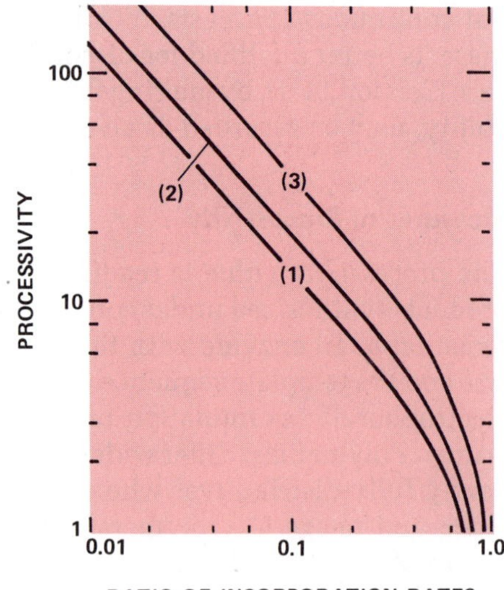

Figure 15-5
Theoretical relationship between experimentally measurable ratio of incorporation rates and processivity. The ratios are of incorporation rates in reactions containing one, two, or three dNTPs compared to the full complement of four dNTPs. Activated thymus DNA is the template-primer.

zation step even with the full complement of dNTPs present. With a processive polymerase, however, a long replication run exploits every initiation, provided all four dNTPs are available.

A precise quantitative method for assessing processivity is based on a comparison of polymerization rates with one, two, or three versus four dNTPs (Fig. 15-5).[48] Based on this assay, pol I is processive for about 20 residues on an activated (nicked and gapped) thymus DNA template (Table 4-6); a value near 200 with poly d(AT) is drastically reduced at low temperature and high ionic strength. In contrast with pol I, eukaryotic pol β remains distributive under a variety of conditions.

Template challenge affords a less quantitative, but often useful, measure of processivity. With two polymers of distinctive composition, such as poly dA·oligo dT and poly dC·oligo dG, switching from one to the other can be identified by the incorporation of dTTP versus dGTP. With natural template-primers (e.g., plasmids), their distinctive size, detected by gel electrophoresis, can afford a measure of whether a polymerase, replicating one template-primer, can be challenged to switch to another primed template.

15-9 Fidelity of Replication[49]

Preservation of the genome of a species is entrusted to a replication process with an error frequency—in *E. coli* cells—of 10^{-10} or less per base pair. These rare replication errors are the principal source of the so-called spontaneous mutations. The frequency of these mutations is increased enormously, up to 10^5-fold,

48. Bambara RA, Uyemura D, Lehman IR (1976) *JBC 251*:4090; Bambara RA, Uyemura D, Choi T (1978) *JBC 253*:413.
49. Schaaper RM, Danforth BN, Glickman BW (1986) *JMB 189*:273; Maki H, Akiyama M, Horiuchi T, Sekiguchi M (1990) in *Mutagenesis and Anticarcinogenesis Mechanisms II* (Kuroda Y, Shankel DM, Waters MD, eds.). Plenum Press, New York; Horiuchi T, Maki H, Sekiguchi M (1989) *Bull Inst Pasteur 87*:309.

by alterations in a few genes [*mutator* (mut) *genes*] responsible for the replication proteins that ensure the high fidelity of the process.

Impaired fidelity, leading to faulty or distorted proteins and aberrant metabolism, can cause cancer and other diseases and may contribute significantly to the aging process. At the same time, mutability provides the opportunities for the selective forces of evolution to increase the frequency of novel genetic combinations.

Measurement of Replication Errors

Assays of replication error frequency in vitro are generally of two kinds: (1) incorporation of a nucleotide mismatched with a homopolymeric template (e.g., incorporation of dGTP with a poly dT template) and (2) change of a phenotype (e.g., mutant to wild type) by incorporation of a mismatched nucleotide at the mutated site. Extrapolations from these assays to the natural occurrence of replication errors have limitations: an unnatural template in the first case, and the interposition of an in vivo expression of an in vitro mismatch in the other. In addition, numerous parameters, including the structural state of the template-primer, the concentration and ratios of the dNTPs, the ionic strength of the reaction mixture, and the presence of polyamines, can have profound effects. Aside from these factors, polymerases may differ by 1000-fold in fidelity under nearly identical conditions.

In vitro studies of mutability not only have offered insights into the mechanisms of replication and distinctions among the polymerases, but also have been of practical importance in detecting the mutagenic effects of environmental agents. Mutagenesis assays, in bacteria and cell cultures, of drugs, dietary additives, radiations, and occupational agents have been valuable guides to the hazards they pose for carcinogenesis.

Mechanisms for Ensuring Fidelity

Mechanisms for fidelity are manifested at five or more stages.

(1) *Balanced levels of the dNTPs.* These are provided by fine regulation of key biosynthetic steps (Section 2-15). Aberrantly high levels of a dNTP favor its misincorporation; low levels of a dNTP invite the incorporation of an incorrect dNTP present at a higher concentration. Incorporation of uracil in place of thymine (Section 10) is minimized by the action of dUTPase and the maintenance of dTTP levels. If uracil is misincorporated or generated in DNA by cytosine deamination, it is removed in a repair reaction initiated by uracil N-glycosylase.

(2) *Complementary base pairing of a dNTP to the template.* On the basis of hydrogen-bonding forces, the error frequency at this stage is estimated at 10^{-3} to 10^{-4} per base pair. At this stage and the next, mispairing may be provoked by an unnatural base[50] or the lack of a base[51] in the template.

(3) *"Induced fits" of polymerase and DNA.*[52] The polymerase active site adapts to the size and shape of a correct base pair and the conformational features of the DNA template, adjusted by bending and base stacking. The DNA

---

50. Heinrich M, Krauss G (1989) *JBC 264*:119; Preston BD, Singer B, Loeb LA (1986) *PNAS 83*:8501.
51. Randall SK, Eritja R, Kaplan BE, Petruska J, Goodman MF (1987) *JBC 262*:6864.
52. Sloane DL, Goodman MF, Echols H (1988) *NAR 16*:6465.

sequence ("context"), that is, the nearest-neighbor effects of base stacking, influences fidelity.[53] At this stage the error frequency is reduced to about 10^{-5} to 10^{-6} per base pair.

(4) *Proofreading by 3′→5′ exonuclease.* Discrimination by the polymerase against a mismatched primer terminus is manifested by removal of the mismatch by 3′→5′ exonuclease action[54] or by the reduced rate of extension of a mismatched terminus.[55] These actions may further reduce the error frequency by 10^2 or 10^3 per base pair. Transient misalignment of polymerase on the template, in effect a "dislocation" error,[56] can contribute to mutations undetected by proofreading.[57]

(5) *Mismatch correction.* At least three different systems operate at a final postreplication stage (Section 21-3) to bring the error frequency down to the level observed in vivo, near 10^{-10} per base pair.

Mutator Genes and Proteins[58]

Some of the *mut* genes were discovered by searching for mutations that greatly increase the frequency of reversion of a mutant allele (e.g., gal^- to gal^+) or that induce resistance to a drug (e.g., rifampicin). Other genes which increase mutation frequency were recognized because of alterations in proteins known to operate in replication or in postreplicational repair (Table 15-6).

mutT. The very first mutator gene discovered in *E. coli* was *mutT*. Its mutations enhance the transversion of an AT base pair to CG by 1000-fold or more.[59] The wild-type MutT protein (14.9 kDa)[60] hydrolyzes a form of dGTP (perhaps the minor *syn* form) that can pair with a tautomer of A in the template. The mutant protein, by failing to prevent the mismatching of G to A, allows the transversion to a CG base pair in the next round of replication. The *mutT* mutation when present for many generations enhances the GC frequency in DNA and may be one of the ways in which the GC content has diverged widely (ranging from 25% to 75%) in microbial genomes.

Mutations in Polymerases and Their Domains. The genes that encode polymerases and their 3′→5′ exonuclease (proofreading) domains or subunits represent another important class of mutator genes. Phage T4 polymerase becomes a powerful mutator when its 3′→5′ exonuclease activity is diminished by mutation. Alterations in T4 polymerase that increase the exonuclease activity[61] cause an increase in fidelity, an antimutator action; antimutator *(amu)* genes have also been observed in the *E. coli* genome.[62]

53. Mendelman LV, Boosalis MS, Petruska J, Goodman MF (1989) *JBC 264*:14415.
54. Kunkel TA (1988) *Cell 53*:837; Bialek G, Nasheuer H-P, Goetz H, Grosse F (1989) *EMBO J 8*:1833.
55. Perrino FW, Loeb LA (1989) *JBC 264*:2898.
56. Kunkel TA, Alexander PS (1986) *JBC 261*:160.
57. Boosalis MS, Mosbaugh DW, Hamatake R, Sugino A, Kunkel TA, Goodman MF (1989) *JBC 264*:11360; Papanicolaou C, Ripley LS (1989) *JMB 207*:335.
58. Maki H, Akiyama M, Horiuchi T, Sekiguchi M (1990) in *Mutagenesis and Anticarcinogenesis Mechanisms II* (Kuroda Y, Shankel DM, Waters MD, eds.). Plenum Press, New York; Horiuchi T, Maki H, Sekiguchi M (1989) *Bull Inst Pasteur 87*:309.
59. Cox EC (1976) *ARG 10*:135; Schaaper RM, Bond BI, Fowler GR (1990) *MGG 219*:256.
60. Akiyama M, Horiuchi T, Sekiguchi M (1987) *MGG 206*:9; Bhatnagar SK, Bessman MJ (1988) *JBC 263*:8953; Akiyama M, Maki H, Sekiguchi M, Horiuchi T (1989) *PNAS 86*:3949.
61. Muzyczka N, Poland RL, Bessman MJ (1972) *JBC 247*:7116; Bessman MJ, Muzyczka N, Goodman MF, Schnaar RL (1974) *JMB 88*:409.
62. Cox EC (1976) *ARG 10*:135.

Table 15-6
Mutator genes in *E. coli*

Gene	Map location (minutes)	Product
dnaE (*polC*)	4	α subunit of pol III holoenzyme
dnaQ (*mutD*)	5	ϵ subunit of pol III holoenzyme
polA	87	pol I
mutH	81	endonuclease (mismatch repair)
mutL	93	complex with *mutH* and *mutS* products
mutM		
mutR	59	
mutS	62	mismatch recognition
mutT	2	dGTPase
mutY (*micA*)	64	A/G endonuclease (adenine glycosylase); analogous to *mutB* product in *S. typhimurium*
uvrD (*mutU*)	84	helicase II
lig	52	ligase
ssb	92	SSB
dut	82	dUTPase
ung	56	uracil N-glycosylase
recA	58	RecA protein
lexA	92	LexA protein (repressor)

The *mutD* gene, the source of a nearly 10^5-fold increase in mutation frequency per base pair, proved to be an allele of *dnaQ*, the gene for the ϵ *subunit*, the 3'→5' exonuclease of pol III.[63] For a *mutD5* strain to display its highest mutation frequency, the cells are grown in a rich medium or in a minimal medium supplemented with thymidine, and must also bear a deficiency in methyl-directed mismatch repair.[64]

Mutations in the α *subunit* (*dnaE* gene) of the pol III holoenzyme have various effects on replication fidelity depending on whether the defect is in the polymerase domain or in the region that interacts with the ϵ subunit. Although ϵ alone contains the exonuclease activity, its proofreading activity is elevated 50-fold or more through its interaction with α. With one particular *dnaE* allele, a 10^3- to 10^4-fold increase in mutation frequency appears to be due to misalignments on the template during chain elongation, rather than to effects on the ϵ subunit function.[65] Combined structural and mutational analyses of an increasing number of polymerases (e.g., pol I; T7, T4, and ϕ29 polymerases; pol α) have begun to reveal the domains of the proteins responsible for fidelity at the several stages of polymerization.

Comparison of polymerases from diverse sources, with due allowances for the influence of assay conditions, reveals roughly three categories based on their error frequencies: (1) Low, near 10^{-8}, for enzymes possessing 3'→5' exonuclease proofreading activity; *E. coli* pol I, pol III holoenzyme, and T4 and T7 polymerases are examples. (2) Moderate, near 10^{-6}, for those that lack an overt proof-

63. Scheuermann R, Echols H (1984) *PNAS 81*:7747; Maruyama M, Horiuchi T, Maki H, Sekiguchi M (1983) *JMB 167*:757; Di Francesco R, Bhatnagar SK, Brown A, Bessman MJ (1984) *JBC 259*:5567.
64. Schaaper RM (1988) *PNAS 85*:8126.
65. Maki H, personal communication.

reading activity; eukaryotic pol α is an example. (3) High, near 10^{-4}, for enzymes that lack both a proofreading activity and a refined polymerase discrimination; retroviral reverse transcriptases, particularly that of HIV, are examples.

Among the mutator genes and their products listed in Table 15-6, some, such as *ssb* and *lig*, show slight but significant effects on fidelity, some function in postreplicational mismatch repair *(mutH, L, M, S, Y, uvrD)* (Section 21-2), and some *(lexA, recA)* are key elements in mutagenic repair by the SOS system. The *dut* gene product, dUTPase, prevents the incorporation of uracil in DNA, while mutations in the *ung* gene allow its product, uracil N-glycosylase, to persist (Section 10); the mutagenic consequences of the spontaneous deamination of cytosine to uracil are averted by the removal of uracil.

Metal Ion Substitution

A divalent metal ion, commonly Mg^{2+}, plays several essential roles in replication: chelation of the dNTPs, polymerization, and proofreading. Substitution of Mn^{2+} decreases the fidelity of polymerization by 100-fold or more.[66] Mn^{2+} eliminates the discrimination by T7 polymerase against dideoxy-NTPs and sharply reduces that of *E. coli* pol I.[67] How the metal affects the furanose ring structure and alters the nucleotide as a substrate is unknown, as are the reasons for the inadequacy of various metals (e.g., Be^{2+}, Cd^{2+}, Co^{2+}) to substitute for Mg^{2+} at several stages of the replication process.

15-10 Uracil Incorporation in Replication[68]

E. coli DNA polymerases, and probably others as well, show little discrimination between dUTP and dTTP. Nor is uracil significantly different from thymine as a template for replication and transcription. With uracil replacing one-third of the thymines in *E. coli*, or replacing all thymines in certain phages, no gross defects are apparent. Thus, the incorporation of uracil is unlike the mismatch misincorporations responsible for spontaneous mutations. Nevertheless, uracil incorporation both in vivo and in vitro can have consequences that deserve attention.

Replication in Vivo

The ratio of concentrations of dUTP and dTTP governs the frequency of uracil incorporation; the effectiveness of the uracil N-glycosylase system determines its persistence in DNA. The dUTP level in *E. coli* depends on the balance of its synthesis (three-fourths from dCTP and the remainder from dUDP) against its removal by dUTPase.[69] Hydrolysis of dUTP produces dUMP, as does dCMP deaminase. Thus, these enzymes produce the precursor for the de novo synthe-

---

66. El-Deiry WS, Downey KM, So AG (1984) *PNAS 81*:7378; El-Deiry WS, So AG, Downey KM (1988) *Biochemistry 27*:546.
67. Tabor S, Richardson CC (1989) *PNAS 86*:4076.
68. Hochhauser S, Weiss B (1977) *J Bact 134*:157; Tye B-K, Chien J, Lehman IR, Duncan BK, Warner HR (1978) *PNAS 75*:233; Tamanoi F, Okazaki T (1978) *PNAS 75*:2195; Brynolf K, Eliasson R, Reichard P (1978) *Cell 13*:573; Grafstrom RH, Tseng BY, Goulian M (1978) *Cell 15*:131; Olivera BM, Manlapaz-Ramos P, Warner HR, Duncan BK (1979) *JMB 128*:265.
69. Shlomai J, Kornberg A (1978) *JBC 253*:3305.

sis of dTTP. In sum, the dTTP level is determined by how well the synthesis of dTMP (from either dUMP or by salvage from thymine or thymidine) keeps up with its use in DNA synthesis. At an estimated steady-state level of 1 dUTP per 300 dTTP in *E. coli*, uracil incorporation into DNA would occur at a frequency of about 1 per 1200 nucleotides. This frequency would be increased by agents which depress the dTTP level, such as methotrexate (Section 14-2).

The intricate traffic patterns leading to and from dUTP and dTTP are in delicate balance (Section 2-8). Under certain conditions in which dUTPase levels are rate-limiting, as in *dut* mutants,[70] there is a profound increase in the production of small replication fragments due to removal of incorporated uracil and the slow rate of repair. Thus, the nature and level of precursors in DNA labeling experiments and the genetic makeup of strains can critically affect the rate of synthesis and the structures of DNA intermediates (Section 2-12).

DNA fragments produced by the excision of uracil from DNA in postreplication repair in *E. coli* can be distinguished from nascent, RNA-primed replication fragments by the retention of a 5′-P terminus after alkaline hydrolysis.[71] The resistance of nascent fragments to degradation by spleen exonuclease has provided an additional demonstration that uracil-derived fragments are not significant in number except in dUTPase mutants.

Besides the replicative incorporation of uracil, the spontaneous deamination of DNA cytosines to uracil occurs at a significant rate, and the removal of these uracil residues is essential to avoid mutations.

Replication in Vitro

Uracil incorporation is as common and unavoidable in vitro as in vivo.[72] The same strategies govern its incorporation and excision in vitro and in vivo, but the tactical details may differ. A substrate mixture of the four dNTPs is likely to contain trace amounts of dUTP. As little as 1 dUTP per 1000 dTTP (e.g., 10^{-8} M versus 10^{-5} M) has caused the incorporation of about 1 uracil in the synthesis of a viral circle of some 5000 ntd. Even if contamination by dUTP could be avoided initially, generation of the molecule from dCTP, either spontaneously or by the action of deaminases, would be difficult to prevent. The best insurance against intrusion of dUTP is to provide dUTPase for its removal.

Synthesis of DNA with purified enzymes freed of dUTPase, but still contaminated with uracil N-glycosylase and excision nucleases, leads to fragmented products. Including purified dUTPase as a "sanitizing agent" in the incubation mixtures allows intact uracil-free products to be synthesized.[73]

In semiconservative replication in *E. coli* lysates on cellophane disks, the size and number of replication fragments have been drastically altered by using *dut*$^-$ and *ung*$^-$ cells as sources and by varying the levels of dUTP, dCTP, and dTTP as substrates.[74] Manipulating these substrate levels has similar effects in the analysis of polyoma and animal cell replication intermediates.[75]

---

70. Hochhauser S, Weiss B (1977) *J Bact 134*:157; Warner HR, Duncan BK (1978) *Nature 272*:32.
71. Machida Y, Okazaki T, Miyake T, Ohtsuka E, Ikehara M (1981) *NAR 9*:4755.
72. Olivera BM (1978) *PNAS 75*:238; Shlomai J, Kornberg A (1978) *JBC 253*:3305; Tye B-K, Nyman P-O, Lehman IR (1978) *BBRC 82*:434.
73. Olivera BM (1978) *PNAS 75*:238; Shlomai J, Kornberg A (1978) *JBC 253*:3305; Tye B-K, Nyman P-O, Lehman IR (1978) *BBRC 82*:434.
74. Olivera BM (1978) *PNAS 75*:238.
75. Brynolf K, Eliasson R, Reichard P (1978) *Cell 13*:573; Grafstrom RH, Tseng BY, Goulian M (1978) *Cell 15*:131.

15-11 Rolling-Circle Replication

Fork movement around a circular template, referred to as *rolling-circle replication*, can yield multiple copies and multigenome-length products.

Multiple Single-Stranded Circles as Products

Conversion of the circular duplex replicative intermediate (RF) of phage ϕX174 to many copies of single-stranded viral (+) circles provided early evidence for the rolling-circle mechanism in vivo (Section 17-4). Lariat-shaped (σ-shaped) molecules, in which a single-stranded limb of the (+) strand protrudes from a duplex circle, are seen in the electron microscope. Participating in the multienzyme assembly at the replication fork are the gene *A* protein (gpA), a single molecule of Rep protein, and pol III holoenzyme, each operating processively, with SSB available to coat the synthesized viral strand. With the essential host proteins (Rep, pol III holoenzyme, and SSB) and the phage-encoded gpA in hand, the RF→SS mechanism (Section 17-4) can be reconstituted with high efficiency.

Cleavage by gpA of the (+) strand of a supercoiled ϕX174 RF between specific residues provides the 3′-OH primer terminus for the polymerase (Section 8-7). Covalent linkage of gpA to the 5′-P at the cleaved bond orients Rep protein for its helicase action. Upon separation of the SSB-coated viral (+) strand, the complementary (−) strand becomes available as a template. A remarkable feature of the replication complex is the *looped rolling-circle* structure (Fig. 8-13 and Section 17-4), in which gpA with the 5′-P end attached appears locked to the template as the fork progresses around the circle.

When the circuit of the template is completed and the regenerated origin sequence is reached, it is recognized by gpA, which trails the replication fork clamped to the template. The protein, possessing twin cleavage–ligation sites, now cleaves the regenerated origin, ligates the completed (+) strand, and rolls around the template once again. The 5386-ntd ϕX174 circle is synthesized in about 10 seconds, and 20 or more copies are released from a single rolling-circle intermediate.

In the similar replication cycle of Ff phage (Section 17-3), the generation of viral circles also starts with an RF cleavage by an analogous gene *2* protein. However, for lack of covalent linkage of the protein to the cleaved DNA, the rolling-circle replication is limited to a single round.

Multigenome-Length Single Strands as Products

Under some circumstances in vitro, and possibly in vivo also, the replication of a primed single-stranded circle, upon reaching completion, may enter a rolling-circle stage. The viral strand of ϕX174 or Ff phages, primed by RNA (or DNA), may fail to stop and await closure by ligase when the complementary strand is completed. Instead, with a helicase (e.g., dnaB protein, T4 gp41, or T7 gp4) displacing the synthesized strand, the polymerase can keep going 20 or more times around the circular template. Entry into a rolling-circle mode is facilitated by a model template-primer possessing a 5′ primer end that is displaced from the template because it is unmatched for several residues. When rolling-circle replication is coupled with primase action, the displaced multigenome-length single strand can be converted to the double-strand form.

Concatemeric Duplexes as Products

Replication of circular duplex phage λ proceeds by a theta mechanism (Section 16-4) initially, then enters a rolling-circle stage for production of the phage genomes. The unidirectional rolling-circle fork replaces the bidirectional mode, and multigenome-length concatemers are produced for subsequent division into the genome lengths that fill the phage head (Section 17-7). It is still not known how the switch is made from the theta to the rolling-circle stage.

Other Duplex Products

Additional instances of a rolling-circle mechanism appear in the replication of the 2μ plasmid of yeast (Section 18-10) and in the amplification of ribosomal RNA genes during the development of *Xenopus* oocytes. More examples will very likely be discovered with further studies of replication under sundry circumstances.

15-12 Termination of Replication[76]

Completion of the two daughter chromosomes encounters problems inherent in the structural features of the genome and the mechanism of its replication. With some linear duplexes, gaps at the 5′ ends need to be filled, and with others, telomeric sequences may have to be supplied. With circular genomes, the removal of catenanes is essential, and impeding fork movement at the terminus may be helpful. Beyond that, the devices for the segregation of a pair of chromosomes between the daughter cells in prokaryotic and nucleated organisms depend on cytoskeletal and membranous structures that are not well characterized.

No polymerase is known that can extend a chain from the 5′ end and thus fill the gap created at that end after the RNA primer has been excised (Fig. 15-6). Circularity of the chromosome can solve this problem, because chain growth around a circular template ultimately juxtaposes the 3′ and 5′ ends for union by

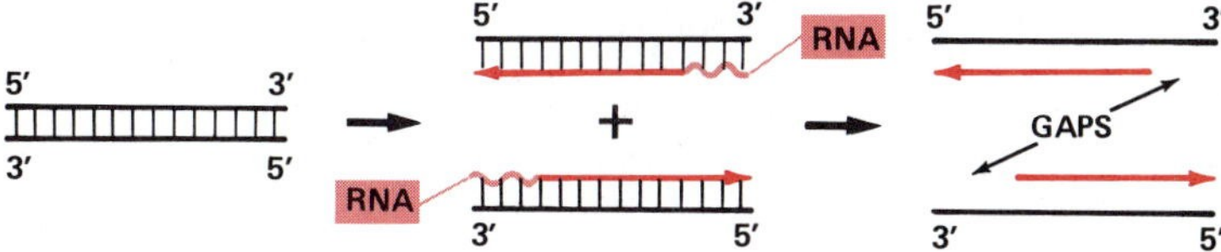

Figure 15-6
Gaps created at 5′ ends of progeny strands by removal of primer RNA.

76. Kuempel PL, Pelletier AJ, Hill TM (1989) *Cell 59*:581; Weaver DT, Fields-Barry SC, De Pamphilis ML (1985) *Cell 41*:565; Fields-Barry SC, De Pamphilis ML (1989) *NAR 17*:3261.

ligase. This may be one of the reasons why the circular form has been adopted by plasmids and a number of viral and bacterial genomes and why it is often an intermediate in the replication of linear chromosomes. As alternatives to circularity, other mechanisms for terminating linear chromosomes depend on special sequences at the ends of the duplex. One mechanism employs hairpin loops formed from palindromic sequences to fill the incomplete 5′ ends; another relies on terminal redundancies and their potential to form cohesive joints and concatemers.

Palindromic Hairpin Loops to Complete the 5′ End[77]

The symmetry of a palindromic base sequence permits it to snap back and form a self-complementary hairpin loop (Fig. 15-7 and Section 1-11). Thus, the unreplicated, single-stranded 3′ end can fold back on itself to form a terminal base-paired loop. The gap at the 5′ end of new DNA can then be filled by a sequence of reactions: extension of the 3′ end by polymerase, sealing by ligase, cleavage by endonuclease at the 5′ end of the palindrome, unfolding of the hairpin loop, and extension by polymerase to fill the new gap at the 3′ end.

The scheme employing terminal palindromic sequences is refined in the

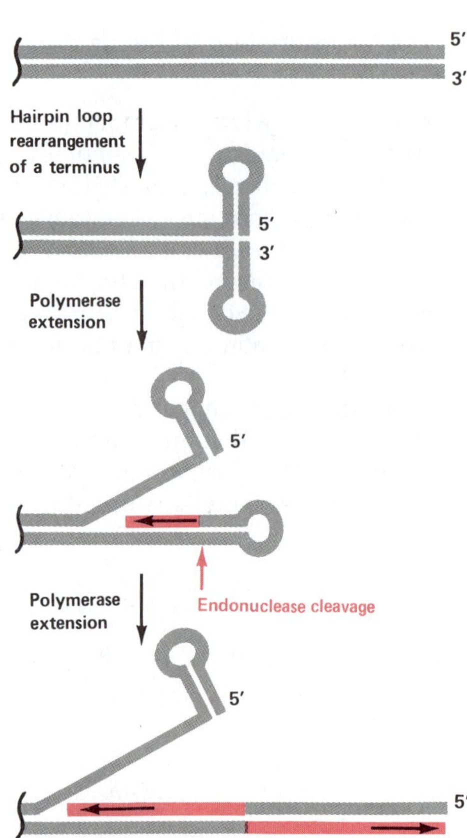

Figure 15-7
Model for replication of eukaryotic chromosomes based on the use of terminal palindromes. [Adapted from Tattersall P, Ward DC (1976) *Nature* *263*:106]

77. Cavalier-Smith T (1974) *Nature 250*:467; Tattersall P, Ward DC (1976) *Nature 263*:106.

rolling-hairpin model. Direct initiation of replication from each end of a linear duplex avoids RNA priming and the consequent complications of incomplete 5′ ends. These hairpin mechanisms have three requirements: terminal palindromic sequences (which need not be perfectly base-paired in the looped end of the potential hairpin), proteins to catalyze the formation and melting of the hairpins, and a specific endonuclease to make a nick at the 5′ end of the palindromic sequence.

Redundant Termini to Form Concatemers to Complete the 5′ End[78]

T7 phage DNA is an example of a nonpermuted linear duplex with a stretch of several hundred nucleotides repeated at each end. Limited removal of 3′ terminal regions by exonuclease III leaves 5′ cohesive ("sticky") tails that link intramolecularly to form circles, or, at higher DNA concentrations, link intermolecularly to form chains or concatemers (Fig. 17-23). In an analogous manner, incomplete 5′ terminal regions in new DNA would leave 3′ tails free for making concatemers by similar lap joints. As indicated in Figure 17-23, if both tails are longer than the redundant zone, the concatemer is completed by gap filling; if the tails cannot include the entire redundant zone, the extra lengths can be removed by exonuclease. Gap-filling by polymerase and sealing by ligase ensure that the concatemer is covalently intact. When packaging of DNA into virions is called for, specific staggered endonuclease nicking of both strands produces unit-length genomes whose ends can then be filled out by polymerase (Fig. 15-8).

Addition of Telomeric Sequences[79]

The termini of duplex linear chromosomes *(telomeres)*, in organisms ranging from yeast and ciliates to plants and mammals, are made up of highly conserved sequences, tandemly repeated 20 to 100 times. The *Tetrahymena* polymerase synthesizes $(TTGGGG)_n$ [the human is $(TTAGGG)_n$][80] at the 3′ end of its telomere, but does not use a DNA template. Rather, synthesis follows the instructions of a 3′-AACCCCAAC (ribonucleotide) sequence contained within an essential RNA (about 160 ntd long) in the polymerase ribonucleoprotein.

The proposed mechanism (Fig. 15-9) entails hybridization of a long protruding 3′ end of the DNA chain to the RNA sequence to permit an elongation by TTG; then, translocation back on this RNA template permits a further elongation by GGGTTG. Thus, growth of the 3′ end of the telomere is by 6-ntd lengths of GGGTTG. By this scheme a 3′ end terminating at any nucleotide within the TTGGGG sequence can be elongated to yield the perfect tandem repeats. The G residues in three or four of the tandem repeats can be paired to form a unique four-stranded structure, enabling them to be recognized as primers.[81] Synthesis of the $(AACCCC)_n$ complementary chain to form the duplex telomere may be primed by a primase.[82] The familiar problem of filling the gap at the 5′ end

78. Watson JD (1972) *NNB 239*:197.
79. Greider CW, Blackburn EH (1989) *Nature 337*:331; Morin GB (1989) *Cell 59*:521; Shippen-Lentz D, Blackburn EH (1990) *Science 247*:546; Yu G-L, Bradley JD, Attardi LD, Blackburn EH (1990) *Nature 344*:126.
80. de Lange T, Shiue L, Myers RM, Cox DR, Naylor SL, Killery AM, Varmus HE (1990) *Mol Cell Biol 10*:518.
81. Sundquist WI, Klug A (1989) *Nature 342*:825; Panyutin IG, Kovalsky OI, Budowsky EI, Dickerson RE, Rikhirev ME, Lipanov AA (1990) *NAS 87*:867.
82. Zahler AM, Prescott DM (1989) *NAR 17*:6299.

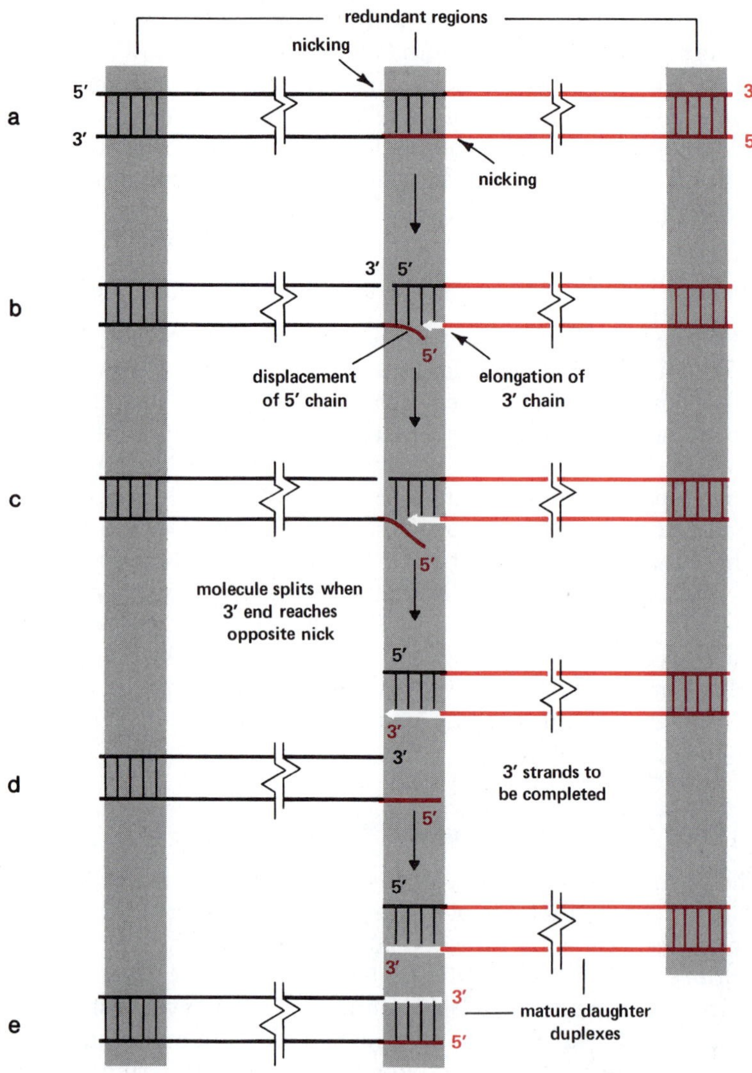

Figure 15-8
Scheme for maturation of T7 DNA from the concatemeric intermediate. (a) The concatemer is nicked at specific, staggered points. (b) The 3′-hydroxyl ends created by nicking are elongated by polymerase action, displacing the 5′ ends. (c) The concatemer comes apart when extension of the 3′-hydroxyl end reaches the nicked region on the template strand. (d) Molecules with incomplete 3′ ends are available for elongation by polymerase action. (e) Unit-length T7 molecules with redundant terminal regions are ready for packaging in phage heads. [From Watson JD (1972) *NNB 239*:197]

created by removal of the RNA primer is solved by the enormous redundancy of the telomeric tandem repeats.

Separation of Progeny Catenanes[83]

The separation of progeny duplexes—long linear chromosomes (e.g., yeast) as well as circles—offers topological problems, as judged by progeny that emerge

83. Weaver DT, Fields-Barry SC, De Pamphilis ML (1985) *Cell 41*:565; Fields-Barry SC, De Pamphilis ML (1989) *NAR 17*:3261.

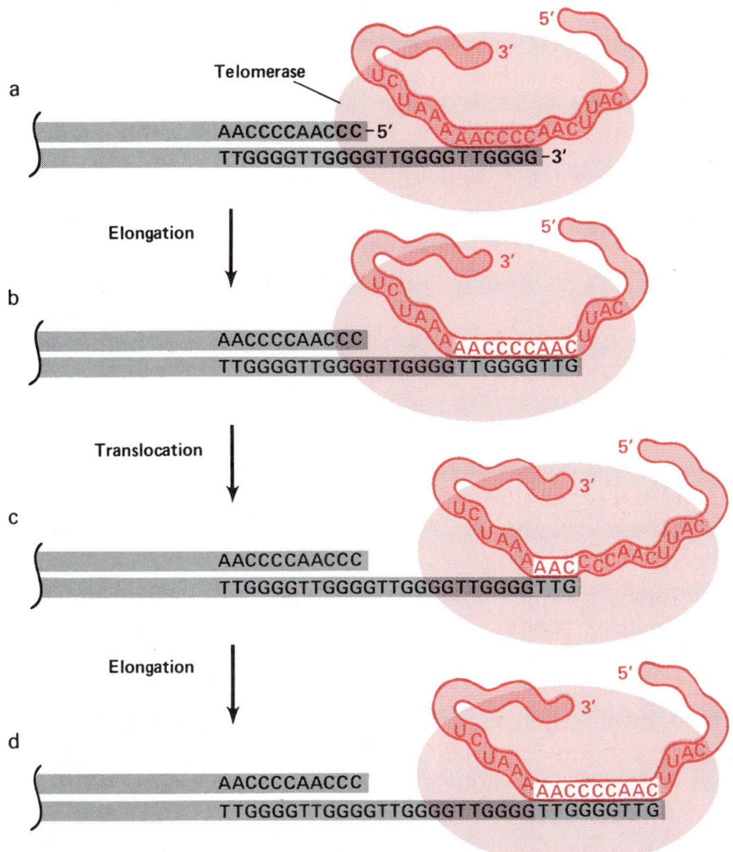

Figure 15-9
Model for elongation of telomeres by telomerase. (a) After recognition of the TTGGGG strand of the *Tetrahymena* telomere by telomerase, the 3′-most nucleotides hybridize to the complementary sequence in the RNA. (b) TTG is added one nucleotide at a time. (c) Telomerase translocates and hybridizes to the 3′-most TTG nucleotides. (d) Elongation completes a new sequence, permitting another translocation. Oligonucleotides with 3′ ends terminating at any nucleotide within the sequence TTGGGG are correctly elongated to yield perfect tandem repeats.

interlocked as catenanes. Significant pausing in the decatenation of daughter molecules just short of their completion has been observed in the replication in vitro of SV40 (Section 19-2) and in the reconstituted systems for *oriC* and ColE1 plasmids (Section 18-2) and for phage λ (Section 17-7). The lacking component for this stage near completion may in some instances be topoisomerase III, which seems well suited for this decatenation function and was without an assigned role for many years after its discovery.

Analysis of the segregation of replicated SV40 DNA into mature supercoiled forms may have general significance.[84] Bidirectional replication around the 5.2-kb circle pauses 100 to 200 bp short of completion. DNA synthesis continues to fill in the gaps of the "Siamese-dimer" daughter circles, sealing one and then the other to give two supercoiled circles that are held together simply by topological linkage. Topoisomerase II activity is inferred as the means of reducing the catemers to monomeric supercoils. Interruption of segregation by placing the infected cells in hypertonic medium does not affect replication, suggesting

84. Sundin O, Varshavsky A (1980) *Cell 21*:103; Sundin O, Varshavsky A (1981) *Cell 25*:659.

that these processes are under independent controls that may be employed for some physiologic function.

Impeding Fork Movement at a Terminus[85]

Opposite the origin of the circular genome of *E. coli* (*oriC*) is a terminus (*ter*) region where the bidirectionally moving forks meet and terminate.[86] The *ter* sequence and its specific binding protein have been further defined by studies of analogous *ter* regions and mechanisms in *E. coli* plasmids (e.g., R6K)[87] and in *B. subtilis*.[88]

A pair of *ter* sequences of about 20 bp are separated from each other by a comparable length of DNA and are inverted in such a way that one orientation blocks the fork moving clockwise and the opposite orientation blocks the fork moving counterclockwise. (The *ter* region of *E. coli* is actually made up of two pairs of inverted repeats of the *ter* sequence.) The *ter* sequences of a pair are virtually identical, and the sequences in *E. coli* and in the R6K plasmid show only small deviations; the *ter* sequence of *B. subtilis* is considerably longer and bears no resemblance to that of *E. coli*.

The *ter* sequence impedes fork movement only when bound by a *ter*-binding protein (TBP or Tus; 36 kDa), product of the *tus* (termination utilization substance) gene (also called *tau*). TBP, purified from overproducing cells on the basis of binding the *ter* sequence, blocks replication fork movements in vitro as it does in vivo. In the replication by a purified enzyme system of *oriC* plasmids containing a *ter* sequence and TBP to bind it, one *ter* orientation blocks only the clockwise movement of the replication fork while the opposite orientation blocks only the counterclockwise movement[89] (Fig. 15-10), just as observed in vivo. Replication of an *oriC* plasmid with *ter* sequences on both sides of the origin is limited to the origin region (Fig. 15-11). Inasmuch as blockage of *oriC* replication in one direction still permits the fork to move around the circle in the other direction, replication of an *oriC* plasmid with one *ter* sequence can produce a genome-length product. However, the rolling-circle mode of replication that follows the single-genome-length production by this enzyme system is severely inhibited by a *ter* sequence in either orientation. In this instance, the *ter*–TBP complex limits replication to making one copy and prevents uncontrolled amplification by the rolling-circle mechanism (Section 11).

Of the enzymes contributing to the progress of the replication fork, the helicases are most affected. Depending on the polarity of the helicase (Section 11-1), one orientation of the *ter*–TBP complex prevents its action in separating the duplex strands while the other orientation does not. The 5′→3′ translocation of dnaB protein is blocked only by one *ter*–TBP orientation, while the 3′→5′

---

85. Kuempel PL, Pelletier AJ, Hill TM (1989) *Cell 59*:581; Smith MT, Wake RG (1989) *J Bact 170*:4083; Lewis PH, Wake RG (1989) *J Bact 171*:1402; Lewis PJ, Smith MT, Wake RG (1988) *J Bact 171*:3564; Williams NK, Wake RG (1989) *NAR 17*:9947; Sista PR, Mukherjee S, Patel P, Khatri GS, Bastia D (1989) *PNAS 86*:3026; Lee EH, Kornberg A, Hidaka M, Kobayashi T, Horiuchi T (1989) *PNAS 86*:9104.
86. de Massy B, Bejar S, Louarn SB, Louarn JM, Bouché JP (1987) *PNAS 84*:1759; Pelletier AJ, Hill TM, Kuempel PL (1988) *J Bact 170*:4293; Hidaka M, Akiyama M, Horiuchi T (1988) *Cell 55*:467.
87. Sista PR, Mukherjee S, Patel P, Khatri GS, Bastia D (1989) *PNAS 86*:3026.
88. Smith MT, Wake RG (1989) *J Bact 170*:4083; Lewis PH, Wake RG (1989) *J Bact 171*:1402; Lewis PJ, Smith MT, Wake RG (1989) *J Bact 171*:3564; Williams NK, Wake RG (1989) *NAR 17*:9947; Lewis PJ, Ralston GB, Christopherson RI, Wake RG (1990) *JMB 214*:73.
89. Lee EH, Kornberg A, Hidaka M, Kobayashi T, Horiuchi T (1989) *PNAS 86*:9104.

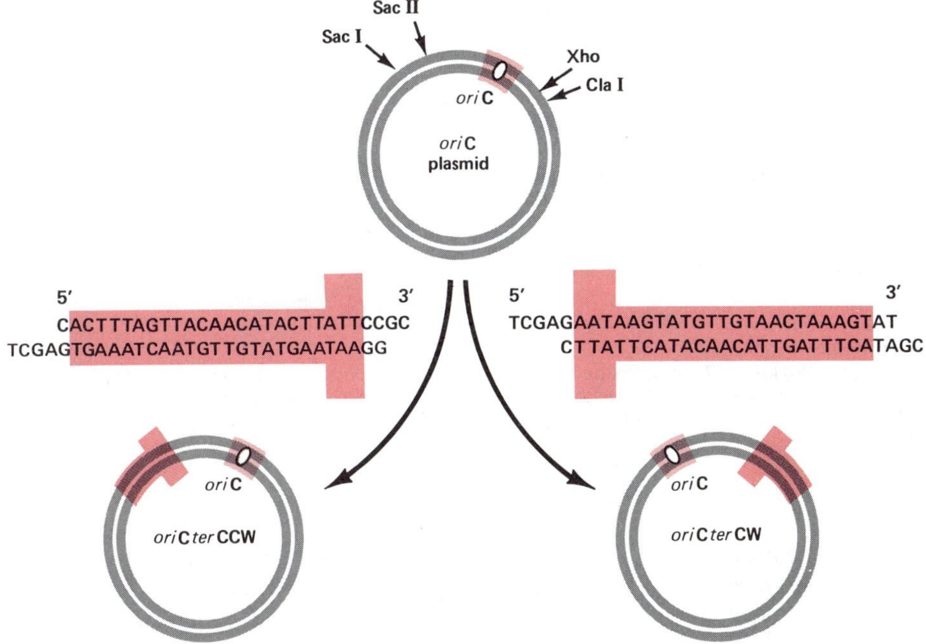

Figure 15-10
Construction of *oriC terCCW* and *oriC terCW* plasmids. Two complementary synthetic oligomers of 30 and 28 nucleotides were annealed to generate a 26-bp duplex containing the 23-bp *ter* consensus sequence and cohesive ends for restriction nuclease SacI and SacII. The duplex was inserted into *oriC* plasmid DNA (pCM959, 4012 bp) cleaved by SacI and SacII. In the resultant *oriC terCCW* plasmid circle (3812 bp; lower left), the *ter* sequence is oriented to block counterclockwise DNA replication. In the *oriC terCW* plasmid (4028 bp; lower right), generated in a similar way with Xho and ClaI, the *ter* sequence is oriented to block the clockwise DNA replication fork.

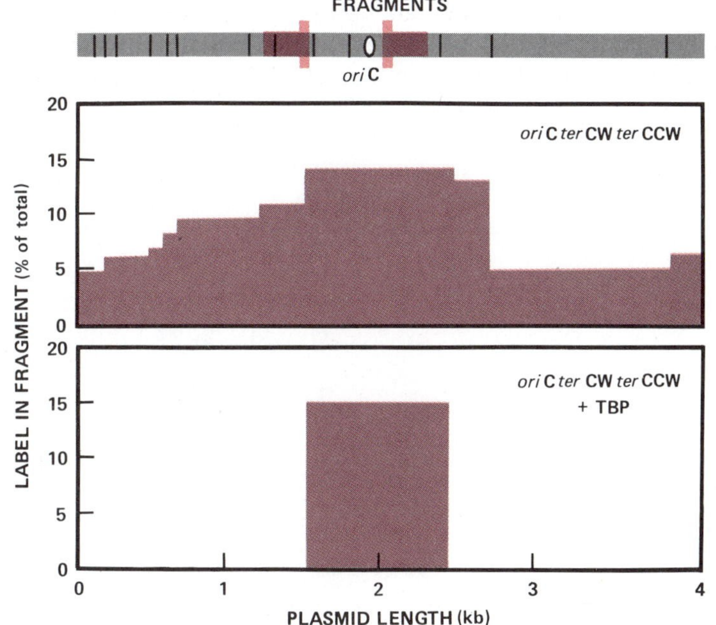

Figure 15-11
Demonstration that TBP-*ter* complexes block the progress of the replication fork in both directions. DNA fragments of the products of a short-term replication initiated at *oriC* were generated by EcoRV and HaeIII digestion and analyzed by polyacrylamide gel electrophoresis and autoradiography. The positions of the digestion fragments, minimal *oriC*, and *ter* (T-shaped area) are shown for plasmid *oriC terCW terCCW*.

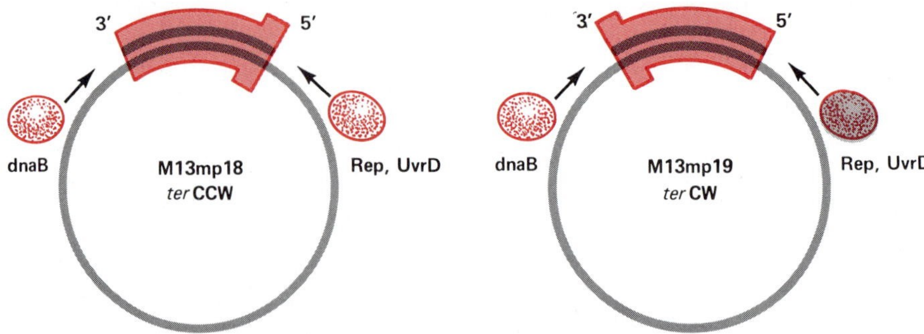

Figure 15-12
The effect of the *ter*–TBP complex on helicases. In M13mp18 *ter*CCW, the *ter* sequence (T-shaped area) is oriented to block counterclockwise replication, thus blocking Rep and UvrD; in M13mp19 *ter*CW it is oriented to block clockwise replication, thus blocking dnaB.

translocations of Rep and UvrD (helicase II) proteins are blocked only by the opposite orientation (Fig. 15-12).[90] When TBP binds the R6K *ter* sequence (a weak terminator in vivo), it does so with less than one-thousandth the affinity it has for the *E. coli ter* sequence. In this instance, the movement of dnaB helicase only is blocked, but not that of Rep and helicase II protein, which move with the opposite polarity.[91] The asymmetry of TBP binding imposed by the *ter* orientation appears to make the complex relatively resistant to disruption and consequent helicase action when encountered by a helicase coming from one direction but susceptible when approached from the other. How TBP binding to a *ter* sequence produces an orientation-specific blockage of a helicase that approaches from one direction and not from the other has yet to be examined.

The physiologic importance of a termination mechanism is uncertain. Absence of the mechanism in a number of genomes is tolerated and, in *E. coli*, deletion of the *ter* sequence or mutational inactivation of TBP has no apparent effect. Yet the pausing of replication forks as they near completion of a round may afford an interval for topological adjustments and may forestall undesirable consequences, such as a runaway rolling circle.

---

90. Lee EH, Kornberg A, Hidaka M, Kobayashi T, Horiuchi T (1989) *PNAS 86*:9104.
91. Khatri GS, MacAllister T, Sista PR, Bastia D (1989) *Cell 59*:667.

CHAPTER 16

Genome Origins

16-1 Identification of Origins

The initiation of replication takes place at a particular sequence in the parental DNA, designated the *origin*. The event is governed by factors that recognize the sequence and either direct the start of a chain de novo or introduce a nick which can be covalently elongated. The isolation of replication origins therefore assists in the identification and purification of initiation factors. The more accessible replicons of plasmids and viruses have been studied as models, and the origins of bacterial and yeast chromosomes have been transplanted into small plasmids to facilitate their analysis.

Two approaches to identifying origins are commonly used. The initiation sites can be (1) physically mapped on replication intermediates by a variety of electron microscopy and gel electrophoresis techniques or (2) functionally mapped by determining the *cis*-acting sequences required for initiation of replication on a DNA molecule. Though often identical, the sequences identified as origins by these physical and functional criteria do not necessarily coincide. In some cases, sequences essential for initiation are displaced by hundreds of base pairs from the actual site where DNA synthesis begins.

Physical Mapping of Origins

Replication Intermediates. The replication fork marks the advance of DNA synthesis within a replication intermediate of a duplex chromosome. Common intermediates include (Fig. 16-1) *theta structures*,[1] resulting from strand separa-

1. Cairns J (1963) *JMB* 6:208.

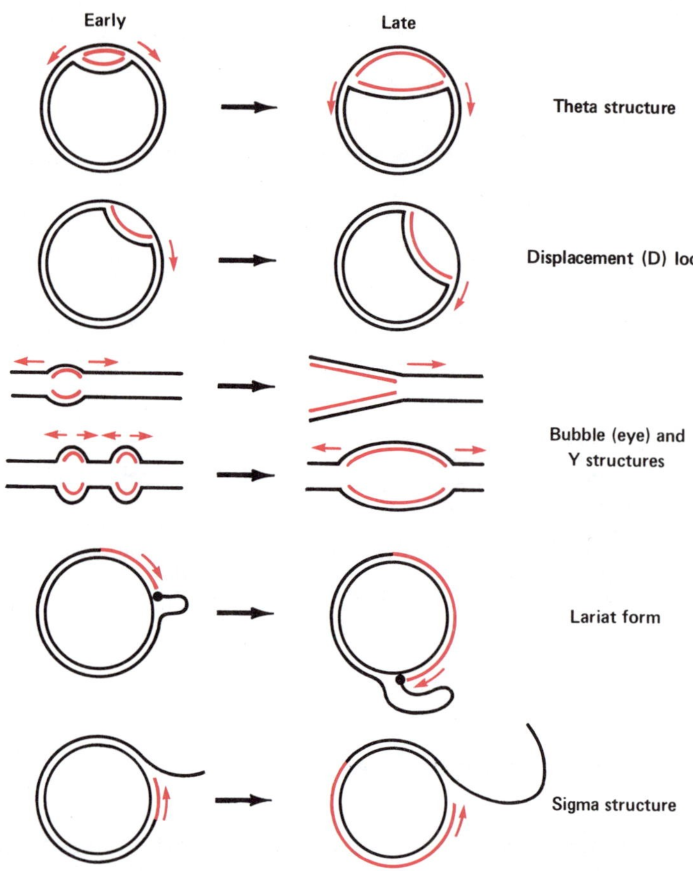

Figure 16-1
Structures of some common replication intermediates.

tion in a specific region of a circular chromosome (see below); *displacement (D) loops*, resulting from synthesis of one daughter strand in a circular or linear chromosome, with synthesis initially in one direction on one template strand; *bubble (eye)* and *Y structures*, resulting from initiation on a linear chromosome (see below); and *lariat (sigma) forms*, resulting from covalent extension of a parental strand in a circular chromosome, as in a rolling circle (Sections 8-7 and 15-11).

Techniques for Physical Mapping. Theta and other bubble structures are the predominant forms in duplex DNA replication and have been studied extensively. Several physical methods have been used to identify the origins and the direction of fork movement on these molecules. The different techniques have diverse advantages and will be described briefly.

(1) *Fiber autoradiography and electron microscopy* are commonly used techniques. An early insight into the structure of the replicating chromosome came from the use of autoradiography,[2] in which the emissions from [^{3}H]thymidine-DNA produce an image of the labeled structure on a photosensitive emulsion. When the *E. coli* chromosome is labeled by an interval of cell growth in [^{3}H]thy-

2. Cairns J (1963) *JMB* 6:208.

midine medium and is then displayed in an autoradiograph, the remarkable picture of a replicating circular structure emerges (Fig. 16-2).

Replicating chromosomes contain three segments, two of which make up the bubble and the third the parental duplex. Portions of the chromosome in which both strands were synthesized during the labeling period (and therefore are radioactive) have twice the grain density of those portions in which only one strand was newly synthesized and the other is the unlabeled, parental strand. The intermediate in this picture of the replicating chromosome, in planar projection, resembles the Greek letter θ and is thus called a *theta structure.*

The two forks in the theta structure can arise from initiation at a fixed point advancing in either one or both directions. Labeling of one fork indicates unidirectional growth; labeling of both forks, bidirectional growth (Fig. 16-2). The distance between the growing fork and the fork at the origin measures the extent of synthesis and locates the initiation site on unidirectionally replicating molecules. In contrast, the origin is found in the center of the bubble when replication is bidirectional and the rate of fork movement is equal in both directions (Fig. 16-3).

While autoradiography distinguishes between unidirectional and bidirectional replication, mapping the location of the origin requires that the molecules contain a fixed marker, visible in the electron microscope. Limited denaturation[3] of DNA molecules generates melted (denatured) loops in regions of high AT content; these loops serve as markers on the chromosome to determine the position of the origin and the direction of replication (Fig. 16-4). If replication is unidirectional, one fork always remains at the origin, and the distance between it and a marker remains constant throughout replication. If bidirectional, the distance between the marker and both forks varies with the extent of replication (Fig. 16-4).

The end of a molecule generated by a restriction enzyme can also provide a fixed marker for mapping replicating structures in the electron microscope.

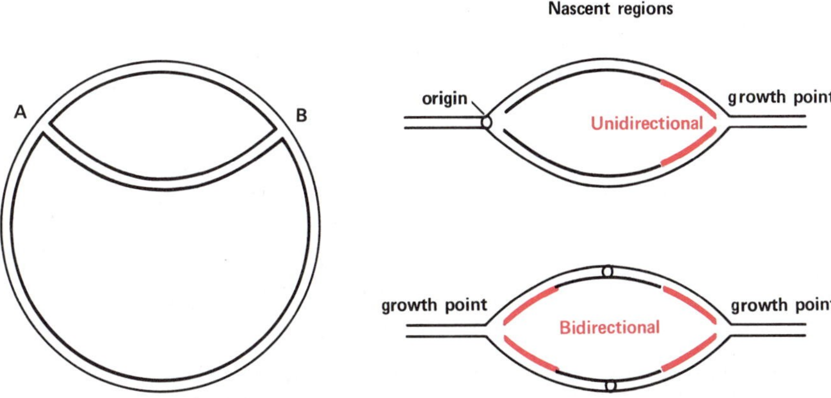

Figure 16-2
Nascent strands within the replicating region of a chromosome. Grain densities at A and B in autoradiographs show whether growth is unidirectional or bidirectional.

3. Inman RB (1966) *JMB* 18:464.

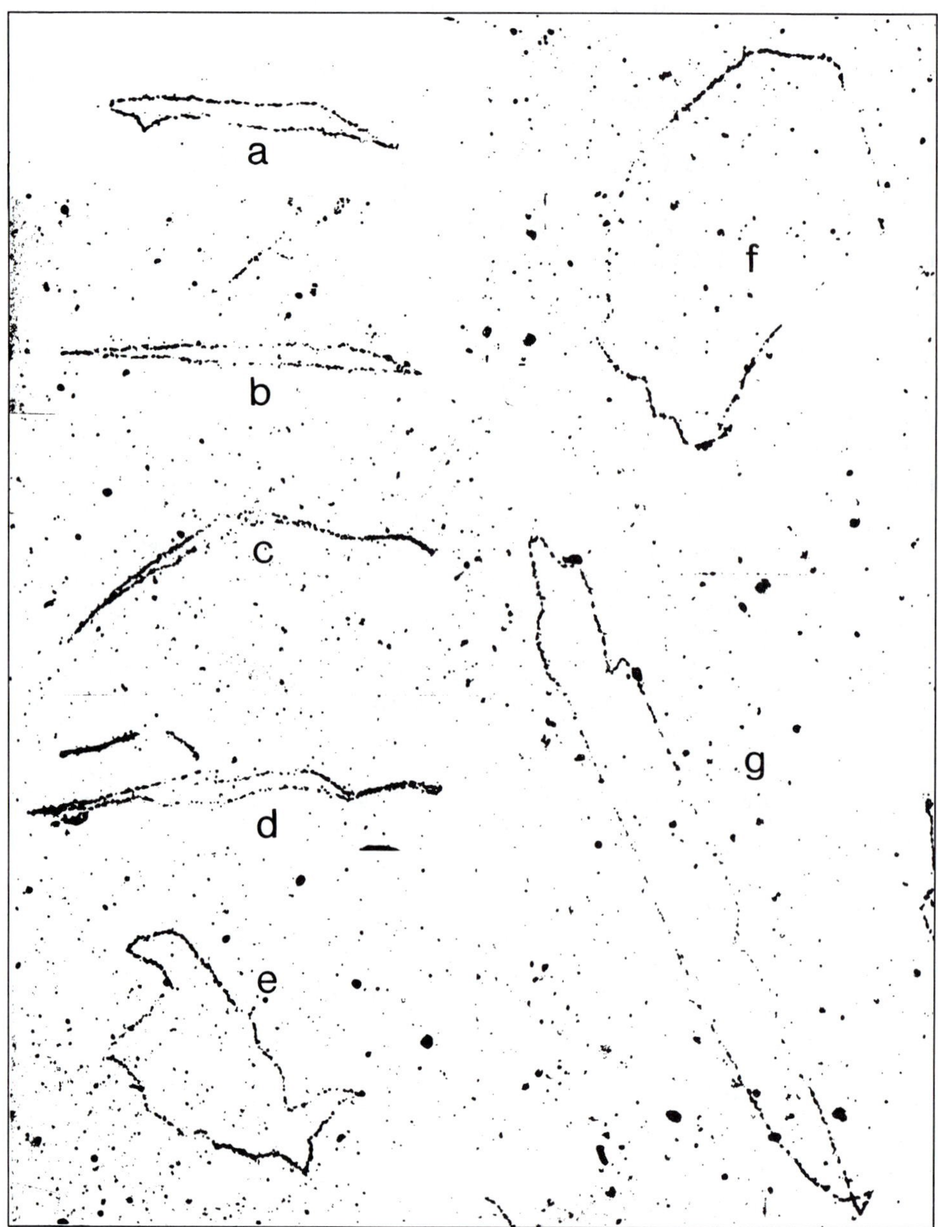

Figure 16-3
Autoradiographs of chromosomal loops in *B. subtilis* (germinating spores) indicating bidirectional growth. (a) Loop with even grain density. (b–g) Loops with heavy grain density at both ends or forks. [From Gyurasits EB, Wake RG (1973) *JMB 73*:55; courtesy of Professor RG Wake]

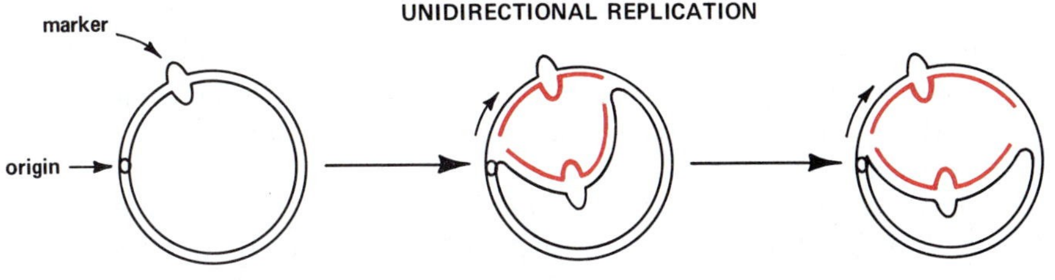

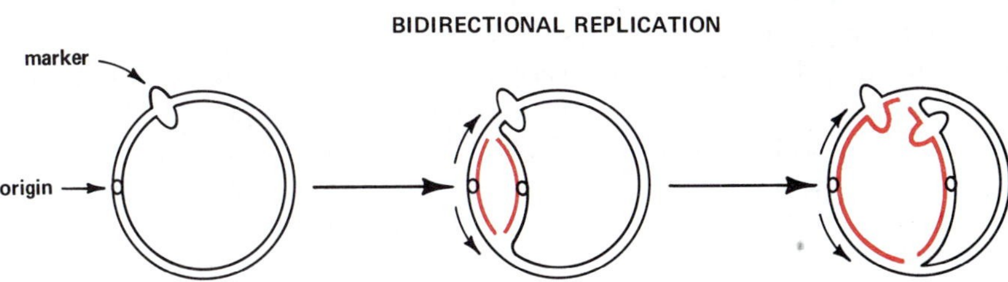

Figure 16-4
Denaturation mapping. Such mapping locates markers for determining the direction of chromosome growth.

When the positions of the replication forks with respect to the ends (located at a known position on the physical map) are plotted against the extent of DNA synthesis, two lines, representing the two forks, are generated (Fig. 16-5). If these lines have similar slopes that are opposite in sign, replication is bidirectional, at a similar rate in each direction. If one of the lines is horizontal, one fork is not moving during synthesis, indicating that replication is unidirectional. Extrapolation of the fork-position lines to the ordinate (the point of 0% replication) locates the origin.

These electron microscopy techniques have shown that replication of the circular genomes of λ and SV40, as well as of the linear phage T7, is bidirectional from a single, uniquely located origin. In contrast, the ColE1 theta structures show initiation from a specific site to be in only one direction.[4] Fiber autoradiography and electron microscopy reveal that bidirectional replication prevails

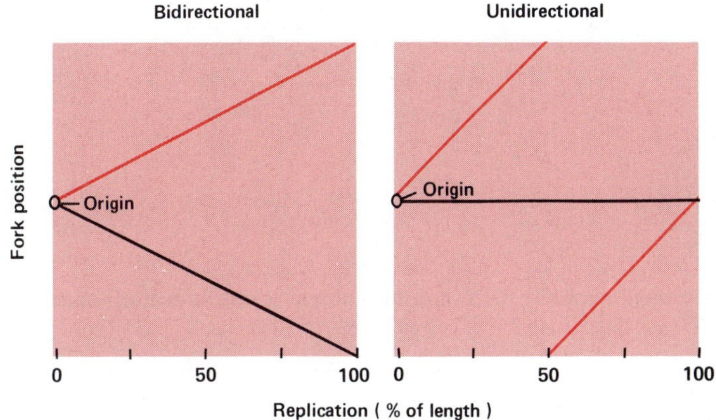

Figure 16-5
Idealized graphs of the position of replication forks at increasing extents of replication. Y-axis intercept (point of 0% replication) locates the origin. Left, bidirectional replication; right, unidirectional replication.

4. Lovett MA, Katz L, Helinski DR (1974) *Nature* *251*:337.

in the multiple and more complex chromosomes of yeast,[5] *Physarum*,[6] *Drosophila*,[7] and mammals.[8] The many starts (replicating units or replicons) form loops which are seen as bubbles or "eyes" along the DNA fiber in electron micrographs (Fig. 16-6). The sequences where these many initiations occur are currently being sought (Section 6).

(2) *Two-dimensional (2D) gel electrophoresis* is an effective way to map initiation sites, including those of eukaryotic genomes replicating in vivo. Two techniques have been developed that are based on the different migration properties of replication intermediates. Molecules are detected by hybridization with labeled probes from various regions of the DNA, making labeling of the replicating DNA unnecessary and allowing analysis of specific regions of the chromosome.

One technique[9] depends on the separation of restriction fragments by mass in one dimension and by the unique mobilities of DNA molecules that contain fork or bubble shapes in the second dimension. Examples of the electrophoretic patterns of such molecules are shown in Figure 16-7. Fragments that form an arc shape characteristic of a bubble are likely to contain an origin, while fragments exhibiting a simple Y pattern should lack one. Double-Y–shaped fragments are judged to contain the termination sites where bubbles coalesce. An asymmetrically located bubble (Fig. 16-7) provides the most information about the position of the origin: the point at which the arc shape changes from that of a bubble to that of a simple Y should mark the distance between the origin and the end of the fragment (i.e., the restriction cleavage site) closest to the origin. By analyzing the behavior of fragments generated by several differently located restriction sites, one can map the location of the origin to within a few hundred base pairs.

This technique, first used to analyze ARS elements on yeast plasmids (Section

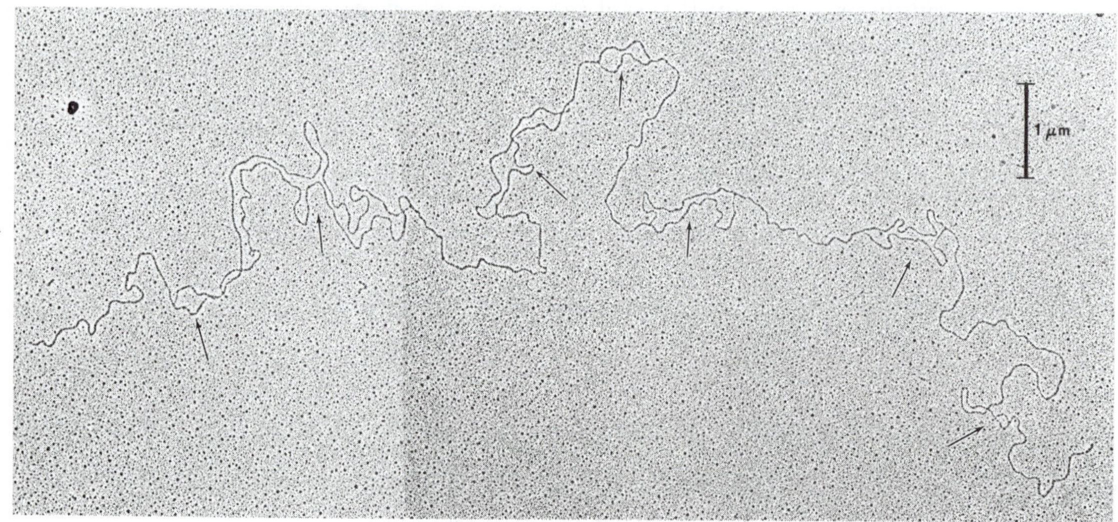

Figure 16-6
Electron micrograph of replicating *Drosophila* DNA showing multiple "eyes" (bubbles). (Courtesy of Dr H Kriegstein)

---

5. Callan HG (1973) *CSHS 38*:195.
6. Newton CS, Petes TD, Hereford LM, Fangman WL (1974) *Nature 247*:32; Petes TD, Williamson DH (1975) *Exp Cell Res 95*:103.
7. Kriegstein HJ, Hogness DS (1974) *PNAS 71*:135.
8. Huberman JA, Riggs AD (1968) *JMB 32*:327; Huberman JA, Tsai A (1973) *JMB 75*:5.
9. Brewer BJ, Fangman WL (1987) *Cell 51*:463.

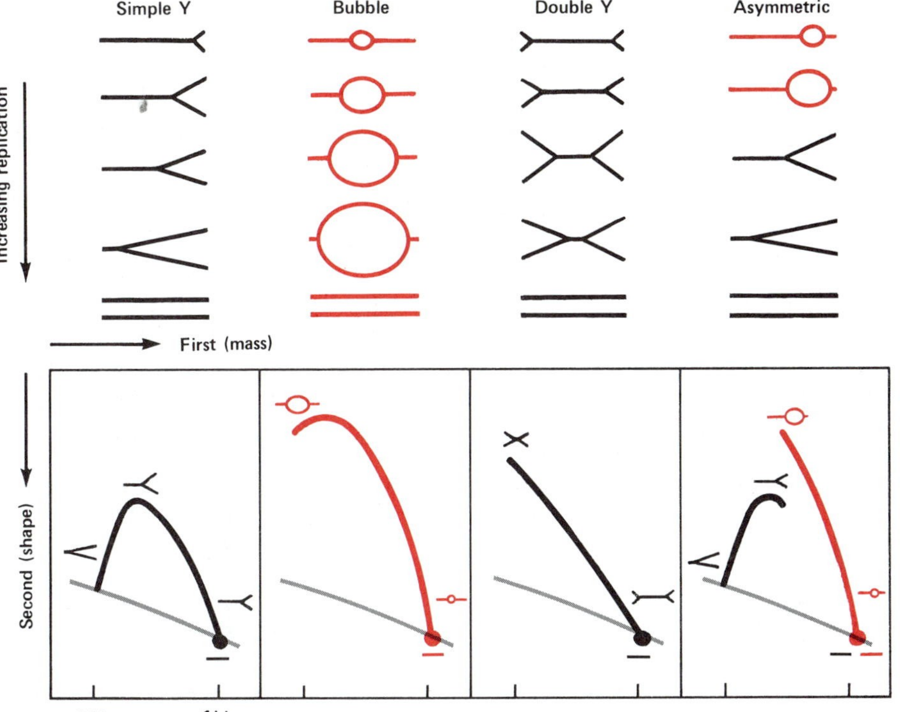

Figure 16-7
Types of replication intermediates generated by replication origins on an arbitrary 1-kb fragment. The lower panels depict the expected migration in two-dimensional agarose gels of the four types of replication intermediates shown; light lines indicate locations of linear molecules of various masses between 1 and 2 kb. [From Brewer BJ, Fangman WL (1987) *Cell 51*:463]

6), has since been applied to eukaryotic viral genomes and even to cellular chromosomes.

A second 2D electrophoresis method[10] (Fig. 16-8) separates the fragments of replicating molecules by mass under neutral conditions in the first dimension and by the size of single DNA strands at alkaline pH in the second dimension. The nascent strands form an arc on the gel, separated from the parental strands and from nonreplicating forms. The resulting pattern can be probed with labeled sequences from known parts of the fragment. Probes from very near the origin detect the entire arc; probes far from the origin hybridize to only the long chains. With multiple probes from different regions, the directions of fork movement can be determined. Analysis of multiple fragments localizes the origin and termination regions and distinguishes uni- from bidirectional replication.

This second technique is similar in sensitivity and accuracy to the one that depends on the shape of the molecules. Although both techniques employ 2D electrophoresis and yield similar information, they depend on totally different characteristics of the replicating molecules. Consequently, the application of both techniques to a given DNA molecule provides the clearest description of its replication pattern.[11]

---

10. Huberman JA, Spotila LD, Nawotka KA, El-Assouli SM, Davis LR (1987) *Cell 51*:473; Nawotka KA, Huberman JA (1988) *Mol Cell Biol 8*:1408.
11. Linskens MHK, Huberman JA (1990) *NAR 18*:647.

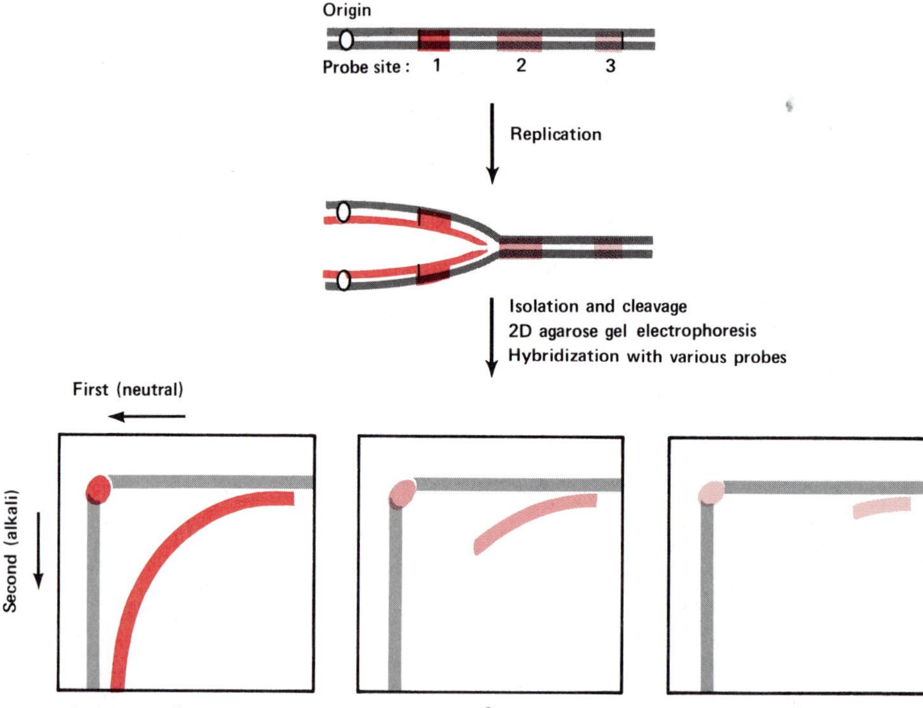

Figure 16-8
Outline of replicon-mapping by neutral–alkaline agarose gel electrophoresis. Details are given in the text. 1, 2, and 3 = cloned sequences to be used as hybridization probes. The lower panels represent autoradiograms of the same gel hybridized with the indicated probes. Each spot at upper left represents nonreplicating DNA, with randomly nicked strands extending downward and the parental strands extending horizontally; colored arcs are nascent strands detected by the indicated probes. [From Nawotka KA, Huberman JA (1988) *Mol Cell Biol* 8:1408]

(3) *One-dimensional gel electrophoresis* can in some cases localize origins, and is especially useful for DNA molecules replicating in vitro. Synchronously replicating molecules labeled with radioactive nucleotides are cleaved into multiple fragments by a restriction endonuclease and analyzed by gel electrophoresis. Early in replication, fragments near the origin are the most strongly labeled; as synthesis proceeds, fragments farther away acquire the label. Within limits imposed by the size of the fragments and the synchrony of replication, the origin can be localized. To improve synchrony in vitro, a preinitiation reaction can be introduced, followed by elongation for various times or with increasing levels of a chain-terminating dideoxynucleotide.

Inasmuch as a population is analyzed, rather than individual molecules (as in electron microscopy), the patterns of labeling may suggest that synthesis is bidirectional when, in fact, half of the molecules replicate in one direction and the other half in the opposite direction.

(4) *The non–cross-linked strand extrusion technique* enriches for nascent chains at origins and can be applied in vivo. DNA strands are cross-linked with psoralen at fairly regular intervals of about 5 kb. After the cross-linking, radio-

active precursors are added for a period to label the newly synthesized DNA.[12] Upon isolation and denaturation of the DNA, the short fragments released are principally nascent strands from the vicinity of replication origins. If strand synthesis has proceeded farther from the origin than the average distance between cross-links, the nascent strand should include at least one cross-link to a parental DNA and thus not be released by denaturation. In addition, the newly synthesized strands remain relatively short due to inhibition of fork movement by the cross-links in the template. Strands isolated in this way can be used as hybridization probes to localize the chromosomal regions likely to contain origins.[13]

(5) *The polymerase chain reaction (PCR)*, an extremely sensitive technique for detecting DNA sequences (Section 4-17), can be applied to mapping initiation sites in chromosomal regions of known sequence[14] (Fig. 16-9). Nascent chains

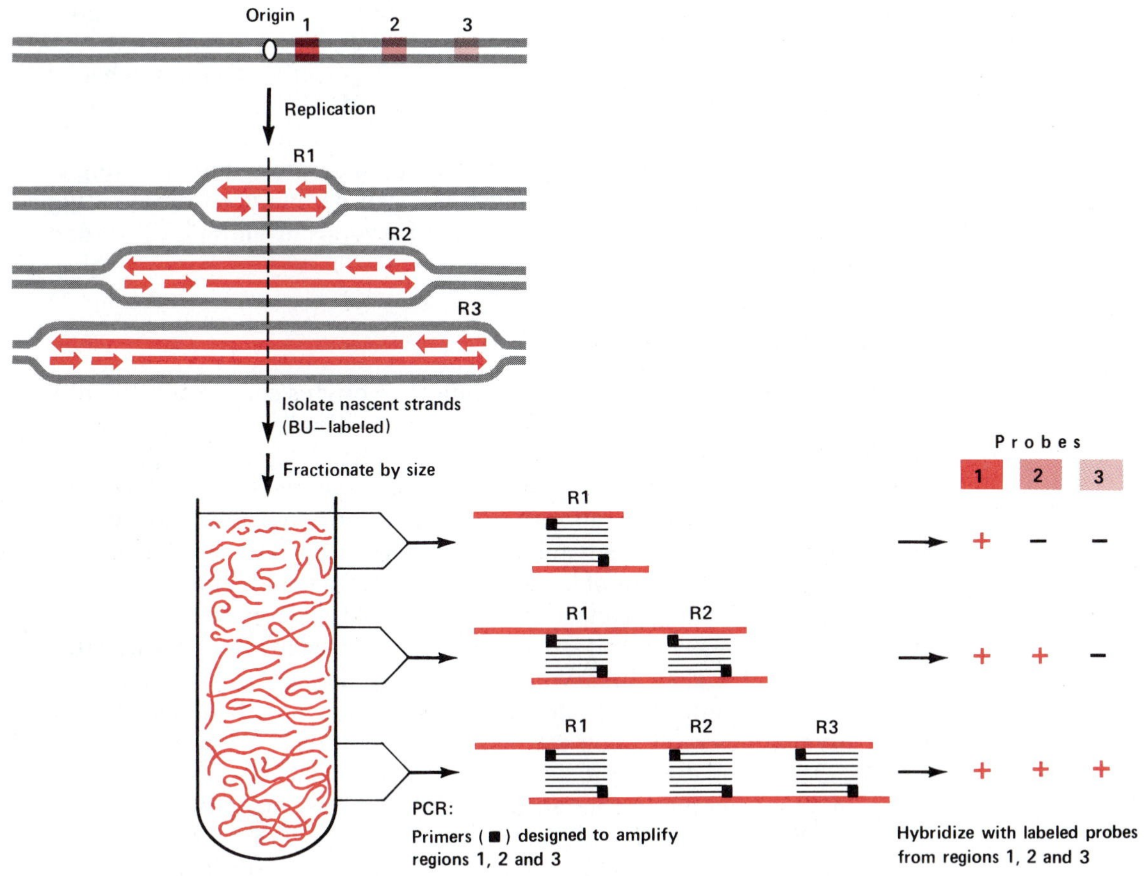

Figure 16-9
Mapping of replication origins by polymerase chain reaction (PCR) amplification of segments on nascent DNA strands. Black lines = parental DNA; colored lines = newly replicated DNA strands. [From Vassilev L, Johnson EM (1989) *NAR 17*:7693]

12. Anachkova B, Hamlin JL (1989) *Mol Cell Biol* 9:532; Russev G, Vassilev L (1982) *JMB 161*:77.
13. Anachkova B, Hamlin JL (1989) *Mol Cell Biol* 9:532; Russev G, Vassilev L (1982) *JMB 161*:77.
14. Vassilev L, Johnson EM (1989) *NAR 17*:7693.

are purified, separated into different classes by size, and amplified by PCR using pairs of primers from different regions. Amplification is successful only when the primers lie within a region of the template that has been replicated and thus is represented in the nascent-fragment pool. Primers near the origin will amplify sequences in all the DNA size classes, while those far away will work only on the long DNA chains. While conceptually similar to the neutral–alkaline 2D gel electrophoresis method, this technique is significantly more sensitive, simplifying its application to single-copy regions within the complex chromosomes of higher eukaryotes.

(6) *Okazaki fragment distribution* can be used to map the direction of replication. Because the template strands for continuous and discontinuous synthesis change abruptly at an origin of bidirectional replication, the polarity of fork movement through a region can establish whether the region is likely to contain an origin. One approach involves labeling newly synthesized chains with bromodeoxyuridine (BUdR).[15] The Okazaki fragments are then purified away from unlabeled parental DNA by immunoprecipitation with anti-BUdR antibodies. Hybridization of this DNA to strand-specific probes allows the direction of fork movement through the region to be determined. A similar technique has used DNA synthesized in the presence of a protein-synthesis inhibitor.[16] Mapping of origins by Okazaki fragment distribution, while not exact, is sensitive, and has been applied to single-copy regions of mammalian chromosomes. Although the replication pattern of large regions of DNA can be analyzed, unidirectional replication origins (about 10% of those in eukaryotic chromosomes) would not be detected.

Functional Mapping of Origins

On plasmid, phage, and viral chromosomes, the origins can often be identified by complementation assays in which proteins required for replication are provided *in trans*. However, the availability of multiple origins or pathways of initiation may complicate this approach. As an example, the multiple mechanisms of initiating phage T4 replication (Section 17-6) have thwarted attempts to identify the origin, despite advanced knowledge of T4 DNA synthesis.

The chromosomal origins of *E. coli* (including its plasmids and phages), yeast, and animal viruses have been cloned by their ability to promote the replication of linked sequences. This strategy has unfortunately been less effective with the chromosomes of higher eukaryotes (Section 6). Sequences identified in this manner are often called *ARS* (autonomous replication sequence) *elements*. They direct autonomous replication and impart a high transformation frequency to the linked DNA, allowing it to be maintained as an episome, rather than integrated into the genome. Although the genetic and metabolic requirements for ARS plasmid replication should be the same as those of the chromosome from which the ARS was derived, these plasmids are often lost at a high frequency at cell division for lack of a specific partitioning signal, such as a centromere.

A replication origin can be rendered conditionally essential, and thus more easily studied, by construction of a chimeric replicon containing two origins which function under different circumstances. In *E. coli*, the ColE1 (or pBR322)

---

15. Burhans WC, Vassilev LT, Caddle MS, Heintz NH, DePamphilis ML (1990) *Cell* 62:955.
16. Handeli S, Klar A, Meuth M, Cedar H (1989) *Cell* 57:909.

origin is especially useful because it is dependent on the host *polA* (pol I) function (Section 18-2). In a *polA*^{ts} strain, at the permissive temperature, a chimeric plasmid containing the ColE1 origin can be maintained. Upon a shift to the nonpermissive temperature, this origin can no longer function and plasmid replication must depend on the presence, effectiveness, and behavior of the other origin, derived from the chromosome, another plasmid, or a phage.

Once a DNA fragment containing an origin is cloned, mutagenesis can be used to delineate the sequences essential for origin function. Physical criteria must then be applied to establish that DNA synthesis is in fact initiated within the identified sequence. Sequences that stimulate DNA transformation can have several alternative functions; for example, they may provide binding sites for a topoisomerase or for the nuclear scaffold, or may function as a partition locus. Physical mapping has confirmed that in many instances, including the origins of the *E. coli* and SV40 chromosomes and yeast ARS elements, the *cis*-acting sequence required for replication is very near the site of initiation. Yet there are discrepancies, as with plasmid R6K (Section 18-5) and the ACE sequences required for amplification of the *Drosophila* chorion genes (Section 6). Furthermore, only part of the essential sequence in a bipartite origin, such as those of Epstein-Barr virus (Section 5) and phage P4 (Section 17-7), may correspond to the site of replication initiation.

16-2 The *E. coli* Chromosomal Origin *(oriC)*

Extensive analysis of the initiation of *E. coli* replication in vivo and in vitro has allowed the unique chromosomal origin *(oriC)* to be identified and its structure characterized. The pathway of initiation at *oriC* (Section 3) will serve as a basis for comparison of the structure and mechanism of initiation at the origins of other prokaryotic (Section 4) and eukaryotic (Sections 5 and 6) genomes.

Identification of *oriC* in Vivo

The single origin of the *E. coli* chromosome was detected and localized by several different strategies. DNA from this region is the first to be replicated after a return to the permissive temperature of *dnaA* or *dnaC* ts mutants, which are blocked at initiation but can complete an ongoing round of replication. Because the genes around the origin are replicated first, they are present in twice the abundance of those farther away. In fast-growing *E. coli*, repeated initiation before a round of replication is completed magnifies this bias (Section 20-2). These gene dosage experiments localize the origin to about 84 minutes on the genetic map.[17] Fragments from this region allow the replication of an F plasmid in a strain (Hfr) that carries an integrated copy of the plasmid,[18] indicating the presence of an origin insensitive to the stringent copy number controls of F.

The *oriC* region is essential. It can be deleted only when there is a secondary origin, provided by an integrated copy of a plasmid or phage (Section 18-1), or when cells are deficient in RNase H, allowing initiation from several distinct sites, known as *oriK* (Section 20-3).

---

17. Bird RE, Louarn JM, Martuscelli J, Caro L (1972) *JMB 70*:549; Master M, Broda P (1971) *NNB 232*:137.
18. Hiraga S (1976) *PNAS 73*:198.

Properties of *oriC* Plasmids

Recombinant plasmids dependent on *oriC* for their autonomous replication have been obtained by several methods. In one method (Fig. 16-10), the region responsible for autonomous replication of a drug-resistance plasmid was excised, and the remaining sequences were combined with fragmented *E. coli* DNA to generate a recombinant plasmid capable of self-replication. The plasmid was traced by its drug-resistance marker. The *oriC* region could be introduced either from the entire fragmented chromosome (the "shotgun" approach; Fig. 16-10) or from a phage in which *oriC* and nearby markers had first been selected and thereby enriched. No essential genes are found within the 2.2 kb region between *oriC* and *asnA* (asparagine synthase) to the right and within the 8.5 kb region between *oriC* and *unc* or *atp* (*ATP*-synthase operon) to the left (Fig. 20-7). Regions near *oriC* encode two nonessential proteins with unknown functions: a 70-kDa protein (the *gidA* gene) to the left and a 16-kDa protein (the *mioC* gene) to the right, with *oriC* contained within the intergenic space.

Replication of *oriC* plasmids requires the same functions as replication of the chromosome, including RNA synthesis; protein synthesis; and dnaA, dnaC, and gyrase functions.[19] Replication is bidirectional, initiating within the *oriC* sequence or slightly to its left (the *gidA* side).[20] Initiation of replication occurs at the same time within the cell cycle for both the plasmid and the chromosome.[21]

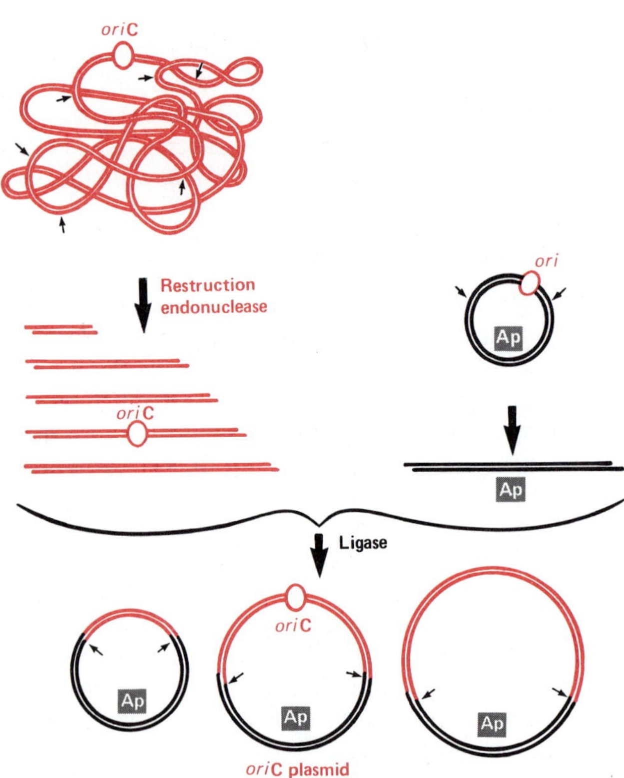

Figure 16-10
Construction of a plasmid, utilizing the *E. coli* replication origin. The origin region, released from an *E. coli* chromosome fragmented by a restriction endonuclease, is selected for the capacity to sustain autonomous replication of a plasmid carrying a gene (*Ap*) for drug (ampicillin) resistance. (Courtesy of Dr S Yasuda)

19. Von Meyenburg K, Hasen FG, Riise E, Bergmans HE, Meijer M, Messer W (1979) *CSHS 43*:121.
20. Meijer M, Messer W (1980) *J Bact 143*:1049.
21. Helmstetter CE, Leonard AC (1987) *J Bact 169*:3489; Leonard AC, Helmstetter CE (1986) *PNAS 83*:5101.

Thus, *oriC* plasmids are faithful minichromosome models for the behavior of the chromosomal origin. Even the presence of high copy numbers of *oriC* plasmids has little deleterious effect on host cell replication or division. Incompatibility between the plamid-borne origins and the chromosomal copies is seen only under special circumstances, such as reduced activity of dnaA protein (Section 20-3). The *oriC* plasmids are, however, unstable, for lack of a mechanism to ensure equal partitioning at cell division; they can be stabilized by the partition functions of some naturally occurring low-copy-number replicons, such as F plasmid[22] (Section 18-6).

Sequence Organization of *oriC*

The functional regions of *oriC* have been determined by using a chimeric ColE1–*oriC* replicons. Plasmid maintenance is assayed by comparing the function of a linked gene (e.g., an antibiotic-resistance determinant) in a *polA*$^+$ and a *polA*$^-$ host. Deletions, insertions, substitutions of one or more base pairs, and biochemical studies reveal several remarkable features within *oriC*[23] (Fig. 16-11).

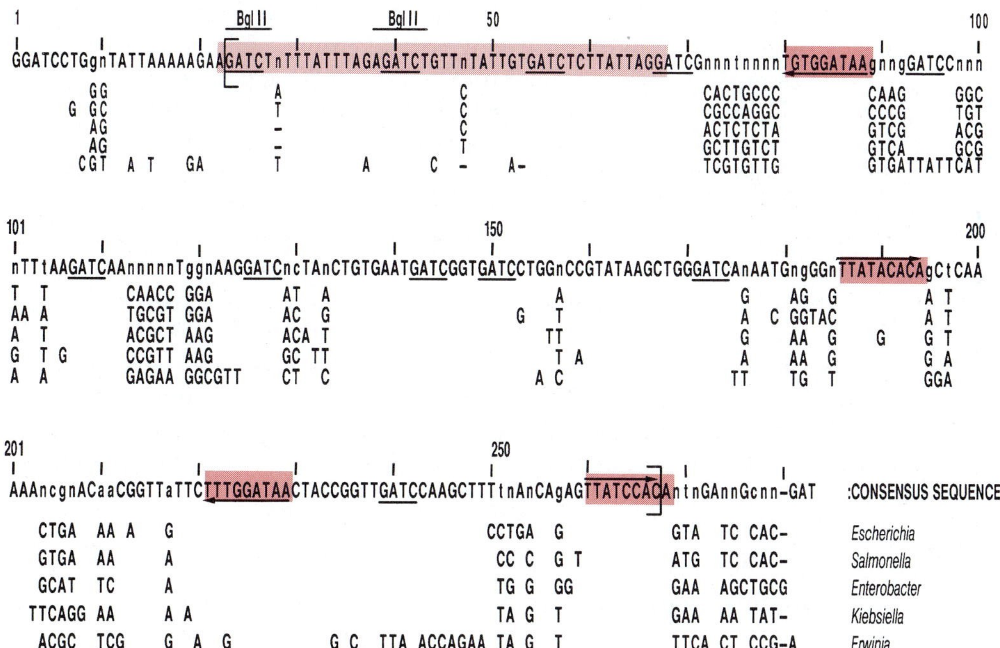

Figure 16-11
Minimal origin of enteric bacterial chromosomes. The numbering of the nucleotide positions in the consensus sequence is the one first used for *E. coli*, and the upper left end is the 5′ end. Capital letter = the same nucleotide in all five origins; lowercase letter = the same nucleotide in three of the five bacterial origins, with only two different nucleotides at that site; n = any of the four possible nucleotides, or a deletion (—); underline = *dam* methylation sites; colored box = 13-mer; box with arrow = dnaA binding sequence. The 245-bp minimal origin is in brackets. (Courtesy of Dr J Zyskind)

22. Løbner-Olesen A, Atlung T, Rasmussen KV (1987) *J Bact 169*:2835.
23. Oka A, Sasaki H, Sugimoto K, Takanami M (1984) *JMB 176*:443; Oka A, Sugimoto K, Sasaki H, Takanami M (1982) *Gene 19*:59; Asada K, Sugimoto K, Oka A, Takanami M, Hirota Y (1982) *NAR 10*:3745.

(1) The minimal required region is 245 bp.[24] The efficiency of initiation is, however, influenced by flanking sequences and by the position and orientation of *oriC* within the host plasmid.

(2) Five sequence blocks (15 to 20 bp) are highly conserved among the *oriC* regions of enteric bacteria (*Escherichia, Salmonella, Enterobacter, Klebsiella, Erwinia*).[25] Elsewhere in the genomes of these organisms, some separated by 100 million or more years of evolution, the sequences are randomized.

(3) In the conserved regions, most substitutions are not tolerated.

(4) Regions of largely random sequence but of fixed length separate the conserved regions and may serve as spacers to confer a unique conformation on *oriC*.

(5) Four 9-mers are the specific recognition sites for binding dnaA, the initiator protein (Section 3). The consensus sequence [5′-TTAT(C/A)CA(C/A)] also occurs in the origins of many plasmids and in the promoter regions of several genes, including *dnaA*.

(6) Three 13-mers at the left edge of *oriC* are the sites for opening the duplex for initiation and loading of the dnaB helicase (Sections 3 and 11-2). This sequence has a high propensity for melting in supercoiled DNA in the absence of any protein factors,[26] as well as for interacting with dnaA protein (Section 3).

(7) Fourteen GATC sequences are located in the 300 bp of *oriC* and neighboring regions, instead of the one or two expected in a 300-bp interval. GATC is a Dam methylase site, but the relatively normal growth of *dam* mutants argues against an essential methyl acceptor role for the GATC sequences. Hemimethylated *oriC* plasmids are, however, unable to replicate in vivo,[27] and specifically bind to membrane fractions in vitro.[28] The time required to methylate the origin after passage of the replication fork is about ten times longer than that observed for other regions of the chromosome.[29] The accuracy of timing of the firing of multiple origins present in the same cell is altered[30] in *dam* mutants, and initiation occurs at random times in the cell cycle.[31] All of these observations suggest that methylation may play an important role in *oriC* function during replication or cell division (Sections 20-3, 20-4, and 20-5).

(8) Two back-to-back promoters may be loci for RNA polymerase action during the initiation of DNA synthesis rather than being connected with mRNA synthesis.[32] Several potential secondary structures exist in the *oriC* sequence,[33] but their significance is uncertain in view of equally attractive branched and stem-and-loop structures outside the *oriC* region.

16-3 Initiation from the *E. coli* Chromosomal Origin *(oriC)*

The establishment of a cell-free system that replicates *oriC* plasmids opened the way to identifying the factors involved in the process and understanding how

---

24. Oka A, Sugimoto K, Takanami M, Hirota Y (1980) *MGG 178*:9.
25. Zyskind JHW, Cleary JM, Brusilow WS, Harding NE, Smith DW (1983) *PNAS 80*:1164.
26. Kowalski D, Eddy MJ (1989) *EMBO J 8*:4335.
27. Russell DW, Zinder ND (1987) *Cell 50*:1071.
28. Ogden GB, Pratt MJ, Schaechter M (1988) *Cell 54*:127.
29. Campbell JL, Kleckner N (1990) *Cell 62*:967.
30. Boye E, Løbner-Olesen A, Skarstad K (1988) *BBA 951*:359; Bakker A, Smith DW (1989) *J Bact 171*:5738.
31. Koppes L, Nordstrom K (1986) *Cell 44*:117.
32. Lother H, Messer W (1981) *Nature 294*:376.
33. Hirota Y, Yamada M, Nishimura A, Oka A, Sugimoto K, Asada K, Takanami M (1981) *Prog N A Res 26*:33.

they initiate replication. This system, in its crude and purified states, has clarified the steps in initiation at *oriC* and has served as a model for studies of initiation of phages λ and Mu and several plasmids in *E. coli* and SV40 in primate cells. The detailed description of the *oriC* system that follows assumes that its patterns may be as prototypical for chromosome initiation as *E. coli* DNA polymerase I has been for chain growth.

Replication in Crude Extracts[34]

The *oriC* plasmid and the small phage–*oriC* chimera (Section 2) were essential for the discovery of the *oriC* enzyme system. An ammonium sulfate fraction (fraction II) prepared from cell lysates that can replicate such *oriC* templates has many of the physiologically relevant features. The soluble enzyme system shows the following characteristics:

(1) It depends completely on exogenously furnished, supercoiled *oriC* plasmids that contain the intact origin sequence.
(2) Replication initiates within or near the *oriC* sequence and proceeds bidirectionally.[35]
(3) The enzyme system depends on RNA polymerase and gyrase, as indicated by total inhibition by rifampicin or nalidixate.
(4) The system depends on certain replication proteins (e.g., dnaB protein, dnaC protein, SSB) but not others (e.g., DNA pol I, RecA protein), as judged by inhibition with specific antibodies.
(5) It operates independently from protein synthesis.
(6) It depends on dnaA activity, as shown by the inactivity of enzyme fractions from each of two *dnaA*^{ts} mutant strains and complementation by homogeneous dnaA protein.[36]

The reasons why this potent system for the recognition of *oriC* remained elusive for so many years became apparent only when the many factors that activate and inhibit it were identified and characterized. Several features are noteworthy:

(1) Cell lysates are uniformly inactive.[37] Only a rather narrow ammonium sulfate fraction (fraction II) derived from the cell lysate is active. Sufficient salt is needed to precipitate and concentrate the many required proteins; more salt precipitates inhibitory factors.
(2) The optimal concentrations of the supercoiled *oriC* plasmid and of Mg^{2+} fall within narrow ranges.
(3) A strong ATP-regenerating system is essential to maintain ATP at 2 mM.
(4) A flexible, hydrophilic polymer, such as polyethylene glycol, polyvinyl alcohol, or methylcellulose, is absolutely required. The polymer increases the effective concentrations of macromolecular reactants by an "excluded volume" effect,[38] in essence providing the "macromolecular crowding" characteristic of intact cells.

---

34. Fuller RS, Kaguni JM, Kornberg A (1981) *PNAS 78*:7370.
35. Kaguni JM, Fuller RS, Kornberg A (1982) *Nature 296*:623.
36. Fuller RS, Kornberg A (1983) *PNAS 80*:5817.
37. Fuller RS, Kaguni JM, Kornberg A (1981) *PNAS 78*:7370.
38. Zimmerman SB, Trach SO (1988) *BBA 949*:297; Zimmerman SB, Harrison B (1987) *PNAS 84*:1871.

DnaA Protein: The Initiation Factor[39]

The in vitro system for the replication of *oriC* plasmids made it possible to purify the dnaA protein by complementing the replication defect of an extract made from *dnaA* mutant cells.[40] Several characteristics of *dnaA* mutants indicated that this protein is a central player in the initiation of chromosomal replication. All *dnaA* alleles are of the slow-stop phenotype, indicating involvement of the protein in initiation (Section 15-4). The *dnaA* gene is also unique among those involved in replication in that the absence of fast-stop alleles indicates that the protein function is limited to initiation. Although many phage and plasmid replicons appeared to be independent of dnaA protein, in being able to replicate in ts mutants at the nonpermissive temperature, more stringent *dnaA* null alleles and in vitro replication systems reveal a role for dnaA protein in the replication of most plasmids (Chapter 18) and some phages (e.g., P1; Section 17-7). In these instances, dnaA protein appears to play an accessory role, rather than being the primary initiator protein as it is for *oriC* (Section 4).

The dnaA protein is a monomer of 52 kDa that tends to aggregate into large complexes during purification. The protein binds ATP tightly ($K_d = 0.03\ \mu M$) and hydrolyzes it very slowly to ADP in a DNA-dependent manner with no notable sequence preference.[41] Low-affinity ATP binding is also required. The amino acid sequence contains a region homologous to the adenine nucleotide–binding consensus sequence.[42] The region of the protein involved in site-specific DNA binding is not known, and there is no evidence for DNA-binding motifs, such as helix–turn–helix or zinc fingers (Section 10-9). A region related to the hydrophobic coil–coil motif is present,[43] and may be involved in interactions between subunits.

The dnaA protein binds specifically to the 9-bp dnaA boxes in the *oriC* sequence, in various plasmid origins, and in the control regions of several genes, including its own.[44] Binding to *oriC* is cooperative, and a region encompassing about 200 bp is protected from DNase I digestion by the bound protein. This complex is visible by electron microscopy and is estimated to contain about 30 monomers of the protein, with the DNA wrapped around the periphery of the complex (Fig. 16-12 and Plate 15). Both the ATP- and ADP-bound forms of the protein bind effectively to *oriC*.[45] However, the structures differ,[46] and the ADP form is unable to progress to the next stages of initiation.[47]

The dnaA protein has three distinct functions in initiation at *oriC* in vitro. It (1) binds to the origin, recognizing it as the site for initiation to take place,[48] (2) promotes melting of the DNA strands,[49] and (3) functions in loading the dnaB helicase at *oriC*.[50] This initiation pathway, and the involvement of dnaA protein, is discussed in more detail below.

---

39. Fuller RS, Kornberg A (1983) *PNAS 80*:5817; Fuller RS, Funnell BE, Kornberg A (1984) *Cell 38*:889; Sekimizu K, Bramhill D, Kornberg A (1987) *Cell 50*:259.
40. Fuller RS, Kornberg A (1983) *PNAS 80*:5817.
41. Sekimizu K, Bramhill D, Kornberg A (1987) *Cell 50*:259.
42. Sekimizu K, Bramhill D, Kornberg A (1987) *Cell 50*:259.
43. Kim P, personal communication.
44. Fuller RS, Funnell BE, Kornberg A (1984) *Cell 38*:889.
45. Sekimizu K, Bramhill D, Kornberg A (1987) *Cell 50*:259.
46. Baker TA, Crooke E, Kornberg A, Griffith JA, unpublished.
47. Sekimizu K, Bramhill D, Kornberg A (1987) *Cell 50*:259; Bramhill D, Kornberg A (1988) *Cell 52*:743.
48. Fuller RS, Funnell BE, Kornberg A (1984) *Cell 38*:889.
49. Bramhill D, Kornberg A (1988) *Cell 52*:743.
50. Bramhill D, Kornberg A (1988) *Cell 52*:743; Baker TA, Sekimizu K, Funnell BE, Kornberg A (1986) *Cell 45*:53.

Replication with Purified Proteins[51]

At least 13 proteins participate in *oriC*-directed DNA replication, as judged by their activity in the reconstituted replication reaction. These factors can be divided into four functional classes:

(1) *Prepriming proteins* act early in initiation to generate an isolatable intermediate that is competent for priming and initiation of DNA synthesis. These factors include[52] the dnaA protein, dnaB and dnaC proteins (Section 8-4), the dsDNA-binding protein HU (Section 10-8), and, in some instances, RNA polymerase (Chapter 7).[53]

(2) *Specificity proteins* are required to maintain the dependence of the replication reaction on the origin sequence and dnaA protein when RNA polymerase is present. These factors apparently abort initiations that occur elsewhere than *oriC*. The most potent specificity factor is RNase H[54] (Section 13-10), which degrades the RNA portion of RNA–DNA hybrids. Mutants deficient in this activity *(rnh⁻)* allow initiation at locations other than *oriC* and have a "stable DNA replication" phenotype (Section 20-3). Topoisomerase I,[55] which relaxes negatively supercoiled DNA, and HU protein also have specificity activity.

(3) *Elongation proteins* are involved in the priming of DNA chains, in fork movement, and in DNA synthesis. Included are the dnaB helicase (Section 11-2), primase (Section 8-3), DNA pol III holoenzyme (Section 5-6), gyrase (Section 12-3), and SSB (Section 10-2).

(4) *Processing proteins,* acting late in replication, convert the nearly completed daughter plasmids into intact monomeric supercoils. DNA pol I (Chapter 4), RNase H, DNA ligase (Section 9-2), and gyrase all participate in this process.[56]

With these 13 purified proteins, negatively supercoiled plasmids containing *oriC* can be efficiently replicated. Initiation is within or near *oriC* and replication proceeds bidirectionally.[57] Daughter plasmids with the expected structure are generated in good yield.[58] This defined system makes it possible to analyze the mechanisms of initiation and replication fork movement and to search for additional proteins that may regulate and modify their operation. There might also be a role for some of the primosomal components (PriA, PriB, PriC, and dnaT proteins) essential for ϕX174 initiation (Section 8-4) and for other helicases (Sections 11-2 and 11-3). Factors may act directly on *oriC* in much the same way as the numerous factors that activate and silence promoter regions for transcription. Clearly important, but not included in the reconstituted system, are the factors that control the methylation and membrane orientation of *oriC*. Significant evidence of a role for Dam methylation of *oriC* has been found in vivo (Sections 2, 20-3, and 20-4), and indications of attachment of the origin and dnaA protein to the cell membrane are too numerous and persuasive to be ignored (Section 20-4).

---

51. Kaguni JM, Kornberg A (1984) *Cell 38*:183.
52. Funnell BE, Baker TA, Kornberg A (1987) *JBC 262*:10327.
53. van der Ende A, Baker TA, Ogawa T, Kornberg A (1985) *PNAS 82*:3954; Ogawa T, Baker TA, van der Ende A, Kornberg A (1985) *PNAS 82*:3562.
54. Ogawa T, Pickett GG, Kogoma T, Kornberg A (1984) *PNAS 81*:1040; Ogawa T, Baker TA, van der Ende A, Kornberg A (1985) *PNAS 82*:3562.
55. Kaguni JM, Kornberg A (1984) *JBC 259*:8578.
56. Funnell BE, Baker TA, Kornberg A (1986) *JBC 261*:5616.
57. Funnell BE, Baker TA, Kornberg A (1987) *JBC 262*:10327.
58. Funnell BE, Baker TA, Kornberg A (1986) *JBC 261*:5616.

Stages in Initiation of Replication from *oriC*[59]

Initiation proceeds through a series of stages identified by the proteins, nucleotides, and reaction conditions required, and by the changes observed in the DNA (Plate 15).

Initial Complex Formation. The first stage involves the cooperative binding of dnaA protein to its several binding sites in the origin, primarily to the four 9-bp dnaA boxes. Nearly 200 base pairs, including all but the three 13-mers in the leftmost part of the *oriC* sequence, are protected against DNase I attack. The complex appears nearly spherical by electron microscopy (Fig. 16-12), with the DNA wrapped around the outside of the protein.[60] Although the exact stoichiometry is not known, 20 to 40 dnaA protein monomers are involved, as judged by microscopy and binding studies. Complex formation occurs at lower temperatures than are required for initiation and does not require a supercoiled template. ATP is not essential.

Open Complex Formation. This depends on a structural change in the dnaA·*oriC* complex and results in localized melting of the DNA within the 13-bp repeats at the leftmost edge of *oriC*.[61] Opening, detected by sensitivity of this region to the single-strand–specific endonuclease P1 (Section 13-5), requires that at least a fraction of the dnaA protein bind ATP. The template must be negatively supercoiled ($\sigma = -0.04$ or more underwound),[62] and an elevated reaction temperature is required (at least 21°, depending on the exact reaction conditions). Small amounts of HU protein, or the related IHF, stimulate this reaction, probably by stabilizing a desirable bend in the DNA.[63]

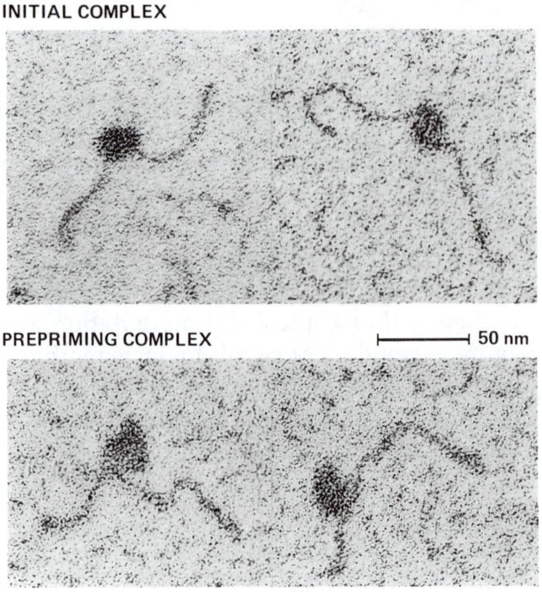

INITIAL COMPLEX

PREPRIMING COMPLEX — 50 nm

Figure 16-12
Electron micrographs of protein complexes at *oriC*. Initial complexes (above) were formed on a supercoiled *oriC* plasmid with dnaA protein only. Prepriming complexes (below) were formed with dnaA, dnaB, dnaC, and HU proteins. Complexes were cross-linked and the DNA was cut with a restriction endonuclease. Protein complexes are seen at the *oriC* site asymmetrically situated on the DNA fragment. [From Funnell BE, Baker TA, Kornberg A (1987) *JBC 262*:10334]

59. Bramhill D, Kornberg A (1988) *Cell 52*:743; Bramhill D, Kornberg A (1988) *Cell 54*:915; Baker TA, Sekimizu K, Funnell BE, Kornberg A (1986) *Cell 45*:53.
60. Fuller RS, Funnell BE, Kornberg A (1984) *Cell 38*:889.
61. Bramhill D, Kornberg A (1988) *Cell 52*:743.
62. Baker TA, Kornberg A (1988) *Cell 55*:113.
63. Skarstad K, Baker TA, Kornberg A (1990) *EMBO J 9*:2341.

Open complex formation probably involves a specific interaction between dnaA protein and one or more of the 13-mers, although this binding is not revealed by DNase I footprinting. The reduced stability of dnaA · *oriC* complexes on templates from which the 13-mers are deleted,[64] competition by a synthetic 13-mer oligonucleotide with dnaA protein binding to *oriC*, and the presence of 13-mer sequences near dnaA boxes in other origins[65] (Section 4) and in certain promoters, all indicate some specificity in the interaction between 13-mers and the protein. One 13-mer is sufficient for open complex formation, but all three are required to form the prepriming complex, a further indication that the three 13-mers form a functional unit. The thermodynamic disposition of this sequence to be melted is manifested by its susceptibility to cleavage by single-strand–specific endonucleases, in the absence of any proteins, under conditions that exploit the available energy of supercoiling.[66] A regulatory role in initiation may be performed by a protein (IciA) that binds the 13-mers and prevents dnaA protein from forming the open complex (Section 20-3).

Prepriming Complex Formation. The association of dnaB protein with *oriC* in the prepriming complex represents the first step at which binding of dnaA protein to the origin results in the assembly of replication enzymes for the initiation of DNA synthesis.[67] DnaB protein is donated from a complex with dnaC protein, formed in solution with ATP (Section 8-4). Although critical to assembly, dnaC protein is not stably maintained in the complex,[68] in which it inhibits the helicase activity of dnaB protein and so must be ejected for replication to proceed.

The prepriming intermediate, containing dnaA and dnaB proteins and probably HU protein bound to *oriC*, is stable, isolatable, and resistant to low temperature. It is larger and more asymmetric in shape than the initial complex (Fig. 16-12) and encompasses the additional 50 bp at the left edge of *oriC*, the location of the 13-mer repeats;[69] the 13-mers are melted in the complex, as judged by extended sensitivity to P1 nuclease. Two restriction enzyme cleavage sites within the 13-mer region are also protected, further indicating that this DNA is altered by melting, or by protein binding, or both.[70]

Evidence for a direct role of dnaA protein in the loading of dnaB protein, in addition to opening the duplex at the 13-mers, is of several kinds.

(1) DnaB protein binds only poorly to ssDNA, and the levels of dnaB protein present in the *oriC* replication reaction are probably inadequate to support its association unaided.

(2) Normally, dnaB protein is unable to bind to SSB-coated ssDNA. If dnaA protein is allowed to associate with the DNA prior to the addition of SSB, dnaB protein, with the aid of dnaC protein, is able to utilize this DNA as a template.[71]

(3) A hairpin region isolated from the plasmid R6K contains the sequence of a dnaA box in the stem and binds dnaA protein. Assembly of dnaB onto this

64. Yung B, Kornberg A (1989) *JBC 264*:6146.
65. Bramhill D, Kornberg A (1988) *Cell 52*:743.
66. Kowalski D, Eddy MJ (1989) *EMBO J 8*:4335.
67. Baker TA, Sekimizu K, Funnell BE, Kornberg A (1986) *Cell 45*:53; Baker TA, Funnell BE, Kornberg A (1987) *JBC 262*:6877; Funnell BE, Baker TA, Kornberg A (1987) *JBC 262*:10327.
68. Baker TA, Allen GC Jr, Kornberg A, unpublished.
69. Funnell BE, Baker TA, Kornberg A (1987) *JBC 262*:10327.
70. Bramhill D, Kornberg A (1988) *Cell 52*:743.
71. Wahle E, Lasken R, Kornberg A (1989) *JBC 264*:2469.

template occurs with the assistance of dnaC protein.[72] This sequence substitutes for a ϕX-type primosome assembly site in plasmid replication of ColE1-type plasmids (Section 8-4).

(4) DnaA boxes and primosome assembly sites are functionally equivalent for the initiation of discontinuous strand synthesis during plasmid replication, both serving to load dnaB protein (Section 8-4).

Thus, dnaA protein helps dnaB protein to bind DNA under circumstances in which neither formation of the large protein structure assembled at *oriC* nor strand opening is involved. The dnaC protein is always required, indicating that the protein–protein interaction may be between dnaA protein and either dnaB or dnaC, or the complex of both proteins.

Unwinding and Priming. Melting of the DNA duplex proceeds bidirectionally from *oriC* as the dnaB helicase migrates from the dnaA · *oriC* complex to provide a template for the priming and replication enzymes (Fig. 16-13).[73] Priming and DNA synthesis start promptly upon the addition of these enzymes to a prepriming complex. Primase, through its interaction with dnaB protein (Section 8-4), is responsible for priming both continuous and discontinuous replication. Primase and DNA pol III holoenzyme have no apparent preference for *oriC* but initiate wherever dnaB protein provides a template.

Replication Fork Movement. The helicase action of dnaB protein melts the duplex for replication of the entire template.[74] The dnaB–primase interaction probably generates a mobile primosome, needed for the multiple primings of discontinuous-strand synthesis. The 5′→3′ polarity of the dnaB helicase (Section 11-2) places it on the template for the discontinuous strand, moving in the direction of the replication fork. The lengths of the nascent fragments depend on the primase concentration; at a high primase concentration, short fragments (500 to 2000 ntd) are generated in addition to the longer chains, indicating both continuous and discontinuous replication occur.[75] Gyrase and SSB are required during fork movement, gyrase providing the swivel that allows the melted parental strands to become unlinked (Section 12-1). DNA pol III holoenzyme is responsible for DNA synthesis.

Processing of the Nascent Strands. This stage generates covalently closed duplex products.[76] Pol I, assisted by RNase H, removes the primers and then fills the resulting gaps to enable the short fragments to be joined by *E. coli* ligase. As with many circular replicons, catenated dimers are prominent intermediates late in replication, and gyrase both decatenates and supercoils them to generate monomeric circles.

RNA Polymerase Function:[77] Transcriptional Activation

RNA polymerase (RNAP) action is strongly implicated in the initiation of replication from *oriC*. The kinetics of inhibition of replication by rifampicin in vivo

72. Masai H, Nomura N, Arai K (1990) *JBC*, *265*:15134.
73. Baker TA, Funnell BE, Kornberg A (1987) *JBC 262*:6877.
74. Baker TA, Funnell BE, Kornberg A (1987) *JBC 262*:6877; Baker TA, Sekimizu K, Funnell BE, Kornberg A (1986) *Cell 45*:53.
75. Fradkin LG, personal communication.
76. Funnell BE, Baker TA, Kornberg A (1986) *JBC 261*:5616.
77. Baker TA, Kornberg A (1988) *Cell 55*:113; Ogawa T, Baker TA, van der Ende A, Kornberg A (1985) *PNAS 82*:3562.

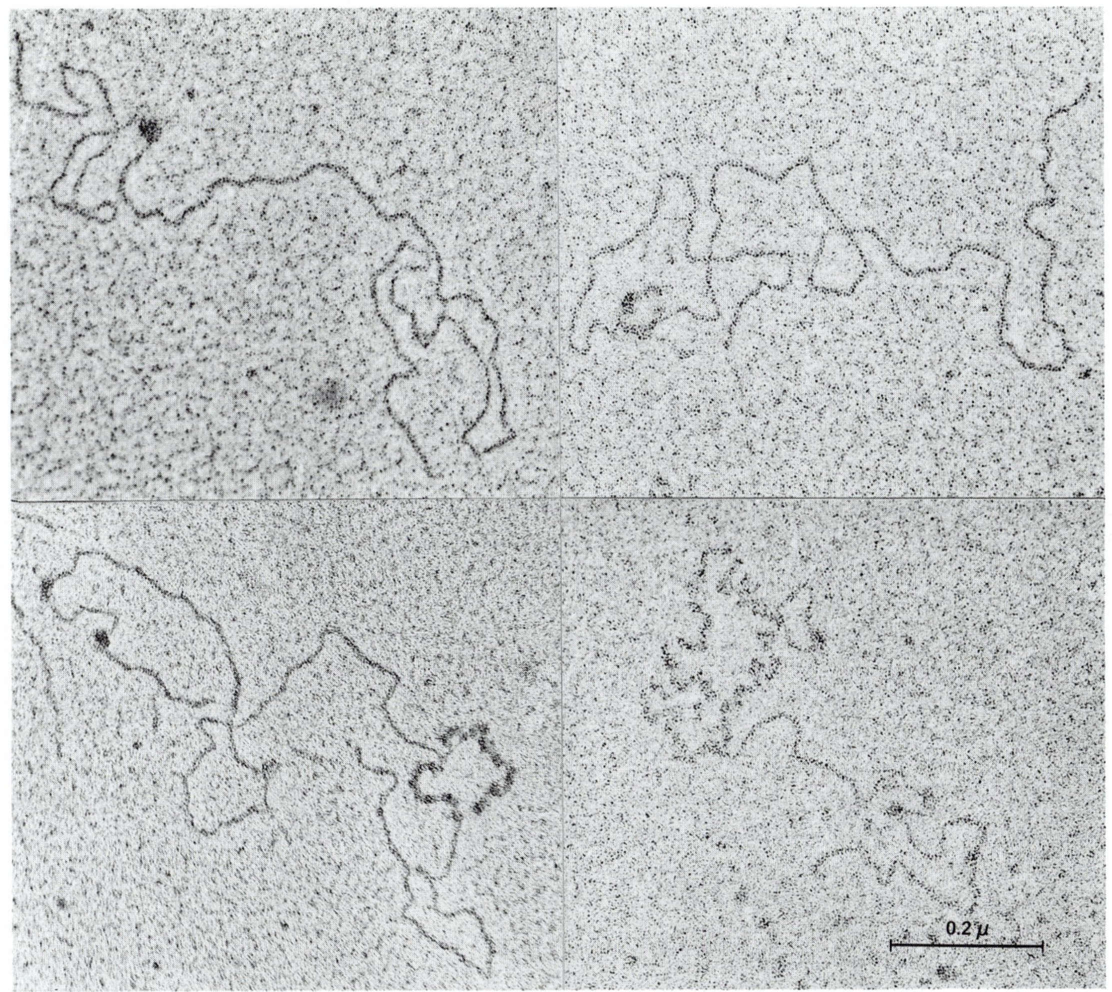

Figure 16-13
Electron micrographs of protein–DNA complexes during unwinding by dnaB helicase and gyrase of a double-stranded *oriC* plasmid DNA. Molecules were fixed and cleaved with a restriction endonuclease. Upper left, initiation complex at *oriC* before addition of gyrase. Upper right, small bubble before addition of gyrase. Lower left and right, larger bubbles after 1 or 2 minutes of unwinding after addition of gyrase. Small proteins distributed along the DNA are presumed to be gyrase.

indicate that RNAP, independent of mRNA synthesis, is required specifically for the initiation of chromosome replication. Mutations in *rpoB*, which encodes the β subunit of RNAP, suppress the ts phenotype of *dnaA* mutant cells;[78] mutations affecting the β and β' subunits increase the copy nmber of *oriC* plasmids and the chromosome.[79]

During the replication of *oriC* plasmids in vitro, RNAP, while not essential, is clearly involved under certain reaction conditions.[80] Several factors that inhibit *oriC* initiation reveal a dependence on transcription (Table 16-1). Open complex formation is the stage affected. The inhibition appears to be caused by structural

78. Atlung T (1984) *MGG 197*:125; Bagdasarian MM, Izakowska M, Bagdasarian M (1977) *J Bact 130*:577.
79. Rasmussen KV, Atlung T, Kerszman G, Hansen GE, Hansen FG (1983) *J Bact 154*:443.
80. Baker TA, Kornberg A (1988) *Cell 55*:113; Ogawa T, Baker TA, van der Ende A, Kornberg A (1985) *PNAS 82*:3562.

Table 16-1
Conditions that elicit a dependence on RNA polymerase by inhibiting open complex formation

Condition	Mechanism of action
High levels of HU protein ($>1/3$ coating of the template)	constrains negative supercoils, increases DNA melting temperature
Absence of HU protein	possible lack of a bend at *oriC*
Reduced negative superhelicity ($\sigma < -0.04$)[a]	insufficient superhelicity for DNA melting by dnaA protein
Reduced reaction temperature ($< 23°$)[a]	insufficient thermal energy for DNA melting by dnaA protein

[a] The superhelicity and temperature values at which the reaction requires transcriptional activation depend on the level of HU present.

changes in the DNA that prevent dnaB protein from joining dnaA protein and forming the prepriming complex. Transcription by RNAP can restore the capacity to form the prepriming complex, for which the ATP-bound form of dnaA protein and the *oriC* sequence, including all three 13-mers, are still required.

The stimulation of initiation by RNAP does not involve the priming of DNA synthesis, as judged by several criteria.[81] (1) Open complex and prepriming complex formation, the stages at which RNAP acts, precede priming. (2) Full activation occurs even when the RNA product is terminated with 3′-dATP and thus lacks the 3′-OH primer group. (3) The activating transcript does not need to cross the *oriC* sequence to be effective; DNA synthesis initiates within or near *oriC* rather than in the transcribed region. (4) Primase is sufficient for priming all DNA chains and replication depends upon it even when RNAP is present; the replication efficiency is reduced tenfold when primase is omitted from reactions containing RNAP and the transcripts it generates must then serve as primers.[82]

In vivo, numerous RNA–DNA junctions, representing the switch from primer to DNA synthesis, occur within and around the origin.[83] The sequence upstream of most of these junctions is consistent with the preferred tri- and tetranucleotide sites recognized by primase (Section 8-3), indicating that primase is responsible for priming DNA synthesis. However, these sites also coincide with the 3′ ends of transcripts initiating from a promoter (Secton 20-3) immediately to the right of *oriC*,[84] indicating that RNAP could also prime DNA synthesis within *oriC*. The promoter can, however, be deleted without any profound effect on replication or cell growth.[85]

The minimal requirement for activation is an RNA–DNA hybrid, or R-loop, generated during transcription; once synthesized, the R-loop can activate without the direct participation of RNAP. This R-loop stimulates initiation by altering the template structure and facilitating melting of the DNA strands at *oriC*. Transcription can precede all other initiation stages. A protein–protein interaction between dnaA protein and RNAP, inferred from genetic suppression of *dnaA*[ts] by *rpoB* mutants, is not responsible for this activation. The RNAPs to

81. Baker TA, Kornberg A (1988) *Cell 55*:113.
82. Ogawa T, Baker TA, van der Ende A, Kornberg A (1985) *PNAS 82*:3562.
83. Hirose S, Hiraga S, Okazaki T (1983) *MGG 189*:422.
84. Rokeach LA, Zyskind JW (1986) *Cell 46*:763; Rokeach LA, Kassavetis GA, Zyskind JW (1987) *JBC 262*:7264.
85. Løbner-Olesen A, Boye E, personal communication.

phages T7 and T3, which are structurally unrelated to the *E. coli* enzyme, are potent activators of *oriC* templates that contain a promoter for these enzymes. The activating RNA–DNA hybrid can work over substantial distances; its activity is modulated by the length and sequence content of the DNA between the R-loop and the origin. Sequences of less than about 200 bp that lack any runs of 10 or more GC base pairs allow full activation, while longer intervening regions and those containing longer stretches of GC base pairs are less effective in activation.[86]

The formation, at *oriC*, of an open complex, a key intermediate in the assembly of the replication complex, is exquisitely sensitive to conditions that either permit or inhibit melting of the duplex. Under several different conditions where dnaA protein alone is unable to open the DNA, an R-loop, located even a substantial distance away, provides a large enough change in the template structure to tip the balance in favor of initiation. Similar roles of transcription are indicated in other prokaryotic origins. In the replication of phage λ, transcription of the template is required for initiation in vivo, although it does not need to physically cross the origin and is not likely to be involved in priming of DNA synthesis.[87] Furthermore, RNAP overcomes the inhibition by HU protein of prepriming complex formation at *oriλ* in vitro[88] (Section 17-7). During ColE1 replication under some conditions, the formation of an RNA–DNA hybrid opens the strands for loading of the discontinuous replication complex[89] (Section 18-2).

16-4 Other Prokaryotic Origins

Aside from the initiator endonuclease replication mechanism (see below), initiation of replication of a duplex chromosome involves three stages: (1) recognition of the origin, (2) initial melting of the duplex, and (3) loading of the helicase, permitting more extensive melting of the duplex and assembly of the replication fork. In *E. coli*, these stages are carried out principally by dnaA protein (Section 3). Recognition is through interaction between dnaA protein and its four boxes, opening of the duplex is through interaction with the nearby 13-mers, and loading of the dnaB helicase is probably accomplished by an interaction between dnaA protein and the dnaB–dnaC protein complex. Analysis of the origin structures and the mechanisms of initiation of other genomes suggests that a sequence of three such steps is general. In some instances, as with the lambdoid phages, the parallels are especially striking. In other cases, as with plasmid R1, variations are played on this basic theme.

Other Bacterial Chromosomal Origins

The consensus derived from the *oriC* sequences of five highly diverse species of enteric bacteria (Section 2 and Fig. 16-11) emphasizes their strong relationship.[90]

86. Skarstad K, Baker TA, Kornberg A (1990) *EMBO J 9*:2341.
87. Furth ME, Dove WF, Meyer BJ (1982) *JMB 154*:65.
88. Mensa-Wilmot K, Carroll K, McMacken R (1989) *EMBO J 8*:2393.
89. Masukata H, Dasgupta S, Tomizawa J (1987) *Cell 51*:1123; Parada CA, Marians KJ (1989) *JBC 264*:15120.
90. Zyskind JW, Cleary JM, Brusilow WSA, Harding NE, Smith DW (1983) *PNAS 80*:1164.

Inasmuch as these origins provide autonomous replication to plasmids in *E. coli*, the sequences must all be recognized by the *E. coli* replication proteins. In addition, the dnaA proteins from *Salmonella typhimurium* and *Serratia marcescens* function in *E. coli*, as manifested by their ability to complement a *dnaA*^{ts} mutation.[91] Their homology to the *E. coli* dnaA protein is strong, with 63 amino acids at the N terminus and 333 at the C terminus being identical. It follows that the mechanism of replication initiation in enteric bacteria is exceptionally well conserved.

The origin sequences from the more distantly related Gram-negative bacteria *Pseudomonas aeruginosa*and *Pseudomonas putida*[92] and the Gram-positive *B. subtilis*,[93] although at a different chromosomal location (Fig. 20-7), are structurally related to the *E. coli* sequence (Fig. 16-14). The relationship is not as great as in the enteric bacteria and the origins fail to support autonomous replication in *E. coli*. However, several dnaA boxes are found within a 200-bp sequence and the consensus (TTATCCACA),[94] derived from comparing the 9-mer sequences from these different species, is identical to that found in *E. coli*. The locations of the first and last boxes with respect to each other and the adjacent repeat regions are also conserved; iterated sequences at the left edge of the origins are related to the *E. coli* 13-mers. *B. subtilis oriC* has three repeats of a 16-bp sequence, identical at eight positions to the *E. coli* 13-mer. The 12-bp motif in the analogous position in the *Pseudomonas* origins is identical at five nucleotides.[95] *B. subtilis*, *P. aeruginosa*, and *P. putida* also encode proteins homologous to *E. coli* dnaA protein. The C-terminal domains, which contain the ATP-binding motif and the sites of several mutations that inactivate the protein for replication, exhibit greater than 50% identity.[96]

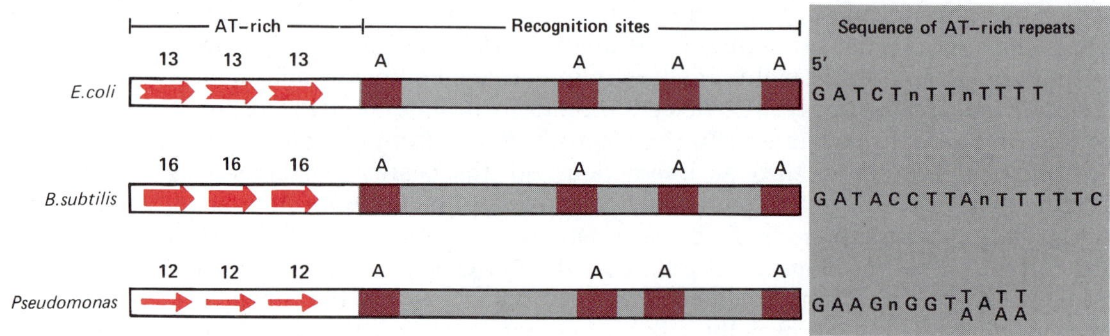

Figure 16-14
Maps of the *E. coli*, *B. subtilis*, and *Pseudomonas* origins. "A" box = 9-mer recognition sequence that directs tight binding of dnaA protein; arrows = AT-rich repeats; n = any nucleotide. [Modified from Bramhill D, Kornberg A (1988) *Cell 54*:915]

91. Skovgaard O, Hansen FG (1987) *J Bact 169*:3976.
92. Yee TW, Smith DW (1990) *PNAS 87*:1278.
93. Ogasawara N, Moriya S, von Meyenburg K, Hansen FG, Yohikawa H (1985) *EMBO J 4*:3345; Moriya S, Fukuoka T, Ogasawara N, Yoshikawa H (1988) *EMBO J 7*:2911.
94. Fujita MQ, Yoshikawa H, Ogasawara N (1989) *MGG 215*:381.
95. Yee TW, Smith DW (1990) *PNAS 87*:1278.
96. Fujita MQ, Yoshikawa H, Ogasawara N (1989) *MGG 215*:381.

The Lambdoid Phage Origins

The lambdoid phages, which provide further examples of the *oriC*-like initiation pathway, are especially noteworthy because the origin sequence and initiator proteins are not related to the *oriC* and dnaA protein of *E. coli*.

In phage λ, origin recognition is through the tight binding of the λ-encoded O protein to four 19-bp elements within the sequence (Fig. 16-15).[97] O protein, thus bound, melts the flanking AT-rich region,[98] which contains three repeats of an 11-bp sequence, and interacts with the λ P protein.[99] The latter, a dnaC protein analog, binds dnaB protein and thus recruits this helicase to the origin for initiation.[100] Subsequent replication requires the same proteins as in *oriC*-dependent synthesis, indicating that identical replication forks are assembled.[101]

While only the replication initiation of λ has been studied in detail, the origins of other lambdoid phages also implicate two classes of repeated elements in interactions with their O proteins. In ϕ82, the 11-mer sequences in the AT-rich region are identical to those in *ori*λ, and the O proteins are closely related (Fig. 16-15).[102] The origin of the more distantly related ϕ80 has a different AT-rich repeat motif,[103] and the N terminus of the O protein, responsible for origin recognition,[104] is also distinctive.

Origins That Use Both a Rep Protein and DnaA Protein

Several *E. coli* plasmids encode their own origin-binding proteins (Rep proteins) which, with varying degrees of assistance by dnaA protein, perform the key steps of recognition, origin melting, and loading of the replication machinery.

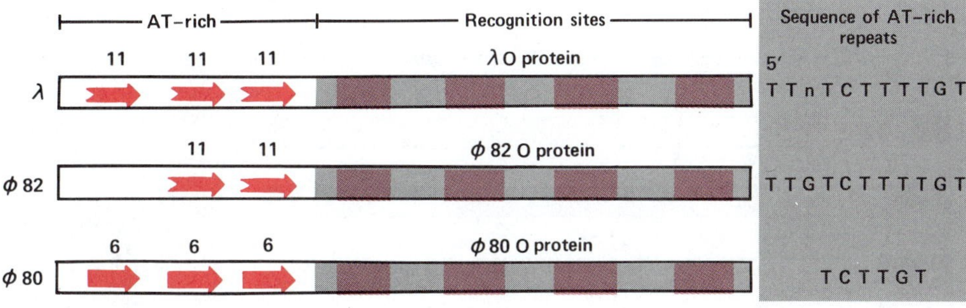

Figure 16-15
Maps of the origins of the lambdoid phages. Boxes = phage O protein binding sites; arrows = AT-rich repeats; n = any nucleotide. [Modified from Bramhill D, Kornberg A (1988) *Cell 54*:915]

97. Dodson M, Roberts J, McMacken R, Echols H (1985) *PNAS 82*:4678.
98. Schnös M, Zahn K, Inman RB, Blattner FR (1988) *Cell 52*:385.
99. Zylicz M, Gorska I, Taylor K, Georgopoulos C (1984) *MGG 196*:401.
100. Dodson M, Echols H, Wickner S, Alfano C, Mensa-Wilmot K, Gomes B, LeBowitz J, Roberts JD, McMacken R (1986) *PNAS 83*:7638.
101. Mensa-Wilmot K, Carroll K, McMacken R (1989) *EMBO J 8*:2393.
102. Moore DD, Denniston KJ, Blattner FR (1981) *Gene 14*:91.
103. Grosschedl R, Hobom G (1979) *Nature 277*:621.
104. Furth M, Yates J (1978) *JMB 126*:227.

Their replication requires dnaB and dnaC proteins, primase, and DNA pol III holoenzyme; the assembled forks must be closely related to those that function during replication from *oriC*.

The origins of the several plasmids (Table 16-2 and Fig. 16-16) all contain (1) reiterated sequences *(iterons)* that direct the binding of the plasmid-encoded initiator protein, (2) one or two dnaA boxes, (3) an AT-rich sequence, which may contain a 13-mer and a binding sequence for IHF, the host histone-like protein, (4) an additional class of iterons within the AT-rich region (pSC101 plasmid

Table 16-2
***E. coli* plasmid origins that use both dnaA protein and their own Rep proteins**

Plasmid	dnaA boxes	13-mers	Rep binding sites[a]	AT-rich repeats[b]	GATC sites	IHF binding
pSC101	1	2	five 18-mers	none	1	yes
P1 *oriR*	2	—	five 19-mers	five 7-mers	5	no
F *oriS*	2	1	four 19-mers	four 8-mers	1	nd[c,d]
ColV-K30	2	—	five 19-mers	five 11-mers	6	nd[c]
R1	1	—	exact sequence not known	three 9-mers	—	yes[e]
R6K *oriγ*	1	1	seven 22-mers	three 10-mers	1	yes

[a] Each plasmid has a distinct origin-specific binding protein, commonly called Rep.
[b] Other than the *oriC*-type 13-mer.
[c] nd = not determined.
[d] HU required in vivo.
[e] Not essential for replication in vivo.

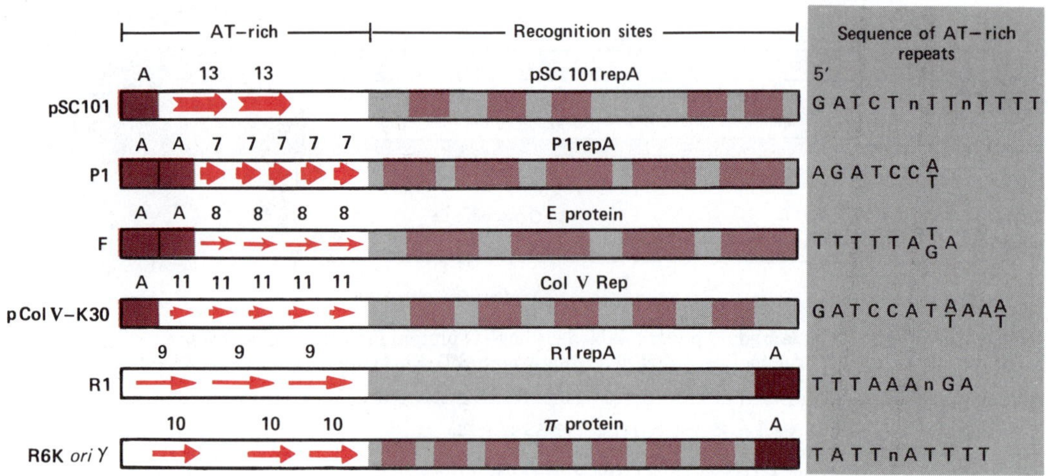

Figure 16-16
Maps of prokaryotic plasmid origins, aligned to emphasize their related structures. Each plasmid (except pSC101) encodes an origin-binding protein which recognizes the iterons within the origin. "A" box = 9-mer recognition sequence that directs tight binding of dnaA protein; arrows = AT-rich repeats; n = any nucleotide. [Modified from Bramhill D, Kornberg A (1988) *Cell 54*:915; pColV-K30 sequence from Perez-Casa JF, Gammie AE, Crosa JH (1989) *J Bact 171*:2195]

excepted), and, commonly, (5) GATC sequences, the sites of methylation by the *E. coli* Dam methylase (Section 21-10).

Genetically, pSC101 has a strong dependence on dnaA protein (Section 18-3), similar to that of the *E. coli* chromosome.[105] For the other plasmids, *dnaA* null alleles, deletions of the dnaA boxes in the origin, or an in vitro analysis is required to reveal the involvement of dnaA.[106] In these plasmids, probably only a subset of the activities of dnaA protein is required for initiation.

Biochemical analysis of replication is most advanced for the plasmid R1 and prophage P1 (Sections 18-4 and 17-7). Their Rep proteins and dnaA protein bind specifically to their respective origins. In R1, a complex is formed at the origin with 40 to 50 monomers of RepA, to which one to two molecules of dnaA protein are added.[107] Analysis by nuclease protection and sensitivity indicates that about 100 bp of the origin DNA is wrapped around the proteins and that the AT-rich region is melted. Clearly, dnaA and the plasmid-specific protein act jointly in the stages of recognition and melting.

For both R1 and P1 initiation, the dnaA protein probably functions in loading the dnaB protein from a dnaB·dnaC protein complex. Consistent with the less stringent requirement for dnaA function in vivo, both the ATP- and the ADP-bound forms of the protein can function in R1 and P1 replication in vitro, and much less dnaA is required than in *oriC* initiation.[108] Whereas the ATP form of dnaA protein is absolutely required for opening of the duplex at *oriC*, this stage in R1 and P1 initiation may be provided largely by the Rep proteins, with dnaA protein serving only to load the other replication enzymes.

ColE1 and Phage T7 Origins: The Same Three Initiation Stages without an Initiator Protein

Initiation of the ColE1 and T7 replicons may not appear to follow the *oriC* pathway in detail, but the three key initiation steps are still evident. Origin recognition and template melting are provided through transcription by RNAP rather than by an origin-binding protein (Sections 17-5 and 18-2). For ColE1, the host RNAP, and for T7 the phage polymerase, initiate RNA synthesis at a particular promoter. RNA–DNA hybrid formation opens the duplex DNA, allowing the replication enzymes to be loaded. Why certain promoters and transcribed regions are used as replication origins while others are not is not known. In ColE1, the formation of a specific RNA secondary structure is required for hybrid formation, and the template sequence at the origin affects the efficiency with which this hybrid is utilized for initiation.

In both ColE1 and T7, the 3′ end of the initiating transcript also primes replication of the continuous strand.[109] The helicase–primase complex, needed to melt the duplex template for continuous-strand synthesis and to prime the fragments of the discontinuous strand, is loaded subsequently on the discontinuous-strand template. Loading on ColE1 plasmids occurs at a primosome assembly site[110] or

105. Hasunuma K, Skiguchi M (1977) *MGG 154*:225.
106. Hansen EB, Yarmolinsky MB (1986) *PNAS 83*:4423; Masai H, Arai K (1987) *PNAS 84*:4781.
107. Masai H, Arai K (1987) *PNAS 84*:4781; Masai H, Arai K (1988) in *DNA Replication and Mutagenesis* (Moses RE, Summers WC, eds.). Am Soc Microbiol, Washington DC, p. 113.
108. Wickner S, Hoskins J, Chattoraj D, McKenney K (1990) *JBC 265*:11622; Masai H, personal communication.
109. Itoh T, Tomizawa J (1980) *PNAS 77*:2450.
110. Minden JS, Marians KJ (1985) *JBC 260*:9316.

dnaA box;[111] the mechanism of positioning T7 gp4 (the helicase–primase) on the T7 template is unknown. Transcription thus provides an alternative to an origin-binding protein in melting the DNA for initiation. For *oriC* and phage λ, advantage may be taken of this capacity by using RNAP to assist the dnaA and λ O proteins in melting the origins (Sections 3 and 17-7).

Unidirectional versus Bidirectional Replication

Origins promote either unidirectional or bidirectional replication (Table 16-3), depending on whether one or two replication complexes are assembled during initiation. There may also be two independent initiation events, with distinctive origin sequences and mechanisms, to assemble the two fork complexes. The relative advantages of uni- and bidirectional replication, and why different genomes choose one mode or the other, remain unknown.

Coupled Bidirectional Replication. In *oriC*-dependent initiation, both replication forks appear to be established by a concerted mechanism.[112] Two dnaB helicase–primase complexes, one on each strand, appear to be loaded at the same time (Fig. 16-17). Preference for one direction of unwinding or of DNA synthesis has not been observed. Replication of both the continuous and discontinuous strands may be primed by the same mechanism.

Unidirectional Initiation. Only one helicase is required to propagate the single fork during unidirectional replication (Fig. 16-18). Discontinuous-strand synthesis needs a primosome, while the continuous strand may be replicated for the entire length of the molecule from a single primer.

The R1 and ColE1 plasmids differ in the way that unidirectional replication is established (Fig. 16-18). In R1 replication, dnaB protein is loaded by the RepA·dnaA protein complex at the origin. As a helicase, dnaB protein moves away from this complex in the 5′→3′ direction on the strand to which it is bound. Primers are synthesized on this strand by the action of primase. Thus, the discontinuous strand is initiated before the continuous strand. The specific

Table 16-3
Unidirectional and bidirectional replication in prokaryotic replicons

Origin	Mode	Comments
E. coli oriC	coupled bidirectional	
Phage λ	coupled bidirectional	partial blocking of the leftward fork gives replication a rightward bias
ColE1; pBR322	unidirectional	leading strand is initiated first; discontinuous-strand synthesis is terminated at the origin
R1	unidirectional	discontinuous strand is initiated first and is terminated at the origin
R6K *ori*α[a]	unidirectional	mechanism similar to that of R1
F *oriS*[a]; R6K *ori*γ[a]	sequentially bidirectional	one strand is initiated at the origin; the second at a separately located *pas*

[a] Mode of replication is suggested by the structure of the origin and location of priming and primosome assembly sequences; the mechanism in vitro and in vivo is not known.

111. Seufert M, Messer W (1987) *Cell 48*:73.
112. Baker TA, Funnell BE, Kornberg A (1987) *JBC 262*:6877.

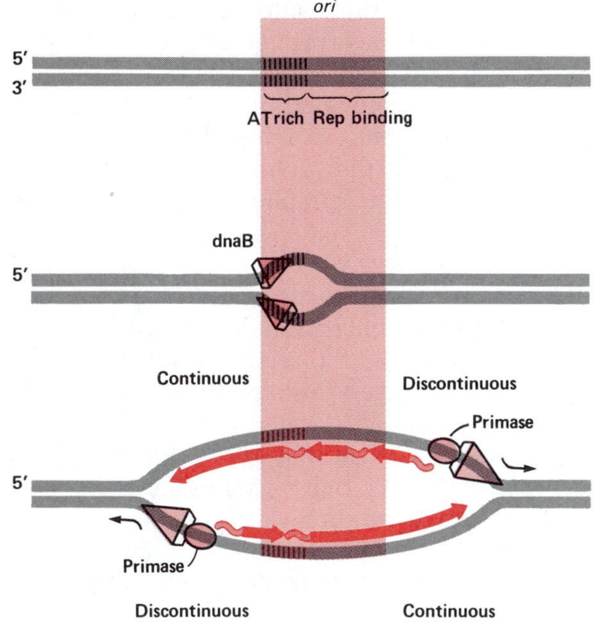

Figure 16-17
Coupled bidirectional replication: two helicase–primase complexes are loaded at once, as at *oriC*.

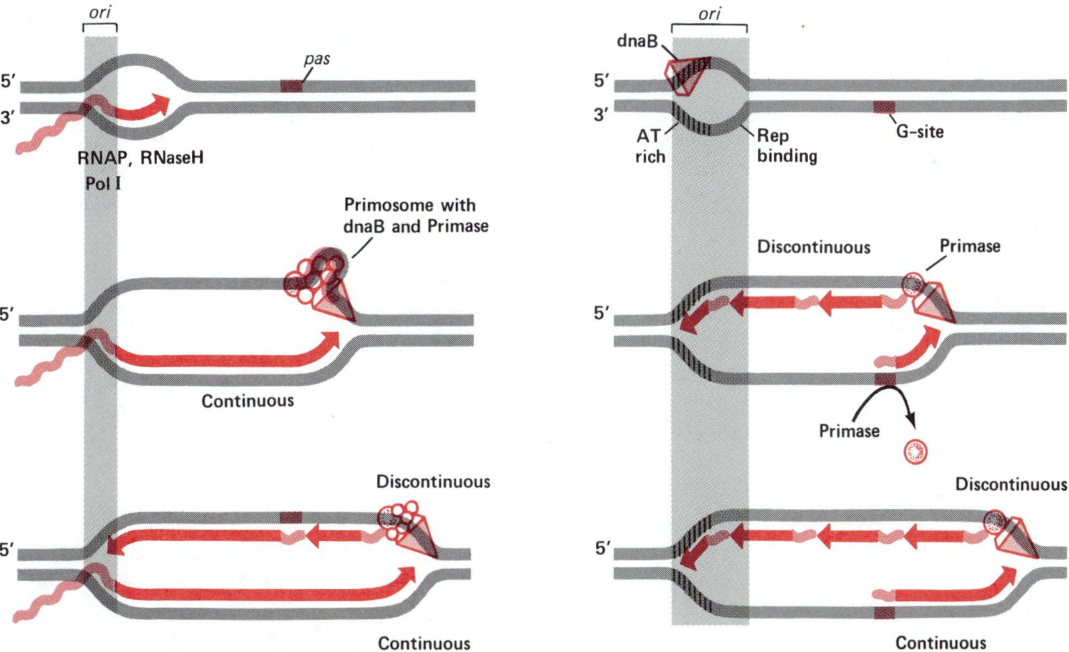

Figure 16-18
Unidirectional replication: only one helicase–primase is loaded; a second mechanism is required to prime the other strand. Left: Continuous-strand synthesis is initiated first (as in ColE1). Right: The discontinuous strand is started at *orir* before initiation of continuous synthesis of the other strand (as in R1). G-site = primase recognition site; *pas* = primosome assembly site.

recognition site for primase (G-site in Fig. 16-18) (Section 8-3), located about 400 bp downstream of the origin, provides the single primer required for continuous-strand synthesis.[113] In contrast, in ColE1 replication, continuous-strand synthesis is initiated first, using the primer generated by RNAP (Fig. 16-18). A single dnaB protein, as part of the primosome, is subsequently loaded on the template for the discontinuous strand at the primosome assembly site, which is rendered single-stranded and available by replication of the opposite strand.

Sequentially Bidirectional Initiation. The mechanisms used to load a single primosome—an origin-binding protein (R1) or a primosome assembly site (ColE1)—can be combined to give sequentially bidirectional initiation. The structures of F plasmid *oriS* and of R6K *oriγ* suggest this type of pathway.[114] In F *oriS*, two φX-type primosome assembly sites are located about 70 bp to the left of the AT-rich repeated elements in the origin (Fig. 16-19). Plausibly, the entry of a single dnaB helicase, loaded at the AT-rich region in the origin, can result in unidirectional replication on the discontinuous-strand template, followed by exposure of the single-stranded *pas* sequences for primosome assembly on the continuous-strand template, to create bidirectional replication.

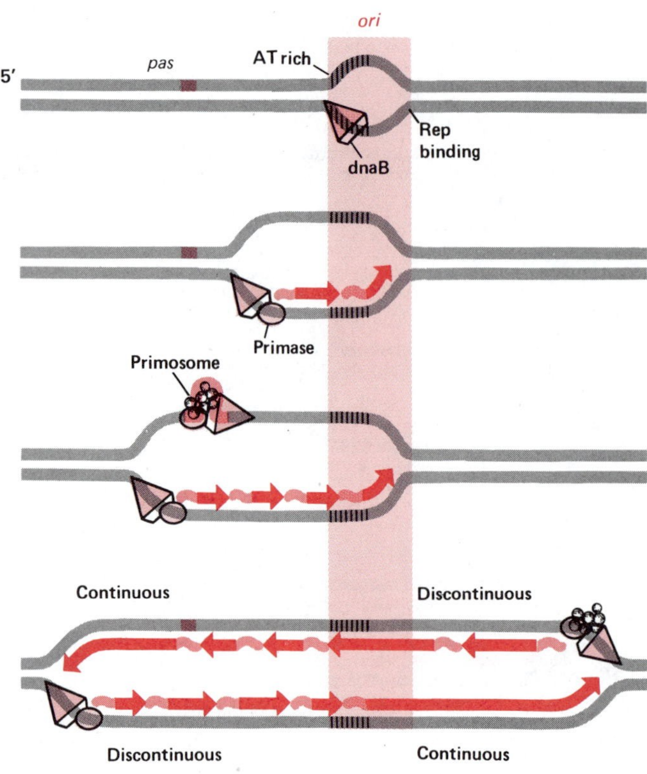

Figure 16-19
A hypothetical scheme for sequential bidirectional initiation, as might occur at F plasmid *oriS*.
pas = primosome assembly site.

113. Masai H, Arai K (1989) *JBC 264*:8082.
114. Masai H, Nomura N, Kubota Y, Arai K (1990) *JBC 265*:15124.

Effect of Termination Signals on Direction. In addition to the intrinsic properties of origins that promote one or two replication forks, termination signals, which stall or dissociate replication forks (Section 15-12), can bias the directional pattern. Some unidirectionally replicating plasmids (e.g., ColE1, R1) have such sequences that stall replication. R6K, with a complicated replication pattern of three origin sequences, has a specific termination signal, located asymmetrically relative to the origins (Section 18-5). Phage λ replication has a rightward bias, especially in vitro; the initiator protein tightly bound to the origin may inhibit the leftward fork (Section 17-7). Even the *oriC* sequence, which nearly always promotes bidirectional replication, will conform to a unidirectional mode if a termination sequence (normally positioned diametrically opposite to the origin) is placed close to either side of the origin (Section 15-12).[115]

Origins Cleaved by Initiator Endonucleases

Replication of most plasmids of Gram-positive bacteria and viral strand synthesis of several single-stranded phages is initiated by site-specific cleavage of the origin to generate a 3′-OH terminus which is elongated for continuous rolling-circle replication (Sections 8-7, 15-11, 17-2, and 17-3). These origins have some structural similarities to θ-type origins, and initiation has several common steps. In both ϕX and Ff phage replication, rolling-circle synthesis generates the viral single strand from an RF intermediate. Initiation involves (1) recognition of the origin sequence by an initiator endonuclease, (2) localized melting and cleavage of the viral DNA strand, also by the endonuclease, and (3) loading of a helicase to sustain extensive replication. Termination also takes place at the origin, and specific sequences in this region are involved in the process.

The ϕX (+) strand origin of 30 bp contains three regions with distinctive functions: a specific binding sequence for the gene *A* protein (gpA), an AT-rich spacer region, and an 8-bp sequence cleaved by gpA at a specific diester bond (Fig. 16-20). On ssDNA, the cleavage site alone is sufficient for cutting by the protein, but on the normal duplex substrate the other sequences are also required. GpA binds first at the binding site, melts the AT-rich sequence, and then binds the cleavage site prior to nicking the viral strand. The distinctive binding and melting activities of gpA are reminiscent of some dnaA protein actions at *oriC*. Also, gpA interacts with the *E. coli* Rep protein, the DNA helicase responsible for melting the duplex in this stage of viral replication (Sections 8-7 and 11-3).

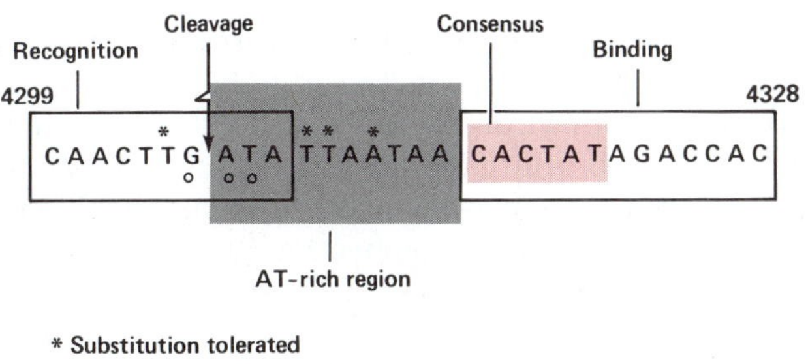

Figure 16-20
Sequence of the origin of ϕX174 DNA for viral strand synthesis. The sequence includes a region essential for binding gpA, separated by an AT-rich spacer from another region which is recognized for cleavage. Residue substitutions affecting replicative activity are marked. [Adapted from Heidekamp F, Baas PD, van Boom JH, Veeneman GH, Zipursky SL, Jansz HS (1981) *NAR* 9:3335]

115. Lee EH, Kornberg A, Hidaka M, Kobayashi T, Horiuchi T (1989) *PNAS* 86:9104.

Ff phage (+) strand origins are also made up of an essential region containing the sites for binding and cleavage by gp2, flanked by an AT-rich region that enhances origin activity about 100-fold. This AT-rich region, like many of those for the θ replication origins, binds the host IHF protein (Section 10-8). Gp2, like ϕX gpA, recruits the Rep helicase for action on this template.

The initiation of rolling-circle replication on the *Staphylococcus* plasmid pT181 requires a 43-bp sequence; an additional 27 bp stimulates activity when there is competition for the plasmid-encoded RepC initiator protein. The sequence is relatively GC-rich (50%) compared to most origins and contains a bend and a dyad symmetry element that can form a cruciform structure. Adoption of this secondary structure, rather than melting of the region (as in ϕX), may provide the optimal DNA conformation for cleavage. Whether RepC protein also has a role in recruiting the replication machinery to the plasmid origin is unknown.

Origins of Single-Stranded Phages

During the initiation of replication on single-stranded phages, melting of the duplex and loading of a DNA helicase are not necessary because the template is already in the single-stranded form, accessible to the replication enzymes. The origin sequences are therefore primarily concerned with priming DNA synthesis or the loading of priming machinery. The varied capacities of these origin structures to promote different modes of priming have been discussed (Chapter 8). They all have extensive regions of secondary structure, which probably stabilize them against melting by SSB.

Single-stranded phage origins fall into three classes: (1) the Ff sequence, which forms a specific structure recognized by *E. coli* RNAP for synthesis of a unique primer RNA (Section 8-2); (2) the G4 origin, recognized specifically by primase for synthesis of a unique primer (Section 8-3); and (3) the ϕX174 origin, recognized as a *pas* sequence by PriA protein, which promotes assembly of the seven-component primosome, a mobile complex that allows synthesis of multiple primers on the single-stranded template (Section 8-4). Primer sites, like that of G4, and primosome assembly sites, like that of ϕX, also appear as components of the complex origins of duplex DNA genomes (see above).

16-5 Eukaryotic Virus and Organelle Origins[116]

Several of the best studied replication origins of eukaryotic viruses resemble, in structure (Fig. 16-21) and in mechanism of initiation, the origins of prokaryotic genomes. Apparent among the eukaryotic viral origins (see below) are iterated elements that direct the binding of specific initiator proteins (Table 16-4). Melting of a nearby AT-rich or palindromic sequence element has been either demonstrated or implied as a region for entry of the replication machinery. The initiation of replication may be strongly influenced by transcriptional enhancers, silencers, and other promoter elements through the transcription fac-

---

116. DePamphilis ML (1988) *Cell* 52:635; Mecsas J, Sugden B (1987) *Ann Rev Cell Biol* 3:87; Challberg MD, Kelly TJ (1989) *ARB* 58:671.

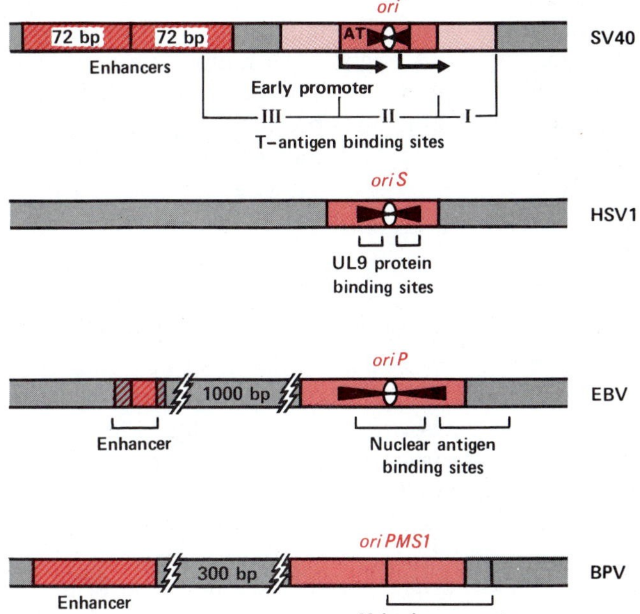

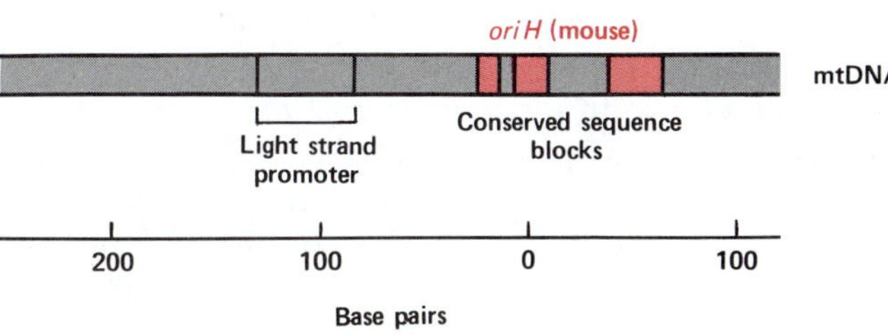

Figure 16-21
Eukaryotic virus and mitochondria origins. SV40 = simian virus 40; HSV-1 = herpes simplex virus 1; EBV = Epstein-Barr virus; BPV = bovine papilloma virus; mtDNA = mitochondrial DNA; PMS = plasmid maintenance sequence; AT = AT-rich sequence; arrowheads = palindrome; colored box = minimal *ori* core. [From DePamphilis ML (1988) *Cell 52*:635; HSV-1 from Elias P, Lehman IR (1988) *PNAS 85*:2959]

Table 16-4
Characteristics of eukaryotic viral and mitochondrial replication origins[a]

Genome	Origin elements	Binding (initiator) protein	Melted region	Transcription effects
SV40, polyoma	four 5-bp repeats; AT region; palindrome	T antigen	palindrome; AT region	nearby promoter stimulates SV40; essential for polyoma
HSV-1	two 8-bp repeats; AT region in center of palindrome	gpUL9	AT region likely	
EBV				
oriP	four 30-bp repeats, two forming part of the palindrome; enhancer of twenty 30-bp repeats	EBNA-1		enhancer essential
oriLyt	two noncontiguous regions			one element is an enhancer; the other overlaps a promoter
BPV	two noncontiguous regions	probably product of E1 open reading frame		as for *oriLyt* of EBV
mtDNA				
H-strand	promoter and RNA processing sites	RNAP; RNase	D-loop	promoter essential for initiation
L-strand	palindrome	primase		

[a] BPV = bovine papilloma virus; EBNA-1 = Epstein-Barr nuclear antigen 1; EBV = Epstein-Barr virus; HSV-1 = herpes simplex 1 virus; mtDNA = mitochondrial DNA; RNAP = RNA polymerase; SV40 = simian virus 40.

tors that recognize them. Transcriptional control elements may activate or even create an origin. For the mitochondrial DNA origin, RNA synthesis has a direct role in initiation, while transcriptional factors may alter nucleosomal packing at the SV40 origin to provide access for the T antigen or promote origin unwinding. For EBV and BPV, the distant location of enhancers makes their contribution to the initiation of replication less clear.

SV40 and Polyoma Origins[117]

Initiation from the SV40 origin has been reconstituted with purified components (Section 19-2). The single origin lies within a 450-bp region that also controls viral transcription. The minimal essential sequence is 64 bp. Flanking regions that include additional binding sites for the large T antigen (the initiator protein) and transcriptional control signals stimulate replication (Fig. 16-21). The minimal or "core" origin consists of three components: (1) four repeats of a 5-bp sequence, which direct the binding of T antigen,[118] (2) 17 AT base pairs, which may participate in origin melting,[119] and (3) a 15-bp palindrome, which is the first region melted during initiation.[120]

The action of T antigen on the SV40 origin is reminiscent of that of dnaA protein at *oriC*. Binding to the four pentameric repeats forms a large nucleoprotein complex.[121] The presence of ATP causes a conformational change in the complex and melting of the duplex.[122] Both the AT region and the T antigen binding sites are required for this step to occur, and melting is localized to the palindrome.[123]

Whether T antigen plays a direct role in loading the rest of the replication machinery, analogous to the loading of dnaB protein by dnaA protein, is unclear. T antigen is itself a helicase and can unwind the duplex enough to make an additional helicase unnecessary for initiation.[124] A complex ormed by T antigen with the host polymerase α–primase complex (Section 6-2) may contribute to the assembly of the replication machinery at the origin.[125] Unwinding and DNA synthesis are bidirectional, evidence that two replication complexes are assembled at the origin at nearly the same time. The discontinuous-strand primers also appear to prime continuous-strand synthesis in the opposite direction, as in the idealized bidirectional replication pattern (Fig. 16-17).[126]

The stimulation of initiation by the promoter elements neighboring the origin appears to be mediated by the binding of specific transcription factors rather than by RNA synthesis. While these elements clearly stimulate replication in vivo, their involvement in vitro critically depends on the reaction conditions. Transcription factors bound to the promoter elements may stimulate the initial melting of the origin DNA by T antigen.[127] Their presence may in addition modify the packing of nucleosomes on the template, resulting in a more exposed chromatin structure in the origin region.[128]

---

117. Challberg MD, Kelly TJ (1989) *ARB 58*:671.
118. Jones KA, Tjian R (1984) *Cell 36*:155.
119. Deb S, Delucia AL, Koff A, Tsui S, Tegtmeyer P (1986) *Mol Cell Biol 6*:4578.
120. Borowiec JA, Hurwitz J (1988) *EMBO J 7*:3149; Borowiec JA, Dean FB, Bullock PA, Hurwitz J (1990) *Cell 60*:181.
121. Mastrangelo IA, Hough PVC, Wilson VG, Wall JS, Hainfeld JF, Tegtmeyer P (1985) *PNAS 82*:3626; Dean FB, Dodson M, Echols H, Hurwitz J (1987) *PNAS 84*:8981.
122. Borowiec JA, Hurwitz J (1988) *PNAS 85*:64.
123. Dean FB, Borowiec JA, Ishimi Y, Deb S, Tegtmeyer P, Hurwitz J (1987) *PNAS 84*:8267.
124. Stahl H, Droge P, Knippers R (1986) *EMBO J 5*:1939.
125. Smale ST, Tjian R (1986) *Mol Cell Biol 6*:4077; Gannon JV, Lane DP (1987) *Nature 329*:456.
126. Hay RT, DePamphilis ML (1982) *Cell 28*:767.
127. Guo ZS, Gutierrez C, Heine U, Sogo JM, DePamphilis ML (1989) *Mol Cell Biol 9*:3593.
128. Cheng L, Kelly TJ (1989) *Cell 59*:541.

The structure of the polyoma replication origin is similar to that of SV40, although homology at the nucleotide-sequence leel is not remarkable.[129] The polyoma antigen binds to recognition sequences and probably activates initiation by the same mechanism as the SV40 protein. The polyoma promoter elements are required for initiation, rather than being only stimulatory as they are for SV40.

Herpes Simplex, Epstein-Barr, and Bovine Papilloma Virus Origins

Herpes simplex virus (HSV-1), Epstein-Barr virus (EBV), and bovine papilloma virus (BPV) origins may function by mechanisms similar to that of SV40.

Herpes Simplex (HSV-1). The HSV-1 genome (Section 19-5) has three origins, *oriL* (large) and two copies of *oriS* (small), which are related in sequence and function by the same mechanism;[130] all three origins are not essential for viral replication but the presence of two may be required.[131]

The *oriS* sequence has been the most studied because the larger palindrome in *oriL* makes plasmids containing it difficult to propagate. A minimum of 60 to 70 bp, containing a large inverted repeat that can form a cruciform, is required for *oriS* function.[132] The arms of the cruciform contain binding sites for the origin-binding protein, while the central 18 bp are all A and T and make up a region that is likely to be melted during initiation.

The product of the *UL9* gene, required for replication in vivo, binds specifically to the origin and thus is probably the initiator protein.[133] In addition, UL9 protein binds ATP and is a DNA-dependent ATPase and DNA helicase.[134] Although in vitro replication from the herpesvirus origins has not yet been achieved, several properties of UL9 protein suggest functions analogous to those of SV40 T antigen.

Epstein-Barr Virus (EBV).[135] This herpesvirus has two distinct replication origins. One, *oriP*, is responsible for maintaining the genome in a latent extrachromosomal state,[136] and the other, *oriLyt*, is responsible for amplification of the genome during lytic growth.[137] The two origins are located separately on the genome, require distinct viral products for their activities, and promote different modes of replication. DNA synthesis from *oriP* is circle-to-circle, through θ intermediates; during lytic replication, long concatemers are generated, indicative of a rolling-circle mechanism.

The latent-state origin consists of two elements, the core origin and an enhancer sequence, normally separated by about 1 kb, which function independently of position and orientation with respect to one another. The core origin contains four copies of a binding sequence for the essential virus-encoded plasmid maintenance protein, EBNA-1 (*E*pstein-*B*arr *n*uclear *a*ntigen 1).[138] Two of

129. DePamphilis ML (1988) *Cell 52*:635.
130. Murchie MJ, McGeoch DJ (1982) *J Gen Virol 62*:1; Knopf CW, Spies B, Kaerner HC (1986) *NAR 14*:8655.
131. Polvino-Bodnar M, Orberg PK, Schaffer PA (1987) *J Virol 61*:3528.
132. Lockshon D, Galloway D (1988) *Mol Cell Biol 8*:4018.
133. Elias P, Lehman IR (1988) *PNAS 85*:2959.
134. Bruckner RC, Crute JJ, Dodson MS, Lehman IR (1991) *JBC 266*:2669; Challberg M, personal communication.
135. Mecsas J, Sugden B (1987) *Ann Rev Cell Biol 3*:87.
136. Yates J, Warren N, Reisman D, Sugden B (1984) *PNAS 81*:3806; Reisman D, Sugden B (1986) *Mol Cell Biol 6*:3838.
137. Hammerschmidt W, Sugden B (1988) *Cell 55*:427.
138. Rawlins DR, Milman G, Hayward SD, Hayward GS (1985) *Cell 42*:859.

these binding sites form part of the 65-bp dyad, similar in structure to the protein binding sites in the SV40 and HSV-1 origins. Plasmid replication starts within or near this sequence.[139] The second origin element consists of 20 direct copies of the same EBNA-1 binding sequence that make up the core origin.[140] These direct repeats bind EBNA-1 and function as a transcriptional enhancer as well as an essential element in replication initiation. Replication is initiated bidirectionally from the core origin but one fork is blocked at the enhancer, so that most of the genome is replicated unidirectionally.[141]

The lytic replication origin is essential for amplified DNA synthesis and, as with *oriP*, is made up of two noncontiguous elements, one of which functions as a transcriptional enhancer and can be substituted by a heterologous enhancer. The structure of *oriLyt* has not been analyzed in detail, nor has the binding protein for it been identified. Replication depends on the viral DNA polymerase, which might be recruited by the origin-binding protein.

Bovine Papilloma Virus (BPV). Like EBV, this virus is maintained as an extrachromosomal plasmid and requires an origin of replication for this state. Two sequences have been identified as *cis*-acting elements that promote autonomous replication.[142] One of these, plasmid maintenance sequence 1 (PMS-1), lies within the region mapped as the start site of replication by electron microscopy[143] and thus is likely to be the primary origin.[144] The required sequence can be divided into two noncontiguous segments that function independently of the distance between them. One can be substituted by a transcriptional enhancer; the second, probably the true origin, overlaps a promoter.

Mutations in the 3′ end of an open reading frame (*E1*) specifically inhibit plasmid replication. The protein product predicted by the *E1* sequence has some homology to SV40 T antigen,[145] and is likely to be the initiator protein. Regulation of BPV replication, which imposes a limit of one initiation per cell cycle, is directed both by *cis*-acting replication signals and *trans*-acting factors (Section 20-7).[146]

Other Viral Replication Origins

Some eukaryotic viruses use distinctive replication strategies and thus have differently structured origins. As examples, the initiation of DNA synthesis depends on priming by a terminal protein in adenoviruses (Sections 8-8 and 19-4), by a specific tRNA in retroviruses (Section 19-7), and by terminal hairpin structures in parvoviruses (Sections 15-12 and 19-3).

Origins of Organelle Genomes

Replication of the supercoiled circular genomes of mammalian mitochondria (Section 18-11) is asymmetric: synthesis of the H-strand is nearly complete before the L-strand is started.[147] The distinctive initiations of the two strands are

---

139. Gahn TA, Schildkraut CL (1989) *Cell 58*:527.
140. Reisman D, Sugden B (1986) *Mol Cell Biol 6*:3838.
141. Gahn TA, Schildkraut CL (1989) *Cell 58*:527.
142. Stenlund A, Bream GL, Botchan MR (1987) *Science 236*:1666.
143. Waldeck W, Rösl F, Zentgraf H (1984) *EMBO J 3*:2173.
144. Lusky M, Botchan MR (1986) *PNAS 83*:3609.
145. Clertant P, Seif I (1984) *Nature 311*:276.
146. Roberts JM, Weintraub H (1988) *Cell 52*:397.
147. Clayton DA (1982) *Cell 28*:693.

also manifested in reliance on different sequences and enzymatic mechanisms.

The H-strand origin and its mechanism of initiation resemble those of the *E. coli* ColE1-type plasmids and phage T7 (Section 4). Origin recognition and melting are provided by transcription rather than binding by an origin-specific protein.[148] RNA synthesis by the mtRNA polymerase is initiated about 100 bp upstream of the origin. Cleavage of this transcript by a specific processing RNase provides a primer for leading-strand synthesis. Assembly of the replication machinery near this region takes advantage of the 3′-OH primer terminus. Continuous synthesis of the H-strand uncovers a hairpin on the L-strand template that serves as the initiation signal for priming by a primase.[149]

The genomes of chloroplasts (Section 18-11) and the minicircle DNA of the kinetoplast (Section 18-11), the single mitochondrion of trypanosomes, possess specific origins of replication, but little is yet known about how they operate.

16-6 Eukaryotic Chromosome Origins[150]

The existence of unique replication origins in all prokaryotic genomes and in eukaryotic viral and organelle genomes suggests that the initiation of cellular chromosomes is unlikely to be random. Finding specific initiation sequences has been difficult, however, due to the complexity of eukaryotic genomes and the limitations of assays to detect the initiation sites. The clearest evidence for specific replication origins is found in yeast. In animal cells, the identification of specific initiation sequences is less advanced, but new techniques (Section 1) and attention to repetitive or amplified regions of the genome offer promise. In some cells, it appears that efficient, properly regulated replication may be independent of specific sequence elements, indicating that specific origins are not universally essential for eukaryotic chromosomal replication.

In assessing approaches to identifying the elusive origins of eukaryotic cellular chromosomes, some features of prokaryotic, viral, and organelle genome origins may be useful guides.

(1) Existence of alternative sequences, such as the primosomal assembly sequences for starting the chains for discontinuous replication (Section 8-4).

(2) Multiple, noncontiguous elements, as in EBV (enhancer and a core origin) or in mitochondria (promoter, RNA processing site, and L-strand priming site).

(3) Lack of coincidence between the essential initiation sequence and the replication start site.

(4) Novel signals, like those suggested for the maintenance and partitioning of episomal (plasmid) genomes of EBV and BPV.[151]

Yeast ARS Elements

Autonomous replication sequences, isolated from chromosomes and naturally occurring plasmids in yeast, mediate the replication of DNA in which they are present. In some cases at least, these ARS elements act as origins of replication.

---

148. Chang DD, Clayton DA (1985) *PNAS 82*:351.
149. Wong TW, Clayton DA (1985) *Cell 42*:951.
150. Umek RM, Linskens MHK, Kowalski D, Huberman JA (1989) *BBA 1007*:1.
151. Mecsas J, Sugden B (1987) *Ann Rev Cell Biol 3*:87.

DNA replication of an ARS plasmid occurs in the nucleus during S phase and initiates only once per cell cycle, as expected of a minichromosome with a unique replication origin. ARS plasmids are, however, unstable during cell division and require a centromere or other type of partition mechanism to ensure their proper partitioning to both daughter cells at division. About 400 ARS elements are estimated to be present in the yeast genome, enough to support replicons of about 36 kb, the average size observed by electron microscopy.

ARS elements are made up of two functional domains. (1) Domain A, always present, consists of an 11-bp sequence that has the consensus (A/T) T T T A T (A/G) T T T (A/T). Mutations that alter this core sequence destroy ARS function. Analogous to other replication origins, this sequence is probably the binding site of an initiator protein; however, such a factor is yet to be identified. (2) Domain B is AT-rich and extends 50 to 100 bp to the 3′ side of the core. Like the core, it is essential to ARS function. In contrast to the core, a unique nucleotide sequence has not been identified in this region. Its general AT-rich character is thought to provide a region of DNA melting.[152]

Two-dimensional gel electrophoresis has established that replication initiates within or near the single ARS element of the 2μ plasmid (Section 18-10); the same is true of an artificial plasmid that contains ARS1, derived from chromosome IV.[153] However, not all sequences that function as an ARS on a plasmid are used as replication origins on the chromosome. For example, there is an ARS element adjacent to each of the tandemly repeated rRNA genes of yeast. Two-dimensional gel electrophoresis shows that one ARS is responsible for the replication of several adjacent repeat units.[154] Thus, most of the ribosomal ARS elements present in the flanking repeats are not used as replication origins within a given S phase. Other chromosomal regions, such as the left end of chromosome III, also contain some ARS elements which do not function as origins in the chromosome.[155] DNA sequence context and chromosomal structure probably play roles in determining which ARS sequences are used for initiation.

In spite of the uncertainty about when they are used on the chromosome, ARS elements are clearly potential replication origins, and thus an understanding of how they function should provide insight into the mechanism of eukaryotic DNA replication. Unfortunately, no efficient in vitro replication system for an ARS plasmid has been established. ARS-specific DNA-binding proteins have been isolated,[156] but none with a convincing initiator function has yet been identified. One such factor, OBF1, binds to domain B and can act as an enhancer of ARS function, perhaps in a role analogous to that of promoter elements and transcription factors in eukaryotic viral origins (Section 5).[157] However, transcription into an ARS sequence seems to inhibit its function.[158]

---

152. Umek RM, Kowalski D (1990) *PNAS 87*:2486.
153. Brewer BJ, Fangman WL (1987) *Cell 51*:463; Huberman JA, Spotila LD, Nawotka KA, El-Assouli SM, Davis LR (1987) *Cell 51*:473.
154. Linskens MHK, Huberman JA (1988) *Mol Cell Biol 8*:4927.
155. Umek RM, Linskens MHK, Kowalski D, Huberman JA (1989) *BBA 1007*:1.
156. Buchman AR, Kimmerly WJ, Rine J, Kornberg RD (1988) *MCB 8*:210; Shore D, Stillman D, Brand AH, Nasmyth K (1987) *EMBO J 6*:461; Diffley JFX, Stillman B (1988) *PNAS 85*:2120; Eisenberg S, Civalier C, Tye BK (1988) *PNAS 85*:743; Biswas EE, Stefanec MJ, Biswas SB (1990) *PNAS 87*:6689.
157. Walker S, Francesconi SC, Eisenberg S (1990) *PNAS 87*:4665.
158. Snyder M, Sapolsky RJ, Davis RW (1988) *Mol Cell Biol 8*:2184.

ARS Elements from Other Organisms

Selection for sequences that promote a high efficiency of transformation and maintenance of DNA in an extrachromosomal state has identified replication origins of prokaryotes and a number of fungal species, including *Ustilago*,[159] *Neurospora*,[160] *Aspergillus*,[161] and *Rhodosporidium*.[162] However, the cloning in yeast of DNA from mammalian or insect sources, in hopes of identifying chromosomal origins, has not proved useful. Sequences obtained in this manner contain close matches to the yeast ARS core element rather than bona fide origins that function in the original organism. Considering that ARS sequences from various fungi and even from other yeasts lack homologous sequences and are frequently not functional in *S. cerevisiae*, this failure is not surprising. The absence of conservation within eukaryotic replication initiation signals contrasts with the homologies in prokaryotes and with a number of eukaryotic transcriptional control elements that remain relatively unchanged throughout evolution.

The transformation of circularized chromosomal DNA fragments into vertebrate tissue culture cells to discover ARS elements has also had only limited success. *Xenopus* egg injections indicate that specific sequences are not required (see below). Sequences isolated from other cell types, while sometimes exhibiting intriguing ARS activity,[163] have not yet been convincingly shown to be functional replication origins. Even the sequences strongly implicated as the replication origins of the amplified dihydrofolate reductase (DHRF) region in Chinese hamster (CHO) cells (see below) have failed to support the replication of episomal DNA.[164]

Gene Amplification: Control Elements and Replication Origins

Amplified genes—chromosomal regions present in multiple copies, including the naturally redundant RNA genes—have been popular objects in the search for replication origins. In none of these efforts has a replication origin been uniquely identified, yet preferences for certain initiation sites are clearly evident.

Two chromosomal locations in *Drosophila* that encode the chorion (eggshell) proteins are amplified in a developmentally specific and tissue-specific manner.[165] Several genes on the X chromosome are amplified 16-fold, while a second region, on chromosome 3, is amplified 60-fold (Fig. 16-22). Overreplication, apparently initiating within the gene cluster and proceeding bidirectionally, is responsible for generating these multiple copies. Four and eight additional rounds of initiation would account for the observed gene levels and the "onion-skin" replication structures visualized by electron microscopy.[166]

---

159. Tsukuda T, Carleton S, Fotheringhan S, Holloman W (1988) *Mol Cell Biol 8*:3703.
160. Grant DM, Lambowitz AM, Rambosek JA, Kinsey JA (1984) *Mol Cell Biol 4*:2041.
161. Timberlake WE, Boylan MT, Cooley MB, Mirabito PM, O'Hara EB, Willet CE (1985) *Exp Mycol 9*:351.
162. Tully M, Gilbert HJ (1985) *Gene 36*:235.
163. McWhinney C, Leffak M (1990) *NAR 18*:1233; Ariga H, Itani T, Iguchi-Ariga SMM (1987) *Mol Cell Biol 7*:1; Frappier L, Zannis-Hadjopoulos M (1987) *PNAS 84*:6668; Krysan PJ, Haase SB, Calos MP (1989) *Mol Cell Biol 9*:1026.
164. Burhans WC, Vassilev LT, Caddle MS, Heintz NH, DePamphilis ML (1990) *Cell, 62*:955; Vassilev, LT, Burhans, WC, DePamphilis ML (1990) *Mol Cell Biol 10*:4685.
165. Spradling A, Orr-Weaver T (1987) *ARG 21*:373.
166. Osheim YN, Miller OL, Beyer AL (1988) *Mol Cell Biol 8*:2811.

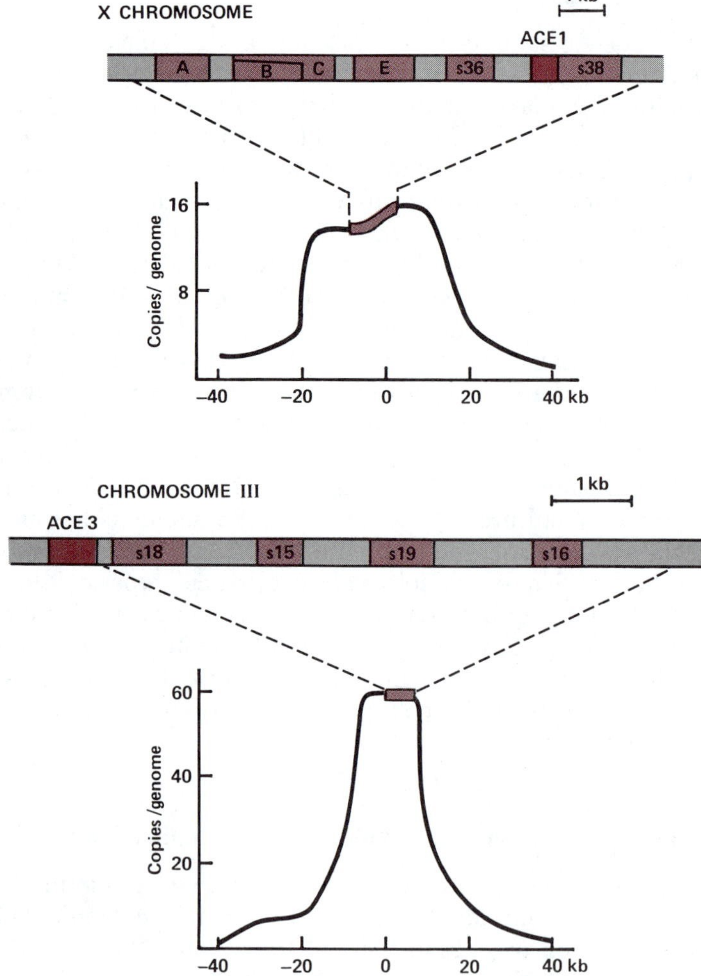

Figure 16-22
Amplification of chorion genes in *Drosophila*. The 80-kb DNA amplification gradients of the X chromosomal domain (upper) and the third chromosomal locus (lower) are graphed, with maps of the relative positions of the chorion genes (shaded) and the regions essential *in cis* for amplification (ACE). ACE = amplification control element. [From Osheim YN, Miller OL Jr, Beyer AL (1988) *MCB* 8:2811]

The *cis*-acting elements involved in amplification of these regions have been isolated on P-element transposons.[167] The elements (known as ACE, for *a*mplification *c*ontrol *e*lements) are sufficient to promote amplification of linked DNA in follicle tissues with the correct timing. The efficiency of amplification depends on the location of the insertion. The ACE region is localized to about 500 bp internal to each gene cluster, inside the region of maximum copy number. Thus the ACE region was thought to contain the replication origin, and the multiple peripheral elements to enhance its activity.[168]

The replication pattern of the ACE region revealed by two-dimensional gel electrophoresis is surprisingly devoid of origin activity.[169] The nearest origin, made up of two or more closely spaced initiation sites, maps about 1 kb away from the ACE site, within a region that enhances amplification but is not essential. Thus, the sequences required for amplification fail to identify the origin. However, initiation does occur at specific sites, which will very likely be refined

167. Orr-Weaver TL, Spradling AC (1986) *Mol Cell Biol* 6:4624.
168. Delidakis C, Kafatos FC (1989) *EMBO J* 8:891.
169. Delidakis C, Kafatos FC (1989) *EMBO J* 8:891; Heck MMS, Spradling AC (1990) *Mol Cell Biol* 110:903.

by further mapping. Perhaps the ACE regions serve as replication enhancers to activate neighboring origins, as for the bipartite origins of EBV and BPV (Section 5); in this instance, the enhancer sequence may be more important in the initiation process than the start site. Whether the same origins are used in normal, nonamplified DNA replication is unknown.

Amplification is not a programmed developmental event in mammalian cells. However, gene amplification occurs spontaneously in transformed cells. It was first detected as an amplification of drug-resistance genes in cells exposed to antineoplastic drugs[170] (Section 20-7). The mechanism by which these multiple copies of a region develop is not well understood and may well involve recombinational events early in the process[171] (Section 20-7). Extrachromosomal elements which contain the amplified gene and appear during the early stages may, after multiple rounds of duplication, integrate back into the chromosome, either in their original locus or elsewhere. Amplification via an extrachromosomal element indicates that autonomously functioning replication origins should be present within the amplified domains.

An extreme example of amplification is seen with the dihydrofolate reductase gene *(DHRF)* in Chinese hamster cells that have become resistant to methotrexate.[172] The gene, and as much as 200 kb of flanking DNA, is amplified approximately 1000-fold.[173] This high level of amplification facilitates the mapping of replication intermediates. By identification of the restriction fragments which are replicated earlest during S phase and by strand extrusion (Section 1), initiation has been localized to a 28-kb region containing two apparent origins, separated by about 22 kb.[174] Mapping of the directions of replication near the *DHFR* gene in normal cells (i.e., cells without the gene amplification) suggests that the same initiation sites are used during normal replication.[175] The distribution of Okazaki fragments indicates that replication initiates within a 450-bp region, making this sequence the best localized mammalian chromosomal origin. However, two-dimensional gel electrophoresis suggests that initiation can take place at multiple sites throughout the 28-kb region, indicating that a chromosomal domain, rather than specific, highly localized origin sequences, is activated for initiation.[176]

Functional criteria are now needed to assess whether the sequences required for initiation coincide with the replication start site. Since the 450-bp sequence fails to support replication of an autonomous plasmid, movement of this DNA fragment to a new chromosomal location, as has been done for the ACE elements of *Drosophila* (see above), will be necessary to define the functionally important sequence elements.

Replication Origins in Ribosomal DNA

The repetitive organization of rRNA genes allows the replication bubbles to be mapped by electron microscopy. In sea urchin, *Tetrahymena*, and *Physarum*, as

170. Stark GR, Debatisse M, Giulotto E, Wahl GM (1989) *Cell 57*:901.
171. Smith KA, Gorman PA, Stark MB, Groves R, Stark GR (1990) *Cell, 63*:1219.
172. Schimke RT (1989) *Bioassays 11*:69.
173. Ma C, Looney JE, Leu T-H, Hamlin JL (1988) *Mol Cell Biol 8*:2316.
174. Leu T-H, Hamlin JL (1989) *Mol Cell Biol* 9:523; Anachkova B, Hamlin JL (1989) *Mol Cell Biol* 9:532.
175. Handeli S, Klar A, Meuth M, Cedar H (1989) *Cell* 57:909; Burhans WC, Vassilev LT, Caddle MS, Heintz NH, De Pamphilis ML (1990) *Cell 62*:955.
176. Vaughn JP, Leu T-H, Hamlin JL (1990) in *Molecular Mechanisms in DNA Replication and Recombination* (Richardson CC, Lehman IR, eds.). *UCLA*, Vol 127. A R Liss, New York, p. 237;Vaughn JP, Dijkwel PA, Hamlin JL (1990) *Cell 61*:1075.

well as in yeast, the origins map within the nontranscribed spacer region.[177] An ARS element is present in this region in yeast, although only about one sequence in five of the repeats functions as an origin during each round of replication (see above).

In *Tetrahymena*, the ribosomal DNA is amplified to 10^4 copies per cell and is maintained as extrachromosomal linear molecules in the somatic macronucleus. The replication origin has been localized to about 700 bp within the nontranscribed spacer. Sequence changes in this region affect the ability of the linked genes to be amplified, thus providing genetic evidence for its importance in replication. Such mutations lie within a highly conserved rRNA promoter element, implicating transcription or transcriptional factors in initiation.[178]

Dispensability of Replication Origins

DNA from any source is semiconservatively replicated upon injection into *Xenopus* eggs[179] and adheres to cell-cycle control. Thus, no unique sequence is needed for initiation or for the regulation of replication in these cells. Analysis of replicating structures provides no evidence for specific sites of initiation; replication bubbles are distributed nearly at random along the length of the DNA.

Why the sequence requirement for initiation is so relaxed in eggs compared to other cells, such as yeast, is unknown. Perhaps the need for rapid replication in early embryonic development has led to such an abundance of initiation factors that virtually any sequence can be responsive and serve as an origin. Furthermore, the difficulty experienced to date in finding the replication origins of eukaryotic chromosomes may indicate that they are not as distinct as their prokaryotic and viral counterparts. Multiple important elements may be spread over a large distance of DNA such that one critical conserved sequence element cannot be found. Sequences that mediate contact with structural elements to organize the DNA within the nucleus or on the chromosome scaffold may be as important as the sequences for binding soluble replication factors.

---

177. Saffer LD, Miller LO (1986) *Mol Cell Biol* 6:1148; Botchan PM, Dayton AI (1982) *Nature* 299:453; Truett MA, Gall JG (1977) *Chromosoma* 64:295; Cech TR, Brehm SL (1981) *NAR* 9:3531; Vogt VM, Braun R (1977) *EJB* 80:557.
178. Larson DD, Blackburn EH, Yaeger PC, Orias E (1986) *Cell* 47:229.
179. Méchali M, Kearsey S (1984) *Cell* 38:55.

CHAPTER 17

Bacterial DNA Viruses

17-1 Viral Windows on Cellular Replication[1]

Because a virus depends on a host cell for its development, it must either use the machinery of this cell to replicate itself or bring in the genetic information to tailor-make its own machinery—or do some of both. In any case, understanding the replication of the relatively small, well-defined viral chromosome illuminates the replication process of the cell itself. The claim that molecular biology originated largely in bacteriophage research[2] is arguable, but no vigorous objections should be raised to emphasizing the contributions that phages have made to our understanding of DNA synthesis. Many of the still missing parts of the replication puzzle in both animal and bacterial cells will be seen through windows opened by research with viruses.

The degree to which the virus relies on the host's replicative apparatus depends on its size; genome sizes of *E. coli* phages vary in length from about 5 kb to over 150 kb (Table 17-1). In the smallest DNA bacteriophages, Ff and ϕX174, with only eight to ten genes, almost all of the chromosomal information is devoted to defining the viral coat and its assembly; the viruses rely almost completely on the machinery of the host cell to replicate their DNA. The small chromosome, only one-thousandth the size of the host chromosome, is then a

1. Calendar R (ed.) (1988) *The Bacteriophages.* Plenum Press, New York, Vols. 1 & 2. (These volumes are excellent sources for both the general and particular properties of the phages.)
2. Cairns J, Stent GS, Watson JD (eds.) (1966) *Phage and the Origins of Molecular Biology.* CSHL; Stent GS, Calendar R (1978) *Molecular Genetics,* 2d ed. Freeman, San Francisco; Echols H (1978) in *The Bacteria* (Sokatch JR, Ornston LN, eds.). Academic Press, New York, Vol. 7, p. 487.

Table 17-1
Genomes of representative *E. coli* phages

Phage	Related phages	Particle shape	DNA Size (kb)	DNA Shape[a]	Comments
ϕX174	S13, G4	icosahedron	5.4	ss circular	duplex replicative form
M13	fd, f1	filament	6.4	ss circular	duplex replicative form
P4		head, tail	11.6	ds linear	cohesive ends; satellite phage
T7	T3	head, short tail	40.0	ds linear	terminal redundancy
λ	ϕ80, 434, 186	head, tail	48.6	ds linear	cohesive ends form replicative circles; lysogenic
P1	P7	head, tail	88	ds linear	terminal redundancy; permuted
T5		head, tail	113	ds linear	one strand interrupted
T4	T2, T6	head, tail	166	ds linear	terminal redundancy; permuted
R17	MS2, f2, Qβ	icosahedron	3.6	ss linear	RNA (not DNA)

[a] ds = double-stranded; ss = single-stranded.

promising instrument for identifying the host replicative enzymes and their functions. Although Ff and ϕX174 are similar, they use different replicative assemblies in the host cell for the initial stage, the conversion of the single-stranded viral genome to the duplex form.

Some phages of intermediate size, such as λ, induce a few enzymes for their own replication but depend primarily on host enzymes. Others, such as T7, devote a significant fraction of their genome to an autonomous DNA replication apparatus. The large phages like T4 contain 20 or more genes that direct the synthesis of precursors and a relatively complete and independent replicative apparatus. Even so, it is easier to grapple with the T4 genome than with the E. *coli* chromosome, which is 20 times larger.

The chromosomes of most temperate phages, such as λ, can be inserted into that of the host, and thus can be used to discover how such recombination takes place and what factors determine whether the course of infection will be virulent or temperate. Infection of healthy *E. coli* with phage λ can take either course with roughly equal probability.The *lytic response* of a *virulent* infection resembles T-phage infections (Sections 5 and 6). In infection by a *temperate* phage (Section 7), the phage DNA is integrated into the host chromosome and replicated with it, without detectable effect on the cell. The integrated phage genome (*prophage*) remains dormant for many cell generations; the cell which carries the prophage is said to be *lysogenic*. However, when the cell undergoes certain stresses, such as UV irradiation or metabolic alterations, the regulatory system that holds the prophage in check responds to the stressful stimulus, and the lysogenic state ends. The prophage is excised from the host chromosome, and unrestrained multiplication of the extrachromosomal phage DNA leads to a typically virulent and lytic infection.

The temperate phage P1 maintains itself as an extrachromosomal plasmid rather than integrating into the host chromosome. Maintained, like the host chromosome, at one or a few copies per cell, P1 encodes several functions to ensure its faithful partitioning into the two daughter cells during cell division.

P1 plasmid replication is thus instructive in the way replication is regulated and how chromosomal copy number is maintained.

The replication of virulent as well as temperate phages has been studied by means of host mutants with defects that impede development of the phage but may not interfere seriously with growth of the host. New insights into replication have been obtained by identifying these defects and by discovering, in turn, phage mutations that overcome them.

Animal cell viruses (Chapter 19) can be useful in understanding animal cell DNA replication, just as bacteriophages can be for the bacterial cell. Reference has already been made to the DNA polymerases encoded by large animal viruses (Chapter 6). The small DNA animal viruses (polyoma and SV40), similar to phages Ff and ϕX174 in size, produce virulent infections, relying almost entirely on host replication enzymes. Occasionally, the chromosome of these small viruses may become integrated into the host DNA and provoke uncontrolled proliferative growth in the host cells and tissues, an event of great significance for cancer biology.

In addition to characterizing the enzymes of DNA synthesis, we need to know their localization in the cell and the means by which their activities are controlled. Once again, elucidation of spatial and temporal factors in the viral life cycle may reveal basic facts about the regulation of replication and expression of the DNA of the host.

17-2 Stages in the Viral Life Cycle

To understand how the viral DNA is synthesized, we must know how the DNA is organized within the virus, delivered into the cell, multiplied, packaged, and finally released from the cell. These stages, each containing several steps, are designated as follows:

(1) *Adsorption* of the virus to a specific receptor on the cell surface.

(2) *Penetration* of the DNA through the cell envelope and membrane layers to the interior. Uncoating of the DNA from its protein shell may be coupled to its replication.

(3) *Recognition* of the template by host metabolic enzymes. The initial recognition can be by one of four enzyme systems, depending on the virus: *replication* to create a DNA duplex (in the ssDNA viruses), *reverse transcription* to make a DNA copy of the viral RNA (in retroviruses), *transcription* of a particular region of the DNA (in the dsDNA viruses), or *translation* of a special region of the RNA (in RNA phages and animal viruses).

(4) *Multiplication* of the DNA through various intermediates, leading ultimately to many copies of a mature form appropriate for packaging.

(5) *Assembly and packaging* of DNA condensed into a preformed protein shell.

(6) *Release* of virus particles as they are packaged, either by progressive budding through the cell surface without major damage to the cell (Ff phage), or by rupture of the cell after particles have accumulated inside in large numbers and a lysis-inducing protein has been produced (ϕX and λ phages).

In reviewing what is known of the chemistry of the viral life cycle, one question remains central. How is the viral DNA recognized and handled with a specificity that ensures its rapid and error-free duplication in competition with an extraordinary excess of cellular DNA? In the case of Ff, the competition would be for the cellular RNA polymerase (RNAP) and DNA polymerase holoenzyme, whereas ϕX174 must immediately appropriate the primosome assembly (Section 8-4). When a duplex viral DNA such as T7 invades the cell, certain genes must be promptly transcribed by the host RNAP and translated so that early proteins, including phage-specific RNA and DNA polymerases, are made available. Recognition of the viral nucleic acid by the host systems for replication, transcription, and translation may depend on powerful promoters in the DNA or strong ribosomal binding sites in the RNA. Viral adsorption proteins may play an additional role.

Possibly, an adsorption protein, which may also be a *pilot protein*,[3] guides the viral DNA to the specific cell surface receptor. By strong interaction with a viral DNA region, the pilot protein orients that region toward the cell membrane. For a polyelectrolyte such as DNA to penetrate the lipid bilayer of the membrane, a pore or sheath is essential. (Passage of even small ions through the plasma membrane requires a protein channel.) An amphipathic pilot protein may provide the hydrophobic exterior required for interaction with membrane lipids and the hydrophilic inner surface needed for passage of the DNA. Spheroplasts, which lack cell walls, can be infected by free DNA (transfected), but this infection is less efficient, by several orders of magnitude, than introduction of the same DNA molecule via phage infection into intact cells.

The pilot protein may have the capacity to direct a phage ssDNA to a replicative assembly, a phage dsDNA to RNAP for transcription of a particular gene, and a phage RNA to a ribosome for synthesis of a phage-specific protein. In each instance, for the tiny bit of invading nucleic acid to be expressed by replication, transcription, or translation, it must overcome competition with enormous excesses of endogenous DNA and RNA.

Viruses have developed many ways of redirecting the host cell's biosynthetic machinery to their own use (Table 17-2). Aside from outcompeting the host DNA for the cellular enzymes and encoding their own specific enzymes such as DNA and RNA polymerases, numerous phages encode proteins designed to inhibit or destroy crucial cellular functions. The A* protein of ϕX174 shuts off host DNA synthesis; T7 encodes nucleases that degrade the host DNA, as well as two proteins that inhibit the host RNAP after the phage-specific enzyme has been synthesized; and T4 directs the synthesis of a novel dNTP, while degrading a related dNTP of the host.

A most remarkable feature of viral multiplication is that so many viruses are produced in so short a time—several hundred coliphages in 20 minutes. The awesome speed and efficiency of the processes do not necessarily make them simpler for the biochemist to fathom, but they are at least easier to detect above the background of normal cellular activity. Despite the extraordinary complexity of the viral life cycle, its dissection and eventual reconstruction with chemi-

3. Jazwinski SM, Lindberg AA, Kornberg A (1975) *Virology* 66:283; Jazwinski SM, Marco R, Kornberg A (1975) *Virology* 66:294.

Table 17-2
Strategies for redirecting and augmenting host functions for the benefit of phage growth

Method	Examples	
	Phage	Proteins[a]
Modification of transcription by host RNAP	T7	gp0.7: protein kinase gp2: RNAP-binding protein
	T4	Mod and Alt: ADP-ribosylation gp55: alternate σ factor
	SPO1	gp28, gp33, and gp34: σ factors
Modification of transcription by phage-encoded RNAP	T7	gp1: RNAP
Degradation of host DNA	T7	gp3: exonuclease gp6: endonuclease
	T4	DenA: endonuclease II DenB: endonuclease IV
Synthesis and use of altered nucleotides	T4	gp42: dCMP hydroxymethylase; gp1: HM·dCMP kinase
	SPO1	gp29: dUMP hydroxymethylase gp23: HM·dUMP kinase
Degradation of host deoxynucleotides	T4	gp56: dCTP/dCDPase
	SPO1	dTTPase
	T5	5′ deoxynucleotidase
Inhibition of host DNA synthesis	ϕX174	gene *A** protein
	T4, N4	products unknown
Use of a pilot protein	ϕX174	gpH; adsorption protein
Capture of host dnaB protein	λ	gpP: dnaC analog

[a] HM·dCMP, HM·dUMP = hydroxymethyl cytosine and uracil deoxynucleoside monophosphates; RNAP = RNA polymerase.

cally defined components in a cell-free system, from start to finish, remains an attractive prospect.

17-3 Small Filamentous (Ff) Phages: M13, fd, f1[4]

E. coli cells bearing hair-like F pili (Section 18-7) are hosts for filamentous viruses, such as M13, fd, and f1. These Ff (*F* pili, *f*ilamentous) viruses are nearly identical in sequence and behavior. A related *Pseudomonas* phage, Pf1, is twice as long as M13 but has about the same DNA content, packaged in a more extended form. Replication of these viruses depends on many of the same enzymes used by the cell to replicate plasmids and its own chromosome.

The Ff phages have recently gained wide popularity as cloning vectors because they have no physical constraints limiting the length of DNA that can be packaged and because they allow the easy purification of ssDNA. The many unusual features of the Ff viruses and their life cycles (see Fig. 17-4) recommend them for intensive study of broader aspects of DNA synthesis.

4. Model P, Russel M (1988) in *The Bacteriophages* (Calendar R, ed.). Plenum Press, New York, Vol. 2, p. 375; Rasched I, Oberer E (1986) *Microbiol Rev* 50:401.

Table 17-3
Characteristics of Ff virions and DNA

Virion		DNA (single-stranded)	
Shape	filament	Shape	circle
Mass	16.4 MDa	Mass	2.11 MDa
Weight of 10^{14}	2.7 mg	Weight of 10^{14}	0.35 mg
Length	895 nm	No. of nucleotides	6407[a]
Diameter, outer	6 nm	Base composition (%)	
Proteins, and number of copies		A	24.6
Major coat		T	34.7
gp8	2710	G	20.5
Adsorption end		C	20.2
gp3	5		
gp6	5		
Packaging end			
gp7	5		
gp9	5		

[a] The genomes of M13 and f1 each have 6407 nucleotides, while fd has 6408.

Virion

The thread-like Ff viruses are 895 nm long and only 6 nm in diameter (Table 17-3; Fig. 17-1).[5] The ssDNA is contained within a protein tube,made up of about 2700 subunits of gp8 in shingle-like arrangement[6] (Fig. 17-2), with an inner diameter of 2.5 nm. Cross-linking the DNA strands with psoralen[7] has disclosed a hairpin loop, approximately 100 ntd long, at one end of the phage.[8] This hairpin contains the packaging signal and lies within the intergenic region along with the replication origins for each DNA strand (Sections 8-2 and 8-7; see also Fig. 17-7). The virion also contains about five copies each of the products of genes *3*, *6*, *7*, and *9*. The proteins encoded by genes *3* and *6* (A and D proteins) make up the adsorption complex at one end of the phage, while gp7 and gp9 (C protein) are at the other end.[9] Surprisingly, the DNA is oriented with the region containing the complementary-strand origin at the end of the virion opposite the adsorption complex.[10]

Coat. Ff can be obtained in large quantities, about 1 gram per 10 liters of culture, and is, thus far, the only bacterial virus suitable for x-ray diffraction analysis. The mature form of the major coat protein (gp8) is 50 residues long[11]

5. Banner DW, Nave C, Marvin DA (1981) *Nature 289*:814.
6. Marvin DA, Wachtel EJ (1976) *Philos Trans R S Lond [Biol] 276*:81.
7. Shen CK, Hearst JE (1976) *PNAS 73*:2649.
8. Shen CK, Ikoku A, Hearst JE (1979) *JMB 127*:163.
9. Simons GFM, Konings RNH, Schoenmakers JGG (1979) *FEBS Lett 106*:8; Simons GFM, Konings RNH, Schoenmakers JGG (1981) *PNAS 78*:4194; Lin T-C, Webster RE, Konigsberg W (1980) *JBC 255*:10331; Grant RA, Lin T-C, Konigsberg W, Webster RE (1981) *JBC 256*:539; Gray CW, Brown RS, Marvin DA (1981) *JMB 146*:621.
10. Webster RE, Grant RA, Hamilton LAW (1981) *JMB 152*:357.
11. Snell DT, Offord RE (1972) *BJ 127*:167; Nakashima Y, Konigsberg W (1974) *JMB 88*:598; Hagen DS, Weiner JH, Sykes BD (1979) *Biochemistry 18*:2007.

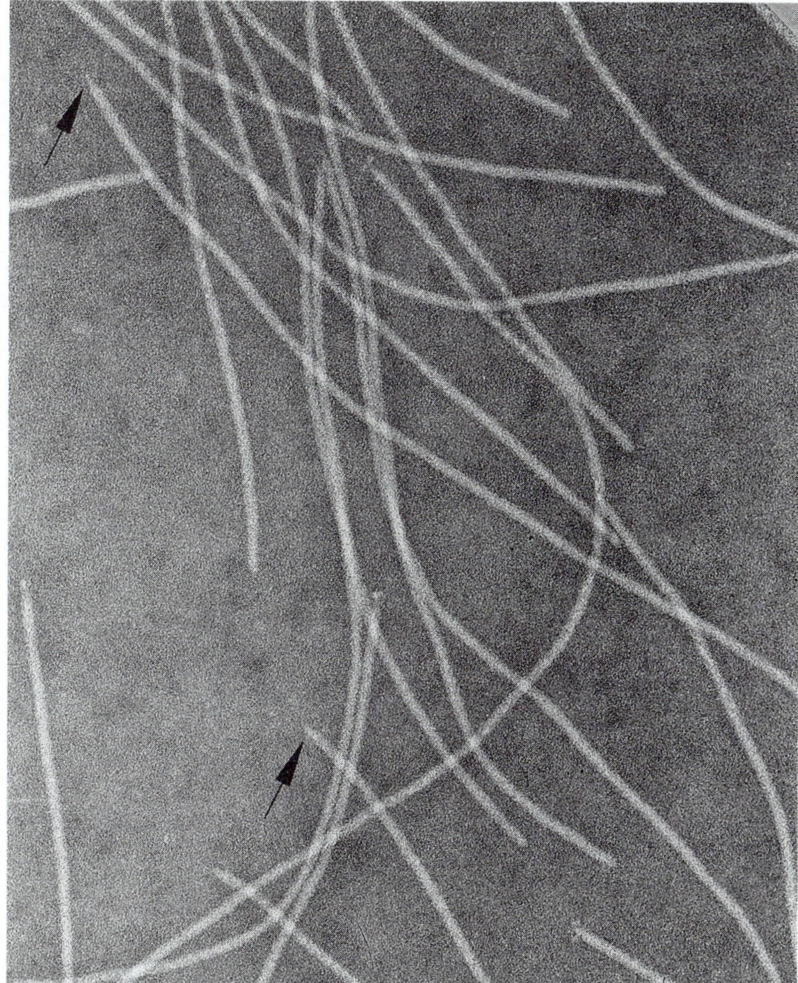

Figure 17-1
Electron micrograph of M13 (×200,000). Some filaments are pointed at one end (arrows). (Courtesy of Professor RC Williams)

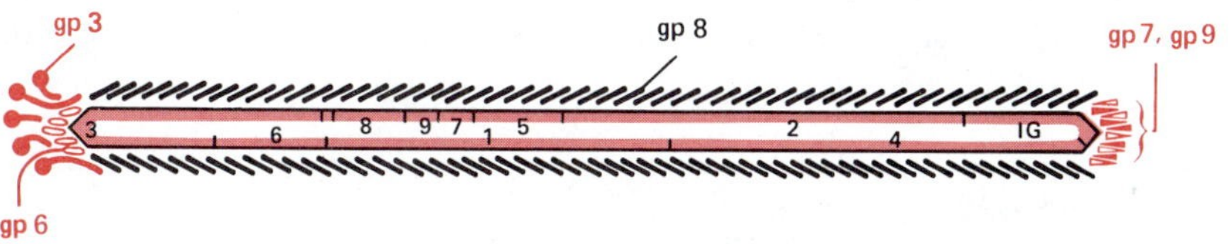

Figure 17-2
Schematic representation of the coat proteins and genetic map of the ssDNA a filamentous (Ff) phage. Numbers = phage genes and their locations; IG = intergenic sequence containing the complementary-strand origin. (Courtesy of Professor RE Webster)

and contains no histidine, cysteine, or arginine. An acidic region at the N terminus (residues 1 to 20) is followed by a hydrophobic interior (residues 21 to 39) and a basic region at the C terminus (residues 40 to 50).

```
1         10         20         30         40         50
N-AEGDDPAKAAFDSLQASATEYIGYAWMVVVVVVIVGATIGIKLFKKFTSKAS-C
       Acidic              Hydrophobic            Basic
```

These regions correspond to functional domains in the protein: the acidic region forms the phage surface, the basic region is internal, in contact with the DNA, and the hydrophobic core provides for the gp8 subunit interactions.[12] This phage, a simple nucleoprotein, readily altered by site-specific mutagenesis, is a model for probing the organization of a protein–DNA complex.

Gp8 is stored in the internal membrane before phage assembly. The α-helix content of the membrane-bound form is about 50%, while in the phage gp8 is almost entirely α-helical.[13] Translation produces a precursor called the *precoat* (or *procoat*). Maturation requires cleavage of 23 amino acids from the N terminus by a membrane-associated leader peptidase.[14] Processing of the signal peptide is essential; mutations that block cleavage are lethal to both host and phage. Membrane insertion of coat protein is independent of the host *secA* and *secY* functions, which are involved in the export of many (but not all) proteins.[15] These genetic characteristics, and the ability of coat protein to integrate into membranes in vitro, have made it a powerful model for investigating membrane assembly.[16]

Genome. The small size of the Ff genome speaks for its relative simplicity. The circular DNA, of known sequence, 6407 nucleotides long, contains ten genes (Table 17-4). The arrangement of the genome differs from that of icosahedral phages (Section 4) in the relative absence of overlapping genes and the prominence of intergenic spaces that encode signals for DNA replication, packaging, and gene expression. The evolution of sequences in Ff phages has been influenced by the lack of constraint in the size of the genome that can be packaged into a phage particle.[17]

A particularly important intergenic (IG) region between genes *2* and *4* (Fig. 17-3; also see Fig. 17-7)[18] comprises 8% of the genome (508 ntd). Stop codons in each of the three translational reading frames would limit any gene product to only 16 amino acids. Thus the function of this region clearly is not to encode a protein product. This region (see below) contains the origins of complementary (−) and viral (+) strand synthesis as well as the signal for packaging the DNA into the phage. The IG region has the potential for forming secondary structures and is preserved in miniphages (see below), on which it confers a selective growth advantage.[19]

---

12. Armstrong J, Hewitt J, Perham R (1983) *EMBO J 2*:1641.
13. Cross T, Opella SJ (1985) *JMB 182*:367; Schneider D, Valentine K, Opella SJ (1985) *Biophys J 47*:398a.
14. Silver P, Watts C, Wickner W (1981) *Cell 25*:341; Wolfe PB, Silver P, Wickner W (1982) *JBC 257*:7898.
15. Wolfe PB, Rice M, Wickner W (1985) *JBC 260*:1836; Kuhn A, Kreil G, Wickner W (1987) *EMBO J 6*:501.
16. Wickner W (1989) *TIBS 14*:280.
17. Hermann R, Neugebauer K, Zentgraf H, Schaller H (1978) *MGG 159*:171; Ohsumi M, Vovis GF, Zinder ND (1978) *Virology 89*:438.
18. Zinder ND, Horiuchi K (1985) *Microbiol Rev 49*:101.
19. Chen T-C, Ray DS (1978) *J Virol 28*:679.

Table 17-4
Ff genes and functions

Gene	Function[a]	Protein product: Mass (kDa)	No. of copies in virion	No. of copies in cell[b]
1	morphogenesis	35	0	—
2	RF replication; nicking	46	0	1,000
3	adsorption protein	43	5	—
4	morphogenesis	50	0	—
5	SSB	9.8	0	150,000
6	minor coat protein (adsorption complex)	12	5	—
7	minor coat protein (morphogenic end)	3.5	5	—
8	major coat protein	5.2	2710	—
9	minor coat protein (morphogenic end)	3.3	5	—
X	switch from RF→RF to RF→SS	12	0	500

[a] RF = replicative form; SS = viral single strands; SSB = single-strand binding protein.
[b] About 10^3 virions are produced per cell, per generation. Morphogenic and DNA replication proteins are present in the cell during phage growth.

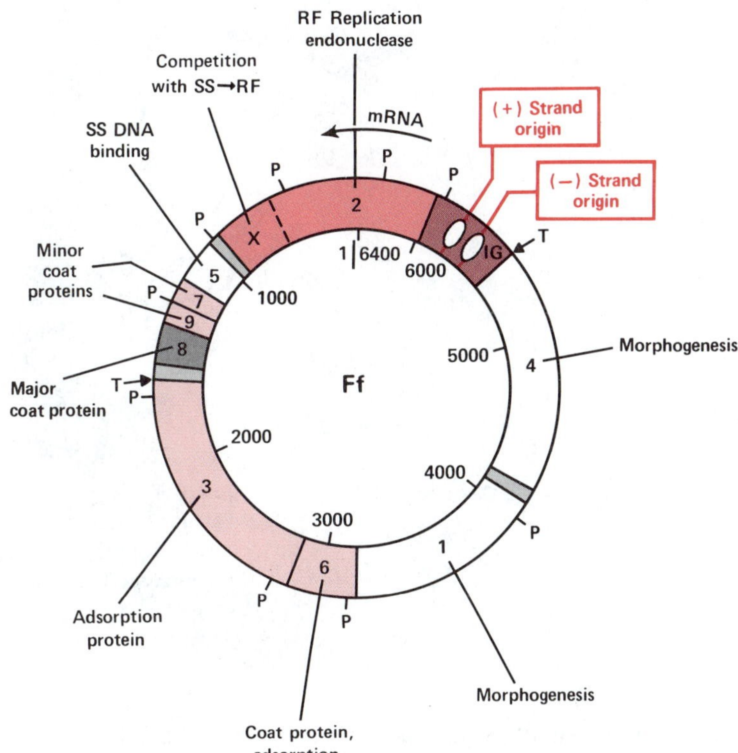

Figure 17-3
Genetic map of an Ff phage with functions of gene products. DNA replication origins are in an intergenic region; P = promoters; T = transcription terminator.

Infection

The virus attaches to the cell at or near the F (sex, or *f*ertility, factor) pilus. The ensuing infection results in extrusion of 1000 or more viral particles in each cell generation (Fig. 17-4). Reduction in the rate of host cell growth by only about one-third sustains phage development and allows the assay of phage particles by their formation of turbid plaques.

Adsorption. Because Ff viruses will infect only strains bearing F pili and have been photographed moored to pili ends (Fig. 17-5), this structure is presumably a means by which the Ff phage is delivered into the cell.[20] F pili are also associated with transferring plasmid and chromosomal DNA to cells that lack F pili and with infection by RNA phages.[21]

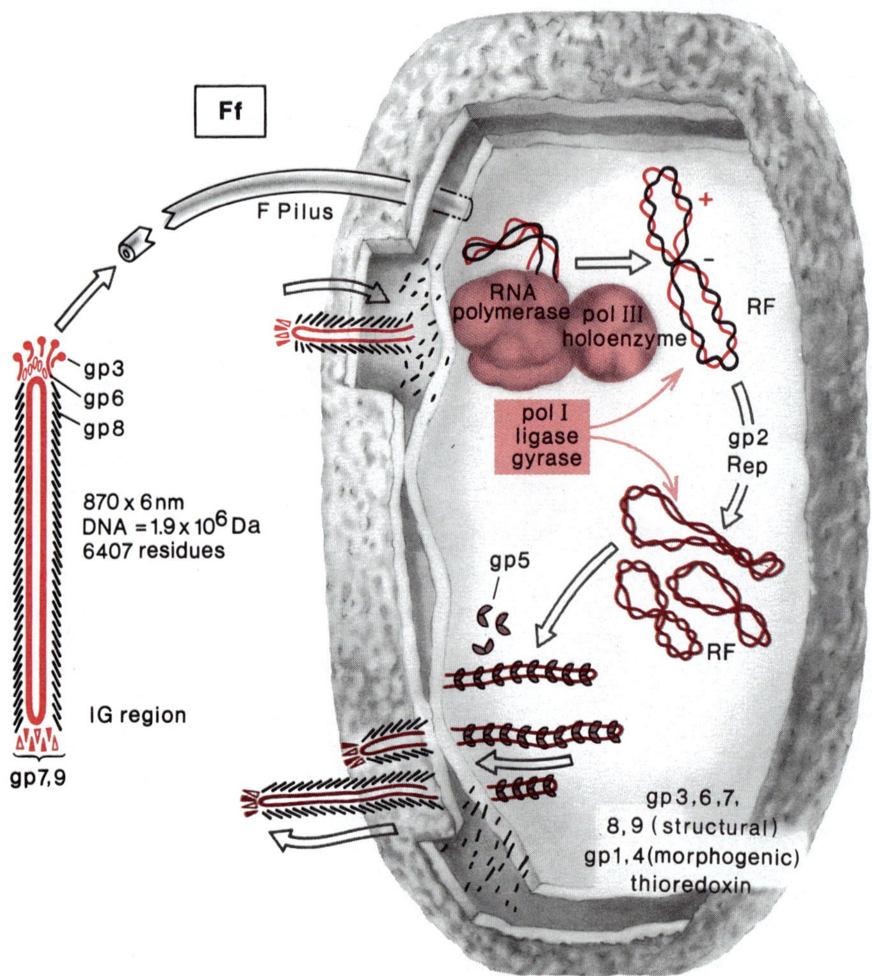

Figure 17-4
Scheme for the life cycle of a filamentous (Ff) phage. See Figure 17-9 for comparison with polyhedral phage ϕX174.

20. Caro LG, Schnös M (1966) *PNAS 56*:126; Jacobson A (1972) *J Virol 10*:835.
21. Willetts N, Skurray R (1980) *ARG 14*:41.

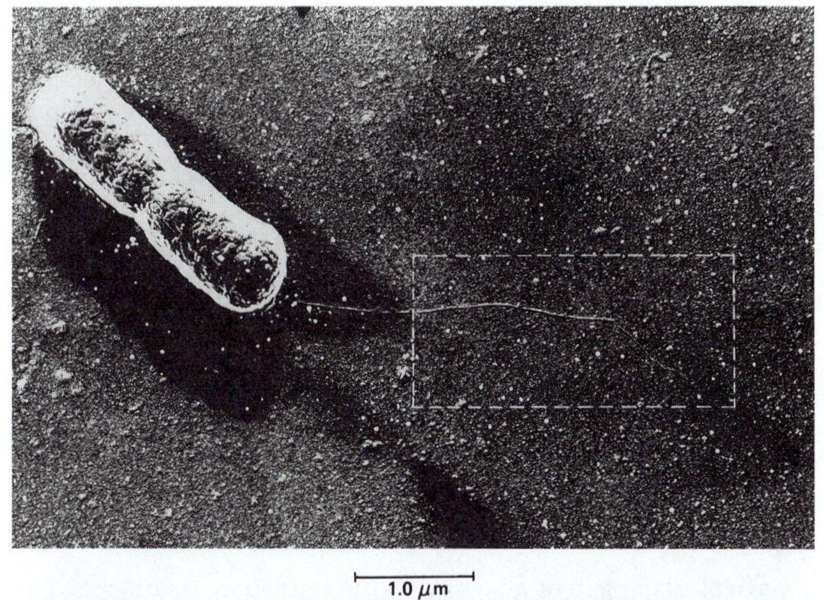

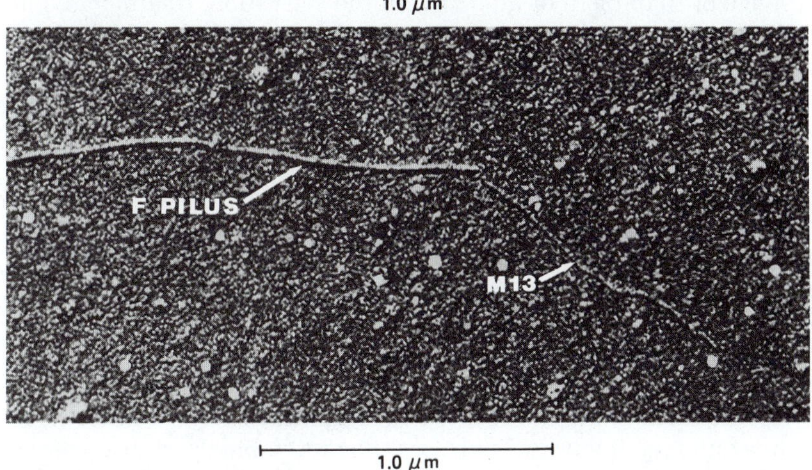

Figure 17-5
Electron micrograph of an F pilus of *E. coli* with an M13 phage attached to its tip. Despite this demonstrable interaction between pilus and phage, the direct role of the pilus in adsorption is uncertain. (Courtesy of Dr J Griffith)

One model of Ff adsorption proposes that the phage nucleoprotein is conducted along a groove of the F pilus to the cell surface. Another adsorption mechanism suggests progressive depolymerization of the pilus that retracts the tip, with the phage attached, down to the base of the pilus at the cell surface.[22] An alternative possibility is that the phage is guided by the pilus to a membrane receptor at its base, exposed when the pilus is shed. Such receptors must also be present in cells lacking pili, inasmuch as such cells when partially disrupted are efficient hosts for Ff phages.[23] Attachment to F pili is not an absolute requirement for infection, but facilitates it enormously. Sensitive assays for infection

22. Marvin DA, Hohn B (1969) *Bact Rev* 33:172.
23. Marco R, Jazwinski SM, Kornberg A (1974) *Virology* 62:209.

using phage–plasmid hybrids carrying drug resistance markers[24] reveal that infection of F^- cells, compared to F^+, have an efficiency of only about 10^{-6}. The *tol* genes, involved in import of colicins, are also required for Ff phage infection, even in cells that lack pili. The products of these genes are thus candidates for a cell-surface receptor.[25]

The gene *3* protein is responsible for adsorption to the pili. By electron microscopy, the adsorption complex appears as a knob and stem structure at one end of the phage. Attachment is via the N-terminal globular domain of gp3;[26] removal of this domain by proteolysis destroys infectivity. The adsorption end of the phage also contains gp6, required for phage infectivity and stability in an unidentified way.

Penetration. Both the phage coat protein and DNA become associated with the host cell upon infection; the coat protein deposited in the inner cell membrane is salvaged for coating the progeny DNA.[27] The notable experiment with T2 phage indicating that DNA is the genetic substance because so much of the DNA and so little of the protein entered the cell would have been less definitive had Ff been used instead.[28]

In the presence of rifampicin or after thymidine starvation, treatments that block the initiation of replication, the phage adsorbs to the cell but the DNA is apparently not uncoated.[29] Decapsidation of the phage thus appears to be coupled to DNA replication. Inasmuch as the complementary-strand origin is located at the phage end opposite to the gp3 adsorption complex, it is not likely to be the first DNA to enter the cell.[30] Furthermore, the complementary-strand origin can function when moved to other locations on the template, and the (−) strand origin of G4 can substitute for that of M13 when located even 1 kb from the normal origin site.[31] Thus, much of the DNA must be accessible to the replication enzymes upon infection, making a tight coupling between uncoating and replication unlikely.

Ff as Parasitic Phages. Ff phages alone among the bacterial viruses do not produce a lytic infection.[32] Rather, the infected cell produces and secretes virus particles without undergoing lysis. Ff viruses thus resemble the myxoviruses, which are filamentous animal viruses (influenza, mumps, Sendai, simian virus 5) composed of nucleoprotein units that are extruded through the cell surface with relatively few pathologic effects. A block in the maturation of myxoviral particles at the cell membrane, however, results in the accumulation of nucleoprotein units, cellular fusion, disintegration, and death. Similarly, interruption of Ff development at any stage beyond RF replication results in accumulation of aborted intermediates, development of large intracellular whorls of the plasma membrane, and loss of cell viability.[33] Both the cellular pathology in abortive infections and the maintenance of a commensal relationship with a cell may

24. Russel M, Whirlow H, Sun T-P, Webster RE (1988) *J Bact 170*:5312.
25. Sun T-P, Webster RE (1986) *J Bact 165*:107; Sun T-P, Webster RE (1987) *J Bact 169*:2667.
26. Nelson FK, Friedman SM, Smith GP (1981) *Virology 108*:338.
27. Smilowitz H (1974) *J Virol 13*:94
28. Hershey AD (1953) *CSHS 18*:135.
29. Brutlag D, Schekman R, Kornberg A (1971) *PNAS 68*:2826.
30. Webster RE, Grant RA, Hamilton LAW (1981) *JMB 152*:357.
31. Kaguni J, Ray DS (1979) *JMB 135*:863.
32. Marvin DA, Hohn B (1969) *Bact Rev 33*:172.
33. Schwartz FM, Zinder ND (1968) *Virology 34*:352.

depend critically on the regulation of biosynthesis and assembly of proteins and lipids in cell membranes.[34]

Replication: Stages in the Cycle

Successive events in the replication of Ff DNA may be divided into three stages (Fig. 17-6). The first stage (SS→RF), conversion of viral DNA (SS) to the parental replicative form (RF), requires no synthesis of phage-encoded proteins and relies on a host replication system. The second stage (RF→RF), multiplication of the replicative form, is initiated by the phage-encoded gp2, and produces enough templates to support adequate levels of transcription. The final stage (RF→SS), synthesis of viral single strands, entails coating of the intracellular DNA by gp5 (Section 10-4) in a form suitable for phage assembly in the membrane. The gp5 is displaced when the progeny DNA is packaged into phage filaments.

SS→RF. Replication of the viral single-stranded circle to a circular duplex provided the first insight into RNA priming[35] (Section 8-1). RNA polymerase (RNAP) containing σ^{70} (Section 7-2) recognizes a specific sequence [(−) strand *ori*; Section 8-2] on the (+) strand in the presence of SSB and produces a short RNA segment[36] (the RNA primer) that is extended by DNA pol III holoenzyme.

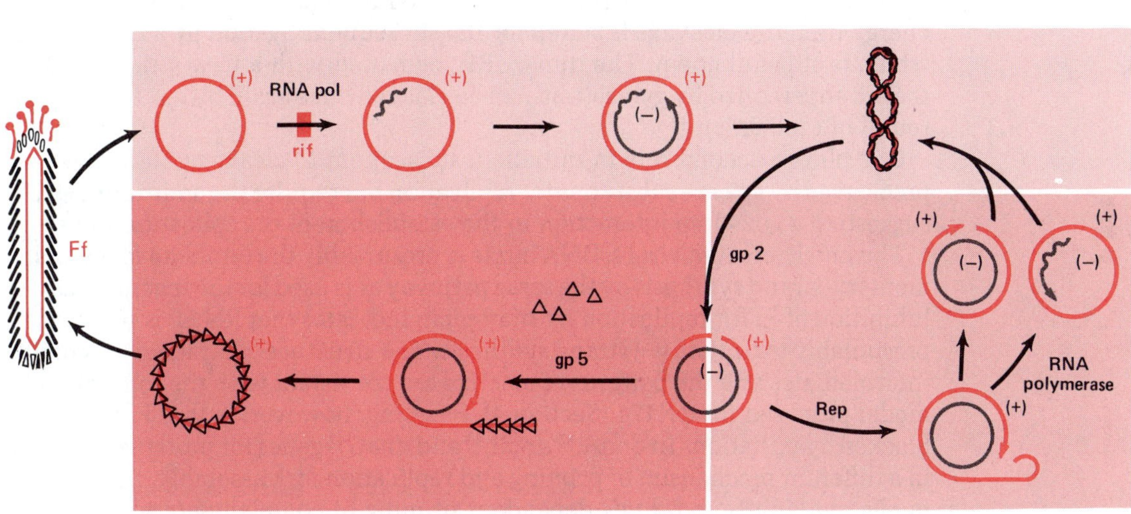

Figure 17-6
Scheme for Ff DNA replication in three successive stages: (1) SS → RF, (2) RF → RF, (3) RF → SS. Notice that during the three stages, only two basic mechanisms are used: synthesis of a (−) strand on a viral single strand and synthesis of a viral (+) strand on the double-stranded circle with concomitant displacement of single strands. RNAP = RNA polymerase; rif = rifampicin.

34. Chamberlain BK, Webster RE (1976) *JBC 251*:7739.
35. Brutlag D, Schekman R, Kornberg A (1971) *PNAS 68*:2826.
36. Kaguni JM, Kornberg A (1982) *JBC 257*:5437.

Although the sequence recognized by RNAP has been identified and is in a region of secondary structure, it contains no homology to typical *E. coli* promoters.

The product of this stage, RFII, is a duplex circle with a small gap. It contains the intact viral DNA and a nearly full-length synthetic complementary strand with the RNA primer covalently attached to the 5′ end.[37] The 5′→3′exonuclease action of DNA pol I is essential for removing the primer.[38] *E. coli* ligase seals and gyrase supercoils the product, making RFI.

Priming of DNA synthesis by RNAP, used for converting Ff viral DNA to RF, is also used for replicating some plasmids, such as ColE1 (Section 18-2). Complementary-strand synthesis on phage φX174, as will be seen in Section 4, is primed instead by *E. coli* primase assisted by a complex primosome.

RF→RF. Synthesis is initiated by the introduction of a specific cleavage in the supercoiled parental RF by the action of gp2 endonuclease. Replication proceeds by a rolling-circle mechanism very similar to that of φX174 (Section 15-11). Gp2 encoded by Ff is a 46-kDa endonuclease (Section 8-7) that cleaves the viral strand of supercoiled RF at the (+) strand origin (Fig. 17-3).[39] The 3′-OH created by this cleavage is elongated by DNA pol III holoenzyme as Rep helicase unwinds the duplex. *E. coli* SSB binds the displaced viral strand.[40]

After synthesis has proceeded around the circle, recreating the duplex origin, the viral strand is liberated by a second gp2 cleavage, and the two ends are ligated into a ssDNA circle.[41] In contrast to the cleavage of φX174 by gpA (Section 8-7), covalent linkage between gp2 and the 5′-P end of the DNA has not been observed. Gp2, like gpA, requires no high-energy cofactor of activity. Thus, the energy from the cleavage is stored for the subsequent ligation by a mechanism which is still unknown. The duplex RF, recreated with a new viral strand, is sealed, supercoiled by gyrase, and can be cleaved once again by gp2 to initiate a round of replication.[42]

Gp2 plays a second role in initiation, subsequent to cleavage. RF molecules precleaved by gp2 are substrates for replication but require the addition of gp2,[43] suggesting a gp2–Rep interaction in the establishment of replication forks.

The newly created viral ssDNA circle is presumably the template for complementary-strand synthesis by the same pathway as is used for the incoming DNA. Inhibition of Ff RF replication by rifampicin indicates that RNAP is required,[44] presumably to initiate (−) strand synthesis. The structural characteristics of the intermediates and the patterns of labeling are consistent with the rolling-circle model proposed for φX174 (Section 4). Puzzling, however, is that Ff growth is blocked by mutations in *dnaA*,[45] *dnaB*,[46] and *dnaG*[47] genes for proteins involved in a different mechansim of priming and replication fork assembly. An attempt to observe *dnaB*- and *dnaG*-dependent priming of complementary strands in

37. Westergaard O, Brutlag D, Kornberg A (1973) *JBC 248*:1361; Dasgupta S, Mitra S (1978) *PNAS 75*:153.
38. Dasgupta S, Mitra S (1977) *BBRC 78*:1108; Dasgupta S, Mitra S (1978) *PNAS 75*:153.
39. Meyer TF, Geider K (1979) *JBC 254*:12636; Meyer TF, Geider K (1979) *JBC 254*:12642.
40. Meyer TF, Geider K (1982) *Nature 296*:828.
41. Geider K, Baumel I, Meyer TF (1982) *JBC 257*:6488.
42. Horiuchi K, Zinder ND (1976) *PNAS 73*:2341; Horiuchi K, Ravetch JV, Zinder ND (1979) *CSHS 43*:389.
43. Meyer TF, Geider K (1982) *Nature 296*:828
44. Brutlag D, Schekman R, Kornberg A (1971) *PNAS 68*:2826.
45. Bouvier F, Zinder ND (1974) *Virology 60*:139.
46. Olsen WL, Staudenbauer WL, Hofschneider PH (1972) *PNAS 69*:2570.
47. Ray DS, Dueber J, Suggs S (1975) *J Virol 16*:348; Dasgupta S, Mitra S (1976) *EJB 67*:47.

vitro was unsuccessful, suggesting that the effect of these mutations on Ff growth may be indirect.[48]

RF→SS. Viral-strand synthesis to make DNA for packaging into phage proceeds by the same mechanism as used during RF→RF synthesis, in which gp2 cleavage initiates rolling-circle replication. Gp5 blocks complementary-strand synthesis by coating the displaced viral strand to form the nucleoprotein precursors for packaging.[49]

The relative levels of gp2, gpX, and gp5 determine whether RF→RF synthesis continues or viral-strand synthesis takes over.[50] When gp5 predominates, replication shifts to viral-strand synthesis. Temperature-sensitive gene *2* mutants, when held at a restrictive temperature, accumulate gp5; then, upon shift to a permissive temperature, only ssDNA is synthesized. Mutants in the C-terminal third of gp2, lacking only gpX, are defective in synthesis of viral single strands.[51] GpX is apparently involved in the switch to viral-strand synthesis, perhaps by assisting gp5 in inhibiting complementary-strand synthesis (SS→RF). The ability of gp5 to repress the translation of gp2 and gpX may also play a role in regulating the switch between the replication pathways.

Gp5 binds tightly and cooperatively to ssDNA and facilitates the unwinding of duplex DNA.[52] As a dimer of 9.8-kDa subunits, it can bind eight nucleotides of ssDNA (Section 10-4). The protein composition of the gp5 · ssDNA complex includes, in addition to 1300 to 1600 molecules of gp5, one to three molecules of *E. coli* SSB and a variable amount of an 11-kDa host protein.[53] In the absence of coat protein or other gene products essential for phage assembly, these complexes accumulate in large numbers, whereas cells infected with gene 5 mutants accumulate RF.

The Intergenic Region[54]

The 508-ntd IG sequence between genes *2* and *4* contains the complementary- and viral-strand origins and the DNA packaging signal (Fig. 17-7). This region has a large degree of secondary structure: five palindromes, which form hairpins (numbered A to E), and a 150-ntd AT-rich region.

The complementary-strand origin is the site of RNA priming. The RNA starts at a unique location and, in the absence of DNA pol III holoenzyme, is 30 ntd long. RNAP specifically protects about 125 ntd, including hairpins B and C, from nuclease digestion.[55] This origin is not absolutely required for phage growth: phage lacking this region make very small plaques but are viable, indicating the existence of a secondary pathway for complementary-strand synthesis.[56] Whether RNAP is used in this alternate pathway and where synthesis is initiated are unknown.

48. Bayne ML, Dumas LB (1979) *J Virol 29*:1014.
49. Salstrom JS, Pratt D (1971) *JMB 61*:489, Pratt D, Laws P, Griffith J (1974) *JMB 82*:425.
50. Fulford W, Model P (1988) *JMB 203*:39; Fulford W, Model P (1988) *JMB 203*:49.
51. Fulford W, Model P (1984) *JMB 178*:137.
52. McPherson A, Brayer GD (1985) in *Biological Molecules and Assemblies; Vol. 2: Nucleic Acids and Interactive Proteins* (Jurnak FA, McPherson A, eds). Wiley, New York, p. 325.
53. Grant RA, Webster RE (1984) *Virology 133*:315.
54. Zinder ND, Horiuchi K (1985) *Microbiol Rev 49*:101.
55. Gray CP, Sommer R, Polke C, Beck E, Schaller H (1978) *PNAS 75*:50.
56. Kim MH, Hines JC, Ray DS (1981) *PNAS 78*:6784.

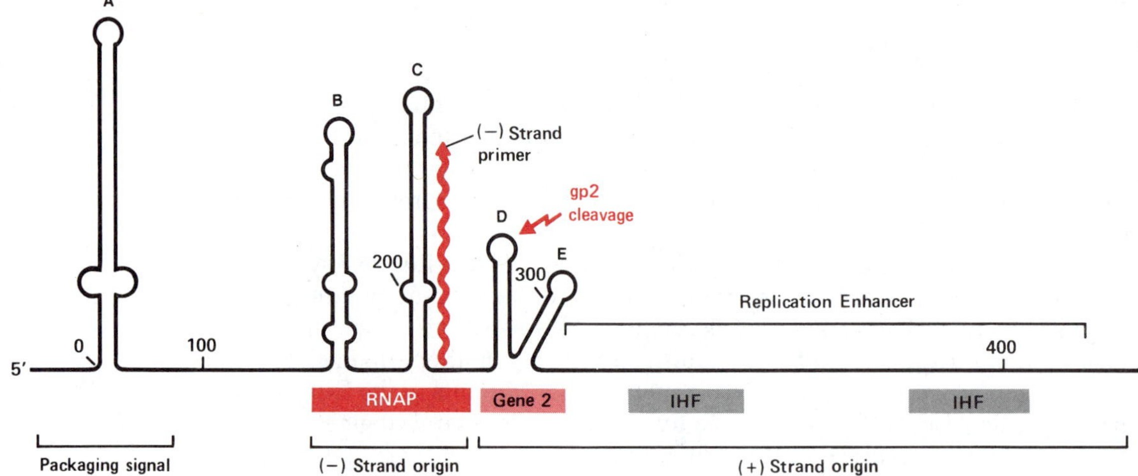

Figure 17-7
The intergenic (IG) region of Ff phage. Vertical elements are self-complementary sequences (hairpins) in the secondary structure of the viral (+) strand. Boxes = protein-binding sites; numbers = approximate nucleotides from the 5′ end of the IG; IHF = integration host factor; A–E = hairpins. [After Model P, Russel M (1988) in *The Bacteriophages* (Calendar R, ed.). Plenum Press, New York, Vol. 2, p. 375]

Gp2 cleaves the (+) strand origin to initiate and terminate viral-strand synthesis. The (+) strand origin (including hairpins D and E and the 3′ AT-rich sequences) is, however, more complex than a simple cleavage site. Distinct and overlapping sequence elements of the (+) strand origin responsible for gp2 cleavage, initiation, and termination of replication have been identified.

The initiation sequences include an essential core region and an enhancer that stimulates replication 100-fold. Four monomers of gp2 bind specifically to the core and, on supercoiled DNA, nick the cleavage site.[57] The 3′ AT-rich region contains the enhancer. Insertions or deletions within the enhancer region reduce phage replication to a few percent of the wild-type level.[58] Phages that overproduce gp2 (or encode an altered form) can compensate for defects in the enhancer, but not for mutations in the core, indicating that gp2 interacts with the enhancer as well as with the cleavage site.[59]

The *E. coli* integration host factor (IHF) (Section 10-8) binds specifically to the enhancer region.[60] IHF is a dimer of small basic proteins that bind DNA and are involved in many processes: site-specific recombination, plasmid replication, and regulation of gene expression.[61] IHF binds to a preferred nucleotide sequence on duplex DNA and often introduces DNA bends. Mutants defective in IHF support Ff phage growth at 3% of the wild-type rate. Phages insensitive to the enhancer mutations also suppress the replication defect of IHF mutants, evidence that the enhancer is the target of IHF action.

Palindrome A, the DNA packaging signal, is important for obtaining phage

57. Horiuchi K (1986) *JMB 188*:215; Greenstein D, Horiuchi K (1987) *JMB 197*:157.
58. Johnston S, Ray DS (1984) *JMB 177*:685; Dotto GP, Horiuchi K, Zinder ND (1984) *JMB 172*:507.
59. Dotto GP, Zinder ND (1984) *PNAS 81*:1336; Dotto GP, Zinder ND (1984) *Nature 311*:279; Kim MH, Ray DS (1985) *J Virol 53*:871.
60. Greenstein D, Zinder ND, Horiuchi K (1988) *PNAS 85*:6262.
61. Nash HA, Robertson CA (1981) *JBC 256*:9246; Friedman DI (1988) *Cell 55*:545; Drlica K, Rouviere-Yaniv J (1987) *Microbiol Rev 51*:301.

particles in high yield but is not required for DNA replication.[62] It is separable from the origins and, in the correct orientation, can function anywhere in the genome. The packaging signal emerges from the cell first and is presumably involved in the initiation of budding.[63]

Mini (Defective) Ff Phage

Filaments only about one-third the standard length were discovered in stocks of phages obtained after many passages beyond an initial single-plaque isolation.[64] These particles vary from 0.2 to 0.5 of a unit length and show deletions of large portions of the genome. Miniphage particles contain no intact genes and can be reproduced only upon coinfection with a helper phage. Partial duplications of the IG region are common, as they provide a growth advantage. Duplications of the (−) strand origin allow the miniphage to outcompete the wild-type for RNAP.[65] However, more than one fully functional (+) strand origin cannot be maintained during rolling-circle replication and is never observed. Possibly, the first appearance of mini-Ff DNA is an uncommon event in the second stage of replication or is the result of recombination. Subsequently the mini-DNA may be more readily replicated and favored over the wild-type in multiplication.

Ff Phages as Cloning Vectors[66]

Ff phages have gained popularity as cloning vectors because the capsid imposes no constraints on size, allowing a wide variety of DNA lengths to be packaged. The extensive use of dideoxynucleotide DNA chain termination reactions for DNA sequencing has made the purification of large quantities of ssDNA very desirable. Furthermore, sequences cloned in Ff phage vectors can be transferred easily between different bacterial strains by transduction.

Vectors have been constructed by inserting multiple cloning sites into the otherwise unmodified phages or by cloning the IG region into a plasmid vector. Single-stranded plasmid DNA is synthesized and packaged into phage particles after infection of the plasmid-containing cells with a helper phage stock. The orientation of the origin and the packaging signal determine which of the two single strands will be packaged; vectors with cloning sites in either orientation with respect to these signals are widely available. One elegant construct contains IG regions from both f1–M13 and Ike (a related phage), cloned in opposite orientations, allowing either strand to be isolated depending on the type of helper phage used.[67]

Gene Expression[68]

The level of expression of the ten Ff phage proteins varies widely: from 10^5 or 10^6 molecules per cell of the major structural proteins (gp8 and gp5) to a few

62. Dotto GP, Zinder ND (1983) *Virology 130*:252.
63. Lopez J, Webster RE (1983) *Virology 127*:177.
64. Griffith J, Kornberg A (1974) *Virology 59*:139; Hewitt JA (1975) *J Gen Virol 26*:87; Enea V, Zinder ND (1975) *Virology 68*:105.
65. Horiuchi K (1983) *JMB 169*:389.
66. Geider K (1986) *J Gen Virol 67*:2287; Smith GP (1988) in *Vectors: A Survey of Molecular Cloning Vectors and Their Uses* (Rodriguez RL, Denhardt DT, eds.). Butterworth, Boston, p. 61; Zinder ND, Boeke JD (1982) *Gene 19*:1.
67. Peeters BPH, Schoenmakers JGG, Konings RNH (1986) *Gene 41*:39; Konings RNH, Luiten RGM, Peeters BPH (1986) *Gene 46*:269.
68. Model P, Russel M (1988) in *The Bacteriophages* (Calendar R, ed.). Plenum Press, New York, Vol. 2, p. 375.

hundred copies of proteins involved catalytically in morphogenesis. Transcription initiates from several promoters (Fig. 17-3) and uses only the complementary strand as a template.[69] Ff is unlike most phages, which have a distinct transcriptional program: all the Ff promoters appear to be active throughout the life cycle, and variable mRNA levels are achieved through different promoter strengths and RNA stabilities.[70] Transcription from IG counterclockwise through gene *8* is efficient and stops at the central terminator at the end of gene *8*. Downstream of this terminator, transcription is much less efficient; a second terminator just within IG blocks transcription through the origin region. The correlation between the amount of RF and the levels of transcription indicates that all RF molecules are used as transcription templates.

Translation efficiency of the genes also varies.[71] As expected, genes *8* and *5* are the most efficiently translated. Translational repression of gene *2* and gene *X* by high levels of gp5 appears to be important in maintaining a balance between RF and viral-strand synthesis.[72] The target in the mRNA for repression of these two overlapping genes appears to be distinct; alteration of the control site for gene *2* still leaves gene *X* repressible by gp5.[73]

Phage Assembly[74]

No mature Ff phage is found in the cytoplasm of the infected cell; viruses are assembled at the membrane and extruded into the media without disrupting cell growth. Assembly occurs at adhesion zones where the inner and outer membranes contact one another. The increase in the number of these zones during phage infection depends on the action of gp1, required for morphogenesis.[75] Gp4 also participates catalytically in assembly, but in an unknown way.

During phage assembly (Fig. 17-8), the gp5 enveloping the viral single strand is replaced by gp8. The packaging signal bound by the minor coat proteins, gp7 and gp9, is the first DNA to emerge from the cell. The adsorption proteins bind to the opposite end of the phage and thus leave the cell last. The high incidence of abnormally long phage particles due to mutations in genes *3* or *6* indicates that these proteins help to terminate assembly.[76]

Among host mutants defective in phage assembly, those affected in thioredoxin and thioredoxin reductase have been studied in some detail.[77] Thioredoxin is a cofactor in a variety of oxidation–reduction reactions,[78] yet replacement of the cysteines required for these reactions has no effect on phage assembly. The conformation of the reduced form of the protein, rather than its reducing power, is essential. Phage mutations that suppress the morphogenesis defect in thioredoxin mutants implicate an interaction between gp1 and thioredoxin that creats assembly sites within the host membrane.

---

69. La Farina M, Model P (1983) *JMB 164*:377; Smits MA, Jansen J, Konings RNH, Schoenmakers JGG (1984) *NAR 12*:4071.
70. Blumer KJ, Steege DA (1984) *NAR 12*:1847.
71. Blumer KJ, Ivey MR, Steege DA (1987) *JMB 197*:439.
72. Fulford W, Model P (1988) *JMB 203*:49; Fulford W, Model P (1988) *JMB 203*:39.
73. Fulford W, Model P (1988) *JMB 203*:49; Fulford W, Model P (1988) *JMB 203*:39.
74. Webster RE, Lopez J (1985) in *Virus Structure and Assembly* (Casjens S, ed.). Jones and Barlet, Boston, p. 235.
75. Lopez J, Webster RE (1985) *J Bact 163*:1270.
76. Grant RA, Webster RE (1984) *Virology 133*:315; Grant RA, Webster RE (1984) *Virology 133*:329; Lopez J, Webster RE (1983) *Virology 127*:177.
77. Russel M, Model P (1984) *J Bact 159*: 1034; Russel M, Model P (1985) *PNAS 82*:29; Russel M, Model P (1985) *J Bact 163*:238; Russel M, Model P (1986) *JBC 261*:14997.
78. Holmgren A (1985) *ARB 54*:237.

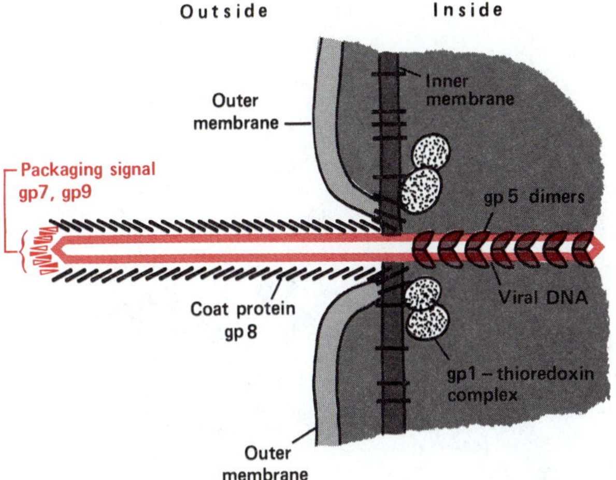

Figure 17-8
Assembly of Ff at the cell surface. The gp5 bound to the viral ssDNA is displaced by membrane-bound gp8 as phage is extruded through adhesions between the inner and outer membranes.

17-4 Small Polyhedral Phages: ϕX174, S13, G4[79]

An early reward for studying the very small ϕX174 phage was the discovery of a chromosome made of a single-stranded circular DNA.[80] Since then, genetic, physical, and biochemical studies of polyhedral ssDNA coliphages and their life cycles (Fig. 17-9), as well as those of Ff phages (Section 3), have advanced our understanding of the molecular biology of DNA replication. ϕX174 has been a remarkably effective probe for elucidating how the *E. coli* chromosome, 1000 times its size, is replicated.

Phage S13 is similar to ϕX174.[81] Also closely related are ϕA, ϕC, ϕK, ϕR, 6SR, BR2, α3, St-1, and the series of G phages. Of special interest, because their replication mechanism differs from that of ϕX174, are phages G4, St-1, ϕK, and α3 (see below; see also Section 8-3).

Virion

The ϕX174 particle (Table 17-5 and Fig. 17-10) is an icosahedron with a knob at each vertex. Its mass is 6.2 MDa, its diameter 25 nm.[82]

Coat. The phage capsid is composed of 60 molecules of gpF and has 12 vertices of fivefold symmetry (Fig. 17-10). Fitted into the capsid at each vertex is a knob (spike) composed of five gpG molecules and one gpH. A fourth virion protein,[83] assigned to gene *J*, is an internal polypeptide, only 37 residues long, 12 of which are lysine and arginine; gpJ probably serves to condense the DNA.

---

79. Hayashi M, Aoyama A, Richardson DL, Hayashi MN (1988) in *The Bacteriophages* (Calendar R, ed.). Plenum Press, New York, Vol. 2, p. 1.
80. Sinsheimer RL (1970) *Harvey Lect 64*:69.
81. Baker R, Tessman E (1967) *PNAS 58*:1438; Godson GN (1973) *JMB 77*:467; Godson GN (1976) *Virology 75*:263; Harbers B, Delaney AD, Harbers K, Spencer JH (1976) *Biochemistry 15*:407.
82. Edgell MH, Hutchison CA III, Sinsheimer RL (1969) *JMB 42*:547.
83. Shank PR, Hutchison CA III, Edgell MH (1977) *Biochemistry 16*:4545; Freymeyer DK II, Shank PR, Edgell MH, Hutchison CA III, Vanaman TC (1977) *Biochemistry 16*:4550.

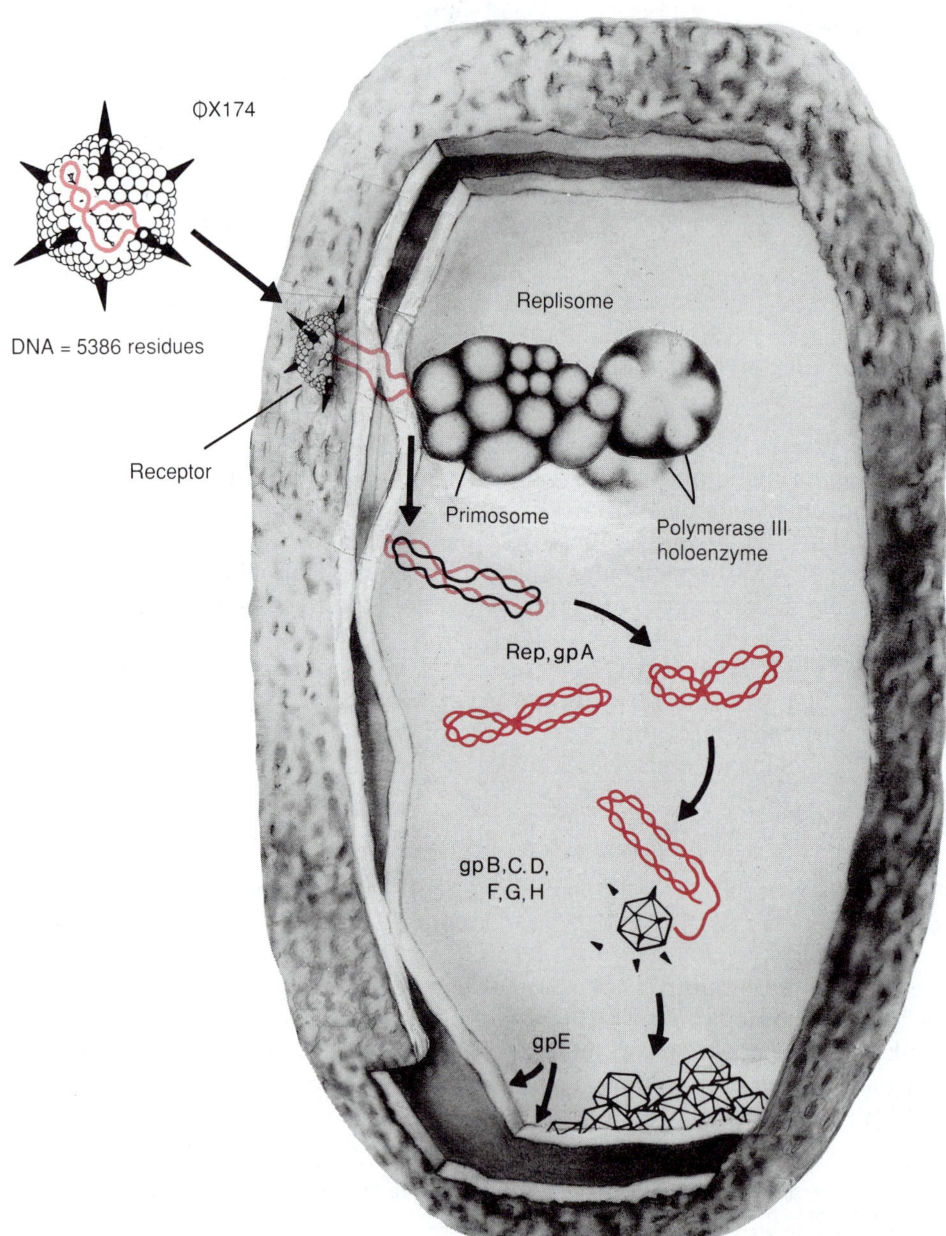

Figure 17-9
Scheme for the life cycle of ϕX174. See Figure 17-4 for comparison with filamentous phages.

Table 17-5
Characteristics of φX174 virion and DNA

Virion		DNA (single-stranded)	
Shape	icosahedron	Shape	circle
Mass	6.2 MDa	Mass	1.8 MDa
Weight of 10^{14}	1 mg	Weight of 10^{14}	0.30 mg
Diameter	25 nm	No. of nucleotides	5386
Major capsid protein	gpF	Base composition (%)	
Major spike protein	gpG	A	23.9
Minor spike protein	gpH	T	31.3
Core protein	gpJ	G	23.3
		C	21.5

Genome. The φX174 viral strand contains 5386 nucleotides. When this chromosome was discovered, its single-stranded character appeared to pose an exception to the rule that replication requires a duplex chromosome. Elucidation of the replication mechanism used by φX174, however, proved to be a potent confirmation of the generality of the semiconservative model. The viral strand (+) has a base composition in which the purines are not equivalent to pyrimidines; A does not match T, nor does G match C (Table 17-5). However, when the virus enters the host, replication of the (+) strand to produce a complementary (−) strand yields a duplex RF with base pairing characteristic of all natural DNAs.

Of 11 proteins encoded by the compact φX174 genome (Table 17-6 and Fig. 17-11), only gpK appears to be nonessential, although *K* mutants have a reduced

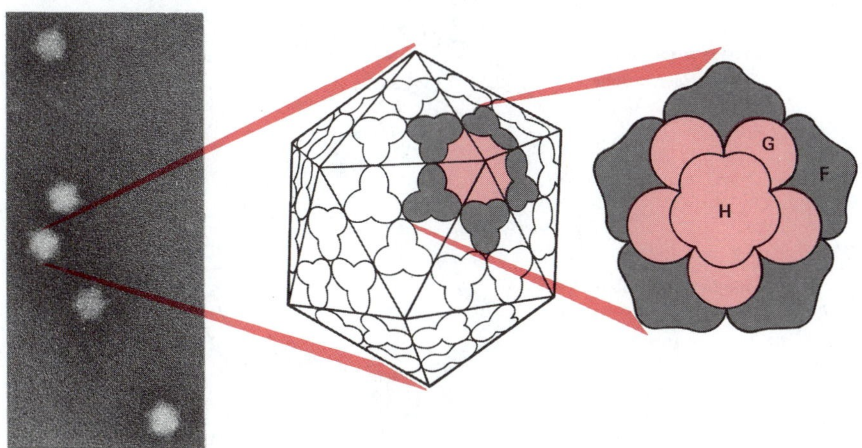

Figure 17-10
Virion of phage φX174. Left, electron micrograph (×200,000). (Courtesy of Professor RC Williams) Right, scheme of φX174 subunits in the polyhedron (middle) and in a spike region (far right), suggesting orientations of gpH, gpG, and the major coat protein (gpF). [From Edgell MH, Hutchison CA, Sinsheimer RL (1969) *JMB* 42:547]

Table 17-6
ϕX174 genes and functions[a]

Gene	Function	Protein mass[b] (kDa)	
		SDS–PAGE	Sequence
A	RF replication; viral strand synthesis	60	58.7
*A**	shutoff of host DNA synthesis	37	38.7
B	capsid morphogenesis	20	13.8
C	DNA maturation	6	10.0
D	capsid morphogenesis and assembly	14	16.9
E	host cell lysis	10	10.4
F	major coat protein	50	48.4
G	major spike protein	20	19.0
H	minor spike protein, adsorption	37	34.4
J	core protein; DNA condensation	4	4.2
K	stimulation of phage production	8	6.4

[a] Hayashi M, Aoyama A, Richardson DL, Hayashi MN (1988) in *The Bacteriophages* (Calendar R, ed.). Plenum Press, New York, Vol. 2, p. 1.
[b] As determined by SDS–PAGE (sodium dodecyl sulfate–polyacrylamide gel electrophoresis) or direct sequence analysis.

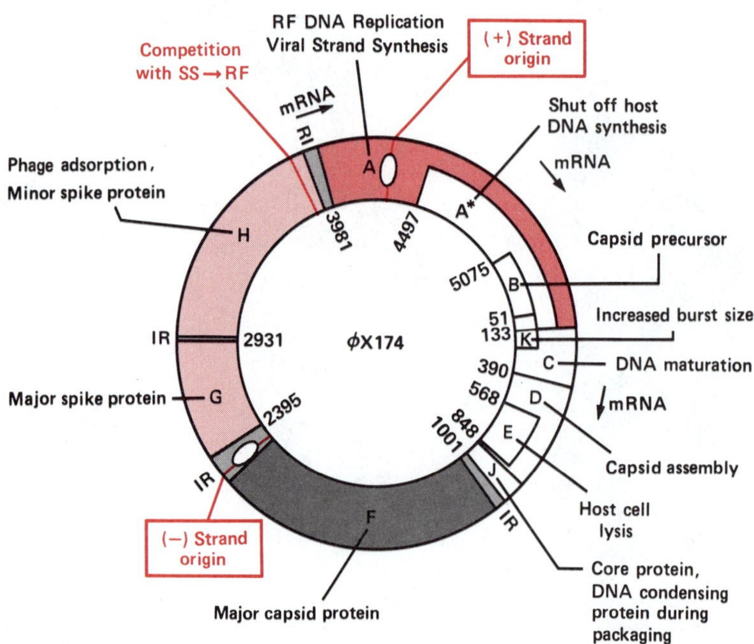

Figure 17-11
Genetic map of phage ϕX174 with suggested functions of gene products. IR = intergenic region; arrow = mRNA promoter and direction of transcription. (Courtesy of Dr F Heidekamp)

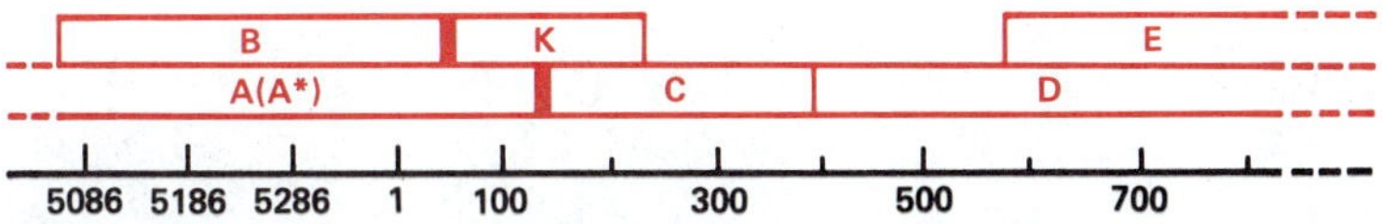

Figure 17-12
Portion of genetic map of phage G4 showing overlapping sequences. Colored regions represent overlap between genes *B* and *K* of one nucleotide and between genes *A*(*A**) and *C* of two nucleotides. Gene *B* is contained within gene *A*(*A**); gene *K* overlaps genes *A*(*A**) and *C* as well as gene *B*. Nucleotide numbering is based on the *Providentia stuartii* 164 (PstI) restriction cleavage map for ϕX174. (Courtesy of Professor GN Godson)

burst size.[84] The combined molecular weights assigned to the gene products exceed the limit imposed by the available DNA. This is explained by the capacity of certain DNA sequences to encode more than one protein (Fig. 17-12). The region originally assigned only to gene *A* also contains a gene (*A**) for a small protein, formed by reinitiation of translation in the same frame,[85] and all of gene *B*, translated in a different reading frame.[86] Gene *K* is read in yet another reading frame at the end of gene *A* and the start of gene *C*,[87] so that for five nucleotides, where genes *A* and *C* overlap, the sequence is read in all three translation frames. Thus, five different proteins derive some or all of their primary structural information from a shared DNA sequence. In addition, farther downstream, gpE is encoded entirely within gene *D*.[88] Use of the ϕX174 chromosome is astonishingly efficient.

Comparisons of the nucleotide sequences of ϕX174 and G4 (5577 ntd) make it possible to observe evolutionary development at both nucleotide and amino acid sequence levels.[89] A change of 22% in the nucleotide sequence of overlapping genes is near the value of 33% for the change in nonoverlapping genes, suggesting that simultaneous use of a coding region in two reading frames is not as restrictive for evolutionary development as had been generally assumed. The origin sequence for viral-strand synthesis, located within gene *A* (Fig. 17-11), is strongly conserved in this family of icosahedral phages that differ considerably in the size and sequence of their genomes.

Infection

Adsorption and Penetration. The spikes, tiny tail-like structures, may function as primitive attachment organs. The observed transfer of gpH with the DNA into the cell[90] suggests a pilot-protein function (Section 2).

The phage adsorbs irreversibly to lipopolysaccharide in the outer membrane layer of rough strains of *E. coli* and *Salmonella typhimurium* (Fig. 17-13).[91] These

84. Tessman ES, Tessman I, Pollock TJ (1980) *J Virol 33*:557; Gillam S, Atkinson T, Markham A, Smith M (1985) *J Virol 53*:708.
85. Linney E, Hayashi M (1974) *Nature 249*:345.
86. Smith M, Brown NL, Air GM, Barrell BG, Coulson AR, Hutchison CA III, Sanger F (1977) *Nature 265*:702.
87. Shaw DC, Walker JE, Northrop RD, Barrell BG, Godson GN, Fiddes JC (1978) *Nature 272*:510.
88. Barrell BG, Air GM, Hutchison CA III (1976) *Nature 264*:34.
89. Godson GM (1978) *TIBS 3*:249; Smiley BL, Warner RC (1979) *NAR 6*:1979.
90. Jazwinski SM, Lindberg AA, Kornberg A (1975) *Virology 66*:283; Jazwinski SM, Marco R, Kornberg A (1975) *Virology 66*:294.
91. Jazwinski SM, Lindberg AA, Kornberg A (1975) *Virology 66*:268.

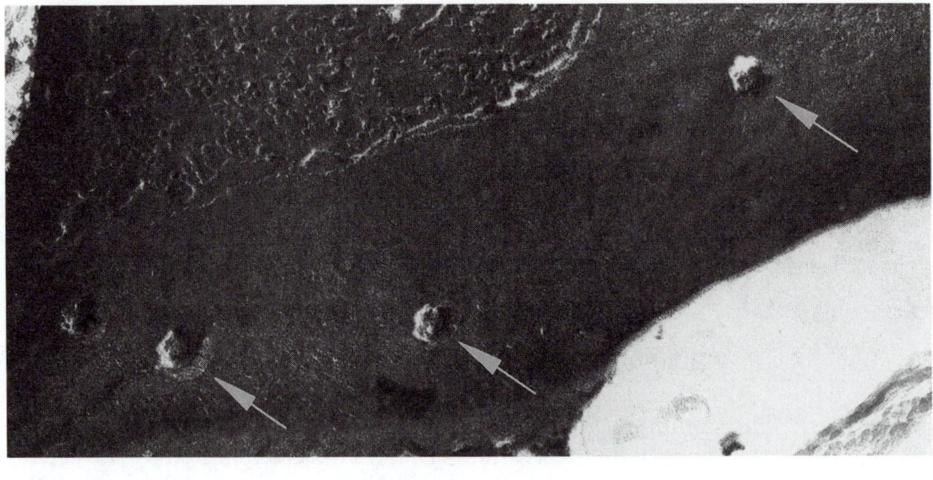

1.0 μm

Figure 17-13
Adsorption of ϕX174 (arrows) to outer membrane of *E. coli* as seen in a freeze-etch electron micrograph. (Courtesy of Dr J Griffith)

strains lack all the O antigen outer chains but retain the core and lipid A. The phage binding site in S. *typhimurium* has the structure

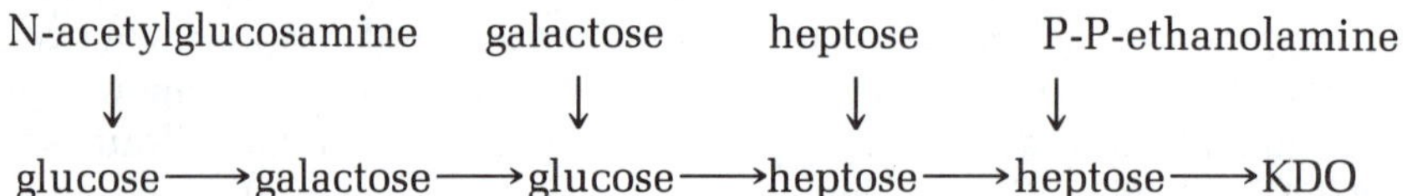

(Arrows show the glycosyl linkages; KDO is 2-keto-3-deoxyoctanoate; and P is phosphate.) N-acetylglucosamine at the nonreducing end of the chain is required for binding ϕX174 but not S13. The extensive structural and serologic studies devoted to *Salmonella* are due to its medical importance; the structure of the *E. coli* surface receptor is not yet known.

Adsorption includes a reversible stage in which the phage has not been detectably altered and remains infectious,[92] followed by an irreversible stage called "*eclipse*."[93] Eclipsed phages have lost infectivity for fresh cells, and their DNA has become susceptible to attack by DNase. Intact lipid A is required for adsorption to progress to the eclipse stage, but not for the earlier reversible stage. Whereas the addition of lipopolysaccharide to phage is sufficient to observe eclipse in vitro, cell-envelope fractions are more active. A heat-labile factor, extractable by detergents, is responsible for this stimulation, suggesting that a host membrane protein participates in eclipse and penetration.[94] Electron microscopy of these eclipsed phage complexes reveals that the DNA is extruded

---

92. Incardona NL, Tuech JK, Murti G (1985) *Biochemistry 24*:6439.
93. Incardona NL, Selvidge L (1973) *J Virol 11*:775.
94. Mano Y, Kawabe T, Komamo T, Yazaki K (1982) *Agric Biol Chem 46*:2041; Mano Y, Kawabe Y, Obata K, Yoshimura T, Komano T (1982) *Agric Biol Chem 46*:631.

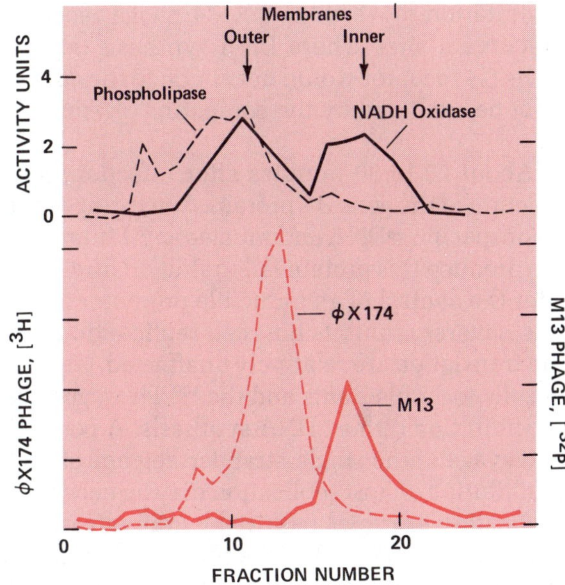

Figure 17-14
ϕX174 and M13 phages and their replicative forms attached at distinctive membrane fractions in a simultaneous infection of *E. coli*. Less dense fraction, marked by peak of NADH oxidase, contains inner plasma membrane; denser fraction, marked by phospholipase A, contains lipopolysaccharide outer membrane. NADH oxidase in outer membrane fraction suggests regions of adhesion between inner and outer membranes in these cells.

through a spike. When the envelope fractions are used, the DNA is associated with a membrane fragment.

The parental RF, like the phage particle, appears to be attached to the outer cell membrane. By this localization, ϕX174 contrasts with M13, which is associated with the inner cell membrane (Fig. 17-14). The distribution of ϕX174 among membrane fractions of *E. coli* simultaneously infected with ϕX174 and M13 is also consistent with binding to zones of adhesion between inner and outer membranes. Such regions of fusion seen in the electron microscope (Section 20-4) are regarded as preferred sites for ϕX174 adsorption and Ff phage assembly (Section 3). A distinctive phospholipase A1 may also be localized in these regions.[95] Since the folded *E. coli* chromosome also appears to be attached to a membranous site characterized by outer membrane markers (Section 20-4), there is an intriguing possibility that the ϕX174 receptor, by its location in this site, gives infecting DNA direct access to chromosomal replicative apparatus fixed to this surface.

A region of the ϕX174 genome between genes *A* and *H*, known as the "reduction sequence," inhibits ϕX174 growth when present on a plasmid. Adsorption of the phage is not affected, but all stages of replication are inhibited. This sequence appears to compete with the infecting phage for a limiting host component required for replication and may be the DNA site responsible for attachment of the genome to the cell membrane.[96] A candidate for the host component is the primosome (Section. 8-4): ssDNA-transducing particles that depend on a primosome assembly site (*pas*) for conversion to the duplex form are much more sensitive to the reduction sequence than are those that depend on the G4 SS→RF origin, which requires only primase. Furthermore, ϕX174 DNA pack-

95. Scandella CJ, Kornberg A (1971) *Biochemistry 10*:4447.
96. van der Avoort HG, van Arkel GA, Weisbeek PJ (1982) *J Virol 42*:1.

aged in G4 capsids escapes the inhibition.[97] Were the ϕX174 phage receptor connected to the membrane attachment sites where DNA synthesis occurs, phage particles entering through the G4 receptor would not attach to the ϕX174 membrane site and thus would not be excluded by the competing plasmid.

Shutoff of Host DNA Synthesis. About 20 to 30 minutes after infection, host DNA synthesis ceases, due to the action of the gene *A** protein. Consisting of the C-terminal half of gpA, gpA* is a nonspecific ssDNA endonuclease.[98] Mutants in the distal part of gene *A* that fail to produce this protein fail to shut off host DNA synthesis.[99] Gene *A**, cloned under the control of an inducible promoter in the absence of any other ϕX174 genes, severely inhibits host cell replication upon induction, while transcription and translation are relatively unaffected. Growth continues for a few hours as the cells form filaments and die.[100] Thus, gpA* is both necessary and sufficient for inhibition of host DNA synthesis. A possible mechanism of this inhibition is cleavage of the single-stranded regions of host (but not phage) replication forks. Inhibition of host replication may be needed to release the limited number of replication enzymes required by the phage. For example, the 30 or so rolling-cirlce intermediates that are present 20 minutes after infection would appropriate all the DNA pol III holoenzyme molecules estimated to be in the cell.[101]

Distinctions between ϕX174 and Ff. The ϕX174 and Ff phage genomes, basically similar in size, structure, and organization, and very likely linked in their evolution, nevertheless present many interesting biological differences in host range, virulence, and pathways of replication (Table 17-7).

Replication

The replication cycle of ϕX174 DNA (Table 17-8) includes three stages: (1) conversion of the viral single strand to the duplex replicative form (SS→RF); (2) multiplication of RF (RF→RF); and (3) rolling-circle replication to generate new viral single strands for packaging into virus particles (RF→SS). Stages in this pathway, very similar to those of the Ff cycle (Fig. 17-6), have been examined both in vivo[102] and in vitro.

SS→RF. Converson of the viral (+) circle to the covalently closed duplex form (Section 8-4) relies entirely on host enzymes. The first demonstration of the enzymatic synthesis of a biologically active DNA employed ϕX174 DNA and was accomplished simply with pol I and DNA ligase, using a small annealed fragment of random DNA as primer.[103] Tailoring by the 3′→5′proofreading and 5′→3′ excision exonucleases of pol I ensured that the primer in the complemen-

---

97. van der Avoort HG, van Arkel GA, Weisbeek PJ (1982) *J Virol 42*:1.
98. Langeveld SA, van Mansfeld ADM, De Winter JM, Weisbeek PJ (1979) *NAR 7*:2177; Langeveld SA, van Mansfeld AD, van der Ende A, van de Pol JH, van Arkel GA, Weisbeek PJ (1981) *NAR 9*:545.
99. Martin DF, Godson GN (1975) *BBRC 65*:323.
100. Colasanti J, Denhardt DT (1985) *J Virol 53*:805.
101. Dressler D, Hourcade D, Koths K, Sims J (1978) in *The Single-Stranded DNA Phages* (Denhardt DT, Dressler D, Ray DS, eds.). CSHL, p. 187.
102. Baas PD (1985) *BBA 825*:111; Baas P, Jansz H (1978) in *The Single-Stranded DNA Phages* (Denhardt DT, Dressler D, Ray DS, eds.). CSHL, p. 215; Dressler D, Hourcade D, Koths K, Sims J (1978) in *The Single-Stranded DNA Phages* (Denhardt DT, Dressler D, Ray DS, eds.). CSHL, p. 187.
103. Goulian M, Kornberg A (1967) *PNAS 58*:1723; Goulian M, Kornberg A, Sinsheimer RL (1967) *PNAS 58*:2321.

Table 17-7
Distinctions between Ff and ϕX174[a]

Property	Ff	ϕX174
Structure of virion	filament	polyhedron
Adsorption receptor	F pilus; inner membrane	lipopolysaccharide; outer membrane
Host strain	male	rough
Salvage of coat protein	yes	no
Replication system: SS→RF	RNA polymerase	primosome
Covalent initiator of protein-DNA complex	no	yes
Viral strand synthesis:		
Binding protein	gp5	unknown
Coupled to encapsidation	no	yes
Pool of progeny DNA	yes	no
New DNA in parental coat	yes	no
Parental DNA in new coat	yes	no
Coat assembly location	membrane	cytoplasm
Virus particles in cell	no	yes
Mode of release	budding	cell lysis
Infection	parasitic	virulent

[a] No detectable homology (< 10%) between Ff and ϕX174 DNA.

tary strand was correctly base-paired and eventually eliminated. Both the synthetic complementary strand and its synthetic viral strand copy were fully infectious. Physiologically, the infecting ϕX174 viral DNA strand is primed and replicated by complex machinery made up of host replication enzymes. In vitro, at least 20 polypeptides are needed for the apparently simple operation of converting a ϕX174 viral circle to supercoiled RF.

The six substages of replication — prepriming (Section 8-4), priming (Section 8-4), elongation (Section 4-2), gap filling (Section 4-2), ligation (Section 9-5), and supercoiling (Section 12-3), — have been reviewed.[104] The complex prepriming substage is dispensed with by phages G4, St-1, ϕK, and α3. Instead, primase

Table 17-8
Replication cycle of ϕX174

	Stage	Time, (minutes, at 33°)	Events
1	SS→RF	0-1	adsorption and penetration; viral SS→parental RF; transcription of RF
2	RF→RF	1-20	parental RF→~60 progeny RF
		25	RF multiplication stops; host DNA synthesis stops
3	RF→SS	20-30	~35 rolling circles→~500 viral SS→phage particles
		40	cell lysis

104. Kornberg A (1978) CSHS 43:1; Meyer RR, Shlomai J, Kobori J, Bates D, Rowen L, McMacken R, Ueda K, Kornberg A (1978) CSHS 43:289; Eisenberg S, Scott JF, Kornberg A (1978) CSHS 43:295.

directly recognizes a unique sequence in the viral DNA and synthesizes a short RNA transcript that primes elongation (Section 8-3). This difference in the enzymes used to prime at the complementary (−) strand origins of ϕX174 and G4 is reflected in the products of the reactions. Whereas primase recognizes and synthesizes a primer on the G4 (−) strand origin, the (−) strand origin on ϕX174 is the assembly site of a mobile primosome (Section 8-4), which translocates around the template DNA synthesizing primers in many places. Thus, while both ϕX174 and G4 have unique sequences required for initiation of (−) strand synthesis, only G4 has a uniquely located primer.

The primosome, either captured whole from the host replication complex or assembled at the (−) strand origin, appears to remain associated with the template throughout subsequent replication stages. The RF bound to the primosome is a superior substrate for cleavage by gpA, which initiates the next stage of replication.[105]

Reconstitution of the multienzyme SS→RF pathway is still incomplete. The physiologic roles of the PriB and PriC prepriming proteins (Section 8-4) remain to be validated. The ϕX174 DNA templates, prepared from phage particles by phenol or alkaline extraction, hardly resemble the viral DNA, whose penetration from an eclipsed particle into the cell is coupled to its replication. The soluble enzyme systems do not reflect the evidence that replicative intermediates are probably membrane-bound. Finally, the pilot-protein function of gpH, the spike protein, suggested by its association with parental RF in vivo, has not been duplicated in vitro. Nevertheless, partial reconstitution in vitro of the first-stage replicative events, starting with phage particles, has been achieved[106] and invites further studies.

Complementary-strand synthesis of ϕX174 has been studied intensively as a model for the discontinuous synthesis of lagging strands. Assembly of the primosome at *pas* sequences is a key initiation step in the replication of some plasmids, such as ColE1 (Sections 8-4 and 18-2). However, many phages and plasmids (including *oriC* plasmids, which bear the origin of the *E. coli* chromosome) replicate in vitro independent of PriA protein and lack *pas* sequences in their origins, suggesting that PriA-mediated primosome assembly is not necessary for discontinuous-strand synthesis (Section 16-3). Mobile helicase–primase protein complexes, similar to the ϕX174 primosome, are clearly required for replication of these templates, but their assembly is directed by origin-binding proteins (other than PriA), such as dnaA or λ O protein (Sections 7,8-4, 16-3, and 16-4). Whereas the primosome, discovered in studies of ϕX174 replication, has wide conceptual significance, its composition and assembly vary with different replicons and circumstances.

RF→SS (+) →RF. Initiation of in vitro multiplication of the duplex circle (Section 8-7) appears simple compared to SS→RF replication. Two proteins that do not participate in SS→RF are required to generate a single-stranded circle from the duplex RF.[107] One is the phage-encoded gpA, whose function is to initiate and conclude the reaction; the other is a DNA helicase, the product of the *E. coli* *rep* gene (Section 11-2). The single (+) strand "peeled off" the duplex by these

105. Low RL, Arai K, Kornberg A (1981) *PNAS 78*:1436.
106. Jazwinski SM, Kornberg A (1975) *PNAS 72*:3863.
107. Ikeda J, Yudelevich A, Hurwitz J (1976) *PNAS 73*:2669; Eisenberg S, Scott JF, Kornberg A (1976) *PNAS 73*:1594.

proteins is converted to a duplex RF by the same SS→RF pathway used for the infecting viral DNA.

Gene A protein (Section 8-7), a 60-kDa endonuclease specific for the ϕX174 (+) strand origin, has been purified on the basis of its ability to initiate replication on covalently closed duplex templates.[108] The Rep helicase (65 kDa; Section 11-2) has been purified by complementing the defect in ϕX174 replication in extracts from *rep* mutants. These two pure proteins, gpA and Rep, acting with only SSB and pol III holoenzyme, use supercoiled RFI catalytically to synthesize viral (+) circles at a physiologic rate. The product, a full-length circle identical to DNA extracted from phage particles, serves as template for the multiprotein SS→RF system to form RF. Whether synthesis of the complementary strand awaits completion of viral-strand synthesis or is coupled to it is not known. Presumably, synthesis of viral strands at the next stage [RF→SS (+); see below] employs the same RF→RF mechanism.

The two mechanisms for (+) and (−) strand synthesis, used separately or in conjunction, can account for all stages in the phage replication cycle (Fig. 17-15).[109] Synthesis of the (+) strand, once initiated, is *continuous*, whereas (−) strand synthesis is *discontinuous*, requiring a fresh initiation on each circle. The parental RF, with its conserved primosome, may be the only source of new (+) strands and duplex RF molecules.

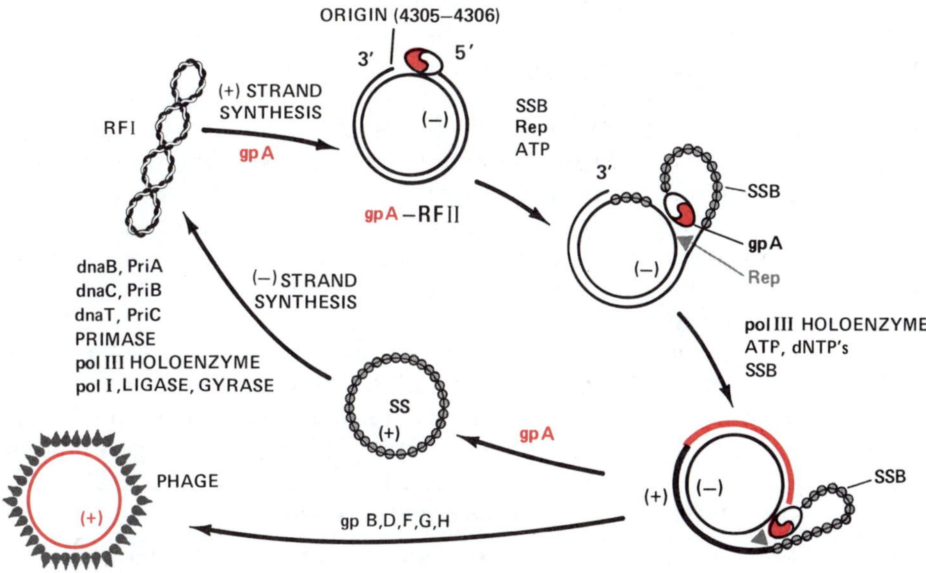

Figure 17-15
Scheme for ϕX174 RF replication in two stages: Continuous replication initiated by gpA cleavage generates viral (+) circles, and discontinuous replication of the viral circles by the SS → RF system produces RF. In the presence of phage-encoded maturation and capsid proteins (gpB, D, F, G, and H), viral circles are encapsidated rather than replicated.

108. Eisenberg S, Kornberg A (1979) *JBC 254*:5328.
109. Eisenberg S, Scott JF, Kornberg A (1976) *PNAS 73*:3151.

Pulse-labeling studies and quantitative electron microscopy[110] of intracellular viral forms show that the number of duplex DNA circles increases to about 60 per cell during the first 25 minutes of infection (Fig. 17-16). In addition, there are about 35 rolling circles per cell, each giving rise to an encapsidated single-stranded circle every 30 seconds (about 170 ntd per second). Analysis of ϕX174 replication in vivo is generally in good agreement with the model based on studies in vitro,[111] except for a few discordant observations. D-loop structures, rather than rolling circles (Section 16-1), were observed as the predominant RF intermediates,[112] and discontinuous synthesis of viral strands was deduced from some pulse-labeling experiments.[113] Initiation at secondary sequences, rather than the gpA cleavage site, might be involved, but functional *pas* sequences have not been detected on the complementary strand.[114]

GpA is remarkably multifunctional (Section 8-7; Fig. 17-15). The protein splits the phosphodiester bond between residues 4305 and 4306 and forms a covalent bond with the 5′-P group at the nick. Complexed with Rep protein, gpA participates in unwinding the duplex. In the absence of DNA synthesis, the gpA · Rep protein complex proceeds around the circle on the (−) strand, unwinding the duplex. The Rep protein catalytically separates duplex strands, using two ATPs per base pair melted (Section 11-2); hydrolysis of ATP provides the energy for strand separation at the fork. SSB traps the single strands as they are unwound. When the gpA · Rep protein complex, as part of a looped rolling circle, reaches the 3′-OH end of the open (+) strand, the linear (+) and circular (−) strands are released.

When unwinding is coupled to replication, the 3′-OH group at the nick is covalently extended by DNA pol III holoenzyme, creating a (+) strand longer than unit length and regenerating a duplex origin. As the replication fork traverses the full length of the template circle, the gpA cleaves the regenerated origin and establishes a fresh covalent bond with the new 5′ end. This second cleavage by gpA liberates unit-length ϕX174 DNA whose 3′-OH group can attack the gpA – 5′-P bond. The DNA ends of the displaced viral strand are thus ligated to form a circle. GpA is covalently linked to the 5′ end of the DNA via an ester linkage with tyrosine in the active site.[115] Each molecule of gpA contains two active tyrosines which alternately carry out the successive cleavage reactions.[116] In each nicking event, gpA is linked anew to a 5′-P end and its catalytic role is preserved (Section 8-7).

The (+) strand origin consists of the sequences required for gpA cleavage of RFI and initiation of rolling-circle replication (Section 16-4). This 30-bp sequence is within gene *A*. Three elements are required:[117] the 8-bp cleavage sequence recognized by gpA,[118] an AT-rich spacer region, and a gpA binding

110. Dressler D (1970) *PNAS 67*:1934; Koths K, Dressler D (1978) *PNAS 75*:605.
111. Godson GN (1977) *JMB 117*:337; Baas PD, Teertstra WR, Jansz JS (1978) *JMB 125*:167; Baas PD, Teertstra WR, van der Ende A, Jansz JS (1980) *JMB 137*:283; Keegstra W, Baas PD, Jansz JS (1979) *JMB 135*:69.
112. Godson GN (1977) *JMB 117*:353.
113. Machida Y, Okazaki T, Okazaki R (1977) *PNAS 74*:2776; Matthes M, Denhardt DT (1980) *JMB 136*:45; Matthes M, Denhardt DT (1982) *J Virol 42*:12; Strathearn MD, Low RL, Ray DS (1984) *J Virol 49*:178.
114. Matthes M, Weisbeek PJ, Denhardt DT (1982) *J Virol 42*:301.
115. van Mansfeld AD, Baas PD, Jansz HS (1984) *Adv Exp Med Biol 179*:221; van Mansfeld AD, van Teeffelen HA, Baas PD, Veeneman GH, van Boom JH, Jansz JS (1984) *FEBS Lett 173*:351; Roth MJ, Brown DR, Kurwitz J (1984) *JBC 259*:10556; Sanhueza S, Eisenberg S (1985) *J Virol 53*:695.
116. van Mansfeld AD, van Teeffelen HA, Baas PD, Jansz HS (1986) *NAR 14*:4229.
117. Baas PD, Heidekamp F, van Mansfeld, ADM, Jansz HS, Langeveld SA, van der Martel GA, Veeneman GH, van Boom JH (1981) in *The Initiation of DNA Replication* (Ray DS, ed.). Academic Press, New York, p. 195; Fluit AC, Baas PD, van Boom JH, Veeneman GH, Jansz HS (1984) *NAR 12*:64434; Fluit AC, Baas PD, Jansz HS, Veeneman GH, van Boom JH (1984) *Adv Exp Med Biol 179*:231.
118. van Mansfeld AD, Baas PD, Jansz HS (1984) *Adv Exp Med Biol 179*:221.

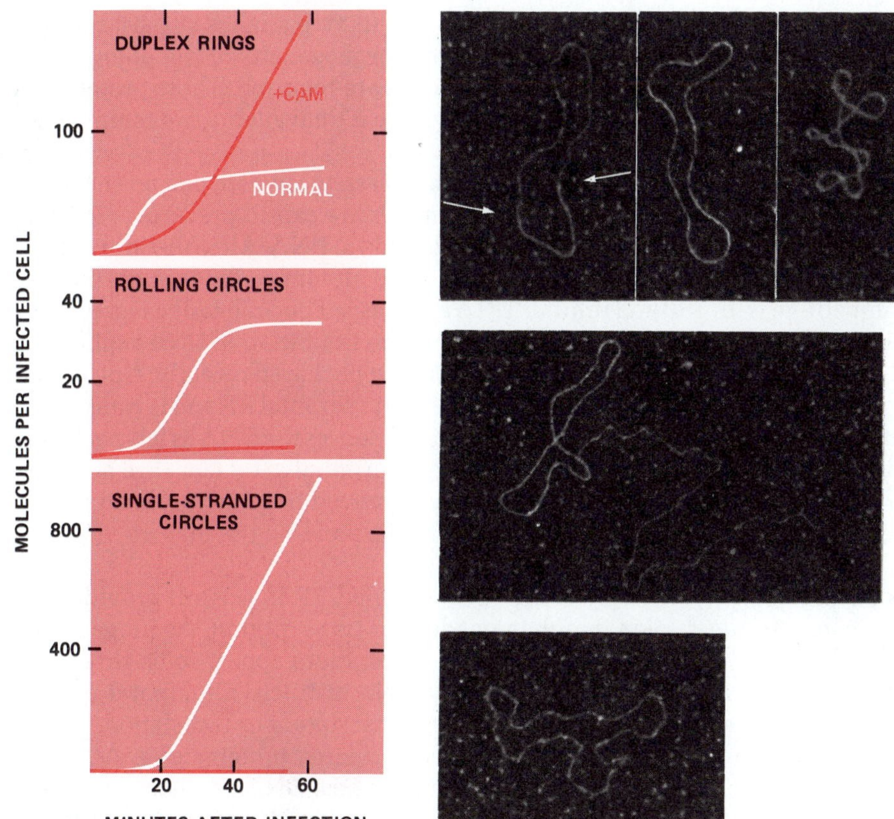

Figure 17-16
Forms of ϕX174 DNA during normal infection and in the presence of chloramphenicol (CAM, 35 mg/ml) assayed with the electron microscope. Single-stranded DNA appears thinner and less rigid than duplex DNA. Electron micrographs are as follows: Top, duplex rings; at left, partial duplex (arrows show extent of duplex region); middle, relaxed; right, supercoiled. Center, rolling circle. Bottom, single-stranded circle. (Courtesy of Professors D Dressler and K Koths)

sequence. The recognition sequence is sufficient to direct ssDNA cleavage by gpA, but additional elements are required for RF cleavage. GpA first binds to the binding sequence, then melts the AT-rich spacer and the cleavage-recognition sequence. Once melted, the recognition site is efficiently nicked. The sequences required for termination of a round of synthesis are the same as those required for initiation, implying that the first and second cleavages proceed by the same mechanism. The first and the last three nucleotides of the 30-bp (+) strand origin are not essential for cleavage and must be involved in some later replication step, perhaps the binding of phage proteins involved in coupling (+) strand synthesis to DNA packaging (see below).[119]

In vivo, gpA acts preferentially *in cis*; the protein is ineffective in complementing gene *A*–defective phage present in the same cell. This behavior is not easily explained, considering that the infected cell yields about 3000 copies of the soluble protein, only one of which is needed to form and sustain a rolling-circle intermediate. The *cis* action may be due to competition between the full-length protein and the truncated gpA*.

119. Brown DR, Schmidt-Glenewinkel T, Reinberg D, Hurwitz J (1983) *JBC 258*:8402; Brown DR, Reinberg D, Schmidt-Glenewinkel T, Roth M, Zipursky SL, Hurwitz J (1983) *CSHS 47*:701.

RF→SS (+) and Phage Assembly. In vitro and in vivo studies suggest that the accumulation of viral strands is coupled to morphogenesis of the phage particle[119a] (Fig. 17-17); eight virus-encoded proteins are required. These proteins act to condense and encapsidate the DNA and make it unavailable as a template for (−) strand synthesis and conversion to RF.

The switch from RF→RF synthesis to phage assembly depends on the accumulation of proheads, gpC, and gpJ.[120] *Proheads* are capsid precursors that are assembled from gpF, G, H, B, and D in the absence of DNA. After the initiation of unwinding,[121] gpC binds to the RFII · gpA · Rep protein complex. This binding is competitive with SSB and inhibits DNA synthesis. The prohead recognizes the initiation complex containing gpC, relieving the inhibition of DNA replication and allowing viral strands to be synthesized and packaged directly. GpJ, a nonspecific DNA-binding protein, associates with the parental RF and is transferred to the viral DNA during replication where it condenses the DNA as it is packaged in the prohead.[122] Rolling-circle replication is terminated by cleavage and ligation of the viral strand ends by gpA. GpB is removed from the prohead during replication and gpD is removed after packaging, yielding the mature virion.

Gene Expression.[123] Multiplication of supercoiled RF in stage II provides the many copies of the duplex template needed for transcription. As judged from RNA polymerase binding sites and mRNA sizes, there appear to be three promoters (Fig 17-11) and several terminators. No distinctive temporal control mechanisms have been observed for either transcription or translation. Translation starts when several genes produce abbreviated proteins:[124] gpA*, gpF*, and gpG*. The function of gpA* in shutting off host DNA synthesis has been mentioned, but the significance of the other shortened gene products is unknown.

Expression of the ϕX174 genome, so remarkable for multiple reading frames within a sequence and for multiple translational starts in the same frame, is a fascinating object for study. Although the *A* and *B* genes overlap, each has its own promoter, allowing their transcription and translation to be unlinked. In contrast, the *E* gene, within the *D* gene but translated in a different frame, does not have its own promoter; thus the two proteins must be translated from the same mRNA. In the highly active translation of the *D* gene, the binding of ribosomes in the *D* frame largely excludes their occupation of the gene *E* ribosome binding site. The initiation of *E* translation thus requires that ribosomes translating in the *D* frame must shift frames and terminate prematurely, in order to free the 3′ portion of the mRNA.[125] This frame shifting is mediated by the misdecoding of certain alanine codons by a class of *E. coli* serine tRNA to provide sufficient gene *E* expression for cell lysis. This strategy of frameshift and termination, allowing expression of the downstream gene in a shared mRNA, first discovered in RNA phages,[126] may apply to other systems as well.

119a. Kornberg A (1978) *CSHS 43*:1; Meyer RR, Shlomai J, Kobori J, Bates D, Rowen L, McMacken R, Ueda K, Kornberg A (1978) *CSHS 43*:289; Eisenberg S, Scott JF, Kornberg A (1978) *CSHS 43*:295.
120. Aoyama A, Hamatake RK, Hayashi M (1983) *PNAS 80*:4195.
121. Aoyama A, Hayashi M (1986) *Cell 47*:99.
122. Hamatake RK, Aoyama A, Hayashi M (1985) *J Virol 54*:345.
123. Hayashi M, Fujimura FK, Hayashi M (1976) *PNAS 73*:3519; Pollock TJ, Tessman I, Tessman ES (1978) *JMB 124*:147.
124. Pollock TJ, Tessman I, Tessman ES (1978) *J Virol 30*:543.
125. Buckley KJ, Hayashi M (1987) *JMB 198*:599.
126. Kastelein RA, Remaut E, Fiers W, van Duin J (1982) *Nature 295*:35.

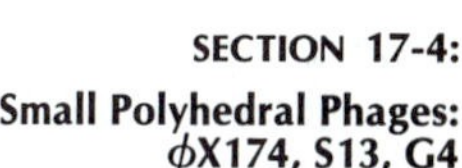

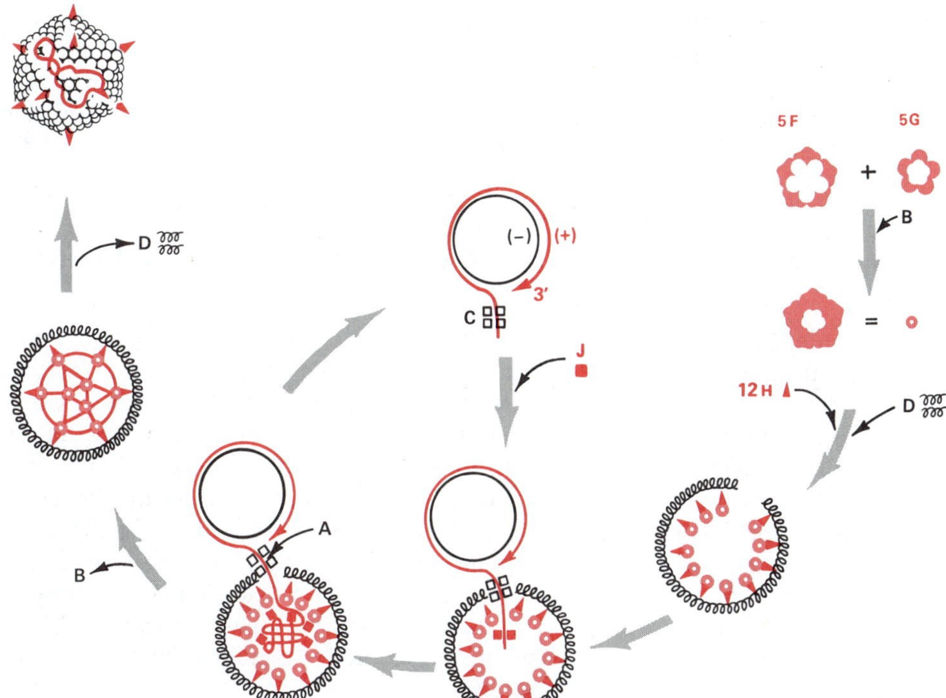

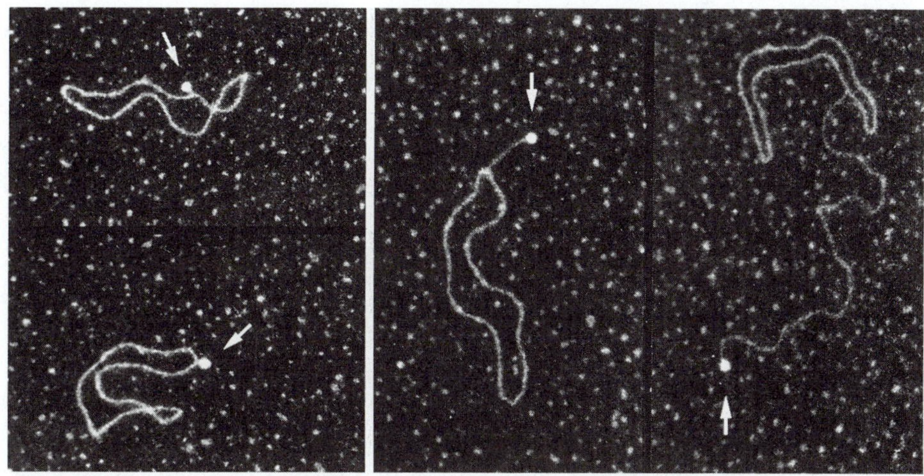

Figure 17-17
Assembly of ϕX174 phage. At top is a model for assembly. Letters indicate ϕX174 proteins that participate in morphogenesis. GpB catalyzes aggregation of five molecules of gpF (major capsid protein) with five of gpG (spike protein) to form a 12S particle. GpD may provide a scaffolding function to form the 108S capsomere; gpH is added at the same time as gpD. GpC facilitates encapsidation of the DNA, and gpA measures unit length of genome and forms the circle. GpJ protein is the least well understood. (Courtesy of Professor M Hayashi). At bottom are electron micrographs of the complex of the rolling circle with the capsid (arrows). In a low-temperature spread (left), single-stranded tails are inside the capsid and not evident, whereas in a room-temperature spread (right) the tail can be seen outside the capsid. (Courtesy of Professors D Dressler and K Koths)

Total Enzymatic Synthesis of a Polyhedral Virus. With the isolated and characterized proteins needed for ϕX174 phage replication and assembly available, and with the host gene-expression system in hand, only the clarification of the viral adsorption process remains for an attempt to reconstitute virus multiplication with molecularly defined components from start to finish.

17-5 Medium-Sized Phages: T7 and Other T-Odd Phages (T1, T3, T5)[127]

T7, a virulent *E. coli* phage with a genome of 40 kb, is more complex than Ff or ϕX174 but far smaller and simpler than the T-even phages. It has attracted interest for several reasons. The entire nucleotide sequence of the 40-kb genome has been determined,[128] and a full understanding of its replication and regulated transcription seems attainable. In its rapid replication and transcription, novel phage and host functions are deftly balanced. As a linear chromosome, its replication poses special problems whose solutions should have general significance.

The genomes of all four T-odd phages are linear duplexes with terminal redundancies. T1, T3, and T7 are near 40 kb; T5 is 121 kb. Phage T3 closely resembles T7. They cross-react serologically, cross-hybridize in DNA annealing, and recombine genetically; yet they induce distinctive enough enzyme patterns to preclude successful complementation in a mixed infection. Two of these distinctions will be mentioned below.

Phage T1 has not been studied in detail. However, it probably replicates by a pathway substantially different from those of T7 and T3, in that T1 depends on the host's DNA polymerase rather than a phage-encoded one.[129] The variable terminally redundant region of T1 is about 2.8 kb, nearly 20 times that of T7. Because its head can accommodate the near 50-kb phage λ DNA and equally large pieces of host DNA, the T1 phage is an attractive vehicle for transduction.[130] Phage T5, because of numerous distinctions from the other T-odd phages, is discussed at the end of this section.

T7 Virion

The T7 icosahedral particle (Fig. 17-18, top) has a short, cone-like tail. Its linear DNA is terminated at each end by 5′-P and 3′-OH groups with an invariant sequence (Fig. 17-18, bottom).[131] Structural, genetic, and biochemical analyses show that every T7 DNA molecule is identical.

Terminal Redundancy. The T7 chromosome, unlike that of phage λ (Section 7), does not go through a circular stage during replication. Problems faced by a linear template include susceptibility to exonuclease action and the need for a mechanism to replicate the 3′ end of each template strand (Section 15-12). RNA

---

127. Hausmann R (1988) in *The Bacteriophages* (Calendar R, ed.). Plenum Press, New York, Vol. 1, p. 259; Richardson CC, Beauchamp BB, Huber HE, Ikeda RA, Meyers JA, Nakai H, Rabkin SD, Tabor S, White J (1987) in *DNA Replication and Recombination* (McMacken R, Kelly TJ, eds.). *UCLA*, Vol. 47. AR Liss, New York, 151; Dunn JJ, Studier FW (1983) *JMB 166*:477.
128. Dunn JJ, Studier FW (1981) *JMB 148*:303; Dunn JJ, Studier FW (1983) *JMB 166*:477.
129. Drexler H (1988) in *The Bacteriophages* (Calendar R, ed.). Plenum Press, New York, Vol. 1, p. 235.
130. Drexler H, Christensen JR (1979) *J Virol 30*:543.
131. Price SS, Schwing JM, Englund PT (1973) *JBC 248*:7001.

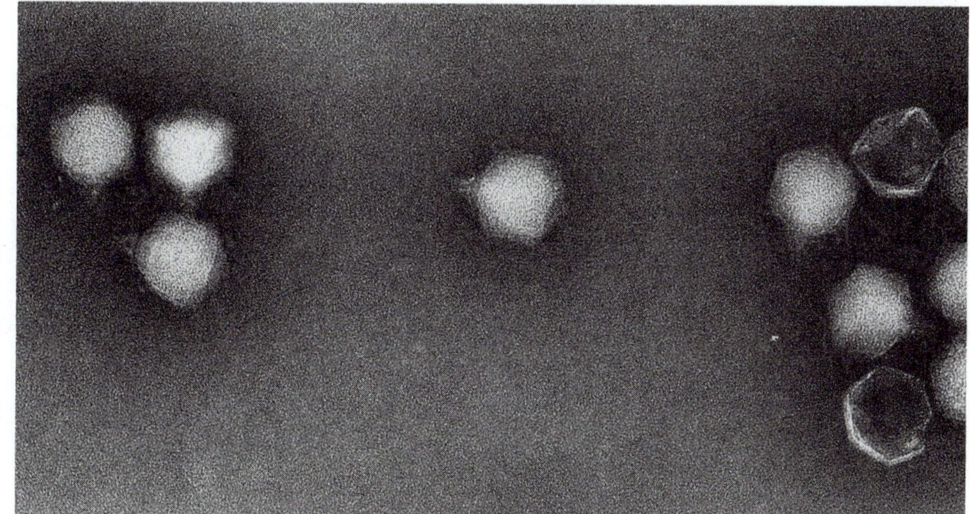

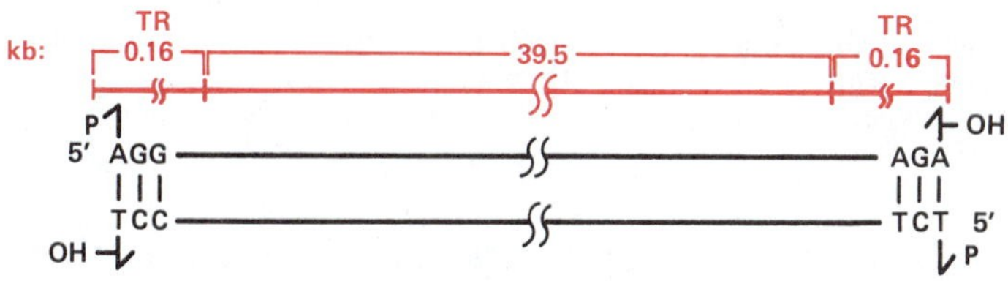

Figure 17-18
Structure of phage T7. Top, electron micrograph (×200,000). (Courtesy of Professor RC Williams) Bottom, terminal sequences of phage T7 molecules showing that 40-kb duplexes are complete and contain, at each end, 5′-P and 3′-OH groups on a unique sequence in a terminal redundancy (TR) of 160 base pairs.

priming leaves an RNA segment at the 5′ end of newly synthesized strands that must later be excised and replaced with DNA. For lack of a mechanism to fill this gap with DNA chain growth, the T7 chromosome depends on a 160-bp perfect direct repeat, or *terminal redundancy*, to replicate its ends. How terminal redundancy and concatemer formation achieve the replication of the 3′ ends has been considered (Section 15-12) and will be discussed further below.

Genetic Map.[132] The completely sequenced 39,936-bp T7 genome (Fig. 17-19) is used very efficiently; 92% of the sequence is involved in coding its 50 genes. The noncoding regions include the terminal repeats and signals that regulate replication and gene expression. The genes were designated by integers in the original mapping. Subsequently discovered genes have required decimal notations to identify their locations.[133]

The genes of T7, like those of T4 and λ, are clustered according to function and order of expression. They are expressed in three groups: Class I genes, at the far left of the genome, are transcribed by host RNA polymerase (RNAP) and encode

132. Studier FW, Dunn JJ (1983) *CSHS 47*:999.
133. Studier FW (1969) *Virology 39*:562.

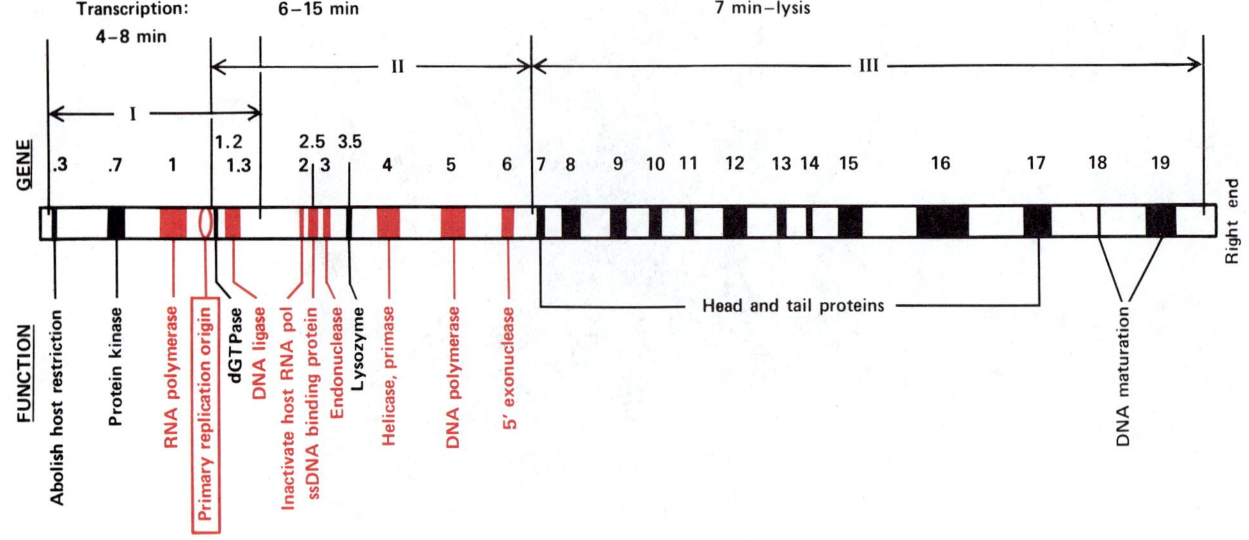

Figure 17-19
Genetic map of phage T7. Gene numbers and their estimated sizes and functions are indicated; genes essential for DNA replication are in color. Regions transcribed in the early (I) and subsequent (II, III) stages of the phage cycle, and the duration of these stages, are also shown.

functions that redirect cellular metabolism to the phage. These include a unique RNAP, inhibitors of the host restriction system, and a protein kinase that inactivates the host RNAP.[134] Class II genes are involved mainly in phage DNA replication; class III genes encode the morphogenetic and capsid proteins. Class II and III genes are transcribed by the T7 RNAP.

Twenty of the T7 genes are known to be essential; of these, seven are required for replication. These include genes for a DNA polymerase (gp5), DNA helicase–primase (gp4), an inhibitor of the host RNAP (gp2), T7 RNAP (gp1),[135] an endonuclease (gp3), an exonuclease (gp6), and an SSB (gp2.5). Some gene *2.5* mutants remain viable by utilizing the host SSB.

Several of the T7 gene products are required only in certain defective host strains. Gp1.3, an ATP-dependent ligase,[136] is normally not needed, but is required in cells of low ligase content; this gene is absent from phage T3. Mutants defective in gene *1.2* fail to grow in the *optA1* mutant of *E. coli*:[137] T7 DNA replication stops early, and the newly synthesized DNA is rapidly degraded. The *optA1* cells overproduce a host-encoded dGTPase[138] that reduces the intracellular dGTP concentration to one-fifth that of wild-type cells.[139] Gp1.2 specifically inhibits this dGTPase by forming a complex with it,[140] allowing the dGTP pool to rise 200-fold upon T7 infection, to a level effective in supporting phage replication. Extracts from *optA1* cells, defective in the DNA synthesis required for packaging T7 DNA, can be complemented with either gp1.2 or dGTP.[141]

---

134. Chamberlin M, McGrath J, Waskell L (1970) *Nature 228*:227.
135. Chamberlin M, Ring J (1973) *JBC 248*:2235; Chamberlin M, Ring J (1973) *JBC 248*:2245.
136. Masamune Y, Frenkel GD, Richardson CC (1971) *JBC 246*:6874.
137. Saito H, Richardson CC (1981) *J Virol 37*:343.
138. Beauchamp BB, Richardson CC (1988) *PNAS 85*:2563; Seto D, Bhatnagar SK, Bessman MJ (1988) *JBC 263*:1494.
139. Myers JA, Beauchamp BB, Richardson CC (1987) *JBC 262*:5288.
140. Huber HE, Beauchamp BB, Richardson CC (1988) *JBC 263*:13549.
141. Myers JA, Beauchamp BB, White JH, Richardson CC (1987) *JBC 262*:5280.

Refined control of transcription characterizes the onset and progress of T7 replication. One of three promoters for host RNAP, located near the left end of the duplex (which enters the cell first), defines the start of transcription.[142] Transcription then proceeds for 20% of the genome to a Rho protein–dependent terminator.[143] The polycistronic message from this "early" region is then processed by host RNase III, an RNA-processing enzyme, into five messages. One of them encodes T7 RNAP (Section 7-13),[144] which transcribes the rest of the genome to provide the DNA replication enzymes and the morphogenetic and coat proteins.

Specific and rapid transcription by the T7 RNAP exerts a positive control for phage functions; shutoff of host RNAP by gp0.7 and gp2 exerts a negative control over the early region, as well as over host functions.[145] The mechanism for switching from the T7 class II to class III genes, both transcribed by T7 RNAP, is unclear. One possibility is that DNA entry and transcription are coupled:[146] the linear genome enters the cell progressively and becomes available for transcription in stages,[147] as has been described for T5 (see below).

Linear entry of DNA may explain the remarkable resistance of T7 DNA to cleavage by the *E. coli* restriction system. The entering DNA survives destruction during the time needed to produce the inhibitor; gene *0.3*, at the end of the genome that enters the cell first, is spared because it lacks restriction sites. Further entry of the genome is delayed until gp0.3 is produced and prevents host restriction by forming a tight complex with the EcoB and EcoK restriction and modification enzymes.[148] (Section 13-7). T3 phage achieves the same result by encoding an enzyme that hydrolyzes S-adenosylmethionine, needed by the restriction system.[149]

The stability of T7 mRNAs implies that the regulation of translation may also function in switching among the gene classes. RNase III processing of T7 mRNAs is common; at least ten of the transcripts are processed, although the function of this processing is unknown.[150] Each of the T7 genes appears to have its own ribosome binding site and is translated independently from a polycistronic message.[151]

A fascinating and unsolved question is the failure of T7, but not T3, to develop in cells bearing the F plasmid (male cells; Section 18-6); the infection aborts after about 8 minutes, when the late genes are being expressed. Membrane damage is observed, along with a generalized breakdown of macromolecular synthesis.[152] The block appears to be at the stage when the DNA carrying the late genes enters into the cell. The products of gene *1.1* (unknown function) and gene *1.2* (dGTPase inhibitor) are required for T3 to escape inhibition by F plasmid.[153] The combined effect of a mutation in gene *1.2* and two alterations in gene *10* (major

142. Dunn JJ, Studier FW (1973) *PNAS 70*:1559; Minkley EG, Pribnow D (1973) *JMB 77*:255.
143. Dunn JJ, Studier FW (1980) *NAR 8*:2119.
144. Dunn JJ, Studier FW (1973) *PNAS 70*:1559.
145. Hesselbach BA, Nakada D (1977) *J Virol 24*:736; Hesselbach BA, Nakada D (1977) *J Virol 24*:346.
146. Zavriev SK, Shemyakin MF (1982) *NAR 10*:1635.
147. McAllister WT, Morris C, Rosenberg AH, Studier FW (1981) *JMB 153*:527.
148. Bandyopadhyay PK, Studier FW, Hamilton DL, Yuan R (1985) *JMB 182*:567; Mark K-K, Studier FW (1981) *JBC 256*:2573.
149. Spoerel N, Herrlich P, Bickle TA (1979) *Nature 278*:30; Spoerel N, Herrlich P (1979) *EBJ 95*:227.
150. Dunn JJ, Studier FW (1973) *PNAS 70*:1559; Dunn JJ, Studier FW (1983) *JMB 166*:477.
151. Dunn JJ, Studier FW (1983) *JMB 166*:477.
152. Yamada Y, Nakada D (1975) *J Virol 16*:1483; Blumberg DD, Mabie CT, Malamy MH (1976) *J Virol 17*:94.
153. Molineux IJ, Spence JL (1984) *PNAS 81*:1465.

capsid protein) are required for T7 to grow normally on male strains.[154] Production of either gp1.2 or gp10 is sufficient to cause toxicity in cells containing F plasmid; how their presence causes the abortive T7 infection and cell death is unknown.

T7 Replication[155]

Host DNA continues to be synthesized for the first 5 minutes after infection and then, with the onset of T7 DNA replication, begins to be degraded. Degradation continues for 10 to 15 minutes, ceasing at about the time of lysis, 25 to 30 minutes after infection at 30°.

In Vivo Replication. Electron microscopy indicates that T7 DNA replication generally starts at a specific point, giving rise to a replication bubble located 17% of the distance from the genetic left end of the molecule, and proceeds in both directions. The resultant bubble gives the initial replicating intermediate an eye-shaped appearance (Section 16-1; Fig. 17-20),[156] and then generates a Y-shaped structure when replication has reached the left end; the Y structure then expands. Replication proceeds nearly to the ends of the linear molecule. Concatemers are then formed, allowing synthesis to the very 5′ ends (see below).

A second round of replication may be initiated before the first is completed, giving rise to multiple eyes and forks. Mutants with a deleted replication origin may replicate[157] by using a secondary origin.[158] Nascent short chains contain RNA primers, the majority of them tetranucleotides of the same sequence as

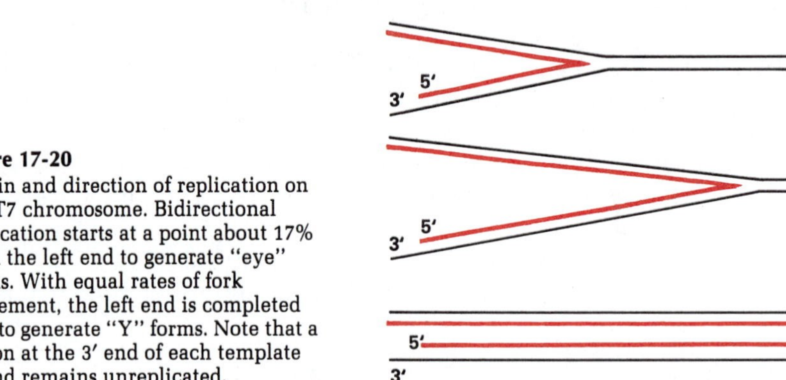

Figure 17-20
Origin and direction of replication on the T7 chromosome. Bidirectional replication starts at a point about 17% from the left end to generate "eye" forms. With equal rates of fork movement, the left end is completed first to generate "Y" forms. Note that a region at the 3′ end of each template strand remains unreplicated.

154. Molineux IJ, Schmitt CK, Condreau JP (1989) *JMB 207*:563.
155. Richardson CC, Beauchamp BB, Huber HE, Ikeda RA, Meyers JA, Nakai H, Rabkin SD, Tabor S, White J (1987) in *DNA Replication and Recombination* (McMacken R, Kelly TJ, eds.). *UCLA*, Vol. 47. AR Liss, New York, p. 151.
156. Wolfson J, Dressler D, Magazin M (1972) *PNAS 69*:499; Dressler D, Wolfson J, Magazin M (1972) *PNAS 69*:998; Wolfson J, Dressler D (1972) *PNAS 69*:2682.
157. Simon MN, Studier FW (1973) *JMB 79*:249.
158. Tamanoi F, Saito H, Richardson CC (1980) *PNAS 77*:2656.

synthesized by gp4 in vitro[159] (see below). The frequency of these RNA-linked nascent DNAs is greatly increased with phages defective in gp6,[160] a 5′→3′exonuclease.

Replication Proteins. The products of seven T7 genes are required for replication.

(1) *DNA polymerase (gp5).* The 80-kDa gp5 contains both DNA polymerase and 3′→5′ exonuclease activities.[161] This protein makes a 1:1 complex with host thioredoxin that is the highly processive form of polymerase, responsible for replication of the T7 genome.[162] Thioredoxin stimulates the processivity of gp5 several hundred-fold by stabilizing its interaction with the primer-template[163] (Section 5-9).

(2) *Helicase and primase (gp4).* The protein exists in two molecular-weight forms (63 and 56 kDa)[164] and provides both the primase and helicase activity for T7 replication forks (Section 11-2). Properties of gp4 will be described below.

(3) *RNA polymerase (gp1).* T7 RNAP (98 kDa) is required for initiation at the T7 primary and secondary origins, in addition to transcribing all the other essential replication proteins (Section 7-13).

(4) *Host RNAP inhibitor (gp2).* The 9-kDa protein binds host RNAP and shuts off its activity.[165] It is also required for the synthesis of concatemeric T7 DNA[166] and for packaging of DNA into phage heads in vitro.[167]

(5) *SSB (gp2.5).* The 25-kDa protein (Section 10-4) specifically stimulates T7 DNA polymerase on single-stranded templates[168] and the priming activity of gp4;[169] it is only partially replaceable in vivo by *E. coli* SSB.[170]

(6) *Exonuclease (gp6).* The 40-kDa protein is a 5′→3′ exonuclease[171] (Section 13-3) resembling the excision nuclease of host DNA pol I. It requires a double strand, attacks a 5′ terminus whether it contains a phosphate or a hydroxyl group, releases oligonucleotides as well as mononucleotides, and has RNase H activity.

(7) *Endonuclease (gp3).* The 17-kDa protein and gp6 exonuclease together extensively degrade the host DNA, furnishing most of the nucleotides for T7 DNA synthesis. The endonuclease, despite a strong preference for ssDNA, also converts duplex DNA to oligonucleotides with 3′-OH and 5′-P termini. The gp3 and gp6 nucleases not only convert host DNA to precursors for T7 DNA replication, but also are involved in recombination and in the formation and utilization of concatemers.

---

159. Okazaki T, Kurosawa Y, Ogawa Y, Seki T, Shinozaki K, Hirose S, Fujiyama A, Kohara Y, Machida Y, Tamanoi F, Hozumi T (1978) *CSHS 43*:203.
160. Shinozaki K, Okazaki T (1977) *MGG 154*:263.
161. Adler S, Modrich P (1979) *JBC 254*:11605; Hori K, Mark DF, Richardson CC (1979) *JBC 254*:11591; Hori K, Mark DF, Richardson CC (1979) *JBC 254*:11598.
162. Modrich P, Richardson CC (1975) *JBC 259*:5515.
163. Huber HE, Tabor S, Richardson CC (1987) *JBC 262*:16224; Tabor S, Huber HE, Richardson CC (1987) *JBC 262*:16212.
164. Dunn JJ, Studier FW (1983) *JMB 166*:477.
165. Hesselbach BA, Nakada D (1977) *J Virol 24*:736; Hesselbach BA, Nakada D (1977) *J Virol 24*:746.
166. Center MS (1975) *J Virol 16*:94.
167. LeClerc JE, Richardson CC (1979) *PNAS 76*:4582.
168. Reuben RC, Gefter ML (1973) *PNAS 70*:1846; Scherzinger E, Litfen R, Jost E (1973) *MGG 123*:247.
169. Nakai H, Richardson CC (1988) *JBC 263*:9831.
170. Araki H, Ogawa H (1981) *MGG 183*:66; Studier FW, personal communication.
171. Kerr C, Sadowski PD (1972) *JBC 247*:305; Kerr C, Sadowski PD (1972) *JBC 247*:311; Shinozaki K, Okazaki T (1978) *NAR 5*:4245.

In Vitro Replication.[172] Reconstitution of site-specific initiation requires only T7 RNAP, DNA polymerase, and a DNA template containing the primary origin.[173] Located 15% from the left end of the T7 map, this origin consists of two T7 RNAP promoters, ϕ1.1A and ϕ1.1B, and a 61-bp AT-rich region.[174] Initiation of replication occurs when T7 RNAP starts synthesis at one of the promoters and DNA polymerase displaces the RNAP and elongates the 3′-OH end of the transcript (Fig. 17-21). RNA–DNA junctions occur in several locations within the ϕ1.1B promoter and the AT-rich region, both in vivo and in vitro;[175] 10 to 60 ntd of RNA are covalently attached to the 5′ end of the newly synthesized DNA strands. Bidirectional propagation of DNA replication (continuous synthesis of the leading strand and discontinuous synthesis of the lagging strand) requires two additional proteins: the gene 4 helicase–primase and an SSB.[176] Gp4 has dual functions: the separation of duplex strands at the fork and the synthesis of primers for the discontinuous strand.

During initiation, gp4 binds the DNA strand displaced by the polymerases at the origin. The association of gp4 with ssDNA requires a nucleoside triphosphate (dTTP is preferred),[177] which fuels gp4 translocation in the 5′→3′ direction. The movement of gp4 melts the DNA duplex,[178] allowing the DNA polymerase to synthesize the continuous strand. This helicase activity of gp4 is

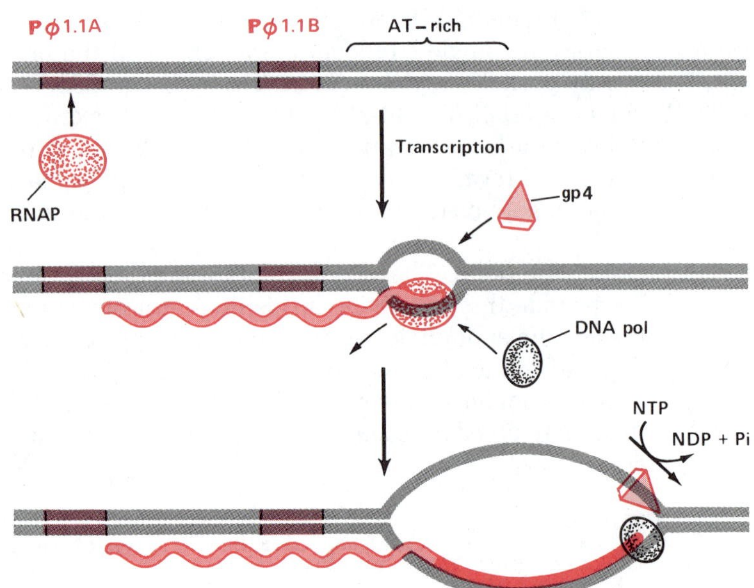

Figure 17-21
Stages of initiation at the T7 primary origin. T7 RNAP (RNA polymerase) initiates transcription at either promoter (ϕ1.1A or ϕ1.1B) and proceeds through an AT-rich region. T7 DNA polymerase displaces RNAP and gp4 binds to the displaced DNA strand. T7 DNA polymerase elongates the 3′ end of the transcript, and helicase action of gp4 fuels replication fork movement.

172. Nakai H, Beauchamp BB, Bernstein J, Huber HE, Tabor S, Richardson CC (1988) in *DNA Replication and Mutagenesis.* Am Soc Microbiol, Washington DC (Moses RE, Summers WC, eds.), p. 85; Richardson CC, Beauchamp BB, Huber HE, Ikeda RA, Meyers JA, Nakai H, Rabkin SD, Tabor S, White J (1987) in *DNA Replication and Recombination* (McMacken R, Kelly TJ, eds.). UCLA, Vol. 47. AR Liss, New York, p. 151.
173. Fuller CW, Richardson CC (1985) *JBC 260*:3185.
174. Saito H, Tabor S, Tamanoi F, Richardson CC (1980) *PNAS 77*:3917; Tamanoi F, Saito H, Richardson CC (1980) *PNAS 77*:2656.
175. Sugimoto K, Kohara Y, Okazaki T (1987) *PNAS 84*:3977; Fuller CW, Richardson CC (1985) *JBC 260*:3185.
176. Fuller CW, Richardson CC (1985) *JBC 260*:3197.
177. Matson SW, Richardson CC (1985) *JBC 260*:2281.
178. Matson SW, Richardson CC (1983) *JBC 258*:14009; Matson SW, Tabor S, Richardson CC (1983) *JBC 258*:14017.

specific for T7 DNA polymerse; no stimulation of *E. coli* pol I, II, or III or of T4 DNA polymerase is observed.[179]

Uncoupled from primer formation in the absence of rNTPs, DNA polymerase and gp4 can produce exceedingly long continuous strands from a circular template[180] (Fig. 17-22). Gp4 exists in two polypeptide forms: 63 kDa and 56 kDa, the latter generated by restarting translation in the same frame used to synthesize the former.[181] Both forms have helicase activity (Section 11-2) and thus stimulate displacement synthesis by T7 polymerase;[182] the small form lacks primase activity (Section 8-5).[183] In the presence of rNTPs and SSB (*E. coli* SSB or T7 gp2.5), the 63-kDa form of gp4 synthesizes RNA primers on the displaced strand. Although gp4 is both helicase and primase at T7 replication forks, more than one

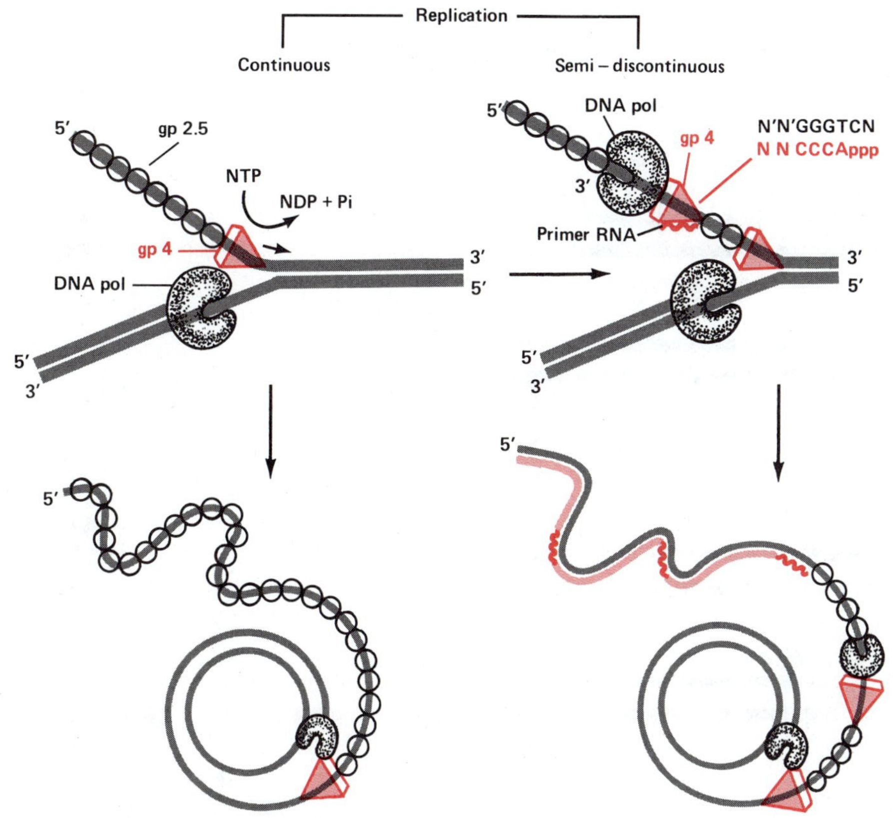

Figure 17-22
T7 replication forks and products. (Left) Continuous-strand synthesis (in the absence of rNTPs). (Right) Coupled continuous- and discontinuous-strand synthesis.

179. Kolodner R, Richardson CC (1977) *PNAS 74*:1525; Lechner RL, Richardson CC (1983) *JBC 258*:11185; Kolodner R, Masamune Y, LeClerc JE, Richardson CC (1978) *JBC 253*:566.
180. Kolodner R, Richardson CC (1978) *JBC 253*:574.
181. Dunn JJ, Studier FW (1983) *JMB 166*:535.
182. Nakai H, Richardson CC (1988) *JBC 263*:9818.
183. Bernstein J, Richardson CC (1988) *PNAS 85*:396.

molecule is required to execute these functions.[184] The helicase activity is highly processive, unwinding as much as 40 kb of duplex DNA without dissociating from the template. In contrast, the priming activity, on the same template DNA molecules, is sensitive to both dilution and challenge DNA. Two types of gp4 must act together to replicate the continuous and discontinuous strands, one a processive helicase and the other a primase that repeatedly dissociates from the template and reassociates with it.

Termination and Maturation. Asymmetrically positioned single-stranded regions are seen in the electron microscope in one of the two growing areas of the T7 replicating fork, consistent with semidiscontinuous DNA synthesis[185] (Section 15-3). These gaps in the duplex must eventually be filled; either T7 or *E. coli* functions can do this (Section 4-16). However, two gaps will remain where the primer RNA has been removed from the 5′ ends of the replicated strand, opposite the 3′ ends of parental template strands (Fig. 17-23). Filling these gaps would require a novel 3′→5′chain growth (Section 3-3), unknown in any DNA polymerase. A model[186] proposed to solve this dilemma suggests that concatemers, developed by homologous recombination within the terminal repeats of the T7

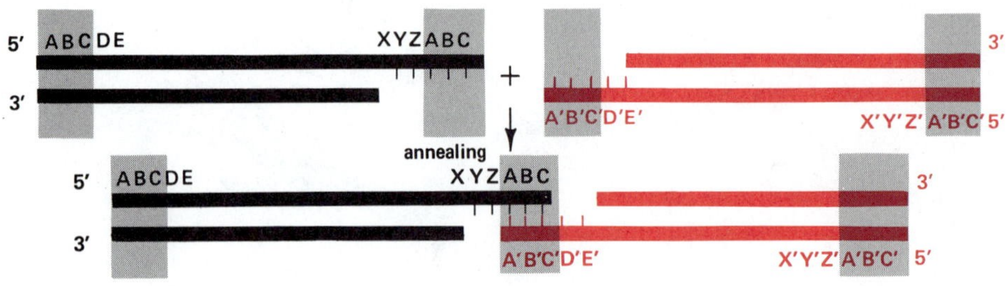

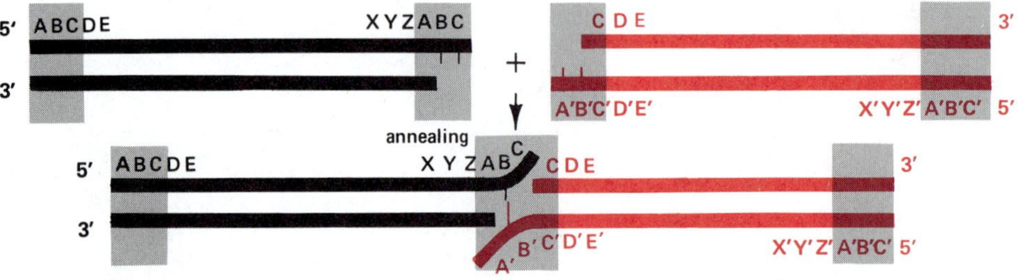

Figure 17-23
Model for using concatemers as intermediates in replication of T7 DNA. (Top) When the unreplicated 3′ tail region extends beyond the redundant region (shaded), the annealing of two molecules at their redundant regions leaves gaps to be filled. (Bottom) When the unreplicated 3′ tail region is only within the redundant region, annealing of two molecules at their redundant regions leaves unmatched lengths of ssDNA which require excision. [From Watson JD (1972) *NNB 239*:197]

184. Nakai H, Richardson CC (1988) *JBC 263*:9818.
185. Wolfson J, Dressler D, Magazin M (1972) *PNAS 69*:499; Dressler D, Wolfson J, Magazin M (1972) *PNAS 69*:998, Wolfson J, Dressler D (1972) *PNAS 69*:2682.
186. Watson JD (1972) *NNB 239*:197.

molecule, are cleaved by nucleases to generate ends with 5′ single-stranded overhangs, which can be readily filled in by a polymerase.

Linear concatemers, containing multiple copies of the genome linked head to tail, are in fact intermediates in T7 replication.[187] These chains are generated through pairing of the unreplicated 3′ ends of individual chromosomes, complementary in sequence due to terminal redundancy.[188] Replication and pairing of dimers in turn creates tetramers. Models for processing propose that the concatemers are specifically nicked at the junctions of the terminal redundancy. Strand displacement synthesis by DNA polymerase from the 3′ ends through the redundant region would yield full-length duplex ends. However, evidence for this site-directed cleavage of the T7 concatemer junction is still lacking and this model, though attractive, has yet to be reconstituted with isolated enzymes.

Several T7 gene products are required for the formation and processing of concatemers. Strains mutant in capsid proteins accumulate concatemers but fail to process the DNA. Coupling of DNA maturation and packaging[189] can also be observed in vitro; newly generated chromosome ends are protected from nuclease degradation, indicating that they are packaged.[190] The minor capsid protein (gp8) and the scaffolding protein (gp9) are required, as are gp18 and gp19, proteins that bind capsid intermediates during packaging. Their absence from mature virions indicates that they play a direct role in DNA processing.[191] Gp18 appears to act in a complex with a host protein.[192] The endonuclease, gp3, is required for processing of concatemers in vivo, but not for maturation of a defined concatemer junction in vitro. Gp3 cleaves X- and Y-shaped DNA branches (Section 21-5),[193] much like T4 endonuclease VII (Section 6), and thus may be involved in resolving complex structures prior to the specific processing of the chromosome ends.

Mutants in gene *2* and gene *6* process concatemers aberrantly, giving rise to short, incomplete chromosomes,[194] but their roles are unknown. The maturation defect in gene *2* mutants, defective in the shutoff of host RNAP, can be overcome by addition of the RNAP inhibitor rifampicin.[195] Why active RNAP blocks concatemer formation is also mysterious.

Involvement of the replication enzymes in concatemer processing can be assayed only in vitro[196] because concatemer substrates fail to accumulate in replicaton-deficient mutants. The gene *4* helicase–primase is not required for the displacement synthesis needed to complete the chromosomal ends. T7 DNA polymerase stimulates the synthesis of full-length ends, but can be replaced by a host DNA polymerase. In extracts lacking polymerase (or gp2 or gp6), full-length right-hand ends are produced but the left ends lack at least 160 bp of the terminal redundancy, an observation in conflict with the symmetric processing model in which both ends are synthesized by the same mechanism. Alternative models in which the right end is created by a double-stranded break have been proposed.

---

187. Kelly TJ, Thomas CA (1969) *JMB 44*:459.
188. Serwer P, Greenhaw GA, Allen JL (1982) *Virology 123*:474; White JH, Richardson CC (1987) *JBC 262*:8851.
189. Hausmann R, LaRue K (1969) *J Virol 3*:278; Serwer P, Watson RH (1981) *Virology 108*:164.
190. White JH, Richardson CC (1987) *JBC 262*:8851.
191. Roeder GS, Sadowski PD (1977) *Virology 76*:263; Serwer P, Watson RH (1981) *Virology 108*:164.
192. White JH, Richardson CC (1987) *JBC 262*:8845.
193. De Massy B, Weisberg RA, Studier FW (1987) *JMB 193*:359.
194. Center MS (1975) *J Virol 16*:94; Miller RC, Lee M, Scraba DG, Paetku Y (1976) *JMB 101*:223.
195. Ontell MP, Nakada D (1980) *J Virol 34*:438.
196. White JH, Richardson CC (1987) *JBC 262*:8851.

T5 Phage[197]

The phage T5 genome is notable for interruptions (nicks) at defined intervals on one of its strands and for transfer of its DNA into the cell in two discrete steps. The viral chromosome, a 121-kb linear duplex with 10-kb terminal repeats,[198] falls between T7 and T4 in size and complexity.

Interrupted Genome. The five major ligase-repairable nicks are in the "light" strand of the duplex at locations near 8%, 18%, 33%, 65%, and 99% from its 3′-OH end.[199] Nicks are repaired early in infection and do not appear in the free, concatemeric progeny DNA before it is packaged. The nicks correlate grossly with the division of the genetic map of mutations into linkage groups, and are thought to function in the transcription of "late" genes. However, the nicks do not appear to be essential for phage replication; mutants producing covalently intact progeny are viable.[200]

Two-Step Entry of T5 DNA. After 8% of the left end of the genome has entered the cell (first-step transfer), there is a pause of about 4 minutes during which several genes are expressed from this region. Among them are genes governing breakdown of the host chromosome, modification of host RNAP, inhibition of detrimental host functions, and transfer of the remainder of the phage genome. The second step of DNA entry can take place even when attached phage heads have been sheared off. A phage-encoded product inhibits the host restriction nuclease, protecting the T5 genome.[201] As with T7 DNA, the first-step transfer DNA of T5 contains no host restriction sites, allowing time for the synthesis of one or more protection proteins before the entry of the vulnerable (restrictable) DNA.

Replication. Unlike T7, but like T4, phage T5 encodes a number of enzymes that ensure an abundant supply of DNA precursors. The unique choice by T5 not to use host DNA precursors is carried to the point of deploying a T5-encoded 5′-deoxynucleotidase[202] to remove nucleotides that arise from the breakdown of host DNA. This enzyme, induced immediately upon infection, is no longer detectable by 15 minutes, when phage DNA synthesis commences. Phage T5, like T4, achieves transcriptional control through modifications of host RNAP, rather than encoding its own enzyme as T7 and T3 do.

T5 DNA polymerase (Section 5-10), encoded by genes *D7*, *D8*, and *D9*,[203] has not been purified to homogeneity and is not adequately characterized. GpD5, required for DNA synthesis and late gene expression, is a DNA-binding protein.[204] The products of the four other genes essential for DNA synthesis have not been identified, nor have the gene products that maintain normal rates of DNA synthesis, avoid delay or arrest, and promote maturation. The D15 protein,

197. McCorquodale DJ, Warner HR (1988) in *The Bacteriophages* (Calendar R, ed.). Plenum Press, New York, Vol. 1, p. 439.
198. Rhoades M (1982) *J Virol 43*:566.
199. Rhoades M (1977) *J Virol 23*:725; Rhoades M (1977) *J Virol 23*:737.
200. Rogers SG, Godwin EA, Shinosky ES, Rhoades M (1979) *J Virol 29*:716.
201. Davison J, Brunel F (1979) *J Virol 29*:11; Davison J, Brunel F (1979) *J Virol 29*:17.
202. Mozer TJ, Warner HR (1977) *J Virol 24*:635; Mozer TJ, Thompson RB, Berget SM, Warner HR (1977) *J Virol 24*:624.
203. Fujmura RK, Tavtigian SV, Choy RL, Roop BC (1985) *J Virol 53*:495.
204. McCorquodale DJ, Gossling J, Benzinger R, Chesney R, Lawhorne L, Moyer RW (1979) *J Virol 29*:322.

known to be a 5′→3′exonuclease and essential for late transcription, has endonucleolytic functions that may contribute to the introduction of the specific nicks found in the progeny DNA.[205]

17-6 Large Phages:[206] T4 and Other T-Even Phages (T2, T6)

A unique base in the DNA of T-even phages and the abrupt switch from host to phage metabolism that T-even phages induce—unusual even in virulent infections—made this phage family an early and attractive subject for study.[207] The fundamental discoveries of mRNA, virus-induced enzyme functions, temporal transcriptional control, modification of macromolecules, and patterns of organelle morphogenesis arose from examining these phages. Much is now known about their structure and their life cycles (Fig. 17-24), but many accessible questions about DNA entry, replication, transcription, and assembly of the virion still remain unanswered.

Intensive analysis of the genetics of T4 and the availability of T4 mutants that overproduce phage-encoded enzymes made this phage the favored member of the T-even family. Because of the close similarity of the T-even phages, indi-

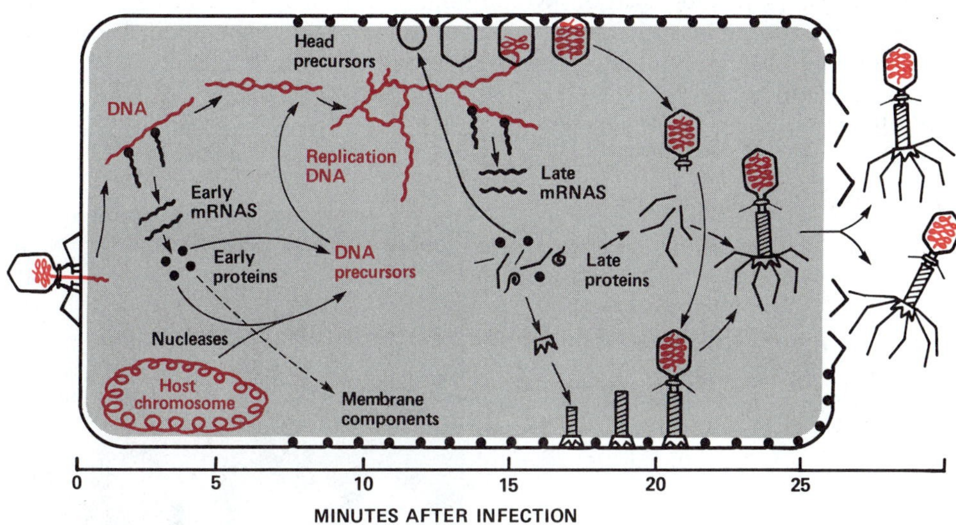

Figure 17-24
Scheme for the life cycle of phage T4. Noteworthy features: Nucleolytic action on host chromosome furnishes DNA precursors; replicating DNA is much longer than virion DNA; several phage-coded proteins become associated with the host membrane; maturation of phage head occurs at a membrane site. (Courtesy of Professor CK Mathews)

205. Moyer RW, Rothe CT (1977) *J Virol 24*:177; Everett RD (1981) *J Gen Virol 52*:25.
206. Mathews CK, Kutter EM, Mosig G, Berget PB (eds.) (1983) *Bacteriophage T4*. Am Soc Microbiol, Washington DC; Mosig G, Eiserling F (1988) in *The Bacteriophages* (Calendar R, ed.). Plenum Press, New York, Vol. 2, p. 521.
207. Cohen SS (1968) *Virus-Induced Enzymes*. Columbia University Press, New York; Edgar RS, Epstein RH (1965) *Sci Amer 212*:70; Wood WB, Edgar RS (1967) *Sci Amer 217*:60.

cated by sequence homology of over 85%, comparative studies of these phages will be particularly instructive about how replication strategies have evolved and how gene action is regulated.

Virion

The anatomy of the long-tailed T4 phage (Fig. 17-25) and the syringe-like injection of its DNA through the *E. coli* surface (Fig. 17-24) have been illustrated again and again. The polyhedral head is packed with DNA and a variety of proteins, including three small core proteins and a few phage-encoded enzymes, as well as with polyamine and Mg^{2+} counterions for the DNA.

The T4 genome is 166 kb long. As with T7, each T4 chromosome has terminally redundant ends a few percent of the genome in length, making the packaged DNA a 171-kb duplex. The nucleotide sequence of T4 DNA is circularly permuted; the ends of the chromosomes in a phage population can be situated anywhere throughout the sequence. Circular permutation (Fig. 17-26), first suggested by the circular genetic map[208] (Fig. 17-27), can be demonstrated by denaturing and then reannealing the DNA;[209] hydrogen-bonded circles with single-stranded tails are generated in a high percentage of the molecules. The importance of terminal redundancy in replicating the ends of a linear chromosome has been discussed for T7 replication, and the additional circular permutation of T4 DNA may contribute to its replication and maturation.

In T4 DNA, cytosine is replaced by glucosylated, hydroxymethylated cytosine (glu-HMC). (This modification, which renders the DNA insensitive to most restriction endonucleases, was initially an impediment to the application of recombinant DNA technology to the T4 genome.) Modification can be bypassed in phage mutants yielding T4 DNA with the standard cytosine instead of glu-HMC.[210] However, several enzymes, notably T4 topoisomerase, primase, and

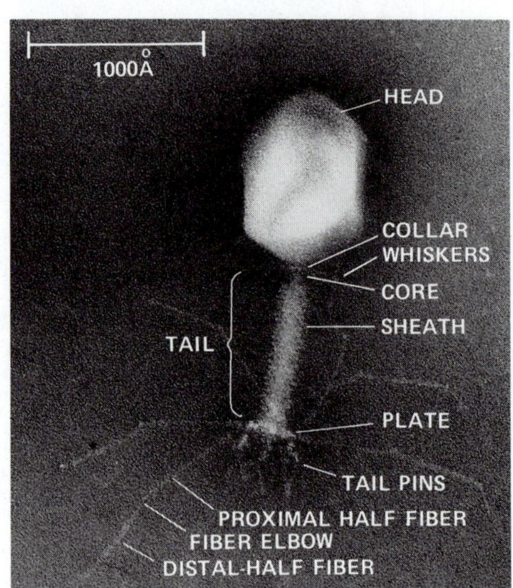

Figure 17-25
Electron micrograph of phage T4 (×270,000). (Courtesy of Professor RC Williams)

208. Streisinger G, Emrich J, Stahl MM (1967) *PNAS* 57:292.
209. MacHattie LA, Ritchie DA Jr, Thormas CA, Richardson CC (1967) *JMB* 23:355.
210. Kutter E, Snyder L (1983) in *Bacteriophage T4* (Mathews CK, Kutter EM, Mosig G, Berget PB, eds.). Am Soc Microbiol, Washington DC, p. 56.

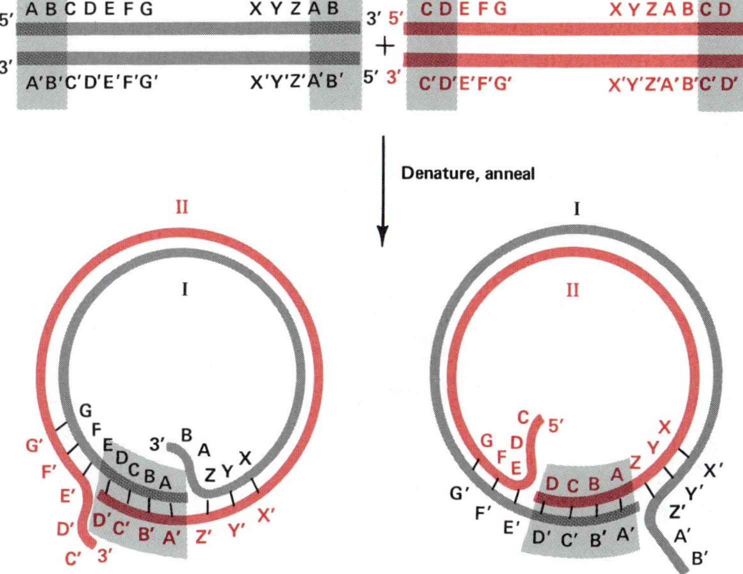

Figure 17-26
Scheme illustrating how circularly permuted, redundant regions at ends of phage T4 DNA can be detected by conversion of linear to circular forms. Shaded areas indicate redundant regions.

T4-modified RNAP, recognize unmodified DNA somewhat differently from glu-HMC DNA, thereby complicating studies with unmodified DNA.

Mutations now define about 130 genes on the T4 map (Fig.17-27).[211] DNA sequencing of about 60% of the T4 genome has revealed approximately 70 more open reading frames of yet unidentified function, most of them encoding proteins of 15 kDa or smaller. The entire genome may contain as many as 250 genes.

The known genes may be classified according to their function in metabolism and phage assembly. Of the 82 metabolic genes, only 22, related to DNA replication, are essential. An enormous fraction of the genome, one-third or more, is devoted to seemingly nonessential functions; this is not unique to T-even phages but has been observed in other mapped genomes, such as phage λ, *E. coli*, and , to a lesser extent, T7. What might appear to be an unnecessary genetic burden undoubtedly extends host range, provides an advantage in growth rate, and adds a capacity to cope with exigencies not imposed by standard culture conditions.

Infection

Adsorption and Penetration. Receptors in the outer *E. coli* membrane are recognized by the tips of the phage tail fibers. The phage then moves over the surface until it becomes fixed by short tail pins to the heptose residues in the lipopolysaccharide layer.[212] Then follow steps of base-plate elongation, tail contraction, core penetration through the cell envelope, unplugging of the core, and finally injection of DNA.

211. Kutter E, Guttman B, Rüger W, Tomaschewski J, Mosig G (1987) in *Genetic Maps 1987* (O'Brien SJ, ed.). CSHL, New York, Vol. 4, p. 22.
212. Zorzopulos J, DeLong S, Chapman V, Kozloff LM (1982) *Virology 120*:33; Riede I, Drexler K, Schwarz H, Henning U (1987) *JMB 194*:23.

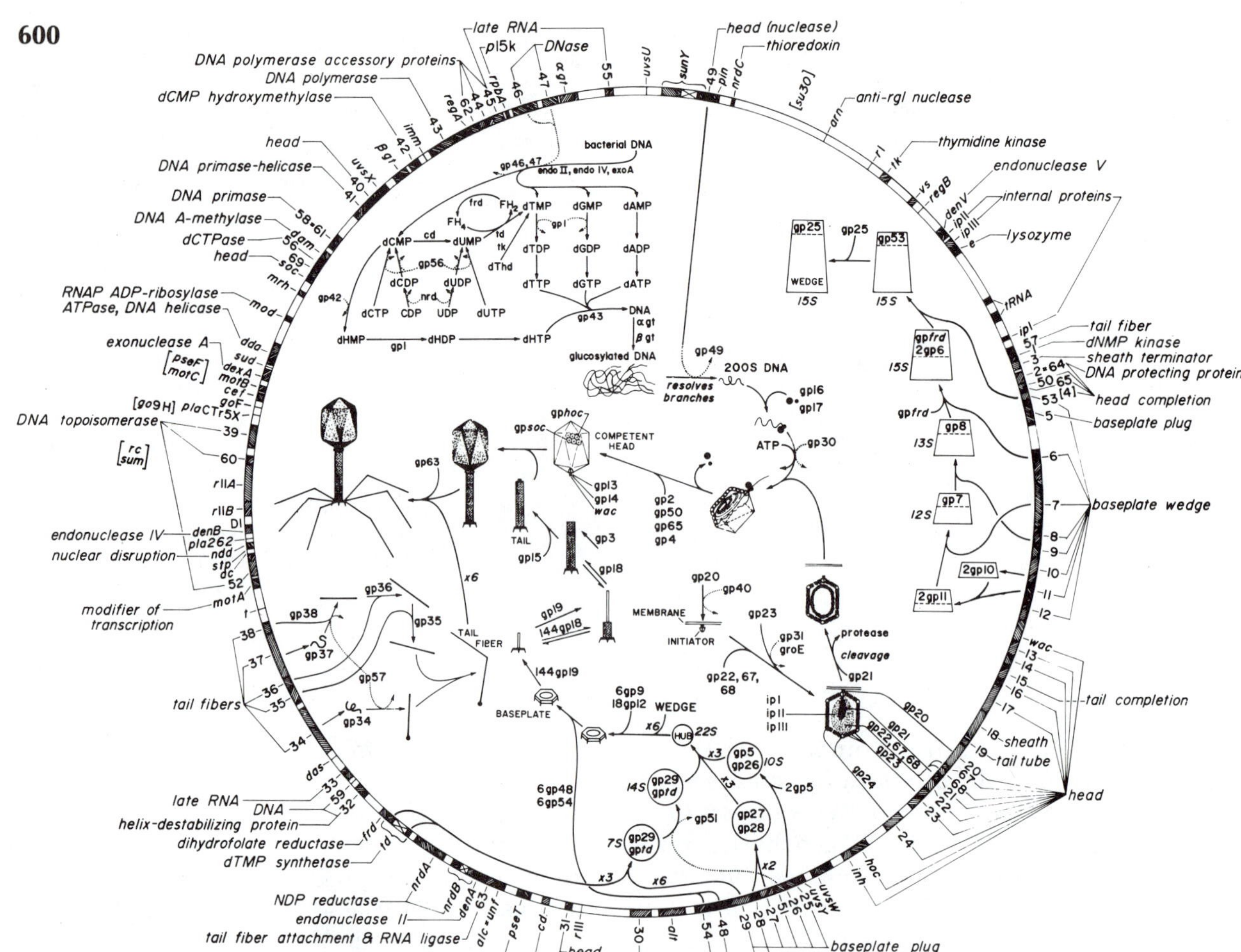

Figure 17-27
Genetic map of T4. (Courtesy of Dr E Kutter)

Shutoff of Host Synthesis.[213] Within one minute after adsorption, synthesis of host macromolecules has virtually ceased and transcription of certain phage genes has been initiated; within 4 minutes, replication of phage DNA is under way. Despite all that is known about T4 infection, there is still no explanation of why host macromolecular synthesis stops so promptly and completely.

One of numerous suggestions is that an effect on the membrane by phage penetration secondarily influences the replication and transcription of the host chromosome tied to the membrane. Other phage-driven modifications of the host chromosome include unfolding of the supercoils,[214] disruption of the nu-

213. Snustad DP, Snyder L, Kutter E (1983) in *Bacteriophage T4* (Mathews CK, Kutter EM, Mosig G, Berget PB, eds.). Am Soc Microbiol, Washington DC, p. 40; Warner HR, Snustad DP (1983) in *Bacteriophage T4* (Mathews CK, Kutter EM, Mosig G, Berget PB, eds.). Am Soc Microbiol, Washington DC, p. 103.

214. Tigges MA, Bursch CJH, Snustad DP (1977) *J Virol 24*:775.

cleoid,[215] its translocation to membrane sites, and, finally, degradation to nucleotides. However, cells infected with mutants in each of the genes responsible for these modifications still shut off the synthesis of host RNA, DNA, and protein rather promptly.

Solution of this problem, beyond clarifying a major event in phage infection, might provide insights into the spatial organization and control of the *E. coli* chromosome and thus claim another phage coup in molecular biology.

Metabolic Functions.[216] Directing the cell's economy to express and replicate the T4 genome entails the introduction of transcriptional and translational controls, as well as membrane modifications. Specificity and programming changes are imposed on the host RNAP by ADP-ribosylation and the addition of subunits (Section 7-13). Various phage-encoded proteins exert further influences on membrane structure and function.

To maintain a rate of replication ten times the host level and to produce a unique DNA, a variety of phage-encoded enzymes are brought into play (Section 2-13). In addition, T4 introduces several enzymes for the repair and recombination of its DNA.

Transcription

T4 depends on the host RNAP throughout its life cycle, but encodes several factors which redirect the enzyme to its own developmental pattern. Transcription can be divided into four stages: immediate early, delayed early, middle, and late. Except for the late transcribed genes, these stages are distinguished more by their unique requirements than by the transcription pattern, inasmuch as many genes are expressed during more than one of the stages. Transcription is regulated by the chemical modification of RNAP, by binding of protein factors to the RNAP, and by the induction of promoter-specific positive activators.

The *immediate early genes* are transcribed by the host σ^{70} RNAP. The T4 *alt* gene product, injected with the DNA during infection, immediately ADP-ribosylates one of the α subunits of RNAP. By 4 to 5 minutes after infection, ADP-ribosylation of both α subunits is completed by the *mod* gene product. This modification, while not essential, reduces the affinity of the core polymerase for σ^{70} and also increases the read-through of transcription terminators.

Delayed early gene transcription is distinguished from the immediate early stage in its dependence on phage-directed protein synthesis after infection. The RNAP-binding proteins produced by the *rpbA* and *rpbB* genes may affect the read-through of transcription terminators or the recognition of new promoters.

Middle gene transcription depends on the *motA* gene product, which is thought to activate transcription by binding to a specific sequence in middle gene promoters.

Late gene transcription depends on extensive modification of the core RNAP and on concurrent DNA replication. Late in infection, *E. coli* σ^{70} is replaced by

---

215. Koerner JF, Thies SK, Snustad DP (1979) *J Virol 31*:506.
216. Rabussay D (1983) in *Bacteriophage T4* (Mathews CK, Kutter EM, Mosig G, Berget PB, eds.). Am Soc Microbiol, Washington DC, p. 167; Rabussay D, Hall DH (1983) in *Bacteriophage T4* (Mathews CK, Kutter EM, Mosig G, Berget PB, eds.). Am Soc Microbiol, Washington DC, p. 174; Geiduschek EP, Elliott T, Kassavetis GA (1983) in *Bacteriophage T4* (Mathews CK, Kutter EM, Mosig G, Berget PB, eds.). Am Soc Microbiol, Washington DC, p. 189; Geiduschek EP, Kassavetis GA (1988) in *The Bacteriophages* (Calendar R, ed.). Plenum Press, New York, Vol. 1, p. 93.

gp55, a T4-specific sigma factor (σ^{gp55}) that redirects the core RNAP to late promoters; σ^{gp55}, like σ^{70}, is required only during promoter recognition and recycles during transcription. Competition of σ^{gp55} with the more abundant σ^{70} is not achieved by its higher affinity for the core enzyme. Rather, gp33 and a 10-kDa product of the *rpbB* gene appear to inhibit σ^{70} binding and thus favor its replacement by σ^{gp55}.

The genes for the structural proteins of T4 phage are transcribed only at late times. By coupling late transcription directly to DNA replication, the phage ensures that its DNA will begin to be packaged only after extensive replication has occurred. Studies in vitro with purified proteins offer clues to a complex mechanism for this coupling.[217]

The T4 late promoters are unusually simple, having an AT-rich −10 sequence, but no −35 sequence. The *E. coli* core RNAP, combined with purified σ^{gp55}, binds to these promoters but cannot form an open promoter complex (Section 7-5) and initiate RNA synthesis unless the DNA template is under negative superhelical tension. T4 DNA is not supercoiled in vivo and the requirement for negative superhelicity in vitro can be replaced by the binding of the DNA polymerase accessory proteins (gp44/62 and gp45) to the DNA template. Since these proteins can exert their effects either upstream or downstream from the promoter and from distances more than 1000 bp away, they seem to stimulate open complex formation by an "enhancer-like" mechanism. This effect requires gp33, which binds to the gp55 form of RNAP and seems to act as a bridge between the RNAP and gp45 in the DNA polymerase accessory protein complex. (Gp33, gp55, and gp45 have also been identified in vivo as essential for late transcription.)

The DNA polymerase accessory proteins need to hydrolyze ATP in order both to exert their stimulatory effect on initiation of late transcription and to maintain the accessory protein complex with the polymerase at a replication fork. Thus, a requirement for the active form of the accessory protein complex can explain why late transcription stops in a T4-infected cell as soon as DNA synthesis is blocked at an immobile replication fork.

Translation

Ordered expression of T4-encoded proteins depends on translational controls as well as altered transcription. The large T4 repertoire for regulating translation includes modification of host ribosomes,[218] alternative RNA secondary structures, translational repressors (gp32 and RegA protein),[219] novel tRNAs,[220] and enzymatic cleavage of certain host tRNAs to assemble a set of tRNAs with a codon usage more suited to T4 than to *E. coli*.

The T4 genome is unique among prokaryotes in that a few mRNAs contain introns. Those of the *td* (thymidylate synthase)[221] and *nrdB* (nucleotide reduc-

217. Herendeen DR, Williams KP, Kassavetis GA, Geiduschek EP (1990) *Science 248*:573; Herendeen DR, Kassavetis GA, Barry J, Alberts BM, Geiduschek EP (1990) *Science 245*:952.
218. Wiberg JS, Karam JD (1983) in *Bacteriophage T4* (Mathews CK, Kutter EM, Mosig G, Berget PB, eds.). Am Soc Microbiol, Washington DC, p. 193.
219. Gold L, Inman M, Miller E, Pribnow D, Schneider TD, Shinedling S, Stormo G (1984) in *Gene Expression: The Translational Step and Its Control* (Clark BFC, Petersen HU, eds.). Munksgaard, Copenhagen, p. 379.
220. Schmidt F, Apirion D (1983) in *Bacteriophage T4* (Mathews CK, Kutter EM, Mosig G, Berget PB, eds.). Am Soc Microbiol, Washington DC, p. 208.
221. Chu FK, Maley GF, Maley F, Belfort M (1984) *PNAS 81*:3049; Chu RK, Maley GF, West DK, Belfort M, Maley F (1986) *Cell 45*:157; Chu FK, Maley F, Martinez J, Maley GF (1987) *J Bact 169*:4368; Belfort M, Pedersin-Lane J, West D, Ehrenman K, Maley G, Chu F, Maley F (1985) *Cell 41*:375.

tase)[222] genes can self-excise; the mRNA tertiary structures presumably provide the information needed for the cleavages. Gene *60*, which encodes a subunit of T4 topoisomerase, contains an intron that is not excised but simply skipped during translation;[223] ribosomes may ignore this sequence by its folding into a stable hairpin. These introns in phage T4, easily manipulated genetically and biochemically, may contribute to the understanding of RNA processing and its relation to translational regulation.

Replication Genes[224] and Proteins

Mutations in the clustered T4 replication genes (Fig. 17-27) are manifested by one of the following phenotypes: complete absence of DNA synthesis, arrest of DNA synthesis, delayed onset of DNA synthesis, or defective DNA maturation (Table 17-9). The numerous phage-encoded replication proteins are involved in nucleotide biosynthesis, DNA synthesis, and DNA modification. Because replication beyond the early stages depends on recombination, mutants deficient in recombination have a "DNA-arrest" phenotype which can be suppressed if the early initiation mode is allowed to continue.

Nucleotide Biosynthesis Enzymes.[225] To ensure an abundant supply of deoxynucleotides beyond those furnished by the host, the phage encodes key biosynthetic enzymes, thereby amplifying the level of dNTPs many-fold. The phage-encoded enzymes (and their genes) include thioredoxin (*nrdC*), nucleoside diphosphate reductase (*nrdA* and *nrdB*), dihydrofolate reductase (*frd*), dCMP deaminase (*cd*), thymidylate synthase (*td*), and deoxynucleotide kinase (gene *1*). The host biosynthetic enzymes, (deoxy)adenylate kinase and nucleoside diphosphate kinase, normally present at high levels, are relied upon completely.

To exclude cytosine from progeny DNA and substitute glu-HMC, phage-encoded enzymes hydrolyze dCTP and dCDP to dCMP (gp56), hydroxymethylate the dCMP (gp42), and phosphorylate the hydroxymethylated dCMP (gp1) (Section 2-13).

The phage-encoded enzymes for nucleotide biosynthesis form an organelle-like complex which may include key host enzymes.[226] The dNTPs synthesized by the complex approximate the ratios present in T4 DNA, which has a GC content of only 33% compared to 50% in host DNA. Whether this complex funnels the precursor nucleotides directly into DNA remains uncertain (Section 2-13).[227]

In permeabilized T4-infected cells, added deoxynucleoside monophosphates are incorporated into DNA more efficiently than are the triphosphates, the true substrates,[228] arguing for compartmentalization of DNA precursors.

---

222. Sjöberg B-M, Hahne S, Mathews CZ, Mathews CK, Rand DN, Gait MJ (1986) *EMBO J 5*:2031; Gott J, Shub D, Belfort M (1986) *Cell 47*:81.
223. Huang WM, Ao S-Z, Casjens S, Orlandi R, Zeikus R, Weiss R, Winge D, Fang M (1988) *Science 239*:1005.
224. Mosig G (1987) *ARG 21*:347; Mosig G, Eiserling F (1988) in *The Bacteriophages* (Calendar R, ed.). Plenum Press, New York, Vol. 2, p. 521.
225. Mathews CK, Allen JR (1983) in *Bacteriophage T4* (Mathews CK, Kutter EM, Mosig G, Berget PG, eds.). Am Soc Microbiol, Washington DC, p. 59.
226. Mathews CK, Moen LK, Wang Y, Sargent RG (1988) *TIBS 13*:394.
227. Flanegan JB, Greenberg GR (1977) *JBC 252*:3019; Flanegan JB, Chiu CS, Greenberg GR (1977) *JBC 252*:6031.
228. Reddy GPV, Mathews CK (1978) *JBC 253*:3461; Mathews CK, Allen JR (1983) in *Bacteriophage T4* (Mathews CK, Kutter EM, Mosig G, Berget PG, eds.). Am Soc Microbiol, Washington DC, p. 59.

Table 17-9
T4 genes involved in DNA replication

Phenotype of mutant	Gene	Function[a]
No DNA synthesis	*1*	deoxynucleotide kinase
	32	SSB
	41	helicase; stimulates primase (gp61)
	42	dCMP hydroxymethyl transferase
	43	DNA polymerase
	44 *45* *62* }	polymerase accessory protein
	56	dCTPase, dUTPase, dCDPase, dUDPase
Arrested DNA synthesis	*30*	DNA ligase
	46 *47* }	exonuclease; recombination
	uvsX	RecA-like; recombination
	uvsY	accessory protein to UvsX
	uvsW	affects recombination
	59	links replication and recombination
Delayed DNA synthesis	*39* *52* *60* }	DNA topoisomerase; affects initiation
	61	primase; stimulates helicase (gp41)
Maturation defective	*16* *17* }	head filling
	33	binds late form of RNA P
	49	endonuclease VII; resolves Holliday structures
	αgt	HMC α-glucosyl transferase
	βgt	HMC β-glucosyl transferase
	55	σ factor of late form of RNAP
Nucleotide synthesis defective	*nrdA* *nrdB* }	nucleoside diphosphate reductase
	nrdC	thioredoxin
	td	thymidylate synthase
	frd	dihydrofolate reductase
	denA	endonuclease II; degrades host DNA
	denB	endonuclease IV; degrades host DNA
	cd	dCMP deaminase
Nonessential: suggestive role in replication	*dda*	helicase
	dexA	exonuclease A; host DNA breakdown
	dam	DNA adenine methyltransferase
	tk	thymidine kinase
	denV	endonuclease V; class I N-glycosylase that excises bulky lesions
	psetT	polynucleotide kinase
	das	suppressor of genes *46*,*47*
	ndd	nucleoid disruption

[a] HMC = hydroxymethylcytosine; RNAP = RNA polymerase; SSB = single-strand binding protein.

DNA Synthesis Enzymes.[229] The protein products of nearly all of the "no-DNA-synthesis" and "DNA-delay" mutants have been identified, purified to homogeneity, and studied in vitro. Seven have a direct role in replication as judged by their participation in DNA synthesis in vitro, including DNA polymerase (gp43) and the accessory proteins gp44, gp62, and gp45, which impart high processivity to the polymerase. (Together these four proteins constitute the T4 DNA polymerase holoenzyme.) Two proteins, gp41 and gp61, together form a helicase–primase complex. Gp32, an SSB (or helix-destabilizing protein), functions in distinctive ways in several reactions and cannot be replaced by *E. coli* SSB.

Additional proteins involved in replication, for which an essential function in vitro has not been proved, include gp39, gp52, and gp60, the subunits of T4 topoisomerase (Section 12-7). This type II enzyme requires ATP and appears to play some role in initiation of replication. As infection proceeds, initiation of replication becomes dependent on recombination (see below): mutations in the genes for the recombination enzymes cause a defect in replication. These T4 recombination proteins include the *uvsX* gene product, similar in activity to the *E. coli* RecA protein (promoting strand exchange and branch migration between homologous DNA molecules; Section 21-5), and the products of genes *uvsY*, *uvsW*, *59*, *46*, and *47*.

DNA Modification Enzymes.[230] Glycosylation of the HMC residues by α- and β-glucosyl transferases (gpαgt and gpβgt) produces a pattern distinctive for each of the T-even phages. Enzymologic analysis of how these unique glycosylation patterns are produced led to the first demonstration of macromolecular modification after polymerization. T4 and T2 also induce a methyl transferase that methylates N6 of adenine in GATC sequences. Unlike the host Dam methyltransferase (Section 21-10), the phage enzyme is not inhibited by the glu-HMC modification.

Nucleases.[231] Of the numerous phage-encoded nucleases, only a few have been assigned meaningful roles. Endo II (DenA) is a nicking enzyme; endo IV (DenB) is a single-strand–specific enzyme. By their initial recognition of cytosine residues, they exercise a specificity for host DNA. (This capacity of endo IV was exploited in early advances in DNA sequencing.) The host DNA initially digested by these enzymes is further degraded by the gp46/47 exonuclease to provide nucleotides for phage DNA synthesis.[232] The gp46/47 exonuclease also has an essential role in recombination. Endo V (DenV) is the key incision enzyme in repair of UV lesions (Sections 13-5 and 21-2). Endo VII (gp49) efficiently cleaves branched DNA at Holliday junctions (Section 21-5) and acts on a rapidly sedimenting form of T4 DNA, processing it to a substrate for the packaging machinery.[233] Several other endo- and exonucleases have been identified, but their distinctive features and functions are still unknown.

229. Nossal NG, Alberts BM (1983) in *Bacteriophage T4* (Mathews CK, Kutter EM, Mosig G, Berget PG, eds.). Am Soc Microbiol, Washington DC, p. 71.
230. Hattman S (1983) in *Bacteriophage T4* (Mathews CK, Kutter EM, Mosig G, Berget PG, eds.). Am Soc Microbiol, Washington DC, p. 152; Revel HR (1983) in *Bacteriophage T4* (Mathews CK, Kuter EM, Mosig G, Berget PG, eds.). Am Soc Microbiol, Washington DC, p. 156.
231. Warner HR, Snustad DP (1983) in *Bacteriophage T4* (Mathews CK, Kutter EM, Mosig G, Berget PG, eds.). Am Soc Microbiol, Washington DC, p. 103.
232. Carlson K, Øvervatn A (1986) *Genetics 114*:669; Carlson K, Wiberg JS (1983) *J Virol 48*:18.
233. Minigawa T, Ryo Y (1978) *Virology 91*:2220.

Initiation of Replication[234]

Three independent and possibly redundant mechanisms for initiating DNA replication have complicated the genetic and molecular analysis of this key event. The mechanisms were named primary, secondary, and tertiary modes in order of their discovery, but these designations can be misleading, inasmuch as both primary and tertiary initiations occur before the onset of secondary initiations. The two early replication modes, primary and tertiary, are distinguished by their different requirements for RNAP and several phage-encoded enzymes.

Primary Initiation. This mode appears to require the recognition of T4 early gene promoters (*E. coli* promoters on the T4 DNA) by the unmodified host RNAP, as initiation is sensitive to rifampicin and is shut off when polymerase modifications (see above) make the enzyme unable to recognize early promoters. T4 topoisomerase is also implicated in this pathway.

Attempts to identify the precise origins by recombinant DNA methods have failed. In vivo data suggest that several potential primary origins are exploited under different growth conditions. The best characterized of the putative primary origins is in gene *69* and is called *oriA*.[235]

Tertiary Initiation. This mode is distinguished from the primary by its resistance to rifampicin, and from the secondary by its independence from the gp46/47 recombination exonuclease. Two tertiary origins which support replication after T4 infection have been cloned and extensively characterized.[236] Replication of plasmids bearing these sequences generates long concatemers of glu-HMC DNA which can be packaged into T4 heads. In contrast to recombination-dependent plasmid replication, the replication of these origin plasmids does not require the *uvsX*, *uvsY*, and *46/47* recombination genes nor DNA sequence homology between the infecting phage and the plasmid.

The two tertiary origins map to genes *23* and *uvsY*. These regions represent two of the three major recombination "hot spots" on the T4 genome. Since replication from these origins does not require recombination proteins, the stimulation of recombination is thought to be the result rather than the cause of initiation.

Both tertiary origins contain a *motA*-dependent T4 middle-gene promoter and an additional 50-bp sequence downstream that is not required for transcription. The MotA protein, thought to be a sequence-specific DNA-binding protein, positively activates the initiation of transcription on promoters that contain its recognition sequence (a MotA box). The presence of a promoter in a tertiary origin is surprising because transcription of T4 middle-gene promoters is sensitive to rifampicin, while initiation at these origins is rifampicin-resistant. Perhaps the MotA protein or MotA and RNAP act in initiation by binding the promoter rather than by transcription. In this way, MotA protein may function analogously to transcription factors in the initiation of eukaryotic viral DNA replication (Section 16-5).

---

234. Kozinski AW (1983) in *Bacteriophage T4* (Mathews CK, Kutter EM, Mosig G, Berget PG, eds.). Am Soc Microbiol, Washington DC, p. 111; Mosig G (1987) *ARG 21*:347; Kreuzer KN, Alberts BM (1986) *JMB 188*:185.
235. Macdonald PM, Mosig G (1984) *EMBO J 3*:2863.
236. Kreuzer KN, Engman HW, Yap WY (1988) *JBC 263*:11348; Menkens AE, Kreuzer KN (1988) *JBC 263*:11358.

Secondary Initiation. This mode, required for replication in the latter stages of T4 development, is independent of RNAP and dependent on the T4 recombination proteins: gp46/47 exonuclease, UvsX recombinase, gp32 (SSB), gp59, and UvsY protein. Secondary replication is also facilitated by the T4 topoisomerase, which is not essential for recombination. After the redirection of RNAP to late promoters, recombination becomes the major initiation pathway. In mutants which fail to modify RNAP, the DNA-arrest phenotype of the recombination-deficient mutants is not observed because the other modes of initiation continue.

Plasmids carrying essentially any T4 sequence replicate in a cell infected with T4 phage by a mechanism resembling secondary initiation.[237] Replication of the plasmid depends on genes required for secondary initiation and on a region of DNA sequence homology between the plasmid and the infecting phage.

The incompletely replicated 3′ ends of the linear T4 genome may initiate recombination by invading duplex molecules and, in the process, prime DNA synthesis. Elongation of these invasive 3′ ends provides a mechanism for completing them, an operation required during the replication of linear duplex molecules. Because the T4 genome is circularly permuted, a sequence at the end of one genome is also found in many locations on others, allowing recombination to prime DNA synthesis in many places when many phage genomes are present in the same cell. The actions of T4 recombination enzymes in DNA synthesis in vitro are in keeping with the interrelationship of these processes in vivo (see below).

Replication with Purified Enzymes[238]

Interactions among Enzymes. A multienzyme system of highly purified T4 proteins (Table 17-10) sustains rapid, extensive, high-fidelity replication. The polymerase and 3′→5′ proofreading exonuclease activities of gp43 are aided by three accessory proteins, gp44, gp62, and gp45. Gp45 stimulates the (d)ATPase

Table 17-10
T4 proteins required for DNA replication fork movement in vitro

Protein	Mass (kDa)	Function in vivo	Activity in vitro
gp41	53.8	priming, fork movement	DNA helicase; ATPase
gp61	39.5	priming	primase
gp32	34.5	replication, recombination	ssDNA binding
gp43	103.5	DNA synthesis	polymerase; 3′→5′ exonuclease
gp44	35.9	accessory to polymerase	ATPase; completion of polymerase holoenzyme
gp62	21.4		
gp45	24.7		

237. Kreuzer KN, Alberts BM (1986) *JMB 188*:185; Kreuzer KN, Yap WY, Menkens AE, Engman HW (1988) *JBC 263*:11366.
238. Richardson RW, Nossal NG (1989) *JBC 264*:4725; Richardson RW, Nossal NG (1989) *JBC 264*:4732; Bedinger P, Munn M, Alberts BM (1989) *JBC 264*:16880; Cha TA, Alberts BM (1986) *JBC 261*:7001; Cha TA, Alberts BM (1989) *JBC 264*:12220; Alberts BM (1987) *Philos Trans R S Lond [Biol] 317*:395; Nossal NG, Hinton DM (1987) *JBC 262*:10879.

activity of the 44/62 protein complex and together the three proteins hydrolyze ATP, increasing the rate and processivity of the polymerase. The four-protein complex (holoenzyme) synthesizes long chains at 600 ntd per second, nearly the rate attained in vivo.

Replication of duplex DNA requires additional proteins: gp32, gp41, and gp61. Gp32, the T4 SSB, stimulates polymerase action on ssDNA by opening short stretches of secondary structure. Whereas the replication of duplex DNA in vitro by the DNA polymerase holoenzyme requires high levels of gp32, physiologic levels suffice if coupled with catalytic unwinding by the 5′→3′ helicase activity of gp41 (Section 11-2).

Gp41 (Section 11-2) and gp61 (Section 8-5) together unwind the template and prime DNA synthesis on the discontinuous strand. The helicase activity of gp41 is stimulated by gp61. More remarkable is the effect of gp41 on the priming activity of gp61. T4 primers are pentamers: 5′-pppApCpNpNpN-3′. The 41/61 complex recognizes the template sequence 3′-TTG-5′, in which the first T is not used as a template. The sequence 3′-TCG-5′ is also used in DNA containing dCMP rather than glu-HMC. At high levels, gp61 alone synthesizes some primers, but only at TCG, a sequence absent from normal T4 DNA. Synthesis by gp61 alone results mainly in dimers, whereas the polymerase requires at least a tetramer as a primer. Thus, gp61, despite possessing the active site for polymerizing rNTPs, must rely on the complex with gp41 to prime T4 discontinuous-strand initiation correctly.

Several interactions among the replication proteins suggest their organization in the operations at a replication fork. Beyond the association of the polymerase with accessory proteins, and the helicase with primase, are interactions between these two complexes. The polymerase accessory proteins stimulate the helicase activity of gp41 and allow the 41/61 complex to function on gp32-coated DNA. Gp41 with an altered C terminus interacts normally with gp61 but fails to stimulate DNA polymerase, thus identifying a domain involved in coupling the T4 primosome and polymerase. Complexes of gp32 with both gp43 and gp45 may account for some of its effects on T4 replication. Finally, the ability of actively replicating molecules to withstand dilution, with no change in the size of Okazaki fragments or in the rate of fork movement, suggests that the polymerases involved in continuous- and discontinuous-strand synthesis are physically linked,[239] allowing the DNA polymerase synthesizing the discontinous strand to recycle repeatedly. The T4 replication forks with their assembled proteins may reveal some of the fundamental architectural features of replicating complexes.

The replication complex also interacts with the transcription apparatus. Beyond the activating contributions of transcription and transcription factors in primary and tertiary initiations, transcription can also inhibit replication. In vitro, the inhibition is greatest when the transcription complex and a replication fork collide head on. The *dda* gene product, a 5′ → 3′ DNA helicase (Section 11-3), can overcome this inhibition by allowing the replication complex to bypass RNA polymerase.[240] The helicase can also permit the replication fork to move through lactose repressor molecules bound to a template. Late transcrip-

---

239. Selick HE, Barry J, Cha T-A, Munn M, Nakanishi M, Wong ML, Alberts BM (1987) in *DNA Replication and Recombination* (McMacken R, Kelly TJ, eds.). UCLA, Vol. 47. AR Liss, New York, p. 183.
240. Bedinger P, Hochstrasser M, Jongeneel CV, Alberts BM (1983) *Cell 34*:115.

tion requires concurrent replication of the template in vivo, and studies in vitro point to ATP hydrolysis by the combined actions of gp44/62 and gp45 at an active replication fork as the signal that replication is in progress.[241]

Recombination-Dependent Replication in Vitro.[242] Genetic recombination and DNA replication are intertwined late in the T4 life cycle. The dependence of replication on T4 recombination proteins is also demonstrated in vitro. Both duplex and single-stranded linear DNAs with sequence homology are required. The UvsX protein, the T4 analog of *E. coli* RecA protein (Section 21-5), binds cooperatively to the single strand and catalyzes pairing and strand exchange between the ssDNA and the complementary strand of the duplex, creating a D-loop. The UvsX activity is assisted by UvsY protein. T4 DNA polymerase holoenzyme recognizes and elongates the 3′ end of the linear ssDNA. The gp41 helicase or gp41/61 primosome is loaded on the ssDNA within these recombination intermediates through the action of gp59, generating complete replication forks.[243] Gp59 (26 kDa) binds both ssDNA and gp32, and appears to interact specifically with gp41. Loading of the T4 primosome onto DNA is probably assisted by gp59 under some circumstances; however, this action is essential in recombination-dependent replication, probably because the ssDNA at the fork is coated with gp32 and UvsX protein, which prohibit the association of gp41.

DNA Maturation[244]

Recombination-dependent replication most likely generates the large networks of T4 DNA observed late in infection. Some evidence for circular forms in T4 replication also exists, compatible with the development of long concatemers, as occurs in phage λ replication (Section 7). Still, a T7-like mechanism of concatemer formation seems more likely. Unreplicated single-stranded 3′ tails can invade homologous duplexes by the action of UvsX protein and then start a replication fork, perhaps assisted by the prior trimming and tailoring actions of the gp46/47 exonuclease (Table 17-9).

Endo VII (gp49) probably functions to release DNA from the multimeric complexes in a form suitable for packaging. Endo VII cleaves Holliday structures created during recombination (Section 21-5), and gene *49* mutants are defective in DNA packaging. A 171-kb linear duplex is packaged into phage heads. Packaging must generate molecules slightly larger than one genome in length to yield the terminally redundant, circularly permuted chromosomes. A double-stranded scission (Fig. 17-28) that provides a "headful"-sized piece does not require a sequence-specific cleaving activity, as is necessary for T7 or λ, but rather a mechanism that measures the desired length. Genes *16* and *17*[245] encode the terminase protein, a multimeric structure that hydrolyzes ATP and acts continually throughout packaging; T4 ligase and topoisomerase are also involved.

---

241. Herendeen DR, Kassavetis GA, Barry J, Alberts BM, Geiduschek EP (1989) Science 245:952.
242. Formosa T, Alberts BM (1986) *Cell* 47:793.
243. Barry J, Alberts BM, personal communication.
244. Mosig G, Eiserling F (1988) in *The Bacteriophages* (Calendar R, ed.). Plenum Press, New York, Vol. 2, p. 521.
245. Berget PB, King J (1983) in *Bacteriophage T4* (Mathews CK, Kutter EM, Mosig G, Berget PG, eds.). Am Soc Microbiol, Washington DC, p. 246.

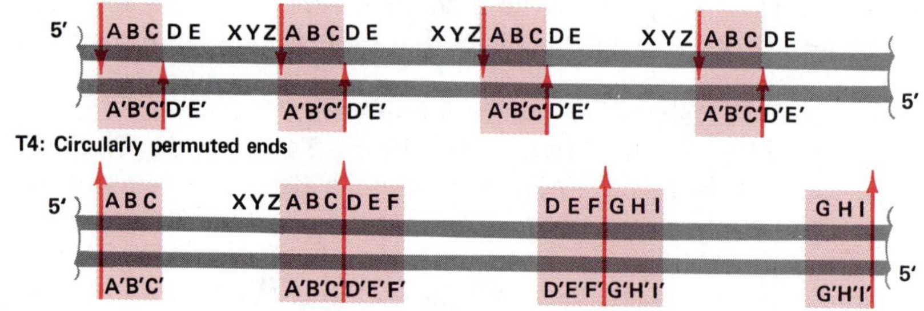

Figure 17-28
Scheme for producing redundant, uniform ends by staggered double-strand scissions of a concatemer at specific sites (T7) and redundant, circularly permuted ends by directly opposite double-strand scissions after a measured length of DNA (T4). Shaded areas indicate redundant regions.

17-7 Temperate Phages:[246] λ, P22, P2, P4, P1, Mu

The temperature phages (Table 17-11) choose between two alternative life-styles. They can grow *lytically*, much like the T phages, or convert *E. coli* to a *lysogenic* state, in which the majority of the phage functions are repressed and the phage genome remains quiescent as a *prophage* for many generations. The prophage genome is either integrated into the host chromosome (λ and Mu) or maintained as a plasmid (P1). An infection with λ seems poised, at several stages in development, either to pursue the *virulent* (lytic) course of λ DNA multiplication, leading to cell lysis, or to insert λ DNA into the host chromosome, repressing further development and autonomous replication. Phage λ is representative of a number of related temperate phages of *E. coli* (21, φ80, 186, and 434).

Several important features associated with temperate phages are related to their integration into the host chromosome: (1) they may provide an alternative origin of replication, integratively suppressing mutations in the host origin; (2) occasionally they package host DNA upon excision, and subsequently transfer these genes to another host (*specialized transduction*); and (3) they may cause mutations by integrating within a host gene (especially phage Mu).

Table 17-11
Representative temperate *E. coli* phages

Phage	DNA (kb)	Comments
λ	48	best studied temperate coliphage; specialized transduction
P22	43	lysogenic in *Salmonella;* generalized transduction vector
P2	33	transactivation with P4
P4	11.5	satellite of P2
P1	88	generalized transduction; maintained in plasmid state; low copy number
Mu	38	multiple integration sites; replicates via transposition; invertible sequence

246. Hendrix RW, Roberts JW, Stahl FW, Weisberg RA (eds.) (1983) *Lambda II.* CSHL; Hershey AD (ed.) (1971) *The Bacteriophage Lambda.* CSHL; Calendar R (ed.) (1988) *The Bacteriophages.* Plenum Press, New York, Vols. 1 & 2.

λ Phage[247]

Phage λ, of medium size, possesses an icosahedral head and a long tail terminating in a single fiber. λ DNA is 48 kb with a 12-ntd single-stranded terminus at each 5′ end; these termini are complementary and therefore cohesive (Fig. 17-29).[248] Phages λ, φ80, φ82, 21, and 434 have identical cohesive termini which differ[249] from the 19-ntd ends of phages 186, P2, and 299.

The genetic map[250] (in the prophage form, Figure 17-30) includes about 50 cistrons determined by recombination frequencies and deletion mapping. It is colinear with a physical map prepared by heteroduplex analysis in the electron microscope. The entire DNA sequence has been determined,[251] revealing 12 additional open reading frames that presumably encode proteins not previously identified genetically.[252] Related functions are clustered: head and tail assembly, integration–excision, generalized recombination, regulation, replication, and lysis.

Infection

Adsorption and Penetration. The fiber protruding from the tail, responsible for host-range specificity, consists of three molecules of gpJ.[253] The fate of gpJ after adsorption and its possible function as a pilot protein for the entering DNA have not been explored. Located at the tip of the tail assembly, gpH* may be injected into the cell with the DNA, suggesting it as another candidate for a pilot-protein function.[254] The λ receptor on *E. coli* (product of the *lamB* gene) is a membrane

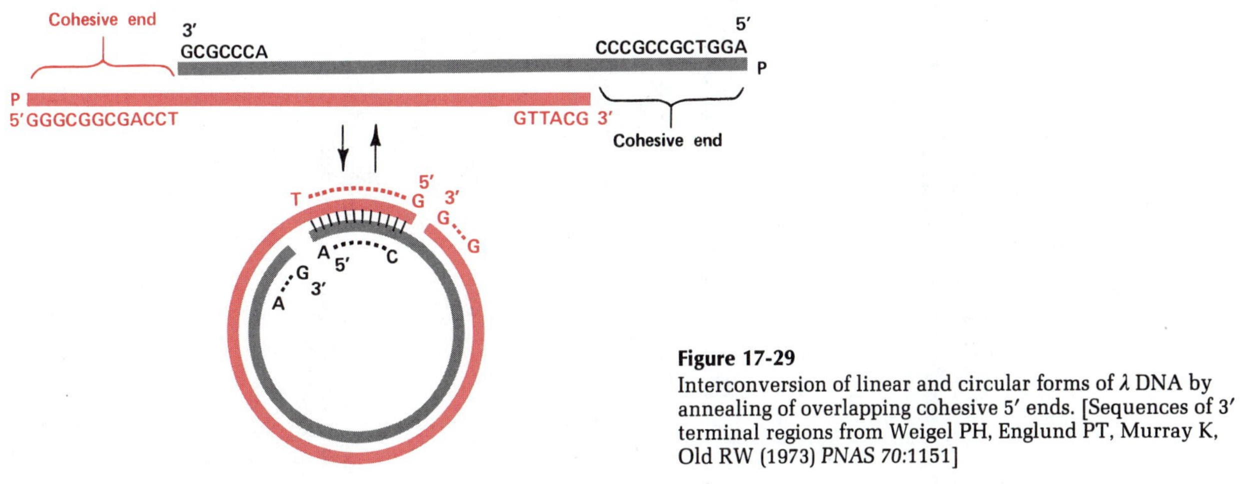

Figure 17-29
Interconversion of linear and circular forms of λ DNA by annealing of overlapping cohesive 5′ ends. [Sequences of 3′ terminal regions from Weigel PH, Englund PT, Murray K, Old RW (1973) *PNAS 70*:1151]

247. Hendrix RW, Roberts JW, Stahl FW, Weisberg RA (eds.) (1983) *Lambda II*. CSHL.
248. Wu R, Taylor E (1971) *JMB 57*:491.
249. Murray K, Murray NE (1973) *NNB 243*:134.
250. Echols H, Murialdo H (1978) *Microbiol Rev 42*:577.
251. Daniels DL, Schroeder JL, Szybalski W, Sanger F, Coulson AR, Hong GF, Hill DF, Petersen GB, Blattner FR (1983) in *Lambda II* (Hendrix RW, Roberts JW, Stahl FW, Weisberg RA, eds.). CSHL, p. 519.
252. Daniels DL, Schroeder JL, Szybalski W, Sanger F, Blattner FR (1983) in *Lambda II* (Hendrix RW, Roberts JW, Stahl FW, Weisberg RA, eds.). CSHL, p. 469.
253. Ruerst CR, Bingham H (1978) *Virology 87*:437.
254. Roa M (1981) *FEMS Microbiol Lett 11*:257.

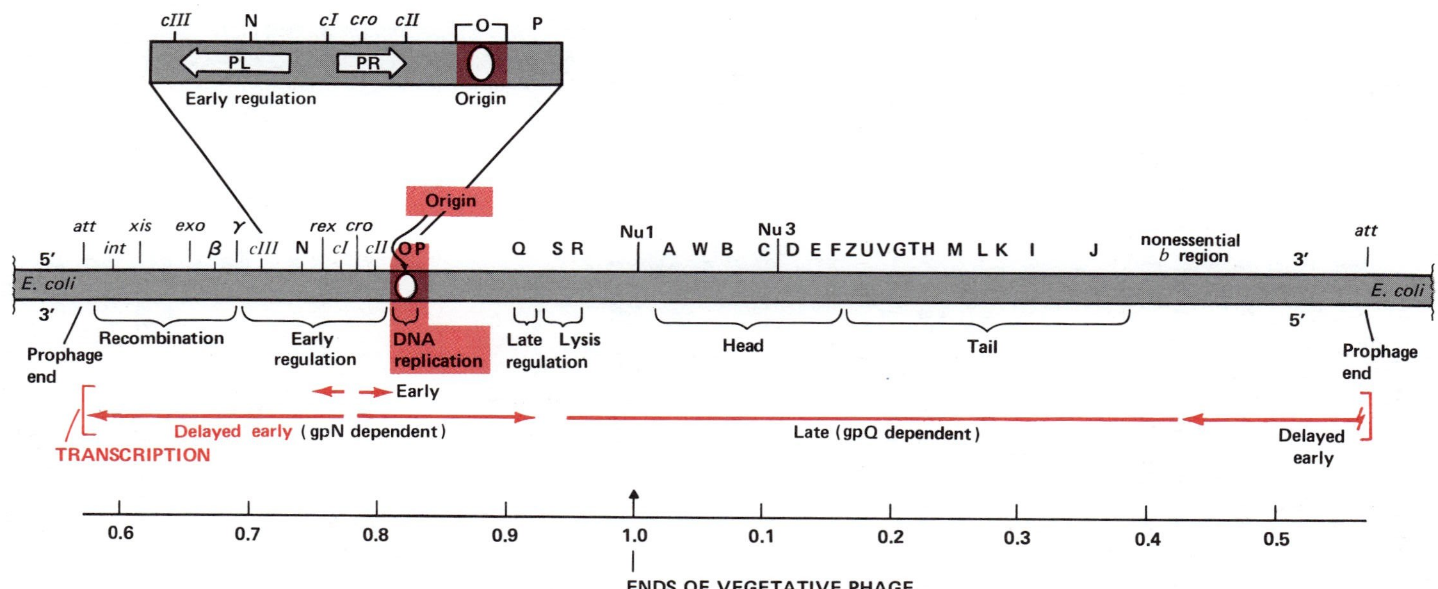

Figure 17-30
Genetic map of prophage form of phage λ. Linear vegetative map reads from *A* rightward to *R*; prophage map is presented because it simplifies the representation of transcription. [From Szybalski W (1971) in *The Bacteriophage Lambda* (Hershey AD, ed.). CSHL, p. 779; Berg D (1974) *JMB 86*:59]

protein also employed in maltose transport. DNA injection requires the participation of the host *pel* gene (penetration of lambda), which is allelic with *ptsM*, one of the genes involved in the phosphotransferase system for sugar uptake.[255]

After the DNA enters the cell, cohesion of the complementary 5′ ssDNA ends converts the linear duplex to a circular form (Fig. 17-29). Hydrogen-bonding of the 12-ntd stretches is completed within a minute in vivo, catalyzed by an unknown mechanism. Covalent closure of the opposed ends by DNA ligase and negative supercoiling by gyrase prepare the DNA for transcription, replication, recombination, or integration.

Lysis or Lysogeny.[256] The choice between lytic infection, producing several hundred mature, infectious particles, and lysogenic integration into the host chromosome is determined by numerous host and phage factors. These factors add up to an intricately balanced system for regulating gene expression.

Gene expression in λ depends entirely on initiation by the host RNAP from promoters having the standard sequence recognized by σ^{70} (Section 7-4). λ achieves transcriptional control of its development by utilizing a combination of repressors, activators,[257] and antiterminators.[258] The use of antitermination catalyzed by gpN and gpQ is especially impressive (Fig. 17-30). Transcription from the same promoters can synthesize (1) only the early regulatory proteins, when both gpN and gpQ are absent, (2) the replication and recombination proteins, as well as the regulatory proteins, when gpN is present, or (3) essentially the entire genome when gpQ is also present.

Two proteins, cI and Int, directly establish the lysogenic state. The cI protein (also called the λ repressor) shuts down the expression of almost all genes, and the *int* gene product catalyzes the insertion of the λ genome into the host chromosome. Repression of P_R (a promoter for developmental genes) by cI protein limits the production of genes required for further development, including the amount of the antiterminator, gpQ. The level of cII protein is most critical in the decision between lysogeny and lytic growth. An activator of transcription, cII participates in the decision by stimulating expression of the *cI* and *int* genes, thereby favoring lysogenic growth. The abundance of cII, an unstable protein, is finely tuned to the physiologic state of the host cell via interactions with other phage and host products.[259]

The integration of supercoiled λ DNA takes place when expression of the *int* gene has generated an adequate supply of Int protein. Once λ is integrated as a prophage, a low level of cI repressor is adequate to repress all lytic functions and maintain the lysogenic state; this level is regulated by cI repressor itself, which stimulates its own synthesis when present at low levels and inhibits it when abundant. Bacteria carrying a λ prophage are also immune to further infection,

---

255. Elliott J, Arber W (1978) *MGG 161*:1.
256. Echols H (1986) *Trends Genet 168*:1; Herskowitz I, Hagen D (1980) *ARG 14*:399; Wulff D, Rosenberg M (1983) in *Lambda II* (Hendrix RW, Roberts JW, Stahl FW, Weisberg RA, eds.). CSHL, p. 53.
257. Ptashne M (1986) *A Genetic Switch: Gene Control and Phage λ.* Cell Press (Blackwell), Cambridge, Mass.
258. Friedman DI (1988) in *The Bacteriophages* (Calendar R, ed.). Plenum Press, New York, Vol. 2, p. 263; Friedman DI, Gottesman M (1983) in *Lambda II* (Hendrix RW, Roberts JW, Stahl FW, Weisberg RA, eds.). CSHL, p. 21.
259. Hoyt MA, Knight DM, Das A, Miller HI, Echols H (1982) *Cell 31*:565; Banuett F, Hoyt MA, McFarlane L, Echols H, Herskowitz I (1986) *JMB 187*:213.

due to the constant intracellular presence of cI repressor. Molecular details of integration and excision of λ are considered in the discussion of site-specific recombination (Section 21-6).

The λ prophage is replicated regularly as part of the host DNA. Following integration, immediate shutdown of prophage-specific DNA synthesis is essential to the stabilization of the lysogenic state because such replication would be deleterious to the host cell. The lysogenic state persists until agents that cause DNA damage, or otherwise inhibit host DNA synthesis, trigger prophage induction by causing RecA-mediated cleavage of the cI repressor[260] (Section 21-3). UV irradiation, thymine starvation, or mitomycin cross-linking are examples of agents which activate the host RecA protein. Activated RecA protein stimulates the autocatalytic cleavage and destruction of the host LexA protein, a repressor of *E. coli* repair and mutagenesis genes (Section 21-3). The λ cI repressor resembles the LexA protein in its capacity for self-cleavage, stimulated by the activated RecA protein, and thus cI keys λ into the cellular SOS response to DNA damage.

Induction of the lytic cycle depends on the relative levels of the two repressors cI and Cro (Table 17-12). Both proteins bind to the same three operators, O_{R1}, O_{R2}, and O_{R3} [which regulate P_R and P_{RM} (the cI promoter)], but they bind in a different order, generating a switch. The cI repressor has the highest affinity for O_{R1} and lowest for O_{R3}; a high cI to Cro ratio inhibits P_R, shutting down lytic development. In contrast, Cro binds first to O_{R3}, inhibiting repressor synthesis from P_{RM}, while allowing production of the lytic genes. Lytic gene expression leads to excision of the prophage as a covalently closed circle; phage replication enables the virus to escape the endangered cell.

λdv: Defective Virus in a Plasmid State.[261] λdv plasmids are supercoiled circles produced by deletion of most of the λ genome. The minimal requirements for

Table 17-12
Relationship between repressor level and transcription pattern in the P_{RM} and P_R regulatory region

Repressor	Binding occupancy	Transcription pattern	
		cro gene	*cI* gene
cI	+ + +	off	off
	+ +	off	stimulated
	cI ← O_{R3} O_{R2} O_{R1} → Cro; P_{RM} P_R		
Cro	+	on	off
	+ +	off	off
	+ + +	off	off
None		on	basal

260. Roberts JW, Devoret R (1983) in *Lambda II* (Hendrix RW, Roberts JW, Stahl FW, Weisberg RA, eds.). CSHL, p. 123.
261. Matsubara K (1976) *JMB 102*:427.

λdv are (1) the presence of the P_R promoter and the *O* and *P* genes, (2) the absence of an active cI repressor, and (3) the presence of at least a partially active Cro protein to dampen transcription from P_R in order to prevent lethal overreplication. The plasmids, numbering about 50 per cell, range in size from about 5% to 25% of λ and are usually found as oligomers of this fundamental unit. Usually connected head to tail, these units in certain instances may instead be linked head to head as inverted repetitions. Whether the plasmids are generated by recombination or by an aberration of replication is still unknown.

Replication[262]

Phage λ replication, like that of the single-stranded phages, Ff and φX174, depends almost entirely on the host replication machinery. The phage encodes only two proteins directly involved in replication, the products of the *O* and *P* genes. Both proteins are probably involved exclusively in initiation, appropriating the host enzymes and redirecting them to λ DNA. In contrast to the T-even phages, λ relies solely on "pirating" proteins in order to reprogram host metabolism to that of the phage. The λ *N* and *cro* genes which encode transcription regulators are also required for normal λ replication, although their effect is indirect.[263]

Early and late stages of replication are characterized by differently structured intermediates[264] (see Fig. 17-32). Circular products predominate early after the virus initiates replication; late in a productive infection, rolling-circle intermediates predominate, giving rise to long concatemeric forms.

Early-Stage Replication. All the λ elements needed for early replication — the *O* and *P* genes, the origin sequence, and the P_R promoter — are localized in a 2500-bp region. Electron microscopy reveals that replication of the supercoiled λ circle starts at a point within the *O* gene, 81% from the left end of the linear genome[265] (Section 16-1). DNA synthesis is bidirectional about 75% of the time; the other molecules replicate unidirectionally to the left or right.

As defined by electron microscopy, by mutations, and by the cloning of the essential region in plasmids, the sequence of *ori*λ contains four nearly identical repeats of a 19-bp palindromic sequence, an approximately 40-bp AT-rich region just to the right, and, further to the right, a 28-bp palindrome (Fig. 17-31). In vitro, the minimal *ori*λ is an abbreviated 85-bp sequence, two of the 19-mers and the AT-rich region providing the essential elements.[266] The sequence contains oligo dA tracts that impart a stable bend to the DNA (Section 1-10), magnified by the binding by O protein.[267] This bendability may be important for opening the duplex at the origin during initiation.

---

262. Furth ME, Wickner SE (1983) in *Lambda II* (Hendrix RW, Roberts JW, Stahl FW, Weisberg RA, eds.). CSHL, p. 145.
263. Dove W, Inokuchi H, Stevens W (1971) in *The Bacteriophage Lambda* (Hershey AD, ed.). CSHL, p. 747; Matsubara K (1981) *Plasmid 5*:32.
264. Sogo JM, Greenstein M, Skalka A (1976) *JMB 103*:537; Reuben RC, Skalka A (1977) *J Virol 21*:673; Better M, Freifelder D (1983) *Virology 126*:168.
265. Schnös M, Inman R (1970) *JMB 51*:61.
266. Wickner S, McKenney K (1987) *JBC 262*:13163; Tsurimoto T, Matsubara K, (1982) *PNAS 79*:7369; Tsurimoto T, Kougara H, Matsubara K (1984) in *Plasmids in Bacteria* (Helinski D, Cohn S, Clewell D, eds.). Plenum Press, New York, p. 151.
267. Zahn K, Blattner FR (1987) *Science 236*:416; Zahn K, Blattner FR (1985) *Nature 317*:451; Zahn H, Blattner FR (1985) *EMBO J 4*:3605.

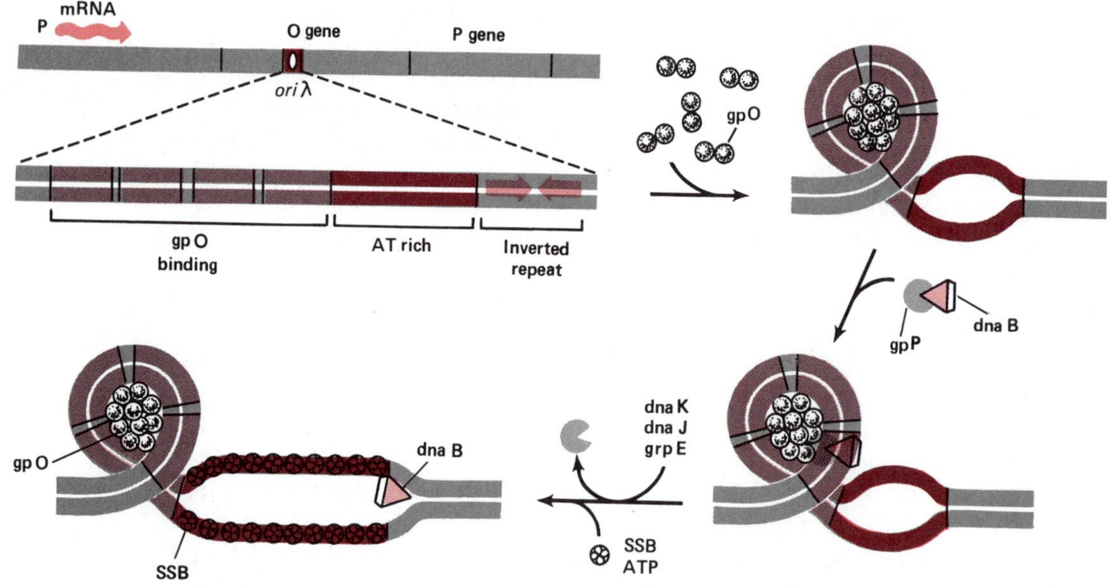

Figure 17-31
Diagrams of the *ori*λ region and the prepriming stages of λ replication. After dnaB protein initiates unwinding, primase and pol III holoenzyme initiate DNA synthesis. Replication is primarily unidirectional in vitro. The locations of *ori*λ within the *O* gene, the *O* and *P* genes, and the major rightward promoter (P) are shown.

The start of λ DNA replication[268] (Fig. 17-31) requires the participation of gpO and gpP, respective counterparts of the host dnaA and dnaC proteins (Section 16-3), which are dispensable for λ replication.[269] The replication machinery is that of the host: dnaB helicase, SSB, primase, gyrase, and pol III holoenzyme. In addition, host-encoded heat-shock proteins, products of the *dnaK*, *dnaJ*, and *grpE* genes, play an essential role (see below) in activating the proteins complexed at *ori*λ prior to the initiation of DNA synthesis.

Initiation of replication at *ori*λ has been reconstituted with purified proteins.[270] Both gpO and gpP have been purified to homogeneity,[271] although the instability of gpO, with a half-life of approximately 1.5 minutes in vivo,[272] hampers its isolation. GpO is a dimer of 34-kDa subunits that binds the 19-bp

268. Keppel F, Fayet O, Georgopoulos C (1988) in *The Bacteriophages.* (Calendar R, ed.). Plenum Press, New York, Vol. 2, p. 145.
269. Furth ME, Wickner SE (1983) in *Lambda II* (Hendrix RW, Roberts JW, Stahl FW, Weisberg RA, eds.). CSHL, p. 145.
270. McMacken R, Mensa-Wilmot K, Alfano C, Seaby R, Carroll K, Gomes B, Stephens K (1988) in *Cancer Cells* (Kelly T, Stillman B, eds.). CSHL, Vol. 6, p. 25; Mensa-Wilmot K, Seaby R, Alfano C, Wold MS, Gomes B, McMacken R (1989) *JBC 264*:2853; McMacken R, Alfano C, Gomes B, LeBowitz JH, Mensa-Wilmot K, Roberts JD, Wold M (1987) in *DNA Replication and Recombination* (McMacken R, Kelly TJ, eds.). UCLA, Vol. 47. AR Liss, New York, p. 227; Zylicz M, Ang D, Liberek K, Georgopoulos C (1989) *EMBO J 8*:1601.
271. Wickner SH, Zahn K (1986) *JBC 261*:7537; Gottesman S, Gottesman M, Shaw JE, Pearson ML (1981) *Cell 24*:225; McMacken R, Wold MS, LeBowitz JH, Roberts JD, Mallory JB, Wilkinson JAK, Loehrlein C (1983) in *Mechanisms of DNA Replication and Recombination* (Cozzarelli NR, ed.). ICN-UCLA Symposia on Molecular and Cellular Biology, Vol. 10. Academic Press, New York, p. 819; Roberts JD, McMacken R (1984) *NAR 11*:7435; LeBowitz JH, Zylicz M, Georgopoulos C, McMacken R (1985) *PNAS 82*:3988.
272. Wyatt WM, Inokuchi H (1974) *Virology 58*:313.

repeats in the origin.[273] Specificity for the origin sequence resides in gpO alone: O proteins of different lambdoid phages are specific for their cognate origins, but the P proteins are not.[274] The N terminus of gpO binds DNA and the C terminus interacts with the 26.5-kDa gpP, a dimer[275] that also binds dnaB protein.[276]

The sequence of initiation events at *oriλ* (Fig. 17-31) has been gleaned largely from studies in vitro. The DNA is bent around gpO in a complex, presumably one gpO dimer for each 19-mer,[277] and the AT-rich region becomes sensitive to nuclease attack, indicating melting of the DNA duplex.[278] This melted DNA bubble probably serves as the entry site for the replication complex. GpO directs the host replication complex to *oriλ* by interacting with the gpP·dnaB protein complex.[279] The strong binding of dnaB protein by gpP competes favorably with dnaC in the dnaB·dnaC complex[280] that functions in the analogous replication fork assembly reactions at *oriC* (Section 16-3). The dnaB protein in the dnaB·gpP complex is inactive.[281] Three heat-shock proteins (*dnaK*, *dnaJ*, and *grpE* gene products) dissociate the *oriλ*·O·P·dnaB complex to liberate dnaB protein.[282] The dnaB protein, as a helicase (Section 11-2), initiates unwinding of the DNA duplex, presumably at the site of strand opening by gpO. Chain initiations by primase (Section 8-3) and elongation by pol III holoenzyme then follow. The sequences of numerous RNA–DNA transition points, observed both in vivo and in vitro, are consistent with a scheme in which both strands are primed by a dnaB–primase primosome.[283]

Transcriptional Activation of Initiation. Initiation of replication at *oriλ* requires RNAP action directly.[284] Normally, transcription through *oriλ* from P_R is required. Mutants that bypass the requirement for transcription from P_R contain new promoters that reinstate transcription in the origin region. Located as far as 95 bp downstream of *oriλ*, these promoters restore replication even though RNA synthesis does not cross the origin, suggesting that transcription "activates" initiation rather than supplies primers for replication.[285]

Replication in vitro depends on transcription only when high levels of HU protein, an *E. coli* histone-like protein (Section 10-8), are present.[286] HU protein

273. Zylicz M, Gorska I, Taylor K, Georgopoulos C (1984) *MGG 196*:401; Dodson M, Roberts J, McMacken R, Echols J (1985) *PNAS 82*:4678; Zahn K, Blattner RFR (1985) *EMBO J 4*:3605; Tsurimoto T, Matsubara K (1981) *NAR 9*:1789.
274. Furth M, McLeester C, Dove W (1978) *JMB 126*:195; Furth M, Yates J (1978) *JMB 126*:227.
275. Zylicz M, Gorska I, Taylor K, Georgopoulos C (1984) *MGG 196*:401.
276. Wickner S (1978) *ARB 47*:1163; Klein A, Landa E, Schuster E (1980) *EJB 105*:1; Tsurimoto T, Matsubara K (1982) *PNAS 79*:7639; McMacken R, Wold NS, LeBowitz JH, Roberts JD, Mallory JB, Wilkinson JAK, Loehrlein C (1983) in *Mechanisms of DNA Replication and Recombination* (Cozzarelli NR, ed.). ICN-UCLA Symposia on Molecular and Cellular Biology, Vol. 10. Academic Press, New York, p. 819.
277. Zylicz M, Gorska I, Taylor K, Georgopoulos C (1984) *MGG 196*:401.
278. Schnös M, Zahn K, Inman RB, Blattner FR (1988) *Cell 52*:385.
279. Dodson M, Roberts J, McMacken R, Echols H (1985) *PNAS 82*:4678; Dodson M, McMacken R, Echols H (1989) *JBC 264*:10719; Wickner SH, Zahn K (1986) *JBC 261*:7537; Alfano C, McMacken R (1989) *JBC 264*:10699.
280. McMacken R, Wold NS, Le Bowitz JH, Roberts JD, Mallory JB, Wilkinson JAK, Loehrleim C (1983) in *Mechanisms of DNA Replication and Recombination* (Cozzarelli NR, ed.). ICN-UCLA Symposia on Molecular and Cellular Biology, Vol. 10. Academic Press, New York, p. 819.
281. Klein A, Lanka E, Schuster E (1980) *EJB 105*:1.
282. Dodson M, Echols H, Wickner S, Alfano C, Mensa-Wilmot K, Gomes B, LeBowitz J, Roberts JD, McMacken R (1986) *PNAS 83*:7638; Liberek K, Georgopoulos C, Zylicz M (1988) *PNAS 85*:6632; Alfano C, McMacken R (1989) *JBC 264*:10709; Dodson M, McMacken R, Echols H (1989) *JBC 264*:10719.
283. Yoda K, Yasuda J, Jiang X-W, Okazaki T (1988) *NAR 16*:6531; Tsurimoto T, Matsubara K (1984) *PNAS 81*:7402.
284. Dove WR, Inokuchi K, Stevens WF (1971) in *The Bacteriophage Lambda* (Hershey AD, ed.). CSHL p. 747; Hobom G, Grosschedl R, Lusky M, Scherer G, Schwartz E, Kossel H (1978) *CSHS 43*:165.
285. Furth ME, Dove WF, Meyer BJ (1982) *JMB 154*:65.
286. McMacken R, Mensa-Wilmot K, Alfano C, Seaby R, Carroll K, Gomes B, Stephens K (1988) in *Cancer Cells* (Kelly T, Stillman B, eds.). CSHL, Vol. 6, p. 25.

apparently blocks initiation by inhibiting *ori*λ·O·P complex formation and DNA duplex unwinding. An analogous situation is observed in replication from *oriC* in vitro (Section 16-3). The inhibitory effects of HU protein, or of other factors that inhibit melting of the origin DNA, such as topoisomerase I, are overcome by transcription near, but not necessarily over, the origin. Transcription counteracts the deleterious effects of these factors on the structure of the template, allowing melting of the DNA duplex and the subsequent protein assembly reactions to proceed.

The control of transcriptional activation plays an important role in λ replication, enabling replication to be shut off very quickly during lysogenization. Immediate inhibition of initiation upon integration is essential, as prophage replication is lethal to the host cell. Activation by transcription from P_R puts initiation directly under the influence of the cI repressor. Mutants in which *ori*λ is divorced from P_R control kill the cell upon integration, showing that other mechanisms that shut down replication, such as the swift degradation of gpO, are insufficiently rapid.[287]

Multiplication of λ circles, by way of theta-form (θ) intermediates (Section 16-1), continues for 5 to 15 minutes after infection (Fig. 17-32). Reinitiation at *ori*λ does not occur while a round of replication is in progress, perhaps because the topologically constrained state of the DNA required for initiation is not reestablished until late in the replication process.[288] This requirement for a negatively supercoiled origin region is relaxed after induction of the SOS response, allowing *ori*λ to refire before a round of replication is completed.[289] The monomeric circular products of early replication are not incorporated as such into virions. Late replication or recombination is essential to produce the multimers, which are substrates for packaging DNA into virions (see below).

The Heat-Shock Proteins in λ Initiation. Host proteins involved in λ replication were first uncovered by the isolation of *E. coli* mutants, called *gro*, which failed to support λ growth.[290] Phage mutants able to propagate on the defective host strains were then isolated to identify interactions between the host and λ functions. Several classes of *gro* mutants were suppressed by alterations in the λ *P* gene, placing their role in the initiation of DNA replication.[291] One class of *groP* mutations in the *dnaB* gene are accounted for by the interaction between dnaB protein and gpP. Other *groP* mutants were in the *dnaK*, *dnaJ*,[292] and *grpE*[293] genes, which encode very abundant proteins induced by an increase in temperature and other types of environmental stress.

The heat-shock proteins function in vitro specifically in an initiation stage, as predicted by their phenotype in vivo. The dnaK and dnaJ proteins help dissociate the gpP and dnaB proteins from the O·P·dnaB complex assembled at *ori*λ, to unmask the helicase and priming functions of dnaB protein. The dnaK and dnaJ proteins act together and require NTP hydrolysis by dnaK protein.[294] When

287. Ohashi M, Dove W (1976) *Virology 72*:299.
288. Inman RB, Schnös M (1987) *JMB 193*:377.
289. Schnös M, Inman RB (1987) *Virology 158*:294.
290. Friedman DI, Olson ER, Georgopoulos C, Tilly K, Herskowitz I, Banuett F (1984) *Microbiol Rev 48*:299.
291. Georgopoulos CP (1971) in *The Bacteriophage Lambda* (Hershey AD, ed.). CSHL, p. 639; Georgopoulos CP, Herskowitz I (1971) in *The Bacteriophage Lambda* (Hershey AD, ed.). CSHL, p. 553.
292. Saito H, Uchida H (1977) *JMB 113*:1; Saito H, Uchida H (1978) *MGG 164*:1.
293. Ang D, Chandrasekhar GN, Zylicz M, Georgopoulos C (1986) *J Bact 167*:25.
294. Liberek K, Georgopoulos C, Zylicz M (1988) *PNAS 85*:6632; Alfano C, McMacken R (1989) *JBC 264*:10709; Dodson M, McMacken R, Echols H (1989) *JBC 264*:10719.

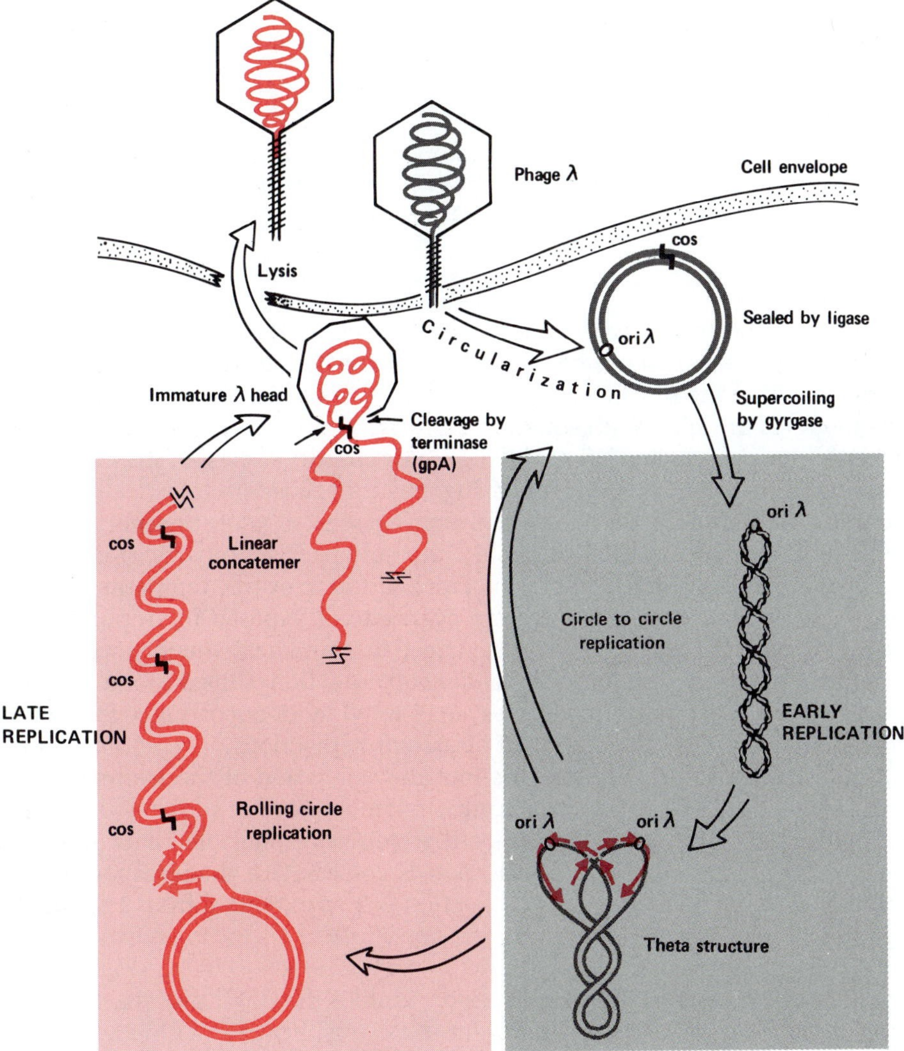

Figure 17-32
Scheme for replication of phage λ. Early stages involve doubly forked, superhelical, circular forms. In later stages, concatemeric forms support DNA maturation. *cos* = cohesive-end sites. [After Enquist LW, Skalka A (1973) *JMB 75*:185; Skalka A (1978) *TIBS 3*:279]

duplex melting is blocked, as in instances when the prepriming complex is formed on DNA lacking negative superhelicity, the heat-shock proteins simply disassemble the complex.[295] GpO, gpP, and dnaB protein appear unmodified in this process and are active upon transfer to a new template. The GrpE protein is dispensable in vitro, but reduces by tenfold or more the amount of dnaK protein required, and thus appears to act synergistically with the dnaK and dnaJ proteins.[296]

295. Alfano C, McMacken R (1989) *JBC 264*:10709.
296. Alfano C, McMacken R (1989) *JBC 264*:10709; Zylicz M, Ang D, Liberek K, Georgopoulos C (1989) *EMBO J 8*:1601.

Table 17-13
Properties of heat-shock proteins involved in λ replication

Gene	Mass (kDa)	Native form	Properties
dnaK	69.1	monomer	ATPase; autophosphorylates; forms tight complex with grpE protein
dnaJ	41.1	dimer	Binds ss- and dsDNA nonspecificially; interacts with gpP
grpE	25	monomer	Binds tightly to dnaK protein; autophosphorylates

Although the stage at which these proteins act has been clearly established, the intimate mechanism remains uncertain. The characteristics and activities (Table 17-13) of the purified proteins provide a few clues. The dnaK protein is 48% identical in amino acid sequence to the Hsp 70 proteins (heat-shock protein, 70 kDa) from humans and *Drosophila*, a highly conserved class of stress-induced proteins.[297] Accumulating evidence suggests that those proteins, called "chaperones," assemble and disassemble protein aggregates, using the energy of NTP hydrolysis to break apart hydrophobic interactions,[298] a property displayed by the dnaK, dnaJ, and GrpE proteins in λ replication. Elevated temperatures may cause cellular proteins to unfold and aggregate at exposed hydrophobic surfaces, thus accounting for the increased need for these heat-shock proteins. The λ replication system, absolutely dependent on the action of these proteins to take apart the dnaB·gpP protein complex, may provide an opportunity for a more refined analysis of the specificity of disassembly reactions.

Studies of λ also provide insights into the regulation of the heat-shock response. Infection, like an increase in growth temperature, stimulates expression of the heat-shock proteins. The viral cIII protein is responsible. In its presence, the half-life of the σ^{32} protein, an alternate *E. coli* RNAP σ factor responsible for expression of the heat-shock genes (Section 7-4), is extended from 2 to about 8 minutes.

Heat-Shock Proteins in Host Replication. Unlike the dnaB protein, the heat-shock proteins appear to serve different roles in replication of λ than in that of the host. Whereas the *dnaK, dnaJ*, and *grpE* genes are essential for *E. coli* only at high temperatures, they are required for λ replication at all temperatures. Mutants in any of the three genes have nearly identical phenotypes at high temperature. The defects are highly pleiotropic, including inhibition of both RNA and DNA synthesis.[299] Deletion derivatives of *dnaK* and *dnaJ* are viable at low temperature but grow slowly and accumulate multiple compensatory mutations, allowing them to grow at higher temperatures. The primary defect does not appear to be in chromosomal DNA replication. Furthermore, the release of dnaC protein from the dnaB·dnaC complex in vitro—the reaction in *oriC* replication analogous to dissociation of the gpP·dnaB complex in λ replication—needs no assistance from heat-shock proteins. There may still be a role for the heat-shock proteins in host replication. Inasmuch as their functions appear to be redundant,

297. Bardwell JCA, Craig E (1984) *PNAS 81*:848.
298. Pelham HRB (1986) *Cell 46*:959.
299. Ang K, Chandrasekhar GN, Zylicz M, Georgopoulos C (1986) *J Bact 167*:25; Saito H, Uchida H (1978) *MGG 164*:1; Itikawa H, Ryu J (1979) *J Bact 138*:339.

a mutational defect in one of the proteins may be compensated by the others. Furthermore, an allele of *dnaK* appears to be specifically defective in initiation of *E. coli* DNA replication.[300]

Late-Stage Replication.[301] By 15 minutes, the θ stage of replication is over and rolling circles [sigma (σ) forms] predominate (Fig. 17-32). A defined switch, controlling the change from the θ to the σ stage, has been assumed, but no gene controlling this process has been described. Studies of replication intermediates in vivo suggest that σ replication occurs throughout the λ life cycle and that the θ mode is inhibited by 15 minutes.

How rolling-circle replication is initiated is largely unknown. GpO and gpP are probably required, although reduced amounts of gpP may suffice. Of the rolling-circle intermediates observed, at least 40% have initiated at *ori*λ, evidence that initiations of the θ and σ forms of replication employ a similar mechanism. An enzyme that specifically nicks the origin, as gpA does in ϕX174 (Section 4), has not been identified. One possibility is that σ intermediates arise from a unidirectional replication fork initiated at *ori*λ. Representing as much as 25% of the population of replication intermediates in vivo, the unidirectional fork colliding with an unresolved initiation structure at the origin[302] may give rise to a rolling circle. Thus, molecules that initiate bidirectional replication remain as simple circles, but unidirectionally replicating molecules may be converted to the rolling-circle mode. Once started, rolling circles keep going even after initiation has been turned off.

Inactivation of a temperature-sensitive gpP has no effect on the late stage of replication, but a shift of an O^{ts} mutant to restrictive temperature stops replication immediately,[303] suggesting that gpO, in addition to its function in initiation, is an essential part of the elongation complex. However, in vitro, there is no evidence that gpO is involved in fork movement. Both the isolation of active replication intermediates and the use of antibodies restrict the involvement of O protein exclusively to initiation.[304]

Monomeric λ DNA is not packaged into phage particles; the synthesis of multimeric DNA is thus essential for a productive infection. The formation of multimers can be achieved either by rolling-circle replication, which produces head-to-tail linear concatemers, or by homologous recombination to yield multimeric circles[305] (Fig. 17-33).

The *E. coli* exonuclease V, product of the *recBCD* genes (Section 21-5), inhibits rolling-circle formation. During infection, exo V is normally inhibited by the λ *gam* (*gamma*) gene product. In gam^- mutants, with rolling-circle replication blocked, λ growth becomes dependent on recombination. Homologous recombination systems of either the host (*recA*-dependent) or phage (*red*-dependent) can generate multimeric circles for packaging. Alternatively, destruction of exo V restores rolling-circle replication and removes the requirement for recombination.

---

300. Sakakibara Y (1988) *J Bact 170*:972.
301. Better M, Freifelder D (1982) *Virology 119*:159.
302. Dodson M, Echols H, Wickner S, Alfano C, Mensa-Wilmot K, Gomes B, LeBowitz J, Roberts JD, McMacken R (1986) *PNAS 83*:7638.
303. Klinkert J, Klein A (1978) *J Virol 25*:730.
304. Alfano C, McMacken R (1989) *JBC 264*:10709; Alfano C, McMacken R (1989) *JBC 264*:10699.
305. Smith GR (1983) in *Lambda II* (Hendrix RW, Roberts JW, Stahl FW, Weisberg RA, eds.). CSHL, p. 175.

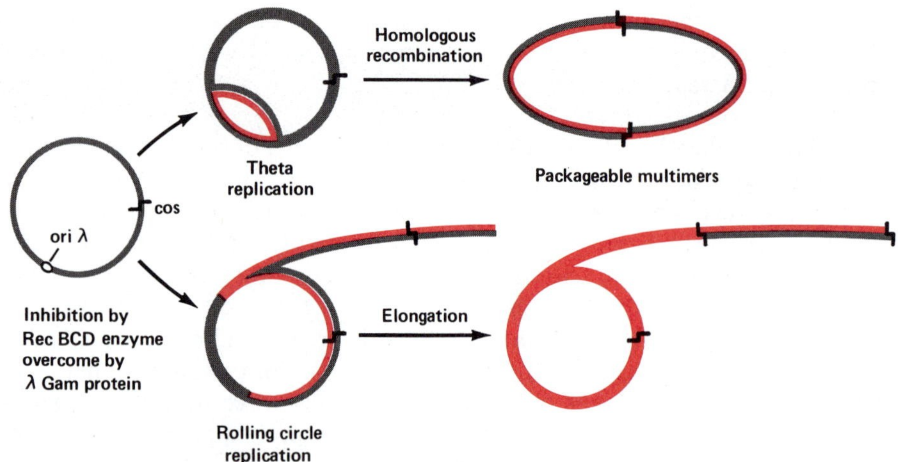

Figure 17-33
Alternative mechanisms for generating multimers of the λ genome for packaging. *cos* = cohesive-end sites. [After Wyman AR, Wertman KF (1987) *Meth Enz 152*:173]

Maturation and Packaging.[306] Maturation of late DNA to unit-length linear duplexes with cohesive ends requires that staggered double-stranded scissions be made precisely at the cohesive-end sites (*cos*).[307] Cleavage is by the terminase enzyme, product of the *A* (74 kDa) and *Nu1* (21 kDa) genes. Terminase has several biochemical activities: binding to the *cos* site, introduction of the 12-bp staggered cut, binding to proheads, and DNA-dependent ATPase. The 200-bp *cos* sequence, required for packaging, contains regions directing both the binding and cleavage by terminase. Absolutely required for DNA maturation in vivo is the presence of five head proteins (products of genes *B, C, Nu3, D,* and *E*) that define the framework of the phage head. In the absence of any one of these factors, heads are not assembled and DNA without cohesive ends and several times the unit length accumulates. Cleavage at *cos* by terminase in vitro does not require proheads. Terminase–DNA and terminase–DNA–prohead complexes can be formed in vitro or isolated from infected cells.

Packaging requires IHF protein (the small, basic, dsDNA-binding protein; Section 10-8) or THF (*t*ermination *h*ost *f*actor, product of an unknown gene), provided by the host cell. DNA maturation is blocked in host *groE* mutants,[308] which fail to support phage head assembly. GpE is the main capsid protein of λ; certain E mutants can overcome the *groE* defect. In many *groE* mutants, head assembly of other phages (T4 and ϕ80) and tail assembly of T5 and 186 are also blocked. The *groE* locus is made up of two closely linked genes with the same phenotype, *groEL* (large, encoding a 65-kDa protein) and *groES* (small, 15-kDa protein). The GroEL protein is extremely abundant, comprising about 2% of total cell protein. The *groE* genes, like *dnaK*, *dnaJ*, and *grpE*, are under heat-

306. Feiss M, Becker A (1983) in *Lambda II* (Hendrix RW, Roberts JW, Stahl FW, Weisberg RA, eds.). CSHL, p. 305; Black LW (1988) in *The Bacteriophages* (Calendar R, ed.). Plenum Press, New York, Vol. 2, p. 321.
307. Nichols BP, Donelson JE (1978) *J Virol 26*:429.
308. Georgopoulos CP, Hendrix RW, Kaiser AD, Wood WB (1972) *NNB 239*:38; Friedman DI, Olson ER, Georgopoulos C, Tilly K, Herskowitz I, Banuett F (1984) *Microbiol Rev 48*:29.

shock regulation and probably represent further examples of proteins that influence macromolecular assemblies.

After encapsidation of the phage DNA,[309] terminase recognizes a second *cos* site and cleaves it, generating a unit-length chromosome. Sequential unidirectional packaging of two to three genomes in a concatemer is usual for each initiation event. The size constraint on packaging is rather strict; only DNAs of 75% to 105% the length of wild-type λ can be packaged. Packaging of a λ derivative carrying a large deletion can be restored in the presence of a foreign (helper) DNA; such phage have become favored vectors for the initial cloning of DNA fragments.[310]

P22 Phage[311]

Phage P22, a temperate phage of *Salmonella typhimurium*, is closely related to λ in structure and in many functions. From studies of P22, along with those of λ and T4, have come many major concepts of molecular biology. P22 is also notable as a convenient vector for generalized transduction as well as for λ-like specialized transduction.

The genome of P22 is a linear, duplex DNA of 41.8 kb, circularly permuted, with 1.7-kb terminal redundancies. Thus, like T4 but unlike λ, P22 has a circular genetic map. Upon entry into the cell, the genome is circularized by homologous recombination between the terminal repeats, catalyzed by either the phage or the host general recombination system.

Two P22 proteins are required for replication, the products of genes *12* and *18*. Gp18 is homologous to the λ and ϕ82 O proteins and thus is likely to be an origin-binding protein.[312] By contrast, gp12 is not homologous to the λ P protein but rather to dnaB protein. P22 phage, unlike λ, is independent of the host *dnaB* and *dnaC* genes.[313] In addition to sequence homology, there is strong biochemical evidence that gp12 is a functional analog of the dnaB protein:[314] it is an ssDNA-dependent ATPase, allows primase to synthesize primers on any ssDNA, and substitutes for the dnaB protein in ϕX174 replication in vitro without dependence on dnaC protein.

Replication of DNA is generally similar in P22 and λ. Bidirectional replication from a unique origin generates circular, θ-form intermediates. The circles give way, presumably by a rolling-circle mechanism, to concatemeric forms in the later stages. The concatemers are measured and cut to fill the phage head ("headful packaging"). The initial cut is made at a specific site, the *pac* sequence. Subsequent cuts are simply determined by measuring the physical length of a "headful," about 43.5 kb.[315] *Generalized transduction* results from packaging headfuls of host DNA into defective particles,[316] starting from the integrated P22 *pac* site or at some homologous nucleotide sequence in the host chromosome.

---

309. Yamagishi H, Okamoto M (1978) *PNAS 75*:3206.
310. Meissner PS, Sisk WP, Berman ML (1987) *PNAS 84*:4171; Dunn IS, Blattner FR (1987) *NAR 15*:2677; Young RA, Davis RW (1983) *PNAS 80*:1194.
311. Susskind MM, Botstein D (1978) *Microbiol Rev 42*:385; Poteete AR (1988) in *The Bacteriophages* (Calendar R, ed.). Plenum Press, New York, Vol. 2, p. 647.
312. Backhaus H, Petri JB (1984) *Gene 32*:289.
313. Schanda-Mulfinger UEM, Schmieger J (1980) *J Bact 143*:1042; Botstein D, Herskowitz I (1974) *Nature 251*:584.
314. Wickner S (1984) *JBC 259*:14038; Wickner S (1984) *JBC 259*:14044.
315. Casjens S, Huang WM, Hayden M, Parr R (1987) *JMB 194*:411.
316. Schmieger H (1982) *MGG 187*:516.

P2 and P4 Phages[317]

The commensal relationship between temperate phage P2 and its small satellite P4 is most instructive as a model for refined transcriptional control and structural interactions. Phage P2 provided the first clear evidence for unidirectional replication from a unique origin and a persuasive example of integrative suppression (Sections 16-3 and 18-1). The P2 and P4 replication patterns contrast sharply in mechanism and in reliance on host proteins.

P4 relies on P2 as a helper phage for all 17 late morphogenetic and capsid proteins and the lytic enzyme. As a result, the phages have identical tails and heads, except that the P4 head is only one-third the size of the P2 head. P4 induces, *transactivates*, the late P2 functions even when a failure of P2 replication has failed to elicit them. P2, in turn, transactivates the late P4 genes. In a host strain that is not a P2-lysogen, P4 replicates, lysogenizes by integration, or is maintained as a plasmid, but forms no infectious particles.

P2 and P4 have duplex DNA genomes of 33 and 11.6 kb, respectively, each with the same 19-ntd, single-stranded cohesive ends (as expected, since P4 is packaged by P2 gene products). Otherwise they share no homology and use very different replication strategies. Replication of P2 proceeds by a rolling-circle mechanism[318] and depends on host replication proteins: dnaB protein, primase, pol III holoenzyme, and Rep protein. P2 encodes two replication proteins, the products of genes *A* and *B*. GpA, the initiation protein, introduces a nick into the template to start rolling-circle replication. In this function, and its preferential *cis* action, it is like ϕX174 gpA (Section 4). The gpB is required for discontinuous-strand synthesis but details of its function are unknown.[319]

There are about ten sites in the *E. coli* chromosome where P2 integrates to form a prophage. Integration at sites near the chromosomal origin, but not near the terminus, allows chromosomal replication to become dependent on P2 rather than on *E. coli* initiation functions, including genes *A* and *rep* (an *E. coli* gene, but normally not essential for chromosome replication), and independent of *oriC* and dnaA protein. In this circumstance, *oriC* and *dnaA* mutations are not lethal, being *integratively suppressed* by the P2 genome. Although fork movement is unidirectional in autonomous replication, it is bidirectional when integrated.

Replication of P4 differs from that of P2 in being bidirectional and generating θ intermediates. It is not dependent on P2 or on host replication proteins other than pol III holoenzyme and SSB. It relies on its own gpα, an RNA polymerase that, in vitro, synthesizes poly rG from a poly dG·polydC template and has primase activity on single-stranded templates.[320] Presumably, the protein participates in RNA priming of replication.[321]

The P4 origin of replication, mapped by electron microscopy, is 2.6 kb from the right end.[322] Deletion analysis has revealed another 300-bp sequence, required *in cis* for replication (called *crr*, *cis* replication region) and located 4.7 kb to the left.[323] The *crr* consists of a 120-bp direct repeat, containing six copies of the 8-bp sequence that also occurs six times in the origin. The *crr* sequence can act at variable distances from the origin and in either orientation. How *crr*

---

317. Bertani LE, Six EW (1988) in *The Bacteriophages* (Calendar R, ed.). Plenum Press, New York, Vol. 2, p. 73.
318. Chattoraj DK (1978) *PNAS 75*:1685.
319. Funnell BE, Inman RB (1983) *JMB 167*:311.
320. Krevolin MD, Calendar R (1985) *JMB 182*:509.
321. Kahn M, Hanawalt P (1979) *JMB 128*:501.
322. Krevolin MD, Inman RB, Roof D, Kahn M, Calendar R (1985) *JMB 182*:519.
323. Flensburg J, Calendar R (1987) *JMB 195*:439.

functions in replication is still unclear, but it is an attractive example of the interaction of distant sequences in the initiation of replication, a recurring and poorly understood motif in replication origins. The gpα may bind specifically to both the *ori* and *crr* sequences.[324]

P1 Phage[325]

The moderately large phage P1 lysogenizes by plasmid formation; the prophage is perpetuated extrachromosomally in about the same number of copies as the chromosome. The phage genome, about 100 kb, resembles that of P22 in that it is circularly permuted, with a 9% to 12% terminal redundancy. Upon entry into the cell, the DNA is circularized by the *lox–cre*, site-specific recombination system, encoded by the phage. The *cre* gene product catalyzes exchange between two *loxP* sites;[326] phages carrying *loxP* within the terminal redundancy (about 25% to 30% of the population) are circularized. The *lox–cre* system also plays an important role in stabilizing the prophage by resolving plasmid multimers[327] and produces a linear genetic map for P1, unusual for a phage with a circularly permuted genome. P1 has a second site-specific recombination system, *cin–cix*, which inverts the C segment (homologous to the G segment of Mu; see below), allowing expression of alternate forms of the tail fibers and thereby determining host range.[328]

Because P1 is maintained as a plasmid (Section 18-6), instead of being integrated into the host chromosome, the prophage cannot rely on passive replication. Thus, in contrast to most temperate phages, replication functions are not repressed in the prophage. However, lytic and lysogenic replication depend on two separate replication pathways.

Prophage Replication. Three phage-encoded elements are involved in prophage replication: an origin sequence (*oriR*),[329] an origin-binding protein (the *repA* gene product),[330] and a locus responsible for copy number control.[331] The origin structure is classic. The 250-bp *oriR* sequence contains five copies of a RepA-binding sequence, a 60-bp AT-rich region, and two copies of the host dnaA-binding sequence (Sections 16-3 and 16-4). The AT-rich region contains five GATC sequences that are methylated by the host DNA adenine methyltransferase; methylation of these sequences is essential for replication.[332] However, completely unmethylated DNA does not normally occur, and hemimethylated DNA, the state that would occur for a period after initiation of replication, appears to replicate normally. Thus, the role of methylation in the normal replication cycle remains unclear.

---

324. Strack B, Christian R, Calendar R, Lanka E, personal communications.
325. Yarmolinsky M, Sternberg N (1988) in *The Bacteriophages* (Calendar R, ed.). Plenum Press, New York, Vol. 1, p. 291.
326. Sternberg N, Sauer B, Hoess R, Abremski K (1986) *JMB 187*:197; Segev N, Cohen G (1981) *Virology 114*:333; Hochman L, Segev N, Sternberg N, Cohen G (1983) *Virology 131*:11.
327. Sternberg N, Hamilton D, Austin S, Yarmolinsky M, Hoess R (1980) *CSHS 45*:297; Austin S, Ziese M, Sternberg N (1981) *Cell 25*:729.
328. Iida S, Meyer J, Kennedy KE, Arber W (1982) *EMBO J 1*:1445; Iida S, Huber H, Hiestand-Nauer R, Meyer J, Bickle TA, Arber W (1984) *CSHS 49*:769.
329. Chattoraj DK, Abeles AL, Yarmolinsky MB (1985) in *Plasmids in Bacteria* (Helinski DR, Cohen SN, Clewell DB, Jackson DA, Hollaender A, eds.). Plenum Press, New York, p. 355; Chattoraj DK, Snyder KM, Abeles AL (1985) *PNAS 82*:2588.
330. Abeles AL, Snyder KM, Chattoraj DK (1984) *JMB 173*:307.
331. Chattoraj D, Cordes K, Abeles A (1984) *PNAS 81*:6456; Pal SK, Chattoraj DK (1987) in *DNA Replication and Recombination* (McMacken R, Kelly TS, eds.). UCLA, Vol. 47. A R Liss, New York, p. 441.
332. Abeles AL, Austin SJ (1987) *EMBO J 6*:3185.

Initiation requires binding of both the RepA and dnaA proteins to the origin. Only a subset of the functions of dnaA involved in host replicaton are essential for P1. In vivo, the requirement for *dnaA* is demonstrable only with null alleles; *dnaA*^{ts} mutations, lethal to the host, are not defective for P1 replication.[333] In vitro, dnaA protein complexed with either ATP or ADP functions in P1 replication,[334] whereas only the ATP form supports *oriC* plasmid replication (Section 16-3). Most likely, some of the functions of dnaA protein are fulfilled by the RepA protein for P1 replication, and the two proteins together assemble the replication complex at the origin. The replication machinery is that of the host: RNAP, dnaB and dnaC proteins, primase, and pol III holoenzyme are required.

A P1 sequence, *incA*, regulates prophage replication, ensuring that the number of plasmids in the cell remains low. Because of its low intracellular abundance, P1 also encodes functions directing the equal partition and stable inheritance of plasmids during cell division. Both these aspects of P1, also common to other plasmids maintained at a few copies per cell (as with the F factor), are discussed in Section 18-6.

Lytic Replication. This replication mode starts about 5 minutes after induction or infection. As with λ replication, for the first 30 minutes there are about an equal number of θ and σ intermediates, while at late times σ replication predominates.[335] Although σ forms appear to arise as a result of replication rather than recombination, the host *recA* function is involved. The origin of lytic replication has not been analyzed in detail. However, a 1-kb region required for lytic, but not plasmid, replication has been identified.[336] This region contains a *cis*-acting origin sequence, *oriL*, and a gene, *repL*, which is required for its function; a promoter, under control of the phage repressor, is also required.

P1 encodes two replication proteins that replace *E. coli* functions: a dnaB protein analog, product of the *ban* (*B analog*) gene,[337] and an SSB.[338] The Ban protein, normally repressed in the prophage state, allows P1 to grow lytically on *dnaB*^{ts} host strains. Derivatives of P1 which constitutively express *ban* suppress the replication defect of *dnaB* mutants;[339] the Ban and dnaB proteins form a heteromultimer which supports host replication.[340] In contrast to the dnaB analog of P22, which bypasses the need for both the *dnaB* and *dnaC* genes, *dnaC* is required for P1 lytic replication, suggesting that the Ban and dnaC proteins interact.[341] As with plasmid replication, primase and pol III holoenzyme are required for lytic replication. Packaging is by the headful mechanism described for P22.

Phage Mu[342]

Named for its *mu*tator properties, this temperate phage integrates nearly randomly throughout the chromosome, disrupting genes upon insertion. The Mu

---

333. Hansen EB, Yarmolinsky MB (1986) *PNAS 83*:4423.
334. Wickner S, personal communication.
335. Cohen G (1983) *Virology 131*:159; Segev N, Laub A, Cohen G (1980) *Virology 114*:333.
336. Hansen EB (1989) *JMB 207*:135; Cohen G, Sternberg N (1989) *JMB 207*:99.
337. Lanka E, Schuster H (1970) *MGG 106*:279.
338. Johnson BF (1982) *MGG 186*:122.
339. D'Ari R, Jaffe-Brachet A, Touati-Schwartz D, Yarmolinsky M (1975) *JMB 94*:341.
340. Lanka E, Mikolajczyk M, Schlicht M, Schuster H (1978) *JBC 253*:5847.
341. Hay J, Cohen G (1983) *Virology 131*:193.
342. Symonds N, Toussaint A, Van de Putte P, Howe MM (eds.) (1987) *Phage Mu*. CSHL; Harshey RM (1988) in *The Bacteriophages* (Calendar R, ed.). Plenum Press, New York, Vol. 1, p. 193.

genome, a linear dsDNA of 37 kb, is covalently associated with host DNA throughout its life cycle. In the virion, a 50- to 150-bp segment of host DNA is attached to the left end and a 1- to 2-kb segment is attached to the right. Arising from the DNA flanking the integration sites, these host sequences are variable, differing in each phage, and give rise to "split ends" on the DNA after heteroduplex analysis[343] (Fig. 17-34). The genome also contains a 3-kb invertible segment (the G region), analogous to the C region of phage P1, encoding alternate forms of the tail-fiber proteins.[344] The orientation of the G segment thus determines the bacterial strains to which Mu can adsorb. Among particles grown by infection, 99% have the same G orientation, while in phage produced by induction of lysogenic cells, both orientations are equally represented. Apparently, inversion of G is rare enough that randomization of the orientation is achieved only after many generations of growth.[345]

Mu encodes two proteins committed to protecting its DNA from attack. The product of the *gam* gene (or *sot*, for stimulation of transformation) protects the linear DNA from digestion by the host RecBCD enzyme (exo V). Gam protein inhibits digestion by binding to the DNA, rather than to the enzyme as is done by its λ counterpart.[346] The *mom* gene (modification of Mu) directs a unique glycinamide modification of about 15% of the adenine residues (at the 6-amino group), rendering the DNA resistant to many restriction systems.[347] The chemical donor of this unusual adduct is unknown, but is presumably of host origin, inasmuch as *mom* is the only Mu gene critical for modification. Transcription of *mom* requires methylation by the host DNA adenine methyltransferase (the *dam* gene product) of three GATC sites upstream of the coding sequence. Thus, in the absence of Dam methylation, *mom* modification is also blocked.[348]

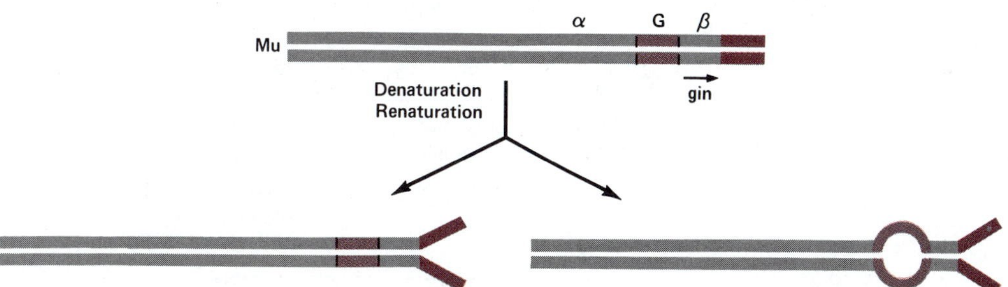

Figure 17-34
Heteroduplexes of Mu DNA. Two G regions may be in the same orientation (left) or in the opposite orientation (right) as observed in electron micrographs of heteroduplexes. α, G, β = DNA regions seen by electron microscopy; gin = recombinase gene. Bacterial sequences at both ends of Mu DNA remain single-stranded in heteroduplexes; only those at the right end are long enough to be seen in electron micrographs.

343. Howe MM (1988) in *Phage Mu* (Symonds N, Toussaint A, Van de Putte P, Howe MM, eds.). CSHL, p. 25.
344. Koch C, Mertens G, Rudt F, Kahmann R, Kanaar R, Plasterk R, van de Putte P, Sandulache R, Kamp D (1988) in *Phage Mu* (Symonds N, Toussaint A, Van de Putte P, Howe MM, eds.). CSHL, p. 75; Grundy FJ, Howe MM (1984) *Virology 134*:296.
345. Symonds N, Coelho A (1978) *Nature 271*:573.
346. Paolozzi L, Symonds N (1988) in *Phage Mu* (Symonds N, Toussaint A, Van de Putte P, Howe MM, eds.). CSHL, p. 53; Akroyd J, Symonds N (1986) *Gene 49*:273.
347. Kahmann R, Hattman S (1988) in *Phage Mu* (Symonds N, Toussaint A, Van de Putte P, Howe MM, eds.). CSHL, p. 93; Swinton D, Hattman S, Crain PF, Cheng C-S, Smith DL, McCloskey JA (1983) *PNAS 80*:7400.
348. Seiler A, Blocker H, Frank R, Kahmann R (1986) *EMBO J 5*:2719; Hattman S, Ives J (1984) *Gene 29*:185.

Circularization and Integration.[349] The linear Mu genome, with its attached host sequences, has neither cohesive ends nor terminal repetitions, and thus no obvious way of circularizing. Yet circularization is apparently a prerequisite to integration; in vitro, integration requires the Mu DNA to be part of a covalently closed supercoiled circle (see below). Although closed circles of Mu DNA have not been detected, the linear genome is circularized by complexing with the 64-kDa gpN shortly after infection.[350] This circular complex transforms cells much more efficiently than deproteinized Mu DNA, arguing for its importance. Whether these protein-linked circles fulfill the requirement for constrained DNA during integration in vitro remains to be established.

Upon infection, the Mu genome must integrate into the host chromosome prior to either lytic or lysogenic growth (Fig. 17-35). Mu integrates by nonhomologous recombination, termed *transposition* (Section 21-7). Integration is by

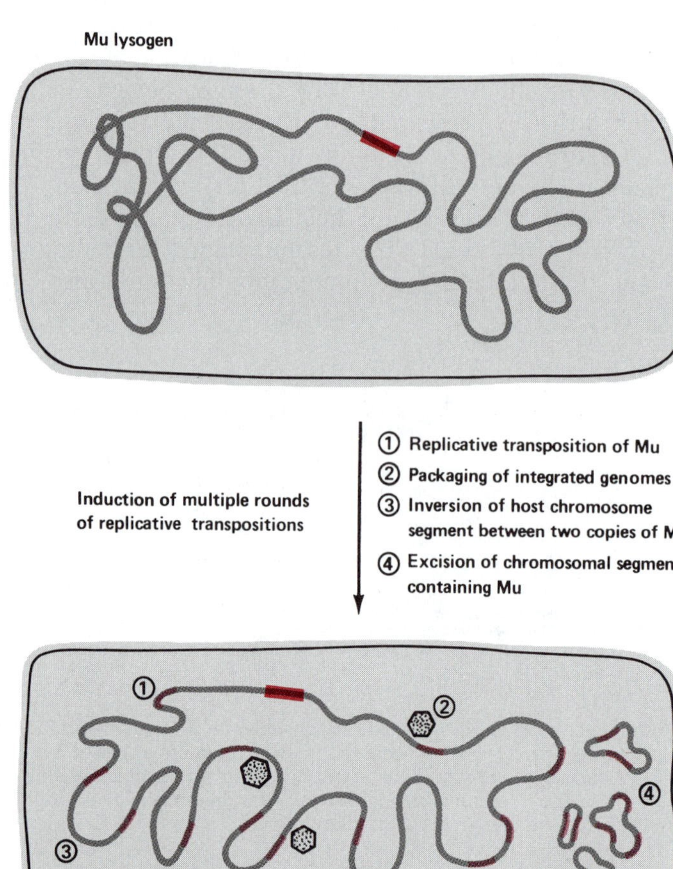

Figure 17-35
Host chromosome rearrangements by Mu phage. (Top) Mu lysogen with a single copy of Mu integrated at an essentially random location in the host chromosome. (Bottom) Multiple rounds of transposition during lytic growth inactivate host genes by insertion and generate inversions and deletions.

349. Harshey RM (1986) in *Phage Mu* (Symonds N, Toussaint A, Van de Putte P, Howe MM, eds.). CSHL, p. 111.
350. Gloor G, Chaconas G (1986) *JBC 261*:16682; Harshey RM, Bukhari AI (1983) *JMB 167*:427.

nonreplicative transposition (Fig. 17-36); the genome is inserted into the host chromosome without extensive replication and the old flanking host DNA is lost in the process[351] (see beow). Repression of further gene expression following integration establishes a stable prophage.

Lytic Growth. Unlike λ, Mu does not excise from the chromosome during induction of the lytic cycle.[352] Instead, multiple rounds of *replicative transposition* commence (Fig. 17-37). During the following 100 or so cycles of replicative transposition, DNA synthesis is semiconservative and restricted to the Mu sequences; the host sequences flanking Mu remain essentially unreplicated. Each daughter copy of Mu DNA is attached at one end to the original flanking DNA and at the other end to a new flanking DNA. The target site of the transposition is cleaved to generate a 5-bp stagger; thus, a 5-bp sequence at the target site is duplicated along with the Mu chromosome. Each round of replicative transposition is accompanied by either inversion or deletion of the host chromosome segment between the replicated copies of Mu DNA. Deletion events generate two DNA circles, each of which contains a replicated copy of Mu DNA. Thus, some of the replicating Mu DNA remains associated with the bulk of the host chromosome, while other copies are found in extrachromosomal supercoiled circles of various sizes called *heterogeneous circular DNA* (HcDNA).

Packaging is by a headful mechanism.[353] The genome size is smaller than a headful, resulting in the packaging of host sequences attached to the ends of the Mu chromosome. Although requiring a site, *pac* (*pac*kaging), near the left end,[354] cleavage and packaging initiate approximately 100 bp away from *pac* within the variable flanking host sequences.

Mechanism of Transposition and Replication.[355] Mu transposition, obligatory for replication, requires the two ends of the linear genome, about 200 bp at the left end (*attL*, *att*achment site *l*eft) and 100 bp at the right (*attR*). The essential phage protein, Mu transposase, the 75-kDa product of the *A* gene, binds specifically to the *att* sites. The replication "origin," that is, the *cis*-acting sequence required for initiation of DNA synthesis, is indistinguishable from the *att* sites. Replication is coupled to transposition and starts at the ends of the genome.

Transposase (gpA) catalyzes the strand cleavages and ligations vital for transposition. A second Mu protein, gpB, stimulates integrative (nonreplicative) transposition 10-fold to 100-fold and is essential for replication. Mu relies on host enzymes for all of the steps in transposition and replication not promoted by these two proteins.

Analysis of Mu transposition in vitro has revealed important aspects of the mechanism. Integrative (Fig. 17-36) and replicative (Fig. 17-37) transposition arise from a common recombination intermediate.[356] Formation of this strand-

---

351. Harshey RM (1983) *Nature 311*:580; Liebaart JC, Ghelardini P, Paolozzi L (1982) *PNAS 79*:4362; Chaconas G, Kennedy DL, Evans D (1983) *Virology 128*:48.
352. Ljungquist E, Bukhari AI (1977) *PNAS 74*:3143.
353. Howe MM (1986) in *Phage Mu* (Symonds N, Toussaint A, van de Putte P, Howe MM, eds.). CSHL, p. 63.
354. Groenen MAM, van de Putte P (1985) *Virology 144*:520.
355. Mizuuchi K, Craigie R (1986) *ARG 20*:385; Chaconas G (1986) in *Phage Mu* (Symonds N, Toussaint A, Van de Putte P, Howe MM, eds.). CSHL, p. 137; Mizuuchi K, Higgins NP (1986) in *Phage Mu* (Symonds N, Toussaint A, van de Putte P, Howe MM, eds.). CSHL, p. 159.
356. Craigie R, Mizuuchi K (1985) *Cell 41*:867.

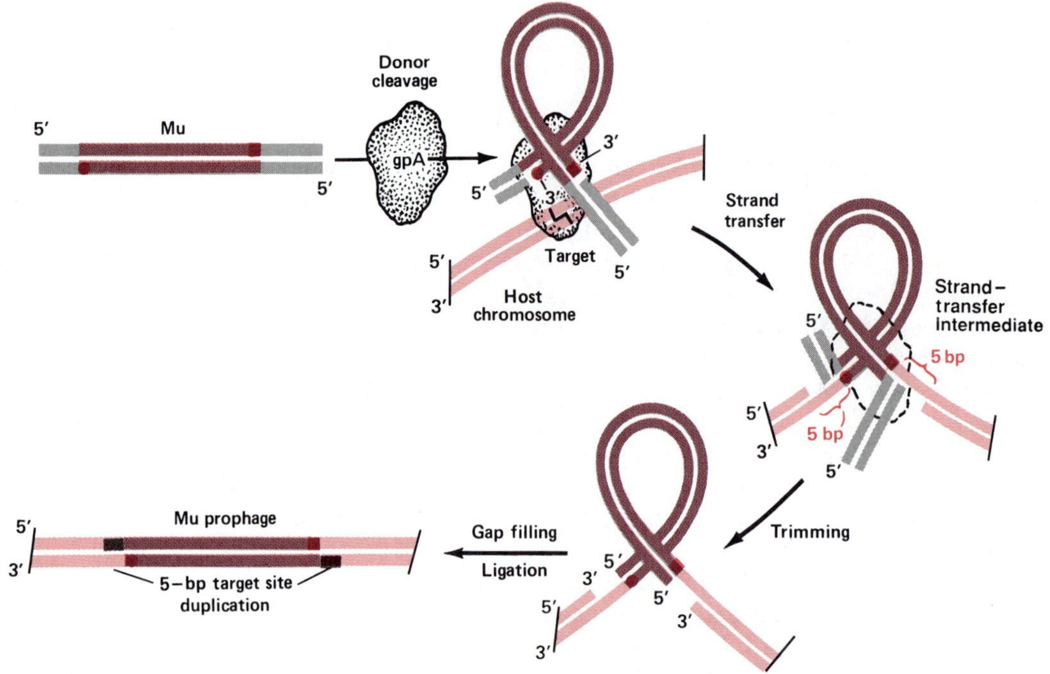

Figure 17-36
Mechanism of Mu integrative (conservative) transposition. All cleavages and ligations are by gpA.

transfer intermediate can be distinguished from its resolution by repair (as during integration) or replication.

The transposase binds to the two *att* sites and nicks the Mu genome at the junction between Mu and host DNA, generating a 3′-OH at each end.[357] HU, a host histone-like protein (Section 10-8), is required for cleavage of the Mu DNA by gpA; an important role of HU for Mu growth has been confirmed genetically. GpA is also responsible for concerted cutting of the target DNA with a 5-bp stagger and ligation of the 5′-P ends to the 3′ ends of the Mu DNA. This step, stimulated by gpB and ATP, produces the strand-transfer intermediate. GpA and HU together can catalyze strand transfer in the absence of any cofactor energy source (e.g., ATP); however, the Mu DNA must be present on a negatively supercoiled template for the reaction to proceed.

An additional transposase-binding site, distinct in sequence from the *att* sites and located about 950 bp from the left end, is essential for efficient transposition.[358] This internal activation sequence, recognized by a domain of gpA, overlaps a region involved in regulation of early gene expression and is bound by the Mu repressor during the establishment of lysogeny. Repressor binding excludes gpA from this sequence, directly inhibiting further transpositions which would endanger the viability of the host cell and thus be undesirable for the phage as well.

357. Mizuuchi K (1984) *Cell* 39:395; Craigie R, Mizuuchi K (1987) *Cell* 51:493.
358. Mizuuchi M, Mizuuchi K (1989) *Cell* 58:399.

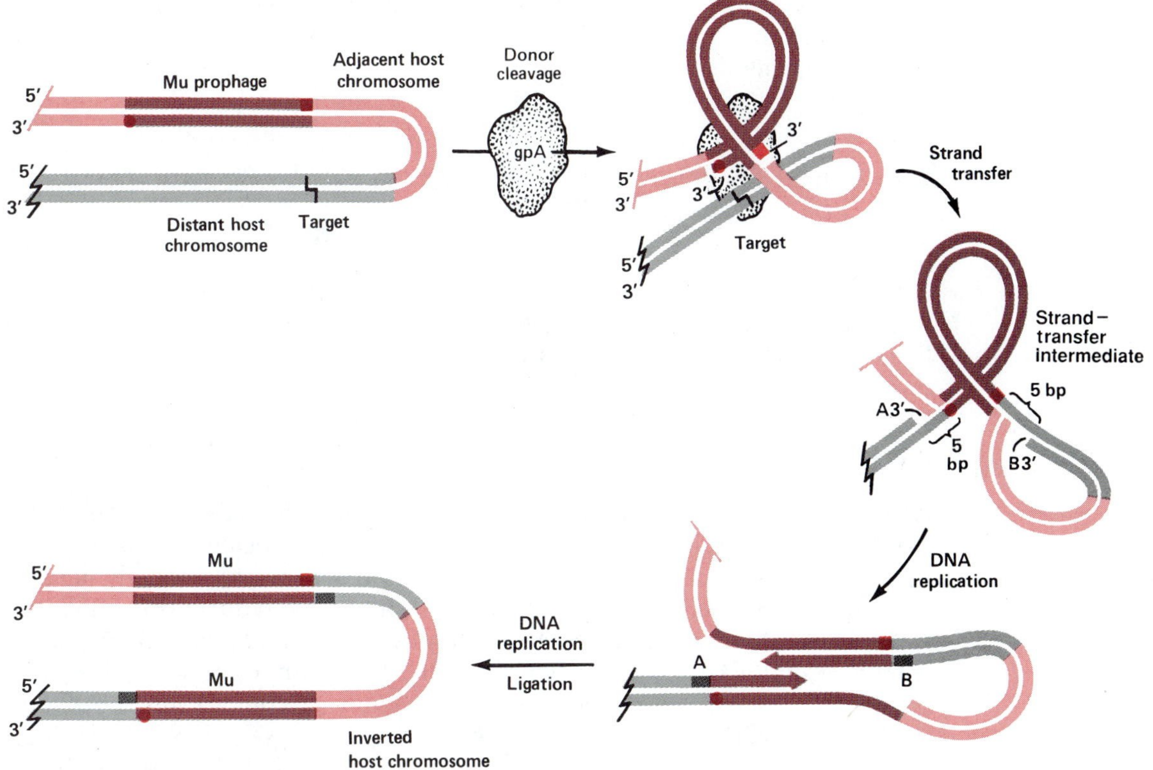

Figure 17-37
Mechanism of Mu replicative transposition. The Mu genome and target DNA are assembled into a complex protein–DNA structure during recombination.

Mu gpB, an ATPase, stimulates strand transfer[359] and dramatically influences target-site selection. The gpA preferentially attacks DNA bound by gpB when cleaving the target site. GpB binds DNA nonspecifically, but will not bind near a gpA · *att* site complex. Thus, in the presence of gpB, intermolecular targets are favored, whereas intramolecular sites are the rule in its absence. By activating intermolecular targets, gpB could prevent Mu from transposing into itself during lytic replication.[360]

How replication and repair enzymes are recruited to the strand-transfer intermediate and how the choice is made between replication and nonreplicative integration are unknown. GpA or gpB, both of which stay bound to the strand-transfer intermediate, may help redirect the host replication enzymes to the Mu template. Replication requires the host dnaB and dnaC proteins, gyrase, pol III holoenzyme, and probably primase.[361] Free 3′-OH groups remain at the junctions between the host and Mu DNA in the strand-transfer intermediate and may serve as primers for the continuous strand. Covalent extension of these

359. Maxwell A, Craigie R, Mizuuchi K (1987) *PNAS 84*:699.
360. Adzuma K, Mizuuchi K (1989) *Cell 57*:41; Adzuma K, Mizuuchi K (1988) *Cell 53*:257.
361. Toussaint A, Faelen M (1974) *MGG 131*:209.

ends generated during transposition would explain why synthesis is restricted to the Mu DNA. The structure of the replication product in vitro is consistent with this mechanism.

17-8 Other Phages: N4, PM2, PR4, and the *Bacillus* Phages (SPO1, PBS1, PBS2, ϕ29, SPP1)

Although the major classes of *E. coli* phages have been considered in preceding sections, other important coliphages, *Pseudomonas* phages, and particularly the *B. subtilis* phages also demand attention. The novelty of their genomes and replication mechanisms has enlarged our understanding of replication.

N4 Phage[362]

An intriguing feature of the coliphage N4 is its use of three distinct RNA polymerases (RNAPs) during its life cycle (Section 7-13);[363] encapsidation of one of them is unique among the phages.[364] This 320-kDa enzyme, one of the largest polypeptides known, is injected into the cell with the DNA during infection and is responsible for rifampicin-resistant, early transcription. Middle transcription is by a second N4-encoded enzyme. N4 RNAP II, a dimer of 30- and 40-kDa polypeptides,[365] is one of the smallest RNAPs known, in remarkable contrast to the virion enzyme. The N4 late transcripts are synthesized by *E. coli* RNAP. The developmental strategy of N4 is clearly different from that of any other phage. Characteristically, a phage depends on the host RNAP for early transcription and then either modifies the enzyme (e.g., T4) or synthesizes its own (e.g., T7) in order to redirect transcription exclusively to the phage DNA.

The N4 genome is a 74-kb linear molecule with a 390- to 440-bp terminal redundancy and a 3′ extension on both ends.[366] The phage encodes virtually all of the replication proteins it needs. Only the host gyrase and ligase are required.[367]

Five N4 gene products are essential for replication: a DNA polymerase with a potent 3′→5′ exonuclease, an SSB, a 5′→3′ exonuclease, a 78-kDa protein of unknown function, and the virion RNAP. Although N4 replication is independent of host helicase and primase, no such enzymes specific for N4 have been identified. Possibly, the virion RNAP may serve in replication as a primase. Host DNA synthesis is inhibited by an N4-encoded factor.

N4 replication probably initiates from both ends of the genome; double Y intermediates have been observed, but no eye structures. Packaging, presumably of concatemeric intermediates,[368] is initiated at specific sequences, rather

---

362. Kiino DR, Rothman-Denes LB (1988) in *The Bacteriophages* (Calendar R, ed.). Plenum Press, New York, Vol. 2, p. 457.
363. Zivin R, Zehring W, Rothman-Denes LB (1981) *JMB 152*:335.
364. Falco SC, VanderLann K, Rothman-Denes LB (1977) *PNAS 74*:520; Falco SC, Zehring WA, Rothman-Denes LB (1980) *JBC 255*:4339.
365. Zehring WA, Rothman-Denes LB (1983) *JBC 258*:8074; Zehring WA, Falco SC, Malone D, Rothman-Denes LB (1983) *Virology 126*:678.
366. Zivin R, Malone C, Rothman-Denes LB (1980) *Virology 104*:205; Ohmori L, Laynes LL, Rothman-Denes LB (1988) *JMB 202*:1.
367. Guinta D, Stambouly J, Falco SC, Rist JK, Rothman-Denes (1986) *Virology 150*:33.
368. Lindberg GK, Pearle M, Rothman-Denes LB (1988) in *DNA Replication and Mutagenesis* (Moses R, Summers W, eds.). Am Soc Microbiol, Washington, DC, p. 130.

than by a headful mechanism. How the 3′ single-strand extension at the ends of the molecule are generated and what function they perform are still mysteries.

PM2 and Other Lipid-Containing Phages[369]

PM2, which infects a marine *Pseudomonas* species, attracts interest for several of its unusual properties: a lipid bilayered envelope, a small 9-kb circular duplex genome condensed in a nucleocapsid, and the lack of the tail that other duplex-DNA phages employ for injection. The envelope has proved to be an attractive model for studies of membrane structure and biosynthesis, and the ease of obtaining large amounts of supercoiled PM2 DNA has made it a popular reagent. Unfortunately, little is known about the replication cycle and its mechanism.

Other lipid-containing phages have been studied as model systems for exploring membrane structure and assembly.[370] These include the *Pseudomonas* phage ϕ6, with a segmented RNA genome, and PRD1 and PR4, representatives of a large family of *E. coli* and *Salmonella* phages. In phages of this family, the linear genomes (about 15 kb) are inside a membrane vesicle within the protein capsid. The phospholipid of the membrane, derived from the host cell, has a similar composition, although deficient in phosphatidylglycerol. The membrane is protein-rich and contains most of the 17 structural proteins. The terminal protein, bound to the ends of the linear genome, is within the membrane vesicle. Proteins responsible for filling the vesicle with DNA during assembly have been identified but details of their operations remain to be explored. Replication is initiated by protein priming, as will be discussed for ϕ29 (see below). Surprisingly, genetic analysis indicates that most of the *E. coli* replication proteins are required for formation of the terminal protein initiation complex.[371]

B. subtilis Phages: SPO1, PBS1, PBS2, ϕ29, SPP1

Much interest in *Bacillus* species has derived from properties that could not be studied in *E. coli*, namely, sporulation and, for a time, the ease of transformation or transfection by DNA (Section 21-8). *B. subtilis* phages (Table 17-14) have not

Table 17-14
***Bacillus subtilis* phages**

Phage	Related phages	DNA (kb)	Thymine or analog[a]	Comments
SPO1	SP8, SP82G, ϕe, 2C	140–195	HMU	phage-encoded DNA polymerase
PBS1	PBS2	300	U	phage-encoded RNA and DNA polymerases
ϕ29	ϕ15, PZA	19	T	sequence complete; covalently bound protein at 5′ termini
SPP1		43	T	separated strands easily isolated; uses host DNA polymerase

[a] HMU = 5-hydroxymethyluracil; T = thymine; U = uracil.

369. Mindich L (1978) in *Comprehensive Virology* (Fraenkel-Conrat H, Wagner RR, eds.). Plenum Press, New York, Vol. 2, p. 271; Brewer GJ (1980) *Int Rev Cytol 68*:53.
370. Mindich L, Bamford DH (1988) in *The Bacteriophages* (Calendar R, ed.). Plenum Press, New York, Vol. 2, p. 475.
371. Banford DH, Mindich L (1984) *J Virol 50*:309.

been studied as extensively as coliphages, but those that have been studied are remarkably interesting. They resemble coliphages morphologically and biologically, including virulent, temperate,[372] and defective classes. The genomes are generally linear, duplex, and nonpermuted. Some are notable for substitution of hydroxymethyluracil or uracil for thymine, and others for protein priming of the initiation of replication. Phage DNA synthesis is insensitive to HP uracil, an agent that inhibits host DNA replication by blocking DNA polymerase III (Section 5-7). Phage-encoded DNA polymerases are responsible for DNA synthesis, as demonstrated for SPO1, PBS1, and ϕ29.

SPO1 Phage.[373] SPO1 and closely related phages (SP82G, ϕe) have the size and complexity of the T-even phages, but the injection of DNA is slower, consuming several minutes. The genes controlling DNA synthesis, located at one end of the duplex genome, enter and are transcribed first.

Biosynthetic pathways are altered in five steps, including the action of a dTTPase involved in the substitution of hydroxymethyl dUTP for dTTP for incorporation into phage DNA (Section 2-13). However, inclusion of some thymine residues (up to 20% of hydroxymethyluracil) is tolerated. Ten phage genes are required for replication,[374] including those directing the synthesis of HM·dUTP, a DNA polymerase, and two genes specifically involved in initiation. Inhibitor studies indicate that the host gyrase is essential;[375] a phage mutant that imparts resistance to nalidixate, the gyrase inhibitor, indicates that SPO1 may modify this enzyme. Analysis of the structure of replication intermediates in vivo indicates that SPO1 has at least two origins of DNA replication.[376] Head-to-tail concatemers of up to 20 genome lengths have been observed and probably are responsible for replicating the ends of the molecule, as described for phage T7[377] (Section 5).

Like T4, these phages regulate transcription by modifying host RNAP (Section 7-4). The phage encodes two alternate σ factors that replace those of the host during development. The ϕe genome has attracted attention because its transcription is inhibited in sporulating cells. This failure, ascribed to sporulation-induced changes in host RNAP, has been used to assay the switch from vegetative metabolism to spore development.

PBS1 and PBS2 Phages. Uracil is substituted for thymine in the very large PBS1 and PBS2 phages. Four phage-encoded enzymes provide the nucleotide biosynthetic pathway for ensuring that uracil rather than thymine is incorporated into PBS1 DNA (Section 2-13). Transcription control is exercised by a phage RNAP (Section 7-13) that resembles the host enzyme in structural complexity. A novel DNA polymerase, probably encoded by the phage, has a mass in the range of 145 to 195 kDa and is unique among phage polymerases for its multisubunit structure.[378] A 3′→5′ exonuclease with a preference for single strands is associated with the polymerase.

---

372. Zahler SA (1988) in *The Bacteriophages* (Calendar R, ed.). Plenum Press, New York, Vol. 1, p. 559.
373. Stewart C (1988) in *The Bacteriophages* (Calendar R, ed.). Plenum Press, New York, Vol. 1, p. 477.
374. Glassberg J, Slomiany RA, Stewart DR (1977) *J Virol 21*:54; Glassberg J, Franck M, Stewart DR (1977) *J Virol 21*:147.
375. Alonso JC, Sarachu AN, Grau O (1981) *J Virol 39*:855.
376. Glassberg J, Franck M, Stewart CR (1977) *Virology 78*:433; Curran JR, Stewart DR (1985) *Virology 142*:78.
377. Levner MH, Cozzarelli NR (1972) *Virology 48*:402.
378. Hitzeman RA, Price AR (1978) *JBC 253*:8526.

ϕ29 Phage.[379] The small ϕ29 phage is one of the most thoroughly studied and best understood among all the phages. It has a linear, double-stranded DNA, 19.3 kb in length.[380] Its most remarkable feature is the covalent attachment of a protein molecule to each end of the genome[381] (Fig. 17-38). This 31-kDa terminal protein, gp3, is linked to each 5′ end by a phosphodiester bond between the OH group of an internal serine and 5′ dAMP, the terminal nucleotide. This covalent attachment of gp3 with dAMP primes the synthesis of both strands. ϕ29 was the earliest example of protein-priming in prokaryotes (Section 8-8), now known to occur in other phages, such as PRD1, and in some fungal and plant plasmids. The mechanism strongly resembles that of the DNA adenoviruses (Section 19-4) and the RNA poliovirus.

In addition to gp3, five other ϕ29 proteins are required for replication: the DNA polymerase (gp2), the dsDNA-binding protein (gp6), the ssDNA-binding protein (gp5), and two of unknown function (gp1 and gp17).[382] During initiation (Fig. 17-38), gp3 binds to the ends of the genome by association with the parental

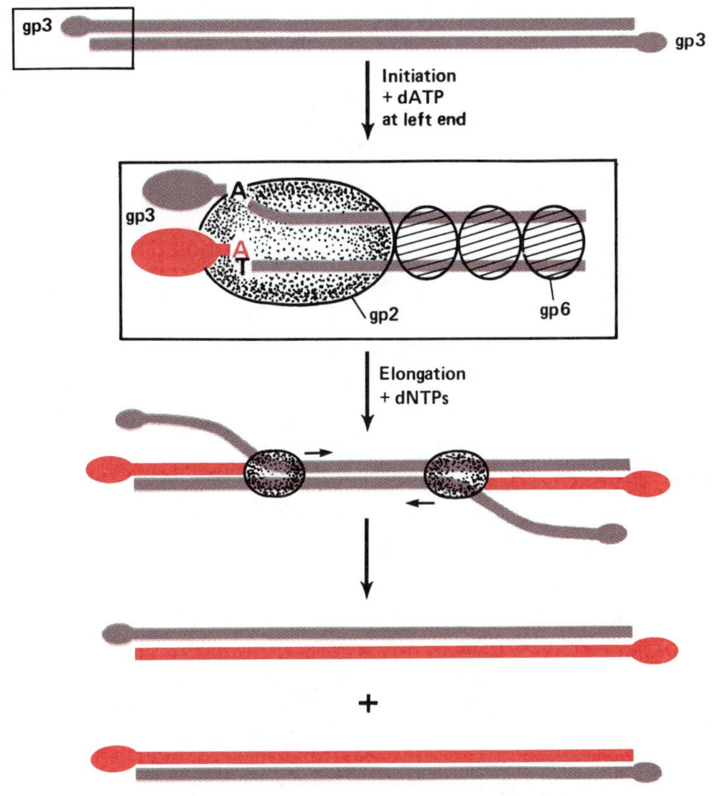

Figure 17-38
DNA replication of ϕ29 phage. Initiation by terminal protein (gp3), DNA polymerase (gp2), and dsDNA-binding protein (gp6) is detailed in the enlarged view.

379. Salas M (1988) in *The Bacteriophages* (Calendar R, ed.). Plenum Press, New York, Vol. 1, p. 169.
380. Yoshikawa H, Ito J (1982) *Gene 17*:323; Garvey KJ, Yoshikawa H, Ito J (1985) *Gene 40*:301; Vlcek C, Paces V (1986) *Gene 46*:215.
381. Hermoso JM, Méndez E, Soriano F, Salas M (1985) *NAR 13*:7715.
382. Carrascosa JL, Camacho A, Moreno F, Jiménez F, Mellado RP, Viñuela E, Salas M (1976) *EJB 66*:229.

gp3 or by direct interaction with the 5′ terminal sequence.[383] Gp2, in the presence of dATP, catalyzes the formation of the gp3·dAMP covalent complex, providing a primer for extensive DNA synthesis by the same gp2 polymerase.[384] Gp6 stimulates the formation of the terminal protein initiation complex by binding to the ends of the genome and altering the DNA structure.[385] Replication of ϕ29 starts at both ends of the genome, non-simultaneously; two continuous strands are synthesized in opposite directions and no discontinuous replication is required.

SPP1.[386] The genome of SPP1 is a linear duplex, terminally repetitious and circularly permuted,[387] like the genome of P22. The two strands are asymmetric in composition, making it simple to separate them and follow their fates. For this reason, SPP1 has been employed in studies of transfection, repair, recombination, and replication. Replication, initiated at a unique origin, depends on host DNA pol III.[388]

---

383. Blanco L, Prieto I, Gutiérrez J, Bernad A, Lázaro JM, Hermoso JM, Salas M (1987) *J Virol 61*:3983; Gutierrez J, Vinós J, Prieto I, Méndez E, Hermoso JM, Salas M (1986) *Virology 135*:474.
384. Blanco L, Salas M (1985) *PNAS 82*:6404; Blanco L, Bernad A, Salas M (1988) in *DNA Replication and Mutagenesis* (Moses RE, Summers WC, eds.). Am Soc Microbiol, Washington DC, p. 122.
385. Blanco L, Gutiérrez J, Lázaro JM, Bernad A, Salas M (1986) *NAR 14*:4923; Prieto I, Serrano M, Lázaro JM, Salas M, Hermoso JM (1988) *PNAS 85*:314.
386. Riva S, Polsinelli M, Falaschi A (1968) *JMB 35*:347.
387. McIntosh PK, Dunker R, Mulder C, Brown MC (1978) *J Virol 28*:865; Morelli G, Fisseau C, Behrens B, Trautner TA, Luh J, Ratcliff SW, Allison DP, Ganesan AT (1979) *MGG 168*:153.
388. Rowley SD, Brown NC (1977) *J Virol 21*:493.

CHAPTER 18

Plasmids and Organelles

18-1 Bacterial Plasmids

The surge of interest in plasmids and advances in understanding of their replication account for the expansion of a modest section in the first edition of *DNA Replication* to this rather large chapter only a decade later. Progress in plasmid research stems largely from the use of recombinant DNA technology, discoveries of in vitro replication systems, and facile methods for synthesizing and sequencing DNA.

Plasmids are extrachromosomal, autonomously replicating, double-stranded DNA circles, found in most bacteria (Table 18-1). In essence, plasmids, like viruses, are clusters of genes with a capacity to replicate. Lacking the information to make a viral coat, they fail to exist outside a cell. These episomal elements are usually identified with genes for drug resistance (R plasmids), self-transmissibility (F plasmid), and antibiotics known as bacteriocins (colicinogenic substances). They are also responsible for resistance to heavy metals and irradiation, for production of restriction endonucleases, and for the metabolism of various compounds such as camphor and toluene. Plasmids also induce tumors in plants. The crown gall tumor in dicotyledons results from the integration of a small region from a plasmid carried by the bacterium *Agrobacterium tumefaciens* into plant chromosomes.

In addition to those that occur naturally, plasmids have been constructed in vitro with a functional origin, and are maintained when introduced into cells. Plasmids that utilize bacterial chromosomal origins (*oriC* plasmids; Section 16-3), the phage λ origin (λdv plasmids; Section 17-7), and even yeast ARS elements (Section 16-6) are useful in the study of replication due to their manageable size and dispensability for cell growth.

Table 18-1
Plasmids of Gram-negative bacteria[a]

Plasmid	Length (kb)	Self-transmissibility	Copy number	Features[b]
ColE1	6	−	10–20	confers immunity to colicinE1
R300B	9	−	10–20	
pSC101	9	−	4–6	Tc^R
F	~90	+	1–2	F pilus
ColV	~90	+	1–2	
EntP30	~90	+	1–2	
R1, R100[c]	90	+	1–2	Cm^R,Sm^R,Su^R,Tc^R
R6K	38	+	10–20	Ap^R,Sm^R
RK2	56	+	5–8	broad host range
Agrobacterium plasmids	140–500	+		tumor induction in plants
Rhizobium plasmids	40–50	+		

[a] Courtesy of Dr DR Helinski.
[b] Superscript R = resistance to the designated drug: Ap = ampicillin, Cm = chloramphenicol, Sm = streptomycin, Su = sulfonamides, Tc = tetracycline.
[c] R1 and R100 differ in their drug resistance.

Other mobile genetic elements, namely insertion sequences and transposons (Table 18-2), play a large role in the evolution of plasmids and promote the exchange of genetic information from one genome to another. *Insertion sequences* (IS elements) are small genetic elements (<2 kb) that encode the capacity to insert themselves nearly randomly into any DNA molecule via transposition (Section 21-7). Genome rearrangements, such as deletions and inversions, are a common side effect of transposition. When insertion elements flank a gene, they can form a *composite transposon*, capable of moving as a unit. Transposons carrying antibiotic-resistance genes have had a large impact on both medical microbiology and bacterial molecular genetics. Insertion sequences and transposons are important components of bacterial plasmids; R plasmids carry several different transposons and insertion sequences (Fig. 18-1), providing, in addition to the resistance genes, regions of homology with other plasmids and bacterial chromosomes. These patches of homologous sequence promote recombination between the different genomes, important for several plasmid-promoted reactions (e.g., mobilization and integrative suppression; see below and Section 4).

Table 18-2
Mobile genetic elements

Element	Example	Size (kb)	Lifestyle
Phage	Mu, λ	~50	self-replicating; generalized (Mu) or specialized (λ) insertion
Plasmid	R, F, ColE1	2–200	self-replicating; insertion via homology between insertion sequence elements
Insertion sequence	IS1, IS2, IS3	0.8–1.4	inserted widely in host and plasmid genomes
Transposon	Tn3, Tn5, Tn10	~4–12	consists of selectable genes flanked by insertion sequences

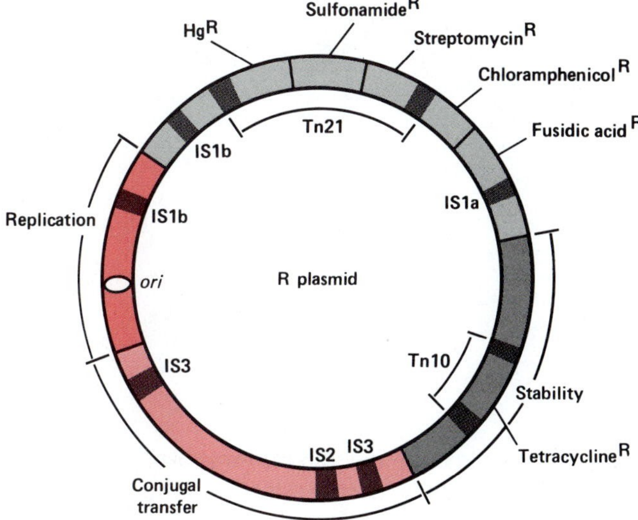

Figure 18-1
Components of R plasmids. Segments are responsible for replication, stability, transfer, and resistance (R) to drugs and mercuric ion. Multiple insertion sequences (IS) and composite transposons (Tn) are common. [Based on the map of NR1 from Womble DD, Rownd RH (1988) *Microbiol Rev 52*:433]

Like viruses, plasmids provide important windows into the replication mechanisms of the host cell. Dispensable for cell survival, these small, easily manipulatable templates are attractive alternatives to the chromosome for the study of DNA replication. The most commonly studied plasmids of Gram-negative bacteria depend on host proteins for priming and DNA synthesis (Table 18-3), but contribute their own factors to the initiation process. Except for ColE1, each plasmid encodes at least one replication protein specifically responsible for recognizing the plasmid origin (Table 18-4). RSF1010 is unusual in encoding three replication proteins which make it independent of host functions except for DNA polymerase. Surprisingly perhaps, the host *dnaA* gene, long considered the specific initiator of the *E. coli* chromosome, is also involved in the replication

Table 18-3
Host proteins involved in plasmid replication[a]

	Host protein							
Plasmid	RNAP[b]	pol I	dnaA	dnaB	dnaC	dnaG	dnaE	dnaT
ColE1, pBR322	+	+	±[c]	+	+	+	+	+[f]
pSC101		−	+	+	+	+	+	
R1, R100	−[d]	−	+[e]	+	+	+	+	−[f]
R6K	+	−	+[e]	+	+	+	+	
F	+[g]	−	+[e]	+	+	+	+	
RK2	−		+	+		+	+	
RSF1010	−	−	−	−	−	−	+	

[a] Blank spaces represent uncertainty due to lack of information.
[b] RNAP = RNA polymerase; usually determined by sensitivity to rifampicin.
[c] Optional.
[d] Required only for protein synthesis.
[e] Replication defect demonstrated clearly only with null mutants of *dnaA*; plasmid replication supported by most missense alleles.
[f] In vitro replication resistant to anti–dnaT protein antibody.
[g] The rare σ^{32} factor is required.

Table 18-4
Plasmid-encoded initiator proteins

Plasmid	Gene	Protein	Monomer mass (kDa)
pSC101	*repA*	RepA	38
R1	*repA*	RepA	33
R100	*repA1*	RepA1	33
R6K	*pir*	π	36
F	*repE*	RepE	29
RK2	*trfA*	TrfA	32 and 43
RSF1010	*repC*	RepC	31
pT181	*repC*	RepC	38

of many plasmids in *E. coli*. How the plasmid-specific initiator proteins work in concert with dnaA protein should provide further insights into the mechanism of initiation on duplex DNA as well as into the multiple activities of dnaA protein.

Plasmids provide opportunities to address several questions not easily accessible with viruses. In contrast to the genomes of phages, plasmids are maintained at a fixed number of copies per cell by controls that regulate the frequency of initiation of replication. Constant monitoring by the plasmid of its intracellular concentration, or *copy number*, is linked to responses in the replication program. Two general mechanisms of negative regulation are used by *E. coli* plasmids to maintain a stable copy number. ColE1 and R1 employ specific inhibitors to limit the rate of production of a critical initiation factor (e.g., the first primer or the initiator protein). Plasmids pSC101, F, and P1 employ another mechanism: multiple binding sites compete with the origin for the initiator protein, decreasing the initiation frequency.

The isolation of mutants and the construction of new plasmids carrying specific elements provide methods for identifying important plasmid-encoded functions. Mutations that interfere with the normal replication pathway invariably lower the copy number, while mutants that disrupt the regulatory circuitry elevate the copy number. Regulatory elements can also be identified by the *incompatibility* phenotypes they impart. Plasmids are incompatible if they cannot be maintained in the same cell. Closely related plasmids are naturally incompatible because they are indistinguishable to the replication machinery. Artificially constructed plasmids carrying elements that inhibit replication exclude natural plasmids when introduced into the same cell. Thus, elements that express incompatibility provide insights into normal plasmid replication and maintenance.

Low-copy-number plasmids encode a wide variety of different functions that ensure the inheritance of the plasmid during cell division. These functions can (1) act directly in partitioning in a way comparable to the centromeres of eukaryotic chromosomes, (2) resolve multimeric forms of plasmids, and (3) kill or prevent the division of cells which have lost the plasmid.

Plasmids also provide vectors for genetic exchange between bacteria during *conjugation* (Section 7). Some large plasmids encode multiple functions that allow them to invade new cells and thus are *self-transmissible*. The transfer functions of these self-transmissible plasmids encode organelles (pili), replication enzymes (helicases and primases), and special DNA sites for initiation of transfer. Smaller *mobilizable* plasmids carry elements allowing their transfer in the presence of a self-transmissible plasmid. Plasmid transfer during conjugation is of enormous medical importance, being responsible for the rapid propagation of drug-resistance determinants throughout bacterial populations.

From the outset of the recombinant DNA revolution, plasmids have been key actors and still occupy center stage. Commonly, the young scientist takes on a "plasmid prep" as the first assignment upon entering a laboratory. With these molecules, virtually any DNA fragment can be amplified and functionally expressed. An appreciation of their replication, regulation, and inheritance is imperative for their effective use.

18-2 ColE1

ColE1 is a 6.6-kb, high-copy-number (about 20 copies per cell) *E. coli* plasmid encoding resistance to colicin E1. In contrast to the host chromosome and many other plasmids, ColE1 does not require de novo protein synthesis for initiation of replication. This independence allows the plasmid to be amplified relative to the host chromosome in the presence of protein-synthesis inhibitors (e.g., chloramphenicol). Most of the popular cloning vectors, including pBR322 and the pUC plasmids, use the ColE1 origin (or a related one[1]) and thus replicate by the same mechanism. Interesting differences between the naturally occurring plasmid and the constructed derivatives do exist, however, and will be mentioned throughout this section.

ColE1 replicates unidirectionally from a unique origin[2] and relies entirely on host-encoded proteins. The minimal replicon is contained on a 600-bp fragment. The synthesis of the two strands is nearly simultaneous,[3] although initiated by different mechanisms which can be experimentally manipulated independently of one another. ColE1 thus provides a system for examining the coordination of events during semidiscontinuous DNA replication. Among several pathways available for initiation of ColE1 replication, the predominant model will be described first; the relative contributions in vivo of the alternative modes observed in vitro are difficult to assess.

Initiation of Replication

Initiation of Continuous-Strand Synthesis. Three host enzymes are required to initiate continuous-strand synthesis on a negatively supercoiled plasmid in vitro: RNA polymerase (RNAP) holoenzyme, DNA pol I, and RNase H.[4] In vivo, initiation is preceded by a rifampicin-sensitive event, indicating that RNAP is essential.[5] Replication is blocked under most circumstances in *polA* mutant strains.[6] The effect of an RNase H deficiency on ColE1 replication is complicated (see below).

1. Sutcliffe JG (1978) *CSHS 43*:77.
2. Tomizawa J-I, Sakakibara Y, Kakfuda T (1974) *PNAS 71*:2260.
3. Inselburg J (1974) *PNAS 71*:2256.
4. Itoh T, Tomizawa J (1980) *PNAS 77*:2450; Itoh T, Tomizawa J (1982) *NAR 10*:5949.
5. Sakakibara Y, Tomizawa J (1974) *PNAS 71*:802; Sakakibara Y, Tomizawa J (1974) *PNAS 71*:1403.
6. Kingsbury ST, Helinski DK (1970) *BBRC 41*:1538.

RNAP synthesizes the primer for DNA synthesis (Fig. 18-2). Transcription of the primer (RNA II) is initiated 555 bp upstream from the transition point between RNA and DNA,[7] the point commonly regarded as the origin.[8] RNA II spontaneously forms an RNA–DNA hybrid with the template in the origin region.[9] Hybrid formation depends critically on the secondary structure of the transcript, a series of stems and loops. When transcription of RNA II leads to proper hybrid formation, initiation of replication can follow. Mutations that disrupt RNA folding are defective in primer synthesis because they fail to form the hybrid. Modulation of the RNA II secondary structure regulates replication (see below). RNase H, specific for cleavage of RNA in RNA–DNA hybrids, generates 3′-OH ends in the origin that serve as primers for DNA synthesis by pol I.[10] Uncleaved transcripts are poor primers and the proper RNA secondary structure is crucial for efficient cleavage by RNase H and for primer utilization.

Pol III holoenzyme, with its superior processivity, replaces pol I after the synthesis of about 400 ntd.[11] This switch may be associated with initiation of discontinuous replication on the other strand, which could provide a helicase for the continuous-strand replication fork, but these details are still unknown.

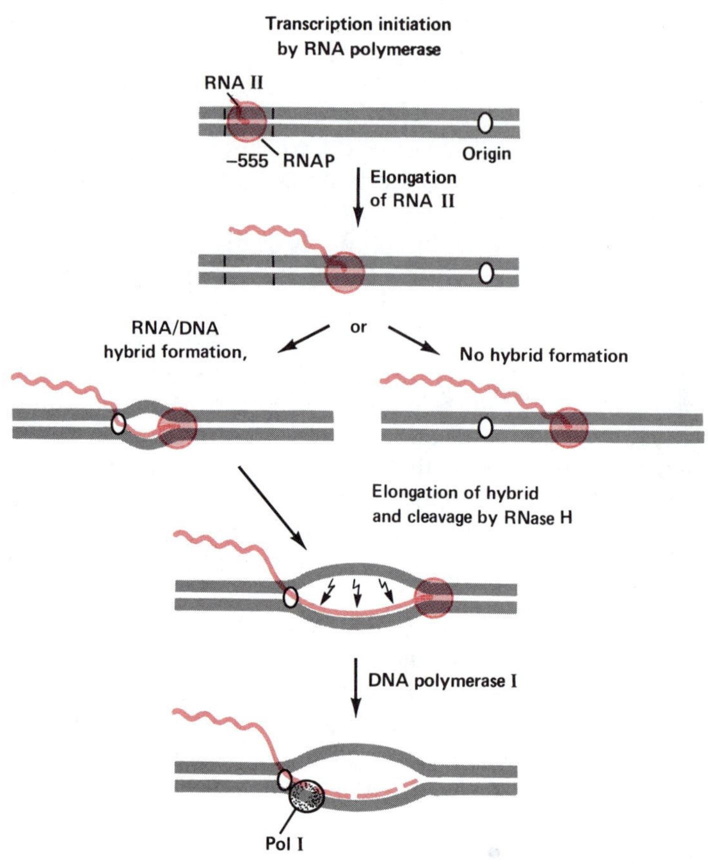

Figure 18-2
Initiation of synthesis of the ColE1 continuous strand by RNA polymerase (RNAP), RNase H, and pol I.

7. Itoh T, Tomizawa J (1980) *PNAS 77*:2450.
8. Tomizawa J-I, Ohmore H, Bird RE (1977) *PNAS 74*:1865.
9. Selzer G, Tomizawa J (1982) *PNAS 79*:7082.
10. Itoh T, Tomizawa J (1978) *CSHS 43*:409.
11. Staudenbauer WL (1977) *MGG 156*:27.

Initiation of Discontinuous-Strand Synthesis. ColE1 contains two sequences that can act as origins of complementary-strand synthesis on single-stranded phage DNA.[12] These sequences are functional analogs of the complementary-strand origin of ϕX174 DNA, and are called primosome assembly sites (*pas* sequences), also known as n′ sites or *rri* (rifampicin-resistant initiation) sequences (Section 8-4). The two *pas* sequences are on opposite strands near the origin. The site on the continuous-strand template is about 400 bp downstream, approximately where the switch from pol I to pol III holoenzyme occurs, suggesting its involvement in this process. The *pas* on the discontinuous-strand template is 150 bp downstream of the origin and functions in initiation on this strand.

The synthesis by pol I could activate the *pas* on the discontinuous-strand template by rendering it single-stranded (Fig. 18-3). In vitro, using purified proteins, initiation of discontinuous replication requires the presence of the seven *E. coli* proteins essential for primosome assembly: PriA, PriB, PriC, dnaB, dnaC, dnaT, and primase (Sections 8-3 and 8-4), as well as the continuous-strand initiation proteins and SSB.[13]

The superhelical state of the plasmid template is critical during initiation. Delicate balancing between gyrase and topoisomerase I activities is vital in vitro

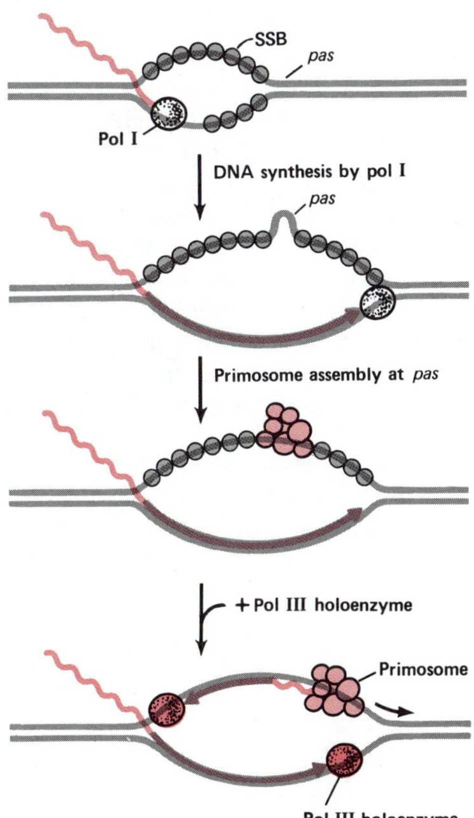

Figure 18-3
Initiation of synthesis of the ColE1 discontinuous strand at the primosome assembly site (*pas*).

12. Zipursky SL, Marians KJ (1980) *PNAS 77*:6521; Zipursky SL, Marians KJ (1981) *PNAS 78*:6111.
13. Minden JS, Marians KJ (1985) *JBC 260*:9316.

to maintain specificity of initiation for the origin and *pas* sequence.[14] This sensitivity to plasmid topology indicates that a major objective of the initiation events is the opening of the DNA duplex, thereby providing a single-stranded template for priming and replication.

Alternative Pathways for Initiation of Discontinuous Synthesis. Although ColE1 replication does not require the host dnaA protein, a dnaA binding site within the origin region very near the *pas*[15] on the template for discontinuous synthesis provides an alternative pathway for initiation of replication on this strand.[16]

In a crude cell extract, dnaA protein stimulates pBR322 replication twofold; replication still requires pol I, dnaB protein, primase, and the synthesis of RNA II. However, in the presence of excess dnaA protein, the primosomal protein dnaT, normally essential for pBR322 replication,[17] is no longer required. The *pas* on the template for discontinuous synthesis and the dnaA binding site are functionally equivalent. Replication requires dnaA protein absolutely and is independent of dnaT protein if the *pas* is deleted. Thus, two pathways are available for the initiation of synthesis on this strand, one dependent on the *pas* and dnaT protein, the other dependent on dnaA protein and its binding site. Either pathway leads to extensive replication; in the absence of both pathways, DNA synthesis is very inefficient. The equivalence of the dnaA binding site and the *pas* emphasizes that an essential role of site-specific DNA-binding proteins in initiation of replication is to direct the helicase and primase to the template; neither of these proteins shows any preference for the replication origin.

The dnaA binding site, like the *pas*, functions only when activated by RNAP and pol I during the initiation of continuous-strand synthesis. DnaA binding sites from five different sources are functional on pBR322, indicating that primosome loading requires no sequence other than the 9-bp dnaA box[18] on the discontinuous-strand template. Location and orientation of the box with respect to the R-loop primer generated by RNAP is important; a dnaA box located 16 but not 457 bp downstream of the original location is functional, and the natural orientation is more active than its inverse. The direction of replication is, however, not affected by this orientation. How the formation of the R-loop activates initiation of discontinuous replication at the dnaA box is unclear, inasmuch as the functional form of the dnaA box, in contrast to a *pas* sequence, is presumably duplex DNA.

Although their replication is inefficient, plasmids in which both the *pas* and the dnaA box are deleted can still be maintained in vivo. Templates lacking both sites can be replicated in vitro, although intermediates accumulate due to inefficient discontinuous-strand synthesis. The low level of synthesis on this strand is absolutely dependent on the presence of dnaT protein in vitro, and replication of the deletion plasmids is inhibited in *dnaT* mutant strains. Thus this third mode of replication depends on dnaT protein while being independent of both the *pas* and the dnaA box.

Transcriptional Activation in the Absence of RNase H. ColE1 can be maintained in strains lacking RNase H, in apparent contradiction to its requirement for

---

14. Minden JS, Marians KJ (1985) *JBC 260*:9316; Marians KJ, Minden JS, Parada C (1986) *Prog N A Res 33*:111.
15. Fuller RS, Funnell BE, Kornberg A (1984) *Cell 38*:889.
16. Seufert W, Messer W (1987) *Cell 48*:73.
17. Masai H, Arai K (1988) *JBC 263*:15016.
18. Seufert W, Dobrinski B, Lurz R, Messer W (1988) *JBC 263*:2719.

initiation in vitro.[19] When RNase H is absent, the normally essential pol I is not needed,[20] indicating that initiation is by a substantially different mechanism. Surprisingly, formation of the RNA–DNA hybrid by RNA II is still required for replication, although its major role is no longer priming of DNA synthesis.[21] Regulation of RNA II hybrid formation by RNA I, which modulates plasmid copy number (see below), is still intact.[22]

In the absence of RNase H and pol I, primarily discontinuous initiation occurs; if pol I is present, some continuous-strand synthesis apparently arises from elongation of the uncleaved RNA II. Neither the *pas* sequences nor the dnaA box appear to be required for replication on transcriptionally activated templates. The presence of the RNA II–DNA hybrid, 30 to 40 ntd in length, allows assembly of the discontinuous-strand replication machinery through the action of proteins and mechanisms that are still unclear.

Replication Control: Regulation of Hybrid Formation by RNA II[23]

ColE1 is maintained at a constant copy number. ColE1-like plasmids with the same origin, but with other distinguishable features, exclude each other. Thus ColE1 contains elements for both copy number control and incompatibility. Both properties derive from the same replication control mechanism, the modulation of hybrid formation between RNA II and the template.

The replication control region is 200 to 600 bp upstream of the origin, within the sequence transcribed to generate RNA II. The regulatory species (RNA I) is a 108-ntd molecule transcribed from the strand opposite that used to produce the RNA II primer. This antisense RNA, complementary to the 5′ end of RNA II,[24] is initiated 445 bp upstream of the origin and is terminated near the initiation site for RNA II. RNA I inhibits primer formation by RNA II on all plasmids which are genetically incompatible.[25] Thus, RNA I is a *trans*-acting regulator of ColE1 replication.

RNA I, by complexing with RNA II, blocks its folding into the conformation required for hybridization with the template. RNA I is made up of three stem-and-loop structures and a single-stranded 5′ end,[26] corresponding to similar structures in RNA II.[27] Mutations in the loop regions dramatically affect association between the two RNAs, without perturbing the secondary structures, indicating the importance of the loops in the interaction.[28] RNA pairing occurs in three stages (Fig. 18-4): (1) the initial loop–loop interaction, termed "kissing," (2) pairing of the 5′ end of RNA I with RNA II, and finally, (3) propagation of pairing from the 5′ end to the 3′ end of RNA I to yield a stable, base-paired dsRNA. RNA I dramatically influences the secondary structure of RNA II even at distances remote from the paired region. Inasmuch as the folded state of RNA II is critical to its ability to hybridize with the template, disruption of the structure by RNA I blocks hybrid formation.

---

19. Naito S, Kitani T, Ogawa T, Okazaki T, Uchida H (1984) *PNAS 81*:550.
20. Kogoma T (1984) *PNAS 81*:7845.
21. Dasgupta S, Masukata H, Tomizawa J (1987) *Cell 51*:1113; Masukata H, Dasgupta S, Tomizawa J (1987) *Cell 51*:1123.
22. Ohmori H, Murakami Y, Nagata T (1987) *JMB 198*:223.
23. Tomizawa J (1987) in *New Perspectives on the Molecular Biology of RNA*, (Inouye M, Dubock BS, eds.). Academic Press, New York, p. 249; Polisky B (1989) *Cell 55*:929.
24. Itoh T, Tomizawa J (1982) *NAR 10*:5949.
25. Tomizawa J, Itoh T (1981) *PNAS 78*:6096; Tomizawa J, Itoh T, Selzer G, Som T (1981) *PNAS 78*:1421.
26. Tamm J, Polisky B (1983) *NAR 11*:6381; Morita M, Oka A (1979) *EJB 97*:435.
27. Masukata H, Tomizawa J (1984) *Cell 36*:513; Masukata H, Tomizawa J (1986) *Cell 44*:125.
28. Tomizawa J (1984) *Cell 38*:861.

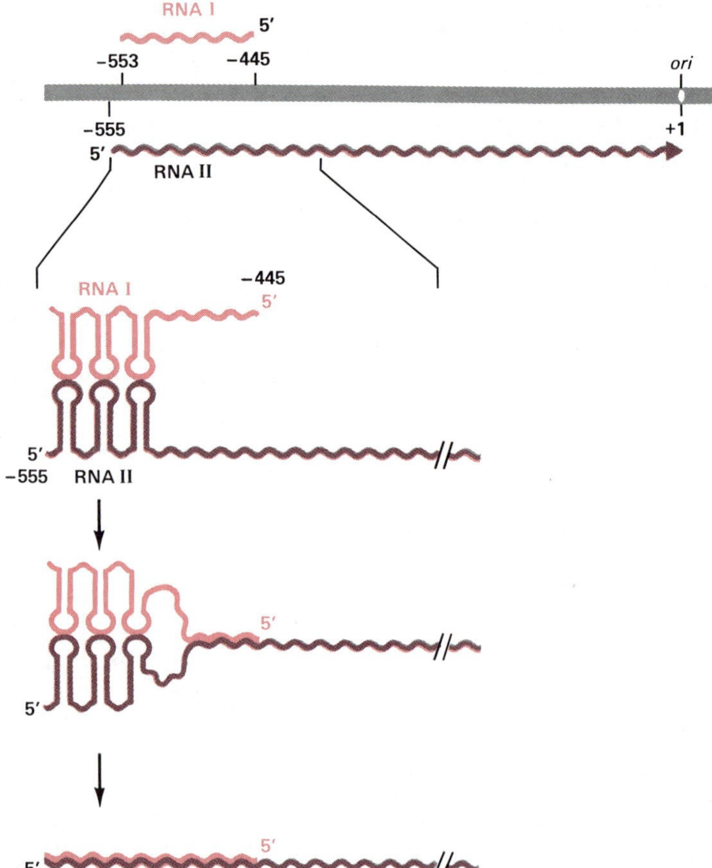

Figure 18-4
Stages in pairing of RNA I and RNA II. (Modified from a figure provided by Dr J-I Tomizawa)

ColE1 encodes a protein that enhances the inhibitory effect of RNA I. This 63-amino-acid protein, called Rom (*R*NA *o*ne inhibition *m*odulator) or Rop (*r*epressor *o*f *p*rimer), increases the rate of binding of RNA I to RNA II.[29] Rom protein is not essential for inhibition by RNA I and is absent from some common plasmids, including the pUC series. Deletion of Rom causes a fourfold to sixfold increase in plasmid copy number. Rom has no effect on plasmids in the absence of RNA I, confirming that the major replication regulator is the RNA and that Rom only magnifies its effect.

Inhibition by RNA I can occur only when RNA II is between 100 and 360 ntd long.[30] Once the RNA–DNA hybrid is formed, RNA I has no further effect. The rate of complex formation is thus critical to regulation. Rom protein, by increasing the rate of interaction, helps to ensure that it occurs during this "window of opportunity." Among the ColE1 plasmid population, templates for replication are chosen at random. No mechanism is known that imposes a delay of initiation on recently replicated plasmids, nor is there a device that provides for each

29. Tomizawa J, Som T (1984) *Cell 38*:871; Lacatena RM, Banner DW, Castagnoli L, Casareni G (1984) *Cell 37*:1009; Tomizawa J (1986) *Cell 47*:89.
30. Tomizawa J (1986) *Cell 47*:89.

plasmid's being replicated at least once per cell cycle. Thus, some plasmids may be replicated many times while others are not used as templates during a given cell cycle. The key factors that regulate copy number are the concentrations of RNA I and Rom protein.

Elongation, Termination, Segregation, and Partitioning

Semidiscontinuous synthesis of the two DNA strands occurs nearly simultaneously to advance the fork. Helicase activity of the primosome, probably by dnaB protein, separates the double helix;[31] gyrase removes the topological strain generated during unwinding. Synthesis of the discontinuous strand proceeds toward the origin, stopping at a distinct location, called *terH*, by an unknown mechanism.[32]

The topological problems associated with the late stages of replication on circular DNA molecules have been studied in some depth with pBR322.[33] During replication, all the helical intertwines between the two parental strands must be removed before the daughter molecules can be completed and segregated. As described earlier (Section 12-1), these helical turns can become interlinks between the duplex regions of the molecule, giving rise to multiply interlinked catenated dimers. In vitro, the predominant products of replication of pBR322, as well as of other circular templates (e.g., *oriC* plasmids, λdv, and SV40), are such catenated dimers.

Among the enzymes capable of resolving dimers generated during pBR322 replication, DNA gyrase is less effective than topoisomerase I. The latter decatenates duplex circles that contain single-stranded regions like those found in the late replication intermediates of pBR322, yet topo I is not required for pBR322 replication in vivo. An activity in cell extracts was found to resolve the catenated replication intermediates efficiently; it proved upon purification to be the poorly characterized and neglected topoisomerase III (Section 12-2).[34] Whereas the importance of topo III for plasmid and chromosome replication in vivo remains to be established, this enzyme is probably one of a variety of topoisomerases available for coping with the assorted topological problems arising in replication and other DNA transactions.

ColE1 plasmids are partitioned randomly at cell division. Stability of inheritance is therefore proportional to the number of plasmids present in the cell.[35] Inasmuch as plasmid multimers reduce the number of units available for partitioning, ColE1 (but not most cloning vectors derived from it) carries a site-specific recombination system (cer system; Section 21-6) that converts multimers to monomers. This recombination system is independent of the host RecA, RecF, and RecE recombination pathways (Section 21-5) but is dependent on three host-encoded proteins, including the products of the *argR* and *pepA* genes. Monomerization helps to keep the number of partitionable units at a level sufficient for random partitioning to maintain the plasmid level for hundreds of generations.

---

31. Minden JS, Marians KJ (1985) *JBC 260*:9316; Masai H, Arai K (1988) *JBC 263*:15016.
32. Minden JS, Marians KJ (1985) *JBC 260*:9316; Dasgupta S, Masukata H, Tomizawa J (1987) *Cell 51*:1113.
33. Marians KJ, Minden JS, Parada C (1986) *Prog N A Res 33*:111.
34. DiGate RJ, Marians KJ (1988) *JBC 263*:13366.
35. Summers DK, Sherratt DJ (1984) *Cell 36*:1097; Stirling CJ, Colloms SD, Collins JF, Szatmari G, Sherratt DJ (1989) *EMBO J 8*:1623; Stirling CJ, Szatmari G, Stewart G, Smith MC, Sherratt DJ, (1988) *EMBO J 7*:4389.

18-3 pSC101

The 9.4-kb pSC101 plasmid,[36] isolated originally from *Salmonella panama*, encodes resistance to tetracycline and is maintained at an intermediate copy number of 4 to 6 per host chromosome. In addition to replication functions, it encodes a region directing stable inheritance.[37] Replication of pSC101 has received significant attention, largely because it was initially the only plasmid known to depend on the host *dnaA* gene,[38] the essential initiator of chromosomal replication. The discovery of dnaA boxes near plasmid origins and the use of *dnaA*-null alleles[39] and in vitro replication reactions[40] have revealed partial or total dependence on dnaA protein in the replication of other plasmids (e.g., ColE1, R1, F, and P1). The sensitivity of pSC101 to numerous *dnaA* alleles suggests that this plasmid exploits more activities of dnaA protein than do other plasmids; pSC101 also appears to lack the redundant initiation mechanisms of ColE1 that make dnaA protein dispensable. How dnaA protein functions in cooperation with the plasmid-encoded origin-binding protein should help to clarify some of the multiple activities of dnaA protein in chromosomal initiation.

Replication

The replication of pSC101 is unidirectional from a unique origin.[41] The minimal replication region consists of a 1.3-kb fragment containing the origin sequence and the gene, *repA*, for the plasmid-specific initiator protein (Fig. 18-5).[42] RepA protein is the only plasmid-encoded protein required for replication. Unlike the ColE1 family of plasmids, pSC101 does not require pol I.[43]

RepA Protein. The 37.5-kDa RepA protein has been overproduced and purified as a fusion protein with *E. coli* β-galactosidase.[44] RepA binds cooperatively to repeated sequences in the origin and, by binding to its own promoter region,

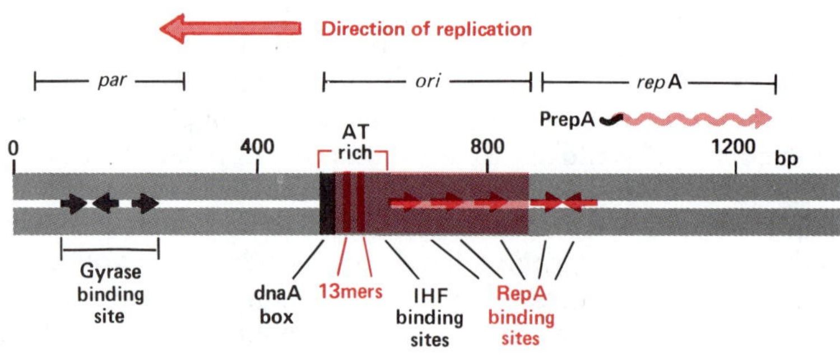

Figure 18-5
Replication region of pSC101, showing elements of the origin, partition locus (*par*), and *repA* gene.

36. Bernardi A, Bernardi F (1984) *NAR 12*:9415.
37. Meacock PA, Cohen SN (1980) *Cell 20*:529.
38. Hasunuma K, Sekiguchi M (1977) *MGG 154*:225.
39. Hansen EB, Yarmolinsky MB (1986) *PNAS 83*:4423.
40. Masai H, Arai K (1987) *PNAS 84*:4781; Seufert W, Messer W (1987) *Cell 48*:73.
41. Cabello F, Timmis K, Cohen SN (1976) *Nature 259*:285.
42. Vocke C, Bastia D (1983) *PNAS 80*:6557.
43. Cabello F, Timmis K, Cohen SN (1976) *Nature 259*:285.
44. Vocke C, Bastia D (1983) *Cell 35*:495.

regulates transcription. Both replication-defective and high-copy-number mutants of *repA* have been described. Thus RepA protein can both activate and inhibit replication.[45]

The Origin Region. The *cis*-acting sequence required for replication is about 400 bp and can be divided into three domains (Fig. 18-5): (1) leftmost, a dnaA box;[46] (2) middle, an 80-bp region of 84% AT content containing a DNA bend and two 13-bp repeats, homologous to those in *oriC* melted during initiation (Section 16-3);[47] and (3) rightmost, five iterated RepA-binding sites (*iterons*)—three as 18-bp direct repeats, and a pair as inverted repeats.[48] The latter overlap the promoter for the *repA* gene (just to the right of the origin) and regulate transcription rather than affect the initiation of replication directly.[49]

In addition to the binding of RepA protein to its iterons, two host proteins, dnaA and IHF (*i*ntegration *h*ost *f*actor; Section 10-7), interact with the origin. DnaA protein binds to the dnaA box, and perhaps to the 13-bp repeats as well, while IHF, required for pSC101 maintenance in vivo,[50] binds the AT-rich region.[51] IHF is a small, basic, histone-like protein that influences numerous replication, transcription, and recombination reactions by altering the DNA conformation upon binding (Section 10-8). Inasmuch as a host mutation in topo I allows pSC101 to replicate in the absence of IHF, modification of DNA structure by IHF is implied.[52] In vitro, the binding of IHF enhances the bent character of the AT-rich region. Alteration of the binding sites of IHF destroys its effect on initiation at the origin.

Regulation of Replication

Mutants in *repA* and *dnaA* reveal that both plasmid and host elements regulate pSC101 replication. Some *repA* mutants have an increased copy number, indicating that RepA protein activity can be rate-limiting for initiation.[53] Extra copies of the three tandem RepA binding sites inhibit pSC101 replication, presumably by sequestering RepA protein away from the origin.[54] The modulation of initiation by multiple iterons is a regulatory motif common to several plasmids and will be discussed in more detail for F plasmid (Section 6). Replication of pSC101 also responds to the activity of dnaA protein. Upon shift of a *dnaA*^{cs} (cold-sensitive) strain to a low temperature, both pSC101 and the *E. coli* chromosome initiate excessively.[55] Unfortunately, there is still no replication system for pSC101 with which to analyze the interplay between the dnaA and RepA proteins in vitro.

---

45. Armstrong KA, Acosta R, Ledner E, Machida Y, Pancotto M, McCormick M, Ohtsubo H, Ohtsubo E (1984) *JMB 175*:331.
46. Fuller RS, Funnell BE, Kornberg A (1984) *Cell 38*:889.
47. Bramhill D, Kornberg A (1988) *Cell 52*:743; Bramhill D, Kornberg A (1988) *Cell 54*:915.
48. Vocke C, Bastia D (1983) *Cell 35*:495.
49. Linder P, Churchward G, Xia GX, Yu YY, Caro L (1985) *JMB 181*:383.
50. Gamas P, Burger AC, Churchward G, Caro L, Galas D, Chandler M (1986) *MGG 204*:85; Biek DP, Cohen SN (1989) *J Bact 171*:2056; Bieck DP, Cohen SN (1989) *J Bact 171*:2066.
51. Stenzel TT, Patel P, Bastia D (1987) *Cell 49*:709.
52. Biek DP, Cohen SN (1989) *J Bact 171*:2056; Bieck DP, Cohen SN (1989) *J Bact 171*:2066.
53. Armstrong KA, Acosta R, Ledner E, Machida Y, Pancotto M, McCormick M, Ohtsubo H, Ohtsubo E (1984) *JMB 175*:331.
54. Linder P, Churchward G, Caro L (1983) *JMB 170*:287.
55. Frey J, Chandler M, Caro L (1984) *JMB 179*:171.

The Partition Locus

Adjacent to the pSC101 replication origin is the *par* (*par*tition) locus (Fig. 18-5), a 270-bp sequence responsible for the stable inheritance of plasmids during cell division.[56] This region, nonessential for replication, has no effect on plasmid copy number or incompatibility. Plasmids lacking *par* are lost at a much higher frequency than predicted for the random segregation of plasmids with this copy number. Apparently *par* allows individual plasmids to be counted and distributed; in the absence of *par* the plasmid population is treated as a single unit.

The mechanism of stabilization by *par* is unknown. Resolution of multimers, as has been described for the *cer* function of ColE1 (Section 2), is not the basis of stability. The *par* sequence acts only *in cis*, encodes no proteins, and stabilizes unrelated plasmids. Thus *par* action depends exclusively on host functions.[57] Plasmids containing a partial deletion of *par* will segregate faithfully, but are excluded from cells containing a plasmid with the wild-type sequence. Apparently partial inactivation of *par* renders plasmids unable to compete with the wild type for replication.

The *par* sequence is a specific binding site for DNA gyrase,[58] which has a 40-fold higher affinity for *par* than for random DNA. A 4-bp deletion within *par*, which renders the plasmid par^-, reduces gyrase-binding affinity. However, the functional relationship between gyrase binding and partitioning remains uncertain. The change in plasmid superhelicity resulting from *par* deletions seems too modest to explain the mutant phenotype. Furthermore, there is no indication of an accumulation of catenated plasmids, which could be resolved by gyrase, in the absence of *par*. Some feature of gyrase other than topoisomerase activity, such as membrane binding, may explain its contribution to plasmid maintenance.[59]

18-4 R Plasmids: R1 and R100

The 100-kb R plasmids are responsible for the rapid propagation of multiple drug-resistance genes in bacterial populations. They are maintained at a copy number of 1 to 2 per chromosome and are self-transmissible by conjugation. The high homology of the transfer genes to those of F plasmids suggests a common ancestry.[60] The mechanisms of F plasmid inheritance and transfer by conjugation are presented in Sections 6 and 7.

The extensive use of antibiotics beginning in the 1950s provided strong selective pressures for generating and spreading the R plasmids. Yet replication and transfer genes homologous to those of present-day R plasmids have been identified in plasmid populations isolated prior to the widespread use of antibiotics.[61] The drug-resistance genes encoded on transposons (Section 21-7) seem to have accumulated in preexisting replicons, thereby generating the R plasmids.

56. Meacock PA, Cohen SN (1980) *Cell 20*:529.
57. Tucker WT, Miller CA, Cohen SN (1984) *Cell 38*:191; Miller CA, Tucker WT, Meacock PA, Gustafsson P, Cohen SN (1983) *Gene 24*:309.
58. Wahle E, Kornberg A (1988) *EMBO J 7*:1889.
59. Gustafsson P, Wolf-Watz H, Lind L, Johansson K-E, Nordström K (1983) *EMBO J 2*:27; Miller CA, Beaucage SL, Cohen SN (1990) *Cell 62*:127.
60. Womble DD, Rownd RH (1988) *Microbiol Rev 52*:433.
61. Datta N (1985) in *Plasmids in Bacteria* (Helinski DR, Cohen SN, Clewell DB, Jackson DA, Hollaender A, eds.). Plenum Press, New York, p. 3.

Transposons and the related insertion sequences, powerful promoters of genetic change, are responsible for the rapid evolution of these replicons.

The two most widely studied members of this class are the homologous, incompatible plasmids R1 and R100 (also called NR1).[62] Both R1 and R100 encode resistance to chloramphenicol, fusidic acid, streptomycin, spectinomycin, and sulfonamides; in addition, R1 also imparts resistance to ampicillin and kanamycin, and R100 also to tetracycline and mercuric ion (Fig. 18-1). In view of the similarity of R1 and R100 genes for replication, partition, and transfer, the strategies employed for these processes must also be closely related. Since most in vitro replication studies have employed R1 rather than R100, the R1 terminology for the replication functions will be used here.

Replication

Replication is unidirectional from a single origin.[63] The replication functions have been identified by constructing autonomously replicating "miniplasmids" (mini R1 or mini R100) or by determining the regions required for an integrated plasmid to suppress defects in host DNA replication. A 2.5-kb region contains the genes and control elements required for properly regulated replication (Fig. 18-6). These include the origin (*oriR*), the *repA* gene (encoding the RepA initiator protein), and the *copA* and *copB* elements that regulate replication by controlling the expression of *repA*.

The 149-bp *oriR* sequence[64] is located 188 bp downstream of the *repA* gene; it lacks iterated elements common to replication origins. The functional units are a 90-bp segment to which RepA protein binds, and a 70-bp region of 80% AT content which contains a binding site for the host IHF. A dnaA box lies just upstream from the 90-bp RepA binding site.

RepA Protein, the *Cis*-Acting Initiator. The *repA* gene encodes a 33-kDa protein that activates initiation at the origin. In vitro, RepA protein binds specifically to *oriR*, covering a 90-bp region. It also binds, less tightly, to sequences in the 3′ end of its structural gene. In contrast to other initiator proteins, RepA acts almost exclusively *in cis*; plasmids defective in *repA* cannot be complemented by a *repA*$^+$ plasmid present in the same cell.[65]

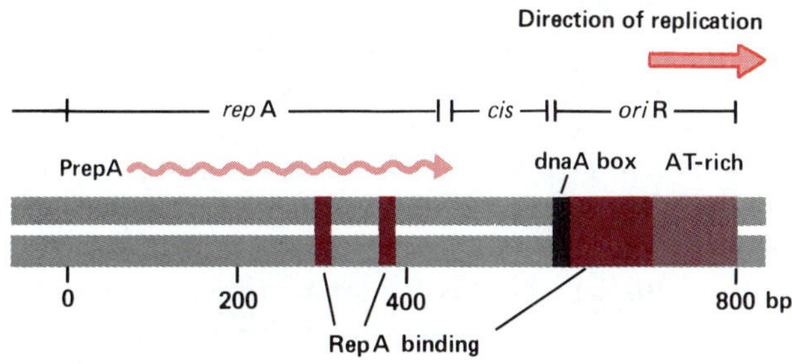

Figure 18-6
Replication region of R1, with multiple binding sites for RepA in *oriR* and its vicinity. The region labeled *cis* is responsible for *cis*-specific action of the RepA protein. Wavy line = transcript of *repA*, with the promoter (P_{repA}) and terminator (arrowhead).

62. Womble DD, Rownd RH (1988) *Microbiol Rev 52*:433; Ryder TB, Davison DB, Rosen JI, Ohtsubo E, Ohtsubo H (1982) *Gene 17*:299.
63. Ohtsubo E, Feingold J, Ohtsubo H, Mickel S, Bauer W (1977) *Plasmid 1*:18; Miyazaki C, Kawai Y, Ohtsubo H, Ohtsubo E (1988) *JMB 204*:331.
64. Masai H, Kaziro Y, Arai K (1983) *PNAS 80*:6814; Ohtsubo H, Ryder TB, Maeda Y, Armstrong K, Ohtsubo E (1986) *Adv Biophys 21*:115.
65. Miki T, Easton AM, Rownd RH (1980) *J Bact 141*:87.

The RepA binding sites near the 3′ end of its gene and the region between *repA* and *oriR* mediate the *cis* action of the protein. This intergenic region (CIS) contains a static DNA bend and a Rho protein–dependent transcription terminator, and appears to impart the *cis*-acting behavior to Rep protein.[66] Perhaps, pausing at the terminator allows the almost completely translated protein to bind to its recognition sequences at the 3′ end of the gene. By this device, the protein would never leave the template from which it is transcribed and a pool of free RepA would never accumulate. The importance of this CIS-sequence is manifested by the decrease in transformation efficiency and plasmid copy number seen in its absence.

In addition to being funneled directly to the template during transcription, RepA protein is inactivated by interacting with the origin; this inability of RepA protein to cycle to a second template further contributes to its *cis*-specific behavior. RepA synthesized from one plasmid can activate replication of another only if the encoding plasmid lacks a functional origin.[67] Thus, RepA protein can work *in trans* in vitro to activate a *repA*$^-$ *oriR*$^+$ plasmid, provided the protein-donor plasmid is *repA*$^+$ *oriR*$^-$. Active RepA protein has been purified from cells which overproduce it in the absence of *oriR*.

Other instances in which proteins act preferentially *in cis*[68] (e.g., gpA of ϕX174 and the Tn10 transposase) are even less well understood. Analysis of the *cis* action of RepA in vitro and in vivo provides an opportunity to analyze a phenomenon of considerable biochemical interest and regulatory importance.

Mechanism of Initiation

The RepA protein and *oriR* are the only plasmid-encoded elements required for replication. RepA has two demonstrable activities: sequence-specific DNA binding and topoisomerase (type I) activity. Binding to *oriR* is clearly significant for initiation, but no role for the topoisomerase has been demonstrated. The host-encoded dnaA, dnaB, and dnaC proteins, primase, pol III holoenzyme, gyrase, and SSB are all involved in R1 replication.[69] RNAP is not essential except when RepA protein is provided by coupled transcription and translation from a plasmid carrying *repA*.[70]

Binding of RepA to *oriR* causes conformational changes that enable dnaA protein to bind to the origin; specific binding of dnaA protein has not been observed in the absence of RepA.[71] Together, RepA and dnaA protein form a large complex, involving 40 to 50 monomers of RepA, one to two molecules of dnaA protein, and about 100 bp of *oriR* DNA. Nuclease-protection and sensitivity studies suggest that the bound DNA winds about the sphere of proteins, as in a nucleosome, and that the DNA in the AT-rich region is melted or bent.[72] These conformational changes may allow the loading of the replication machinery, specifically the dnaB helicase.

The evidence for involvement of dnaA protein in R1 replication shows some inconsistencies. R1 plasmids suppress the host replication defect of *dnaA*^{ts} and

66. Masai H, Arai K (1988) *NAR 16*:6493; Dong S, Rouillard KP, Womble DD, Rownd RH (1989) *J Bact 171*:703.
67. Masai H, Kaziro Y, Arai K (1983) *PNAS 80*:6814; Dong X, Womble DD, Rownd RH (1988) *JMB 202*:495.
68. McFall E (1986) *J Bact 167*:429.
69. Masai H, Kaziro Y, Arai K (1983) *PNAS 80*:6814; Ortega S, Lanka E, Siaz R (1986) *NAR 14*:4865.
70. Masai H, Kaziro Y, Arai K (1983) *PNAS 80*:6814; Dong X, Womble DD, Rownd RH (1988) *JMB 202*:495.
71. Masai H, Arai K (1987) *PNAS 84*:4781.
72. Masai H, Arai K (1988) in *DNA Replication and Mutagenesis* (Moses RE, Summers WC, eds.). Am Soc Microbiol, Washington DC, p. 113.

dnaA-null cells by integrating into the chromosome.[73] The reliance on the R1 origin for chromosomal replication in these mutant strains indicates that R1 replication is independent of *dnaA* function. However, integration of R100-1 was found to be incapable of suppressing strains completely deficient in *dnaA* function (*dnaA*-null), indicating that residual *dnaA* activity is required for integrative suppression of $dnaA^{ts}$ alleles.[74] Mini-R100 plasmids showed a decrease in copy number rather than a total absence of replication in the *dnaA*-null mutant.[75] In contrast, extracts from many $dnaA^{ts}$ strains failed to replicate R1 DNA and were activated by addition of purified dnaA protein.[76] Clearly, *dnaA* performs an important role in R1 replication, but the degree to which *dnaA* is necessary may depend on several unidentified factors. Replication of R1 in $dnaA^{ts}$ mutants suggests that only a subset of the protein's functions required in chromosomal replication is exploited by R1.

DNA synthesis on the continuous strand is initiated at a specific site located about 400 bp downstream of the origin.[77] This sequence can function in a single-stranded form as a template for the synthesis of a unique primer by host primase. The trinucleotide CAG occurs 10 to 12 ntd upstream of the RNA–DNA transition, consistent with the sequence preference of primase.[78] Homology to the phage G4 complementary strand origin (Section 8-3), recognized by primase, is also evident. Thus, continuous-strand synthesis on R1 appears to be primed by primase alone.

The priming sequence is neither a part of *oriR* nor essential, but its presence stimulates replication in vitro and increases plasmid copy number in vivo. The 5′ ends of continuous strands are heterogeneous on plasmids lacking this signal and must arise by an alternate priming mechanism. Regions of sequence homology with the priming site have been found on plasmids F and R6K, indicating the generality of primase recognition sequences.

The discontinuous-strand products have heterogeneous 5′ ends but terminate at a specific site very near the origin.[79] Although the mechanism is unknown, specific termination ensures unidirectional replication by blocking the discontinuous strand from becoming a continuous strand in the opposite direction. A sequence homologous to the R6K replication terminus (Sections 5 and 15-12) is present in this region.

In a scheme for the assembly of replication forks on R1[80] (Fig. 18-7), the combined action of RepA and dnaA proteins at *oriR* assemble the dnaB helicase–primase mobile primosome on the discontinuous-strand template. The dnaB helicase, donated from a dnaB·dnaC complex, unwinds the DNA duplex unidirectionally, uncovering in an active single-stranded form the priming site on the other parental strand, 400 bp downstream of *oriR*. A primer for the continuous strand is synthesized by primase at this site and elongated by pol III holoenzyme, while the discontinuous-strand dnaB protein provides the helicase for continuous-strand synthesis. Multiple primers are synthesized by dnaB protein and primase on the discontinuous strand, which pol III holoenzyme elongates toward the origin.

---

73. Molin S, Nordström K (1980) *J Bact 141*:111; Kline BC, Logoma T, Tam JE, Shields MS (1986) *J Bact 168*:440; Nishimura A, Nishimura Y, Caro L (1973) *J Bact 116*:1107.
74. Nagata T, Murakami Y, Imai M (1988) *MGG 213*:163.
75. Tang XB, Womble DD, Rownd RH (1989) *J Bact 171*:5290.
76. Ortega S, Lanka E, Diaz R (1986) *NAR 14*:4865; Masai H, Arai K (1987) *PNAS 84*:4781.
77. Masai H, Arai K (1989) *JBC 264*:8082; Miyazaki C, Kawai Y, Ohtsubo H, Ohtsubo E (1988) *JMB 204*:331.
78. Yoda K, Yasuda J, Jiang X-W, Okazaki T (1988) *NAR 16*:6531.
79. Masai H, Arai K (1989) *JBC 264*:8082; Miyazaki C, Kawai Y, Ohtsubo H, Ohtsubo E (1988) *JMB 204*:331.
80. Masai H, Arai K (1989) *JBC* 264:8082.

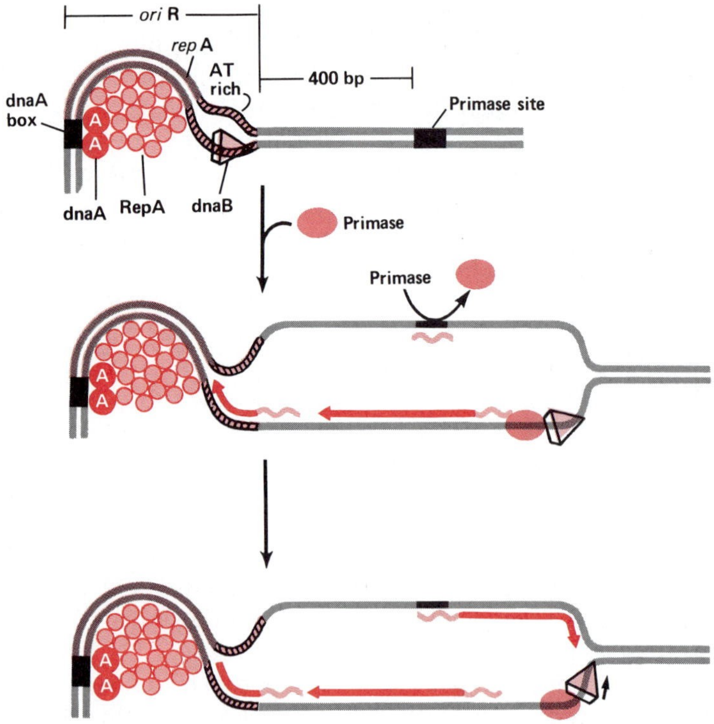

Figure 18-7
Scheme for initiation of continuous- and discontinuous-strand synthesis on R1. The discontinuous-strand replication complex is initiated first, as RepA and dnaA proteins add dnaB protein at the AT-rich region. The primase site for the continuous strand is uncovered by the action of dnaB helicase. Primase synthesizes a unique primer specifically at this G site. Pol III holoenzyme elongates the continuous-strand primer; unidirectional fork movement is fueled by the dnaB helicase on the discontinuous strand.

Regulation of Replication: Regulation of RepA Synthesis[81]

RepA synthesis, required for initiation of replication, is controlled by two *trans*-acting regulators (products of *copA* and *copB*) which maintain the plasmid at a stable, low copy number. The *copA* gene encodes a 91-ntd RNA, opposite in sense to the leader portion of the *repA* message[82] (Fig. 18-8). Identical in R1 and R100, *copA* RNA is the major regulator of copy number and incompatibility and acts independently of *copB*. The *copA* RNA hybridizes with the *repA* mRNA, thereby inhibiting translation. The duplex secondary structure of the RNA probably causes the ribosome binding site and the GUG start codon to become inaccessible for initiation of translation.[83]

The abundance of the free *copA* RNA, transcribed from the promoter P_{copA}, determines plasmid copy number. When the plasmid level rises, *copA* RNA and *repA* mRNA are more abundant and interact more frequently to repress *repA* translation and thus replication. When the plasmid level drops, the resulting decrease in both RNAs reduces the chance of their pairing and thus increases

81. Rownd RH, Womble DD, Dong X (1986) in *Antibiotic Resistance Genes: Ecology, Transfer, and Expression.* Banbury Report 24, CSHL, p. 179; Nordström KS, Molin S, Light J (1984) *Plasmid 12*:71; Womble DD, Rownd RH (1986) *JMB 192*:259.
82. Light J, Molin S (1983) *EMBO J 2*:93; Womble DD, Dong X, Wu RP, Luckow VA, Martinez AF, Rownd RH (1984) *J Bact 160*:28.
83. Dong X, Womble DD, Rownd RH (1987) *J Bact 169*:5353; Rownd RH, Womble DD, Dong X, Luckow VA, Wu R-P (1985) in *Plasmids in Bacteria* (Helinski D, Cohen SN, Clewell D, Jackson D, Hollaender A, eds.). Plenum Press, New York, p. 335; Womble DD, Dong X, Rownd RH (1987) in *New Perspectives on the Molecular Biology of RNA* (Inouye M, Dudock BS, eds.). Academic Press, New York, p. 225.

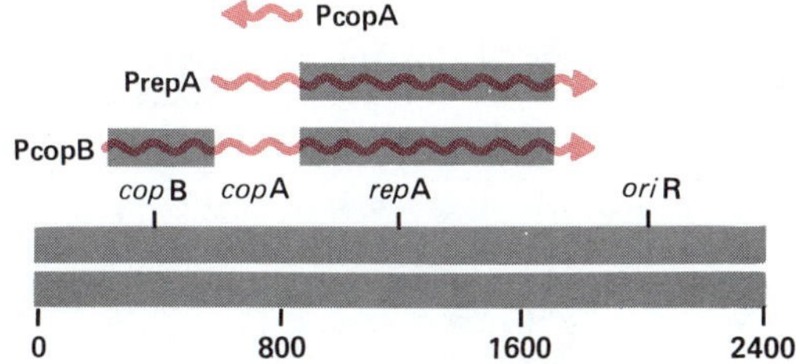

Figure 18-8
R1 copy number regulatory region. Wavy lines = transcripts with promoters (P_{copA}, P_{copB}, P_{repA}), direction, terminators (arrowheads), and translated regions (color areas).

the frequency of *repA* translation, resulting in a constant level of RepA production that is independent of plasmid copy number and can compensate for any random fluctuations that may occur.

The *copB* gene encodes a 10-kDa protein which represses transcription from the *repA* promoter. The *repA* gene has two promoters: P_{copB} is upstream of *copB* and is constitutively expressed; P_{repA} is downstream of, and regulated by, *copB*[84] (Fig. 18-8). While deletion of *copB* causes a 10- to 15-fold increase in copy number, excess *copB* has only a minimal inhibitory effect on replication, indicating that P_{repA} is already mostly repressed by CopB protein when the plasmid is present at the normal copy number. Regulation by *copB* most likely functions by sensing a drop in the plasmid concentration below a critical level; at this low level, the gene dosage of *copB* is also low, allowing an increase in transcription from P_{repA}. Thus, *copB* action leads to a transient increase in replication frequency until the normal plasmid concentration is reestablished.

18-5 R6K

Information about many aspects of replication, from initiation to termination, has come from studies of R6K. A naturally occurring *E. coli* plasmid, R6K encodes resistance to streptomycin and ampicillin and is maintained at about 15 copies per chromosome, in contrast to the copy number of 1 or 2 for most R plasmids. By encoding the functions required for conjugation, R6K is self-transmissible to bacteria that lack the plasmid.

The structure of in vivo replication intermediates immediately reveals several curious facets of R6K replication. As defined by the locations of bubbles in early intermediates, R6K has three origins of replication within a 4-kb region.[85] Most molecules, however, use only one origin in a replication cycle. Thus, during initiation, R6K must choose which origin will fire. The three origins, α, β, and γ, are used at the relative frequencies of 48%, 34%, and 18%, respectively. R6K also has a defined replication terminus, located asymmetrically from the origins (Fig. 18-9). Replication proceeds unidirectionally until it reaches the terminus,[86] after which initiation occurs from the same origin in the opposite direction.

---

84. Light J, Molin S (1982) *MGG 187*:486; Liu C-P, Churchward G, Caro L (1983) *Plasmid 10*:148; Rownd RH, Womble DD, Dong X, Luckow VA, Wu R-P (1985) in *Plasmids in Bacteria* (Helinski D, Cohen SN, Clewell D, Jackson D, Hollaender A, eds.). Plenum Press, New York, p. 335; Dong X, Womble DD, Luckow VA, Rownd RH (1985) *J Bact 161*:544.
85. Crosa JH (1980) *JBC 255*:11075; Inuzuka N, Inuzuka M, Helinski DR (1980) *JBC 255*:11071.
86. Crosa JH, Luttrop LK, Falkow S (1976) *J Bact 129*:454; Lovett MA, Sprads RB, Helinski D (1975) *PNAS 72*:2905.

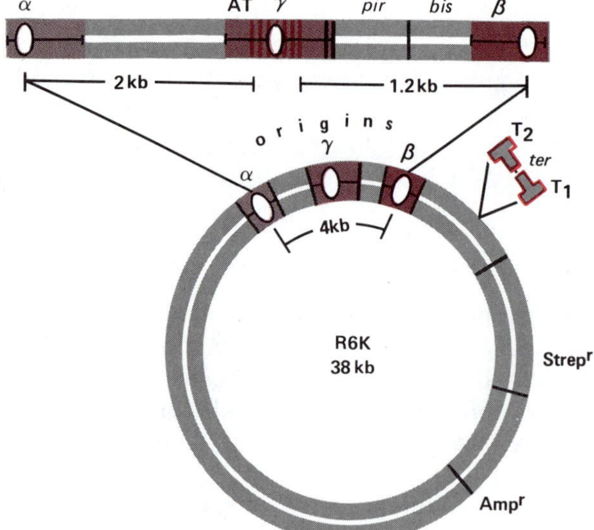

Figure 18-9
Map of R6K, showing the three clustered replication origins and the asymmetric location of the terminus (*ter*). AT = AT-rich sequence.

R6K replication has been analyzed both in vivo and in vitro. The plasmid-encoded π protein is essential for initiation from all three origins, and a second protein, Bis, is specifically involved in initiation from *ori*β.[87] RNA synthesis and the host *dnaA*,[88] *dnaB*, *dnaC*, and *dnaE* genes are required;[89] pol I is dispensable.

The Origin Region

The three replication origins and genes for the two replication proteins, *pir* (encoding π protein) and *bis*, are contained on a 4-kb fragment (Fig. 18-9). Although the α and β origins are used most frequently, the γ origin is required *in cis* for initiation at these other two sites. Thus, if the replication origin is defined as the *cis*-acting sequence essential for initiation, rather than the location of a replication bubble, γ would be a component of both the α and β origins. However, if only the site of assembly of the replication complex is specified, γ alone would be considered the R6K origin even though it is only rarely the site of initiation of DNA synthesis (see below).

The organization of the γ origin is very similar to that of the single origin of pSC101 and is classic for plasmid origins (Section 16-4). The *ori*γ sequence contains seven 22-bp direct repeats that govern the binding of π protein (Fig. 18-10). To the left of the repeats is a 90-bp 80% AT-rich sequence;[90] to the right, a dnaA box. The AT-rich region contains a sequence homologous to the *E. coli oriC* 13-bp repeats (Section 16-2) and binds IHF protein in vitro.[91] In vivo, IHF is required for maintenance of plasmids that depend only on the γ origin, but not for those relying on α or β.

87. Mukhopadhyay P, Filutowicz M, Helinski DR (1986) *JBC 261*:9534.
88. Kline BC (1988) *Can J Microbiol 34*:526.
89. Arai T, Clowes R (1975) in *Microbiology 1974* (Schelessinger D, ed.). Am Soc Microbiol, Washington DC; p. 141.
90. Stalker DM, Kolter R, Helinski DR (1979) *PNAS 76*:1150.
91. Filutowicz M, Appelt K (1988) *NAR 16*:3829.

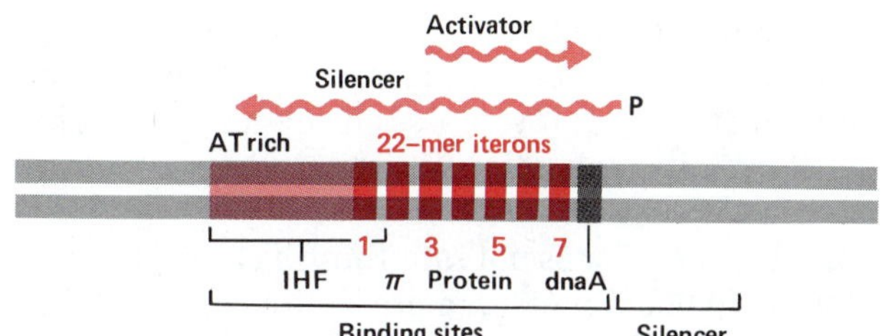

Figure 18-10
Structure of R6K *oriγ*. Silencer activity resides in a promoter located to the right of the dnaA box, just outside the origin.

Mutagenesis of the γ origin has revealed the sequence requirements crucial for initiation.[92] A single base change in the first or the sixth of the 22-bp repeats destroys the binding of π protein in vitro and the origin activity in vivo. However, deletion of only the altered repeat restores replication activity; thus six repeats are sufficient. An essential sequence lies just outside the repeats; it has no effect on π binding but greatly reduces plasmid copy number and may be a primase initiation site, similar to that in R1[93] (Sections 4 and 8-3).

In spite of its classic structure, the γ origin is only rarely the site of initiation because of an adjacent inhibitory sequence (Fig. 18-10). This "silencer" sequence acts *in cis* and contains a weak promoter directing transcription into *oriγ*.[94] The mechanism of inhibition is not yet known. Possibly an "activator RNA" is blocked by the antisense "silencer RNA" much as the primer RNA II of ColE1 is inhibited by the antisense RNA I (Section 2). A transcript within *oriγ* with polarity opposite to that of the silencer has been identified,[95] but its role in replication is unknown; π protein stimulates association of these antisense RNAs in vitro. However, this mechanism does not explain why the silencer inhibits *oriγ* only *in cis*, when other antisense RNAs act *in trans*. Disruption of the origin structure by transcription is also a possibility.

Mechanism of Initiation: Activation of *oriβ* by *oriγ*

Oriγ, used rarely due to the action of the silencer, is required for initiation from the α and β origins. The π protein, which binds strongly only to *oriγ*, is required for all origins. A possible model for the activation of *oriβ* by *oriγ* is that the replication complex assembled by π protein at *oriγ* is transferred to *oriβ*, where initiation begins.[96] Some evidence in support of this model has been obtained.

Binding of π to *oriβ*, or to *oriα*, in the absence of *oriγ*, has not been detected. However, a sequence homologous to half of a 22-bp π binding site exists within *oriβ* and is required for activation of β by γ.[97] Formation of a loop in the DNA between *oriγ* and *oriβ*, 2000 bp apart, allows binding to this weak site in *oriβ* in vitro. The AT-rich sequences are also required for activation, suggesting that proteins in addition to π are assembled at γ and transferred to β.

Regulation of Replication: Two Functions of the π Protein

The π protein and its reiterated binding sites are the two elements controlling plasmid replication. The iterons both inhibit replication and mediate incompat-

92. McEachern MJ, Filutowicz M, Helinski DR (1985) *PNAS 82*:1480.
93. Masai H, Arai K (1989) *JBC 264*:8082.
94. Patel I, Bastia D (1986) *Cell 47*:785.
95. Patel I, Bastia D (1987) *Cell 51*:455.
96. Mukherjee S, Erickson H, Bastia D (1988) *Cell 52*:375.
97. Mukherjee S, Erickson H, Bastia D (1988) *Cell 52*:375.

ibility between plasmids. Thus, competition between the multiple binding sites for π protein inhibits replication. However, π protein, an activator of initiation at all three origins, has two additional effects: (1) it represses its own synthesis by binding to its promoter,[98] and (2) it inhibits replication initiation when present at high levels.

In contrast to the Rep proteins of R1 and F, π is not rate-limiting for initiation. Large amounts of π protein (3500 to 10,000 dimers) are present in cells containing R6K.[99] Reduction of π protein to 0.1% of its normal level supports R6K replication at the standard copy number, while overproduction of π results in a substantial reduction. Autoregulation of π protein, by repressing its own promoter, may serve to ensure that the levels do not reach the point of inhibiting DNA synthesis.

Both the π protein and the repeat sequences are required for initiation, yet an excess of either component inhibits replication. To explain these positive and negative effects, the requirement for two distinct forms of π protein, an activating form and a repressor, has been proposed. The same conclusion has been drawn from studies of the control of F plasmid replication. However, the biochemical basis for the two forms is unexplored.

The R6K Terminus

R6K contains a sequence that arrests replication forks moving in either direction. This *terminus* (*ter*) is located asymmetrically from the origins (Fig. 18-9); its activity is readily detected by the structure of the replication intermediates. The *ter* sequence functions on heterologous plasmids, such as ColE1 and pSC101,[100] encodes no proteins, and is active in soluble extracts made from plasmid-free cells.[101] The required protein (TBP, for *t*erminus *b*inding *p*rotein) is encoded by the host *tus* (terminus utilization substance) or *tau* gene (Section 15-12). When present on a unidirectionally replicating plasmid, *ter* causes a dramatic inhibition of fork movement, but replication is eventually completed by synthesis in the same direction.[102]

The minimal termination sequence contains two 22-bp elements, the orientation of which impedes fork movement in one or the other direction.[103] Nearly identical sequences occur in R1 and R100, plasmids unrelated to R6K, and in the terminus region of the *E. coli* chromosome (Section 15-12). This sequence is the binding site for TBP, the *tus* product.[104] The terminus sequences of R6K and *E. coli* are not essential; their absence causes no striking defects in replication or cell division.

18-6 F Plasmid[105]

F, the first plasmid discovered, held center stage in the development of *E. coli* molecular genetics. As the prototypic "fertility factor," F is capable of transmit-

98. Kelley W, Bastia D (1985) *PNAS 82*:2574.
99. Filutowicz M, McEachern MJ, Helinski DR (1986) *PNAS 83*:9645.
100. Kolter R, Helinski DR (1978) *JMB 124*:425; Bastia D, Germino J, Crosa JH, Hale P (1981) *Gene 14*:81.
101. Germino J, Bastia D (1981) *Cell 23*:681.
102. Kolter R, Helinski DR (1978) *JMB 124*:425.
103. Horiuchi T, Hidaka M (1988) *Cell 54*:515.
104. Hill TM, Kopp BJ, Kuempel PL (1988) *J Bact 170*:662.
105. Willetts N, Skurray R (1987) in *Escherichia coli and Salmonella typhimurium* (Neidhardt FC, ed.). Am Soc Microbiol, Washington DC, Vol. 2, p. 1110.

ting itself, and any attached DNA, including the entire host chromosome, to a new cell. Of the 60 known genes on the 94.5-kb chromosome (Fig. 18-11), about 20, the *tra* (for *transfer*) genes, are involved in conjugation, the mechanism of which is presented in Section 7.

The remnants of three independent replication systems, or *replicons*, are present on F (Fig. 18-11).[106] The RepF1A replicon, most resembling the complete F plasmid in copy number, stability, and incompatibility, is of principal importance. Relatives of F lacking RepF1A depend entirely on RepF1B, but this system is probably not the key to survival. Plasmids depending on RepF1B are maintained, but are much less stable than their RepF1A counterparts. The third system, RepF1C, rendered nonfunctional on F by insertion of Tn1000 but active in some members of the F family, is homologous to the replication region of plasmid R1.

Mini-F is a 9-kb plasmid derived from the RepF1A region (Fig. 18-11). It encodes seven proteins involved in replication and plasmid maintenance, and two origins. First identified by electron microscopy, the bidirectional *oriV* (also called *ori1*) was found to be nonessential by deletion analysis, leading to the discovery of *oriS* (*ori2*). The *oriS* sequence controls unidirectional synthesis and is essential for all mini-F replication, including initiation from *oriV*. Although the relative contributions of *oriV* and *oriS* to plasmid replication have not been assigned, the autonomous and essential nature of *oriS* is quite clear.

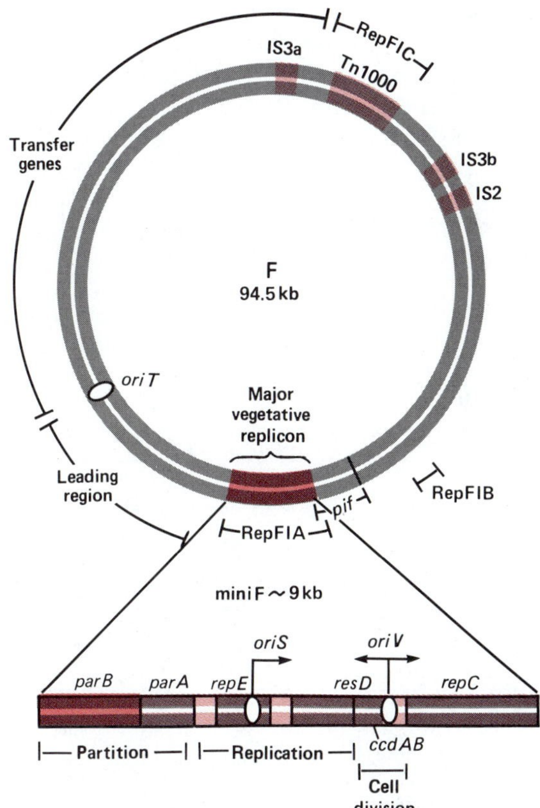

Figure 18-11
Structure of F plasmid, showing the three replicons, the transfer genes, and the structure of mini-F plasmid. [Based on Willetts N, Skurray R (1987) in *Escherichia coli and Salmonella typhimurium*, (Neidhardt FC, ed.). Am Soc Microbiol, Washington DC, Vol. 2, p. 1110]

106. Couturier M, Bex F, Bergquist PL, Maas WK (1988) *Microbiol Rev* 52:375.

The *oriS* sequence and the *repE* gene, encoding the RepE protein, an origin-binding protein, are the two F components absolutely essential for replication (Fig. 18-12). The 217-bp *oriS* sequence contains two dnaA boxes, a 46-bp AT-rich region which includes a 13-bp sequence homologous to that melted at *oriC* by dnaA protein (Sections 16-2 and 16-3), and four 19-bp RepE protein binding sites. To the right of *oriS* is *repE*, followed by five more 19-bp repeats which comprise the *incC* (*inc*ompatibility) locus. Properly regulated initiation requires *incC* in addition to *oriS* and *repE*. In its regulatory organization, the F replication region resembles those of P1, pSC101, and R6K (Section 16-4).

Replication from *oriV* requires, in addition to the RepE protein, the *repC* and *ccdB* products; *repC*, also known as *pifC* (for *p*hage *i*nhibition by *F*), is responsible for the exclusion of certain phages, such as T7 (Section 17-5). The 34-kDa RepC protein binds to *oriV*[107] and to its own promoter,[108] repressing transcription. The *ccdB* gene product is involved in plasmid maintenance (see below); how RepC and CcdB proteins activate *oriV* is still a mystery.

Mechanism of Replication

The mechanism of replication of F has received much less attention than has its regulation. Of the few molecular details available, it is known that RepE protein[109] is a dimer of 29-kDa subunits and binds (1) the 19-bp repeats in *oriS* and *incC*[110] and (2) an operator overlapping its promoter.

F replication depends on the host *dnaB*, *dnaC*, *dnaG* (primase), and *dnaE* (pol III) genes. Primosome assembly sites (*pas* sequences) are found near both *oriS* and *oriV*, but their contribution and those of the primosomal proteins (PriA and dnaT; Section 8-4) are unknown. A possible mechanism of initiation, based on the structure of the origin, has been proposed (Section 16-4).

The involvement of dnaA protein in F replication was long uncertain. Like R1, F was known to integratively suppress the replication defect of *dnaA*^{ts} strains. The *dnaA*-null alleles and analysis of suppressed *dnaA*^{ts} strains have now established that *dnaA* is essential for F replication.[111] The integrity of at least one dnaA box in *oriS* is required for plasmid maintenance.[112] That *dnaA* mutants

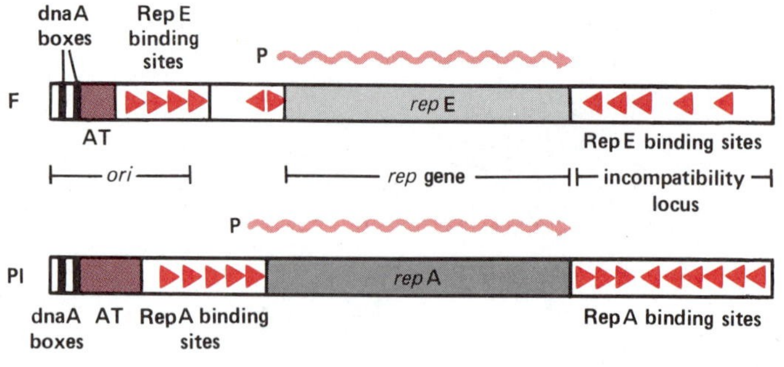

Figure 18-12
Replication control regions of plasmids F and P1. Although the DNA sequences are not homologous, they are nearly identically organized. The Rep protein binding sites responsible for autoregulation of *rep* transcription overlap the origin in P1 but are independent from the origin sites in F.

107. Caughey PA, de Feyter R, Lane HED (1986) *NAR 14*:9699.
108. Miller JF, Malamy MH (1986) *PNAS 83*:1433.
109. Masson L, Ray DS (1988) *NAR 16*:413; Tokino T, Murotsu T, Matsubara K (1986) *PNAS 83*:4109.
110. Masson L, Ray DS (1986) *NAR 14*:5693.
111. Kline BC, Kogoma T, Tam JE, Shields M (1986) *J Bact 168*:440; Kogoma T, Kline BC (1987) *MGG 210*:262; Murakami Y, Ohmori H, Yura T, Nagata T (1987) *J Bact 169*:1724.
112. Murakami Y, Ohmori H, Yura T, Nagata T (1987) *J Bact 169*:1724.

have more drastic effects on replication of the *E. coli* chromosome than of F indicates that F plasmid and the chromosome require dnaA protein in different ways. Mini-F mutants independent of *dnaA* should offer insights into these multiple activities of dnaA protein.

Replication of *oriS* plasmids is supported by a crude extract from plasmid-free cells supplemented with purified RepE protein.[113] The 217-bp *oriS* fragment, required in vivo, is also active in vitro. Rifampicin and novobiocin prevent replication, indicating that RNAP and gyrase are required. Chloramphenicol has no effect on replication; in contrast to the RepA protein of R1, RepE protein need not be freshly synthesized and acts *in trans*. Replication is supported by 100 molecules of RepE protein per template, similar to the amount required to fill all the binding sites on mini-F.

An unusual form of the host RNAP is involved in F replication. The *htpR* (*rpoH*) gene encodes σ^{32}, the rare RNAP sigma factor responsible for expression of heat-shock genes (Table 7-3). Mutations in *htpR* block the replication[114] and maintenance of F plasmids. Suppressors of *htpR*, which produce heat-shock proteins in the absence of σ^{32}, do not overcome the F replication defect, indicating that the heat-shock sigma factor is directly involved in F replication. Transcription of the *repE* gene by RNAP containing σ^{32} is required.

Regulation of Replication and Copy Number[115]

Replication of F is perhaps the most strictly regulated of all plasmids. Should the plasmid number exceed the normal 1 to 2 copies per chromosome, plasmid replication is completely repressed until segregation at cell division reestablishes the normal level. The regulatory elements of F are analogous to those of the P1 prophage[116] (Fig. 18-12): *oriS* is equivalent to *oriR* of P1; *repE*, like P1 *repA*, encodes the initiator protein and is autoregulated; and *incC*, like P1 *incA*, consists of multiple initiator-protein binding sites. The terms *ori*, *rep*, and *inc* are used below to describe both the F and P1 systems together. Although the F and P1 replicons are unrelated and possess virtually no sequence homology, they have equivalent regions for regulating replication and partition.

The negative control of both F and P1 replication is manifested by regulatory mutants with increased copy numbers. The mutants are defective in the autoregulation of *rep* or suffer disruptions of the *inc* repeats.

Changes in Rep protein that destroy autoregulation raise the levels of Rep protein and stimulate plasmid replication. The mutant proteins are deficient in operator binding but are able to bind the origin and promote initiation, a surprising result in that the operator and *ori* repeats have similar nucleotide sequences. However, the footprinting patterns of wild-type Rep protein binding to the operator and to *ori* sequences are distinctive.[117] Further, wild-type Rep binds the operator more strongly than *ori*, whereas the mutant protein favors *ori*.

Deletion of the *inc* region causes a fivefold to eightfold increase in copy number.[118] Additional copies of the *inc* repeats have the opposite effect, inhibit-

113. Muraiso K, Tokino T, Murotsu T, Matsubara K (1987) *MGG 206*:519.
114. Wada C, Imai M, Yura T (1987) *PNAS 84*:8849.
115. Kline BC (1988) *Can J Microbiol 34*:526; Kline BC (1985) *Plasmid 14*:1.
116. Chattoraj DK, Abeles AL, Yarmolinsky M (1985) in *Plasmids in Bacteria* (Helinski DR, Cohen SN, Clewell DB, Jackson D, Hollaender, A, eds.). Plenum Press, New York, p. 335.
117. Masson L, Ray DS (1986) *NAR 14*:5693.
118. Pal S, Mason RJ, Chattoraj DK (1986) *JMB 192*:275.

ing replication and excluding *ori* plasmids. Thus, the initiation frequency depends on the number of *inc* elements present; *inc* appears to compete with *ori* for initiator molecules.

These two regulatory loops operate jointly to determine initiation frequency: Rep protein, by repressing its own synthesis, limits its availability for replication, and the *inc*-binding sites compete with *ori* for the limited amount of Rep. Yet it seems paradoxical that competition between *ori* and *inc* can regulate replication if Rep is autoregulated. Binding of Rep to these sequences should derepress the *rep* gene and thus allow synthesis to increase until all the binding sites are adequately filled. Several models that attempt to explain this paradox will be discussed briefly.

Simplest is that Rep has different affinities for its distinct binding sites, so that far lower concentrations are required to repress the *rep* gene than to fill the *inc* and *ori* sites. In fact, Rep does bind more tightly to the operator than to *ori*. However, levels of *inc* 50 to 100 times higher than normal fail to derepress the *rep* promoter, yet one extra copy of *inc* inhibits replication. This model is thus inadequate.

The possibility of two forms of Rep, one an abundant repressor and the other a rare initiator, has been proposed.[119] Binding of the initiator form to *ori* or *inc* would not affect occupancy of the operator because it would not change the concentration of the repressor form. A computer model based on these ideas faithfully achieves a stable plasmid copy number and corrects alterations generated by mistakes in replication or division.[120] The behavior of copy number mutants can also be reconciled. However, there is no biochemical evidence for two forms of Rep; homogeneous Rep preparations bind both operator and *inc* sequences efficiently in vitro.

A third possibility is that Rep can bind to *inc* and operator simultaneously to maintain repression.[121] Physical evidence for such protein–protein complexes has been observed. Yet this model fails to explain how plasmids with the *inc* region deleted have an increased but *stable* copy number. Runaway replication would be expected were *inc* required to maintain repression while Rep binds *ori*, but this is not observed. A final complicating factor, not considered by the regulatory models, is that the inhibition of replication by *inc* may not be due merely to the sequestering of Rep protein away from the origin.[122] Reversal of the negative *inc* effects by high levels of Rep has not been demonstrated.

Plasmid Maintenance by Host Killing: The *ccd* Genes

The F plasmid possesses the remarkable capacity to destroy those cells which lose the plasmid by errors in replication or cell division. First discovered and most extensively characterized in F plasmids, these killing functions are also found in R1, R100, and P1. The mechanism may have evolved in these low-copy-number plasmids because there is no leeway for a drop in copy number below one plasmid per cell.

The *ccdA* and *ccdB* (coupled cell division) genes are responsible for killing the

119. Trawick JD, Kline BC (1985) *Plasmid 13*:59.
120. Womble DD, Rownd RH (1987) *JMB 195*:99.
121. Chattoraj DK, Masson RJ, Wickner SH (1988) *Cell 52*:551.
122. Pal SK, Chattoraj DK (1988) *J Bact 170*:3554.

F^- daughters of F^+ cells within three generations of the plasmid loss (Fig. 18-13). The *ccdB* gene product (11.7 kDa) is the inhibitor of cell growth;[123] it is counteracted by *ccdA*, which encodes an 8.7-kDa protein.[124] The instability of the CcdA factor ensures that if the plasmid concentration drops below a critical level, loss of the factor by dilution or inactivation unleashes the inhibitory activity of the CcdB protein in the plasmid-free cell, causing the cell to form filaments, produce progeny lacking chromosomal DNA, and die. Plasmid-containing cells in the same population are unaffected because they maintain adequate levels of CcdA, and thus they grow preferentially.

Exactly how the *CcdB* protein exerts its effects is not yet known. The *ccd* genes induce the host SOS response (Section 21-3), causing a variety of physiologic changes, including induction of λ prophages. However, the inhibition of cell growth by *ccd* is independent of the host *recA* gene and thus does not depend on SOS induction.[125] Among *E. coli* mutants resistant to killing by *CcdB*, some mutations are located in the *groES* gene, which encodes abundant heat-shock proteins.[126] The mechanism by which these mutants suppress killing may provide insights into the killing by *ccd* genes and the mechanisms of normal cell division.

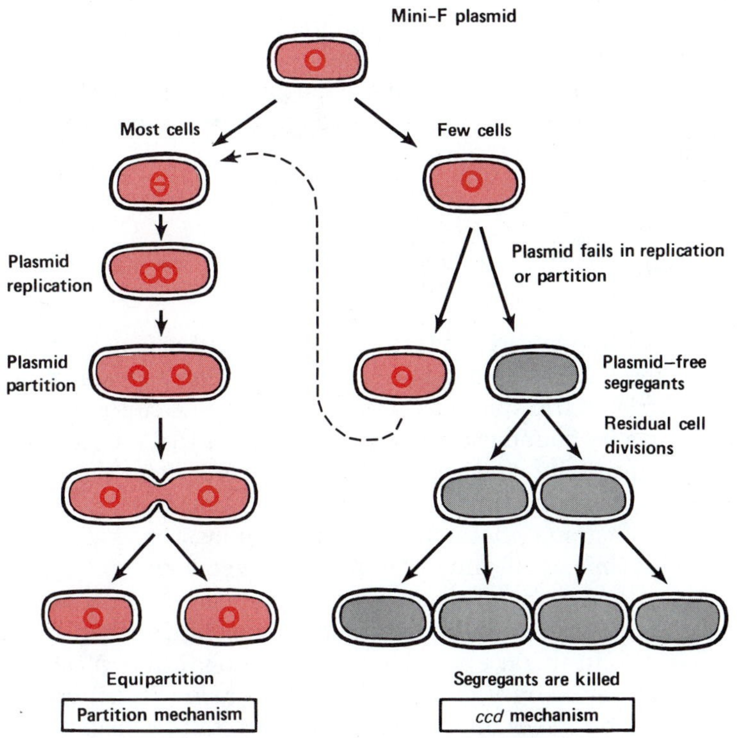

Figure 18-13
Scheme of stable inheritance of mini-F plasmid showing how the partition and *ccd* mechanisms work together to maintain the plasmid within the population of growing cells. (Modified from a figure provided by Dr S Hiraga)

123. Jaffé A, Ogura T, Hiraga S (1985) *J Bact 163*:841; Karoui H, Bex F, Drèze P, Couturier M (1983) *EMBO J 2*:1863; Miki T, Chang ZT, Horiuchi T (1984) *JMB 174*:627; Miki T, Yoshioka K, Horiuchi T (1984) *JMB 174*:605; Mori H, Ogura T, Hiraga S (1984) *MGG 196*:185.
124. Ogura T, Hiraga S (1983) *PNAS 80*:4784.
125. Jaffé A, Ogura T, Hiraga S (1985) *J Bact 163*:841; Hiraga S, Jaffé A, Ogura T, Mori H, Takahashi H (1986) *J Bact 166*:100.
126. Miki T, Orita T, Furun M, Horiuchi T (1988) *JMB 210*:327.

Partition Functions

F plasmid encodes three elements which function to ensure that both daughter cells receive a copy of the plasmid during cell division. These are the *sop* genes (stability of plasmid): *sopA* and *sopB* encode *trans*-acting proteins (43.7 and 35.4 kDa, respectively) that stabilize plasmids containing the *sopC* sequence,[127] which is the functional analog of the centromere of a eukaryotic chromosome. The partition genes *parA*, *parB*, and *parS* of plasmid P1 are homologous to *sopA*, *sopB*, and *sopC*.[128]

The *sopC* region contains a 43-bp sequence, repeated 12 times in the same orientation.[129] Each element contains a 7-bp inverted repeat to which the sopB protein binds.[130] Genetic and biochemical evidence supports a SopB–SopC (or ParB–ParS) interaction for partition. The proteins specifically bind their respective sequences in vitro,[131] and overexpression of SopB (or ParB) protein specifically destabilizes plasmids containing the *sopC* (or *parS*) sequence. In the P1 system, maximal binding of ParB to ParS requires assistance by IHF, the host DNA-binding protein,[132] which is also involved in plasmid partition in vivo.

How the partition elements function to ensure plasmid inheritance needs to be examined. Presumably, the proteins affix the plasmids to a protein on the host membrane or septum by interactions of the centromere-like sequence complexed with the Sop (or Par) proteins. Much interest has focused on identifying host mutants defective in plasmid partitioning in the hope of uncovering such interactions. Mutations in the β subunit of IHF (Section 10-8) belong to this class, and certain gyrase mutants (Section 12-3) also appear to have a specific defect in F partitioning.

18-7 Bacterial Conjugation[133]

Conjugation, the process whereby DNA is transferred from cell to cell, is common in Gram-positive and Gram-negative bacteria. Conjugation appears to be the major mechanism for DNA transfer in nature, both within a bacterial population and even across species boundaries. The many plasmids that encode the capacity to promote conjugation include F, R1, R100, and R6K. These *conjugative*, or *self-transmissible*, plasmids have highly related transfer genes, almost certainly arising from a common ancestor. F plasmid conjugation has been analyzed most thoroughly, but significant differences between F and more distantly related self-transmissible plasmids (e.g., RK2) deserve attention as well.

In addition to its own genome, a conjugative plasmid can *mobilize* other DNA for transfer to another cell. In this cooperative action, a mobilizable plasmid with a *cis* sequence for initiating transfer (the *oriT* site; see below) is activated both by its own mobilization genes and by factors from a self-transmissible

127. Ogura T, Hiraga S (1983) *Cell 32*:351.
128. Abeles AL, Snyder KM, Chattoraj DK (1984) *JMB 173*:307.
129. Mori H, Kondo A, Ohshima A, Ogura T, Hiraga S (1986) *JMB 192*:1.
130. Hayakawa Y, Murotsu T, Matsubara K (1985) *J Bact 163*:349.
131. Funnell BE (1988) *J Bact 170*:954; Kusukawa N, Mori H, Kondo A, Hiraga S (1987) *MGG 208*:365.
132. Funnell BE (1988) *PNAS 85*:6657.
133. Willetts N, Skurray R (1987) in *Escherichia coli and Salmonella typhimurium* (Neidhardt FC, ed.). Am Soc Microbiol, Washington DC, Vol. 2, p. 1110; Willetts N, Wilkins B (1984) *Microbiol Rev 48*:24; Willetts N, Skurray R (1980) *ARG 14*:41.

plasmid. Still other genomes, including the *E. coli* chromosome, recombine with a self-transmissible plasmid at a region of homology provided by an insertion sequence, and are transferred with the plasmid.

About one-third of the F genome is devoted to conjugation (Table 18-5). At least 22 genes (the *tra* family) are involved in *transfer*, and two elements regulate *tra* gene expression. All but two of the *tra* genes are expressed as part of a 32-kb operon and are thus coordinately controlled. Included in the *tra* genes are functions for (1) constructing the *pilus*, an extracellular organelle responsible for initiating cell-to-cell contact, (2) processing the DNA during transfer, and (3) inhibiting conjugation with another cell that already contains a plasmid. The conjugation pathway (Fig. 18-14) is summarized as follows:

(1) The pilus on the F^+ donor cell initiates a connection with a plasmid-free (F^-) recipient. This proteinaceous, flexible, hair-like structure contacts the F^- cell surface to form a mating pair. Contraction of the pilus stabilizes the pair and establishes the direct cell-to-cell contact.

(2) DNA transfer is initiated by the introduction of a specific nick in the plasmid. The nicked strand is then peeled off and, in some unknown way, is transferred, as ssDNA, through the membranes and into the recipient cell. The pilus itself is not directly involved in the transfer.

Table 18-5
The *tra* genes of the F plasmid

	Mass (kDa)			
Gene[a]	SDS–PAGE	DNA sequence	Gene product location[b]	Function
traA	13–14	13.2	IM	major pilus subunit, pilin
traB	55–64		OM	
traC	78–98		IM	
traD	77–90		IM,OM	
traE	19–22	21.2	IM	
traF	25		OM	
traG	110–116		IM	N-glycosylation of pilin (?); stabilization of mating pairs
traH	39–45		IM	
traI	174–180		CY	DNA helicase
traJ	23.5–27	27	CY	regulation of *tra* operon
traK	24		OM	
traL	11	10.3	OM	
traM	13–14	14.5	IM	initiation of transfer
traN	66			stabilization of mating pairs
traP	23.5		OM	
traQ	12.5	10.9		processing of pilin
traS	18	16.9	IM	surface exclusion
traT	24–28	26	OM	surface exclusion
traU	20			
traV	21			
traY	16	13.8	IM or OM	nicking at *oriT*
traZ	78–87		CY	fragment of TraI protein

[a] Listed alphabetically.
[b] IM = inner membrane; OM = outer membrane; CY = cytoplasm.

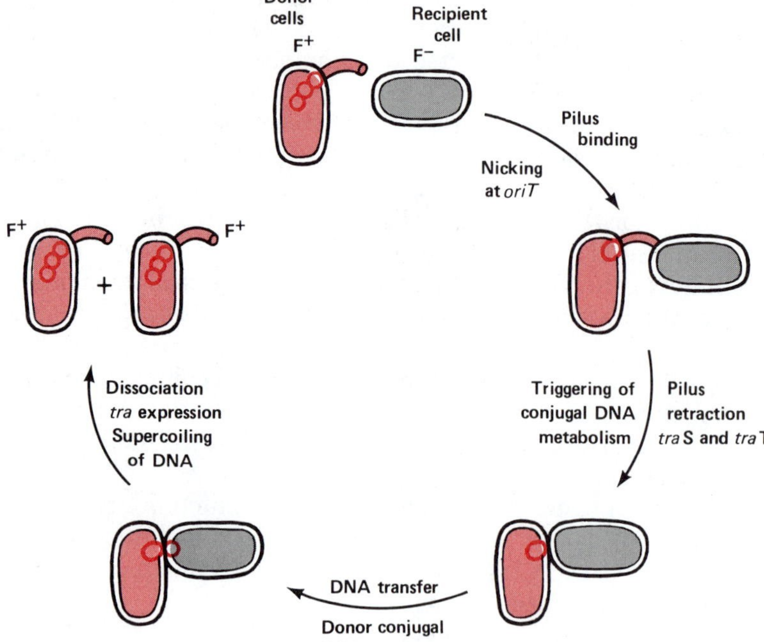

Figure 18-14
A model for the *E. coli* mating cycle. F plasmid DNA appears as superhelical and nicked circular forms. Donor conjugal DNA synthesis is not dependent upon stabilization of mating pairs and may not be dependent upon pilus retraction. [Modified from Willetts N, Skurray R (1980) *ARG 14*:41]

(3) DNA synthesis in both the donor and the recipient cell generates a duplex copy of the plasmid by copying both the transferred and the retained single strands.

(4) After conjugation, the mating pairs separate. The recipient cell expresses the plasmid-encoded genes, making it phenotypically F^+ and able to participate in subsequent matings as a donor cell.

The Pilus

F^+ cells have one to three pili on the cell surface. Pili are assembled from pilin, the product of the *traA* gene. An inner-membrane protein, pilin is expressed as a 13-kDa peptide, cleaved to 7 kDa,[134] and then acetylated, phosphorylated, and glucosylated. Ten genes, in addition to *traA*, are required for modification of pilin and assembly of the pilus. The *traQ* gene has been implicated in the proteolytic processing[135] and *traG* in the acetylation.[136] These functions are not F-specific and can be provided by a number of transfer-proficient plasmids.

In the structure of the pilus analyzed by x-ray diffraction, about 10^4 molecules of pilin are in a helical arrangement, 2 to 3 μm long and 8 nm in diameter, with a 2-nm pore.[137] No data are available concerning the base structure from which the pilus emerges from the cell membrane.

Once cell-to-cell contact is made (Fig. 18-14), the pilus contracts or depolymerizes to bring the two cells into direct association.[138] The signal that initiates

134. Frost LS, Paranchych W, Willetts NS (1984) *J Bact 160*:395.
135. Moore D, Sowa BA, Ippen-Ihler K (1981) *MGG 184*:260; Moore D, Sowa BA, Ippen-Ihler K (1982) *MGG 188*:459.
136. Creeger E, Schulet T, Rothfield L (1984) *JBC 259*:3064.
137. Folkhard W, Leonard KR, Malsey S, Marvin DA, Dubochet J, Engel A, Achtman M, Helmuth R (1979) *JMB 130*:145.
138. Achtman M, Morelli G, Schwuchow S (1978) *J Bact 135*:1053.

contraction is not known, but may be the same as the signal that makes pili retract upon attachment of DNA and RNA phages.[139] The *traN* and *traG* gene products, required for stable mating-pair formation, are apparently involved in establishing a direct channel between mating cells.[140] The TraD protein may facilitate DNA passage through the cell membranes and is ssential for the entry of certain RNA phage genomes.[141]

DNA Transfer

Cleavage at *oriT*. DNA transfer is initiated at a specific plasmid sequence called *oriT*. Plasmids with this sequence can be mobilized *in trans* by a tra^+ plasmid. In F plasmids, *oriT* has been localized to a 373-bp fragment at the edge of the transfer operon[142] (Fig. 18-11). Fragments of similar size that contain *oriT* sequences have been isolated from other conjugative plasmids (R1, R100, RK2, RP1) and from several small, mobilizable plasmids (e.g., ColE1, CloDF13, and RSF1010[143]). A nick made at a specific phosphodiester bond in one DNA strand initiates transfer. All *oriT* sequences include regions of dyad symmetry and a highly AT-rich region. Inverted repeats in the F *oriT* are important for nicking and transfer.[144]

Nicking at *oriT* can be identified by isolating plasmid–protein complexes which are cleaved in vitro upon addition of alkali or detergent.[145] These "relaxation complexes," used to map the nick sites in several plasmids, have been most extensively analyzed using ColE1. One of the three proteins present in these complexes becomes covalently bound to the 5′ end of the nicked DNA.[146] Phage λ carrying a cloned copy of *oriT* is specifically nicked when grown in F^+, but not F^-, cells[147] and provides a means of identifying both the sequences and the proteins needed for the cleavage event. The *traY* gene encodes a 17-kDa membrane protein, the putative endonuclease responsible for nicking at *oriT*. A product of *traI*, the gene for helicase I (Section 11-3), also may be involved, although purified TraY and TraI proteins together are not sufficient to promote nicking of *oriT* in vitro.[148]

In contrast to most of the transfer genes, the *traI*, *traY*, and *oriT* functions are plasmid-specific. The differences in nucleotide sequence between the *oriT* of F and that of R1 preclude complementation by the respective TraY proteins.[149] In other plasmid pairs, the transfer specificity lies within the *traI* gene.[150] Since helicase I shows no template specificity in vitro, a TraY–TraI protein interaction may be inferred. The function of the *traM* gene product is also restricted to certain plasmids, presumably in the binding to *oriT* and in the initiation of unwinding by helicase I.

The *oriT* sequences of small mobilizable plasmids differ from those of the plasmids that activate their transfer. For example, ColE1 is efficiently mobilized

139. Jacobson A (1972) *J Virol 10*:835.
140. Manning PA, Morelli G, Achtman M (1981) *PNAS 78*:7487.
141. Paranchych W (1975) in *RNA Phages* (Zinder ND, ed.). CSHL, p. 85.
142. Everett R, Willetts M (1982) *EMBO J 1*:747.
143. Willetts N, Wilkins B (1984) *Microbiol Rev 48*:24.
144. Thompson R, Taylor L, Kelly K, Everett R, Willetts N (1984) *EMBO J 3*:1175.
145. Clewell DB, Helinski DR (1969) *PNAS 62*:1159; Clewell DB, Helinski DR (1970) *Biochemistry 9*:4428.
146. Blair DG, Helinski DR (1975) *JBC 250*:8785; Guiney DG, Helinski DR (1975) *JBC 250*:8796.
147. Everett R, Willetts N (1980) *JMB 136*:129.
148. Abdel-Monem M, Taucher-Scholz G, Klinkert M-Q (1983) *PNAS 80*:4659; Lahue EE, Matson SW (1990) *J Bact 172*:1385.
149. Everett R, Willetts N (1980) *JMB 136*:129.
150. Willetts NS, Maule J (1979) *MGG 169*:325; Willetts NS, Maule J (1985) *Genet Res 47*:1.

by both F and RK2, which have different *oriT* sequences and cannot activate the transfer of one another. It follows that the *oriT*-specific *traY* and *traI* genes are not required for ColE1 mobilization.[151] Rather, the plasmid-specific mobilization functions of the little-known *mob* (*mob*ilization) genes of ColE1 substitute for these essential activities.

Unwinding and Transfer. Unwinding of the nicked plasmid is coupled to DNA transfer and is initiated only upon formation of mating pairs. DNA is transferred as a single strand; neither replication to replace the transferred strand in the donor cell nor generation of the complementary strand in the recipient is obligatory during transfer. However, in an unperturbed mating, transfer and replication are simultaneous (Fig. 18–15).

The 5′ end of the strand nicked at *oriT* enters the recipient cell first.[152] Helicase I probably catalyzes the displacement of this strand, inasmuch as TraI protein is required specifically for transfer during conjugation and, in contrast to nearly all *tra* gene products, is located in the cytoplasm. Helicase I (Section 11-3) unwinds DNA with a 5′ → 3′ polarity at about 1000 ntd per second, a direction and speed appropriate for unwinding the DNA during transfer.[153]

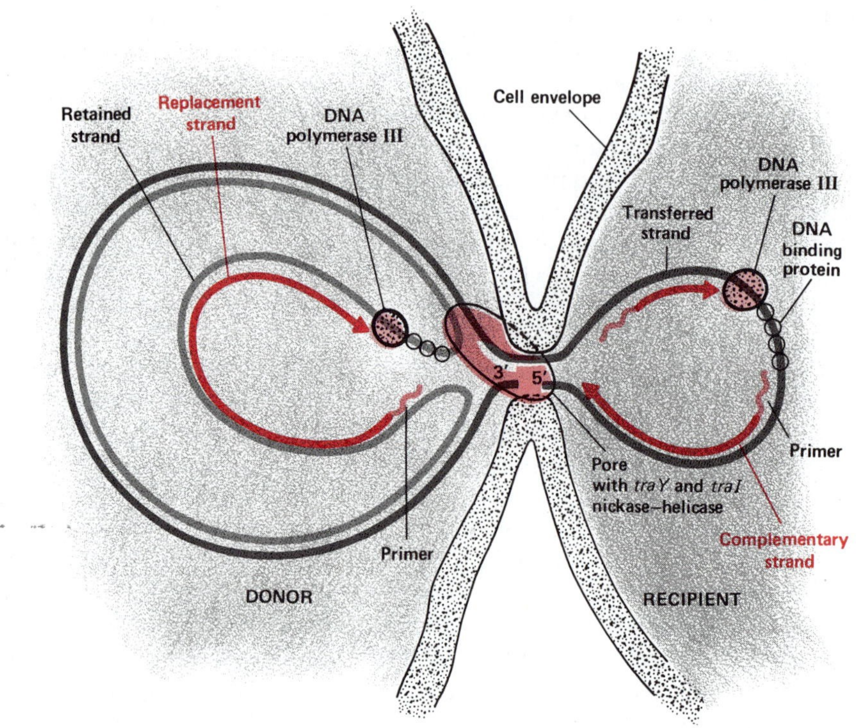

Figure 18-15
Model for conjugative transfer and replacement-strand synthesis in donor and recipient cell. [From Willetts N, Skurray R (1987) in *Escherichia coli and Salmonella typhimurium* (Neidhardt FC, ed.). Am Soc Microbiol, Washington DC, Vol. 2, p. 1110]

151. Everett R, Willetts N (1980) *JMB 136*:129; Willetts N, Maule J (1979) *MGG 169*:325; Willetts N (1980) *MGG 180*:213.
152. Ihler G, Rupp WD (1969) *PNAS 63*:138; Ohki M, Tomizawa M (1968) *CSHS 33*:651.
153. Abdeo-Monem M, Lauppe HF, Kartenbeck J, Durwald H, Hoffmann-Berling H (1977) *JMB 110*:667.

SSB proteins, required during unwinding, protect the transferred single strand from nuclease attack and prepare it for replication. Nearly all conjugative plasmids, including F, encode an SSB homologous to the *E. coli* protein.[154] Whether this protein is essential for transfer, in the absence of the host protein, has not been established.

DNA Synthesis. DNA synthesis regenerates the double-stranded plasmid genome for both donor and recipient cells. Defective conjugative synthesis in *dnaE*^{ts} mutants indicates that DNA pol III holoenzyme is required in both cells.[155] Because the DNA transfer is led by the 5′ end, synthesis of the complementary strand in the recipient cell is most likely discontinuous (Fig. 18-15). Synthesis of the complementary strand in the donor cell may be continuous, by elongation from either the 3′ end of the nick or from a single primer.

Initiation of replication in the donor cell requires the host RNA polymerase, evidence against both covalent extension of the 3′ end of the nick and a rolling-circle mechanism. The *dnaB* and *dnaG* genes are not essential; thus, involvement of the primosome is unlikely.[156] In contrast, priming of F DNA in the recipient cell can be performed by either the primosome or RNA polymerase; mobilized ColE1 DNA is probably primed by the primosome, initiated at the primosome assembly site.

Many conjugative plasmids, although not F, encode their own primase.[157] On these other plasmid templates, complementary-strand synthesis is independent of the host priming systems and relies on the plasmid enzyme transferred into the new cell along with the DNA. The primase genes are dispensable for plasmid maintenance, but are required for transfer into certain bacterial hosts[158] with inadequate primase activity. Often these plasmids can suppress *dnaG* mutants, indicating that the plasmid-encoded primases are sufficiently similar to the host enzyme to support chromosomal replication.[159] In recipient cells lacking primase and treated with rifampicin to block the expression of plasmid-borne genes, both plasmid and host DNA synthesis is observed, evidence for the transfer of primase with the entering DNA.

Termination of Conjugation. Although the final stages of DNA synthesis and processing that generate covalently closed supercoiled plasmids are not known in detail, it is assumed that replication precedes circularization of the transferred strand.[160] The enzymes required for ligation are probably plasmid-encoded, inasmuch as the *oriT* sequence is required. A mechanism similar to the ligation of the displaced strand during ϕX174 replication may be operating. The endonuclease could remain at the surface between the two cells and religate the nicked strand upon reaching the 3′ end.

Mating pairs dissociate upon completion of DNA transfer.[161] The recipient cell, upon expressing the plasmid-encoded genes, acquires pili and the surface exclusion proteins (encoded by *traS* and *traT*) that inhibit mating between two

154. Golub EI, Low KB (1985) *J Bact 162*:235.
155. Kingsman A, Willetts N (1978) *JMB 122*:287; Wilkins BM, Hollom SE (1974) *MGG 134*:143.
156. Kingsman A, Willetts N (1978) *JMB 122*:287; Wilkins BM, Hollom SE (1974) *MGG 134*:143.
157. Lanka E, Barth PT (1981) *J Bact 148*:769.
158. Lanka E, Barth PT (1981) *J Bact 148*:769; Wilkins BN (1975) *J Bact 122*:899; Wilkins BM, Boulnois GJ, Lanka E (1981) *Nature 290*:217.
159. Willetts N, Wilkins B (1984) *Microbiol Rev 48*:24.
160. Willetts N, Wilkins B (1984) *Microbiol Rev 48*:24.
161. Achtman M, Morelli G, Schwuchow S (1978) *J Bact 135*:1053.

F^+ cells.[162] The *traT* product, an outer-membrane lipoprotein, may block the interaction between the pilus and its receptor. The *traS* product, an inner-membrane protein, blocks mating by a mechanism still unknown. Various plasmids have different surface exclusion properties.[163] Even though the transfer genes of F and R1 are very homologous, *traT* and *traS* are sufficiently different to permit conjugation between F- and R1-bearing cells.

Regulation of Fertility[164]

Conjugative plasmids encode three functions (*traJ*, *finO*, and *finP*) that regulate expression of the large *tra* operon, and thus regulate fertility. The 27-kDa product of the *traJ* gene activates transcription;[165] in its absence nearly all stages of conjugation are inhibited (i.e., pilus formation, surface exclusion, and nicking at *oriT*). Genes *traM*, *finO*, *finP*, and *traJ* itself escape from *traJ* regulation; this is due to their location outside the *tra* operon. The *traI* and *traT* genes, internal to the operon, also escape by having their own promoters.[166]

The *finO* and *finP* (*f*ertility *in*hibition) genes repress *traJ* expression.[167] When *finO* is disrupted by insertion of IS3,[168] *tra* gene expression becomes high compared to that of most naturally occurring plasmids (including R1 and R100). When the *finO* product is now supplied by a *finO*$^+$ plasmid, repression of the F *traJ* gene brings F fertility down to the level normal for *finO*$^+$ plasmids.[169] R100-1, a mutant of R100 with a 1000-fold increased transfer efficiency, is also defective in *finO*. The product of the *finO* gene has not been characterized, nor has the mechanism by which it represses *traJ*.

The other regulator of *traJ* is *finP*, which encodes a 78-ntd RNA, complementary to the 5′ end of the *traJ* message.[170] The RNA is thought to repress *traJ* expression by hybridizing to the end of the message by a mechanism analogous to the copy-number control of R1 and pT181 (see below).

By the action of the *fin* genes, the *tra* functions of most self-transmissible plasmids (although not those of F, which is naturally *finO*$^-$) are normally repressed: cells lack pili, are unable to mate, and become fertile only during the rare periods of *tra* gene induction.[171] Whether such induction occurs spontaneously, or in response to some environmental signal, is unknown. Nonetheless, efficient transfer of plasmids throughout a population is achieved because, once a successful mating occurs, the *tra* genes are expressed in the recipient cell for several hours before repression by the *fin* system is established. Regulated, rather than constitutive, expression of the *tra* genes may be advantageous.

---

162. Achtman M, Kennedy N, Skurray R (1977) *PNAS 74*:5104: Achtman M, Manning PA, Kusecek B, Schwuchow S, Willetts N (1980) *JMB 138*:779.
163. Willetts N, Skurray R (1980) *ARG 14*:41.
164. Willetts N, Skurray R (1987) in *Escherichia coli and Salmonella typhimurium* (Neidhardt FC, ed.). Am Soc Microbiol, Washington DC, Vol. 2, p. 1110.
165. Willetts NS (1977) *JMB 112*:141.
166. Mullineaux P, Willetts N (1985) in *Plasmids in Bacteria* (Helinski DR, Cohen SN, Clewell DB, Jackson DA, Holleander A, eds.). Plenum Press, New York, p. 605.
167. Finnegan D, Willetts N (1973) *MGG 127*:307; Finnegan D, Willetts N (1971) *MGG 111*:256; Willetts NS (1977) *JMB 112*:141.
168. Cheah KC, Skurray R (1986) *J Gen Microbiol 132*:3269.
169. Egawa R, Hirota Y (1962) *Jpn J Genet 37*:66.
170. Mullineaux P, Willetts N (1985) in *Plasmids in Bacteria* (Helinski DR, Cohen SN, Clewell DB, Jackson DA, Holleander A, eds.). Plenum Press, New York, p. 605.
171. Willetts N, Skurray R (1980) *ARG 14*:41.

Otherwise, the constant presence of pili would render the cell sensitive to pili-specific phages (such as the Ff phages; Section 17-3), even when no mating opportunities are available.

18-8 Broad-Host-Range Plasmids

Most plasmids grow only in a restricted group of bacterial hosts. However, two classes of plasmids (RK2 and RSF1010) can be transferred and maintained in almost all species of Gram-negative bacteria. Their strategies for replication and the regulation of gene expression, essential for proper replication, must have evolved in order that these plasmids could flourish in rather diverse metabolic environments.

Broad-host-range plasmids are of technical importance because they make it possible for recombinant DNA technology to exploit organisms in which the standard ColE1-based vectors cannot replicate. Their transfer properties make them excellent "shuttle vectors," suitable for use both in *E. coli* and in other bacteria, and permitting the introduction of cloned DNA into a cell that is not easily transformed by naked DNA.

RK2

One class of broad-host-range plasmids is exemplified by RK2, closely related or identical to RP1 and RP4. This self-transmissible plasmid is relatively large (~60 kb) and encodes multiple drug-resistance determinants. Replication requires two plasmid-encoded functions: the origin (*oriV*) sequence and the *trfA* (trans replication *f*actor) gene. The TrfA protein exists in 32- and 43-kDa forms, arising from translation of the same reading frame from two different start sites.[172] In *E. coli*, RK2 replicates by a mechanism similar to that of F plasmid. The TrfA protein activates initiation by binding to repeated elements in the origin; the amount of available TrfA regulates the initiation frequency. Initiation also requires dnaA protein and two dnaA boxes that are found within *oriV*. The dnaB protein, primase, DNA gyrase, and pol III holoenzyme are required for replication in crude *E. coli* extracts.[173]

Several observations suggest that, in bacteria other than *E. coli*, initiation may occur by a different mechanism. An insertion mutation in *oriV*, between the TrfA binding sites and one of the dnaA boxes, abolishes replication activity in *E. coli* but not in *Pseudomonas aeruginosa*.[174] In *E. coli*, the small TrfA protein supports replication, while the large form is required in *P. aeruginosa*.[175] The plasmid-encoded primase, dispensable for vegetative replication, is essential for conjugation in some species.[176]

RK2 depends almost exclusively on host-encoded replication proteins, yet is able to survive in a wide variety of bacterial hosts where these proteins are expected to vary significantly. Thus, the RK2 initiation system must be uncom-

---

172. Shingler V, Thomas CM (1984) *JMB 175*:229; Smith CA, Thomas CM (1984) *JMB 175*:251.
173. Pinkney M, Diaz R, Lanka E, Thomas CM (1988) *JMB 203*:927.
174. Cross MA, Warne SR, Thomas CM (1986) *Plasmid 15*:132; Shingler V, Thomas CM (1989) *BBA 1007*:301.
175. Shingler V, Thomas CM (1989) *BBA 1007*:301.
176. Lanka E, Barth PT (1981) *J Bact 148*:769.

mon in its ability to assemble distinct replication complexes, a capacity which should provide insight into the evolution of replication fork assembly.

RSF1010

A second class of broad-host-range plasmids, typified by RSF1010, encode enough of their own replication proteins to be independent of much of the host machinery.[177]

The 8.7-kb RSF1010, originally isolated from *E. coli*, encodes resistance to streptomycin and sulfonamides and is mobilizable, although not self-transmissible. The plasmid carries an origin sequence, *oriV*, and genes for three replication proteins.[178] The *repC* gene encodes an origin-binding protein required for initiation, *repB* encodes a primase, and *repA* a DNA helicase. The *repB* products are two related proteins, 38 and 70 kDa; the smaller form has been purified and supports plasmid replication; the larger may be involved in mobilization. By virtue of these proteins, RSF1010 replication is independent of the host *dnaA*, *dnaB*, *dnaC*, and *dnaG* (primase) genes,[179] thereby providing an adaptive advantage for survival in varied intracellular environments.

The RSF1010 replication proteins are plasmid-specific, inasmuch as they fail to complement defects in the analogous host genes. This contrasts with the primase and SSB genes of conjugative plasmids, which often suppress the defect in host replication in *dnaG* and *ssb* mutants, as well as acting during plasmid transfer. The RSF1010 primase appears to recognize a specific sequence on each strand of the plasmid.

RSF1010 is efficiently mobilized by certain tra^+ plasmids, including RK2, for transfer into almost all bacteria. At least three *mob* genes and an *oriT* sequence are required. RSF1010 can even be transferred from *Agrobacterium tumefaciens* into plant cells in the presence of its cognate *mob* and the RK2 *tra* functions.[180] This observation is of technical importance in facilitating the introduction of DNA into plants. Furthermore, since the same DNA sequence and proteins required for bacterial conjugation also function in DNA transfer from bacteria into plants, the mechanistic information available on bacterial conjugation can be generalized to DNA transfer between kingdoms in nature.[181]

18-9 Plasmids of Gram-Positive Bacteria

As in *E. coli*, small, circular plasmids are found in Gram-positive bacteria, where they carry drug-resistance genes and are maintained at a fixed number of copies per cell. However, the mechanism of replication is strikingly different. A rolling-circle mode, involving the covalent extension of a nicked template and a

---

177. Frey J, Bagdasarian M (1989) in *Promiscuous Plasmids of Gram-Negative Bacteria* (Thomas CM, ed.). Academic Press, New York, p. 79.
178. Scherzinger E, Bagdasarian MM, Scholz P, Lurz R, Rückert B, Bagdasarian M (1984) *PNAS 81*:654.
179. Scherzinger E, Bagdasarian MM, Scholz P, Lurz R, Rückert B, Bagdasarian M (1984) *PNAS 81*:654; Scholz P, Haring V, Scherzinger E, Lurz R, Bagdasarian MM, Schuster H, Bagdasarian M (1985) in *Plasmids in Bacteria* (Helinski DR, Cohen SN, Clewell DB, Jackson DA, Hollaender A, eds.). Plenum Press, New York, p. 243.
180. Buchanan-Wollaston V, Passiatore JE, Cannon F (1987) *Nature 328*:172.
181. Stachel SE, Zambryski PC (1986) *Cell 47*:155; Lichtenstein C (1986) *Nature 322*:682.

single-stranded circular intermediate, is responsible for the duplication of plasmid genomes (Fig. 18-16). This replication program resembles that used by single-stranded phages (Ff and ϕX174; Sections 17-3 and 17-4) rather than the theta replication employed by *E. coli* plasmids.

The prototypic plasmid of Gram-positive bacteria is the 4.4-kb pT181. Originally isolated from *Staphylococcus aureus*, it encodes resistance to tetracycline[182] and is maintained at about 20 to 25 copies per cell. Studies of its replication in vivo and in vitro established the rolling-circle mechanism of replication.

First-Strand Synthesis

Replication initiates from a single origin, located within the *repC* gene, and proceeds unidirectionally.[183] Essential for initiation, *repC* encodes a 38-kDa endonuclease which specifically cleaves the plasmid DNA within the origin (Section 8-7).[184] A unique nick, with a 3′-OH and a blocked 5′-P, is introduced into one strand to start rolling-circle replication, with the cleavage protein, presumably, covalently bound to the 5′ end. The capacity of RepC protein to religate the nicked DNA qualifies it, like ϕX174 gpA, as a sequence-specific type I topoisomerase.[185]

In contrast to the sequences from plasmids of Gram-negative bacteria, the pT181 origin lacks repeated iterons and is relatively GC-rich (50%).[186] A 43-bp fragment functions as an origin in vivo and in vitro, although an additional 27

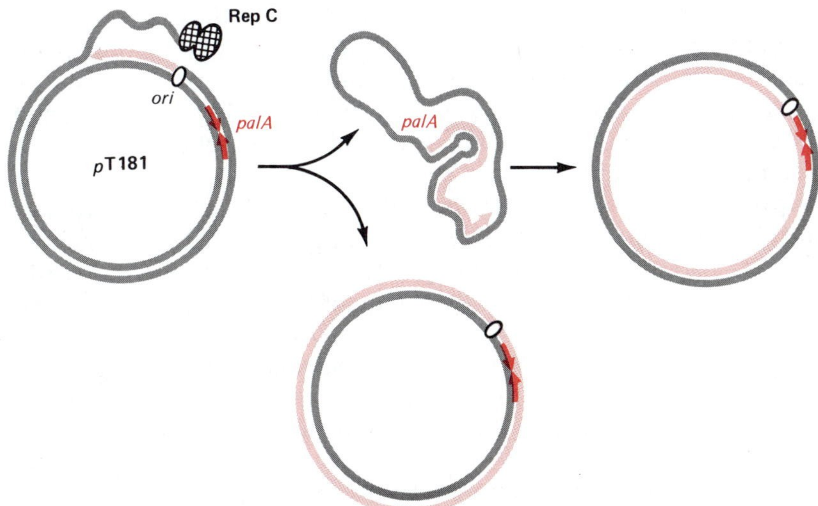

Figure 18-16
Asymmetric rolling-circle replication of pT181. *ori* = site of cleavage by RepC protein; *palA* = second-strand initiation signal. In the absence of *palA*, single-stranded circles accumulate. When *palA* is present, initiation of replication on the displaced strand occurs either coupled or uncoupled to first-strand synthesis. (Modified from a figure provided by Dr R Novick)

---

182. Khan SA, Novick RP (1983) *Plasmid 10*:251.
183. Khan SA, Adler GK, Novick RP (1982) *PNAS 79*:4580.
184. Koepsel RR, Khan SA (1986) Science *233*:1316.
185. Koepsel RR, Murray RW, Rosenblum WD, Khan SA (1985) *PNAS 82*:6845.
186. Koepsel RR, Murray RW, Khan SA (1986) *PNAS 83*:5484.

bp, containing part of the RepC binding site, is required for it to compete with the wild-type origin for RepC. The sequence contains a static, RepC-enhanced DNA bend and a region of dyad symmetry that can form a secondary structure both in vivo and in vitro.[187]

Covalent extension of the 3′-OH generated by RepC cleavage occurs in extracts of *S. aureus*, but not of *E. coli*;[188] neither RNA nor protein synthesis is required if sufficient RepC protein is present to cleave all the plasmids. DNA synthesis proceeds around the circle, displacing a single strand.[189] Once a round is completed, RepC protein ends a round of replication by again cleaving the origin and joining the ends of the displaced strand into a single-stranded circle.[190] Plasmids with two origin sequences generate two single-stranded circles by initiation at one origin and termination at the second.[191]

Second-Strand Replication

The initiation signal for complementary-strand synthesis is located upstream of the first-strand nick site and thus the signal is converted to its active single-stranded form only shortly before completion of first-strand replication.[192] The two strands are therefore replicated sequentially, rather than simultaneously. Single-stranded circular replication intermediates are observed both in vivo and in vitro.

The origin of second-strand synthesis is a large (~160 ntd) *pal*indromic sequence, *palA*.[193] Highly related *palA* sequences have been identified in eight different *Staphylococcus* plasmids. Deletion of *palA* results in a large reduction in plasmid copy number and an increased accumulation of the ssDNA products of first-strand replication. The *palA* sequence is not absolutely essential, however, a finding indicative of secondary mechanisms of initiation on the displaced strand. A cell-free extract from *S. aureus* can convert the single-strand form to duplex DNA in vitro,[194] but the mechanism of initiation and the enzymes involved have not been identified.

One group of plasmids isolated from Gram-positive bacteria have the remarkable property of also replicating in Gram-negative cells; pLS1, derived by a small deletion from the streptococcal plasmid pMV158, is an example. Replication follows the same strategy as with pT181: rolling-circle synthesis initiates at a nick introduced at the origin, generating circular ssDNA.[195] A signal responsible for initiation of the second strand has been isolated. Like *palA* of pT181, it is a palindrome, but unlike the pT181 sequence, it is functional in *Streptococcus pneumoniae*, *B. subtilis*, and *E. coli*. Sequence comparisons reveal that the pLS1 signal is 50% homologous to the complementary-strand origin of ϕX174, and 48% to that of M13. Thus, it is substantially more related to these *E. coli* phage sequences than to *palA* of the *Staphylococcus* plasmids (35%).[196] The remark-

---

187. Noirot P, Novick RP, personal communication.
188. Khan SA, Carleton SM, Novick RP (1981) *PNAS 78*:4902; Koepsel RR, Murray RW, Rosenblum WD, Khan SA (1985) *JBC 260*:8571.
189. Murray RW, Koepsel RR, Khan SA (1989) *JBC 264*:1051.
190. Iordanescu S, Projan SJ (1988) *J Bact 170*:3427.
191. Murray RW, Koepsel RR, Khan SA (1989) *JBC 264*:1051; Iordanescu S, Progan SJ (1988) *J Bact 170*:33427.
192. Gruss AD, Ross HF, Novick RP (1987) *PNAS 84*:2165.
193. Gennaro ML, Iordanescu S, Novick RP, Murray RW, Steck TR, Khan SA (1989) *JMB 205*:355.
194. Khan S, personal communication.
195. Puyet A, del Solar GH, Espinosa M (1988) *NAR 16*:115.
196. del Solar GH, Puyet A, Espinosa M (1987) *NAR 15*:5561.

able host range of pLS1 may result from its carrying a second-strand origin recognizable by the host machinery of many organisms.

Regulation of Plasmid Replication

Two small antisense RNAs, homologous to the untranslated 5′ end of the *repC* mRNA, regulate *repC* expression and thus plasmid replication.[197] The small RNA molecules are thought to interact with a loop in the leader region of the message. RNA–RNA hybrid formation changes the mRNA secondary structure in such a way that a stable stem-and-loop structure is formed. This structure resembles a Rho-independent terminator and causes transcription to stop prematurely.[198] Thus, the antisense RNA may regulate by transcriptional attenuation, as in the control of many biosynthetic operons (Sections 7-7 and 7-8), rather than by inhibition of translation. Cis-acting DNA sequences also play a role in regulation of plasmid replication. The pT181 plasmid contains *cmp* (competition), a sequence of less than 70 bp, that acts as an enhancer of origin function from a distant location.[199] Enhancement appears to require the participation of a host-encoded factor.[200]

18-10 Yeast Plasmids: The 2μ Circle and Others

The 2μ circle is a 6.3-kb, circular extrachromosomal element present in the nucleus of most *Saccharomyces cerevisiae* strains, although it imparts no apparent selective advantage. Maintained at 50 to 100 copies per haploid genome, the plasmid is efficiently transmitted to the newly budded cell during mitosis and to all four spores in meiosis.[201] Covered with nucleosomes and relying on the host machinery for replication, the 2μ genome responds to cellular control, replicating once per S phase. Thus, the 2μ system is attractive for studying the mechanism and regulation of replication in a eukaryote highly amenable to genetic and biochemical analysis.

Plasmid Structure

The 2μ genome contains two large unique regions between two copies of a 599-bp sequence in inverted orientation.[202] Embedded in the inverted repeats are recognition sequences for the plasmid-directed, site-specific recombination system [the FLP ("flip") system; Section 21-6]. Flip-mediated recombination between these sites inverts the unique regions with respect to one another.[203] Thus, the 2μ circle occurs in two isomeric forms (Fig. 18-17) with approximately equal frequency. This inversion is necessary for plasmid amplification (see below).

---

197. Carleton SM, Projan SJ, Highlander SK, Moghazeh SM, Novick RP (1984) *EMBO J 3*:2407; Kumar C, Novick R (1985) *PNAS 82*:638.
198. Novick RP, Iordanescu S, Projan SJ, Edelman I, personal communication.
199. Gennaro ML, Novick R (1986) *J Bact 168*:160; Gennaro ML, Novick R (1988) *J Bact 170*:5709.
200. Gennaro ML, personal communication.
201. Livingston DM (1977) *Genetics 86*:73.
202. Hartley JL, Donelson JE (1980) *Nature 286*:860.
203. Broach JR, Hicks JB (1980) *Cell 21*:501; Broach JR, Guarascio VR, Jayaram M (1982) *Cell 29*:227.

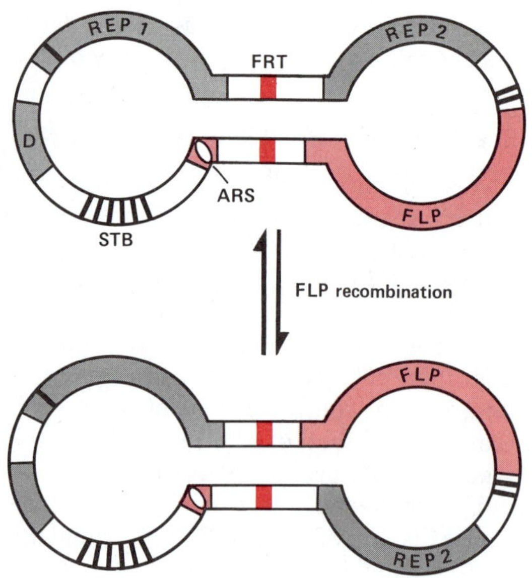

Figure 18-17
Flip-mediated recombination isomers of the 2μ plasmid, drawn with the large inverted repeats aligned. The *FLP* (for *flip*) gene encodes the recombinase. *FRT* = *f*lip *r*ecombination *t*arget; STB = *stab*ility site repeats; ARS = *a*utonomous *r*eplication *s*equence; black bands = 9-bp repeat sequences.

Four proteins involved in replication, maintenance, and copy number control are encoded by the 2μ plasmid. The 48-kDa recombinase is the product of the largest gene, *FLP* (*flip*). The *REP1* and *REP2* gene products are required for stable plasmid inheritance and regulate copy number by repressing expression of the plasmid genes. The D protein, product of the fourth gene, also regulates gene expression by antagonizing the repressor activity of the REP1 and REP2 proteins. In addition to the flip recombination sequences, the plasmid contains a sequence responsible for autonomous replication [the ARS (*a*utonomous *r*eplication *s*equence) element] and a *stab*ility locus (STB), both of which are required *in cis*, the former for replication and the latter for partitioning.

Replication

The 2μ circle replicates early in S phase and relies entirely on host-encoded factors.[204] Replication initiates near the ARS element (within 200 bp, as defined by deletions), proceeds bidirectionally, and terminates diametrically opposite the site of initiation.[205] The coincidence between the location of an ARS element and an initiation bubble was first observed with the 2μ circle. ARS elements impart replication competence to plasmids and occur in the genome at the frequency expected for origins. Establishing the ARS sequence as an origin of replication favors the thesis that initiation of replication occurs at specific sequences on eukaryotic chromosomes (Section 16-6).

In synchronized cells, replication intermediates appear within 10 minutes of entering S phase and disappear by about 40 minutes.[206] Intermediates are primarily theta (θ) forms and are multiply intertwined, catenated dimers (as in the late stages of replication of many circular templates; Section 12-1). Among the

204. Livingston DM, Kupfer DM (1977) *JMB 16*:249; Zakian VA, Brewer BJ, Fangman WL (1979) *Cell 17*:923.
205. Brewer BJ, Fangman WL (1987) *Cell 51*:463; Huberman JA, Spotila LD, Nawotka KA, el-Assouli SM, Davis LR (1987) *Cell 51*:473.
206. Brewer BJ, Fangman WL (1987) *Cell 51*:463.

intermediates, the two isomers of 2μ are equally represented. Decatenation occurs efficiently during S phase, yielding covalently closed, supercoiled circular products.

Amplification of Plasmid Copy Number

Flip-directed, site-specific recombination (Section 21-6) allows amplification of the plasmid copy number without violating the eukaryotic replication rule of only one initiation per cell cycle. According to one model, when replication has been initiated but has not yet reached the second flip site, recombination inverts the bidirectionally moving replication forks, so that they both move in the same direction[207] (Fig. 18-18). A θ-replication intermediate is thus converted into a "double rolling circle." Continuing replication generates long concatemeric DNA, which can be monomerized by site-specific recombination. Many copies of the plasmid thus result from a single initiation. Termination of rolling-circle replication occurs when a second inversion causes the replication forks again to travel toward each other until they collide.

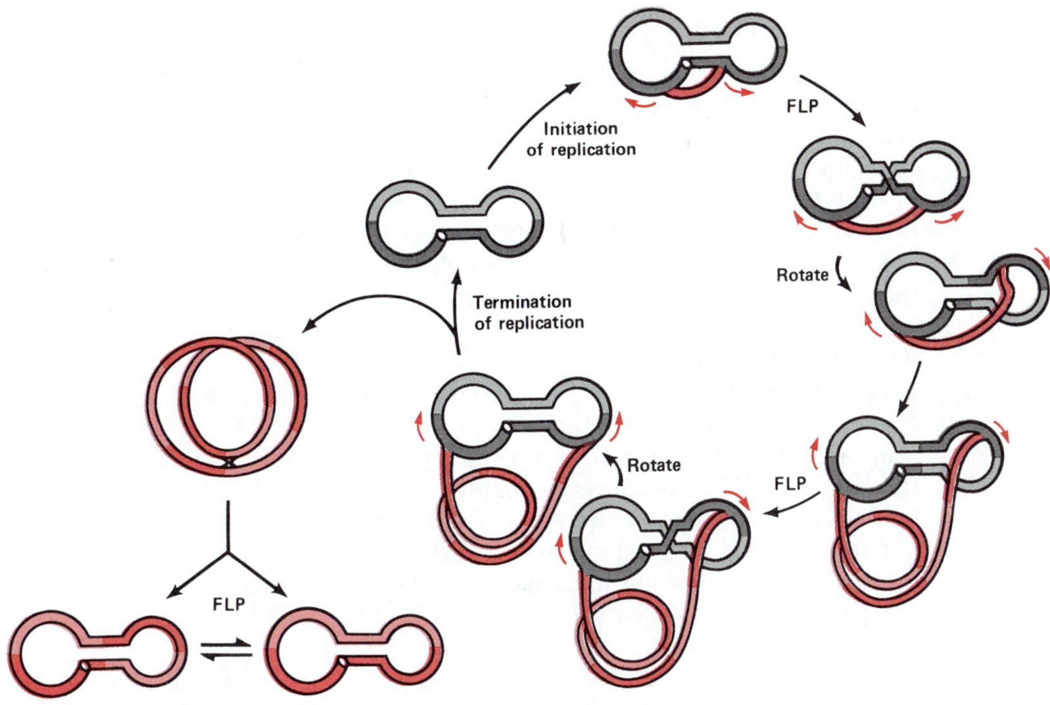

Figure 18-18
Model for recombinational amplification of the 2μ plasmid. Semiconservative DNA replication proceeds bidirectionally from the plasmid origin. Colored arrows = direction of replication fork movement. *FLP*-mediated recombination reorients the forks so that they no longer converge, and continuing replication yields a multimeric replication intermediate. Another *FLP* recombination event restores the converging orientation of the replication forks. Completion of replication yields a multimer and the original 2μ-circle monomer. Further *FLP*-mediated or general recombination resolves the multimer into monomers.

207. Futcher AB (1986) *J Theor Biol 119*:197.

Evidence for this recombination-dependent model of amplification is as follows: (1) Amplification from a single copy to the normal high copy number occurs only in recombination-proficient plasmids; both the recombination sites (*FRT*, for *flip* recombination target) and the FLP protein are required.[208] (2) Long replication intermediates are observed during amplification, in contrast to the monomeric θ intermediates normally observed. (3) Finally, the ratio of the two plasmid isomers is highly biased after amplification synthesis, indicating that many of the daughter plasmids arose from replication of a single template.[209]

Recombination-dependent amplification may be used during replication of other eukaryotic genomes. There are yeast plasmids, unrelated to the 2μ circle, that have the same genome organization—two unique regions and a large inverted repeat—and encode site-specific recombination systems, suggesting replication by a similar pathway.[210] The same genome structure is shared by the chloroplast chromosomes of many plants.[211] In *Chlamydomonas* also, recombination between the repeats has been demonstrated,[212] and the replication origin lies very near one repeat, thereby maximizing the chance that inversion will occur during DNA synthesis.[213] Both θ and rolling-circle intermediates have been observed in chloroplast DNA replication[214] (Section 11).

Regulation of Plasmid Copy Number[215]

In contrast to the control of bacterial plasmid replication, adjusted by the frequency of initiation, the 2μ copy number is regulated by adjusting *FLP* expression and thus the amount of synthesis by amplification. The *REP1* and *REP2* gene products together repress the *FLP* and *REP1* genes. Thus *REP1* is autoregulated, while *REP2* is constitutively expressed. The D protein antagonizes the repression of *FLP* by the REP1–REP2 repressor.[216]

The circuit that controls copy number can be formulated by the operation of these components. When the plasmid concentration is normal, REP1 and REP2 proteins repress the *FLP* gene and inhibit amplification synthesis. However, when the copy number falls, the level of REP1 and REP2 also falls, inducing *FLP* expression and triggering amplification synthesis. The D protein magnifies the effect on *FLP* expression of the drop in REP1–REP2 by antagonizing their repressor activity. The increased gene dosage resulting from amplification raises the level of REP1–REP2, reestablishing repression of *FLP* (and *REP1*), thus halting the burst of replication.

This regulatory network suggests that the REP1–REP2 repressor should bind to three regions of the plasmid: the *FLP* and *REP1* promoters and the STB locus (involved in partition; see below). A 9-bp sequence is found once in the *REP1*

---

208. Volkert FC, Broach JR (1986) *Cell 46*:541.
209. Armstrong KA, Som T, Volkert RC, Broach JR (1988) in *Cancer Cells*, Vol. 6 (Kelly T, Stillman B, eds.). CSHL, p. 213.
210. Araki H, Jearnpipatkul A, Tatsumi H, Sakurai T, Ushio K, Muta T, Oshima Y (1985) *JMB 182*:191; Murray JAH, Cesareni G, Argos P (1988) *JMB 200*:601.
211. Rochaix JD (1978) *JMB 126*:596.
212. Palmer JD (1983) *Nature 301*:92.
213. Wu M, Lou JK, Chang DY, Chang CH, Nie ZQ (1986) *PNAS 83*:6761.
214. Kolodner RD, Tewari KK (1975) *Nature 256*:708.
215. Som T, Armstrong KA, Volkert FC, Broach JR (1988) *Cell 52*:27; Armstrong KA, Som T, Volkert RC, Broach JR (1988) in *Cancer Cells*, Vol. 6 (Kelly T, Stillman B, eds.). CSHL, p. 213.
216. Murray JAH, Scarpa M, Rossi N, Cesareni G (1987) *EMBO J 6*:4205.

promoter, twice upstream of the *FLP* gene, and six times in the STB locus, and is thus an attractive candidate for the REP1–REP2 recognition sequence.[217]

Partitioning at Cell Division

In the absence of a partition system, yeast plasmids are retained in the mother cell during budding or division.[218] The primary function of the *REP1, REP2*, and STB (also called *REP3*) loci is to overcome this segregational bias, allowing inheritance of the plasmid by both daughter cells.[219] These functions also operate in meiosis to transmit the plasmid to all four spores.

The mechanism of partitioning is still unknown. However, since the REP1 and REP2 proteins act together as transcriptional regulators, at least one is likely to be a DNA-binding protein. STB, the site on the plasmid required for stabilization by REP1–REP2, contains five direct repeats of a 62-bp sequence, probably involved in REP1–REP2 binding.[220] A second domain of STB contains a transcription terminator. Transcription through the STB repeats and the ARS appears to inhibit the function of these regions.

The REP1 protein is found exclusively in the nucleus, and fractionates with the karyoskeleton.[221] By DNA sequence, the carboxyl terminus of REP1 is homologous to the fibrous proteins: the myosin heavy chain, vimentin, and the nuclear lamins. Thus, REP1 protein may function in partitioning by simultaneously intercalating into the nuclear matrix and (with REP2 protein) binding to the plasmid DNA. By anchoring the DNA to this structural component, REP1 protein could ensure partitioning to the budded cell.

Artificial Yeast Plasmids[222]

Plasmids constructed in vitro from chromosomal constituents replicate and are stably maintained in yeast cells. Studies of the behavior in vivo of such plasmids have facilitated the identification and dissection of ARS elements, centromeres, and telomeres. Plasmid replication depends on the *cdc* (cell-division cycle) genes and is inhibited by factors, such as the mating pheromone α-factor, that block entrance into S phase. A similar control obtains for nuclear DNA but not for the mitochondrial chromosome.

ARS plasmids are present at 100 to 200 copies, but are inefficiently partitioned into the bud during mitosis.[223] The addition of a centromere stabilizes the plasmids enormously and lowers their copy number to approximately one per cell.[224] Centromeres, presumably the site of spindle attachment to the chromosome, appear to be deleterious when present at high levels. The addition of telomeres converts plasmids into artificial, linear "minichromosomes." In early studies, fragments of *Tetrahymena* ribosomal DNA were used to provide the

217. Armstrong KA, Som T, Volkert RC, Broach JR (1988) in *Cancer Cells*, Vol. 6 (Kelly T, Stillman B, eds.). CSHL, p. 213.
218. Murray AW, Szostak JW (1983) *Cell 34*:961.
219. Cashmore AM, Albury MS, Hadfield G, Meacock PA (1986) *MGG 203*:154; Kikuchi Y (1983) *Cell 35*:487.
220. Murray JAH, Cesareni G (1986) *EMBO J* 5:3391.
221. Wu L-C C, Fisher PA, Broach JR (1987) *JBC 262*:883.
222. Williamson DH (1985) *Yeast 1*:1; Blackburn EH, Szostak JW (1984) *ARB 53*:163; Walmsley RM (1987) *Yeast* 3:139.
223. Murray AW, Szostak JW (1983) *Cell 34*:961.
224. Dani GM, Zakian VA (1983) *PNAS 80*:3406; Murray AW, Szostak JW (1983) *Nature 305*:189.

telomere function. When these sequences are introduced into yeast, the termini acquire repeated sequences, identical to those found at the ends of the yeast chromosomes.[225] Their physiologic behavior, coupled with the ease of experimental manipulation, make minichromosomes attractive objects for genetic and enzymologic analysis.

18-11 Mitochondria, Kinetoplasts, and Chloroplasts

The origin of mitochondria and chloroplasts by the parasitic prokaryote invasion of larger cells is a favored theory that encompasses the origin of eukaryotic cells as well. Evidence in its favor is the composition of these organelles: a small circular genome, ribosomes that resemble those of prokaryotes, and membranes more like those of bacteria than of the cell in which they reside.

Reasons for the retention of independent genomes by these organelles remain a mystery. Most of the proteins required for their assembly and function are encoded in the nucleus, translated in the cytoplasm, and then transported into the organelle. Of the five proteins specifically involved in mitochondrial DNA replication, all appear encoded by nuclear genes (see below). The organelle genomes do, however, direct the synthesis of some vital molecules, including tRNAs that decode mRNA by rules different from those used in the nucleus. Besides this surprising departure from the "universal" genetic code, organelle inheritance differs from that of nuclear genes: inheritance of organelle DNA, by virtue of its cytoplasmic location, is strictly maternal. These variations from the usual mechanisms have made the changes in DNA sequence of organelle genomes a favored object in studies of evolution.

Cells contain many mitochondria (100 to 1000 in the average mammalian cell), and plant cells possess many chloroplasts as well. Each organelle contains several copies of the genome (5 to 10 in mitochondria, 20 to 80 in chloroplasts). Despite their minute size compared to the nuclear genome, the numerous organelle chromosomes make up from 1% to 15% of the total cellular DNA. Mitochondria and chloroplasts grow and divide throughout all phases of the cell cycle, doubling in number, on the average, before each cell division. DNA replication is not limited to the nuclear S phase but appears to occur randomly throughout the cell cycle. Replication of mitochondrial DNA relies on a special DNA polymerase and a distinctive mechanism, which may prove instructive for both eukaryotic and prokaryotic replication.

Mitochondria[226]

Genome. Mitochondrial DNA (mtDNA) of animal cells is a supercoiled, circular duplex of 16 kb. The complete nucleotide sequence of the mouse and human mtDNAs has been determined.[227] Most molecules (>75% in mouse L cells)

---

225. Shampay J, Szostak JW, Blackburn EH (1983) *Nature 310*:154; Walmsley RW, Chan CSM, Tye B-K, Petes TD (1984) *Nature 310*:157.
226. Clayton DA (1982) *Cell 28*:693; Clayton DA (1984) *ARB 53*:573; Topper JN, Chang DD, Fisher RP, Clayton DA (1988) in *Cancer Cells*, Vol. 6 (Stillman B, Kelly TJ, eds.). CSHL, p. 207.
227. Anderson S, Bankier AT, Barrell BG, de Bruijn MHL, Coulson AR, Drouin J, Eperon IC, Nierlich DP, Roe BA, Sanger R, Schreier PH, Smith AJH, Staden R, Young IG (1981) *Nature 290*:457; Bibb MJ, Van Etten RA, Wright CT, Walberg MW, Clayton DA (1981) *Cell 26*:167.

contain a D-loop at a unique location (Fig. 18-19). Catenated dimers, circular dimers, and higher oligomers are found at levels of several percent of the DNA molecules in many tissues and cultured cells.[228] The most extreme complexity is found in the networks of kinetoplast DNA of flagellated organisms (see below).

Mammalian mtDNA encodes 2 rRNAs, 22 tRNAs, and 13 proteins.[229] The remaining mitochondrial proteins are supplied by nuclear genes and translated by cytoplasmic ribosomes. The only substantial stretch of noncoding sequence is in the D-loop region, which contains the signals for initiation of transcription and replication.

Mitochondrial DNA of yeast (*Saccharomyces*) is five times as long as mammalian mtDNA. Yet the number of genes is about the same, because much of the genome is composed of noncoding sequences. The 75-kb molecule, with a GC mole fraction of only 18%, is one of the most AT-rich functional DNAs in nature. Yeast provides an attractive system for the study of mitochondrial biosynthesis. When grown on glucose, yeast can survive without mitochondria, making it easier to identify the nuclear and mitochondrial genes required for organelle function. One class of mt-deficient mutants, the "cytoplasmic petites," lacks most of the mitochondrial genome. The remainder of the molecule, amplified by tandem duplication, is faithfully replicated and maintained at the same cellular concentration as wild-type mtDNA. The sequences maintained must therefore include the signals required for DNA replication. Morphologically normal but nonfunctional promitochondria are formed, attesting to the importance of nuclear genes in mitochondrial assembly.

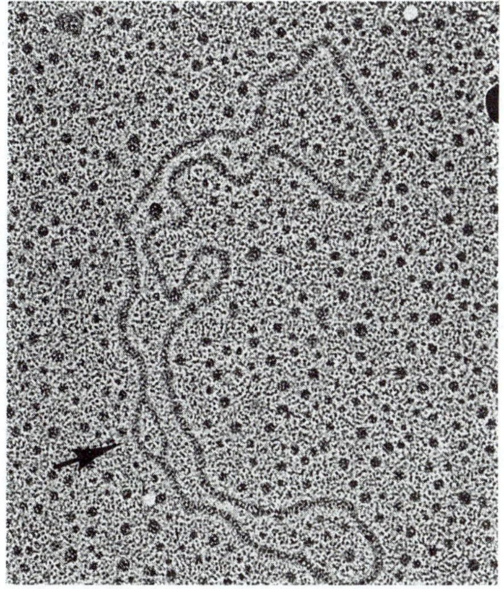

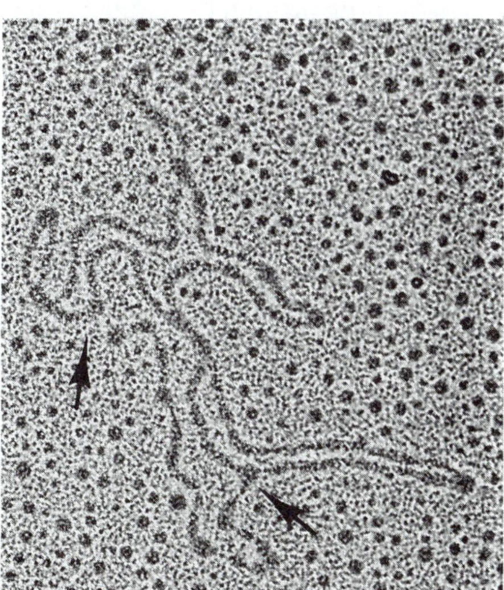

Figure 18-19
Electron micrographs of D-loops of replicating mitochondrial DNA. (Left) small D-loop; (right) expanding D-loop with extent of the displaced single strand marked by arrows. DNA contour length, 5 μm. [Courtesy of Professor D Clayton]

228. Clayton DA, Smith CA (1975) *Int Rev Exp Pathol* 14:1.
229. Clayton DA (1984) *ARB* 53:573.

Sensitivity of animal-cell mtDNA to cleavage by alkali and RNase H is attributable to interspersed ribonucleotides.[230] Approximately ten residues are included in the 16-kb DNA, very likely vestiges of priming events in replication. A significant fraction of these ribonucleotides has been located near the replication starts of the two mtDNA strands[231] (see below). Mitochondria appear to lack the capacity to repair DNA damage, such as that resulting from UV irradiation, and apparently fail to excise incorporated ribonucleotides as well (Section 13-8).

Replication. Mitochondrial DNA and its replication have been studied in rat liver, cultures of human and mouse cells, yeast, and other sources. Mouse L cells are convenient because culture lines lacking cytoplasmic thymidine kinase but retaining the mitochondrial enzyme are available, allowing specific radioactive labeling of mtDNA.[232] In this cell line, mtDNA replication continues throughout the cell cycle, doubling at the same rate as the cell. Mitochondrial DNA molecules are selected at random for replication, with some molecules replicating twice and others not at all. Observation of in vivo replication intermediates reveals that synthesis of the two complementary strands [heavy (H) and light (L)] is relatively asynchronous (Fig. 18-20).

H-Strand Synthesis. Initiation of H-strand synthesis, the first replication event, occurs within the D-loop region near the two divergent promoters that direct transcription of the template strands.[233] RNA initiated at the L-strand promoter, located about 100 bp upstream of the H-strand origin, is responsible for priming DNA synthesis and transcribing the entire L-strand. Both mitochondrial RNA polymerase (mtRNAP) and transcription factor (mtTF), a sequence-specific DNA-binding protein that imparts specificity to the polymerase, are essential.[234] Transcripts initiating at the L-strand promoter cross the origin region and are either processed to be used as primers[235] or continue to be extended. Cleavage by the site-specific mtRNA processing enzyme is thus the committed step for initiation of replication.

RNA cleavage occurs at three specific locations within a 90-bp region, coinciding with the three conserved sequence blocks (CSB I, II, and III) found in all vertebrate mtDNAs. The RNA processing enzyme, purified from mouse cells, cleaves RNA from either a mouse or a human template. An activity from human mitochondria that is capable of processing both RNAs has also been identified. Since homology between the two templates is confined to the CSBs, these sequences are considered to be involved in enzyme recognition. The processing enzyme contains a 136-ntd RNA, required for activity, with a 10-ntd stretch complementary to CSBII.[236] The CSBII sequence is located in the substrate RNA immediately downstream of the cleavage site, implicating a RNA–RNA pairing in the specificity of processing.

Cleavage yields 3′-OH and 5′-P termini, consistent with its role in converting a

---

230. Wong-Staal F, Mendelsohn J, Goulian M (1973) *BBRC 53*:140; Grossman LI, Watson R, Vinograd J (1973) *PNAS 70*:3339.
231. Brennicke A, Clayton DA (1981) *JBC 256*:10613.
232. Berk AJ, Clayton DA (1973) *JBC 248*:2722.
233. Chang DD, Clayton DA (1984) *Cell 36*:635.
234. Fisher RP, Clayton DA (1985) *JBC 260*:11330; Fisher RP, Topper JN, Clayton DA (1987) *Cell 50*:247.
235. Chang DD, Clayton DA (1987) *EMBO J 6*:409.
236. Chang DD, Clayton DA (1989) *Cell 56*:131.

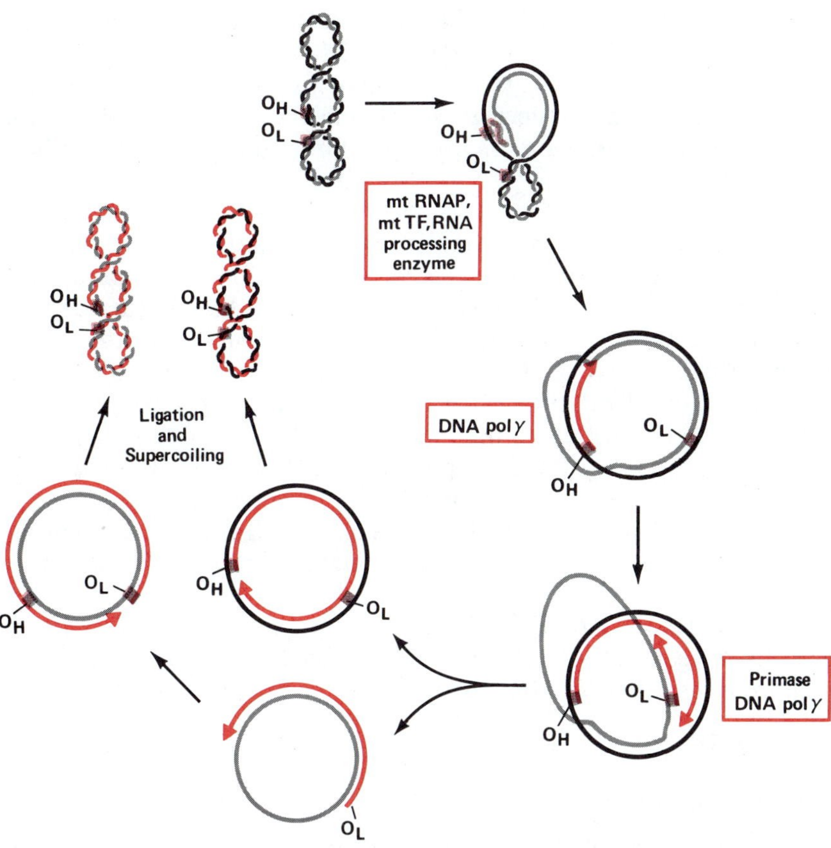

Figure 18-20
Scheme for asymmetric, asynchronous replication of mammalian mitochondrial DNA. O_H = heavy-strand origin; O_L = light-strand origin; mtRNAP = mitochondrial RNA polymerase; mtTF = mitochondrial transcription factor. [Modified from Clayton DA (1982) *Cell 28*:693]

nascent transcript into a primer. The enzyme is active against the free RNA as well.[237] The RNA component of the processing enzyme, as well as the polypeptide, is encoded by a nuclear gene,[238] and thus must be transported into the mitochondria, challenging a long-held belief that nucleic acids are unable to traverse the mitochondrial membrane.

L-Strand Synthesis. Continuous unidirectional synthesis of about 70% of the H-strand uncovers the L-strand origin in its active single-stranded form. The origin is a 30-ntd stem-and-loop structure, preserved in the mouse and in humans despite substantial divergence in the primary sequence. Apparently, it is the secondary structure of the L-strand that is essential for origin function.[239]

An enzyme fraction containing primase, DNA pol γ, and RNase H from human mitochondria can initiate replication at the L-strand origin.[240] Primase specifically recognizes the origin and synthesizes primers of varying length depending

237. Chang DD, Clayton DA (1987) *EMBO J 6*:409.
238. Chang DD, Clayton DA (1987) *Science 235*:1178.
239. Clayton DA (1982) *Cell 28*:693.
240. Wong TW, Clayton DA (1985) *Cell 42*:951.

on the availability of rNTPs. Transition from RNA to DNA occurs at the base of a stem-and-loop structure, in agreement with the DNA starts in the replication intermediates made in vivo; elongation proceeds unidirectionally. Whether RNase H or any other factors in the enzyme fraction are required is not known. Some components seem to function in providing for origin specificity of initiation.

Elongation and Maturation. Asymmetric replication allows DNA synthesis to be continuous on both strands. The H-strand is completed before the L-strand, with the gap between the two origins being the last segment to be replicated. Complete replication takes about one hour; the average rate of polymerization is less than 10 ntd per second, a rate 100 times slower than that of phage and bacterial DNA synthesis. Much of the time during mtDNA synthesis is consumed in pauses at topological constraints in the molecule. Helicase and topoisomerase activities are presumably required for replication, but have not been identified. The DNA product is initially relaxed; another 40 minutes elapses before the supercoiled form, with 100 superhelical turns, is produced. The D-loops, synthesized on the supercoiled templates, are continually destroyed and regenerated.[241]

Unlike replication of the circular DNA of animal mitochondria, replication of the linear DNA of *Tetrahymena* includes bidirectional fork movement from a central origin and repetitive terminal sequences for completion by a telomerase (Section 6-9).[242]

Kinetoplast DNA Networks[243]

The single mitochondrion of trypanosomes, called a *kinetoplast*, contains an enormous network of 1- to 2-kb catenated DNA circles (minicircles) in addition to the 20 to 40 larger molecules (maxicircles) of conventional mtDNA (Fig. 18-21). These 5000 to 10,000 minicircles make up close to 20% of the total cellular DNA, yet serve no known function. Unlike mammalian mtDNA, kinetoplast DNA (kDNA) replicates synchronously with nuclear DNA.[244] Replication is highly ordered; the entire network is duplicated before it is divided at mitosis. Kinetoplast replication requires an understanding of how DNA is organized within a cell and may lead to a better appreciation of coordinately regulated replication in eukaryotes.

Composition of Minicircles. Minicircles from different organisms differ in DNA sequence; in some species, circles with highly heterogeneous sequences are present within a single cell, while in others, all the molecules are essentially identical. The minicircles from a given species usually share a 100- to 200-bp region which contains the origins of replication for both strands. There is little sequence homology between organisms. The only sequence common to all circles, the 12-ntd "universal minicircle sequence" (UMS), is the site of priming of DNA replication.[245]

---

241. Bogenhagen D, Clayton DA (1978) *JMB 119*:49.
242. Goldbach RW, Bollen-Deboer JE, Van Bruggen ERJ, Borst P (1979) *BBA 562*:400.
243. Ray DS (1987) *Plasmid 17*:177; Simpson L (1987) *Ann Rev Microbiol 41*:363.
244. Cosgrove WB, Skeen MJ (1970) *J Protozool 17*:172; Simpson L, Braly P (1970) *J Protozool 17*:511.
245. Ntambi JM, Shapiro TA, Ryan KA, Englund PT (1986) *JBC 261*:11890.

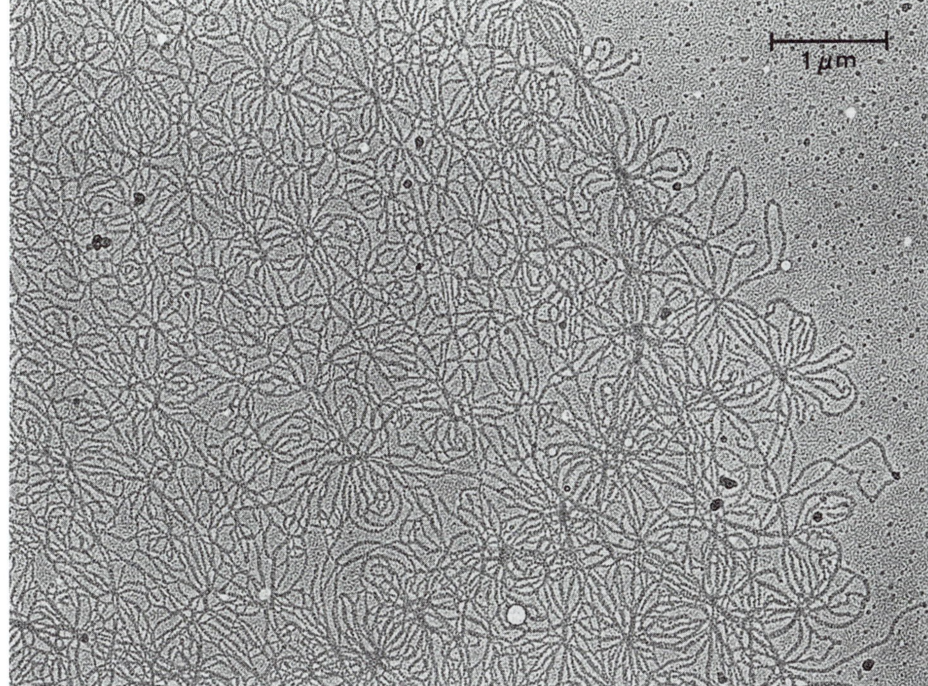

Figure 18-21
Electron micrograph of a portion of the kinetoplast DNA network (kDNA) from *Crithidia fasciculata*. (Courtesy of Professor D Ray and C Sheline)

Minicircle DNA is generally considered as lacking in any coding capacity; transcription has rarely been detected and open reading frames are restricted to short stretches. One study, using antibodies raised against minicircle-encoded peptides generated by cloning DNA fragments into an expression vector, suggests that minicircle information does appear in proteins,[246] a finding that warrants attention, especially considering the posttranscriptional editing by the addition of uridine residues into RNA encoded by the maxicircles.[247]

A stable DNA bend (Section 1-10) is present on most minicircles.[248] The bend, due to runs of 4 to 6 adenine residues phased with the DNA helix repeat, is located at places distinct from the replication origins. The function of this bend remains unknown. Possibly, it is involved in directing the attachment and removal of minicircles from the network. Inasmuch as covalently closed circles are removed from the network while gapped circles are joined to it (see below), a significantly different conformation of the bend in these two DNA forms could identify which molecules are destined for assembly or disassembly.

246. Shlomai J, Zadok A (1984) *NAR 12*:8017.
247. Shaw JM, Feagin JE, Stuart K, Simpson L (1988) *Cell 53*:401; van der Spek H, Van den Burg J, Sloof P, Benn R (1988) *EMBO J 7*:2509; Van der Spek H, Speijer D, Arts G-J, Van den Burg J, Van Steeg H, Sloof P, Benne R (1990) *EMBO J 9*:257.
248. Kitchin PA, Klein VA, Ryan KA, Gann KL, Rauch CA, Kang DS, Wells RD, Englund PT (1986) *JBC 261*:11302; Ray DS, Hines JC, Sugisaki H, Sheline C (1986) *NAR 14*:7953.

Replication of the kDNA Network.[249] Ordered removal of minicircles from the network for replication ensures that each circle is copied once per cycle. Covalently closed circles are removed from the center of the network, replicated as free molecules, and rejoined to the periphery. The final gap-filling and ligation stages of replication are postponed until the entire network is replicated. As a result of this cycling, networks contain primarily covalently closed circles in the middle, and newly replicated gapped molecules at the edges. Once the structure has been duplicated, it is divided by an unknown mechanism into two networks, one for each daughter cell.

Free minicircles replicate by a mechanism similar to that for mtDNA, with each strand possessing a distinct replication origin. In contrast to mtDNA, replication is semidiscontinuous, suggesting that it is essentially simultaneous for both strands.[250] Initiation at both origins depends on RNA priming; gaps, containing residual ribonucleotides, are found in the regions of the conserved nucleotide sequences where initiations occur.[251] In *Crithidia fasciculata* minicircles, the H- and L-strand origins are separated by about 100 bp, each in a region of conserved sequence. There are two such regions and origins per circle, and initiations occur with equal frequency at each of them.[252]

Most nicks and gaps on the lagging strand are rapidly ligated; however, discontinuities persist on both strands until after the replicated molecule has been reconnected to the network.[253] Excision of RNA primers, gap-filling, and ligation can all occur following reattachment. Closure of the molecules may be postponed until replication of the network is complete in order to mark the minicircles that have already been duplicated. Maxicircle replication has not been studied in detail, but a rolling-circle mechanism has been suggested.

Topoisomerases are almost certainly involved in replication of the minicircle network (Section 12-7). The properties of two topoisomerases suggest their involvement. A 60-kDa, type II enzyme, purified from *C. fasciculata*, efficiently removes covalently closed minicircles from the network, but not those that are specifically nicked.[254] A decatenation activity, specific for closed circles, is attractive, in keeping with the replication pathway as described above. A second type II enzyme (a homodimer of 132-kDa subunits) catenates and decatenates minicircles, as well as relaxing DNA.[255] This protein, like the smaller enzyme, has been localized specifically to the mitochondria.[256] Immunovisualization locates the protein primarily at the periphery of the kinetoplast DNA network, suggesting its involvement in joining minicircles to the network following their replication.

Chloroplasts

The DNA of chloroplasts[257] (cpDNA) from higher plants (pea, corn) and from algae (*Euglena, Chlamydomonas*) is a circular duplex of 130 to 180 kb. The

249. Englund PT (1978) *Cell 14*:157; Englund PT, Hajduk SL, Marini JC (1982) *ARB 51*:695.
250. Birkenmeyer L, Sugisaki H, Ray DS (1985) *NAR 13*:7107.
251. Birkenmeyer L, Sugisaki H, Ray DS (1985) *NAR 13*:7107; Ntambi JM, Englund PT (1985) *JBC 260*:5574.
252. Birkenmeyer L, Ray DS (1986) *JBC 261*:2362.
253. Kitchin PA, Klein VA, Englund PT (1985) *JBC 260*:3844.
254. Shlomai J, Zadok A (1983) *NAR 11*:4019; Shlomai J, Linial M (1986) *JBC 261*:16219.
255. Melendy T, Ray DS (1989) *JBC 264*:1870.
256. Melendy T, Sheline C, Ray DS (1988) *Cell 55*:1083.
257. Jope DA, Hirai A, Widman SG (1978) *J Cell Biol 79*:631; Kolodner RD, Tewari KK (1975) *JBC 250*:4888; Kolodner R, Warner RC, Tewari KK (1975) *JBC 250*:7020; Kolodner RD, Tewari KK (1979) *PNAS 76*:41.

replicative intermediates seen in the electron microscope suggest that replication begins with the formation of two displacement loops, which expand toward each other to form a theta-type structure. Upon completion of one round, further replication is by a rolling-circle mechanism. The large inverted repeats in the genome, reminiscent of the organization of the yeast 2μ circle, suggest that site-specific recombination may be involved in the switch to rolling-circle replication (Section 10).

Several strategies have been used to identify the replication origins of cpDNA, including the isolation of fragments that replicate in yeast, in *Chlamydomonas*, and in crude chloroplast extracts. A D-loop region from *Chlamydomonas* cpDNA has been cloned and characterized.[258] This region melts readily in vitro; replication in crude algal fractions is stimulated by this sequence and initiates near it, a finding indicative of its function as an origin of replication.[259] A fragment from maize cpDNA, which promotes replication in an in vitro system prepared from pea chloroplasts, shows a 67% homology to the *Chlamydomonas* sequence, a remarkable conservation among cpDNA replication origins.[260] The successful introduction of exogenous DNA into chloroplasts by using tungsten microprojectiles[261] holds much promise for identifying cpDNA origins by observing their activity in vivo directly.

---

258. Waddell J, Wang W, Wu M (1984) *NAR 12*:3843; Wang W, Chang CH, Waddell J. Wu M (1984) *NAR 12*:3857.
259. Wu M, Lou JH, Chang DY, Chang CH, Nie ZQ (1986) *PNAS 83*:6761.
260. Gold B, Carrillo N, Tewari KK, Bogarad L (1987) *PNAS 84*:194.
261. Boynton JE, Gillham NW, Harris EH, Hosler JP, Johnson AM, Jones AR, Randolph-Anderson BL, Robertson D, Klein TM, Shark KB, Sanford JC (1988) *Science 240*:1534.

CHAPTER 19

Animal DNA Viruses and Retroviruses

19-1 Utility of Animal Viruses[1]

The many and various DNA animal viruses (Table 19-1) provide windows on animal cell DNA replication, as the phages (Chapter 17) do for bacteria. Each offers distinctive features susceptible to genetic manipulation and analysis in cell-free systems.

At one extreme, the very small parvoviruses and papovaviruses, containing a genome of 5000 bp or fewer, rely almost entirely on host replication systems for their multiplication. Very large viral genomes, such as the herpes and vaccinia (pox) genomes, containing 100 to 300 kb, resemble the T-even phage DNAs in complexity. They contain many genes coding for new DNA replication proteins to synthesize their DNA. Herpesviruses multiply in the nucleus, while poxviruses are assembled in the cytoplasm, each type offering clues to the subcellular distribution of replication processes.

In addition to disclosing the replication mechanisms of animal cells, the viruses illuminate many other facets of cellular structure and function. Tumor viruses employ recombination processes for integration into the host chromosome, where they may provoke cellular transformation and proliferative growth. Chromosome organization, control of gene expression, aspects of carbohydrate, protein, and lipid metabolism, bioenergetics, and patterns of organelle development may all be studied by focusing on the viral life cycle.

1. Fields BN, Knipe DM, Chanock RM, Hirsch MS, Melnick JL, Monath TP, Roizman B (eds.) (1990) *Virology*. 2nd ed. Raven Press, New York; Campbell JL (1986) *ARB* 55:733; Hay RT, Russell WC (1989) *BJ* 258:3.

Table 19-1
Sizes of virions and genomes of representative DNA animal viruses and retroviruses

Classification	Examples	Virion: Shape	Virion: Size (nm)	Genome: Single-strand or duplex	Genome: Size (kb)
Parvovirus		icosahedron	20	ss	1.8–2.7
Nondefective	H-1, H-3, RV, MVM				
Defective (satellite)	adenovirus-associated				
Papovavirus		icosahedron	30–50	ds, circle	4.5–7.5
Papilloma	rabbit, human, etc. papilloma viruses				
Polyoma	mouse polyoma virus				
Vacuolating	SV40, BK				
Adenovirus	human (e.g., Ad2), simian, avian adenoviruses	icosahedron, spikes	80–90	ds	30–40
Herpesvirus	Herpes simplex viruses (e.g., HSV-1) Epstein-Barr virus cytomegalovirus pseudorabies virus Marek's disease virus	icosahedron, envelope	100–150	ds	80–140
Poxvirus	vaccinia virus animal poxviruses (cow, sheep, fowl)	brick-shaped	300 × 200 × 100	ds	240–300
Retrovirus (leukovirus, oncornavirus, lentivirus)	murine leukemia viruses avian leukosis virus Rous sarcoma virus mouse mammary tumor virus human immunodeficiency virus	spherical, envelope	100–120	ss, RNA	~8.5

19-2 Papovaviruses:[2] Simian Virus 40 (SV40), Polyoma Virus, and Bovine Papilloma Virus (BPV)[3]

The papovavirus family includes *papilloma*, *polyoma*, and *vacuolating* (e.g., SV40) viruses and the human viruses BK[4] and JC. These small viruses are notable for producing cancer in animals and humans; bovine papilloma virus (BPV) is the major cause of cervical cancer, the most common malignancy in the developing world. Papovaviruses attract interest for additional reasons:

(1) Their small size has made it possible to map their genes, determine their complete nucleotide sequence, analyze their gene expression, and select a large

2. Eckhart W (1990) in *Virology* (Fields BN et al., eds.) 2nd ed. Raven Press, New York, p. 1593; Shah KV (1990) in *Virology* (Fields BN et al., eds.) 2nd ed. Raven Press, New York, p. 1609; DePamphilis ML, Bradley MK (1986) in *The Papovaviridae* (Salzman N, ed.). Plenum Press, New York, p. 99. Kelly TJ (1988) *JBC 263*:17889; Kelly TJ, Wold MS, Li J (1988) *Adv Virus Res 34*:1; Challberg MD, Kelly TJ (1989) *ARB 58*:671; Howley PM (1990) in *Virology* (Fields BN et al., eds.) 2nd ed. Raven Press, New York, p. 1625; Shah KV, Howley PM (1980) in *Virology* (Fields BN et al., eds.) 2nd ed. Raven Press, New York, p. 1651; Stillman B (1989) *Ann Rev Cell Biol 5*:197.
3. Challberg MD, Kelly TJ (1989) *ARB 58*:671.
4. Kelly TJ (1988) *JBC 263*:17889; Kelly TJ, Wold MS, Li J (1988) *Adv Virus Res 34*:1.

variety of mutants. This detailed genetic and biochemical knowledge has made them useful vectors for foreign genes and DNA segments.

(2) Their reliance on host systems of replication and transcription provides a means for probing the basic operation of these systems. The resolution of cell-free replication systems and purification of the responsible proteins has furnished the most revealing details of eukaryotic replication.

(3) These viruses recombine in vivo with other viruses (e.g., adenovirus) and form complementing hybrids.

(4) Integration of viral DNA into the host chromosome results in cell transformation.

(5) Alternative routes of viral replication following infection lead to cellular death and lysis in permissive cells but to a proliferative response in nonpermissive cells. These alternatives offer an opportunity to analyze the basis for this important difference.

(6) BPV, maintained as a plasmid at a fixed copy number and replicated only once per cell cycle, offers an avenue for exploring mechanisms for the strict regulation of replication in a eukaryotic cell.

The avalanche of interest in the papovaviruses has built a body of information which, as in the case of phage λ, grows exponentially and exposes still more refined and fundamental questions for investigation.

Virion[5]

SV40 and polyoma are particles about 50 nm in diameter, with icosahedral symmetry. The capsid of SV40 consists of three virus-encoded polypeptides: the major capsid protein, VP1, 39.7 kDa; and the two minor capsid proteins, VP2 and VP3, 38.5 and 27 kDa, respectively. The DNA genome within the capsid is complexed with histones, indistinguishable from the chromatin of cellular nucleosomes (Section 10-7). Polyoma is similar to SV40 in structure but differs in its capsid proteins.

Genome and Origin

Papovavirus DNA is a supercoiled, circular duplex of about 5 kb. Despite the physical and biological similarity of SV40 and polyoma, homology in their DNA sequences is weak. Genetic loci, the origin and termini of replication, beginnings and ends of mRNA transcripts, and distinctive structural features are expressed in map units (linear dimensions) relative to the EcoRI cleavage site (Fig. 19-1).

The DNA replication origin (Section 16-5) in the papovaviruses is in a small, highly conserved region within a noncoding sequence of about 450 bp which contains the regulatory signals for transcription (Fig. 19-2); a 64-bp core is sufficient for initiation of SV40 replication, but the sequences nearby influence the efficiency, especially in vivo.

The origin is complex and consists of at least three essential domains. (1) At the center of the core are four copies of a pentameric motif (GAGGC), organized

5. DePamphilis ML, Bradley MK (1986) in *The Papovaviridae* (Salzman N, ed.). Plenum Press, New York, p. 99.

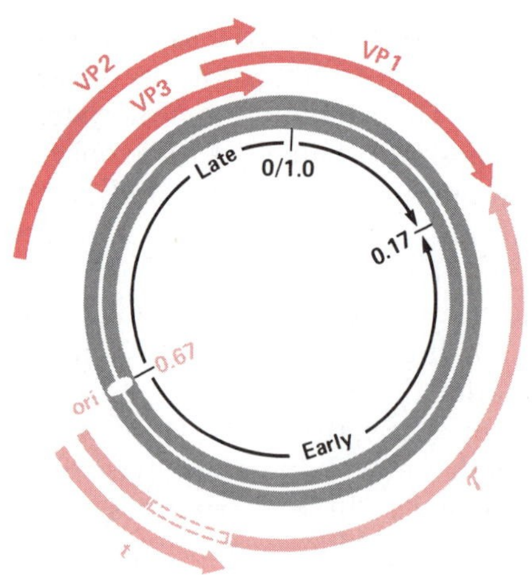

Figure 19-1
Genetic map of SV40. The map is oriented by the site of cleavage by EcoRI restriction nuclease, set at 0 or 1.0. The replication origin, near 0.67, orients transcriptional starts of early and late functions in permissive cells; proteins = colored arrows. Large T antigen is translated from an mRNA processed to remove the intervening sequence (dashed region); mRNA for small t antigen includes this sequence. Viral capsid proteins (VP1, VP2, and VP3) are translated from late mRNAs.

as an inverted repeat, which provide the prime binding site for the hexameric T antigen (see below). (2) To the right of the core, a region of 17 AT base pairs is most likely the initial site for opening the duplex. (3) To the left is a 15-bp imperfect palindrome.

Adjacent to the core region, a second T antigen binding site increases replication efficiency several-fold in vivo and in vitro. Sequence elements, such as enhancers, near the origin, are binding sites for transcriptional factors and

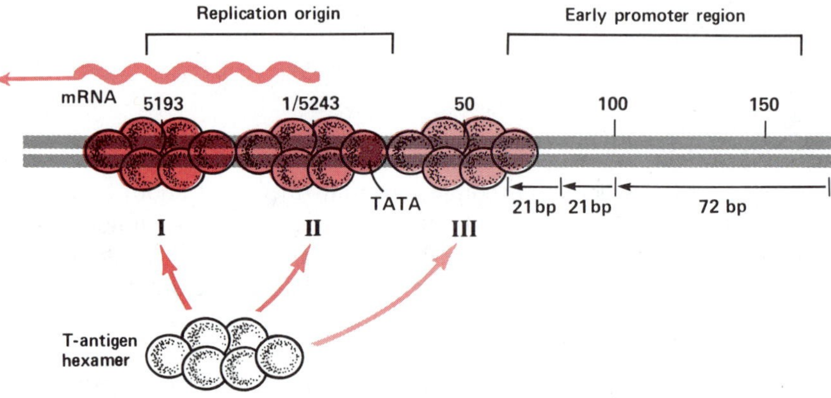

Figure 19-2
Model for the sequential binding of T antigen to SV40 DNA in viral DNA replication and the control of early transcription. In the region of the SV40 genome containing the three T antigen binding sites (I, II, and III), the nucleotides are numbered above the center of each binding site, with the *Bgl*I restriction site as 1/5243. Horizontal arrows depict 21-bp and 72-bp direct repeat sequences. At early times after viral infection, when the binding sites are free of T antigen, RNA polymerase II can initiate transcription at the early promoter region. A hexamer of T antigen binds at site I and coordinates binding of a second hexamer at site II, allowing DNA replication to begin. As the infection progresses, the T antigen concentration increases to a level that allows binding of a third hexamer to site III. At this level of T antigen, all classes of early transcripts would be repressed. (Courtesy of Professor R Tjian)

enhance replication in vivo tenfold or more.[6] Transcriptional activation of the initiation of replication (Section 16-5) is also observed in other eukaryotic viruses besides SV40, polyoma, and BPV (e.g., Epstein-Barr virus and adenovirus), in *Drosophila*, *Tetrahymena*, and yeast, and in prokaryotes (e.g., *E. coli oriC* and phage λ *ori*λ).

Infection

The fate of infecting particles is difficult to determine with isotopic tracers and chemical analysis because radioactive labeling of the virus is not sufficiently intense and because the multiplicity needed for infection is large, about 100 particles or more per plaque-forming unit. As a result, less is known about the early biochemical events of SV40 and polyoma replication cycles than of phages M13, ϕX174, and λ. However, certain important features have been clarified.

Lytic Cycle and the T Antigen.[7] Virus particles are transported to the host nucleus. The viral genome, complexed with histones, is in effect a minichromosome. Replicative (productive) infection of SV40 in permissive monkey cells proceeds with transcription of the early region in a counterclockwise direction (Fig. 19-1). For historical reasons, two early proteins are called the T (tumor) antigens: large T, 82 kDa, and small t, 17 kDa. Polyoma, unlike SV40, also encodes a third early protein, the middle t antigen. Large T antigen merits special attention since it appears to act as a primary inducer of transformation and forms a complex with a cellular tyrosine-specific protein kinase ($pp60^{c\text{-}src}$) that in other instances has been linked to transforming activity (Section 7).

T antigen is required for initiation of replication and for the regulation of transcription of both early and late viral genes. The protein binds to key sites in the origin domain and, by its helicase action, linked to ATPase, participates in opening of the duplex at the origin. T antigen also binds DNA polymerase α and recruits cellular replication proteins to the origin. This interaction is inhibited by murine p53, the product of a cellular anti-oncogene[8] (see below).

Binding of T antigen is modulated by phosphorylation[9] of threonine and serine residues located in clusters in the N- and C-terminal regions. T antigen produced in *E. coli* lacks phosphate and is deficient in both DNA-binding specificity and the support of replication in vitro.[10] However, phosphorylation of a single threonine residue (amino acid 124) of this form of T antigen by the yeast (*S. pombé*) CDC2 protein kinase enables it to bind its primary site in the origin and to promote replication.[11]

Yet dephosphorylation of T antigen at residues other than threonine 124 may be activating. RF-C (RP-C), a protein purified as a factor essential for the initial stages of SV40 DNA replication, is identical to the catalytic subunit of the cellular protein phosphatase 2A (PP2A).[12] This protein, which forms a tight and

---

6. de Villiers J, Schaffner W, Tyndall C, Lupton S, Kamen R (1984) *Nature 312*:242; DePamphilis ML (1988) *Cell 52*:635; Cheng L, Kelly TJ (1989) *Cell 59*:541.
7. Deb S, Tsui S, Koff A, De Lucia AL, Parsons R, Tegtmeyer P (1987) *J Virol 61*:2143; Deb S, Tegtmeyer P (1987) *J Virol 61*:3649; Mastrangelo IA, Hough PVC, Wall JS, Dodson M, Dean FB, Hurwitz J (1989) *Nature 338*:658; Borowiec JA, Dean FB, Bullock PA, Hurwitz J (1990) *Cell 60*:181.
8. Gannon JV, Lane DP (1987) *Nature 329*:456; Gannon JV, Lane DP (1990) *New Biologist 2*:84.
9. Prives C (1990) *Cell 61*:735.
10. Mohr IJ, Gluzman Y, Fairman MP, Strauss M, McVey D, Stillman S, Gerard RD (1989) *PNAS 86*:6479.
11. McVey D, Brizuela L, Mohr I, Marshak DR, Gluzman Y, Beach D (1989) *Nature 341*:503.
12. Virshup DM, Kelly TJ (1989) *PNAS 86*:3584.

specific complex with T antigen, can specifically remove the phosphate from one or two particular serine residues of the phosphorylated form of T antigen. A striking stimulation of SV40 DNA replication by PP2A added to extracts from early G_1-phase but not S-phase cells may be due to its action on T antigen.[13] Such cell-cycle–dependent activities targeted to T antigen have strong implications for an analogous regulation of host chromosomal replication.

Apart from its direct involvement in viral replication, T antigen also performs in key regulatory roles by perturbing a quiescent cell, inducing the cell to express many of the proteins needed for proliferation of its own genome. Negative regulators of cell proliferation, such as p53 and retinoblastoma protein, may be antagonized by T antigen binding. At an elevated level, T antigen binds to its promoter located within or adjacent to the minimal origin and represses transcription.

Infection induces increased levels of the host replication enzymes, among them thymidine kinase, dTMP kinase, dCMP deaminase, and DNA polymerase α. The mechanism of this effect and the viral factors that produce it are obscure. With the onset of DNA replication, transcription of the late region proceeds clockwise from near the origin to 0.16 on the L (late) strand, leading to the production of capsid proteins.

Translation requires processing of the transcripts to remove noncoding intervening sequences. The sequence of VP3 is made up of the distal 70% of VP2; their termini overlap the start of VP1 by 22 nucleotides. With the synthesis of the capsid proteins, progeny DNA is packaged into virions, after which cell death and lysis ensue.

Transformation.[14] Infection in nonpermissive cells (rodent, for SV40) is accompanied by the same early transcription and synthesis of large T and small t antigens. But DNA replication, late-strand transcription, and capsid protein synthesis are blocked. Cells are induced to replicate DNA and undergo mitosis, but over 90% revert to a quiescent state with no evidence of prior infection. A small fraction become genetically transformed. Part or all of several copies of the genome are covalently integrated into chromosomal DNA. The transformed cells, their morphology altered, sustain DNA replication and cell division under conditions that suppress the growth of normal cells. Inoculation of transformed cells into appropriate animals produces tumors.

In most murine cells, some of the SV40 T antigen is bound to the cellular phosphoprotein, p53, known to affect the proliferation of cells.[15] Whereas the wild-type (nononcogenic) p53 profoundly interferes with T antigen in its support of SV40 DNA replication, a mutant (oncogenic) p53 has no effect. These and related findings support a normal role for p53 as an anti-oncogene, in this instance blocking the replication and transforming activity of a tumor virus. Whether these findings can be generalized to other systems remains to be determined.

The SV40 genome is integrated at many chromosomal sites, and the SV40 sequence at which integration occurs is not unique. Large T and small t antigens are required for transformation, but the mechanism of integration is not known.

---

13. Virshup DM, Kaufmann MG, Kelly TJ (1989) *EMBO J 8*:3891.
14. Tooze J (ed.) (1980) *Molecular Biology of Tumor Viruses: DNA Tumor Viruses.* Part 2. CSHL; Fried M, Prives C (1986) *Cancer Cells 4*:1.
15. Oren M (1985) *BBA 823*:67; Gannon JV, Lane DP (1987) *Nature 329*:456; Gannon JV, Lane DP (1990) *New Biologist 2*:84; Wang EH, Friedman PN, Prives C (1989) *Cell 57*:379; Friedman PN, Kern SE, Vogelstein B, Prives C (1990) *PNAS 87*:9275.

Nor is it understood how small t antigen, which does not bind SV40 DNA, contributes to maintaining the transformed state; a complex with PP2A phosphatase (see above) may be involved.

The integration of papovavirus DNA into the host chromosome superficially resembles the integration of phage λ (Section 17-7), but the differences are basic. Phage λ integration is "deliberate," is promoted by phage genes, takes place at specific sites on the host chromosome, and occurs with a frequency near 50%. Papovavirus integration appears less deliberate, depends on host genes, and takes place at random sites and at a frequency near 0.001%.

Rescue of an Integrated SV40 Genome by Saltatory Replication.[16] Transformed rodent cells have no free SV40 genomes; less than one per 50 cells is detected. However, SV40 genomes are produced when these nonpermissive cells are fused with permissive simian cells; half the heterokaryons form infectious centers. Presumably, monkey cells provide replicative functions not present in rodent cells, making it possible for SV40 DNA replication, activated by a T antigen system, to proceed in situ.

The model in Figure 19-3 suggests how repeated rounds of replication of an

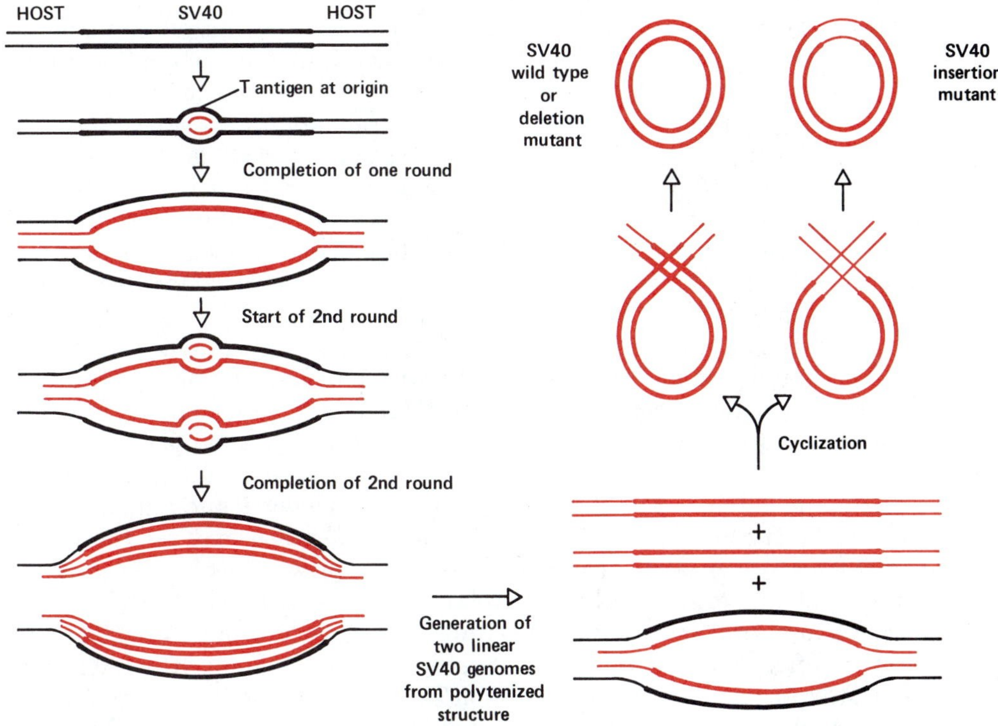

Figure 19-3
Model for excising integrated SV40 genome by selective replication. Two rounds of replication, each terminated at limits of the viral genome, can generate two linear genomes. Depending on where cyclization takes place, progeny may be wild type, or mutants with inserted host DNA, or mutants with deletions of viral DNA. (Courtesy of Dr M Botchan)

16. Botchan M, Topp W, Sambrook J (1978) *CSHS 43*:709.

SV40 genome, in effect a polytenization of that region, may liberate a linear viral DNA, which is then free to recombine with itself to form a circle. In cell lines with tandem duplication of integrated SV40 DNA, the efficiency of infectious virus formation is increased 100-fold to 1000-fold, very likely because of the favorable effect of terminal duplication on recombination. In the absence of tandem duplication, cyclization by nonhomologous recombination may form mutants with deletions of virus DNA or insertions of host DNA. Because viral genomes may be integrated in several places in host chromosomes, their replication entails jumping from one to another of these regions—a saltatory feature rather than the progressive form of replication.

Replication in Vivo[17]

Important features of SV40 DNA replication are known from studies of replicating DNA in cultures of monkey cells. (Similar patterns are found in murine cells infected with polyoma.) Replicating DNA is isolated following a pulse of [^{3}H]thymidine applied to cell cultures infected for 24 to 48 hours. Since host DNA is also labeled after infection, the replicating viral DNA is separated by a procedure that exploits the small, compact size of the viral replicative forms. The following features of SV40 DNA are notable:

(1) Replication begins within the 100-bp origin and proceeds semidiscontinuously and bidirectionally at about equal speeds, terminating at a point on the circle nearly opposite the origin. In a pattern not unlike that of the linear host chromosome with its multiple origins, replication is completed when the two advancing replication forks meet and merge. Termination and segregation of the progeny molecules are not determined by a specific nucleotide sequence.

(2) Completion of progeny molecules as twisted circles (form I, Fig. 19-4) takes 5 to 10 minutes. Although this might suggest fork movement of only 0.5 kb per minute, it is likely that initiation and completion of new molecules, rather than chain growth, limit the rate. For comparison, the overall rate of fork movement is 3 kb per minute for a mammalian chromosome and 40 kb for *E. coli.*

(3) Replicative intermediates (Fig. 19-4) contain intact parental strands; the growing strands include small fragments with a sedimentation coefficient of only 4S. Replication is semidiscontinuous.[18] Intermediates contain two loops of equal length attached to a supercoiled region of unreplicated parental molecules (Fig. 19-4). Unlike the replicative forms of phage λ, superhelical turns persist in the circular parental duplexes of SV40, polyoma, and mitochondrial (Section 8-11) replicative intermediates. This difference may be due to the nucleosomal organization of SV40 and the tenfold larger size of λ DNA, with its consequent greater susceptibility to relaxation by nicking.

(4) Mature viral DNA normally occurs as a chain of about 20 nucleosomal beads (minichromosome)[19] (Fig. 19-5; see also Section 10-7). It is likely that replicative intermediates are also complexed with histones. In the higher-order structure of SV40 chromatin,[20] a stretch of 75 to 400 bp long, which includes the

---

17. DePamphilis ML, Bradley MK (1986) in *The Papovaviridae* (Salzman N, ed.). Plenum Press, New York, p. 99. Challberg MD, Kelly TJ (1989) *ARB 58*:671; Howley PM (1990) in *Virology* (Fields BN et al., eds.) 2nd ed. Raven Press, New York, p. 1625; Shah KV, Howley PM (1990) in *Virology* (Fields BN et al., eds.) 2nd ed. Raven Press, New York, p. 1651.
18. Perlman D, Huberman JW (1977) *Cell 12*:1029.
19. Griffith JD (1975) *Science 187*:1202.
20. Müller U, Zentgraf H, Eicken I, Keller W (1978) *Science 201*:406.

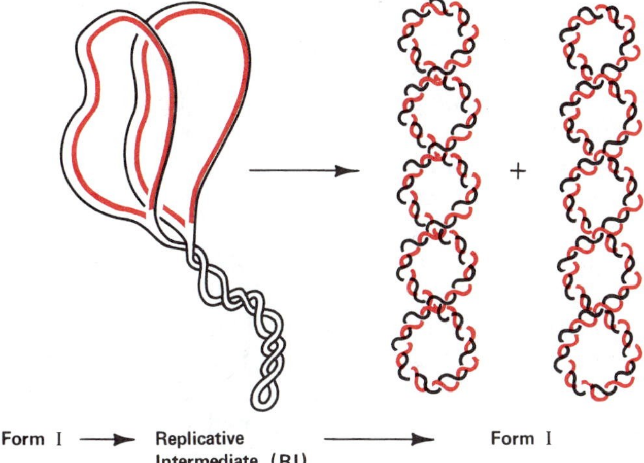

Figure 19-4
Intermediates in replication of supercoiled papovaviral DNA (form I). Replicative intermediates (RIs) retain unreplicated template strands in superhelical form and progeny strands as linear structures not attached covalently to template strands. [From Winnacker E-L, Magnusson G, Reichard P (1972) *JMB 72*:523; Winnacker E-L, Magnusson G, Reichard P (1972) *JMB 72*:539; Fareed GC, McKerlie ML, Salzman NP (1973) *JMB 74*:95]

origin of replication, is specifically uncoated, as indicated by sensitivity to endonuclease action.[21]

(5) Torque developed during replication is probably relieved by a topoisomerase (Section 12-6).

(6) Chains of greater than unit length have been found as part of sigma forms (Section 15-11) in cells treated with the inhibitor 2′-deoxy-2′-azidocytidine[22] (Section 2-7), but their significance as replication intermediates seems doubtful.

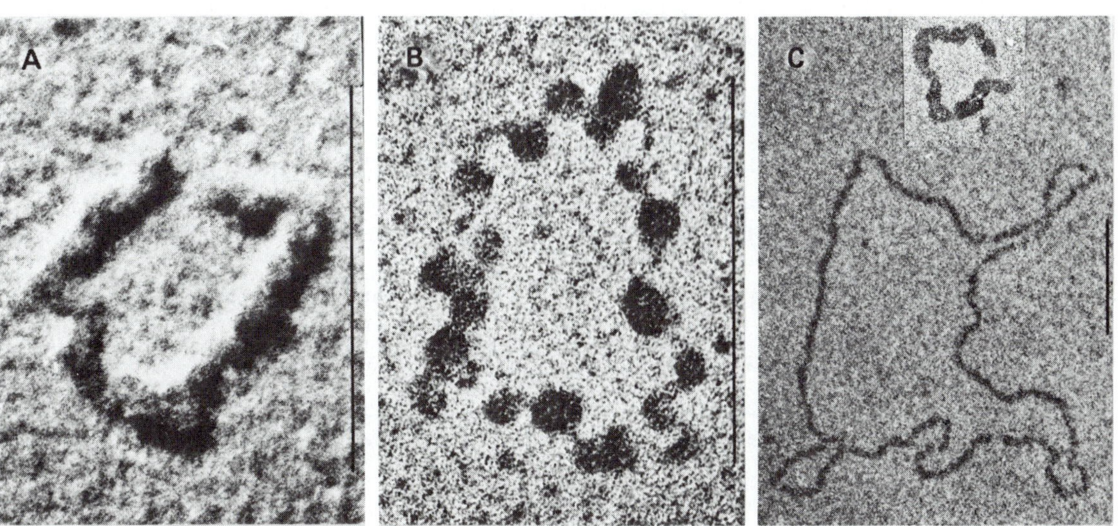

Figure 19-5
SV40 minichromosomes in various states: (A) native, (B) beaded by dilution in water, and (C) deproteinized and released. Inset in (C) compares native chromosome (A) with its DNA (C) at the same magnification; vertical bars = 100 nm. (Courtesy of Professor J Griffith)

21. Varshavsky AJ, Sundin OH, Bohn MJ (1978) *NAR 5*:3469; Waldeck W, Föhring B, Chowdhury K, Gruss P, Sauer G (1978) *PNAS 75*:5964.
22. Bjursell G (1978) *J Virol 9*:136.

(7) Large amounts of SV40 DNA (30,000 to 220,000 genomes per cell) are isolated with the large-molecular-weight fraction of cell DNA late in infection.[23] The SV40 DNA is composed of tandem, head-to-tail repeats, few if any of them covalently joined to the host DNA.[24] This polymeric form is highly infectious and is efficiently cleaved into supercoiled, monomer-length genomes.

(8) After injection of SV40 or polyoma DNA into unfertilized *Xenopus* eggs, the viral DNA is replicated in phase with the cell cycle during maturation of the oocyte,[25] but replication does not depend on the origin region.

(9) Origins are also random in the low level of SV40 DNA replication that remains when a functional large T antigen is absent in infections at a restrictive temperature with a temperature-sensitive T mutant.[26] Thus, T antigen may represent a class of proteins that enables a virus to initiate many rounds of replication without restraint by the normal controls of cellular DNA replication. Yet a chimera of SV40 and BPV (see below) is maintained at a stable copy number even when T antigen is in excess.[27] Evidently, a *cis*-acting negative control mechanism can limit this episome to a single replication per cell cycle.

Replication in Vitro[28]

Many insights had been gleaned from studies of intact and disrupted nuclei but not until the breakthrough of obtaining a vigorous soluble extract[29] was it possible to examine initiation at the origin and to identify the numerous specific proteins and events needed for efficient replication. As with the discoveries of the cell-free replication of phage ϕX174 and the *oriC* minichromosome, the SV40 system has illuminated significant aspects of the host replication machinery, as well as the subtle features of T antigen, the sole functional element encoded by the virus. Also emerging from these enzymologic studies has been the revelation of the basic similarities in design between the prokaryotic and eukaryotic devices for initiating and propagating the replication fork of a genome.

The principal replication proteins include T antigen, DNA polymerase δ [dependent on proliferating cell nuclear antigen (PCNA)], DNA polymerase α–primase (Section 6-2), replication factors (or replication proteins) RF-A (or RP-A, an SSB; see Section 10-6) and RF-C (or RP-C; see below), topoisomerase I or II, and DNA ligase. These proteins contribute to replication at four stages:

(1) *Opening the origin.* The hexameric T antigen recognizes and binds its repeated elements and then effects the opening of an adjacent AT-rich region by its ATP-driven conformational changes[30] and helicase activity.[31] The appropri-

23. Hirai K, Defendi V (1972) *J Virol 9*:705; Hölzel F, Sokol F (1974) *JMB 84*:423.
24. Rigby PWJ, Berg P (1978) *J Virol 28*:475.
25. Harland RM, Laskey RA (1980) *Cell 21*:761.
26. Martin RG, Stelow VP (1980) *Cell 20*:381.
27. Roberts JM, Weintraub H (1988) *Cell 52*:397.
28. Challberg MD, Kelly TJ (1989) *ARB 58*:671; Howley PM (1990) in *Virology* (Fields BN et al., eds.) 2nd ed. Raven Press, New York, p. 1625; Shah KV, Howley PM (1990) in *Virology* (Fields BN et al., eds.) 2nd ed. Raven Press, New York, p. 1651; Stillman B (1989) *Ann Rev Cell Biol* 5:197; Lee S-H, Eki T, Hurwitz J (1989) *PNAS 86*:7361.
29. Li JJ, Kelly TJ (1984) *PNAS 81*:6973.
30. Roberts JM (1989) *PNAS 86*:3939; Scheffner M, Wessel R, Stahl H (1989) *NAR 17*:93; Bullock PA, Seo YS, Hurwitz J (1989) *PNAS 86*:3944; Tsurimoto T, Fairman MP, Stillman BW (1989) *Mol Cell Biol 9*:3839.
31. Stahl H, Droge P, Knippers R (1986) *EMBO J 5*:1939; Wiekowski M, Droge P, Stahl H (1987) *J Virol 61*:411; Dean FB, Bullock PA, Murakami Y, Wobbe CR, Weissbach L, Hurwitz J (1987) *PNAS 84*:16.

ate phosphorylated form of T antigen may depend on RF-C,[32] discovered to be a subunit of protein phosphatase 2A. A series of ensuing alterations in the origin, not unlike those effected by *E. coli* dnaA protein at *oriC* (Section 16-3), prepare the ground for the two replication forks. Participation by RF-A, a three-subunit phosphoprotein, is essential in this stage. The structurally and functionally related yeast replication factor A could replace the human host protein in melting the origin but is relatively feeble in supporting the replication reaction.[33] Still other cellular proteins may be needed. Transcriptional activators that bind sequence elements adjacent to the origin core (e.g., NFI) greatly enhance accessibility to the origin.

(2) *Initiation.* Pol α–primase presumably lays down a primer on each strand in the opened origin.

(3) *Elongation.* Pol δ with PCNA extends each primer processively and may coordinate nicely with pol α–primase to advance the replication forks bidirectionally, the pol δ with PCNA being responsible for the continuous-strand synthesis and pol α-primase for discontinuous-strand synthesis.[34] Either topo I or topo II can provide the swivel (Section 12-1). A cellular helicase that translocates $5' \rightarrow 3'$ on the discontinuous-strand template may complement or substitute for T antigen, which has a $3' \rightarrow 5'$ polarity that would orient it toward the fork on the opposite template strand. Fork movement, estimated at about 20 bp per second, is probably limited by the rate of helicase melting. RF-A may also contribute to elongation.

(4) *Termination and segregation.* At or near completion, the multiply intertwined daughter molecules need to be separated, most likely by the action of topo II, which is uniquely equipped for decatenation.

Bovine Papilloma Virus (BPV)[35]

The papilloma viruses are responsible for epithelial tumors in a variety of animal species. Most of the cells in these tumors contain no infectious virus. Rather, they contain many copies of the viral genome in a latent state which become virus particles only in a terminally differentiated cell, a keratinocyte. One subset of papilloma viruses, BPV-1, is amenable to study because it can transform rodent cells in culture and is maintained as a plasmid at a fixed copy number. This feature makes it an attractive object for the study of *cis*- and *trans*-acting elements responsible for the intricate and orderly regulation of chromosome replication in eukaryotic cells.

The BPV-1 genome is a circular duplex of 7945 bp, of which a 5.4-kb segment is sufficient for transformation of cells in vitro. Cells transformed by virus or DNA carry the genome as a plasmid at about 200 copies per cell. Intensive genetic analysis has identified certain *cis*- and *trans*-acting elements responsible for maintaining this copy number and ensuring that, as with the host chromosome, each copy is replicated only once in a cell cycle.

To one side of the eight open reading frames (ORFs) that occupy 4.5 kb of the

---

32. Virshup DM, Kelly TJ (1989) *PNAS 86*:3584.
33. Brill SJ, Stillman B (1989) *Nature 342*:92.
34. Kelly TJ (1988) *JBC 263*:17889; Kelly TJ, Wold MS, Li J (1988) *Adv Virus Res 34*:1; Lee S-H, Eki T, Hurwitz J (1989) *PNAS 86*:7361; Tsurimoto T, Stillman B (1989) *EMBO J 8*:3883.
35. Challberg MD, Kelly TJ (1989) *ARB 58*:671; Howley PM (1990) in *Virology* (Fields BN et al., eds.) 2nd ed. Raven Press, New York, p. 1625; Shah KV, Howley PM (1990) in *Virology* (Fields BN et al., eds.) 2nd ed. Raven Press, New York, p. 1651.

minimal genome is a 1-kb long control region (LCR) that includes transcriptional controls and the replication origin. Interactions between host factors and the several transcriptional elements may generate structural changes in the LCR that activate the origin for replication.[36]

Viral gene products have yet to be isolated and characterized. Two ORFs (E1 and E2) are essential and may suffice for viral DNA synthesis. Mutations in certain ORFs affect copy number and integration, as well as replication. Of two *cis*-acting plasmid maintenance sequences (PMSs),[37] the one contained in a 521-bp segment within the LCR may well be associated with the origin of replication (Section 16-5) and is also identified with plasmid maintenance and suppression of integration. A second 140-bp PMS nearby may also function as an origin. A BPV chimera created with an SV40 origin replicates from that origin but responds to the BPV regulatory systems and is maintained at the fixed copy number.[38]

19-3 Parvoviruses: Autonomous and Helper-Dependent Viruses[39]

The smallest of the DNA viruses are included under the generic designation *parvovirus* (Latin *parvus*, small). The linear, single-stranded genomes of the parvoviruses invoke novel replication features that may apply to large duplex chromosomes.

Parvoviruses fall into two principal groups: the autonomous and the helper-dependent. The latter, also known as adeno-associated viruses (AAVs) and dependoviruses, usually depend on help from a concurrent adenovirus or herpesvirus infection; yet certain cells under appropriate conditions can dispense with the helper virus.[40] The autonomous parvoviruses, such as the minute virus of mice (MVM), H-1, and the Kilham rat virus, reproduce on their own in appropriate host cells; even so, the host range may be extended by a helper virus infection. Despite apparent differences between autonomous viruses and AAVs, the similarities in their genomes and replication mechanisms are most impressive.

Virion and Genome

The icosahedral particles, 20 to 25 nm in diameter, are remarkably resistant to heating, low pH, organic solvents, and enzymes that inactivate most other vertebrate viruses.

Parvoviral DNAs are linear single strands of 4.7 to 5.2 kb with a hairpin duplex at each end. As with other small viruses, overlapping reading frames in the genome encode the three coat polypeptides and nonstructural proteins. The hairpins at the 5′ termini of autonomous viruses range from 207 to 245 bases; the 3′ palindromic sequences are approximately 115 bases. Sequence and symmetry within the T-shaped palindromic termini of the autonomous viruses are responsible for the cross-arm structure.[41] In most autonomous viruses, the 3′- and 5′-terminal sequences are unrelated, but those of the human B19 are the same.

36. Lusky M, Botchan MR (1986) *PNAS 83*:3609.
37. Berg L, Lusky M, Stenlund A, Botchan MR (1986) *Cell 46*:753.
38. Roberts JM, Weintraub H (1988) *Cell 52*:397.
39. Cotmore SF, Tattersall P (1987) *Adv Virus Res 33*:91; Berns KI (1990) in *Virology* (Fields BN et al., eds.) 2nd ed. Raven Press, New York, p. 1743.
40. Yakobson B, Koch T, Winocour E (1987) *J Virol 61*:972.
41. Bohenzky RA, LeFebvre RB, Berns KI (1988) *Virology 166*:316; Bohenzky RA, Berns KI (1989) *JMB 206*:91.

Generally, the autonomous viruses encapsidate the strand complementary to mRNA, while AAVs package either the (+) or (−) strand with equal frequency in separate particles. Upon release from the virion, complementary strands can anneal to form a linear duplex. Strand ends of parvoviral DNA have inverted terminal repetitions that can form a variety of hairpin structures upon self-annealing (Fig. 19-6).

Infection

Parvoviruses enter actively dividing cells and migrate to the nucleus by unknown means. They depend on host replication functions expressed in the G_1

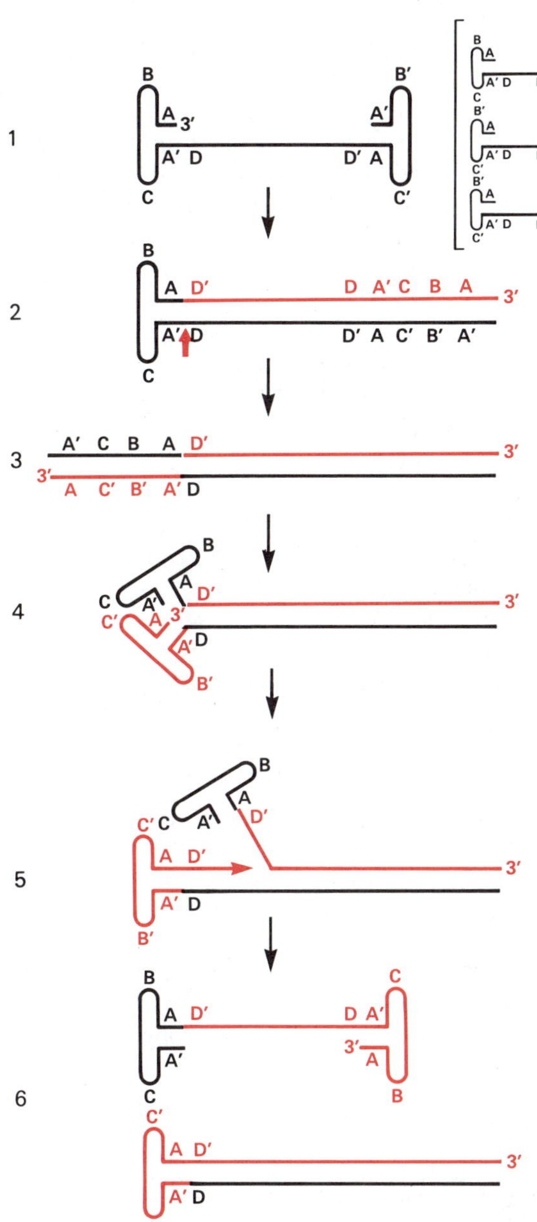

Figure 19-6
Proposed model for replication of AAV. The letters A, A′, B, B′, etc. refer to terminal repeated sequences. (1) Viral strand. Of the four possible terminal configurations for a strand, three are shown in brackets. (2) A hairpin loop has primed complementary-strand synthesis to form the duplex replicative form. (Note that the distance between C and B in the linear segment on the right should be the same as in the folded form on the left.) Nicking (arrow) takes place opposite the 3′ terminus of the parental strand. (3) The nicking, and chain extension at the nick, has resulted in a "hairpin transfer." (4) Reformation of terminal hairpin loops, equally likely at either end, has generated a 3′ primer terminus. (5) Synthesis from the 3′ primer terminus has displaced a unit length of genome. (6) Completion of one round of displacement synthesis has resulted in a displaced single strand (ready to start synthesis of a complementary strand, as in step 1) and a duplex molecule that can undergo another round of hairpin transfer and displacement synthesis (beginning at step 2). [After Berns KI, Hauswirth WW (1982) in *Organization and Replication of Viral DNA* (Kaplan A, ed.). CRC Press, Cleveland, p. 25]

and S phases of cell growth. Multiple viral mRNAs are generated from a variety of promoters and alternative splicings. The left half of the genome is the source of regulatory proteins involved in gene expression and DNA replication; the right half encodes the coat proteins. The adenovirus helper effect for AAVs is targeted to the regulation of gene expression. Without this intervention, an AAV encodes functions which can inhibit its own genes and DNA replication and force it into latency. Thus, the adenoviral presence enables the AAV regulatory proteins to activate the expression of AAV genes and promote replication. The cytopathic effect of autonomous parvovirus infections is the rapid lysis of suitable host cells.

Replication in Vivo

The patterns of replication, as represented by the rolling-hairpin scheme (Section 15-7), apply in a general way to both the autonomous and AAV groups. Three features are notable: (1) Terminal hairpin sequences, used as primers, dispense with the need for RNA or protein priming. (2) Elongation is by single-strand displacement in replicative intermediates, as observed in adenoviral replication (Section 4). (3) Site-specific cleavage of a replicative intermediate by a viral-encoded enzyme with covalent attachment of the enzyme to the cleavage site resembles the actions of gene A protein in phage ϕX174 replication (Sections 8-7, and 17-4).

The terminal sequences provide the basis for hairpin priming and hairpin-loop transfers[42] (Fig. 19-6). The four possible combinations of terminal sequences for each strand of an AAV are found in equal proportions, showing that there is no bias for a particular orientation of the terminal sequences. Inasmuch as the inversion does not occur at the 3′ end of the virion AAV strand, a complex model of replication may be inferred.

Terminally cross-linked molecules, in both parental and progeny intracellular replicative intermediates,[43] are in accord with terminal hairpin strand initiation and a hairpin-loop transfer mechanism to replicate 5′-terminal sequences. Synthesis of both strands takes place by continuous elongation in the $5' \rightarrow 3'$ direction by a strand-displacement mechanism. The displaced single strands are either encapsidated or recycled back to a replicative form by synthesis of the complementary strand, as in the initial synthesis of the parental replicative form from infecting viral single strands (Fig. 19-6).

Site-specific cleavage of a replicative intermediate is crucial at a particular stage (Fig. 19-6). Autonomous viruses encode two nonstructural proteins (NS-1 and NS-2) that are later phosphorylated.[44] Protein covalently linked to the 5′ termini that persist in the virion corresponds to NS-1 in the case of MVM; the AAV Rep proteins (see below) are similarly implicated as site-specific incision enzymes.[45]

Negative regulation of AAV replication is indicated in the behavior of a chimera of an AAV with the SV40 origin region.[46] In cells that would normally support replication of an episome bearing this SV40 sequence, replication of the

42. Lusby E, Fife KH, Berns KI (1980) *J Virol 34*:402; Lusby E, Bohenzky R, Berns KI (1981) *J Virol 37*:1083.
43. Hauswirth WW, Berns KI (1979) *Virology 93*:57.
44. Cotmore SF, Tattersall P (1986) *Virus Res 4*:243.
45. Cotmore SF, Tattersall P (1988) *J Virol 62*:851; Snyder RO, Samulski RJ, Muzyczka N (1990) *Cell 60*:105.
46. Labow MA, Berns KI (1988) *J Virol 62*:1705; Berns KI, Kotin RM, Labow MA (1988) *BBA 951*:425.

chimera is inhibited. Presumably, without an adenoviral helper, products of the *rep* gene of AAVs act on a target sequence in the AAV terminal repeat to inhibit replication of the chimeric episome.

When the host cell is not permissive for AAV replication, as in the instance when adenoviral helper functions are lacking, AAVs can establish a latent infection by integrating into cellular DNA. Human AAV is integrated through junctions between the ends of the viral genome and host DNA at a specific site in chromosome 19.[47] In the use of these unique terminal viral sequences, integration of an AAV differs from the nonspecific selection of SV40 sequences in its integration.[48] Emergence of integrated AAV, facilitated by adenovirus superinfection, depends on adenoviral tumor antigen(s) and might involve the AAV termini in a manner resembling their use in replication.

Replication in Vitro

A reconstituted system with the potency and features anticipated from parvoviral replication in vivo has not been achieved. Inasmuch as viral replication relies almost entirely on the host polymerases and related proteins, and because the palindromic 3′-terminal hairpins are such effective primers, there are limited opportunities to identify the enzymatic events unique to the parvoviral replicative cycle.

The specificity of nicking by the autonomous viral NS proteins and the AAV Rep proteins, and their covalent attachments to their respective target sequences, may provide the most reliable touchstones for tracking the viral replicative events. Of the four viral Rep (nonstructural) proteins (with theoretical masses of 71, 61, 45, and 35 kDa), the two larger Rep proteins bind the AAV terminal repeat and only in the hairpin conformation.[49] The purified Rep68 (corresponding to the 61-kDa protein) has at least three activities:[50] (1) it is an ATP-dependent, site-specific, strand-specific endonuclease that cuts at the terminal resolution site, (2) it binds covalently to the 5′ end at the cut site, and (3) it has an ATP-dependent DNA helicase activity. These properties are in keeping with what is required for resolution of the termini and thus Rep68 performs in vitro in ways that might be anticipated from the binding of NS-1 to the autonomous viral DNA in vivo. Evidence has also been obtained for a ligation step at some stage in the replication of an autonomous virus.[51]

19-4 Adenoviruses[52]

The adenoviruses, which transform cells and produce cancer in certain animals, have features similar to those that make phage T7 experimentally attractive

47. Kotin RM, Siniscaleo M, Samulski RJ, Zhu X, Hunter L, Laughlin CA, McLaughlin S, Muzyczka N, Rocchi M, Berns KI (1990) *PNAS 87*:2211.
48. Weinberg RA (1980) *ARB 49*:197; Stringer JR (1981) *J Virol 38*:671.
49. Ashktorab H, Srivastava A (1989) *J Virol 63*:3034; Im D-S, Muzyczka N (1989) *J Virol 63*:3095.
50. Im D-S, Muzyczka N (1990) *Cell 61*:447.
51. Horwitz MS (1990) in *Virology* (Fields BN et al., eds.) 2nd ed. Raven Press, New York, p. 1679; Horwitz MS (1990) in *Virology* (Fields BN et al., eds.) 2nd ed. Raven Press, New York, p. 1723.
52. Kelly TJ (1988) *JBC 263*:17889; Kelly TJ, Wold MS, Li J (1988) *Adv Virus Res 34*:1; Stillman B (1989) *Ann Rev Cell Biol 5*:197; Challberg MD, Kelly TJ (1989) *ARB 58*:671; Van Hille B, Duponchel N, Salomé N, Spruyt N, Cotmore SF, Tattersall P, Cornelis JJ, Rommelaere J (1989) *Virology 171*:89. Van der Vliet PC, Claessens J, De Vries E, Leegwater PAJ, Pruijn GJM, Van Driel W, Van Miltenburg RT (1988) *Cancer Cells 6*:61.

(Section 17-5). Their size, intermediate between the tiny papovaviruses and the large herpes and pox viruses, places them within reach of complete chemical and genetic analysis, while endowing them with sufficient complexity to provide novel proteins and replication mechanisms. Discovery of an in vitro system[53] and its subsequent resolution has disclosed a protein-priming mechanism for initiation of replication and the appropriation of cellular transcription factors into the viral replication machinery, features with general significance in prokaryotic as well as eukaryotic replication. Some adenoviral functions complement defective parvoviruses (Section 3), as suggested by their hybridization with SV40.

More than 80 serologically distinct adenovirus strains have been isolated from humans, simians, and other animals. Adenoviruses are found in several human tissues, often associated with acute respiratory disease. The 31 human strains can be sorted into three groups based on oncogenicity, cross-reactivity with T antigen, and DNA homology. Viruses in group A (e.g., Ad12) produce tumors in newborn hamsters within a few months; those in group B (e.g., Ad7) are weakly oncogenic; members of group C (e.g., Ad2 and Ad5) produce no tumors but do transform rat cells in culture. Easy cultivation of Ad2 and of the very closely related Ad5 provides large quantities of virus and infected cells for study and has made them favored strains for experimental work.

Virion

Adenoviruses are icosahedrons, about 80 nm in diameter, made up of 252 sections, or *capsomers*. Of these, 240 face six neighbors and are called *hexons*; the others, located at each of the 12 vertices, face five neighbors and are called *pentons*. Projecting from each penton is a fiber with a knob at the end. As many as 15 structural polypeptides have been identified in Ad2, ranging in size from 5 to 120 kDa. The nucleocapsid, or core, is formed from DNA and four polypeptides. Three are arginine-rich: (1) the major core polypeptide VII (18.5 kDa, 1070 copies), which complexes with the DNA; (2) polypeptide V (48.5 kDa, 180 copies), which organizes the complex into a multilobular form and binds it to each of the 12 pentons directly or through interaction with other capsid proteins; and (3) polypeptide X or μ (4 kDa, 125 copies). Of special interest is the 55-kDa protein present in two copies, covalently bound to the 5′ ends of the genome, one molecule at each end (see below). Despite the absence of histones, the DNA is organized in a chromatin-like structure.

Genome. The DNA is a linear, blunt-ended duplex of 36 kb with a unique (nonpermuted) sequence. Lacking the terminal redundancy of T phages or the cohesive ends of phage λ, Ad2 possesses instead an inverted repetition of 103 bp at the ends and additional palindromes 180 bp from each terminus (Fig. 19-7).

Denaturing and reannealing the viral DNA in vitro produces circular single strands, presumably via panhandle structures, from both the *r* and *l* strands (Fig. 19-7); their terminal sequences are identical. Removal of the 3′ end by exonuclease III enables the internal palindrome in Ad2 DNA to form a snapback structure (hairpin) if it lacks homology with the terminal palindrome (case 2, Fig. 19-7).

---

53. Challberg MD, Kelly TJ (1979) *JMB* 135:499; Challberg MD, Kelly TJ (1979) *PNAS* 76:655.

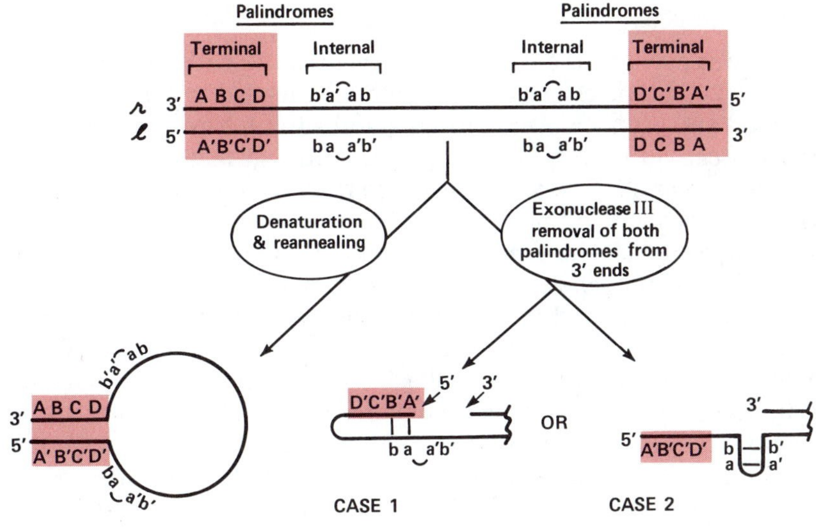

Figure 19-7
Structure of adenovirus 2 (Ad2) genome. Terminal inverted repeats (palindromes) lead to circular structures ("panhandles") upon denaturation and reannealing. Hairpin structures are formed upon removal of 3′-terminal palindromes: they are hairpin-loop terminals if there is homology between terminal and internal palindromes (case 1), or internal hairpins if there is no such homology (case 2). Designations r and l stand for rightward and leftward transcription of strands. [Adapted from Wold WSM, Green M, Büttner W (1978) in *Molecular Biology of Animal Viruses* (Nayak DP, ed.). Dekker, New York, p. 673]

The terminal 55-kDa protein of the virion attached to the 5′ end of each strand is derived from an 80-kDa precursor by cleavage late in the infective cycle. The protein is linked by a phosphodiester bond from serine to the 5′-terminal deoxycytidylate residue of each strand. In the virion of a protease-deficient mutant (Ads *tsl*) grown at a nonpermissive temperature, the terminal protein is 80 kDa. Circular structures formed by cohesion of the two 5′-terminal 55-kDa protein molecules observed in the disrupted virion have also been seen among the replicative forms of adenovirus in infected HeLa cell nuclei.[54]

Origins. The identical origins at each end of the genome consist of domains A, B, and C. Domain A (nucleotides 1 to 18) is essential and can serve as a minimal origin; all serotypes share the 10-bp consensus sequence ATAATATACC. B (nucleotides 19 to 39), with the consensus sequence TTGGCN$_5$GCCAA, stimulates 30-fold or more. C (nucleotides 29 to 51), with the consensus sequence TATGATAAT, stimulates threefold. Domains B and C are bound by cellular proteins identified independently as transcriptional activators. The spacing of the domains as well as their composition is critical to their functioning in initiation.

54. Weber J (1976) *J Virol* 17:462.

Infection

The fate of infecting particles is difficult to trace because, as with papovaviruses (Section 2), 10 to 200 particles per cell are generally needed to ensure that all cells are infected and produce plaque-forming units (pfu). Nevertheless, early events in the replicative cycle are becoming clearer through in vitro studies, and later viral stages are easier to examine because they dominate cell activities. Final virus yields are about 10^4 pfu (10^5 to 10^6 genomes) or more per cell.

In the replicative cycle,[55] the virus is adsorbed through its fibers (spikes) to cell-surface receptors and penetrates the plasma membrane. Uncoating of the virion, transport of the core to the nucleus, and disassembly of the core rely on host enzymes. Transcription of early genes by an α-amanitin–sensitive host enzyme, presumably RNA polymerase II, begins within 2 to 3 hours after Ad2 infection (Fig. 19-8). The extent of the genome represented by early mRNAs is uncertain. Values range from 10% to 50% of the genome, depending on whether only abundant transcripts are detected or more sensitive techniques are used. In any case, early genes are not localized as in phage T7 but are distributed throughout the map.

The most abundant of early gene products, Ad DBP (Section 10-6), is the SSB needed for DNA replication. It is detectable as early as 4 hours after Ad2 infection. At about 6 hours, genes coding largely for virion structural proteins begin to be expressed, and virions appear between 13 and 24 hours. DNA synthesis at this stage (about 18 hours) is almost exclusively viral. Mechanisms for the switch from early to late gene expression and the nature and location of signals for

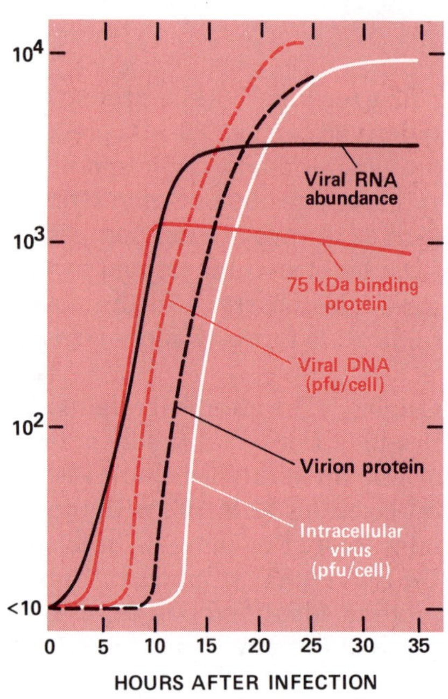

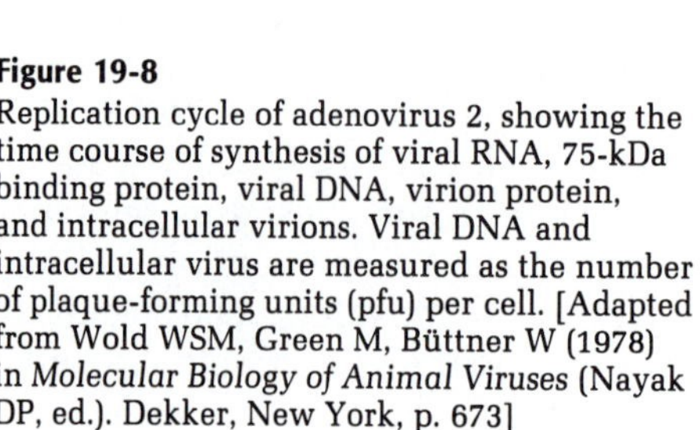

Figure 19-8
Replication cycle of adenovirus 2, showing the time course of synthesis of viral RNA, 75-kDa binding protein, viral DNA, virion protein, and intracellular virions. Viral DNA and intracellular virus are measured as the number of plaque-forming units (pfu) per cell. [Adapted from Wold WSM, Green M, Büttner W (1978) in *Molecular Biology of Animal Viruses* (Nayak DP, ed.). Dekker, New York, p. 673]

55. Girard M, Bouché JP, Marty L, Revet B, Berthelot N (1977) *Virology* 83:34.

transcription have been examined. How host biosynthetic functions are turned off during the latter part of the viral productive cycle is unclear, but a contributing factor may be a specific block to the nuclear export of cellular, as opposed to viral, mRNA.

Replication in Vivo

Replicative forms observed in the electron microscope and the location of newly synthesized DNA (determined by hybridization with defined viral segments) are best interpreted by knowledge of the in vitro replication mechanism (Fig. 19-9). Replication takes place in two stages. In the first, initiation and processive synthesis from either end replicates one strand of the template to produce a duplex and displaces the other parental strand as a single strand. In the next stage, the displaced single strand is in turn replicated to produce the second daughter duplex. In more detail:

(1) Replication is unidirectional, starting at each end of Ad DNA and forming a duplex molecule with displaced single-stranded tails (Type I forms, Fig. 19-9). The origin of each template strand is at its 3′ end and the terminus is at the 5′ end, as in semiconservative replication.

(2) The terminal 55-kDa protein of the virion is produced early in infection, and each 5′ terminus of progeny DNA that ends up in virions has one molecule of the protein covalently linked. Virtual identity of the peptide map in two viral strains of low homology shows that the protein is highly conserved in composition among adenoviruses. The deoxycytidylate that initiates the chain is cova-

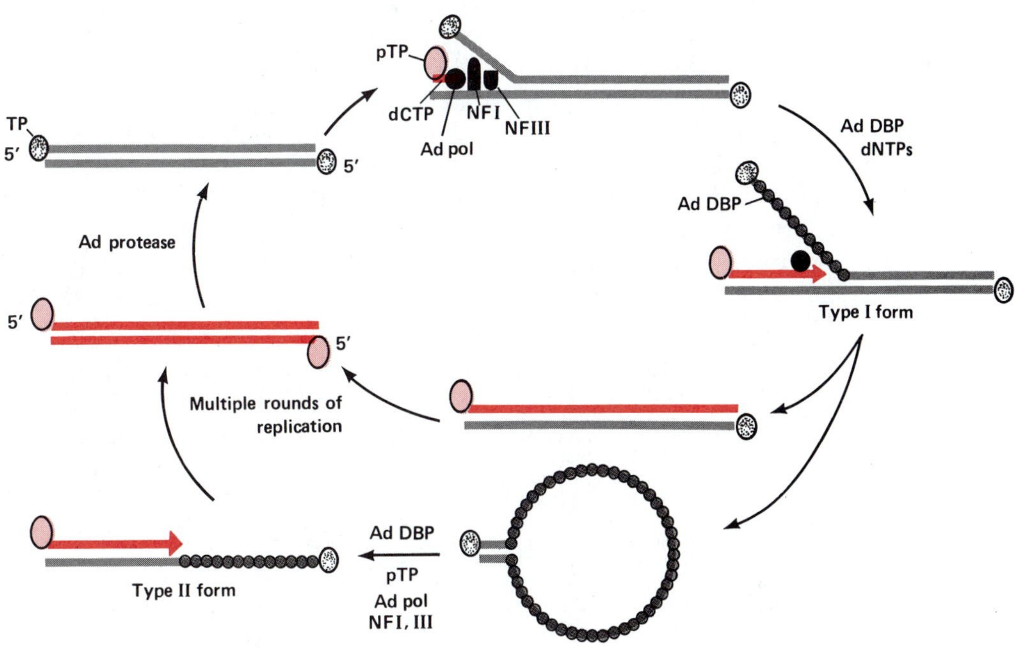

Figure 19-9
Replication pathway of adenovirus DNA. (From Stillman B (1989) *Ann Rev Cell Biol* 5:197]

lently attached to the protein before the protein recognizes the template origin. Related examples are an analogous protein bound to the 5′ deoxyadenylate ends of DNA phage ϕ29 (Section 17-8) and a 6-kDa protein linked by tyrosine to the 5′ terminus of nascent poliovirus RNA.

(3) Completion of the first stage of replication releases the displaced single strands. They may assume a panhandle form due to their terminal palindromes.

(4) The duplex end of the panhandle has the same structure as the ends of the duplex viral genome and is probably recognized, initiated, and replicated in the same way. Intermediates in this stage appear as partially duplex molecules (Type II forms, Fig. 19-9).

(5) Circular forms, formed by cohesion of the two 5′-terminal 55-kDa protein molecules, have been observed in the disrupted virion but have not been found as yet among replicative forms in the cell.

(6) Single-stranded DNA is prominent initially as a displaced strand and later as a template, explaining the need for large quantities of the SSB produced before the start of replication. The SSB may be needed to protect the DNA from nucleases and endow it with a favorable template conformation.

Replication in Vitro[56]

Three virus-encoded proteins are essential (Table 19-2): terminal protein precursor (pTP), DNA polymerase (Ad pol),[57] and the SSB (Ad DBP).[58] The mRNAs for these proteins account for about 25% of the coding capacity of the genome and are differentially spliced from a common precursor. Also required are three host proteins (nuclear factors): two transcriptional activators (NFI,[59] NFIII[60]) and a topoisomerase (NFII[61]). A complex of equal numbers of pTP, Ad pol, NFI, and NFIII, isolated from infected cells, suffices for initiation.[62]

Initiation of replication at one or the other end of the linear duplex entails (1) displacement of the 55-kDa viral terminal protein (processed pTP) covalently linked to the 5′ strand end, (2) orientation of the 80-kDa pTP at the 3′ end of the complementary strand, and (3) linkage of deoxycytidylate to a serine residue of the pTP, presumably by Ad pol. Processive replication by the same polymerase follows. Binding at the terminus by NFI to domain B and NFIII to domain C is needed for initiation, but the contributions of AdDBP and NFIII are made only after the fork has penetrated some 50 or more nucleotides downstream.

The requirement for ATP in initiation can be replaced by nonhydrolyzable analogs. The locus of this effector function has not been determined. Strand

---

56. Challberg MD, Kelly TJ (1989) *ARB 58*:671; Howley PM (1990) in *Virology* (Fields BN et al., eds.) 2nd ed. Raven Press, New York, p. 1625; Shah KV, Howley PM, (1990) in *Virology* (Fields BN et al., eds.) 2nd ed. Raven Press, New York, p. 1651; Stillman B (1989) *Ann Rev Cell Biol 5*:197.
57. Tamanoi F, Stillman BW (1982) *PNAS 79*:2221; Challberg MD, Ostrove J, Kelly TJ (1982) *J Virol 41*:265; Kenny MK, Balogh LA, Hurwitz J (1988) *JBC 263*:9801; Salas M (1988) *Curr Top Microbiol Immunol 136*:71.
58. Lindenbaum JO, Field J, Hurwitz J (1986) *JBC 261*:10218.
59. Challberg MD, Desiderio SV, Kelly TJ (1980) *PNAS 77*:5105; Enomoto T, Lichy JH, Ikeda J-E, Hurwitz J (1981) *PNAS 78*:6779.
60. Rosenfeld PJ, Kelly TJ (1986) *JBC 261*:1398; Santoro C, Mermod N, Andrews PC, Tjian R (1988): *Nature 334*:218; de Vries E, van Driel W, van den Heuvel SJL, van der Vliet PC (1987) *EMBO J 6*:161.
61. Pruijn GJM, van Driel W, van der Vliet PC (1986) *Nature 322*:656; Rosenfeld PJ, O'Neill EA, Wides RJ, Kelly TJ (1987) *Mol Cell Biol 7*:875; O'Neill EA, Kelly TJ (1988) *JBC 263*:931; van der Vliet PC, Claessens J, de Vries E, Leegwater PAJ, Pruijn GJM, van Driel W, van Mittenburg RT (1988) in *Cancer Cells 6* (Kelly TJ, Stillman BW, eds.). CSHL, p. 61.
62. Nagata K, Guggenheimer RA, Hurwitz J (1983) *PNAS 80*:6177; Pruijn GJM, van Driel W, van der Vliet PC (1986) *Nature 322*:656; O'Neill EA, Fletcher C, Burrow CR, Heintz N, Roeder RG, Kelly TJ (1988) *Science 241*:1210.

Table 19-2
Viral and host proteins in adenoviral replication

Protein	Mass (kDa)	Function
Viral		
Terminal protein precursor (pTP)	80	initiation of strands
Terminal protein (TP)	55	processed pTP; linked to 5′ viral ends
DNA polymerase (Ad pol)	140	synthesis of strand
SSB (AdDBP)	59[a]	coating of strands
Host		
Nuclear factor I (NFI)[b]	52–66[c]	initiation, cellular transcription
Nuclear factor II (NFII)	30	topoisomerase
Nuclear factor III (NFIII)[b]	92	initiation, cellular transcription

[a] Apparent size in SDS–polyacrylamide gel electrophoresis is 72 kDa.
[b] NFI is also called CTF (for cellular transcription factor); NFIII is also called OTF-1 (for octamer transcription factor).
[c] Three closely related proteins in this size range are each active.

displacement during elongation does not require helicase action. Fork movement can be sustained by polymerization of the dNTPs and the coating of the displaced single strand by AdDBP.

With the purified enzyme system, elongation of a DNA chain reaches only about one-quarter genome length (9 kb) and requires the intervention of the topoisomerase activity of NFII. A structural impediment at this stage of replication may explain the rather tardy appearance of the need for this activity. Upon completion of one daughter duplex (Fig. 19-9), the displaced single strand, probably in a panhandle conformation, provides the same terminus for initiation that the parental duplex does and is replicated in a like manner.

19-5 Herpesviruses[63]

The size and complexity of herpesviruses, comparable to the T-even phages (Section 17-6), endow them with considerable independence from cell replicative systems. The novel devices for DNA replication in nuclei introduced by these viruses are of basic interest. Of practical as well as academic importance is their influence on cellular DNA replication, reflected by the widespread latent and recurrent diseases they produce and by their association with human cancer. Examples of the 80 or more herpesviruses that infect a variety of animal species are the herpes simplex viruses (HSV-1 and HSV-2), which cause cold sores and genital tract infections in humans; Epstein-Barr virus (EBV), associated with infectious mononucleosis and with two-human cancers, Burkitt's lymphoma and nasopharyngeal carcinoma; cytomegalovirus (CMV), causing animal and human diseases, particularly in immunodeficient individuals; varicella-zoster virus (VZV), producing chickenpox in children and shingles in adults; and Marek's herpesvirus, which causes malignant avian lymphoma.

63. Challberg MD, Kelly TJ (1989) *ARB 58*:671; Wagner EK (1985) in *The Herpesviruses* (Roizman B, ed.). Plenum Press, New York, Vol. 3 p. 104; Roisman B, Sears AE (1990) in *Virology* (Fields B et al., eds.) 2nd ed. Raven Press, New York, p. 1795.

Virion

Herpesviruses are seen in infected cells as icosahedral particles, about 100 nm in diameter, budding through an inner nuclear membrane. The particles are usually found in the perinuclear space, in the cisternae of the endoplasmic reticulum, and in extracellular fluids. The large virion has a nucleocapsid core surrounded by concentric layers of a capsid composed of 162 capsomers, a tegument of unidentified composition, and a membranous envelope with projecting spikes. Over 30 polypeptides, ranging in size from under 25 to over 200 kDa, are present in many copies per particle, but no host-coded proteins have been identified among them.

HSV Genome[64]

The completely sequenced HSV-1 is a linear duplex of 153 kb, sufficient to code for 100 different 55-kDa polypeptides (human CMV is 220 kb). The GC content is generally high, reaching 67% and 69% in HSV-1 and HSV-2, compared to 41% to 42% in the cellular DNA. Interruptions in the phosphodiester backbone are most likely due to nicks or gaps rather than cleavages at intermittent ribose residues. Most remarkable is the regular arrangement of repeated sequences. Short (S) and long (L) unique regions are flanked by inverted repeats (Fig. 19-10). A brief terminal sequence, shared by the inverted repeats, makes the genome terminally redundant. Commonly, during replication, intramolecular crossovers (inversions of the S and L sequences) by breakage and reunion yield equal numbers of the four isomeric arrangements, all infectious; mutant viruses incapable of this isomerization produce progeny at normal rates in cell culture.[65]

Origin sequences have been located in two places:[66] *oriL* in the middle of the L region and two copies of *oriS* in the inverted repeats flanking the S region (Fig. 19-10). The closely related origin sequences seem to be functionally equivalent; either *oriL* or one copy of *oriS* is dispensable. The origin contains a central, long, AT-rich hairpin (palindrome), presumably the site of opening the duplex, and arms on either side recognized by the origin-binding protein (see below). Dele-

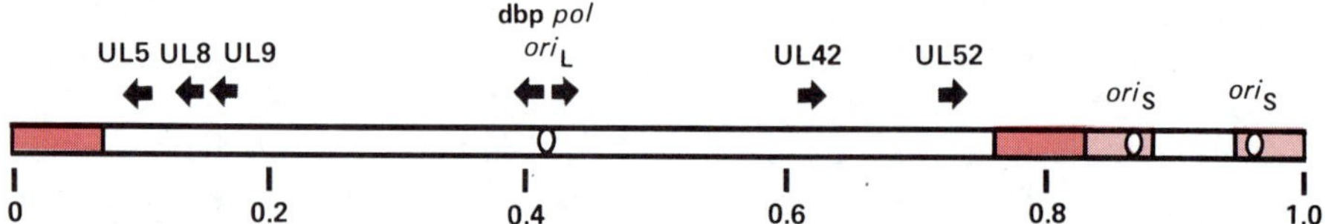

Figure 19-10
Organization of the herpes simplex virus (HSV) genome. Open areas = unique regions of the genome; dark and light color areas = inverted repeats flanking the long unique region and short unique region, respectively; arrows = open reading frames (orfs) for genes encoding proteins involved in DNA replication (see Table 19-3). [Adapted from Challberg MD, Kelly TJ (1989) *ARB* *58*:671]

64. McGeoch DJ, Dalrymple MA, Davison AJ, Dolan A, Frame MC, McNab D, Perry LJ, Scott JE, Taylor P (1988) *J Gen Virol 69*:1531; Chou J, Roizman B (1986) *J Virol 57*:629.
65. Jenkins F, Roizman B (1986) *J Virol 59*:494.
66. Deiss LP, Chou J, Frenkel N (1986) *J Virol 59*:605; Varmuza SL, Smiley JR (1985) *Cell 41*:793; Lockshon D, Galloway D (1988) *Mol Cell Biol 8*:4018.

tion of the palindrome or replacement of the AT-rich region by GC base pairs blocks DNA replication. An essential region, in addition to the origins, holds the genome termini where cleavage of "endless intermediates" is needed for packaging into virions.

The genome, in which introns are uncommon, encodes at least 73 proteins.[67] Of these, seven are necessary and sufficient for maintaining an autonomously replicating minigenome and for replication of the virus.[68] Although HSV appears to be independent of cellular replication functions, some HSV proteins may be important in replacing them, and some host enzymes, such as topoisomerases, are surely needed.

HSV Replication in Vivo[69]

The virus replicates lytically in epithelial cells in individuals and in cell cultures. Like other herpesviruses, HSV can be latent for many years in the infected individual, commonly in neurons of sensory ganglia. Little is known of the extent, if any, of its replication in this state.

Depending on the virus and the host cell, the infection may be productive or latent. The basis for cellular permissiveness that supports a productive infection or the relative lack of it that leads to latency or transformation is not understood. Reproductive cycles with HSV-1 are "fast," consuming only 13 to 19 hours and yielding 10^4 to 10^5 particles per cell; they are "slow" with EBV or CMV.

The temporal stages of HSV-1 gene expression resemble those of a phage T4 infection. Immediate early (α) gene products are followed by the early (β), which include the replication proteins, and finally by the late (γ) proteins. By 4 to 7 hours after HSV-1 infection, host DNA synthesis is completely replaced by viral DNA synthesis. Replication is not restricted, as it is with parvoviruses, to the DNA-synthesis (S) phase of the cell cycle, implying that the virus operates under its own controls. Shutoff of virtually all host macromolecular biosyntheses and gross morphologic changes in organelles and membranes lead to death of the cell.

Mutations in any one of seven HSV genes[70] (recognized by complementation studies and as the open reading frame for *UL42*) block the replication of plasmids dependent on *oriL* or *oriS*. The protein products of each of these genes have been identified (see below). Best analyzed of the phenotypes are (1) the temperature-sensitive mutants with amino acid substitutions in the DNA polymerase (Section 6-8), the principal target of antiviral drugs (Section 14-6), and (2) mutants in the SSB [known as DBP or infected-cell polypeptide 8 (ICP8); Section 10-6]. Mutants in ICP8 display pleiotropic effects which include altered regulation of gene expression, modified sensitivity to antiviral drugs, and defects in nuclear localization and replication.

In addition to the proteins encoded by the seven essential replicative genes, other HSV proteins probably augment DNA metabolic functions in the infected

---

67. McGeoch DJ, Dalrymple MA, Davison AJ, Dolan A, Frame MC, McNab D, Perry LJ, Scott JE, Taylor P (1988) *J Gen Virol 69*:1531; Chou J, Roizman B (1986) *J Virol 57*:629.
68. Weller SK, Aschman DP, Sacks WR, Coen DM, Schafer PA (1983) *Virology 130*:290; Marchetti ME, Smith CA, Schaffer PA (1988) *J Virol 62*:715.
69. Wagner EK (1985) in *The Herpesviruses* (Roizman B, ed.). Plenum Press, New York, Vol. 3 p. 104; Roizman B, Sears AE (1990) in *Virology* (Fields B et al., eds.) 2nd ed. Raven Press, New York, p. 1795.
70. Weller SK, Aschman DP, Sacks WR, Coen DM, Schafer PA (1983) *Virology 130*:290; Marchetti ME, Smith CA, Schaffer PA (1988) *J Virol 62*:715.

cell.[71] These include thymidine kinase with an associated thymidylate kinase, deoxycytidine deaminase, ribonucleotide reductase, uracil N-glycosylase, and an alkaline DNase (exonuclease).

Biosynthesis of deoxynucleotides by the viral-encoded enzymes is exempt from the allosteric controls that inhibit the corresponding host enzymes. The DNases, in contrast to the endonucleolytic cellular DNases, can act exonucleolytically on both native and denatured DNA (host and viral) to release 5′-mononucleotides, mainly from 3′ termini. If nuclease function is to scavenge host DNA for viral replication, it is not clear how the viral DNA, which is not known to be modified, is spared from attack.

At least two viral gene products are needed to shut off host functions. Thus far, too little is known from studies with cells, washed nuclei, chromatin, and soluble enzyme preparations to delineate the outline, let alone the details, of the replication mechanism. Because of terminal redundancies, the genome might, by appropriate exonuclease action, develop cohesive ends with which to form circular or concatemeric structures, as phages T4 and λ do (Sections 17-6, and 17-7). The large genome size, isomerizations, and recombinations complicate interpretations of the pattern of replication. Multigenome-sized intermediates are found which lack defined termini and suggest rolling circles, concatemers of tandem repeats, and unresolved recombinant forms.

Cell Transformation

Only part of the HSV genome is needed to transform susceptible, nonpermissive cells. Fragments produced by shearing or restriction-nuclease cleavage are effective and in some instances can be selected by linkage with the thymidine kinase gene. To what extent viral genes and genomes are integrated into host chromosomes is still uncertain. Many copies of EBV are associated with cellular DNA in Burkitt's lymphoma and lymphoblastoid cells and may be present as plasmids or in unconventional linkages.

EBV is unique in the efficiency with which it can immortalize human B lymphocytes.[72] From 10% to 100% of infected peripheral B cells yield progeny that proliferate indefinitely with EBV maintained in a latent plasmid state. Among several genes implicated in the immortalization reaction, the EBV nuclear antigen 1 (EBNA-1) is required for replication of the EBV plasmid and serves as a transcriptional activator. Vectors with the EBV origin for plasmid (*oriP*) and lytic (*oriLyt*) DNA replication[73] (Section 16-5) should encourage more intensive replication studies comparable to those of HSV-1.

Replication Proteins[74]

The sequences of the products of the seven replication genes (Fig. 19-10; Table 19-3) have been determined for HSV and two other human herpesviruses (VZV

71. Challberg MD, Kelly TJ (1989) *ARB 58*:671; Howley PM (1990) in *Virology* (Fields BN et al., eds.) 2nd ed. Raven Press, New York, p. 1625; Shah KV, Howley PM (1990) in *Virology* (Fields BN et al., eds.) 2nd ed. Raven Press, New York, p. 1651.
72. Sugden B (1989) *Cell 57*:5.
73. Hammerschmidt W, Sugden B (1988) *Cell 55*:427.
74. Challberg MD, Kelly TJ (1989) *ARB 58*:671; Howley PM (1990) in *Virology* (Fields BN et al., eds.) 2nd ed. Raven Press, New York, p. 1625; Shah KV, Howley PM (1990) in *Virology* (Fields BN et al., eds.) 2nd ed. Raven Press, New York, p. 1651.

Table 19-3
HSV replicative genes and proteins

Gene	Protein mass (kDa)	Protein function
UL9	94	origin-binding
pol	136	DNA polymerase
UL42	51	dsDNA-binding; polymerase accessory
UL5	99	helicase–primase
UL8	80	unknown
UL52	114	helicase–primase
dbp (ICP8)	130	ssDNA-binding

and EBV). Even though EBV is only distantly related to the others, the conservation of these genes and proteins, except for the origin-binding protein, indicates that replication patterns during lytic growth of herpesviruses are basically alike. Some features of these key proteins, also considered in other chapters, are especially notable.

The *origin-binding protein* of HSV (gpUL9)[75] behaves as a dimer and binds *oriS* cooperatively to two nearly identical 8-bp boxes adjacent to the palindrome, binding ten times more strongly to the left box than to the right one. The DNA-binding domain is localized in the C-terminal region. Like SV40 T antigen and *E. coli* dnaA protein, gpUL9 binds and hydrolyzes ATP and can be expected to participate like a helicase in opening the origin and directing the entry of other replication proteins.

In the three-subunit *helicase–primase* complex[76] of gpUL5, gpUL8 and gpUL52, several features suggest that gpUL5 and gpUL52 serve together as the helicase, using ATP or GTP for unwinding; gpUL8 appears to be dispensable. Translocation in the 5′ → 3′ direction resembles that of dnaB protein in its primosomal function in discontinuous-strand synthesis. The purified *DNA polymerase*[77] is found to be complexed with gpUL42, the *dsDNA-binding protein*. In an accessory polymerase function, gpUL42 enhances processivity when ICP8 is present. The 5′ → 3′ exonuclease intrinsic to the polymerase has the same RNase H capacity as does *E. coli* pol I. The abundant ICP8[78] (DBP) resembles the *SSBs* of phage T4 (gp32) and *E. coli* in its highly cooperative binding to ssDNA; yet its contributions to template use by primase and DNA polymerase remain to be examined.

19-6 Poxviruses: Vaccinia[79]

An unusually large particle and genome size, replication in the cytoplasm, and major medical importance have attracted attention to the poxviruses for over 50

75. Elias P, O'Donnell ME, Mocarski ES, Lehman IR (1986) *PNAS 83*:6322; Olivo PD, Nelson NJ, Challberg MD (1988) *PNAS 85*:5414; Koff A, Tegtmeyer P (1988) *J Virol 62*:4096.
76. Crute JJ, Tsurumi T, Zhu L, Weller SK, Olivo PD, Challberg MD, Mocarski ES, Lehman IR (1989) *PNAS 86*:2186; Dodson MS, Crute JJ, Bruckner RC, Lehman IR (1989) *JBC 264*:20835; Calder JM, Stow ND (1990) *NAR 18*:3573.
77. Crute JJ, Lehman IR (1989) *JBC 264*:19266; Hernandez TR, Lehman IR (1990) *JBC 265*:11227.
78. Ruyechan WT (1983) *J Virol 46*:661; O'Donnell ME, Elias P, Funnell B, Lehman IR (1987) *JBC 262*:4260; de Bruyn Kops A, Knipe DM (1988) *Cell 55*:857; Gao M, Bouchey J, Curtin K, Knipe DM (1988) *Virology 163*:319.
79. Moss B (1990) in *Virology* (Fields BN et al., eds.) 2nd ed. Raven Press, New York, p. 2079.

years. To the virologist, the severe cell pathology of variola (smallpox) and vaccinia (cowpox) infections and the malignant or benign tumors in rabbits caused by the myxoma–fibroma group are of compelling interest. To the biotechnologist, a genetically attenuated poxvirus is a handy vector for the delivery of 25 kb or more of DNA for genetic therapy of disease, for vaccination (e.g., hepatitis B and influenza antigens), and for production of foreign proteins.[80] To the biochemist, a virion packed with an assortment of enzymes, a genome with a unique DNA structure, a vigorous DNA replication cycle in the cytoplasm, and the synchronous development of a readily isolated organelle offer almost unlimited opportunities.

Virion and Genome

These largest of all animal viruses are brick-shaped or oval, with dimensions in the range of 300 × 200 × 100 nm. They are easily sedimented and purified and are available in quantity for analysis. The nucleocapsid core is surrounded by several concentric shells of membranes and proteins. Over 30 polypeptides, small and large, have been identified.

The virion is replete with enzymes for the synthesis and processing of RNA. Permeabilized virions afford an in vitro system in which transcripts are guanylylated (capped), methylated, and polyadenylated for translation by a protein-synthesizing system. The RNA polymerase has been purified extensively,[81] and it is now possible to take a closer look at the fascinating dependence of transcription on ATP hydrolysis. Action of phosphohydrolase II of the virion may be coupled to transcription in a manner analogous to involvement of the DNA-dependent ATPases in replication. The multisubunit RNA polymerase of the poxvirus, as the first example of an enzyme within a virion used immediately on entry to express the viral genome, has had a major impact on virology; the reverse transcriptases in RNA tumor viruses (Section 7) were recognized later on.

Poxvirus DNAs (180 to 300 kb) are among the largest viral nucleic acids known; the vaccinia virus genome is 186 kb. The GC content, near 35%, is distinctly lower than that of animal cells (about 42%). The most remarkable feature of the DNA is its continuity ("cross-linkage") at the ends that enables a giant, single-stranded circle to be formed upon denaturation. At each end is a 10-kb repetition with direct tandem repeats (Fig. 19-11); this region is hypervariable within a viral strain. Hairpin terminal loops can be formed from palindromic sequences at each end of the duplex, as in the replicative forms of autonomous parvoviruses (Section 3).

Infection and Replication

Vaccinia particles enter the cytoplasm of cells in only a few minutes and are uncoated. The RNA polymerase of the virion transcribes an early region (14%) of the genome. Viral DNA synthesis begins at 1.5 hours, peaks about an hour later, and declines sharply thereafter. The maximal rate is several times that of nor-

---

80. Karacostas V, Nagashima K, Gonda MA, Moss B (1989) *PNAS 86*:8964.
81. Morrison DK, Moyer RW (1986) *Cell 44*:587.

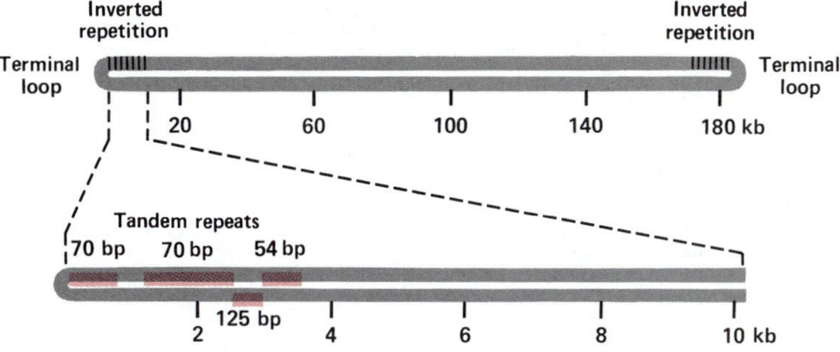

Figure 19-11
Structure of vaccinia virus DNA. A representation of the entire genome and an expansion of the 10-kb inverted terminal repetition are shown. The tandem repeats are clustered as indicated. (From Bernard M (1985) in *Virology* (Fields BN, Knipe DM, Chanock RM, Hirsch MS, Melnick JL, Monath TP, Roizman B, eds.). Raven Press, New York, p. 685]

mal host DNA synthesis and produces viral DNA to a level 50 percent that of cellular DNA. The viral DNA is distributed in discrete cytosolic factory areas called "virosomes" and provides the focus around which each virion develops. The infection yields 10^4 particles per cell between 12 and 15 hours after infection, following an eclipse period of 4 to 6 hours. Shutoff of host functions is part of the extensive cytopathology that causes cell death.

The requirement for biosynthetic and replicative enzymes is more stringent for the cytoplasmic poxviruses because of the general inaccessibility of the host enzymes in the nucleus. Although the virus can be replicated in enucleated cells, some nuclear components, such as the largest subunit of RNA polymerase II are recruited.[82] For synthesis of the mRNA, the virion packages the RNA polymerase, capping and methylating enzymes, and the poly A polymerase which adds a poly A tail to the mRNA.

Also in the virion are polynucleotide kinase, a nicking–joining enzyme, DNA-dependent ATPase, endonuclease, exonuclease, and topoisomerase. For DNA synthesis the virus encodes ribonucleotide reductase,[83] thymidine kinase, thymidylate kinase,[84] topoisomerase,[85] DNA polymerase (Section 6-8), DNA ligase,[86] and an SSB. Virion enzymes with potential but undefined replicative functions include an endonucleolytic DNase, two DNA-dependent ATPases, and a protein kinase. The DNA polymerase[87] is notable for its sensitivity to aphidicolin and for a $3' \rightarrow 5'$ exonuclease activity.

A scheme[88] proposed for the replication of vaccinia DNA (Fig. 19-12) is based on electron micrographs of viral DNA structures in infected cells and on plausible functions for enzymes associated with the virus before and after infection. The origin of replication has not been defined, nor is it known whether RNA priming initiates replication. Fork movement appears to be unidirectional.

82. Moyer RW (1987) *Virus Res 8*:173.
83. Shuman S, Moss B (1987) *PNAS 84*:7478.
84. Smith GL, de Carlos A, Chan YS (1989) *NAR 17*:7581.
85. Slabaugh M, Roseman N, Davis R, Mathews C (1988) *J Virol 62*:519; Tengelsen LA, Slabaugh MB, Bibler JK, Hruby DE (1988) *Virology 164*:121.
86. Kerr SM, Smith GL (1989) *NAR 17*:9039.
87. Earl PL, Jones EV, Moss B (1986) *PNAS 83*:3659.
88. Baroudy BM, Vankatesan S, Moss B (1982) *Cell 28*:315.

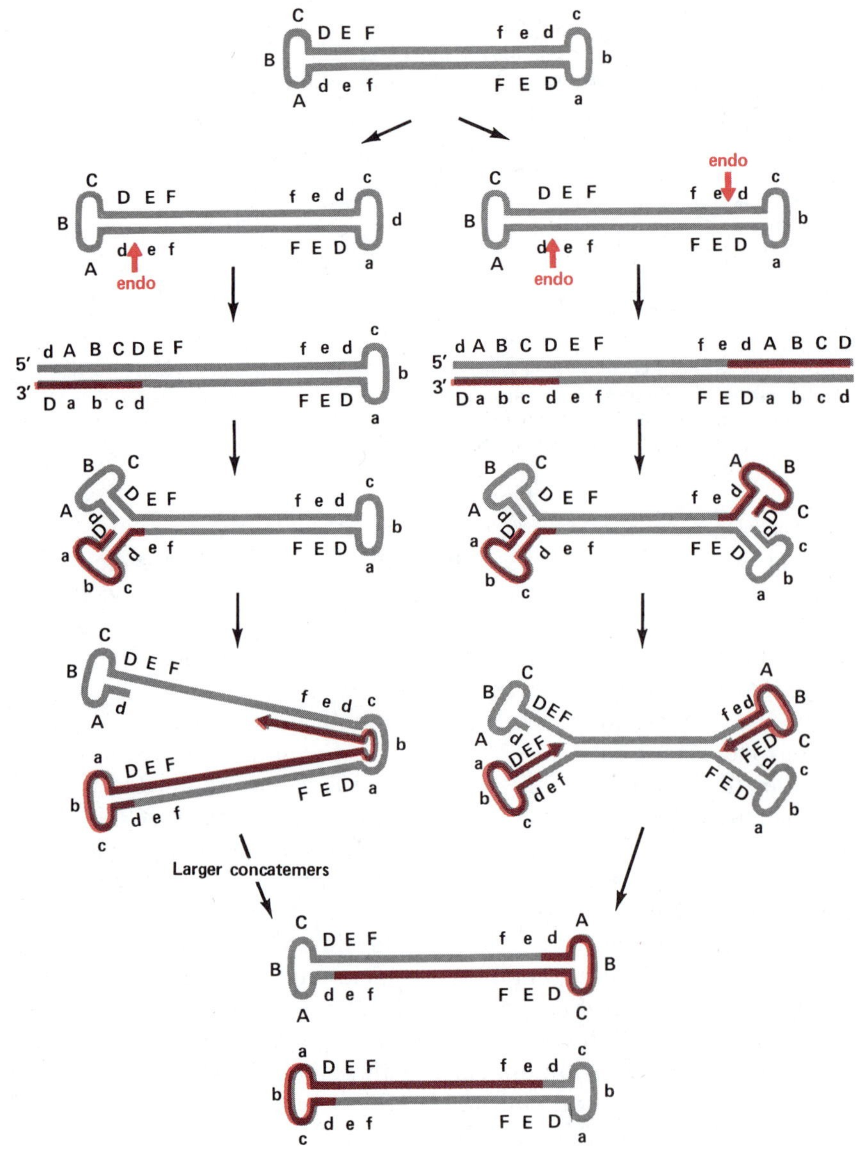

Figure 19-12
Model for replication of vaccinia virus DNA. Replication initiated at the looped region at one end progresses unidirectionally to produce concatemers from which linear genomes are released and cyclized by a hairpin-loop mechanism (left pathway). Replication initiated at both ends produces linear genomes directly (right pathway). [After Moss B (1990) in *Virology* (Fields BN, Knipe DM, eds.). Raven Press, New York, p. 2079]

Opening of the template circle at some location at an appropriate time is needed to produce a full duplex. Presumably, palindromic sequences fold back and produce hairpins that finally yield circular viral progeny. The nicking–joining enzyme may be responsible for the intramolecular linking that completes the genome.[89]

89. Reddy MK, Bauer WR (1989) *JBC 264*:443.

The unique replicative feature of these small, single-stranded RNA genomes is that multiplication requires their integration as DNA duplexes into the host chromosomes. The reverse (retro) transcription of RNA to DNA sets these RNA viruses apart from all other RNA viruses and focuses attention on the DNA polymerase, so remarkable for its synthesis of a complete DNA genome tailored for integration and for an RNase H of extraordinary specificity.

The retroviruses have been of long-standing biological and medical interest. They multiply in cells without interrupting growth or division and are transmitted "vertically" as a provirus from one cell generation to the next. On occasion, they cause a variety of diseases, most prominently several kinds of cancer (leukemias and sarcomas). (The oncogenic retroviruses are also called *oncornaviruses.*) Retroviruses may cause anemia, arthritis, and immunodeficiency states. The latter disease in humans—human immunodeficiency disease (known as acquired immunodeficiency syndrome, or AIDS)—has made the causative HIV retrovirus the target of the most massive crusade in modern microbial history.

Before the focus on HIV, the most intensively studied of the retroviruses were those that produce cancer in chickens (Rous sarcoma virus) and in monkeys; leukemia in birds (avian myeloblastosis virus, AMV), mice (Moloney murine leukemia virus, M-MuLV), and cats; and mammary tumors in mice and monkeys.

The avian sarcoma virus became favored for study because cell transformation can be detected by a simple, highly efficient in vitro assay. Sarcomas are produced readily at the site of injection in vivo. Mammalian cell lines can be transformed at low frequency (about 10^{-5} that of chick cells), but do not sustain viral replication. A variety of deletion and temperature-sensitive mutants that affect viral transformation or replication have clarified the genomic organization and specific functions.

Reverse transcription of RNA into DNA has also been recognized in the life cycles of other viruses and genetic elements.[91] These include (1) hepatitis B and cauliflower mosaic viruses (Sections 8 and 9), whose DNA genomes are first converted to RNA in the course of their replication; (2) a class of transposons in eukaryotic chromosomes (retrotransposons), which proliferate by synthesizing transposon RNA and, after reverse transcription, reintegrate at new chromosomal sites, never acquiring an extracellular or viral form; (3) processed pseudogenes that, by inference, arise by reverse transcription of cellular mRNAs; and (4) the multicopy, single-stranded (ms) DNAs of *E. coli* (Section 5-7).

Virion

Retrovirus particles are spheres, 100 nm in diameter, covered by an envelope derived from the plasma membrane of the host cell. Glycoprotein spikes pro-

---

90. Varmus HE, Swanstrom R (1985) in *Molecular Biology of Tumor Viruses;* Vol. 2: *RNA Tumor Viruses* (Weiss R, Teich N, Varmus H, Coffin J, eds.). CSH, p. 75; Varmus HE, Brown PO (1989) in *Mobile DNA* (Howe M, Berg D, eds.). Am Soc Microbiol, Washington DC, p. 53; Varmus HE (1988) *Science 240:*1427; Varmus HE (1989) *Harvey Lect 83:*35; Coffin JM (1990) in *Virology* (Fields BN et al., eds.) 2nd ed. Raven Press, New York, p. 1437.
91. Boeke JD, Garfinkel DJ, Styles CA, Fink GR (1985) *Cell 40:*491.

truding from the envelope are specified by the virus; the spikes recognize receptors on susceptible cells. Within the envelope is a "core shell" composed of several virus-coded proteins, which in turn contains the ribonucleoprotein core. In addition to the genome and primer tRNAs, the core contains reverse transcriptase (Section 6-9), protease, integrase, and small RNA-binding and *gag* gene proteins encoded by the virus. The presumed functions of the several binding proteins in organizing the viral genome for packaging and replication may be modulated by phosphorylation.

Genome

Two identical linear RNA strands, each near 10 kb long, are joined near their 5′ ends by hydrogen-bonding. Like typical mRNAs, each strand is capped at the 5′ end and polyadenylated (about 200 ntd) at the 3′ end (Fig. 19-13). A short terminal repetition or redundancy of 13 to 250 ntd (R) is at each end, along with one unique sequence (U_5) near the 5′ end (80 to 200 ntd) and another (U_3) near the 3′ end (170 to 1200 ntd). Together these terminal sequences form the long terminal repeats (LTRs), organized as 5′-U_3-R-U_5-3′. The genomes fall into four classes (Fig. 19-14):

(1) The most common retroviral genome contains the *gag* gene, which encodes the structural proteins of the viral core, the *pol* enzyme genes, which

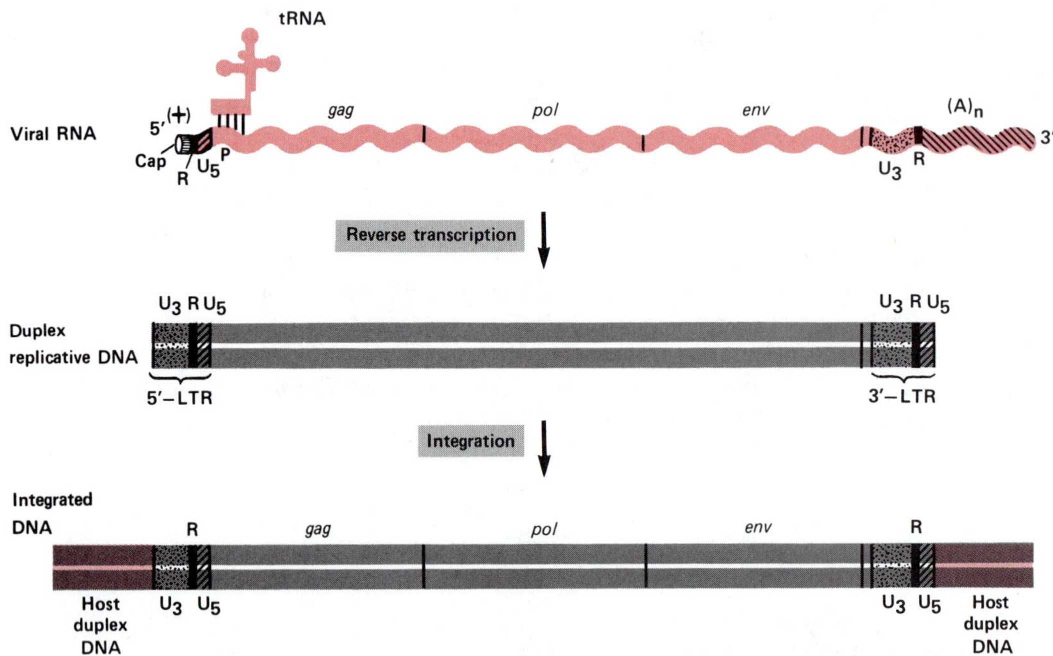

Figure 19-13
Structure of retroviral genomic RNA, duplex forms of replicative DNA, and integrated viral DNA. P = binding site for the primer tRNA. The integrated viral DNA and its unintegrated precursor are several hundred nucleotides longer than genomic RNA by virtue of having two LTRs. [Adapted from a figure from Professor JM Bishop and Dr IM Verma]

Simple

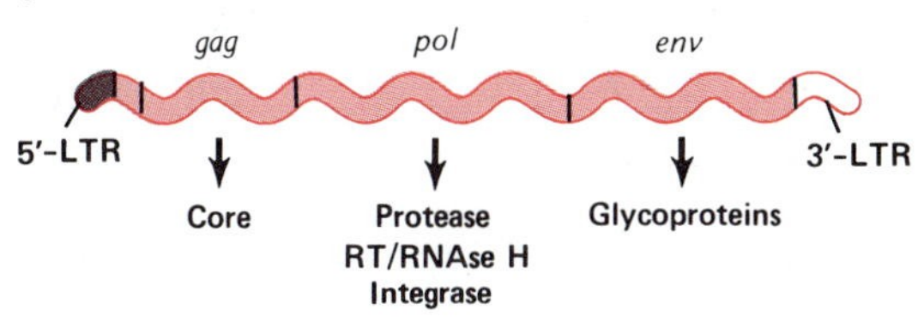

Natural recombinants

Sophisticated

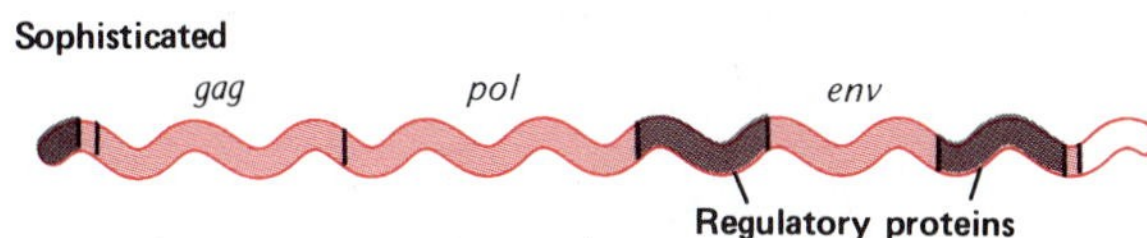

Redesigned as vectors for genes

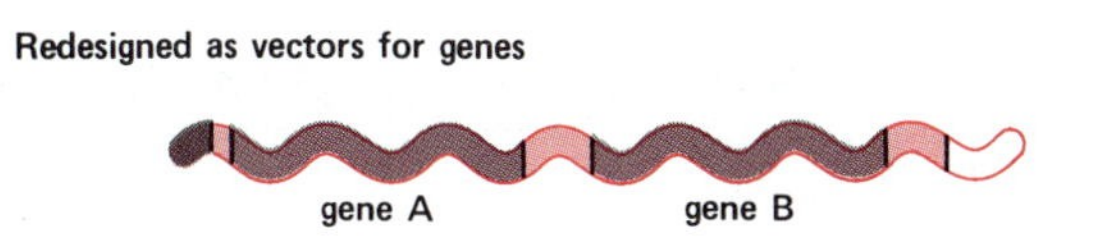

Figure 19-14
The four general types of retroviral genomes. Each diagram represents one molecule of viral RNA in simplified, schematic form with the 5′ end on the left. Regulatory proteins are usually encoded in overlapping reading frames. [Adapted from Varmus HE (1989) *Harvey Lect 83*:35]

include protease, reverse transcriptase, and integrase, and the *env* gene, which encodes the glycoproteins of the viral envelope.

(2) Less commonly, a retroviral genome has acquired part or all of the coding sequence of a cellular gene, usually one involved in growth control. This *protooncogene*, when aberrantly expressed or mutated, becomes an *oncogene*, capable of producing cancerous change. Fifty or more such viral oncogenes (v-*oncs*) have been characterized. The *src* oncogene of avian sarcoma (v-*src*), for example, encodes a 60-kDa phosphoprotein ($pp60^{v\text{-}src}$) with protein kinase activity resembling that of the c-*src* host cell counterpart ($pp60^{c\text{-}src}$). Because the oncogene usually displaces part of the normal viral genome (Fig. 19-14), the virus is defective and must be grown in the presence of a helper virus.

(3) A far more sophisticated class of retroviral genomes is endowed with additional genes that are reponsible for striking activations and repressions of gene expression; the genomes still remain replication-competent. Examples are the foamy viruses, human T-cell leukemia viruses (HTLVs), and the lentiviruses, including HIV.[92]

(4) Rapidly increasing in number is a class of genomes engineered as vectors for the integration of altered or foreign genes.[93]

Life Cycle

Interaction of viral glycoproteins with the host cell-surface receptor is followed by fusion of the viral envelope with the cell membrane (Fig. 19-15). Except for the specific interaction of HIV with the T-lymphocyte CD4 protein, the molecular details of viral entry are unknown. Uncoating of the viral envelope frees the complex nucleoprotein core, which becomes the substrate and enzyme source

92. Wong-Staal F (1990) in *Virology* (Fields BN et al., eds.) 2nd ed. Raven Press, New York, p. 1529.
93. Coffin J (1985) in *Molecular Biology of Tumor Viruses*; Vol. 2: *RNA Tumor Viruses* (Weiss R, Teich N, Varmus H, Coffin J, eds.). CSH, p. 17; Roux P, Jeanteur P, Piechaczyk M (1989) *PNAS 86*:9079.

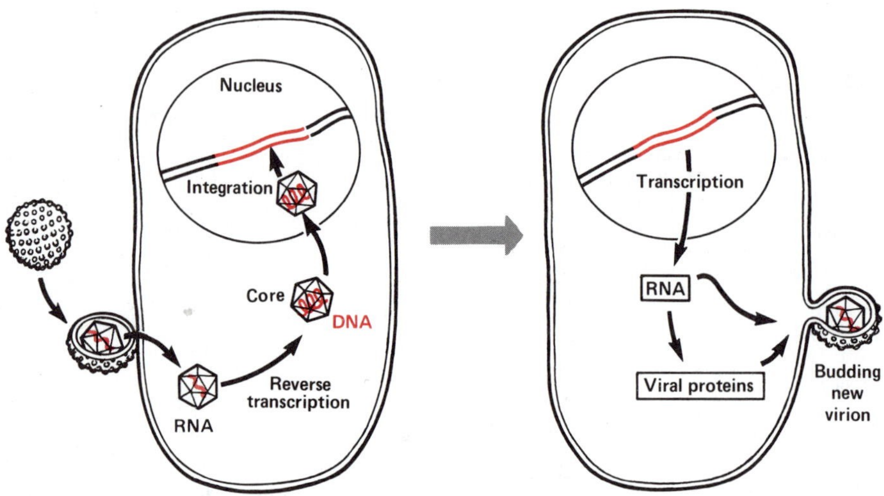

Figure 19-15
Life cycle of a retrovirus. Reverse transcription of entering RNA produces duplex DNA, which is integrated into the host chromosome. Transcription of the proviral genome produces RNAs for synthesis of virions and, in some cases, a transforming protein.

for subsequent conversion of the RNA genome to duplex DNA, for transport of the DNA to the nucleus, and for integration of the DNA into the chromosome. After the proviral genome has been transcribed back into mRNA and its genes expressed, the primary transcript is packaged in the viral core. Budding of the virions through the cell membrane is at sites enriched for viral glycoproteins.

Retroviruses use an effective strategy to simplify the daunting problem of assembling seven or more distinctive proteins in precise stoichiometry. The viral core is assembled from just two protein components, Gag protein and the Gag–Pol polyprotein. After budding, the proteins are proteolytically processed into their native forms by the viral protease. Thus, the individual core components need not possess the information to identify them for packaging and assembly.

Replication[94]

The succession of steps in the conversion of the retroviral genome to duplex DNA suitable for integration has been inferred from the intermediates produced in vivo, from the reverse transcription within purified permeabilized virions that produces infectious viral DNA, and from the actions of pure reverse transcriptase on defined templates. This scheme (Fig. 19-16) accounts for the genesis of LTRs, the replication of both ends of a linear template without sacrifice of genetic content, the initiation and propagation of both strands of viral DNA, and the role of the terminal redundancy (R) in viral DNA synthesis.

---

94. Varmus HE, Swanstrom R (1985) in *Molecular Biology of Tumor Viruses*; Vol 2: *RNA Tumor Viruses* (Weiss R, Teich N, Varmus H, Coffin J, eds.). CSH, p. 75; Varmus HE, Brown PO (1989) in *Mobile DNA* (Howe M, Berg D, eds.). Am Soc Microbiol, Washington, DC, p. 53.

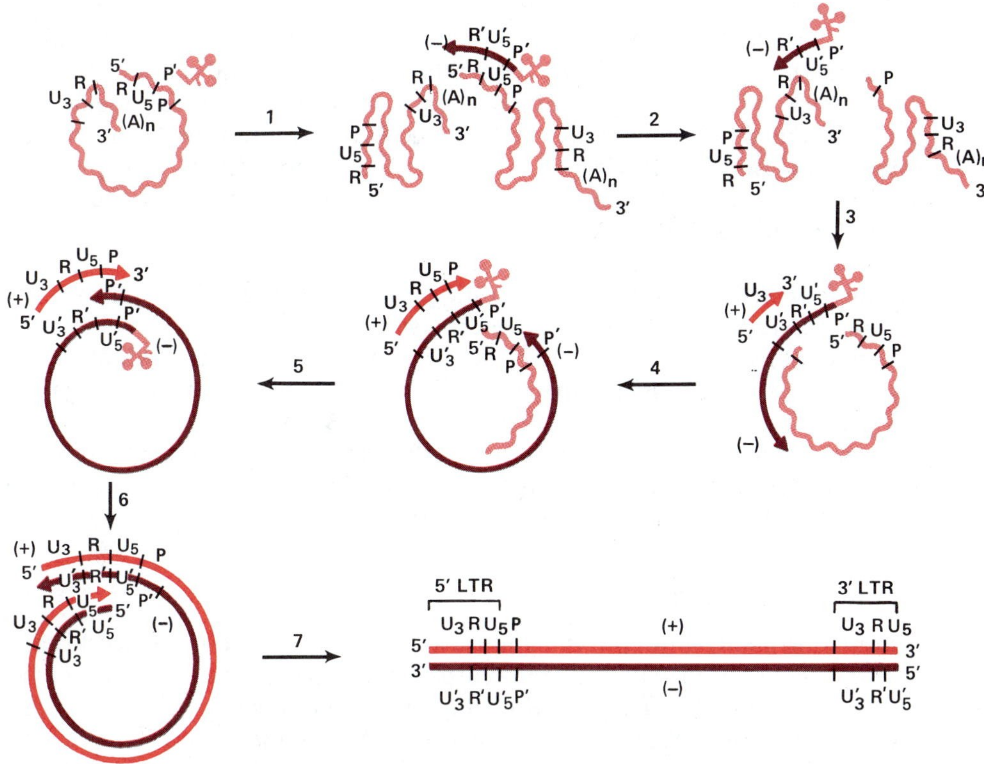

Figure 19-16
Model for reverse transcription of a viral RNA genome to a linear duplex replicative DNA form. Step 1: Initiation of minus strand (cDNA). Step 2: Minus strand transfer ("jump"). Step 3: Initiation of plus strand. Step 4: Further DNA synthesis and removal of viral RNA. Step 5: Plus-strand transfer. Step 6: Continued synthesis of plus-strand DNA and removal of the tRNA primer. Step 7: Completion of linear duplex DNA. Precisely how and when in these and preceding steps the genomic RNA is removed, presumably by the RNase H activity of the reverse transcriptase, is still uncertain. The final duplex DNA, having complete LTRs, is longer than the genomic RNA. Cloverleaf = tRNA, wavy line = viral genomic RNA, smooth line with arrowhead = DNA with direction of synthesis, letter and primed letter = complementary sequences. [Based on Bilboa E, Mitra SW, Goff S, Baltimore D (1979) *Cell 18*:93]

It remains uncertain whether the template for (−) DNA strand synthesis is furnished by one of the two copies of the (+) strand genome in the viral particle or whether both copies are required. If both copies participate, as claimed,[95] an *inter*strand transfer ("jump") would take place (as shown in Fig. 19-16), in contrast to an intrastrand jump with replication confined to a single template.

Synthesis of the (−) DNA Strand. The primer is a specific host tRNA with its 3′ stem region annealed to a site (P), about 18 ntd long, located next to U_5 on the

95. Panganiban AT, Fiore D (1988) *Science 241*:1064.

viral RNA. The particular tRNA incorporated in the virion depends on the virus: $tRNA^{trp}$ for AMV, $tRNA^{pro}$ for M-MuLV, $tRNA^{lys}$ for murine mammary tumor virus. The tRNA is extended by DNA synthesis to the very end of the 5′ R region, whereupon removal of the viral RNA template for this synthesis makes a 5′ single-stranded tail of the newly synthesized DNA. This tail can then make an interstrand jump to pair with the R region at the 3′ end of the other copy of the viral RNA. Elongation from this point generates a complete (−) DNA strand, hybridized to the viral RNA.

Synthesis of the (+) DNA Strand. Priming at a highly conserved polypurine tract is made possible by a remarkably specific cleavage of the viral RNA by RNase H. A persistent hybridized fragment is extended by the polymerase, which copies the (−) DNA strand template up to the tRNA attached to its 5′ end. Once again, a DNA tail is created by removal of the vestigial RNA. Pairing of the (−) strand by a jump to this tail (intrastrand) provides the remainder of the (−) strand as a template for synthesis of a complete (+) strand DNA and the final duplex DNA copy of the viral genome, bounded by LTRs.

Integration and the Proviral State[96]

Information about retroviral integration and the proviral state gleaned from in vivo studies has been expanded and sharpened by in vitro studies examining the process mediated by nucleoprotein complexes from infected cells.[97] The complexes contain the DNA duplex derived from a viral genome engineered to have in its LTR an *E. coli* gene for an amber suppressor tRNA. The target for integration of this viral DNA is a tandem array of phage λ DNAs with multiple amber mutations. A successful integration event is recorded when a plaque forms on nonsuppressor host cells, indicating the suppression of the λ amber mutations by an integrated retroviral genome.

Salient facts about integration (Fig. 19-17) are as follows:

(1) Two viral components are essential: the integrase encoded at the 3′ end of the *pol* gene, and short sequences at the ends of the LTRs.

(2) The integrase in the nucleoprotein complex recognizes a specific site at the ends of the viral (donor) DNA and removes one or two nucleotides from the 3′ end, exposing the 3′-OH that will be used for joining to the chromosomal target DNA.

(3) The viral 3′ ends are then covalently joined to the staggered 5′ ends resulting from cleavage of the target DNA molecules (e.g., λ DNA in the in vitro system). Inasmuch as this joining does not require an input of energy from ATP, the reaction must be tightly coupled to the target DNA cleavage.

(4) Following the initial joining step (strand transfer step), the gaps flanking the provirus are filled in, resulting in a short duplication of the target DNA, the length of which (4 to 6 bp) is determined by the virus species (e.g., 4 bp for MLV; 5

---

96. Varmus HE, Brown PO (1989) in *Mobile DNA* (Howe M, Berg D, eds.). Am Soc Microbiol, Washington DC, p. 53; Varmus HE (1988) *Science 240*:1427; Varmus HE (1989) *Harvey Lect 83*:35; Grandgenett DP, Mumm SR (1990) *Cell 60*:3.
97. Brown PO, Bowerman B, Varmus HE, Bishop JM (1989) *PNAS 86*:2525; Fujiwara T, Mizuuchi K (1988) *Cell 54*:497; Fujiwara T, Craigie R (1989) *PNAS 86*:3065; Bowerman B, Brown PO, Bishop JM, Varmus HE (1989) *Genes Dev 3*:469; Ellison V, Abrams H, Roe T, Lifson J, Brown P (1990) *J Virol 64*:2711.

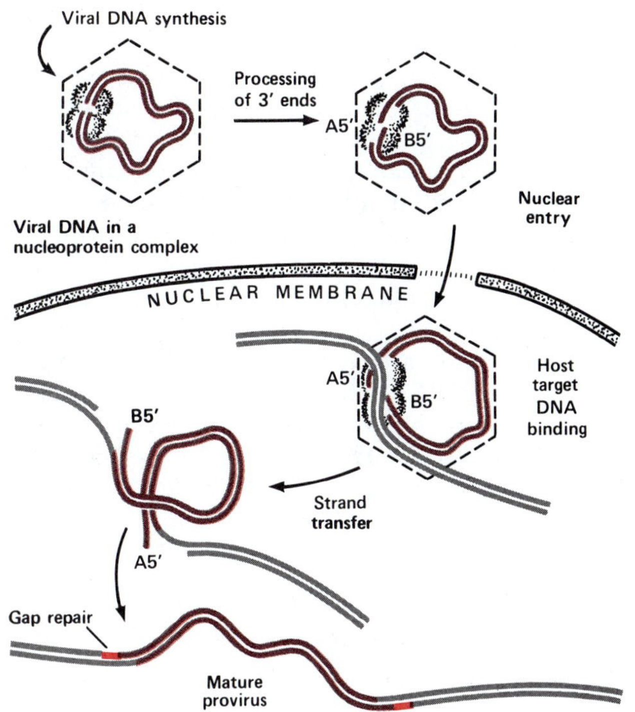

Figure 19-17
Mechanism of MLV integration. (Courtesy of Professor P Brown)

bp for HIV). The duplication reflects the separation of the staggered cleavage of the target DNA chains.

(5) Integration sites in the target DNA are numerous, but not completely random; there is no requirement for homology between the interacting sites on viral and target DNA.

(6) The integrated provirus has an invariant structure: it is joined to the target DNA with the loss of 1 or 2 bp from what had been the 3′ ends of the linear DNA;[98] proviruses of diverse strains all terminate with characteristic dinucleotides (5′-TpG . . . CpA-3′) and are flanked by the short duplication of the target DNA. The provirus resembles the DNA of yeast transposase,[99] and the integration reaction mechanism is remarkably similar to the transposition of phage Mu (Section 17-7). Yet there is no evidence that proviruses can transpose *directly* from one location in the host DNA to another.

(7) The nucleoprotein complex that carries out the integration reaction is a 160S particle with some associated capsid proteins. Resolution of the components[100] and their reconstitution into an active integration complex will answer questions about the detailed mechanism of this recombination event.

(8) Proviral integration generally does no harm to the host except as excess baggage. Infrequently, the virus may be inserted within a gene and inactivate it, may amplify expression of a nearby gene that leads to proliferation, or may initiate transduction of a host gene to create a viral oncogene (v-*onc*).

---

98. Roth MJ, Schwartzberg PL, Goff SP (1989) *Cell 58*:47.
99. Eichinger DJ, Boeke JD (1988) *Cell 54*:955; Eichinger DJ, Boeke JD (1990) *Genes Dev*, 4:324.
100. Katz RA, Merkel G, Kulkowsky J, Leis J, Skalka AM (1990) *Cell 63*:87.

Expression of the Provirus

The provirus appears to be a self-contained transcriptional unit, with a characteristic site for initiation of RNA synthesis, canonical signals that direct the splicing of RNA during the genesis of viral mRNAs, and a signal for polyadenylylation of the completed mRNAs; in the case of mouse mammary tumor virus, there is also a domain that mediates the stimulation of viral RNA synthesis by steroid hormones. Cellular DNA adjacent to the provirus may nevertheless modulate or even completely repress the synthesis of viral RNA. As a consequence, the site of integration on the cellular genome can be an important determinant of viral gene expression.

Translation of the *pol* gene depends on a novel (−1) ribosomal frameshift relative to that of the upstream *gag* gene.[101] To produce the catalytic amounts of Pol needed by the virus, compared to the higher levels of Gag protein, a (−1) frameshift once in 20 times produces a Gag–Pol fusion protein which can later be processed proteolytically to the constituent proteins.

19-8 Hepadna Viruses:[102] Hepatitis B Virus (HBV)

Hepadna (*hepatitis DNA*) viruses are distributed worldwide in humans; in some areas 20% of the population is actively infected.[103] The close association of hepatitis B virus (HBV) with chronic hepatitis that progresses to hepatocellular carcinomas in 2% to 4% of patients has focused attention on a better understanding of the replicative cycle of HBV. Finding viruses closely related to HBV in woodchucks, ground squirrels,[104] and ducks is the basis for assembling them in a new hepadna group.

A novel organization of a tiny DNA genome and a virus-associated DNA polymerase pose an interesting problem in replication. Replication of this small strange DNA virus in liver cells, and of the similar cauliflower mosaic virus in plants (Section 9), involves as an intermediate a genomic (+) strand RNA with terminal redundancies (repeats) packaged into a viral nucleocapsid. Similarities with retroviruses (Section 7) are striking: reverse transcription, integration, and terminal repeats in the linearized HBV DNA which resemble the retroviral LTR sequences.

Virion and Genome

The HBV virion, also known as the *Dane particle*, is a 42-nm spherical particle with a lipid envelope dominated by the hepatitis B surface antigen (HB_sAg). A 28-nm nucleocapsid core contains the core antigen (HB_cAg), the viral DNA, a DNA polymerase activity, and probably HB_sAg in a cryptic form. These features and polypeptides characterize the hepadna virus group. In persistent human

101. Jacks T, Power MD, Masiarz FR, Luciw PA, Barr PJ, Varmus HE (1988) *Nature 331*:280; Jacks T, Madhani HD, Masiarz FR, Varmus HE (1988) *Cell 55*:447.
102. Robinson WS (1990) in *Virology* (Fields BN et al., eds.) 2nd ed. Raven Press, New York, p. 2137; Hollinger FB (1990) in *Virology* (Fields BN et al., eds.) 2nd ed. Raven Press, New York, p. 2171; Ganem D, Varmus HE (1987) *ARB 56*:651; Hirsch RC, Lavine JE, Chang L, Varmus HE, Ganem D (1990) *Nature 344*:552.
103. Ayoola EA, Deinhardt F, Gust I (1988) *Bull WHO 66*:443.
104. Marion PL, Oshiro LS, Regnery DC, Scullard GH, Robinson WS (1980) *PNAS 77*:2941.

infections, high concentrations of incomplete viral forms, as well as intact particles, are found in the blood. These include small HB_sAg particles 16 to 25 nm wide and long HB_sAg filaments (22 × 500 nm).

Viral DNA, a partially duplex circle, has been cloned in *E. coli.*[105] The longer (−) strand of 3182 nucleotides is matched in part by a complementary (+) strand, of varied lengths from 1700 to 2800 bases. The long strand is nicked at a point 300 ntd from the 5′ end of the shorter strand. The duplex is held in a circular conformation by base pairing between the 5′ overlapping ends of the nearly complete (−) strand and the half-length (+) strand (Fig. 19-18). The 5′ terminus of the long strand has a covalently attached protein;[106] the 5′ end of the short strand has no protein attached but, rather, a capped oligoribonucleotide.

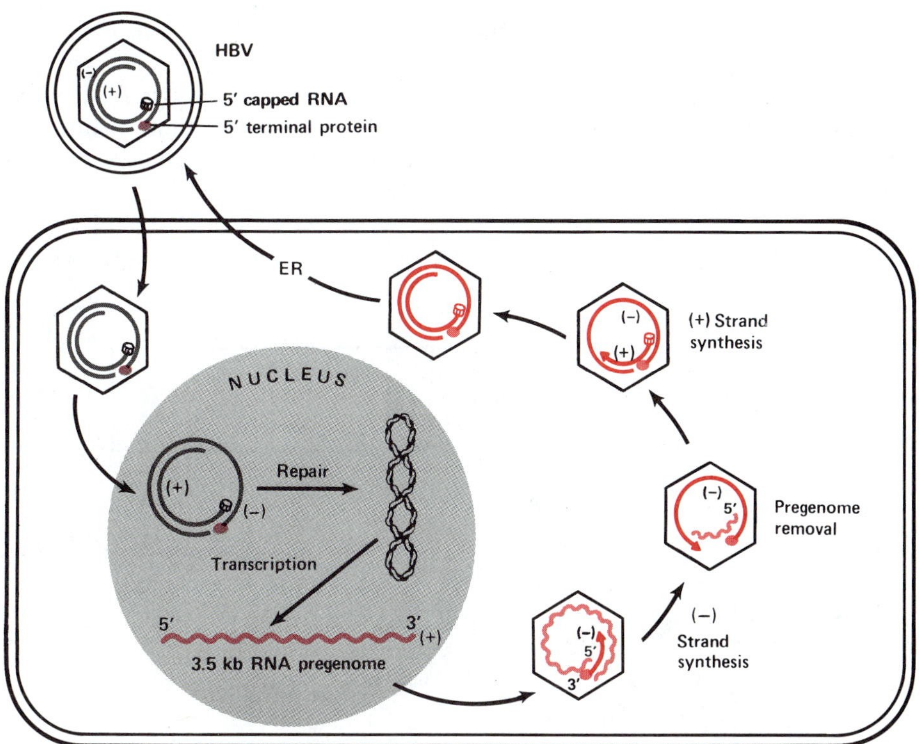

Figure 19-18
Replication cycle of hepatitis B virus (HBV). After entry into the cell and uncoating, the viral genome is transported to the nucleus where it is converted into a supercoiled circular molecule by cellular repair mechanisms. This supercoiled DNA is transcribed into unspliced, capped, polyadenylated mRNA species which are used for the production of viral proteins. In addition, the terminally redundant 3.5-kb RNA functions as an RNA pregenome which is packaged into viral cores, reverse transcribed, and exported by budding through the endoplasmic reticulum (ER). (Courtesy of Dr H Schaller)

---

105. Ono Y, Onda H, Sasada R, Igarishi K, Sugino Y, Nishioka K (1983) *NAR 11*:1747; Fujiyama A, Miyanohara A, Nozaki C, Yoneyama T, Ohtomo N, Matsubara K (1983) *NAR 11*:4601.
106. Bosch, V, Bartenschlager R, Radziwill G, Schaller H (1988) *Virology 166*:475.

Replication[107]

The initial replicative event is completion of the (+) DNA strand by the endogenous viral reverse transcriptase to produce a closed circular form. (This reaction in the particle in vitro provides a useful assay for the enzyme and a marker for the presence of HBV.) Ensuing transcription (probably by the host RNA polymerase II) of the long (−) viral strand produces a greater-than-genome-length (3.5-kb) mRNA (Fig. 19-19). The pregenome (+) mRNA is encapsidated and becomes the template for reverse transcription to form the (−) DNA viral strand. Subsequent synthesis of the (+) DNA strand stops approximately halfway in achieving the mature viral form of the genome. The DNA polymerase and RNase H domains of the hepadna reverse transcriptase resemble those of the retroviruses. Neither the size nor the replicative function of the protein covalently bound at the 5′ end of the (−) strand has been determined.

Despite many similarities to retroviral replication, there are distinctive features of hepadna replication (Fig. 19-19): (1) The reverse transcriptase action takes place in the HBV particle while it develops in the infected cell, whereas in retroviruses, it occurs in the infecting *mature* virus (with its complete RNA genome) after the virus has entered the new host cell. (2) The HBV genome contains a viral protein covalently linked to the 5′ end of its (−) strand, suggesting a priming role of the protein in place of a tRNA; in this event, priming of hepadna replication would resemble that of adenoviruses, phage ϕ29, and the

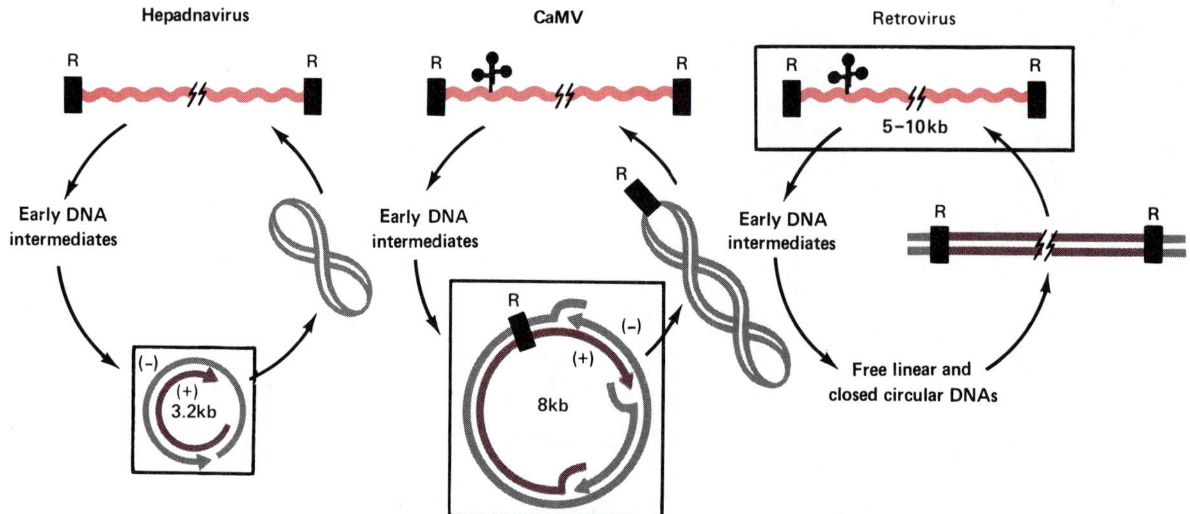

Figure 19-19
Comparison of the central features of the life cycles of three viruses that use reverse transcriptase: hepadnaviruses, cauliflower mosaic virus (CaMV), and retroviruses. The forms of the viral genome packaged into extracellular particles are boxed. R = terminal repeats; cloverleaf = tRNA. (From Ganem D, Varmus HE (1987) *ARB 56*:651]

107. Ganem D, Varmus HE (1987) *ARB 56*:651; Hirsch RC, Lavine JE, Chang L, Varmus HE, Ganem D (1990) *Nature 344*:552; Will H, Reiser W, Weimer T, Pfaff E, Büscher M, Sprengel R, Cattaneo R, Schaller H (1987) *J Virol 61*:904; Radziwill G, Schaller H, Bosch V (1988) *Virology 163*:123.

picorna viruses. (3) Integration of viral DNA into the host chromosome is a regular part of the life cycle of retroviruses but not of HBV. (4) Expression of retroviral genes is from the integrated, proviral template; in contrast, the template for HBV transcription is episomal. Cauliflower mosaic virus replication (Fig. 19-19) combines some features of both retroviral and HBV replication.

Many obscure aspects of the biology of the life cycle may be clarified in future with studies of HBV and the duck hepadna virus in differentiated human hepatoma (hepatoblastoma) cell lines that can express the HBV genes and produce infectious virus with cloned HBV DNA.

19-9 Plant[108] and Insect[109] Viruses

Among the numerous DNA viruses of plants and insects, some have attracted attention by their potential as gene vectors. As autonomously replicating, multicopy vectors, they can vie with bacterial and yeast plasmids. In the plant world, they offer an alternative to agents that integrate into the host genome, as in the case of the T1 plasmid of *Agrobacterium tumefaciens.* Of comparable interest are unique features of their life cycles and replication. Best characterized of the plant viruses are the cauliflower mosaic virus (a member of the caulimovirus group) and the geminiviruses; of the insect viruses, the baculovirus of silkworms is the most intensively studied.

Cauliflower Mosaic Virus (CaMV)[110]

The small isometric viruses of the caulimovirus group have fairly restricted host ranges, but often can be experimentally transmitted to other species. CaMV, for example, can also infect other *Brassica* species (e.g., turnip) and some *Datura* and *Nicotiana* species. In nature, CaMV is spread by aphids, but it is easily transmitted mechanically by rubbing it on the leaves of susceptible plants. Disease in infected plants varies in severity from extreme stunting and necrosis of leaves to few symptoms without effect on growth.

Virion and Genome. The icosahedral particle has a diameter of about 50 nm and is coated by phosphorylated and glycosylated capsid proteins. Internal viral proteins have DNA polymerase and ssDNA-binding activities.

The CaMV genome (Fig. 19-20) is duplex DNA, near 8 kb long, that contains three discontinuities, as judged by sensitivity to S1 nuclease. Regarded initially as nicks or gaps, these discontinuities are actually triple-stranded overlaps; they are present at a single position in the (−) strand and at two fixed locations in the (+) strand. Complete copies of the genome cloned in *E. coli* do not contain these remarkable structures and are fully infectious. Yet after passage through plants, the progeny acquire the three discontinuities in the precise locations.

Of the eight open reading frames, functional assignments have been suggested for six: a protein kinase, an aphid-acquisition factor, a duplex DNA-binding protein, the capsid protein, a viroplasm (inclusion-body) protein responsible for

---

108. Hohn T, Schell J (eds.) (1987) *Plant DNA Infectious Agents.* Springer-Verlag, New York.
109. Miller LK (1988) *Ann Rev Microbiol 42*:177; Maeda S (1989) *Ann Rev Entomol 34*:351.
110. Gronenborn B (1987) in *Plant DNA Infectious Agents* (Hohn T, Schell J, eds.). Springer-Verlag, New York, p. 1.

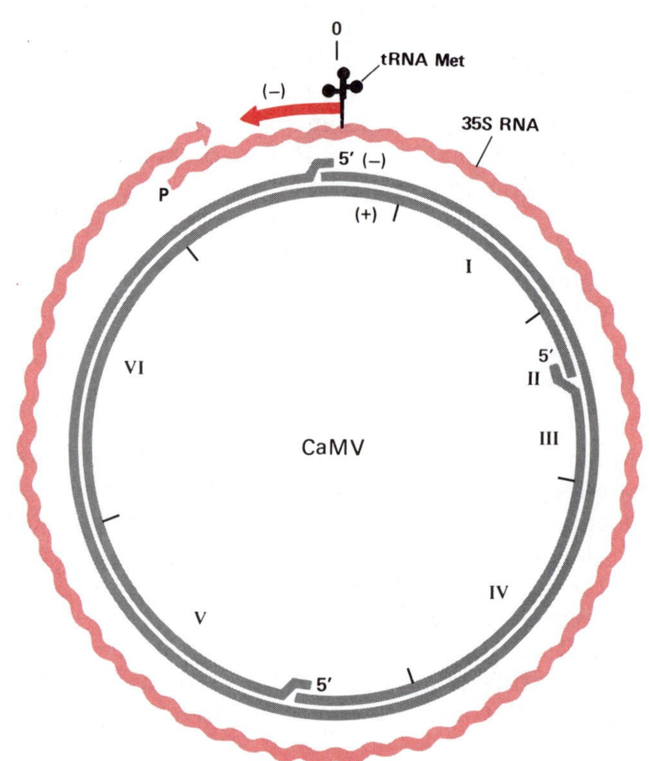

Figure 19-20
The genome organization of cauliflower mosaic virus (CaMV). The genes (open reading frames, orf's) are numbered I through VI, and the 5′ ends of the three short overlap structures (discontinuities) are indicated. The single discontinuity of the (−) DNA strand is defined as the zero point of the map; only the discontinuity in orf V is essential for replication. The synthesis, counterclockwise, of the (−) DNA strand by reverse transcriptase is primed by $tRNA^{Met}$ (cloverleaf) hybridizing to a 35S transcript from a promoter (P) located just after orf VI in a large intergenic region. [From Gronenborn B (1987) in *Plant DNA Infectious Agents* (Hohn T, Schell J, eds.). Springer-Verlag, New York, p. 1]

the cytoplasmic sites of replication, and a reverse transcriptase with considerable amino acid homology to the hepadna and retroviral enzymes. In addition, an RNA-dependent DNA polymerase activity has been partially purified from infected turnip leaves and from bacterial and yeast strains engineered to express the CaMV reverse transcriptase gene.

Replication. The remarkable features of the CaMV genome and the reverse transcriptase step in hepadna replication sparked the revelation of a similar mechanism for CaMV. Replication in the nucleus and cytoplasm of infected plant cells follows this sequence of steps:

(1) The discontinuities are repaired and a supercoiled form is synthesized, probably in the nucleus.

(2) Transcription of the entire (−) viral strand in the nucleus produces the key replicative intermediate, a terminally redundant 35S RNA. Expression of the minichromosome, a key phase in the multiplication cycle, is regulated differentially in various host tissues.[111]

(3) Synthesis of the progeny (−) DNA strand in the cytoplasm, directed by the 35S RNA template, is primed by a methionine initiator-tRNA. In the region of terminal redundancy of the 35S RNA, the template is switched from the 5′ end to the 3′ end of the RNA, and synthesis then proceeds until a full-length cDNA copy is made.

111. Covey SN, Turner DS, Lucy AP, Saunders K (1990) *PNAS 87*:1633.

(4) Initiation of the (+) strand to produce the progeny duplex occurs at a site nearly 180° from the tRNA primer site. The mechanism for this initiation is unknown.

The asymmetric mode of replication suggests possible ways in which the discontinuities (overlaps) might be formed by strand displacement at sites of termination of both (−) and (+) DNA strands. Insights gained from studies of CaMV may also illuminate other retrovirus-like replicative cycles.

Geminiviruses[112]

These small, single-stranded, circular viruses are studied because of the simplicity of their genomes and their potential as gene vectors in a wide range of plant hosts. The favored member of this group is the *cassava latent virus* (CLV), transmitted by white flies to dicotyledonous plants and possessing a *duplicate (twinned) genome*. Representing another class is the *maize streak virus* (MSV), which is transmitted by leafhoppers to monocotyledonous plants (rice, wheat, maize, sorghum, millet, sugar cane) and probably possesses only a *single-copy genome*.

The twin circles of CLV are 2779 and 2724 ntd long. Although the larger is self-replicative, both are needed for infection; the smaller unit may be important for systemic spread of the virus. Twelve open reading frames have been identified in the bipartite genome, along with intergenic regions that occupy 16% of the larger circle and 39% of the smaller. A stem–loop structure, containing the 9-mer TAATATTAC, is shared widely among the geminiviruses and resembles the primosome assembly site of the single-stranded circular phage ϕX174 (Section 17-4).

The genome of MSV resembles the larger CLV circle. Except for the coat protein, none of the gene products has been characterized. As for the replication process, little is known except for its intranuclear location and the almost certain involvement of double-stranded replicative forms as intermediates.

Baculoviruses[113]

The silkworm nuclear polyhedrosis virus belongs to a group of baculoviruses which infect the pupae of the silkworm (*Bombyx mori*). Cell lines of a lepidopteran (*Spodoptera frugiperda*) also support the growth of a nuclear polyhedrosis virus, *Autographa californica*. The capacity of these viruses to produce astonishing quantities of a foreign protein in permissive cells expressed under the powerful polyhedrin promoter has made them a popular vector in genetic engineering.[114]

The mechanism of replication of the circular, double-stranded genome of 120 to 130 kb is largely unknown beyond the induction of an encoded DNA polymerase, the domains of which resemble those of other eukaryotic viral and cellular polymerases (Section 5-11).[115] The polymerase is sensitive to aphidico-

112. Davies JW, Townsend R, Stanley J (1987) in *Plant DNA Infectious Agents* (Hohn T, Schell J, eds.). Springer-Verlag, New York, p. 31.
113. Varmus HE, Brown PO (1989) in *Mobile DNA* (Howe M, Berg D, eds.). Am Soc Microbiol, Washington DC, p. 53.
114. Varmus HE, Brown PO (1989) in *Mobile DNA* (Howe M, Berg D, eds.). Am Soc Microbiol, Washington, DC, p. 53; Iatrou K, Meidinger RG, Goldsmith MR (1989) *PNAS 86*:9129.
115. Tomalski MD, Wu J, Miller LK (1988) *Virology 167*:591.

lin and possesses a 3′ → 5′ exonuclease,[116] as in the host DNA polymerase δ. A further relationship to the replicative pol δ is that a gene in *A. californica* encodes a protein similar to the proliferating cell nuclear antigen (PCNA),[117] the accessory protein of pol δ. The baculovirus PCNA has a 42% amino acid identity with rat PCNA; a mutant lacking the intact gene shows a delay in DNA replication. While partially dispensable for a proliferating cell culture, the PCNA may be essential for quiescent cells, in which the virus is known to establish an S-phase environment and thereby replicate efficiently when the host DNA cannot.

116. Mikhailov VS, Marlyev KA, Ataeva JO, Kullyer PK, Atrazhev AM (1986) *NAR* 14:3841.
117. O'Reilly DR, Crawford AM, Miller LK (1989) *Nature* 337:606.

CHAPTER 20

Regulation of Chromosomal Replication and Cell Division

20-1 Regulation of Replication: Multiple Types of Control

Replication of the chromosome is the central event in the cell division process. Decisions of when and with what frequency to replicate the genome are made in response to a variety of signals. To identify these signals and understand how they are communicated to the replication machinery is one of the frontiers of study in replication.

During steady-state growth, the following distinct types of control are needed:

(1) *Timing of initiation within the cell cycle.* This is needed to coordinate replication with the other events in cell division.

(2) *Coordination of initiation at multiple origins.* In bacteria, all the replication origins in the cell fire nearly synchronously, whereas in eukaryotic cells initiation at the multiple origins follows a distinct temporal program.

(3) *Inhibition of refiring at an origin within a single cell cycle.* This ensures that the genome is duplicated only once. The inhibition is apparently overridden under conditions of gene amplification.

In response to changing conditions, further types of control may become necessary:

(4) *Activation of replication when growth is favorable.* This response is signalled by an abundance of nutrients for a bacterial culture or by the presence of growth factors for eukaryotic cells.

(5) *Inhibition of replication when conditions are unfavorable.* Inhibition occurs, for example, in response to DNA damage, the lack of an essential

nutrient, or the attainment of a terminally differentiated state, as in bacterial spores or cells of a specialized eukaryotic tissue.

Finally, the copy number of extrachromosomal elements is generally limited to a specific intracellular value. *Copy number control* is among the best understood examples of regulation of replication. The mechanisms are, however, distinct from those used to regulate the cellular chromosomes, and are discussed in Chapter 18.

The control of chromosome replication is generally imposed at the stages of initiation. As with other extensive biochemical pathways, the step at which the commitment is made is the most likely target for its regulation. Knowledge of the molecular details of the "on/off switch" for initiation is essential to an understanding of its operation in response to the many signals that may regulate it. Whereas the enzymatic analysis of viral replicative systems has been an indispensable guide to the cellular mechanisms of replication, the uncontrolled nature of these viral cycles limits their influence as regulatory models. Autonomous plasmids with rigid copy number control (Chapter 18), as well as those plasmids engineered to depend on a cellular chromosomal origin, in effect minichromosomes, have proved exceedingly instructive in studies of regulation of the cellular genomes.

20-2 The Bacterial Cell Cycle[1]

Bacteria grow and divide at different rates, depending on culture conditions, while the rate of growth of the DNA chains is relatively inflexible. For rapid division, more replication forks are introduced; for slow division, intervals without replication occur. Thus, initiation is the critical step in the control of chromosomal replication.

Since the rate of chain elongation is essentially fixed, the time elapsed between initiation at the single replication origin and completion of chromosomal replication is invariable.[2] In *E. coli** strains at 37°, growing with a doubling time of 20 to 60 minutes, the replication time is about 40 minutes. (At doubling times of 2 hours or longer, the synthesis period is considerably extended.) The time consumed by cell division following termination of a round of replication is also quite constant, about 20 minutes at these moderate to fast growth rates.

The major variable in the cycle is the time required to initiate a new round of replication. Initiation can occur when previously started replication forks have progressed only halfway around the chromosome, or there may be a long delay after cell division before the start of a new round. Several cycle patterns are shown as examples (Fig. 20-1). The designations are as follows: I = *interinitiation time*, the period between initiations. Because initiation limits the

---

* Most physiologic experiments on cell division have used B/r strains, whereas genetic studies usually use K12 strains. The cell division values here are for B/r at 37°; some of the experiments in Section 3 which employ K12 have somewhat different values.

1. Helmstetter CE (1987) in *Escherichia Coli and Salmonella Typhimurium* (Neidhardt FC, ed.). Am Soc Microbiol, Washington DC, Vol. 2 p. 1594; Cooper S, Helmstetter CE (1968) *JMB 31*:519; Pritchard RH (1974) *Philos Trans R S Lond [Biol] 267*:303; Donachie WD, Begg KJ, Vicente M (1976) *Nature 264*:328.
2. Cooper S, Helmstetter CE (1968) *JMB 31*:519; Pritchard RH (1974) *Philos Trans R S Lond [Biol] 267*:303; Donachie WD, Begg KJ, Vicente M (1976) *Nature 264*:328.

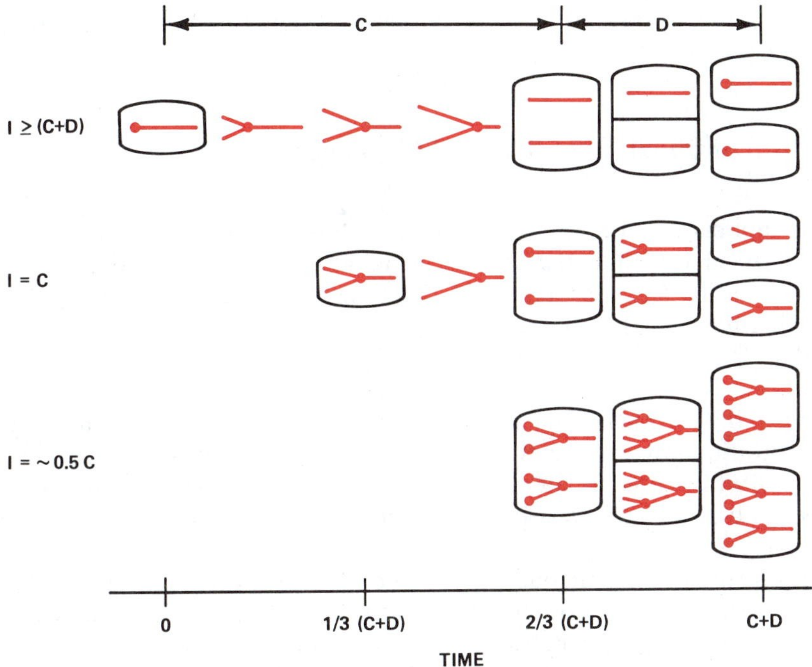

Figure 20-1
Bacterial cell cycle, showing chromosome replication and cell division as a function of the initiation interval, I, the time between initiations. C is chain growth time; D, division time. The circular chromosome is shown in linear form. [After Helmstetter C, Cooper S, Pierucci O, Revelas E (1968) *CSHS 33*:809]

cycle, I is equal to the doubling time during balanced growth. C = *chain growth time*. D = *division time*, the period required after chromosomal replication is completed, before division occurs. *Prereplication time*, the period after cell division and prior to initiation, is only present at slow growth rates.

$I \geq (C + D)$ When doubling of the cell mass requires longer than 60 minutes, for instance 2 hours (a slow growth rate), fresh initiations are made well after the daughter cells have separated.

$I = C$ If the cell mass doubles in 40 minutes (a moderate growth rate), fresh initiations are made when the previously started chains are finished but before cell division is under way.

$I < 0.5\ C$ When the doubling of cell mass requires less time than the C period, such as 20 minutes (a rapid growth rate), initiation occurs when the previously started chains are only about half finished. This replication is *multiforked*, or *dichotomous*.

These examples illustrate that initiation of replication is independent of cell division, but not the reverse. It is commonly assumed that the chromosome must be completed in order to trigger division (Section 5). Yet, despite their usual synchrony, the replication and cell division cycles may not be directly coupled but, rather, each may respond to various signals, some shared and others specific (Section 5).

20-3 Control of Initiation of Bacterial Chromosomes[3]

Initiation of replication in *E. coli* occurs at a unique sequence, *oriC*. Much is known about the factors that operate at this origin to assemble replication forks for duplication of the chromosomal DNA (Sections 16-2 and 16-3). During balanced growth, initiation is regulated at three levels to couple replication to cell growth and division.

(1) *Timing in the cell cycle.* The replication rate is related to cell mass in such a way that initiation always occurs at a fixed ratio of origins to cell mass.[4] This *initiation mass*, while constant for all growth rates, is achieved at different times with respect to cell division, resulting in a specific *timing* of initiation in the division cycle.

(2) *Synchrony of initiation at multiple copies of oriC.* Dichotomous replication in rapidly growing cells (Section 2) results in multiple copies of *oriC* per cell; these origins fire essentially synchronously under normal conditions.

(3) *Inhibition of immediate reinitiation.* Each origin fires only once per cell cycle. The fact that several origins present in a cell may all fire during a single initiation period, but that each fires only once, suggests the presence of a *cis*-acting mechanism to inhibit immediate reinitiation.

In addition to these three distinct controls that operate under balanced growth conditions, other mechanisms are probably needed to respond to an environmental change. For example, inhibition of initiation occurs after DNA damage to allow time for repair, and an increase in initiation frequency accompanies improved growth conditions and must be controlled to prevent runaway replication.

The DNA sequences required for the proper regulation of initiation are largely included within *oriC*, inasmuch as plasmids with the minimal *oriC* initiate synchronously with the chromosome and replicate only once per cell cycle.[5] The effects of specific mutations, inhibitors, and changes in the levels of replication proteins, both in vivo and in vitro, help to identify the *trans*-acting factors that control initiation.

Biosynthetic and Metabolic Processes Required for Initiation

Disruption of a process essential for initiation allows the started replication complexes to complete a cycle, but prevents the onset of a new round.

Protein Synthesis. When a growing culture is deprived of an essential amino acid, protein synthesis stops promptly; increases in RNA and cell mass cease soon afterwards[6] (Fig. 20-2). However, DNA synthesis continues until the level

---

3. von Meyenburg K, Hansen FG (1987) in *Escherichia Coli and Salmonella Typhimurium* (Neidhardt FC, ed.). Am Soc Microbiol, Washington, DC, Vol. 2 p. 1555.
4. Donachie WD (1968) *Nature 219*:1077.
5. Leonard AC, Helmstetter CE (1986) *PNAS 83*:5101; Koppes LJH, von Meyenburg K (1987) *J Bact 169*:430.
6. Schaechter M (1961) *CSHS 26*:53; Maaløe O, Hanawalt PC (1961) *JMB 3*:144.

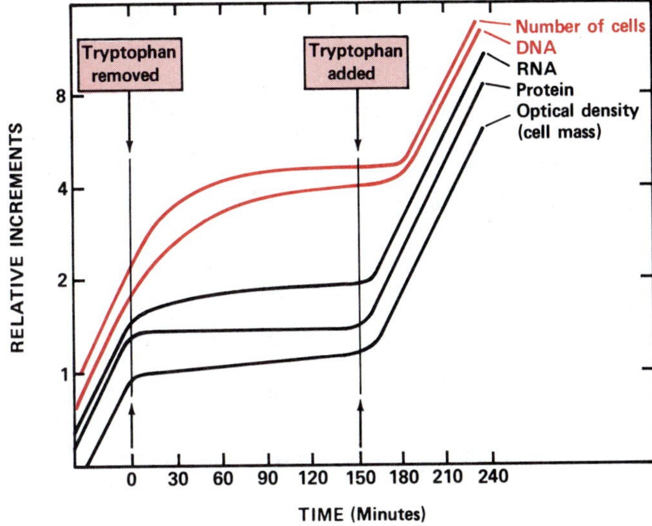

Figure 20-2
Influence of tryptophan starvation on synthesis of DNA and other macromolecules. DNA synthesis persists in starved cells, but a new round of replication is not initiated until tryptophan is restored and an interval of protein synthesis has supervened.

nearly doubles, enabling cell division to ensue upon completion of chromosomal replication (Section 5). After restoration of the missing nutrient, the resumption of protein synthesis, RNA synthesis, and cell growth is immediate, but DNA synthesis starts only after a substantial lag; the increase in cell number trails even further behind. The addition and subsequent removal of an inhibitor of protein synthesis, such as chloramphenicol, has the same effect as deprivation and restoration of an amino acid. Thus, essential substances are consumed in initiation and must be replenished by protein synthesis.

RNA Synthesis. Inactivation of RNA polymerase depresses initiation independently from its effect on protein synthesis, as indicated by the kinetics of inhibition by rifampicin compared to that by chloramphenicol[7] (Fig. 20-3). Whether transcription must be within the origin region or not is unclear; a unique RNA

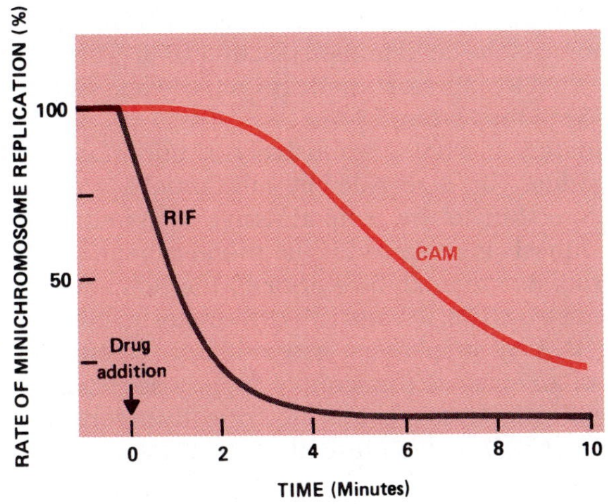

Figure 20-3
Rate of *oriC* plasmid replication after inhibition of RNA synthesis by rifampicin (RIF) or inhibition of protein synthesis by chloramphenicol (CAM). [From von Meyenburg K, Hansen FG, Riise E, Bermans HEN, Meijer M (1979) *CSHS 43*:124]

7. Lark KG (1972) *JMB 64*:47; Messer W (1972) *J Bact 112*:7; Zyskind JW, Deen LT, Smith DW (1977) *J Bact 129*:1466.

essential for initiation has not been identified in spite of the implied role of transcription.

Replication of *oriC* plasmids is *transiently* stimulated by a change to a richer medium (a "nutritional shift up") and by several treatments that inhibit protein synthesis.[8] Initiation in *dnaA*^{ts} mutants at semipermissive temperatures is induced by the same conditions. Transcription of stable RNA is specifically stimulated, an indication that a stringently regulated RNA may be involved in initiation.

DNA Methylation.[9] After passage of a replication fork, DNA is hemimethylated until the DNA adenine methyltransferase (Dam) modifies the newly synthesized strand. Hemimethylation of *oriC* blocks initiation in vivo; the time required to methylate the origin region after its replication is about ten times longer than that needed for most other chromosomal locations. The hemimethylated state thus contributes to the refractory period between sequential rounds of initiation (see below).

Mutations That Specifically Affect Initiation

Several specific proteins are required for initiation (Section 16-3) beyond those needed for protein synthesis, RNA synthesis, and DNA methylation. The dnaA protein is of paramount importance. In addition, genetic and biochemical evidence supports roles for the dnaC and dnaB proteins and gyrase (Table 20-1). Also, certain RNA polymerase mutations (in *rpoB* and *rpoC*) specifically alter chromosomal copy number, but it remains unclear whether the effect of these mutations is direct or part of a global change in gene expression. Mutations in *rpoB* and deletion of *topA* (the gene for topoisomerase I) suppress *dnaA*^{ts} mutations, indicating an influence of transcription and DNA superhelicity on dnaA protein activity in vivo. Roles in initiation for the *dnaK*, *dnaJ*, *dnaT*, and *dnaY* genes are indicated by their "slow-stop" phenotypes (Section 15-4), but their particular contributions are not understood.

Role of dnaA Protein. Three essential initiation functions are carried out by dnaA protein at *oriC*: (1) recognition and binding of the origin, (2) melting of the duplex DNA strands, and (3) loading of the dnaB protein (Section 16-3). Through these steps, dnaB protein is oriented to express its helicase activity and enable primase to prime DNA chains (Sections 11-2 and 8-4).

The role of dnaA during initiation in vivo is exclusively in initiation; the protein is not involved in elongation. This is revealed by substantial DNA synthesis in a *dnaA*^{ts} strain despite a shift to the nonpermissive temperature.[10] However, this residual DNA synthesis is abolished if the round of replication has been completed at the permissive temperature prior to the temperature shift. Furthermore, integrated episomes (P2, R1, and F)[11] correct the inability of a *dnaA*^{ts} mutant to synthesize DNA by supplying a new initiator protein and origin. In contrast, integration of an episome (integrative suppression) cannot relieve mutations in genes for replication enzymes that are critically involved in elongation, such as dnaB protein or DNA polymerase III.

---

8. Orr E, Meacock PA, Pritchard RH (1978) in *DNA Synthesis, NATO*, p. 85.
9. Campbell JL, Kleckner N (1990) *Cell 62*:967; Russell DW, Zinder ND (1987) *Cell 50*:1071.
10. Wechsler JA, Gross JD (1971) *MGG 113*:273.
11. Lindahl G, Hirota Y, Jacob F (1971) *PNAS 68*:2407; Iida S (1977) *MGG 155*:153.

Table 20-1
Genes involved in initiation at *oriC*

Gene	Comments
dnaA	mutants are defective only in initiation; action is required very early
dnaB	most mutants are defective in chain elongation; however, *dnaB252* is altered only in initiation, possibly defective in forming prepriming complex at *oriC*
dnaC	except for *dnaC1* and *dnaC7*, mutants are defective in initiation
dnaK	*dnaK111* allele is defective in initiation
dnaT	slow-stop phenotype; involved in initiation or termination
gyrB	gyrase is involved in fork movement but some *gyrB* alleles are defective specifically in initiation; hence requirements for this enzyme in initiation and elongation are distinguishable
hupA, hupB	encode the two subunits of HU protein. Partially defective in *oriC* plasmid replication; chromosomal replication is less affected
rnh	required to maintain specificity for initiation at *oriC* and dependence on *dnaA*; mutants show constitutive stable DNA replication
rpoB, topA	mutations suppress *dnaA*^{ts} mutations, suggesting that changes in transcription or DNA superhelicity may help partially defective dnaA protein
trx	encodes thioredoxin; mutations suppress *dnaA*^{ts} mutations

Role of dnaC Protein.[12] Mutations in *dnaC* in most instances cause only a delay in the cessation of replication at the restrictive temperature, indicating that, as with *dnaA* mutants, the defect is in chromosome initiation. However, some alleles fail in replication more promptly, implying a role for dnaC protein in chain elongation as well. During initiation at *oriC* in vitro, dnaC protein functions exclusively in initiation by complexing with dnaB protein as a prerequisite to the assembly of the dnaA·dnaB prepriming complex at *oriC* (Section 16-3). The requirement for dnaC protein in the assembly of the primosome on ϕX174 and ColE1 templates, however, further indicates a function beyond *oriC* initiation (Section 8-4).

Relationship between dnaA Protein Level and Timing of Initiation within the Cell Cycle[13]

The effects of varied intracellular levels of dnaA protein and *dnaA* mutants[14] indicate that dnaA protein is rate-limiting for initiation under normal growth conditions. Induction of *dnaA* expression from a *lac* or λP_L promoter (two promoters commonly used to control gene expression experimentally) causes a rapid increase in initiation at *oriC* on the chromosome and on *oriC* plasmids.[15]

Systematic alteration of the steady-state level of dnaA protein from below normal to five times normal (by expression from the *lac* promoter) reveals multiple effects on cell growth and DNA replication.[16] At low dnaA levels, growth is slow and the ratio of DNA to cell mass is unusually small, indicating infrequent initiation. At higher levels, cellular growth rate, replication time,

12. Wechsler JA (1978) *DNA Synthesis, NATO*, p. 49.
13. Løbner-Olesen A, Skarstad K, Hansen FG, von Meyenburg K, Boye E (1989) *Cell 57*:881.
14. Hansen FG, Rasmussen KV (1977) *MGG 155*:219; Kellenberger-Gujer G, Podhajska AJ, Caro L (1978) *MGG 162*:9; Lycett GW, Orr E, Pritchard RH (1980) *MGG 178*:329; Tippe-Schindler R, Zahn G, Messer W (1979) *MGG 168*:185.
15. Atlung T, Løbner-Olesen A, Hansen FG (1987) *MGG 206*:51; Pierucci O, Rickert M, Helmstetter CE (1989) *J Bact 171*:3760; Skarstad K, Løbner-Olesen A, Atlung T, von Meyenburg K, Boye E (1989) *MGG 218*:50.
16. Løbner-Olesen A, Skarstad K, Hansen FG, von Meyenburg K, Boye E (1989) *Cell 57*:881.

DNA/mass ratio, and timing of initiation within the cell cycle are all within the normal range. Further increases in *dnaA* expression cause the DNA/mass ratio to rise, indicating initiations earlier than usual in the cell cycle. Thus the frequency of initiation appears to be coupled to the increase in cell mass by the accumulation of dnaA protein, as opposed to dilution of an inhibitor,[17] another model offered to explain initiation at a fixed ratio of mass to origins.

Suppression of amber mutations in *dnaA* by a temperature-sensitive suppressor tRNA argues that dnaA protein must be newly synthesized for each round of initiation.[18] The requirement for de novo protein synthesis prior to initiation (see above), as well as the coupling of initiation rate to the increase in cell mass, may therefore be attributable to the synthesis of dnaA protein.

Inasmuch as correctly timed initiation can be attained by levels of dnaA protein expressed from the *lac* promoter, cell-cycle regulation of *dnaA* expression does not appear to be critical. Autoregulation of *dnaA* normally allows expression to respond to both the dnaA protein level and the presence of the many dnaA protein binding sites in the cellular DNA.[19] The state of methylation of the GATC sequences within the *dnaA* promoter also affects expression.[20] The *lac* promoter, expressed constantly throughout the cell cycle, is not responsive to either the level of dnaA protein or the state of DNA methylation. The changes in *dnaA* expression caused by fluctuations in the dnaA protein level and the state of methylation may be exploited for regulation of initiation under changing growth conditions, rather than within the normal cell cycle.[21]

With dnaA protein at a high level, the abundance of genes proximal to the origin is increased to three to four times that of genes located farther away,[22] suggesting that some of the newly generated replication forks are aborted as they progress from the origin. The increased time required to complete a round of synthesis when cells are forced to initiate earlier than normal in the cell cycle[23] further indicates that other replication factors become limiting when dnaA protein is present in excess.

Factors That Disrupt the Synchrony of Initiation[24]

Initiation normally occurs simultaneously at all the origins present in a single cell. This synchrony of firing is nicely demonstrated by flow cytometry.[25] When exponentially growing cells are treated with inhibitors of RNA or protein synthesis, new initiation events are blocked, while ongoing rounds are allowed to finish. At the end of this "run-out" period, the cells contain fully replicated chromosomes equal in number to the number of origins present at the time the inhibitory drug was administered (Fig. 20-4). The DNA content of individual cells can be determined during flow cytometry by staining the DNA with a

---

17. Pritchard RH, Barth PT, Collins J (1969): *Symp Soc Gen Microbiol 19*:263.
18. Schau N, O'Day K, Peter W, Wright A (1981) *J Bact 145*:904; Schau N, O'Day K, Wright A (1981) in *The Initiation of DNA Replication* (Ray DS, ed.). ICN–UCLA Symposia on Molecular and Cellular Biology, Vol. 22. Academic Press, New York, p. 315; Kimura M, Yura T, Nagata T (1980) *J Bact 144*:649.
19. Atlung T, Clausen ES, Hansen FG (1985) *MGG 200*:442.
20. Braun RE, Wright A (1986) *MGG 202*:246.
21. Løbner-Olesen A, Skarstad K, Hansen FG, von Meyenburg K, Boye E (1989) *Cell 57*:881.
22. Atlung T, Løbner-Olesen A, Hansen FG (1987) *MGG 206*:51.
23. Løbner-Olesen A, Skarstad K, Hansen FG, von Meyenburg K, Boye E (1989) *Cell 57*:881.
24. Boye E, Løbner-Olesen A, Skarstad K (1988) *BBA 951*:359.
25. Skarstad K, Boye E, Steen HB (1986) *EMBO J 5*:1711.

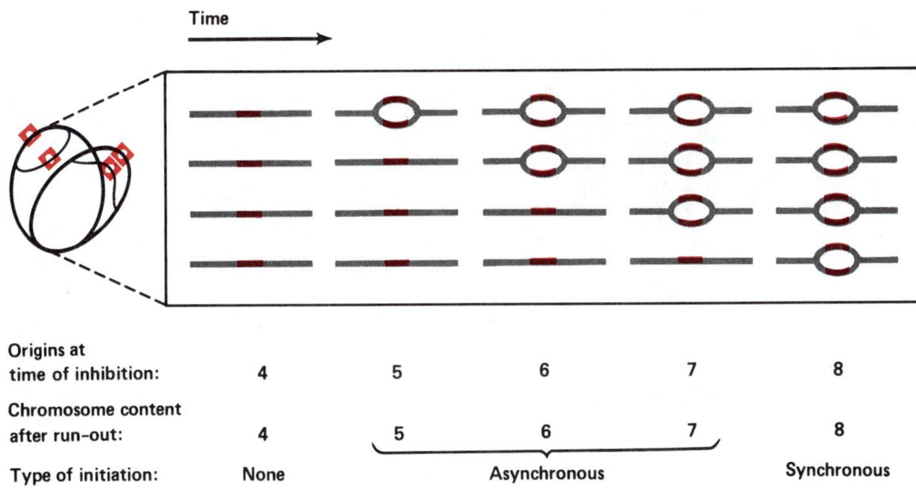

Figure 20-4
Scheme of chromosome configuration in cells initiating replication either synchronously or asynchronously at four origins. [After Skarstad K, Boye E, Steen HB (1986) *EMBO J* 5:1711]

fluorescent dye. Normally growing wild-type cells have chromosome numbers of 2^n ($n = 1, 2, 3, 4$). However if initiation is asynchronous, chromosome numbers of 3, 5, 6, 7, 9, and 10 per cell, as well as 2, 4, 8, and 16 per cell, are observed (see Fig. 20-5).

Mutants. Mutations in several DNA replication genes affect initiation synchrony (Table 20-2 and Fig. 20-5). Some *dnaA* mutations result in uncoordinated initiation, indicating an involvement of dnaA protein in the synchrony as well as the timing of initiation.[26] Not all *dnaA* mutations affect synchrony, however; the most severely affected alleles are clustered, indicating that a central domain of the protein is specifically involved. Asynchronous initiation occurs even at the permissive growth temperature and is dominant over the synchronous behavior of the wild-type dnaA protein. Thus it is the altered form

Table 20-2
Mutations that disrupt initiation synchrony

Mutation	Comments
dam	severe disruption by deletion and missense alleles; level of Dam methylase is critical
dnaA	severe disruption by mutations in the central domain; asynchrony phenotype is dominant
dnaC	increased initiation at even numbers of origins
recA	recombination-defective alleles only
rpoC	increased initiation at even numbers of origins; effect likely through changes in gene expression

26. Skarstad K, von Meyenburg K, Hansen FG, Boye E (1988) *J Bact 170*:852.

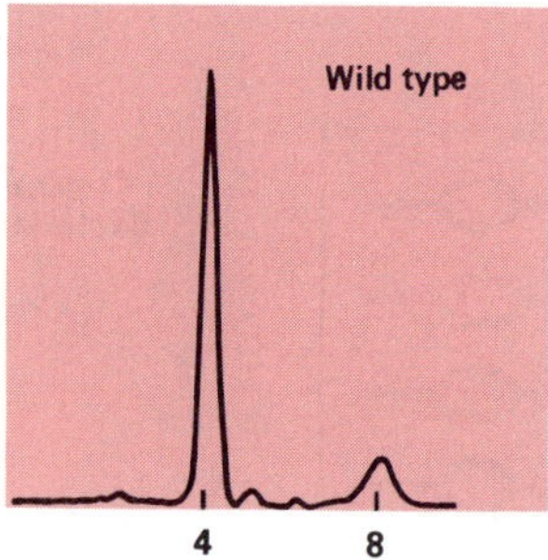

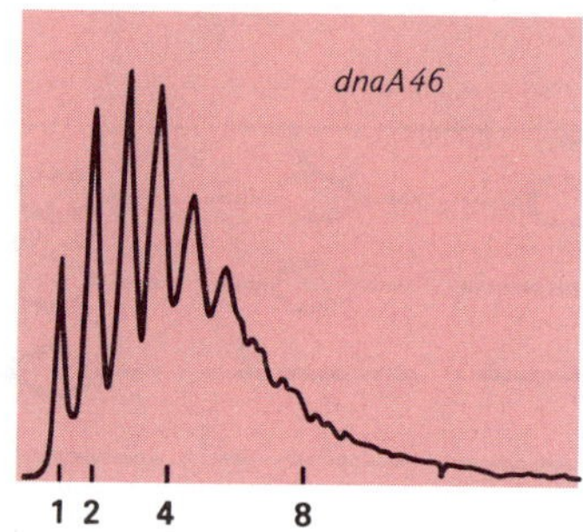

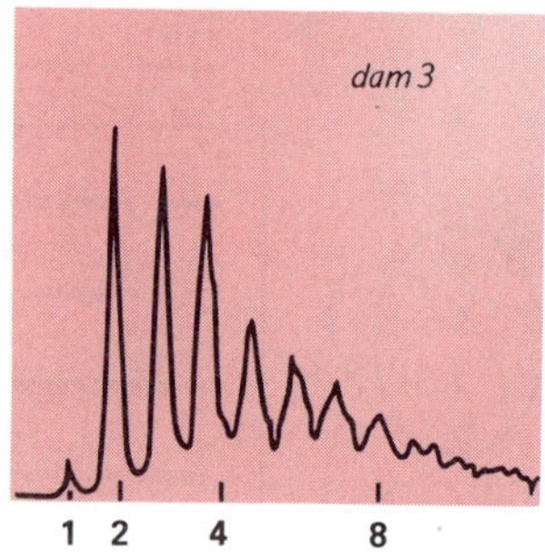

Figure 20-5
Run-out replication in *E. coli* showing the number of chromosomes in wild-type *dnaA46*, and *dam3* cells. [From Boye E, Løbner-Olesen A, Skarstad K (1988) *BBA 951*:359]

of the protein participating in initiation that is responsible for the defect in coordinating the process, rather than a lower level of activity of the mutant protein.

Mutations in *dnaC*, *rpoC*, *recA*, and *dam* also influence the synchrony of initiation.[27] The effect of *dnaC*, consistent with its role early in initiation in vitro, is probably direct. The *rpoC* mutation may involve either a global change in gene expression or the specific interaction of the altered RNA polymerase at *oriC*. Strikingly, in both the asynchronous *dnaC* and *rpoC* mutants, but not the *dnaA* strains, even numbers of chromosomes (i.e., 6, 10, 12) are predominant. This pattern indicates that *oriC* sequences may be organized in pairs during initiation and that dnaA protein may be involved in this association. RecA protein (Section 21-5), which is proficient in recombination but not necessarily in activation of the SOS response (Section 21-3), is required for synchronous initiation.[28] Asynchrony in *recA* mutants can be explained by stalling of newly formed replication forks due to the presence of the mutant RecA protein, rather than by a direct action of the protein in initiation. The *rpoB* suppressor alleles of *dnaA*^{ts} mutations do not influence coordination of initiation, nor do *dnaK* and *dnaJ* mutations.[29]

Methylation. The minimal *oriC* sequence contains 11 GATC sites which direct adenine methylation. Dam methyltransferase probably affects initiation synchrony directly by altering methylation of these sites.[30] In addition to the severe asynchrony in *dam* mutants, altered intracellular levels of Dam methyltransferase outside a narrow range have a dramatic effect on initiation synchrony:[31] both lowered and elevated levels of the enzyme cause asynchrony.

Initiation is prevented in vivo when *oriC* is hemimethylated.[32] The period of hemimethylation appears to be important in determining the refractory time after initiation before the origin can fire again. Rapid methylation effected by elevated levels of methyltransferase may cause asynchrony by allowing refiring

27. Boye E, Løbner-Olesen A, Skarstad K (1988) *BBA 951*:359.
28. Skarstad K, Boye E (1988) *J Bact 170*:2549.
29. Boye E, Løbner-Olesen A, Skarstad K (1988) *BBA 951*:359.
30. Bakker A, Smith DW (1989) *J Bact 171*:5738; Boye E, Løbner-Olesen A (1990) *Cell 62*:981.
31. Boye E, Løbner-Olesen A (1990) *Cell 62*:981; Boye E, Løbner-Olesen A, Skarstad K (1988) *BBA 951*:359.
32. Russel DW, Zinder ND (1987) *Cell 50*:1071.

of an origin within a single initiation window. In contrast, at low enzyme levels, the origin may remain hemimethylated when the second round of replication is scheduled to occur; although most of the origins in the cell will fire, they fail to do so in a tightly coordinated fashion. Regulation of the period of hemimethylation of *oriC* is also indicated by the much longer time required to remethylate this sequence compared to GATC sites located elsewhere in the chromosome.[33]

Hemimethylated *oriC* sequences bind specifically to cellular membrane fractions, suggesting that methylation may play a role in partitioning the daughter chromosomes at cell division (Sections 4 and 5). The replication defect of hemimethylated *oriC* plasmids in vivo has not been observed in either the crude or reconstituted soluble enzyme systems,[34] perhaps due to a need for binding of the origin to the membrane components that inhibit reinitiations at *oriC* in vivo.

Regulation of Origin Copy Number and the Possible Role of Transcription

Upon the overexpression of dnaA protein, the increase in initiation frequency produces a smaller elevation of chromosomes relative to cell mass than of origins to cell mass, indicating that a block downstream from initiation prevents lethal overreplication. Runaway replication is observed in *E. coli* when an R1 plasmid origin, temperature-sensitive for copy number control, is substituted for the *oriC* sequence in the chromosome.[35] Whereas the plasmid origin supports chromosomal replication and normal cell growth at the permissive low temperature, the timing of initiation with respect to cell division is completely randomized. With a shift to the restrictive high temperature, unregulated replication is induced and is lethal.

Certain *rpoB* and *rpoC* mutations, encoding altered β and β' subunits of RNA polymerase, cause a stable high copy number of the chromosome and of *oriC* plasmids. One *rpoC* allele has been analyzed in the most detail: at high temperature it raises the chromosomal content twofold to threefold; upon return to permissive temperature, replication is blocked until the normal ratio of origin to cell mass is reestablished by cell division. Sublethal levels of the RNA polymerase inhibitor streptolydigin have the same effect. Inasmuch as no single mutation, for example in *dnaA*, causes this high-copy-number phenotype, and since only mutations in RNA polymerase were found in screening for chromosomal copy number mutants, a change in gene expression is most likely responsible, rather than a specific interaction of the altered RNA polymerase with the *oriC* sequence during initiation.

In contrast to replication of naturally occurring plasmids, *E. coli* replication is not especially sensitive to the concentration of *oriC* sequences in the cell. Even high levels of *oriC* plasmids (e.g., 50 copies per cell) exert little deleterious effect on chromosomal replication. However, a few examples of incompatibility between *oriC* sequences have been observed.[36] The presence of *oriC* on a high-copy-number plasmid, such as a ColE1–*oriC* chimera, inhibits replication of plasmids that depend exclusively on *oriC*. Competition for dnaA protein, which binds to the plasmid-borne dnaA boxes, is responsible for this inhibition. The

---

33. Campbell JL, Kleckner N (1990) *Cell 62*:967.
34. Landoulsi A, Hughes P, Kern R, Kohiyama M (1989) *MGG 216*:217; Boye E, personal communication.
35. Bernander R, Merryweather A, Nordström K (1989) *J Bact 171*:674.
36. Stuitje AR, de Wind N, van der Spek JC, Pors TH, Meijer M (1986) *NAR 14*:2333; de Wind N, Parren P, Stuitje AR, Meijer M (1987) *NAR 15*:4901.

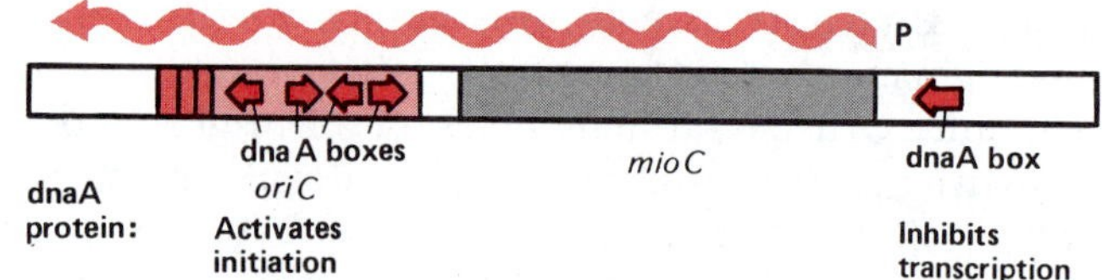

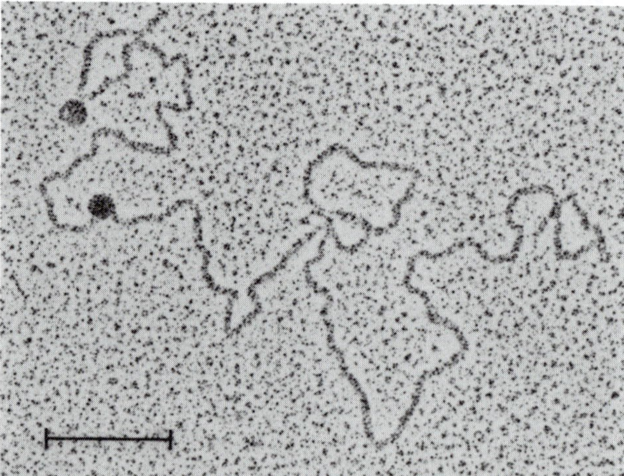

Figure 20-6
Above: Map of the *E. coli oriC* region, showing the dnaA boxes in *oriC* and in the promoter (P) for the adjacent *mioC* gene. Below: Electron micrograph of one dnaA protein complex bound at *oriC* (upper spot) and another at the *mioC* promoter on EcoR1-linearized M13 *ori* C26 DNA. Bar = 0.2μm. (Courtesy of Dr B Funnell)

presence on the high-copy plasmid of the *mioC* promoter, which directs transcription toward *oriC* and is negatively regulated by dnaA protein, magnifies this competition (Fig. 20-6).

Incompatibility between *oriC* plasmids and the chromosome is seen only under very special conditions, indicating that the chromosome is a superior competitor for dnaA protein. However, in strains containing certain *dnaA*^{ts} alleles, and with *oriC* from *E. coli* B/r (instead of K12), cell growth, and presumably replication, is inhibited by *oriC* plasmids. Growth inhibition is reversed either by increasing the level of *dnaA* expression or by an alteration in the *mioC* promoter, converting it to the sequence normally found in K12. Transcription from this promoter again affects the ability of the origin to compete for limiting amounts of dnaA protein.

The *mioC* promoter (P_{mioC}) increases the copy number and stability of *oriC* plasmids.[37] Transcription, rather than the protein product of the *mioC* gene (a 16-kDa protein of unknown function), is required, and regulation by dnaA protein is important. Other promoters do not substitute for P_{mioC}, and high levels of transcription inhibit rather than stimulate replication. Plasmid replication in HU- or IHF-deficient mutants requires P_{mioC}, linking its function to the transcriptional activation of *oriC* characterized in vitro (Section 16-3). P_{mioC} is, however, not essential for chromosomal replication.

In contrast to *E. coli*, the Gram-positive *B. subtilis*, while likely initiating replication by a similar mechanism (Section 16-4), strictly regulates origin copy

37. Løbner-Olesen A, Atlung T, Rasmussen KV (1987) *J Bact 169*:2835; Stuitje AR, Meijer M (1983) *NAR 11*:5775.

number. Plasmids containing *B. subtilis oriC* cannot be maintained unless the dnaA binding sites are mutated, a strong suggestion that origin number is restricted by limiting levels of dnaA protein. Perhaps in *E. coli, dnaA* expression is less tightly regulated, allowing increases to high enough dnaA levels to support even a large number of extrachromosomal copies of *oriC*. Interestingly, what appears to be a chromosomal rearrangement has separated *oriC* from the *dnaA* gene in the evolution of enteric bacteria. Yet in *B. subtilis* and *Pseudomonas*, *oriC* and the *dnaA* gene are closely linked (Fig. 20-7), reminiscent of the organization of origins and initiator proteins of many plasmids. As a result of divorcing of the *dnaA* gene from *oriC*, its expression may be more loosely regulated and less sensitive to the presence of its binding sites. How *E. coli* avoids overreplication, when it allows exceedingly high numbers of episomal origins, is not understood.

Regulation under Changing Growth Conditions

UV irradiation and other DNA-damaging agents affect both the initiation and elongation phases of DNA replication. Infection of UV-irradiated cells with a hybrid phage that depends on *E. coli oriC* for replication suggests that a *trans*-acting factor is produced upon irradiation that specifically inhibits initiation.[38] Replication of undamaged DNA templates in the irradiated cell is inhibited. The

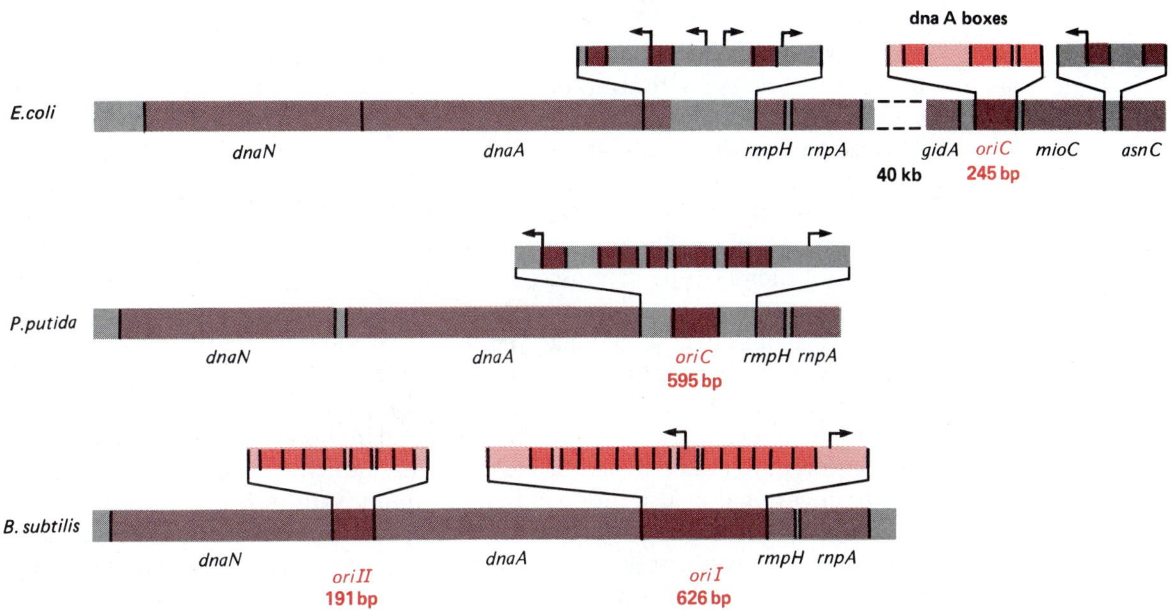

Figure 20-7
Comparison of the chromosomal organization of the origin and *dnaA* regions in *E. coli*, *Pseudomonas putida*, and *B. subtilis*. Arrow = bacterial promoter. Note the long distance separating *oriC* and *dnaA* in *E. coli* compared to the contiguous organization in the other bacteria. (Courtesy of Drs TW Yee and DW Smith)

38. Verma M, Moffet KG, Egan JB (1989) MGG *216*:446.

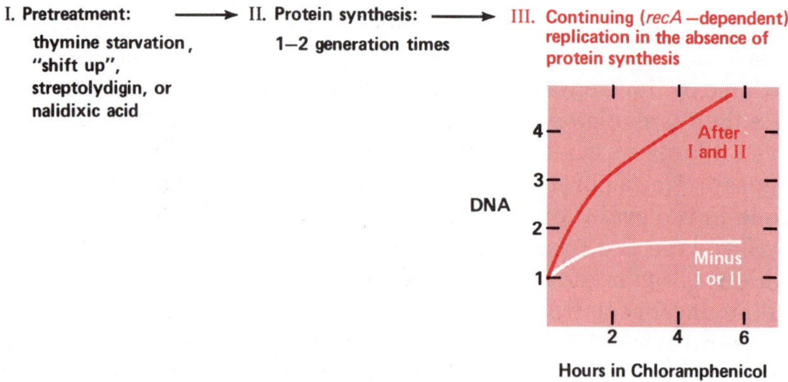

Figure 20-8
Continuing ("stable") replication in the absence of protein synthesis. This ensues in normal *E. coli* after thymine starvation, shift up to a rich medium, or addition of streptolydigin or nalidixic acid. Chloramphenicol inhibits protein synthesis.

blockage of initiation is transient and the protein or gene responsible has yet to be identified. The similar inhibition of phage 186 replication, which depends on the dnaA and dnaC proteins, and the resistance of phage λ, which does not, suggests these proteins as possible targets of the inhibitor.

A form of replication dependent on *recA* follows certain metabolic derangements and enables replication to continue indefinitely, free of dependence on protein synthesis (Fig. 20-8). Cells starved for thymine, or "shifted up" to a rich medium, or treated with streptolydigin or nalidixate, after an interval of protein synthesis, continue replication indefinitely without further protein synthesis.[39] This pathway of *inducible stable DNA replication* requires the functions of the *recA*, *dnaT*, and *dnaC* genes. Cells mutant in any of these genes behave as do normal, unperturbed cells in failing to reinitiate DNA synthesis when protein synthesis is blocked. The requirement for *dnaT* and *dnaC*, both encoding components of the φX-type primosome (Section 8-4), suggests the involvement of this enzyme complex in the pathway. Induction of stable replication is part of the SOS response to DNA damage (Section 21-3) and requires, in addition to the activated form of RecA protein, the RecBCD enzyme[40] (Section 21-5) and at least one product under LexA repressor control. In *rin* mutants, induction of stable DNA replication occurs even in *recA*$^-$ cells, and in this case *recF* (Section 21-5) is also required. Reduced RNase H activity, which uncovers an alternate replication pathway that is also independent of protein synthesis (*constitutive stable DNA replication*; see below), is not responsible for induction after DNA damage. Thus the relationship between these two protein-synthesis–independent replication pathways is unknown.

The scheme in Figure 20-9 proposes that dnaT protein is an integral part of the normal replication complex and contributes to its ultimate destabilization. Protein synthesis is then required to replenish the components for a fresh complex to initiate another round of replication. The direct interaction with dnaC pro-

39. Kogoma T, Lark KG (1975) *JMB 94*:243; Kogoma T, Connaughton MJ, Allzadeh BA (1978) *DNA Synthesis, NATO*, p. 113.
40. Magee TR, Kogoma T (1990) *J Bact 172*:1834.

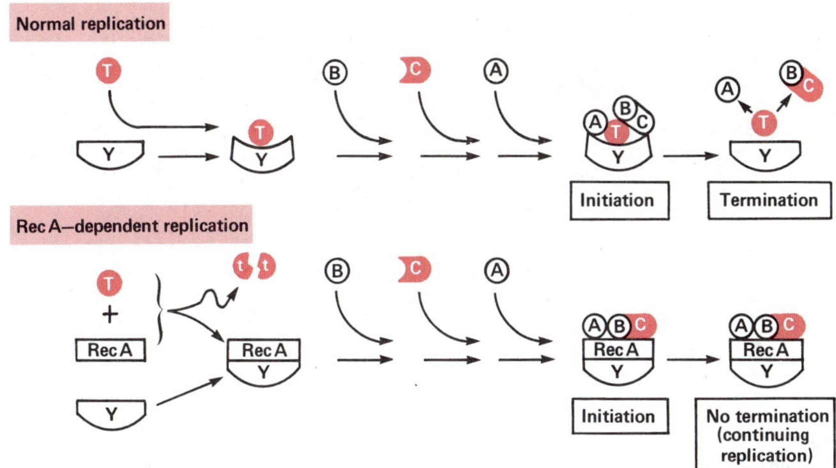

Figure 20-9
Model for multiple functions of dnaT protein. Scheme explains the participation of dnaT protein (T) in initiation and termination complexes, and continuing RecA-dependent replication in its absence. Y represents a protein synthesized early in the formation of the replication complex. The state of Y depends on interactions with RecA, dnaA, dnaB, dnaC, and dnaT proteins and determines whether replication will be normal or continuing (stable). (Courtesy of Professor KG Lark)

tein in this scheme would explain the suppression of mutant dnaC protein by mutant dnaT protein. Under metabolic circumstances that activate RecA protein, dnaT protein may be proteolyzed, with the result that the replicating complex is no longer destabilized upon termination.

Possible Control Mechanisms Suggested from in Vitro Studies

In addition to the adjustment of its level, dnaA protein activity can be modulated in other ways to control initiation. The protein, in vitro, exists in several forms:[41] tightly and loosely bound to ATP, tightly bound to ADP, as a monomer, and as a large aggregate. Of the nucleotide-bound forms, only the ATP-bound dnaA is active in initiation at *oriC*; (Fig. 20-10); the tightly bound ATP is hydrolyzed very slowly to ADP in the presence of DNA. Exchange of bound nucleotide with high external levels of ATP is very feeble. The cellular components which promote nucleotide exchange are acidic phospholipids (Fig. 20-10) in a fluid membrane (Section 4).

About half of the isolated dnaA protein is found in aggregates which contain from 2 to 12 molecules of phospholipid per dnaA molecule. The aggregated protein, relatively inert in supporting replication in the purified in vitro system, can be monomerized and activated by treatment either with dnaK protein and ATP or with phospholipase A2.[42] The dnaA protein isolated from *dnaA46* mutant cells forms aggregates even more readily and can be activated by dnaK protein assisted by unidentified factors.[43]

41. Sekimizu K, Yung BY-Y, Kornberg A (1988) *JBC 263*:7136; Sekimizu K, Kornberg A (1988) *JBC 263*:7131.
42. Hwang DS, Crooke E, Kornberg A (1990) *JBC 265*:19244.
43. Hwang DS, Kaguni JM (1988) *JBC 263*:10633; Hwang DS, Kaguni JM (1988) *JBC 263*:10625.

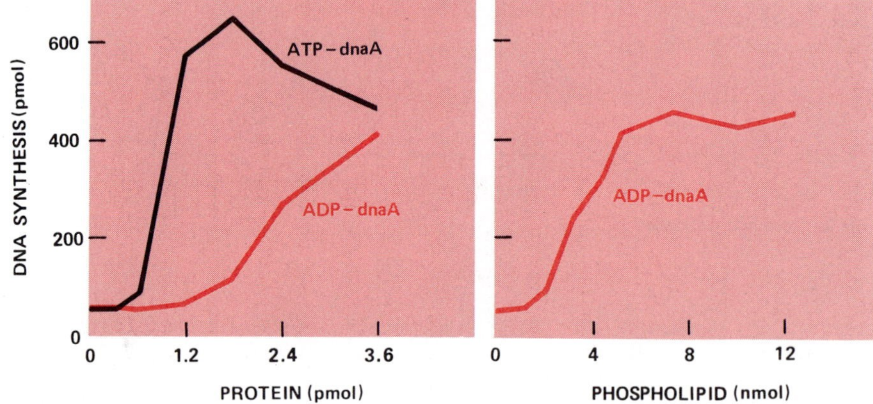

Figure 20-10
Left: Activity of the ATP- and ADP-bound forms of dnaA protein in *oriC* plasmid replication in vitro. Right: Activation in vitro of ADP–dnaA protein (1.7 pmol) by *E. coli* phospholipids in the presence of ATP. (Courtesy of Dr E Crooke)

IciA, a Negative Effector for *oriC*[44]

Formation of the open complex by dnaA protein, an early event in initiation, is blocked by the binding of a novel 33-kDa protein to the 13-mer region; this protein has no effect on the subsequent stages of initiation and replication. About 10 to 20 molecules (likely as 5 to 10 dimers) cover the three tandem 13 mers. The encoding gene named *iciA* (*i*nhibitor of *c*hromosome *i*nitiation) has been sequenced and overexpressed. The effects of mutations in the gene and in the 13-mers point to a negative regulatory role for the IciA protein in vivo.

Bypass of the Normal Pathway: Constitutive Stable DNA Replication[45]

Constitutive stable DNA replication (cSDR) is an alternative initiation pathway available in cells that lack RNase H, the enzyme that degrades RNA in RNA–DNA hybrids. Replication in these *rnh*$^-$ cells is altered in several ways: (1) The dnaA protein is not required; *rnh* mutants suppress both *dnaA*^{ts} and *dnaA*^{null} alleles. (2) The *oriC* sequence can be deleted; initiation occurs at several other sites called *oriK*.[46] (3) Protein synthesis, normally required to initiate a round of replication, is not essential. (4) Initiation is random with respect to cell age and mass. (5) The choice of initiation at one of the several *oriK* sites also appears to be random. (6) Transcription by RNA polymerase is critical for stable DNA replication, as judged by rifampicin-sensitivity of initiation. (7) This transcription is stringently controlled; initiation is inhibited by amino acid starvation in *relA*$^+$ but not in *relA* mutant strains. (8) The RecA function is required.[47]

A model for initiation of cSDR is as follows: transcription from a stringently regulated promoter forms an RNA–DNA hybrid, which persists due to the absence of RNase H. The promoter or hybridization sites correspond to *oriK*.

44. Hwang DS, Kornberg A (1990) *Cell 63*:325; Thöny B, Hwang DS, Fradkin L, Kornberg A (1991) *PNAS*, in press.
45. von Meyenburg K, Boye E, Skarstad K, Koppes L, Kogoma T (1987) *J Bact 169*:2650; Kogoma T (1978) *JMB 121*:55.
46. deMassy B, Fayet O, Kogoma T (1984) *JMB 178*:227.
47. Kogoma T, Skarstad K, Boye E, von Meyenburg K, Steen HB (1985) *J Bact 163*:439.

RecA protein interacts with the displaced strand of the R-loop, enabling the entry and assembly of the replication complex, as achieved by dnaA protein at *oriC*.

Regulation of Replication in *B. subtilis*

Initiation of replication in the Gram-positive *B. subtilis* resembles that of *E. coli* in that it involves (Section 16-4): a related sequence at *oriC*,[48] a protein homologous to dnaA protein, and a requirement for de novo synthesis of both protein and RNA prior to initiation.[49] Yet, there are distinctions. The *B. subtilis oriC* cannot be cloned on an autonomously replicating plasmid because of the pronounced episomal competition for the *B. subtilis* analog of dnaA protein,[50] an effect that indicates a tighter "control loop" for origin copy number than in *E. coli*.

Additional distinctive replication control features in *B. subtilis* include the following:

(1) An RNA–DNA copolymer extends a transcript originating from the *rrnO* operon, located near the origin.[51] Synthesis of this RNA is correlated with the transcriptional requirement for initiation.

(2) Stringent regulation of initiation is more clearly demonstrated in *B. subtilis* than in *E. coli*.[52]

(3) Replication initiates quite synchronously for several rounds during germination of spores.

(4) A key initiation factor, product of the *dnaB* gene (unrelated to *E. coli dnaB*), is a membrane protein.[53] Linkage between the origin, the replication complex, and the cell membrane is inferred from cell fractionation and in vitro replication studies[54] (Section 4).

(5) Overinitiation, occurring after return of $dnaB^{ts}$ mutants to the permissive temperature, causes an increase in the copy number of genes proximal to the origin but not in those located farther away. Stalling of newly formed replication forks may provide a postinitiation control mechanism to ensure against excessive overreplication,[55] as in *E. coli* (see above).

20-4 Bacterial Envelopes and Chromosome–Membrane Association

The cell envelope may play an important role in the regulation of DNA synthesis and cell division. Binding of the chromsome to the cell envelope and spatial organization of the molecular components on the membrane may provide the physical and temporal linkage of replication and cell division. Chemical evidence for such associations is, however, still rather meager.

---

48. Moriya S, Ogasawara N, Yoshikawa H (1985) *NAR 13*:2251.
49. Laurent SJ, Vannier FS (1973) *J Bact 114*:474; Laurent SJ (1973) *J Bact 116*:141; Laurent SJ (1974) *J Bact 117*:329.
50. Moriya S, Fukuoka T, Ogasawara N, Yoshikawa H (1988) *EMBO J 7*:2911.
51. Séror-Laurent SJ, Henckes G (1985) *PNAS 82*:3586.
52. Séror SJ, Vannier F, Levine A, Henckes G (1986) *Nature 321*:709.
53. Hoshino T, McKenzie T, Schmidt S, Tanaka T, Sueoka N (1987) *PNAS 84*:653.
54. Laffan J, Firshein W (1987) *J Bact 169*:2819.
55. Henckes G, Harper R, Levine A, Vannier F, Séror SJ (1989) *PNAS 86*:8660; Watabe K, Forough R (1987) *BBRC 145*:861.

The Cell Envelope

This multilayered structure defines the size, shape, and functional integrity of the cell. It consists in general of four figuratively concentric layers (Fig. 20-11).

(1) *Inner (plasma) membrane.* This structure encircles the entire cell and regulates the active passage of small molecules. It is a lipid bilayer, rich in a variety of proteins, including the enzymes that synthesize the envelope. Phosphatidylethanolamine is the predominant lipid species (~70%), with the acidic phospholipids phosphatidylglycerol (~20%) and cardiolipin (~5%) making up the remainder. The membrane furnishes the interface for intracellular structures and reactions. Complex invaginations of the membrane called *mesosomes* (see below) may be sites of attachment for the chromosome.

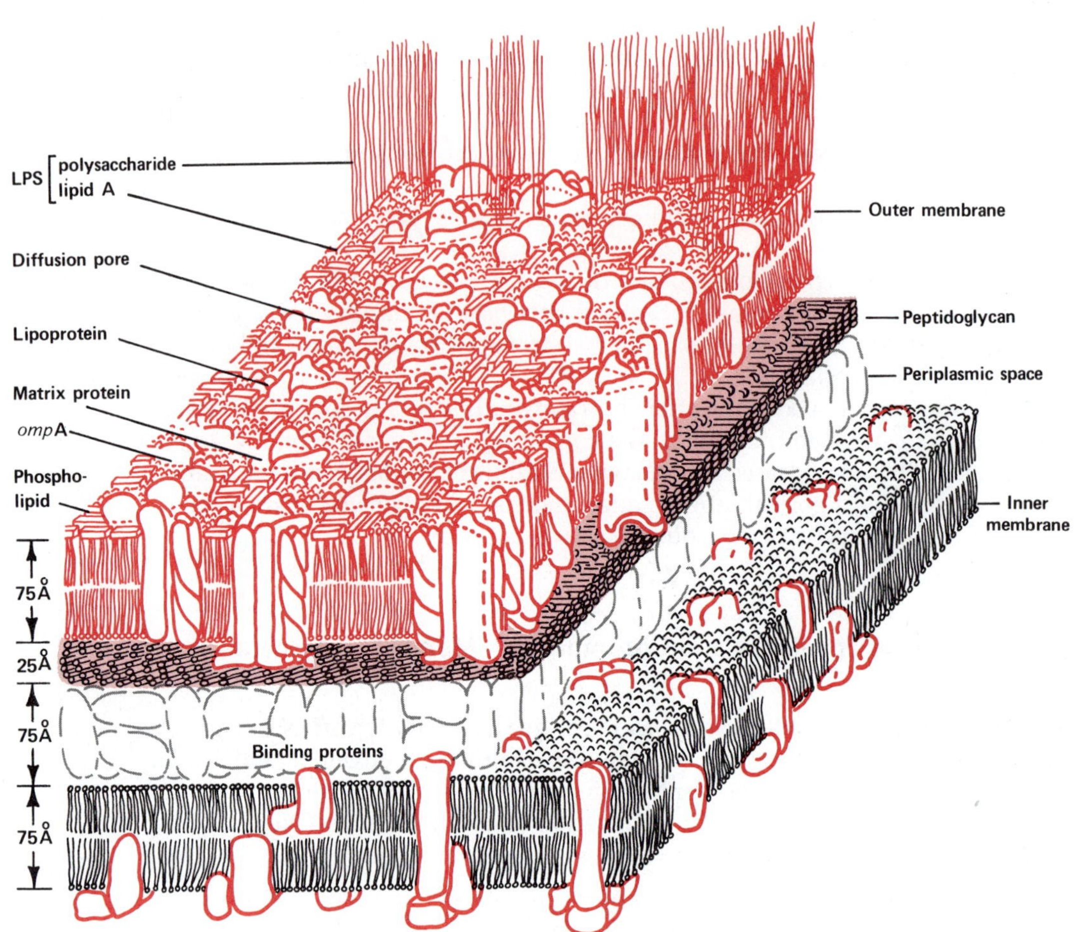

Figure 20-11
Drawing of a portion of the *E. coli* cell envelope. OmpA (outer membrane protein A) is equivalent to TolG, which provides bacteriocin tolerance. LPS = lipopolysaccharide. (Courtesy of Professors JM DiRienzo and M Inouye)

(2) *Periplasm.* The space between the inner membrane and the exoskeleton (cell wall and outer membrane) makes up 20% to 40% of the cell volume. It has a defined polymeric and ionic composition, and is isoosmotic with the cytoplasm. The periplasmic space is in effect a cellular compartment.[56]

(3) *Cell wall* (also called *peptidoglycan layer, sacculus,* or *murein*). Single units of a heteropolysaccharide, about 5×10^6 kDa, cross-linked by short peptide bridges, give the cell its shape and rigidity. Penicillin and other β-lactam antibiotics interfere with enzymes that synthesize these cross-linkages and thus prevent orderly growth and septation;[57] the cells are thereby disposed to misshapen forms and osmotic lysis.

(4) *Outer membrane.* This structure in Gram-negative bacteria, such as *E. coli* and *Salmonella,* is a lipopolysaccharide-rich phospholipid bilayer. Although the percentage of the membrane contributed by protein is similar to that of the inner membrane, there are relatively few (less than 10) abundant protein components. The major proteins include a lipoprotein, some of which is covalently linked to the cell wall, and the porins, which are responsible for selective permeability.[58] As the outermost cellular surface, the outer membrane contains common antigenic determinants, receptors for phages, and the glycocalyx, a coating of fibers that enables cells to adhere to surfaces and to each other. The outer membrane, in essence the social face of the cell, acts also as a passive sieve, permitting the passage of hydrophilic molecules no larger than about 600 Da.

Bacterial Chromosome–Membrane Association[59]

With replication of the bacterial chromosome geared to cell division, it is plausible that segregation of the genomes and their partitioning between daughter cells might be achieved by a linkage of the replication origin or fork to the cell membrane. Concurrent growth of the new cell envelope would separate the developing genomes; septation of the newly synthesized envelope would partition them.

Membranes are phospholipid bilayers in which proteins are in a mosaic arrangement. A membrane protein, if peripherally associated with the bilayer, is removable by high salt concentrations; if integrally associated, its removal requires solubilization of the membrane by detergents. Location of the DNA at a membrane surface must depend on interactions with specific proteins seated in or attached to the membrane.

The fragmentary evidence of nascent DNA–membrane associations is derived from several sources: electron microscopy; cell fractionation; and binding in vitro between origin sequences, membrane components, and proteins; the best characterized of the proteins is the dnaA protein of *E. coli* (see below).

Evidence Derived from Electron Microscopy.[60] Thin sections of bacteria show a *nucleoid,* a pleomorphic central region of condensed DNA. One or more points

56. Stock JB, Rauch B, Roseman S (1977) *JBC 252*:7850.
57. Spratt BG (1977) *EJB 72*:341.
58. Nakae T (1976) *JBC 251*:2176; Nikaido H, Song SA, Shaltiel L, Nurminen H (1977) *BBRC 76*:324.
59. Schaechter M (1990) in *Molecular Mechanisms in DNA Replication and Recombination* (Richardson CC, Lehman IR, eds.). UCLA, Vol. 127. AR Liss, New York, p. 165; Ogden G, Schaechter M (1986) in *Bacterial Chromatin* (Gualerzi CO, Pon CL, eds.). Springer-Verlag, Berlin, p. 49.
60. Siegel PJ, Schaechter M (1973) *Ann Rev Microbiol 27*:261; Leibowitz PJ, Schaechter M (1975) *Int Rev Cytol 41*:1.

on the nucleoid seem to touch a *mesosome*,[61] a lamellar invagination of the inner membrane. As daughter cells separate during cell division, the mesosome of each cell retains contact with the DNA. In protoplasts, which lack mesosomes, there nevertheless seems to be a point of contact between DNA and membrane. Although the electron microscope image has pointed out the proximity of DNA to membranes, the existence or nature of any chemical linkage between them has remained invisible.

Rapidly Sedimenting Complexes.[62] In cellular fractionation, the chromosomal DNA of *B. subtilis* and *E. coli* associates with membranes at three unique sites: the replication fork, the origin, and the terminus of replication. Cells lysed following a brief pulse of [^{3}H]thymidine yield a rapidly sedimenting membrane fraction many-fold enriched in labeled DNA relative to total DNA. This enrichment for nascent DNA can be reduced by a chase with unlabeled thymidine. Genetic markers at the origin and terminus of the *B. subtilis* chromosome[63] remain associated with the membrane fraction more often than do other genes. Enrichment for genetic markers near the origin is even greater in rapidly growing cells undergoing dichotomous replication and possessing more initiated chromosomes per cell. Similar observations[64] are obtained in *E. coli* using radioactive rather than genetic markers. Yet disrupting the association of the origin with the membrane—by inhibiting RNA or protein synthesis—does not prevent initiation of chromosome replication.[65]

Replicating DNAs of infecting phages[66] (P22, λ, T4, T7, φX174, and M13) at several stages in their life cycles are also found in fractions that contain host cell membrane. Maturation and assembly of various phages involve the cell envelope, in which components of the phage coat in various stages of development are found.

Phospholipid Complexes.[67] The nascent DNA of cells lysed with Sarkosyl detergent has been found associated with crystals containing the detergent and magnesium, which form a sharp band, the M band, in sucrose gradients. Appearance of the DNA in the M band depends on phospholipid in this membrane fraction. Since 20-fold fragmentation of the DNA by x-rays does not decrease binding to the M band, multiple points of attachment may be inferred. Reduction of the binding by rifampicin, an inhibitor of RNA polymerase, suggests that this attachment of nascent DNA depends on a transcriptional complex rather than a replicative one.

Folded-Chromosome Complex. Lysis of *E. coli* spheroplasts with nonionic detergents in high salt or spermidine, at or below 10°, yields the complete, folded

---

61. Ryter A (1968) *Bact Rev 32*:39.
62. Ogden G, Schaechter M (1986) in *Bacterial Chromatin* (Gualerzi CO, Pon CL, eds.). Springer-Verlag, Berlin, p. 49; Siegel PJ, Schaechter M (1973) *Ann Rev Microbiol 27*:261; Leibowitz PJ, Schaechter M (1975) *Int Rev Cytol 41*:1.
63. Yamaguchi K, Yoshikawa H (1975) *J Bact 124*:1030; Yamaguchi K, Yoshikawa H (1977) *JMB 110*:219; Toyoda H, Yamaguchi K, Yoshikawa H (1978) *J Bact 136*:799.
64. Craine BL, Rupert CS (1978) *J Bact 134*:193; Nicolaidis AA, Holland IB (1978) *J Bact 138*:178.
65. Craine BL, Rupert CS (1979) *J Bact 137*:740.
66. Siegel PJ, Schaechter M (1973) *Ann Rev Microbiol 27*:261; Leibowitz PJ, Schaechter M (1975) *Int Rev Cytol 41*:1.
67. Siegel PJ, Schaechter M (1973) *Ann Rev Microbiol 27*:261; Leibowitz PJ, Schaechter M (1975) *Int Rev Cytol 41*:1.

chromosome attached to cell envelope fragments.[68] At higher temperatures or upon dissolution of the membranes by ionic detergents, the chromosome is released. Outer-membrane markers are most prominent among the envelope fragments. The chromosome attachment correlates with initiation of a round of replication and depends on the synthesis of specific proteins.[69]

Origin–Membrane Binding and Specific Membrane Proteins

Hemimethylation of *oriC* and Membrane Binding.[70] The *oriC* sequence specifically binds to fractions that appear to be from the outer membrane of *E. coli* cells. When total chromosomal DNA is cleaved into fragments, the *oriC* fragment is far more tightly bound than other segments, indicating that a rather specific membrane–*oriC* complex is formed. To form the complex requires the *oriC* sequence from growing cells, and especially from synchronized cells shortly after initiation of replication. The increased affinity of *oriC* is due to its hemimethylated state during the interval after replication and before modification of the newly synthesized strand by the methyltransferase.

Binding studies can be performed with a 460-bp fragment containing the minimal *oriC* sequence and some flanking DNA. Within this minimal sequence two binding sites are apparent: one at 38 to 155 bp and one at 155 to 244 bp. Of the 11 GATC sites in *oriC*, it is uncertain which are involved in binding. The longer time required to methylate *oriC* after passage of the replication fork, compared to other regions of the chromosome, suggests that *oriC* acts as a domain with respect to methylation; two GATC sites from different parts of the origin have been analyzed in detail; both show similar slow methylation kinetics.[71]

These methylation results indicate that hemimethylation of *oriC* may be involved in chromosome segregation as well as serving to prevent refiring of the origin immediately after initiation (Section 3). The proteins responsible for membrane association are still undefined, and how the lack of a hemimethylated state in *dam* mutant strains is bypassed as a signal for chromosome replication and segregation is also unknown.

Membrane Proteins That Bind *oriC*. Two distinct proteins have been purified from membrane fractions on the basis of their affinity for binding the *oriC* sequence: B′, about 60 kDa, binds two sites in and near the minimal origin on ssDNA.[72] MOB2 (*m*embrane *o*rigin-*b*inding protein 2), about 10 kDa, binds to a fragment +244 to +417 bp at the right edge of the origin and prefers supercoiled DNA.[73] Antibodies to MOB2 indicate that 80% to 90% of the protein is localized to the inner membrane. Under some conditions, B′ and MOB2 apparently form a complex, but mutations in the genes encoding these proteins are needed to assess their physiologic roles.

---

68. Kavenoff R, Ryder OA (1976) *Chromosoma 55*:15; Drlica K, Burgi E, Worcel A (1978) *J Bact 134*:1108.
69. Worcel A, Burgi E (1974) *JMB 82*:91.
70. Ogden GB, Pratt MJ, Schaechter M (1988) *Cell 54*:127; Hendrickson WG, Kusano T, Yamaki H, Balakrishnan R, King M, Murchie J, Schaechter M (1982) *Cell 30*:915; Kusano T, Steinmetz D, Hendrickson WG, Murchie J, King M, Benson A, Schaechter M (1984) *J Bact 158*:313.
71. Campbell JL, Kleckner N (1990) *Cell 62*:967.
72. Jacq A, Kohiyama M, Lother H, Messer W (1983) *MGG 191*:460.
73. Jacq A, Kern R, Tsugita A, Kohiyama M (1989) *J Bact 171*:1409.

Membrane Association of dnaA Protein. Although dnaA is not a classic integral membrane protein, several lines of evidence argue for the presence and importance of dnaA–membrane interactions.

(1) About half of the dnaA protein isolated from overproducing cells is in an aggregated form, associated with phospholipids.[74]

(2) Purified dnaA protein binds acidic phospholipids (i.e., cardiolipin and phosphatidylglycerol).[75] This binding can be demonstrated by co-sedimentation of dnaA with membranes in sucrose gradients, and it can profoundly influence dnaA protein activity in initiation of replication.[76]

(3) The binding of phospholipids destabilizes ATP or ADP tightly bound to dnaA protein.[77] Thus, dnaA rendered inert for replication by virtue of tight binding to ADP can be rejuvenated in the presence of acidic phospholipids. In addition to the acidity of the phospholipid head groups, the fluidity of the membranes in which the phospholipids are presented to the protein affects the activation reaction, thus indicating that a membrane–protein interaction is involved.

(4) 3-Decynoyl-N-acetylcysteamine, a transition-state analog that inhibits β-hydroxythioester dehydrase, the enzyme responsible for introducing the double bond to form oleic acid, also prevents the initiation of a cycle of replication in *E. coli*. This inhibition of replication, relieved by providing the cells with oleic acid, can be ascribed to diminished membrane fluidity due to the lack of an unsaturated fatty acid in the lipid bilayer.[78] Neither phospholipids isolated from inhibited cells nor saturated phospholipids are able to displace ADP bound to dnaA protein; this inability nicely correlates the reactivation of dnaA protein for initiation observed in vitro, with in vivo phenomena.[79]

In membrane complexes with the *B. subtilis* origin, the endogenous DNA continues to be replicated in vitro.[80] The *dnaB* gene product and key initiation factors appear to be membrane proteins.[81] Another membrane protein, identified as binding tightly to the origin region, may inhibit initiation.[82]

Summary of DNA–Membrane Relationships

Despite its plausibility and the morphologic and biochemical evidence in its favor, the DNA–membrane relationship remains chemically ill-defined and tenuous. The difficulty in capturing this complex may be inherent in the noncovalent associations that characterize the structure of membranes. Contiguity in electron microscope pictures, co-sedimentation in velocity gradients, and metabolic relationships—all fail to clarify the chemical nature of DNA attachment to the membrane. Nevertheless, such an attachment continues to be an attractive hypothesis.

---

74. Sekimizu K, Yung BY-Y, Kornberg A (1988) *JBC 263*:7136.
75. Sekimizu K, Kornberg A (1988) *JBC 263*:7131.
76. Sekimizu K, Yung BY-Y, Kornberg A (1988) *JBC 263*:7136.
77. Sekimizu K, Kornberg A (1988) *JBC 263*:7131.
78. Fralick JA, Lark KG (1973) *JMB 80*:459.
79. Yung BY-Y, Kornberg A (1988) *PNAS 85*:7202.
80. Laffan J, Firshein W (1987) *J Bact 169*:2819.
81. Hoshino T, McKenzie T, Schmidt S, Tanaka T, Sueoka N (1987) *PNAS 84*:653.
82. Laffan JF, Firshein W (1988) *PNAS 85*:7452.

20-5 Bacterial Cell Division[83]

The rod-shaped *E. coli* cell increases in length as it grows; when it is twice the length of a newborn cell, a septum is introduced at the center. Production of a new peptidoglycan layer and invagination of both cell membranes are required before the daughter cells can separate. For division to be an orderly process, (1) replication of the chromosome must be complete, ensuring a full complement of DNA for each daughter cell; (2) division must be coupled to cell growth so that the cells are of a proper age and size to divide; (3) the septum must be located in the middle of the cell; and (4) critical cellular components must be partitioned into both of the developing cell compartments.

Genetics of Cell Division

Certain mutations identify genes that are specifically required for septum formation and cell separation, but are not essential for cell growth (Table 20-3). Genes in which conditional mutations cause the cells to filament are called *fts* genes (*f*ilamentation *t*emperature-*s*ensitive). Long filaments form and usually

Table 20-3
Some genes involved in *E. coli* cell division

Stage	Gene	Map position	Function
Initiation of septation	*ftsH*	69	required only during initiation
	ftsZ	2	rate-limiting factor for division; required throughout septum formation; target for inhibition by *sfiA* product
Septum formation	*ftsA*	2	partial septum formed; product interacts with *ftsI* product
	ftsI (pbpB)	2	penicillin-binding protein 3 (PBP3) transpeptidase, transglycosidase; required for synthesis of septal form of peptidoglycan
	ftsF	82	septum formation; specific function unknown
	ftsE	76	ATP binding; membrane ion transport?
	ftsQ	2	septum formation; specific function unknown
Cell separation	*envA*	2	separation of peptidoglycan layer between sister cells; N-acetylmuramoyl-L-alanine amidase levels depressed in mutant
Inactivation of septation sites	*minB*	26	locus comprising *minC*, *D*, and *E*;
	minC	26	*minC* and *D* together form inhibitor of septation
	minD	26	
	minE	26	prevents *minC–D* inhibition at central septation sites but not at cell poles
Regulation of septation	*sfiA (sulA)*	22	inhibitor of cell division produced during SOS response; interacts with *ftsZ*
	lon	10	degrades *sfiA* protein, allowing septation to resume after DNA repair is complete

83. Donachie WD, Robinson AC (1987) in *Escherichia Coli and Salmonella Typhimurium* (Neidhardt FC, ed.). Am Soc Microbiol, Washington, DC, Vol. 2 p. 1578; Lutkenhaus J (1990): *Trends Genet* 6:22; Donachie WD, Begg KJ, Sullivan NF (1984) in *Microbial Development* (Losick R, Shapiro L, eds.). CSHL, p. 27.

contain multiple nucleoids, spaced at proper intervals. Thus, septum formation can be blocked without disrupting cell growth, DNA replication, and chromosome partitioning.

Mutations which result in filamentous cells that contain no constrictions (such as *ftsZ*) identify genes whose products act early in the division process, while those encoding products that act later (such as *ftsA*) result in filaments that contain partial septa. The *envA* gene acts late, during separation of the peptidoglycan layer between the sister cells, and *envA* mutants appear as chains of completely divided, but unseparated, cells; synchronous division at the proper locations along the filament occurs after the mutants are returned to the permissive temperature. The morphology of mutant strains, the effects of double mutants, and the sensitivity to lysis by β-lactam antibiotics indicate a pathway for the ordered functioning of the various cell-division genes (Fig. 20-12).

Initiation of Septation. The *ftsZ* gene acts early and is rate-limiting in septum formation.[84] Overproduction of the *ftsZ* product increases the frequency of cell division. During a single cell cycle, septation may occur at both the centrally located division site and the cell poles (usually inactivated by the action of the *minB* gene products; see below). The uncoupling of division from the replication

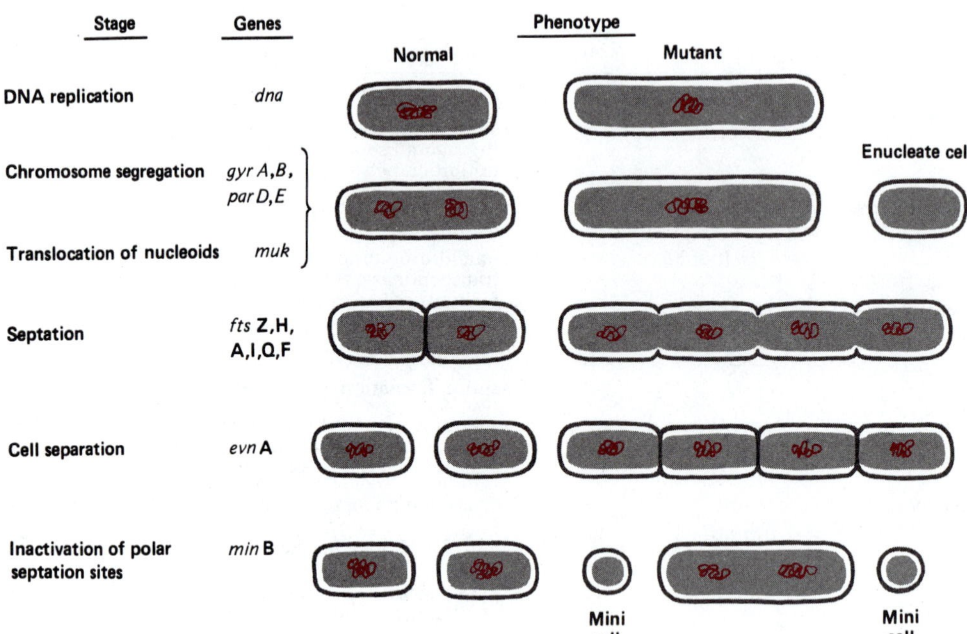

Figure 20-12
Genes that function in chromosome partition and cell division and the effect of the deficient mutation on cell morphology. Left, the shapes of normal cells when the gene product acts; right, the appearance of a mutant cell in which the gene has been inactivated.

84. Ward JE, Lutkenhaus JF (1985) *Cell 42*:941; Bi E, Lutkenhaus L (1990) *J Bact 172*:2765.

cycle, and the use of polar division sites, results in "minicells" that lack chromosomal DNA. Overproduction of *E. coli* FtsZ protein also causes minicell formation in *Salmonella, Shigella,* and *Klebsiella;* a related protein has been detected in several Gram-positive bacteria as well. In addition to its function in cell division under balanced growth conditions, FtsZ protein is also the target of the cell-division inhibitor SfiA protein, induced during the SOS response to DNA damage[85] (see below).

Septum Synthesis. The *ftsI* gene encodes penicillin-binding protein 3 (PBP3).[86] This enzyme has both transpeptidase and transglycosidase activity on peptidoglycan substrates. Present at about 50 copies in the cell inner membrane, it is the target of the β-lactam antibiotic cephalexin, which specifically blocks septation without affecting cell elongation. Thus it seems that PBP3 synthesizes the septum peptidoglycan layer.[87] The *ftsA* product interacts with PBP3, implicating it in peptidoglycan synthesis at the septum as well.[88] A fraction of the PBP3 is covalently modified by attachment of lipid.

The *ftsE* product is a 25-kDa protein with homology to a family of membrane-bound nucleotide-binding proteins involved in sugar, amino acid, and vitamin transport, as well as in other processes. These proteins may function by coupling ATP binding or hydrolysis at the membrane to some cellular event.[89] The FtsE protein is associated with the inner membrane and may couple ATP hydrolysis to some aspect of cell division.[90] However, the filamentous phenotype of *ftsE* mutants can be partially suppressed by conditions that cause slow growth, suggesting that the gene is not obligatorily involved in division.[91]

Cell Separation. The *envA* gene is required for separation of the daughter cells.[92] In the *envA* mutant, chains of cells accumulate that have completely pinched-off inner cell membranes but still retain a double thickness of peptidoglycan, enough for two cell poles. Apparently, the *envA* mutant is defective in cleavage of the peptidoglycan layer, leaving the daughter cells covalently attached. The depressed levels of N-acetylmuramoyl-L-alanine amidase activity in these mutant cells are in keeping with this defect, but it is not known whether this enzyme is the *envA* gene product.[93]

Regulation of Septum Formation. The three proteins encoded by the *minB* locus (MinC, MinD, and MinE) are required to inactivate septum sites used during previous rounds of cell division, thereby restricting division to the center of the cell.[94] The MinC (25 kDa) and MinD (30 kDa) proteins together form an inhibitor of cell division. Their production in the absence of MinE causes filamentation. In contrast, the absence of either MinC or MinD allows division at the cell poles

---

85. Jones CA, Holland IB (1985) *PNAS 82*:6045; Lutkenhaus JF (1983) *J Bact 154*:1339.
86. Spratt BC (1977) *EJB 72*:341.
87. Ishino F, Matsuhashi M (1981) *BBRC 101*:905.
88. Tormo A, Ayala JA, de Pedro MA, Aldea M, Vicente M (1986) *J Bact 166*:985.
89. Higgins CF, Hiles ID, Salmond GPC, Gill DR, Downie JA, Evans IJ, Holland IB, Gray L, Buckel SD, Bell AW, Hermodson MA (1986) *Nature 323*:448.
90. Gill DR, Salmond GPC (1987) *MGG 210*:504.
91. Taschner PEM, Huls PG, Pas E, Woldringh CL (1988) *J Bact 170*:1533.
92. Normark S, Bowman HG, Matsson E (1969) *J Bact 97*:1334.
93. Wolf-Watz H, Normark S (1976) *J Bact 128*:580; Lutkenhaus JF, Wu HC (1980) *J Bact 143*:1281.
94. de Boer PAJ, Crossley RE, Rothfield LI (1989) *Cell 56*:641.

and formation of minicells. The MinE protein (10 kDa) is the topological specificity factor that restricts inhibition by MinC–MinD to the cell poles. These polar sites, which were at the cell center prior to division (Fig. 20-13), are still recognizable as sites for septum formation in the absence of the *minB* factors. Thus, properly positioned division requires the balanced production of the three *minB* products. Although normally involved in the division process, the *minB* products are not essential.

Periseptal Annuli: Domains of the Envelope at the Division Sites[95]

Long before the septum can be seen, periseptal annuli appear as the first signs of development of a division site. The annuli consist of two concentric rings around the circumference of the cell where the inner membrane, peptidoglycan layer, and outer membrane are closely associated (Fig. 20-13). The region of the cell envelope between the two rings is the site of the future septum.

Newborn cells already contain a pair of annuli at their center, formed in the previous generation, prior to cell division. Predivisional cells contain, in addition to the central annuli, pairs of annuli located at $\frac{1}{4}$ and $\frac{3}{4}$ of the cell length (Fig. 20-13); after division, these sites will be located at the centers of the newborn cells. New annuli first appear near the original midcell location and are translocated to the $\frac{1}{4}$ and $\frac{3}{4}$ locations during cell growth. Translocation may occur through progressive insertion of new cell envelope material between the midcell and the newly formed annuli.

Annuli are synthesized and positioned normally in *ftsA* mutants. Filaments that accumulate at the restrictive temperature have regularly spaced annuli, between which septation occurs after a return to the permissive temperature. The annuli may function in cell division as boundaries in the envelope, allowing the cell to selectively adjust the composition of the compartment between the two adhesion rings. Annuli can act as physical barriers to protein movement within the periplasmic space.

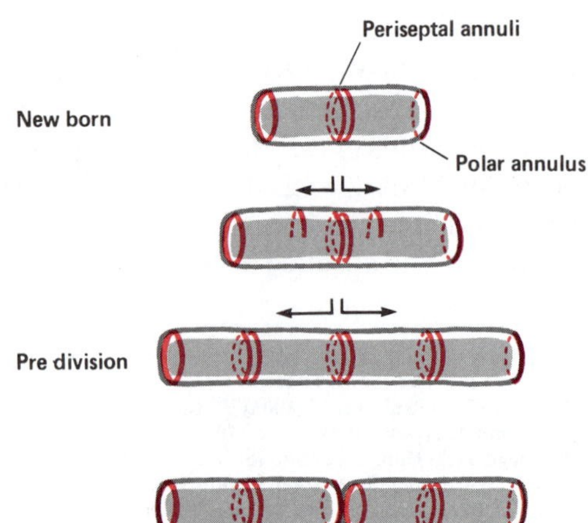

Figure 20-13
Development of periseptal annuli during cell growth. (Modified from a figure provided by Dr L Rothfield)

95. Cook WR, Kepes F, Joseleau-Petit D, MacAlister TJ, Rothfield LI (1987) *PNAS* 84:7144; Rothfield LI, Cook WR (1988) *Microbiol Sci* 5:182.

Cells divide 20 minutes after the completion of a round of replication. However, the time required to synthesize the chromosome and to build the division sites, including construction of periseptal annuli, exceeds the generation time when cells are growing at maximal rates. Thus in rapidly growing cells, early events such as initiation of replication and synthesis of annuli must occur in the grandmother cell, to ensure that daughter cells are properly formed two division cycles later.

Although the replication and cell-division cycles are clearly linked through some signalling devices, they are not obligatorily coupled. The clearest example of interaction between the two cycles is the rapid, efficient inhibition of cell division in response to disruption of chromosomal replication.

DNA damage, or other inducers of the SOS response, transiently inhibit septation until repair is complete and repression of the SOS regulon (Section 21-3) is reestablished. The inhibitor responsible for blocking division is the product of *sfiA* (suppressor of *fi*lamentation; also called *sulA*, suppressor of *lon*), which is under LexA repressor control. Expression of *sfiA* is induced by DNA damage, which activates RecA-mediated cleavage of LexA; SfiA protein inhibits FtsZ protein, blocking septation. Induction of *sfiA*, independent of DNA damage or any other component of the SOS system, is sufficient to block division.[96] Inhibition of septation by *SfiA* can be overcome by increasing the level of FtsZ, indicating that a stoichiometric interaction between FtsZ and SfiA proteins is involved in blocking division. Mutations in *ftsZ*, called *sfiB* or *sulB*, render cells insensitive to SfiA inhibition.[97]

The *sfiA* division block is temporary, lasting only as long as the RecA protein remains activated. Division is restored by the degradation of SfiA protein by *lon*-encoded protease.[98] Because SfiA protein has a half-life of only 1.2 minutes in lon^+ cells it disappears quickly once its synthesis is repressed, except in *lon* mutants, in which its half-life reaches 19 minutes. Thus, lon^- cells filament after inhibition of DNA synthesis and are hypersensitive to UV light and other DNA-damaging agents.

The *sfiA* gene is nonessential and is not involved in coupling the division and replication cycles under normal conditions.[99] Its function is restricted to blocking septation in response to DNA damage. A second SOS-induced division inhibitor, product of the *sfiC* gene present in some *E. coli* strains, is encoded by a cryptic genetic element. Like *sfiA*, its target is *ftsZ*.

In cells unable to prevent division by the SOS pathway, division is still inhibited by DNA damage or other blocks in chromosomal replication. Possibly, synthesis of a product required for division is triggered by replication of the chromosomal terminus. Synthesis of FtsA protein follows this pattern, but the mechanism of coupling is unclear as *ftsA* maps at 2 minutes, a position far from the terminus.[100] The *cfcA* (control *f*requency of *c*ell division) mutation prevents

---

96. Huisman O, D'Ari R, Gottesman S (1984) *PNAS 81*:4490.
97. Lutkenhaus JF (1983) *J Bact 154*:1339; Jones CA, Holland IB (1985) *PNAS 82*:6045; Jones CA, Holland IB (1984) *EMBO J 3*:1181.
98. Mizusawa S, Gottesman S (1983) *PNAS 80*:358; Schoemaker JM, Gayda RC, Markovitz A (1984) *J Bact 158*:551.
99. Huisman O, D'Ari R (1981) *Nature 290*:797; Huisman O, Jacques M, D'Ari R, Caro L (1983) *J Bact 153*:1072.
100. Tormo A, Dopazo A, de la Campa AG, Aldea M, Vicente M (1985) *J Bact 164*:950; Tormo A, Fernández-Cabrera C, Vicente M (1985) *J Gen Microbiol 131*:239; Masters M, Paterson T, Popplewell AG, Owen-Hughes T, Pringle JH, Begg KJ (1989) *MGG 216*:475.

both the SOS-dependent and SOS-independent pathways of division inhibition, and these cells divide more frequently than normal when replication is undisturbed.[101] This mutation therefore may provide further clues to the pathway of SOS-independent coupling of replication and division.

In the absence of SOS, inhibition of division after a replication block is not completely effective. In a normal strain, fewer than 1% of the cells lack chromosomal DNA; by contrast, in SOS-defective mutants, 10% to 17% of the cells have no DNA after inhibition of DNA synthesis. The size and shape of these cells are unaltered, indicating that normal division can occur even when chromosomal replication has not been completed.

Interdependence of the Replication and Division Cycles Summarized

During balanced growth, division occurs 20 minutes after completion of a round of replication, which requires 40 minutes (Section 2). The cyclic events of replication and cell division are thus superimposed in an ordered way which ensures that newborn cells will have a completed chromosome. Yet the cycles can be uncoupled. Some rules are apparent that describe the response of the replication cycle to changes in the division cycle, and the reverse:

(1) *Inhibition of replication inhibits cell division.* This takes place at the stage of septation, principally by SOS induction of the *sfiA* inhibitor of *ftsZ;* additional pathways are also evident.

(2) *Inhibition of division does not inhibit replication.* Thus in *fts* mutants, replication and chromosome segregation continue, producing filaments with a normal DNA content and with nucleoids regularly spaced along their length.

(3) *Neither the initiation nor the termination of replication is sufficient to trigger cell division.* Chromosomal replication frequency can be increased and initiation advanced in the cell cycle without causing a corresponding increase in the frequency of division. Examples include induction of initiation by overproducing the dnaA protein, or by initiating chromosomal replication from a temperature-inducible plasmid origin.[102]

(4) *The division rate does not determine the rate of replication.* The cell division frequency can be increased by overproduction of FtsZ protein without causing a corresponding increase in replication. The result is a minicell without DNA.

Replication and cell division are regulated, at least partially, by independent mechanisms. Both cycles may be triggered by common signals related to cell growth, such as the rate of protein synthesis, the stringent response to amino acid starvation, changes in growth temperature (the heat-shock response;[103] Section 7-8), and carbon availability via the CAP · cAMP complex (Section 10-9). For cell division to occur, two completed chromosomes should be available and cell growth should have reached a certain minimal length; under most growth conditions, these requirements are fulfilled at about the same time.

101. Nishimura A (1989) *MGG 215*:286.
102. Bernander R, Nordström K (1990) *Cell 60*:365.
103. Tsuchido T, VanBogelen RA, Neidhardt FC (1986) *PNAS 83*:6959.

Prior to division, replication of the chromosome must have been completed and the two nucleoids partitioned into the two halves of the maternal cell. Little is known about the mechanism of this partitioning. The required chromosomal signals do not appear to be within or near *oriC*, inasmuch as *oriC* plasmids are randomly segregated at division. Centromeres and microtubules, responsible for chromosome segregation in eukaryotic cells, have not been identified in bacteria.

In the maternal cell before division, replicated nucleoids, like the periseptal annuli (see above), are located at $\frac{1}{4}$ and $\frac{3}{4}$ of the cell length prior to construction of the septa. This positioning requires protein synthesis. When replication is completed in the absence of protein synthesis, the two completed nucleoids remain in the middle of the cell. After removal of the block to protein synthesis, nucleoids move to the proper location before any significant increase in cell length is achieved, indicating the independence of chromosome positioning from cell growth.

Chromosome segregation is normal in most cell-division mutants; nucleoids are positioned regularly along the filaments. Mutations conditionally defective in partitioning give rise to filaments with centrally located nucleoids. The *parA* and *parD* mutations are alleles of *gyrB* and *gyrA*, respectively, and may encode gyrase subunits especially defective in decatenation (Section 12-3). The *parB* mutation, an allele of *dnaG*, which encodes primase (Section 8-3), is unlikely to be involved in segregation in a direct way. Mutants in *parC*, and *parE* frequently produce DNA-free cells.[104] These genes appear to encode a type II topoisomerase distinct from, but homologous with, gyrase.[105] The *muk* mutant cells produce a large number of "nondisjunction" divisions in which one daughter cell has two nucleoids and the other has none.

20-6 The Eukaryotic Cell Cycle[106]

Eukaryotic cells duplicate their DNA during a distinct synthesis (S) period, preceded by a gap (G_1) phase and separated from mitosis (M) by a second gap (G_2) phase (Fig. 20-14). The duration of each of these four phases varies considerably, but the values are reasonably constant, totalling 24 hours, in animal cells in culture: S phase lasts about 7 hours, G_2 about 4 hours, M about 1 hour, and G_1, the most variable, 12 hours or more. An analogous cycle in yeast has a generation time as brief as 1.4 hours, and in *Xenopus*, embryos divide every 30 minutes for the first 7 hours. In these rapidly dividing cells, the S and M periods are abbreviated and the G_1 and G_2 phases are essentially eliminated.

Despite the enormous importance of control of the replication cycle, and huge effort devoted to its study, the biochemical mechanisms are incompletely understood. The first major control in the replication cycle is the choice between

104. Kato J, Nishimura Y, Yamada M, Suzuki H, Hirota Y (1988) *J Bact 170*:3967; Kato J, Nishimura Y, Imamura R, Niki H, Hiraga S, Suzuki H (1990) *Cell 63*:393.
105. Hiraga S, Niki H, Ogura T, Ichinose C, Mori H, Ezaki B, Jaffé A (1989) *J Bact 171*:1496.
106. Dunphy WG, Newport JW (1988) *Cell 55*:925; Hartwell L, Weinert T (1989) *Science 246*:621; Lee M, Nurse P (1988) *Nature 327*:287; Murray AW, Kirschner MW (1989) *Science 246*:614.

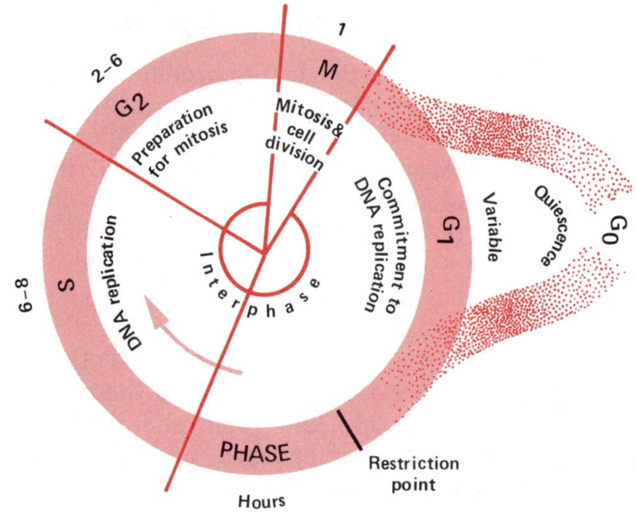

Figure 20-14
Growth cycle of a eukaryotic cell, including a quiescent phase (G_0). Variation in cell cycle time is due largely to variation in the duration of G_1 phase.

remaining quiescent (in G_0) and entering the cycle (G_1). Once committed, the cycle usually progresses to completion.

$G_1 \rightarrow S$: Commitment to the Cell Cycle; Restriction Point or Start[107]

A cell is committed to proliferate once it proceeds past the "restriction point" (R), called "Start" in yeast. If it fails to pass R, a cell becomes committed to an indefinite period of quiescence. Passage of a cell through R depends on extracellular signals. For the cells of multicellular organisms, specific *growth factors*, small proteins that bind to receptors on the cell surface, are essential.[108] Different cell types have different receptors and respond to distinct growth factors. As examples, epithelial cells proliferate in response to platelet-derived growth factor (PDGF), and T cells are stimulated by interleukin 2 (IL2).

Upon binding to its receptor, the growth factor initiates a regulatory cascade. PDGF binding alters the conformation of the cytoplasmic domain of its receptor, activating its intrinsic kinase which then phosphorylates target proteins, triggering a sequence of events required for entrance into S phase. In addition to phosphorylated proteins, second-messenger molecules include Ca^{2+}, calmodulin, cyclic nucleotides, inositol phosphates, and prostaglandins. Control of replication at the G_1–S boundary may also be modulated by the relative activities of a large variety of protein kinases and phosphatases, some of which act directly on the replication enzymes (Section 7). In addition to growth factors, certain nutrients and de novo protein synthesis are prerequisites for the commitment to a round of cell division. For cells in culture, sufficient anchorage sites are also required.

As a unicellular organism, yeast is sensitive to the growth medium rather than to growth factors in its passage through Start.[109] Cells must reach a minimal size

107. Pardee AB (1989) *Science 246*:603.
108. Yarden Y, Ullrich A (1988) *ARB 57*:443.
109. Hartwell L, Weinert T (1989) *Science 246*:621.

to enter the division cycle. The mating pheromones, a- and α-factors, produced and secreted by haploid cells, arrest division of cells of the opposite mating type at this stage. Certain mutations that disrupt the cell cycle identify *cdc* (cell division cycle) genes specifically required to progress through Start. The *CDC28* gene in *Saccharomyces cerevisiae* (*cdc2* in *Schizosaccharomyces pombé*) is required at both $G_0 \rightarrow G_1 \rightarrow S$ and at $G_2 \rightarrow M$; the gene encodes a protein kinase analogous to the catalytic subunit of mitosis-promoting factor (MPF) that regulates entry into mitosis (see below).

$S \rightarrow G_2$

Once started, the remaining stages of the cell cycle occur in order, consuming relatively fixed times irrespective of extracellular signals. In S phase, the genome is replicated exactly once; initiation occurs at multiple origins which fire at different times influenced by transcription and their position on the chromosome (Section 7). While the cycle is predetermined once the restriction point has been passed, a blockade of replication by inhibitors or by DNA-damaging agents can delay the onset of G_2 and M (Section 7). In G_2, the nuclei prepare for division, the replicated chromosomes remain as paired sister chromatids, and further DNA synthesis is prohibited (Section 7).

$G_2 \rightarrow M$: Mitosis-Promoting Factor ($p34^{cdc2}$ and Cyclin)[110]

Entry into mitosis is controlled by the cytoplasmic *mitosis-* (or *maturation-*) *promoting factor* (MPF). Injection of active MPF into cells at any stage in the cycle leads to the breakdown of the nuclear envelope and chromosome condensation that are characteristic of mitosis. MPF consists of two subunits, a 34-kDa serine–threonine protein kinase (called $p34^{cdc2}$, as it is the product of the *cdc2* gene in *Schizosaccharomyces pombé*) and an approximately 50-kDa subunit called *cyclin.* First purified from *Xenopus,* MPF is highly conserved, and promotes mitosis in mammals, yeasts, and the sea urchin. MPF triggers mitosis by its protein kinase activity. Phosphorylation of histone H1 influences chromosome condensation, and some additional regulatory proteins are probably also specifically modified.[111]

The level of activated MPF varies throughout the cell cycle, peaking during mitosis (Fig. 20-15). Activation of MPF appears to involve at least four events: (1) dephosphorylation of certain tyrosine, serine, and threonine residues on the $p34^{cdc2}$ subunit, activating its protein kinase; (2) accumulation of cyclin through new synthesis; (3) association of the kinase and the cyclin subunits; and (4) phosphorylation of cyclin by $p34^{cdc2}$. The protein synthesis requirement for entry into M allows the production of the critical level of cyclin needed to activate MPF.[112] However, once activation is achieved, cyclin is abruptly degraded, lowering the cyclin level and thus reducing the level of active MPF. The kinase activity of the $p34^{cdc2}$ subunit is further inactivated by rephosphorylation, returning the protein to its inactive, interphase status.

---

110. Lee M, Nurse P (1988) *Nature 327*:287; Dunphy WG, Newport JW (1988) *Cell 55*:925; Murray AW, Kirschner MD (1989) *Science 246*:614.
111. Arion D, Meijer L, Brizuela L, Beach D (1988) *Cell 55*:371; Maller JL (1990) *Biochemistry 29*:3157.
112. Murray AW, Kirschner MW (1989) *Nature 339*:287.

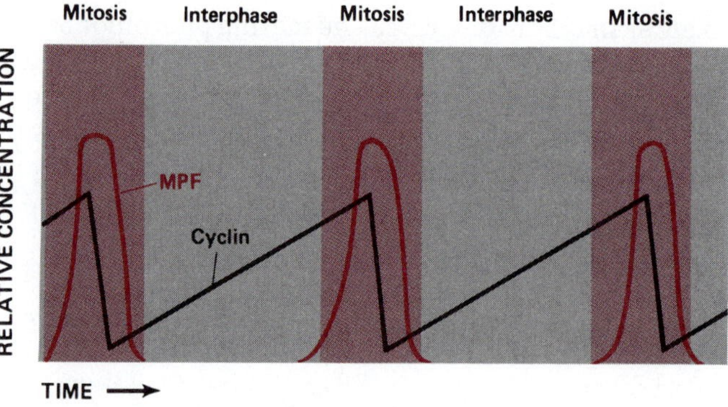

Figure 20-15
Correlation between mitosis and the rise and fall in levels of activated MPF and its cyclin subunit.

$G_1 \rightarrow G_0$: Commitment to Quiescence[113]

Most tissues in an adult organism arrest DNA replication and cell division upon maturity. To understand the biochemical mechanism, a number of experimental means have been used to induce the quiescent phase in tissue culture cells. The effects observed, termed a "negative pleiotypic response," are strongly influenced by the system under study. The most common experimental inducers of G_0 are (1) exhaustion of a nutrient, such as an essential amino acid, glucose, or phosphate, (2) absence of serum growth factors, (3) confluent growth of cells ("contact inhibition" or "density-dependent inhibition"), accompanied by a loss of contact with the substratum, and (4) inhibition of protein synthesis. Mutational loss of a required factor, such as in the *cdc* mutants in yeast, is also useful, as are drugs (e.g., colchicine, vincristine) that arrest the cycle in mitosis or early in G_1.

G_1 cells that fail to find the requirements for passing the restriction point (R) enter the quiescent state of G_0. Once they have entered G_0, much longer times are required to pass R after return of the proper conditions than are needed for continually cycling cells. G_0 cells are distinct from those in G_1 in exhibiting only about one-third the rate of RNA and protein synthesis and in having a smaller cell size due to losses of protein and RNA. The enzymes required for replication must be newly synthesized when cells emerge from G_0 (Section 7).

In cancer cells, the requirements for entry into the cell cycle are relaxed. Transformed cells often show cell-cycle control abnormalities in which the requirement for growth factors is reduced, as are the sensitivity to high cell density and the need for anchorage sites. Among the cellular protooncogenes are those for growth factors and their receptors, for protein kinases in cytoplasmic membrane and cytoskeletal locations, and for a large number of nuclear regulatory proteins. Clearly, the pathway of transduction of an extracellular signal that ultimately leads to quiescence or proliferation provides a large target for factors that can disrupt cell-cycle control.

20-7 Control of Replication in Eukaryotes

Replication of chromosomal DNA is restricted to the S phase by a combination of *trans*-acting factors and *cis*-acting signals. Within S, additional control mecha-

113. Pardee AB (1989) *Science* *246*:603.

nisms ensure that the entire genome is copied only once and that continuation of the division cycle is blocked if replication is incomplete. Also regulated are the number of origins that fire in a given S phase and the timing of replication of different regions in a chromosome.

Replication Block in G_1: Lack of *Trans*-Acting Factors

G_1 nuclei cannot replicate their DNA because essential *trans*-acting factors are missing. If cells containing nuclei at different stages in the cell cycle are fused[114] (Fig. 20-16), the nuclei in G_1 replicate abnormally early when exposed to cytoplasm from S-phase cells. The requirement for protein synthesis before cells can pass from the restriction point into G_1 (Section 6) is another indication that critical proteins must be synthesized at this stage.

The intracellular levels of enzymes required for DNA synthesis, such as thymidine kinase,[115] primase (p49 subunit),[116] and pol α,[117] has been compared in cells induced to enter the cycle from quiescence ($G_0 \rightarrow S$) and in actively proliferating cells. Strikingly increased production of the essential replication proteins accompanies the commitment to DNA synthesis in the induced cells. In contrast, actively cycling cells synchronized by physical methods (such as fractionation by size, rather than by inducing a metabolic block) contain comparatively uniform levels of replication enzymes in the four cycle stages. Therefore, the protein synthesis requirement for cells to pass beyond the restriction point is not attributable to the need for replication enzymes in order to enter S from G_1.

Stimulation of Replication by Phosphatase and Kinase Activities. A clue to the regulation of replication at the G_1–S border is provided by studies of SV40 DNA replication in extracts from human cells. Aside from requiring the viral T antigen, a DNA helicase that binds and melts the SV40 origin (Section 11-4), SV40 replication depends entirely on cellular factors (Section 19-2). Remarkably, replication in extracts from G_1 cells is tenfold less active than in extracts from cells in other stages of the cycle.[118] Separation of the DNA strands at the origin is especially poor in these G_1 extracts. Replication protein 3 (RP-C), which stimulates melting of the origin DNA and initiation of SV40 replication in a partially

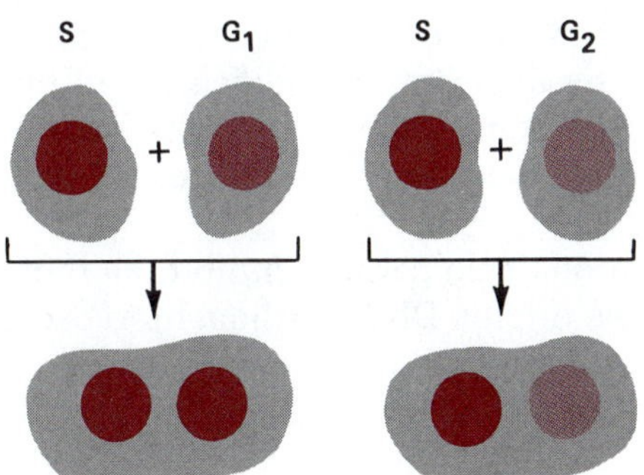

Figure 20-16
Schematic representation of cell-fusion experiments demonstrating the ability of a G_1-phase nucleus, but not a G_2-phase nucleus, to be activated *in trans* by an S-phase cell.

114. Rao PN, Johnson RT (1970) *Nature 225*:159.
115. Sherley JL, Kelly TJ (1988) *JBC 263*:8350.
116. Tseng BY, Prussak CE, Almazan MT (1989) *Mol Cell Biol 9*:1940.
117. Wahl AF, Geis AM, Spain BH, Wong SW, Korn D, Wang TSF (1988) *Mol Cell Biol 8*:5016.
118. Roberts JM, D'Urso G (1988) *Science 241*:1486.

purified system,[119] is implicated as the factor whose activity is depleted in G_1. Identical to the catalytic subunit of protein phosphatase 2A, RP-C specifically removes certain phosphates from T antigen, stimulating its activity[120] (Section 19-2). Extracts of G_1-phase cells are activated eightfold by RP-C, while those from S phase are little affected.

In another study, a protein homologous to the yeast kinase p34^{cdc2} (Section 6) was found to activate SV40 replication in extracts from G_1.[121] Phosphorylation of a protein other than T antigen seems to be required. These data suggest that, as in yeast, MPF functions at both the $G_1 \rightarrow S$ and $G_2 \rightarrow M$ transitions in mammalian cells (Section 6). Thus, a proper balance between kinases and phosphatases may trigger the onset of replication as it does other events in the cell cycle (Section 6).

Replication Block in G_2: *Cis*-Inhibition Relieved by Mitosis

The G_2-phase nucleus of a cell, if fused with one in S phase, is not activated for replication (Fig. 20-16), indicating that in contrast to G_1 nuclei, G_2 nuclei cannot be activated *in trans*.[122] The ability of the fused S-phase nucleus to continue replication argues against the presence of a diffusible G_2 inhibitor. Inasmuch as extracts from G_2-phase cells have abundant replication factors and are active for DNA synthesis in vitro, the inhibition of replication in G_2 probably operates by a *cis*-specific mechanism, relieved by passage through mitosis.

Effect of *Cis*-Acting Sequences on Replication in Bovine Papilloma Virus (BPV).[123] Replication of each of the multiple extra chromosomal copies of this provirus occurs once per S phase in contrast to the process in prokaryotic episomes, in which individual plasmids are chosen at random for replication until the appropriate plasmid concentration is achieved (Chapter 18). In BPV, two sequences have been identified near the origin that are responsible for limiting DNA synthesis to a single round. A *trans*-acting factor encoded by the 5′ portion of the open reading frame E1, called the M (maintenance) factor, also appears to be involved.

Although the mechanism by which the BPV signals control replication is unknown, possibilities have been suggested that may be applicable to the *cis*-inhibition of reinitiation on cellular chromosomes. For example, a chromatin structure generated during replication could preclude the further action of replication enzymes. Chromatin assembly is linked to replication, and the newly assembled fibers could be marked by differential modification of histones. The highly condensed state of chromosomes in M phase also may be sufficient to prevent replication until the decondensation after mitosis removes this barrier.

Pairing of sister chromatids after replication could perhaps inhibit the reinitiation of replication. Association of the two duplex DNAs is thought to occur along their lengths, mediated by the action of *cis*-acting sequences. Separation of sister chromatids during mitosis would automatically eliminate this feature were it a block to replication.

---

119. Virshup DM, Kelly TJ (1989) *PNAS 86*:3584.
120. Virshup DM, Kauffman MG, Kelly TJ (1989) *EMBO J 8*:3891.
121. D'Urso G, Marraccino RL, Marshak DR, Roberts JM (1990) *Science 250*:786.
122. Rao PN, Johnson RT (1970) *Nature 225*:159.
123. Roberts JM, Weintraub H (1988) *Cell 52*:397; Roberts JM, Weintraub H (1986) *Cell 46*:741. Botchan M, Berg L, Reynolds J, Lusky M (1986) in *Papillomaviruses* (Evered D, Clark S, eds.). Wiley, New York. p. 53.

Sequence-Independent Inhibition of Replication Reinitiation in *Xenopus* Embryos.[124] Replication in *Xenopus* embryos and DNA synthesis in extracts from *Xenopus* eggs do not require any specific nucleotide sequences (Section 16-6). Origin-independent replication is efficient and semiconservative, and the entire genome is replicated once. Thus, no specific sequence is required for this remarkable regulation of replication.

The mechanism of limiting replication to one round may depend on the packaging of the DNA into nuclei prior to the onset of replication. In extracts from eggs, replication cannot go beyond a single round in the absence of protein synthesis. Additional replication cycles, coincident with cycles of nuclear breakdown and chromosome condensation that mimic mitosis, depend on protein synthesis or the addition of activated MPF (Section 6). Thus, the inhibition of reinitiation may perhaps be relieved by "mitosis" in vitro much as it is in vivo.

Of the mitotic events that occur upon addition of MPF, breakdown of the nuclear envelope is sufficient to allow reinitiation of replication. Chemical, enzymatic, and physical treatments that permeabilize the membrane can activate a second round of replication. This effect may be explained if a factor required for initiation gains access to the nucleus only when the nuclear membrane is dispersed. This "licensing" factor must be able to promote a single round of replication. While able to explain regulation in *Xenopus* embryos and egg extracts, this mechanism may be inadequate to account for the ordered initiation at relatively fixed chromosomal locations throughout S phase that occurs in other types of cells (see below).

Restriction of Replication to S Phase Summarized

(1) Absence of replication in G_0 can be attributed to an inadequate supply of critical replication proteins constitutively present in proliferating cells. These replication enzymes are synthesized in G_1 when cells enter the cycle from G_0.

(2) DNA replication in G_1 is blocked for lack of *trans*-acting factors. Activation of the replication enzymes to a functional form may depend on the proper adjustment of protein phosphatase and kinase activities at the G_1–S border.

(3) Reinitiation of replication may be prevented by the state of the DNA in the S and G_2 phases. Chromatid pairing, chromatin conformation, and permeability of the nuclear membrane may account for this negative control.

(4) Passage through mitosis restores the G_1 state. Although the *cis*-inhibition of a new round of replication is removed, the *trans*-acting replication potential, apparently present throughout S and G_2, must also be inactivated, possibly by changes in the protein phosphatase and kinase activities that modify the replication proteins.

Inhibition of DNA Synthesis and Block of Mitosis[125]

The arrest of DNA synthesis by inhibitors or mutations in essential replication genes blocks the onset of mitosis. DNA-damaging agents have the same effect, causing the cell cycle to halt in G_2 to provide time for repair. This dependence of mitosis on the completion of replication is clearly seen in yeast. Temperature-sensitive mutations in the *S. cerevisiae* replication enzymes display the cell-

---

124. Blow JJ, Laskey RA (1988) *Nature 332*:546.
125. Hartwell L, Weinert TA (1989) *Science 246*:629.

division-cycle phenotype: pol I = *CDC17*, pol III = *CDC2*, ligase = *CDC9*. Suspension of the cell cycle prior to mitosis characterizes these mutants.

The yeast *RAD9* (*ra*diation sensitivity) gene controls coupling of mitosis to the completion of replication.[126] Cells defective in this gene (*rad9* null mutants) divide despite DNA damage or when replication is blocked by a temperature-sensitive replication mutant. Even in the absence of DNA damage, *rad9* cells lose chromosomes at a 20-fold higher rate than wild-type cells, an indication that the gene normally ensures that replication is complete before allowing cell division. A possible mechanism for the RAD9 protein is to inhibit MPF activity until the completion of replication.

In mammalian cells, as well as in yeast, completion of replication or of DNA repair is necessary prior to the onset of mitosis. The G_2 block imposed by DNA damage or incomplete replication can be bypassed by treating the cells with caffeine.[127] Caffeine may promote mitosis by stabilizing a labile protein essential for the process. Alternatively, caffeine could function by activating phosphatases or inhibiting kinases involved in regulating the entrance into mitosis.[128]

The link between completion of DNA synthesis and initiation of mitosis is also evident in *Xenopus* egg extracts, which are competent for nuclear assembly, DNA synthesis, and mitotic cycling.[129] The cyclic rise and fall of MPF activity and nuclear breakdown and reassembly are mitotic events that normally occur in these extracts. When DNA synthesis is inhibited by the addition of aphidicolin, these mitotic events are blocked; the extracts appear permanently arrested in S phase. The inhibition of initiation of mitosis can be traced directly to the absence of activation of MPF (Section 6). Unreplicated DNA blocks a critical posttranslational modification, required for MPF activation. Production of cyclin and association of the cyclin and kinase subunits of MPF do not appear to be the stages inhibited. As in mammalian cell culture, addition of caffeine reverses the inhibition, allowing mitosis to occur without prior completion of DNA synthesis.

Temporal Regulation of Initiation: Firing of Replication Origins[130]

A cell committed to replication needs additional signals to determine when and where, on the complex array of chromosomal DNA, synthesis is to begin. Regulation operates on both the *number* of origins that will be used in a given S phase and the *timing of firing* of origins in different locations on the genome.

Change in Length of S Phase during Development. The duration of S phase varies markedly during development of an organism. Changes in the replication pattern during *Drosophila* development, for example, are dramatic. During the first divisions of the fertilized embryo, the S phase is completed in 4 minutes, while later in development, the same amount of DNA synthesis requires over 10 hours. In *Xenopus*, a 25-minute synthesis period in the embryo can be compared to many hours in the adult. The length of time consumed by DNA synthesis is determined principally by the average replicon size, which in turn depends on the number of origins activated during S.

---

126. Weinert T, Hartwell L (1988) *Science 241*:317.
127. Schlegel R, Pardee A (1986) *Science 232*:1264; Schlegel R, Croy R, Pardee A (1987) *J Cell Physiol 131*:85.
128. Dasso M, Newport JW (1990) *Cell 61*:811.
129. Dasso M, Newport JW (1990) *Cell 61*:811.
130. Laskey RA, Fairman MP, Blow JJ (1989) *Science 246*:609.

Table 20-4
Changes in S phase during *Drosophila* development[a]

Stage	Length of S phase (hours)	Mean interorigin distance (kb)	Elongation rate (kb/min)
Cleavage nuclei	0.06	7.9	2.6
Cellular blastoderm	0.33	10.2	—
Larval brain	11.9	107	1.0

[a] Modified from Spradling A, Orr-Weaver T (1987) *ARG 21*:373.

Measuring the distance between replication bubbles in electron micrographs (Section 16-1) reveals that origins are much more closely spaced in the rapidly replicating cells than in cells of mature tissues (Table 20-4). Between 10 and 100 times as many origins must fire in each S phase in the early embryo. In contrast to the initiation patterns, the rate of DNA chain elongation is relatively constant throughout development.

Selection of Replication Origins. Little is known about replication origins in eukaryotes other than those in yeast. Yet arguments for preferred initiation sites in most cell types are persuasive (Section 16-6). *Xenopus* eggs are unusually permissive in that specific origin sequences are not required. Early in development (in *Xenopus* and perhaps generally) many sequences can serve as origins that cannot be used later. A possible explanation for this decreased sequence specificity early in development is that the DNA in embryonic cells is not transcriptionally active, whereas in more mature cells replication and transcription are concurrent. Most known eukaryotic replication origins overlap with promoters and transcription regulatory elements (Sections 16-5 and 16-6). Replication timing of many genes also depends on their transcription (see below). Thus, early in development, frequently spaced, transcription-independent origins may be used, whereas later on, coordination of transcription and replication imposes a more stringent selection among replication initiation sites.

Timing of Firing at Origins during S Phase. DNA synthesis in the S phase occurs in an ordered pattern: different regions of the chromosome replicate in a characteristic temporal sequence. In a mammalian chromosome, some regions replicate early in S, while others are late-replicating. As indicated by the length of a coordinately regulated region, some 20 individual origins are probably initiated together in a chromosomal domain.[131] Electron microscopic evidence also suggests that neighboring origins are activated nearly simultaneously.[132]

The timing of replication of different chromosomal regions in yeast suggests a pattern in which centromeric sequences and surrounding regions replicate within the first half of the S phase, while telomeres at the chromosome ends are replicated late. Since telomeres contain putative origins capable of functioning in plasmids, the tardy replication of telomeres suggests that certain origins are programmed to fire last.

131. Crossen PE, Pathak S, Arrighi FE (1975) *Chromosoma 52*:339; Holmquist G, Gray M, Porter T, Jordan J (1982) *Cell 31*:121.
132. Hand R (1975) *JCB 64*:89.

The replication pattern of a 200-kb region of yeast chromosome 3, analyzed in detail, confirms other evidence that the centromere is replicated early and that the telomere is last. Initiation appears to occur principally at three ARS elements, with synthesis proceeding at about 3.6 kb per minute. Fork movement is considerably slowed near the telomere. The three origins appear to fire at about the same time, consistent with their coordinate regulation. Also, several sequences that are known to function as ARS elements do not seem to be used.

When the timing of S-phase replication of specific genes was analyzed in cultured mammalian cells, several rules emerged.[133] (1) Individual genes replicate at a characteristic time in a given cell type. (2) Actively transcribed genes often replicate early. (3) The timing of replication of a certain gene may differ in different cell types. (4) Active expression of a gene often causes it to replicate earlier in S phase than it does in a cell where it is not expressed. The control of replication and transcription is clearly interrelated (Sections 16-5 and 16-6).

Gene Amplification: Overriding the Rules against Reinitiation[134]

In at least two cases, gene amplification is used in terminally differentiated cells to achieve a high level of expression of mRNAs and proteins. In *Drosophila*, genes responsible for the continuous high-level synthesis of the chorion proteins are amplified about 50-fold in follicle cells by selective hyperreplication, resulting in a partially polytene structure extending over about 100 kb[135] (Section 16-6). In developing chick myotubules, the α-actin genes (and probably other muscle-specific genes as well) are amplified about 100-fold, but only during the 2-day period of rapid synthesis of muscle proteins. In this instance, a high turnover of extrachromosomal templates for transcription seems likely. These templates may be circles or linear tandem repeats produced by rolling-circle replication.[136] Novel DNA structures found in the amplified DNA support such a mechanism, which is similar to that used for the high-level synthesis of rRNA in amphibian oocytes.[137]

Gene Amplification Revealed by Drug Resistance. In some cell lines resistant to methotrexate, small episomal "double-minute" chromosomes carry the amplified dihydrofolate reductase genes.[138] Although the detailed structure of double-minute chromosomes is unknown, it seems likely that they comprise a large number of circular or linear tandem repeats of the gene and neighboring sequences derived from the chromosome. Such chromosomal units are also found consistently in several different types of tumors, especially neuroblastomas and retinoblastomas, inviting the speculation that amplification and consequent overproduction of one or more proteins may be related to tumorigenesis in such cases. By contrast, DNA amplification is very rare in normal human cells.[139]

---

133. Brown EH, Ibqal MA, Stuart S, Hatton KS, Schildkraut CL (1987) *MCB 7*:450; Dhar V, Skoultichi AI, Schildkraut CL (1989) *MCB 9*:3524.
134. Schimke RT (ed.) (1982) *Gene Amplification*. CSHL.
135. Spradling AC (1981) *Cell 27*:193.
136. Zimmer WE Jr, Schwartz RJ (1982) in *Gene Amplification* (Schimke RT, ed.). CSHL, p. 137.
137. Brown DD, Dawid IB (1968) *Science 160*:272.
138. Schimke RT, Brown PC, Kaufman RJ, McGrogan M, Slate DL (1980) *CSHS 45*:785; Haber DA, Schimke RT (1981) *Cell 26*:355.
139. Wright JA, Smith HS, Watt FM, Hancock MC, Hudson DL, Stark GR (1990) *PNAS 87*:1791.

In other examples of resistance to methotrexate,[140] to N-phosphonacetyl-L-aspartate (PALA),[141] and to cadmium ions, the highly amplified genes for the respectively overproduced dihydrofolate reductase, CAD (first three enzymes of pyrimidine biosynthesis), and metallothionein I are arranged linearly over a considerable length of chromosomal DNA, with 100 to 500 kb of the amplified DNA present per gene. In one case, these sequences were observed to replicate synchronously, early in S phase.[142] Much of the amplified DNA surrounding the CAD gene has been cloned, and novel junctions with the neighboring sequences are present. How the amplified structures arise is unknown, but recombination, rather than replication, appears to be the initial event.[143] Circular episomal elements may then serve as intermediates in the amplification process (Section 16-6).[144]

140. Biedler JL, Spengler BA (1976) *Science* 191:185; Mibrandt JD, Heintz NH, White WC, Rothman SM, Hamlin JL (1981) *PNAS* 78:6043.
141. Wahl GM, Vitto L, Padgett RA, Stark GR (1982) *Mol Cell Biol* 2:308.
142. Mibrandt JD, Heintz NH, White WC, Rothman SM, Hamlin JL (1981) *PNAS* 78:6043.
143. Smith KA, Gorman PA, Stark MB, Groves RP, Stark GR (1990) *Cell* 63:1219.
144. Stark GR, Debatisse M, Guilotto E, Wahl GM (1989) *Cell* 57:901; Ruiz JC, Choi KH, von Hoff DD, Robinson IB, Wahl GM (1989) *Mol Cell Biol* 9:109.

CHAPTER 21

Repair, Recombination, Transformation, Restriction, and Modification

21-1 DNA Damage and Mutations[1]

DNA is subject to damage by chemical and physical agents in the environment and by free radicals or alkylating agents generated in metabolism. DNA also suffers errors during its replication and in the course of other metabolic processes. These lesions and errors may be removed by repair (Section 2), replaced by recombination (Section 5), or retained, leading to mutations or cell death. DNA-damaging agents have been applied in the treatment of disease (e.g., cancer) and used experimentally to produce mutations for genetic analyses that probe metabolic mechanisms (Section 14-1).

DNA Lesions

The most common forms of DNA damage are the following:

(1) *Missing base.* Acid and heat remove purines; at physiologic temperatures and pH, roughly 10,000 purines and several hundred pyrimidines are lost per mammalian genome per day. Certain alterations destabilize bases, leading to their elimination (see below). A large variety of altered bases (e.g., uracil) are also excised by specific DNA glycosylases. The resulting apurinic or apyrimidinic sites *(AP sites)* are mutagenic during replication in vivo, especially when present during the SOS-induced mutagenic state (Sections 3 and 5).

1. Lindahl T (1982) *ARB* 51:61; Walker GC (1985) *ARB* 54:425; Friedberg EC (1985) *DNA Repair.* WH Freeman, New York; Sancar A, Sancar GB (1988) *ARB* 57:29; Echols H, Goodman MF (1991) *Mutat Res* 236:301.

(2) *Altered base (or nucleoside).* Ionizing radiation and electrophilic agents (including alkylating agents) modify bases and sugars. Potent carcinogens (e.g., aflatoxin B1, 2-acetyl-aminofluorene, benzo[a]pyrene) introduce *bulky adducts.* In addition, base (or nucleoside) analogs (Section 14-3) may be substituted during replication.

Deaminations that give rise to uracil, hypoxanthine, and xanthine are spontaneous, because of inherent chemical instability, or are inducible by nitrous acid. Mutations due to deamination are exclusively *transitions,* rather than transversions. That is, one pyrimidine replaces the other, or one purine replaces the other; for example, a TA base pair becomes CG. (In *transversions* a purine replaces a pyrimidine, a pyrimidine replaces a purine; e.g., TA becomes AT.)

(3) *Incorrect base.* A wrong base pair may result from an incorporation error during replication. Mutator genes, which increase the frequency of such "spontaneous mutations," encode defective polymerases or nucleases or other proteins essential for the normal high fidelity of replication (Section 15-9).

(4) *Bulge due to deletion or insertion.* Intercalating agents (e.g., acridines; Section 14-4), inserted between the planar rings of the bases, cause omission of a nucleotide or addition of extra nucleotides during replication or recombination. These errors produce *frameshift* (phase-shift) *mutations* which result in a shifting of the translation reading frame out of phase.

(5) *Linked pyrimidines.* UV irradiation (254 nm) produces cyclobutyl dimers between adjacent pyrimidines, most prominently *thymine dimers* (Fig. 21-1). The *(6–4) lesion,* the linkage of C6 of a cytosine (or thymine) to C4 of an adjacent (3′) cytosine, is another biologically significant photoproduct. Normal base pairing is prevented by the presence of these dimers.

(6) *Strand breaks.* Either directly by attack at a sugar residue or as a consequence of base alterations, phosphodiester bond breaks (single or double) follow ionizing radiation (Section 14-7) or action of certain chemical agents (e.g., bleomycin; Section 14-4).

(7) *3′-Deoxyribose fragments.* Rupture of deoxyribose by many free radical–generating agents (e.g., bleomycin and ionizing radiation) leaves strand breaks terminated with 3′-sugar products [e.g., phosphoglycolate].[2]

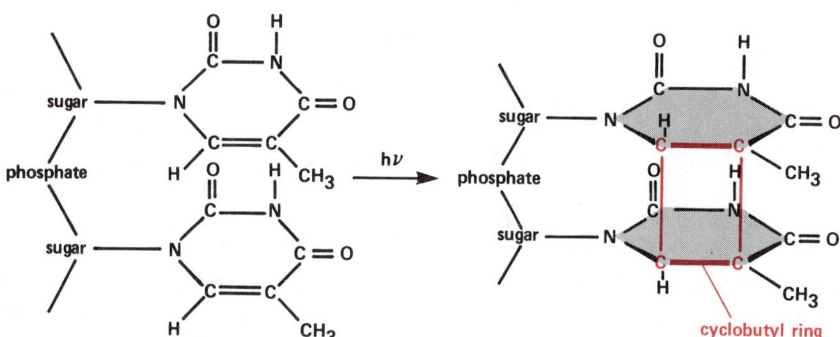

Figure 21-1
Structure of a cyclobutylthymine dimer. Adjacent thymine residues in a DNA chain join upon ultraviolet irradiation (hν).

2. Hutchinson F (1985) *Prog N A Res 32*:115; Levin JD, Johnson AW, Demple B (1988) *JBC 263*:8066; Johnson AW, Demple B (1988) *JBC 263*:18009; Johnson AW, Demple B (1988) *JBC 263*:18017.

(8) *Cross-linkage of strands.* Bifunctional aklylating agents (Section 14-4), mitomycin C, and photoactivated psoralens can form covalent bridges between the two DNA strands, preventing their separation.

Although the lesions and mutations produced by some agents can be predicted, in most instances the result is determined by the exact atom attacked, and is also affected by the chemistry of the cell at the time the agent is acting and by the particular replication or repair responses the cell may make to the injury. In rare instances, the lesion may lead to changes in the DNA that improve function and provide a selective advantage for evolution.

Oxygen Radical Toxicity[3]

Of the many sources of DNA damage, oxidative stresses are common, serious, and among the least appreciated. In bacterial and eukaryotic cells, hyperbaric oxygen, γ and UV radiation, ozone, peroxides, and redox-cycling drugs all exert adverse effects through some species of oxygen. The sequence of univalent reductions of O_2 is as follows:

$$O_2 \longrightarrow \underset{\text{superoxide}}{O_2^{-}\cdot} \longrightarrow \underset{\text{peroxide}}{H_2O_2} \longrightarrow \underset{\text{hydroxyl radical}}{HO\cdot} \longrightarrow H_2O$$

Damage to DNA by oxygen radicals (oxyradicals) is mediated by metal ions (Fig. 21-2), especially iron, as in the classic Fenton reaction:

$$Fe^{2+} + H_2O_2 + H^+ \longrightarrow Fe^{3+} + HO\cdot + H_2O$$

Thus, chelators, oxygen-radical scavengers, and increased levels of certain enzymes (superoxide dismutase, catalase, peroxidase) reduce oxygen toxicity both in vivo and in vitro.

Detection of Altered Bases

A sensitive and applicable test[4] detects the presence in DNA of bases altered by a large variety of chemical agents from the simplest oxyradical or base-alkylation products to complex bulky adducts. DNA digested enzymatically to 3′-deoxynucleotides is phosphorylated with [γ-^{32}P]ATP by phage T4 polynucleotide kinase; high-resolution chromatography is then used and can detect a single adduct in 10^5 nucleotides as an extra and distinctive ^{32}P-nucleoside diphosphate. Screening many chemicals (e.g., electrophilic agents) for their capacity to bind and alter DNA is therefore simple and does not require having each of them in a highly radioactive form to begin with. AP sites can also be detected with a similar ^{32}P-postlabeling assay[5] by exploiting the specificity of the kinase to phosphorylate ApS (where S represents the abasic sugar group) but not SpA; ApS

3. Imlay J, Linn S (1988) Science *240*:1302.
4. Randerath K, Reddy MV, Gupta RC (1981) PNAS *78*:6126.
5. Weinfeld M, Liuzzi M, Paterson MC (1990) *Biochemistry 29*:1737.

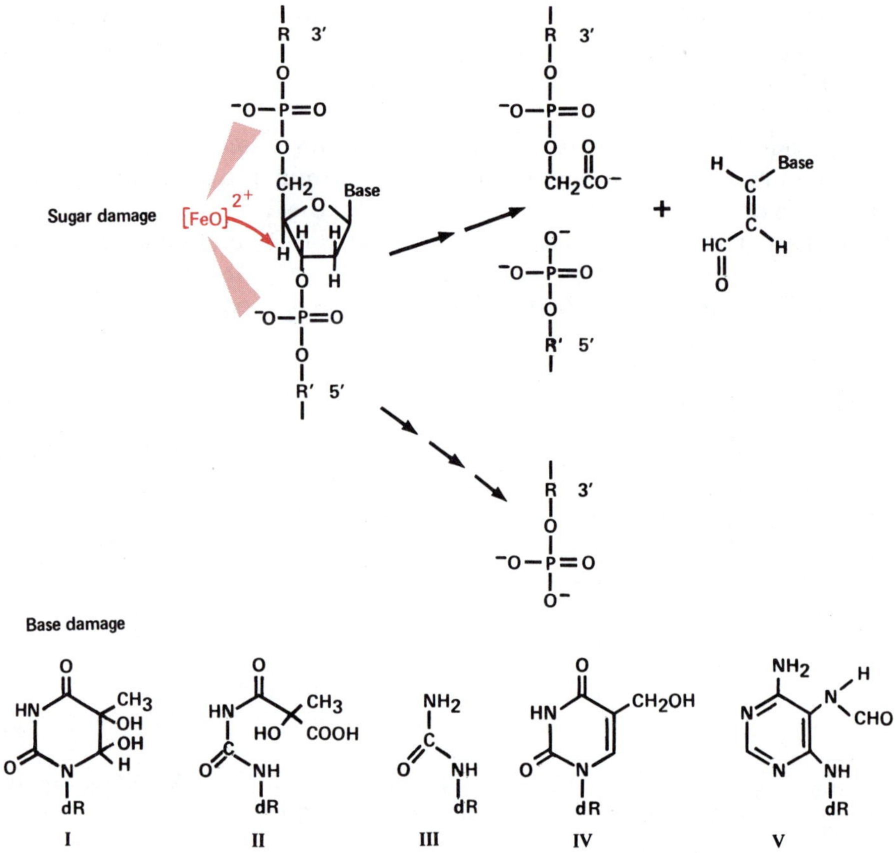

Sugar damage

Base damage

Figure 21-2
DNA damage products formed by oxygen radicals. Sugar damage: Phosphoglycolate and 3′-phosphates are removed by *E. coli* exonuclease III or endonuclease IV. Base damage: The structures are I, thymine glycol (*cis* or *trans*); II, methyltartronylurea; III, urea; IV, 5-hydroxymethyluracil; V, 4,6-diamino-5-formamidopyrimidine. I to IV are formed from thymine. I to III are removed by endonuclease III (bacteria) or γ endonuclease (mammalian cells). IV is apparently excised enzymatically in mammalian but not bacterial cells. V is formed from adenine and is excised in both cell types. [From Imlay JA, Linn S (1988) *Science 240*:1302]

is the isomer generated by treatment of AP-containing DNA with snake venom phosphodiesterase and calf alkaline phosphatase.

Oxidative damage to DNA can be assessed by the presence of 8-hydroxyguanine[6] (one of 20 oxidative DNA-damage products), and its appearance in the urine indicates repair of the lesion.

Probing Sites of DNA Damage in Vivo

Defined DNA sequences are used to probe the spectrum and distribution of damage produced by various agents. The pattern of damage in highly reiterated genomic sequences demonstrates that the distribution of UV-induced cyclobutyl dimers in intact chromatin in cells is similar to that in irradiated naked DNA.[7] Furthermore, this approach revealed a new type of photodamage [(6–4) lesions; see above] at cytidine positions to the 3′ side of another pyrimidine.

6. Fraga CG, Shigenaga MK, Park JW, Degan P, Ames BN (1990) *PNAS 87*:4533.
7. Lippke JA, Gordon LK, Brash DE, Haseltine WA (1981) *PNAS 78*:3388.

DNA is the only molecule which, when altered or damaged, is repaired by the cell. In no other instance is the integrity of a single molecule so vital to the survival of the cell. A bacterial gene has a 50% chance of remaining unaltered even after having been duplicated 100 million times. This remarkable stability of DNA in all cells is in considerable measure due to a variety of devices (Fig. 21-3) for preserving its integrity and repairing virtually any lesion it may sustain.

The repair systems are extraordinarily diverse and effective. Perhaps 100 loci in *E. coli* are involved in DNA repair and related functions (Table 21-1). The duplex complementary structure of DNA ensures that information lost upon excision of a lesion from one strand can be retrieved from the other. (When the strands are separated, as at the growing fork, repair may be more difficult.) However, some errors that extend across both strands can also be repaired. Deletions or insertions of nucleotides, or cross-linkage of strands, may be corrected by replacement of the region through recombination (Section 5). An improper base pair resulting from a proofreading failure, seemingly irremediable because of ambiguity about which of the bases, parental or progeny, is correct, can be identified by the mismatch repair system (see below). Even a double-strand break, perhaps the most damaging of lesions, may be sealed by ligases or repaired by recombination.

The molecular mechanisms for repair can be divided into (1) those that reverse the damage (i.e., photoreactivation, dealkylation), and (2) those that excise and replace the damaged unit by replication, recombination, or the mismatch repair pathways (see below). When repair fails, continuity of the genome may be preserved through "error-prone" replication, in which bypass of the lesion permits replication to proceed.

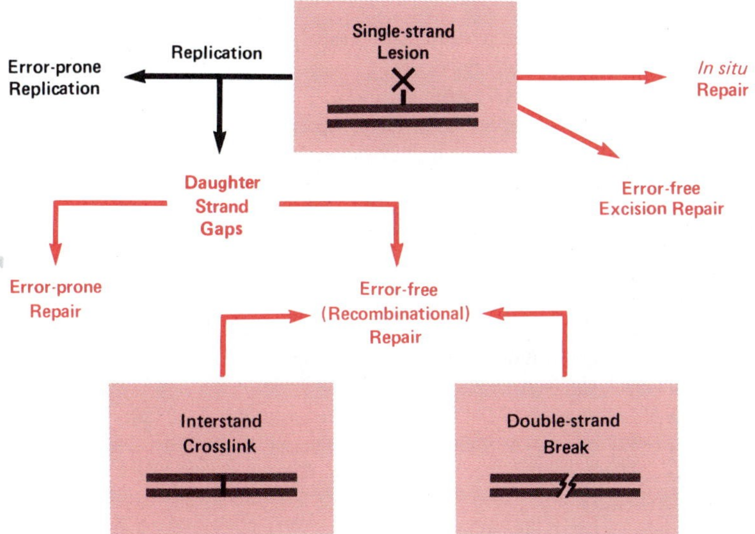

Figure 21-3
Variety of repair mechanisms. (Courtesy of Professor S Linn)

8. Friedberg EC (1985) *DNA Repair.* WH Freeman, New York; Sancar A, Sancar GB (1988) *ARB* 57:29; Sibghat-Ullah, Sancar A (1990) *Biochemistry* 29:5711.

Table 21-1
Loci and functions of representative DNA repair, recombination, and related genes in the linkage map of *E. coli* K12[a]

Gene	Map locus (min.)	Function or comment
dam	74	DNA adenine methylation
dcm	43	DNA cytosine methylation
dinF	92	induced by UV and mitomycin C
dnaQ (*mutD*)	5	DNA polymerase III, ϵ subunit
endA	64	DNA-specific endonuclease I
fis	72	site-specific DNA inversion
ftsE, Q, S, X, Y	76	cell division
ftsH	69	cell division
ftsM, Z	2	cell division
hsdM	99	host modification; DNA methylase M
hsdR	99	host restriction; endonuclease R
hsdS	99	specificity determinant for *hsdM* and *hsdR* activities
lexA (*tsl, spr*)	92	regulatory gene for SOS operon
lig	52	DNA ligase
lon	10	ATP-dependent protease La
minB	26	segregation of minute cells containing no DNA; complex locus
mutH (*mutR, prv*)	61	methyl-directed mismatch repair
mutL (*mut-25*)	95	methyl-directed mismatch repair
mutM	82	$G \cdot C \longrightarrow T \cdot A$ transversions
mutS	59	methyl-directed mismatch repair
mutT	2	$A \cdot T \longrightarrow C \cdot G$ transversions
mutY	64	$G \cdot C \longrightarrow T \cdot A$ transversions
phr	16	photolyase
polA	87	DNA polymerase I
polB	2	DNA polymerase II
rac	30	defective prophage rac
recA (*lexB, tif, zab*)	58	general recombination; SOS repair
recB, C, D	61	exonuclease V subunits
recE	30	exonuclease VIII
recF (*uvrF*)	83	general recombination; postreplication repair of UV damage
recJ	62	recombination and DNA repair
recN	57	recombination and DNA repair
recO	56	conjugational recombination and DNA repair
recQ	86	conjugational recombination and DNA repair
ruvA, B	41	filamentation; sensitivity to UV irradiation
sbcB (*xonA*)	44	exonuclease I (*xonA*); suppresses *recBC*$^-$
sbcC	9	suppressor of *recBC*$^-$
sfiC	26	cell division inhibition
ssb (*lexC*)	92	single-strand DNA binding protein
sulA (*sfiA*)	22	suppressor of *lon*
umuC, D	26	error-prone repair; induction of mutations by UV
ung	56	uracil–DNA glycosylase
uvrA	92	repair of UV damage to DNA; excision nuclease
uvrB	18	repair of UV damage to DNA; excision nuclease
uvrC	42	repair of UV damage to DNA; excision nuclease
uvrD (*mutU, pdeB, recL, UvrE*)	86	DNA helicase II
xseA	54	exonuclease VII, large subunit
xseB	10	exonuclease VII, small subunit
xthA	38	exonuclease III (endonuclease for AP sites)

[a] Data from Bachmann BJ (1990) *Microbiol Rev* 54:130.

Photoreactivation by Photolyases[9]

The energy of visible light (300 to 600 nm) is used by photolyases to break the cyclobutyl pyrimidine dimer ring in situ, restoring the bases to their monomeric form. These enzymes operate in cells from many sources, ranging from mycoplasmas, the tiniest of bacteria, to animal cells, such as the human leukocyte; yeast has two distinct enzymes. The sizes, action spectra, and other features vary among the photolyases. The fact that mycoplasmas, with only a few hundred genes, devote one of them to an enzyme for photoreactivation illustrates the importance of DNA repair.

Photoreactivation (Fig. 21-4), as an alternative to excision, repairs any UV-induced cyclobutyl dimer. In the first stage of photoreactivation, the enzyme recognizes and binds specifically to the dimer in the dark. UV-irradiated oligothymidylate is a good substrate;[10] oligo $(dT)_{18}$ with an average of 3.5 dimers is bound tightly by two molecules of enzyme. All photolyases contain two chromophores, one of which is 1,5-dihydroflavin adenine dinucleotide ($FADH_2$). The nature of the second chromophore divides photolyases into two classes; the yeast and *E. coli* enzymes belong to one which employs a pterin (folate coenzyme) and the other class uses a deazaflavin.[11] When the lesion absorbs light (of a wavelength characteristic for the chromophore), the energy is used by the stable enzyme–DNA complex to convert the cyclobutyl dimer to the pyrimidine monomers. The enzyme then dissociates from the DNA.

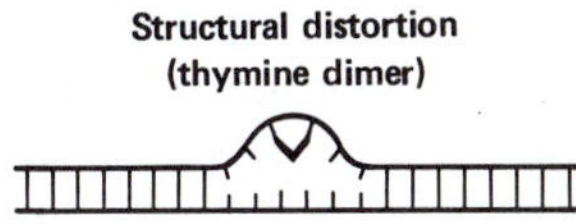

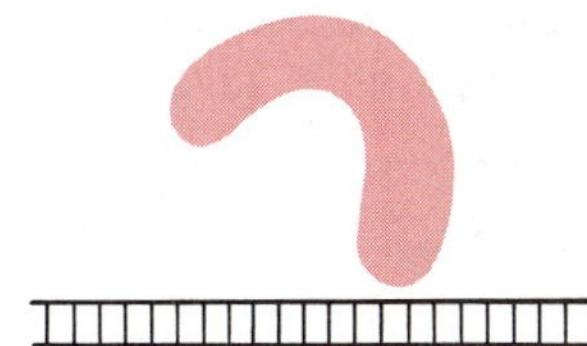

Figure 21-4
Scheme for enzymatic photoreactivation of a pyrimidine dimer. (Courtesy of Professor PC Hanawalt)

9. Walker GC (1985) *ARB 54*:425; Jorns MS (1989) in *Chemistry and Biochemistry of Flavoenzymes* (Muller F, ed.). CRC Press, Boca Raton, Florida, Vol. 2; Jorns MS, Wang B, Jordan SP, Chanderkar LP (1990) *Biochemistry 29*:552; Li YK, Sancar A (1990) *Biochemistry 29*:5698; Payne G, Wills M, Walsh C, Sancar A (1990) *Biochemistry 29*:5706.
10. Weinfeld M, Liuzzi M, Paterson MC (1990) *Biochemistry 29*:1737.
11. Sancar GB, Sancar A (1987) *TIBS 12*:259; Wang B, Jorns MS (1989) *Biochemistry 28*:1148.

Repair by Direct Dealkylation[12]

Another example of direct reversal of damage is the transfer of a methyl group from the precarcinogenic O^6-methylguanine (or an ethyl group from O^6-ethylguanine)[13] to a cysteine residue of an O^6-methylguanine–DNA methyltransferase (MTase). The enzyme also removes an alkyl group from a phosphotriester by alkylation of another cysteine residue. Unlike a typical enzyme, the acceptor protein is irreversibly inactivated upon alkylation. MTases are present in *E. coli*,[14] yeast,[15] and human cells.[16]

Synthesis of the 39-kDa *E. coli* MTase, product of the *ada* gene, is induced[17] by exposure to low levels of an alkylating agent such as N-methyl-N′-N-nitrosoguanidine. The protein, modified by removal of an alkyl group from a phosphotriester, positively regulates genes involved in alkylation repair, including its own. The alkylated MTase binds the *ada* box ($A_3N_3A_3$GCGCA) upstream of *ada* regulon promoters, activating transcription. Resistance to high levels of alkylating agents is mediated through the large increases in intracellular levels of this protein. Another transferase of *E. coli*, the noninducible 19-kDa Ogt protein, is homologous to the MTase of human cells,[18] elevated levels of which are correlated with resistance of certain tumors to alkylating drugs.[19]

Excision Repair[20]

Systems for the excision of DNA lesions depend largely on the nucleases described in Chapter 13. Repair of the great variety of damaged and modified bases is achieved by excisions (Fig. 21-5) that remove (1) the damaged base, creating an AP site, (2) a fragment containing an AP site, (3) the nucleotide lesion and neighboring region of DNA, or (4) an interstrand cross-link. The gap generated by excision is filled by DNA polymerase (pol I in *E. coli*) and then covalently joined by DNA ligase.

Base Excision. Base excision to remove the lesion, thereby creating an AP site, is carried out by one of several N-glycosylases (Section 13-9) that recognize a deaminated or altered base, or a helical deformation caused by the lesion, and then hydrolyze the bond linking the base to the sugar.

Excision of an AP Region. This is achieved by either of two ways (Fig. 21-5; see also Section 13-9 and Fig. 13-6). By one pathway, a class II endonuclease makes the initial incision next to the AP site, after which the fragment is removed by an exonuclease. By the second route, a class I AP endonuclease, which possesses both N-glycosylase and AP endonuclease domains, uses the latter for an incision of the phosphodiester backbone to the 3′ side of the AP site. Removal of the AP fragment is then carried out by a class II endonuclease, which incises the backbone to the 5′ side of the AP site. Deoxyribose fragments at 3′ termini are

---

12. Lindahl T, Sedgwick B, Sekiguchi M, Nakabeppu Y (1988) *ARB 57*:133.
13. Ellison KS, Dogliotti E, Connors TD, Basu AK, Essigmann JM (1989) *PNAS 86*:8620.
14. Lindahl T, Sedgwick B, Sekiguchi M, Nakabeppu Y (1988) *ARB 57*:133.
15. Sassanfar M, Samson L (1990) *JBC 265*:20.
16. Hayakawa H, Koike G, Sekiguchi M (1990) *JMB 213*:739; Koike G, Maki H, Takeya H, Hayakawa H, Sekiguchi M (1990) *JBC 265*:14754; Tano K, Shiota S, Collier J, Foote RS, Mitra S (1990) *PNAS 87*:686.
17. Teo I, Sedgwick B, Kilpatrick MW, McCarthy TV, Lindahl T (1986) *Cell 45*:315; Sedgwick B, Robins P, Totty N, Lindahl T (1988) *JBC 263*:4430; Sekiguchi M, Nakabeppu Y (1987) *Trends Genet 3*:51.
18. Hayakawa H, Koike G, Sekiguchi M (1990) *JMB 213*:739; Lindahl T, Sedgwick B, Sekiguchi M, Nakabeppu Y (1988) *ARB 57*:133.
19. Gonzaga PE, Brent TP (1989) *NAR 17*:6581.
20. Walker GC (1985) *ARB 54*:425; Friedberg EC (1985) *DNA Repair*. WH Freeman, New York.

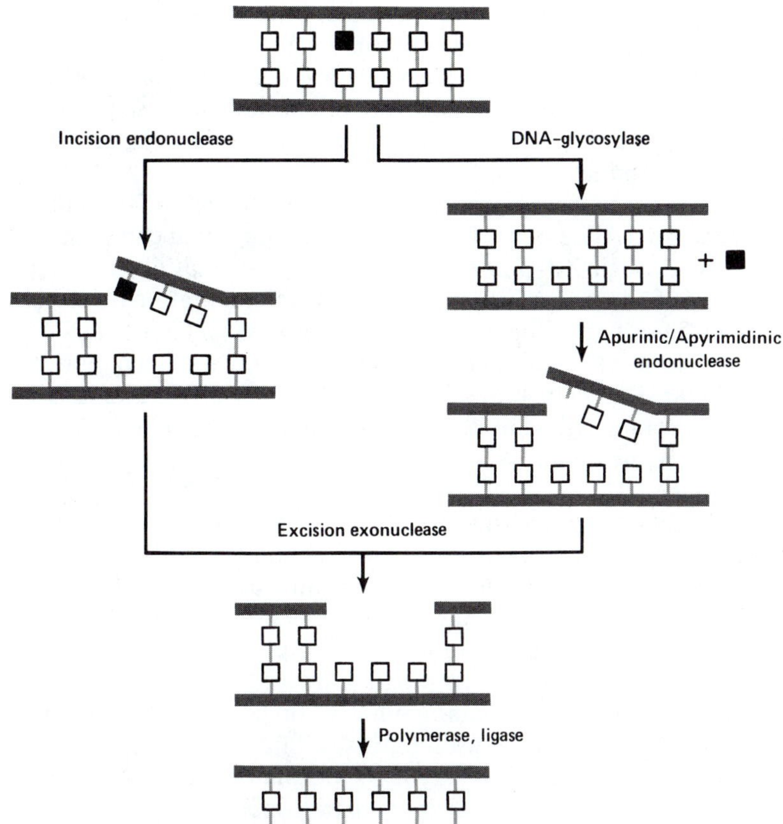

Figure 21-5
Two pathways of excision repair. (Courtesy of Professor S Linn)

removed in *E. coli* largely (95%) by exonuclease III (endonuclease II); endonuclease IV and some novel activities account for the remainder.[21]

Phage T4 endonuclease V (Sections 13-5 and 13-9) is a potent enzyme for incising next to pyrimidine dimers. The enzyme has been used in vitro to complement deficiencies in *E. coli uvr* mutants[22] and to allow repair synthesis in extracts of cells from xeroderma pigmentosum patients (Section 4). Exonucleases encoded by T4 (Sections 13-2 and 13-3) follow up the enzymatic incision. The remarkable capacity of phage T4 polynucleotide kinase (Section 9-7) to hydrolyze phosphate from 3′ ends as well as to phosphorylate 5′-OH ends may be a device to prepare strand breaks for ligation.

Oligonucleotide Excision. Excision[23] that includes the region containing a UV dimer, bulky adduct, or interstrand cross-link (see below) is performed by an excinuclease, such as the UvrABC complex of *E. coli* (Section 13-9), which cuts out a 12- to 13-ntd fragment from one strand (Fig. 13-10). The resulting gap is filled in by DNA pol I and sealed by ligase. The hydrolytic incisions are made at the eighth phosphodiester bond 5′ to the lesion and the fourth or fifth bond 3′ to it. While not required for incision, pol I and helicase II are essential for the removal of the oligonucleotide and recycling of the excinuclease. Similar func-

21. Bernelot-Moens C, Demple B (1989) *NAR 17*:587.
22. Shimizu K, Sekiguchi M (1979) *MGG 168*:37.
23. Sancar A, Sancar GB (1988) *ARB 57*:29; Van Houten B (1990) *Microbiol Rev 54*:18.

tional features are found in the UV endonuclease α of *Dinococcus (Micrococcus) radiodurans*, the most UV-resistant organism known.[24]

The pathways of excision repair in eukaryotes are not well characterized. Five *RAD* genes in yeast,[25] essential for repair, have been cloned and sequenced. Except for knowing that the RAD3 protein is a helicase, the biochemical functions of the other gene products are still unclear.

In humans, studies of xeroderma pigmentosum (XP) (Section 4) show the basis for the defects in repair of UV lesions to be distributed among eight complementation groups (XP-A through XP-G and an XP variant). Several of the corresponding proteins, purified from wild-type cells by complementation of XP cell extracts, have been identified as having UV-repair functions. One is XP-D, a DNA glycosylase/class I AP endonuclease which acts on oxidized pyrimidines.[26] Another, the XP-E protein, binds damaged DNA and seems functionally analogous to a yeast protein linked to photolyase.[27] In addition to light-dependent repair, the photolyases from *E. coli* and yeast appear to participate in excision repair.[28]

Cell extracts of XP cells defective in complementation groups A, C, D, and G are complemented in DNA repair synthesis by the *E. coli* UvrABC proteins,[29] implying that these cells are deficient in incision but not in the subsequent stages of repair. The gene for group A has been cloned and encodes a protein which contains a zinc-finger domain and is therefore presumed to be a DNA binding protein.[30] The gene for group B, a repair gene (ERCC-3), has also been cloned and encodes a protein which contains a helix–turn–helix DNA binding domain and a set of motifs characteristic of helicases.[31] Interestingly, the single patient in group B also suffered from Cockayne's syndrome (Section 4), suggesting that the defective protein is involved in two separate biochemical pathways.

Further complexity is indicated by still another human gene (*ERCC*-1),[32] which complements a UV-sensitive rodent-cell mutant but none of the XP mutations. The N-terminal half of the *ERCC*-1 open reading frame is homologous to *RAD10* of yeast, and the C-terminal region resembles *E. coli uvrC*.

Excision of an Interstrand Cross-Link. Such cross-links are produced by psoralens and by many carcinogenic and chemotherapeutic agents (mitomycin C, nitrogen and sulfur mustards, cisplatin); excision requires repair of both strands (Fig. 21-6). In *E. coli* and mammalian cells, cross-links are repaired by a combination of nucleotide excision and recombinational repair.[33] In *E. coli*, the requirement for the Uvr excinuclease[34] and RecA recombinase[35] and the assist-

24. Evans DM, Moseley BEB (1985) *Mutat Res 145*:119.
25. Naumovski L, Chu G, Berg P, Friedberg EC (1985) *Mol Cell Biol 5*:17; Reynolds P, Higgins DR, Prakash L, Prakash S (1985) *NAR 13*:2357; Sung P, Prakash L, Weber S, Prakash S (1987) *PNAS 84*:6045; Bardwell L, Burtscher H, Weiss WA, Nicolet CM, Friedberg EC (1990) *Biochemistry 29*:3119.
26. Keeney S, Linn S (1990) *Mutation Res 236*:239.
27. Chu G, Chang, E (1988) *Science 242*:564; Patterson M, Chu G (1989) *Mol Cell Biol 9*:5105; Chu G, Chang E (1990) *PNAS 87*:3324.
28. Sancar GB (1990) *Mutation Res 236*:147.
29. Hansson J, Grossman L, Lindahl T, Wood RD (1990) *NAR 18*:35.
30. Tanaka K, Miura N, Satokata I, Miyamoto I, Yoshida MC, Satoh Y, Kondo S, Yasui A, Okayama H, Okada Y (1990) *Nature 348*:73.
31. Weeda G, van Ham R, Vermeulen W, Bottsma D, van der Eb A-J, Hoeijmakers JH (1990) *Cell 62*:777.
32. van Duin M, de Wit J, Odijk H, Westerveld A, Yasui A, Koken HM, Hoeijmakers JHJ, Bootsma D (1986) *Cell 44*:913.
33. Sancar A, Sancar GB (1988) *ARB 57*:29.
34. Jones BK, Yeung AT (1988) *PNAS 85*:8410; Sladek FM, Melian A, Howard-Flanders P (1989) *PNAS 86*:3982.
35. Sinden RR, Cole R (1978) *PNAS 75*:2373.

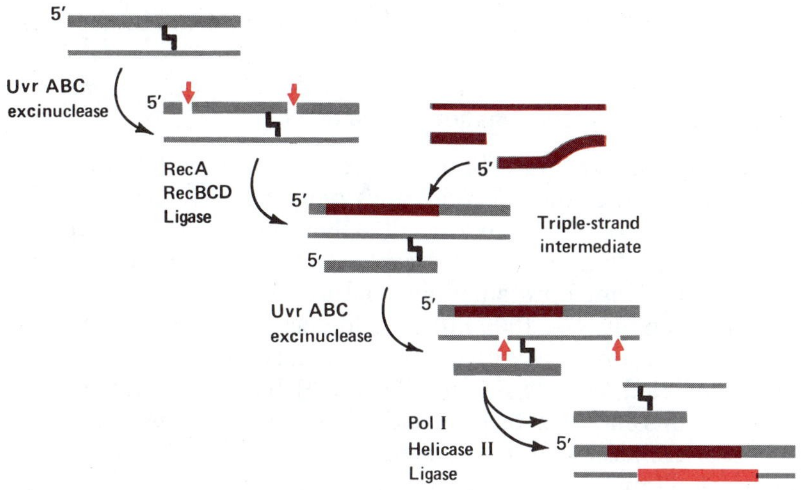

Figure 21-6
Model for repair of a psoralen cross-link in *E. coli*. [Based on Sancar A, Sancar GB (1988) *ARB 57*:29]

ance of the RecBCD nuclease may account for the repair products.[36] Excision of an oligonucleotide from one strand and its replacement by a RecA-mediated recombination is followed by excision of the corresponding oligonucleotide from the other strand and its replacement by gap-filling synthesis.

Excision–Repair Patches.[37] The excision–repair patch in *E. coli* produced by pol I is normally about 20 ntd long. Although these "short patches" are longer in *polA* mutants deficient in 5′ → 3′ exonuclease function, repair is still efficient, owing perhaps to exonuclease VI and RecBCD nuclease. In *polA* mutants deficient in the polymerase function, repair patches are more extended. These "long patches" may be several thousand nucleotides long under some undefined circumstances. Presumably, when pol II and pol III (Sections 5-2 and 5-3) substitute for pol I, they do not bind at the incision nick but require a sizable gap to accommodate them. Enlarging the discontinuity at the lesion requires the $recA^+$ phenotype[38] and thus may represent still another SOS repair function (Section 3).

In mammalian cells, the short patch is only 3 or 4 ntd long and the long patch is about 35 ntd. While patch size in bacteria is determined by distinctive enzymes, patch size in mammalian cells is more closely related to the nature of the lesion. Thus, damage from ionizing radiation and from alkylation is repaired in short patches; long patches repair DNA distortion produced by UV and strand cross-linkages.

Preferential Repair.[39] Repair of transcriptionally active (and presumably more accessible) genes has been observed in mammalian cells, yeast, and *E. coli*.[40]

36. Cimino GD, Gamper HB, Isaacs ST, Hearst JE (1985) *ARB 54*:1151.
37. Hanawalt PC, Cooper PK, Ganesan AK, Smith CA (1979) *ARB 48*:783.
38. Cooper PK, Hanawalt PC (1972) *PNAS 69*:1156.
39. Hanawalt PC (1987) *Environ Health Perspect 76*:9; Bohr VA, Wassermann K (1988) *TIBS 13*:429.
40. Terleth C, van Sluis CA, van de Putte P (1989) *NAR 17*:4433; Mellon I, Hanawalt P (1989) *Nature 342*:95.

Survival of cells after UV damage may depend on the prompt and preferential repair of specific loci essential for recovery.

Translesion Replication and Recombinational Repair

When a DNA polymerase encounters a pyrimidine dimer or certain other lesions in the template, either one of two things happens (Fig. 21-7): (1) The polymerase, as part of the SOS response, fills that spot by non-template-directed ("error-prone," "noninstructed") replication across the lesion, employing the UmuC and UmuD proteins[41] and pol III holoenzyme. (2) The polymerase stops, and then resumes 1000 ntd or so downstream; the discontinuity or postreplication gap may be filled with a complementary strand from the sister duplex by RecA-mediated recombination[42] (Section 5). By either pathway (Fig. 21-7), the lesion remains and must be removed subsequently by one or another of the direct-repair or excision-repair systems.

Both the translesion (mutagenic) replication and the recombinational bypass mechanisms are governed by products of the SOS response (see below and Section 3). Polymerase action is modified in some undefined ways; the modification relaxes the rigorous requirement of polymerase for template direction during a stage of synthesis or proofreading and thus increases the frequency of mutagenic errors at the site of a lesion, at an AP site, or elsewhere along the template. Recombination is stimulated by the highly elevated levels of RecA protein.

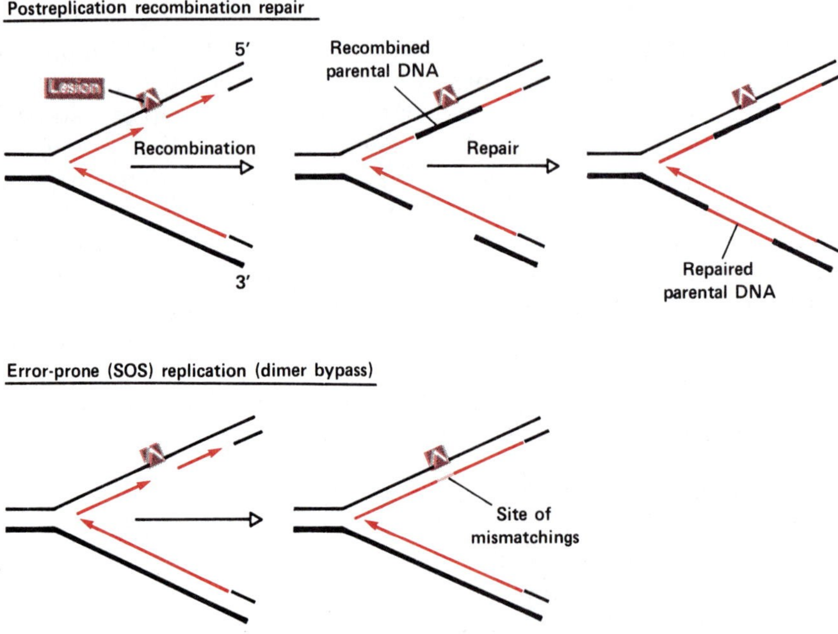

Figure 21-7
Schemes for repair, bypassing removal of the lesion, by error-prone replication of the sequence containing the lesion, or by recombination.

41. Bridges BA, Woodgate R (1985) *PNAS 82*:4193; Lu C, Echols H (1987) *JMB 196*:497.
42. Livneh Z, Lehman IR (1982) *PNAS 79*:3171; West SC, Korner A (1985) *PNAS 82*:6445; Symington, LS, Kolodner R (1985) *PNAS 82*:7247.

Mismatch Repair[43]

Recognition and correction of base-pairing errors in DNA is known in *E. coli*,[44] yeast,[45] and mammalian cells.[46] In *E. coli*, three distinctive enzyme systems are employed to remove (1) errors in replication, (2) mismatches within recombination intermediates, and (3) thymine of a GT base pair produced by spontaneous deamination of 5-methylcytosine. Heteroduplex intermediates in genetic recombination with one or few mismatches are generally aborted. The importance of mismatch repair in monitoring recombination intermediates is indicated by the common failure of recombinations between *E. coli* and *Salmonella typhimurium*, which are 20% divergent in DNA sequence. This barrier to recombination is breached in mutants deficient in mismatch repair.[47]

The Methyl-Directed Pathway.[48] For an interval of several minutes, the newly synthesized DNA strand of a duplex can be identified because its GATC sites have not been methylated. Should any of the eight possible base–base mispairs (except for C–C) be discovered by this repair system, then that section of the unmethylated (i.e., new) strand containing the mispaired base is removed. Mismatch repair depends on seven proteins: MutS (97 kDa), MutL (70 kDa), MutH (25 kDa), MutU (helicase II, UvrD helicase), exonuclease I, SSB, and pol III holoenzyme.

The sequence of events is as follows: MutS forms a complex with the heteroduplex and is then joined by MutL and MutH. The S·L·H complex is then translocated along the DNA, in either direction, for several thousand base pairs, if necessary, until it encounters an unmethylated GATC sequence in one strand of the duplex. The latent endonuclease function of MutH, in the presence of ATP, incises that strand at the GATC. (A GATT sequence, for example, is ignored.) Excision of the incised strand, from the break up to and including the mismatch, depends on exonuclease I, helicase II, and SSB and the concerted replacement of the DNA by pol III holoenzyme synthesis.[49]

Remarkably, MutS can recognize the slight helical distortion of an incorrect base pair and the DNA can be tracked for great distances to identify which of the strands is newly synthesized in order to remove and replace the entire section, all with high efficiency.

Very Short Patch Repair.[50] This system is dependent on MutS, MutL, Dcm (DNA cytosine methylase), and Pol I, but not on MutH. As defined by genetic analysis, the system excises a short tract that contains the GT mismatch resulting from the deamination of 5-methylcytosine and replaces it with a GC base pair. This system may also account for the repair of other mismatches in the absence of a GATC-site signal.[51]

---

43. Modrich P (1987) *ARB 56*:435; Lahue RS, Au KG, Modrich P (1989) *Science 245*:160; Modrich P (1989) *JBC 264*:6597.
44. Modrich P (1987) *ARB 56*:435; Lahue RS, Au KG, Modrich P (1989) *Science 245*:160, Modrich P (1989) *JBC 264*:6597.
45. Bishop DK, Andersen J, Kolodner RD (1989) *PNAS 86*:3713; Kramer B, Kramer W, Williamson MS, Fogel S (1989) *Mol Cell Biol 9*:4432; Kramer W, Kramer B, Williamson MS, Fogel S (1989) *J Bact 171*:5339.
46. Holmes J Jr, Clark S, Modrich P (1990) *PNAS 87*:5837.
47. Rayssiguier C, Thaler DS, Radman M (1989) *Nature 342*:396.
48. Modrich P (1987) *ARB 56*:435; Lahue RS, Au KG, Modrich P (1989) *Science 245*:160.
49. Au KG, Clark S, Miller JH, Modrich P (1989) *PNAS 86*:8877; Lu A-L, Cuipa MJ, Ip MS, Shanabruch WG (1990) *J Bact 172*:1232.
50. Lieb M, Allen E, Read D (1986) *Genetics 114*:1041; Lieb M (1987) *J Bact 169*:5241; Jones M, Wagner R, Radman M (1987) *JMB 194*:155; Dzidic S, Radman M (1989) *MGG 217*:254.
51. Claverys J-P, Méjean V (1988) *MGG 214*:574.

MutY Repair System.[52] This system for correcting the GA mismatch is entirely independent of the methyl-directed pathway. MutY (MicA) of *E. coli* is a 39-kDa adenine glycosylase, an iron–sulfur protein with homology to endonuclease III. The enzyme recognizes the GA mismatch in the duplex and excises the adenine. The resulting apurinic site is then incised and removed by an AP endonuclease (Section 13-9), and the lesion is replaced by a proper GC base pair.

Other Mismatch Repair Systems. A mismatch repair system identified in yeast is more efficient in removing some mismatches than others.[53] Extracts of human cells possess a protein that binds the GT mismatch and another that binds AC, TC, and TT mismatches, actions presumably involved in the repair of these mismatches.[54]

21-3 Repair Responses and Regulation

Poly(ADP-Ribose) and Repair[55]

The ADP-ribose moiety, derived by cleavage of the ribose–nicotinamide glycosylic bond of NAD^+, is polymerized into long chains by poly(ADP-ribose) synthase (polymerase). Stimulation of poly(ADP-ribose) synthesis is correlated with DNA strand breaks, the basis for which is still obscure. Activation of the synthase in vitro requires linear DNA with nicks or termini with 5′-OH groups.[56] A zinc-finger domain in the enzyme determines the specificity for single-stranded breaks in the DNA.[57] Inhibitors of the synthase, such as 5-methylnicotinamide and 3-aminobenzamide,[58] increase the toxicity and chromosomal aberrations caused by DNA-damaging agents. Although the action of ADP-ribose appears to be at a late stage of excision repair, inferences about its role in ligation are weakened by the disparity between its involvement in repair of damage by alkylating agents but not by UV irradiation.[59]

Regulation of Repair

The relatively scarce repair enzymes (20 to 100 molecules per *E. coli* cell) are encoded by genes modulated by at least four operons: (1) SOS (e.g., UvrA, UvrB, UvrC; RecA; UmuC, UmuD); (2) *ada* (e.g., methyltransferases; N-glycosylase); (3) *soxR* (e.g., endonuclease IV[60]); and (4) *katF*[61] (e.g., exonuclease III; catalase II). Additional regulons (i.e., operons) respond to oxidative stress[62] and to heat shock, the former including the effects of H_2O_2.

---

52. Michaels ML, Pham L, Nghiem Y, Cruz C, Miller JH (1990) *NAR 18*:3841.
53. Bishop DK, Andersen J, Kolodner RD (1989) *PNAS 86*:3713.
54. Wiebauer K, Jiricny J (1989) *Nature 339*:234; Stephenson C, Karran P (1989) *JBC 264*:21177.
55. Ueda K, Hayaishi O (1985) *ARB 54*:73; Althaus FR, Richter CR (1987) *Mol Biol Biochem Biophys 37*:1.
56. Benjamin RC, Gill DM (1980) *JBC 255*:10493; Benjamin RC, Gill DM (1980) *JBC 255*:10502; Shall S (1981) in *Chromosome Damage and Repair* (Seeberg E, Kleppe K, eds.). Plenum, New York, p. 477.
57. Gradwohl G, Menissier de Murcia JM, Molinete M, Simonin F, Koken M, Hoeijmakers JHJ, De Murcia G (1990) *PNAS 87*:2990.
58. Morgan WF, Cleaver JE (1983) *Canc Res 43*:3104.
59. Cleaver JE, Bodell WJ, Morgan WF, Zelle B (1983) *JBC 258*:9059.
60. Chan E, Weiss B (1987) *PNAS 84*:3189; Greenberg JT, Demple B (1989) *J Bact 171*:3933.
61. Sak BD, Eisenstark A, Touati D (1989) *PNAS 86*:3271; Mulvey MR, Loewen PC (1989) *NAR 17*:9979.
62. Chan E, Weiss B (1987) *PNAS 84*:3189; Greenberg JT, Demple B (1989) *J Bact 171*:3933.

The SOS Response.[63] Induced by DNA damage that blocks replication (e.g., damage from UV, mitomycin, or nalidixate), the SOS response (Fig. 21-8) restores replication and converts the lesion to an error-prone site. The response in *E. coli* includes a large number of well-studied phenomena observed when DNA damage turns on some 20 SOS-response genes, negatively controlled by the LexA repressor. The latter binds the SOS box (5′-CTGN$_{10}$CAG) that overlaps the promoters of the SOS genes, including both *recA* and *lexA* itself.

The signal for the SOS response[64] is ssDNA at a replication fork stalled by UV damage or generated by repair of other lesions. Both the ssDNA and UV-irradiated dsDNA[65] are bound by RecA protein. The activated RecA complex in turn binds LexA protein, inducing a conformational change that promotes LexA self-cleavage and loss of its repressor activity.[66] RecA activation also results in the self-cleavage of two other proteins: the λ repressor[67] (enabling the prophage to be induced and escape the damaged cell) and UmuD protein[68] (enabling it to function in mutagenic bypass). A complex of the activated UmuD protein with UmuC and RecA proteins and DNA polymerase III, a "mutasome" localized at the lesion, may permit the polymerase to replicate past it.

Removal of the LexA repressor turns on the expression of genes involved in several functions: *lexA* and *recA* (which serve as key regulators), *uvrA*, *uvrB*, and *uvrD* (which act in excision repair), *umuC* and *umuD* (serving in mutagenic bypass), *sfiA* and *sfiB* (cell division inhibitors), *recN*, *recQ*, *uvrD*, *himA*, and *ruv*,[69] as well as *recA* (all involved in recombination), and several *din* (damage-inducible) genes of known (Section 5-2) and unknown functions. Cell survival is promoted by enhanced repair and recombination over an interval extended by the postponement of cell division. Upon the disappearance of the ssDNA signal and the activated form of RecA, the autodigestion of LexA ceases and expression of SOS genes returns to the low constitutive levels.

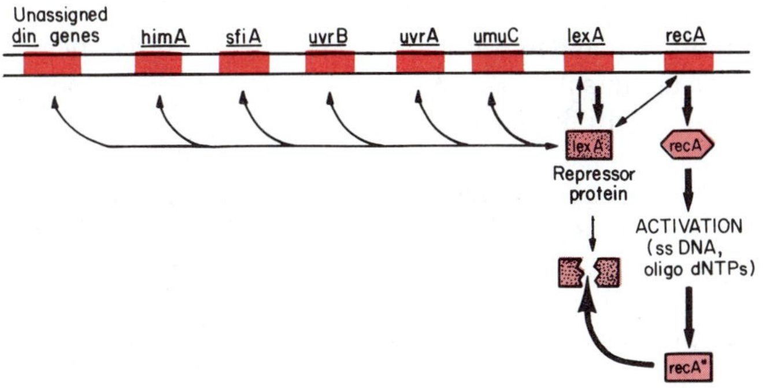

Figure 21-8
The SOS regulatory system of *E. coli*. The sequence of genes does not represent their order in the chromosome. The genes are repressed by LexA protein during normal growth and induced when LexA is cleaved during DNA damage or inhibition of DNA synthesis. (Courtesy of Professor PC Hanawalt)

63. Walker GC (1984) *Microbiol Rev 48*:60.
64. Sassanfar M, Roberts JW (1990) *JMB 212*:79.
65. Lu C, Echols H (1987) *JMB 196*:497.
66. Little JW (1984) *PNAS 81*:1375; Slilaty SN, Rupley JA, Little JW (1986) *Biochemistry 25*:6866; Slilaty SN, Little JW (1987) *PNAS 84*:3987.
67. Craig NL, Roberts JW (1980) *Nature 283*:26; Roberts JW, Devoret R (1983) in *Lambda II* (Hendrix RW, Roberts JW, Stahl FW, Weisberg RA, eds.). CSHL, p. 123.
68. Burckhardt SE, Woodgate R, Scheuermann RH, Echols H (1988) *PNAS 85*:1811; Woodgate R, Rajagopalan M, Lu C, Echols H (1989) *PNAS 86*:7301.
69. Iwasaki H, Shiba T, Nakata A, Shinagawa H (1989) *MGG 219*:328; Sharples GJ, Benson FE, Illing GT, Lloyd RG (1990) *MGG 221*:219.

Some responses to DNA damage depend on the components of the SOS regulon, but in a less direct way. For example, an increased expression of *dnaN* and *dnaQ* (encoding the β and ϵ subunits, respectively, of DNA polymerase III holoenzyme), induced by methyl methanesulfonate alkylation or by mitomycin C, is abolished in *lexA* and *recA* mutants.[70] Compared to the typical SOS functions, the subunit responses are delayed, the genes lack *lexA* binding sites, and they fail to show a constitutive derepression. Nevertheless, this link to the SOS system may provide a response to an increased need for the polymerase following DNA damage. Enhanced synthesis of the ϵ subunit sharply reduces mutagenesis.[71]

Response to Oxidative Stress. Exposure to H_2O_2 or other oxidative stress provokes in *E. coli* the production of ApppA and related alarmones that may in turn induce expression of some 30 proteins.[72] Within the *oxyR* regulon[73] of *S. typhimurium* are genes for catalase–peroxidase, superoxide dismutase, glutathione reductase, and an alkyl hydroperoxide reductase; endonuclease IV is part of another regulon[74] induced by superoxide-generating agents, but not by H_2O_2.

Assessment of Repair Systems[75]

The variety of lesions, the complexity of DNA, and the many repair pathways make it difficult to identify the systems for repair of damage caused experimentally or naturally. Principal methods, to be discussed below, include determining the viability of cells or organisms bearing various mutations; examining the state of the DNA (for breaks, gaps, residual lesions, or patches) after specific treatments; and using damaged viruses and plasmids as probes for repair of lesions.

Viability of Mutants. Mutants selected for increased sensitivity (or resistance) to DNA-damaging agents have been isolated, and in some instances the responsible gene products have been identified (Table 21-1).

E. coli mutants in DNA polymerase I show reduced survival after UV irradiation (Fig. 21-9) but fare better than a cell deficient in one of the UvrA, B, or C proteins or RecA protein. A cell lacking both the UvrA and recA proteins is helpless to repair the lesions or even circumvent them; one pyrimidine dimer per genome is lethal.

Other repair pathways have been inferred from the behavior of *E. coli* mutants with lesions in *recB* and *recC* (Section 13-9), *recF* and *uvrD*, and *polC*. At the concluding stages of virtually all repair pathways, ligase is predictably essential. Animal-cell DNA-repair mutants are not so well studied as the prokaryotic mutants, but recognition of an increasing number of inborn errors of DNA repair in humans will greatly advance the work (Section 4).

---

70. Quiñones A, Kaasch J, Kaasch M, Messer W (1989) *EMBO J 8*:587; Kaasch M, Kaasch J, Quiñones A (1990) *MGG 219*:187.
71. Ciesla Z, Jonczyk P, Fijalkowska I (1990) *MGG 221*:251.
72. Bochner BR, Lee PC, Wilson SW, Cutler CW, Ames BN (1984) *Cell 37*:225.
73. Tartaglia LA, Storz G, Ames BN (1989) *JMB 210*:709.
74. Chan E, Weiss B (1987) *PNAS 84*:3189.
75. Friedberg EC (1985) *DNA Repair*. WH Freeman, New York.

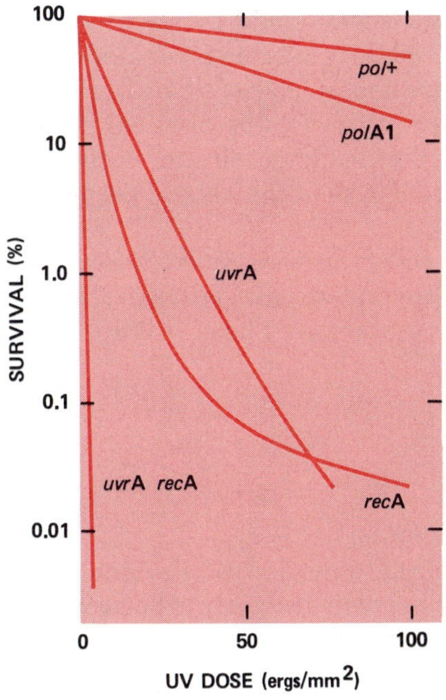

Figure 21-9
Comparative sensitivities of *E. coli* strains to UV light. Strain *pol*$^+$ maintains normal repair, whereas the *polA1* mutant, with reduced DNA polymerase I, is partially deficient in excision–repair. The *uvrA* strain, which lacks damage-specific endonuclease, is totally deficient in this function. The *recA* strain is completely deficient in recombination repair; double mutant *uvrA,recA* thus lacks both dark-repair mechanisms. [From Hanawalt PC (1972) *Endeavor* *31*:83]

State of the DNA. The specificity of UV lesions and the precision with which the radiation can be administered have made excision–repair of these lesions the most amenable to study. Several ways can be used to monitor the repair.

(1) *Loss of dimers* from DNA is measured by comparing the abundance of [^{3}H]thymine dimers (chromatographically analyzed) in the acid-soluble fraction with that remaining in DNA. The distribution of dimers can be detected by the single-strand breaks in the DNA after treating it in vitro with T4 endonuclease V.

(2) *Single-strand breaks* and alkali-labile sites are determined by sedimentation of DNA in alkaline sucrose gradients.

(3) *Conservative repair DNA synthesis* can be distinguished from semiconservative replicative DNA synthesis by 5-bromouracil (BU) or bromodeoxyuridine (BUdR) density labeling. The DNA is labeled with [^{14}C]thymidine in the growth medium before irradiation and then with [^{3}H]BU after irradiation. The buoyant density of extracted, sheared DNA is analyzed in a cesium chloride density gradient. BU incorporated in semiconservative replication bands at a hybrid density, whereas BU patches introduced in repair are too small in amount to shift the buoyant density significantly. Further analysis can determine the repair-patch size and furnish information about the pathway employed.

Damaged Viruses and Plasmids as Probes.[76] The well-defined genomes of small viruses and plasmids can be traced, isolated, and analyzed, offering the advan-

76. Lehmann A, Domen A (1985) *NAR* *13*:2087; Drinkwater NR, Klinedinst DK (1986) *PNAS* *83*:3402; DuBridge RB, Tang P, Hsia HC, Leong PM, Miller JH, Calos MP (1987) *Mol Cell Biol* *7*:379; Chu G, Berg P (1987) *Miol Biol Med* *4*:277; Dixon K, Roilides E, Miskin R, Levine AS (1988) in *DNA Replication and Mutagenesis* (Moses RE, Summers WC, eds.). Am Soc Microbiol, Washington DC, p. 472.

tages exploited in studies of replication and recombination. Lesions can be introduced into isolated DNA or virion DNA while it is being replicated in the cell or after integration into the host chromosome. Repair systems can be assessed by infectivity and mutation frequency, as well as by physical criteria applied to the host chromosome. However, the repair of "shuttle vectors" is highly mutagenic and may thus be unrepresentative of the natural process.

Unscheduled DNA Synthesis. In eukaryotic cells, this is assumed to represent repair replication. The incorporation of [^{3}H]thymidine into repaired DNA can be observed in single-cell autoradiographs as grains over nuclei that are not in the S phase of the cell-division cycle.

21-4 Repair in Disease[77] and Aging[78]

Knowing that mutagenic agents are often carcinogens adds to the interest in DNA repair. While the repair processes in animal cells are likely to be as diverse and complex as in *E. coli*, their study is facilitated by a variety of human inborn errors uncovered in clinical practice. Cells cultured from patients with DNA-repair disorders generally show the repair defect.

Diseases with DNA-Repair Deficiencies

With facile assays for measuring excision–repair in cells (Sections 2 and 3), several diseases are now known to be associated with a deficiency in the repair of UV-induced lesions (Table 21-2). These diseases are generally autosomal and recessive. Patients often manifest photosensitivity, a low tolerance to other DNA-damaging agents, chromosome aberrations, and a high incidence of cancers.

Table 21-2
Human diseases with DNA-repair defects[a]

Disease	Sensitivity	Cancer susceptibility	Symptoms
Ataxia telangiectasia	γ irradiation	lymphomas	ataxia, dilation of blood vessels in skin and eyes, chromosome aberrations, immune dysfunction
Bloom's syndrome	mild alkylating agents	carcinomas, leukemias, lymphomas	photosensitivity, facial telangiectases, chromosome alterations
Cockayne's syndrome	ultraviolet		dwarfism, retinal atrophy, photosensitivity, progeria, deafness, trisomy 10
Fanconi's anemia	cross-linking agents	leukemias	hypoplastic pancytopenia, congenital anomalies
Xeroderma pigmentosum	ultraviolet, chemical mutagens	skin carcinomas and melanomas	skin and eye photosensitivity, keratoses

[a] Other human hereditary disorders that may be related to DNA-repair defects include dyskeratosis congenita (Zinsser-Cole-Engman syndrome), progeria (Hutchinson-Gilford syndrome), and trichothiodystrophy.

77. Friedberg EC (1985) *DNA Repair.* WH Freeman, New York.
78. Cutler RG (1976) in *Aging, Carcinogenesis and Radiation Biology* (Smith KC, ed.). Plenum Press, New York, p. 443; Hart RW (1976) in *Aging, Carcinogenesis and Radiation Biology* (Smith KC, ed.). Plenum, New York, p. 537.

Xeroderma Pigmentosum.[79] Patients with this rare skin disease are extremely sensitive to sunlight and commonly develop skin cancer. Fibroblasts cultured from affected individuals are specifically defective in excision–repair of cyclobutyl dimers. This defect results in a high frequency of UV-induced somatic mutations that lead to skin cancer. Included in this disease group are at least eight genetically distinct complementation groups (Section 2). Although excision–repair in cells cultured from most or all of these groups is strongly stimulated in vitro by phage T4 endonuclease V, a class I repair endonuclease (Section 13-9), viability of these cells is not restored. Possibly, a multienzyme complex is responsible for the early steps in excision–repair in human skin, and in this disease both access of the complex to the DNA lesion and the repair steps are deficient.

Other Defective-Repair Diseases. *Bloom's syndrome* is an autosomal recessively inherited disease with immune system defects, an increased incidence of early cancer, chromosomal breaks, and sister-chromatid exchanges. Impairment of DNA repair is responsible for the fivefold to tenfold elevation in spontaneous mutation rates in fibroblasts from patients. Bloom's syndrome cells are defective in DNA ligase I (Section 9-5), but proof that this is the primary cause of the disease is still pending.[80]

In *ataxia telangiectasia,* a hypersensitivity to ionizing radiation is accompanied by a degenerative ataxia, dilated blood vessels, chromosome aberrations, and lymphomas. *Cockayne's syndrome,* characterized by dwarfism and features of premature aging, is associated with a defect in the repair of UV-induced damage in transcriptionally active DNA.[81] The list of diseases with repair deficiencies (Table 21-2) is likely to grow to include additional genetically and functionally distinguishable disorders. Even though these congenital disorders are rare, their study furnishes important guides to understanding DNA metabolism and other fundamental aspects of human biology.

Aging

What is aging and how can it be measured? The complexity of the problem seems even greater than that of cancer, known to be multifactorial (requiring many independent events) in its origin and to assume a hundred guises. Despite their enigmatic origin, the phenomena of aging can be observed in the decline and death of cells and animals.

It seems virtually certain that numerous factors underlie the variety of changes associated with an aging cell or organism, ranging from programmed differentiation to the ravages of environmental agents. Even though the fraction of aging effects due to inadequate DNA repair remains controversial, it is clear that an accumulation of DNA lesions contributes to the loss of cell function and viability, and the effects can be catastrophic.

A striking correlation has been noted, in animals from seven species with a 20-fold range in life span, between a capacity to repair UV lesions and longevity[82] (Fig. 21-10). Furthermore, the error frequencies of DNA polymerases α and

79. Cleaver JE, Kraemer KH (1989) in *The Metabolic Basis of Inherited Disease* (Scriver CS, Beaudet AL, Sly WS, Valle D, eds.). 6th ed. McGraw-Hill, New York, p 2949; Kraemer KH, Lee MM, Scotto J (1987) *Arch Dermatol 123*:241.
80. Lehmann A, Norris P (1990) *Sem Dermatology* 9:55.
81. Venema J, Mullenders LHF, Natarajan AT, van Zeeland AA, Mayne LV (1990) *PNAS 87*:4707.
82. Hart RW, Setlow RB (1974) *PNAS* 71:2169.

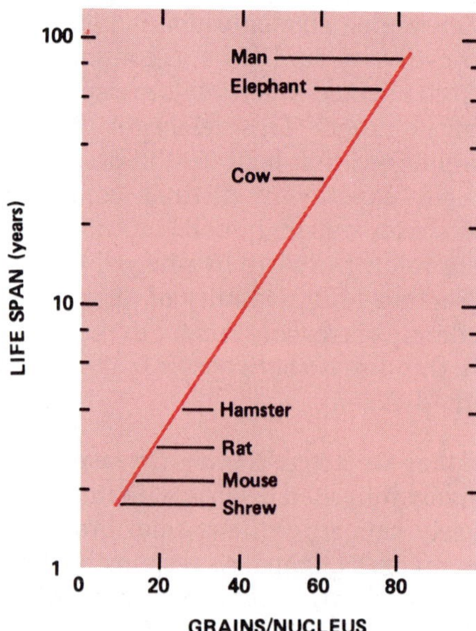

Figure 21-10
Correlation between the amount of unscheduled (repair) DNA synthesis in young fibroblasts and estimated life span of the species. Synthesis was measured at 13 hours after exposure to UV irradiation (20 Jm^{-2}). [After Hart RW, Setlow RB (1974) *PNAS 71*:2169.]

γ from senescent human fibroblasts are significantly greater than are those from "young" cells.[83] An accumulation in DNA of 8-hydroxyguanine, a gauge of oxidative damage, increases in some tissues of aged rats,[84] possibly due to the loss of repair activity.

On the other hand, increased deficiencies in replication are not apparent in peripheral lymphocytes from older people.[85] No difference was detected in excision–repair capacity in three cold-blooded vertebrates with life spans ranging from 3 to over 118 years,[86] nor were significant correlations between repair capacity and aging seen among 34 species in 11 orders of mammals.[87] Excision–repair capacity also does not differ significantly in UV-irradiated human skin cells from newborns and from donors up to 88 years of age.[88]

Overall genomic DNA repair activity may be a poor indicator of cell sensitivity to DNA-damaging agents. Instead, subtle age-related changes in particular classes of genes repaired during differentiation may be part of a programmed maturation process.[89] For example, the repair levels of some terminally differentiated cells may be barely detectable, while selected, transcribed genetic domains are actively repaired.

An interesting hypothesis holds that damage by free-radical reactions is the main cause of aging and age-related disorders,[90] and its prevention by antioxidants is suggested to be of prime importance. If DNA damage proves to be one of the primary consequences of free radicals, then their effect on DNA and repair of the resulting lesions deserves considerable attention.

83. Murray V, Holliday R (1981) *JMB 146*:55; Krauss SW, Linn S (1982) *Biochemistry 21*:1002.
84. Fraga CG, Shigenaga MK, Park JW, Degan P, Ames BN (1990) *PNAS 87*:4533.
85. Agarwal SS, Tuffner M, Loeb LA (1978) *J Cell Physiol 96*:235.
86. Woodhead AD, Satlow RB, Grist E (1980) *Exp Gerontol 15*:301.
87. Kato H, Harada M, Tsuchiya K, Moriwaki K (1980) *Jpn J Genet 55*:99.
88. Liu S, Hanawalt PC, personal communication.
89. Hanawalt PC, Gee P, Ho L (1990) in *Molecular Biology of Aging* (Finch CE, Johnson TE, eds.). UCLA, Vol. 123. AR Liss, New York, p. 45.
90. Harman D (1981) *PNAS 78*:7124.

The aging diseases progeria and Werner's syndrome seem to be associated with defects in one or a few genetic loci. However, it is not likely that one enzyme or enzyme complex is the primary locus of the changes associated with aging in general. A defect in DNA repair may be especially catastrophic inasmuch as the increased mutation rate amplifies other metabolic failures, including the enzymatic machinery for repair itself. Integrity of DNA and fidelity of its transcription and replication are of basic importance. Failure to repair incessant damage to DNA must be responsible for many cellular disorders, including some of the symptoms or diseases of aging.

21-5 Homologous Recombination

Assortment of genes by recombination in meiosis and the duplication, deletion, inversion, and translocation of genes were recognized phenomena even before DNA was identified as the genetic material. The repair of DNA (Section 2) may also employ recombination mechanisms in addition to excision–repair. By the latter device, the faulty segment is excised by incisions on both sides of the lesion and replaced by replication directed by the intact complementary strand. An alternative is postreplication recombination, in which a segment supplied by the parental DNA is used to match the defective segment that could not be properly replicated. Of the numerous ways by which DNA molecules recombine to redistribute their genetic contents, four categories are now recognized.

(1) In *homologous recombination*, exchange between homologous segments takes place anywhere along the length of the DNA molecule. Initiation appears to depend on a homology search, involving DNA–DNA pairing between the two recombination partners. The process in *E. coli* depends on RecA protein (see below) employed in a multiplicity of pathways.

(2) In *site-specific recombination* (Section 6), exchange occurs at a specific site, as in the integration of phage λ DNA into the *E. coli* chromosome. Recombination of λ DNA depends on both phage-encoded and host proteins but is independent of the RecA protein. Initiation involves protein–DNA interactions at the site of recombination, rather than pairing of the DNAs by homologous sequences.

(3) *Transposition* (Section 7) is the recombination event by which certain DNA elements, called *transposons*, catalyze their own movement to a new chromosomal location. Transposition is mechanistically related to site-specific recombination, in involving protein–DNA interactions rather than DNA homology for pairing the recombining DNAs. In contrast to the site-specific mechanisms, however, a specific protein–DNA complex is required on only one of the recombination partners, with recognition of the second DNA usually being relatively nonspecific.

(4) *Illegitimate recombination* includes events that do not fall into the above classes. The integration of transfected DNA at nonhomologous sites and certain genome rearrangements such as translocations and deletions are examples. The mechanisms by which these events arise are largely unknown.

In view of the many kinds of recombination and the variety of genetic and physiologic backgrounds in which recombination takes place, we can expect considerable variation among the biochemical mechanisms and pathways even within a single cell. Nevertheless, a recurring motif in recombination in both prokaryotes and eukaryotes is the breakage and reunion of DNA segments.

Mechanisms of Homologous Recombination

In the process of exchanging sequences between homologous DNA molecules at essentially any site, the mechanisms used are largely inferred from the structure of the products. Recombination is *reciprocal* when the exchange of sequences between the two molecules results in no loss of information. *Nonreciprocal* recombination, or *gene conversion*, occurs when the two partners end up with the same sequence information, rather than exchanging it. Although the outcome may be different, both reciprocal recombination and gene conversion probably operate through closely related mechanisms (see below).

Despite the isolation of many recombination-defective mutants (especially in phage, *E. coli*, and yeast) and the availability of several purified and characterized protein products, homologous recombination has not been reconstituted in vitro, and the detailed pathway is therefore not known.

Models for Strand Exchange[91]

Initiation of recombination is thought to involve the introduction of a single- or double-stranded break in the DNA of the recombining partners. Models for recombination via the single-strand invasion (Fig. 21-11) or double-stranded gap repair mechanisms (see Fig. 21-13) are presented.

Single-Strand – Initiated Exchange. This entails:[92]

(1) *Alignment* of the homologous duplexes.

(2) *Invasion* of the cleaved strands into the duplex of the recombination partner. Ligation of these strands to their homologs generates a *Holliday junction* intermediate[93] in which the two duplexes are associated by a single-stranded crossover (Fig. 21-12). Apparent in structural studies is the complete base-pairing of the DNA strands in these intermediates (Section 1-13). Strands from both parental duplexes may invade at the same point (Fig. 21-11, left), or only one strand may invade, in which case replication may fill the gap left by the invading strand (Fig. 21-11, right).

(3) *Branch migration* of the junction, increasing the length of the exchanged region.

(4) *Resolution* of the joint molecule. As in initiation, this requires cleavage of the strands; the location of these cleavages determines the outcome of recombination. If the terminating cleavages are on the same strands as those initiating the reaction, the products are merely patched, whereas cleavage of the opposite strands generates spliced (recombined) products (Fig. 21-12).

---

91. Whitehouse HLK (1982) *Genetic Recombination: Understanding the Mechanisms.* Wiley, New York.
92. Meselson M, Radding CM (1975) *PNAS* 72:356.
93. Holliday R (1964) *Genet Res* 5:282.

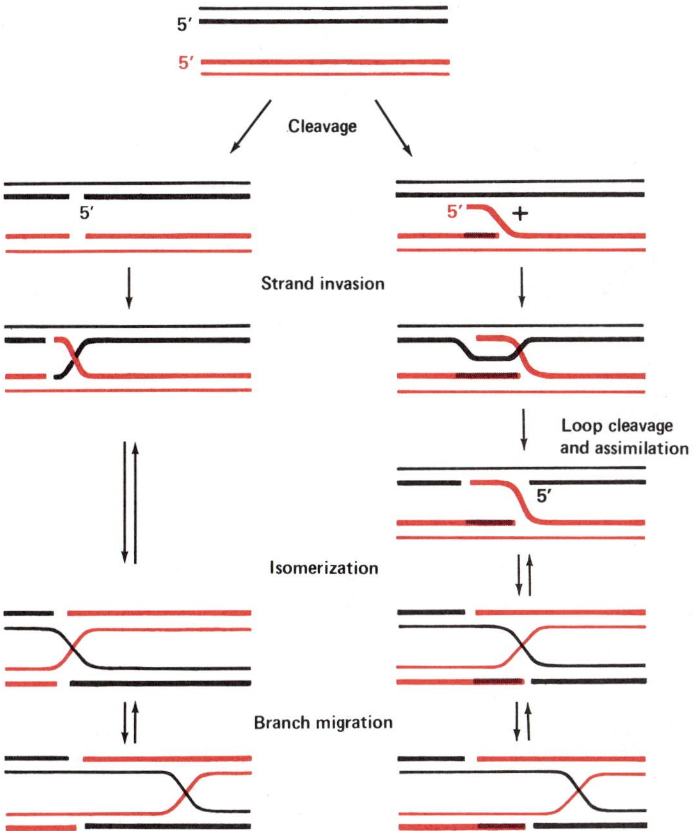

Figure 21-11
Models for single-strand–initiated homologous recombination: (left) no DNA synthesis; (right) DNA synthesis required. [Left, modified from Lewis B (1990) *Genes IV*, Oxford University Press, New York, p 628; right, from Radding CM (1978) *ARB 47*:847]

Double-Stranded Gap Repair.[94] As a recombination mechanism in phages λ[95] and T4, in yeast,[96] and in some *E. coli* pathways, this form of recombination (Fig. 21-13) entails:

(1) *Alignment* of the two homologous DNA molecules.

(2) *Exonuclease digestion* from the break to make a gap with single-strand extensions.

(3) *Invasion* of one of the single-stranded ends into the intact duplex to generate a D-loop.

(4) *DNA synthesis* initiated at the 3′-OH of the invading strand. This replaces the sequence lost by exonuclease digestion and covers the site of the initial

94. Szostak JW, Orr-Weaver TL, Rothstein RJ, Stahl FW (1983) *Cell 33*:25; Thaler DS, Stahl FW (1988) *ARG 22*:169.
95. Takahashi N, Kobayashi I (1990) *PNAS 87*:2790.
96. Nicolas A, Treco D, Schultes NP, Szostak JW (1989) *Nature 338*:35; Sun H, Treco D, Schultes NP, Szostak JW (1989) *Nature 338*:87.

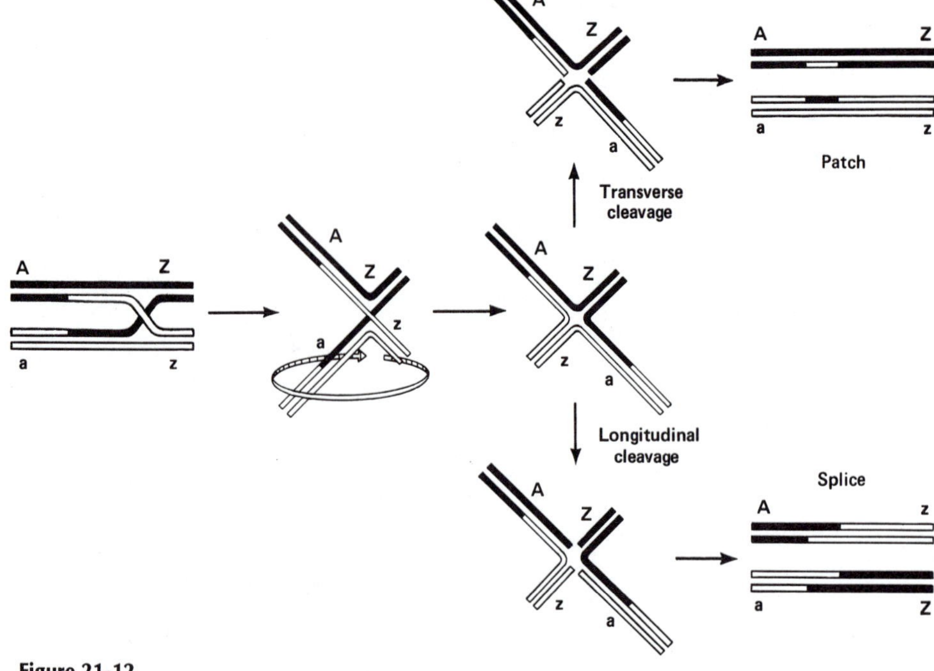

Figure 21-12
Isomerization and cleavage of Holliday junctions to give alternative recombination products. [After Potter H, Dressler D (1978) *PNAS 75*:3698]

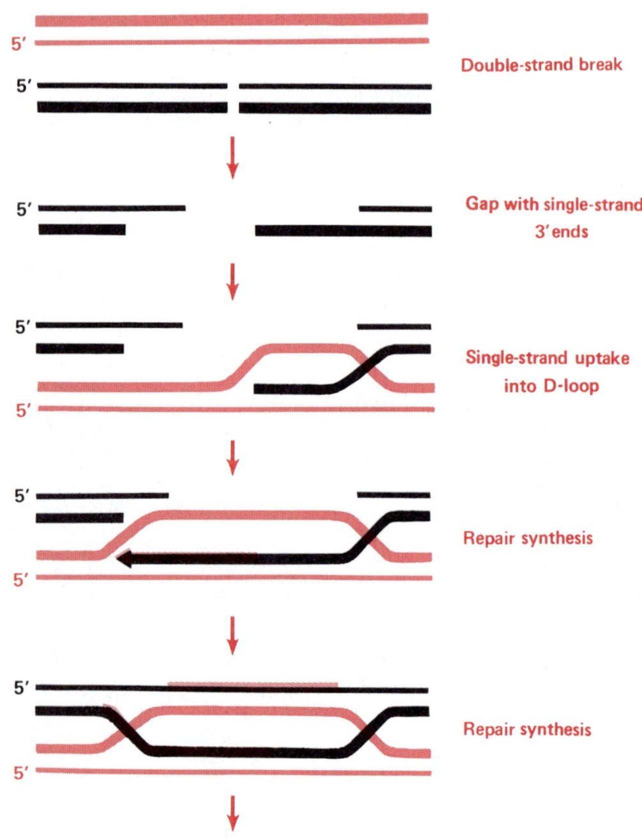

Figure 21-13
Double-strand gap repair model for initiation of homologous recombination. [Modified from Lewin B (1990) *Genes IV*, Oxford University Press, New York, p 630]

break. The strand displaced by D-loop formation can pair with sequences from the other side of the break, forming a primer for repair DNA synthesis of the other strand.

(5) *Resolution* of the Holliday junctions, generating two recombinant molecules, in which the initial break is completely healed.

These two models, and the many variations of them that have been suggested, explain much of the available data about the structure of recombinant molecules and thus provide a useful framework with which to approach the enzymology of the process. Further, these models suggest that recombination may involve considerable gap-filling by DNA polymerase and the action of endo- and exonucleases and DNA ligase. In addition to some of the enzymes already considered in replication and repair, genetic studies implicate others essential for recombination. Many of the recombination-defective mutants are also defective in repair and in the DNA replication of some phages, emphasizing the interrelationship between these processes.

Multiple Recombination Pathways in *E. coli*[97]

Genetic analyses in *E. coli* have revealed many mutations that affect the efficiency of genetic exchange. These mutants and the various in vivo recombination assays (conjugation, P1 transduction, recombination between plasmids or λ phage) reveal a multiplicity of recombination pathways (Table 21-3). The *recA* gene, encoding the strand-exchange protein which pairs homologous DNA molecules (see below), is central to all of these pathways.* Enzymes unique for each of the different pathways are probably responsible for generating the substrates used by RecA protein and resolving the joint molecules into the recombinant products.

Table 21-3
Major recombination pathways in *E. coli*[a]

Property	Pathway		
	RecBCD	RecF	RecE
Rec proteins required[b]	RecA, B, C, D	RecA, F, J, N, O, Q; Ruv	RecA, E, F, J, O, Q
Activating mutation	none	*sbcB, C*	*sbcA*
Responsible for conjugal recombination in wild-type cells	+++	±	–
Promotes plasmid recombination in wild-type cells	–	+++	–
Requires double-strand DNA break	yes	no	no
Activated by *chi* sites	yes	no	no

[a] Modified from Smith GR (1989) *Cell 58*:807.
[b] Assayed by conjugal recombination; when other recombination assays are employed, the genetic requirements may vary. In addition to these recombination factors, other gene products are required. The RecBCD pathway depends on SSB, gyrase, pol I, and ligase; the other pathways are likely to have similar requirements.

* Although RecA protein is required for all the homologous recombination pathways listed, in certain cases it may not be essential. For example, recombination between plasmids by the RecE pathway appears not to require RecA, while in conjugative recombination by the same pathway, RecA is clearly involved.

97. Smith GR (1989) *Cell 58*:807; Smith GR (1988) *Microbiol Rev 52*:1; Mahajan SK (1988) in *Genetic Recombination* (Kucherlapati R, Smith GR, eds.). Am Soc Microbiol, Washington DC, p. 87.

The *RecBCD pathway*[98] is largely responsible for the genetic exchange assayed by conjugation or transduction in wild-type *E. coli.* The RecA, RecB, RecC, and RecD proteins and SSB, pol I, DNA ligase, and gyrase are required. The RecA strand-exchange protein, the RecBCD nuclease–helicase[99] (see below), and the replication proteins have all been purified and characterized, making it possible to pursue the reconstitution of this recombination pathway in vitro (see below).[100]

The *RecF pathway* becomes detectable in *recB* or *recC* mutants and is responsible for plasmid–plasmid recombination in wild-type cells. The recombination deficiency of *recB* or *recC* mutants can be restored by changes in the *sbcB* and *sbcC* genes (suppressor of rec *BC*), which activate the RecF pathway. The *sbcB* allele inactivates exonuclease I, a 3′→5′ nuclease[101] (Section 13-2); the absence of this enzyme may allow DNA molecules with recombinogenic 3′ single-stranded tails to persist long enough to be utilized in recombination. The basis of the *sbcC* effect is unknown.[102]

Among the genes required for the RecF pathway, *recJ* encodes a single-strand–specific exonuclease with a preference for 5′ termini.[103] The RecJ function also becomes essential for conjugative recombination in *recD* mutants which are *recB*$^+$ and *recC*$^+$. The inactivated exonuclease of the RecBCD enzyme in these mutants may be substituted by the RecJ activity. The RecQ protein appears to be a DNA helicase with properties similar to those of helicase II[104] (Section 11-3). Several additional genes of unknown function are also required (Table 21-3).

The *RecE pathway* is activated in *recB*- or *recC*-defective strains by the *sbcA* mutation which turns on expression of exonuclease VIII, encoded by *recE*.[105] This 5′→3′ exonuclease, specific for duplex DNA,[106] may function by digesting DNA from double-strand breaks, generating recombinogenic 3′ single-stranded tails. Encoded by the *rac* cryptic prophage, the *recE* product resembles the λ exonuclease; recombination by the RecE pathway is mechanistically similar to that of the λ Red pathway (Section 17-7), a homologous recombination system that functions in λ-infected *E. coli.* The Red pathway depends on the phage recombination genes *exo* and *β*, but is independent of *recA*.

RecE recombination requires most of the same genes as does the RecF pathway (see Table 21-3). The mutations that activate RecE and RecF recombination both appear to increase the availability of DNA molecules with single-stranded tails, and recombination is probably mediated by related mechanisms.

In addition to the major recombination pathways present in *E. coli,* many phages encode their own homologous recombination systems. These phage recombination enzymes, in addition to being responsible for genetic exchange, are often required for efficient DNA replication and DNA maturation. Among the best studied phage recombination systems are those of T7 (Section 17-5), T4[107] (Section 17-6), and the λ Red pathway[108] (Section 17-7).

98. Taylor AF, Schultz DW, Ponticelli AS, Smith GR (1985) *Cell 41*:153.
99. Taylor AR (1988) in *Genetic Recombination* (Kucherlapati R, Smith GR, eds.). Am Soc Microbiol, Washington DC, p. 232.
100. Roman LJ, Kowalczykowski SC (1989) *JBC 264*:18340; Wang T-C, Smith KC (1989) *MGG 216*:315.
101. Kushner SR, Nagaishi H, Clark AJ (1972) *PNAS 69*:1366; Phillips GJ, Kushner SR (1987) *JBC 262*:455.
102. Lloyd RG, Buckman C (1985) *J Bact 164*:836.
103. Lovett ST, Kolodner RD (1989) *PNAS 86*:2627.
104. Umezu K, Kakayama K, Nakayama H (1990) *PNAS 87*:5363.
105. Kushner SR, Nagaishi H, Clark AJ (1974) *PNAS 71*:3593; Gillen JR, Willis DK, Clark AJ (1981) *J Bact 145*:521.
106. Joseph JW, Kolodner R (1983) *JBC 258*:10411; Joseph JW, Kolodner R (1983) *JBC 258*:10418.
107. Yonesaki T, Minagawa T (1989) *JBC 264*:7814.
108. Smith GR (1988) *Microbiol Rev 52*:1; Takahashi N, Kobayshi I (1987) *PNAS* 87:2790.

RecA. The *E. coli* RecA protein is the prototype of a class of strand-exchange proteins. Mutants defective in the *recA* gene do not execute homologous recombination, postreplication repair (Section 2), or any of the induced (SOS) functions. The properties of the purified protein provide insights into the central role of this enzyme in nearly all modes of homologous recombination.

Purified RecA, a 38-kDa monomer, binds cooperatively and stoichiometrically to ssDNA, generating a nucleoprotein complex that is the active species in DNA strand exchange (Fig. 21-14). Other activities of RecA, including ATP hydrolysis and repressor cleavage (Section 3), are also activated by the binding of RecA to ssDNA.[110] The filament contains about one monomer of RecA to each 4 to 6 ntd of DNA in a stable protein–DNA complex with a half-life of 30 minutes. ATP is not required for filament formation but the addition of ATP or ADP increases the rate of protein dissociation. RecA also binds to duplex DNA, although with much slower kinetics than in its binding to ssDNA; association is facilitated by low pH (i.e., pH ~ 6) or single-stranded regions, and the DNA helix is partially unwound by the protein.

Binding of RecA to ssDNA is polar, proceeding $5'\rightarrow3'$ in coating the DNA. As a consequence, the RecA molecules migrate to the 3′ end of the linear DNA molecule and collect there. Under conditions required for recombination in vitro, SSB helps RecA overcome secondary structure, allowing it to coat the ssDNA completely.[111] SSB is also important for maintaining the stability and activity of RecA·ssDNA complexes throughout the strand-exchange reaction.[112]

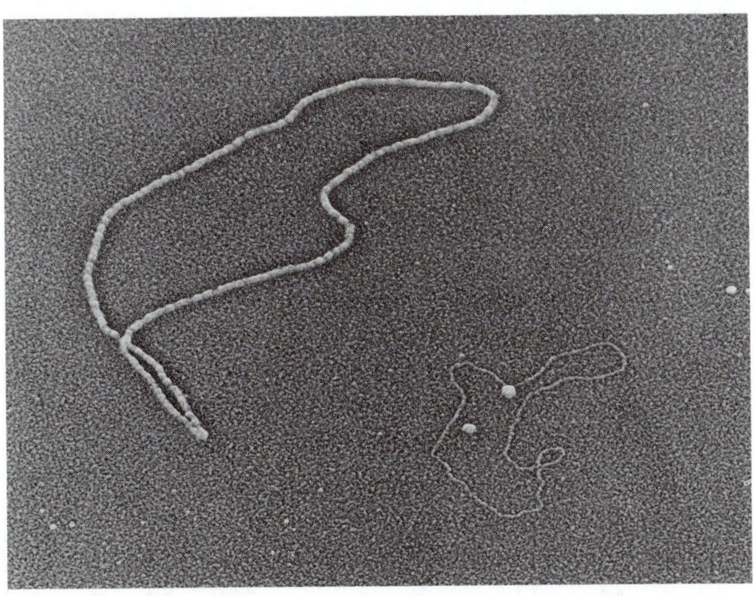

Figure 21-14
Electron micrograph of two M13 ssDNA circles; one bound by RecA (left), one free (right). Note the extended form of the RecA–DNA complex, 1.5 to 2 times the length of the free DNA. [Courtesy of Dr J Griffith]

---

109. Cox MM, Lehman IR (1987) *ARB 56*:229; Radding CM (1982) *ARG 16*:405; Griffith J, Harris L (1988) *Crit Rev B 23*:S43; Radding CM (1989) *BBA 1008*:131.
110. Roberts JW, Robert CW, Craig NL (1978) *PNAS 75*:4714.
111. Cox MM, Lehman IR (1982) *JBC 257*:8523; Wu AM, Kahn R, DasGupta C, Radding CM (1982) *Cell 30*:37; West SC, Cassuto E, Howard-Flanders P (1982) *MGG 186*:333; Cox MM, Lehman IR (1981) *PNAS 78*:3433.
112. Morrical SW, Cox MM (1990) *Biochemistry 29*:837.

Renaturation of complementary single strands by RecA probably depends on some of its recombinase properties. In contrast to the stimulation of renaturation by simple SSBs (such as *E. coli* SSB and T4 gp32; Sections 10-2, 10-3, and 10-4), the RecA-catalyzed reaction requires subsaturating amounts of protein and is stimulated by ATP. Kinetics of renaturation are first-order with respect to DNA concentration, indicating that an intermediate complex is quickly formed between the two strands prior to formation of the DNA duplex. Several lines of evidence suggest that RecA has two modes of DNA binding involved in the pairing of the two strands.[113]

The Strand-Exchange Reaction. RecA catalyzes strand exchange between a large variety of homologous DNA molecules with different conformations (Fig. 21-15). The structural requirements for a DNA substrate are (1) a region of

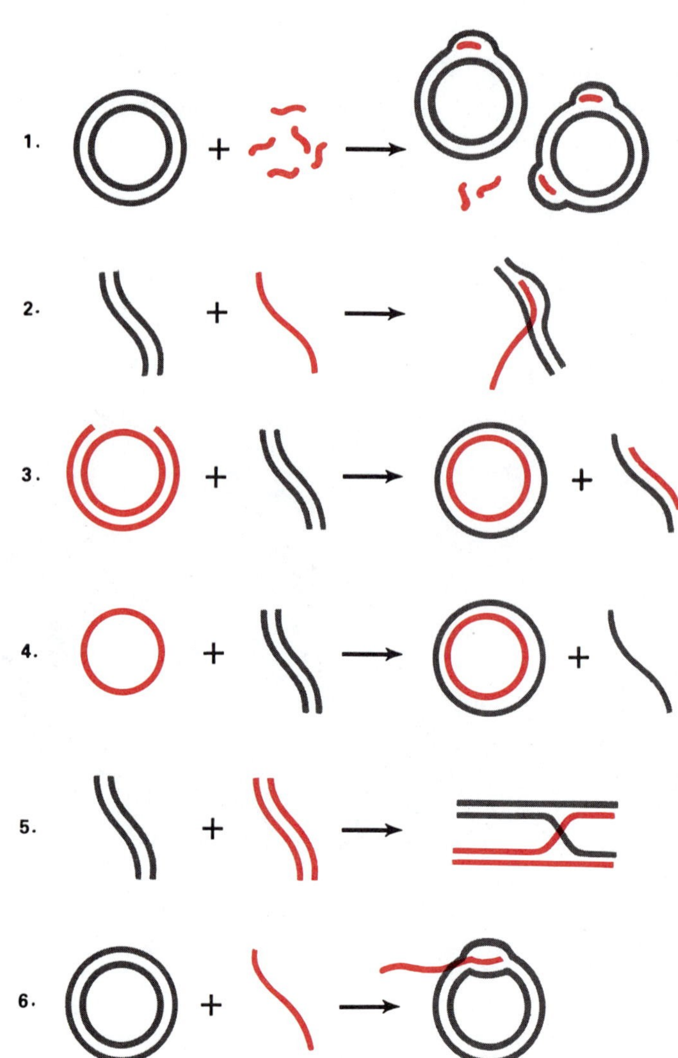

Figure 21-15
Strand-exchange reactions promoted by RecA. Molecule contributing the single-stranded partner is shown in color in each case.

113. Menetski JP, Kowalczykowski SC (1989) *Biochemistry 28*:5871.

ssDNA for assembly of a RecA filament, (2) DNA–DNA homology between the two partners, and (3) a free end, within the region of homology, to allow the strands to intertwine. This final requirement can be circumvented in the presence of a topoisomerase to allow the strands of covalently closed molecules to be interlinked (Section 12-1). The stages of the strand-exchange reaction are as follows:[114]

(1) Assembly of the RecA nucleoprotein filament on the ssDNA partner (see above). Binding is principally to the DNA backbone, shows no sequence specificity, and extends the DNA strand prior to strand pairing.

(2) DNA–DNA aggregation mediated by RecA. This association does not require homology, and brings the partners together to facilitate the search for homologous sequences.

(3) Association of homologous DNA segments between the ssDNA contained in the nucleoprotein filament and its complementary sequence in the duplex DNA. This association initially does not require a free DNA end within the region of homology.

(4) Conversion of the initial homology-based joints to stable D-loops (*plectonemic joints*) when a free DNA end or topoisomerase is available.

(5) *Strand exchange* by elongating the heteroduplex. The previously paired strand is displaced from the duplex by the new strand. Branch migration is unidirectional, and is not impeded by pyrimidine dimers and short regions of nonhomology. Exchange can go to completion, provided it is not limited by topological constraints or extensive nonhomologous sequences; it may even proceed through deletions or insertions hundreds of base pairs in length.

Branch migration appears to proceed 5′→3′ (with respect to the incoming ssDNA) on several types of DNA substrates.[115] However, the preference for strand exchange between the 5′ end of a linear ssDNA and a circular supercoiled partner is not easily reconciled with this polarity.[116] Furthermore, the fact that both binding of RecA to the ssDNA and branch migration are polar reactions means that the basis of the overall polarity of exchange may originate from more than one source.

The strand-exchange reaction establishes that RecA has the intrinsic capacity to (1) associate DNA molecules, (2) assist in the recognition between homologous DNAs, (3) exchange base-paired DNA strands, and (4) catalyze extensive branch migration. Although most experiments on RecA-catalyzed strand exchange have used one single-stranded and one duplex partner, RecA protein can also mediate strand exchange between two homologous DNA duplexes provided one has a single-stranded tail for the initial assembly of RecA. This reaction forms a classic Holliday junction; the central role of RecA in recombination is thus clearly established both in vivo and in vitro.

Other Strand-Exchange Proteins. Recognition of the key contributions of RecA protein to homologous recombination has helped in the search for similar proteins from other sources (Table 21-4), using the strand-exchange assay. The best characterized strand-exchange protein is UvsX from phage T4; Rec1 has been

114. Kowalczykowski SC (1987) *TIBS 12*:141.
115. Cox MM, Lehman IR (1981) *PNAS 78*:6018.
116. Konforti BB, Davis RW (1987) *PNAS 84*:690; Konforti BB, Davis RW (1990) *JBC 265*:6916.

Table 21-4
Purified strand-exchange proteins

Source	Protein	Mass (kDa)	Stoichiometric	ATP-Dependent
E. coli[a]	RecA	38	yes	yes
Phage T4[b]	UvsX	40	yes	yes
Ustilago[c]	Rec1			yes
Yeast				
Mitotic cells[d]		120	yes	no
Vegetative cells[e]		132	yes	no
Meiotic cells[f]			no	no
Drosophila embryos[g]		105	no	no

[a] See text for references.
[b] Yonesaki T, Ryo Y, Minagawa T, Takahashi H (1985) *EJB 148*:127; Formosa TY, Alberts BM (1985) *JBC 261*:6107.
[c] Kmiec E, Holloman WK (1982) *Cell 29*:367; Kmiec EB, Holloman WK (1984) *Cell 36*:593.
[d] Halbrook J, McEntee K (1989) *JBC 264*:21403.
[e] Heyer W-D, Evans DH, Kolodner RD (1988) *JBC 263*:15189.
[f] Sugino A, Nitiss J, Resnick MA (1988) *PNAS 85*:3683.
[g] Lowenhaupt K, Sander M, Hauser C, Rich A (1989) *JBC 264*:20568; Eisen A, Camerini-Otero RD (1988) *PNAS 85*:7481.

reported in the fungus *Ustilago*. Both proteins are similar biochemically to RecA; mutations in their genes disrupt recombination and repair. Among proteins isolated from other cell types are some interesting mechanistic differences. The eukaryotic strand-exchange proteins, with the exception of Rec1, function without ATP (or nucleotide hydrolysis) and fail to form the stoichiometric protein–DNA filaments characteristic of RecA. Their involvement in recombination remains to be proven.

Recombination Nucleases

The models for recombination and the genetic evidence for the essential factors indicate an important role for nucleases in homologous recombination. Nucleases (Section 13-9) function in both the initiation and the resolution stages. However, the multiple roles that nucleases play in recombination have complicated the analysis of their contributions. Some nucleases that are needed to generate or process one type of recombination intermediate may destroy the intermediates made by a competing pathway.

The multifunctional RecBCD enzyme,[117] with its exo- and endonuclease (Section 13-9) and helicase (Section 12-2) activities, is the principal nuclease in recombination in wild-type *E. coli*. Specific cleavage in vitro of *chi* sequences,[118] which are recombination hot-spots in vivo, confirms the importance of this enzyme. However, the residual recombination in RecBCD mutants, and the ability of suppressor mutations to activate secondary recombination pathways, indicates that other exonucleases can substitute at least partially for this enzyme (Table 21-5). Conditions that activate the alternative pathways either

117. Taylor AF (1988) in *Genetic Recombination* (Kucherlapati R, Smith GR, eds.). Am Soc Microbiol, Washington DC, p. 232.
118. Ponticelli AS, Schultz DW, Taylor AF, Smith GR (1985) *Cell 41*:145.

Table 21-5
Nucleases involved in homologous recombination in *E. coli*[a]

Enzyme	Gene	Substrate	Polarity	Role in recombination
RecBCD, Exo V	*recB*, *recC*, *recD*	ds or ss; also ss endo	either	required in WT; multiple functions
RecJ exo	*recJ*	ss-specific	5′→3′	required for RecE and RecF pathways; substitutes for RecBCD exo
Exo I	*sbcB*	ss-specific	3′→5′	inhibits RecE and RecF pathways; inactivated in *sbcB* mutant.
Exo VIII	*recE*	ds-specific	5′→3′	required for RecE pathway; activated by *sbcA* mutant.

[a] ds = double-stranded DNA; endo = endonuclease; exo = exonuclease; ss = single-stranded DNA; WT = wild-type cells.

generate or stabilize 3′ ends in ssDNA, consistent with the preference for 3′ ends during RecA-promoted strand exchange in vitro.

Enzymes That Cleave Holliday Junctions[119]

Specific cleavage of DNA with structures resembling Holliday junctions points to involvement of the cleaving enzymes in the late stages of homologous strand exchange (see models above). Examples include phage T4 endonuclease VII,[120] phage T7 endonuclease I,[121] and a protein purified from mitotic yeast[122] (Table 21-6). Specific cleavage of junctions has also been detected in extracts of human cells.[123]

The cleavage specificities of these enzymes indicate that they recognize the forked structure but are relatively indifferent to the sequence in or around the junction.[124] The enzymes introduce a pair of nicks near the base of the four-way

Table 21-6
Enzymes that cleave Holliday junctions

Enzyme	Cleavage				Sequence specificity
	X junction	Y junction	Distance[a]	Position	
T7 endo I (gp3)	yes	yes	1	5′	some
T4 endo VII (gp49)	yes	yes	1 to 5	3′	some
Yeast endo	yes	no	4 to 8	5′	none

[a] Nucleotides from the four-way junction.

119. West SC (1989) in *Nucleic Acids and Molecular Biology* (Eckstein F, Lilley DMJ, eds.). Springer-Verlag, Berlin, Vol. 3, p. 44.
120. Mizuuchi K, Kemper B, Hays J, Weisberg R (1982) *Cell 29*:357.
121. deMassy B, Weisber RA, Studier FW (1987) *JMB 193*:359.
122. West SC, Körner A (1985) *PNAS 82*:6445; Symington LS, Kolodner R (1985) *PNAS* 82:7247.
123. Waldman AS, Liskay RM (1988) *NAR 16*:10249.
124. Mueller JE, Kemper B, Cunningham RP, Kallenbach NR, Seeman NC (1988) *PNAS 85*:9441; Dickie P, McFadden G, Morgan AR (1987) *JBC 262*:14826; Duckett DR, Murchie AIH, Diekmann S, von Kitzing E, Kemper B, Lilley DMJ (1988) *Cell 55*:79; Picksley SM, Parsons CA, Kemper B, West SC (1990) *JMB 212*:723; Parsons CA, Murchie AI, Lilley DMJ, West SC (1989) *EMBO J 8*:239.

branch, resolving the molecule into two linear pieces. The T4 and T7 enzymes also cleave three-way junctions (Y-shaped molecules).[125]

A significant feature of T4 endo VII and the yeast enzyme is that cleavage of the two homologous arms of a four-armed structure is at the same sequence in the two strands of the same polarity.[126] The enzyme, after recognizing the junction, must be able to recognize the two homologous arms of the molecule. As a result of the symmetric cleavage, the two linear DNAs generated are directly ligatable without the need for gap-filling or repair.

The T7 and T4 enzymes, while implicated in recombination, most likely function in other DNA maturation processes as well. T7 gene *3*, which encodes T7 endo I, is required for phage recombination and the breakdown of host DNA; mutants of T4 gene *49* (which codes for T4 endo VII) fail to process the large DNA networks formed during recombination-dependent replication late in phage growth. Although the yeast enzyme properties are suitable for a role in homologous recombination, mutations to evaluate this role in vivo are not yet known. While Holliday junctions are invariably included in recombination models, resolvases for these junctions have not been identified within the known *E. coli* recombination proteins. The RecBCD enzyme has been suggested to provide this function, but does not appear to cleave preexisting Holliday junctions into recombinant products in vitro.[127] An activity capable of cleaving Holliday junctions generated by RecA protein in vitro has been detected and partially purified and is distinct from RecBCD.[128] Whether this enzyme plays a role in recombination in vivo remains to be established.

Recombination by the RecBCD Pathway

A model based on the properties of the RecA and RecBCD enzymes has been proposed for homologous recombination[129] (Fig. 21-16). The RecBCD enzyme complex enters one of the recombination partners at the double-strand break and melts the strands by progressive helicase action. Upon its reaching a *chi* sequence, the endonuclease is activated, generating an ssDNA tail with a 3′-OH. RecA protein assembles on this strand, using it to form a D-loop and initiate strand exchange with the homologous DNA duplex. How this Holliday junction is resolved after strand exchange is unknown. T4 endo VII can function in this capacity in vitro,[130] as does an activity present in *E. coli* cell extracts.[131]

Alternative models for the participation of RecBCD enzyme in recombination have been proposed to explain how *recD* mutants, which produce a RecBCD enzyme lacking all known enzymatic activities, demonstrate increased rather than decreased rates of recombination in vivo.[132]

Meiotic and Mitotic Recombination[133]

Recombination in eukaryotic cells occurs chiefly in meiosis, at rates 100 to 1000 times those in mitosis. Crossing-over and gene conversion in meiosis achieve the

125. Jensch F, Kemper B (1986) *EMBO J* 5:181; deMassy B, Weisber RA, Studier FW (1987) *JMB* 193:359.
126. Mizuuchi K, Kemper B, Hays J, Weisberg R (1982) *Cell* 29:357; Parsons CA, West SC (1988) *Cell* 52:621.
127. Taylor AF, Smith GR (1990) *JMB* 211:117.
128. Connolly B, West SC (1990) *PNAS* 87:8476.
129. Smith GR (1988) *Microbiol Rev* 52:1.
130. Müller B, Jones C, Kemper B, West SC (1990) *Cell* 60:329.
131. Connolly B, West SC (1990) *PNAS* 87:8476.
132. Thaler DS, Stahl FW (1988) *ARG* 22:169.
133. Giroux CN (1988) in *Genetic Recombination* (Kucherlapati R, Smith GR, eds.). Am Soc Microbiol, Washington DC, p. 465; Roeder GS, Stewart SE (1988) *Trends Genet* 4:263.

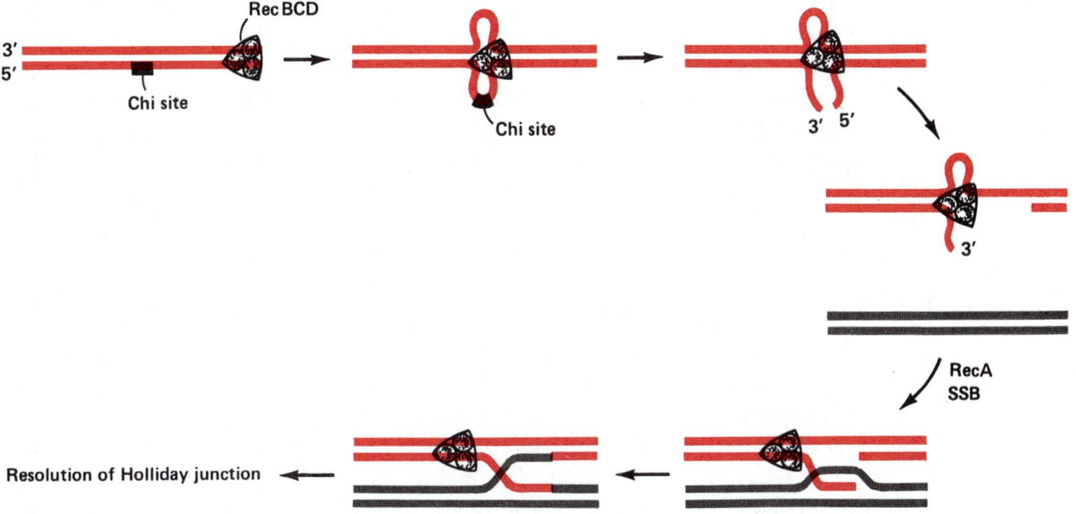

Figure 21-16
Model for homologous recombination promoted by RecA and RecBCD enzymes and *chi* sites. RecBCD enzyme cuts and unwinds one DNA strand near the *chi* sequence to generate an ssDNA tail. Aided by RecA and SSB proteins, the tail invades homologous dsDNA of a second parent, which may be circular and supercoiled. The displaced strand, a D-loop, anneals with the gap in the first parental DNA to form a Holliday junction. [Modified from Smith GR (1988) *Microbiol Rev 52*:1]

relatively independent assortment of markers from the parental to progeny chromosomes. Recombination occurs at the four-strand stage, after DNA synthesis and chromosome pairing and prior to the first meiotic division. Recombination is associated with the *synaptonemal complex,* a structure visible by microscopy in which the homologous chromosomes are physically aligned. All organisms with this complex show high levels of recombination between homologs during meiosis, while mutants which lack it [*rad50, spo11, hop1* in yeast; *C(3)G* in *Drosophila*][134] have reduced levels of homologous exchange in meiosis but retain normal levels of mitotic recombination. Still other mutations (such as yeast *rad52*) affect both pathways, indicating that some enzymes are common to both. Mitotic recombination is associated with DNA repair and is induced by DNA-damaging agents which introduce double-strand breaks and interstrand cross-links. Mitotic recombination occurs at both the two- and four-strand stages of chromosome pairing.

Gene Conversion

Gene conversion is nonreciprocal recombination between alleles in which there is a net gain of one of the alleles among the DNA strands, at the expense of the other allele, as a result of the recombination event. Double-strand gap repair is an example of an event that can lead to gene conversion because the sequence originally on the broken DNA duplex is "converted" to the sequence found in the homologous recombination partner. Single-strand invasion accompanied by

134. Farnet C, Padmore R, Cao L, Raymond W, Alani E, Kleckner N (1989) in *Mechanisms and Consequences of DNA Damage Processing* (Friedberg EC, Hanawalt PC, eds.). UCLA Vol. 83. AR Liss, New York, p. 201; Klapholz S, Waddell CS, Esposito RE (1985) *Genetics 110*:187; Giroux CN, Tiano HF, Dresser ME (1986) *Yeast 2*:133; Hollingsworth NM, Byers B (1989) *Genetics 121*:445.

repair DNA synthesis can also cause gene conversion, as can mismatch repair within the heteroduplex of a strand-exchange intermediate.

Gene conversion not only is important in both meiotic and mitotic recombination between homologous chromosomes, it also plays a role in exchange between copies of repeated sequences within the genome, and between expressed and silent alleles contributing to the control of gene expression in several organisms. Thus there are a number of specialized gene rearrangement events that use this mechanism. These events may be called *specialized gene conversions* in contrast to the events associated with general homologous recombination.

The classic example of genetic control by gene conversion is *mating-type switching* in yeast.[135] Although the yeast genome contains three copies of the mating-type information, only one copy, that at the *MAT* locus, is expressed. Replacement of the copy at *MAT*, with alternate alleles stored as nonexpressed "silent" genes by unidirectional gene conversion, is responsible for the switching of yeast cells between the two alternative mating types. Conversion, or mating-type switching, is initiated by a specific double-strand break at *MAT*, introduced by the HO endonuclease; recombination is thus thought to occur by the double-strand gap-repair mechanism (see above). Cell-cycle–specific and mother-cell–specific control of expression of *HO* regulates switching in a predictable pattern.

In other examples of gene conversion between expressed and silent alleles,[136] the choice of one allele out of several to more than 1000 possible alleles accounts for the surface antigen of trypanosomes, for antigenic variation in bacterial pathogens (*Borrelia hermsii* and *Neisseria gonorrhoeae*), and for the diversity of immunoglobin light-chain genes in chickens.

Gene Targeting[137]

Homologous recombination between DNA sequences located on the chromosome and experimentally introduced cloned sequences enables the genome of an organism to be modified. Through the technique of gene targeting, essentially any desired mutation can be constructed and analyzed. The ability to target genetic modifications in yeast successfully has contributed significantly to this organism's usefulness and popularity.[138] Modification of the genome of mammalian pluripotent stem cells (i.e., embryo-derived stem cells) in turn allows the alterations to be introduced into the germ line of a living organism, as is now routinely done to make *transgenic mice*.

The application of gene replacement to mammalian cells has, however, proved difficult because, in contrast to the situation in yeast, most DNA introduced into mammalian cells integrates into the genome by nonhomologous recombination at seemingly random sites. When DNA is introduced into mammalian cells, usually less than a few percent (often as low as 1 in 10^4) of the stably integrated DNA settles in at a homologous site. The large background of nonhomologous events necessitates the use of selection and screening methods to identify the cells carrying the targeted genetic change.

---

135. Strathern JN (1988) in *Genetic Recombination* (Kucherlapati R, Smith GR, eds.). Am Soc Microbiol, Washington DC, p. 445.
136. Borst P, Greaves ER (1987) *Science 235*:658.
137. Capecchi MR (1989) *Trends Genet 5*:70; Capecchi MR (1989) *Science 244*:1288.
138. Botstein D, Fink GR (1988) *Science 240*:1439.

The mouse hypoxanthine phosphoribosyl transferase gene (*hprt*) has been a favorite target for gene replacement. It has three distinct advantages: (1) its location on the X chromosome means that in male cells only one disruption event is required to generate a $hprt^-$ phenotype; (2) the mutant phenotype can be selected directly because $hprt^-$, but not $hprt^+$, cells can grow in the presence of the base analog 6-thioguanine; and (3) the gene is expressed in the embryo-derived stem cells.

Several strategies have been developed, largely from the study of insertion into *hprt*, to select and screen for properly targeted recombination.

(1) *Selection for expression of the introduced copy of the gene.* This method employs replacement vectors made in such a way that expression of the drug-resistance gene, used to select for stable transformants, depends on homologous recombination to supply a missing enhancer or promoter.[139] Enrichment of homologous events in the resulting transformants is about 100-fold.

(2) *Selection against insertion at random sites.*[140] In this case the integration vector is constructed so that if integration occurs by a nonhomologous pathway, a gene will be expressed in the vector that can be selected against in cell culture. In contrast, insertion via homologous recombination will result in loss of this detrimental sequence. Enrichments of 2000-fold in the recovery of properly targeted insertions have been reported.

(3) *Screening by the polymerase chain reaction (PCR).* In the absence of selection, PCR (Section 4-17) is the method of choice to screen for properly integrated recombinants.[141] If appropriate amplification primers are chosen, one from the insertion vector and one from the DNA flanking the desired insertion site, amplification will occur only if the primer sites are brought in close proximity as a result of homologous recombination.

Several factors seem to affect the frequency of homologous recombination. One of these is the length of DNA shared between the introduced gene and the chromosomal site. Increasing the length of homologous sequences about five-fold, from 3 to 15 kb, appears to increase the frequency of homologous recombination about 100-fold.[142] The length of the disruption in the gene to be introduced may also have a significant influence, with longer disruptions being less efficiently recombined. The frequency of integration of introduced DNA into expressed regions may be higher than that into transcriptionally silent chromosomal locations. Finally, whether the new DNA is introduced into the cells by microinjection or by electroporation may affect the frequency and outcome of integration events.

Additional Associations of Recombination and Replication

The obligatory association of recombination with postreplication repair, lesion bypass, and healing of double-strand breaks has been mentioned. There are additional interactions between DNA replication and general recombination: (1) the requirement for homologous recombination for replication during the

139. Jasin M, Berg P (1988) *Genes Dev 2*:1353; Sedivy JM, Sharp PA (1989) *PNAS 86*:227.
140. Mansour SL, Thomas KR, Capecchi MR (1988) *Nature 336*:348.
141. Kim HS, Smithies O (1988) *NAR 16*:8887.
142. Capecchi MR (1989) *Science 244*:1288.

latter growth phases of phage T4, in which the 3′ ends of recombination intermediates serve as primers for extensive DNA synthesis[143] (Section 17-6); (2) the function of RecA in the initiation of stable DNA replication and possibly in normal initiation as well (Section 20-3); (3) recombination in phage λ (Section 6), which is stimulated by replication; and (4) the "copy-choice" DNA synthesis (template switching) in viral duplication[144] which may also lead to recombinant progeny molecules.

21-6 Site-Specific Recombination[145]

In contrast to homologous recombination, these DNA transactions do not require extensive regions of DNA homology or the homologous recombination machinery, such as the RecA function. Site-specific recombination depends on only one or a few proteins. At least one of these proteins interacts with short recognition sites, and initiation of recombination depends on protein–DNA interactions rather than DNA–DNA interactions. In typical reactions of this class, DNA cleavages and ligations at defined locations produce uniform, exact recombinant molecules. Site-specific recombination is widely applied to achieve alternative DNA arrangements (Table 21-7).

Table 21-7
Site-specific recombination systems

Function	System	Comments
Phage integration and excision	λ Int	best-characterized system; other lambdoid phages (e.g., φ80) use similar systems, as do P22, P2, and P4
Resolution of circular multimers	Tn3 and γδ resolvase	processing of cointegrate products of transposition
	P1 *lox–cre*	circle formation upon phage infection; resolution of multimers increases plasmid stability
	ColE1 Cer	resolution of multimers increases plasmid stability by increasing the number of partitionable units; three host proteins required[a]
	2μ FLP	inversion of segment allows plasmid amplification by forming a double rolling-circle intermediate, subsequently reduced to monomers
Inversions for expression of alternate genes	Hin	*Salmonella* flagella genes
	Gin	Mu tail fibers
	Pin	P1 tail fibers
	Cin	*E. coli* cryptic element; function unknown
Assembly of genes during development	*B. subtilis* mother-cell–specific σ factor[b]	
	Anabaena nitrogen-fixation genes[c]	
	immunoglobulin and T-cell receptor genes in mammals[d]	

[a] Stirling CJ, Szatmari G, Stewart G, Smith MCM, Sheratt DJ (1988) *EMBO J 7*:4389.
[b] Strager P, Kunkel B, Kroos L, Losick R (1989) *Science 243*:507.
[c] Golden JW, Robinson SJ, Haselkorn R (1985) *Nature 314*:419; Golden JW, Wiest DR (1988) *Science 242*:1421.
[d] Although included in this table as a site-specific reaction, the mechanism by which these genes are assembled is clearly distinct from the rest of the reactions listed.

143. Mosig G (1987) *ARG 21*:347.
144. Kirkegaard K, Baltimore D (1986) *Cell 47*:433.
145. Craig NL (1988) *ARG 22*:77; Sadowski P (1986) *J Bact 165*:341.

The recombination reactions which have been analyzed in detail share several characteristics:

(1) Recombination is conservative, DNA synthesis is not required, and sequences are neither lost nor gained in the reaction.
(2) Exchange occurs between relatively small DNA sites of nearly identical sequence. Some reactions also require additional *cis*-acting DNA sequences for high efficiency.
(3) One recombinase protein is chiefly responsible for recognizing the recombination sites and for cleaving and ligating the DNA. Transient covalent protein–DNA intermediates conserve the energy of cleavage for the ligations, thus avoiding the need for nucleotide cofactors. Additional DNA-binding proteins often assist in assembly of the recombinase at the recombination sites.
(4) Reactions can be either *intermolecular*, such as the integration of the λ genome in the host chromosome, or *intramolecular*, as in the resolution of circular multimers and in gene inversions (Fig. 21-17).
(5) Intramolecular reactions are either *inversions*, in which the orientation of the DNA segment between the recombination sites is flipped with respect to the flanking DNA, or *deletions* that result in excision of this piece of DNA (Fig. 21-17). The products depend on the orientation of the recombination sites: deletions result from recombination between direct repeats, whereas inversions occur when the orientation of the sites is inverted. Certain recombinases catalyze only one of these events, whereas others can carry out both (Table 21-8).

Recombinases fall into two distinct classes of structurally related proteins:

(1) The *integrase family* includes the P1 Cre and 2μ FLP proteins, λ Int protein, and several other phage integrases.[146] These proteins make staggered breaks in the DNA with 5′ overhangs of 6 to 8 ntd and form a covalent DNA–

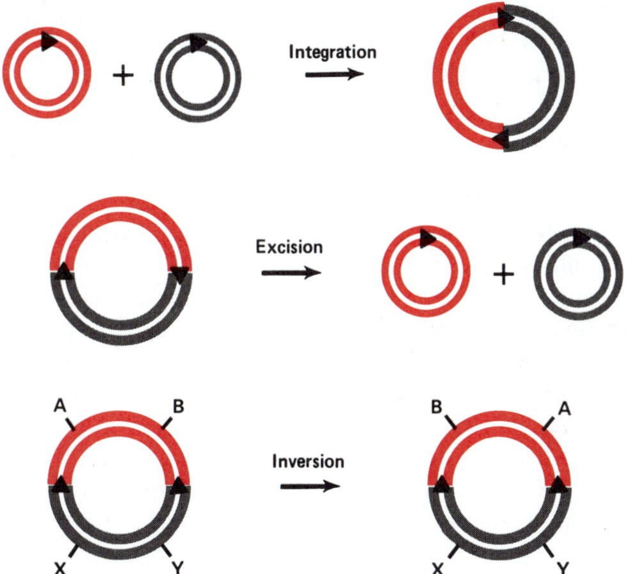

Figure 21-17
Physical consequences of site-specific recombination.

146. Argos P, Landy A, Abremski K, Egan JB, Haggard-Ljungquist E, Hoess RH, Kahn ML, Kalionis B, Narayana SVL, Pierson LS III, Sternberg N, Leong JM (1986) *EMBO J* 5:433.

Table 21-8
Properties of various site-specific recombination systems[a]

Property	Recombinase: Int	Recombinase: FLP	Resolvase	Invertase
Reaction types[b]	inter deletion or inversion	inter deletion or inversion	intra deletion only	intra inversion only
Site structure	*attP*: complex *attB*: simple	simple	complex	simple + enhancer
Superhelicity required	for *attP*	no	yes	yes
Auxiliary proteins	Xis, IHF, Fis	none	none	Fis, HU
Size and type of cleavage overhang	7 ntd; 5′	8 ntd; 5′	2 ntd; 3′	2 ntd; 3′
Covalent protein–DNA attachment	3′P–tyr	3′P–tyr	5′P–ser	5′P–ser
Strand exchange	2 sequential single strands	2 sequential single strands	double strand	double strand

[a] Modified from Sadowski P (1986) *J Bact 165*:341.
[b] intra = intramolecular recombination; inter = intermolecular recombination.

protein linkage via a tyrosine to the 3′-P. They have in common a 40-amino-acid region near the C terminus in which histidine, arginine, and tyrosine residues, probably part of the active center, are completely conserved. For λ Int protein, the tyrosine has been shown to form the covalent protein–DNA bond.[147] The N-terminal domains are unrelated, and probably account for the specific DNA binding.

(2) The *resolvase family* includes the Tn3 and γδ resolvases and the Hin family of invertases. Covalent protein–DNA attachment is by a serine–5′-P bond; the 3′ ends of the DNA at the cleavage site form a 2-ntd overhang. In this protein family, a 13% identity in amino acid sequence, especially near the N termini, includes the domain involved in strand exchange. The C-terminal regions are responsible for specific DNA binding and contain an amino acid sequence of the helix–turn–helix motif (Section 10-9).

λ Integration and Excision[148]

Integration of the λ genome into that of the host cell during lysogenization is the first and most studied example of site-specific recombination. Int protein (40 kDa) recognizes the recombination sites, and carries out the cleavage and ligation reactions. In addition to Int protein, both integration and excision require IHF (*i*ntegration *h*ost *f*actor; Section 10-8), a small basic *E. coli* protein which binds to preferred sequences and bends the DNA. Excision involves the phage excision protein (Xis, 8.6 kDa) and is stimulated by the *E. coli* Fis protein (*f*actor for *i*nversion *s*timulation; Section 10-8) under some conditions.[149]

Integrative recombination occurs between the phage (*attP*) and bacterial (*attB*) attachment sites, whereas excision takes place between the left and right (*attL* and *attR*) junctions of the prophage (Fig. 21-18). The 235-bp *attP* sequence is

147. Pargellis CA, Nunes-Duby SE, Moitoso de Vargas L, Landy A (1988) *JBC 263*:7678.
148. Landy A (1989) *ARB 58*:913; Weisberg RA, Landy A (1983) in *Lambda II* (Hendrix RW, Roberts JW, Stahl FW, Weisberg RA, eds.). CSHL, p. 211.
149. Thompson JF, Moitoso de Vargas L, Koch C, Kahmann R, Landy A (1987) *Cell 50*:901.

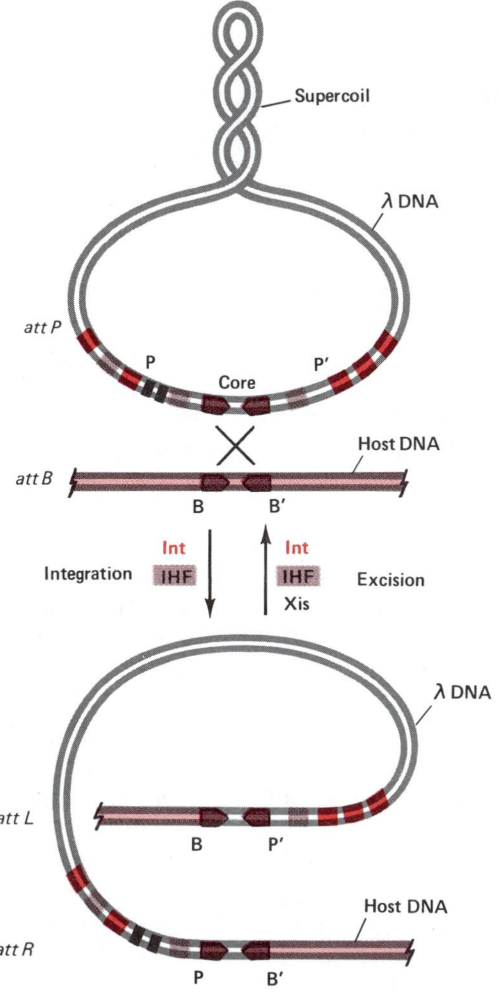

Figure 21-18
The components of integrative and excisive recombination of phage λ. Supercoiled *attP* integrates into linear *attB* in the presence of Int and IHF to yield the prophage products *attL* and *attR*; the excision reaction additionally requires Xis. The inverted core-type Int binding sites are indicated by colored arrows in the core. The phage P and P′ arms carry five arm-type Int, three IHF and two Xis sites, color coded to the named proteins. [From Nunes-Düby SE, Matsumoto L, Landy A (1987) *Cell 50*:779]

complex, containing seven Int binding sites and three IHF binding sites. Additional sequences, bound by Xis and Fis, are in *attR* after recombination and are involved in excision. Within *attP*, the Int protein binds the core elements at the cleavage sites and the "arm" sequences in the DNA flanking the recombination junctions.[150] An unusual feature of Int protein is its possession of two DNA-binding domains that recognize these two distinct DNA sequence elements.[151] In contrast to *attP*, the *attB* sequence is only 30 bp and contains two core Int binding sites in inverted orientation. Between these sites are seven base pairs, called the overlap region, which can be essentially any sequence but must be identical in the two recombining DNAs.

Int Protein Assembly at *attP*. This first step in recombination involves binding of Int protein to its multiple sites.[152] Affinity for the arm sequences, being stronger than that for the core elements, facilitates filling of the central sites. Multiple

150. Ross W, Landy A (1982) *PNAS 79*:7724; Ross W, Landy A (1983) *Cell 33*:261.
151. Moitoso de Vargas L, Pargellis CA, Hasan NM, Bushman EW, Landy A (1988) *Cell 54*:923.
152. Richet E, Abcarian P, Nash HA (1986) *Cell 46*:1011; Thompson JF, Moitoso de Vargas L, Skinner SE, Landy A (1987) *JMB 195*:481.

protein–DNA and protein–protein interactions are involved, and correct spacing of the protein binding sites is critical.[153] IHF stimulates complex formation by bending the DNA; at least one of the IHF sites can be replaced by an intrinsically bent DNA sequence.[154] The DNA carrying *attP* must be negatively supercoiled in a structure with the DNA wrapped around the protein core, as in a nucleosome; this *attP*–Int protein complex is called the *intasome.*

Synapses of the *attP* and *attB* Sites.[155] The *attP* intasome binds to the *attB* sequence (devoid of proteins) to associate the two recombination sites prior to strand exchange. The bivalent DNA-binding capacity of Int protein allows molecules bound to the arm sites of *attP* to simultaneously bind the core elements in *attB.* The two *att* sites associate by random collision requiring no specific sequence orientation. Synapses depend entirely on protein–DNA and protein–protein interactions rather than DNA–DNA pairing; sequence homology between the recombination partners is not required until a subsequent stage. The stable, highly ordered synaptic complex fixes the DNA sites in the proper conformation for strand exchange.

Strand Exchange by Int Protein. The exchange involves two sequential single-strand cleavages and ligations (Fig. 21-19) within the core region.[156] One pair of strands is cleaved first, and strand exchange at this site forms a Holliday intermediate (Section 5). Branch migration of this junction through the overlap region progresses to the second core Int site, explaining the need for identical sequences (although no specific sequence) between the Int sites of the two recombination partners. Exchange of the second pair of strands occurs at the second Int site, resulting in a 7-bp splice in the two strands at the junction.

Excision. Integration divides *attP* so that a normal intasome cannot form at either *attR* or *attL.* Thus excision, although regenerating the starting DNA molecules, is not simply a reversal of the integration pathway. The phage Xis protein has no known enzymatic activity and probably functions, as does IHF, to alter DNA conformation. By binding to two adjacent sites, Xis protein induces a bend of greater than 140°, facilitating formation of an intasome at *attR.* Protein assembly at *attR,* in contrast to that at *attP,* is stimulated by supercoiling but does not require it; lower levels of Int protein are needed and IHF binds to a different set of recognition sites.

At low levels of Xis protein, the host Fis protein (Section 10-8) stimulates excision 20-fold; in vivo the dependence on Fis for excision is even greater.[157] Fis protein may allow λ to sense the physiologic state of the host cell, inasmuch as Fis levels drop 70-fold when cells enter the stationary phase.

Secondary Integrations and Excisions. Less frequent, by about 100-fold, are integrations and excisions at secondary sites, presumably regions of weaker homology with *attP.* These events are related to transposition reactions (Section

153. Snyder UK, Thompson JF, Landy A (1989) *Nature 341*:255.
154. Goodman SD, Nash HA (1989) *Nature 341*:251.
155. Richet E, Abcarian P, Nash HA (1988) *Cell 52*:9; Griffith JD, Nash HA (1985) *PNAS 82*:3124.
156. Kitts PA, Nash HA (1987) *Nature 329*:346; Nash HA, Robertson CA (1989) *EMBO J 8*:3523; Nunes-Düby SE, Matsumoto L, Landy A (1987) *Cell 50*:779.
157. Thompson JF, Moitoso de Vargas L, Koch C, Kahmann R, Landy A (1987) *Cell 50*:901.

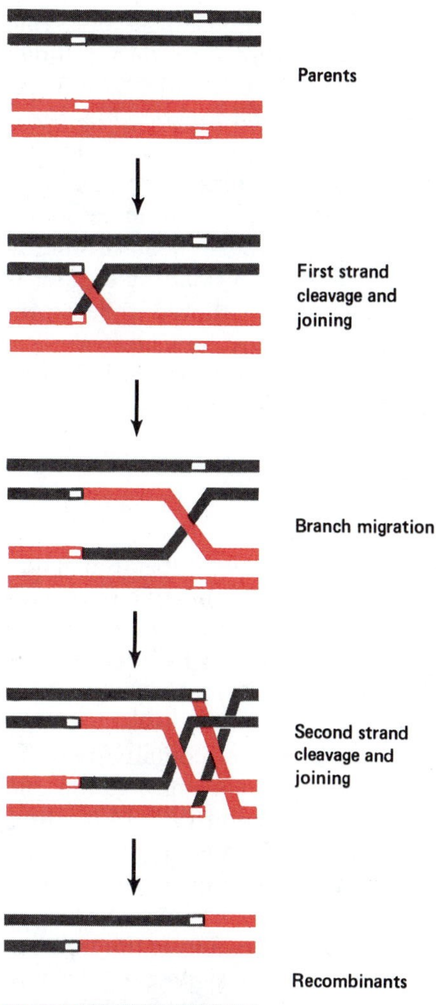

Figure 21-19
Sequential single-strand cleavage and ligation mechanism for site-specific recombination. [From Craig NL (1988) *ARG* 22:77.]

7), in which very specific sequences are needed on only one of the two recombining molecules. In fact there is a class of transposons which appear to use a mechanism analogous to that of λ Int (Section 7).

Flip (FLP) Recombination of the Yeast 2μ Plasmid

The FLP recombinase of the yeast 2μ plasmid generates two DNA isomers that differ in the orientation of the unique sequences with respect to each other (Section 18-10). Inversion during replication generates a type of rolling-circle intermediate that allows the synthesis of multiple copies of the plasmid from a single initiation event[158] (Section 18-10).

---

158. Volkert FC, Broach JR (1986) *Cell* 46:541.

As a member of the integrase family, the FLP recombinase uses the same chemical mechanism as λ Int protein, and strand exchange occurs via a Holliday intermediate.[159] Recombination takes place between two identical 48-bp sequences [*FRT* (*flip* recombination *target*) sites], each containing two FLP binding sites in inverted orientation, with an 8-bp overlap region in the middle. A third FLP binding site is adjacent on one side, but is not required for recombination. Both *FRT* sites are similar in structure to λ *attB*. Complex sequences flanking the recombination site, negative superhelicity, and accessory DNA-binding protein are not required for FLP recombination, but DNA bending by FLP is indicated. Inasmuch as the *FRT* sites are identical, and not altered by recombination, the forward and reverse reactions use the same mechanism.

The *lox* – *cre* recombination system[160] of phage P1 is similar to the FLP system. Both use a simplified integrase reaction, without the involvement of accessory proteins or complex recombination sites.

Tn3 and γδ Resolvases[161]

Resolution systems are encoded by some transposable elements (Section 7) that use a two-stage propagation mechanism (see Fig. 21-24). Replicative transposition results in fusion of the donor and target DNAs, forming a *cointegrate* containing two copies of the transposon. Recombination between special sequences (*res* sites) "resolves" the cointegrate, regenerating the donor molecule and leaving a copy of the transposon in the target.

Resolvase, the only protein required for recombination, contains two domains. The N-terminal region is involved in DNA cleavage and ligation and supplies the protein – protein interactions needed for cooperative DNA binding. Serine 10 appears to be the residue covalently bound to DNA during the recombination reaction. The C-terminal domain contains the sequence-specific DNA recognition determinants.

Recombination occurs between the *res* sites in the two copies of the transposon in the cointegrate. Each 120-bp site consists of three sequences, each of which is bound by a resolvase dimer. Binding bends the DNA, facilitating protein – protein interactions between the neighboring complexes. A dramatic "kink" is induced at the cleavage site which may strain the helix, assisting breakage of the DNA strand; the two base pairs at this site are not recognized by resolvase, but must be "kinkable" for efficient recombination.

Synapsis of *res* sites is by protein – protein interactions. A tight, highly ordered structure, favored by negative superhelicity, is formed. In contrast to the λ Int and FLP systems, resolvase operates only on intramolecular sites in direct orientation; resolution events (deletions) are catalyzed, but inversions or integrations are not. This specificity is derived from a special site alignment during synapsis that is required for a productive reaction. Strand exchange appears to take place by a concerted double-strand break and ligation mechanism (Fig. 21-20), rather than by the sequential single-strand exchanges used by λ Int and FLP.

159. Meyer-Leon L, Huang LC, Umlauf SW, Cox MM, Inman RB (1988) *Mol Cell Biol 8*:3784; Meyer-Leon L, Inman RB, Cox MM (1990) *Mol Cell Biol 10*:235; Gronostajski RM, Sadowski PD (1985) *Mol Cell Biol 5*:3274.
160. Hoess RH, Abremski K (1985) *JMB 181*:351.
161. Dröge P, Cozzarelli NR (1989) *PNAS 86*:6062; Benjamin HW, Cozzarelli NR (1988) *EMBO J 7*:1897.

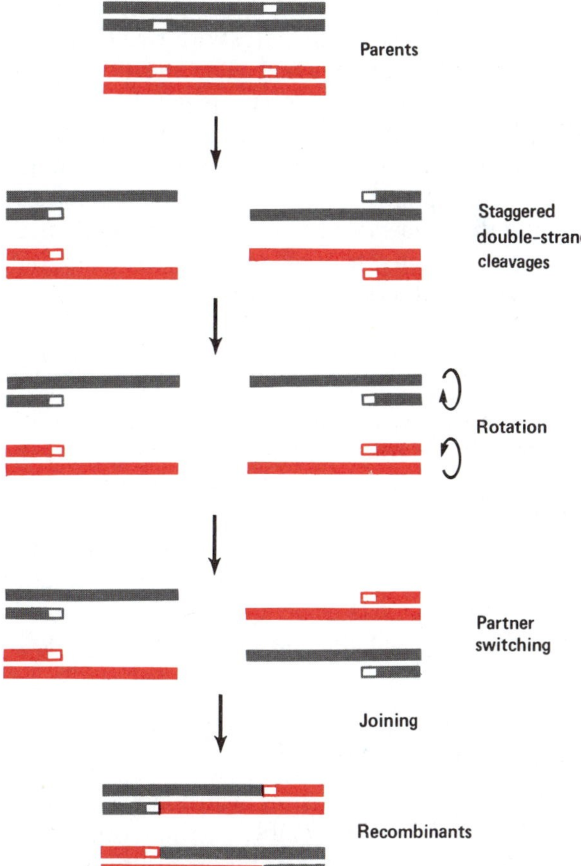

Figure 21-20
Concerted double-strand break and ligation mechanism for site-specific recombination. [From Craig NL (1988) *ARG* *22*:77.]

Invertases[162]

The invertases, mechanistically similar to resolvase, catalyze inversions of a DNA segment rather than deletions, and are unable to promote intermolecular recombination. The *Salmonella* Hin system inverts DNA containing the promoter for the flagella *H1* gene and the repressor of *H2*, causing expression of alternate protein forms *(phase variation)*. In other examples, the Gin system of phage Mu and the Cin system of P1 flip the phage tail-fiber genes, resulting in expression of alternate copies and a change in the host range of these viruses. These three invertases, and the *E. coli* Pin protein, encoded by a cryptic genetic element, have similar amino acid sequences (60% related) and are functionally interchangeable. With a 13% identity to resolvase, the structures, the chemistry of cleavage, and the mechanism of strand exchange of this class of proteins are all rather similar.

162. Craig NL (1985) *Cell 41*:649; Glasgow AC, Simon MI (1989) in *Mobile DNA* (Berg DE, Howe MM, eds.). Am Soc Microbiol, Washington DC, p. 211.

The recombination sites are simple: about 25 bp long, they consist of a 12-bp imperfect inverted repeat of the invertase recognition site, with two base pairs in the center which form the 3′ overhang upon cleavage. The additional presence of a 60-bp enhancer sequence stimulates inversion between the recombination sites 20- to 200-fold. This enhancer, which functions in either orientation and in many places on the DNA molecule, consists of two Fis protein binding sites. To bring the three sequences together, proper synapsis appears to require interaction between Fis protein and the invertases bound at the recombination sites. HU protein (Section 10-8) stimulates the rate of recombination by facilitating bending of the DNA segments required to form this structure.[163]

Gene Assembly during Development

Site-specific recombination can activate genes at a precise time in development. In *B. subtilis,* the gene for an RNA polymerase σ factor is assembled in the mother cell during sporulation by specific deletion of a region of the genome; an adjacent gene, required for recombination, encodes a protein homologous to the resolvase family.[164] In the filamentous cyanobacterium, *Anabaena,* the genes required for nitrogen fixation are interrupted by an 11-kb segment that must be removed during the development of heterocysts (specialized cells responsible for fixing nitrogen); the *xisA* gene probably encodes the site-specific recombinase required for this excision.[165]

Immunoglobulin and T-Cell Receptor Gene Assembly[166]

The extensive repertoire of different binding specificities of antibody and T-cell receptor molecules is generated largely by the assembly of their genes from widely separated gene segments. The germ-line DNA contains multiple copies of gene segments, called *V* (variable), *D* (diversity), *J* (joining), and *C* (constant regions (Fig. 21-21), which are assembled during B cell and T cell development. The diversity to values of 10^7 or greater is provided by the large numbers of these segments, multiplied by the many different combinations in which they can be assembled. The "error-prone" nature of assembly contributes further to the diversity (see below).

The gene segments are flanked by conserved sequences containing the recombination signals, highly conserved heptamer and nonamer elements, separated by spacers composed of either 12 or 23 bp of nonconserved nucleotide sequence (Fig. 21-21). The spacer length is critical: during gene assembly, a signal with a 12-bp spacer is always joined to one with a 23-bp sequence. The two recombination signals are fused to make a *signal joint* and the two gene segments, now contiguous, form a *coding joint* (Fig. 21-22).

Similarities and differences between the VDJ recombination reaction and the prokaryotic systems are indicated by the structures of the products. As in the prokaryotic examples, the recombination signals appear to be sites for specific

163. Johnson RC, Bruist MF, Simon MI (1986) *Cell 46*:531; Johnson RC, Glasgow AC, Simon MI (1987) *Nature 329*:462; Kanaar R, van de Putte P, Cozzarelli NR (1989) *Cell 58*:147.
164. Sato T, Samori Y, Kobayashi Y (1990) *J Bact 172*:1092.
165. Golden JW, Robinson SJ, Haselkorn R (1985) *Nature 314*:419; Golden JW, Wiest DR (1988) *Science 242*:1421; Lammers PJ, Golden JW, Haselkorn R (1986) *Cell 44*:905.
166. Lewis S, Gellert M (1989) *Cell 59*:585; Blackwell TK, Alt FW (1989) *JBC 264*:10327.

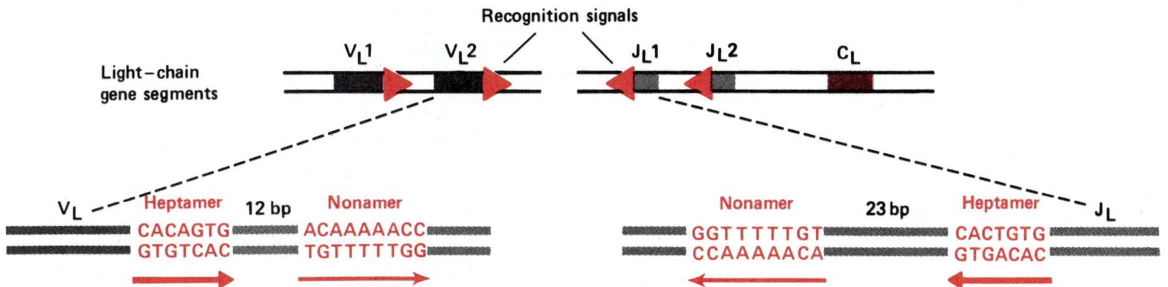

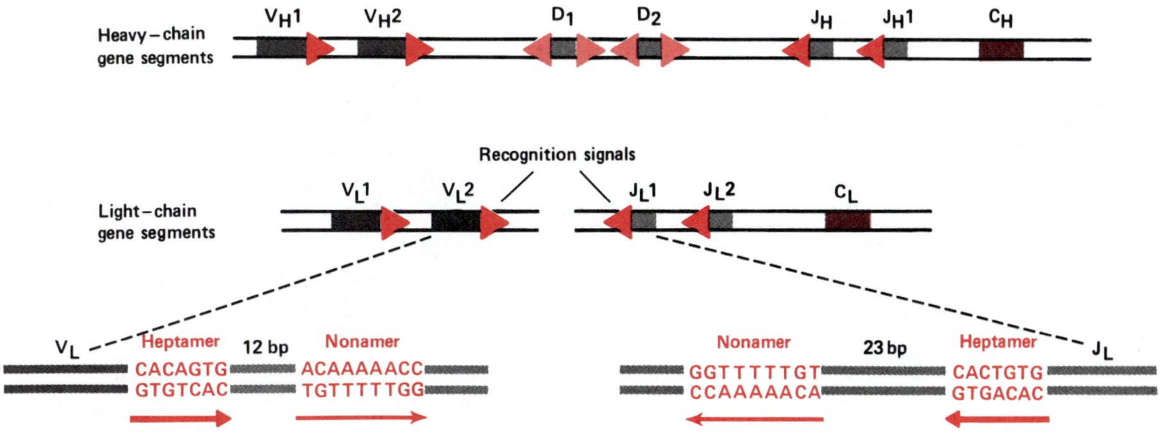

Figure 21-21
Structure of assembly recognition signals of immunoglobulin genes. H = heavy chain, L = light chain, V = variable region, D = diversity region, J = joining region, C = constant region. [Redrawn from Alberts B, Bray D, Lewis J, Raff M, Roberts K, Watson JD (1989) *Molecular Biology of the Cell*, 2 ed, Garland Publishing, NY, p 1026]

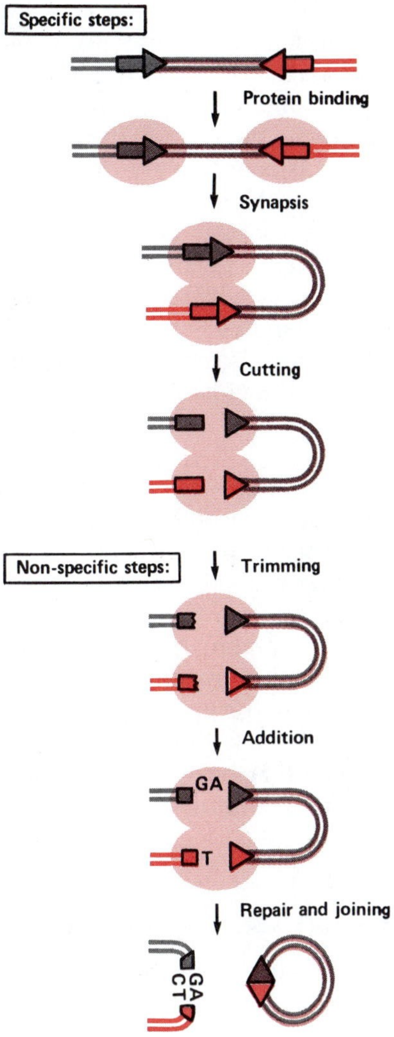

Figure 21-22
Proposed steps in immunoglobulin and T-cell receptor gene assembly. [Modified from Lewis S, Gellert M (1989) *Cell* 59:585]

protein recognition; synapsis probably occurs via protein–protein interactions rather than by DNA–DNA pairing, and cleavage is thought to occur at a specific site at the edge of the recombination signal. In contrast, concerted cleavage and ligation seems unlikely in the VDJ system. Deletions or additions, or both, of several nucleotides at the recombination site (especially at the coding junction) often accompany gene assembly. Thus, after cleavage, the DNA ends must be available for enzymatic alteration prior to ligation, making the involvement of a protein–DNA covalent intermediate unlikely.

Terminal deoxynucleotidyl transferase (Section 6-10) is implicated in the addition of non–germ-line nucleotides at the coding junction.[167] Limited exonuclease digestion probably produces the deletions. This uncoupling of cleavage and ligation, and the use of additional modification enzymes, argue that the "VDJ recombinase" will be substantially different from the prokaryotic enzymes. Separate activities for site pairing, cleavage, deletion, insertion, and ligation may be involved. Ligation of the joints must depend on an external energy source, inasmuch as the energy from the cleavage cannot be conserved.

A single recombination machinery in B cells and T cells is indicated by the similarities in the reactions and their ability to recombine the same artificial substrates. A pair of closely linked genes, *RAG-1* and *RAG-2* (recombination-activating genes 1 and 2), have been identified which synergistically activate VDJ recombination in fibroblast cells, where such activity is normally lacking.[168] *RAG-1* encodes a 119-kDa open reading frame conserved between human and mouse and expressed in cell types that are recombination-proficient. Thus, it is an attractive candidate for an important component of the recombination machinery. *RAG-2* codes for a smaller protein and is coordinately expressed with *RAG-1*. In addition, several proteins have been purified by their ability to bind the recombination signals or cut the DNA near these sites. However, their relation to VDJ recombination remains unclear. Finally, mice with a severe combined immunodeficiency (*scid* mice) are defective in VDJ recombination and thus cannot develop an immune system. Further analysis of the biochemical and genetic defects in these *scid* mutant cells, which also appear to have an altered ability to repair DNA damage, should assist in the identification of the components of the recombination machinery.

Class-Switching.[169] In response to antigenic stimulation, immunoglobulin-expressing lymphocytes can undergo a distinct type of recombination called "class-switching." By this mechanism, cells expressing specific heavy-chain variable and constant regions can switch to producing antibodies with the same specificity but with a different heavy-chain constant region, and therefore of a different immunoglobulin class. Thus, by switch recombination, a cell can change from producing IgM to IgG, while maintaining the antigenic specificity. Class-switching involves recombination between switch signals upstream of sequential constant-region segments. The first signal lies within an intron between the assembled variable region and the μ constant region (for IgM). Recombination can occur at numerous positions within these short ($<$80 bp) switch

---

167. Landau N, Schatz D, Rosa M, Baltimore D (1987) *Mol Cell Biol 7*:3237; Lieber M, Hesse J, Mizuuchi K, Gellert M (1988) *PNAS 85*:8588.
168. Schatz DG, Oettinger MA, Baltimore D (1989) *Cell 59*:1035; Oettinger MA, Schatz DG, Gorka C, Baltimore D (1990) *Science 248*:1517.
169. Marcu KB (1982) *Cell 29*:719; Radbruch A, Burger C, Klein S, Muller W (1986) *Immunol Rev 89*:69.

signal regions, which are made up of simple pentameric sequence repeats. A role for DNA replication in the class-switching process has been suggested.[170]

Macronucleus Assembly in Ciliated Protozoa[171]

Certain ciliated protozoa have two types of nuclei. The *micronucleus* is the germ-line nucleus containing the whole genome in high-molecular-weight chromosomes that are inactive for transcription. The *macronucleus* contains only a fraction of the genomic DNA, rearranged, amplified, and transcriptionally active. After a sexual cycle, construction of a new macronuclear genome involves massive rearrangement of the micronuclear DNA to generate the 20,000 individual gene-sized pieces, 0.5 to 1.5 kb in length. The process includes (1) fragmentation of the micronuclear chromosomes into the small segments; (2) deletion of intron-like sequences, called *internal eliminated sequences*, to assemble the genes; (3) rearrangement of gene segments in some instances; (4) addition of telomeres at the ends of the linear "minichromosomes"; and (5) DNA synthesis to amplify their copy number approximately 1000-fold.

Both removal of the internal eliminated sequences and rearrangement of the gene segments undoubtedly involves recombination. In *Euplotes crassus*, some of the internal eliminated sequences are of a repetitive DNA class resembling a transposable element[172] (Section 7). Massive excision of these elements as circles accompanies macronuclear development;[173] their removal splices together coding segments by a mechanism that may resemble the excision of a phage or transposon.

A dramatic example of rearrangement during gene assembly is the splicing together of the actin genes for the macronucleus in *Oxytricha nova*.[174] The nine coding segments (exons) for this gene in the micronuclear chromosome must be assembled and rearranged (Fig. 21-23) in such a way that their order, if 1 to 9, say, in the micronucleus, is 8-7-1-2-4-3-5-9-6 in the macronucleus. The order of seven out of the nine segments is switched and, in addition, the orientation of segment 7 is flipped. Short ($<$14 bp) inverted repeats flank the gene segments and appear to be signals for the process inasmuch as matching up of the like sequence elements is needed to generate the correct configuration (Fig. 21-23).

21-7 Transposition[175]

A specific DNA sequence moves to a new chromosomal location by transposition. DNA elements which catalyze their own movement are called *transposons, transposable elements,* or *insertion sequences* (Section 18-1). Widespread in nature, transposons are found in bacteria, fungi, plants, insects, and animals

170. Dunnick W, Wilson M, Stavnezer J (1989) *Mol Cell Biol 9*:1850.
171. Yao M-C (1989) in *Mobile DNA* (Berg DE, Howe MM, eds.). Am Soc Microbiol, Washington DC, p. 715; Blackburn EH, Karrer KM (1986) *ARG 20*:501.
172. Baird SE, Fino GM, Tausta SL, Klobutcher LA (1989) *Mol Cell Biol* 9:3793.
173. Tausta SL, Klobutcher LA (1989) *Cell 59*:1019.
174. Greslin AF, Prescott DM, Oka Y, Loukin S, Chappell JC (1989) *PNAS 86*:6264.
175. Berg DE, Howe MM (eds.) (1989) *Mobile DNA*. Am Soc Microbiol, Washington DC; Mizuuchi K, Craigie R (1986) *ARG 20*:385; Craig NL, Kleckner N (1987) in *Escherichia Coli and Salmonella* Typhimurium (Neidhardt FC, ed.). Am Soc Microbiol, Washington DC, Vol. 2, p. 1054; Grindley NDF, Reed RR (1985) *ARB 54*:863.

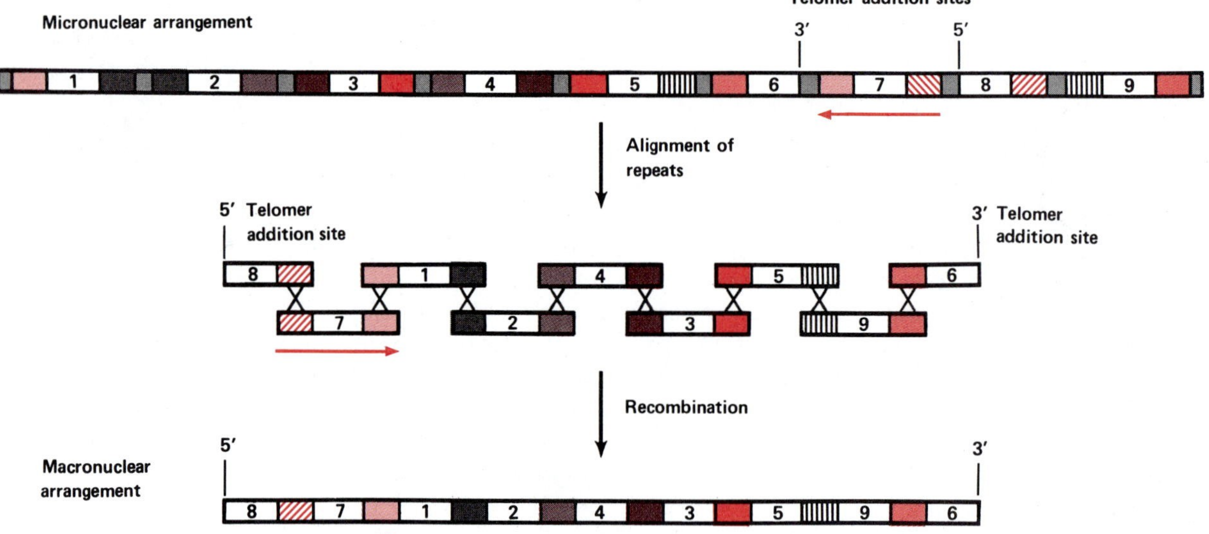

Figure 21-23
Assembly of the macronuclear actin gene from gene segments in the micronucleus. Top: Arrangement of direct and inverted repeats (blocks of various patterns), exons, and introns of the micronucleus. Middle: All exons can be arranged in the order of the macronuclear gene by aligning pairs of repeats. Bottom: Homologous crossing-over yields a complete macronuclear actin gene (except for telomeres) with the introns removed. The polarity inversion of exon 7 is shown by a directional arrow. [From Greslin AF, Prescott DM, Oka Y, Loukin SH, Chappell JC (1989) *PNAS 86*:6264]

and function as important vectors for genetic change; insertions, deletions, and genome rearrangements accompany their movement. They both activate and inactivate genes by inserting near or within them. As repetitive elements, transposons also provide regions of homology throughout the genome, upon which the homologous recombination systems can act.

Transposition events are of three general types: (1) In *nonreplicative transposition*, the element is "cut out" of the donor DNA and "pasted" into the target DNA molecule. (2) In *replicative transposition*, the DNA is duplicated, leaving a copy of itself in the original site while inserting itself into a new location via formation of a cointegrate (Fig. 21-24). (3) In *retrotransposition*, movement is through an RNA intermediate followed by reverse transcription and insertion of the cDNA into a new chromosomal location. This third class of transposition departs mechanistically from the other two and will be discussed separately below.

Nonreplicative and Replicative Transposition

Transposons that translocate via a DNA intermediate, be they nonreplicative or replicative (Table 21-9), share several features:

(1) The two ends of the element carry nearly the same sequence in inverted orientation, necessary *in cis* for transposition.

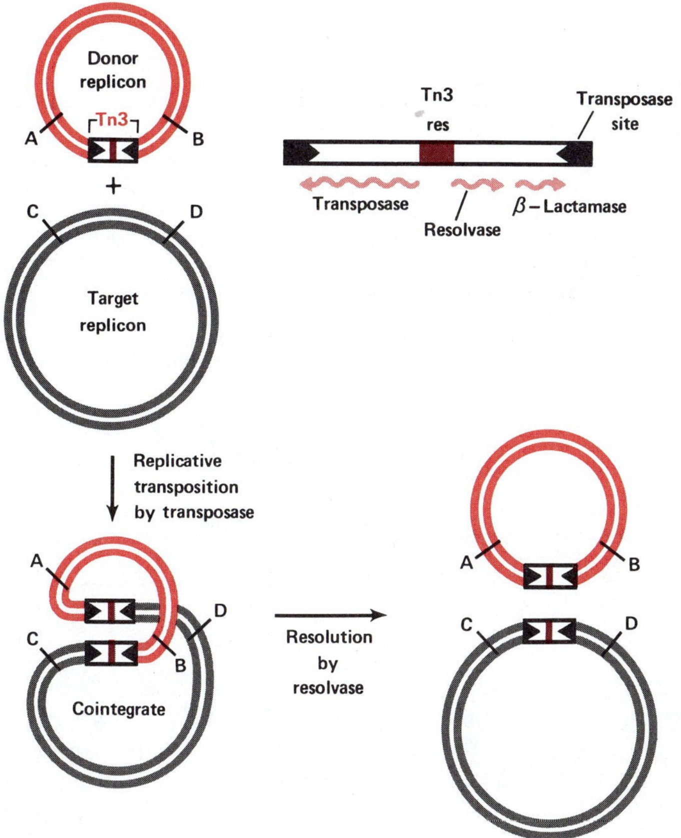

Figure 21-24
Scheme for replicative transposition of the transposon Tn3. A, B = donor replicon genes; C, D = target replicon genes; res = resolution site.

(2) The transposons encode at least one protein, the transposase, required for movement that specifically binds and cleaves the end sequences.

(3) The transposons generate a short (≤12 bp) duplication of the DNA at the target site during transposition. The length of this sequence is characteristic and invariant for a given element and is generated by the staggered cleavage of the target DNA by the transposase.

Mechanism of Transposition. This has been worked out in most detail for phage Mu, a prokaryotic virus which propagates chiefly via multiple rounds of replicative transposition (Section 17-7); occasionally, Mu also undergoes nonreplicative transposition. A common intermediate can be resolved by repair of the transposon–target-site junctions, to give a simple insertion, or by replication of the transposon, to yield a cointegrate. The steps in formation of the intermediate are, in brief, as follows (see also Section 17-7, Fig. 17-37):

(1) Recognition and pairing of the two ends of the transposon by the transposase forms a specialized protein–DNA structure. Complex formation is stimulated by the host HU and IHF proteins, as is common in site-specific recombination (Section 6). An additional *cis*-acting sequence is involved. Mu transposase, like λ Int protein (Section 6), has two independent DNA-binding domains.

Table 21-9
Characteristics of some DNA-mediated transposons

Element	Mechanism	Ends	Comments
Bacteria			
Tn10 (IS10)	nonreplicative	22-bp inverted repeats	encodes transposase and *trans*-acting RNA inhibitor
Tn5 (IS50)	nonreplicative	9-bp inverted repeats	encodes transposase and truncated form that inhibits transposition; requires host dnaA protein
Tn7	nonreplicative	asymmetric	encodes 5 proteins involved in transposition, 2 of which affect target site selection; one predominant integration site in *E. coli*
Tn3	replicative	38-bp inverted repeats	encodes transposase and resolvase; resolvase also regulates transposition
Mu	replicative	asymmetric	encodes transposase and protein that stimulates replicative pathway and affects target site selection; no resolvase function
Drosophila			
P element	unknown	31-bp inverted repeats	encoded transposition protein binds internally; host protein binds to ends; transposition in germ-line cells only; truncated transposase inhibits transposition
Maize			
En/Spm[a]	nonreplicative	13-bp inverted repeats	encodes 1 or 2 proteins; preferentially moves to locations near original insertion site

[a] Gerl A, Saedler H (1989) *ARG 23*:71.

(2) Nicking by transposase generates a 3′-OH at each end of Mu.

(3) Cleavage of the target site, by the transposase within the Mu DNA complex, generates a staggered cut with 5′-P overhangs of 5 ntd. This staggered cleavage gives rise to the 5-bp target-site duplication characteristic of Mu insertions.

(4) Ligation of the 5′-P of the target to the 3′-OH of the element gives the strand-transfer intermediate. Cleavage of the target is energetically coupled to the ligation, although, in contrast to the site-specific recombination reactions, coupling does not involve a protein–DNA covalent intermediate.

Tn10, which moves exclusively via nonreplicative transposition, appears to use a similar mechanism.[176] The polarity of cleavage of the element and the targetsite are the same as for Mu, although the target-site break has a 9-ntd 5′ overhang. Initial cleavage of the transposon ends appears to break both strands. Double-strand rather than single-strand cleavage of the ends of the element may be the hallmark differentiating nonreplicative and replicative transposition. Host pol I is implicated in healing the transposon junctions after insertion of Tn10 (also Tn5). The donor site from which the transposon has been excised must be repaired by host machinery or else the excision has lethal consequences. Transposition by Tn10 induces the SOS DNA-damage response (Section 3) by generating broken strands at the donor site.

Regulation of Transposition. Transposable elements, in order to be effective, must limit their movements so as not to endanger the viability of the host cell. Several strategies are used to keep the expression of transposase, and thus the rate of transposition, to an acceptably low level. Some elements encode specific

176. Kleckner N (1989) in *Mobile DNA* (Berg DE, Howe MM, eds.). Am Soc Microbiol, Washington DC, p. 221; Benjamin HW, Kleckner N (1989) *Cell 59*:373; Haniford DB, Chelouche A, Kleckner N (1989) *Cell 59*:385.

trans-acting inhibitors. Tn10 makes an RNA that is antisense to the transposase message, which inhibits its production. Tn5 synthesizes a truncated version of transposase that seems to combine with the full-length form and inhibit its action. *Drosophila* P element transposition is restricted to germ-line cells by regulated splicing of the introns in the transposase message; like Tn5, P elements encode a small version of the transposase which inhibits movement of the element.

In addition to exerting some type of copy-number control on their own transposition, several elements impart *transposition immunity* to the DNA in which they are inserted, thereby greatly decreasing the frequency of a second transposition event into the occupied molecule. The mechanism of this *cis* inhibition of transposition is unknown (except for the distinctive case of Mu phage; Section 17-7) but does depend on the presence of the ends of an element and exerts an influence over distances greater than 100 kb.

Some transposons respond to changes in host cell physiology. In the case of Tn10, transposition is sensitive to chromosomal replication via the host Dam methylase system (Section 10). When the element is hemimethylated, as during the short period after passage of the replication fork, expression of transposase and the transposase binding site at the end of the element are both activated. In this way, movement of Tn10 is stimulated by replication, which may in turn provide a new target for transposition. Also sensitive to Dam methylation are Tn5 (IS50) and IS903; the activity of several maize elements appears to be regulated by methylation as well. Transposition of the bacterial transposon Tn501, which encodes mercury resistance, is stimulated by exposure to Hg^{2+}, and Tn917, which imparts erythromycin resistance, is activated by erythromycin.

Retrotransposition and Retroviral Integration[177]

A class of transposons, prominent in yeast (Ty) and *Drosophila (copia)*, and related to retroviruses, move to a new genomic location via an RNA intermediate. The yeast Ty (Transposon, yeast) elements, like retroviruses (Section 19-7), have long terminal repeats, called δ sequences, which flank the internal coding sequence (Fig. 21-25). Encoded are proteins with sequence homology to the retroviral integrase and reverse transcriptase (Fig. 21-25) which are specifically required for transposition.

The pathway for reverse transcription by these elements is analogous to that of retroviruses. In fact, Ty elements are packaged into intracellular virus-like particles containing Ty RNA, reverse transcriptase, and integrase.[178] These isolated particles are capable of synthesizing a double-stranded DNA copy in vitro.

The substrate for integration by retroviruses, and presumably by retrotransposons as well, is the full-length, linear duplex DNA. The integration reaction is reminiscent of insertion of phage Mu. A staggered cut with 5′-P overhangs is introduced at the target site by the integrase protein–donor DNA complex; the 3′-OH ends of the retroviral DNA are ligated to the 5′-P at the cleavage site in a concerted reaction. No external energy source is required. Repair of the staggered cleavage of the target site generates the short target-site duplications that

177. Skalka AM (1988) in *Genetic Recombination* (Kucherlapati R, Smith GR, eds.). Am Soc Microbiol, Washington DC, p. 701; Grandgenett DP, Mumm SR (1990) *Cell 60*:3; Varmus H, Brown P (1989) in *Mobile DNA* (Berg DE, Howe MM, eds.). Am Soc Microbiol, Washington DC, p. 53.
178. Eichinger DJ, Boeke JD (1989) *Cell 54*:955.

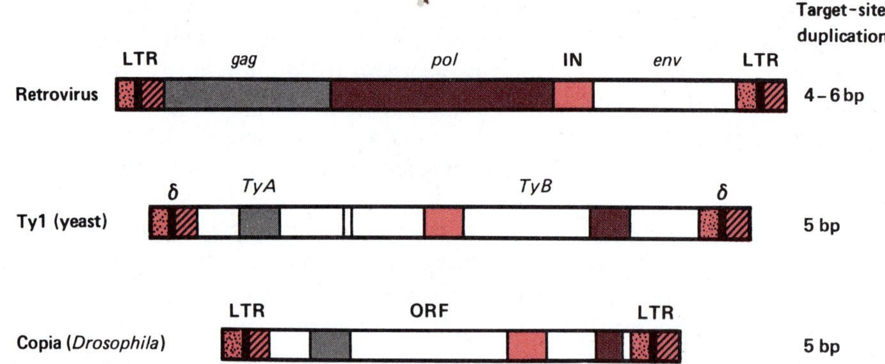

Figure 21-25
Structural similarity between retroviruses and retrotransposons of yeast (Ty) and *Copia* (*Drosophila*). LTR and δ = long terminal repeats; orf = open reading frame; IN = integration protein. Homology between retroviral proteins and those of Tyl and *Copia* are color coded. [Modified from Lewin B (1990) *Genes IV*, Oxford University Press, New York, p 685]

are characteristic of retroviral integration events, as well as those of all known transposable elements.

As a consequence of transposing via RNA, the transposition frequency of Ty elements is stimulated by induction of Ty transcription. Movement of numerous defective elements present in the genome is also stimulated, probably by the increased production of the Ty protein products, resulting in packaging of the defective as well as functional Ty RNA into the virus-like particles. Transposition results in a copy of the element in a new location, while the original donor element remains intact. Excision of the donor element does not accompany the process (as occurs in conservative transposition of DNA elements), nor are cointegrates formed (as in replicative transposition). Retrotransposition thus results in fewer disruptions and rearrangements of the host chromosome. The changes in the genome associated with retrotransposons are principally due to their disruption or activation of genes or to the action of the general recombination machinery on the multiple copies of the element that may be present in the genome.

In addition to the retroviral-type transposons, some repetitive DNA elements in cells probably arose by reverse transcription of cellular tRNAs and structural RNAs (Section 1-5). These elements encode no transposition proteins, and the means by which they were propagated is still unknown.

21-8 Bacterial Transformation and Competence[179]

A bacterium can become genetically transformed by taking up DNA from the medium and assimilating it into its genome. The experimental observation of

179. Smith HO, Danner DB, Deich RA (1981) *ARB 50*:41; Notani NK, Setlow JK (1974) Prog N A Res 14:39; Portoles A, Lopez R, Espinosa M (eds.) (1977) *Modern Trends in Bacterial Transformation and Transfection*. North Holland, Amsterdam; Low KB, Porter RD (1978) *ARG 12*:249.

bacterial transformation,[180] more than any other finding, established DNA as the genetic substance (Section 1-1). Transformation of yeast spheroplasts has become feasible,[181] as has the transfer of chromosomes[182] and genes[183] between animal cells. (*Transformation*, in reference to animal cells, applies to a state of uncontrolled growth, as in a malignant cell, sometimes resulting from integration of viral DNA into the chromosome.) *Transfection*, in which free viral DNA or RNA is taken up into bacterial or animal cells to produce infectious particles, may resemble bacterial transformation in the initial stages.[184]

Transformation has been studied intensively in only a few bacterial species: *Streptococcus* species [e.g., *S. sanguis* and *S. pneumoniae* (colloquially known as pneumococcus)], *Bacillus* species (e.g., *B. subtilis*), and *Hemophilus influenzae* are the best known. Despite intensive study for over 30 years, most molecular details of this form of genetic transfer are still vague. The process is completed within a few minutes; it entails uptake of DNA into a form insusceptible to external DNase. The DNA becomes single-stranded (in *Streptococcus* and *Bacillus*) and then recombines with a homologous region of the cellular chromosome to form a heteroduplex. The efficiency of transformation can be very high. In *Hemophilus* virtually all cells in a culture may be transformed; in *Streptococcus* and *Bacillus* 10% to 20% of the cells may be competent in transformation; as much as 50% of the homologous DNA taken up into *Streptococcus* and *Bacillus* cells may be integrated into the genome.

Hemophilus Influenzae[185]

Competence of the cell for DNA uptake develops when an exponentially growing culture is shifted from a rich medium to one that supports protein synthesis but not cell division. Three changes in the cells occur during development of competence: (1) modification of the cell envelope by synthesis of new proteins, change in the lipopolysaccharide composition, and formation of vesicle-like structures at the cell surface; (2) an increase in the recombination frequency; and (3) appearance of single-strand gaps and "tails" within the genomic DNA.[186] The envelope changes probably prepare the cell for binding and uptake of extracellular DNA, while the other effects may facilitate the incorporation of this DNA into the genome.

Every cell in the culture is induced to synthesize a membrane receptor protein made up of two polypeptides, 28.5 and 52 kDa. Uptake by *Hemophilus*[187] is highly specific for *Hemophilus* DNA, unlike the case with pneumococci[188] and *Bacillus* species,[189] which show no discrimination between homologous and foreign DNA. Distributed through *Hemophilus* DNA at 4-kb intervals are 11-bp uptake sites (5′-AAGTGCGGTCA), recognized by several receptor proteins on

180. Avery OT, MacLeod CM, McCarty M (1944) *J Exp Med 79*:137; Hotchkiss RD (1957) in *Chemical Basis of Heredity* (McElroy WD, Glass B, eds.). Johns Hopkins Press, Baltimore, p. 321.
181. Hinnen A, Hicks JB, Fink GR (1978) *PNAS 75*:1929; Struhl K, Stinchcomb DT, Scherer S, Davis RW (1979) *PNAS 76*:1035.
182. Miller CL, Ruddle FH (1978) *PNAS 75*:3346.
183. Wigler M, Pellicer A, Silverstein S, Axel R (1978) *Cell 14*:725; Wigler M, Pellicer A, Silverstein S, Axel R, Urlaub G, Chasin L (1979) *PNAS 76*:1373.
184. Trautner TA, Spatz H (1973) *Curr Top Microbiol Immunol 62*:61.
185. Kahn ME, Barany F, Smith HO (1983) *PNAS 80*:6927; Kahn ME, Maul G, Goodgal SH (1982) *PNAS 79*:6370; Sedgwick B, Setlow JK (1976) *J Bact 125*:588; Sisco KL, Smith HO (1979) *PNAS 76*:972.
186. McCarthy D, Kupfer DM (1987) *J Bact 169*:565.
187. Scocca JJ, Poland RL, Zoon KC (1974) *J Bact 118*:369.
188. Raina JL, Ravin AW (1978) *J Bact 133*:1212; Raina JL, Metzer E, Ravin AW (1978) *J Bact 133*:1224.
189. Soltyk A, Shugar D, Piechowska M (1975) *J Bact 124*:1429.

the recipient-cell surface.[190] Standard DNA preparations 30 to 50 kb long are likely to have about te uptake sites per molecule, thus ensuring that all chromosomal markers will be transferred. Uptake appears to be independent of the size of DNA on which the site is located.

Binding of the DNA uptake site by the receptor protein is reversible and is followed by the irreversible transport of the DNA into the cell. The internalized DNA remains duplex, and escapes restriction and degradation by cellular enzymes. Specialized membrane extensions, called *transformosomes*, are responsible for uptake and protection.[191] These structures are located at points of fusion between the inner and outer membranes and have a composition similar to that of the outer membrane; they are five times more abundant on the surface of competent cells than on their noncompetent counterparts. The DNA-binding activity of the cell surface resides exclusively in the transformosomes. Upon addition of homologous DNA, the structures are efficiently internalized; the DNA enters the cell in the protected state and can be extracted in the form of rapidly sedimenting DNA–membrane complexes. After about 5 minutes, the newly internalized sequences become associated with the chromosomal DNA. While closed circular DNA is taken up by cells, it enters the unprotected state much less efficiently than does linear DNA and is thus less effective in transforming *Hemophilus*.

Transformation in *Neisseria*, another organism of clinical importance, may employ a similar system;[192] specific sequences on the DNA are required. The silent copies of the pilin genes, which are involved in the ability of these organisms to change their surface antigens ("antigenic switching"), can be integrated into the expressed locus after they enter a cell through transformation.[193]

Streptococcus Species[194]

Like *Hemophilus*, streptococci must be competent to adsorb and take up DNA for transformation. This competence is triggered by the accumulation of a critical concentration of a secreted protein called *competence factor* (CF),[195] produced by growing cultures. CF is a 10-kDa basic protein which does not bind DNA but apparently interacts with the cell surface. Its action causes a global change in protein synthesis, activating the production of several competence-specific products. An inhibitor of CF action may be responsible for the transient nature of the competent state. A locus controlling competence encodes two proteins (ComA, 77 kDa, and ComB, 49 kDa).[196] Cells mutant in *comA* fail to produce CF, but respond normally to exogenously added CF.

Unlike the case with *Hemophilus*, the surface receptors for DNA have not been identified in streptococci, and there is little discrimination between homospecific and foreign DNA. Subsequent events leading to assimilation of homologous DNA are not known in detail except that the DNA is single-stranded and in a complexed and protected form before it becomes associated with the

190. Danner DB, Deich RA, Sisco KL, Smith HO (1980) *Gene 11*:311.
191. Kahn ME, Barany R, Smith HO (1983) *PNAS 80*:6927.
192. Dougherty TJ, Asmus A, Tomasz A (1979) *BBRC 86*:97.
193. Seifert HS, Ajioka RS, Marchal C, Sparling PF, So M (1988) *Nature 336*:392.
194. Chandler MS, Morrison DA (1988) *J Bact* 170:3136; Fujii T, Naka D, Toyoda N, Seto H (1987) J Bact *169*:4901; Seto H, Tomasz A (1974) *PNAS 71*:1493; Lacks S, Greenberg B, Neuberger M (1974) *PNAS 71*:2305; Lacks S, Greenberg B (1976) *JMB 101*:255; Lacks S (1979) *J Bact 138*:404.
195. Tomasz A (1966) *J Bact 91*:1050.
196. Chandler MS, Morrison DA (1988) *J Bact 170*:3136; Chandler MS, Morrison DA (1987) *J Bact 169*:2005.

recipient chromosome. A competence-specific nickase is implicated in DNA uptake.[197] Specific degradation of one strand of the incoming DNA, coupled to transport, has been suggested (as in *B. subtilis*; see below).

Bacillus Species[198]

B. subtilis cells become competent upon entry into stationary phase in media containing glucose;[199] several environmental conditions affect the development of competence, including inhibition by glutamine, suggesting that nitrogen starvation may be a key signal.[200] Mutants which uncouple expression of the competence genes from these conditions have been identified, and should assist in the description of the regulatory pathway.[201] Competent cells become smaller and lighter, separable from their noncompetent counterparts by centrifugation.

Any DNA sequence is bound to the surface of competent cells, cleaved at random sites into pieces about 20 kb long, and then converted into a single-stranded form, inaccessible to nuclease. About half of the DNA is degraded during uptake, suggesting that one strand of the duplex is specifically digested during DNA entry. This degradation may fuel transport of the DNA across the cell membranes, although maintenance of the proton motive force is also required. The single-stranded fragments can be isolated from the cells, and are rapidly incorporated into the chromosome by homologous recombination.

Several genes specifically involved in the development of competence have been identified. The *comA* and *comB* genes act very early, most likely in regulation of the process. The sequence of the *comA* gene is similar to that of the effector partner of several known prokaryotic signal transducers; however, an associated sensor protein has not been identified.

Of the competence-specific proteins, the functions of a few are suggested (Table 21-10). A specific SSB has been partially purified, and may function in protecting the ssDNA from degradation once it is internalized.[202] A 75-kDa nuclease complex, made up of multiple 17-kDa and 18-kDa subunits, is found associated with membranes and may be involved in DNA binding and uptake.[203] The 17-kDa protein has the nuclease activity, whereas the 18-kDa subunit appears to limit its action. Disruption of these genes reduces transformation to 5% of the wild-type level, indicating that this complex is important for transformation, but that other DNA entry pathways must also exist.

Escherichia coli

The relative insusceptibility of coliform organisms to transformation has been overcome in *E. coli* by exposing cells to calcium chloride and low temperature, followed by a brief period of warming. A modification of the $CaCl_2$ treatment[204] of recipient cells has raised the efficiency so that 20% of the viable cells can be

---

197. Fujii T, Naka D, Toyoda N, Seto H (1987) *J Bact 169*:4901.
198. Dubnau D (1989) in *Regulation of Procaryotic Development* (Smith I, Slepecky R, Setlow P, eds.). Am Soc Microbiol, Washington DC, p. 147.
199. Anagnostopoulos C, Spizizen J (1961) *J Bact 81*:741.
200. Albano M, Hahn J, Dubnau D (1987) *J Bact 169*:3110.
201. Dubnau D (1989) in *Regulation of Procaryotic Development* (Smith I, Slepecky R, Setlow P, eds.). Am Soc Microbiol, Washington DC, p. 147.
202. Eisenstadt E, Lange R, Willecke K (1975) *PNAS 72*:323.
203. Vosman B, Kuiken G, Kooistra J, Venema G (1988) *J Bact 170*:3703; Smith H, Wiersma K, Bron S, Venema G (1983) *J Bact 156*:101; Smith H, Wiersma K, Venema G, Bron S (1984) *J Bact 157*:733.
204. Dagert M, Ehrlich SD (1979) *Gene 6*:23; Chang S, Cohen SN (1979) *MGG 168*:111.

Table 21-10
***B. subtilis* genes involved in transformation**

Gene	Transformation (% of wild-type)	Function or comment
comA[a]		similar in sequence to effectors of bacterial two-component regulatory systems; putative helix-turn-helix DNA binding motif
comB[a]		
comI	5	encodes nuclease (17-kDa) subunit of 75-kDa membrane-DNA binding complex
comJ	25	encodes 18-kDa subunit of 75-kDa membrane complex; may limit digestion by nuclease
comN	0.02	DNA binding and uptake are two-thirds of normal in mutant cells
comO	10^{-4}	DNA binding to cells is blocked in mutant
comG[b]	10^{-5} to 10^{-6}	complex locus with seven orfs
orf1		sequence contains an ATP-binding motif; similar to *virB* of Ti plasmids, which encodes a membrane ATPase
orfs 3, 4, 5		sequence homology to bacterial pilins; may be involved in pore formation

[a] blocked very early, cells fail to develop competence.
[b] orf = open reading frame.

transformed (2×10^7 transformants per microgram of plasmid DNA). In contrast to the DNA specificity of "naturally" transformable bacteria, the efficiencies for plasmid and viral DNA transformations are higher than for chromosomal DNA. The mechanisms of the calcium effect and DNA uptake in the artificial system are still obscure. Contrary to the DNA utilized in the natural systems, the donor DNA can be nonhomologous (unlike the case in other Gram-negative bacteria) and remains double-stranded during uptake (unlike that in Gram-positive bacteria).

Poly-β-Hydroxybutyrate in Membranes of Competent Bacteria

In several naturally transformable bacteria, including *Azotobacter vinelandii*, *B. subtilis*, and *H. influenzae*, as well as in the artificially induced competence of *E. coli*, conditions that produce the competent state also stimulate production of the neutral lipid D(−)-poly-β-hydroxybutyrate (PHB).[205] Coincident with its production is a change in the membrane state of the bacteria, characterized by an alteration in the high-temperature-induced lipid-phase transition. In *E. coli*, this membrane alteration cannot be correlated with a change in the lipid composition and is thus attributed to insertion of PHB into the inner membrane.[206] Blockage of PHB synthesis prevents the development of the competent state, but inhibition of respiration, ATP synthesis, or protein synthesis does not.

The membrane-bound PHB structures in *E. coli* and *A. vinelandii* have been visualized by freeze-fracture electron microscopy[207] and isolated by chloroform extraction.[208] The isolated structures contain PHB, polyphosphate, and calcium

205. Reusch RN, Sadoff HL (1983) *J Bact 156*:778.
206. Reusch RN, Hiske TW, Sadoff HL (1986) *J Bact 168*:553.
207. Reusch R, Hiske T, Sadoff HL, Harris R, Beveridge T (1987) *Can J Microbiol 33*:435.
208. Reusch RN, Sadoff HL (1988) *PNAS 85*:4176.

(Ca^{2+}) in a molar ratio of about 1 : 1 : 0.5. In a structure proposed for this complex a helical arrangement of the PHB has the hydrophobic methyl groups facing the fatty acid chains and the hydrophilic ester groups coordinated with an inner cylinder of calcium and polyphosphate. Whether this structure might serve as a transmembrane conduit for DNA remains to be determined.

21-9 Restriction and Modification

Restriction – Modification Systems

Bacteria have a remarkable immunity to foreign DNA. Enzymatic mechanisms modify the cell's own DNA in a characteristic pattern and degrade or restrict DNA molecules that lack the distinctive modification pattern. A *restriction–modification (R–M) system* [also called *host-specificity (hs) systems*] might prevent foreign DNA from functioning in the cell, but it might also facilitate recombination. This is especially true of the R–M systems directed by plasmids that carry and transfer drug-resistance genes.

The R–M system has two essential features: A cell with a restriction specificity, r_x, not only degrades entering DNA from cells lacking corresponding modification specificity, m_x, but also has an m_x activity that protects its own DNA from the r_x activity.

Two kinds of R–M systems are found in some strains of *E. coli:* The allelic host system, distinctive for strains B and K, is expressed by a type I combined restriction endonuclease–methylase (Section 13-7), and the system encoded in plasmids is expressed by a type II restriction endonuclease (Section 13-8) separate from its corresponding methylase. Two R–M systems comparable to those in *E. coli* are also found in *H. influenzae*, but no R–M system of any kind has been discovered as yet in animal cells.

Modification usually consists of methylation of the 6-amino group of adenine residues in a specific DNA sequence; in the type II systems the C5 of cytosine residues may be methylated.[209] Restriction depends on an endonuclease (Section 13-7) that recognizes the same specific DNA sequence, provided that at least one strand is unmodified (some enzymes only cleave the completely unmethylated sequence, whereas others recognize the hemimethylated). The endonuclease introduces successive cleavages in each strand, resulting in a double-strand break located at that sequence or, in some cases, elsewhere in the DNA.[210] A double-strand break may mark the death of a DNA molecule as an integral unit when repair of the break cannot take place. Extensive and nonspecific breakdown of the restricted DNA by other nucleases then follows.[211]

R–M systems were first clearly recognized and defined in studies of phage infections of *E. coli* strains.[212] The efficiency of phage infectivity, as judged by the number of plaques produced on a bacterial lawn, provides a measure of the susceptibility of the invading phage DNA to the DNA-immunity system of the host strain. As indicated in Table 21-11, phage λ grown on *E. coli* strain C, which

---

209. Hattman S, Keister T, Bottehrer A (1978) *JMB 124*:701.
210. Horiuchi K, Zinder ND (1972) *PNAS 69*:3220; Horiuchi K, Vovis GF, Zinder ND (1974) *JBC 249*:543.
211. Simmon VF, Lederberg S (1972) *J Bact 112*:161.
212. Arber W, Linn S (1969) *ARB 38*:467.

Table 21-11
Host-specificity (hs) or restriction–modification (R–M) systems in *E. coli*, analyzed by efficiency of plating λ phage[a,b]

E. coli strain on which parental phage was grown	*E. coli* strain for plating progeny phage					
	C	K	B	C(P1)	K(P1)	Kr⁻
C	1	< 0.001	< 0.001	< 0.0001	< 0.0001	1
K	1	1	< 0.001	< 0.0001	< 0.0001	1
B	1	< 0.001	1	< 0.0001	< 0.0001	1
C(P1)	1	< 0.001	< 0.001	1	< 0.001	1
K(P1)	1	1	< 0.001	1	1	1

[a] *E. coli* C has no known R–M system. A lysogenized strain, K(P1) (i.e., strain K carrying phage P1), displays the R–M system of the phage as well as its own. Strain C(P1) carries only the P1 restriction system. Kr⁻ is a mutant strain lacking restriction capacity. The plating efficiency values in the table are relative to those for *E. coli* C.

[b] From Arber W, Linn S (1969) *ARB 38*:467.

lacks any known modification system, fails to "plate" (i.e., multiply) on all strains except C or a mutant of strain K that lacks the restriction system. However, phage λ, when grown on strain K and carrying that modification pattern, does plate efficiently on strain K as well as on C. The genes for an R–M system in *E. coli* can also be provided by a lysogenic phage or by plasmids and, when present, operate independently of the host system. As seen for strain C lysogenized with phage P1 (Table 21-11), all entering DNA is restricted, except that which develops in such a lysogenized cell. When two R–M systems are present in the same cell, restriction of foreign DNA is more efficient and surviving phage DNA acquires both modification patterns.

Modification Requirements. Maintenance of a characteristic modification pattern of DNA thus requires (1) persistence of the specific DNA sequence recognized for methylation, and (2) replication in a host that has the methylation system and lacks a restriction system that would recognize the DNA as foreign. Thus, modified DNA is an unstable genetic characteristic because it is lost within two generations when replicated in a host that lacks the proper system.

Genetic information for the *hsd system*, as the R–M system of *E. coli* is designated, is contained in three adjacent genes: *hsdR*, *hsdM*, and *hsdS* (Section 13-7). The *hsdR* gene specifies the endonuclease function; the *hsdM*, a polypeptide required for both endonuclease and methylase functions; and the *hsdS* gene, the sequence recognition that determines the specificity of the entire system. Whereas *hsdR* mutants fail to restrict but do modify, *hsdM* and *hsdS* mutants neither restrict nor modify.

Genetic complementation studies suggest that in *E. coli* B the three closely linked genes code for three diffusible products: polypeptides for restriction, modification, and specificity of site recognition. Support for this model has come from the isolation of *E. coli* B and K *restriction* enzymes (Section 13-6), each of which is in fact composed of three nonidentical subunits that are products of the *hsdR*, *hsdM*, and *hsdS* genes. The B modification enzyme has been shown to have two of these three subunits.[213]

Double-stranded DNA is sensitive to modification and restriction. Single-stranded phage DNAs of M13 and ϕX174 are less readily modified or restricted

213. Lautenberger JA, Linn S (1972) *JBC 247*:6176.

than the duplex replicative forms.[214] Modification of one strand is sufficient to confer immunity to restriction; a heteroduplex prepared from one modified and one unmodified strand is not restricted. For this reason, DNA, as it is produced during semiconservative replication, is protected by the modification on the template strand.

Modification must be nearly error-free if the cell is to survive. Restriction may have this accuracy in distinguishing the correct site sequence but may miss some sequences; unmodified phages are known to escape restriction more frequently than can be explained by mutations. Such tolerance for foreign DNA may be of some evolutionary value. In the competition between modification and restriction of DNA entering the cell, additional features of the host and the DNA may influence the outcome. For example, enzymatic activities may be affected by the physiologic as well as the genetic state of the host. Also, very short chromosomes, such as M13, with at most a few cleavage sites for a particular restriction enzyme, may enjoy a special survival advantage.

Restriction–Modification Methylases.[215] *Type I restriction endonuclease* (Section 13-7), or R–M enzyme, methylates DNA at a specific site. An unmethylated site directs the restriction endonuclease component to cleave the DNA at a nonspecific distant location. S-Adenosylmethionine is the methyl donor and duplex DNA the acceptor, to produce a pair of methylated adenines (or cytosines) symmetrically oriented in the specific R–M sequence; ATP and Mg^{2+} are required.

Mutation in the modification gene *(hsdM)* of the *E. coli* K restriction enzyme results in a loss of all enzyme activities, including methylation. Failure of the mutant enzyme to bind S-adenosylmethionine prevents the conformational changes essential for recognition and binding of the unmodified DNA substrate. Inasmuch as both the restriction and modification activities of the intact enzyme are expressed in vivo, there must be a means for determining whether binding to an unmodified recognition sequence is followed by cleavage or by protective methylation.

EcoP1 methylase (Section 13-7), an R–M enzyme encoded by phage P1 (Section 17-7),[216] methylates the central A in the sequence AGACC. Each methylation site is associated with cleavage at a nonspecific site nearby. Unlike the type I enzymes, the P1 methylase does not require ATP and carries out restriction in the absence of S-adenosylmethionine.

The methylases corresponding to the *type II restriction endonucleases* (Section 13-8) are entirely separate entities. They transfer methyl groups one at a time, dissociating from the DNA after each transfer, a behavior in sharp contrast to that of the restriction endonucleases (see below).

EcoRI methylase[217] recognizes a hexanucleotide sequence with twofold symmetry in duplex DNA and transfers methyl groups to the adenine residues adjacent to the symmetry axis:

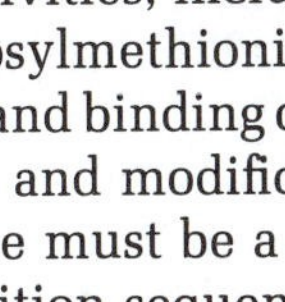

```
↓ *¦
GAA¦TTC
CTT¦AAG
   ¦* ↑
```

214. Blakesley RW, Dodgson JB, Nes IF, Wells RC (1977) *JBC 252*:7300.
215. Haverman A, Heywood J, Meselson M (1972) *PNAS 69*:3138; Vovis GF, Zinder ND (1975) *JMB 95*:557.
216. Reiser J, Yuan R (1977) *JBC 252*:451; Bachi B, Reiser J, Pirrotta V (1979) *JMB 128*:143.
217. Rubin RA, Modrich P (1977) *JBC 252*:7265; Modrich P, Rubin RA (1977) *JBC 252*:7273.

Cleavage by EcoRI endonuclease, a dimer of 28.5-kDa subunits, is at the linkages indicated by the arrows. Both strands of the recognition sequence are cleaved in a single binding event. The methylase, by contrast, is a single polypeptide chain of 39.5 kDa, and dissociates from the DNA with each methyl group transfer. The asymmetry of methylase action contrasts with that of the nuclease. A double-strand break by the nuclease can serve either recombination or degradation. For the methylase, however, the in vivo substrate is presumably a duplex of a methylated parental template strand and the nascent unmethylated progeny.

Two distinct cytosine methylases in *Bacillus amyloliquefaciens* strain H (Bam) and two adenine methylases in *Bacillus brevis* (Bbv) methylate specific residues in the sequences recognized by the corresponding restriction endonucleases.[218] Methylation of GGATCC at the subterminal C residue renders the sequence resistant to hydrolysis by BamI endonuclease, but methylation of the A or the other C residue does not.

E. coli Adenine and Cytosine Methyltransferases

Methylation of adenine and cytosine is widespread in all DNAs. Yet an adequate appreciation of the functions of this major form of decoration is lacking. In prokaryotes, including their phages and plasmids, one clear role is the destruction of foreign DNA (see above) by marking self. An additional role of methylation may be to distinguish parental DNA from newly synthesized strands that may require mismatch correction. In eukaryotes, which lack restriction systems, the principal role of methylation appears to be in the regulation of gene expression (see below).

The methylated nucleotides are not incorporated into DNA directly. Specific DNA methylases (methyltransferases) transfer the methyl group from S-adenosylmethionine to adenine and cytosine residues at specific sites in the genome.[219] These methyltransferases in *E. coli* are distinct from the R–M system methylases (see above). The *DNA adenine methylase*, product of the *dam* gene, specifically recognizes the symmetric tetranucleotide d(pGATC) in duplex DNA and methylates the adenine in each strand. Unlike cytosine methylation by the *dcm* system, which has yet to be given a physiologic role, several functions are served by Dam methylation:

(1) *Repair.* A delay in methylation of the daughter strand distinguishes it from the parental strand and identifies it as the target for methyl-directed mismatch DNA repair (Section 13-9). In the absence of Dam methylase, both strands are recognized and cleaved by the *mutSLH* enzymes (leading to the mutator phenotype of *dam* mutants) to generate a double-strand break.

(2) *Replication.* Lack of synchrony of initiation of replication from *oriC* in *dam* mutants may be due to a needed role for hemimethylated DNA in the inhibition of reinitiation (Section 20-3). Membrane binding of *oriC* DNA also depends on its hemimethylated state (Section 20-4).

(3) *Transcription.* Methylation regulates the expression of some genes, such as *dnaA* and *mioC* (Section 20-3) and the *mom* gene of phage Mu (Section 17-7).

(4) *Transposition.* Transposition by certain transposons (Tn10, Tn5) is regulated by *Dam* methylation (Section 7).

---

218. Hattman S, Keister T, Gottehre A (1978) *JMB 124*:701.
219. Hattman S, Brooks JE, Masurekar M (1978) *JMB 126*:367.

(5) *Initiation of replication and packaging of phage P1.* These functions depend on the methylation of specific sequences (Section 17-7).

Despite the contributions of Dam methylation to so many aspects of DNA metabolism, this function is still dispensable for cell growth. Furthermore, this methylation system is limited to enteric bacteria. Other species, despite having DNA repair, replication, and transcriptional functions operating by mechanisms similar to those in *E. coli*, have neither the Dam methylation system nor an apparent substitute.

Phage T4 encodes its own Dam methylase, which not only recognizes the target sequence of the *E. coli* enzyme, but also can operate on phage DNA which is heavily glycosylated and contains hydroxymethyl cytosine, rendering it resistant to cellular nucleases (Section 17-6). In phage Mu DNA, about 15% of the adenine residues are modified in a novel way by the Mu *mom* gene product, produced under control of *E. coli* Dam methylation (Section 17-7). This modification appears to be there to protect Mu DNA against restriction.

Methylation of Eukaryotic DNA

Methylation of DNA in eukaryotes differs from that in prokaryotes in several ways:

(1) The principal modified base in eukaryotes is 5-methylcytosine except in some unicellular organisms in which methyladenine predominates. The only modified base in mammalian DNA is 5-methylcytosine, 90% of which is in the symmetrical sequence CpG. Only 50% to 70% of the mammalian CpG sequences are modified, depending on the species and tissue.

(2) Viral DNA is not methylated.

(3) For lack of known R–M systems in eukaryotes, it would appear that methylation does not contribute to protection against foreign DNA.

Methylation of mammalian DNA is noteworthy as a device for gene regulation and cellular differentiation. The correlations between hypomethylation and gene activation, and conversely between extensive methylation and transcriptional inactivity[220] are impressive. Yet causal relationships are not conclusive. Furthermore, the absence of methyl groups in *Drosophila* DNA indicates that a highly differentiated organism can manage without a methylation mechanism.

Several aspects of mammalian DNA methylation suggest it as a mechanism for gene regulation.

(1) Methylation can alter the state of DNA and depress the affinity of a repressor for an operator site.[221]

(2) Methylation is symmetrical in both DNA strands, and methylation patterns are somatically (clonally) heritable.[222] Methyltransferase action on a CpG in a newly synthesized strand may require that the opposite CpG in the template strand be methylated. Thus, unmethylated sites in the parent will remain unmethylated in the progeny, just as sites on hemimethylated DNA signal methylation of specific sites on the progeny.

220. Sobieski DA, Eden FC (1981) *NAR* 9:6001; Naveh-Many T, Cedar H (1981) *PNAS* 78:4246.
221. Fisher EF, Caruthers MH (1979) *NAR* 7:401.
222. Stein R, Gruenbaum Y, Pollack Y, Razin A, Cedar H (1982) *PNAS* 79:61.

(3) Methylation patterns differ between tissues.[223] The hypermethylated state of a germ cell or undifferentiated stem cell can, as a result of hypomethylation during subsequent replication stages, lead to progressive lack of methylation.

(4) 5-Azacytidine, a cytidine analog, causes fibroblasts to differentiate into muscle cells, presumably by activating gene expression through inhibition of DNA methylation.[224] The dramatic reversal by 5-azacytidine of X chromosome inactivation during normal mammalian differentiation[225] is another instance suggesting DNA methylation to be the basis for inhibition of gene expression.

Other DNA Modifications

Although methylation of adenine and cytosine residues are the principal forms of DNA modification, other unusual nucleotides occur especially in certain phage DNAs. In the T-even phages (Section 17-6), dCMP is replaced by hydroxymethylcytosine by phage-directed changes in nucleotide biosynthesis (Section 2-13); these residues, after incorporation into DNA, are glycosylated by the phage-encoded β–glycosyl transferases. In addition to rendering the phage DNA resistant to host restriction, these modifications enable the phage to recognize its own DNA as distinct from that of the host and allow the preferential degradation of the bacterial chromosome. In phage Mu DNA, a unique glycinamide moiety modifies about 15% of the adenine residues (Section 17-7). Modification is directed by the Mu *mom* gene and occurs only when the host Dam methyltransferase is active (see above). The glycinamide adducts protect the phage DNA from restriction. Several *B. subtilis* phages (Section 17-8) substitute uracil or hydroxymethyluracil for thymidine. Inasmuch as the alternate nucleotides are synthesized and incorporated directly into the DNA, modification of nucleotide biosynthesis and the removal of TTP by degradation (Section 2-13) achieve the unusual DNA without postreplication modifications as observed in T-even phages.

---

223. Sobieski DA, Eden FC (1981) *NAR 9*:6001; Naveh-Many T, Cedar H (1981) *PNAS 78*:4246.
224. Jones PA, Taylor SM (1980) *Cell 20*: 85; Jones PA, Taylor SM (1981) *NAR 9*:2933; Creusot F, Acs G, Christman JK (1982) *JBC 257*:2041.
225. Liskay RM, Evans RJ (1980) *PNAS 77*:4895; Mohandas T, Sparkes RS, Shapiro LJ (1981) *Science 211*:393.

CHAPTER 22

Synthesis of Genes and Chromosomes

22-1 Origin of DNA on Earth[1]

Leaving aside the proposal that life originated on another planet and was planted on Earth by rockets (directed panspermia),[2] and the controversy as to whether prebiotic chemistry on Earth was conditioned by a reducing atmosphere, a view seriously challenged by some geologists, there must still be earnest interest in nonenzymatic mechanisms of RNA and DNA synthesis. Speculations about the evolution of DNA from prebiotic molecules have been bolstered by a number of facile reactions emerging from experimental studies that make credible steps in a progression from simple gas molecules to a self-replicating system.

Fossils show that biochemically advanced microbial life was present on earth over 3 billion years ago. The evolution of protocells[3] must have been preceded by the slow development of macromolecules. The emergence of nucleic acids on the primitive earth can be considered simply as progressing from the synthesis of purines, pyrimidines, and sugars, to the synthesis of nucleotides and oligonucleotides, and finally to the template-directed assembly of polynucleotides. Whether polypeptides participated in these synthetic reactions is an important but difficult question.

Conjectures that RNA preceded DNA in evolution[4] are supported by the catalytic capacities of RNA in the splicing and processing of RNA transcripts, the priming by RNA of DNA chain initiations, the template direction by RNA of the synthesis of DNA telomeres, and by the interposition of an RNA stage in the life cycle of certain DNA viruses. The selection of DNA as the virtually exclusive

1. Orgel LE, Lohrman R (1974) *Accts Chem Res* 7:368; Oro J (1976) in *Reflections on Biochemistry* (Kornberg A, Horecker BL, Cornudella L, Oro J, eds.). Pergamon Press, New York, p. 423.
2. Crick F (1981) *Life Itself, Its Origin and Nature.* Simon and Schuster, New York.
3. Woese CR, Fox GE (1977) *J Mol Evol* 10:1.
4. Wintersberger U, Wintersberger E (1987) *Trends Genet* 3:198; Weiner AM (1988) *Cell* 52:155.

hereditary substance is ascribed to its greater stability and, thus, its superior ability to protect the sequence of the genome.

It is hard to assess the importance of kinetic and thermodynamic aspects of one prebiotic reaction versus another. Through the power of natural selection, the feeble reaction may have been favored over the more facile; huge amounts of prebiotic time were available to permit exceedingly fine distinctions to be exploited.

Nonenzymatic Synthesis of Purines, Pyrimidines, and Sugars

Complex organic compounds might have been brought to Earth by comets or asteroids.[5] Alternatively, a large number of organic reagents can be derived by applying energy to a reducing atmosphere containing N_2, NH_4, CO, CO_2, CH_4, and H_2O. Heat, electrical discharges, and UV irradiation are examples of energy sources. Prominent products are amino acids, formaldehyde, and hydrogen cyanide, vast amounts of which occur in extraterrestrial dust clouds.

Adenine can be obtained in 0.5% yield by refluxing a solution of ammonium cyanide.[6] [The empirical formula of adenine, $C_5H_5N_5$, can also be written $(HCN)_5$.] Oligomerization of HCN in the presence of formaldehyde produces 8-hydroxymethyladenine in far greater amounts than adenine.[7] Nucleoside analogs may have been produced in complex mixtures with other aldehydes on the prebiotic Earth.

Several pathways that include cyanide, cyanate, cyanogen, and carbonitriles yield adenine, hypoxanthine, guanine, and diaminopurine. Cyanoacetylene, which has also been identified in interstellar dust clouds, could be a precursor of pyrimidines. Ribose is one of the sugars in the complex mixture produced from formaldehyde by alkaline catalysts. Thus, the synthesis of the elementary building blocks of nucleic acids (purines, pyrimidines, sugars, and amino acids) can be readily demonstrated in aqueous solutions of simple molecules known to be abundant in outer space.

Nonenzymatic Synthesis of Nucleotides

Purines or pyrimidines with ribose or deoxyribose have not been condensed to the β-nucleosides to a significant extent under the mild conditions that form the precursor compounds. Nucleosides have been phosphorylated with molecules such as cyanamide serving as condensing agents in aqueous solutions. Dry inorganic phosphates condensed at high temperatures into polyphosphates also make effective phosphorylating agents. However, organic catalysts are more plausible prebiotic agents; urea, for example, can convert inorganic phosphates and nucleosides to inorganic polyphosphates and nucleoside 5′-mono- and 5′-polyphosphates.

Nonenzymatic Synthesis of Oligonucleotides

Activated nucleotides condensed into oligonucleotides that can act as templates are components of a plausible sequence leading to the development of nucleic

---

5. Chyba CF, Thomas PJ, Brookshaw L, Sagan C (1990) *Science* *249*:366.
6. Oro J (1976) in *Reflections on Biochemistry* (Kornberg A, Horecker BL, Cornudella L, Oro J, eds.). Pergamon Press, New York, p. 423.
7. Schwartz AW, Bakker CG (1989) *Science* *245*:1102.

acids. However, nucleoside 5′-di- and 5′-triphosphates and cyclic nucleoside 2′,3′-phosphates, which are effective substrates for polymerization by enzymes, are not used effectively for oligonucleotide synthesis in solutions simulating prebiotic conditions. On the other hand, solid-state reactions of adenosine cyclic 2′,3′-phosphate with a number of amine catalysts have yielded substantial amounts of dimers in which the natural 3′,5′ linkage predominates.[8]

Minerals or polypeptides might once have served as polymerase catalysts, and zinc salts, zinc–amino acid complexes, imidazoles, and polylysine have been tried. Zinc chloride catalyzes oligonucleotide synthesis from adenosine 5′-phosphoroimidazolide, but over 80% of the dimeric product has the 2′,5′ linkage.[9] Whenever in the universe template-directed synthesis of polynucleotides did take place, it is helpful to know the important catalytic role that metals, such as Zn^{2+} and Pb^{2+}, could have played.

Zinc is of special interest because it is an essential component of many of the DNA and RNA polymerases. With a poly C template, Zn^{2+} acting alone can catalyze the assembly of an activated GMP derivative (guanosine 5′-phosphoroimidazolide) into poly G chains of 30 to 40 residues in the natural 3′, 5′ linkage.[10] This regiospecificity is remarkable in view of the far greater reactivity of the 2′-OH compared with the 3′-OH. Zinc action is probably directed at changing the geometry of the polynucleotide helix rather than activating the 3′-OH group. Compared to GMP, the other nucleotides (AMP, UMP, and CMP) are selected against by a factor of 200. Although Pb^{2+} is an efficient catalyst,[11] the oligomers it produces are 2′, 5′-linked, and the fidelity to template direction is far less than with Zn^{2+}.

Prebiotic synthesis of polynucleotides may have been influenced by interaction with short peptides. This possibility is supported by the effects that rather minor changes in the imidazole substituent have on efficiency and regiospecificity of poly G synthesis, even in the absence of Zn^{2+}. From the 2-methylimidazole GMP derivative, for example, in the absence of Zn^{2+}, 89% of the products are oligomers, four or more residues long, connected predominantly by 3′, 5′ diester bonds.[12]

Nonenzymatic Synthesis of Polynucleotides by Template Direction[13]

Helical structures formed by hydrogen-bonding AMP with polyuridylate can direct the synthesis of oligoadenylate. However, this template-catalyzed, nonenzymatic condensation requires a water-soluble carbodiimide, a rather implausible ingredient of a prebiotic soup. Imidazolides of the nucleoside 5′-phosphates, readily formed under mild conditions, can substitute as efficient condensing agents.

A 14-mer oligodeoxynucleotide, $(C_3G)_3C_2$, directs the synthesis of smaller 3′, 5′-linked oligonucleotides using the 2-methylimidazolides of guanosine and cytidine.[14] The predominant products, complementary to the template, range in length from GGC to the 14-mer $G_2(G_3C)_3$. In these simple nonenzymatic reactions, substantial information is transferred from template to products.

8. Verlander MS, Lohrmann R, Orgel LE (1973) *J Mol Evol 2*:303.
9. Sawai H, Orgel LE (1975) *JACS 97*:3532; Sawai H (1976) *JACS 98*:7037.
10. Lohrmann R, Bridson PK, Orgel LE (1980) Science *208*:1464; Bridson PK, Orgel LE (1980) *JMB 144*:567.
11. Lohrmann R, Orgel LE (1980) *JMB 142*:555; van Roode JHG, Orgel LE (1980) *JMB 144*:579.
12. Inoue T, Orgel LE (1981) *JACS 103*:7666.
13. Orgel LE, Lohrmann R (1974) *Accts Chem Res 7*:368.
14. Acevedo OL, Orgel LE (1987) *JMB 197*:187.

Two major problems in attempts at nonenzymatic template-directed synthesis of polynucleotides have been the failure to polymerize pyrimidines and the predominant formation of the unnatural 2′, 5′ internucleotide bond. The difficulty in polymerizing pyrimidines may be due to their weaker stacking interactions compared to purines. The assembly of purine–pyrimidine dinucleotides (e.g., pGpU) directed by an alternating copolymer [e.g., poly(pApC)] produces repetitive duplexes (Section 3) that sustain facile reiterative self-replication (Section 4-14).

To circumvent the feebler stacking of pyrimidines, AI base pairing[15] and other all-purine duplexes can be envisioned as an ancestral nucleic acid,[16] later modified to include the pyrimidines. The standard N9-sugar-linked purine hydrogen-bonded to a postulated N3-sugar-linked xanthine or isoguanine possesses the same geometry and charge as the Watson-Crick base pairs. Biosynthesis of the novel N3-linked purines might take place at a step in the de novo pathway (Section 2-3) when a tautomer of the formylglycinamide ribotide could give rise to the N3 instead of the N9 linkage.

The 2′, 5′ phosphodiester linkage is more readily formed than the 3′, 5′ bond, presumably because the 2′-OH group is more reactive. However, a 3′,5′-linked polynucleotide, once it becomes part of a helical structure, is more resistant to hydrolysis than the 2′,5′ isomer.[17] Thus, natural selection of the 3′,5′-nucleic acid structure may have been based on the survival of the more stable helix (Section 1-3). Specific catalysis by one of the transition metals, such as zinc or lead, serving as a primitive polymerase, is another possibility. Lead salts in low concentration catalyze poly U–directed synthesis of poly A chains up to 10 residues long, preferentially linked by the natural 3′, 5′ phosphodiester bond.

Evolution of Genomes

No clue to the nature of the primordial polymerase is evident among the structures of the families of DNA and RNA polymerases. Understanding the mechanisms of reiterative replication responsible for "de novo" synthesis of repetitive polymers (Section 4-14) might clarify the prebiotic polymerizations.

In the transition from repetitive polymers to the intricate genomes of phages, bacteria, and eukaryotic cells, the reiteration of sequences and genetic segments is a common theme. Highly repetitive sequences characterize heterochromatin in eukaryotic cells (Section 1-5). Duplications are evident in the sequences of proteins whose evolution can be traced.[18] The major peptides in the precursor of bovine corticotropin-β-lipoprotein show little resemblance in amino acid sequence, although homologies in the coding regions suggest they evolved by a series of genetic duplications.[19]

The accretion of DNA in genomes can be seen both as tandem[20] and nontandem (insertional) duplications of genetic segments. Most of the 20 or more genes in the *E. coli* chromosome that are responsible for glucose catabolism are contained in four clusters. The spacing of these clusters, 90° or 180° apart on the circular map, suggests evolutionary divergence of an ancestral glucolytic gene after two doublings of a primitive genome.[21] Mechanisms for the duplication of

15. Crick FHC (1968) *JMB 38*:367.
16. Wachterschäuser G (1988) *PNAS 85*:1134.
17. Usher DA, McHale AH (1976) *PNAS 73*:1149.
18. Wu TT, Fitch WM, Margoliash E (1974) *ARB 43*:539; McLachlan AD (1979) *JMB 128*:49.
19. Nakanishi S, Inoue A, Kita T, Numa S, Chang ACY, Cohen SN, Nunberg J, Schimke RT (1978) *PNAS 75*:6021.
20. Anderson RP, Roth JR (1977) *Ann Rev Microbiol 31*:473.
21. Riley M, Anilionis A (1978) *Ann Rev Microbiol 32*:519.

animal genes also appear to be widespread, as indicated by drug resistance [e.g., to methotrexate or N-phosphonacetyl-L-aspartate (PALA); Section 14-2] acquired through progressive and extensive amplification of the gene and the enzyme that the drug antagonizes.[22]

22-2 Determination of DNA Sequence[23]

It is hard to imagine now how slow and difficult it once was to determine a DNA sequence compared to that of protein and even RNA. In 1974, lacking small, defined DNA molecules and specific chemical and enzymatic reagents to degrade them, the sequencing of a 40-bp fragment was a notable achievement. Analyses are now done easily and rapidly because (1) site-specific restriction nucleases cleave large DNA molecules into small fragments, (2) molecular cloning can separate and amplify fragments from 0.1 to 50 kb, (3) intense radioactive labeling can be used to mark the terminal residues of the fragments, and (4) multiple procedures are available for rapid and accurate analysis of the sequences of the fragments.[24]

Currently, the sequence of a segment several kilobases in length is routinely completed in most laboratories in a few weeks. Most recently, sequencing the 3-billion-bp human genome has been undertaken; storage and retrieval of the information looms as one of the bottlenecks in the organization and direction of this monumental effort.

The power of DNA sequencing technology continues to expand, producing novel insights into structures that affect many aspects of genetic evolution and function. As examples, analyses of mitochondrial DNAs revealed differences in the genetic code, anomalous tRNAs, additional codon–anticodon interactions, and a different initiation mechanism for protein synthesis.[25] Comparison of the germ-line and somatic-cell sequences of the immunoglobulins identified somatic recombination and mutations that help to account for the structures and diversity of antibodies[26] (Section 21-6). The amino acid sequence of a protein, inferred from an open reading frame in the DNA sequence, can guide the chemical synthesis of oligopeptides, which can then serve as immunogens to elicit antibodies; these in turn can identify and isolate a previously unknown protein from a cell or tissue.[27]

The remarkable accuracy of DNA sequencing rests on the cross-checking afforded by analysis of both strands of the duplex, on the availability of numerous restriction enzymes that allow the generation of multiple overlapping sequences, and on the presence of an open reading frame in cistronic regions. Both the chemical and the enzymatic replication methods for sequence analysis have been assisted by modifications in gel electrophoretic separations[28] that achieve high resolution with automated reading and data recording.

---

22. Schimke RT, Kaufman RJ, Alt FW, Kellems RF (1978) *Science 202*:1051; Wahl GM, Padgett RA, Stark GR (1979) *JBC 254*:8679.
23. Barrell BG (1978) *Int Rev Biochem 17*:125; Wu R (1978) *ARB 47*:607; Air GM (1979) *Crit Rev B 6*:1.
24. Sambrook J, Fritsch EF, Maniatis T (1989) *Molecular Cloning: A Laboratory Manual*. 2nd ed. CSHL; Ausubel FM, Brent R, Kingston RE, Moore DD, Seidman JG, Smith JA, Struhl K (1987 et seq.) *Current Protocols in Molecular Biology*. Wiley, New York.
25. Sanger F (1981) *Science 214*:1205.
26. Tonewaga S (1979–80) *Harvey Lect 75*:61; Ravetch JV, Siebenlist U, Korsmeyer S, Waldman T, Leder P (1981) *Cell 27*:583; Kim S, Davis M, Sinn E, Patten P, Hood L (1981) *Cell 27*:573.
27. Lerner RA, Sutcliffe JG, Shinnick TM (1981) *Cell 23*:309.
28. Sanger F, Coulson AR (1978) *FEBS Lett 87*:107.

Isolation of Homogeneous DNA

A DNA chain can be severed by nucleases into fragments that can then be separated by agarose gel electrophoresis. With even the modest-sized *E. coli* chromosome, 4 million base pairs long, the yield of a unique 200-ntd fragment would be vanishingly small. Fortunately, by molecular cloning (Section 5), segments of DNA 100 to 40,000 bp long can be amplified and isolated in pure form. By using additional nucleases to generate fragments of overlapping sequences, along with various hybridization techniques, a map of the arrangement of the DNA segments can be constructed. Amplification of a scant amount of DNA can be achieved by the polymerase chain reaction (PCR) technique (Section 4-17). Only a very brief sequence (e.g., 20 bp) must be known in order to be used as a primer in the production of large quantities of the adjacent DNA.

Labeling of the DNA

All sequencing methods depend on labeling the DNA in vitro with ^{32}P, ^{35}S, or fluorescent nucleotides and determining the size and composition of fragments. In the *chemical method* of sequence analysis, the terminal nucleotide of a fragment is labeled and then the fragment is hydrolyzed by base-specific reactions. Very high specific radioactivities can be obtained, and the labeled terminus is a fixed reference point for the digestion products of the fragment. The nucleotide sequences of the fragment can then be read directly from the gel mobilities of the ladder of digestion products. In the *enzymatic replication method* of sequence analysis, the DNA fragments are used to prime the synthesis of new DNA chains from specific sites on a template DNA molecule. Labeled deoxynucleotides are incorporated internally during polymerization.

Labeling the 5′-Terminal Nucleotide. Polynucleotide kinase[29] (Section 9-7) transfers ^{32}P from [$\gamma^{32}P$]ATP to the 5′-OH end of ribonucleotide and deoxyribonucleotide chains of all sizes, down to and including a 3′-mononucleotide. Phosphatase may be needed to generate a free 5′-OH terminus. By using ATP at a specific radioactivity in excess of 10^{11} counts per minute per micromole, subpicomolar quantities of fragments can be easily analyzed. By labeling the 5′ end of a chain of known mass with ATP of known specific radioactivity, chain length can be accurately determined. Complete nuclease digestion of the DNA permits direct identification of the terminal nucleotide.

Labeling the 3′-Terminal Nucleotide. When phosphorylation of the 5′ end proves difficult, the 3′ end can be labeled by enzymatic addition of a nucleotide, in either of two ways. One method uses [α-^{32}P]NTP and terminal deoxynucleotidyl transferase (Section 6-10) to add an alkali-labile ribonucleotide. The other method uses turnover of the 3′ chain end by T4 DNA polymerase.[30]

Unique End-Labeling of Fragments. When either or both ends (5′ or 3′) of a duplex fragment have been labeled, the strands can be separated by cleavage with a restriction nuclease into unequal-sized fragments that are then distin-

29. Weiss B, Richardson CC (1967) *JMB 23*:405; Murray K (1973) *BJ 131*:569; Bernardi A (1974) *Anal Biochem* 59:501.
30. Englund PT (1971) *JBC 246*:3269; Englund PT, Price SS, Weigel PH (1974) *Meth Enz 29*:273.

guished by gel electrophoresis. Alternatively, the duplex can be denatured and the strands separated on gels, by density centrifugation, or by hybridization to specific RNA.

Chemical Method for Sequence Analysis[31]

The chemical sequencing method is advantageous for its ready applicability to almost any DNA fragment.[32] Chemical reactions used in sequence studies break a DNA chain (with a terminal label) preferentially at one of the four nucleotides under conditions adjusted so that, on the average, only one break is made per chain. A family of fragments is produced, and their sizes are found by determining the position of the break in the chain relative to the same labeled end. Gel electrophoresis resolves the fragments by size and measures their exact length by comparison with standards.

Four distinctive reactions are applied to cleave DNA either at guanines alone (G), at adenines as well as at guanines (A + G), at cytosines and thymines equally (C + T), or at cytosines alone (C). G and A methylated by dimethyl sulfate are distinguishable by different conditions for chain cleavage. Hydrazinolysis affects T and C, or C alone, depending on ionic strength; piperidine then breaks the chain at the affected pyrimidine sites. When the four families of fragments (G, A + G, C + T, C) are analyzed side by side on a polyacrylamide gel, the DNA sequence for 300 ntd or more can be read directly from the pattern of radioactive bands on an autoradiograph.

Enzymatic Replication Methods for Sequence Analysis

Enzymatic methods rely on a DNA polymerase to copy a particular sequence of ssDNA, starting from the terminus of an annealed primer. Generally the primer is a chemically synthesized oligonucleotide, complementary to a region upstream of that to be sequenced. Cleavage by one of several devices then generates four families of synthetic DNA chains terminated at A, T, G, or C and representative of all possible lengths. Nucleotide sequences of 10 to 400 ntd from the primer 3′ end can be analyzed rapidly and accurately by the various replication methods.

Chain-Terminating Inhibitors.[33] Chain-terminating inhibition has rapidly become the method of choice because of its simplicity and accuracy. The 2′, 3′-dideoxy analogs and the arabinose analogs of the dNTPs, when incorporated, block further chain growth.[34] With appropriate ratios of analog to natural dNTP, a uniform spread of chain lengths is obtained rapidly and simply. Wherever an analog is inserted during replication, chains will terminate; a pile-up of chains will eventually be observed at each position of that nucleotide in the sequence. By using analogs of each of the four dNTPs in turn, the total sequence can be read from four adjacent channels in a polyacrylamide gel. A T7 DNA polymerase (Sequenase), modified to minimize pausing at template secondary structures and mismatches, helps to sharpen the gel patterns.[35]

---

31. Maxam AM, Gilbert W (1977) *PNAS 74*:560.
32. Gilbert W (1982) *Science 214*:1305.
33. Sanger F, Nicklen S, Coulson AR (1977) *PNAS 74*:5463; Maat J, Smith AJH (1978) *NAR 5*:4537.
34. Atkinson MR, Deutscher MP, Kornberg A, Russel AF, Moffatt JG (1969) *Biochemistry 8*:4897.
35. Tabor S, Richardson CC (1987) *PNAS 84*:4767.

Procedures. The dideoxy method (now used exclusively among the enzymatic methods) is effective for determining extensive, genome-length sequences. The requirement for ssDNA, a limitation of enzymatic methods, has been circumvented in two ways. One is by degradation of a duplex fragment by exonuclease III ($3' \rightarrow 5'$) or phage λ exonuclease ($5' \rightarrow 3'$) to generate long stretches of single-stranded template, an ideal method for primed-synthesis sequence analysis.[36] The other approach, termed *"shotgun" DNA sequencing*, is to clone random fragments 200 to 400 ntd long (produced by pancreatic DNase I in the presence of Mn^{2+}) in a single-stranded phage vector such as M13.[37] A β-galactosidase gene with an EcoRI restriction site is included in the M13 genome and provides both a convenient site for insertion of the fragments and the means for identifying recombinant phage with the use of a suitable indicator system.

In the shotgun procedure, a collection of phage clones, each with a random fragment, encompasses the entire genome to be sequenced; a chemically synthesized oligonucleotide, complementary to the sequence flanking the cloning site, serves as a universal primer on ssDNA extracted from the phage. Once again, simple cloning replaces the laborious and inexact fractionations of restriction digests. The only serious limitation of the shotgun method is that as the sequencing approaches completion, new information diminishes asymptotically: analysis of 20 random clones, with an average insert of 227 residues, yielded 3 kb of a 4.3-kb sequence, but analysis of another 20 to 30 clones was needed to complete the sequence.[38]

The sequences of unexplored chromosomal regions of special interest have been obtained by "walking" from a nearby region as a point of reference. With clones containing overlapping sequences, the 100-kb bithorax region of *Drosophila* that determines its segmental patterns has been reached after a jaunt of 300 or so kb.[39] Analyses of the sequences define transcriptional boundaries and show that most of the spontaneous mutations in this vast locus are the result of transpositional events.

22-3 Chemical Synthesis of Oligodeoxyribonucleotides and Genes[40]

The forward leap in chemical synthesis of DNA rivals that of DNA sequencing. Synthesis of a 14-ntd sequence that previously consumed a year and could be undertaken by only a few laboratories in the world can now be done in hours by a trained technician almost anywhere. The combination of an automated synthetic procedure that adds a nucleotide in a 30-minute cycle, practical procedures for isolation and characterization of the product, and reliable methods for enzymatic repair and ligation of the oligonucleotides produces a gene of 500 bp or more long in about the time it takes to isolate the gene from natural sources.

Beyond the use of oligodeoxyribonucleotides as the building blocks of genes, their value is impressive in many other ways: as binding sites for a variety of regulatory and catalytic proteins; as linkers in splicing genes into chromosomes; as probes to isolate a genetic sequence from cells or tissues; as primers to synthe-

---

36. Smith AJH (1979) *NAR 6*:831; Zain BS, Roberts RJ (1979) *JMB 131*:341; Guo L-H, Wu R (1982) *NAR 10*:2065.
37. Orgel LE, Lohrmann R (1974) *Accts Chem Res 7*:368.
38. Anderson S (1981) *NAR 9*:3015.
39. Bender W, Spierer P, Hogness DS (1979) *J Supramol Struct [Suppl] 3*:32.
40. Khorana HG (1968) *Harvey Lect 62*:79; Kössel H, Seliger H (1975) Prog Chem Org *Nat Prod 32*:298; Wu R, Bahl CP, Narang SA (1978) Prog *N A Res 21*:102.

size or sequence DNA (or RNA); as altered primers for site-directed mutagenesis; and as the source material for x-ray crystallography and other refined studies of DNA structure.

Chemical syntheses of di-, tri-, and tetranucleotides set the stage for the synthesis of biologically active genetic segments (Section 4). Although oligonucleotide synthesis is now a highly automated, solid-phase procedure, some of the basic synthetic features remain noteworthy. Oligo- and polynucleotide synthesis has three prerequisites: (1) protecting groups for the 3′-OH and 5′-OH chain ends, for the amino substituents of A, G, and C, and for the phosphate moiety itself; (2) an effective condensing agent; and (3) a good yield of pure intermediates.

Protecting Groups

The protecting groups must be added and removed under conditions that do not damage the intermediates or products (Fig. 22-1). Selective removal of the

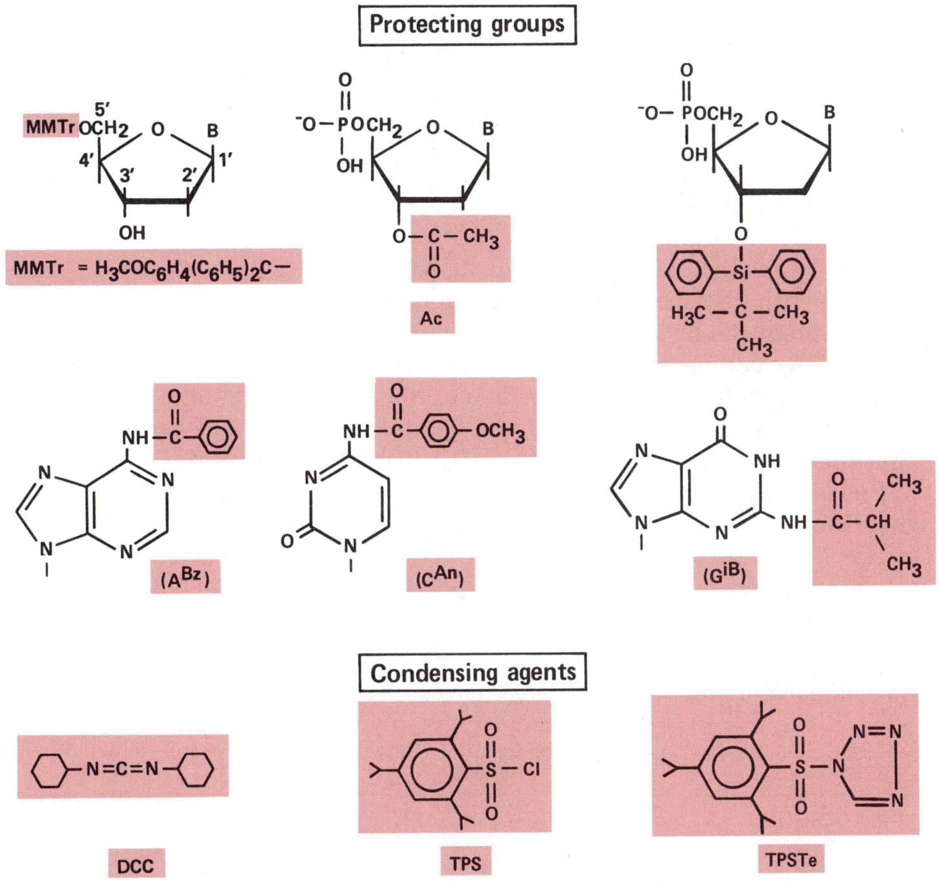

Protecting groups

MMTr = $H_3COC_6H_4(C_6H_5)_2C-$

Ac

(A^{Bz})

(C^{An})

(G^{iB})

Condensing agents

DCC

TPS

TPSTe

Figure 22-1

Protecting groups (in color blocks) for deoxynucleosides and deoxynucleotides, and condensing agents (in color blocks) used in chemical synthesis of polydeoxynucleotide chains. Abbreviations: MMTr, monomethoxytrityl; Ac, acetyl; Bz, benzoyl; An, anisoyl; iB, isobutyryl; DCC, dicyclohexylcarbodiimide; TPS, triisopropylbenzenesulfonyl chloride; TPSTe, triisipropylbenzenesulfonyl tetrazolide. The protecting group in the upper right is *tert*-butyldiphenylsilyl.

groups at the 3′ and 5′ termini is required for repeated condensations in the synthesis of longer chains.

Among the many available ways of supplying protecting groups, two may be cited. One involves a choice between the phosphotriester approach, in which the internucleotidic phosphate is protected by a β-cyanoethyl or related group, and the now classic phosphodiester method, in which the internucleotidic phosphate is not protected. Although longer reaction times may handicap the phosphotriester method, separation of the uncharged triester permits comparatively large-scale operations and excellent yields (see below). Solid-phase syntheses employ the triesters, and are capable of automation. A second example uses a *tert*-butyldiphenylsilyl group to protect the 3′-OH of mono- and oligodeoxyribonucleotides.[41] The strong lipophilic character of the protecting group facilitates rapid solvent extraction of the silylated intermediates, which are readily isolated by high-pressure liquid chromatography.[42] The silyl group can be removed by fluoride ion at room temperature.

Synthesis of Repetitive DNA Duplexes

With the phosphodiester method, di-, tri-, and tetranucleotides have been synthesized and then condensed as oligonucleotide blocks into chains containing the di-, tri-, or tetranucleotides repeated two or three times. For each such polymer chain a polynucleotide was prepared, with a sequence that exactly matched it, by base pairing in an antiparallel orientation.

These synthetic repetitive DNA duplexes (Table 22-1) helped to elucidate the genetic code and other biochemical and genetic aspects of protein synthesis.[43] The short polymers, serving as template-primers, were extended by DNA polymerase with the four dNTPs to macromolecular size. The minute quantity of polymer produced by the original organic synthesis was thus expanded manyfold to a large polymer, and this could serve as a template to direct the synthesis of an mRNA by RNA polymerase. One or the other strand of the DNA duplex could be transcribed, depending on which of the four rNTPs were supplied. Thus the mRNA produced from $d(AAC)_n \cdot d(GTT)_n$ in the presence of UTP and GTP was $r(GUU)_n$; in the presence of CTP and ATP it was $r(AAC)_n$. Upon translation of $r(GUU)_n$ in each of the three possible reading frames, three homopolypeptides were obtained: polyvaline, polyleucine, and polycysteine; $r(AAC)_n$ led to the synthesis of polyasparagine, polythreonine, and polyglutamine.

Table 22-1
Synthetic polydeoxynucleotides with repeating nucleotide sequences

Repeating dinucleotide sequences	Repeating trinucleotide sequences	Repeating tetranucleotide sequences
$d(TC)_n \cdot d(GA)_n$	$d(TTC)_n \cdot d(GAA)_n$	$d(TTAC)_n \cdot d(GTAA)_n$
$d(TG)_n \cdot d(CA)_n$	$d(TTG)_n \cdot d(CAA)_n$	$d(TATC)_n \cdot d(GATA)_n$
	$d(TAC)_n \cdot d(GTA)_n$	
	$d(ATC)_n \cdot d(GAT)_n$	

41. Jones RA, Fritz H-J, Khorana HG (1978) *Biochemistry* 17:1268.
42. Fritz H-J, Belagaje R, Brown EL, Fritz RH, Jones RA, Lees RG, Khorana HG (1978) *Biochemistry* 17:1257.
43. Khorana HG (1968) *Harvey Lect* 62:79.

From the nature of the homo- and copolypeptides obtained in translating a variety of repeating di-, tri-, and tetranucleotide messengers, the triplet nature of the genetic code and the identity of most of the codons could be deduced or confirmed.

Synthesis of Codon Triplets[44] and the Enzymatic Synthesis of Oligodeoxyribonucleotides[45]

The phosphotriester method has allowed the rapid synthesis of fully protected trideoxyribonucleotide blocks representing virtually all 64 codons, in yields of 50% or greater relative to the starting 5′-OH monomer.

Polynucleotide phosphorylase provides a remarkably convenient and highly specific way to synthesize chains up to 13 residues long. Substituting a dNDP for the normal rNDP substrate, in the presence of Mn^{2+}, permits the controlled addition of one or more nucleotides to the 3′ end of primers with chain lengths of about 3 to 12 residues. Oligonucleotides synthesized by the repeated addition of a single residue are used as defined primers, mutated sequences, and heteroduplexes.

Solid-Phase Synthesis

Synthesis on a solid-phase support of cellulose,[46] silica gel,[47] or resins of polydimethylacrylamide[48] or polystyrene,[49] rather than in solution, has become the method of choice because of its speed, efficiency, and capacity for automation. In each cycle of nucleotide addition, unreacted reagents are quickly removed and an unreacted 5′-OH group can be masked against further additions. The choice of suitably protected mono-, di-, or trinucleotides as building blocks is dictated by several considerations. Only four types of mononucleotides are needed, but 16 dimers and 64 trimers are required. However, the trimers are readily synthesized from monomers and dimers and can be stored safely.

The use of trimers not only saves time and effort in synthesizing long chains but also makes the separation of product from impurities more decisive because of the larger differences in size and charge. Oligomers of 37 and 34 nucleotides, synthesized by successive addition of trimers, have been annealed to generate a duplex with 5′ tails.[50] A flush-ended 55-bp duplex (containing the PriA protein recognition site at the origin of phage ϕX174 DNA replication; Section 8-4) has been formed by filling in these tails with DNA polymerase I.[51]

Current procedures use phosphites or phosphotriesters and phosphorylating reagents such as methyl phosphodichloridites or aryl phosphodichloridates and their triazolide derivatives.[52] Others find advantage in using condensing re-

---

44. Hirose T, Crea R, Itakura K (1978) *Tetrahedron Lett 28*:2449.
45. Gillam S, Jahnke P, Smith M (1978) *JBC 253*:2532; Trip EM, Smith M (1978) *NAR 5*:1529.
46. Crea R, Horn T (1980) *NAR 8*:2331.
47. Alvarado-Urbina G, Sathe GM, Liu W-C, Gillen MF, Duck PD, Bender R, Ogilvie KK (1981) *Science 214*:270; Matteucci MD, Caruthers MH (1981) *JACS 103*:3185.
48. Markham AF, Edge MD, Atkinson TC, Greene AR, Heathcliffe GR, Newton CR, Scanlon D (1980) *NAR 8*:5193.
49. Miyoshi K, Arentzen R, Huang T, Itakura K (1980) *NAR 8*:5507.
50. Crea R, Vasser MP, Struble ME, personal communication.
51. Crea R, Vasser MP, Struble ME, personal communication.
52. Alvarado-Urbina G, Sathe GM, Liu W-C, Gillen MF, Duck PD, Bender R, Ogilvie KK (1981) *Science 214*:270; Matteucci MD, Caruthers MH (1981) *JACS 103*:3185; Broka C, Hozumi T, Arentzen R, Itakura K (1980) *NAR 8*:5461.

agents such as triisopropylbenzenesulfonyl tetrazolide or mesitylenesulfonyl 3-nitrotriazolide to form the internucleotide linkage.[53]

Synthesis of Genes

Synthesis of the genes for small hormones, such as somatostatin, insulin, and thymosin, was carried out in solution. With solid-phase support, the total synthesis of a 514-bp duplex containing the human leukocyte interferon gene has been achieved.[54] Engineered into the duplex are initiation and termination signals for expression in *E. coli* and appropriate restriction enzyme sites for insertion into plasmid vectors. The strategy for synthesis was based on the preparation of 66 duplex oligonucleotides ranging in size from 14 to 21 residues and designed with overlapping ends for use in enzymatic ligation after annealing of complementary fragments.

The native sequence of the interferon gene contains many repeated heptamers and octamers, all of which may be troublesome in ligation of synthetic fragments. These potential difficulties were taken into account, as were the preferred codon usages in *E. coli*, by making 71 substitutions within the constraints of the genetic code. *E. coli* clones containing the anticipated nucleotide sequence were obtained.

The most impressive advantage of chemical synthesis over the isolation of natural genes (or the cDNA copies of processed mRNAs) is the ease and precision of altering every residue in the chain. The striking effects of even a single base substitution in the regulation of gene expression by attenuation[55] (Section 7-8), let alone the profound influence of an amino acid change on the activity of a protein, demonstrate the power of creating analogs for basic studies and clinical applications. Also, the ability to introduce restriction endonuclease sites by manipulation of the third nucleotide position of codons (or the second position, or both) makes for ease of genetic manipulation of the synthetic gene versus the natural one.

22-4 Assembling Genes into Chromosomes: Recombinant DNA[56]

Developments in a major, worldwide research industry reported in daily newspapers, in monographs, and in every issue of hundreds of journals can hardly be summarized here in a few pages. In effect, much of all current biological and medical scientific effort is based on improved cloning devices, superior techniques for transforming bacterial, plant, and animal cells with DNA, and advanced techniques of DNA sequencing and chemical synthesis. These advances have opened up vast opportunities for exploring gene evolution, structure, and function, and virtually every facet of cellular architecture and development.

---

53. Edge MD, Greene AR, Heathcliffe GR, Meacock PA, Schuch W, Scanlon DB, Atkinson TC, Newton CR, Markham AF (1981) *Nature* *292*:756.
54. Edge MD, Greene AR, Heathcliffe GR, Meacock PA, Schuch W, Scanlon DB, Atkinson TC, Newton CR, Markham AF (1981) *Nature* *292*:756.
55. Yanofsky C (1981) *Nature* *289*:751.
56. Sambrook J, Fritsch EF, Maniatis T (1989) *Molecular Cloning: A Laboratory Manual*. 2nd ed. CSHL; Ausubel FM, Brent R, Kingston RE, Moore DD, Seidman JG, Smith JA, Struhl K (1987 et seq.) *Current Protocols in Molecular Biology*. Wiley, New York.

Only a few of the recent advances in genetic engineering and related technologies with broad practical applications need be cited as illustrative examples.

The success of recombinant DNA work rests on restriction nucleases, cloning vehicles, DNA synthesis and ligation, and hosts for cloning. The current focus on these agents and operations may overlook the role of nucleic acid enzymology at each stage in the development of recombinant DNA.[57] The first molecules were fashioned with poly dA and poly dT tails added by terminal deoxynucleotidyl transferase after treatment with λ exonuclease.[58] Covalent closures required exonuclease III, DNA polymerase I, and ligase.[59] Restriction endonucleases were discovered in enzymatic studies of restriction–modification systems (Section 13-6). These and other enzymes, such as alkaline phosphatase, S1 nuclease, reverse transcriptase, exonuclease VII, and RNA polymerase, are used to make recombinant DNAs and to analyze their structure and function.

Restriction Endonucleases

DNA can be fragmented randomly by mechanical shearing or specifically by one or several of the large variety of available restriction endonucleases (Section 13-7). Some nucleases produce duplex fragments with blunt, or flush, ends; some produce fragments with cohesive ends. The latter are extended single-stranded termini, two or more nucleotides long, at each of the 5′ or 3′ ends (depending on the enzyme), which cohere readily because of their complementary nature. Fragments produced from diverse DNAs by a particular nuclease can be joined by ligase at these lap joints and cleaved subsequently at this very linkage by a nuclease with the same specificity.

Cloning Vehicles

Plasmids and viruses are the vectors for transferring genes to organisms in which the genes can be cloned and amplified. The vector is self-replicating. Plasmids carry an identifying genetic marker, and recombinants in viral vectors may be identified by the formation of plaques.

Numerous plasmids are in use, and new ones are constantly being fashioned by additions and deletions to fit the particular needs of experiments. For example, ColE1, the plasmid conferring immunity to colicin, can be enlarged to include a gene for resistance to drugs (e.g., Tc^R, tetracycline resistance), or it can be miniaturized by removing nonessential regions, including some that limit its copy number (Section 18-2). However, the self-replicating capacity of a plasmid must remain intact. The number of copies of ColE1 per cell, near 20 in general, can be raised to several hundred by an inhibitor of protein synthesis (e.g., chloramphenicol) that prevents replication of the host chromosome, so that the concentrations of ColE1 and host DNAs become comparable.

Segments of phages, particularly λ, that are not needed for lytic infection or integration can be removed. The number of restriction sites for a particular nuclease can be reduced to two (e.g., by mutation) and confined to a nonessential region. Such deletions can then be replaced by a foreign DNA "passenger." Phage λ DNA cannot be packaged if it is less than 75% or more than 102% of the

---

57. Singer M (1979) in *Genetic Engineering* (Setlow J, Hollaender A, eds.). Plenum Press, New York, p. 1.
58. Lobban PE, Kaiser AD (1973) *JMB 78*:453.
59. Jackson DA, Symons RH, Berg P (1972) *PNAS 69*:2904.

natural unit length. With a phage DNA constructed below the minimum length, only those particles incorporating foreign DNA will give rise to progeny and produce plaques. The carrying limit for these λ vehicles is about 17 kb, but further manipulations allow virtually the entire genome to be replaced, raising this limit to over 40 kb. Adaptability in length of the filamentous phage M13 (Section 17-3) makes it an attractive vehicle.

A eukaryotic virus such as SV40 can likewise be adapted by replacing part of the genome with foreign DNA. The deleted viral function—for example, the coding region for a capsid protein—can be furnished by a "helper" SV40 genome containing a temperature-sensitive defect in a gene other than the one deleted from the vector chromosome.

Assembly of Recombinant DNA

Foreign DNA is prepared for the cloning vehicle by fragmenting the genome (by shearing or with a restriction nuclease) or by reverse transcription of mRNA to its double-stranded complementary DNA (cDNA). The duplex DNA fragment is then covalently joined to the cloning vehicle by ligase through one of several means: (1) cohesion of overlapping ends generated by certain restriction nucleases; (2) cohesion of oligo dA and oligo dT tails or oligo dG and oligo dC tails (12 to 200 ntd long) added to the 3′ flush ends by terminal deoxynucleotidyl transferase; (3) joining of flush ends directly; or (4) cohesion of chemically synthesized overlapping ends: linkers which constitute a restriction nuclease recognition sequence are added by flush-end joining and are later cleaved by the nuclease.

Although the products of recombining DNA fragments are formidably numerous, careful adjustment of the vector/insert ratio and phosphatase treatment of the vector to prevent self-ligation can minimize complexity. By selecting the size and shape of intermediates and products, using gel electrophoresis or other separation methods, the yield of the desired recombinant can be further refined. The most definitive and powerful of all purification techniques is selection through cloning guided by observation of phenotypic markers for the presence of the desired insert.

Cellular Hosts for Amplification and Cloning

DNA is introduced into *E. coli* by transformation with plasmids (Section 21-8), by infection with λ phage packaged in vitro, or by direct transfection with phage DNA. Each transformed cell or infective center is grown as a separate colony or plaque. The low frequency of transformants makes it improbable that a recipient cell will acquire more than one recombinant molecule. In this manner, a genome as complex as that of *Drosophila* can be divided into many thousands of fragments, and each fragment can then be isolated as a separate clone.

Selecting the desired clone is based on the presence of both vehicle (vector) markers and passenger markers (inserts). For example, in cloning fragmented *Drosophila* DNA in some ColE1-derived plasmids, immunity to colicin and resistance to tetracycline establish the presence of the plasmid. Hybridization of an *E. coli* colony with histone mRNA identifies which one, among the many millions of *E. coli* cells, harbors a plasmid with the histone genes. In this manner, several thousand colonies or λ phages can constitute a reference library in which all genes of the genome of *E. coli*[60] or yeast or *Drosophila*[61] are represented at least once.

A clone can be identified, isolated, and then extensively amplified by growing in a large-scale culture. The passenger genetic segment can then be separated by excising it from the cloning vehicle with the restriction nuclease used originally to introduce it. The DNA segment can be isolated in pure form and in milligram amounts even when present in only one part per million in the genome.

Applications

Recombinant DNA techniques now permeate every facet of genetic research, from analysis of complex chromosomes to industrial synthesis of known or novel gene products. A few examples suffice to illustrate the power of these techniques.

Mapping a Chromosome. By determining the sequence of genes in a cloned fragment and then matching it on either side with overlapping neighboring fragments from other clones, large stretches of a genome can be delineated. The approximate location of an isolated gene and the number of copies can be identified by in situ hybridization of the cloned gene to a polytene chromosome.

Gene Expression. The direction of transcription, the start and stop signals, the origins of replication, and genes and gene clusters are analyzed in vitro and in vivo with isolated gene segments.

Superior promoters of gene expression can be engineered. Genes under the control of the *trp* or *lac* promoters can be induced or repressed by nutrients or nutrient analogs in the medium; transcriptional promoters provided in several tandem copies or in special locations can outcompete other cell promoters. Similarly, superior ribosomal binding sites may be fashioned to enhance translation.

The permissiveness of a bacterium (e.g., *E. coli*) in responding to a strongly promoted gene enables protein overproduction to attain bizarre levels; human insulin, *E. coli* RecA protein, or *E. coli* tryptophan synthetase can make up 50% or more of the cellular protein. The bacterium, with its proteolytic disposal mechanisms overwhelmed, can sequester a huge deposit of a foreign protein (or one of its own proteins normally present in a trace amount), so that a simple cell lysate may yield the protein in a nearly pure form.

Fused Genes. Promoters and other transcriptional control elements can be analyzed by fusing them to a gene (e.g., for β-galactosidase or a drug-resistance enzyme) whose expression is very easily measured. Fusing the DNA sequence for the signal peptide of a membrane or secretory protein to the gene for a cytoplasmic protein can translocate the latter to a membranous or extracellular location.

---

60. Kohara Y, Akiyama K, Isono K (1987) *Cell 50*:495.
61. Sambrook J, Fritsch EF, Maniatis T (1989) *Molecular Cloning: A Laboratory Manual.* 2nd ed. CSHL; Ausubel FM, Brent R, Kingston RE, Moore DD, Seidman JG, Smith JA, Struhl K (1987 et seq.) *Current Protocols in Molecular Biology*. Wiley, New York.

Amplifying Genes and Gene Products. An increase in gene copy number may elevate the level of the gene product 100-fold or more. Overproduction of DNA ligase (Section 9-2) and DNA polymerase I (Section 4-1), and use of the *E. coli lac* operon are among the many examples in which gene amplification has advanced enzyme chemistry and studies of gene regulation.

Vectors with mutations in copy-number control can be induced to levels of 1000 per cell and thus exceed the amount of chromosomal DNA; the level of a protein encoded by a gene in the "runaway" plasmid can be amplified as much or more.

Correcting a Mutation and Site-Directed Mutagenesis. Substituting one or a few bases in a DNA sequence may create mutations or repair them, may convert a gene of one species to the same gene of another species, and may create or delete certain regions, such as origins, promoters, and restriction sites. For example, the *am3* mutation of phage ϕX714 can be corrected by hybridizing to that region of the viral strand a short sequence of wild-type DNA prepared by enzymatic or chemical synthesis. After introduction of this DNA hybrid into cells, a large percentage of the resulting phage particles possess the wild-type sequence. Mutagenesis of the active sites of enzymes, hormones, receptors, and antibodies is a favored means for analyzing structure–function relationships.

Inserting Eukaryotic Genes into Prokaryotes. *E. coli* can replicate and express eukaryotic genes. A yeast gene for histidine biosynthesis introduced into *E. coli* by a plasmid corrects the deficiency in a mutant defective at that step in the pathway. Chicken ovalbumin is synthesized and secreted by *E. coli.*[62] Syntheses of animal hormones such as somatostatin, insulin,[63] and growth hormone[64] in *E. coli* are examples of industrial production of scarce and needed therapeutic agents.

Inserting Novel Genes into Eukaryotes. Segments of the yeast chromosome can be fashioned into autonomous vehicles (Section 18-10) for introducing genes from a variety of sources into yeast. The thoroughly analyzed chromosome of SV40 (Section 19-2) has been used as a vehicle for genes whose expression in monkey kidney cells converts them into factories to produce rabbit β-globin or *Drosophila* histones.[65]

Shuttle Vectors (Chimeric Plasmids). It is often useful to have a vector that can be grown and expressed in two different cellular habitats. Such shuttle vectors replicate autonomously and support gene expression in both a bacterial and a yeast cell, or in a bacterial and an animal cell, or in two different bacterial species (e.g., *E. coli* and *B. subtilis*).

Minichromosomes. Chromosomal origins introduced into plasmids can generate autonomously replicating genomes that behave like the intact chromosomes

---

62. Fraser TH, Bruce BJ (1978) *PNAS 75*:5936.
63. Goeddel DV, Kleid DG, Bolivar F, Heyneker HL, Yansura DG, Crea R, Hirose T, Kraszewski A, Itakura K, Riggs AD (1979) *PNAS 76*:5936.
64. Martial JA, Hallewell RA, Baxter JD, Goodman HM (1979) *Science 205*:602.
65. Mulligan RC, Howard BH, Berg P (1979) *Nature 277*:108.

Table 22-2
Enzymatic tools of genetic engineering

Class	Enzyme[a]	Class	Enzyme[a]
Nucleases	restriction endonucleases	Kinases	T4 polynucleotide kinase
	S1 and mung bean endonuclease		DNA kinase
	DNase I	Polymerases	DNA polymerase I
	exonuclease III		T4 DNA polymerase
	λ exonuclease		reverse transcriptase
	Bal31 exonuclease		terminal deoxynucleotidyl transferase
	RNase H		T7 polymerase (Sequenase®)
Ligases	T4 DNA ligase		*Thermus aquaticus* (Taq) polymerase
	T4 RNA ligase	Phosphatases	*E. coli* alkaline phosphatase
	E. coli ligase		calf alkaline phosphatase

[a] From Singer, M (1979) in *Genetic Engineering* (Setlow JK, Hollaender A, eds.). Plenum Press, New York, p. 1.

(Section 16-6). Introduction of a centromeric sequence into a yeast plasmid endows it with the status of a stable, functional yeast chromosome.

22-5 Homage to Enzymes; an Epilogue

The replication of DNA, as well as other DNA transactions—repair, recombination, restriction, and transcription—all are carried out by the enzymes that form the main subject of this book. Having these enzymes in hand and knowing what they do has made it possible to dissect, analyze, multiply, and rearrange segments of DNA, genes, and chromosomes. These enzymes form the cast that is largely responsible for the drama of genetic engineering (Table 22-2).

The relative simplicity of recombinant DNA techniques and the commercial availability of enzymes and equipment have led to rapid expansion and ingenious applications. In most instances, the DNA manipulations are carried out with packaged kits in which the enzymes are generally as faceless as the buffers and salts. The results of the manipulations are usually monitored in intact cells or organisms in which the enzymatic functions are totally obscure. Clearly, the ease of analyzing and creating DNA has diverted the spotlight from the laborious isolation and characterization of enzymes. Despite the insights gained from the sequences of genomes, including the primary structures of encoded proteins, the last word about cellular form and function will not be said without an intimate knowledge of the protein products of the DNA blueprint.

Author Index

Subject Index

Page numbers in **bold** type denote definitions and structures.

Structures, Symbols, and Abbreviations

am	amber mutation: a nonsense mutation containing the chain-terminating codon UAG
AMP, GMP, IMP, UMP, CMP	5′-phosphates of ribonucleosides of adenine, guanine, hypoxanthine, uracil, cytosine
2′-AMP, 3′-AMP, etc.	2′- and 3′-phosphates of the nucleosides
Ap, Gp, etc.	3′-AMP, 3′-GMP, etc.
AP sites	abasic (i.e., apurinic or apyrimidinic) sites
AT, GC	base pairs of adenine with thymine, and of guanine with cytosine, or their respective deoxyribonucleotides
bp	base pairs
capsid	a coat protein unit of a virus
cDNA	complementary DNA obtained by reverse transcription of RNA
$(CG)_n$	a sequence consisting of *n* repeats of CG base pairs
cis-acting sequence	a sequence that affects its own DNA molecule
concatemer	a chain of an unstated number of repeated units of duplex DNA
$(dA)_{40}$, etc.	homopolymer chain of about 40 deoxyriboadenylate (etc.) residues
$dA_{4000} \cdot dT_{200}$	a poly dA·poly dT_n DNA duplex
$d(AT)_{12}$	copolymer chain of 24 residues consisting of alternating A and T, 12 residues each; associated with another such chain by base pairing
$d(pT)_4$	a chain of 4 thymidylates
$d(pC)_4(pT)_{200}$	a polymer with 4 C residues and 200 T residues
dAMP, etc.	5′-phosphates of 2′-deoxyriboadenine, etc.
(d)CMP, etc.	cytidine (etc.) monophosphate with either ribose or deoxyribose
ddC	2′-3′-dideoxycytidine
ddNTPs	dideoxyribonucleoside triphosphates
Δ	abasic lesion in DNA
D-loop	displacement loop: a DNA replicative intermediate in which only one chain of a duplex region is replicated, leaving the other single strand displaced
dnaA box	a 9-bp DNA consensus sequence to which dnaA protein (the *E. coli* replication initiator) specifically binds
dna^{ts}	genetic locus for a replication protein in a temperature-sensitive (conditionally lethal) mutant
dNMP	5′-phosphate of a 2′-deoxyribonucleoside
dsDNA	double-stranded DNA
EBNA-1	Epstein-Barr nuclear antigen 1
E. coli	*Escherichia coli*
EDTA	ethylenediamine tetraacetic acid
gp	gene product
holoenzyme	complete enzyme including cofactors
kb	kilobases: number of base pairs in thousands
kDa	kilodalton (1000 Da)
LTR	long terminal repeat
MDa	megadalton (10^6 Da; approx. 3 kb of DNA)
msDNA	multicopy single-stranded DNA
mtRNA	mitochondrial RNA
NMR	nuclear magnetic resonance
ntd	nucleotide
oligo dA, etc.	homopolymer chain of deoxyriboadenylate (etc.) of undefined length, generally under several hundred residues
pA, pG, etc.	AMP, GMP, etc.
$(pA)_{40}$, etc.	homopolymer chain of about 40 riboadenylate (etc.) residues
P_i; PP_i	orthophosphate; inorganic pyrophosphate
phosphodiester	$R{-}O{-}\overset{\overset{\displaystyle O}{\|}}{\underset{\underset{\displaystyle O^-}{\|}}{P}}{-}O{-}R$
pol I	*E. coli* DNA polymerase I
poly dA, etc.	deoxyriboadenylate (etc.) homopolymer chain of undefined length, generally over 1000 residues
poly dA·poly dT	homopolymer chains of poly dA and poly dT associated by base pairing
poly d(AT)	copolymer chain of alternating A and T residues; associated with another such chain by base pairing
poly rA·poly rU	homopolymer chains of riboadenylate and ribouridylate associated by base pairing
processive	repetitive enzyme action without dissociation between steps; nondistributive
prophage	provirus stage of a temperate phage
provirus	state of a virus integrated into the host cell chromosome
RF	double-stranded circular replicative form of DNA
RFI	covalently closed RF
RFII	RF with discontinuity in at least one strand
retrovirus	RNA virus possessing a reverse transcriptase
RNase	ribonuclease
rRNA	ribosomal RNA
σ	sigma subunit of RNA polymerase; also the lariat-shaped DNA replicative intermediate in a rolling-circle mechanism
snRNA	small nuclear RNA
SSBs	single-strand binding proteins
ssDNA	single-stranded DNA
SS → RF	conversion of circular viral ssDNA to the duplex replicative form
suppressor mutation	mutation that restores a function lost by another mutation at another genetic site
θ	theta structure, a DNA replicative intermediate
T_m	thermal midpoint of melting
Taq	*Thermus aquaticus*
trans-acting sequence	a sequence that can affect any DNA molecule
ts	temperature-sensitive
UV	ultraviolet
V_{max}	maximal enzyme velocity
vegetative phage	free state of a phage
virion	a free virus particle